QUANTUM FIELD THEORY

QUANTUM FIELD THEORY

Claude Itzykson and
Jean-Bernard Zuber
Commissariat à l'Energie Atomique
Centre d'Etudes Nucléaires de Saclay

DOVER PUBLICATIONS
Garden City, New York

Bibliographical Note

This Dover edition, first published in 2005, is a republication of the work originally published by McGraw-Hill, Inc., New York, in 1980. A new preface to the Dover Edition and list of errata have been added.

International Standard Book Number

ISBN-13: 978-0-486-44568-7
ISBN-10: 0-486-44568-2

Manufactured in the United States of America
44568212 2022
www.doverpublications.com

CONTENTS

PREFACE TO THE DOVER EDITION

It is a great pleasure to see this book, first published a quarter of a century ago by McGraw-Hill, now being reprinted by Dover Publications, Inc. Of course, in the meantime, other textbooks of quantum field theory have appeared and the reader has a wide choice among several excellent and modern references. It is, however, my sincere hope and secret ambition that the present book will continue to be useful to many readers.

The present edition is a mere reprint of the original McGraw-Hill edition. Aside from a list of errata, that the reader will find beginning on page xxv, we have not attempted to revise the text. Obviously, there are several sections which we would write differently today. Perhaps the most dramatic obsolescence in this book is the discussion of non-abelian anomalies in chapter XI, which has been completely outdated by the progress in its understanding made in the early eighties. The reader is invited to consult the literature to fill this gap.

The new printing takes place in a special year, since 2005 marks the tenth anniversary of the death of Claude Itzykson in May 1995. It is with particular emotion that I devote this new edition to the memory of a dear friend and a great physicist, who initiated the lectures which led to this book and was the main driving force in the whole project.

Saclay/Jussieu, August 2005
Jean-Bernard Zuber

PREFACE

Quantum field theory has remained throughout the years one of the most important tools in understanding the microscopic world. The recent years have seen a blossoming of developments and applications which go far beyond the original scope.

Our attempt at presenting a pedagogical survey of this subject arose from lectures given in Orsay and Saclay, and first materialized in a set of notes written in French. As with any such endeavor it reflects to a large extent our prejudices and enthusiasms, even though we have tried to be as thorough as possible and to keep a balance between the formalism and practical examples of calculations.

In its present version this book addresses itself both to students and researchers in field theory, particle physics, and related areas. It presupposes a general background in quantum mechanics, electrodynamics, and relativity, and assumes some familiarity with classical calculus including group theory and complex analysis. To avoid one-sided views, we have respected the historical perspective and given equal weight to the operator formulation, propagator approach, and more synthetic path integral representation. Nevertheless, this is by no means a complete treatise. We list some of the main omissions. For lack of space and because we felt incompetent we do not treat axiomatic field theory and critical phenomena in statistical mechanics. Specific topics such as a detailed study of the Poincaré group and its representations, higher spin fields, a thorough discussion of infrared problems, gravitational interactions, two-dimensional models, etc., could not be discussed here.

The book is roughly divided into two parts. The first eight chapters cover the standard quantization of electrodynamics and culminate in the perturbative renormalization. The remaining five chapters are devoted to functional methods, relativistic bound states, broken symmetries, nonabelian gauge fields, and asymptotic behavior.

Sections of the text in smaller type contain exercises, comments, and suggestions. It was thought inappropriate to have a separate set of problems. The latter often consist in completing derivations and extending or applying some methods, and are therefore more usefully related directly with the main discussion—the more so as one progresses toward the recent and richer aspects of the theory.

To give some perspective, provide our sources, and suggest some further reading, each chapter closes with a section entitled "notes," which collects a number of references. Needless to say, they are not meant to be complete in any way. We present our apologies in advance to all of those who will not find their work properly quoted.

We owe, of course, a great deal to many of the excellent textbooks on the subject. Some overlap was unavoidable. This is also true of several review articles and lecture notes.

We hope that the reader will not suffer too much from our insufficient mastery of the English language. In the process of translation we have undoubtedly lost some accuracy in style that only belongs to one who expresses himself in his mother tongue.

The stimulating atmosphere of the Service de Physique Théorique in Saclay has been essential in developing our interest in field theory, and we have a great debt toward our friends and colleagues past and present. To R. Stora our teacher and example, to M. Bander, A. Carroll, G. Mahoux, G. Parisi, N. Salingaros, and V. Ganapathi who read the manuscript, and to the students and physicists who pointed out misprints, made comments and suggestions, and who cannot all be quoted here, we express our warmest thanks. We are also grateful to T. Kinoshita who was kind enough to provide us with a review of recent results in quantum electrodynamics.

This enterprise would have been impossible without the help of our laboratory. We thank above all Brigitte Bunel, Danielle Lechaton, and Jeanine Rigou, who carried the burden of typing the manuscript, Madeleine Porneuf who supervised the project, and the Commissariat à l'Energie Atomique and its staff who supported it.

Claude Itzykson
Jean-Bernard Zuber

Saclay
March, 1978

GENERAL REFERENCES

Adler, S., and R. Dashen: "Current Algebras," Benjamin, New York, 1968.

Akhiezer, A., and V. B. Berestetskii: "Quantum Electrodynamics," Interscience, New York, 1965.

Alfaro, V. de, S. Fubini, G. Furlan, and C. Rossetti: "Currents in Hadron Physics," North-Holland, Amsterdam, 1973.

Berestetskii, V. B., E. M. Lifshitz, and L. P. Pitaevskii: "Relativistic Quantum Theory," Pergamon Press, Oxford, 1971.

Berezin, F. A.: "The Method of Second Quantization," Academic Press, New York, 1966.

Bethe, H. A., and E. E. Salpeter: "Quantum Mechanics of One and Two Electron Atoms," Springer-Verlag, Berlin, 1957.

Bjorken, J. D., and S. D. Drell: "Relativistic Quantum Mechanics," and "Relativistic Quantum Fields," McGraw-Hill, New York, 1964 and 1965.

Bogoliubov, N. N., and D. V. Shirkov: "Introduction to the Theory of Quantized Fields," Interscience, New York, 1959.

Eden, R. J., P. V. Landshoff, D. I. Olive, and J. C. Polkinghorne: "The Analytic *S*-Matrix," Cambridge University Press, 1966.

Feynman, R. P.: "Quantum Electrodynamics," Benjamin, New York, 1973.

Feynman, R. P., and A. R. Hibbs: "Quantum Mechanics and Path Integrals," McGraw-Hill, New York, 1965.

Gell-Mann, M., and Y. Ne'eman: "The Eightfold Way," Benjamin, New York, 1964.

Hepp, K.: "Théorie de la Renormalisation," Springer-Verlag, Berlin, 1969.

Jost, R.: "The General Theory of Quantized Fields," AMS, Providence, 1965.

Källen, G.: "Quantum Electrodynamics," Springer-Verlag, Berlin, 1972.

Landau, L. D., and E. M. Lifshitz: "The Classical Theory of Fields," Pergamon Press, Oxford, 1975.

Nakanishi, N.: "Graph Theory and Feynman Integrals," Gordon and Breach, New York, 1970.

Schwinger, J. (ed.): "Quantum Electrodynamics," Dover, New York, 1958.

Schwinger, J.: "Particles, Sources and Fields," vols. 1 and 2, Addison-Wesley, Reading, 1970 and 1973.

Streater, R. F., and A. S. Wightman: "*PCT*, Spin and Statistics, and All That," Benjamin, New York, 1964.

Thirring, W.: "Principles of Quantum Electrodynamics," Academic Press, New York, 1958.

Taylor, J. C.: "Gauge Theories of Weak Interactions," Cambridge University Press, 1976.

Todorov, I. T.: "Analytic Properties of Feynman Diagrams in Quantum Field Theory," Pergamon Press, Oxford, 1971.

ERRATA

CHAPTER I

Page 16	Line 1	after "force law", add "it follows that"
Page 16	Line 3	for: we learn ... read: This expression holds for a weak magnetic field when we average over the motion of the orbiting particle. It amounts to dropping total time derivatives, and shows ...
Page 20	Eq. (1-88)	for: $\frac{\partial I}{\partial t_2} - \frac{\partial I}{\partial t_1} = 0$ read: $\frac{\partial I}{\partial t_2} + \frac{\partial I}{\partial t_1} = 0$
Page 23	Eq. (1-97)	for: $\widetilde{\Theta}^{\mu\nu} = \frac{\partial \mathcal{L}}{\partial(\partial_\mu \phi_i)} \partial_\nu \phi_i - g_{\mu\nu} \mathcal{L}$ read: $\widetilde{\Theta}^{\mu\nu} = \frac{\partial \mathcal{L}}{\partial(\partial_\mu \phi_i)} \partial^\nu \phi_i - g^{\mu\nu} \mathcal{L}$
Page 27	Eq. (1-131)	for: $\pi(\mathbf{x}, t) = \frac{\partial L}{\partial \partial_0 \varphi(\mathbf{x}, t)} = \cdots$ read: $\pi(\mathbf{x}, t) = \frac{\delta L}{\delta \partial_0 \varphi(\mathbf{x}, t)} = \cdots$
Page 31	Eq. (1-158)	for: $J^\mu = j^\mu + \frac{\partial \mathcal{L}_{\text{int}}}{\partial[\partial_\mu \alpha(x)]} = \cdots$ read: $J^\nu = j^\nu + \frac{\partial \mathcal{L}_{\text{int}}}{\partial[\partial_\nu \alpha(x)]} = \cdots$
Page 33	Eq. (1-169)	for: $\widetilde{G}_{\substack{\text{ret} \\ \text{adv}}}(p) = \frac{-1}{(p_0 \pm i\varepsilon) - \mathbf{p}^2 - m^2}$ read: $\widetilde{G}_{\substack{\text{ret} \\ \text{adv}}}(p) = \frac{-1}{(p_0 \pm i\varepsilon)^2 - \mathbf{p}^2 - m^2}$
Page 34	Eq. (1-172)	for: $e^{\pm i\omega_p x_0 + \cdots}$ read: $e^{\mp i\omega_p x_0 + \cdots}$

Page 35	Line –1	for: Sec. 1-2 read: Sec. 1-1-2
Page 41	Eq. (1-212)	for: $\cdots = -j.\widetilde{j}^*$ read: $\cdots = -\widetilde{j}.\widetilde{j}^*$

CHAPTER II

Page 47	Eq. (2-3)	for: $\left(\frac{\partial^2}{\partial t^2} - \boldsymbol{\nabla}^2 - m^2\right)\psi(\mathbf{x},t) = 0$ read: $\left(\frac{\partial^2}{\partial t^2} - \boldsymbol{\nabla}^2 + m^2\right)\psi(\mathbf{x},t) = 0$
Page 59	Eq. (2-50)	for: $P(n_k)\Lambda_\pm(k) = \left(I \pm \frac{\boldsymbol{\Sigma}.\mathbf{k}}{\lvert\mathbf{k}\rvert}\right)\Lambda_\pm(k)$ read: $P(n_k)\Lambda_\pm(k) = \frac{1}{2}\left(I \pm \frac{\boldsymbol{\Sigma}.\mathbf{k}}{\lvert\mathbf{k}\rvert}\right)\Lambda_\pm(k)$
Page 60	Eq. (2-51) and	for: $\frac{\not k+m}{2}$ read: $\frac{\not k+m}{2m}$ for: $\frac{-\not k+m}{2}$ read: $\frac{-\not k+m}{2m}$
Page 61	Line –9	for: b/d^* read: d^*/b
Page 66	Lines 5, 7, 12	for: $\frac{e\hbar}{2mc}$ read: $\frac{e\hbar}{2m}$
Page 66	Line 7	for: $\frac{e\hbar}{mc}$ read: $\frac{e\hbar}{m}$
Page 82	Line 13	for: $\frac{mZ^4\alpha^5}{n^3}$ read: $\frac{mZ^4\alpha^5}{\pi n^3}$
Page 84	Line 1	for: $E_{\rm tot} = m + M - \frac{\alpha}{2n^2}\frac{mM}{m+M} + \cdots$ read: $E_{\rm tot} = m + M - \frac{\alpha^2}{2n^2}\frac{mM}{m+M} + \cdots$
Page 93	Eq. (2-118)	for: $e\int d^4x_2 \not A(x_2) S_A(x_2,x_1)$ read: $e\int d^4x_2\, S_F(x_3,x_2)\not A(x_2) S_A(x_2,x_1)$

CHAPTER III

Page 109	Eq. (3-16)	for: $[a_k, a_{k'}] = \delta(k-k')$ read: $[a_k, a_{k'}^\dagger] = \delta(k-k')$
Page 117	Eq. (3-56)	for: $\Delta(x-y) = \cdots$ read: $\Delta(x) = \cdots$
Page 119	Line 2	for: $D(\eta_2)\lvert\eta_1) = \lvert\eta_1 + \eta_2)$ read: $D(\eta_2)\lvert\eta_1) = e^{i\int d\widetilde{k}\,\mathrm{Im}(\eta_1^*\eta_2)}\lvert\eta_1 + \eta_2)$
Page 119	Eq. (3-67)	for: $\widetilde{f}_n(k) = \int d^3x\, e^{i\mathbf{k}.\mathbf{x}} F_n(\mathbf{x}) = \widetilde{F}_n(\mathbf{k})$ read: $\widetilde{f}_n(k) = \int d^3x\, e^{-i\mathbf{k}.\mathbf{x}} F_n(\mathbf{x}) = \widetilde{F}_n(\mathbf{k})$

Page 120	Eq. (3-69)	for: $\varphi_c(\mathbf{x}) = \sum_n \varphi_{n,c} \int \widetilde{dk}\, e^{-i\mathbf{k}.\mathbf{x}} F_n^*(\mathbf{k})$ read: $\varphi_c(\mathbf{x}) = \sum_n \varphi_{n,c} \int \widetilde{dk}\, e^{-i\mathbf{k}.\mathbf{x}} \widetilde{F}_n^*(\mathbf{k})$
Page 135	Eq. (3-136)	for: $k^0 = \sqrt{k^2 + \mu^2}$ read: $k^0 = \sqrt{\mathbf{k}^2 + \mu^2}$
Page 135	Line –6	for: $\varepsilon^{(\lambda)}(k).\varepsilon^{(\lambda')}(k) = \delta_{\lambda\lambda'}$ read: $\varepsilon^{(\lambda)}(k).\varepsilon^{(\lambda')}(k) = -\delta_{\lambda\lambda'}$
Page 140	Line 11	for: $u = a^2k^2/\pi$ read: $u = a^2k^2/\pi^2$
Page 149	Eq. (3-169)	for: $\{\psi_\xi(x), \bar{\psi}_{\xi'}(x')\} = \cdots$ read: $\{\psi_\xi(x), \bar{\psi}_{\xi'}(y)\} = \cdots$
Page 153	Line 5	for: $C = i\gamma^0\gamma^2$ read: $C = i\gamma^2\gamma^0$
Page 156	Line 17	for: $\gamma^5 C u^*(\hat{k}, \varepsilon) = \cdots$ read: $\gamma^5 C u^*(\tilde{k}, \varepsilon) = \cdots$
Page 159	Line –6	for: $\gamma^0 O^\mu(p', p)^\dagger \gamma^0 = O^\mu(p', p)$ read: $\gamma^0 O^\mu(p', p)^\dagger \gamma^0 = O^\mu(p, p')$
Page 160	Line 6	In the bracket of Eq. (3.203), add a fourth term $+\gamma^5(q^2\gamma^\mu - \not{q}q^\mu)F_4(q^2)$.
Page 160	Lines 10–11	for: $F_3(q^2) = -F_3(q^2)$. Thus parity conservation alone yields $F_3 = 0$. read: $F_3(q^2) = -F_3(q^2)$ and $F_4(q^2) = -F_4(q^2)$. Thus parity conservation alone yields $F_3 = F_4 = 0$.
Page 160	Line 14	for: $\epsilon_1 = \epsilon_2 = -\epsilon_3 = 1$. read: $\epsilon_1 = \epsilon_2 = -\epsilon_3 = \epsilon_4 = 1$.
Page 161	Line 2	for: $\sigma^{\mu\nu} = \frac{i}{2}[{\scriptstyle\gamma^\mu}, \gamma^\nu]$ read: $\sigma^{\mu\nu} = \frac{i}{2}[\gamma^\mu, \gamma^\nu]$

CHAPTER IV

Page 179	Line 17	for: $\sum\limits_{1\leq k<l\leq N}$ read: $\sum\limits_{1\leq l<k\leq N}$
Page 188	Line 15	for: $\mathcal{F}(A)$ read: $\mathcal{T}(A)$
Page 193	Eq. (4-113)	for: $(P - eA)^2 - m^2 = \cdots$ read: $(P - eA)^2 = \cdots$

CHAPTER V

Page 207	Eq. (5-28)	for: $\langle p_1, \ldots, p_n, \text{in}\vert S \vert q_1, \ldots, q_l, \text{out}\rangle$ read: $\langle p_1, \ldots, p_n, \text{in}\vert S \vert q_1, \ldots, q_l, \text{in}\rangle$

Page 209	Line 4	for: $\cdots = \frac{a^{\dagger m}\|0\rangle\langle 0\|a^n}{m!n!}\cdots$ read: $\cdots = \sum_{m,n=0}^{\infty} \frac{a^{\dagger m}\|0\rangle\langle 0\|a^n}{m!n!}\cdots$
Page 215	Line –5	for: $\cdots - iZ^{-1/2} \int d^3x \cdots$ read: $\cdots - iZ^{-1/2} \int d^4x \cdots$
Page 222	Eq. (5-88)	for: $\varepsilon_i.j(x)$ read: $\varepsilon_i.j(y)$
Page 223	Line 12	Delete: ", using the fact that $W_M(x-y)$ depends only on $(x-y)^2$".
Page 225	Line –6	for: $\cdots b^\dagger(q_i,\alpha_i)\|0\rangle$ read: $\cdots b^\dagger(p_i,\alpha_i)\|0\rangle$
Page 228	Line –3	for: $= 8k_i.p_i\, k_f.p_f\, [\cdots$ read: $= 8k_i.p_i\, k_f.p_i\, [\cdots$
Page 232	Eq. (5-120)	for: $-4(\varepsilon_1.\varepsilon_2)$ read: $-4(\varepsilon_1.\varepsilon_2)^2$
Page 239	Line –3	for: $\cdots(\not p_f + \not k + m)\gamma^0(\not p_f + \not k + m)\cdots$ read: $\cdots(\not p_f + \not k + m)\gamma^0(\not p_i + m)\gamma^0(\not p_f + \not k + m)\cdots$
Page 240	Line 3	for: $\cdots p_i.k\, p_j.k]$ read: $\cdots p_i.k\, p_f.k]$
Page 240	Eq. (5-150)	for: $2\omega^2 \frac{p_i^2 \sin^2\theta_i + p_f^2 \sin^2\theta_f}{(E_f - p_f\cos\theta_f)(E_i - p_i \sin\theta_i)}$ read: $2\omega^2 \frac{p_i^2 \sin^2\theta_i + p_f^2 \sin^2\theta_f}{(E_f - p_f\cos\theta_f)(E_i - p_i \cos\theta_i)}$
Page 246	Line 9	for: ... complex field φ to describe ... read: ... complex field φ, creating A and annihilating $\overline{A}$, to describe ...

CHAPTER VI

Page 278	Eq. (6-39)	for: $\frac{m^2 e^4}{4E^2(2\pi)^2}$ read: $\frac{m^4 e^4}{4E^2(2\pi)^2}$
Page 281	Eq. (6-47)	for: $\frac{\alpha}{2E^2}\left[\frac{5}{4} - \frac{8E^4 - m^4}{E^2(E^2 - m^2)(1-\cos\theta)} + \cdots\right.$ read: $\frac{\alpha^2}{2E^2}\left[\frac{5}{4} - \frac{8E^4 - m^4}{4E^2(E^2 - m^2)(1-\cos\theta)} + \cdots\right.$
Page 286	Line 18	for: $\|T\|^2 = 4[\cdots]$ read: $\|T\|^2 = 4[\cdots]^2$
Page 290	Line –7	for: $i\frac{\delta}{\partial\varphi_c(x)}$ read: $i\frac{\delta}{\delta\varphi_c(x)}$
Page 294	Line 10	for: Sec. 4-3 read: Sec. 4-2-2

Page 298	Line –2	for: $2L - 4V + 4 < 0$ read: $2I - 4V + 4 < 0$
Page 303	Line –4	for: $z_i = z_j^0$ read: $z_j = z_j^0$
Page 311	Line 21	for: after elimination of $\alpha_1 = 1-\alpha_2-\alpha_3$, yield, for $\alpha = \alpha_2 = \alpha_3$ $(0 < \alpha < \frac{1}{2}$ since $0 < \alpha_1 < 1)$ read: after elimination of $\alpha_3 = 1 - \alpha_1 - \alpha_2$, yield, for $\alpha = \alpha_1 = \alpha_2$ $(0 < \alpha < \frac{1}{2}$ since $0 < \alpha_3 < 1$)

CHAPTER VII

Page 326	Line 10	for: $\log(\Lambda^2/m^2)$ read: $\ln(\Lambda^2/m^2)$
Page 328	Line 17	for: Sec. 6-3-2 read: Sec. 7-3-2
Page 333	Line –10	for: ... ultraviolet divergent. read: ... ultraviolet convergent.
Page 345	Line –11	for: Sec. 3-3-4 read: Sec. 4-3-4
Page 354	Line –2	for: $\varphi \tanh\varphi + (1 - \varphi\tanh\varphi)$ read: $\frac{1}{2}\varphi\tanh\varphi + (1 - \varphi\coth\varphi)$
Page 355	Line 8	for: contributions read: contribution
Page 361	Line –10	for: $B = \dfrac{\alpha}{4\pi^2}\cdots$ read: $B = \dfrac{\alpha}{2\pi^2}\cdots$
Page 370	Line –5	for: vol. 75, p. 1912 read: vol. 75, p. 898,

CHAPTER VIII

Page 377	Line 9	for: $\dfrac{k_\mu k_\nu \delta^{\mu\nu} - m_2^2}{[(p-k)^2 - m_1^2]^n (k^2 - m_2^2)^p}$ read: $\dfrac{k_\mu k_\nu \delta^{\mu\nu} + m_2^2}{[(p-k)^2 + m_1^2]^n (k^2 + m_2^2)^p}$
	and	for: $\dfrac{1}{[(p-k)^2 - m_1^2]^n (k^2 - m_2^2)^{p-1}}$ read: $\dfrac{1}{[(p-k)^2 + m_1^2]^n (k^2 + m_2^2)^{p-1}}$
Page 380	Line 17	for: Therefore ω_v is read: Therefore $\hat{\omega}_v$ is
Page 381	Line 8	for: at least an ... read: at least one ...
Page 383	Eq. (8-23)	for: $\cdots = 2L_l - 4I_l$ read: $\cdots = 4L_l - 2I_l$

Page 408	Eq. (8-78)	for: $A_\sigma(x)\|0\rangle$ read: $A_\sigma(0)\|0\rangle$
Page 409	Line 6	for: $-i\frac{M^2}{\mu^2(k^2-M^2)}$ read: $-i\frac{M^2k_\sigma}{\mu^2(k^2-M^2)}$
Page 409	Line 13	for: $B(k^2)=\lambda\frac{k^2-M^2}{k^2}$ read: $B(k^2)=-\lambda\frac{k^2-M^2}{k^2}$
Page 410	Line 5	for: Contracting Eq. (8-83) with k_σ and ... read: Contracting Eq. (8-83) with q_σ and ...
Page 410	Line –1	for: p_1, p_2, p_3, p_4 read: k_1, k_2, k_3, k_4
Page 413	Line –6	for: $+eZ_1\bar{\psi}\not{A}\psi$ read: $-eZ_1\bar{\psi}\not{A}\psi$
Page 421	Eq. (8-124)	for: C_2 read: C_1
Page 421	Line –7	for: $[\cdots]^{d/2-4}$ read: $[\cdots]^{d-4}$

CHAPTER IX

Page 430	Line –3	for: $M_{k,k+1}=M_{k-1,k}=\cdots \quad 1\le k\le n-1$ read: $M_{k,k+1}=M_{k+1,k}=\cdots \quad 0\le k\le n-1$
Page 437	Eq. (9-46)	for: $A=\sum_{n,m}A_{n,m}\cdots$ read: $A=\sum_{n,m}\frac{A_{n,m}}{\sqrt{n!m!}}\cdots$
Page 438	Line 2	for: $\exp\left[\bar{z}_n z_{n-1}-\sum_1^{n-1}\bar{z}_k z_k+\bar{z}_1 z_0-i\cdots\right]$ read: $\exp\left[\sum_0^{n-1}\bar{z}_{k+1}z_k-\sum_1^{n-1}\bar{z}_k z_k-i\cdots\right]$
Page 438	Line 10	for: which requires to read: which requires us to
Page 442	Eq. (9-77)	for: $\frac{\bar{\eta}\dot{\eta}-\dot{\bar{\eta}}\eta}{2i}$ read: $\frac{\dot{\bar{\eta}}\eta-\bar{\eta}\dot{\eta}}{2i}$
Page 459	Line 21	for: A canonical ... read: Requiring that the g's have vanishing Poisson brackets, a canonical ...
Page 471	Line 8	for: integrate the normalized version of $\dot{\tilde{\varphi}}$ in the subspace orthogonal to ψ. read: integrate in the subspace orthogonal to ψ, the normalized version of $\dot{\tilde{\varphi}}$.

CHAPTER X

Page 480		Interchange figures 10-3 and 10-4. Captions remain the same.

Page 480	Eq. (10-24)	for: $S(x_1, y_1; x_2, y_2; J)\|_{J=0}$ read: $S(x_1, x_2; y_1, y_2; J)\|_{J=0}$
Page 492	Eq. (10-72)	for: $D^{-3}\chi(P)$ read: $D^{-3}\chi(p)$
Page 497	Line 9	for: $p_0 = \mp E/2 + \omega - i\varepsilon$ read: $p_0 = \mp E/2 - \omega - i\varepsilon$
Page 499	Line -7	for: $(E' - E) \int d^3p\, \eta^\dagger(\mathbf{p})\varphi(\mathbf{p})$ read: $(E' - E) \int d^3p\, \eta^\dagger(\mathbf{p})\varphi'(\mathbf{p})$
Page 499	Eq. (10-107)	for: $\eta^*(\mathbf{p})$ read: $\eta^\dagger(\mathbf{p})$
Page 504	Eq. (10-120)	for: $(k^0 + m^2) - \cdots$ read: $(k^0 + m)^2 - \cdots$

CHAPTER XI

Page 523 Lines 4–6

for: the lagrangian reads ... decoupled massless fields,
read: the lagrangian has the form

$$\begin{aligned}\mathcal{L} &= \tfrac{1}{2}(\partial\rho)^2 + \tfrac{1}{2}(\partial\boldsymbol{\xi})^2 \left(1 + \tfrac{\rho}{v}\right)^2 \left(1 + f\left(\tfrac{\boldsymbol{\xi}}{v}\right)\right) \\ &- \frac{\mu^2}{2}(v + \rho)^2 - \frac{\lambda}{4}(v + \rho)^4\end{aligned}$$

with some function f. From this we see that the $\boldsymbol{\xi}$ correspond to $(n - 1)$ massless fields,

Page 547 Eq. (11-178)

for: $\cdots = \delta^4(x - y)\, T^\alpha_{kl}[\phi_l(y) + v_l]$

read: $\cdots = -i\delta^4(x - y)\, T^\alpha_{kl}[\phi_l(y) + v_l]$

Page 551 Lines 1–5

Replace with:

At $k_1^2 = k_2^2 = 0$, hence $k_1.q = k_2.q = k_1.k_2$, only two independent tensors are consistent with these requirements:

$$\mathcal{B}_{1\,\mu\nu\rho} = \varepsilon_{\mu\nu\sigma\tau} k_1^\sigma k_2^\tau q_\rho$$

and

$$\mathcal{B}_{2\,\mu\nu\rho} = (\varepsilon_{\mu\rho\sigma\tau} k_{1\,\nu} - \varepsilon_{\nu\rho\sigma\tau} k_{2\,\mu})\, k_1^\sigma k_2^\tau - \varepsilon_{\mu\nu\rho\sigma}(k_1^\sigma - k_2^\sigma)\, k_1.k_2\,.$$

A third possible tensor $\mathcal{B}_{3\mu\nu\rho} = (\varepsilon_{\mu\,\rho\sigma\tau} k_{2\nu} - \varepsilon_{\nu\rho\sigma\tau} k_{1\,\mu}) k_1^\sigma k_2^\tau$ is actually not independent, because of the identity $q_\sigma \varepsilon_{\tau\mu\nu\rho} + q_\rho \varepsilon_{\sigma\tau\mu\nu}$

$+q_\nu \varepsilon_{\rho\sigma\tau\mu} + q_\mu \varepsilon_{\nu\rho\sigma\tau} + q_\tau \varepsilon_{\mu\nu\rho\sigma} = 0$. This expresses that a totally antisymmetric rank 5 tensor vanishes in four dimensions. Contracted with $k_1^\sigma k_2^\tau$, this yields $\mathcal{B}_{1\,\mu\nu\rho} = \mathcal{B}_{2\,\mu\nu\rho} + \mathcal{B}_{3\,\mu\nu\rho}$. We therefore write the following expression when $k_1^2 = k_2^2 = 0$

$$\begin{aligned} T_{\mu\nu\rho}(k_1,k_2) &= \varepsilon_{\mu\nu\sigma\tau} k_1^\sigma k_2^\tau q_\rho T_1(q^2) \\ &\quad + \left[\left(\varepsilon_{\mu\rho\sigma\tau} k_{1\,\nu} - \varepsilon_{\nu\rho\sigma\tau} k_{2\,\mu} \right) k_1^\sigma k_2^\tau \right. \\ &\quad \left. - \varepsilon_{\mu\nu\rho\sigma} (k_1^\sigma - k_2^\sigma)\, k_1.k_2 \right] T_2(q^2) \,. \end{aligned} \qquad (11-188)$$

Consequently,

$$q^\rho T_{\mu\nu\rho} = \varepsilon_{\mu\nu\sigma\tau} k_1^\sigma k_2^\tau q^2 \left[T_1(q^2) + T_2(q^2) \right] \qquad (11\text{–}189)$$

Page 553 Line 7 — An updated value is $\Gamma^{\exp} = (7.75 \pm 0.5)\,\mathrm{eV}$.

CHAPTER XII

Page 564 Line 6 — for: $s(0) = x_1, s(1) = x_2$ read: $x(0) = x_1, x(1) = x_2$

Page 574 Line −13 — for: the latter read: Γ

Page 577 Line 12 — for: $\mathcal{M}_0 = -\Delta\delta_{ij} \cdots$ read: $\mathcal{M}_0 = -\Delta\delta_{ab} \cdots$

Page 595 Lines 1–7 — Replace with:

To be precise we write $\det \mathcal{M}_\mathcal{F}(A) \equiv \Delta_\mathcal{F}(A, \mathcal{F}(A))$ according to the definition

$$\Delta_\mathcal{F}^{-1}(A,C) = \int \mathcal{D}(g)\, \delta(\mathcal{F}({}^gA) - C)$$

For a gauge transformation independent of A we have obviously

$$\Delta_\mathcal{F}\,({}^gA, C) = \Delta_\mathcal{F}(A, C)$$

due to the group invariance of the measure $\mathcal{D}(g)$. In the present case however where

$$A' = {}^{g(A)}A \qquad \mathcal{F}(A) = \mathcal{F}'(A') = \mathcal{F}'\left({}^{g(A)}A\right)$$

the gauge transformation depends on the potential. We shall show that the jacobians in $\mathcal{D}(A)$ and $\Delta_\mathcal{F}$ conspire to cancel. Consider

$$\int \mathcal{D}(A)\Delta_{\mathcal{F}}(A,\mathcal{F}(A))$$
$$= \int \mathcal{D}(A)\Delta_{\mathcal{F}}(A,\mathcal{F}(A))\left[\int \mathcal{D}(A')\,\delta\left(A' - {}^{g(A)}A\right)\right]$$
$$\times \left[\Delta_{\mathcal{F}'}(A,\mathcal{F}(A)) \int \mathcal{D}(g)\,\delta\left(\mathcal{F}'\left({}^{g}A\right) - \mathcal{F}(A)\right)\right]$$

Both terms between brackets on the right hand side are equal to unity. The argument of the last δ-function vanishes for $g = g(A)$, we can therefore substitute the generic g for $g(A)$ in the first δ-function. We then set $A =^{g^{-1}} B$, recognize that $\mathcal{D}(A) = \mathcal{D}(B)$ and integrate over B. Thus, using the invariance of Δ under potential independent gauge transformations

$$\int \mathcal{D}(A)\,\Delta_{\mathcal{F}}(A,\mathcal{F}(A)) =$$
$$= \int \mathcal{D}(A')\mathcal{D}(g)\,\Delta_{\mathcal{F}}\left(A',\mathcal{F}\left({}^{g^{-1}}A'\right)\right)$$
$$\Delta_{\mathcal{F}'}\left(A',\mathcal{F}\left({}^{g^{-1}}A'\right)\right)\delta\left(\mathcal{F}'(A') - \mathcal{F}\left({}^{g^{-1}}A'\right)\right)$$
$$= \int \mathcal{D}(A')\mathcal{D}\,(g)\;\Delta_{\mathcal{F}}\left(A',\mathcal{F}'\left(A'\right)\right)$$
$$\Delta_{\mathcal{F}'}\left(A',\mathcal{F}'\left(A'\right)\right)\delta\left(\mathcal{F}'(A') - \mathcal{F}\left({}^{g^{-1}}A'\right)\right)$$
$$= \int \mathcal{D}\left(A'\right)\,\Delta_{\mathcal{F}'}\left(A',\mathcal{F}'\left(A'\right)\right)$$

The last equality follows from the integration over g and translates in precise terms Eq. (12-129).

Page 596	Eq. (12-137)	Delete the first 0=.
Page 600	Line −10	for: $\widetilde{I}_1 = I - \cdots$ read: $\widetilde{I}_1 = \widetilde{I} - \cdots$
Page 603	Line -9	for: involve no λ_5 matrix read: involve no γ_5 matrix
Page 611	Line −10	for: $(k^2 - m^2)^{-1}$ read: $(k^2 - M^2)^{-1}$
Page 625	Line 8	An updated value is $\sin^2\theta_W \simeq 0.23$.

CHAPTER XIII

Page 635	Line 11	for: the derivative $(\partial/\partial x)d^{as}(\alpha_1, x)$

		read: the derivative $(\partial/\partial x)d^{as}(\alpha, x)$
Page 648	Eq. (13-61)	for: $\cdots = m_0 \frac{\partial}{\partial m_0} \cdots$ read: $\cdots = \frac{1}{2} m_0 \frac{\partial}{\partial m_0} \cdots$
Page 649	Eq.(13-65)	for: $\frac{\partial m_0(\Lambda/m, g)}{\partial m}$ read: $\frac{\partial m_0(\Lambda/m, g_0)}{\partial m}$
Page 653	Line –3	for: ... Migdal read: ... Migdal, as well as the three-loop one worked out by Tarasov, Vladimirov and Zharkov
Page 653	Line –2	for: $+O(g^7)$ read: $+\frac{g^7}{(4\pi)^6}\Big(-\frac{2857}{54}C^3 + \frac{1415}{27}C^2T_f - \frac{158}{27}CT_f^2 + \frac{205}{9}CC_fT_f - \frac{44}{9}C_fT_f^2 - 2C_f^2T_f\Big) + O(g^9)$
Page 660	Line 14	for: . A mean read: . A means
Page 662	Line –3	for: "Experimental conditions, $\sin^2(\theta/2) \ll 1$, read: Experimental limitations on the accessible range of $\sin^2(\theta/2)$
Page 670	Eq. (13-124)	for: $m_\mu^2/q^2 \to \infty$ read: $m_\mu^2/q^2 \to 0$
Page 676	Eq. (13-144)	for: $\int \frac{d^4p}{(2\pi)^4}$ read: $\int \frac{d^4q}{(2\pi)^4}$
Page 683	Line 4	for: which play read: which plays
Page 683	Line –5	for: $W(q^2, -\omega) = -W(q^2, -\omega)$ read: $W(q^2, -\omega) = -W(q^2, \omega)$
Page 689	Line 9	add: For the three-loop calculation, see O.V. Tarasov, A.A. Vladimirov and A.Yu. Zharkov, *Phys. Lett.*, ser. B, vol. 93, p. 429, 1980.

APPENDIX

Page 696	Eq. (A-40)	for: $(2\pi^3)$ read: $(2\pi)^3$

INDEX

Page 702	Line –6	for: paramagnetic representation read: parametric representation
Page 702	Line 26	for: Gell-Mann low function read: Gell-Mann Low function

QUANTUM FIELD THEORY

CHAPTER

ONE

CLASSICAL THEORY

In this chapter we survey the lagrangian formalism in classical mechanics and its extension to systems with infinitely many degrees of freedom, emphasizing symmetries and conservation laws in their local form. Using electromagnetism as an example, we introduce Green functions and propagators. Elementary radiation problems are presented, and we close the chapter by studying the inconsistencies of self-interaction. As classical field configurations play a growing role in modern developments we will return to this subject in later chapters.

1-1 PRINCIPLE OF LEAST ACTION

1-1-1 Classical Motion

In classical mechanics the equations of motion follow from a principle of least action. If q stands for a finite collection of configuration variables, $q \equiv \{q_1, q_2, \ldots, q_N\}$, with $\dot{q}$ their velocities at time t, the action is defined as

$$I = \int_{t_1}^{t_2} dt\, L[q(t), \dot{q}(t)] \tag{1-1}$$

The Lagrange function L depends on the positions, velocities, and sometimes also explicitly on time for open systems, i.e., systems subjected to given external forces. The principle of least action states that among all trajectories $q(t)$ which join q_1 at time t_1 to q_2 at time t_2, the physical one yields a stationary value for the action. This stationary value is a unique minimum if q_1, t_1 and q_2, t_2 are close enough. The action is therefore to be considered as a functional of all regular functions $q(t)$ satisfying the boundary conditions $q(t_1) = q_1$, $q(t_2) = q_2$. If $Q(t)$ is the actual trajectory, a nearby one is written $q(t) = Q(t) + \delta q(t)$. Expanding the action in powers of δq as

$$I(q) = I(Q) + \int_{t_1}^{t_2} dt \, \frac{\delta I}{\delta q(t)}(Q)\, \delta q(t) + \cdots \tag{1-2}$$

we express the principle by setting

$$\frac{\delta I}{\delta q(t)}(Q) = 0 \tag{1-3}$$

To compare this expression with the Euler-Lagrange equations we note that since

$$\delta \dot{q}(t) = \frac{d}{dt}\, \delta q(t) \qquad \delta q(t_1) = \delta q(t_2) = 0 \tag{1-4}$$

the variation of (1-1) reads

$$\delta I = \int_{t_1}^{t_2} dt \, \frac{\delta I}{\delta q(t)}\, \delta q(t) = \int_{t_1}^{t_2} dt \left[\frac{\partial L}{\partial q(t)}\, \delta q(t) + \frac{\partial L}{\partial \dot{q}(t)} \frac{d}{dt}\, \delta q(t) \right]$$

An integration by parts of the last term, taking into account boundary conditions, yields the familiar form

$$\frac{\delta I}{\delta q(t)} \equiv \frac{\partial L}{\partial q(t)} - \frac{d}{dt} \frac{\partial L}{\partial \dot{q}(t)} = 0 \tag{1-5}$$

which is to be interpreted as a vector equation in the case of several degrees of freedom. In the simplest cases L is the difference between a kinetic part quadratic in the velocities and a potential part.

The equations are unchanged if we add to L a total derivative with respect to time, and the action is only modified by contributions depending on the boundary conditions. This observation shows that the Lagrange function is not an intrinsic object to determine the motion and leads to a more abstract formalism, developed in particular by E. Cartan.

The hamiltonian formulation is obtained through the definition of conjugate momenta

$$p_i = \frac{\partial L}{\partial \dot{q}_i}(q, \dot{q}) \tag{1-6}$$

Assume that we are able to invert this equation, i.e., to express the velocities in terms of the momenta and the positions. We shall see later what happens if the jacobian of this transformation vanishes. Hamilton's function is then obtained by a Legendre transformation

$$H(p, q) = p_i \dot{q}_i(p, q) - L[q, \dot{q}(p, q)] \tag{1-7}$$

where summation over dummy indices is understood. By differentiation

$$dH = \left[\dot{q}_i + \frac{\partial \dot{q}_j}{\partial p_i}\left(p_j - \frac{\partial L}{\partial \dot{q}_j}\right)\right] dp_i + \left[-\frac{\partial L}{\partial q_i} - \frac{\partial \dot{q}_j}{\partial q_i}\left(\frac{\partial L}{\partial \dot{q}_j} - p_j\right)\right] dq_i$$

We see, using the expression of conjugate momenta, that the Euler-Lagrange equations take the form

$$\dot{q}_i = \frac{\partial H}{\partial p_i} \qquad \dot{p}_i = -\frac{\partial H}{\partial q_i} \tag{1-8}$$

More generally the variation of a function defined on phase space (the space of the $2N$ variables p, q) is

$$\frac{df}{dt} = \frac{\partial f}{\partial t} + \frac{\partial H}{\partial p_i}\frac{\partial f}{\partial q_i} - \frac{\partial H}{\partial q_i}\frac{\partial f}{\partial p_i} = \frac{\partial f}{\partial t} + \{H, f\} \tag{1-9}$$

We have introduced the Poisson bracket notation

$$\{f, g\} = \frac{\partial f}{\partial p_i}\frac{\partial g}{\partial q_i} - \frac{\partial f}{\partial q_i}\frac{\partial g}{\partial p_i} \tag{1-10}$$

It follows from (1-9) that a function without an explicit dependence on time, and such that its Poisson bracket with H vanishes, is a constant of the motion.

It is remarkable that Hamilton's Eqs. (1-8) also follow from a principle of stationarity. Indeed, let us insert

$$L(q, \dot{q})\, dt = p\, dq - H(p, q)\, dt$$

in (1-1) so that the action reads

$$I = \int_{t_1}^{t_2} p\, dq - H\, dt \tag{1-11}$$

and can be thought of as a functional of $2N$ independent functions $q(t)$ and $p(t)$. Assume again that $q(t_1) = q_1$, $q(t_2) = q_2$ without any restriction on $p(t_1)$ or $p(t_2)$. We then have

$$\delta I = \int_{t_1}^{t_2} \left[\delta p\left(\dot{q} - \frac{\partial H}{\partial p}\right) + \left(p\frac{d}{dt}\delta q - \frac{\partial H}{\partial q}\delta q\right)\right] dt$$

Integrating by parts the $p(d/dt)\,\delta q$ term we obtain

$$\frac{\delta I}{\delta p(t)} = \dot{q} - \frac{\partial H}{\partial p} \qquad -\frac{\delta I}{\delta q(t)} = \dot{p} + \frac{\partial H}{\partial q}$$

By setting the variation $\delta I = 0$ we therefore recover Eqs. (1-8). Note that the action I can also be written

$$I = p_2 q_2 - p_1 q_1 - \int_{t_1}^{t_2} q\, dp + H\, dt$$

Hence p's and q's play a similar role in the action.

Let us now compute the action I along the stationary trajectory. We assume (q_2, t_2) close enough to (q_1, t_1) so that the latter is unique and use the notation $I(q_2, t_2; q_1, t_1)$ or even the shorter form $I(2, 1)$.

It is by no means infrequent to find several trajectories joining two points in a given time. For an harmonic oscillator, for instance, infinitely many trajectories return to the origin after half a period.

We can easily verify that

$$\delta I\,(q_2, t_2; q_1, t_1) = (p_2\, \delta q_2 - H_2\, \delta t_2) - (p_1\, \delta q_1 - H_1\, \delta t_1) \tag{1-12}$$

where

$$p_2 = \frac{\partial I(2,1)}{\partial q_2} \qquad H_2 = -\frac{\partial I(2,1)}{\partial t_2}$$

If the Lagrange function depends explicitly on some parameter α, we have, furthermore,

$$\frac{d}{d\alpha} I(q_2, t_2; q_1, t_1) = \int_{t_1}^{t_2} dt\, \frac{\partial L}{\partial \alpha}\,[\alpha, Q(t, \alpha), \dot{Q}(t, \alpha)] \tag{1-13}$$

where the derivative on the right-hand side is taken on the explicit dependence of L on α. As an application of this result, let us add to a given Lagrange function a term $qF(t)$ which will contribute an external driving force $F(t)$ to the equations of motion. The action will become a functional of $F(t)$, and for $t_1 < t < t_2$ we obtain from (1-13)

$$\frac{\delta I(2,1)}{\delta F(t)} = Q(t) \tag{1-14}$$

where $Q(t)$ stands for the trajectory in the presence of F (we can also evaluate both sides for $F = 0$). This is not a priori the best method to solve the equations of motion, but it illustrates the method of generating functions, a procedure occurring frequently in the following.

Consider, for instance, a particle of mass m constrained to move along a one-dimensional axis under the action of a given force $F(t)$. Its Lagrange function is

$$L(q, \dot{q}) = \frac{m\dot{q}^2}{2} + qF(t) \tag{1-15}$$

The solution of the equation of motion can be written

$$Q(t) = \frac{q_1(t_2 - t) + q_2(t - t_1)}{t_2 - t_1} + \int_{t_1}^{t_2} dt'\, G(t, t')\, \frac{F(t')}{m} \tag{1-16}$$

where the Green function $G(t, t')$, symmetric in the interchange of t and t' and vanishing at $t = t_1$ and $t = t_2$, is a solution of

$$\frac{\partial^2}{\partial t^2} G(t, t') = \delta(t - t') \tag{1-17}$$

Here $\delta(t - t')$ is Dirac's point distribution, the derivative of the step function $\theta(t - t')$. We find

$$G(t, t') = -\frac{1}{t_2 - t_1}\,[(t_2 - t)(t' - t_1)\theta(t - t') + (t_2 - t')(t - t_1)\theta(t' - t)] \tag{1-18}$$

The value of the action along the trajectory reads

$$I(q_2, t_2; q_1, t_1) = \frac{m}{2}\frac{(q_2 - q_1)^2}{t_2 - t_1} + \int_{t_1}^{t_2} dt\, \frac{q_1(t_2 - t) + q_2(t - t_1)}{t_2 - t_1} F(t)$$

$$+ \frac{1}{2m}\int_{t_1}^{t_2} dt \int_{t_1}^{t_2} dt'\, F(t)G(t, t')F(t') \tag{1-19}$$

We verify that the functional derivative with respect to F yields the trajectory. If we set $F = 0$ we also recover

$$\frac{\partial I(2, 1)}{\partial q_2} = m\frac{q_2 - q_1}{t_2 - t_1} = p_2 \qquad \frac{\partial I(2, 1)}{\partial t_2} = -\frac{m}{2}\left(\frac{q_2 - q_1}{t_2 - t_1}\right)^2 = -H(p_2, q_2) \tag{1-20}$$

As a second example let us recall how this formalism applies to a free relativistic particle. Let $\mathbf{x}$ and $\mathbf{v}$ be the position and velocity of the particle written q and $\dot{q}$ up to now. The space-time interval reads

$$\Delta s^2 = c^2\Delta t^2 - \Delta\mathbf{x}^2 = c^2\Delta\tau^2 \tag{1-21}$$

where τ is the proper time along the trajectory $d\tau = (1 - \mathbf{v}^2/c^2)^{1/2}\, dt = \gamma^{-1} dt$ and c is the velocity of light. In four-vector notations

$$x^\mu \equiv (ct, \mathbf{x}) \qquad \mu = 0, 1, 2, 3$$

and

$$\Delta s^2 = g_{\mu\nu}\,\Delta x^\mu\,\Delta x^\nu = \Delta x_\mu\,\Delta x^\mu$$

with the metric tensor

$$g_{\mu\nu} = g^{\mu\nu} = \begin{pmatrix} 1 & 0 & 0 & 0 \\ 0 & -1 & 0 & 0 \\ 0 & 0 & -1 & 0 \\ 0 & 0 & 0 & -1 \end{pmatrix} \qquad g_\mu{}^\nu = \delta_\mu{}^\nu$$

used to lower or raise the Lorentz indices. The interval Δs^2 is invariant under Poincaré transformations, combining translations (a) and homogeneous Lorentz transformations (Λ):

$$x^\mu \to x'^\mu = \Lambda^\mu{}_\nu x^\nu + a^\mu \qquad \Lambda^T g\Lambda = g \tag{1-22}$$

A Lorentz transformation can be factorized as a product of an ordinary rotation and a boost along a direction $\mathbf{n}$:

$$\mathbf{x}' = \mathbf{x} + \mathbf{n}[(\cosh\alpha - 1)\mathbf{x}\cdot\mathbf{n} - ct\sinh\alpha] \tag{1-23}$$

$$ct' = ct\cosh\alpha - \mathbf{n}\cdot\mathbf{x}\sinh\alpha$$

where $\mathbf{v} = c\tanh\alpha\mathbf{n}$ is the velocity of the moving frame. The four-velocity

$$u^\mu = \frac{dx^\mu}{d\tau} = \left(c\frac{dt}{d\tau}, \frac{d\mathbf{x}}{d\tau}\right) = \frac{dt}{d\tau}(c, \mathbf{v}) \tag{1-24}$$

is a time-like vector of constant length c. Its derivative with respect to proper time $d^2x^\mu/d\tau^2 = du^\mu/d\tau$, orthogonal to u^μ, is therefore space-like. For a free particle the four-momentum satisfies

$$p^\mu = mu^\mu = \left(\frac{E}{c}, \mathbf{p}\right) = \frac{m}{\sqrt{1 - \mathbf{v}^2/c^2}}(c, \mathbf{v})$$

$$p^2 = m^2c^2 = \frac{E^2}{c^2} - \mathbf{p}^2 \tag{1-25}$$

Note that $\mathbf{p} = E\mathbf{v}/c^2$. Let us now find the corresponding Lagrange function using as principles (1) the relativistic invariance of the action and (2) the occurrence of only first derivatives of the position in L. The last requirement is, of course, borrowed from nonrelativistic mechanics and means that position and velocity should be sufficient to prescribe the motion. It may be relaxed under more general circumstances to comply with the finite velocity of signals in a relativistic context. In the case of a free particle the irrelevance of the origin of space and time coordinates implies, furthermore, that L should only depend on first derivatives and that in the limit $|\mathbf{v}| \ll c$ we should recover $L = \frac{1}{2}m\mathbf{v}^2$ up to a total derivative in time.

It is clear that $L = a\, ds/dt$ satisfies (1) and (2) with the constant a given by the nonrelativistic limit. We thus have

$$L = -mc^2 \frac{d\tau}{dt} = -mc^2 \sqrt{1 - \frac{\mathbf{v}^2}{c^2}}$$
$$I = \int_{t_1}^{t_2} dt\, L = -mc \int_{s_1}^{s_2} ds \tag{1-26}$$

Momentum and energy follow from the general expressions

$$\mathbf{p} = \frac{\partial L}{\partial \mathbf{v}} = \frac{m\mathbf{v}}{\sqrt{1 - \mathbf{v}^2/c^2}} \qquad E = \mathbf{p}\cdot\mathbf{v} - L = \frac{mc^2}{\sqrt{1 - \mathbf{v}^2/c^2}}$$

and agree, of course, with (1-25). The free equation of motion is $d\mathbf{p}/dt = 0$, in such a way that the generalized force $f^\mu = dp^\mu/d\tau$ vanishes. The action along the trajectory

$$I(2,1) = -mc(s_2 - s_1) = -mc^2[c^2(t_2 - t_1)^2 - (\mathbf{x}_2 - \mathbf{x}_1)^2]^{1/2}$$

is a Lorentz scalar. The standard relations

$$-\frac{\partial I(2,1)}{\partial t_2} = E_2 = E \qquad \frac{\partial I(2,1)}{\partial \mathbf{x}_2} = \mathbf{p}_2 = \mathbf{p}$$

mean that the relativistic momentum $-\partial I/\partial x_\mu = p^\mu$ is a four-vector.

An alternative form of the action may be used:

$$I = -\int_1^2 d\tau \frac{m}{2}\left[\frac{dx(\tau)}{d\tau}\right]^2 \tag{1-27}$$

where τ is a priori an arbitrary parameter along the trajectory. The equation of motion obtained by requiring I to be stationary, $m\, d^2x^\mu/d\tau^2 = 0$, implies that τ is proportional to proper time and can be chosen such that $u^2 = (dx/d\tau)^2 = c^2$. In the following we shall set $c = 1$, unless otherwise stated.

One of the virtues of the Lagrange-Hamilton formalism is to suggest a wide class of transformations leaving the structure of the equations of motion invariant. These go beyond simple reparametrizations of the configuration space since they naturally mix positions and momenta. A transformation $(p, q) \leftrightarrow (p', q')$ (possibly time dependent) is canonical if there exists a function H' of p', q' (and perhaps t) such that in terms of the new variables the equations of motion also read

$$\dot{q}' = \frac{\partial H'}{\partial p'} \qquad \dot{p}' = -\frac{\partial H'}{\partial q'}$$

where indices have again been suppressed. A sufficient condition for this property is suggested by the least-action principle. We require that the differential forms $p'\, dq' - H'\, dt$ and $p\, dq - H\, dt$ differ at most by a total differential. In turn this means the equality of the external derivatives of these forms:

$$\sum_i dp_i \wedge dq_i - dH \wedge dt = \sum_i dp'_i \wedge dq'_i - dH' \wedge dt \tag{1-28}$$

This condition implies that functions on phase space have equal Poisson brackets expressed in terms of old (q, p) or new (q', p') variables:

$$\{f, g\} = \sum_i \left(\frac{\partial f}{\partial p_i} \frac{\partial g}{\partial q_i} - \frac{\partial f}{\partial q_i} \frac{\partial g}{\partial p_i} \right) = \sum_i \left(\frac{\partial f}{\partial p'_i} \frac{\partial g}{\partial q'_i} - \frac{\partial f}{\partial q'_i} \frac{\partial g}{\partial p'_i} \right) \tag{1-29}$$

From (1-28) we can also derive equations for H' which turn out to form an integrable system by virtue of (1-29):

$$\begin{aligned} \frac{\partial H'}{\partial p'_i} &= \sum_k \left[\frac{\partial p_k}{\partial p'_i} \left(\frac{\partial H}{\partial p_k} - \frac{\partial q_k}{\partial t} \right) + \frac{\partial q_k}{\partial p'_i} \left(\frac{\partial H}{\partial q_k} + \frac{\partial p_k}{\partial t} \right) \right] \\ \frac{\partial H'}{\partial q'_i} &= \sum_k \left[\frac{\partial p_k}{\partial q'_i} \left(\frac{\partial H}{\partial q_k} - \frac{\partial q_k}{\partial t} \right) + \frac{\partial q_k}{\partial q'_i} \left(\frac{\partial H}{\partial q_k} + \frac{\partial p_k}{\partial t} \right) \right] \end{aligned} \tag{1-30}$$

A glance at (1-29) and (1-30) enables us to recognize that a typical example of canonical transformation is given by the solution of the equations of motion where (q', p') correspond to initial data at time t_0 and (q, p) to the phase space position at time t. It is clear that H' vanishes in this case and that Poisson brackets are left invariant.

Assume that H is time independent. The $2N$ functions q', p' expressed in terms of q, p, and t are constants of motion. Eliminating t one obtains $2N - 1$ (local) constants of motion. Only a small number of them can be extended as well-defined functions over all phase space. A theorem due to Poincaré states that the latter are associated with symmetries of the motion.

Canonical transformations are, of course, not limited to solutions of the equations of motion. Their set generates an infinite group. Finally we remark that restricting (1-28) at fixed time and taking the Nth external power of both sides shows that canonical transformations preserve the measure $\Pi_i\, dp_i \wedge dq_i$ on phase space; this is Liouville's theorem.

1-1-2 Electromagnetic Field as an Infinite Dynamical System

Systems with infinitely many degrees of freedom are familiar in fluid mechanics, electromagnetism, solid-state physics, etc. We shall discuss here how the lagrangian formalism extends to the electromagnetic field.

To obtain the Lagrange function we start from the field equations in the presence of fixed external sources. We use Heaviside's units (with the Coulomb force given by $QQ'/4\pi r^2$) and take $c = 1$. In terms of the electric field $\mathbf{E}$, magnetic induction $\mathbf{B}$, charge and current density ρ and $\mathbf{j}$, Maxwell's equations read

$$\begin{aligned} &(a)\quad \operatorname{div} \mathbf{E} = \rho && (c)\quad \operatorname{div} \mathbf{B} = 0 \\ &(b)\quad \operatorname{curl} \mathbf{B} - \frac{\partial \mathbf{E}}{\partial t} = \mathbf{j} && (d)\quad \operatorname{curl} \mathbf{E} + \frac{\partial \mathbf{B}}{\partial t} = 0 \end{aligned} \tag{1-31}$$

Local charge conservation is expressed as

$$\frac{\partial \rho}{\partial t} + \operatorname{div} \mathbf{j} = 0 \tag{1-32}$$

Let us recall the physical interpretation of these equations in integral form:

(*a*) $\int_S \mathbf{E} \cdot d\mathbf{S} = \int_V \rho d^3x$; the flux of **E** through a closed surface equals the enclosed charge (Gauss' law).

(*b*) $\oint_C \mathbf{B} \cdot d\mathbf{x} = \int_S \left(\mathbf{j} + \frac{\partial \mathbf{E}}{\partial t}\right) d\mathbf{S}$; the circulation of **B** along a closed curve C bordering a surface S equals the flux through S of the sum of the usual current plus Maxwell's displacement current ($\partial \mathbf{E}/\partial t$).

(*c*) $\int_S \mathbf{B} \cdot d\mathbf{S} = 0$; the flux of **B** through a closed surface vanishes—magnetic charges (monopoles) are absent.

(*d*) $-\oint_C \mathbf{E} \cdot d\mathbf{x} = \int_S d\mathbf{S} \cdot \frac{\partial \mathbf{B}}{\partial t}$; a varying magnetic flux generates an electromotive force (Faraday's induction law).

It is convenient to use a compact notation for vector, tensor, etc., fields which exhibits relativistic covariance. With greek indices running from zero to three:

$$x^\mu = (t, \mathbf{x}) \qquad j^\mu = (\rho, \mathbf{j})$$

we write

$$F^{\mu\nu} = -F^{\nu\mu} = \begin{pmatrix} 0 & -E^1 & -E^2 & -E^3 \\ E^1 & 0 & -B^3 & B^2 \\ E^2 & B^3 & 0 & -B^1 \\ E^3 & -B^2 & B^1 & 0 \end{pmatrix} \tag{1-33}$$

while derivatives are abbreviated as $\partial/\partial x^\mu \equiv \partial_\mu$. We also define Levi-Civita's antisymmetric symbol $\varepsilon^{\mu\nu\rho\sigma}$ to be equal to 1 or -1 according to whether $(\mu\nu\rho\sigma)$ is an even or odd permutation of (0, 1, 2, 3) and zero otherwise. Note that $\varepsilon_{\mu\nu\rho\sigma} = -\varepsilon^{\mu\nu\rho\sigma}$. This symbol is used to transform an antisymmetric tensor into its dual, as for instance

$$\tilde{F}^{\mu\nu} = -\tilde{F}^{\nu\mu} = \tfrac{1}{2}\varepsilon^{\mu\nu\rho\sigma}F_{\rho\sigma} = \begin{pmatrix} 0 & -B^1 & -B^2 & -B^3 \\ B^1 & 0 & E^3 & -E^2 \\ B^2 & -E^3 & 0 & E^1 \\ B^3 & E^2 & -E^1 & 0 \end{pmatrix} \tag{1-34}$$

In words, $\tilde{F}$ is obtained from F by substituting $E \to B$, $B \to -E$, and

$$\tfrac{1}{2}\varepsilon^{\mu\nu\rho\sigma}\tilde{F}_{\rho\sigma} = -F^{\mu\nu}$$

Under Lorentz transformations F and $\tilde{F}$ transform as antisymmetric tensors. In particular, if we perform the boost (1-23) we obtain ($c = 1$)

$$
\begin{aligned}
\mathbf{E}' &= (\mathbf{E}\cdot\mathbf{n})\mathbf{n} + \frac{\mathbf{n}\times(\mathbf{E}\times\mathbf{n}) + \mathbf{v}\times\mathbf{B}}{\sqrt{1-\mathbf{v}^2}} \\
\mathbf{B}' &= (\mathbf{B}\cdot\mathbf{n})\mathbf{n} + \frac{\mathbf{n}\times(\mathbf{B}\times\mathbf{n}) - \mathbf{v}\times\mathbf{E}}{\sqrt{1-\mathbf{v}^2}}
\end{aligned}
\tag{1-35}
$$

where

$$\mathbf{n} = \frac{\mathbf{v}}{|\mathbf{v}|}$$

These explicit forms are not always useful. For instance, to find the relativistic invariants constructed from **E** and **B** it is easier to use tensor notations and observe that they can only be combinations of

$$
\begin{aligned}
F_{\mu\nu}F^{\mu\nu} &= -\tilde{F}_{\mu\nu}\tilde{F}^{\mu\nu} = -2(\mathbf{E}^2 - \mathbf{B}^2) \\
F_{\mu\nu}\tilde{F}^{\mu\nu} &= -4\mathbf{E}\cdot\mathbf{B}
\end{aligned}
\tag{1-36}
$$

Note the following identities:

$$\tilde{F}_{\mu\lambda}\tilde{F}^{\lambda\nu} = F_{\mu\lambda}F^{\lambda\nu} + \tfrac{1}{2}g_\mu{}^\nu F_{\alpha\beta}F^{\alpha\beta}$$

$$F^{\mu\nu}\tilde{F}_{\nu\rho} = g^\mu{}_\rho \mathbf{E}\cdot\mathbf{B}$$

With these notations and Einstein's summation convention, Maxwell's equations take the compact form

$$(a)\quad \partial_\mu F^{\mu\nu} = j^\nu \qquad (b)\quad \partial_\mu \tilde{F}^{\mu\nu} = 0 \tag{1-37}$$

while current conservation appears as a natural compatibility condition

$$\partial_\mu j^\mu = 0 \tag{1-38}$$

To proceed, we have to identify the space coordinate $\mathbf{x}$ with the index i of the various degrees of freedom, this correspondence being such that $\sum_i \to \int d^3x$. However, we are at first slightly embarrassed since only first-order derivatives with respect to time occur in Maxwell's equations if we are to identify $\mathbf{E}(\mathbf{x}, t)$ and $\mathbf{B}(\mathbf{x}, t)$ with the configuration variables. Furthermore, Lorentz invariance does not appear naturally in this formulation.

To overcome these difficulties let us first transform the equations into equivalent second-order ones. The trick is to introduce the four-potential A^μ by recognizing that the homogeneous set of equations (1-37b) is precisely the condition enabling one to write

$$F^{\mu\nu} = \partial^\mu A^\nu - \partial^\nu A^\mu \tag{1-39}$$

that is,

$$\mathbf{E} = -\nabla A^0 - \frac{\partial \mathbf{A}}{\partial t} \qquad \mathbf{B} = \operatorname{curl}\mathbf{A}$$

This is in general a local statement with a particular solution, in the vicinity of a point taken as origin, of the form

$$A^\mu(x) = -\int_0^1 d\lambda\, \lambda F^{\mu\nu}(\lambda x) x_\nu$$

Such a potential is, however, not uniquely defined by (1-39). It can be modified by the addition of a four-gradient. This is called a gauge transformation

$$A^\mu(x) \to A^\mu(x) + \partial^\mu \phi \tag{1-40}$$

For regular fields satisfying (1-37) $A^\mu(x)$ can in fact be defined over all space-time, leading to the equivalent form of Maxwell's equation

$$\Box A^\mu - \partial^\mu(\partial_\nu A^\nu) = j^\mu \tag{1-41}$$

where $\Box$ stands for the d'alembertian: $\Box \equiv \partial_\mu \partial^\mu \equiv (\partial/\partial t)^2 - \Delta$. Equation (1-41) is clearly not affected by a gauge transformation. Gauge arbitrariness of electrodynamics may appear sometimes annoying and sometimes a deep and far-reaching principle. Clever choices of gauge lead to interesting simplifications but can also destroy manifest covariance.

Our first goal has been reached in the sense that we now have second-order equations. Compact notations may, however, be misleading. Let us therefore have a closer look at these equations, which read explicitly

$$\rho = \Box A^0 - \frac{\partial}{\partial t}\left(\frac{\partial A^0}{\partial t} + \operatorname{div} \mathbf{A}\right) = -\Delta A^0 - \frac{\partial}{\partial t} \operatorname{div} \mathbf{A}$$

$$\mathbf{j} = \Box \mathbf{A} + \boldsymbol{\nabla}\left(\frac{\partial}{\partial t} A^0 + \operatorname{div} \mathbf{A}\right)$$

For some gauge-fixing conditions, such as $\operatorname{div} \mathbf{A} = 0$ (Coulomb gauge), no time derivatives appear in the first of these relations (Poisson's equation), which plays the role of a constraint. One may then solve for A^0:

$$A^0(t, \mathbf{x}) = \int \frac{d^3x'}{4\pi} \frac{\rho(t, x')}{|\mathbf{x} - \mathbf{x}'|}$$

The vector potential is now given by

$$\Box \mathbf{A} = \mathbf{j} - \boldsymbol{\nabla} \int \frac{d^3x'}{4\pi} \frac{(\partial \rho/\partial t)(t, x')}{|\mathbf{x} - \mathbf{x}'|}$$

The divergence of the right-hand side vanishes, of course, due to current conservation.

The above choice is sometimes used in order to get a lagrangian formulation as a first step toward quantization. It has the obvious drawback of breaking manifest covariance even though the underlying physical picture, corresponding to the elimination of some irrelevant degrees of freedom, may be appealing in specific instances such as bound-state problems.

We want, however, to write an action using a Lagrange function preserving the local character of the theory. This is why we shall not use this gauge with its instantaneous Coulomb potential. Locality is a deeply rooted physical principle emerging from the nineteenth century formulation of field theory. Its implications underlie most of the developments of relativistic field theory, and its verification down to very small space-time intervals (via dispersion relations, for instance) is unquestioned up to now.

The Lagrange function should therefore be expressed as a space integral over a density, the so-called lagrangian $\mathscr{L}(x)$, which in turn should depend only on the fields and on finitely many of their derivatives. We do not really claim that these fields are directly measurable quantities; this is obviously not the case for the gauge-dependent potential A_μ. Locally measurable quantities should, however, be expressed in terms of local combinations of the dynamical variables. We write

$$I = \int d^4x \mathscr{L}(x) \tag{1-42}$$

Most of the time we shall not specify the boundary conditions of this integral. Let us assume that it extends throughout space and that the fields vanish sufficiently fast at infinity to justify integration by parts.

We assume $A^\mu(x)$ to transform as a four-vector field and the lagrangian as a scalar density in order for the action to be a Lorentz invariant. We require, furthermore, that $\mathscr{L}$ depends only on the fields and their first derivatives and be modified at most by a divergence under a gauge transformation (1-40). This is a natural generalization of the analogous property of the Lagrange function and ensures the gauge invariance of the equations of motion.

In general, if a lagrangian depends on fields $\varphi_i(x)$ and their gradients $\partial_\mu \varphi_i(x)$, then under an infinitesimal variation $\delta\varphi_i(x)$ of the fields the action will change according to

$$\begin{aligned}\delta I &= \int d^4x \left\{ \frac{\partial \mathscr{L}(x)}{\partial \varphi_i(x)} \delta\varphi_i(x) + \frac{\partial \mathscr{L}(x)}{\partial[\partial_\mu \varphi_i(x)]} \delta[\partial_\mu \varphi_i(x)] \right\} \\ &= \int d^4x \left\{ \frac{\partial \mathscr{L}(x)}{\partial \varphi_i(x)} - \partial_\mu \frac{\partial \mathscr{L}}{\partial[\partial_\mu \varphi_i(x)]} \right\} \delta\varphi_i(x)\end{aligned} \tag{1-43}$$

Requiring the action to be stationary leads to the generalized Euler-Lagrange equations

$$\frac{\delta I}{\delta \varphi_i(x)} \equiv \frac{\partial \mathscr{L}(x)}{\partial \varphi_i(x)} - \partial_\mu \frac{\partial \mathscr{L}(x)}{\partial[\partial_\mu \varphi_i(x)]} = 0 \tag{1-44}$$

According to the rules set before, we have thus in the electromagnetic case to adjust the coefficients a, b, c, d, e in the lagrangian

$$\mathscr{L}(x) = \tfrac{1}{2}[a\partial_\mu A^\nu \partial^\mu A_\nu + b\partial_\mu A^\nu \partial_\nu A^\mu + c(\partial_\mu A^\mu)^2 + dA_\mu A^\mu + eA_\mu j^\mu]$$

in such a way that Eqs. (1-44) will coincide with (1-41). Up to an overall coefficient, a simple calculation yields

$$\mathscr{L}(x) = -\tfrac{1}{4}F_{\mu\nu}F^{\mu\nu} - j_\mu A^\mu + \frac{c}{2}[(\partial_\mu A^\mu)^2 - \partial_\mu A^\nu \partial_\nu A^\mu]$$

where $F_{\mu\nu}$ is used here as a shorthand notation for the expression of the electromagnetic tensor in terms of the potential. The coefficient c remains arbitrary and multiplies a divergence as was expected:

$$(\partial_\mu A^\mu)^2 - \partial_\mu A^\nu \partial_\nu A^\mu = \partial_\mu[A_\nu(g^{\mu\nu}\partial_\rho A^\rho - \partial^\nu A^\mu)]$$

This last term can therefore be omitted and we are led to the action

$$I = -\int d^4x(\tfrac{1}{4}F^2 + j \cdot A) = \int d^4x\left(\frac{\mathbf{E}^2 - \mathbf{B}^2}{2} - \rho A^0 + \mathbf{j} \cdot \mathbf{A}\right) \tag{1-45}$$

In the presence of a given external current this lagrangian is not gauge invariant. Under a gauge transformation (1-40) it picks an additional contribution which is a divergence by virtue of current conservation:

$$j_\mu \partial^\mu \phi = \partial^\mu(\phi j_\mu)$$

This fact explains the invariance of Maxwell's equations. We conclude that current conservation is a necessary and sufficient condition for the gauge invariance of the theory.

Since this question is of the utmost importance in the quantum case, it is interesting to present another point of view. Up to now we did not restrict the arbitrariness on A^μ. The structure of Eq. (1-41) naturally suggests the Lorentz condition

$$\partial_\mu A^\mu(x) = 0 \tag{1-46}$$

in which case (1-41) reduces to

$$\Box A^\mu = j^\mu \tag{1-47}$$

This is, of course, compatible with (1-46) and restricts severely the gauge arbitrariness, since now only those transformations $A^\mu \to A^\mu + \partial^\mu \phi$ with

$$\Box \phi = 0 \tag{1-48}$$

are allowed.

The Lorentz constraint can be incorporated in the formalism using a Lagrange multiplier λ to add a term $\lambda \int d^4x \frac{1}{2}(\partial \cdot A)^2$ to the action. In this way Maxwell's equations are modified and the new equations read

$$\Box A^\mu + (\lambda - 1)\partial^\mu \partial \cdot A = j^\mu \tag{1-49}$$

Taking the divergence of both sides we see that, for $\lambda \neq 0$,

$$\Box \partial_\mu A^\mu = 0 \tag{1-50}$$

Hence if $\partial_\mu A^\mu$ vanishes for large $|t|$ it will be guaranteed to vanish at all times, and (1-49) will therefore be equivalent to Maxwell's equations in the Lorentz gauge.

Let us summarize the logic:

1. $\mathscr{L} = -\frac{1}{4}F^2 - j \cdot A$. Gauge invariance $\leftrightarrow$ current conservation. Least-action principle $\to$ Maxwell's equations.

2. $\mathscr{L} = -\frac{1}{4}F^2 + \frac{1}{2}\lambda(\partial \cdot A)^2 - j \cdot A$ and $\partial \cdot j = 0$. Least-action principle $\rightarrow$ $\Box \partial_\mu A^\mu = 0$. Boundary conditions $\rightarrow$ Lorentz gauge $\rightarrow$ Maxwell's equations.

Simple remarks enable one to remember the signs in (1-45) irrespective of the conventions. The "kinetic" terms in $(\partial \mathbf{A}/\partial t)^2$ should appear with a factor $+\frac{1}{2}$. Moreover, since $\rho^0 A^0$ is a potential energy it has to occur with a minus sign. The rest follows from Lorentz invariance. Note also that in rationalized units $\int d^3x(\mathbf{E}^2/2)$ is an energy and therefore $\int d^4x(\mathbf{E}^2/2)$ has indeed the dimension of an action: energy $\times$ time.

1-1-3 Electromagnetic Interaction of a Point Particle

The four-vector current of a charged point particle is localized on its space-time trajectory $x^\mu(\tau)$. Therefore

$$j^\mu(\mathbf{y}, t) = e \frac{dx^\mu}{dt} \delta^3[\mathbf{y} - \mathbf{x}(\tau)]\,|_{t=x^0(\tau)} = e \int d\tau \frac{dx^\mu}{d\tau} \delta^4[y - x(\tau)] \tag{1-51}$$

with e denoting the conserved charged $\int d^3y\, j^0(t, \mathbf{y}) = e$. The current j^μ as expressed by (1-51) fulfills the continuity equation $\partial_\mu j^\mu = 0$.

For the combined system particle + field we simply add the actions

$$I = -\int d^4x\left(\frac{F^2}{4} + j \cdot A\right) - m \int ds = -\int d^4x \frac{F^2}{4} - e \int dx_\mu A^\mu[x(\tau)] - m \int ds$$

$$= \int d^4x\, \mathscr{L}_{\text{em}}(x) + \int dt(-m\sqrt{1 - \mathbf{v}^2} + e\mathbf{A} \cdot \mathbf{v} - eA^0) \tag{1-52}$$

Any charged particle acts as a source, thus modifying the surrounding field. As a first approximation we shall neglect this effect and assume A_μ fixed by external conditions. This allows us to drop the contribution of $\mathscr{L}_{\text{em}}$ in (1-52). Observe that we take in this case an opposite view to the one followed at the beginning of this discussion. Genuinely coupled systems, whether classical or quantum mechanical, are obviously difficult to study. It is a fair practice to first investigate simplified models before looking at the more realistic ones. In Sec. 1-3-2, we shall briefly investigate a classical coupled system. We are thus led to study the motion of a particle under the action of a given external field. Its Lagrange function is

$$L = -m\sqrt{1 - \mathbf{v}^2} + e(\mathbf{A} \cdot \mathbf{v} - A^0) \tag{1-53}$$

We see that its conjugate momentum $\mathbf{p}$ differs from the free value $m\mathbf{v}/\sqrt{1 - \mathbf{v}^2}$, since

$$\mathbf{p} = \frac{\partial L}{\partial \mathbf{v}} = \frac{m\mathbf{v}}{\sqrt{1 - \mathbf{v}^2}} + e\mathbf{A} \tag{1-54}$$

Hamilton's function is expressed as

$$H = \mathbf{v} \cdot \frac{\partial L}{\partial \mathbf{v}} - L = \frac{m}{\sqrt{1 - \mathbf{v}^2}} + eA^0 = [m^2 + (\mathbf{p} - e\mathbf{A})^2]^{1/2} + eA^0 \tag{1-55}$$

while the equation of motion $(d/dt)(\partial L/\partial \mathbf{v}) = \partial L/\partial \mathbf{x}$ takes the form

$$\frac{d}{dt}\frac{m\mathbf{v}}{\sqrt{1-\mathbf{v}^2}} = -e\frac{\partial \mathbf{A}}{\partial t} - e\nabla A^0 + e\mathbf{v}\times \operatorname{curl}\mathbf{A} = e(\mathbf{E}+\mathbf{v}\times\mathbf{B}) \tag{1-56}$$

which is nothing but the Lorentz force law. Finally the variation of energy $\mathscr{E}$ is

$$\frac{d\mathscr{E}}{dt} = \frac{d}{dt}\left(\frac{m}{\sqrt{1-\mathbf{v}^2}}\right) = \mathbf{v}\frac{d}{dt}\frac{m\mathbf{v}}{\sqrt{1-\mathbf{v}^2}} = e\mathbf{E}\cdot\mathbf{v} \tag{1-57}$$

a well-known result expressing the fact that only electric fields do work. Equations (1-56) and (1-57) can be written compactly as

$$m\frac{du^\mu}{d\tau} = eF^{\mu\nu}u_\nu \tag{1-58}$$

with $u^\mu = dx^\mu/d\tau$ being the four-velocity.

The reader can verify that the variational principle applied to the action $I = -\int d\tau\, m(u^2/2) - \int eA\cdot dx$ leads directly to (1-58).

Let us now study three simple examples:

1. *Motion in a constant uniform field*
Let $F_{\mu\nu}$ be independent of x. Equations (1-56) and (1-57) lead at once to

$$\frac{m\mathbf{v}}{\sqrt{1-v^2}} - \frac{m\mathbf{v}_0}{\sqrt{1-v_0^2}} = e(\mathbf{E}t + \mathbf{x}\times\mathbf{B}) \qquad \mathscr{E}-\mathscr{E}_0 = e\mathbf{E}\cdot\mathbf{x} \tag{1-59}$$

We can also solve directly the covariant Eq. (1-58) in matrix form:

$$u^\mu(\tau) = \left(\exp\frac{e\tau}{m}F\right)^\mu{}_\nu u^\nu(0)$$

Finally it is instructive to use a spinorial representation. For a four-vector u we introduce the 2×2 matrix:

$$\underset{\sim}{u} = u_0 I + \mathbf{u}\cdot\boldsymbol{\sigma}$$

where $\boldsymbol{\sigma}$ stand for the Pauli matrices. Using the identity $\boldsymbol{\sigma}\cdot\mathbf{a}\,\boldsymbol{\sigma}\cdot\mathbf{b} = (\mathbf{a}\cdot\mathbf{b})I + i\boldsymbol{\sigma}\cdot(\mathbf{a}\times\mathbf{b})$ we rewrite (1-58) as

$$\frac{d}{d\tau}\underset{\sim}{u} = \frac{e}{m}\left(\frac{\mathbf{E}+i\mathbf{B}}{2}\cdot\boldsymbol{\sigma}\underset{\sim}{u} + \underset{\sim}{u}\frac{\mathbf{E}-i\mathbf{B}}{2}\cdot\boldsymbol{\sigma}\right) \tag{1-60}$$

from which it follows that

$$\underset{\sim}{u}(\tau) = \exp\left(\frac{e\tau}{m}\frac{\mathbf{E}+i\mathbf{B}}{2}\cdot\boldsymbol{\sigma}\right)\underset{\sim}{u}(0)\exp\left(\frac{e\tau}{m}\frac{\mathbf{E}-i\mathbf{B}}{2}\cdot\boldsymbol{\sigma}\right)$$

If $\mathbf{n}$ stands for the complex three-vector $\mathbf{E}+i\mathbf{B}$ and $a = (e/2m)(\mathbf{n}^2)^{1/2}$ we find that

$$\exp\left[\frac{e\tau}{2m}(\mathbf{E}+i\mathbf{B})\cdot\boldsymbol{\sigma}\right] = \cosh(a\tau)I + \sinh(a\tau)\frac{\mathbf{n}\cdot\boldsymbol{\sigma}}{(\mathbf{n}^2)^{1/2}}$$

In the frame where $\underset{\sim}{u}(0) = I$ and $\underset{\sim}{x}(0) = 0$ we obtain in the case where $\mathbf{E}\cdot\mathbf{B} = 0$, i.e., when the parameter a is real,

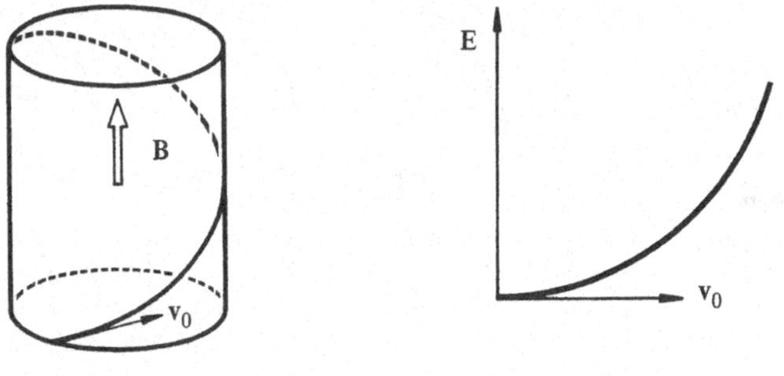

Figure 1-1 Motion in a constant uniform field. (*a*) $\mathbf{E} = 0$. (*b*) $\mathbf{B} = 0$, $\mathbf{E} \cdot \mathbf{v}_0 = 0$.

$$\underset{\sim}{u}(\tau) = \left[\cosh^2(a\tau) + \sinh^2(a\tau)\frac{\mathbf{E}^2 + \mathbf{B}^2}{\mathbf{n}^2}\right]I + \left[2\sinh(a\tau)\cosh(a\tau)\frac{\mathbf{E}}{(\mathbf{n}^2)^{1/2}} + 2\sinh^2(a\tau)\frac{\mathbf{E}\times\mathbf{B}}{\mathbf{n}^2}\right]\cdot\boldsymbol{\sigma}$$

$$\underset{\sim}{x}(\tau) = \left[\tau + \left(\frac{\sinh(2a\tau)}{4a} - \frac{\tau}{2}\right)\left(1 + \frac{\mathbf{E}^2 + \mathbf{B}^2}{\mathbf{n}^2}\right)\right]I + \left[\frac{\cosh(2a\tau) - 1}{2a}\frac{\mathbf{E}}{(\mathbf{n}^2)^{1/2}} + \frac{\sinh(2a\tau) - 2a\tau}{2a}\frac{\mathbf{E}\times\mathbf{B}}{\mathbf{n}^2}\right]\cdot\boldsymbol{\sigma} \tag{1-61}$$

In the limiting case where $\mathbf{n}^2 = 0$, that is, $\mathbf{E}\cdot\mathbf{B} = 0$, $\mathbf{E}^2 - \mathbf{B}^2 = 0$, we find

$$u_0(\tau) = 1 + \frac{e^2\mathbf{E}^2}{m^2}\frac{\tau^2}{2} \qquad \mathbf{u}(\tau) = \tau\frac{e\mathbf{E}}{m} + \frac{e^2}{m^2}\frac{\tau^2}{2}\,\mathbf{E}\times\mathbf{B}$$

Note that the velocity grows faster in the $\mathbf{E}\times\mathbf{B}$ direction. If $\mathbf{E}\cdot\mathbf{B}$ vanishes but $\mathbf{E}^2 - \mathbf{B}^2 \neq 0$ we can in fact pick a frame where either $\mathbf{E}$ or $\mathbf{B}$ vanishes. If $\mathbf{E} = 0$ the particle describes a helix with an axis parallel to $\mathbf{B}$, at constant angular velocity $\omega = (eB/m)\sqrt{1 - \mathbf{v}^2}$. If $\mathbf{B}$ vanishes and $\mathbf{v}_0\cdot\mathbf{E} = 0$ the trajectory is a "catenary" in the $(\mathbf{v}_0, \mathbf{E})$ plane with its concavity in the $e\mathbf{E}$ direction (this catenary reduces to a parabola in the nonrelativistic limit) (see Fig. 1-1).

2. *Gyromagnetic ratio. Thomas precession. Bargmann-Michel-Telegdi equation*

We introduce the concept of intrinsic magnetic moment and gyromagnetic ratio in this classical picture. This is a subject of controversy if one requires a consistent theory. Suffice it to say that we simply consider here a useful limiting situation of a complete quantum treatment.

Recall that an elementary current loop is equivalent to a magnetic moment $d\boldsymbol{\mu} = i\,d\boldsymbol{\Sigma}$, where i is the current and $d\boldsymbol{\Sigma} = \mathbf{n}\,d\Sigma$ stands for a vector normal to the plane of the loop of magnitude equal to the area. When the current is generated by a nonrelativistic orbiting charge, $i = ev/2\pi r$:

$$\boldsymbol{\mu} = \frac{e}{2m}\mathbf{L} \tag{1-62}$$

where $\mathbf{L} = \mathbf{r}\times\mathbf{p}$ is the orbital angular momentum. Equivalently in a homogeneous external magnetic induction $\mathbf{B}$ the interaction part of the Lagrange function can be given the form

$$L_{\text{int}} = e\mathbf{A}\cdot\mathbf{v} = \frac{e}{2}(\mathbf{B}\times\mathbf{r})\cdot\mathbf{v} = \boldsymbol{\mu}\cdot\mathbf{B} \tag{1-63}$$

From the Lorentz force law

$$\frac{d\mathbf{L}}{dt} = \mathbf{r} \times (e\mathbf{v} \times \mathbf{B}) = \frac{e}{2m}\mathbf{L} \times \mathbf{B} \tag{1-64}$$

we learn that both the angular momentum and the magnetic moment precess around the magnetic field with the classical Larmor frequency $\omega = eB/2m$.

In 1926 Uhlenbeck and Goudsmit introduced the idea of electron spin (intrinsic angular momentum) and magnetic moment, with $\boldsymbol{\mu} = g(e/2m)\mathbf{S}\,(|\mathbf{S}| = \hbar/2$; in modern language, spin $\frac{1}{2}$) and g (reduced gyromagnetic ratio) equal to 2, to reproduce the Zeeman splitting. Unfortunately this same value, $g = 2$, seemed to produce, for an electron moving in a central potential, a spin-orbit coupling twice as large as the one required by the fine structure of hydrogen. Let us assume that

$$\frac{d\mathbf{S}}{dt} = \boldsymbol{\mu} \times \mathbf{B}_{\text{rest}} \tag{1-65}$$

holds in the rest frame of the electron. If $\mathbf{E}$ and $\mathbf{B}$ are the fields in the laboratory frame where the electron has velocity $\mathbf{v}$, then from (1-35) $\mathbf{B}_{\text{rest}} = \mathbf{B} - \mathbf{v} \times \mathbf{E} + O(v^2)$. Therefore, in the small velocity approximation the magnetic interaction energy due to spin would be

$$U' = -\boldsymbol{\mu}\cdot(\mathbf{B} - \mathbf{v} \times \mathbf{E}) \tag{1-66}$$

If the electric field of an atomic nucleus is taken to derive from a spherical mean potential

$$e\mathbf{E} = -\frac{\mathbf{r}}{r}\frac{dV}{dr} \tag{1-67}$$

the above energy reads

$$U' = -\frac{ge}{2m}\mathbf{S}\cdot\mathbf{B} + \frac{g}{2m^2}\mathbf{S}\cdot\mathbf{L}\frac{1}{r}\frac{dV}{dr} \tag{1-68}$$

The catch, as first shown by Thomas, comes from the incorrect treatment of Lorentz transformations due to rotational motion. In other words, a pure boost from the laboratory inertial frame leads to a frame rotating with angular velocity ω_T so that the correct energy must be written

$$U = U' - \mathbf{S}\cdot\boldsymbol{\omega}_T \tag{1-69}$$

This is a typical relativistic effect which can be obtained by considering the product of the two boosts of velocities $-\mathbf{v}$ and $\mathbf{v} + \delta\mathbf{v}$. Here $\delta\mathbf{v} = \dot{\mathbf{v}}\,\delta t$. As any Lorentz transformation this product can be factorized into a boost of velocity $\Delta\mathbf{v}$ given by

$$\Delta\mathbf{v} = \frac{1}{\sqrt{1 - \mathbf{v}^2}}\left[\delta\mathbf{v} + \left(\frac{1}{\sqrt{1 - \mathbf{v}^2}} - 1\right)\mathbf{v}\frac{\mathbf{v}\cdot\delta\mathbf{v}}{\mathbf{v}^2}\right]$$

and a three-dimensional rotation

$$\Delta\boldsymbol{\omega} = \boldsymbol{\omega}_T\,\delta t = \left(\frac{1}{\sqrt{1 - \mathbf{v}^2}} - 1\right)\frac{\mathbf{v} \times \delta\mathbf{v}}{\mathbf{v}^2} \simeq \tfrac{1}{2}\mathbf{v} \times \dot{\mathbf{v}}\,\delta t$$

In the present case

$$\dot{\mathbf{v}} \simeq -\frac{\mathbf{r}}{mr}\frac{dV}{dr} \qquad \boldsymbol{\omega}_T \simeq \frac{\mathbf{L}}{2m^2}\frac{1}{r}\frac{dV}{dr} \tag{1-70}$$

and

$$U = -\frac{ge}{2m}\mathbf{S}\cdot\mathbf{B} + \left(\frac{g-1}{2m^2}\right)\mathbf{S}\cdot\mathbf{L}\frac{1}{r}\frac{dV}{dr}$$

If $g = 2$ this effect indeed reduces by half the spin-orbit coupling in agreement with observation. Dirac's theory (Chap. 2) leads naturally to this g value for an electron or muon. But of course other spin $\frac{1}{2}$ particles, such as the proton ($g = 5.59$) or the neutron ($g = -3.83$), have very different g values due to internal structure.

Let us return to the motion of a spinning particle in a constant magnetic field at small velocities. The velocity precesses with an angular frequency eB/m while for spin the corresponding frequency is $geB/2m$. In one period the relative phase will therefore be $2\pi(g/2 - 1)eB/m$. Bargmann, Michel, and Telegdi have obtained a relativistic description of this classical motion of spin valid in slowly varying external fields.

We represent the spin degrees of freedom by a three-vector $\mathbf{S}$ in the rest frame of the particle. In covariant notations it is therefore described by a space-like four-vector S^μ orthogonal to the velocity u^μ. We want to generalize the equation $d\mathbf{S}/dt = g(e/2m)\mathbf{S} \times \mathbf{B}$ valid in the rest frame where $S^\mu = (0, \mathbf{S})$. To preserve the condition $S \cdot u = 0$, we should have $S \cdot \dot{u} + \dot{S} \cdot u = 0$ using dots for derivatives with respect to proper time. In the instantaneous rest frame, $u = (1, \mathbf{0})$; therefore $\dot{S}^0 = \mathbf{S} \cdot \dot{\mathbf{u}}$ and $\dot{S}^\mu = [\mathbf{S} \cdot \dot{\mathbf{u}}, (ge/2m)\mathbf{S} \times \mathbf{B}]$.

To express this in any frame, we observe that $F^\mu{}_\nu S^\nu$ reduces to $(\mathbf{E} \cdot \mathbf{S}, \mathbf{S} \times \mathbf{B})$ in the rest frame. Therefore the required equation is

$$\dot{S}^\mu = \frac{g}{2}\frac{e}{m} F^\mu{}_\nu S^\nu + \frac{e}{m}\left(\frac{g}{2} - 1\right) u^\mu (S^\alpha F_{\alpha\beta} u^\beta) \tag{1-71}$$

From (1-71) it is clear that if $g = 2$, S and u move rigidly. This is not the case when $g \neq 2$. Thus we derive a method to measure the particle's magnetic anomaly $g/2 - 1$.

Define in the laboratory plane with time axis $\hat{t} = (1, \mathbf{0})$ the four-vector L with $\mathbf{L}$ along $\mathbf{u}$ and $L \cdot u = 0$, that is, $L = (u \cosh \varphi - \hat{t})/\sinh \varphi$, $\cosh \varphi = u \cdot \hat{t} = u^0$. Furthermore, let M be a unit four-vector in the (S, L) plane, orthogonal to u and L. We set $S = L \cos \theta + M \sin \theta$ (see Fig. 1-2). Let us find from (1-71) the equation of motion for θ:

$$\begin{aligned}\dot{S} &= \dot{\theta}(-L \sin \theta + M \cos \theta) + \dot{L} \cos \theta + \dot{M} \sin \theta \\ &= \frac{g}{2}\frac{e}{m}(F \cdot L \cos \theta + F \cdot M \sin \theta) + \frac{e}{m}\left(\frac{g}{2} - 1\right) u(L \cdot F \cdot u \cos \theta + M \cdot F \cdot u \sin \theta)\end{aligned} \tag{1-72}$$

Taking the scalar product with M and using $L \cdot M = u \cdot L = u \cdot M = 0$ we find

$$-\dot{\theta} + M \cdot \dot{L} = \frac{g}{2}\frac{e}{m} M \cdot F \cdot L$$

Now $M \cdot \dot{L} = \coth \varphi \dot{u} \cdot M = (e/m) \coth \varphi M \cdot F \cdot u$, leading to

$$\dot{\theta} = \frac{e}{m}\left(\coth \varphi M \cdot F \cdot u - \frac{g}{2} M \cdot F \cdot L\right) \tag{1-73}$$

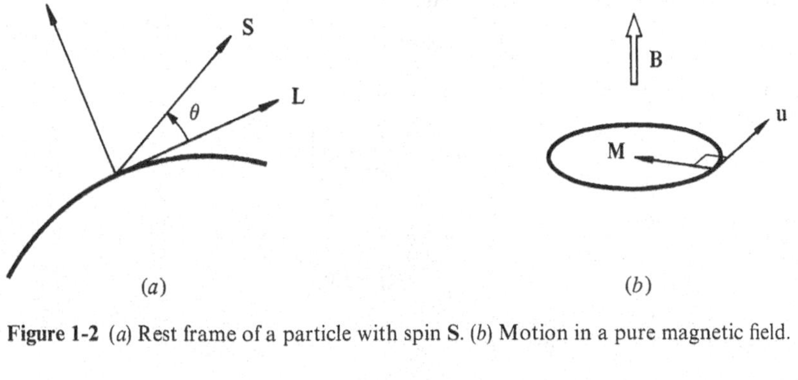

Figure 1-2 (*a*) Rest frame of a particle with spin **S**. (*b*) Motion in a pure magnetic field.

or explicitly

$$\dot{\theta} = \frac{e}{m}\left[\left(\frac{g}{2}\sinh^2\varphi - \cosh^2\varphi\right)\frac{\mathbf{E}\cdot\mathbf{M}}{\sinh\varphi} + \left(\frac{g}{2}-1\right)\coth\varphi(\mathbf{M}\times\mathbf{B})\cdot\mathbf{u}\right] \tag{1-74}$$

In the case of a pure magnetic field with the particle orbiting on a circle at the (proper) Larmor frequency $\omega = eB/m$, the mixed product $(\mathbf{u}\times\mathbf{M}\cdot\mathbf{B})$ equals $\sinh\varphi B$. Thus

$$\dot{\theta} = \left(\frac{g}{2}-1\right)\omega\cosh\varphi$$

$$\theta - \theta_0 = \left(\frac{g}{2}-1\right)\omega\tau\cosh\varphi$$

For a period $\Delta\tau = 2\pi/\omega$,

$$\Delta\theta\big|_{\text{period}} = \left(\frac{g}{2}-1\right)2\pi\cosh\varphi = \left(\frac{g}{2}-1\right)2\pi\frac{\mathscr{E}}{mc^2} \tag{1-75}$$

3. *Motion in a plane wave*

Finally, we study the motion of a charged spinless particle in the electromagnetic field of a plane wave, which we assume linearly polarized for simplicity. The wave is characterized by its light-like propagation vector n^μ and polarization ε^μ. These two vectors are such that

$$n^2 = \varepsilon\cdot n = 0 \qquad \varepsilon^2 = -1 \tag{1-76}$$

The potential depends on an arbitrary function of the variable $\xi = n\cdot x$:

$$A^\mu(x) = \varepsilon^\mu f(\xi) \tag{1-77}$$

From (1-76) it follows that $\partial\cdot A = 0$, while the field

$$F^{\mu\nu}(x) = (n^\mu\varepsilon^\nu - n^\nu\varepsilon^\mu)f'(\xi) \tag{1-78}$$

satisfies

$$F^\mu{}_\lambda F^{\lambda\nu} = \tilde{F}^\mu{}_\lambda\tilde{F}^{\lambda\nu} = n^\mu n^\nu f'(\xi)^2$$

$$F^\mu{}_\nu\tilde{F}^{\lambda\nu} = 0$$

$$F^2 = 0$$

Therefore $|\mathbf{E}| = |\mathbf{B}|$ and $\mathbf{E}\cdot\mathbf{B} = 0$. Since $n_\mu F^{\mu\nu} = n_\mu\tilde{F}^{\mu\nu} = 0$, the classical Lorentz equation

$$m\frac{du^\mu}{d\tau} = eF^\mu{}_\nu u^\nu \tag{1-79}$$

leads to $u\cdot n =$ constant. Choosing coordinates such that at $\tau = 0$, $x(0) = 0$,

$$\xi = x\cdot n = u(0)\cdot n\tau$$

The variable ξ can be used in place of τ in (1-79) in such a way that

$$\frac{du^\mu}{d\xi} = \frac{e}{m}f'(\xi)\left[n^\mu\frac{\varepsilon\cdot u}{n\cdot u(0)} - \varepsilon^\mu\right]$$

Multiplying by ε and integrating gives

$$u\cdot\varepsilon = u(0)\cdot\varepsilon + \frac{e}{m}[f(\xi) - f(0)]$$

which can be inserted in $du^\mu/d\xi$, leading to

$$\frac{du^\mu}{d\xi} = \frac{e}{m} f'(\xi)\left\{n_\mu \frac{\varepsilon \cdot u(0) + (e/m)[f(\xi) - f(0)]}{n \cdot u(0)} - \varepsilon_\mu\right\} \tag{1-80}$$

A last integration gives $u(\xi)$ as

$$u^\mu(\xi) = u^\mu(0) + \frac{e}{m}[f(\xi) - f(0)]\left[n^\mu \frac{\varepsilon \cdot u(0)}{n \cdot u(0)} - \varepsilon^\mu\right] + \frac{e^2}{2m^2}[f(\xi) - f(0)]^2 \frac{n^\mu}{n \cdot u(0)} \tag{1-81}$$

If we assume that the wave is damped at large ξ, we find

$$u^\mu(\infty) = u^\mu(0) + \frac{e}{m} f(0)\varepsilon^\mu - \left[\frac{e}{m} f(0) \frac{\varepsilon \cdot u(0)}{n \cdot u(0)} - \frac{e^2}{2m^2} \frac{f^2(0)}{n \cdot u(0)}\right] n^\mu \tag{1-82}$$

We note the appearance of nonlinear effects (terms in f^2).

For a monochromatic plane wave such as $f(\xi) = a \sin \xi$, for instance, we can compute x^μ in a straightforward way:

$$x^\mu(\xi) = u^\mu(0)\tau - \frac{2e}{m} a \frac{\sin^2 \xi/2}{n \cdot u(0)} \varepsilon^\mu + \frac{e}{m} \frac{n^\mu}{[n \cdot u(0)]^2}\left[2a\varepsilon \cdot u(0) \sin^2 \frac{\xi}{2} + \frac{ea^2}{4m}(\xi - \tfrac{1}{2} \sin 2\xi)\right] \tag{1-83}$$

The action is expressed as

$$\begin{aligned} I(\xi) &= -\int_0^\tau d\tau'(m + eA \cdot u) \\ &= -m\tau - \frac{e}{n \cdot u(0)} \int_0^\xi d\xi' \, f(\xi')\left\{\varepsilon \cdot u(0) + \frac{e}{m}[f(\xi') - f(0)]\right\} \\ &= -mu(\infty) \cdot x - \int_0^\xi \frac{d\xi'}{n \cdot u}\left[eA \cdot u(\infty) - \frac{e^2 A^2}{2m}\right] \end{aligned} \tag{1-84}$$

In the last form we have used $u(\infty)$ defined by (1-82) under the assumption that $f(\xi)$ vanishes at infinity (recall that $n \cdot u$ is conserved). From (1-84) we can obtain the conjugate momentum as

$$\pi^\mu = -\frac{\partial I(\xi)}{\partial x_\mu} = mu^\mu(\infty) + \frac{n^\mu}{n \cdot u}\left[eA \cdot u(\infty) - \frac{e^2}{2m} A^2\right] \tag{1-85}$$

For certain potentials it is meaningful to define averages such that $\overline{A} = 0$ [for instance if $f(\xi) = e^{-\eta|\xi|} g(\xi)$ where g is periodic and $\eta \to 0$]. We then derive from (1-85) that

$$\overline{\pi^2} = m^2 - e^2\overline{A^2} = m^2 + e^2\,\overline{|A^2|} = m^2{}_{\text{eff}} \tag{1-86}$$

This last formula is to be interpreted with care because of the anisotropies arising from the plane wave. It shows, however, that a particle in a strong periodic field will respond to external perturbations with a larger inertia.

Linear accelerators provide a typical instance of such a motion of electrons in a traveling wave.

1-2 SYMMETRIES AND CONSERVATION LAWS

1-2-1 Fundamental Invariants

We return to systems with finitely many degrees of freedom. The main task is, of course, to solve the equations of motion, with appropriate boundary conditions. General properties of the motion such as symmetries are helpful since they simplify the calculations. They can also be used to restrict the class of dynamical models. We have already seen examples of this type with Lorentz invariance.

Symmetries may play a double role. They enable one to generate families of solutions from a given one if some transformations leave the dynamical equations invariant. Or they lead to the conservation of quantities such as charge, energy, momentum, etc. The deep connection between these two aspects is our present subject. We start from a very simple example. A nonrelativistic point particle moves in a force field deriving from a time-independent potential. The position and velocity at time t given initial conditions at time zero are, of course, the same as those at time $t + \tau$ if the same initial conditions were given at time τ. The problem is invariant under time translation. We also know that in such a case energy (i.e. the value of Hamilton's function) is conserved. Let us see how the two properties are related. Any function on phase space varies along the motion according to

$$\frac{df}{dt} = \frac{\partial f}{\partial t} + \{H, f\}$$

Time-translation invariance is equivalent to the statement $\partial H/\partial t = 0$. Since one also has $\{H, H\} = 0$ it indeed follows that

$$\frac{dH}{dt} = 0 \tag{1-87}$$

and energy is conserved. This simple remark is sufficient, for instance, to find the motion explicitly if the particle is restricted to move in one dimension.

Alternatively, consider the action computed along the stationary trajectory leading from (q_1, t_1) to (q_2, t_2). Invariance under time translation means

$$I(q_2, t_2 + \tau; q_1, t_1 + \tau) = I(q_2, t_2; q_1, t_1)$$

or in differential form

$$\frac{\partial I}{\partial t_2} - \frac{\partial I}{\partial t_1} = 0 \tag{1-88}$$

Taking into account Eq. (1-12) this is indeed

$$H_2 = H_1 \tag{1-89}$$

The conservation law clearly follows from the existence of a continuous invariance group. Similarly, for space translations

$$I(\{q_n + a\}_2, t_2; \{q_n + a\}_1, t_1) = I(\{q_n\}_2, t_2; \{q_n\}_1, t_1)$$

Recalling again Eq. (1-12) and differentiating with respect to a, we get the conservation law of total momentum

$$(\sum_n p_n)_2 = (\sum_n p_n)_1 \tag{1-90}$$

The previous examples can be thought of as special cases of the formula giving the variation of the stationary action when an external parameter α is changed, Eq. (1-13). Indeed, α could be chosen as the parameter characterizing the transformation.

If we also have rotational invariance, consider an infinitesimal transformation of angle $\delta\alpha$ around the axis $\mathbf{n}$:

$$\mathbf{q} \to \mathbf{q} + \delta\alpha\,\mathbf{n} \times \mathbf{q}$$

The very same formula $\delta I(2, 1) = p\,\delta q - H\,\delta t\,|_1^2$ leads to

$$\sum_n \mathbf{p}_n \cdot (\mathbf{n} \times \mathbf{q}_n)\,|_2 = \sum_n \mathbf{p}_n \cdot (\mathbf{n} \times \mathbf{q}_n)\,|_1$$

Since $\mathbf{n}$ is arbitrary we therefore find the conservation of total angular momentum as

$$\sum_n \mathbf{q}_n \times \mathbf{p}_n\,|_2 = \sum_n \mathbf{q}_n \times \mathbf{p}_n\,|_1 \tag{1-91}$$

Of course, if we only have rotational invariance around an axis the corresponding component of angular momentum will be the only one conserved.

In summary, when the dynamical problem admits a symmetry the stationary actions $I(2, 1)$ and $I(2', 1')$ are equal, where primes denote transformed boundary conditions. Whenever the transformations form a continuous group we obtain a conservation law by differentiating with respect to the group parameters.

However, symmetries need not be continuous. Such is the case for parity, time reversal, etc. For instance, in the latter case we have $I(q_2, t_2; q_1, t_1) = I(q_1, -t_1; q_2, -t_2)$ where boundary conditions are interchanged and time is reversed, corresponding to $p_2 \leftrightarrow -p_1$. Of course, this invariance does not lead to a conservation law.

We finally cast the above considerations in a form suitable for later generalizations. To be specific we return, for instance, to rotational invariance. Under an infinitesimal rotation $\mathbf{q}$ goes into $R\mathbf{q} = \mathbf{q} + \delta\alpha\,\mathbf{n} \times \mathbf{q}$. The Lagrange function is assumed invariant when R is time independent. However, the least-action principle allows us to choose variations around the stationary path $Q(t)$ of the form

$$\delta\mathbf{q}(t) = \delta\alpha(t)\,\mathbf{n} \times \mathbf{Q}(t) \tag{1-92}$$

provided $\delta\alpha(t_2) = \delta\alpha(t_1) = 0$. Using the invariance of L when $\delta\alpha$ is constant it follows that

$$\frac{\delta I(Q)}{\delta\alpha(t)} = -\frac{d}{dt}\frac{\partial L}{\partial\delta\dot{\alpha}(t)} = 0$$

Hence the conserved quantity is proportional to

$$\frac{\partial L}{\partial\delta\dot{\alpha}} = \sum_i \frac{\partial L}{\partial\dot{q}_i}\frac{\partial\delta\dot{q}_i}{\partial\delta\dot{\alpha}(t)} = \sum_i p_i \frac{\partial\delta q_i}{\partial\alpha} \tag{1-93}$$

since $\partial\dot{q}_i/\partial\delta\dot{\alpha} = \partial q_i/\partial\delta\alpha$. In the present case this is, of course, the component of angular momentum along the direction $\mathbf{n}$.

For time translations the above method is to be used with care. If the infinitesimal parameter $\delta\alpha$ is constant, initial and final times are displaced. We set $\delta q = \delta\alpha(t)\dot{q}$, $\delta\dot{q} = \delta\alpha\ddot{q} + \delta\dot{\alpha}\dot{q}$ with $\delta\alpha(t_1) = \delta\alpha(t_2) = 0$. In the neighborhood of the real trajectory

$$O = \delta I = \int_1^2 dt \left[\delta\alpha\left(\dot{q}\frac{\partial L}{\partial q} + \ddot{q}\frac{\partial L}{\partial \dot{q}}\right) + \delta\dot{\alpha}\frac{\partial L}{\partial \dot{q}}\dot{q}\right]$$

$$= \int_1^2 dt\, \delta\alpha \left[\frac{d}{dt}\left(L - \dot{q}\frac{\partial L}{\partial \dot{q}}\right) - \frac{\partial L}{\partial t}\right]$$

Invariance means generally that $\partial L/\partial t$ vanishes, in which case energy $H = p\dot{q} - L$ is conserved; more generally $(d/dt)H = -\partial L/\partial t$.

It may occur that the equations of motion are invariant but not the Lagrange function. Under an infinitesimal time-independent transformation, L is modified through the addition of a total derivative in time. In other words, $\partial L/\partial\delta\alpha = (d/dt)\phi$ and $(d/dt)(\partial L/\partial\delta\dot{\alpha} - \phi) = 0$ for a time-dependent $\delta\alpha$. Thus we find a conserved quantity which is not $\partial L/\partial\delta\dot{\alpha}$ any more and explicitly depends on time.

As an example, the dynamics of a particle moving under a constant force is translation invariant but momentum is, of course, not conserved. The Lagrange function is $L = \frac{1}{2}m\dot{\mathbf{q}}^2 + \mathbf{q}\cdot\mathbf{F}(t)$. Under a translation $\delta\mathbf{a}(t)$, $\delta L = \mathbf{F}\cdot\delta\mathbf{a} = (d/dt)\int^t dt'\, \mathbf{F}(t')\cdot\delta\mathbf{a}(t')$. The integral of the motion is therefore $m\dot{\mathbf{q}} - \int_{t_0}^t dt'\, \mathbf{F}(t') =$ constant in agreement with naive expectations. Note that even for a constant $\mathbf{F}$ this is explicitly time dependent. Physically it is, of course, impossible to create such a force throughout all space as required by translational invariance.

Extending these relations to infinite systems will not create difficulties. We shall distinguish two types of symmetries. The first one will correspond to geometrical transformations of space and time under which the lagrangian $\mathscr{L}(x)$ will go into $\mathscr{L}(x')$, where x' is the transformed point (Sec. 1-2-2). The second type will leave the lagrangian invariant and will be called internal (Sec. 1-2-3). Symmetries play such a fundamental role that we shall devote Chap. 11 to a deeper study.

1-2-2 Energy Momentum Tensor

For an infinite system we assume a lagrangian depending on the space-time coordinates x only through fields and their gradients. Under a translation we therefore have

$$\mathscr{L}(x + a) \equiv \mathscr{L}[\phi_i(x + a), \partial_\mu\phi_i(x + a)] \tag{1-94}$$

Consider an infinitesimal x-dependent transformation

$$\begin{aligned} \delta\phi_i &= \delta a^\mu(x)\partial_\mu\phi_i(x) \\ \delta\partial_\mu\phi_i(x) &= \delta a^\nu\partial_\nu\partial_\mu\phi_i(x) + \partial_\mu[\delta a^\nu(x)]\partial_\nu\phi_i(x) \end{aligned} \tag{1-95}$$

After an integration by parts the corresponding variation of the action is

$$\delta I = \int d^4x \left\{\partial_\nu\mathscr{L} - \partial_\mu\left[\frac{\partial\mathscr{L}}{\partial(\partial_\mu\phi_i)}\partial_\nu\phi_i(x)\right]\right\}\delta a^\nu(x) \tag{1-96}$$

We obtain in this way a generalization of our previous discussion of energy momentum in a completely local form. From the vanishing of δI for arbitrary $\delta a^\nu(x)$, we deduce that the energy momentum flow described by the canonical tensor

$$\tilde{\Theta}^{\mu\nu} = \frac{\partial \mathscr{L}}{\partial(\partial_\mu \phi_i)} \partial_\nu \phi_i - g_{\mu\nu} \mathscr{L} \tag{1-97}$$

satisfies the conservation law

$$\partial_\mu \tilde{\Theta}^{\mu\nu} = 0 \tag{1-98}$$

(From this derivation the two indices of $\tilde{\Theta}^{\mu\nu}$ play a dissimilar role.) It follows that the four quantities P^ν corresponding to total energy ($\nu = 0$) and three-momentum ($\nu = 1, 2, 3$):

$$P^\nu = \int d^3x \, \tilde{\Theta}^{0\nu}(\mathbf{x}, t) \tag{1-99}$$

are time independent, since

$$\dot{P}^\nu = \int d^3x \, \partial_0 \tilde{\Theta}^{0\nu}(\mathbf{x}, t) = -\int d^3x \sum_{i=1}^{3} \partial_i \tilde{\Theta}^{i\nu}(\mathbf{x}, t) = 0$$

provided that the fields vanish sufficiently rapidly for large arguments, i.e., no energy or momentum escapes at infinity.

This result is a typical case of Noether's theorem. The latter states that to any continuous one-parameter set of invariances of the lagrangian is associated a local conserved current. Integrating the fourth component of this current over three-space generates a conserved "charge." Invariance of the lagrangian means in this geometrical context that we also allow the possible transformation of the space-time argument as appears, for instance, in (1-94). Furthermore, the space integral defining the charge can be performed on a space-like surface σ with a surface element $d\sigma_\mu$ without affecting the result:

$$P^\nu = \int d\sigma_\mu \, \tilde{\Theta}^{\mu\nu} \tag{1-100}$$

It can, of course, happen that the lagrangian also depends explicitly on the coordinates x, in which case (1-94) is no longer valid and Eq. (1-98) is replaced by

$$\partial_\mu \tilde{\Theta}^{\mu\nu} = -\partial^\nu_{\text{explicit}} \mathscr{L} \tag{1-101}$$

Such is the case if $\mathscr{L}$ is a sum of an invariant part $\mathscr{L}_0$ plus a coupling term to external sources $j_i(x)$ of the form $\mathscr{L}_1 = \Sigma_i \, \phi_i(x) j_i(x)$. The energy momentum tensor gets a contribution $\tilde{\Theta}_0^{\mu\nu}$ from $\mathscr{L}_0$ plus an added term from $\mathscr{L}_1$ and reads

$$\tilde{\Theta}^{\mu\nu} = \tilde{\Theta}_0^{\mu\nu} - g^{\mu\nu} \mathscr{L}_1 \tag{1-102}$$

with

$$\partial_\mu \tilde{\Theta}^{\mu\nu} = -\sum_i \phi_i \partial^\nu j_i$$

according to (1-101). This last equation can be rewritten as

$$\partial_\mu \tilde{\Theta}_0^{\mu\nu} = \sum_i j_i(x)\, \partial^\nu \phi_i \tag{1-103}$$

These two ways of expressing the local variation of energy and momentum differ according to whether the interaction energy $-\int d^3x\, \mathscr{L}_1$ is included or not in the system.

Let us apply this to the electromagnetic field coupled to an external conserved current j where the lagrangian is given by

$$\mathscr{L} = -\tfrac{1}{4}F^2 - j \cdot A \tag{1-104}$$

and

$$\begin{aligned} \tilde{\Theta}^{\mu\nu} &= \tilde{\Theta}_0^{\mu\nu} + g^{\mu\nu} j \cdot A \\ \tilde{\Theta}_0^{\mu\nu} &= -F^{\mu\rho}\, \partial^\nu A_\rho + \tfrac{1}{4} g^{\mu\nu} F^2 \end{aligned} \tag{1-105}$$

The tensor $\tilde{\Theta}$ is not gauge invariant even if j is zero! Under a transformation $A \to A + \partial\phi$:

$$\tilde{\Theta}^{\mu\nu} \to \tilde{\Theta}^{\mu\nu} + \partial_\rho (g^{\mu\nu} j^\rho \phi - F^{\mu\rho}\, \partial^\nu \phi) - j^\mu\, \partial^\nu \phi \tag{1-106}$$

In the absence of external sources the added term is a divergence and will not contribute to the value of the total energy momentum if the fields vanish at infinity.

The energy momentum density is in principle measurable and, moreover, is coupled to the gravitational field. It is therefore very unpleasant to find a gauge-dependent expression. In addition, the antisymmetric part of $\tilde{\Theta}^{\mu\nu}$ is nonvanishing:

$$\tilde{\Theta}^{\mu\nu} - \tilde{\Theta}^{\nu\mu} = F^{\nu\rho}\, \partial^\mu A_\rho - F^{\mu\rho}\, \partial^\nu A_\rho \tag{1-107}$$

We know that the lagrangian is not entirely determined by the equations of motion. Correspondingly, the energy momentum tensor admits some arbitrariness. Generally speaking, let $k^\mu(x)$ denote a conserved current. Then the charge $K = \int d^3x\, k^0(\mathbf{x}, t)$ and the local conservation law are unchanged if k^μ goes into

$$k^\mu \to k^\mu + \Delta k^\mu \tag{1-108}$$

provided that

$$\partial_\mu\, \Delta k^\mu = 0 \tag{1-109}$$

and

$$\int d^3x\, \Delta k^0(\mathbf{x}, t) = 0 \tag{1-110}$$

A solution to these constraints is given by

$$\Delta k^\mu = \partial_\rho k^{\mu\rho} \tag{1-111}$$

where $k^{\mu\rho}$ is antisymmetric and depends locally on the fields. Indeed, (1-109) is verified and so is (1-110), since

$$\int d^3x\, \partial_\rho k^{0\rho} \equiv \int d^3x \sum_i^3 \partial_i k^{0i} = 0 \tag{1-112}$$

Returning to the electromagnetic energy momentum tensor, we see that we can add to the canonical $\tilde{\Theta}^{\mu\nu}$ a piece

$$\Delta\tilde{\Theta}^{\mu\nu} = \partial_\rho \Theta^{\mu\rho,\nu} \tag{1-113}$$

where $\Theta^{\mu\rho,\nu}$ is antisymmetric in (μ, ρ) and local in the fields. The discussion of the gauge dependence where

$$\Delta_{\text{gauge}}\tilde{\Theta}^{\mu\nu} = -\partial_\rho(F^{\mu\rho}\,\partial^\nu\phi) \tag{1-114}$$

suggests the compensating term

$$\Delta\tilde{\Theta}^{\mu\nu} = \partial_\rho(F^{\mu\rho}A^\nu) \tag{1-115}$$

We therefore set

$$\Theta^{\mu\nu} = \tilde{\Theta}^{\mu\nu} + \partial_\rho(F^{\mu\rho}A^\nu) = g^{\mu\nu}(\tfrac{1}{4}F^2 + j\cdot A) - F^{\mu\rho}\,\partial^\nu A_\rho + \partial_\rho(F^{\mu\rho}A^\nu) \tag{1-116}$$

Using Maxwell's equations, we get $\partial_\rho F^{\mu\rho} = -j^\mu$, so that an alternative expression for $\Theta^{\mu\nu}$ is

$$\Theta^{\mu\nu} = \tfrac{1}{4}g^{\mu\nu}F^2 + F^{\mu\rho}F_\rho{}^\nu + g^{\mu\nu}j\cdot A - j^\mu A^\nu \tag{1-117}$$

In the absence of sources this energy momentum density enjoys the following properties: it is gauge invariant, conserved, symmetric, and traceless. According to the general framework, if j is nonvanishing then

$$\partial_\mu\Theta^{\mu\nu} = \partial_\mu\tilde{\Theta}^{\mu\nu} = A_\rho\,\partial^\nu j^\rho \tag{1-118}$$

This can be also stated as

$$\partial_\mu\Theta_0^{\mu\nu} = \partial_\mu(\tfrac{1}{4}g^{\mu\nu}F^2 + F^{\mu\rho}F_\rho{}^\nu) = j_\rho F^{\rho\nu} \tag{1-119}$$

The new tensor Θ_0 can thus be identified with the pure electromagnetic contribution. The energy density $\Theta_0^{00} = \frac{1}{2}(\mathbf{E}^2 + \mathbf{B}^2)$ is positive and the momentum density $\Theta_0^{i0} = (\mathbf{E}\times\mathbf{B})^i$ is the familiar Poynting vector. We recover the known result that

$$-\frac{d}{dt}\int_V d^3x\,\Theta_0^{00} = \int_S d\mathbf{S}\cdot(\mathbf{E}\times\mathbf{B}) + \int d^3x\,\mathbf{E}\cdot\mathbf{j}$$

After these manipulations $\Theta^{\mu\nu}$ is still gauge dependent in the presence of sources. This should not be too surprising because the system is an open one. To see what happens in a closed system, taking the sources into account, consider a set of charged point particles interacting electromagnetically. The total action reads

$$I = -\int \tfrac{1}{4}F^2\,d^4x - \sum_n \int d\tau_n\left[\frac{m_n}{2}\,\dot{x}_n^2(\tau_n) + e_n\dot{x}_n(\tau_n)\cdot A(x_n)\right] \tag{1-120}$$

It is invariant under a space-time translation

$$A(x)\to A(x+a) \qquad x_n(\tau_n)\to x_n(\tau_n) - a \tag{1-121}$$

If we consider infinitesimal variations of the form:

$$\delta A^\mu(x) = \delta a^\nu(x)\,\partial_\nu A^\mu(x) \qquad \delta x_n^\mu = -\delta a^\mu(x_n) \tag{1-122}$$

we find

$$-\frac{\delta I}{\delta a_\nu(x)} = \partial_\mu \tilde{\Theta}^{\mu\nu} = 0 \tag{1-123}$$

where
$$\tilde{\Theta}^{\mu\nu}(x) = \tfrac{1}{4}g^{\mu\nu}F^2 - F^{\mu\rho}\,\partial^\nu A_\rho + \sum_n \int d\tau_n e_n \dot{x}^\mu(\tau_n)\,\delta^4[x - x_n(\tau_n)]\,A^\nu[x_n(\tau_n)]$$
$$+ \sum_n m_n \int d\tau_n \dot{x}_n^\mu(\tau_n)\dot{x}_n^\nu(\tau_n)\,\delta^4[x - x_n(\tau_n)] \tag{1-124}$$

Repeating the steps leading to a gauge-invariant tensor, i.e., adding $\partial_\rho(F^{\mu\rho}A^\nu)$, we get

$$\Theta^{\mu\nu} = \tfrac{1}{4}g^{\mu\nu}F^2 + F^{\mu\rho}F_\rho{}^\nu + \sum_n m_n \int d\tau_n \dot{x}_n^\mu \dot{x}_n^\nu\,\delta^4[x - x_n(\tau_n)] \tag{1-125}$$

All unpleasant terms have disappeared leaving a gauge-invariant, symmetric, and conserved total energy momentum tensor. Observe that even though the dynamical equations couple the material points to the fields, their contributions appear additively in (1-125).

Let us now study the consequences of Lorentz invariance about a fixed point to generalize the conservation of angular momentum. In infinitesimal form

$$\delta x^\mu = \delta\omega^{\mu\nu}x_\nu \tag{1-126}$$

with $\delta\omega^{\mu\nu}$ antisymmetric. We observe that the lagrangian is not, in general, invariant under such transformation on the arguments alone, since they should be accompanied by corresponding transformations on the fields. In other words, the fields carry a representation of the homogeneous Lorentz group. The elements Λ of the latter transform them according to

$${}^\Lambda\varphi(x) = S(\Lambda)\varphi(\Lambda^{-1}x) \tag{1-127}$$

We have collected the fields φ in a column vector and $S(\Lambda)$ is a matrix representation of the group. For Λ close to unity, $\Lambda^{-1}x$ will be equal to $x - \delta x$ with δx given by (1-126), and $S(\Lambda)$ will assume a corresponding form. We leave it as an exercise for the reader to study the case of the electromagnetic field as an example of such a circumstance, and we will limit ourselves to the simpler situation where the fields are scalars $[S(\Lambda) = 1]$. To apply Noether's theorem we have to consider $\delta\omega^{\mu\nu}$ in (1-126) as functions of x and to identify their coefficient in the variation of the action, with due care paid to the fact that $\delta\omega^{\mu\nu} = -\delta\omega^{\nu\mu}$. If $\Theta^{\mu\nu}$ stands for the symmetric energy momentum tensor, one finds the conservation of the generalized angular momentum:

$$\begin{aligned} J^{\mu,\nu\rho} &= \Theta^{\mu\nu}x^\rho - \Theta^{\mu\rho}x^\nu \\ \partial_\mu J^{\mu,\nu\rho} &= 0 \end{aligned} \tag{1-128}$$

This represents only the orbital part in the case where the fields transform according to nontrivial representations of the Lorentz group. Additional parts corresponding to the internal contributions appear in $J^{\mu,\nu\rho}$ to build the conserved quantity. In

Chaps. 2 and 3 we shall deal explicitly with such instances. It is crucial that $\Theta^{\mu\nu}$ be symmetric for (1-128) to yield such conserved densities and time-independent associated charges (six in number):

$$J^{\nu\rho} = \int d^3x \, J^{0,\nu\rho}(\mathbf{x}, t) \tag{1-129}$$

Note that $J^{\nu\rho}$ is not translationally invariant. Under a displacement a^μ of the origin of coordinates, the orbital part varies by an amount $a^\nu P^\rho - a^\rho P^\nu$. To obtain what really deserves to be called the intrinsic angular momentum we construct the Pauli-Lubanski vector

$$W_\alpha = -\tfrac{1}{2}\varepsilon_{\alpha\beta\gamma\delta} \frac{J^{\beta\gamma} P^\delta}{\sqrt{P^2}} \tag{1-130}$$

which reduces in a rest frame ($\mathbf{P} = 0$) to ordinary three-dimensional angular momentum.

Up to now in this relativistic context we have avoided appealing to the hamiltonian formalism or introducing Poisson brackets. The reason is that we did not want time to play a special role. However, nothing prevents us from doing so. At a given time t and for a generic $\varphi(\mathbf{x}, t)$ we can associate the conjugate field

$$\pi(\mathbf{x}, t) = \frac{\partial L}{\partial \partial_0 \varphi(\mathbf{x}, t)} = \frac{\delta}{\delta \partial_0 \varphi(\mathbf{x}, t)} \int d^3y \, \mathscr{L}(\mathbf{y}, t) \tag{1-131}$$

Similarly, we define the Poisson bracket of two functionals L_1 and L_2 of the fields φ and π at time t as follows:

$$\{L_1, L_2\} = -\{L_2, L_1\} = \int d^3x \left[\frac{\delta L_1}{\delta \pi(\mathbf{x}, t)} \frac{\delta L_2}{\delta \varphi(\mathbf{x}, t)} - \frac{\delta L_1}{\delta \varphi(\mathbf{x}, t)} \frac{\delta L_2}{\delta \pi(\mathbf{x}, t)} \right] \tag{1-132}$$

These functional derivatives are to be taken with a grain of salt. For instance, if L is expressed as a space integral over a density involving gradients of the fields, a suitable integration by parts is always understood. In particular,

$$\begin{aligned} \{\pi(\mathbf{x}, t), \varphi(\mathbf{y}, t)\} &= \delta^3(\mathbf{x} - \mathbf{y}) \\ \{\pi(\mathbf{x}, t), \pi(\mathbf{y}, t)\} &= \{\varphi(\mathbf{x}, t), \varphi(\mathbf{y}, t)\} = 0 \end{aligned} \tag{1-133}$$

and, for instance,

$$\{\pi(\mathbf{x}, t), \nabla\varphi(\mathbf{y}, t)\} = -\nabla_x \delta^3(x - y)$$

It is easy to see that the field equations (1-44) take the form:

$$\begin{aligned} \partial_0 \varphi(\mathbf{x}, t) &= \{H, \varphi(\mathbf{x}, t)\} = \frac{\delta}{\delta \pi(\mathbf{x}, t)} H \\ \partial_0 \pi(\mathbf{x}, t) &= \{H, \pi(\mathbf{x}, t)\} = -\frac{\delta}{\delta \varphi(\mathbf{x}, t)} H \end{aligned} \tag{1-134}$$

in complete analogy with nonrelativistic dynamics, when we have identified Hamilton's function H as the integral of the energy density:

$$H = \int d^3x\, \Theta^{00}(\mathbf{x}, t) \tag{1-135}$$

expressed in terms of $\varphi(\mathbf{x}, t)$, $\nabla\varphi(\mathbf{x}, t)$, and $\pi(\mathbf{x}, t)$. Equation (1-134) expresses the fact that H generates time translations of the system. The reader should find it easy to generalize it to space translations and infinitesimal Lorentz transformations.

1-2-3 Internal Symmetries

Systems such as those of particle physics are endowed with internal symmetries which play a prominent part in analyzing their spectrum and interactions. As far as we know most of these symmetries are, however, only approximate, being broken by forces weaker than the ones under consideration. It is nevertheless useful to pretend at first that these are exact invariances and to study the pattern of violations as a second approximation. Let us briefly derive Noether's theorem in this classical context.

Let $\phi_1 \cdots \phi_N$ stand for N interacting fields. Some of them can be real or complex and carry Lorentz indices. For each value of x we consider them as a vector in an N-dimensional space where we are given a representation of a group G. The latter can leave certain subspaces invariant (such will necessarily be the case if we have fields having inequivalent transformation properties under the Lorentz group). If G is a compact Lie group we shall write $T^s (s = 1, \ldots, r)$, the corresponding generators with T^s antihermitian, and

$$[T^{s_1}, T^{s_2}] = C^{s_1 s_2}{}_{s_3} T^{s_3} \tag{1-136}$$

Recall that the structure constants $C^{s_1 s_2}{}_{s_3}$ can be chosen totally antisymmetric for a compact group. For brevity we do not distinguish the group from its linear realization on ϕ.

Let us consider space-time-dependent variations of the fields

$$\begin{aligned} \delta\phi(x) &= \delta\alpha_s(x) T^s \phi(x) \\ \delta\phi^\dagger(x) &= -\delta\alpha_s(x)\phi^\dagger(x) T^s \end{aligned} \tag{1-137}$$

under which we have $\mathscr{L}(\phi) \to \mathscr{L}(\phi + \delta\phi) = \hat{\mathscr{L}}$, where $\hat{\mathscr{L}}$ is a function of $\phi, \partial_\mu\phi$ (and their complex conjugates), $\delta\alpha_s$, and $\partial_\mu\delta\alpha_s$. A variation around the stationary solution will yield

$$O = \delta I = \int d^4x\, \delta\alpha_s(x) \left\{ \frac{\partial\hat{\mathscr{L}}}{\partial\delta\alpha_s(x)} - \partial_\mu \frac{\partial\hat{\mathscr{L}}}{\partial[\partial_\mu\delta\alpha_s(x)]} \right\} \tag{1-138}$$

We define the corresponding currents

$$j_s^\mu(x) = \frac{\partial\hat{\mathscr{L}}}{\partial[\partial_\mu\delta\alpha_s(x)]} \tag{1-139}$$

Only the part of the lagrangian containing field derivatives will contribute to

j_s^μ. Noether's theorem follows from (1-138) and gives the divergence of these currents as

$$\partial_\mu j_s^\mu(x) = \frac{\partial \hat{\mathscr{L}}(x)}{\partial \delta\alpha_s} \tag{1-140}$$

where we have noted that $\partial \hat{\mathscr{L}}(x)/\partial \delta\alpha_s(x)$ can be identified with the same derivative for constant variations $\delta\alpha_s$. A current j_s^μ will be conserved if the corresponding term on the right-hand side of (1-140) vanishes, which means that the original lagrangian is invariant under the one-dimensional subgroup of G generated by T_s. Stated differently, to each such subgroup is associated a conserved charge

$$Q_s = \int d^3x\, j_s^0(\mathbf{x}, t) \qquad \frac{dQ_s}{dt} = \int d^3x\, \partial_\mu j_s^\mu(\mathbf{x}, t) = 0 \tag{1-141}$$

Let us return to the general case without assuming the conservation of the currents. This does not prevent us from defining the charges $Q_s(t)$ at time t. In what follows, t will be fixed and we shall not write it explicitly. If $\mathscr{F}$ stands for a functional of the fields and conjugate momenta at time t, we want to compute the Poisson bracket

$$\{Q_s, \mathscr{F}\} = \int d^3x \sum_a \left(\frac{\delta Q_s}{\delta \pi_a} \frac{\delta \mathscr{F}}{\delta \phi_a} - \frac{\delta Q_s}{\delta \phi_a} \frac{\delta \mathscr{F}}{\delta \pi_a} \right) \tag{1-142}$$

The dependence of $\hat{\mathscr{L}}$ on $\partial_0 \delta\alpha_s(x)$ arises only from the dependence of $\mathscr{L}$ on $\partial_0 \phi_\alpha$. To simplify, let us assume that the fields are real (in which case T_s is real and antisymmetric). If this were not the case, we would consider separately the real and imaginary parts. Then it follows that

$$\begin{aligned} j_s^0(x) &= \frac{\partial \hat{\mathscr{L}}}{\partial[\partial_0 \delta\alpha_s(x)]} = \frac{\partial \mathscr{L}}{\partial[\partial_0 \phi_a(x)]} T_{ab}^s \phi_b(x) = \pi(x) T^s \phi(x) \\ Q^s(x) &= \int d^3x\, \pi(x) T^s \phi(x) \end{aligned} \tag{1-143}$$

We have, therefore,

$$\{Q^s, \mathscr{F}\} = \int d^3x \left(\frac{\delta \mathscr{F}}{\delta \phi} T^s \phi - \pi T^s \frac{\delta \mathscr{F}}{\delta \pi} \right) \tag{1-144}$$

In particular,

$$\{Q^s, \phi(x)\} = T^s \phi(x) \qquad \{Q^s, \pi(x)\} = -\pi(x) T^s = T^s \pi(x) \tag{1-145}$$

which can be rewritten

$$\{\delta\alpha_s Q^s, \phi(x)\} = \delta\phi(x) \qquad \{\delta\alpha_s Q^s, \pi(x)\} = \delta\pi(x)$$

where $\delta\phi(x)$ and $\delta\pi(x)$ are the variations (1-137) of the fields under infinitesimal transformations. In other words, the charges Q^s generate these infinitesimal transformations through Poisson brackets in the same way that the hamiltonian generates time translations. Similarly, we have

$$\{Q^{s_1}, Q^{s_2}\} = -\int d^3x\, \pi(x)[T^{s_1}, T^{s_2}]\phi(x) = -C^{s_1 s_2}{}_{s_3} Q^{s_3} \tag{1-146}$$

as should obviously be the case. The invariance of the lagrangian was not required to derive the above relations. Indeed, if H stands for the hamiltonian we easily derive that

$$\frac{dQ^s}{dt}(t) = \{H(t), Q^s(t)\} = \int d^3x \frac{\partial \hat{\mathscr{L}}}{\partial \delta\alpha_s}(\mathbf{x}, t) \tag{1-147}$$

We can generalize Eq. (1-146) to the equal-time Poisson brackets of the time components of the currents with the following result:

$$\{j_0^{s_1}(\mathbf{x}, t), j_0^{s_2}(\mathbf{y}, t)\} = -C^{s_1 s_2}{}_{s_3}\, \delta^3(\mathbf{x} - \mathbf{y}) j_0^{s_3}(\mathbf{x}, t) \tag{1-148}$$

In their quantized version these relations will give rise to the important developments of current algebra (see Chap. 11).

The reader will find it instructive to study similarly the equal-time Poisson brackets of the energy momentum tensor:

$$\begin{aligned}
\{\Theta^{00}(\mathbf{x}, t), \Theta^{00}(\mathbf{y}, t)\} &= [\Theta^{0k}(\mathbf{x}, t) + \Theta^{0k}(\mathbf{y}, t)]\, \partial_k^x \delta^3(\mathbf{x} - \mathbf{y}) \\
\{\Theta^{00}(\mathbf{x}, t), \Theta^{0k}(\mathbf{y}, t)\} &= [\Theta^{kl}(\mathbf{x}, t) - g^{kl}\Theta^{00}(\mathbf{y}, t)]\, \partial_l^x \delta^3(\mathbf{x} - \mathbf{y}) \\
\{\Theta^{0k}(\mathbf{x}, t), \Theta^{0l}(\mathbf{y}, t)\} &= [\Theta^{0k}(\mathbf{y}, t)\, \partial_l^x + \Theta^{0l}(\mathbf{x}, t)\, \partial_k^x]\, \delta^3(\mathbf{x} - \mathbf{y})
\end{aligned} \tag{1-149}$$

from which we recover the Lie algebra of the Poincaré group:

$$\begin{aligned}
&\{P^\mu, P^\nu\} = 0 \qquad \{J^{\mu\nu}, P^\sigma\} = P^\mu g^{\nu\sigma} - P^\nu g^{\mu\sigma} \\
&\{J^{\mu\nu}, J^{\bar{\mu}\bar{\nu}}\} = g^{\mu\bar{\nu}} J^{\nu\bar{\mu}} + g^{\nu\bar{\mu}} J^{\mu\bar{\nu}} - g^{\mu\bar{\mu}} J^{\nu\bar{\nu}} - g^{\nu\bar{\nu}} J^{\mu\bar{\mu}}
\end{aligned} \tag{1-150}$$

To close this section let us study a simple lagrangian involving two real fields ϕ_1 and ϕ_2, which we shall describe using the complex (independent) quantities

$$\phi = \frac{\phi_1 + i\phi_2}{\sqrt{2}} \qquad \phi^* = \frac{\phi_1 - i\phi_2}{\sqrt{2}} \tag{1-151}$$

Let us assume the dynamics invariant under internal rotations in the $(1, 2)$ space or, equivalently, under

$$\phi(x) \to e^{-ie\alpha}\phi(x) \qquad \phi^*(x) \to e^{ie\alpha}\phi^*(x) \tag{1-152}$$

The constant e appearing here will be identified with the elementary coupling constant to the electromagnetic field, i.e., the electric charge. For instance, $\mathscr{L}$ assumes the form:

$$\mathscr{L}_{\text{scalar}} = (\partial_\mu\phi)^*(\partial^\mu\phi) - m^2\phi^*\phi - V(\phi^*\phi) \tag{1-153}$$

with V an arbitrary smooth function, for instance, a polynomial. Noether's theorem yields a conserved current

$$j_\mu(x) = \frac{\partial \mathscr{L}}{\partial[\partial^\mu\alpha(x)]} = ie[\phi^*\partial_\mu\phi - (\partial_\mu\phi^*)\phi] \equiv ie\phi^* \overleftrightarrow{\partial_\mu}\phi \tag{1-154}$$

We can, of course, directly verify this conservation using the equations of motion:

$$(\Box + m^2)\phi = -\frac{\partial V}{\partial \phi^*} \qquad (\Box + m^2)\phi^* = -\frac{\partial V}{\partial \phi} \tag{1-155}$$

The charge

$$Q = \int d^3x\, j^0(\mathbf{x}, t) = ie \int d^3x\, \phi^*(\mathbf{x}, t)\overleftrightarrow{\partial_0}\phi(\mathbf{x}, t) \tag{1-156}$$

is therefore time independent. We now wish to couple this system to the electromagnetic field. Since we are given a conserved current we might think of using it on the right-hand side of Maxwell's equations. But some care must be exercised since the coupling of ϕ to the vector potential A_μ might modify the structure of the current itself. We seek therefore a total lagrangian given as a sum of three pieces—$\mathscr{L}_{\text{em}} = -\frac{1}{4}F^2$, $\mathscr{L}_{\text{scalar}}$ given by (1.153), and an interaction part $\mathscr{L}_{\text{int}}(\phi, \phi^*, A_\mu)$:

$$\mathscr{L} = \mathscr{L}_{\text{em}} + \mathscr{L}_{\text{scalar}} + \mathscr{L}_{\text{int}} \tag{1-157}$$

Let us assume that the current J^μ given by Noether's theorem applied to the full lagrangian (1-157) is indeed the electromagnetic current. In other words, we have

$$J^\mu = j^\mu + \frac{\partial \mathscr{L}_{\text{int}}}{\partial[\partial_\mu \alpha(x)]} = \partial_\mu F^{\mu\nu} = -\frac{\partial \mathscr{L}_{\text{int}}}{\partial A_\nu(x)} \tag{1-158}$$

It is easy to convince oneself that the following interaction:

$$\begin{aligned}\mathscr{L}_{\text{int}} &= -ieA_\mu\phi^* \overleftrightarrow{\partial^\mu}\phi + e^2A^2\phi^*\phi \\ &= -A_\mu j^\mu + e^2A^2\phi^*\phi\end{aligned} \tag{1-159}$$

fulfills these conditions in such a way that

$$J^\mu = j^\mu - 2e^2A^\mu\phi^*\phi \tag{1-160}$$

The full lagrangian is therefore

$$\mathscr{L} = -\tfrac{1}{4}F^2 + [(\partial_\mu + ieA_\mu)\phi]^*[(\partial^\mu + ieA^\mu)\phi] - m^2\phi^*\phi - V(\phi^*\phi) \tag{1-161}$$

leading to the coupled equations

$$\partial_\mu F^{\mu\nu} = J^\nu \qquad [(\partial_\mu + ieA_\mu)(\partial^\mu + ieA^\mu) + m^2]\phi = -\frac{\partial V}{\partial \phi^*} \tag{1-162}$$

We observe that (1-161) follows from a principle of minimal coupling to the electromagnetic field, according to which any derivative ∂_μ acting on a charged field (with charge e) has to be replaced by the covariant derivative $\partial_\mu + ieA_\mu$. As a consequence, the lagrangian is not only invariant under the transformations (1-152) with constant α but under more general x-dependent (i.e., local) gauge transformations:

$$\phi(x) \to e^{-ie\alpha(x)}\phi(x) \qquad \phi^*(x) \to e^{ie\alpha(x)}\phi^*(x) \qquad A_\mu(x) \to A_\mu(x) + \partial_\mu\alpha(x) \tag{1-163}$$

What was initially a trick to derive Noether's theorem has now become a deeper property of electromagnetic couplings which generalizes the gauge invariance already discussed for the free electromagnetic field. This idea suitably extended to noncommutative groups yields very interesting model field theories (see Chap. 12).

1-3 PROPAGATION AND RADIATION

1-3-1 Green Functions

The dynamical equations of field theory are typically of the Klein-Gordon form:

$$(\Box + m^2)\phi(x) = j(x) \tag{1-164}$$

where j may depend on the fields ϕ and extra indices have been omitted. We had already an example with Maxwell's equations for the potential in the Lorentz gauge, where the mass term of (1-164) was absent.

For the time being, let us assume the source $j(x)$ to be given and $m^2 \geq 0$. We are thus dealing with an hyperbolic second-order partial differential equation, which determines ϕ in the neighborhood of a point x in terms of its values together with those of its normal derivative on a space-like surface element passing through x. Characteristic elements are tangent to the light cone, showing that causality is locally obeyed.

In scattering theory one seldom has to tackle the problem in the way just mentioned. Boundary conditions on ϕ are rather imposed along space-like surfaces widely separated by a time-like interval. It is then useful to construct standard solutions to (1-164) where the right-hand side is replaced by a distribution concentrated around a point x'. We shall generically denote $G(x, x')$ the solution of

$$(\Box + m^2)G(x, x') = \delta^4(x - x') \tag{1-165}$$

with an appended suffix to characterize the boundary conditions imposed on G. The latter will most frequently be translationally invariant in such a way that the corresponding Green functions (or propagators) will only depend on the argument $x - x'$. From the superposition principle, solutions to (1-164) will be generated by

$$\phi(x) = \phi^{(0)}(x) + \int d^4x'\, G(x - x')j(x') \tag{1-166}$$

where $\phi^{(0)}(x)$ obeys the homogeneous equation and is chosen in such a way that ϕ satisfies the boundary conditions.

Making further use of translation invariance, (1-165) is solved through a Fourier transformation which replaces it by an algebraic equation. Setting

$$G(x - x') = \frac{1}{(2\pi)^4} \int d^4p\, e^{-ip\cdot(x-x')}\, \tilde{G}(p) \tag{1-167}$$

we get

$$(-p^2 + m^2)\tilde{G}(p) = 1 \tag{1-168}$$

To divide both sides by $-p^2 + m^2$ we have to cope with the zero of this expression on the two-sheeted hyperboloid $p^2 - m^2 = 0$ (or the cone $p^2 = 0$ if $m^2 = 0$). This in turn is equivalent to prescribing in (1-167) a slightly deformed contour of integration. We note that all these choices differ at most by a contribution $g(\mathbf{p}, p_0/|p_0|)\,\delta(p^2 - m^2)$ to $\tilde{G}(p)$, an expression corresponding to the solution of the homogeneous equation. This choice is, of course, related to boundary conditions at infinity.

Let us first define the retarded and advanced Green functions:

$$\tilde{G}_{\substack{\text{ret}\\ \text{adv}}}(p) = \frac{-1}{(p_0 \pm i\varepsilon) - \mathbf{p}^2 - m^2}$$
$$G_{\substack{\text{ret}\\ \text{adv}}}(x) = -\frac{1}{(2\pi)^4}\int d^4p \frac{e^{-ip\cdot x}}{(p_0 \pm i\varepsilon)^2 - \mathbf{p}^2 - m^2} \tag{1-169}$$

The ε prescription is equivalent to a slight contour deformation in the integration over p_0 and shows that $G(x)$ is to be interpreted as a distribution. If we first perform the integral over p_0, in the case of $G_{\text{ret}}(x)$, say, we can close the integration contour in the upper complex plane if $x^0 < 0$, without encountering any singularity. From Cauchy's theorem we conclude that $G_{\text{ret}}(x)$ vanishes for $x_0 < 0$. The opposite conclusion applies to $G_{\text{adv}}(x)$. It can be checked that these distributions are Lorentz invariant so that $G_{\text{ret}}(x)$ vanishes outside the forward light cone while $G_{\text{adv}}(x)$ vanishes outside the backward one. These properties are in agreement with causal propagation. We also note that both Green functions are real, with $G_{\text{adv}}(x) = G_{\text{ret}}(-x)$. When $m^2 = 0$ we recover

$$G_{\substack{\text{ret}\\ \text{adv}}}(x)\big|_{m^2=0} = \frac{1}{2\pi}\,\theta(\pm x_0)\delta(x^2) \tag{1-170}$$

while for $m^2 > 0$ the explicit expressions involving Bessel functions are not too illuminating. However, no matter what m^2 is, the singularity of these Green functions on the light cone remains given by (1-170), a reflection of the fact that the small x^2 behavior is entirely dictated by the differential operator in (1-165). The mass term is then responsible for the fact that the support is not concentrated on the light cone as in (1-170), but also involves signals propagating at a speed smaller than one. Returning to the integral representation (1-169) let us perform the p_0 integral using the decomposition

$$\frac{1}{(p_0 + i\varepsilon)^2 - \mathbf{p}^2 - m^2} = \frac{1}{2\omega_p}\left(\frac{1}{p_0 - \omega_p + i\varepsilon} - \frac{1}{p_0 + \omega_p + i\varepsilon}\right)$$

with $\omega_p = \sqrt{\mathbf{p}^2 + m^2}$. We readily find

$$G_{\text{ret}}(x - x') = i\,\frac{\theta(x_0 - x'_0)}{(2\pi)^3}\int \frac{d^3\mathbf{p}}{2\omega_p}\{e^{-i\omega_p(x_0 - x'_0) + i\mathbf{p}\cdot(\mathbf{x} - \mathbf{x}')} - e^{i\omega_p(x_0 - x'_0) + i\mathbf{p}\cdot(\mathbf{x} - \mathbf{x}')}\}$$

We wish to give a physical interpretation of this formula. To do so, we may imagine the system enclosed in a very large cubic box of size L. The momentum integral is replaced by a Riemann sum, and each component of the momentum is of the form $2\pi n/L$ with n an arbitrary integer, in such a way that $[1/(2\pi)^3]\int d^3p \to \Sigma_n/L^3$.

We then have

$$G_{\text{ret}}(x - x') \to i\theta(x_0 - x'_0)[\Sigma\, \varphi_{+,\mathbf{p}}(x)\varphi^*_{+,\mathbf{p}}(x') - \Sigma\, \varphi_{-,\mathbf{p}}(x)\varphi^*_{-,\mathbf{p}}(x')] \quad (1\text{-}171)$$

The momentum $\mathbf{p}$ assumes only discrete (but very dense) values and the functions

$$\varphi_{\pm,\mathbf{p}}(x) = \frac{1}{L^{3/2}\sqrt{2\omega_p}}\, e^{\pm i\omega_p x_0 + i\mathbf{p}\cdot\mathbf{x}} \quad (1\text{-}172)$$

stand for periodic solutions of the homogeneous Klein-Gordon equation. The plus or minus sign corresponds to the sign of the frequency. The relative minus sign in (1-171) [together with the factor i in front to build a real quantity: $\varphi^*_{+,\mathbf{p}}(x) = \varphi_{-,-\mathbf{p}}(x)$] comes from the fact that the solutions of the equations are to be normalized according to a nondefinite "norm" [see formula (2-5)]. We shall see that while φ_+ has a positive norm, φ_- has a negative one.

In any case the expression (1-171) leads to an interesting interpretation of propagation according to $G_{\text{ret}}(x - x')$. Positive and negative frequencies are respectively propagated forward in time by the first and second term of (1-171). For $G_{\text{adv}}(x - x')$ we have a similar expression with x and x' interchanged.

The difference

$$\begin{aligned} G^{(-)}(x) &= G_{\text{ret}}(x) - G_{\text{adv}}(x) \\ &= \frac{i}{(2\pi)^3}\int d^4p\, e^{-ip\cdot x}\, \varepsilon(p_0)\delta(p^2 - m^2) \\ &= \frac{i}{(2\pi)^3}\int \frac{d^3p}{2\omega_p}\,(e^{-i\omega_p x_0 + i\mathbf{p}\cdot\mathbf{x}} - e^{i\omega_p x_0 + i\mathbf{p}\cdot\mathbf{x}}) \end{aligned} \quad (1\text{-}173)$$

is an odd function vanishing outside the light cone such that

$$\frac{\partial}{\partial x_0} G^{(-)}(x)\big|_{x^0=0} = \delta^3(\mathbf{x}) \quad (1\text{-}174)$$

It clearly satisfies the homogeneous Klein-Gordon equation. When $m^2 = 0$, it reduces to

$$\begin{aligned} G^{(-)}(x)\big|_{m^2=0} &= \frac{1}{2\pi}\,\varepsilon(x_0)\,\delta(x^2) \\ &= \lim_{\eta\to 0}\frac{1}{4\pi^2 i}\left[\frac{1}{(x - i\eta)^2} - \frac{1}{(x + i\eta)^2}\right] \end{aligned} \quad (1\text{-}175)$$

where η is an infinitesimal positive time-like vector.

Finally, the half sum

$$G^{(+)}(x) = \tfrac{1}{2}[G_{\text{ret}}(x) + G_{\text{adv}}(x)] = -\frac{\text{PP}}{(2\pi)^4}\int d^4p\, e^{-ip\cdot x}\,\frac{1}{p^2 - m^2} \quad (1\text{-}176)$$

is an even solution of (1-165). In (1-176) the principal part symbol applies to the p_0 integral, and we recall that

$$\frac{1}{p_0 - \omega_p \pm i\varepsilon} = \text{PP}\frac{1}{p_0 - \omega_p} \mp i\pi\delta(p_0 - \omega_p)$$
$$\delta(p^2 - m^2) = \frac{\delta(p_0 - \omega_p) + \delta(p_0 + \omega_p)}{2\omega_p} \tag{1-177}$$

The above discussion takes place in a purely classical context. We shall, however, encounter in the quantized version another even solution to the very same equation, first introduced by Stueckelberg and Feynman. One reason to explain why it does not naturally appear in classical physics is that it is a complex distribution defined through

$$G_F(x) = -\frac{1}{(2\pi)^4}\int d^4p\, e^{-ip\cdot x}\frac{1}{p^2 - m^2 + i\varepsilon} \tag{1-178}$$

It therefore satisfies

$$(\Box + m^2)G_F(x) = \delta^4(x) \qquad G_F(-x) = G_F(x)$$

If we discretize as above we find the equivalent form:

$$G_F(x - x') \to i[\theta(x_0 - x'_0)\Sigma\, \varphi_{+,\mathbf{p}}(x)\varphi^*_{+,\mathbf{p}}(x') + \theta(x'_0 - x_0)\Sigma\, \varphi_{-,\mathbf{p}}(x)\varphi^*_{-,\mathbf{p}}(x')] \tag{1-179}$$

While the previous Green functions were zero outside the light cone, this is not the case for G_F, which has an exponential tail for negative x^2. Let $x_0 = 0$ and set $r = |\mathbf{x}|$. Then

$$G_F(0, r) = \frac{i}{(2\pi)^2 r}\int_m^\infty dp\,\frac{p}{\sqrt{p^2 - m^2}}e^{-pr} \sim \frac{i\,e^{-mr}}{(2\pi)^2 r^2}\left(\frac{\pi m r}{2}\right)^{1/2} \tag{1-180}$$

From (1-179) we read off that, according to G_F, positive (negative) frequencies propagate forward (backward) in time. This distinction according to frequency explains why G_F is intrinsically complex. Note also that $\tilde{G}_F(p)$ is a meromorphic function of the complex variable p^2 in contradistinction to the previous Green functions.

These expressions can be generalized to more general cases such as the electromagnetic one. Corresponding to the fact that Maxwell's equation alone

$$\Box A^\mu - \partial^\mu(\partial\cdot A) = j^\mu \tag{1-181}$$

is not sufficient to determine the potential A^μ in terms of the conserved current j^μ, we find in momentum space that the matrix

$$L^\mu{}_\nu = p^2 g^\mu{}_\nu - p^\mu p_\nu \tag{1-182}$$

is singular. Its determinant vanishes identically. The operator L is in fact proportional to a projector since

$$L^2 = p^2 L \tag{1-183}$$

To avoid this difficulty, we can either add a small mass term or proceed as in Sec. 1-2 and introduce an extra term $(\lambda/2)(\partial \cdot A)^2$ in the lagrangian. The modified equations of motion are

$$[\Box g^\mu{}_\nu - (1-\lambda)\partial^\mu \partial_\nu]A^\nu = j^\mu \tag{1-184}$$

and the propagator becomes

$$G_F^{\mu\nu}(x,\lambda) = -\frac{1}{(2\pi)^4}\int d^4p\, e^{-ip\cdot x}\frac{g^{\mu\nu} + [(1-\lambda)/\lambda]p^\mu p^\nu/p^2}{p^2 + i\varepsilon} \tag{1-185}$$

The apparent singularity in the numerator plays no role when $G_F^{\mu\nu}$ is convoluted with a conserved current. For $\lambda = 1$ we recover the Feynman propagator $g^{\mu\nu}G_F(x)$. We shall have more to say about this in Chap. 3.

1-3-2 Radiation

As an elementary application of Green functions let us recall the computation of the electromagnetic fields generated by a moving point charge. Let $x^\mu(\tau)$ be its space-time trajectory and $j^\mu(\mathbf{y}, t)$ the associated current given by (1-51). Causality requires to use the retarded Green function to solve for the potential. Moreover, the remarks at the end of the preceding section lead at once to the expression of the potential in the Lorentz gauge:

$$\begin{aligned} A^\mu(y) &= \frac{1}{2\pi}\int d^4z\, \theta(y^0 - z^0)\, \delta[(y-z)^2]j^\mu(z) \\ &= \frac{e}{2\pi}\int_{-\infty}^{\infty} d\tau\, \theta[y^0 - x^0(\tau)]\, \delta\{[y - x(\tau)]^2\}\dot{x}^\mu(\tau) \end{aligned} \tag{1-186}$$

The dots refer to derivatives with respect to proper time. Let $x_+ \equiv x(\tau_+)$, the unique y-dependent retarded point on the trajectory, such that

$$(y - x_+)^2 = 0 \qquad x_+^0 < y^0 \tag{1-187}$$

The following identity holds:

$$\int_{-\infty}^{+\infty} d\tau\, \theta[y^0 - x^0(\tau)]\, \delta\{[y - x(\tau)]^2\} f(\tau) = \frac{f(\tau_+)}{2\dot{x}_+ \cdot (y - x_+)} \tag{1-188}$$

From Eq. (1-187) and the fact that for a physical trajectory $\dot{x}_+$ belongs to the forward light cone, it follows that $\dot{x}_+ \cdot (y - x_+) > 0$. We thus obtain the Lienard-Wiechert retarded potential:

$$A_{\text{ret}}^\mu(y) = \frac{e}{4\pi}\frac{\dot{x}_+^\mu}{\dot{x}_+ \cdot (y - x_+)} \tag{1-189}$$

the relativistic generalization of the Coulomb potential to which it reduces in the frame where $\dot{x}_+^\mu = (1, 0, 0, 0)$. Explicitly, if $\mathbf{r} = \mathbf{y} - \mathbf{x}_+$ and $\mathbf{v}_+ = (d\mathbf{x}/dt)(\tau_+)$:

$$A_{\text{ret}}^0(y) = \frac{e}{4\pi(r - \mathbf{r}\cdot\mathbf{v}_+)} \qquad \mathbf{A}_{\text{ret}}(y) = \frac{e\mathbf{v}_+}{4\pi(r - \mathbf{r}\cdot\mathbf{v}_+)} \tag{1-190}$$

Dropping the + index, the corresponding fields read

$$\mathbf{E} = \frac{e}{4\pi} \frac{(\mathbf{r} - r\mathbf{v})(1 - \mathbf{v}^2) + \mathbf{r} \times [(\mathbf{r} - r\mathbf{v}) \times d\mathbf{v}/dt]}{(r - \mathbf{r} \cdot \mathbf{v})^3}$$

$$\mathbf{B} = \frac{\mathbf{r} \times \mathbf{E}}{r} \tag{1-191}$$

and involve, of course, the retarded point. Both **E** and **B** contain terms in $1/r^2$ which only contribute at short distance and $1/r$ radiated parts. The energy flux across a sphere is given in terms of the Poynting vector as

$$\frac{d\mathscr{E}}{dt} = \int_S d\mathbf{S} \cdot (\mathbf{E} \times \mathbf{B}) = \int_S d\Omega \, (\mathbf{r} \times \mathbf{E})^2 > 0 \tag{1-192}$$

At small velocities the relevant $1/r$ contributions to (1-192) are

$$\mathbf{E}_{\text{rad}} = \frac{e}{4\pi r^3} \left[\mathbf{r} \times \left(\mathbf{r} \times \frac{d\mathbf{v}}{dt} \right) \right] \qquad \mathbf{B}_{\text{rad}} = e \frac{\mathbf{r} \times d\mathbf{v}/dt}{r^2} \tag{1-193}$$

which are the standard dipole fields. In the same limit the radiated power is given by the Larmor formula:

$$\frac{d\mathscr{E}}{dt} = \frac{e^2}{(4\pi)^2} \int \frac{d\Omega}{r^2} \left(\mathbf{r} \times \frac{d\mathbf{v}}{dt} \right)^2 = \frac{e^2}{8\pi} \left(\frac{d\mathbf{v}}{dt} \right)^2 \int d \cos\theta \sin^2\theta = \frac{2}{3} \frac{e^2}{4\pi} \left(\frac{d\mathbf{v}}{dt} \right)^2 \tag{1-194}$$

These results apply to many interesting situations of which we are only going to consider a few.

As an example let a charged particle oscillate nonrelativistically under the influence of a weak electric field of an incident plane wave of low frequency:

$$\frac{d\mathbf{v}}{dt} = \frac{e}{m} \mathbf{E} = \frac{e}{m} \boldsymbol{\varepsilon} E_0 \, e^{-i\omega t + i\mathbf{k} \cdot \mathbf{x}} \tag{1-195}$$

Here $\boldsymbol{\varepsilon}$ is the polarization of the wave, E_0 its amplitude, and **k** its wave vector. According to Larmor's formula the power radiated in the solid angle $d\Omega$ (see

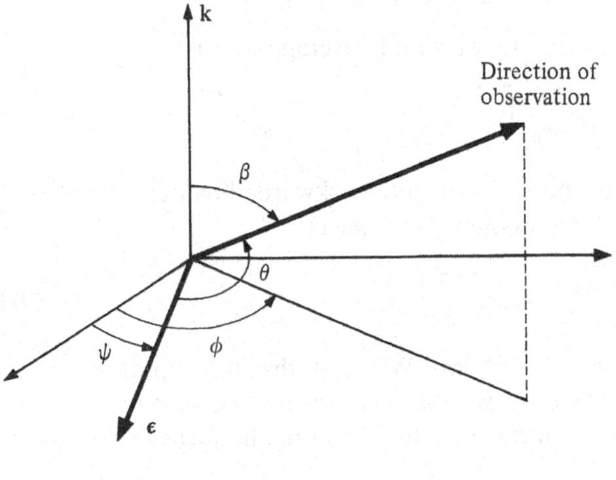

Figure 1-3 Radiation of a charged particle oscillating in a plane wave of wave vector **k** and polarization $\boldsymbol{\varepsilon}$.

Fig. 1-3 for the definition of angles) is given by

$$dP = \frac{e^2}{(4\pi)^2}\left(\mathbf{r}\times\frac{d\mathbf{v}}{dt}\right)^2\frac{d\Omega}{r^2} = \frac{e^2}{(4\pi)^2}\left(\frac{d\mathbf{v}}{dt}\right)^2\sin^2\theta\,d\Omega \tag{1-196}$$

If the particle displacement during a period is only a small fraction of the wavelength of the incident wave, the mean value of $(d\mathbf{v}/dt)^2$ is

$$\left\langle\left(\frac{d\mathbf{v}}{dt}\right)^2\right\rangle = \tfrac{1}{2}\,\mathrm{Re}\,\frac{d\mathbf{v}}{dt}\cdot\frac{d\mathbf{v}^*}{dt} = \frac{e^2}{2m^2}|\,\mathbf{E}_0\,|^2$$

and thus

$$\left\langle\frac{dP}{d\Omega}\right\rangle = \frac{1}{2}\left(\frac{e^2}{4\pi m}\right)^2|\,\mathbf{E}_0\,|^2\sin^2\theta \tag{1-197}$$

The average incident energy flux per unit time and unit area perpendicular to the propagation is given by the average of the incident Poynting vector as $\frac{1}{2}|\,\mathbf{E}_0\,|^2$. We define the differential scattering cross section $d\sigma/d\Omega$ as the ratio of radiated power per unit solid angle to incident power flux across a unit area:

$$\frac{d\sigma}{d\Omega} = \left(\frac{e^2}{4\pi m}\right)^2\sin^2\theta = r_e^2\sin^2\theta \tag{1-198}$$

We have introduced the notation r_e for the classical electromagnetic radius (restoring the velocity of light c):

$$r_e = \frac{e^2}{4\pi mc^2} = \frac{\hbar\alpha}{mc} \tag{1-199}$$

with a value of 2.82×10^{-13} cm for an electron, and α stands for the fine structure constant $\alpha = e^2/4\pi\hbar c \simeq 1/137$. The expression (1-198) is referred to as the Thomson scattering cross section; it is given here for an incident polarized wave. In terms of the angles β, ϕ, and ψ defined on Fig. 1-3, we have

$$\sin^2\theta = 1 - \sin^2\beta\cos^2(\phi-\psi)$$

To obtain the differential cross section, we simply average over ψ:

$$\left.\frac{d\sigma}{d\Omega}\right|_{\text{unpolarized}} = \frac{1}{2\pi}\int_0^{2\pi}d\psi\,\frac{d\sigma}{d\Omega} = r_e^2\,\frac{1+\cos^2\beta}{2} \tag{1-200}$$

Scattering is maximal both in the forward and backward directions. The total Thomson cross section is the integral of (1-200) over Ω:

$$\sigma_{\text{tot}} = \frac{8\pi}{3}r_e^2 \tag{1-201}$$

For an electron this yields 0.66×10^{-24} cm^2. We recall that this expression holds for weak fields and small frequencies $\omega \ll mc^2/\hbar$. When $\hbar\omega \gtrsim mc^2$ or $\lambda \lesssim \hbar/mc$ (the quantum mechanical Compton wavelength), the initial frequency is no longer

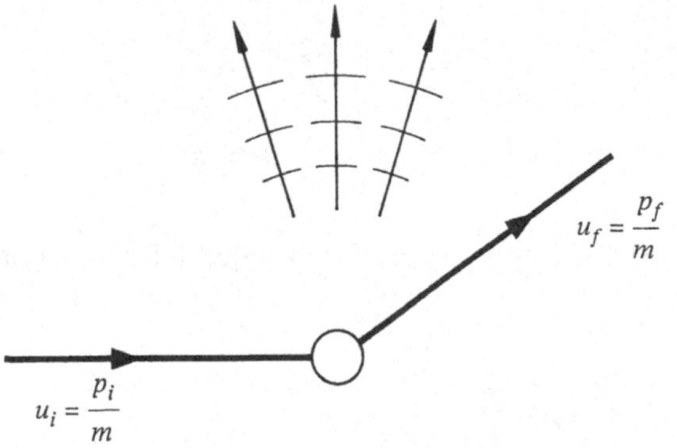

Figure 1-4 Bremsstrahlung.

conserved, we encounter both quantum mechanical and relativistic effects, and the process is then called Compton scattering (see Chap. 5). To a good approximation the cross section is then given by the Klein-Nishina formula [(5-116) and (5-117)]. Let us only quote here the lowest-order correction in $\hbar\omega/mc^2$ to Thomson scattering:

$$\sigma_{\text{tot}} = \frac{8\pi}{3} r_e^2 \left(1 - \frac{2\hbar\omega}{mc^2} + \cdots\right) \tag{1-202}$$

We can study along the same lines classical bremsstrahlung, i.e., radiation by a charge suddenly accelerated. Let $u_i = p_i/m$ and $u_f = p_f/m$ be the initial and final four-velocities (see Fig. 1-4). Choose the origin of coordinates at the space-time point of acceleration, which can be thought of as the idealization of a collision process. The space-time trajectory is parametrized as

$$x(\tau) = \begin{cases} \dfrac{p_i}{m}\tau & \tau < 0 \\ \dfrac{p_f}{m}\tau & \tau > 0 \end{cases} \tag{1-203}$$

while the current

$$\begin{aligned} j_\mu(x) &= e \int d\tau \frac{dx_\mu}{d\tau} \delta^4[x - x(\tau)] \\ &= -\frac{ie}{(2\pi)^4} \int d^4k \, e^{-ik\cdot x} \left(\frac{p_\mu^i}{p^i \cdot k} - \frac{p_\mu^f}{p^f \cdot k}\right) \end{aligned} \tag{1-204}$$

takes after Fourier transformation the simple expression

$$\tilde{j}_\mu(k) = -ie\left(\frac{p_\mu^i}{p^i \cdot k} - \frac{p_\mu^f}{p^f \cdot k}\right) = \tilde{j}_\mu^*(-k)$$

We require that as $t \to -\infty$ the electromagnetic field reduces to the Coulomb field of the incident particle, since current conservation forbids the vanishing of $j_\mu(x)$ when $t \to -\infty$. In the Lorentz gauge we write, using the definition (1-173),

$$A^\mu(x) = \int d^4x'\, G_{\text{ret}}(x - x')j^\mu(x')$$
$$= \int d^4x'\, G^{(-)}(x - x')j^\mu(x') + \int d^4x'\, G_{\text{adv}}(x - x')j^\mu(x') \tag{1-205}$$

We have exhibited the structure of $A^\mu(x)$ as the sum of a radiated field (solution of the homogeneous Maxwell equation) and a Coulomb field attached to the particle. Therefore

$$A^\mu_{\text{rad}}(x) = \int d^4x'\, G^{(-)}(x - x')j^\mu(x')$$
$$= \frac{i}{(2\pi)^3} \int d^4k\, e^{-ik\cdot x}\, \varepsilon(k_0)\delta(k^2)\tilde{j}^\mu(k) \tag{1-206}$$

while

$$F^{\mu\nu}_{\text{rad}}(x) = \frac{1}{(2\pi)^3} \int d^4k\, e^{-ik\cdot x}\, \varepsilon(k_0)\delta(k^2)[k^\mu \tilde{j}^\nu(k) - k^\nu \tilde{j}^\mu(k)] \tag{1-207}$$

The density of radiated energy is given by (1-117):

$$\Theta^{00} = F^{0\mu}F_\mu{}^0 + \tfrac{1}{4}g^{00}F^{\mu\nu}F_{\mu\nu}$$

For a light-like vector $k^0 = |\mathbf{k}|$, we define two orthogonal space-like polarization vectors ε_α satisfying

$$\varepsilon_\alpha^2 = -1 \qquad \varepsilon_1 \cdot \varepsilon_2 = 0 \qquad \varepsilon_\alpha \cdot k = 0 \tag{1-208}$$

We readily derive that the emitted energy $\mathscr{E}$ at a positive time is

$$\mathscr{E} = \int d^3x\, \Theta^{00}(t, x) = \frac{1}{(2\pi)^3} \int \frac{d^3k}{2k_0}\, k_0 \sum_{\alpha=1,2} |\varepsilon_\alpha \cdot \tilde{j}(k)|^2 \tag{1-209}$$

Let us anticipate the interpretation of this radiation in terms of light quanta, i.e., photons of energy momentum $\hbar k$. Henceforth set $\hbar = 1$. Then the energy emitted in a phase space element d^3k will read

$$d\mathscr{E} = \frac{1}{(2\pi)^3}\frac{d^3k}{2}\, e^2 \sum_{\alpha=1,2} \left| \frac{\varepsilon_\alpha \cdot p^i}{k \cdot p^i} - \frac{\varepsilon_\alpha \cdot p^f}{k \cdot p^f} \right|^2 \tag{1-210}$$

This semiclassical calculation enables us to obtain the number of emitted photons of polarization ε by dividing by k_0 the energy of an individual quantum:

$$dN = \frac{d\mathscr{E}}{k^0} = e^2 \left| \frac{\varepsilon \cdot p^i}{k \cdot p^i} - \frac{\varepsilon \cdot p^f}{k \cdot p^f} \right|^2 \frac{d^3k}{2(2\pi)^3 k^0} \tag{1-211}$$

This result agrees, in fact, with the full quantum mechanical treatment (see Secs. 5-2-4 and 7-2-3). By integrating over k (for small k) we note that the total energy is finite but the total number of photons is not. This is the infrared catastrophe to be discussed in greater detail later.

The angular distribution given by (1-211) is of interest. In the frame where $k^\mu \equiv |\mathbf{k}|(1, 1, 0, 0)$, $\varepsilon_1^\mu = (0, 0, 1, 0)$, $\varepsilon_2^\mu = (0, 0, 0, 1)$, taking into account that $k \cdot \tilde{j}(k) = 0$ implies $j^0 = j^1$, we see that

$$\sum_\alpha |\varepsilon_\alpha \cdot \tilde{j}(k)|^2 = -j \cdot \tilde{j}^* \tag{1-212}$$

Therefore,

$$d\mathscr{E} = \frac{e^2}{2(2\pi)^3} d^3k \left[\frac{2p^i \cdot p^f}{(k \cdot p^i)(k \cdot p^f)} - \frac{m^2}{(k \cdot p^i)^2} - \frac{m^2}{(k \cdot p^f)^2} \right]$$

This can be rewritten in terms of the particle initial and final velocities, with $\hat{k} = \mathbf{k}/|\mathbf{k}|$ as

$$d\mathscr{E} = \frac{e^2}{2(2\pi)^3} \frac{d^3k}{|\mathbf{k}|^2} \left[\frac{2(1 - \mathbf{v}_i \cdot \mathbf{v}_f)}{(1 - \hat{k} \cdot \mathbf{v}_i)(1 - \hat{k} \cdot \mathbf{v}_f)} - \frac{m^2}{E_i^2(1 - \hat{k} \cdot \mathbf{v}_i)^2} - \frac{m^2}{E_f^2(1 - \hat{k} \cdot \mathbf{v}_f)^2} \right] \tag{1-213}$$

For the emission of soft quanta ($|\mathbf{k}| \to 0$) the radiation is strongly peaked in the directions of the initial and final velocities, a typical property of bremsstrahlung.

To evaluate the total number of radiated photons, we integrate (1-211) from a lower value $k_{\min}$, needed because of the infrared catastrophe, up to some maximum momentum $k_{\max}$, needed because of the unrealistic sharp angle in the trajectory (1-203). A typical cross section $d\sigma_{\text{coll}}$ will describe the collision process and the total cross section $d\sigma_{\text{brems}}$ will include the final emitted photons. To integrate (1-211) we introduce the notation $q^2 = (p_f - p_i)^2 \simeq -4E^2 \sin^2 \theta/2$ in the ultrarelativistic limit where $E_i \sim E_f = E$ with θ the scattering angle, while in the nonrelativistic limit $|\mathbf{v}_i| \sim |\mathbf{v}_f| = v$. We use Feynman's integral representation:

$$\frac{1}{AB} = \int_0^1 \frac{dx}{[xA + (1 - x)B]^2} \tag{1-214}$$

to compute

$$\int \frac{d^2\hat{k}(1 - \mathbf{v}_i \cdot \mathbf{v}_f)}{(1 - \hat{k} \cdot \mathbf{v}_i)(1 - \hat{k} \cdot \mathbf{v}_f)} = 2\pi \int_0^1 dx \frac{2(1 - \mathbf{v}_i \cdot \mathbf{v}_f)}{1 - [x\mathbf{v}_i + (1 - x)\mathbf{v}_f]^2}$$
$$\simeq \begin{cases} 4\pi\left(1 + \dfrac{4}{3} v^2 \sin^2 \dfrac{\theta}{2}\right) + O(v^4) & v \ll 1, \text{ nonrelativistic} \\ 4\pi \ln\left(-\dfrac{q^2}{m^2}\right) + O\left(\dfrac{m^2}{q^2}\right) & q^2 \gg m^2, \text{ ultrarelativistic} \end{cases}$$

Furthermore,

$$\int d^2\hat{k} \frac{m^2}{E^2(1 - \hat{k} \cdot \mathbf{v})^2} = 4\pi$$

The final expression for the soft photon emission cross section reads

$$\left(\frac{d\sigma}{d\Omega_f}\right)_{\text{brems}} = \left(\frac{d\sigma}{d\Omega_f}\right)_{\text{coll}} \left(\ln \frac{k_{\max}}{k_{\min}}\right) \frac{2\alpha}{\pi} \begin{cases} \frac{4}{3} v^2 \sin^2 (\theta/2) & \text{nonrelativistic} \\ \ln(-q^2/m^2) - 1 & \text{ultrarelativistic} \end{cases} \tag{1-215}$$

In this quick survey of radiation problems we have neglected the interaction of the particle with its own emitted field. The same was true in the discussion of the motion in Sec. 1-3. As a matter of principle this is, of course, wrong, but except in extreme situations it is nonetheless a perfectly justified approximation. Indeed, from Larmor's formula (1-194) the radiated energy $\mathscr{E}_{\text{rad}} \sim \frac{2}{3}(e^2/4\pi) \times (d\mathbf{v}/dt)^2(\Delta t/c^3)$. As long as $\mathscr{E}_{\text{rad}}$ is small as compared to typical energies in the problem, for instance, the particle energy $\mathscr{E}_0 \sim m[(d\mathbf{v}/dt)\Delta t]^2$ for a moving charge, we can neglect classical radiative corrections. This leads to the condition

$$\Delta t \gg \tau_0 = \frac{2}{3}\frac{e^2}{4\pi mc^3} = \frac{2}{3}\frac{r_e}{c} \tag{1-216}$$

where a characteristic radiation time τ_0 has been introduced. On the contrary, this should become relevant if the forces vary appreciably in a time τ_0 or on distances of the order $c\tau_0$.

The classical point charge theory due to Lorentz, including radiative corrections, was the subject of much controversy before the advent of quantum mechanics, which has changed the perspective considerably. Even though in most situations it is not the relevant approximation, it is instructive to understand the limitations of classical mechanics in its most elaborate treatment of point charges. Let us present an heuristic derivation of the correction f_μ to the Lorentz force law to take into account the self-interaction. We write

$$m\dot{u}^\mu \equiv m\frac{du^\mu}{d\tau} = eF^{\mu\nu}_{\text{ext}}u_\nu + f^\mu \tag{1-217}$$

The extra term f^μ should be a four-vector such that for small velocities one obtains from (1-217) the Larmor energy loss

$$\frac{d\mathscr{E}_{\text{rad}}}{dt} = m\frac{du^0}{dt} = -\frac{2}{3}\frac{e^2}{4\pi}\left(\frac{d\mathbf{v}}{dt}\right)^2$$

in the nonrelativistic limit. Furthermore, from translation invariance it should only depend on u and its derivatives with respect to proper time. No new quantity independent of e and m with the dimension of a length should arise if the particle is structureless. Finally, we want to keep the definition of proper time, so that u^2 remains equal to one, from which we deduce that $u \cdot f$ has to vanish.

The four-vector $-(\ddot{u}\cdot u)u^\mu = (\dot{u}^2)u^\mu$ has a fourth component reducing to $-(d\mathbf{v}/dt)^2$ in the nonrelativistic limit. The requirement $f \cdot u = 0$ forces us to construct the combination $\ddot{u}^\mu - (\ddot{u}\cdot u)u^\mu$ orthogonal to u. Hence the above requirements lead to the classical Lorentz-Dirac equation in the form:

$$m\dot{u}^\mu = eF^{\mu\nu}_{\text{ext}}u_\nu + \frac{2}{3}\frac{e^2}{4\pi}\left[\ddot{u}^\mu - (u\cdot\ddot{u})u^\mu\right] \tag{1-218}$$

where F_{ext} represents the contribution of all external charges. The time component of this equation gives a relativistic generalization of the energy balance

$$\frac{d\mathscr{E}}{d\tau} = eF^{0\nu}_{\text{ext}}u_\nu + \frac{2}{3}\frac{e^2}{4\pi}(\dot{u})^2u_0 + m\tau_0\ddot{u}^0 \tag{1-219}$$

The first term on the right-hand side obviously corresponds to the work of external forces; the second is the dissipative ($\dot{u}^2 < 0$) Larmor term. The third one (the so-called Schott's energy term) is a total derivative. It can be neglected when taking averages over (almost) periodic motions or, more generally, when the variation of the acceleration is small during time intervals of order τ_0. The original derivation of (1-218) by Lorentz used a spherical model of the charge and was not free of objections from the relativistic standpoint. Dirac obtained the same equation in a fully relativistic way using local conservation of energy and momentum. Both, however, had to incorporate in the inertial mass an infinite (positive) contribution equal to the electrostatic Coulomb energy created by the charge, a typical renormalization effect. If the observed mass were only electromagnetic in origin we would have to introduce in this Coulomb energy a short-distance cutoff a such that $e^2/4\pi a \sim mc^2$, that is, $a \sim c\tau_0 \sim r_e$. This is therefore the limit on distances where classical physics comes into difficulty. It is much smaller than the Compton wavelength $\lambda = \hbar/mc$ where quantum effects become important ($r_e/\lambda = \alpha \sim 10^{-2}$). Therefore the latter always hides the small-distance classical effects. We shall see that rather than diverging linearly (with inverse length) as here, the "bare" mass of a spin $\frac{1}{2}$ electron diverges only logarithmically.

Even ignoring these infinities, we can expect difficulties at short distances or short times. Thus let us give a closer look at the equation of motion, omitting even dissipative and relativistic effects. We rewrite it in three-dimensional notation as

$$m(\dot{\mathbf{v}} - \tau_0 \ddot{\mathbf{v}}) = \mathbf{F}_{\text{ext}} \tag{1-220}$$

In the absence of external force, apart from free-motion solutions $\mathbf{v}$ = constant, it admits runaway self-accelerated solutions $\dot{\mathbf{v}} = \dot{\mathbf{v}}_0\, e^{t/\tau_0}$. Mathematically, this arises from the occurrence of higher derivatives in the equations which lead to totally unphysical situations. To obtain sensible results we must insist on boundary conditions and replace (1-220) by an integro-differential equation which incorporates these conditions (in particular $\dot{\mathbf{v}} \to 0$ at $t \to \infty$ if $\mathbf{F}_{\text{ext}}$ vanishes in this limit). Then

$$m\dot{\mathbf{v}}(t) = \int_0^\infty ds\, e^{-s}\, \mathbf{F}_{\text{ext}}(t + \tau_0 s) \tag{1-221}$$

In this form runaway solutions are eliminated but another unpleasant feature arises, namely preacceleration. If $\mathbf{F}_{\text{ext}}$ is zero for negative t (Fig. 1-5), then $\dot{\mathbf{v}}$ does not vanish but rather starts increasing at earlier times of order τ_0($\sim 10^{-24}$ s for an electron), which is the time required for light to cross the electromagnetic radius. We can again convince ourselves that quantum mechanics should blur this unwanted feature. Switching on an external field during a time Δt implies an un-

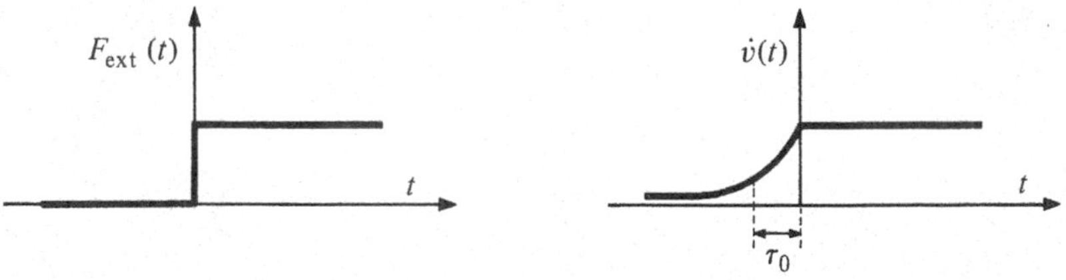

Figure 1-5 Preacceleration of a classical charge.

certainty in energy of order $\Delta E \sim \hbar/\Delta t$. If ΔE is comparable to mc^2, we find $\Delta t \sim \hbar/mc^2 \sim \tau_0/\alpha \sim 137\tau_0$. Classical acausal effects are therefore unobservable.

We have glossed over the difficulties associated with radiative corrections. This should not prevent us, however, from applying these corrections under proper circumstances to study such phenomena as line-width broadening, corrections to scattering, etc.

NOTES

For an account of classical mechanics and its application to field theory see L. D. Landau and E. M. Lifshitz, "Mechanics" and "The Classical Theory of Fields," Pergamon Press, Oxford, 1969 and 1975. A deep treatment of mechanics is given, for instance, in E. Cartan, "Leçons sur les Invariants Intégraux," Hermann, Paris, 1958. Classical electromagnetism and radiation is thoroughly discussed in J. D. Jackson, "Classical Electrodynamics," John Wiley, New York, 1975. Consistency problems of electrodynamics are studied in F. T. Rohrlich, "Classical Charged Particles," Addison-Wesley, Reading, Mass., 1965.

CHAPTER

TWO

THE DIRAC EQUATION

To prepare the construction of a fully relativistic quantum mechanics, we follow the historical path by attempting to build a one-particle theory. We introduce the Klein-Gordon and Dirac equations and discuss their limitations. Applications involve the electromagnetic interactions, the relativistic hydrogen spectrum, and Coulomb scattering. The hole interpretation of negative energy states in terms of antiparticles requires a reformulation, the so-called second quantization, as a many-particle theory.

2-1 TOWARD A RELATIVISTIC WAVE EQUATION

2-1-1 Quantum Mechanics and Relativity

Our first aim will be to try to accommodate the principles of quantum mechanics and of relativistic invariance, namely, to construct a Lorentz covariant wave equation. Along the way, we shall encounter increasing difficulties and inconsistencies, which ultimately will force us to a complete recasting of our physical concepts.

In quantum mechanics, the states of a system are represented by normalized vectors $|\psi\rangle$ (or density matrices $\rho = \Sigma\, p_i |\psi_i\rangle\langle\psi_i|$) of a Hilbert space $\mathscr{H}$: $|\langle\varphi|\psi\rangle|^2$ (or $\langle\varphi|\rho|\varphi\rangle$) is the probability of finding the system in the state $|\varphi\rangle$. Physical observables are identified with self-adjoint, $A = A^\dagger$ (but generally unbounded), operators on the space $\mathscr{H}$. The expectation value of the observable A when the system is in the state $|\psi\rangle$, that is, the average value for many measurements on identically prepared states, is $\langle\psi|A|\psi\rangle$. The time evolution of the system under its self-interaction, or under external forces represented by classical given force fields, is described by the Schrödinger equation:

$$i\hbar\frac{\partial}{\partial t}|\psi(t)\rangle = H|\psi(t)\rangle \tag{2-1}$$

or, equivalently,

$$|\psi(t_2)\rangle = U(t_2, t_1)|\psi(t_1)\rangle$$

H is self-adjoint and U is unitary and satisfies

$$i\hbar\frac{\partial}{\partial t_2}U(t_2, t_1) = H(t_2)U(t_2, t_1)$$

It frequently occurs that a system is invariant under certain symmetries, for instance, symmetries of the external forces. A theorem of Wigner then states that such symmetries are represented by unitary (or antiunitary‡) operators, which map the Hilbert space onto itself, conserve the modulus of scalar products, and commute with H.

On the other hand, special relativity states that the laws of nature are independent of the observer's frame if it belongs to the class of frames—the "galilean frames"—obtained from each other by transformations of the Poincaré group. The latter is generated by the space and time translations, the usual space rotations, and the special Lorentz transformations (or boosts), which relate frames moving with constant relative velocity (see Chap. 1). The speed of light c is an absolute upper bound on the velocity of any signal. Information originating from the space-time point $(\mathbf{x}_0, t_0)$ reaches only points $(\mathbf{x}_1, t_1)$ inside the future cone

$$c^2(t_1 - t_0)^2 - (\mathbf{x}_1 - \mathbf{x}_0)^2 \geq 0 \qquad t_1 - t_0 \geq 0$$

This is the relativistic expression of causality. For typical velocities much smaller than c, galilean mechanics is a reliable approximation.

We may expect some trouble in the search for a relativistic and quantum description of a point particle. Indeed, relativity associates a momentum scale $p = mc$ to a particle of mass m. But the uncertainty relations $\Delta x \cdot \Delta p \sim \hbar$ tell us that for length scales smaller than the Compton wavelength $\lambdabar \equiv \hbar/mc$ ($\lambdabar = 3.8 \times 10^{-11}$ cm for the electron), the concept of a point particle may suffer difficulties. Analyzing the position of the particle with a greater accuracy requires an energy momentum of the same order as the rest mass, thus allowing the creation of new particles. We see that we shall unavoidably be led to the concept of antiparticle. Nevertheless, in an intermediate range quantum relativistic mechanics is worth while and justifies the following developments.

‡ An operator B is said to be antilinear if

$$B(\lambda|\varphi\rangle + \mu|\psi\rangle) = \lambda^* B|\varphi\rangle + \mu^* B|\psi\rangle$$

Defining its adjoint $B^\dagger$ by $\langle B\varphi|\psi\rangle = \langle\varphi|B^\dagger\psi\rangle^* = \langle B^\dagger\psi|\varphi\rangle$ B is antiunitary if $\langle B\varphi|B\psi\rangle = \langle\varphi|B^\dagger B|\psi\rangle = \langle\varphi|\psi\rangle^* = \langle\psi|\varphi\rangle$.

To combine relativistic invariance with quantum mechanics let us return to the correspondence principle. In the usual configuration space representation of quantum mechanics, we associate the operators $i\hbar(\partial/\partial t)$ and $(\hbar/i)\nabla_i = (\hbar/i)(\partial/\partial x^i)$ to the energy E and momentum p^i respectively. For a free massive particle, the energy is given in terms of the momentum by

$$E = \frac{\mathbf{p}^2}{2m} + \text{constant} \qquad \text{in the nonrelativistic picture} \tag{2-2a}$$

$$E^2 = \mathbf{p}^2 c^2 + m^2 c^4 \qquad \text{in the relativistic case} \tag{2-2b}$$

Unless explicitly specified, we shall use the convenient system of units such that $\hbar = c = 1$.

In the same way that the correspondence principle transforms Eq. (2-2a) into the Schrödinger equation for the wave function $\psi(x, t) = \langle x, t|\psi\rangle$:

$$i\frac{\partial}{\partial t}\psi(\mathbf{x}, t) = -\frac{\nabla^2}{2m}\psi(\mathbf{x}, t)$$

it leads, in the relativistic case, from Eq. (2-2b) to the Klein-Gordon equation:

$$\left(\frac{\partial^2}{\partial t^2} - \nabla^2 - m^2\right)\psi(\mathbf{x}, t) = 0 \tag{2-3}$$

Although this equation does not have the Schrödinger-like form (2-1), we may remedy this by casting it in a matrix form. Introducing the notations

$$\psi_{-1} \equiv m\psi \qquad \psi_0 \equiv \frac{\partial\psi}{\partial t} \qquad \psi_i = \frac{\partial\psi}{\partial x^i} \qquad i = 1, 2, 3$$

the vector $\underset{\sim}{\psi} = \{\psi_\alpha\}$ $(\alpha = -1, 0, \ldots, 3)$ satisfies

$$i\frac{\partial\underset{\sim}{\psi}}{\partial t} = \left(m\beta + \frac{1}{i}\boldsymbol{\alpha}\cdot\nabla\right)\underset{\sim}{\psi}$$

for a suitable set of 5×5 hermitian matrices. The reader may explicitly write this set of matrices and obtain an auxiliary condition in order to reproduce a set of equations equivalent to Eq. (2-3).

If we want to interpret ψ as a wave function, we have to find a nonnegative norm, conserved by the time evolution. There does indeed exist a continuity equation:

$$\frac{\partial}{\partial t}\rho + \operatorname{div}\mathbf{j} \equiv \partial_\mu j^\mu = 0 \tag{2-4}$$

where the four-vector $j^\mu \equiv (j^0 = \rho, j^i)$ is defined as

$$\begin{aligned} \rho &= \frac{i}{2m}\left(\psi^*\frac{\partial}{\partial t}\psi - \frac{\partial\psi^*}{\partial t}\psi\right) \\ \mathbf{j} &= \frac{1}{2im}[\psi^*\nabla\psi - (\nabla\psi)^*\psi] \end{aligned} \tag{2-5}$$

In integral form we have equivalently

$$-\frac{\partial}{\partial t}\int_V d^3x\,\rho = \int_S d\mathbf{S}\cdot\mathbf{j}$$

which expresses that the change in the total "charge" inside the volume V corresponds to the flux of $\mathbf{j}$ through the surface S enclosing V. However, the density ρ is not positive definite. Therefore, it may well be considered as the density of a conserved quantity (the electric charge, for instance), but not as a positive probability.

A second problem arises when we realize the existence of negative energy solutions. Any plane wave function

$$\psi(\mathbf{x}, t) = N\, e^{-i(Et-\mathbf{p}\cdot\mathbf{x})}$$

satisfies Eq. (2-3), provided $E^2 = \mathbf{p}^2 + m^2$. Thus negative energies $E = -\sqrt{\mathbf{p}^2 + m^2}$ are on the same footing as the physical ones $E = \sqrt{\mathbf{p}^2 + m^2}$. This is a severe difficulty because the spectrum is no longer bounded from below. It seems that an arbitrarily large amount of energy may be extracted from the system. For a particle initially at rest, this will be the case if an external perturbation allows it to jump over the energy gap $\Delta E = 2m$ between the positive and negative continuum of states. This is clearly a failure of the concept of stable stationary states.

These reasons seemed at a time so overwhelming that they led Dirac to introduce another equation. Although the latter has a positive norm, we shall ultimately have to face the same problems of physical interpretation of the negative energy states. At that stage, we shall come back to the Klein-Gordon equation and recast our relativistic quantum mechanics as a many-body theory, where the negative energy states may be interpreted in terms of antiparticles.

2-1-2 The Dirac Equation

Since the Klein-Gordon equation was found physically unsatisfactory, we shall try to construct a wave equation

$$i\frac{\partial\psi}{\partial t} = \left(\frac{1}{i}\boldsymbol{\alpha}\cdot\nabla + \beta m\right)\psi \equiv H\psi \tag{2-6}$$

where ψ is a vector wave function and $\boldsymbol{\alpha}$, β are hermitian matrices to make H hermitian, such that a positive conserved probability density exists. We now insist on the three following points:

1. The components of ψ must satisfy the Klein-Gordon equation, so that a plane wave with $E^2 = \mathbf{p}^2 + m^2$ is a solution.
2. There exists a four-vector current density which is conserved and whose fourth component is a positive density.
3. The components of ψ do not have to satisfy any auxiliary condition, namely, at a given time they are independent functions of x. We shall also have to verify the relativistic covariance of this formalism.

Dirac proposed that the α_i and β matrices be anticommuting matrices of square equal to one:

$$\begin{aligned} \{\alpha_i, \alpha_k\} &= 0 \qquad \text{for } i \neq k \\ \{\alpha_i, \beta\} &= 0 \\ \alpha_i^2 = \beta^2 &= I \end{aligned} \tag{2-7}$$

with the bracket $\{A, B\}$ of two operators standing for the symmetric combination $AB + BA$, called the anticommutator.

It is easy to verify that condition 1 is fulfilled:

$$\begin{aligned} -\frac{\partial^2}{\partial t^2}\psi &= \left(\frac{1}{i}\boldsymbol{\alpha}\cdot\nabla + \beta m\right)^2 \psi \\ &= (-\nabla^2 + m^2)\psi \end{aligned}$$

Let us introduce the notation γ^μ:

$$\begin{aligned} \gamma^0 &= \beta \\ \gamma^i &= \beta\alpha^i \qquad i = 1, 2, 3 \\ \{\gamma^\mu, \gamma^\nu\} &= 2g^{\mu\nu} \end{aligned} \tag{2-8}$$

and the Feynman "slash":

$$\not{a} \equiv a_\mu \gamma^\mu$$

This enables us to rewrite the Dirac equation as

$$(i\gamma^\mu \partial_\mu - m)\psi \equiv (i\not{\partial} - m)\psi = 0 \tag{2-9}$$

The Klein-Gordon equation is then obtained by multiplying by $(i\not{\partial} + m)$. Four is the smallest dimension in which matrices fulfilling (2-8) can be found.

The matrices α^i and β have eigenvalues equal to ± 1; for $i \neq j$, $\det \alpha^i\alpha^j = \det - \alpha^j\alpha^i = (-1)^d \det \alpha^i\alpha^j$; thus their dimension d must be even. Since for $d = 2$ there exist only three anticommuting hermitian matrices, the Pauli matrices, we have $d \geq 4$.

An explicit representation is provided by

$$\begin{aligned} \gamma^0 &= \begin{pmatrix} I & 0 \\ 0 & -I \end{pmatrix} \qquad \gamma^i = \begin{pmatrix} 0 & \sigma^i \\ -\sigma^i & 0 \end{pmatrix} \\ \beta &= \begin{pmatrix} I & 0 \\ 0 & -I \end{pmatrix} \qquad \alpha^i = \begin{pmatrix} 0 & \sigma^i \\ \sigma^i & 0 \end{pmatrix} \end{aligned} \tag{2-10}$$

in terms of the 2×2 unit I and Pauli σ^i matrices. This representation is useful when discussing the nonrelativistic limit of the Dirac equation.

Among all possible equivalent representations, obtained by a nonsingular transformation: $\gamma \to U\gamma U^{-1}$, the Majorana representation plays a special role. It is designed so as to make the Dirac equation real. This is achieved by interchanging α_2 and β and changing the sign of α_1 and α_3 in the previous representation: $\hat{\alpha}_1 = -\alpha_1, \hat{\alpha}_2 = \beta, \hat{\alpha}_3 = -\alpha_3, \hat{\beta} = \alpha_2$. Then only $\hat{\beta}$ is imaginary and the Dirac equation:

$$\left(\frac{\partial}{\partial t} + \hat{\boldsymbol{\alpha}} \cdot \nabla + i\hat{\beta}m\right)\psi = 0$$

is real; its solutions are linear combinations of real solutions. The matrix U which performs this change of representation and the new form of the γ matrices may be easily determined (see the Appendix).

In the four-dimensional representation (2-10), ψ may be written as a bispinor $\psi = \begin{pmatrix}\varphi \\ \chi\end{pmatrix}$ in terms of two-component spinors φ and χ. For reasons that will soon be clear, φ and χ are referred to as the large and small components respectively. They satisfy

$$\begin{cases} i\dfrac{\partial \varphi}{\partial t} = m\varphi + \dfrac{1}{i}\boldsymbol{\sigma} \cdot \nabla\chi \\[2ex] i\dfrac{\partial \chi}{\partial t} = -m\chi + \dfrac{1}{i}\boldsymbol{\sigma} \cdot \nabla\varphi \end{cases} \tag{2-11}$$

It is interesting to notice the similarity between these equations and two of the four Maxwell equations:

$$\text{rot } \mathbf{E} + \frac{\partial \mathbf{B}}{\partial t} = 0 \qquad \text{rot } \mathbf{B} - \frac{\partial \mathbf{E}}{\partial t} = 0$$

or, explicitly,

$$i\frac{\partial \mathbf{E}}{\partial t} = \frac{1}{i}\mathbf{S} \cdot \nabla(i\mathbf{B}) \qquad i\frac{\partial (i\mathbf{B})}{\partial t} = \frac{1}{i}\mathbf{S} \cdot \nabla(\mathbf{E})$$

where $(S^i)_{jk} \equiv (1/i)\varepsilon_{ijk}$.

The spin matrices S^i play for the spin 1 electromagnetic field the same role as the Pauli matrices $\boldsymbol{\sigma}$ for the spin $\frac{1}{2}$ and $(\mathbf{E}, i\mathbf{B})$ is analogous to (φ, χ).

The main reason for the construction of the Dirac equation was to obtain a positive probability density $\rho = j^0$, together with a continuity equation $\partial_\mu j^\mu = 0$. Since ψ is a complex spinor, ρ has to be of the form $\psi^\dagger \mathscr{R} \psi$ in order to be real and positive. Let us first derive the Dirac equation for $\psi^\dagger$. From (2-9), we deduce

$$\psi^\dagger(i\gamma^{\dagger\mu}\overleftarrow{\partial}_\mu + m) = 0$$

But $\gamma^{\mu\dagger}$ is easily expressed in terms of γ^μ:

$$\gamma^{0\dagger} = \gamma^0 \quad \boldsymbol{\gamma}^\dagger = (\beta\boldsymbol{\alpha})^\dagger = \boldsymbol{\alpha}\beta = \beta(\beta\boldsymbol{\alpha})\beta = \gamma^0\boldsymbol{\gamma}\gamma^0$$

Thus, introducing $\bar{\psi}$:

$$\bar{\psi} \equiv \psi^\dagger\gamma^0 \tag{2-12}$$

we have

$$\bar{\psi}(i\overleftarrow{\not{\partial}} + m) = 0 \tag{2-9a}$$

Combining Eqs. (2-9) and (2-9*a*) leads to

$$\bar{\psi}(\overleftarrow{\not\partial} + \overrightarrow{\not\partial})\psi \equiv \partial_\mu(\bar{\psi}\gamma^\mu\psi) = 0$$

We have, therefore, a candidate for the current

$$j^\mu = \bar{\psi}\gamma^\mu\psi \begin{cases} j^0 = \rho = \bar{\psi}\gamma^0\psi = \psi^\dagger\psi = \varphi^\dagger\varphi + \chi^\dagger\chi \\ \mathbf{j} = \bar{\psi}\boldsymbol{\gamma}\psi = \psi^\dagger\boldsymbol{\alpha}\psi = \varphi^\dagger\boldsymbol{\sigma}\chi + \chi^\dagger\boldsymbol{\sigma}\varphi \end{cases} \tag{2-13}$$

The density ρ is positive. Small and large components contribute equally to ρ whereas $\mathbf{j}$ involves cross terms. We shall see below that j^μ transforms as a Lorentz four-vector.

2-1-3 Relativistic Covariance

According to the relativity principle, we want to verify that the Dirac equation keeps its form in two frames related by a Poincaré transformation. Alternatively, we require that to a system satisfying the equation with certain boundary conditions in a given frame we may associate by Poincaré transformations a family of transformed states satisfying the same equation, with transformed boundary conditions.

Sticking to the first point of view (independence with respect to the observer), we first remark that translation invariance is obvious. Consider now a Lorentz transformation Λ. Let our system be described by the wave function ψ in the first frame and by ψ' in the transformed frame. Both must satisfy the Dirac equation:

$$i\gamma^\mu \frac{\partial}{\partial x^\mu}\psi(x) - m\psi(x) = 0 \tag{2-9b}$$

$$i\gamma^\mu \frac{\partial}{\partial x'^\mu}\psi'(x') - m\psi'(x') = 0 \qquad \text{with } x' = \Lambda x \tag{2-9c}$$

There must be a local relation between ψ and ψ', so that the observer in the second frame may reconstruct ψ' when ψ is given. We assume that this relation is linear:

$$\psi'(x') = S(\Lambda)\psi(x) \tag{2-14}$$

where $S(\Lambda)$ is a nonsingular 4×4 matrix. Equation (2-9*c*) now reads

$$i\gamma^\mu \frac{\partial x^\nu}{\partial x'^\mu}\frac{\partial}{\partial x^\nu}S(\Lambda)\psi(x) - mS(\Lambda)\psi(x) = 0$$

In order that this equation be a consequence of (2-9*b*) for any ψ, and since $\partial x^\nu/\partial x'^\mu = (\Lambda^{-1})^\nu{}_\mu$, we must have

$$S(\Lambda)\gamma^\mu S^{-1}(\Lambda) = (\Lambda^{-1})^\mu{}_\nu\gamma^\nu \tag{2-15}$$

Let us first construct $S(\Lambda)$ for an infinitesimal proper transformation Λ, which may be written as

$$\Lambda^{\mu}{}_{\nu} = g^{\mu}{}_{\nu} + \omega^{\mu}{}_{\nu} \qquad (\Lambda^{-1})^{\mu}{}_{\nu} = g^{\mu}{}_{\nu} - \omega^{\mu}{}_{\nu} + \cdots$$

where the infinitesimal matrix $\omega_{\mu\nu}$ is antisymmetric. We write

$$\begin{aligned} S(\Lambda) &= I - \frac{i}{4}\sigma_{\mu\nu}\omega^{\mu\nu} + \cdots \\ S^{-1}(\Lambda) &= I + \frac{i}{4}\sigma_{\mu\nu}\omega^{\mu\nu} + \cdots \end{aligned} \tag{2-16}$$

where the matrices $\sigma_{\mu\nu}$ are antisymmetric in $\mu\nu$. To first order in ω, Eq. (2-15) yields

$$[\gamma^{\mu}, \sigma_{\alpha\beta}] = 2i(g^{\mu}{}_{\alpha}\gamma_{\beta} - g^{\mu}{}_{\beta}\gamma_{\alpha}) \tag{2-17}$$

A set of matrices $\sigma_{\alpha\beta}$ satisfying this relation is given by

$$\sigma_{\alpha\beta} = \frac{i}{2}[\gamma_{\alpha}, \gamma_{\beta}] \tag{2-18}$$

A finite transformation is of the form

$$S(\Lambda) = e^{-(i/4)\sigma_{\alpha\beta}\omega^{\alpha\beta}} \tag{2-19}$$

where $\omega^{\alpha\beta}$ is now finite.

For spatial rotations S is unitary, whereas it is hermitian for Lorentz boosts.

The form of the finite transformations is most easily derived in the chiral representation for γ matrices:

$$\begin{aligned} &\gamma^0 = \beta = \begin{pmatrix} 0 & -I \\ -I & 0 \end{pmatrix} \qquad \boldsymbol{\alpha} = \begin{pmatrix} \boldsymbol{\sigma} & 0 \\ 0 & -\boldsymbol{\sigma} \end{pmatrix} \qquad \boldsymbol{\gamma} = \begin{pmatrix} 0 & \boldsymbol{\sigma} \\ -\boldsymbol{\sigma} & 0 \end{pmatrix} \\ &\sigma_{0i} = \frac{i}{2}[\gamma_0, \gamma_i] = -i\alpha_i = \frac{1}{i}\begin{pmatrix} \sigma_i & 0 \\ 0 & -\sigma_i \end{pmatrix} \\ &\sigma_{ij} = \frac{i}{2}[\gamma_i, \gamma_j] = -\frac{i}{2}[\alpha_i, \alpha_j] = \varepsilon_{ijk}\begin{pmatrix} \sigma_k & 0 \\ 0 & \sigma_k \end{pmatrix} \end{aligned} \tag{2-20}$$

In this representation, the two Pauli spinors of the decomposition of the bispinor ψ transform independently under rotations and boosts. The representation of the Lorentz group [more exactly, of its covering group $SL(2, C)$] is reducible into a sum of two inequivalent representations: $(\frac{1}{2}, 0) + (0, \frac{1}{2})$. However, we shall see that the representation is irreducible if we include the transformation under parity (space reflection).

We recall that the representations of the Poincaré group are classified according to the values of two Casimir operators P^2 and W^2; P_{μ} is the momentum energy operator, which is the infinitesimal generator of translations, whereas W_{μ} is constructed from the angular momentum operator $J_{\mu\nu}$, the infinitesimal generator of Lorentz transformations, as

$$W_{\mu} = -\tfrac{1}{2}\varepsilon_{\mu\nu\rho\sigma}J^{\nu\rho}P^{\sigma} \tag{2-21}$$

If M^2 denotes the eigenvalue of P^2, W^2 takes only values of the form

$$W^2 = -M^2 S(S+1)$$

where the spin S is integer or half integer.

For the solutions of the Dirac equation, and therefore of the Klein-Gordon equation, $P^2 = -\partial^2$ takes the value m^2, while $J_{\mu\nu}$ is given by

$$\begin{aligned}\psi'(x) &= \left(I - \frac{i}{2} J_{\mu\nu}\omega^{\mu\nu}\right)\psi(x)\\ &= \left(I - \frac{i}{4}\sigma_{\mu\nu}\omega^{\mu\nu}\right)\psi(x^\rho - \omega^\rho{}_\nu x^\nu)\\ &= \left(I - \frac{i}{4}\sigma_{\mu\nu}\omega^{\mu\nu} + x_\mu \omega^{\mu\nu}\partial_\nu\right)\psi(x)\end{aligned}$$

which yields

$$J_{\mu\nu} = \tfrac{1}{2}\sigma_{\mu\nu} + i(x_\mu \partial_\nu - x_\nu \partial_\mu) \tag{2-22}$$

Let us then compute W_μ from Eq. (2-21):

$$W_\mu = -\frac{1}{4i}\varepsilon_{\mu\nu\rho\sigma}\sigma^{\nu\rho}\partial^\sigma \tag{2-21a}$$

The orbital contribution has disappeared, justifying that W_μ corresponds to intrinsic angular momentum. We then use the identity

$$\begin{aligned}\varepsilon_{\mu\alpha\beta\gamma}\varepsilon^{\mu}{}_{\alpha'\beta'\gamma'} &= -\det(g_{\tau\tau'})\\ &= -\sum_P (-1)^P g_{\alpha P_{\alpha'}} g_{\beta P_{\beta'}} g_{\gamma P_{\gamma'}}\end{aligned} \tag{2-23}$$

where in the first expression τ (or τ' respectively) takes the values α, β, γ (α', β', γ'), and in the second one, the sum runs over the permutations P of (α', β', γ'). After some algebra using the Dirac equation, this leads to

$$W^2 = -\tfrac{3}{4}m^2 = -\tfrac{1}{2}(\tfrac{1}{2}+1)m^2 \tag{2-24}$$

Thus the equation describes spin $\frac{1}{2}$ particles.

Finally, we derive the transformation law of the spinor ψ under parity. We have again to find $S(\Lambda)$ satisfying (2-15), where Λ denotes the matrix

$$\Lambda^\mu{}_\nu = \begin{pmatrix}1 & & & \\ & -1 & & \\ & & -1 & \\ & & & -1\end{pmatrix} \tag{2-25}$$

It is easy to see that

$$\psi'(x') = \eta_P \gamma^0 \psi(x) \tag{2-26}$$

is the desired transformation. Here η_P is an arbitrary, unobservable phase. The important point is that the positive and negative energy solutions have relative opposite parities corresponding to the two opposite eigenvalues of γ^0. After the reinterpretation of negative energy solutions, this will mean opposite intrinsic parities for particle and antiparticle.

The various bilinear forms constructed from ψ and $\bar{\psi}$ play an important role in the sequel. The remainder of this section is devoted to the study of their transformation properties under Lorentz transformations. From Eq. (2-14), we deduce that

$$\begin{aligned}\bar{\psi}'(x') &= \bar{\psi}(x)\gamma^0 S(\Lambda)^\dagger \gamma^0 \\ &= \bar{\psi}(x) S^{-1}(\Lambda)\end{aligned}$$

where the second expression is verified using the explicit expressions (2-19) of $S(\Lambda)$ [and (2-26) for parity]. Thus a bilinear product $\bar{\psi}(x)A\psi(x)$ transforms according to

$$\bar{\psi}'(x')A\psi'(x') = \bar{\psi}(x)S^{-1}(\Lambda)AS(\Lambda)\psi(x) \tag{2-27}$$

For instance, from (2-15), we learn that $\bar{\psi}\gamma^\mu\psi$ transforms as a four-vector:

$$\bar{\psi}'(x')\gamma^\mu\psi'(x') = \Lambda^\mu{}_\nu \bar{\psi}(x)\gamma^\nu\psi(x) \tag{2-28}$$

whereas $\bar{\psi}(x)\psi(x)$ is a (nonpositive definite) scalar density.

More generally, any 4×4 matrix may be expanded on a basis of 16 matrices. It may be shown that the algebra generated by the γ matrices—a Clifford algebra for mathematicians—is nothing but the complete algebra of these 4×4 matrices. Let us introduce the notation‡

$$\gamma_5 \equiv \gamma^5 \equiv i\gamma^0\gamma^1\gamma^2\gamma^3 \tag{2-29}$$

In the representation (2-10)

$$\gamma^5 = \begin{pmatrix} 0 & I \\ I & 0 \end{pmatrix} \tag{2-30}$$

The matrix γ^5 satisfies

$$\{\gamma^5, \gamma^\mu\} = 0 \qquad \text{and} \qquad (\gamma^5)^2 = I \tag{2-31}$$

We now consider the 16 matrices:

$$\begin{aligned}\Gamma^S &\equiv I \\ \Gamma^V{}_\mu &\equiv \gamma_\mu \\ \Gamma^T{}_{\mu\nu} &\equiv \sigma_{\mu\nu} = \frac{i}{2}[\gamma_\mu, \gamma_\nu] \\ \Gamma^A_\mu &\equiv \gamma_5\gamma_\mu \\ \Gamma^P &\equiv \gamma_5\end{aligned}$$

‡ A consensus has not been reached on this notation. Some authors do not introduce the factor i (then $\gamma_5^2 = -1$), or use a different sign.

They have the following properties:

1. $(\Gamma^a)^2 = \pm I$.
2. For any $\Gamma^a (\Gamma^a \neq \Gamma^S = I)$, there exists a Γ^b such that $\Gamma^a\Gamma^b = -\Gamma^b\Gamma^a$.
3. Thus the trace of all Γ^a, except Γ^S, vanishes:

$$\operatorname{tr} \Gamma^a (\Gamma^b)^2 = -\operatorname{tr} \Gamma^b \Gamma^a \Gamma^b = -\operatorname{tr} (\Gamma^b)^2 \Gamma^a = 0$$

4. For any couple (Γ^a, Γ^b), $a \neq b$, there exists $\Gamma^c \neq \Gamma^S = I$ such that $\Gamma^a\Gamma^b = \Gamma^c$ up to a factor ± 1 or $\pm i$.
5. From these properties, we deduce the linear independence of our set $\{\Gamma^a\}$. Suppose that

$$\sum_a \lambda_a \Gamma^a = 0$$

Multiplying successively by all the Γ^a and taking the trace leads to $\lambda_a = 0$ for all a.
6. We note the following identities:

$$\begin{aligned} \gamma^\mu \gamma_\mu &= 4 & \gamma^\mu \gamma^\nu \gamma_\mu &= -2\gamma^\nu \\ \gamma^\mu \gamma^\nu \gamma^\rho \gamma_\mu &= 4g^{\nu\rho} & \gamma^\mu \gamma^\nu \gamma^\rho \gamma^\sigma \gamma_\mu &= -2\gamma^\sigma \gamma^\rho \gamma^\nu \end{aligned} \tag{2-32}$$

These and other useful identities have been listed in the Appendix.

Using this basis, we now give the properties of the corresponding bilinears $\bar{\psi} A \psi$ under proper Lorentz or parity transformations:

S:	$\bar{\psi}'(x)\psi'(x') = \bar{\psi}(x)\psi(x)$	scalar
V:	$\bar{\psi}'(x')\gamma^\mu\psi'(x') = \Lambda^\mu{}_\nu \bar{\psi}(x)\gamma^\nu\psi(x)$	vector
T:	$\bar{\psi}'(x')\sigma^{\mu\nu}\psi'(x') = \Lambda^\mu{}_\rho \Lambda^\nu{}_\sigma \bar{\psi}(x)\sigma^{\rho\sigma}\psi(x)$	antisymmetric tensor
A:	$\bar{\psi}'(x')\gamma_5\gamma^\mu\psi'(x') = \det(\Lambda)\Lambda^\mu{}_\nu \bar{\psi}(x)\gamma_5\gamma^\nu\psi(x)$	pseudovector
P:	$\bar{\psi}'(x')\gamma_5\psi'(x') = \det(\Lambda)\bar{\psi}(x)\gamma_5\psi(x)$	pseudoscalar

(2-33)

The prefix "pseudo" refers to parity and x' stands for $x'^\mu = \Lambda^\mu{}_\nu x^\nu$.

2-2 PHYSICAL CONTENT

2-2-1 Plane Wave Solutions and Projectors

We seek plane wave solutions of the free Dirac equation (2-9), i.e., solutions of the form

$$\begin{aligned} \psi^{(+)}(x) &= e^{-ik\cdot x}u(k) \qquad \text{positive energy} \\ \psi^{(-)}(x) &= e^{ik\cdot x}v(k) \qquad \text{negative energy} \end{aligned} \tag{2-34}$$

with the condition that k^0 is positive. To verify the Klein-Gordon equation, we also must have $k^2 = m^2$. The positive time-like four-vector k^μ is nothing but the energy momentum of the particle (with $\hbar$ set equal to one). The Dirac equation implies

$$(\not{k} - m)u(k) = 0 \qquad (\not{k} + m)v(k) = 0 \tag{2-35}$$

Let us assume that the particle is massive, $m \neq 0$. In the rest frame of the particle, $k^\mu = (m, \mathbf{0})$, and Eqs. (2-35) reduce to

$$(\gamma^0 - 1)u(m, \mathbf{0}) = 0$$
$$(\gamma^0 + 1)v(m, \mathbf{0}) = 0$$

There are clearly two linearly independent u solutions, and two v's. In the usual representation (2-10), we denote them as follows:

$$u^{(1)}(m, \mathbf{0}) = \begin{pmatrix} 1 \\ 0 \\ 0 \\ 0 \end{pmatrix} \quad u^{(2)}(m, \mathbf{0}) = \begin{pmatrix} 0 \\ 1 \\ 0 \\ 0 \end{pmatrix} \quad v^{(1)}(m, \mathbf{0}) = \begin{pmatrix} 0 \\ 0 \\ 1 \\ 0 \end{pmatrix} \quad v^{(2)}(m, \mathbf{0}) = \begin{pmatrix} 0 \\ 0 \\ 0 \\ 1 \end{pmatrix} \tag{2-36}$$

We could now boost these solutions from rest up to a velocity $v = |\mathbf{k}|/k^0$ by a pure Lorentz transformation, using Eq. (2-19). It is simpler to observe that

$$(\not{k} - m)(\not{k} + m) = k^2 - m^2 = 0$$

so that we may write

$$\begin{aligned} u^{(\alpha)}(k) &= \frac{\not{k} + m}{\sqrt{2m(m+E)}}\, u^{(\alpha)}(m, \mathbf{0}) = \begin{pmatrix} \left(\dfrac{E+m}{2m}\right)^{1/2} \varphi^{(\alpha)}(m, \mathbf{0}) \\ \dfrac{\boldsymbol{\sigma}\cdot\mathbf{k}}{[2m(m+E)]^{1/2}}\, \varphi^{(\alpha)}(m, \mathbf{0}) \end{pmatrix} \\ v^{(\alpha)}(k) &= \frac{-\not{k} + m}{\sqrt{2m(m+E)}}\, v^{(\alpha)}(m, \mathbf{0}) = \begin{pmatrix} \dfrac{\boldsymbol{\sigma}\cdot\mathbf{k}}{[2m(m+E)]^{1/2}}\, \chi^{(\alpha)}(m, \mathbf{0}) \\ \left(\dfrac{E+m}{2m}\right)^{1/2} \chi^{(\alpha)}(m, \mathbf{0}) \end{pmatrix} \end{aligned} \tag{2-37}$$

Here E denotes the positive quantity: $E \equiv k^0 = (\mathbf{k}^2 + m^2)^{1/2}$; the two component spinors φ and χ are the nonvanishing components of $u(m, \mathbf{0})$ and $v(m, \mathbf{0})$ respectively.

For the conjugate spinors we find

$$\begin{aligned} \bar{u}^{(\alpha)}(k) &= \bar{u}^{(\alpha)}(m, \mathbf{0})\, \frac{\not{k} + m}{\sqrt{2m(m+E)}} \\ \bar{v}^{(\alpha)}(k) &= \bar{v}^{(\alpha)}(m, \mathbf{0})\, \frac{-\not{k} + m}{\sqrt{2m(m+E)}} \end{aligned} \tag{2-38}$$

The normalization factors have been chosen in order that

$$\begin{aligned} \bar{u}^{(\alpha)}(k)u^{(\beta)}(k) &= \delta_{\alpha\beta} \qquad & \bar{u}^{(\alpha)}(k)v^{(\beta)}(k) &= 0 \\ \bar{v}^{(\alpha)}(k)v^{(\beta)}(k) &= -\delta_{\alpha\beta} \qquad & \bar{v}^{(\alpha)}(k)u^{(\beta)}(k) &= 0 \end{aligned} \tag{2-39}$$

Consider now the matrices

$$\begin{aligned} \Lambda_+(k) &\equiv \sum_{\alpha=1,2} u^{(\alpha)}(k) \otimes \bar{u}^{(\alpha)}(k) \\ &= \frac{1}{2m(m+E)} (\not{k} + m) \frac{1+\gamma^0}{2} (\not{k} + m) \\ &= \frac{\not{k} + m}{2m} \end{aligned} \tag{2-40}$$

where use has been made of the identity valid for $k^2 = m^2$:

$$(\not{k} + m)\gamma^0(\not{k} + m) = 2E(\not{k} + m)$$

Similarly, let

$$\begin{aligned} \Lambda_-(k) &\equiv -\sum_{\alpha=1,2} v^{(\alpha)}(k) \otimes \bar{v}^{(\alpha)}(k) \\ &= \frac{1}{2m(m+E)} (\not{k} - m) \frac{1-\gamma^0}{2} (\not{k} - m) \\ &= \frac{-\not{k} + m}{2m} \end{aligned} \tag{2-41}$$

The operators Λ_+ and Λ_- project over the positive and negative energy states respectively. They satisfy

$$\begin{aligned} \Lambda_\pm^2(k) &= \Lambda_\pm(k) \\ \operatorname{tr} \Lambda_\pm(k) &= 2 \\ \Lambda_+(k) + \Lambda_-(k) &= I \end{aligned} \tag{2-42}$$

The normalization in (2-39) is Lorentz invariant. However, the positive definite density per unit volume is $\rho = j^0(k) = \bar{\psi}(k)\gamma^0\psi(k)$. Let us compute it for our plane wave functions

$$\begin{aligned} \bar{\psi}^{(+)(\alpha)}(x)\gamma^0 \psi^{(+)(\beta)}(x) &= \bar{u}^{(\alpha)}(k)\gamma^0 u^{(\beta)}(k) \\ &= \bar{u}^{(\alpha)}(k) \frac{\{\not{k}, \gamma^0\}}{2m} u^{(\beta)}(k) \\ &= \frac{E}{m} \delta_{\alpha\beta} \end{aligned} \tag{2-43a}$$

for positive energy solutions, and

$$\bar{\psi}^{(-)(\alpha)}(x)\gamma^0\psi^{(-)(\beta)}(x) = \bar{v}^{(\alpha)}(k)\gamma^0 v^{(\beta)}(k)$$
$$= -\bar{v}^{(\alpha)}(k)\frac{\{\not{k}, \gamma^0\}}{2m}v^{(\beta)}(k)$$
$$= \frac{E}{m}\delta_{\alpha\beta} \tag{2-43b}$$

for negative ones. The spinors have been normalized at rest; since density times volume has to remain constant, when the latter is reduced by the contraction factor E/m the former must increase by the same amount.

Positive and negative energy states are mutually orthogonal, if we consider states with opposite energies but the same three-momentum:

$$\psi^{(+)(\alpha)}(x) = e^{-i(k^0x^0 - \mathbf{k}\cdot\mathbf{x})}u^{(\alpha)}(k)$$
$$\psi^{(-)(\beta)}(x) = e^{i(k^0x^0 + \mathbf{k}\cdot\mathbf{x})}v^{(\beta)}(\tilde{k}) \qquad \text{with } \tilde{k} \equiv (k^0, -\mathbf{k}) \tag{2-44}$$
$$\bar{\psi}^{(-)(\beta)}(x)\psi^{(+)(\alpha)}(x) = e^{-2ik^0x^0}\bar{v}^{(\beta)}(\tilde{k})\gamma^0 u^{(\alpha)}(k)$$
$$= \tfrac{1}{2}e^{-2ik^0x^0}\bar{v}^{(\beta)}(\tilde{k})\left(-\frac{\tilde{\not{k}}}{m}\gamma^0 + \gamma^0\frac{\not{k}}{m}\right)u^{(\alpha)}(k) = 0$$

The physical meaning of negative energy solutions has yet to be clarified. Also, our construction of plane wave states does not make sense for zero-mass particles. Those will be studied in detail in Sec. 2-4-3.

To characterize the remaining degeneracy of the plane wave solutions u and v, we construct the projectors onto states of definite polarization. For any spacelike normalized four-vector n ($n^2 = -1$) orthogonal to k, we have, from (2-21),

$$W\cdot n = -\tfrac{1}{4}\varepsilon_{\mu\nu\rho\sigma}n^\mu k^\nu\sigma^{\rho\sigma}$$
$$= -\tfrac{1}{2}\gamma_5\not{n}\not{k} \tag{2-45}$$

Therefore, in the rest frame

$$W^0 = 0 \qquad \frac{\mathbf{W}}{m} = \tfrac{1}{2}\gamma^5\gamma^0\boldsymbol{\gamma} \equiv \tfrac{1}{2}\boldsymbol{\Sigma}$$

In the usual basis, $\boldsymbol{\Sigma} = \begin{pmatrix}\boldsymbol{\sigma} & 0\\ 0 & \boldsymbol{\sigma}\end{pmatrix}$. If we choose n along the z axis, $n = n_{(3)} \equiv (0, 0, 0, 1)$, we see that the solutions (2-36) are eigenstates of $-W\cdot n_{(3)}/m = W^3/m$, with eigenvalues $+\frac{1}{2}$ (spin-up) for $u^{(1)}$ and $v^{(1)}$, and $-\frac{1}{2}$ (spin-down) for $u^{(2)}$ and $v^{(2)}$. The projector onto $u^{(1)}(m, \mathbf{0})$ and $v^{(2)}(m, \mathbf{0})$ may thus be written

$$P(n_{(3)}) = \frac{I + \gamma_5\not{n}_{(3)}}{2} = \frac{1}{2}\begin{pmatrix}I + \sigma_3 & 0\\ 0 & I - \sigma_3\end{pmatrix}$$

After a Lorentz transformation, the spinors $u^{(\alpha)}(k)$, $v^{(\alpha)}(k)$ are eigenstates of $-W\cdot n/m$, where n is now the transform of $n_{(3)}$:

$$
\begin{aligned}
-\frac{W\cdot n}{m}u^{(\alpha)}(k) &= \frac{1}{2}\gamma_5 \not{n} u^{(\alpha)}(k) = \pm\frac{1}{2}u^{(\alpha)}(k) \\
-\frac{W\cdot n}{m}v^{(\alpha)}(k) &= -\frac{1}{2}\gamma_5 \not{n} v^{(\alpha)}(k) = \pm\frac{1}{2}v^{(\alpha)}(k)
\end{aligned}
\tag{2-46}
$$

The plus sign refers to $\alpha = 1$, the minus sign to $\alpha = 2$. The projector onto $u^{(1)}(k)$ and $v^{(2)}(k)$ reads

$$P(n) = \tfrac{1}{2}(I + \gamma_5 \not{n}) \tag{2-47}$$

This expression remains valid for an arbitrary normalized vector n, orthogonal to k; $P(n)$ projects onto the state which in its rest frame has a spin $\boldsymbol{\sigma}\cdot\mathbf{n}/2 = \frac{1}{2}$ for a positive energy solution, and a spin $\boldsymbol{\sigma}\cdot\mathbf{n}/2 = -\frac{1}{2}$ for a negative energy one. (Note the signs!!)

We shall denote $u(p, n)$ and $v(p, n)$ the (positive and negative energy respectively) eigenvectors of $P(n)$:

$$
\begin{aligned}
P(n)u(k, n) &= u(k, n) \\
P(n)v(k, n) &= v(k, n)
\end{aligned}
\tag{2-48}
$$

The projector $P(n)$ has the following properties:

$$
\begin{aligned}
[\Lambda_\pm(k), P(n)] &= 0 \\
\Lambda_+(k)P(n) + \Lambda_-(k)P(n) + \Lambda_+(k)P(-n) + \Lambda_-(k)P(-n) &= I \\
\operatorname{tr} \Lambda_\pm(k)P(\pm n) &= 1
\end{aligned}
$$

Relaxing the condition on the norm of n,

$$\rho(n) = \frac{1 + \gamma_5 \not{n}}{2} \qquad -1 < n^2 < 0$$

may be interpreted as a spin density matrix

$$\operatorname{tr} \rho = 2 \qquad \operatorname{tr} \rho^2 < 4$$

There exists a particular choice of n, such that $\mathbf{n}$ is proportional to $\mathbf{k}$ in the reference frame. Let n_k be equal to

$$n_k = \left(\frac{|\mathbf{k}|}{m}, \frac{k^0}{m}\frac{\mathbf{k}}{|\mathbf{k}|}\right) \tag{2-49}$$

This definition of polarization is called helicity, and is such that

$$P(n_k)\Lambda_\pm(k) = \left(I \pm \frac{\boldsymbol{\Sigma}\cdot\mathbf{k}}{|\mathbf{k}|}\right)\Lambda_\pm(k) \tag{2-50}$$

Therefore $P(n_k)$ projects over positive helicity, positive energy and negative helicity, negative energy states.

In the ultrarelativistic limit $m/k_0 \to 0$, $|\mathbf{k}|/k_0 \to 1$, $n_k^\mu \to k^\mu/m$, and

$$P(\pm n_k)\Lambda_+(k) \to \frac{1 \pm \gamma_5}{2} \frac{\not{k} + m}{2}$$
$$P(\pm n_k)\Lambda_-(k) \to \frac{1 \mp \gamma_5}{2} \frac{-\not{k} + m}{2} \tag{2-51}$$

2-2-2 Wave Packets

Let us proceed to the construction of normalizable wave packets. We would like to superpose only positive energy plane waves, the only physically sensible ones at this stage. However, we shall be led to inconsistencies and will have to abandon this requirement. Let $\psi^{(+)}(x)$ be

$$\psi^{(+)}(x) = \int \frac{d^3p}{(2\pi)^3} \frac{m}{E} \sum_{\alpha=1,2} b(p, \alpha) u^{(\alpha)}(p)\, e^{-ip\cdot x} \tag{2-52}$$

The factor $[1/(2\pi)^3](m/E)$ is designed to make the normalization condition simple:

$$\int d^3x\, j^{(+)0}(t, x) = \int \frac{d^3x}{(2\pi)^6} \iint d^3p\, d^3p' \frac{m^2}{EE'} \sum_{\alpha,\alpha'} b^*(p, \alpha) b(p', \alpha')$$
$$\times\, u^{(\alpha)\dagger}(p) u^{(\alpha')}(p')\, e^{i(E-E')t - i(\mathbf{p}-\mathbf{p}')\cdot\mathbf{x}}$$
$$= \sum_\alpha \int \frac{d^3p}{(2\pi)^3} \frac{m}{E} |b(p, \alpha)|^2 = 1 \tag{2-53}$$

where we observe that d^3p/E is a Lorentz invariant measure.

We also compute the total current

$$\mathbf{J}^{(+)} = \int d^3x\, \mathbf{j}^{(+)}(t, \mathbf{x}) = \int \frac{d^3x}{(2\pi)^6} \iint d^3p\, d^3p' \frac{m^2}{EE'} \sum_{\alpha\alpha'} b^*(p, \alpha) b(p', \alpha')$$
$$\times\, u^{(\alpha)\dagger}(p) \boldsymbol{\alpha} u^{(\alpha')}(p')\, e^{i(E-E')t - i(\mathbf{p}-\mathbf{p}')\cdot\mathbf{x}}$$
$$= \int \frac{d^3p}{(2\pi)^3} \sum_{\alpha\alpha'} \frac{m^2}{E^2} b^*(p, \alpha) b(p, \alpha') u^{(\alpha)\dagger}(p) \boldsymbol{\alpha} u^{(\alpha')}(p)$$

We need here the Gordon identity which states that for any two positive energy solutions $u^{(\alpha)}(p)$ and $u^{(\beta)}(q)$ of the Dirac equation, we have

$$\bar{u}^{(\alpha)}(p)\gamma^\mu u^{(\beta)}(q) = \frac{1}{2m} \bar{u}^{(\alpha)}(p)\left[(p+q)^\mu + i\sigma^{\mu\nu}(p-q)_\nu\right] u^{(\beta)}(q) \tag{2-54}$$

Indeed, as a consequence of the Dirac equation

$$0 = \bar{u}^{(\alpha)}(p)\left[\not{a}(\not{q} - m) + (\not{p} - m)\not{a}\right] u^{(\beta)}(q)$$
$$= -2m\bar{u}^{(\alpha)}(p)\not{a}u^{(\beta)}(q) + \bar{u}^{(\alpha)}(p)\left(\left\{\frac{\not{p} + \not{q}}{2}, \not{a}\right\} + \left[\frac{\not{p} - \not{q}}{2}, \not{a}\right]\right) u^{(\beta)}(q)$$

where a is an arbitrary four-vector. Equation (2-54) follows by differentiation with respect to a_μ. Using this identity together with Eq. (2-39), we may write

$$\mathbf{J}^{(+)} = \sum_\alpha \int \frac{d^3p}{(2\pi)^3} \frac{m}{E} |b(p, \alpha)|^2 \frac{\mathbf{p}}{E} = \left\langle \frac{\mathbf{p}}{E} \right\rangle \tag{2-55}$$

Therefore the total current for a superposition of positive energy solutions is just the group velocity. This is analogous to what happens in the Schrödinger theory, and seems satisfactory. However, there is an inconsistency in the assumption of superposition of positive energy solutions only.

To illustrate this point, let us consider the time evolution of a wave packet given at time $t = 0$ by a gaussian distribution of half width d:

$$\psi(0, \mathbf{x}) = \frac{1}{(\pi d^2)^{3/4}} e^{-\mathbf{x}^2/2d^2} w \tag{2-56}$$

where w is some fixed spinor, say $\begin{pmatrix} \varphi \\ 0 \end{pmatrix}$.

The corresponding (normalizable) solution of the Dirac equation has the form

$$\psi(t, \mathbf{x}) = \int \frac{d^3p}{(2\pi)^3} \frac{m}{E} \sum_\alpha [b(p, \alpha)u^{(\alpha)}(p) e^{-ip\cdot x} + d^*(p, \alpha)v^{(\alpha)}(p) e^{ip\cdot x}] \tag{2-57}$$

Since the Fourier transform of a gaussian is a gaussian

$$\int d^3x \, e^{-\mathbf{x}^2/2d^2 - i\mathbf{p}\cdot\mathbf{x}} = (2\pi d^2)^{3/2} e^{-\mathbf{p}^2 d^2/2}$$

we may write

$$(4\pi d^2)^{3/4} e^{-\mathbf{p}^2 d^2/2} w = \frac{m}{E} \sum_\alpha [b(p, \alpha)u^{(\alpha)}(p) + d^*(\tilde{p}, \alpha)v^{(\alpha)}(\tilde{p})]$$

where $\tilde{p}$ stands for $(p^0, -\mathbf{p})$.

From the orthogonality relations (2-43), it follows that

$$\begin{aligned} b(p, \alpha) &= (4\pi d^2)^{3/4} e^{-\mathbf{p}^2 d^2/2} u^{(\alpha)\dagger}(p) w \\ d^*(p, \alpha) &= (4\pi d^2)^{3/4} e^{-\mathbf{p}^2 d^2/2} v^{(\alpha)\dagger}(p) w \end{aligned} \tag{2-58}$$

Using the explicit expressions (2-37), we see that the ratio b/d^* is typically of order $|\mathbf{p}|/(m + E)$ and becomes important when $|\mathbf{p}| \sim m$. If the wave packet is spread out over a distance $d \gg 1/m$, the contribution of momenta $|\mathbf{p}| \sim m \gg 1/d$ is heavily suppressed, and the negative energy components are negligible; the one-particle theory is then consistent. However, if we want to localize the wave packet in a region of space of the same size as the Compton wavelength, that is, $d \lesssim 1/m$, negative energy solutions play an appreciable role. This quantitative discussion is in agreement with the heuristic arguments presented at the beginning of this chapter. For a wave packet with negative energy contributions as in (2-57), we compute as above the normalization condition

$$\int \frac{d^3p}{(2\pi)^3}\frac{m}{E}\sum_\alpha\left(|b(p,\alpha)|^2+|d(p,\alpha)|^2\right)=1 \tag{2-59}$$

and, with the notation $\tilde{p}=(p^0,-\mathbf{p})$, the total current is

$$\begin{aligned} J^i(t)=\int\frac{d^3p}{(2\pi)^3}\frac{m}{E}\Bigg\{&\frac{p^i}{E}\sum_\alpha\left[|b(p,\alpha)|^2+|d(p,\alpha)|^2\right]\\ &+i\sum_{\alpha\alpha'}\left[b^*(\tilde{p},\alpha)\,d^*(p,\alpha')\,e^{2iEt}\,\bar{u}^{(\alpha)}(\tilde{p})\sigma^{i0}v^{(\alpha')}(p)\right.\\ &\left.-\,b(\tilde{p},\alpha)\,d(p,\alpha')\,e^{-2iEt}\,\bar{v}^{(\alpha')}(p)\sigma^{i0}u^{(\alpha)}(\tilde{p})\right]\Bigg\} \end{aligned} \tag{2-60}$$

It is now time dependent. Besides the group velocity term, there is a real, oscillating term. The frequency of these oscillations is very high—larger than

$$2m\frac{c^2}{\hbar}\simeq 2\times 10^{21}\ \mathrm{s}^{-1}$$

This phenomenon, traditionally called zitterbewegung, is an example of the difficulties due to the negative energy states in the framework of a one-particle theory.

A more striking manifestation is the famous Klein paradox. Let us idealize the localization process by a square potential barrier of height V in the half space $z\equiv x^3>0$ (Fig. 2-1). Consider now in the $z<0$ half space an incident positive energy plane wave of momentum $k>0$ along the z axis:

$$\psi_{\text{inc}}(z)=e^{ikz}\begin{pmatrix}1\\0\\ \dfrac{k}{E+m}\\0\end{pmatrix}\qquad\text{(spin-up along the } z \text{ axis)}$$

The reflected wave has the form

$$\psi_{\text{ref}}(z)=a\,e^{-ikz}\begin{pmatrix}1\\0\\ \dfrac{-k}{E+m}\\0\end{pmatrix}+b\,e^{-ikz}\begin{pmatrix}0\\1\\0\\ \dfrac{k}{E+m}\end{pmatrix}$$

(superposition of spin-up and spin-down positive energy solutions). In the $z>0$ half space, i.e., in the presence of the constant potential V, the transmitted wave has a similar form:

$$\psi_{\text{trans}}(z)=c\,e^{iqz}\begin{pmatrix}1\\0\\ \dfrac{q}{E-V+m}\\0\end{pmatrix}+d\,e^{-iqz}\begin{pmatrix}0\\1\\0\\ \dfrac{-q}{E-V+m}\end{pmatrix}$$

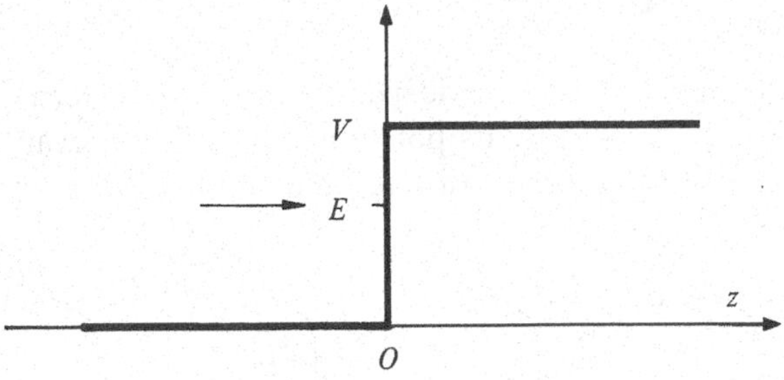

Figure 2-1 Klein's paradox in a square potential.

with an effective momentum q of

$$q = [(E - V)^2 - m^2]^{1/2}$$

Writing down the continuity of the solution at $z = 0$,

$$\psi(z) = \theta(-z)[\psi_{\text{inc}}(z) + \psi_{\text{ref}}(z)] + \theta(z)\psi_{\text{trans}}(z)$$

determines the coefficients $a, \ldots, d$:

$$b = d = 0 \qquad \text{(no spin-flip)}$$

$$1 + a = c$$

$$1 - a = rc \qquad \text{where } r \equiv \frac{q}{k} \frac{E + m}{E - V + m}$$

As long as $|E - V| < m$, q is imaginary and the transmitted wave decays exponentially; beyond a few Compton wavelengths, it is negligible. If we increase V so as to restrict this penetration region, the transmitted wave becomes oscillatory when $V \geq E + m$.

The computation of the transmitted, reflected, and incident currents yields

$$\frac{j_{\text{trans}}}{j_{\text{inc}}} = \frac{4r}{(1 + r)^2} \qquad \frac{j_{\text{ref}}}{j_{\text{inc}}} = \left(\frac{1 - r}{1 + r}\right)^2 = 1 - \frac{j_{\text{trans}}}{j_{\text{inc}}}$$

The conservation of probabilities does indeed look satisfied:

$$j_{\text{inc}} = j_{\text{trans}} + j_{\text{ref}}$$

Unfortunately, since $r < 0$, the reflected flux is larger than the incident one! We are again in trouble when we try to localize the particle within a distance of the order of the Compton wavelength.

In spite of these difficulties, the Dirac equation and its one-particle interpretation are very useful and physically sensible as long as we consider external forces which are slowly varying on a scale of a few Compton wavelengths. They provide us with the first relativistic corrections to the Schrödinger picture. This is what we are going to explore at length in the next sections, before returning to a deeper investigation of the meaning of negative energy states. We now realize that the difficulties which led us to disregard the Klein-Gordon equation have not been really solved. Even though we shall pursue this discussion in the framework of the spin $\frac{1}{2}$ theory because of its important physical implications, we could as well concern ourselves with the scalar case within the same range of validity. This is another instance where important physical theories were constructed for what seems afterwards to be unconvincing motivations.

2-2-3 Electromagnetic Coupling

We wish now to study the interactions of a Dirac particle with an external (classical) electromagnetic field characterized by its potential $A_\mu(x)$. The relevant coupling is obtained from the free Dirac equation through the minimal coupling prescription described in Chap. 1:

$$\partial_\mu \to \partial_\mu + ieA_\mu \tag{2-61}$$

The Dirac equation then reads

$$(i\not{\partial} - e\not{A} - m)\psi(x) = 0 \tag{2-62}$$

This prescription ensures the invariance of the equation under gauge transformations:

$$\begin{cases} \psi(x) \to e^{ie\alpha(x)}\psi(x) \\ A_\mu(x) \to A_\mu(x) - \partial_\mu\alpha(x) \end{cases} \tag{2-63}$$

Here e denotes the charge of the particle; it is negative $e = -|e|$ for the electron. The Lorentz covariance of this equation is clear. If we change our reference frame, the electromagnetic potential transforms as a vector

$$A'_\mu(x' = \Lambda x) = (\Lambda^{-1})_\mu{}^\nu A_\nu(x)$$

and therefore the analysis of Sec. 2-1-3 may be extended to the present case.

Equation (2-62) can be rewritten more explicitly as

$$\begin{aligned} i\frac{\partial\psi}{\partial t} &= \left[\boldsymbol{\alpha}\cdot\left(\frac{1}{i}\nabla - e\mathbf{A}\right) + \beta m + eA^0\right]\psi \\ &= (\boldsymbol{\alpha}\cdot\mathbf{p} + \beta m)\psi + (-e\boldsymbol{\alpha}\cdot\mathbf{A} + eA^0)\psi \\ &= (H_0 + H_{\text{int}})\psi \end{aligned} \tag{2-64}$$

We note a strong resemblance of the interaction part H_{int} with the hamiltonian of a classical particle in an external field $-e\mathbf{v}\cdot\mathbf{A} + eA^0$, in agreement with the interpretation of $\boldsymbol{\alpha}$ as a velocity operator. In the Heisenberg representation, an operator $\mathcal{O}$ satisfies the equation of motion

$$\frac{d}{dt}\mathcal{O}(t) = i[H, \mathcal{O}(t)] + \frac{\partial\mathcal{O}}{\partial t}$$

Thus, here, the position operator $\mathbf{r}$ and the gauge-invariant momentum $\boldsymbol{\pi} \equiv \mathbf{p} - e\mathbf{A}$ satisfy

$$\begin{aligned} \frac{d\mathbf{r}}{dt} &= i[H, \mathbf{r}] = \boldsymbol{\alpha} \\ \frac{d\boldsymbol{\pi}}{dt} &= i[H, \boldsymbol{\pi}] - e\frac{\partial\mathbf{A}}{\partial t} = e(\mathbf{E} + \boldsymbol{\alpha}\times\mathbf{B}) \end{aligned} \tag{2-65}$$

with

$$\mathbf{E} = -\frac{\partial \mathbf{A}}{\partial t} - \nabla A^0$$

$$\mathbf{B} = \text{curl } \mathbf{A}$$

The second equation is the operator version of the Lorentz force equation. In view of the paradoxes encountered in the preceding subsection, the interpretation of $\mathbf{r}$ and $\boldsymbol{\pi}$ as the position and momentum is, however, limited.

To study the physical implications of these equations, we consider their nonrelativistic limit. We write $\psi = \begin{pmatrix} \varphi \\ \chi \end{pmatrix}$ and use the representation $\beta = \begin{pmatrix} I & 0 \\ 0 & -I \end{pmatrix}$, $\boldsymbol{\alpha} = \begin{pmatrix} 0 & \boldsymbol{\sigma} \\ \boldsymbol{\sigma} & 0 \end{pmatrix}$. Equation (2-64) leads to

$$\begin{aligned} i\frac{\partial \varphi}{\partial t} &= \boldsymbol{\sigma} \cdot \boldsymbol{\pi}\chi + eA^0\varphi + m\varphi \\ i\frac{\partial \chi}{\partial t} &= \boldsymbol{\sigma} \cdot \boldsymbol{\pi}\varphi + eA^0\chi - m\chi \end{aligned} \tag{2-66}$$

In the nonrelativistic limit, the large energy m is the driving term in (2-66). We introduce the slowly varying functions of time Φ and X:

$$\begin{aligned} \varphi &= e^{-imt}\Phi \\ \chi &= e^{-imt}X \end{aligned} \tag{2-67}$$

These spinors satisfy

$$\begin{aligned} i\frac{\partial \Phi}{\partial t} &= \boldsymbol{\sigma} \cdot \boldsymbol{\pi}X + eA^0\Phi \\ i\frac{\partial X}{\partial t} &= \boldsymbol{\sigma} \cdot \boldsymbol{\pi}\Phi + eA^0X - 2mX \end{aligned} \tag{2-68}$$

If we assume $eA^0 \ll 2m$ the second equation is solved approximately as

$$X \simeq \frac{\boldsymbol{\sigma} \cdot \boldsymbol{\pi}}{2m}\Phi \ll \Phi$$

and the first one is the Pauli equation

$$i\frac{\partial \Phi}{\partial t} = \left[\frac{(\boldsymbol{\sigma} \cdot \boldsymbol{\pi})^2}{2m} + eA^0\right]\Phi \tag{2-69}$$

This justifies the use of the terms large and small components for φ and χ (or Φ and X respectively). As for Eq. (2-69), it is a generalization to spinors of the Schrödinger equation in an electromagnetic field. After simple algebraic manipulations,

$$(\boldsymbol{\sigma} \cdot \boldsymbol{\pi})^2 = \sigma_i\sigma_j\pi^i\pi^j = \boldsymbol{\pi}^2 + \tfrac{1}{4}[\sigma_i, \sigma_j][\pi^i, \pi^j] = \boldsymbol{\pi}^2 - e\boldsymbol{\sigma} \cdot \mathbf{B}$$

we may rewrite it as

$$i\frac{\partial\Phi}{\partial t}=\left[\frac{(\mathbf{p}-e\mathbf{A})^2}{2m}-\frac{e}{2m}\boldsymbol{\sigma}\cdot\mathbf{B}+eA^0\right]\Phi \tag{2-70}$$

The only spin dependence is through the magnetic interaction $\boldsymbol{\sigma}\cdot\mathbf{B}$. Restoring the factors $\hbar$ and c,

$$H_{\text{magn}}=-\frac{e\hbar}{2mc}\boldsymbol{\sigma}\cdot\mathbf{B}=-\boldsymbol{\mu}\cdot\mathbf{B} \tag{2-71}$$

where the magnetic moment $\boldsymbol{\mu}$ is defined as

$$\boldsymbol{\mu}\equiv\frac{e}{mc}\frac{\hbar\boldsymbol{\sigma}}{2}=2\left(\frac{e}{2mc}\right)\mathbf{S} \tag{2-72}$$

The spin operator $\mathbf{S}$ is $\hbar\boldsymbol{\sigma}/2$. According to the definition of Sec. 1-1-3, the gyromagnetic ratio g is 2. This is a nontrivial prediction of the Dirac theory, derived within the nonrelativistic context of the Pauli equation.

The actual value of $|\boldsymbol{\mu}|$ equals

$$|\boldsymbol{\mu}|=\frac{e\hbar}{2mc}=5.79\times10^{-9}\ \text{eV/G}$$

Radiative corrections affect the experimentally measured value of g by a tiny amount, which we shall study later.

The Pauli equation (2-70) may be further reduced, if we consider a uniform magnetic field $\mathbf{B}=\text{curl}\,\mathbf{A}$, with the choice $\mathbf{A}=\frac{1}{2}\mathbf{B}\times\mathbf{r}$, and neglect the quadratic term in A (weak field approximation). We obtain

$$i\frac{\partial\Phi}{\partial t}=\left[\frac{\mathbf{p}^2}{2m}-\frac{e}{2m}(\mathbf{L}+2\mathbf{S})\cdot\mathbf{B}\right]\Phi$$

where $\mathbf{L}=\mathbf{r}\times\mathbf{p}$ is the orbital angular momentum operator. The reader is invited to derive a complete set of solutions.

The previous study may, in fact, also be performed on the quadratic form of the Dirac equation, i.e., without assuming the nonrelativistic approximation. Starting from (2-62), we multiply it by the operator $(i\not{\partial}-e\not{A}+m)$. This yields

$$[(i\not{\partial}-e\not{A})^2-m^2]\psi=\left\{(i\partial-eA)^2+\frac{1}{2i}\sigma^{\mu\nu}[i\partial_\mu-eA_\mu,i\partial_\nu-eA_\nu]-m^2\right\}\psi$$

$$=\left[(i\partial-eA)^2-\frac{e}{2}\sigma^{\mu\nu}F_{\mu\nu}-m^2\right]\psi=0 \tag{2-73}$$

Therefore the spin-dependent term reads

$$-g\frac{e}{2}\frac{\sigma^{\mu\nu}}{2}F_{\mu\nu}=-g\left(\frac{e}{2}\right)(i\boldsymbol{\alpha}\cdot\mathbf{E}+\boldsymbol{\sigma}\cdot\mathbf{B})$$

in the usual representation, and again corresponds to the value $g = 2$ for the gyromagnetic ratio. Notice that this value is a consequence of the minimal prescription (2-61). We could have written a nonminimal equation:

$$\left[(i\not\partial - e\not A) - m + \frac{\Delta g}{2}\frac{e}{4m}\sigma^{\mu\nu}F_{\mu\nu}\right]\psi = 0 \tag{2-74}$$

which would lead to $g = 2 + \Delta g$. Such an equation has to be used to study the behavior in weak fields of particles with g factors very different from 2.

It is interesting to determine the energy levels in a uniform magnetic field. Assume the field **B** along the z axis. The potential vector A may be chosen such that $A^0 = A^x = A^z = 0$, $A^y = Bx$. For a stationary solution of energy E, $\psi = e^{-iEt}\begin{pmatrix}\varphi\\ \chi\end{pmatrix}$, Eq. (2-66) reads

$$(E - m)\varphi = \boldsymbol{\sigma}\cdot(\mathbf{p} - e\mathbf{A})\chi$$

$$(E + m)\chi = \boldsymbol{\sigma}\cdot(\mathbf{p} - e\mathbf{A})\varphi$$

Eliminating χ yields an equation for φ:

$$(E^2 - m^2)\varphi = [\boldsymbol{\sigma}\cdot(\mathbf{p} - e\mathbf{A})]^2\varphi = [(\mathbf{p} - e\mathbf{A})^2 - e\boldsymbol{\sigma}\cdot\mathbf{B}]\,\varphi$$
$$= [\mathbf{p}^2 + e^2B^2x^2 - eB(\sigma_z + 2xp_y)]\,\varphi$$

This is the hamiltonian of an harmonic oscillator. Since p_y, p_z, and σ_z commute with the right-hand side, we seek solutions of the form

$$\varphi(\mathbf{x}) = e^{i(p_y y + p_z z)}f(x)$$

where $f(x)$ satisfies

$$\left[-\frac{d^2}{dx^2} + (eBx - p_y)^2 - eB\sigma_z\right]f(x) = (E^2 - m^2 - p_z^2)\,f(x) \tag{2-75}$$

Assuming the sign of B such that $eB > 0$, we introduce the auxiliary variables

$$\xi = \sqrt{eB}\left(x - \frac{p_y}{eB}\right)$$

$$a = \frac{E^2 - m^2 - p_z^2}{eB}$$

thereby reducing (2-75) to

$$\left(-\frac{d^2}{d\xi^2} + \xi^2 - \sigma_z\right)f = af$$

If f is an eigenvector of σ_z with eigenvalue $\alpha = \pm 1$,

$$f = \begin{pmatrix}f_1\\ 0\end{pmatrix} \text{ for } \alpha = 1 \qquad f = \begin{pmatrix}0\\ f_{-1}\end{pmatrix} \text{ for } \alpha = -1$$

then f_α satisfies

$$\left(\frac{d^2}{d\xi^2} - \xi^2\right)f_\alpha(\xi) = -(a + \alpha)\,f_\alpha(\xi) \qquad \alpha = \pm 1$$

The solution which vanishes at infinity is expressed in terms of a Hermite polynomial $H_n(\xi)$:

$$f_\alpha = c\, e^{-\xi^2/2} H_n(\xi)$$

provided $a + \alpha = 2n + 1$, n integer, $n = 0, 1, 2, \ldots$. Therefore, the energy levels are given by

$$E^2 = m^2 + p_z^2 + eB(2n + 1 - \alpha) \tag{2-76}$$

and the corresponding wave functions may be easily written. These levels have both a discrete ($n, \alpha = -1$; $n + 1, \alpha = 1$) and a continuous degeneracy (in p_y). The latter may be reduced to a discrete one if we consider a particle in a finite box. Equation (2-76) gives a relativistic generalization of the Landau levels. The value $g = 2$ is such that the spectrum extends down to $E^2 = m^2$.

As a second example let us study the case of a Dirac particle in an electromagnetic plane wave, generalizing thereby the classical treatment of Chap. 1. The plane wave, which is supposed to be linearly polarized, is characterized by its propagation vector n_μ ($n^2 = 0$) and its polarization vector ε_μ ($\varepsilon^2 = -1$, $\varepsilon \cdot n = 0$). We write $A_\mu = \varepsilon_\mu f(\xi)$ where $\xi \equiv n \cdot x$ and $\partial_\mu A_\nu = n_\mu A'_\nu$ with $A'_\nu \equiv \varepsilon_\nu f'(\xi)$. The quadratic equation (2-73) takes the form

$$(-\Box - 2ieA \cdot \partial + e^2 A^2 - m^2 - ie\not{n}\not{A}')\psi = 0 \tag{2-77}$$

Let us exhibit solutions of the form

$$\psi_p(x) = e^{-ip\cdot x}\varphi(\xi) \tag{2-78}$$

with φ a Dirac spinor and p a four-vector which is not orthogonal to n. By adding to p some quantity of the form λn, we may realize the condition

$$p^2 = m^2$$

The interpretation of this four-vector is the following. In the frame where $\varepsilon^0 = 0$, and thus $A^0 = 0$, **E** and **A** along **0x**, **B** along **0y**, and **n** along **0z**, the operators $p_1 = i\partial_1$, $p_2 = i\partial_2$, and $i(\partial_0 + \partial_3) = p_0 + p_3$ commute with the Dirac hamiltonian. The substitution of (2-78) into Eq. (2-77) leads to the condition

$$2in \cdot p\varphi'(\xi) + (e^2 A^2 - 2eA \cdot p - ie\not{n}\not{A}')\varphi(\xi) = 0$$

which is readily integrated as

$$\varphi(\xi) = \left(\frac{m}{p_0}\right)^{1/2} \exp\left\{e\frac{\not{n}\not{A}}{2n \cdot p} - i\int_0^\xi d\xi'\left[e\frac{A(\xi')\cdot p}{n \cdot p} - \frac{e^2 A^2(\xi')}{2n \cdot p}\right]\right\}u \tag{2-79}$$

where u is a constant bispinor. Since $(\not{n}\not{A})^2 = -n^2 A^2 = 0$, we may write

$$\psi_p(x) = \left(1 + \frac{e}{2n \cdot p}\not{n}\not{A}\right)\left(\frac{m}{p_0}\right)^{1/2} u\, e^{iI}$$

where I stands for the action of a classical particle in a plane wave (damped at infinity), with $p = mu(\infty) = p(\infty)$ [compare Eq. (1-84)]:

$$I = -p \cdot x - \int_0^{n\cdot x} d\xi\left[\frac{e}{n \cdot p}A(\xi)\cdot p - \frac{e^2}{2n \cdot p}A^2(\xi)\right]$$

For ψ to satisfy the original Dirac equation and not only the squared equation (2-73), u has to obey some auxiliary condition. After some algebra, we find that

$$(i\not{\partial} - e\not{A} - m)\psi_p(x) = \left(\frac{m}{p_0}\right)^{1/2}\left(1 + \frac{e}{2p \cdot n}\not{n}\not{A}\right)(\not{p} - m)u\, e^{iI}$$

Therefore

$$(\not{p} - m)u = 0$$

and $u = u(p)$ is a solution of the free Dirac equation. The reader will verify that the solution (2-79) has the correct normalization

$$\frac{1}{(2\pi)^3}\int d^3x\, \psi^\dagger_{p'}(t,\mathbf{x})\psi_p(t,\mathbf{x}) = \delta^3(\mathbf{p}-\mathbf{p}')$$

and that the associated current reads

$$j^\mu(x) = \bar\psi_p(x)\gamma^\mu\psi_p(x) = \frac{1}{p^0}\left[p^\mu - eA^\mu + n^\mu\left(e\frac{p\cdot A}{n\cdot p} - \frac{e^2A^2}{2n\cdot p}\right)\right]$$

If $A(\xi)$ is a quasiperiodic function of ξ (slowly damped at infinity), the average value of j^μ is

$$\overline{j^\mu} = \frac{1}{p^0}\left(p^\mu - \frac{e^2}{2n\cdot p}\overline{A^2}n^\mu\right)$$

showing the same phenomenon as its classical counterpart. These expressions were originally derived by Volkow in 1935.

2-2-4 Foldy-Wouthuysen Transformation

In the last subsections, the physical meaning of the Dirac equation has been investigated using a nonrelativistic approximation. It is worth while showing that this may be pursued in a systematic way. This is the purpose of the Foldy-Wouthuysen transformation. To be more explicit, we want to find a unitary transformation

$$\psi = e^{-iS}\psi'$$

which decouples the small and large components and where S may be time dependent.

Let us call odd the operators such as $\boldsymbol{\alpha}$ which couple large and small components, even those which do not (for example, I, $\beta,\ldots$). Since ψ' satisfies the equation

$$i\partial_t\psi' = \left[e^{iS}(H - i\partial_t)\,e^{-iS}\right]\psi' \equiv H'\psi' \tag{2-80}$$

our problem is to find S so as to get rid of the odd operators in H' at a given order in $1/m$. In practice, we shall do it up to terms of order (kinetic energy/m)^3 or (kinetic energy $\times$ potential energy/m^2). This will lead to a further insight in relativistic corrections entailed by the Dirac equation.

In the free case, we can construct S exactly. It is time independent and can be chosen as

$$S = -i\beta\frac{\boldsymbol{\alpha}\cdot\mathbf{p}}{|\mathbf{p}|}\theta = -i\frac{\boldsymbol{\gamma}\cdot\mathbf{p}}{|\mathbf{p}|}\theta$$

Since $(\boldsymbol{\gamma}\cdot\mathbf{p}/|\mathbf{p}|)^2 = -I$, we may compute e^{iS} and H' in a closed form:

$$e^{iS} = e^{(\boldsymbol{\gamma}\cdot\mathbf{p}/|\mathbf{p}|)\theta} = \cos\theta + \beta\frac{\boldsymbol{\alpha}\cdot\mathbf{p}}{|\mathbf{p}|}\sin\theta$$

$$
\begin{aligned}
H' &= e^{iS} H\, e^{-iS} \\
&= \left(\cos\theta + \beta \frac{\boldsymbol{\alpha}\cdot\mathbf{p}}{|\mathbf{p}|}\sin\theta\right)(\boldsymbol{\alpha}\cdot\mathbf{p} + \beta m)\left(\cos\theta - \beta\frac{\boldsymbol{\alpha}\cdot\mathbf{p}}{|\mathbf{p}|}\sin\theta\right) \\
&= \boldsymbol{\alpha}\cdot\mathbf{p}\left(\cos 2\theta - \frac{m}{|\mathbf{p}|}\sin 2\theta\right) + \beta(m\cos 2\theta + |\mathbf{p}|\sin 2\theta)
\end{aligned}
$$

The choice of θ such that

$$
\sin 2\theta = \frac{|\mathbf{p}|}{E} \qquad \cos 2\theta = \frac{m}{E}
$$

eliminates the odd operator $\boldsymbol{\alpha}\cdot\mathbf{p}$ and leaves

$$
H' = \frac{\beta}{E}(\mathbf{p}^2 + m^2) = \beta E = \beta(m^2 + \mathbf{p}^2)^{1/2}
$$

as we would expect. In other words, we have decomposed H into a direct sum of two nonlocal hamiltonians $\pm\sqrt{m^2 + p^2}$. It is clear that these square roots cannot be represented in configuration space by a finite set of differential operators.

In the interacting case, we therefore expect S to be of order m^{-1}, and we expand H' to the desired order:

$$
\begin{aligned}
H' = H + i[S, H] - \frac{1}{2}[S,[S,H]] - \frac{i}{6}[S,[S,[S,H]]] \\
+ \frac{1}{24}[S,[S,[S,[S,H]]]] - \dot{S} - \frac{i}{2}[S,\dot{S}] + \frac{1}{6}[S,[S,\dot{S}]] + \cdots
\end{aligned}
$$

Here use has been made of the general identity

$$
e^A B\, e^{-A} = \sum_{n=0}^{\infty} \frac{1}{n!}[A,[A,[\cdots[A,B]]\cdots]] \tag{2-81}
$$

where, in the generic term of the sum, A appears n times. This identity is easily derived by computing the successive derivatives of $e^{sA} B\, e^{-sA}$ at $s = 0$, and using them in the (formal) Taylor expansion at $s = 1$.

We start from $H = \beta m + \mathcal{O} + \mathcal{E}$ where $\mathcal{O}$ denotes the odd operator $\mathcal{O} = \boldsymbol{\alpha}\cdot(\mathbf{p} - e\mathbf{A})$ and $\mathcal{E}$ the even one eA^0. The solution of the free case suggests that we take $S = -i\beta\mathcal{O}/2$ to first order. Then according to the above expression for H', we calculate

$$
H' = \beta m + \mathcal{O}' + \mathcal{E}'
$$

with $$\mathcal{O}' = \frac{\beta}{2m}[\mathcal{O}, \mathcal{E}] - \frac{\mathcal{O}^3}{3m^2} + i\frac{\beta\dot{\mathcal{O}}}{2m}$$

and $$\mathcal{E}' = \mathcal{E} + \beta\left(\frac{\mathcal{O}^2}{2m} - \frac{\mathcal{O}^4}{8m^3}\right) - \frac{1}{8m^2}[\mathcal{O},[\mathcal{O},\mathcal{E}]] - \frac{i}{8m^2}[\mathcal{O},\dot{\mathcal{O}}]$$

$\mathcal{O}'$ is now of order m^{-1}. We then iterate the process. A second transformation $e^{-iS'}$, with $S' = -i(\beta\mathcal{O}'/2m)$, leads to

$$H'' = \beta m + \mathcal{E}' + \mathcal{O}''$$

where $\mathcal{O}'' = O(m^{-2})$. Finally, a third step with $S'' = -(i\beta\mathcal{O}''/2m)$ eliminates this odd term, leaving the desired hamiltonian

$$H''' = \beta\left[m + \frac{(\mathbf{p} - e\mathbf{A})^2}{2m} - \frac{(\mathbf{p})^4}{8m^3}\right] + eA^0 - \frac{e}{2m}\beta\boldsymbol{\sigma}\cdot\mathbf{B}$$
$$+ \left(-\frac{ie}{8m^2}\boldsymbol{\sigma}\cdot\text{curl }\mathbf{E} - \frac{e}{4m^2}\boldsymbol{\sigma}\cdot\mathbf{E}\times\mathbf{p}\right) - \frac{e}{8m^2}\text{div }\mathbf{E} \qquad \text{(2-82)}$$

The interpretation of the various terms deserves some comments. The term in the bracket is the expansion (to the required order) of $[(\mathbf{p} - e\mathbf{A})^2 + m^2]^{1/2}$. The second term eA^0 is the electrostatic energy of a point-like charge, whereas the third one represents the energy of a magnetic dipole for $g = 2$. The term inside parentheses may be seen to correspond to a spin-orbit (s.o.) interaction. Indeed, for a static spherically symmetric potential, curl $\mathbf{E} = 0$ and $\mathbf{E} = -\nabla A^0$. Therefore

$$\boldsymbol{\sigma}\cdot(\mathbf{E}\times\mathbf{p}) = -\frac{1}{r}\frac{dA^0}{dr}\boldsymbol{\sigma}\cdot(\mathbf{r}\times\mathbf{p}) = -\frac{1}{r}\frac{dA^0}{dr}\boldsymbol{\sigma}\cdot\mathbf{L}$$

and the term in parentheses is

$$H_{\text{s.o.}} = \frac{e}{4m^2 r}\frac{dA^0}{dr}\boldsymbol{\sigma}\cdot\mathbf{L}$$

The magnetic field $\mathbf{B}' = -\mathbf{v}\times\mathbf{E}$ acting on the particle is responsible for this additional magnetic energy. Its interaction with the magnetic moment $\boldsymbol{\mu}$ (2-72) would give

$$H_{\text{s.o.}} = -\frac{e}{2m}\boldsymbol{\sigma}\cdot\mathbf{B}' = -\frac{e}{2m^2}\boldsymbol{\sigma}\cdot(\mathbf{E}\times\mathbf{p})$$

but due to the Thomas precession, this result is reduced by a factor 2.

Finally the last term in Eq. (2-82), referred to as the Darwin term $-(e/8m^2)$ div $\mathbf{E}$, may be traced to the zitterbewegung. The electron position fluctuates by an amount δr such that $\langle\delta r^2\rangle \sim 1/m^2$, and its effective electrostatic energy is the average

$$\langle eA^0(\mathbf{r} + \delta\mathbf{r})\rangle = eA^0(\mathbf{r}) + \frac{e}{2}\frac{\partial^2 A^0(\mathbf{r})}{\partial r^i\,\partial r^j}\langle\delta r^i\,\delta r^j\rangle + \cdots$$

From spherical symmetry, this random fluctuation is

$$\langle\delta r^i\,\delta r^j\rangle = \frac{\delta^{ij}}{3}\langle\delta r^2\rangle \simeq \frac{\delta^{ij}}{3m^2}$$

and the correction to $eA^0(r)$ is

$$\delta(eA^0) = \frac{e}{6m^2}\Delta A^0 = -\frac{e}{6m^2}\operatorname{div}\mathbf{E}$$

in good agreement with the sign and magnitude of the Darwin term.

The reader will have noticed that as the Foldy-Wouthuysen transformation is time dependent, the expectation values of H' in the state ψ' are, in general, different from those of H in the corresponding state ψ.

2-3 HYDROGEN-LIKE ATOMS

An important application of the Dirac equation is the discussion of the fine structure of atomic spectra. This is a field notorious for the successes of quantum mechanics and its relativistic generalizations, including radiative corrections. In view of the intricacy of the Foldy-Wouthuysen method, it is quite remarkable that an exact solution of the Dirac equation in a static Coulomb field exists and leads to an excellent agreement with the observed results on hydrogen-like atoms. The difficulties discussed in Sec. 2-2 are not expected to play a significant role. In atomic physics, the relevant length scale, the Bohr radius $a_0 = \hbar/m_e c\alpha = 0.53$ Å, is 137 times larger than the electron wavelength. For heavy atoms these difficulties do arise and other methods have to be developed.

2-3-1 Nonrelativistic Versus Relativistic Spectrum

Recall the nonrelativistic result obtained from the Schrödinger equation, a triumph for early quantum mechanics. The wave equation is

$$\left[-\frac{\Delta}{2m} - \frac{Ze^2}{4\pi r} - \varepsilon_{n,l}\right]\psi_{n,l}(\mathbf{r}) = 0$$

where

$$-\Delta \equiv -\frac{\partial^2}{\partial r^2} - \frac{2}{r}\frac{\partial}{\partial r} + \frac{L^2}{r^2}$$

$$L^2\psi_{n,l} = l(l+1)\psi_{n,l} \qquad l = 0, 1, \ldots$$

and m is the reduced mass of the electron nucleus system:

$$m^{-1} = m_e^{-1} + m_N^{-1} \simeq m_e^{-1}$$

The quantization condition requires that $n' = n - (l+1)$ must be a nonnegative integer (number of zeros of the wave function), and the energy levels are

$$\varepsilon_{n,l} = -\frac{m(Z\alpha)^2}{2n^2} \qquad n = 1, 2, \ldots \tag{2-83}$$

Numerically, the Rydberg constant $m\alpha^2/2$ is equal to 13.6 eV.

The s-state ($l = 0$) wave function at the origin is

$$\psi(0) = \frac{1}{\pi^{1/2}}\left(\frac{mZ\alpha}{n}\right)^{3/2} \tag{2-84}$$

Since for a given $n \geq 1$, l can take integer values from 0 to $n - 1$, each level has a degeneracy equal to $\sum_0^{n-1}(2l + 1) = n^2$. This is related to a dynamical symmetry of the Coulomb problem according to a group $O(4)$ of rotations in four-dimensional space, which was used by Pauli and Fock in the early days of quantum mechanics to give an algebraic derivation of the Balmer spectrum (2-83). This degeneracy disappears in the relativistic treatment leading to the fine splitting of the spectrum.

Ignoring for a while the effects of spin, let us see the predictions of the Klein-Gordon theory when the electromagnetic coupling is introduced in a minimal way. Let E denote the total energy equal to the rest energy mc^2 plus negative binding energy ε:

$$\left[\left(E + \frac{Z\alpha}{r}\right)^2 + \Delta - m^2\right]\phi = 0$$

or

$$\left[-\frac{\partial^2}{\partial r^2} - \frac{2}{r}\frac{\partial}{\partial r} + \frac{L^2 - Z^2\alpha^2}{r^2} - \frac{2Z\alpha E}{r} - (E^2 - m^2)\right]\phi = 0 \tag{2-85}$$

This equation is formally identical to the Schrödinger equation, after the substitutions

$$L^2 \to L^2 - Z^2\alpha^2 \equiv \lambda(\lambda + 1)$$

$$\alpha \to \alpha\frac{E}{m}$$

$$\varepsilon \to \frac{E^2 - m^2}{2m}$$

The orbital quantum number is shifted by δ_l, $l \to \lambda = l - \delta_l$, where

$$\delta_l = l + \tfrac{1}{2} - [(l + \tfrac{1}{2})^2 - Z^2\alpha^2]^{1/2}$$

$$\simeq \frac{Z^2\alpha^2}{2l + 1} + O(\alpha^4)$$

and the principal quantum number n is similarly displaced by the same amount, since $n' = n - (l + 1)$ must be an integer. Therefore, the energy levels are given by

$$\frac{E_{nl}^2 - m^2}{2m} = -\frac{mZ^2\alpha^2}{2}\frac{E_{nl}^2}{m^2}\frac{1}{(n - \delta_l)^2}$$

or

$$E_{nl} = \frac{m}{\sqrt{1 + [Z^2\alpha^2/(n - \delta_l)^2]}} \tag{2-86}$$

$$= m - \frac{mZ^2\alpha^2}{2n^2} - \frac{mZ^4\alpha^4}{n^3(2l + 1)} + \frac{3}{8}\frac{mZ^4\alpha^4}{n^4} + O(\alpha^6)$$

The second term is the nonrelativistic binding energy and the third one breaks the $O(4)$ degeneracy. We are not going to discuss the pathologies of this case, namely, the singular behavior of the wave function at the origin due to the attractive term $-(Z^2\alpha^2/r^2)$ or the catastrophe which occurs when $Z > 137/2$ (δ_l and thus E_{nl} become complex!). Equation (2-86) is in poor agreement with the experimental situation, meaning that the effects of spin cannot be neglected.

2-3-2 Dirac Theory

We turn now to the predictions of the Dirac equation. Before constructing the wave functions, we first proceed to a simple derivation of the spectrum, as in the Klein-Gordon case. For this purpose, we square the equation, as in Eq. (2-73), and insert the nonvanishing component of the potential $A_0 = -(Ze/4\pi r)$, while $F_{0i} = -F_{i0} = -\partial_i A_0 = E^i$. It is convenient to work in the chiral representation (2-20) for the γ matrices where σ^{0i} is diagonal: $\sigma^{0i} = i\begin{pmatrix} \sigma^i & 0 \\ 0 & -\sigma^i \end{pmatrix}$. Then the spin term in Eq. (2-73) reads

$$\frac{e}{2}\sigma^{\mu\nu}F_{\mu\nu} = \pm ie\boldsymbol{\sigma}\cdot\mathbf{E} = \mp iZ\alpha\frac{\boldsymbol{\sigma}\cdot\hat{r}}{r^2}$$

where $\hat{r}$ is the unit vector $\mathbf{r}/r$. The analog of Eq. (2-84) is an equation for two-component spinors:

$$\left[-\left(\frac{\partial^2}{\partial r^2} + \frac{2}{r}\frac{\partial}{\partial r}\right) + (L^2 - Z^2\alpha^2 \mp iZ\alpha\boldsymbol{\sigma}\cdot\hat{r})\frac{1}{r^2} - \frac{2Z\alpha E}{r} - (E^2 - m^2)\right]\psi_\pm = 0$$

The total angular momentum $\mathbf{J} = \mathbf{L} + \mathbf{S} = \mathbf{L} + \boldsymbol{\sigma}/2$ commutes with the hamiltonian and with L^2. In the subspace where $J^2 = j(j+1)$, $J_z = m$ ($j = \frac{1}{2}, \frac{3}{2}, \ldots$; $-j \le m \le j$), and $L^2 = l(l+1)$, the integer l takes two values: $l = j + \frac{1}{2} \equiv l_+$ and $l = j - \frac{1}{2} \equiv l_-$. Since $\boldsymbol{\sigma}\cdot\hat{r}$ has no diagonal matrix element, $\langle l\pm|\boldsymbol{\sigma}\cdot\hat{r}|l\pm\rangle = 0$, is hermitian and has a square equal to one, the operator $[L^2 - Z^2\alpha^2 \mp iZ\alpha\boldsymbol{\sigma}\cdot\hat{r}]$ assumes in this subspace the following form:

$$L^2 - Z^2\alpha^2 \mp iZ\alpha\boldsymbol{\sigma}\cdot\hat{r} = \begin{pmatrix} (j+\frac{1}{2})(j+\frac{3}{2}) - Z^2\alpha^2 & \mp iZ\alpha \\ \mp iZ\alpha & (j-\frac{1}{2})(j+\frac{1}{2}) - Z^2\alpha^2 \end{pmatrix}$$

Let $\lambda(\lambda+1)$ be its eigenvalues

$$\lambda = [(j+\tfrac{1}{2})^2 - Z^2\alpha^2]^{1/2} \qquad \text{and} \qquad \lambda = [(j+\tfrac{1}{2})^2 - Z^2\alpha^2]^{1/2} - 1$$

which may be written

$$\lambda = (j \pm \tfrac{1}{2}) - \delta_j$$

with
$$\delta_j = j + \frac{1}{2} - \sqrt{\left(j+\frac{1}{2}\right)^2 - Z^2\alpha^2} \simeq \frac{Z^2\alpha^2}{2j+1} + O(Z^4\alpha^4)$$

Again, n is also shifted by δ_j so as to keep $n' = (n - \delta_j) - \lambda - 1$ integer. The condition $n' \geq 0$ restricts j to the range $j \leq n - \frac{3}{2}$ for $\lambda = j + \frac{1}{2} - \delta_j$ and $j \leq n - \frac{1}{2}$ for $\lambda = j - \frac{1}{2} - \delta_j$. Therefore, there is a twofold degeneracy except for the state $j = n - \frac{1}{2}$. The final result is

$$E_{nj} = \frac{m}{\sqrt{1 + [Z^2\alpha^2/(n - \delta_j)^2]}} = m - \frac{mZ^2\alpha^2}{2n^2} - \frac{mZ^4\alpha^4}{n^3(2j+1)} + \frac{3}{8}\frac{mZ^4\alpha^4}{n^4} + O(\alpha^6) \tag{2-87}$$

with $n = 1, 2, \ldots$ and $j = \frac{1}{2}, \frac{3}{2}, \ldots, n - \frac{1}{2}$. A catastrophe occurs now for $Z = 137$; $\delta_{1/2}$ becomes imaginary beyond this value.

The degenerate states may be distinguished by their orbital angular momentum l, which takes the values $j \pm \frac{1}{2}$ (except for $j = n - \frac{1}{2}$ where $l = n - 1$); this is in turn related to the transformation property of the state under parity. The energies of the low-lying states have been represented on Fig. 2-2, where we have used the customary, nonrelativistic spectroscopic notation nl_j.

The new phenomenon is the occurrence of the fine structure, i.e., the difference in energy between levels of different j, for the same value of n. Typically, for $Z = 1$,

$$E(2P_{3/2}) - E(2P_{1/2}) \simeq \frac{m\alpha^4}{32} = 4.53 \times 10^{-5} \text{ eV}$$

$$= 10.9 \text{ GHz}$$

This fine splitting may be seen as a consequence of the spin-orbit coupling in (2-82):

$$\Delta E = \left\langle \frac{Z\alpha}{4m^2} \frac{\boldsymbol{\sigma} \cdot \mathbf{L}}{r^3} \right\rangle$$

This term vanishes for an s wave, whereas for a p wave, $\mathbf{L} \cdot \boldsymbol{\sigma} = (\mathbf{J}^2 - \mathbf{L}^2 - \boldsymbol{\sigma}^2/4)$ takes the values 1 and -2 for $j = 3/2$ and $j = 1/2$ respectively. On the other hand, the expectation value of $1/r^3$ is, on dimensional grounds, of the form $\langle nl|1/r^3|nl\rangle = k_{nl}(mZ\alpha)^3$, where k_{nl} is a pure number. Any textbook on quantum mechanics tells us that $k_{nl} = 8/(2l+1)n^3[(2l+1)^2 - 1]$ $(= \frac{1}{24}$ for $n = 2$ and $l = 1)$:

$$\Delta E_{\text{s.o.}}(2P_{3/2} - 2P_{1/2}) = \frac{m(Z\alpha)^4}{32}$$

in agreement with the previous estimate. (This holds also for higher values of n and l.)

Let us now construct the spinors which are the energy eigenstates of this problem. Returning to the Dirac representation (2-10), we write the bispinors $\psi = \begin{pmatrix} \varphi \\ \chi \end{pmatrix}$ and look for two-component spinors which are eigenstates of J^2, J_z, and L^2 with eigenvalues $j(j+1)$, m, and $l(l+1)$ respectively. Let $\varphi_{j,m}^{(\pm)}$ be the eigenstate for $j = l \pm \frac{1}{2}$.

As eigenstates of L^2, $\varphi_{j,m}^{(\pm)}$ must have the form

$$\varphi_{j,m}^{(\pm)} = \begin{pmatrix} a_\mu & Y_l^\mu \\ b_{\mu'} & Y_l^{\mu'} \end{pmatrix} \qquad \text{(no summation over } \mu \text{ and } \mu')$$

where we use the standard notation Y_l^μ for spherical harmonics. They must also be eigenstates of $J_z = L_z + S_z$ with eigenvalue m and eigenstates of $\mathbf{L}\cdot\boldsymbol{\sigma} = \mathbf{J}^2 - \mathbf{L}^2 - \frac{3}{4}$ with eigenvalue l for $\varphi_{j,m}^{(+)}$ and $-(l+1)$ for $\varphi_{j,m}^{(-)}$. The first condition gives $\mu = m - \frac{1}{2}$, and $\mu' = m + \frac{1}{2}$, the second one, together with the normalization

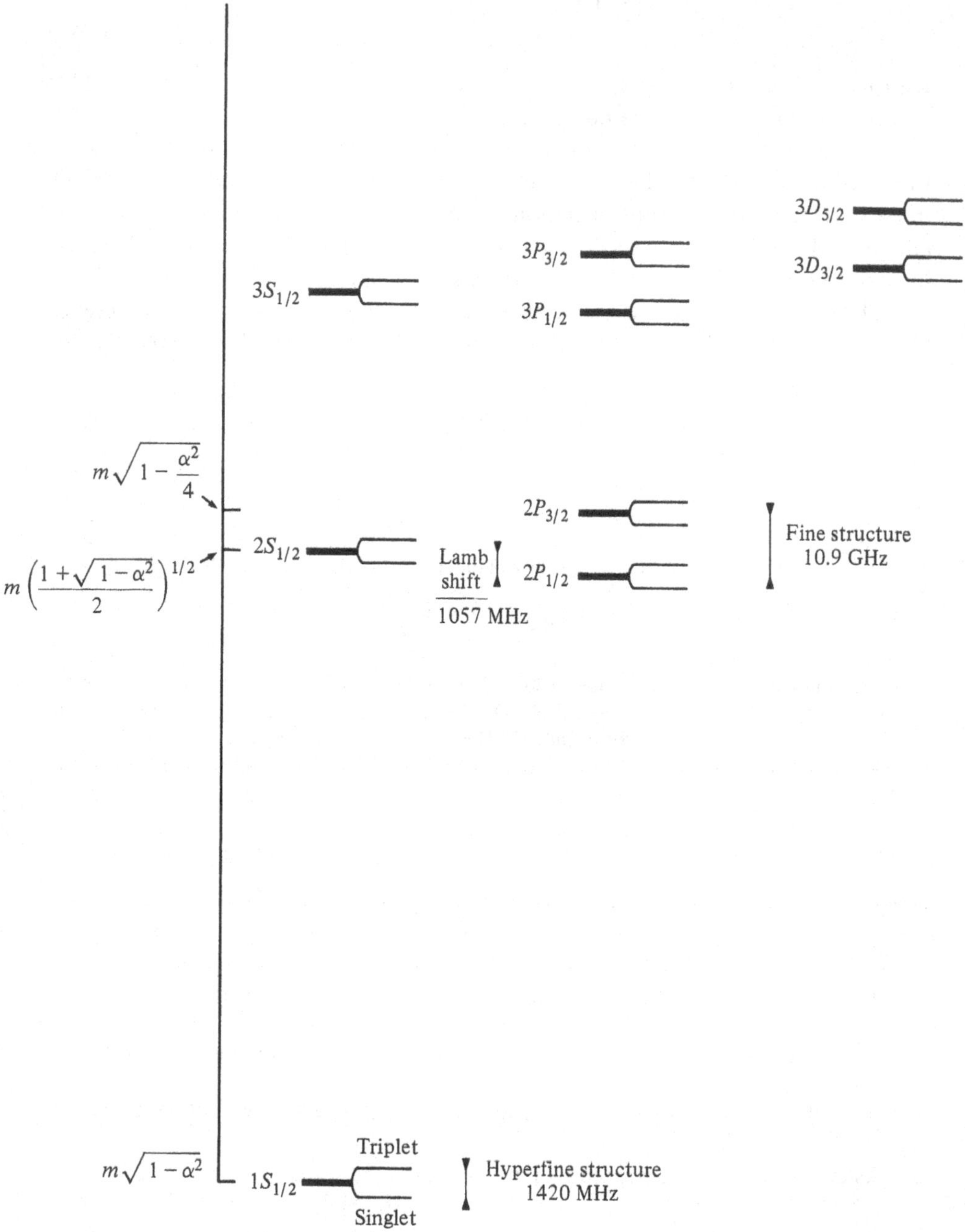

Figure 2-2 Low-lying energy levels of hydrogen.

condition $|a|^2 + |b|^2$, determines a and b. We finally get the spinor harmonics

$$\varphi_{j,m}^{(+)} = (2l+1)^{-1/2}\begin{pmatrix}(l+m+\frac{1}{2})^{1/2} & Y_l^{m-1/2}\\ (l-m+\frac{1}{2})^{1/2} & Y_l^{m+1/2}\end{pmatrix} \qquad j = l+\tfrac{1}{2} \tag{2-88}$$

$$\varphi_{j,m}^{(-)} = (2l+1)^{-1/2}\begin{pmatrix}(l-m+\frac{1}{2})^{1/2} & Y_l^{m-1/2}\\ -(l+m+\frac{1}{2})^{1/2} & Y_l^{m+1/2}\end{pmatrix} \qquad j = l-\tfrac{1}{2}, l>0 \tag{2-89}$$

The phase has been chosen in such a way that

$$\varphi_{j,m}^{(+)} = \boldsymbol{\sigma}\cdot\hat{r}\,\varphi_{j,m}^{(-)} \tag{2-90}$$

Since $\boldsymbol{\sigma}\cdot\hat{r}$ is a pseudoscalar operator, $\varphi_{jm}^{(+)}$ and $\varphi_{jm}^{(-)}$ have opposite parities (also, their angular momenta l differ by one unit). It is convenient to introduce a common notation. Let $\varphi_{j,m}^{l}$ denote $\varphi_{jm}^{(+)}$ if $j = l+\frac{1}{2}$ and $\varphi_{jm}^{(-)}$ if $j = l-\frac{1}{2}$. We verify by inspection that $\varphi_{j,m}^{l}$ has parity $(-1)^l$.

Since the Dirac equation in the Coulomb potential

$$E\psi = \left(\frac{1}{i}\boldsymbol{\alpha}\cdot\boldsymbol{\nabla} + \beta m - \frac{Z\alpha}{r}\right)\psi \equiv H\psi \tag{2-91}$$

is invariant under a space reflection, odd and even eigenstates may be constructed, namely,

$$\beta\psi_{j,m}^{(\pm)}(\tilde{x}) = \pm\psi_{j,m}^{(\pm)}(x)$$

It is clear that the spinors

$$\psi_{jm}^{l} \equiv \begin{bmatrix}\dfrac{iG_{lj}(r)}{r}\,\varphi_{jm}^{l}\\[2ex] \dfrac{F_{lj}(r)}{r}(\boldsymbol{\sigma}\cdot\hat{r})\,\varphi_{jm}^{l}\end{bmatrix} \tag{2-92}$$

have parity $(-1)^l$. In (2-92), the factors i and $1/r$ have been introduced for later convenience. Noting that H in (2-91) reads

$$H = \begin{pmatrix} m - \dfrac{Z\alpha}{r} & \boldsymbol{\sigma}\cdot\mathbf{p}\\[2ex] \boldsymbol{\sigma}\cdot\mathbf{p} & -m - \dfrac{Z\alpha}{r}\end{pmatrix}$$

we do the intermediate calculations

$$\boldsymbol{\sigma}\cdot\mathbf{p}f(r)\varphi_{jm}^{l} = \boldsymbol{\sigma}\cdot\hat{r}(\boldsymbol{\sigma}\cdot\hat{r}\;\boldsymbol{\sigma}\cdot\mathbf{p})f(r)\varphi_{jm}^{l} = \frac{\boldsymbol{\sigma}\cdot\hat{r}}{r}(\mathbf{r}\cdot\mathbf{p} + i\boldsymbol{\sigma}\cdot\mathbf{L})f(r)\varphi_{jm}^{l}$$

$$= -i\,\frac{\boldsymbol{\sigma}\cdot\hat{r}}{r}\left\{r\frac{df(r)}{dr} + [1 \mp (j+1/2)]f(r)\right\}\varphi_{jm}^{l} \qquad \text{for } j = l \pm \tfrac{1}{2}$$

and, similarly,

$$(\sigma \cdot \mathbf{p})(\sigma \cdot \hat{r}) f(r)\varphi^l_{jm} = -\frac{i}{r}\left[r\frac{\partial}{\partial r} + 1 \pm \left(j + \frac{1}{2}\right)\right] f(r)\varphi^l_{jm}$$

Our lengthy separation of variables results in the pair of radial equations

$$\begin{cases} \left(E - m + \dfrac{Z\alpha}{r}\right) G_{lj}(r) = -\dfrac{dF_{lj}}{dr}(r) \mp \left(j + \dfrac{1}{2}\right)\dfrac{F_{lj}(r)}{r} \\ \left(E + m + \dfrac{Z\alpha}{r}\right) F_{lj}(r) = \dfrac{dG_{lj}}{dr}(r) \mp \left(j + \dfrac{1}{2}\right)\dfrac{G_{lj}(r)}{r} \end{cases} \tag{2-93}$$

To solve this pair of equations, we introduce the notation $\lambda = \sqrt{m^2 - E^2}$ (since $E^2 < m^2$), the new variable $\rho = 2\lambda r$, and, for l and j given, the new functions $F_1(\rho)$ and $F_2(\rho)$:

$$G(r) = \left(1 + \frac{E}{m}\right)^{1/2} e^{-\rho/2}(F_1 + F_2)(\rho)$$

$$F(r) = \left(1 - \frac{E}{m}\right)^{1/2} e^{-\rho/2}(F_1 - F_2)(\rho)$$

By eliminating F_2 in favor of F_1, we get a second-order differential equation, the solution of which is

$$F_1 = \frac{\gamma - Z\alpha E/\lambda}{-\lambda + Z\alpha m/\lambda}\rho^\gamma F\left(\gamma + 1 - \frac{Z\alpha E}{\lambda}, 2\gamma + 1; \rho\right)$$

$$F_2 = \rho^\gamma F\left(\gamma - \frac{Z\alpha E}{\lambda}, 2\gamma + 1; \rho\right)$$

with
$$\gamma \equiv [(j + \tfrac{1}{2})^2 - Z^2\alpha^2]^{1/2} = j + \tfrac{1}{2} - \delta_j$$

Here $F(a, b; \rho)$ denotes the degenerate hypergeometric function solution of

$$\left[\rho\frac{d^2}{d\rho^2} + (b - \rho)\frac{d}{d\rho} - a\right] F(a, b; \rho) = 0$$

For large ρ this function behaves as $[\Gamma(b)/\Gamma(a)]\rho^{a-b}e^\rho$. Demanding that F_{lj} and G_{lj} be normalizable, that is, $\int_0^\infty dr(F^2_{lj} + G^2_{lj}) = 1$, implies that $[\Gamma(\gamma - Z\alpha E/\lambda)]^{-1}$ must vanish. This is the desired quantization condition

$$\frac{Z\alpha E}{\lambda} - \gamma = n_r \qquad \text{(nonnegative integer)}$$

$$\equiv n - (j + \tfrac{1}{2})$$

which leads to (2-87):

$$\frac{Z\alpha E}{(m^2 - E^2)^{1/2}} = n - \delta_j$$

Collecting all the factors, it is then possible to write the expression of the normalized solutions F_{lj} and G_{lj}, and therefore of the ψ^l_{jm}.

We only quote here the form of the ground-state wave functions

$$\psi_{n=1,j=1/2,m=1/2} = \frac{(2mZ\alpha)^{3/2}}{(4\pi)^{1/2}} \left(\frac{1+\gamma}{2\Gamma(1+2\gamma)}\right)^{1/2} (2mZ\alpha r)^{\gamma-1} e^{-mZ\alpha r} \begin{pmatrix} 1 \\ 0 \\ \dfrac{i(1-\gamma)}{Z\alpha}\cos\theta \\ \dfrac{i(1-\gamma)}{Z\alpha}\sin\theta\, e^{i\varphi} \end{pmatrix}$$

and (2-94)

$$\psi_{n=1,j=1/2,m=-1/2} = \frac{(2mZ\alpha)^{3/2}}{(4\pi)^{1/2}} \left(\frac{1+\gamma}{2\Gamma(1+2\gamma)}\right)^{1/2} (2mZ\alpha r)^{\gamma-1} e^{-mZ\alpha r} \begin{pmatrix} 0 \\ 1 \\ \dfrac{i(1-\gamma)}{Z\alpha}\sin\theta\, e^{-i\varphi} \\ \dfrac{-i(1-\gamma)}{Z\alpha}\cos\theta \end{pmatrix}$$

Note that $\gamma = \sqrt{1 - Z^2\alpha^2} \sim 1 - (Z^2\alpha^2/2)$. In the nonrelativistic limit $\gamma \to 1$, we recover Schrödinger wave functions multiplied by Pauli spinors. On the other hand, these wave functions are singular at the origin, but this effect is noticeable only for

$$2mZ\alpha r \lesssim e^{-2/Z^2\alpha^2} \approx 10^{-16300/Z^2}$$

i.e., in a rather tiny region!

To compare these results with the experimental levels, several other effects have to be taken into account.

Hyperfine structure As a first approximation, we have neglected the magnetic field induced by the magnetic moment of the nucleus (we consider henceforth the hydrogen atom $Z = 1$). The coupling of the proton spin with the total electron angular momentum splits the levels into doublets. To get an estimate of this effect, let us use a nonrelativistic approximation for s states. The new interaction is

$$H_{\text{hf}} = -\frac{e}{2m}\,\boldsymbol{\sigma}_e \cdot \mathbf{B}$$

where $$\mathbf{B} = \operatorname{curl}\mathbf{A} \qquad \mathbf{A} = -\frac{1}{4\pi}\,\boldsymbol{\mu}_p \times \nabla \frac{1}{r}$$

$\boldsymbol{\sigma}_e$ denotes the electron spin, and $\boldsymbol{\mu}_p$ the proton magnetic moment. We recall that for a current distribution $\mathbf{j}(x)$, the magnetic moment is $\boldsymbol{\mu} = \frac{1}{2}\int \mathbf{r} \times \mathbf{j}(x)\, d^3x$ and $\Delta\mathbf{A} = -\mathbf{j}$; therefore for a moment localized at the origin $\mathbf{j} = -\boldsymbol{\mu} \times \nabla\delta^3(\mathbf{x})$. Hence

$$H_{\text{hf}} = \frac{e}{4\pi}\,\frac{1}{2m}\,\boldsymbol{\sigma}_e \cdot [\nabla \times (\boldsymbol{\mu}_p \times \nabla)]\,\frac{1}{r}$$

$$= \frac{e}{8\pi m}\,\boldsymbol{\sigma}_e \cdot [\boldsymbol{\mu}_p\Delta - (\boldsymbol{\mu}_p \cdot \nabla)\nabla]\,\frac{1}{r}$$

In s states, we only need the angular average of H_{hf}, $\nabla_i \nabla_j$ can be replaced by $\frac{1}{3}\Delta\delta_{ij}$, and

$$\bar{H}_{\text{hf}} = -\frac{e}{3m}\,\boldsymbol{\sigma}_e \cdot \boldsymbol{\mu}_p\, \delta^3(r) \tag{2-95}$$

Introducing the proton gyromagnetic ratio g_p, $\boldsymbol{\mu}_p = -(g_p e/2m_p)(\boldsymbol{\sigma}_p/2)$, we obtain

$$\langle \bar{H}_{\text{hf}} \rangle = \frac{e^2}{12 m_e m_p}\, g_p \boldsymbol{\sigma}_e \cdot \boldsymbol{\sigma}_p |\psi(0)|^2$$

For the ground state, $n = 1$ [compare (2-84)]:

$$|\psi(0)|^2 = \frac{(m_e \alpha)^3}{\pi}$$

and $\boldsymbol{\sigma}_e \cdot \boldsymbol{\sigma}_p$ takes the values 1 in the triplet state and -3 in the singlet one. Finally, we get

$$\Delta E_{\text{hf}\atop n=1,j=1/2} \text{(triplet-singlet)} = \frac{4}{3}\, m_e \alpha^4 \frac{m_e}{m_p}\, g_p = 5.89 \times 10^{-6}\,\text{eV} = 1.42 \times 10^9\,\text{Hz}$$

If we compare this splitting to the fine splitting, we see that the dominant effect is a reduction by a factor $m_e/m_p \sim \mu_p/\mu_e$. The previous estimate may, of course, be extended to higher waves ($l \geq 1$).

Radiative corrections There are several kinds of such corrections. First, the excited states are unstable—they acquire a width—and the atom may undergo a spontaneous transition to a lower state.

In a nonrelativistic dipole approximation, the probability per unit time of radiative transition between two states λ and μ per unit time is

$$W_{\mu \leftarrow \lambda} = \frac{4}{3}\frac{(E_\lambda - E_\mu)^3}{2j_\lambda + 1} |\langle \mu \| D \| \lambda \rangle|^2$$

where $\langle \mu \| D \| \lambda \rangle$ is the reduced matrix element of the dipole operator $\mathbf{D} = e\mathbf{r}$, as defined by the Wigner-Eckart theorem:

$$\langle \mu j_\mu m_\mu | D_q | \lambda j_\lambda m_\lambda \rangle = \frac{1}{(2j_\lambda + 1)^{1/2}} \langle j_\mu m_\mu ; 1q | j_\lambda m_\lambda \rangle \langle \mu \| D \| \lambda \rangle$$

In particular, for a transition $2P \to 1S$, we find

$$W_{1S \leftarrow 2P} = 6.2 \times 10^8\,\text{s}^{-1} = 4.1 \times 10^{-7}\,\text{eV}$$

Second, the charged particles interact with the fluctuations of the quantized electromagnetic field. The latter vanishes only on an average. As a result the levels are slightly shifted. Although a complete and systematic treatment of these effects requires the methods of quantum field theory to be described later, we may give, following Welton, a qualitative description of the main effect: the Lamb shift. It is based upon the same kind of argument as the discussion of the Darwin term in the last section, but here the fluctuation of the position is due to the electromagnetic field, rather than to a relativistic zitterbewegung. Hence the new contribution to the hamiltonian has the form

$$\Delta H_{\text{Lamb}} = \tfrac{1}{6}\langle(\delta\mathbf{r})^2\rangle\,\Delta V$$

where $V = -(Z\alpha/r)$ and therefore, according to Poisson's law, $\Delta V = 4\pi Z\alpha\delta^3(r)$. In this simple picture, treating this perturbation to first order, only s waves are affected and the nth level is shifted by the amount

$$\begin{aligned}\Delta E_{\text{Lamb}}(n) &= \frac{2\pi Z\alpha}{3}\langle(\delta\mathbf{r})^2\rangle|\psi_n(0)|^2 \\ &= \frac{(2mZ\alpha)^3}{12}\frac{\alpha}{n^3}\langle(\delta\mathbf{r})^2\rangle\end{aligned}$$

(For higher waves, the shift is reduced by the vanishing of the wave function at the origin, and is considerably smaller.) The estimate of $\langle(\delta\mathbf{r})^2\rangle$ relies on a classical description of the motion of the electron in the fluctuating field (the nucleus, which is heavier, does not move). The electron oscillates according to

$$m\delta\ddot{\mathbf{r}} = e\mathbf{E}$$

The Fourier component E_ω with frequency ω of the electric field contributes

$$m\delta\mathbf{r}_\omega = -\frac{e}{\omega^2}\mathbf{E}_\omega$$

and assuming that there is no correlation between various modes

$$\langle(\delta\mathbf{r})^2\rangle = \frac{e^2}{m^2}\int_0^\infty \frac{d\omega}{\omega^4}\langle\mathbf{E}_\omega^2\rangle$$

Anticipating the quantum treatment of the electromagnetic field, we suppose that the field is an incoherent superposition of plane waves and remember that the vacuum energy of such a field is the sum of zero point energies

$$\frac{1}{2}\int d^3x\,(\mathbf{E}^2 + \mathbf{B}^2) = \sum_{\lambda=1,2}\sum_k \frac{\omega_{k\lambda}}{2}$$

over all wave numbers k and polarizations λ. In a big box of size L, $\mathbf{k} = (2\pi/L)\mathbf{n}$ (n_x, n_y, and n_z integers) and

$$\frac{1}{2}\int d^3x\,(\mathbf{E}^2 + \mathbf{B}^2) = 2L^3\int\frac{d^3k}{(2\pi)^3}\frac{\omega_k}{2}$$

$$\begin{aligned}\langle\mathbf{E}^2\rangle = \int d\omega\,\langle\mathbf{E}_\omega^2\rangle &= \frac{1}{L^3}\int d^3x\,\mathbf{E}^2 \\ &= \int\frac{d^3k}{(2\pi)^3}\omega_k = \int\frac{d\omega\,\omega^3}{2\pi^2}\end{aligned}$$

Therefore,

$$\langle(\delta \mathbf{r})^2\rangle = \frac{2\alpha}{\pi m^2}\int_0^\infty \frac{d\omega}{\omega}$$

The last integral is divergent both for small and large frequencies. The ultraviolet divergence when $\omega \to \infty$ comes from our poor quantum treatment; there is actually a cutoff for distances of order $\hbar/mc$, that is, frequencies $\omega \sim mc^2/\hbar$. At the other end, the infrared divergence for small frequencies should be cured by a more accurate treatment of the electromagnetic field in the presence of the charges. The large wavelength modes are sensitive to the low-lying electronic states. This suggests an infrared cutoff of the order $\omega \sim c/a = mc^2\alpha/\hbar$. Using these crude estimates we expect a value

$$\langle \delta \mathbf{r}^2\rangle \sim \frac{2\alpha}{\pi m^2}\ln\frac{1}{\alpha}$$

and therefore a Lamb shift

$$\Delta E_{\text{Lamb}}(n, l) \sim \frac{4}{3}\frac{mZ^4\alpha^5}{n^3}\ln\left(\frac{1}{\alpha}\right)\delta_{l0}$$

For the level $n = 2$ of the hydrogen atom the calculated shift $2S_{1/2} - 2P_{1/2}$ is

$$\Delta E_{\text{Lamb}} \simeq 660 \text{ MHz}$$

in rough agreement with the observed shift of 1057 MHz. The comparison of this term with the Darwin term in Eq. (2-82) shows a reduction factor of order $\alpha \ln 1/\alpha$. These approximations will be considerably improved in the sequel (see Chap. 7).

Nuclear effects The nucleus has a finite size and its charge distribution is not concentrated at a point. For the proton this is represented by a form factor. This affects predominantly the *s*-wave states, since higher *l*-wave functions vanish at the origin. An interesting consequence is the isotopic effect, predicted as early as 1932, where different isotopes would have slightly shifted levels.

For light nuclei, the main contribution comes from the mass difference of the various isotopes:

$$\delta E \sim \frac{1}{2}\Delta\left(\frac{1}{m_{\text{nucl}}}\right)\langle \sum_{\text{electrons}} p_i^2\rangle$$

since the reduced mass of the electrons is $1/m = 1/m_e + 1/m_{\text{nucl}}$. For heavy nuclei, however, the finite size of the charge distribution is the leading effect. In a nonrelativistic approximation, we write the correction as

$$\delta E \simeq e\int d^3x\,|\psi(x)|^2\left[V(r) + \frac{Ze}{4\pi r}\right]$$

where $V(r)$ is the true potential and $-(Ze/r)$ its Coulomb approximation.

$$\delta E \simeq e\,|\psi(0)|^2 \int d^3x \left[V(r) + \frac{Ze}{4\pi r} \right]$$

$$\simeq \frac{e}{6}\,|\psi(0)|^2 \int d^3x\, r^2 \Delta V$$

$$\simeq \frac{e}{6}\,|\psi(0)|^2\, Ze\langle r^2 \rangle_{\text{nucleus}}$$

where we have inserted $\Delta r^2 = 6$, integrated by parts, and used the Poisson law: $\Delta V = -\rho$, $\int \rho\, d^3x = -Ze$. According to (2-84), the effect is proportional to Z^4. This effect might be used in practice for isotopic separation. A given isotope is excited to some level by a first laser ray and then ionized by a second laser ray. An electrostatic separation may then be achieved.

A second consequence is that the critical value $Z_c \simeq 137$ beyond which the ground state becomes unstable is pushed to larger values $Z \simeq 175$.

Two-body relativistic corrections A correct treatment should also include the recoil of the nucleus. This is a difficult problem, which may be tackled by a relativistic two-body equation (Chap. 10). An heuristic reasoning may, however, shed some light on this point. From the predictions of the Schrödinger and Klein-Gordon equations [(2-83) and (2-86)], we see that in both cases they can be interpreted by imposing the following condition on the velocity v:

$$\frac{i\alpha}{|\mathbf{v}|} = n = \text{positive integer}$$

Indeed, in the Schrödinger case,

$$E = \frac{m\mathbf{v}^2}{2} = -\frac{m\alpha^2}{2n^2}$$

whereas for the Klein-Gordon equation $\mathbf{v}^2 = \mathbf{p}^2/E^2$,

$$\frac{E^2 - m^2}{E^2} = -\frac{\alpha^2}{n^2} \qquad \text{or} \qquad E = \frac{m}{(1 + \alpha^2/n^2)^{1/2}}$$

and the desired result is obtained by replacing n by $n - \delta_j$ as in (2-86). In a two-body problem, let us assume that the relevant velocity is the relative one, namely, the velocity of one of the particles measured in the rest frame of the other. Then

$$E_{\text{tot}}^2 = (p + P)^2 = m^2 + M^2 + \frac{2mM}{(1 - \mathbf{v}^2)^{1/2}}$$

which is symmetric in the interchange $m \leftrightarrow M$. Using again our empirical rule

$$\mathbf{v}^2 = -\frac{\alpha^2}{(n - \delta_j)^2}$$

we obtain

$$E_{\text{tot}}^2 = m^2 + M^2 + \frac{2mM}{[1 + \alpha^2/(n - \delta_j)^2]^{1/2}}$$

or

$$E_{\text{tot}} = m + M - \frac{\alpha}{2n^2}\frac{mM}{m+M} + \frac{\alpha^4}{8n^4}\frac{mM}{m+M}\left[3 - \frac{mM}{(m+M)^2}\right] - \frac{\alpha^4}{n^3}\frac{mM}{m+M}\frac{1}{2j+1} + O(\alpha^6)$$

This expression reproduces fairly well the recoil effects obtained in a more refined treatment.

2-4 HOLE THEORY AND CHARGE CONJUGATION

2-4-1 Reinterpretation of Negative Energy Solutions

In spite of the successes of the Dirac equation, we must abandon our ostrich policy and face the interpretation of negative energy solutions. As explained previously, their presence is intolerable, since they make all positive energy states unstable in the final analysis.

A solution was proposed by Dirac as early as 1930 in terms of a many-particle theory. Although this shall not be the final standpoint, as it does not apply to scalar particles, for instance, it is instructive to retrace his reasoning. It provides an intuitive physical picture useful in practical instances, and permits fruitful analogies with different situations such as electrons in a metal. Its major assumption is that all the negative energy levels are filled up in the vacuum state. According to the Pauli exclusion principle, this prevents any electron from falling into these negative energy states, and thereby insures the stability of positive energy physical states. In turn, an electron of the negative energy sea may be excited to a positive energy state. It then leaves a hole in the sea. This hole in the negative energy, negatively charged states appears as a positive energy, positively charged particle—the positron. Besides the properties of the positron, its charge $|e| = -e$ and its rest mass m_e, this theory also predicts new observable phenomena:

1. The annihilation of an electron-positron pair. A (positive energy) electron falls into a hole in the negative energy sea with the emission of radiation. From energy momentum conservation at least two photons are emitted, unless a nucleus is present to absorb energy and momentum.
2. Conversely, an electron-positron pair may be created from the vacuum by an incident photon beam in the presence of a target to balance energy and momentum. This is the process mentioned above; a hole is created while the excited electron acquires a positive energy.

Thus the theory predicts the existence of positrons which were in fact observed in 1932. Since positrons and electrons may annihilate, we must abandon the interpretation of the Dirac equation as a wave equation. Also, the reasons for discarding the Klein-Gordon equation no longer hold. As we shall see, it actually describes spinless particles, such as pions. However, the hole interpretation is not satisfactory for bosons, since Fermi statistics plays a crucial role in Dirac's argument.

Even for fermions, the concept of an infinitely charged unobservable sea looks rather queer. We have instead to construct a true many-body theory to accommodate particles and antiparticles in a consistent way. This will be achieved by the "second quantization," i.e., the introduction of quantized fields capable of creating or annihilating particles.

2-4-2 Charge Conjugation

Hole theory implies the existence of electrons and positrons with the same mass and opposite charges which obey the same equation. The Dirac equation must therefore admit a new symmetry corresponding to the interchange particle $\leftrightarrow$ antiparticle. We thus seek a transformation $\psi \to \psi^c$ reversing the charge, i.e., such that

$$(i\not{\partial} - e\not{A} - m)\psi = 0$$

and

$$(i\not{\partial} + e\not{A} - m)\psi^c = 0 \tag{2-96}$$

We demand that this transformation be local and that its square amount at most to multiplying ψ by an unobservable phase. To construct ψ^c we conjugate and transpose the first equation and get

$$[\gamma^{\mu T}(-i\partial_\mu - eA_\mu) - m]\bar{\psi}^T = 0$$

with $\bar{\psi}^T = \gamma^{0T}\psi^*$. In any representation of the γ algebra there must exist a matrix C which satisfies

$$C\gamma_\mu^T C^{-1} = -\gamma_\mu \tag{2-97}$$

For instance, in the representation (2-10), C may be taken as

$$C = i\gamma^2\gamma^0 = \begin{pmatrix} 0 & -i\sigma^2 \\ -i\sigma^2 & 0 \end{pmatrix} \tag{2-97a}$$

$$-C = C^{-1} = C^T = C^\dagger \tag{2-97b}$$

We then identify ψ^c as

$$\psi^c = \eta_c C\bar{\psi}^T \tag{2-98}$$

with η_c an arbitrary unobservable phase, generally taken as being equal to unity. In the present framework charge conjugation is an antilinear transformation. This is consistent with the hole interpretation, since when computing a transition probability the presence of a particle in a certain state will be represented by ψ and its absence by ψ^*. Let us examine more closely the properties of this charge conjugation. We compute ψ^c for ψ describing a spin-down negative energy electron at rest. In the absence of external field

$$\psi = e^{imt}\begin{pmatrix}0\\0\\0\\1\end{pmatrix} \qquad \psi^c = \eta_c C\bar{\psi}^T = \eta_c\, e^{-imt}\begin{pmatrix}1\\0\\0\\0\end{pmatrix}$$

Therefore, the charge conjugate of a negative energy spin-down electron is indeed equivalent to a positive energy spin-up electron.

For an arbitrary solution ψ of energy momentum p polarized along n, we know that

$$\psi = \frac{\varepsilon \not{p} + m}{2m}\,\frac{1+\gamma_5 \not{n}}{2}\,\psi \qquad p^0 > 0$$

where $\varepsilon = \pm 1$ denotes the sign of the energy. Since C commutes with $\gamma_5 = \gamma_5^*$ in the Dirac representation,

$$\begin{aligned}\psi^c = C\bar{\psi}^T &= C\gamma^0\left(\frac{\varepsilon\not{p}+m}{2m}\right)^*\left(\frac{1+\gamma_5\not{n}}{2}\right)^*\psi^* \\ &= \left(\frac{-\varepsilon\not{p}+m}{2m}\right)\left(\frac{1+\gamma_5\not{n}}{2}\right)\psi^c \qquad \text{(2-99)}\end{aligned}$$

ψ^c is described by the same four-vectors p and n, but the sign of the energy has been reversed. Using the notations (2-48), we have

$$\begin{aligned} u(p, n) &= \eta(p, n) v^c(p, n) \\ v(p, n) &= \eta(p, n) u^c(p, n) \end{aligned} \qquad \text{(2-100)}$$

where the phase $\eta(p, n)$ may depend on p and n. We recall that the projector $(1 + \gamma_5 \not{n})/2$ projects onto spin states $\pm\frac{1}{2}$ along n according to the sign of the energy. Thus the spin is reversed by charge conjugation.

We note, furthermore, that under a common transformation on the spinor ψ and the potential A,

$$\begin{aligned} \psi &\to \psi^c = \eta_c C\bar{\psi}^T \\ A_\mu &\to A_\mu^c = -A_\mu \end{aligned} \qquad \text{(2-98}a\text{)}$$

the Dirac equation (2-96) remains unchanged.

The transformation law of the four-vector current under charge conjugation is

$$\begin{aligned} j_\mu = \bar{\psi}\gamma_\mu\psi \to j_\mu^c &= \bar{\psi}^c\gamma_\mu\psi^c \\ &= \psi^T C\gamma_\mu C\bar{\psi}^T = \psi^T\gamma_\mu^T\bar{\psi}^T \end{aligned}$$

We could naively conclude that $j_\mu^c = \bar{\psi}\gamma_\mu\psi = j_\mu$. However, we shall see in the next chapter that ψ and $\bar{\psi}$ have to be considered as anticommuting operators (Fermi-Dirac statistics). Therefore charge conjugation will reverse the sign of j_μ and leave $ej \cdot A$ unchanged.

We shall explore the intricacies of the last discrete symmetry, namely time reversal, in Chap. 3, when we have a satisfactory formulation of particles and antiparticles.

2-4-3 Zero-Mass Particles

In Sec. 2-2-1, while constructing the spinor solutions of the free Dirac equation, we have discarded the massless case $m = 0$. However, the neutrinos are massless spin $\frac{1}{2}$ particles. In addition, we expect that at very high energy massive particles behave as massless. Therefore we reexamine this case. We start from the massless Dirac equation

$$\not{p}\psi = 0 \qquad p_\mu = i\partial_\mu \tag{2-101}$$

Multiplying this equation by $\gamma^5\gamma^0 = -i\gamma^1\gamma^2\gamma^3$ yields

$$\Sigma \cdot \mathbf{p}\psi = \gamma^5 p^0 \psi \tag{2-102}$$

since, for instance, $(\gamma^5\gamma^0)\gamma^1 = i\gamma^2\gamma^3 = \sigma^{23} \equiv \Sigma^1$. The chirality operator γ_5 anticommutes with $\not{p}$. For a positive energy solution

$$\psi(x) = e^{-ik\cdot x}\psi(k)$$

Equation (2-101) requires $k^2 = 0$, thus $k^0 = E = |\mathbf{k}|$, and Eq. (2-102) implies

$$\Sigma \cdot \hat{k}\psi = \gamma^5\psi \tag{2-103}$$

Therefore the chirality equals the helicity (it is opposite for negative energy solutions). Let us label the independent solutions of (2-101) by their chirality:

$$\psi(x) = \begin{cases} e^{-ik\cdot x}u_\pm(k) \\ e^{ik\cdot x}v_\pm(k) \end{cases} \quad \text{with } k^2 = 0,\ k^0 = |\mathbf{k}| > 0$$

$$\gamma_5 u_\pm(k) = \pm u_\pm(k)$$
$$\gamma_5 v_\pm(k) = \pm v_\pm(k)$$

In the usual representation $\gamma_5 = \begin{pmatrix} 0 & I \\ I & 0 \end{pmatrix}$:

$$u_\pm(k) = \frac{1}{\sqrt{2}}\begin{pmatrix} a_\pm(k) \\ \pm a_\pm(k) \end{pmatrix} \qquad v_\pm(k) = \frac{1}{\sqrt{2}}\begin{pmatrix} b_\pm(k) \\ \pm b_\pm(k) \end{pmatrix}$$

and $$a_+(k) = \boldsymbol{\sigma}\cdot\hat{k}a_+(k) = \begin{pmatrix} \cos\dfrac{\theta}{2} \\ \sin\dfrac{\theta}{2}e^{i\varphi} \end{pmatrix} = i\sigma_2 a_-^*(k)$$

$$a_-(k) = -\boldsymbol{\sigma}\cdot\hat{k}a_-(k) = \begin{pmatrix} -\sin\dfrac{\theta}{2}e^{-i\varphi} \\ \cos\dfrac{\theta}{2} \end{pmatrix} = -i\sigma_2 a_+^*(k)$$

where θ and φ are the polar angles of $\hat{k}$. Similarly,

$$v_+(k) = C\bar{u}_-^T(k) = \frac{1}{\sqrt{2}}\begin{pmatrix} b_+(k) = -i\sigma_2 a_-^*(k) = -a_+(k) \\ b_+(k) \end{pmatrix} = -u_+(k)$$

$$v_-(k) = C\bar{u}_+^T(k) = \frac{1}{\sqrt{2}}\begin{pmatrix} b_-(k) = i\sigma_2 a_+^*(k) = -a_-(k) \\ -b_-(k) \end{pmatrix} = -u_-(k)$$

There exists only two independent solutions, for a given k.

Experimental observation shows that only negative chirality neutrinos exist. Neutrinos have the helicity -1, antineutrinos the helicity $+1$. This will be better understood in terms of two-component spinors. Indeed the reason for using four-component spinors no longer holds for the massless Dirac equation

$$i\frac{\partial\psi}{\partial t} = \frac{1}{i}\boldsymbol{\alpha}\cdot\boldsymbol{\nabla}\psi$$

where the algebra

$$\{\alpha_i, \alpha_j\} = 2\delta_{ij}$$

may be realized by the three two-dimensional Pauli matrices. The identification $\alpha_i \to \sigma_i$ leads to positive helicity, positive energy particles, whereas $\alpha_i \to -\sigma_i$ gives negative helicity. Such spinors, initially introduced by H. Weyl, were rejected because they were incompatible with parity conservation (which reverses the sign of helicity). This is not a serious objection any more since neutrinos are involved in weak interactions which do not conserve parity.

We have already introduced in (2-20) the corresponding chiral representations of α matrices:

$$\boldsymbol{\alpha} = \begin{pmatrix} \boldsymbol{\sigma} & 0 \\ 0 & -\boldsymbol{\sigma} \end{pmatrix} \qquad \gamma^0 = \begin{pmatrix} 0 & -I \\ -I & 0 \end{pmatrix} \qquad \gamma^5 = \begin{pmatrix} I & 0 \\ 0 & -I \end{pmatrix}$$

For positive chirality, $\gamma^5 = +1$, $\psi = \begin{pmatrix} \varphi \\ 0 \end{pmatrix}$ and $\gamma\cdot p\psi = 0$ reduces to

$$(-p^0 + \mathbf{p}\cdot\boldsymbol{\sigma})\varphi = 0 \tag{2-104}$$

whereas for $\gamma^5 = -1$, $\psi = \begin{pmatrix} 0 \\ \chi \end{pmatrix}$ and

$$(p^0 + \mathbf{p}\cdot\boldsymbol{\sigma})\chi = 0 \tag{2-105}$$

In both cases, we have a two-component theory, and the Dirac equation is equivalent to the pair of Weyl equations. The so-called charge conjugation C (the neutrinos have no charge!) connects the two chiralities and changes the sign of the energy. There is no C invariance if nature uses only neutrinos of a definite chirality. Actually, since the parity operation P also connects the two types of solutions

$$\psi(t, \mathbf{x}) \to \gamma^0 \psi(t, -\mathbf{x})$$

(γ^0 is antidiagonal), the combined operation CP leaves the Weyl equations invariant. In the new representation, the matrix C of (2-97) reads

$$C = \begin{pmatrix} -i\sigma_2 & \\ & i\sigma_2 \end{pmatrix}$$

Therefore, the CP operation acts according to

$$\psi^{CP}(t, \mathbf{x}) = \eta C \psi^*(t, -\mathbf{x}) = \mp i\eta\sigma_2 \psi^*(t, -\mathbf{x}) \tag{2-106}$$

for chirality $\gamma^5 = \pm 1$ respectively.

We observe that the Lorentz invariant normalizations (2-43*a, b*) of the massive solutions have to be modified in the massless case. Then we shall write

$$\bar{u}^{(\alpha)}(k)\gamma^0 u^{(\beta)}(k) = 2E\delta_{\alpha\beta}$$

$$\bar{v}^{(\alpha)}(k)\gamma^0 v^{(\beta)}(k) = 2E\delta_{\alpha\beta}$$

and leave it to the reader to construct the appropriate plane wave solutions.

2-5 DIRAC PROPAGATOR

2-5-1 Free Propagator

In Chap. 1 we developed the concept of Green functions of a classical scalar field. We will extend it here to spin $\frac{1}{2}$ particles. We consider first free propagation.

Let us try to determine the solution of the Dirac equation at time t_2 as a function of its value at an earlier time t_1. This is possible since we deal with a first-order equation. We thus look for a kernel $K(x_2, x_1)$ such that

$$\psi(t_2, \mathbf{x}_2) = \int d^3x_1 \, K(t_2, \mathbf{x}_2 ; t_1, \mathbf{x}_1)\gamma^0 \psi(t_1, \mathbf{x}_1) \tag{2-107}$$

The introduction of γ^0 will be justified soon. Any solution ψ is a linear superposition of plane wave solutions

$$\psi(t, \mathbf{x}) = \sum_{(\alpha)} \int \frac{d^3k}{(2\pi)^3} \frac{m}{E} [a^{(\alpha)}(k)u^{(\alpha)}(k) \, e^{-ik \cdot x} + b^{(\alpha)*}(k)v^{(\alpha)}(k) \, e^{ik \cdot x}]$$

Owing to the relations (2-43), we may write

$$\int d^3x \, \bar{u}^{(\alpha)}(k) \, e^{-ik \cdot x}\gamma^0 \psi(0, \mathbf{x}) = a^{(\alpha)}(k)$$

and

$$\int d^3x \, \bar{v}^{(\alpha)}(k) \, e^{ik \cdot x}\gamma^0 \psi(0, \mathbf{x}) = b^{(\alpha)*}(k)$$

Therefore,

$$\psi(t_2, \mathbf{x}_2) = \int \frac{d^3k}{(2\pi)^3} \frac{m}{E} \sum_\alpha \int d^3x_1 \, [u^{(\alpha)}(k) \otimes \bar{u}^{(\alpha)}(k) \, e^{-ik\cdot(x_2-x_1)}$$
$$+ v^{(\alpha)}(k) \otimes \bar{v}^{(\alpha)}(k) \, e^{ik\cdot(x_2-x_1)}]\gamma^0 \psi(t_1, \mathbf{x}_1)$$

Interchanging the order of integration, we find the desired kernel

$$K(t_2, \mathbf{x}_2; t_1, \mathbf{x}_1) = \int \frac{d^3k}{(2\pi)^3} \frac{m}{E} \sum_\alpha [u^{(\alpha)}(k) \otimes \bar{u}^{(\alpha)}(k) \, e^{-ik\cdot(x_2-x_1)}$$
$$+ v^{(\alpha)}(k) \otimes \bar{v}^{(\alpha)}(k) \, e^{ik\cdot(x_2-x_1)}] \quad \text{for } t_2 > t_1 \qquad (2\text{-}108)$$

Notice that K depends only on $(x_2 - x_1)$, which is a reflection of the translation invariance of the free equation. We may also use the projectors $\Lambda_\pm(k)$ of Eqs. (2-40) and (2-41) to recast $K(x_2, x_1)$ in a more compact form

$$K(x_2, x_1) = \theta(t_2 - t_1) \int \frac{d^3k}{2E(2\pi)^3} [(\not{k} + m) \, e^{-ik\cdot(x_2-x_1)} + (\not{k} - m) \, e^{ik\cdot(x_2-x_1)}] \qquad (2\text{-}109)$$

We denote this retarded kernel by K_{ret}. Let us show directly that it is a Green function of the Dirac equation. Acting on $K_{\text{ret}}(x_2, x_1)$ with $(i\not{\partial}_2 - m)$ yields

$$(i\not{\partial}_2 - m)K_{\text{ret}}(x_2, x_1) = i\gamma^0 \delta(t_2 - t_1) \int \frac{d^3k}{2E(2\pi)^3} [(\not{k} + m) \, e^{i\mathbf{k}\cdot(\mathbf{x}_2-\mathbf{x}_1)} + (\not{k} - m) \, e^{-i\mathbf{k}\cdot(\mathbf{x}_2-\mathbf{x}_1)}]$$

We may change $\mathbf{k}$ into $-\mathbf{k}$ in the second term of the right-hand side, the coefficient of $(\gamma\cdot\mathbf{k} + m)$ vanishes, and we obtain

$$(i\not{\partial}_2 - m)K_{\text{ret}}(x_2, x_1) = i\delta^4(x_2 - x_1) \qquad (2\text{-}110)$$

$K_{\text{ret}}(x_2, x_1)$ may also be expressed in terms of the scalar retarded Green function $G_{\text{ret}}(x_2 - x_1)$ [Eq. (1-169)] as

$$K_{\text{ret}}(x_2 - x_1) = -i(i\not{\partial}_2 + m)G_{\text{ret}}(x_2 - x_1)$$

Equation (2-110) also follows from the identity (1-165) satisfied by G_{ret}:

$$(\Box_2 + m^2)G_{\text{ret}}(x_2 - x_1) = \delta^4(x_2 - x_1)$$

From

$$G_{\text{ret}}(x) = -\frac{1}{(2\pi)^4} \int d^4k \frac{e^{ik\cdot x}}{(k_0 + i\varepsilon)^2 - \mathbf{k}^2 - m^2}$$

it follows that

$$K_{\text{ret}}(x) = \frac{i}{(2\pi)^4} \int d^4k \, e^{-ik\cdot x} \frac{\not{k} + m}{(k_0 + i\varepsilon)^2 - \mathbf{k}^2 - m^2}$$

where the k_0 integration has to be performed first, along the dashed contour of Fig. 2-3.

The hole theory suggests the introduction of a different Green function, the Feynman propagator already discussed in the last section of Chap. 1. It will

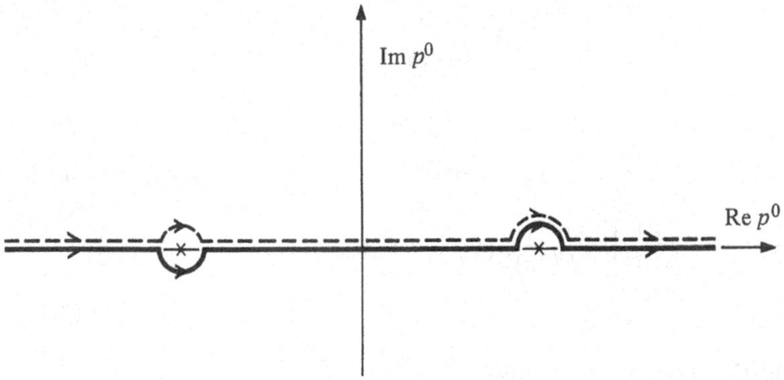

Figure 2-3 Contour of integration for the definition of Green functions. The broken line corresponds to the retarded propagator, the solid line to the Feynman one.

appear in a natural way in the quantized field theory. Nevertheless, let us sketch the ideas that led Feynman and Stueckelberg to its construction.

A Green function may be considered as describing three successive steps:

1. Appearance of an electron at $(t_1, \mathbf{x}_1)$
2. Propagation of the electron from $(t_1, \mathbf{x}_1)$ to $(t_2, \mathbf{x}_2)$
3. Disappearance of the electron at $(t_2, \mathbf{x}_2)$

As long as the electron has a positive energy, this process is physically acceptable for $t_2 > t_1$. On the other hand, if we deal with a negative energy electron, we would like to interpret its vanishing as the appearance of a positron, and vice versa. The second step should be then considered as the propagation of the positron from $(t_2, \mathbf{x}_2)$ to $(t_1, \mathbf{x}_1)$, which makes sense only for $t_2 < t_1$. Therefore in the hole theory, we would like to construct a Green function which propagates the positive energy solutions only for $t_2 > t_1$, the negative energy ones (more precisely the positrons) only for $t_1 > t_2$:

$$S_F(x_2, x_1) = \int \frac{d^3k}{(2\pi)^3 2E} \left[\theta(t_2 - t_1) a(\not{k} + m)\, e^{-ik\cdot(x_2 - x_1)} + \theta(t_1 - t_2) b(\not{k} - m)\, e^{ik\cdot(x_2 - x_1)}\right]$$

The constants a and b are determined by imposing that

$$(i\not{\partial}_2 - m) S_F(x_2, x_1) = \delta^4(x_2 - x_1) \tag{2-111}$$

[Notice the change of normalization with respect to (2-110)]. It follows from a straightforward calculation that

$$(i\not{\partial}_2 - m) S_F(x_2, x_1) = i\delta(t_2 - t_1) \int \frac{d^3k}{2E(2\pi)^3} \gamma^0 \left[a(\not{k} + m)\, e^{i\mathbf{k}\cdot(\mathbf{x}_2 - \mathbf{x}_1)} - b(\not{k} - m)\, e^{-i\mathbf{k}\cdot(\mathbf{x}_2 - \mathbf{x}_1)}\right]$$

$$= i\delta(t_2 - t_1) \int \frac{d^3k}{2E(2\pi)^3} e^{i\mathbf{k}\cdot(\mathbf{x}_2 - \mathbf{x}_1)}$$

$$\times \gamma^0[a(E\gamma^0 - \boldsymbol{\gamma}\cdot\mathbf{k} + m) - b(E\gamma^0 + \boldsymbol{\gamma}\cdot\mathbf{k} - m)]$$

and the condition (2-111) is fulfilled provided $a = -b = 1/i$. The result is thus

$$S_F(x_2 - x_1) = \frac{1}{i}\int \frac{d^3k}{2E(2\pi)^3} [\theta(t_2 - t_1)(\not{k} + m)\, e^{-ik\cdot(x_2 - x_1)}$$

$$-\theta(t_1 - t_2)(\not{k} - m)\, e^{ik\cdot(x_2 - x_1)}] \tag{2-112}$$

$K_{\text{ret}}(x_2 - x_1)$ and $iS_F(x_2 - x_1)$ may differ only by a solution of the homogeneous Dirac equation. This is indeed what is found by a direct calculation:

$$K_{\text{ret}}(x_2 - x_1) - iS_F(x_2 - x_1) = \int \frac{d^3k}{2E(2\pi)^3} (\not{k} - m)\, e^{ik\cdot(x_2 - x_1)}$$

A covariant expression is obtained by means of the integral representation

$$\theta(t) = \lim_{\varepsilon \to 0_+} \int_{-\infty}^{\infty} \frac{d\omega}{2i\pi} \frac{e^{i\omega t}}{\omega - i\varepsilon}$$

The limit $\varepsilon \to 0_+$ will be understood in what follows. The quantities we are dealing with are distributions acting on smooth test functions. After insertion of this expression into (2-112), we get

$$S_F(x) = -\int \frac{d^3k}{(2\pi)^4} \frac{1}{2E} \left[(\not{k} + m)\, e^{-ik\cdot x} \int_{-\infty}^{\infty} \frac{d\omega}{\omega - i\varepsilon} e^{i\omega t} \right.$$

$$\left. - (\not{k} - m)\, e^{ik\cdot x} \int_{-\infty}^{\infty} \frac{d\omega}{\omega - i\varepsilon} e^{-i\omega t} \right]$$

In the first integral we set $p^0 = E - \omega$, $\mathbf{p} = \mathbf{k}[E = k^0 = (\mathbf{k}^2 + m^2)^{1/2}]$ and in the second one $p^0 = \omega - E$, $\mathbf{p} = -\mathbf{k}$:

$$S_F(x) = \int \frac{d^4p}{(2\pi)^4} \frac{e^{-ip\cdot x}}{2E} \left[\frac{E\gamma^0 + \boldsymbol{\gamma}\cdot\mathbf{p} - m}{E + p^0 - i\varepsilon} - \frac{E\gamma^0 - \boldsymbol{\gamma}\cdot\mathbf{p} + m}{E - p^0 - i\varepsilon} \right]$$

For a vanishing positive ε, we may write

$$(E + p^0 - i\varepsilon)(E - p^0 - i\varepsilon) = -(p_0^2 - \mathbf{p}^2 - m^2 + i\varepsilon)$$

Finally,

$$S_F(x) = \int \frac{d^4p}{(2\pi)^4} e^{-ip\cdot x} \frac{\not{p} + m}{p^2 - m^2 + i\varepsilon} \tag{2-113}$$

The $i\varepsilon$ term gives the prescription for the momentum integration. The integration over p^0 is performed first, along the solid contour shown on Fig. 2-3, and then we integrate over $\mathbf{p}$. If we consider m as complex $m_c = m - i\varepsilon$, $\varepsilon \to 0_+$, we have

$$\frac{\not p + m}{p^2 - m^2 + i\varepsilon} = \frac{\not p + m_c}{p^2 - m_c^2} = \frac{1}{\not p - m_c} = \frac{1}{\not p - m + i\varepsilon}$$

Therefore, we may write the Fourier transform of $S_F(x)$ as

$$S_F(p) = \frac{\not p + m}{p^2 - m^2 + i\varepsilon} = \frac{1}{\not p - m + i\varepsilon} \tag{2-114}$$

Notice the relationship between $S_F(x)$ and the Feynman propagator of the scalar field (1-178):

$$S_F(x) = -(i\not\partial + m)G_F(x) \tag{2-115}$$

To summarize, the role of the Feynman propagator is to propagate the positive frequencies toward positive times and the negative ones backward in time.

Let $\psi^{(+)}(t_1, \mathbf{x}_0)$ and $\psi^{(-)}(t_2, \mathbf{x})$ be the positive and negative frequency components of a solution, given for t_1 and t_2 respectively, $t_1 < t_2$ and all $\mathbf{x}$. The propagator S_F allows us to find $\psi(t, \mathbf{x})$ at intermediate times t:

$$\psi(t, \mathbf{x}) = i \int d^3y \, [S_F(t - t_1, \mathbf{x} - \mathbf{y})\gamma^0\psi^{(+)}(t_1, \mathbf{y}) - S_F(t - t_2, \mathbf{x} - \mathbf{y})\gamma^0\psi^{(-)}(t_2, \mathbf{y})] \tag{2-116}$$

2-5-2 Propagation in an Arbitrary External Electromagnetic Field

In practice, we deal with the propagation in the presence of obstacles: diffusion processes, external fields, interactions with other particles. Let us treat the propagation in an external electromagnetic field

$$[i\not\partial_2 - e\not A(x_2) - m]S_A(x_2, x_1) = \delta^4(x_2 - x_1) \tag{2-117}$$

With very few exceptions, we are unable to find a compact expression for S_A. Fortunately, it frequently occurs that the term $e\not A$ is small enough to be treated as a perturbation, and S_A may be expressed as an (asymptotic) expansion in $e\not A$. To derive it, we multiply both sides of (2-117) by $S_F(x_3, x_2)$ and integrate over x_2:

$$\begin{aligned} S_F(x_3, x_1) &= \int d^4x_2 \, S_F(x_3, x_2)[i\overrightarrow{\not\partial}_2 - e\not A(x_2) - m]S_A(x_2, x_1) \\ &= \int d^4x_2 \, S_F(x_3, x_2)[-i\overleftarrow{\not\partial}_2 - e\not A(x_2) - m]S_A(x_2, x_1) \end{aligned}$$

From (2-111), it follows that

$$S_F(x_3, x_2)(-i\overleftarrow{\not\partial}_2 - m) = \delta^4(x_3 - x_2)$$

Therefore, the integral equation which determines S_A reads

$$S_A(x_3, x_1) = S_F(x_3, x_1) + e\int d^4x_2 \, \not A(x_2)S_A(x_2, x_1) \tag{2-118}$$

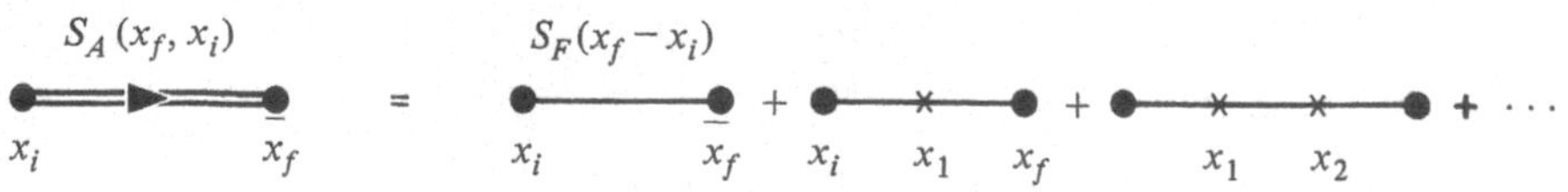

Figure 2-4 Diagrammatic representation of the perturbative expansion (2-119). The solid line between x_k and x_l stands for the propagator $S_F(x_k - x_l)$, the cross for $e\not{A}(x)$.

Equation (2-118) is adapted to a perturbative expansion obtained by iteration:

$$S_A(x_f, x_i) = S_F(x_f, x_i) + \int d^4x_1\, S_F(x_f, x_1)e\not{A}(x_1)S_F(x_1, x_i)$$
$$+ \iint d^4x_1\, d^4x_2\, S_F(x_f, x_1)e\not{A}(x_1)S_F(x_1, x_2)e\not{A}(x_2)S_F(x_2, x_i) + \cdots \tag{2-119}$$

This expansion is depicted diagrammatically on Fig. 2-4.

Introducing the Fourier transforms of S_A and $A(x)$:

$$S_A(x_f, x_i) = \iint \frac{d^4p_i}{(2\pi)^4}\frac{d^4p_f}{(2\pi)^4}\, e^{-i(p_f \cdot x_f - p_i \cdot x_i)}\, S_A(p_f, p_i)$$
$$A_\mu(x) = \int \frac{d^4p}{(2\pi)^4} A_\mu(p)\, e^{-ip \cdot x}$$

(we use the same notation in configuration and momentum space for the sake of simplicity) a perturbative expansion of $S_A(p_f, p_i)$ may also be written. Since $S_F(x_f, x_i) = S_F(x_f - x_i)$ is translation invariant,

$$S_F(p_f, p_i) = (2\pi)^4\delta^4(p_f - p_i)S_F(p_i)$$

where $S_F(p)$ has been given in (2-114) and

$$S_A(p_f, p_i) = S_F(p_f)(2\pi)^4\delta^4(p_f - p_i) + \int d^4p_1\, S_F(p_f)e\not{A}(p_1)S_F(p_i)(2\pi)^4\delta^4(p_f - p_1 - p_i)$$
$$+ \iint d^4p_1\, d^4p_2\, S_F(p_f)e\not{A}(p_1)S_F(p_2 + p_i)e\not{A}(p_2)S_F(p_i)(2\pi)^4\delta^4(p_f - p_1 - p_2 - p_i)$$
$$+ \cdots$$

2-5-3 Application to the Coulomb Scattering

Coulomb scattering will serve as a testing ground of the propagator method. The process under study is the scattering of a charged electron of mass m by a center with charge $-Ze$ and infinite mass. The latter creates a potential $A_0 = -Ze/4\pi r$, $\mathbf{A} = 0$, where $\mathbf{r}$ stands for the vector joining the center to the charge.

In classical nonrelativistic mechanics, the trajectories are hyperbolas. The scattering angle θ is related to the impact parameter b (see Fig. 2-5 for notations) by

1. Geometrical relations

$$\frac{a}{b} = \tan\frac{\theta}{2} \qquad \frac{b}{c} = \cos\frac{\theta}{2}$$

2. Energy conservation

$$\varepsilon = \frac{p_i^2}{2m} = \frac{p_f^2}{2m} = \frac{p_A^2}{2m} - \frac{Z\alpha}{c-a}$$

(p_A is the momentum at point A.)

3. Angular momentum conservation

$$l = p_i b = p_f b = p_A(c-a)$$

After elimination of p_i, p_A, c, and a, we get

$$b = \frac{Z\alpha}{2\varepsilon \tan \theta/2} \tag{2-120}$$

We consider a uniform flux of electrons, of density ρ and incident velocity $v = p_i/m = p_f/m$. The number of scattered particles in the solid angle $d\Omega = 2\pi\, d\cos\theta$ per unit time is equal to the number of incident particles on the ring of area $2\pi b\, db$, namely,

$$\frac{dN}{dt} = \rho v\, 2\pi b\, db = \rho v \left(\frac{Z\alpha}{2\varepsilon}\right)^2 \frac{d\Omega}{\sin^4 \theta/2}$$

Therefore, the differential cross section, defined as the ratio of $dN/dt\, d\Omega$ to the incident flux, is

$$\frac{d\sigma}{d\Omega} = \frac{1}{4\sin^4 \theta/2}\left(\frac{Z\alpha}{2\varepsilon}\right)^2 = \frac{4Z^2\alpha^2 m^2}{|\mathbf{q}|^4} \tag{2-121}$$

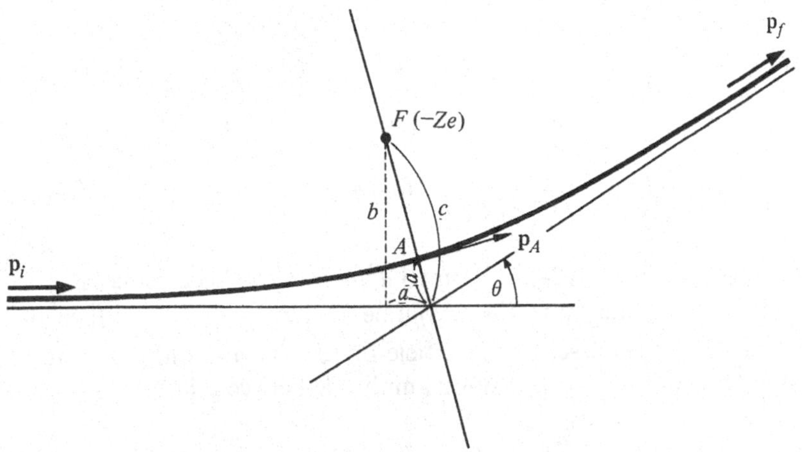

Figure 2-5 Coulomb scattering off a charge located at F.

Here $\mathbf{q} = \mathbf{p}_f - \mathbf{p}_i$ is the momentum transfer and $|\mathbf{q}| = 2p_i \sin\theta/2$. This is the classical Rutherford formula.

We turn to the relativistic quantum case. We shall use the expression derived in (2-116), after substituting S_A for S_F and taking the following boundary conditions: for $t_1 = -\infty$, $\psi^{(+)}(t_1, \mathbf{x})$ is an incident plane wave of positive energy electrons, whereas for $t_2 = +\infty$, $\psi^{(-)}(t_2, \mathbf{x})$ vanishes. Since we do not know S_A, we content ourselves with the first two orders of the perturbative expansion (2-119). Let $\psi_{\text{inc}}(t, \mathbf{x})$ be the solution of the free Dirac equation which reduces to the incident wave when $t \to t_1 = -\infty$. According to (2-119), the perturbative wave function reads

$$\psi(x) = \psi_{\text{inc}}(x) + \psi_{\text{diff}}(x) \tag{2-122}$$

where

$$\psi_{\text{diff}}(x) = \lim_{t_1 \to -\infty} i \int d^3y \int d^4z\, S_F(x-z)e\not{A}(z)S_F(z-y)\gamma^0\psi_{\text{inc}}(t_1, \mathbf{y})$$

and $\qquad y \equiv (t_1, \mathbf{y}) \qquad$ (2-123)

Since for $z^0 > t_1$,

$$i \int d^3y\, S_F(z-y)\gamma^0\psi_{\text{inc}}(t_1, \mathbf{y}) = \psi_{\text{inc}}(z)$$

we have

$$\psi_{\text{diff}}(x) = \int d^4z\, S_F(x-z)e\not{A}(z)\psi_{\text{inc}}(z)$$

As x^0 tends to $+\infty$, ψ_{diff} behaves as a pure positive energy solution of the free Dirac equation. Indeed, using the expression (2-112), we see that only the first term contributes in this limit, and we get

$$\psi_{\text{diff}}(x) = \int \frac{d^3p}{(2\pi)^3}\frac{m}{E}\int d^4z\, \frac{\not{p}+m}{2m} e^{-ip\cdot(x-z)}[-ie\not{A}(z)]\psi_{\text{inc}}(z)$$

$$= \sum_{k_f,\alpha}\left(\frac{m}{VE_f}\right)^{1/2} e^{-ip_f\cdot x}u^{(\alpha)}(p_f)S_{fi} \tag{2-124}$$

where

$$S_{fi} = -ie\int d^4z \left(\frac{m}{VE_f}\right)^{1/2} \bar{u}^{(\alpha)}(p_f)\not{A}(z)\, e^{ip_f\cdot z}\psi_i(z) \tag{2-125}$$

In the second expression of (2-124), the sum $\int d^3p/(2\pi)^3$ has been replaced by a sum over the final states in a finite space volume V: $1/V \sum_{\text{final states } k_f,\alpha}$; then the wave $(m/VE)^{1/2}\, e^{-ip\cdot x}u^{(\alpha)}(p)$ describes a particle of velocity $\mathbf{p}/E$ and polarization α in the volume V. Therefore, the transition amplitude between the state

$$\psi_i(x) = \left(\frac{m}{VE_i}\right)^{1/2} e^{-ip_i\cdot x}u^{(\beta)}(p_i)$$

and an analogous state $\psi_f(x)$ is

$$S_{fi} = -ie \frac{m}{V(E_i E_f)^{1/2}} \int d^4z\, \bar{u}^{(\alpha)}(p_f) A\!\!\!/(z)\, e^{i(p_f - p_i)\cdot z}\, u^{(\beta)}(p_i) \qquad \text{(2-125}a\text{)}$$

For the Coulomb problem, $A \equiv (A_0, \mathbf{0})$ and $A_0 = -Ze/4\pi r$; therefore,

$$S_{fi} = \frac{iZ\alpha}{V} \frac{m}{(E_i E_f)^{1/2}} 2\pi\delta(E_f - E_i) \int d^3r \frac{e^{-i\mathbf{q}\cdot\mathbf{r}}}{r} \bar{u}^{(\alpha)}(p_f)\gamma^0 u^{(\beta)}(p_i) \qquad \text{(2-126)}$$

It is gratifying to recover the energy conservation. We finally need the Fourier transform of the potential

$$\int d^3r \frac{e^{-i\mathbf{q}\cdot\mathbf{r}}}{r} = \frac{4\pi}{|\mathbf{q}|^2} \qquad \mathbf{q} \equiv \mathbf{p}_f - \mathbf{p}_i$$

We are using stationary plane waves instead of wave packets; no wonder, then, that the amplitude (2-126) cannot be squared. This is remedied if according to Fermi's golden rule we consider a finite time interval and replace the δ function by

$$2\pi\delta(E_f - E_i) = \int_{-\infty}^{\infty} dt\, e^{i(E_f - E_i)t} \rightarrow \int_{-T/2}^{T/2} dt\, e^{i(E_f - E_i)t}$$

$$= \frac{2\sin\left[T(E_f - E_i)/2\right]}{E_f - E_i}$$

The square of this expression behaves for large T as $T 2\pi\delta(E_f - E_i)$.

The transition probability between states i and f per unit time and per incident particle is then

$$\frac{dP_{fi}}{dt} = \int \left| \frac{iZ\alpha}{V} \frac{m}{(E_f E_i)^{1/2}} \frac{4\pi}{|\mathbf{q}|^2} \bar{u}^{(\alpha)}(p_f)\gamma^0 u^{(\beta)}(p_i) \right|^2 2\pi\delta(E_f - E_i)\, V \frac{d^3p_f}{(2\pi)^3} \qquad \text{(2-127)}$$

The summation runs over all possible final states, the number of which in the momentum volume element d^3p_f is $V\, d^3p_f/(2\pi)^3$. Dividing by the incident flux $(1/V)(|\mathbf{p}_i|/E_i)$, we obtain the differential cross section

$$d\sigma_{fi} = \int \frac{4Z^2\alpha^2 m^2}{|\mathbf{p}_i| E_f |\mathbf{q}|^4} \delta(E_f - E_i) p_f^2\, dp_f\, d\Omega_f |\bar{u}^{(\alpha)}(p_f)\gamma^0 u^{(\beta)}(p_i)|^2$$

We use $|\mathbf{p}_i| = |\mathbf{p}_f| = p_f$ and $p_f\, dp_f = E_f\, dE_f$ to perform the trivial p_f integration

$$d\sigma_{fi} = \frac{4Z^2\alpha^2 m^2}{|\mathbf{q}|^4} |\bar{u}^{(\alpha)}(p_f)\gamma^0 u^{(\beta)}(p_i)|^2\, d\Omega_f \qquad \text{(2-128)}$$

In the nonrelativistic limit, $\bar{u}^{(\alpha)}\gamma^0 u^{(\beta)}$ is proportional to $\delta^{\alpha\beta}$. If we do not observe the final polarization, a sum over α has to be performed, whereas for an unpolarized incident state we average over the two equally probable polarizations β:

$$\left.\frac{d\sigma_{fi}}{d\Omega}\right|_{\text{unpolarized}} = \frac{4Z^2\alpha^2m^2}{|\mathbf{q}|^4}\sum_\alpha \frac{1}{2}\sum_\beta |\bar{u}^{(\alpha)}(p_f)\gamma^0 u^{(\beta)}(p_i)|^2$$

$$= \frac{4Z^2\alpha^2m^2}{|\mathbf{q}|^4}\frac{1}{2}\operatorname{tr}\left(\gamma^0\frac{\not{p}_i + m}{2m}\gamma^0\frac{\not{p}_f + m}{2m}\right) \tag{2-129}$$

Once again, the expression (2-40) has been used. We now need identities for traces of γ matrices. For any product of an odd number of γ matrices, the traces vanish. For an even number, the following identity may be proved by induction:

$$\operatorname{tr}(\not{a}_1\not{a}_2\cdots\not{a}_{2n}) = a_1 \cdot a_2 \operatorname{tr}(\not{a}_3\cdots\not{a}_{2n}) - a_1 \cdot a_3 \operatorname{tr}(\not{a}_2\not{a}_4\cdots\not{a}_{2n}) + \cdots$$

Here this reduces to

$$\operatorname{tr}\gamma^0\not{p}_i\gamma^0\not{p}_f = 4(E_iE_f - p_i \cdot p_f + E_iE_f)$$

$$\operatorname{tr}\gamma^0\gamma^0 = 4$$

We also need the kinematical relations

$$E_i = E_f = E \qquad p_i \cdot p_f = E^2 - p^2\cos\theta = m^2 + 2E^2\beta^2\sin^2\frac{\theta}{2}$$

where $\beta \equiv v/c = |\mathbf{p}|/E$ is the incoming (or outgoing) velocity and

$$|\mathbf{q}|^4 = 16|\mathbf{p}|^4\sin^4\frac{\theta}{2} = 16\beta^2E^2\mathbf{p}^2\sin^4\frac{\theta}{2}$$

The final expression of the unpolarized cross section (Mott cross section) is

$$\left.\frac{d\sigma}{d\Omega}\right|_{\text{unpolarized}} = \frac{Z^2\alpha^2}{4\mathbf{p}^2\beta^2\sin^4\theta/2}\left(1 - \beta^2\sin^2\frac{\theta}{2}\right) \tag{2-130}$$

When $\beta \to 0$, this reduces to the Rutherford formula. Also notice that the relativistic correction $(1 - \beta^2\sin^2\theta/2)(1 - \beta^2)^{-1}$ affects predominantly the backward scattering.

This result has been derived for incident electrons. Let us discuss briefly the scattering of positrons in the same Coulomb field. The attractive Coulomb force is now replaced by a repulsive one. In classical nonrelativistic mechanics, this leads to the same Rutherford formula (this remarkable result is a peculiar feature of the Coulomb field). In our quantum treatment, we know that the theory is invariant under charge conjugation. The scattering of an electron off a charge $-Ze$ is the same as that of a positron off a charge $+Ze$. On the other hand, to lowest order, the cross section is an even function of Z. Therefore, the Mott cross section is also valid for positrons.

We may verify this by a direct calculation; we shall rather use the hole theory interpretation. An outgoing positron of four-momentum p_f and polarization α is represented by an "incoming" negative energy solution running backward in time:

$$\psi^{(-)}(x) = e^{ip_f\cdot x} v^{(-\alpha)}(p_f) \qquad p_f^0 > 0$$

We thus return to Eq. (2-116), using this time the boundary conditions

$$\psi^{(+)}(t_1, \mathbf{x}) = 0 \qquad t_1 = -\infty$$

$$\psi^{(-)}(t_2, \mathbf{x}) = e^{ip_f\cdot x} v^{(-\alpha)}(p_f) \qquad x^0 = t_2 = +\infty$$

Repeating the steps which led us from (2-122) to (2-125*a*), with due attention paid to the signs, we get

$$S_{fi} = ie \int d^4z \left(\frac{m^2}{V^2 E_i E_f}\right)^{1/2} \bar{v}^{(-\beta)}(p_i)\, e^{i(p_f - p_i)\cdot z} \not{A}(z) v^{(-\alpha)}(p_f)$$

This is in agreement with a direct computation. The incoming positron is described by the wave function

$$\psi_i^c(z) = e^{-ip_i\cdot z} u^{(\beta)}(p_i) = C\, e^{-ip_i\cdot z} \bar{v}^{(-\beta)^T}(p_i)$$

(up to a phase). According to Eq. (2-125), the scattering of these charge $-e$ positrons will be described by

$$\begin{aligned} S_{fi} &= ie \int d^4z\, \bar{\psi}_f^c(z) \not{A}(z) \psi_i^c(z) \\ &= -ie \left(\frac{m^2}{V^2 E_i E_f}\right)^{1/2} \int d^4z\, v^{(-\alpha)^T}(p_f) C^{-1} \not{A} C \bar{v}^{(-\beta)^T}(p_i)\, e^{i(p_f - p_i)\cdot z} \\ &= ie \left(\frac{m^2}{V^2 E_i E_f}\right)^{1/2} \int d^4z\, \bar{v}^{(-\beta)}(p_i) \not{A} v^{(-\alpha)}(p_f)\, e^{i(p_f - p_i)\cdot z} \end{aligned} \qquad (2\text{-}131)$$

It is now clear that this expression leads to the cross section (2-128).

We let the reader discuss the polarization effects or the corrections due to the recoil of the nucleus. We shall simply recall the useful concept of a form factor. Suppose that we study the scattering by a finite charge distribution, a finite radius nucleus, instead of a point-like charge. The spherically symmetric distribution $-Ze\,\rho(r)$ will be normalized according to

$$\int \rho(r)\, d^3r = 1$$

As in (2-126) the cross section to lowest order in α is proportional to $|\tilde{V}(q)|^2$, where $\tilde{V}(q)$ is the Fourier transform of the potential $V(r)$. From the Poisson equation $\Delta V = -\rho$, we learn that $\tilde{V}(q)$ is related to the form factor $\tilde{\rho}(q)$:

$$\tilde{\rho}(q) = \int \rho(r)\, e^{-i\mathbf{q}\cdot\mathbf{r}}\, d^3r$$

by $\tilde{V}(q) = (1/q^2)\tilde{\rho}(q)$. Therefore, the cross section has to be amended to read

$$\frac{d\sigma}{d\Omega} = \left.\frac{d\sigma}{d\Omega}\right|_{\text{Mott}} |\tilde{\rho}(q)|^2 \qquad (2\text{-}132)$$

2-5-4 Fock-Schwinger Proper Time Method

As a complement to the previous study of propagators, we shall present the beautiful method introduced by Fock and Schwinger and use it to give exact expressions of the Dirac propagator in two particular configurations of the external electromagnetic field: constant uniform field and plane wave field which we have already encountered in the classical theory.

Suppose that we seek for the Green function $G(x, x')$ solution of

$$H(x, i\partial_x)G(x, x') = \delta^4(x - x') \tag{2-133}$$

where H is a polynomial in ∂_x. The idea is to consider H as an hamiltonian that describes the proper time evolution of some system. The previous equation is nothing but the definition of the Green function of $H(x, p)$ in the x representation

$$\begin{aligned} \langle x|x'\rangle &= \delta^4(x - x') \\ [x^\mu, p^\nu] &= -ig^{\mu\nu} \qquad p_\mu = i\partial_\mu \end{aligned} \tag{2-134}$$

We introduce the unitary evolution operator $U(x, x'; \tau)$:

$$i\frac{\partial}{\partial\tau} U(x, x'; \tau) = H(x, p)U(x, x'; \tau) \tag{2-135}$$

with the boundary conditions

$$\lim_{\tau\to 0} U(x, x'; \tau) = \delta^4(x - x') \tag{2-136a}$$

$$\lim_{\tau\to -\infty} U(x, x'; \tau) = 0 \tag{2-136b}$$

We have

$$U(x, x'; \tau) = \langle x|\, e^{-iH\tau}|x'\rangle \equiv \langle x|U(\tau)|x'\rangle \tag{2-137}$$

$$G(x, x') = -i\int_{-\infty}^{0} d\tau\, U(x, x'; \tau) \tag{2-138}$$

Equation (2-135) may be rewritten as

$$\begin{aligned} &i\partial_\tau\langle x|U(\tau)|x'\rangle = \langle x|H(x, p)U|x'\rangle = \langle x|U(\tau)U^\dagger(\tau)H(x, p)U(\tau)|x'\rangle \\ \text{or}\quad &i\partial_\tau\langle x(\tau)|x'(0)\rangle = \langle x(\tau)|H\big(x(\tau), p(\tau)\big)|x'(0)\rangle \end{aligned} \tag{2-139}$$

with obvious notations. In favourable instances, we are able to solve for $x(\tau)$ and $p(\tau)$ showing the relation between quantum and classical treatments. We then express $H\big(x(\tau), p(\tau)\big)$ as a function of the suitably ordered operators $x(\tau)$ and $x'(0)$:

$$\langle x(\tau)|H\big(x(\tau), p(\tau)\big)|x'(0)\rangle = F(x, x'; \tau)\langle x(\tau)|x'(0)\rangle$$

Then the equation for $U(x, x'; \tau)$ becomes an ordinary linear differential equation which may be integrated as

$$U(x, x'; \tau) = \exp\left[-i\int^{\tau} d\tau'\, F(x, x'; \tau')\right]C(x, x')$$

where $C(x, x')$ has still to be determined in order that $U(x, x'; \tau)$ satisfies the correct relations between position and momentum; for instance,

$$\begin{aligned} [i\partial^x_\mu - eA_\mu(x)]\,\langle x(\tau)|x'(0)\rangle &= \langle x(\tau)|\pi_\mu(\tau)|x(0)\rangle \\ [-i\partial^{x'}_\mu - eA_\mu(x')]\,\langle x(\tau)|x'(0)\rangle &= \langle x(\tau)|\pi_\mu(0)|x(0)\rangle \end{aligned} \qquad \pi_\mu \equiv p_\mu - eA_\mu \tag{2-140}$$

In the case of a Dirac particle interacting with an external electromagnetic field, we have to solve

$$[i\not{\partial} - e\not{A}(x) - m]S_A(x, x') = \delta^4(x - x')$$

$G_A(x, x')$ defined by

$$S_A(x, x') = [i\not{\partial} - e\not{A}(x) + m]G_A(x, x') \tag{2-141}$$

satisfies

$$HG_A \equiv \left[(i\partial - eA)^2 - m^2 - \frac{e}{2}\sigma_{\mu\nu}F^{\mu\nu}\right]G_A(x, x') = \delta^4(x - x') \tag{2-142}$$

In the Heisenberg representation, the operators $x(\tau) = U^\dagger(\tau)xU(\tau)$ and $\pi(\tau) = U^\dagger(\tau)\pi U(\tau)$ satisfy the Ehrenfest relations

$$\begin{aligned} \frac{dx_\mu(\tau)}{d\tau} &= i[H, x_\mu] = -2\pi_\mu \\ \frac{d\pi_\mu(\tau)}{d\tau} &= i[H, \pi_\mu] = -2eF_{\mu\rho}\pi^\rho - ie\,\partial^\rho F_{\mu\rho} - \frac{e}{2}\partial_\mu F_{\rho\nu}\sigma^{\rho\nu} \end{aligned} \tag{2-143}$$

since $[\pi_\mu, \pi_\nu] = -ieF_{\mu\nu}$.

Consider first a constant field. The previous equations reduce to

$$\frac{dx_\mu}{d\tau} = -2\pi_\mu \qquad \frac{d\pi_\mu}{d\tau} = -2eF_{\mu\rho}\pi^\rho \tag{2-144}$$

These equations are readily integrated, using matrix notations

$$\pi(\tau) = e^{-2eF\tau}\pi(0)$$

$$x(\tau) - x(0) = \left(\frac{e^{-2eF\tau} - 1}{eF}\right)\pi(0)$$

We then compute $\pi(\tau)$ as a function of $x(\tau)$, $x(0)$:

$$\pi(\tau) = -\tfrac{1}{2}eF\,e^{-eF\tau}[\sinh(eF\tau)]^{-1}[x(\tau) - x(0)]$$

and using the antisymmetry of F:

$$\pi^2(\tau) = [x(\tau) - x(0)]K[x(\tau) - x(0)]$$

where $$K \equiv \tfrac{1}{4}e^2F^2[\sinh(eF\tau)]^{-2}$$

Rearranging this expression involves the commutator $[x(\tau), x(0)]$:

$$[x_\mu(\tau), x_\nu(0)] = i\left(\frac{e^{-2eF\tau} - 1}{eF}\right)_{\mu\nu}$$

We finally get H as

$$H = x(\tau)Kx(\tau) - 2x(\tau)Kx(0) + x(0)Kx(0) - \frac{i}{2}\operatorname{tr}[eF\coth(eF\tau)] - \frac{e}{2}\sigma_{\mu\nu}F^{\mu\nu} - m^2 \tag{2-145}$$

and the differential equation satisfied by $U(x, x'; \tau)$ is

$$i\partial_\tau U(x, x'; \tau) = \left\{-\frac{e}{2}\sigma_{\mu\nu}F^{\mu\nu} - m^2 + (x - x')K(x - x') - \frac{i}{2}\operatorname{tr}[eF\operatorname{cotanh}(eF\tau)]\right\}U(x, x'; \tau)$$

The latter may be integrated to read

$$U(x, x'; \tau) = C(x, x')\tau^{-2} \exp\{-\tfrac{1}{2} \operatorname{tr} \ln [(eF\tau)^{-1} \sinh (eF\tau)]\}$$

$$\times \exp\left[\frac{i}{4}(x - x')eF \coth (eF\tau)(x - x') + \frac{i}{2}\sigma_{\mu\nu}F^{\mu\nu}\tau + im^2\tau\right]$$

The function $C(x, x')$ is determined by Eqs. (2-140) which lead to

$$\left[i\partial_\mu^x - eA_\mu(x) - \frac{eF_{\mu\nu}}{2}(x - x')^\nu\right]C(x, x') = 0$$

$$\left[-i\partial_\mu^{x'} - eA_\mu(x') + \frac{eF_{\mu\nu}}{2}(x - x')^\nu\right]C(x, x') = 0$$

The solution has the form

$$C(x, x') = C(x') \exp -ie\int_{x'}^{x} d\xi \cdot [A(\xi) + \tfrac{1}{2}F(\xi - x')]$$

Since $A_\mu(\xi) + \frac{1}{2}F_{\mu\nu}(\xi - x')^\nu$ has a vanishing curl, the integral is independent of the path of integration. Taking a straight line from x' to x, the second term does not contribute, owing to the antisymmetry of $F_{\mu\nu}$, and we may write

$$C(x, x') = C \exp -ie\int_{x'}^{x} d\xi \cdot A(\xi)$$

The constant C is finally determined by (2-136a) as

$$C = -\frac{i}{(4\pi)^2}$$

In summary, the propagator in a constant field reads

$$S_A(x, x') = [i\partial\!\!\!/_x - eA\!\!\!/(x) + m](-i)\int_{-\infty}^{0} d\tau\, U(x, x'; \tau) \tag{2-141a}$$

with

$$U(x, x'; \tau) = \frac{-i}{(4\pi)^2\tau^2} \exp\left\{-ie\int_{x'}^{x} d\xi^\mu A_\mu(\xi) - \frac{1}{2}\operatorname{tr} \ln [(eF\tau)^{-1} \sinh (eF\tau)]\right.$$
$$\left. + \frac{i}{4}(x - x')eF \operatorname{cotanh} (eF\tau)(x - x') + \frac{i}{2}\sigma_{\mu\nu}F^{\mu\nu}\tau + i(m^2 - i\varepsilon)\tau\right\} \tag{2-146}$$

Notice the presence of the $-i\varepsilon$ term [in order to fulfill (2-136b)] and of the phase factor $\exp -ie \int d\xi \cdot A$. Its role is to make U gauge covariant; when $A_\mu(x) \to A_\mu(x) + \partial_\mu\Lambda(x)$,

$$U(x, x'; \tau) \to e^{-ie\Lambda(x)}\, U(x, x'; \tau)\, e^{ie\Lambda(x')}$$

We now turn to the plane wave case. The calculation is quite analogous to the previous one, and we merely sketch the successive steps. We consider a linearly polarized plane wave and use the same notations as in Sec. 2-2-3: $A_\mu = \varepsilon_\mu f(\xi)$ with $\xi = n \cdot x$, $n^2 = 0$, $F_{\mu\nu} = \phi_{\mu\nu}f'(\xi)$ where $\phi_{\mu\nu} = n_\mu\varepsilon_\nu - n_\nu\varepsilon_\mu$. Notice that $\partial^\rho F_{\mu\rho} = 0$. Therefore Eqs. (2-143) take the form

$$\frac{dx_\mu}{d\tau} = -2\pi_\mu$$
$$\frac{d\pi_\mu}{d\tau} = -2e\phi_{\mu\rho}\pi^\rho f'(\xi) - \frac{e}{2}n_\mu\phi_{\rho\nu}\sigma^{\rho\nu}f''(\xi) \tag{2-147}$$

Remarking that $(d/d\tau)(\pi\cdot n) = 0$, $d\xi/d\tau = -2\pi\cdot n$ and $[\xi, d\xi/d\tau] = 0$, we first solve for ξ:

$$\xi(\tau) = n\cdot x(\tau) = \xi(0) - 2\pi\cdot n\tau$$

Then we integrate the equation for $\phi^{\nu\mu}\pi_\mu$ and obtain

$$\phi^{\nu\mu}\pi_\mu = en^\nu f(\xi) + C^\nu$$

where C^ν is a constant operator which commutes with $\pi\cdot n$. Inserting this expression into (2-147) and integrating it, we get

$$-\frac{1}{2}\frac{dx_\mu(\tau)}{d\tau} = \pi_\mu(\tau) = \frac{1}{2\pi\cdot n}\left[2eC_\mu f(\xi) + e^2 n_\mu f^2(\xi) + \frac{e}{2}n_\mu\phi_{\rho\nu}\sigma^{\rho\nu}f'(\xi)\right] + D_\mu$$

Here D_μ is a new constant operator commuting with $\pi\cdot n$. We then calculate $x(\tau) - x(0)$ and eliminate D_μ:

$$\pi_\mu(\tau) = -\frac{x_\mu(\tau) - x_\mu(0)}{2\tau} + \frac{\tau}{[\xi(\tau)-\xi(0)]^2}\int_{\xi(0)}^{\xi(\tau)} d\xi\left[2eC_\mu f(\xi) + e^2 n_\mu f^2(\xi) + \frac{e}{2}n_\mu\phi_{\rho\nu}\sigma^{\rho\nu}f'(\xi)\right]$$
$$-\frac{\tau}{\xi(\tau)-\xi(0)}\left\{2eC_\mu f[\xi(\tau)] + e^2 n_\mu f^2[\xi(\tau)] + \frac{e}{2}n_\mu\phi_{\rho\nu}\sigma^{\rho\nu}f'[\xi(\tau)]\right\}$$

This allows us to express the constant C_μ as

$$C_\mu = -\frac{1}{2\tau}\phi_{\mu\rho}[x^\rho(\tau) - x^\rho(0)] - \frac{en_\mu}{\xi(\tau)-\xi(0)}\int_{\xi(0)}^{\xi(\tau)} d\xi\, f(\xi)$$

After computation of the various commutators

$$[\xi(\tau), x_\mu(0)] = 0$$
$$[\xi(0), x_\mu(\tau)] = 2in_\mu\tau$$
$$[x_\mu(\tau), x^\mu(0)] = -8i\tau$$

we may write the hamiltonian H as

$$H = \frac{1}{4\tau^2}[x^2(\tau) - 2x(\tau)\cdot x(0) + x^2(0)] - \frac{2i}{\tau} - (e^2\langle\delta f^2\rangle + m^2) - \frac{e}{2}\phi_{\rho\nu}\sigma^{\rho\nu}\frac{f[\xi(\tau)] - f[\xi(0)]}{\xi(\tau)-\xi(0)} \tag{2-148}$$

$\langle\delta f^2\rangle$ denotes the quantity

$$\langle\delta f^2\rangle = \int_{\xi(0)}^{\xi(\tau)}\frac{d\xi\, f^2(\xi)}{\xi(\tau)-\xi(0)} - \left[\int_{\xi(0)}^{\xi(\tau)}\frac{d\xi\, f(\xi)}{\xi(\tau)-\xi(0)}\right]^2$$

The evolution operator has the form

$$U(x, x'; \tau) = \frac{C(x, x')}{\tau^2}\exp i\left[\frac{(x-x')^2}{4\tau} + \tau(e^2\langle\delta f^2\rangle + m^2) + e\tau\frac{\phi_{\rho\nu}\sigma^{\rho\nu}}{2}\frac{f(\xi) - f(\xi')}{\xi - \xi'}\right]$$

where the function $C(x, x')$ is again determined by the relations (2-140). We find

$$C(x, x') = C(x')\exp\left(-ie\int_{x'}^{x} dy_\mu\left\{A^\mu(y) - \frac{\phi^{\mu\rho}(y_\rho - x'_\rho)}{n\cdot y - \xi'}\left[\int_{\xi'}^{n\cdot y}\frac{du\, f(u)}{n\cdot y - \xi'} - f(n\cdot y)\right]\right\}\right)$$

where the integral is path independent. For a straight line, the only remaining term in the phase is $\exp -ie\int_{x'}^{x} dy_\mu\, A^\mu(y)$, and we find, finally,

$$U(x, x'; \tau) = -\frac{i}{(4\pi)^2\tau^2}\exp i\left\{\frac{(x-x')^2}{4\tau} + \left[e^2\langle\delta f^2\rangle + m^2 + \frac{e}{2}\phi_{\rho\nu}\sigma^{\rho\nu}\frac{f(\xi)-f(\xi')}{\xi-\xi'} - i\varepsilon\right]\tau\right.$$
$$\left. - e\int_{x'}^{x} dy_\mu\, A^\mu(y)\right\} \tag{2-149}$$

This result recalls the classical result of Sec. 1-1-3. For a periodic function $f(\xi)$, the term proportional to $\phi_{\rho\nu}\sigma^{\rho\nu}$ is smeared out if we average over a few periods. The net effect is a mass shift

$$m_{\text{eff}}^2 = m^2 + e^2\langle \delta f^2 \rangle = m^2 + e^2\overline{f^2}$$

Such a nonlinear effect is hard to detect; very high intensity beams are required, since for a monochromatic plane wave of frequency $\omega/2\pi = c/\lambda$ and of energy density $\mathscr{E} = \overline{E^2} = \overline{f^2}\omega^2 = \rho\hbar\omega$, we have

$$\frac{\Delta m^2}{m^2} = \frac{e^2\overline{f^2}}{m^2c^2} = 4\pi\alpha\bar{\lambda}_e^2\frac{\rho c}{\omega} = 2\alpha\bar{\lambda}_e^2\lambda\rho$$

Here $\bar{\lambda}_e$ is the Compton wavelength of the electron and ρ is the number of photons per unit volume in the incident beam. At present the most powerful laser beams do not enable us to reach a sizable value for this ratio.

NOTES

The material of this chapter is very standard and we could hardly escape rephrasing what is found in many excellent textbooks. See, for instance, J. D. Bjorken and S. D. Drell, "Relativistic Quantum Mechanics," McGraw-Hill, New York, 1964, and also A. Messiah, "Quantum Mechanics," vol. 2, North-Holland, Amsterdam, 1962. Hole theory is beautifully described by P. A. M. Dirac in his Solvay report of 1934, reprinted in "Quantum Electrodynamics," edited by J. Schwinger, Dover, New York, 1958, which contains many of the fundamental papers on relativistic quantum field theory. For the propagator approach see R. P. Feynman, *Phys. Rev.*, vol. 76, pp. 749–769, 1949, and for the proper time method, J. Schwinger, *Phys. Rev.*, vol. 82, p. 664, 1951. The original solution for a Dirac electron in a plane wave is due to D. M. Volkow, *Z. Physik*, vol. 94, p. 25, 1935. For a review on the Klein-Gordon equation see H. Feschbach and F. Villars, *Rev. Mod. Phys.*, vol. 30, p. 24, 1958. The nonrelativistic limit of the Dirac equation is studied in L. L. Foldy and S. A. Wouthuysen, *Phys. Rev.*, vol. 78, p. 29, 1950. An overall view on the bound-state spectrum is given in H. A. Bethe and E. E. Salpeter, "Quantum Mechanics of One and Two Electron Atoms," Springer-Verlag, Berlin, 1957. See also M. E. Rose, "Relativistic Electron Theory," John Wiley, New York, 1961. The interesting semiclassical interpretation of the Lamb shift by T. A. Welton is in *Phys. Rev.*, vol. 74, p. 1157, 1948. For the two-component neutrino theory see R. P. Feynman and M. Gell-Mann, *Phys. Rev.*, vol., 109, p. 193, 1958.

An account of relativistic quantum kinematics is given by P. Moussa and R. Stora in "Analysis of Scattering and Decay," edited by M. Nikolic, Gordon and Breach, New York, 1968. We refer to this work for complementary details on the representations of the Poincaré group and their applications.

CHAPTER

THREE

QUANTIZATION—FREE FIELDS

Quantization is presented in the canonical way and applied to free fields. Starting from a mechanical analog we discuss in turn the neutral and charged scalar field. The Gupta-Bleuler indefinite metric is used in the electromagnetic case with emphasis on gauge invariance and the problems arising from the vanishing photon mass. The vacuum fluctuations are beautifully evidenced by the Casimir effect. For Dirac fields, the stability of the vacuum and the exclusion principle lead to quantization according to anticommutation rules. This implies the connection between spin and statistics and the PCT theorem.

3-1 CANONICAL QUANTIZATION

In the nonrelativistic case we obtain Schrödinger's equation by replacing classical observables by operators and Poisson brackets by commutators. As usual, conjugate momenta are derivatives of the Lagrange function with respect to velocities. We have seen in Chap. 1 that the classical procedure can be extended to infinite systems if discrete indices are replaced by continuous ones and the Kronecker symbol by a Dirac δ function. We shall therefore boldly generalize quantization to this case without worrying too much at first. The only point which we should be cautious about is to preserve Lorentz invariance. In the specific case of electrodynamics the lagrangian is known, at least classically. The precise role of the lagrangian after quantization will have to be clarified. This will be done in Chap. 9, where we shall investigate an alternative and fruitful interpretation of quantization using the methods of path integrals.

It is clear that the correct interpretation of a quantum theory depends on its dynamics (which can lead, for instance, to the appearance of bound states, etc.). Unfortunately, in most cases we cannot solve the equations of motion and we have to treat interactions using approximations. This is a major difficulty which sometimes prevents us from reaching a logical and coherent presentation. Part of this difficulty can be overcome by renormalization which hides some of the interactions by reexpressing the observable quantities in terms of measured masses and coupling constants.

A more consistent but difficult axiomatic approach was pioneered by Wightman. One tries to develop a quantum field theory from a few well-defined axioms obtained from an idealization of physical requirements. To uncover the content of the theory requires sophisticated mathematical developments. Two main lines of attack have been pursued. According to one of them, one studies Green functions in detail: analyticity in momentum space, algebraic properties, discontinuities. At an ultimate stage one may reach a scattering theory after proper identification of the asymptotic states. A more ambitious program involves the explicit construction of the fundamental operators, such as the hamiltonian, in order to study their spectral properties. The latter should provide the particle interpretation, bound states, scattering states, etc.

Our goal will be more modest and follows partly the historical path. It combines approximations, physical intuition, and mathematical deduction. One may hope that these various methods will at some point merge together and that some rigorous basis will be given to the above treatment.

At first sight a field theory does not seem to have an interpretation in terms of microobjects identified as particles. The long story of the wave-particle duality in the theory of light is a testimony of the confusion which may arise and still remains in certain aspects up to this day. One of the triumphs of the quantum theory will be to give a better understanding of this phenomenon. Naive pictures derived from classical experiences have limitations, however, arising in particular from the concept of indistinguishable particles. The structure of the Hilbert space of states constructed for free fields—the Fock space—will reflect this aspect.

We shall only deal here with local theories. The meaning of locality will become clearer as we go along. It assumes an idealization of space-time measurements in arbitrarily small regions. As required by Lorentz invariance and a weak form of causality, measurements separated by a space-like interval cannot influence one another. In other words, (1) local observables exist and (2) local observables relative to space-like separated regions commute. Experimental verifications of this postulate are only indirect since we do not possess any apparatus comparable to nuclear or subnuclear sizes. In this case, measurement necessarily involves objects of the same nature as those under study. Nevertheless, it is a fair statement to say that no violation of this principle has been observed down to distances $\hbar c/\sqrt{s}$ where s is the squared center-of-mass energy of particle collisions in present-day accelerators ($\sqrt{s} \lesssim 60$ GeV at the CERN ISR). These lengths ($\sim 10^{-15}$ cm) are minute fractions of atomic sizes (10^{-8} cm).

Among the striking consequences emerging when combining microcausality with relativistic invariance, let us quote the spin statistics relation (half-integer spin particles are fermions, integer spin ones are bosons) and the existence of a *TCP* invariance. The latter involves a product of time reversal (*T*), parity (*P*), and charge conjugation (*C*). This implies the existence of antiparticles with the same kinematic invariants as their particle counterparts and opposite additive quantum numbers (electric, baryonic, leptonic charges, etc.).

3-1-1 General Formulation

Let $\varphi_\alpha(x)$ be the fields whose dynamics we propose to study. The index α stands for internal characteristics (charges, etc.) or kinematic ones (such as Lorentz indices). For the moment we assume these fields to be free of constraints (which would reduce the number of degrees of freedom). We provisionally set aside half-integer spin fields, the correct treatment of which requires special considerations (Sec. 3-3).

From the classical Lagrange function at a fixed time

$$L(t) = \int d^3x \, \mathscr{L}(\varphi, \partial\varphi) \tag{3-1}$$

we derive the conjugate fields

$$\pi_\alpha(t, \mathbf{x}) = \frac{\delta L(t)}{\delta[\partial_0 \varphi_\alpha(t, \mathbf{x})]} = \frac{\partial \mathscr{L}}{\partial[\partial_0 \varphi_\alpha(t, \mathbf{x})]} \tag{3-2}$$

To construct the hamiltonian operator H we first replace the c-number fields by operators satisfying canonical equal-time commutation rules:

$$[\varphi_\alpha(t, \mathbf{x}), \pi_\beta(t, \mathbf{y})] = i\delta_{\alpha\beta}\delta^3(\mathbf{x} - \mathbf{y}) \tag{3-3}$$

with the commutators $[\varphi, \varphi]$ and $[\pi, \pi]$ vanishing. After inverting (3-2) to give $\partial_0\varphi$ in terms of π and φ we obtain H as

$$H = \int d^3x \left[\sum_\alpha \pi_\alpha(t, \mathbf{x}) \partial_0 \varphi_\alpha(t, \mathbf{x}) - \mathscr{L}(\varphi, \partial\varphi) \right] \tag{3-4}$$

This procedure suffers from the usual drawbacks arising from the operator ordering. Moreover, the multiplication of operator fields at the same point will lead to new difficulties, as we shall soon realize. The two aspects are related.

It must be stressed that when writing (3-3) we have not yet specified in which Hilbert space these operators act. This question has a simple answer in the case of free fields, as we shall see, and hence is bypassed when studying small perturbations around this situation. It is, however, entirely nontrivial in the general case, where its answer requires some knowledge of the dynamics. This is why the latter has some bearing on the very construction of the theory.

To be specific, let us assume that we deal with only one real field φ in the classical picture—hence an hermitian φ in the quantum one—with a lagrangian

$$\mathscr{L} = \tfrac{1}{2}(\partial\varphi)^2 - V(\varphi) \tag{3-5}$$

where V is a smooth function (a polynomial, say). The classical equations of motion read

$$\partial_\mu \frac{\partial \mathscr{L}}{\partial(\partial_\mu \varphi)} - \frac{\partial \mathscr{L}}{\partial \varphi} \equiv \Box\varphi + \frac{dV(\varphi)}{d\varphi} = 0 \tag{3-6}$$

If V reduces to a quadratic term $V(\varphi) = (m^2/2)\varphi^2$,

$$\mathscr{L}(\varphi) = \tfrac{1}{2}(\partial\varphi)^2 - \tfrac{1}{2}m^2\varphi^2 \tag{3-7}$$

from which follows the Klein-Gordon equation

$$(\Box + m^2)\varphi = 0 \tag{3-8}$$

to be interpreted here as a classical field equation and not as a relativistic generalization of Schrödinger's equation. For an arbitrary $V(\varphi)$, that is, for an arbitrary self-interaction without derivatives, the conjugate momentum π is given by

$$\pi = \frac{\partial \mathscr{L}}{\partial(\partial_0 \varphi)} = \partial_0 \varphi \tag{3-9}$$

so that the hamiltonian, expressed in terms of φ and π, is

$$H = \int d^3x \, \{\tfrac{1}{2}[\pi^2 + (\nabla\varphi)^2] + V(\varphi)\} \tag{3-10}$$

and in the simple case of a quadratic $V(\varphi)$ given by (3-7)

$$H = \int d^3x \, \tfrac{1}{2}[\pi^2 + (\nabla\varphi)^2 + m^2\varphi^2] \tag{3-11}$$

We have here no problem of operator ordering. Hence the only source of trouble may come from multiplying operators at the same point. Let us stick for the time being to the case (3-11). We recognize a simple structure of coupled harmonic oscillators. In order to uncover its physical interpretation, let us pretend that space has only one dimension and that instead of assuming continuous values the coordinate x can take only discrete ones which are integral multiples of an elementary length taken as unity. In this case (3-11) would be replaced by

$$\begin{gathered} H = \tfrac{1}{2} \sum_{-\infty}^{+\infty} [\pi_n^2 + (\varphi_n - \varphi_{n-1})^2 + m^2\varphi_n^2] \\ [\varphi_n, \varphi_{n'}] = [\pi_n, \pi_{n'}] = 0 \qquad [\varphi_n, \pi_{n'}] = i\delta_{nn'} \end{gathered} \tag{3-12}$$

A physical model which could be described by (3-12) would be the vibrations of a one-dimensional "crystal," with φ_n standing for the displacement of the nth atom and π_n the conjugate variable. Each individual oscillator with restoring force provided by the $m^2\varphi_n^2$ term would be coupled to its nearest neighbors through the $(\varphi_n - \varphi_{n\pm1})^2$ contributions to the potential energy. To study such a model it is natural to search for the proper modes. Using the discrete translational invariance of H this leads us to introduce Fourier transforms in the form

$$\begin{aligned} \varphi_n &= \int_{-\pi}^{+\pi} \frac{dk}{2\pi}\, e^{ikn}\, \tilde{\varphi}(k) & \pi_n &= \int_{-\pi}^{+\pi} \frac{dk}{2\pi}\, e^{ikn}\, \tilde{\pi}(k) \\ \tilde{\varphi}^\dagger(k) &= \tilde{\varphi}(-k) & \tilde{\pi}^\dagger(k) &= \tilde{\pi}(-k) \end{aligned} \tag{3-13}$$

The commutation rules assume the form

$$\begin{aligned} [\tilde{\varphi}(k), \tilde{\varphi}(k')] &= [\tilde{\pi}(k), \tilde{\pi}(k')] = 0 \\ [\tilde{\varphi}(k), \tilde{\pi}(-k')] &= i \sum_n e^{-i(k-k')n} \end{aligned} \tag{3-14}$$

The last sum is also $2\pi \sum_n \delta(k - k' + 2\pi n)$ which means that if $\tilde{\varphi}$ and $\tilde{\pi}$ are extended as periodic functions of k this reduces to $2\pi\delta(k - k')$. By virtue of (3-13) the hermitian hamiltonian is now expressed as

$$H = \frac{1}{2} \int_{-\pi}^{+\pi} \frac{dk}{2\pi} \{\tilde{\pi}^\dagger(k)\tilde{\pi}(k) + \tilde{\varphi}^\dagger(k)[m^2 + 2(1 - \cos k)]\tilde{\varphi}(k)\} \tag{3-15}$$

To identify this with a sum of individual uncoupled oscillators we set

$$\begin{aligned} \omega_k &= \omega_{-k} = \sqrt{m^2 + 2(1 - \cos k)} \\ a_k &= \frac{1}{\sqrt{4\pi\omega_k}} [\omega_k \tilde{\varphi}(k) + i\tilde{\pi}(k)] \\ a_k^\dagger &= \frac{1}{\sqrt{4\pi\omega_k}} [\omega_k \tilde{\varphi}^\dagger(k) - i\tilde{\pi}^\dagger(k)] \\ [a_k, a_{k'}^\dagger] &= \delta(k - k') \end{aligned} \tag{3-16}$$

The operators $a_k^\dagger$ and a_k create or destroy the mode k with energy ω_k. The hamiltonian H reads

$$H = \frac{1}{2} \int_{-\pi}^{+\pi} dk\, \omega_k (a_k^\dagger a_k + a_k a_k^\dagger) = \int_{-\pi}^{+\pi} dk\, H_k \tag{3-17}$$

while the fields are expressed as

$$\begin{aligned} \varphi_n &= \int_{-\pi}^{+\pi} \frac{dk}{\sqrt{4\pi\omega_k}} (a_k\, e^{ikn} + a_k^\dagger\, e^{-ikn}) \\ \pi_n &= -i \int_{-\pi}^{+\pi} dk \sqrt{\frac{\omega_k}{4\pi}} (a_k\, e^{ikn} - a_k^\dagger\, e^{-ikn}) \\ &= \frac{\partial}{\partial t} \left[\int_{-\pi}^{+\pi} \frac{dk}{\sqrt{4\pi\omega_k}} (a_k\, e^{-i\omega_k t + ikn} + a_k^\dagger\, e^{i\omega_k t - ikn}) \right]_{t=0} \end{aligned} \tag{3-18}$$

The denomination of a_k and $a_k^\dagger$ as destruction and creation operators is justified by the fact that if $|E\rangle$ is an eigenstate of H with energy E we have

$$\begin{aligned} Ha_k|E\rangle &= Ea_k|E\rangle + [H, a_k]|E\rangle = (E - \omega_k)a_k|E\rangle \\ Ha_k^\dagger|E\rangle &= Ea_k^\dagger|E\rangle + [H, a_k^\dagger]|E\rangle = (E + \omega_k)a_k^\dagger|E\rangle \end{aligned} \tag{3-19}$$

The mode k of energy ω_k is interpreted in this mechanical model as a coherent quantized vibration of the lattice atoms, or phonon. Here we understand clearly the relation between particles (phonons) and fields (φ_n is the displacement of the nth atom). Our physical intuition might lead us first to consider states of the crystal characterized by a wave function obtained by diagonalizing the field

$$|y\rangle = |\{y_n\}\rangle \qquad \varphi_n|y\rangle = y_n|y\rangle$$

We shall, rather, choose to generate the states starting from the ground state $|0\rangle$ and its excitations in terms of phonons. We encounter at once a difficulty since the ground-state energy, the lowest eigenvalue of the hamiltonian, is in fact infinite. Each mode k contributes an amount $\frac{1}{2}\omega_k$ to the zero point energy. This is in agreement with the uncertainty relations since each oscillator has a minimum momentum spread due to its potential energy. Since we have a continuum of modes, each slice $(k, k + dk)$ leads to an infinite energy—a fact to be traced to the infinite size of the system. Indeed, since

$$a_k|0\rangle = 0 \qquad \langle 0| a_k^\dagger = 0 \tag{3-20}$$

for any k, we have

$$\langle 0|H|0\rangle = \frac{1}{2}\int_{-\pi}^{+\pi} dk\, \omega_k \langle 0| a_k a_k^\dagger |0\rangle = \frac{1}{2}\int_{-\pi}^{+\pi} dk\, \omega_k \langle 0| [a_k, a_k^\dagger]|0\rangle$$

and the last integral is meaningless since $[a_k, a_k^\dagger] = \delta(0)!$. If, however, the crystal is of finite size $(-N \le n \le N)$ let us take periodic boundary conditions by identifying the sites n and $n + p(2N + 1)$, realizing a circular arrangement. The wave vector k would then be restricted to values $k = [2\pi/(2N + 1)]q$ with q an integer and $-N \le q \le N$, and the zero point energy $(N/2N)\sum_{-N}^{+N}\omega_q$ would be finite. Clearly $(1/2N)\sum_{-N}^{+N}\omega_q$ has a finite limit as N tends to infinity, showing that the above energy is indeed proportional to the size of the system.

By taking a discrete instead of a continuous model, we have introduced a Brillouin zone $-\pi \le k \le \pi$ in momentum space, equivalent to an ultraviolet cutoff in the original model. The reader will recall that the origin of the expression "ultraviolet catastrophe" stems from the divergent contribution to the blackbody thermal radiation of high-frequency modes of the electromagnetic field. A space cutoff allows, further, an unambiguous definition of the hamiltonian operator. This is referred to by saying that we put the system in a box. We can, however, use the following device. Zero point energy is unobservable unless we destroy the crystal. In field theory, the ground state will be interpreted as the vacuum and it will be even harder to destroy! Energy exchanges with the crystal are insensitive to the choice of an origin. We declare by fiat that the ground state has zero energy and we redefine the hamiltonian as

$$\begin{aligned} H &= \frac{1}{2}\int_{-\pi}^{+\pi} dk\, \omega_k [(a_k^\dagger a_k + a_k a_k^\dagger) - \langle 0|(a_k^\dagger a_k + a_k a_k^\dagger)|0\rangle] \\ &= \int_{-\pi}^{+\pi} dk\, \omega_k a_k^\dagger a_k \end{aligned} \tag{3-21}$$

We shall, of course, have to make sure in relativistic theories that this procedure preserves Lorentz covariance.

In this new expression the creation and annihilation operators appear in "normal order," the latter to the right of the former. This is also called Wick's ordering and is denoted by a double-dot symbol:

$$:\tfrac{1}{2}(a_k^\dagger a_k + a_k a_k^\dagger):\, \equiv a_k^\dagger a_k$$

Note that under the normal order sign operators commute and that the definition uses as reference the free-field ground state.

If we restore the lattice spacing a, for values of $|k|$ much smaller than the edge of the Brillouin zone π/a, the frequency ω_k can be approximated by

$$\omega_k = \sqrt{m^2 + \frac{2}{a^2}(1 - \cos ak)} \to \sqrt{m^2 + k^2} \tag{3-22}$$

which is just the form of the dispersion law of relativistic mechanics if k is interpreted as the phonon momentum. In solid-state physics, since it is unaffected by the addition of multiples of $2\pi/a$, k is only a quasimomentum.

In the framework of the harmonic theory with its dynamics dictated by (3-12), phonons are well-defined entities and conserved in the sense that the number operator

$$N = \int_{-\pi}^{+\pi} dk\, a_k^\dagger a_k \tag{3-23}$$

commutes with the hamiltonian.

The superposition of several vibration modes leads to a state where the phonons have only a limited individual status, while the atoms on the various sites can perfectly well be distinguished (it is, for instance, meaningful to introduce an impurity on one of the sites). Indeed, the phonons behave as a boson system. This manifests itself when we add interactions. We can, for instance, consider an extra anharmonic term in the energy or couple the crystal to a thermostat.

The Bose symmetry appears when we generate the states by acting with creation operators on the normalized vacuum $|0\rangle$. Unfortunately, the state $a_k^\dagger|0\rangle$ is not normalizable since $\langle 0|a_k a_k^\dagger|0\rangle = \langle 0|[a_k, a_k^\dagger]|0\rangle = \infty$ due to the infinite size, as before. We can, however, build a wave packet by linear superposition:

$$a_f^\dagger|0\rangle = \int_{-\pi}^{+\pi} dk\, f(k) a_k^\dagger|0\rangle$$

where $f(k)$ can be chosen concentrated around a peak value $\bar{k}$ but such that

$$\begin{aligned}\langle 0|\, a_f a_f^\dagger \,|0\rangle &= \int_{-\pi}^{+\pi} dk_1 \int_{-\pi}^{+\pi} dk_2\, f^*(k_1) f(k_2) \langle 0|\, a_{k_1} a_{k_2}^\dagger \,|0\rangle \\ &= \int_{-\pi}^{+\pi} dk\, |f(k)|^2 < \infty\end{aligned}$$

In other words, $f(k)$ has to be square integrable. The fundamental operators a_k and $a_k^\dagger$ have to be smeared with test functions in order to lead to meaningful expressions; they are called operator-valued distributions. The idealization of a wave of definite momentum k is to be thought of as a simplification when the spread Δk (or $\Delta\omega_k$) is negligible as compared to the momenta (or energies) involved in the problem at hand. In a crystal, for instance, Δk is even limited from below by $(Na)^{-1}$, an effect which can sometimes be observed. Even in a translation-invariant world, the geometry of the experimental apparatus always implies a finite spread.

The state $a_f^\dagger |0\rangle$ is to be interpreted as a one-particle (phonon) state. To construct the full Hilbert space of states generated by vectors such as

$$|r\rangle = a_{f_1}^\dagger a_{f_2}^\dagger \cdots a_{f_r}^\dagger |0\rangle \tag{3-24}$$

we use the standard mathematical procedure of linear superposition and Cauchy's completion. The resulting space is called the Fock space. Consider the action of N and H on the state $|r\rangle$. Let us assume the wave functions f_s to be strongly peaked around the values k_s. Then

$$N|r\rangle = r|r\rangle \qquad H|r\rangle \simeq \left(\sum_1^r \omega_{k_s}\right)|r\rangle \tag{3-25}$$

Hence we have an r phonon state with an almost well-defined energy. However, let us look more closely at its wave function. Using the commutativity of the creation operators we may write

$$\begin{aligned}|r\rangle &= \int_{-\pi}^{+\pi} dk_1 \cdots \int_{-\pi}^{+\pi} dk_r\, f_1(k_1)\cdots f_r(k_r) a_{k_1}^\dagger \cdots a_{k_r}^\dagger |0\rangle \\ &= \int_{-\pi}^{+\pi} dk_1 \cdots \int_{-\pi}^{+\pi} dk_r \frac{1}{r!} \sum_\alpha f_1(k_{\alpha_1})\cdots f_r(k_{\alpha_r}) a_{k_1}^\dagger \cdots a_{k_r}^\dagger |0\rangle\end{aligned} \tag{3-26}$$

where in the last integrand the sum runs over all permutations α of the set of integers $(1, \ldots, r)$. Up to a normalization factor, shall we say that $F_{\text{NS}}(k_1, \ldots, k_r) = f_1(k_1)\cdots f_r(k_r)$ or that

$$F_{\text{S}}(k_1, \ldots, k_r) = \frac{1}{r!} \sum_\alpha f_1(k_{\alpha_1})\cdots f_r(k_{\alpha_r}) \tag{3-27}$$

is the wave function of the state? This question would make sense if there existed an observable with different mean values for one or the other wave function. This is, however, not the case. Indeed, even if we seem to introduce at first the nonsymmetric kernel F_{NS} in the representation of the state $|r\rangle$ the commutation of creation operators symmetrizes it automatically. Phonons are identical particles and obey Bose statistics as a consequence of the basic commutation rules (3-12).

To obtain a normalized r-particle state assume, for instance, that the one-particle wave functions $f_s(k)$ are orthonormalized:

$$\int_{-\pi}^{+\pi} dk\, f_s^*(k) f_t(k) = \delta_{st} \tag{3-28}$$

If $\langle r | r \rangle = 1$ then

$$|r\rangle = \lambda \int_{-\pi}^{+\pi} dk_1 \cdots \int_{-\pi}^{+\pi} dk_r\, f_1(k_1) \cdots f_r(k_r) a_{k_1}^\dagger \cdots a_{k_r}^\dagger |0\rangle$$

Hence

$$1 = \langle r | r \rangle = |\lambda|^2 \int_{-\pi}^{+\pi} dk_1 \cdots \int_{-\pi}^{+\pi} dk_r'\, f_1^*(k_1) \cdots f_r(k_r') \langle 0 | a_{k_r} \cdots a_{k_r'}^\dagger |0\rangle \tag{3-29}$$

Moving each operator a_k to the right and using $a_k a_{k'}^\dagger = a_{k'}^\dagger a_k + \delta(k - k')$ together with (3-28) leads at once to

$$|\lambda|^2 = 1 \tag{3-30}$$

showing that λ can be taken equal to a phase factor. If, however, r_1 of the one-particle wave functions are equal to f_1, r_2 to f_2, and so on, with the f functions still orthonormal, it is readily shown that the correctly normalized state $|r\rangle$ reads

$$|r\rangle \equiv |\{r_1, r_2, \ldots\}\rangle = \frac{1}{\sqrt{r_1! r_2! \cdots}} \int_{-\pi}^{+\pi} dk_1 \cdots \int_{-\pi}^{+\pi} dk_r\, f_1(k_1) \cdots f_1(k_{r_1}) f_2(k_{r_1+1}) \cdots f_2(k_{r_1+r_2}) \cdots a_{k_1}^\dagger \cdots a_{k_r}^\dagger |0\rangle \tag{3-31}$$

More generally, if $F(k_1, \ldots, k_r)$ is a symmetric function then the normalized r-particle state is

$$\begin{aligned} |r\rangle &= \lambda \int_{-\pi}^{+\pi} dk_1 \cdots \int_{-\pi}^{+\pi} dk_r\, F(k_1, \ldots, k_r) a^\dagger{}_{k_1} \cdots a_{k_r}^\dagger |0\rangle \\ |\lambda|^{-2} &= r! \int_{-\pi}^{+\pi} dk_1 \cdots \int_{-\pi}^{+\pi} dk_r\, |F(k_1, \ldots, k_r)|^2 \end{aligned} \tag{3-32}$$

It is an interesting exercise left to the reader to compute the wave function of a state with a fixed number of phonons, and in particular the ground state, in the basis which diagonalizes the field.

Canonical quantization has been performed at a fixed time, say $t = 0$. However, the theory is invariant under time translation. We could therefore have chosen another reference time t and used the operators $\varphi_n(t)$, $\pi_n(t)$ such that

$$\begin{aligned} \dot{\varphi}_n(t) &= i[H, \varphi_n(t)] = \pi_n(t) \qquad & \varphi_n(0) \equiv \varphi_n \\ \dot{\pi}_n(t) &= i[H, \pi_n(t)] \qquad & \pi_n(0) \equiv \pi_n \end{aligned} \tag{3-33}$$

with

$$\begin{aligned} \varphi_n(t) &= e^{iHt} \varphi_n e^{-iHt} = \int_{-\pi}^{+\pi} \frac{dk}{\sqrt{4\pi\omega_k}} (a_k e^{-i\omega_k t + ikn} + a_k^\dagger e^{i\omega_k t - ikn}) \\ \pi_n(t) &= \dot{\varphi}_n(t) \end{aligned} \tag{3-34}$$

Clearly $a_k e^{-i\omega_k t}$, $a_k^\dagger e^{i\omega_k t}$ satisfy canonical commutation rules, and H and N are time independent. The relation between the observables at times t and 0 is a unitarily implemented canonical transformation. The matrix elements representing measurements are independent of the choice of the Heisenberg or Schrödinger picture

$$\langle a|\, e^{iHt}\, A(0)\, e^{-iHt}\, |a\rangle = (\langle a|\, e^{iHt})\, A(0)(e^{-iHt}\, |a\rangle) \qquad \text{(Schrödinger)}$$
$$= \langle a|\, (e^{iHt}\, A(0)\, e^{-iHt})\, |a\rangle \qquad \text{(Heisenberg)}$$

The structure of Eq. (3-34) clarifies the presence of positive and negative frequencies in the field. They correspond to the creation or destruction of phonon modes. The latter have honest positive energies ($\hbar\omega$). Hence φ is not to be understood as a wave function. It is an operator in Fock space even though it is written in terms of the solutions to a wave equation. The superposition is weighted with operator-valued amplitudes. This procedure is referred to as second quantization. The mechanical example of the crystal has prepared the way for the quantum treatment of the relativistic scalar field. The atomic interpretation of the vibrations will disappear but the analogy between phonons and particles remains. We shall have even fewer scruples in setting the vacuum energy equal to zero.

3-1-2 Scalar Field

For the quantized free scalar field with an hamiltonian given by Eq. (3-11) we have essentially to transcribe the previous formulas in three-dimensional continuous space. The three-momenta (an interpretation to be confirmed below) are denoted $\mathbf{k}$, while k^0 also stands for $\omega_k = \omega_{-k} = \sqrt{\mathbf{k}^2 + m^2} > 0$. Throughout this book, the phase space measure for bosons will be noted:

$$d\tilde{k} = \frac{d^3k}{(2\pi)^3 2\omega_k} = \frac{d^4k}{(2\pi)^4}\, 2\pi\delta(k^2 - m^2)\theta(k^0) \tag{3-35}$$

The last expression exhibits clearly the Lorentz invariance of this measure. This may be checked directly on d^3k/ω_k: it is invariant under rotations; a boost of velocity $\tanh\theta$ leaves invariant the transverse components $\mathbf{k}_T$ as well as $\omega_k^2 - k_L^2$. Therefore,

$$dk'_L = \left(\cosh\theta + \sinh\theta\, \frac{k_L}{\omega_k}\right) dk_L$$

$$\omega_{k'} = \left(\cosh\theta + \sinh\theta\, \frac{k_L}{\omega_k}\right)\omega_k$$

proving that $dk'_L/\omega_{k'} = dk_L/\omega_k$.

In terms of the creation and annihilation operators satisfying

$$\begin{aligned} [a(k), a^\dagger(k')] &= (2\pi)^3 2\omega_k \delta^3(k - k') \\ [a(k), a(k')] &= [a^\dagger(k), a^\dagger(k')] = 0 \end{aligned} \tag{3-36}$$

the conjugate fields $\varphi(\mathbf{x})$, $\pi(\mathbf{x})$ at time $t = 0$, fulfilling the basic commutation rules (3.3) are

$$\varphi(\mathbf{x}) = \int d\tilde{k}\,[a(k)\,e^{i\mathbf{k}\cdot\mathbf{x}} + a^\dagger(k)\,e^{-i\mathbf{k}\cdot\mathbf{x}}]$$
$$\pi(\mathbf{x}) = -i\int d\tilde{k}\,\omega_k[a(k)\,e^{i\mathbf{k}\cdot\mathbf{x}} - a^\dagger(k)\,e^{-i\mathbf{k}\cdot\mathbf{x}}] \tag{3-37}$$

The vacuum ground state is defined through

$$a(k)\,|0\rangle = 0 \qquad \langle 0\,|\,0\rangle = 1$$

while the hamiltonian takes the form

$$H = \frac{1}{2}\int d\tilde{k}\,\omega_k : a^\dagger(k)a(k) + a(k)a^\dagger(k) \colon = \int d\tilde{k}\,\omega_k a^\dagger(k)a(k) \tag{3-38}$$

According to Chap. 1 it is natural to expect that the linear momentum operator **P** is given by

$$P^j = \int d^3x\,\Theta^{0j}(\mathbf{x}) = \int d\tilde{k}\,k^j\,\frac{[a^\dagger(k)a(k) + a(k)a^\dagger(k)]}{2} = \int d\tilde{k}\,k^j a^\dagger(k)a(k) \tag{3-39}$$

Unlike the case of energy, no normal ordering is required. The modes **k** and $-\mathbf{k}$ compensate each other, so that the vacuum is translationally invariant, i.e., an eigenstate of **P** with zero eigenvalue. The operators $H \equiv P^0$ and P^j commute, and

$$[P^\mu, a^\dagger(k)] = k^\mu a^\dagger(k) \tag{3-40}$$

showing that $a^\dagger(k)$ acting on a state adds a four-momentum k^μ.

Up to normal ordering the energy momentum tensor density is identical with its classical counterpart

$$\Theta^{\mu\nu}(x) = \partial^\mu\varphi\partial^\nu\varphi - g^{\mu\nu}[\tfrac{1}{2}(\partial\varphi)^2 - \tfrac{1}{2}m^2\varphi^2] \tag{3-41}$$

with $\partial^0\varphi(x)$ replaced by $\pi(x)$.

At time $t = x^0$ the field φ reads

$$\varphi(t, \mathbf{x}) = e^{iHt}\,\varphi(0, \mathbf{x})\,e^{-iHt} = \int d\tilde{k}\,[a(k)\,e^{-ik\cdot x} + a^\dagger(k)\,e^{ik\cdot x}] \tag{3-42}$$

This expression automatically fulfills the Klein-Gordon equation. If we use the symbol $u\overleftrightarrow{\partial}v$ for

$$u\overleftrightarrow{\partial}v = u(\partial v) - (\partial u)v$$

we can write $a(k)$ in terms of $\varphi(0, \mathbf{x})$ and $\pi(0, \mathbf{x})$ as

$$a(k) = \int d^3x\,e^{-i\mathbf{k}\cdot\mathbf{x}}\,[\omega_k\varphi(0, \mathbf{x}) + i\pi(0, \mathbf{x})] = i\int d^3x\,[e^{ik\cdot x}\overleftrightarrow{\partial}_0\varphi(t, \mathbf{x})]_{t=0} \tag{3-43}$$

and the last expression is in fact time independent. Indeed,

$$\partial_0 \int d^3x\, e^{ik\cdot x} \overleftrightarrow{\partial_0} \varphi(x) = \int d^3x \, \{e^{ik\cdot x}\, \partial_0^2 \varphi(x) + [(m^2 - \Delta)\, e^{ik\cdot x}]\varphi(x)\}$$

using the fact that $k_0^2 = m^2 + \mathbf{k}^2$. In all applications we have to consider a normalized superposition of plane waves, in which case the integration by parts of the laplacian operator will be justified. Taking now into account the fact that $\varphi(x)$ satisfies the Klein-Gordon equation, we find that the above expression vanishes. Hence

$$\begin{aligned} a(k) &= i \int d^3x\, e^{ik\cdot x} \overleftrightarrow{\partial_0} \varphi(x) \\ a^\dagger(k) &= -i \int d^3x\, e^{-ik\cdot x} \overleftrightarrow{\partial_0} \varphi(x) \end{aligned} \tag{3-44}$$

A basis in Fock space can be constructed from the normalized r-particle states:

$$|r\rangle = \left(r! \int d\tilde{k}_1 \cdots d\tilde{k}_r |F(k_1, \ldots, k_r)|^2 \right)^{-1/2} \times \int d\tilde{k}_1 \cdots d\tilde{k}_r\, F(k_1, \ldots, k_r) a^\dagger(k_1) \cdots a^\dagger(k_r) |0\rangle \tag{3-45}$$

The wave function $F(k_1, \ldots, k_r)$ is symmetric in the interchange of the four-momenta k_s, all belonging to the upper sheet of the hyperboloid $k^2 = m^2$, the so-called mass shell. A number operator N commuting with P^μ is

$$N = \int d\tilde{k}\, a^\dagger(k) a(k) \tag{3-46}$$

and, of course, acting on the states (3-45):

$$N|r\rangle = r|r\rangle \tag{3-47}$$

It is crucial to verify that relativistic invariance has been maintained. The above formulation uses a particular frame, and we may wonder whether quantization in any other one obtained by a Poincaré transformation would lead to an equivalent theory. If primes refer to the coordinates in this new frame, we want to know whether there exists a quantum canonical transformation relating the fundamental creation and annihilation operators a and $a^\dagger$ to the corresponding a' and $a'^\dagger$ obtained when quantizing on the hyperplane $t' = \text{constant}$. This will be the case if to each transformation $x \to x' = \Lambda x + a$ we may associate a unitary operator $U(a, \Lambda)$ in Fock space, such that

$$U(a, \Lambda)\varphi(x) U^\dagger(a, \Lambda) = \varphi(\Lambda x + a) \tag{3-48}$$

Assuming differentiability in the parameters a and Λ it is sufficient to study infinitesimal transformations of the type

$$x'^\mu = x^\mu + \delta\omega^\mu{}_\nu x^\nu + \delta a^\mu$$

with $\delta\omega_{\mu\nu} + \delta\omega_{\nu\mu} = 0$. Hence we want to find 10 hermitian operators $M^{\mu\nu}$, $(M^{\mu\nu} + M^{\nu\mu} = 0)$, and

P^μ such that

$$U = I - \frac{i}{2}\,\delta\omega_{\mu\nu}M^{\mu\nu} + i\delta a_\mu P^\mu \tag{3-49}$$

and
$$i[\delta a_\mu P^\mu - \tfrac{1}{2}\delta\omega_{\mu\nu}M^{\mu\nu}, \varphi(x)] = \delta a_\mu \partial^\mu \varphi + \tfrac{1}{2}\delta\omega_{\mu\nu}(x^\nu\partial^\mu - x^\mu\partial^\nu)\varphi \tag{3-50}$$

The 10 generators P and M will have to fulfill the geometrical commutation rules

$$\begin{aligned}
&[P^\mu, P^\nu] = 0\\
&[M^{\mu\nu}, P^\lambda] = i(g^{\mu\lambda}P^\nu - g^{\nu\lambda}P^\mu)\\
&[M^{\mu\nu}, M^{\lambda\sigma}] = i(g^{\mu\lambda}M^{\nu\sigma} - g^{\nu\lambda}M^{\mu\sigma} - g^{\mu\sigma}M^{\nu\lambda} + g^{\nu\sigma}M^{\mu\lambda})
\end{aligned} \tag{3-51}$$

where the commutator replaces (up to i) the classical Poisson bracket. As was already anticipated when deriving P^μ, we expect that these generators are given by Noether's theorem

$$P^\mu = \int_t d^3x\, \Theta^{0\mu} \qquad M^{\mu\nu} = \int_t d^3x\, (x^\mu\Theta^{0\nu} - x^\nu\Theta^{0\mu}) \tag{3-52}$$

with $\Theta^{\mu\nu}$ given by Eq. (3-41), or more explicitly

$$\begin{aligned}
&\Theta^{00}(\mathbf{x}) = \tfrac{1}{2}:(\pi^2 + (\nabla\varphi)^2 + m^2\varphi^2):\\
&\Theta^{0j}(\mathbf{x}) = :\pi\partial^j\varphi:
\end{aligned} \tag{3-53}$$

The expressions for P^μ coincide with (3-38) and (3-39) and are indeed such that $i[P^\mu, \varphi(x)] = \partial^\mu\varphi(x)$. As for $M^{\mu\nu}$ we find

$$\begin{aligned}
M_{0j} &= i\int d\tilde{k}\; a^\dagger(k)\left(\omega_k \frac{\partial}{\partial k^j}\right)a(k)\\
M_{jl} &= i\int d\tilde{k}\; a^\dagger(k)\left(k_j\frac{\partial}{\partial k^l} - k_l\frac{\partial}{\partial k^j}\right)a(k)
\end{aligned} \tag{3-54}$$

We verify that Eqs. (3-50) and (3-51) are satisfied. Since $M^{\mu\nu}$ contains only an angular contribution the particles created and destroyed by the field φ have no intrinsic spin.

Having satisfied ourselves with the covariance of the quantized theory we now want to explore the relationship between the description in terms of particles of mass m, spin zero, and the hermitian field φ which could belong to the set of measurable quantities.

Even though at fixed time fields referring to various positions do commute, this is no longer the case when we compare them at different times. From canonical quantization, the commutator of two free fields $\varphi(x)$ and $\varphi(y)$ has, however, a simple structure as a c-number distribution which assumes the form

$$\begin{aligned}
[\varphi(x), \varphi(y)] &= \int d\tilde{k}\,[e^{-ik\cdot(x-y)} - e^{ik\cdot(x-y)}]\\
&= i\Delta(x-y)
\end{aligned} \tag{3-55}$$

The real distribution $\Delta(x)$ can also be written with the help of the function $\varepsilon(u) = u/|u|$:

$$\Delta(x-y) = \frac{1}{i}\int \frac{d^4k}{(2\pi)^3}\,\delta(k^2 - m^2)\varepsilon(k^0)\,e^{-ik\cdot x} \tag{3-56}$$

showing that it is an odd, Lorentz invariant, solution of the Klein-Gordon equation. From $\Delta(0, \mathbf{x}) = 0$ and Lorentz invariance we find, in fact, that $\Delta(x)$ vanishes outside the light cone, i.e., in the region $x^2 < 0$. Measurements at space-like separated points do not interfere, a consequence of locality and causality. Note also that canonical quantization entails

$$\partial_0 \Delta(x)\big|_{x^0=0} = -\delta^3(\mathbf{x}) \tag{3-57}$$

To construct coherent states which diagonalize the positive frequency part of the quantum field—the analogs of the minimal wave packets of the harmonic oscillator—we assume a normalizable function $\eta(k)$ to be given on the mass shell $k^2 = m^2$, $k^0 > 0$:

$$\int d\tilde{k}\, |\eta(k)|^2 < \infty$$

Consider the state $|\eta\rangle$ given by

$$|\eta\rangle = \exp\left[\int d\tilde{k}\, \eta(k) a^\dagger(k)\right] |0\rangle \tag{3-58}$$

which appears in Fock space as a coherent superposition of states with 0, 1, 2, ... particles. It can be used as a generating function to obtain the states with a fixed number of particles if we allow η to remain arbitrary. Let us compute the norm of $|\eta\rangle$ using the identity $e^A e^B = e^{A+B+[A,B]/2}$ valid for two operators A and B commuting with $[A, B]$. We find

$$\langle \eta_1 | \eta_2 \rangle = \exp\left[\int d\tilde{k}\, \eta_1^*(k) \eta_2(k)\right] \tag{3-59}$$

Therefore two such states are in general not orthogonal and the set of coherent states is an over-complete one. If we denote $|\eta)$ as the normalized state corresponding to $|\eta\rangle$:

$$|\eta) = \exp\left[-\frac{1}{2}\int d\tilde{k}\, |\eta(k)|^2\right] |\eta\rangle$$

we have

$$(\eta_1 | \eta_2) = \exp\left[\int d\tilde{k}\, (i\, \mathrm{Im}\, \eta_1^* \eta_2 - \tfrac{1}{2}|\eta_1 - \eta_2|^2)\right]$$

The vacuum state corresponds to $\eta = 0$, and more generally $a(k)|\eta\rangle = \eta(k)|\eta\rangle$, so that we have diagonalized the annihilation (positive frequency) part of the field

$$\varphi^{(+)}(x) = \int d\tilde{k}\, a(k)\, e^{-ik\cdot x} \qquad \varphi^{(+)}(x)|\eta\rangle = \int d\tilde{k}\, \eta(k)\, e^{-ik\cdot x} |\eta\rangle \tag{3-60}$$

Coherent states remain coherent under time evolution. Indeed,

$$\begin{aligned} e^{-iHt}|\eta\rangle &= \exp\left\{\int d\tilde{k}\, \eta(k)[e^{-iHt} a^\dagger(k)\, e^{+iHt}]\right\}|0\rangle \\ &= \exp\left[\int d\tilde{k}\, \eta(k)\, e^{-ik_0 t} a^\dagger(k)\right]|0\rangle \\ &= |\eta_t\rangle \end{aligned} \tag{3-61}$$

where $\eta_t(k) = e^{-ik_0 t}\eta(k)$.

It is also interesting to construct the unitarity operator $D(\eta)$ which realizes the transformation

$$D(\eta_2)|\eta_1) = |\eta_1 + \eta_2)$$

The reader may convince himself that

$$D(\eta) = \exp\left\{-\int d\tilde{k}\,[\eta^*(k)a(k) - \eta(k)a^\dagger(k)]\right\}$$

does the job.

Since $\varphi(x)$ is an operator-valued distribution let us smear it with a test function

$$\varphi_f = \int d^4x\, f(x)\varphi(x) \tag{3-62}$$

When f approaches $\delta(t)g(\mathbf{x})$ or $\delta'(t)g(\mathbf{x})$ we recover a space average of $\varphi(0, \mathbf{x})$ or $\pi(0, \mathbf{x})$. We may study the probability distribution of φ_f in a state corresponding to a fixed number r of particles

$$\rho_r(a) = \langle r|\,\delta(\varphi_f - a)\,|r\rangle = \int \frac{d\alpha}{2\pi}\, e^{-i\alpha a}\,\langle r|\, e^{i\alpha\varphi_f}\,|r\rangle \tag{3-63}$$

To compute this quantity we can use the coherent states as generating functions so the problem is reduced to the evaluation of

$$\langle\eta_1|\, e^{i\alpha\varphi_f}\,|\eta_2\rangle = \exp\left\{\int d\tilde{k}\left[\eta_1^*\eta_2 + i\alpha(\eta_1^*\tilde{f} + \tilde{f}^*\eta_2) - \frac{\alpha^2}{2}|\tilde{f}|^2\right]\right\} \tag{3-64}$$

where $\tilde{f}$ is the Fourier transform of f:

$$\tilde{f}(k) = \int d^4x\, e^{ik\cdot x} f(x)$$

Observe that only the mass shell components of $\tilde{f}$ contribute to Eq. (3-64). Leaving the general case as an exercise to the reader, we derive the vacuum distribution of the field as

$$\begin{aligned}\rho_0(a) &= \int \frac{d\alpha}{2\pi} \exp\left[-i\alpha a - \frac{\alpha^2}{2}\int d\tilde{k}\,|\tilde{f}(k)|^2\right] \\ &= \left[2\pi \int d\tilde{k}\,|\tilde{f}(k)|^2\right]^{-1/2} \exp\left[-\frac{a^2}{2\int d\tilde{k}\,|\tilde{f}(k)|^2}\right]\end{aligned} \tag{3-65}$$

Not surprisingly we obtain a gaussian distribution with mean square fluctuation

$$\sigma = \left[\int d\tilde{k}\,|\tilde{f}(k)|^2\right]^{1/2}$$

around a mean value zero. The field at a fixed space-time point corresponding to $\tilde{f} = 1$ has an infinite fluctuation and is therefore unobservable. But smeared operators are meaningful even when the function f has a support restricted to a fixed time. We note that the vacuum is not a zero field state and we shall soon show that vacuum fluctuations are observable.

Instead of considering $\varphi^{(+)}(x)$ we may also wish to construct a complete set of tested values of $\varphi(0, \mathbf{x})$:

$$\varphi_n = \int d^3x\, \varphi(0, \mathbf{x})F_n(\mathbf{x}) \tag{3-66}$$

corresponding to $f_n(x) = \delta(x^0)F_n(\mathbf{x})$ in the previous notation, with

$$\tilde{f}_n(k) = \int d^3x\, e^{i\mathbf{k}\cdot\mathbf{x}}\, F_n(\mathbf{x}) = \tilde{F}_n(\mathbf{k}) \tag{3-67}$$

normalized according to

$$\int d\tilde{k}\, \tilde{F}_n^*(\mathbf{k})\tilde{F}_{n'}(\mathbf{k}) = \delta_{n,n'}$$

$$\tilde{F}_n(-\mathbf{k}) = \tilde{F}_n^*(\mathbf{k}) \tag{3-68}$$

$$\sum_n \tilde{F}_n(\mathbf{k})\tilde{F}_n^*(\mathbf{k}') \equiv (2\pi)^3 2\omega_k \delta^3(\mathbf{k}-\mathbf{k}')$$

An eigenstate of the operators φ_n with eigenvalues $\varphi_{n,c}$ may, by abuse of language, be called an eigenstate of the field $\varphi(0, \mathbf{x})$ with eigenvalue

$$\varphi_c(\mathbf{x}) = \sum_n \varphi_{n,c} \int d\tilde{k}\, e^{-i\mathbf{k}\cdot\mathbf{x}}\, F_n^*(\mathbf{k}) \tag{3-69}$$

such that

$$\int d^3x\, \varphi_c(\mathbf{x})F_n(\mathbf{x}) = \varphi_{n,c}$$

Let us denote by $|\varphi_c\rangle$ the corresponding state. To obtain its components in the Fock basis it is sufficient to know its scalar product with a coherent state $|\eta\rangle$. This is obtained by solving

$$\langle\eta|\,\varphi_n\,|\varphi_c\rangle = \varphi_{n,c}\langle\eta\,|\varphi_c\rangle \tag{3-70}$$

for every value of n. If the normalization is such that

$$I = \int |\varphi_c\rangle \prod_n \frac{d\varphi_{n,c}}{\sqrt{2\pi}} \langle\varphi_c| \tag{3-71}$$

we find that

$$\langle\eta\,|\varphi_c\rangle = \exp\left\{-\tfrac{1}{4}\sum_n \varphi_{n,c}^2 + \int d\tilde{k}\left[-\tfrac{1}{2}\eta^*(\mathbf{k})\eta^*(-\mathbf{k}) + \eta^*(\mathbf{k})\sum_n \varphi_{n,c}\tilde{F}_n(\mathbf{k})\right]\right\} \tag{3-72}$$

from which we may study the physical content of these states.

3-1-3 Charged Scalar Field

The above description of the hermitian scalar field does not allow a distinction between particles and antiparticles. Particles and antiparticles have to carry some opposite-charge quantum number, whatever the nature of this charge may be. Classically, the minimal coupling prescription requires at least a complex field. In the quantum case let us therefore introduce a doublet of hermitian fields φ_1 and φ_2 represented by the complex quantity

$$\varphi = \frac{\varphi_1 + i\varphi_2}{\sqrt{2}}$$

and its hermitian conjugate. The total lagrangian $\mathscr{L}_{\text{tot}}$ will be the sum of two identical lagrangians of the form (3-7) pertaining to φ_1 and φ_2:

$$\mathscr{L}_{\text{tot}} = \mathscr{L}(\varphi_1) + \mathscr{L}(\varphi_2) = (\partial_\mu\varphi)^\dagger(\partial^\mu\varphi) - m^2\varphi^\dagger\varphi \tag{3-73}$$

The quadratic $\mathscr{L}$ is invariant under any rotation in the internal space characterized by the indices 1 and 2, or equivalently when φ is multiplied by a phase and $\varphi^\dagger$

by the conjugate one. The invariance group is $U(1)$ or $O(2)$. Quantization is performed independently for φ_1 and φ_2, with the states labeled by the numbers n_1 and n_2 of quanta of type 1 or 2 corresponding to the operators

$$N_i = \int d\tilde{k}\; a_i^\dagger(k) a_i(k) \qquad i = 1, 2 \tag{3-74}$$

In terms of the fields φ and $\varphi^\dagger$ and the conjugate momenta

$$\pi = \dot{\varphi}^\dagger = \frac{\dot{\varphi}_1 - i\dot{\varphi}_2}{\sqrt{2}} \qquad \pi^\dagger = \dot{\varphi} = \frac{\dot{\varphi}_1 + i\dot{\varphi}_2}{\sqrt{2}} \tag{3-75}$$

the hamiltonian reads

$$H = \int d^3x : \pi^\dagger \pi + \nabla\varphi^\dagger \nabla\varphi + m^2 \varphi^\dagger \varphi : \tag{3-76}$$

Note the absence of factor $\frac{1}{2}$ in Eqs. (3-73) and (3-76) as compared to the hermitian case.

The commutation rules

$$[\varphi(x), \varphi(y)] = [\varphi^\dagger(x), \varphi^\dagger(y)] = 0 \qquad [\varphi(x), \varphi^\dagger(y)] = i\Delta(x - y) \tag{3-77}$$

reduce at equal times to

$$[\varphi(t, \mathbf{x}), \pi(t, \mathbf{y})] = [\varphi^\dagger(t, \mathbf{x}), \pi^\dagger(t, \mathbf{y})] = i\delta^3(\mathbf{x} - \mathbf{y})$$

The field itself may be written

$$\varphi(x) = \int d\tilde{k}\, [a(k)\, e^{-ik \cdot x} + b^\dagger(k)\, e^{ik \cdot x}]$$

$$\varphi^\dagger(x) = \int d\tilde{k}\, [b(k)\, e^{-ik \cdot x} + a^\dagger(k)\, e^{ik \cdot x}]$$

$$\begin{aligned} a(k) &= \frac{a_1(k) + ia_2(k)}{\sqrt{2}} \\ a^\dagger(k) &= \frac{a_1^\dagger(k) - ia_2^\dagger(k)}{\sqrt{2}} \\ b(k) &= \frac{a_1(k) - ia_2(k)}{\sqrt{2}} \\ b^\dagger(k) &= \frac{a_1^\dagger(k) + ia_2^\dagger(k)}{\sqrt{2}} \end{aligned} \tag{3-78}$$

$$[a(k), a^\dagger(k')] = [b(k), b^\dagger(k')] = (2\pi)^3 2\omega_k \delta^3(\mathbf{k} - \mathbf{k}')$$

with all other commutators vanishing.

We see that φ destroys quanta of type a and creates quanta of type b, and vice versa for $\varphi^\dagger$, corresponding to the two observables

$$N_a = \int d\tilde{k}\; a^\dagger(k)a(k) \qquad N_b = \int d\tilde{k}\; b^\dagger(k)b(k) \tag{3-79}$$

The choice of diagonalizing N_a and N_b rather than N_1 and N_2 will appear natural when discussing the charge below. It is reminiscent of the choice of circular polarization versus linear polarization of light. Of course, $N_a + N_b = N_1 + N_2$. Finally, energy momentum is expressed as

$$P^\mu = \int d\tilde{k}\; k^\mu[a^\dagger(k)a(k) + b^\dagger(k)b(k)] \tag{3-80}$$

while the vacuum is annihilated by all operators $a(k)$ and $b(k)$.

Applying Noether's theorem to the $U(1)$ invariance of the theory, we expect the existence of a conserved current

$$j_\mu = :i\varphi^\dagger \overleftrightarrow{\partial}_\mu \varphi: \tag{3-81}$$

and a corresponding time-independent charge

$$\begin{aligned} Q &= \int d^3x :i(\varphi^\dagger\dot{\varphi} - \dot{\varphi}^\dagger\varphi): \\ &= \int d\tilde{k}\; [a^\dagger(k)a(k) - b^\dagger(k)b(k)] = N_a - N_b \end{aligned} \tag{3-82}$$

Indeed, we verify that

$$\dot{Q} = i[H, Q] = 0 \qquad \partial_\mu j^\mu = 0 \tag{3-83}$$

Therefore a quanta carry a charge $+1$ and b quanta a charge -1. The two types play symmetric roles and to distinguish them physically we have to introduce an external coupling.

It has become traditional to identify one of the types with the particles, the other one with antiparticles. The symmetry of the theory requires $m_a = m_b$ and also the equality of spins, here zero. The discrete symmetry under interchange of particles and antiparticles is called charge conjugation.

Show that one can find a unitary operator $\mathscr{C}$ in Fock space such that

$$\mathscr{C}\varphi(x)\mathscr{C}^\dagger = \eta_c\varphi^\dagger(x) \qquad |\eta_c| = 1 \tag{3-84}$$

which commutes with the hamiltonian, and discuss its properties.

Examples of distinct particle-antiparticle scalar pairs are the mesons $\pi^+\pi^-$, K^+K^-, or $K^0\bar{K}^0$. In the last case the opposite "charge" is strangeness.

The unified description given by the complex field φ which enters as a unique combination in charge conjugation-invariant couplings shows a profound relationship between the dynamical behavior of particles and antiparticles which goes beyond the equality of their kinematical invariants, as we shall see later.

3-1-4 Time-Ordered Product

When acting on a state the operator $\varphi^\dagger(x)$ creates a particle of charge $+1$ or destroys an antiparticle of charge -1. In either case it adds a total charge $+1$. Similarly, if we act on a state with $\varphi(x')$ we destroy a unit charge. We can interpret the combined action of the two operators, leaving the total charge invariant, in two ways according to the sign of $t' - t$. If $t' > t$ we first create a particle at time t with $\varphi^\dagger(x)$ and destroy it at a later time t' with $\varphi(x')$. The corresponding amplitude will be given by the mean value of the operator

$$\theta(t' - t)\varphi(t', \mathbf{x}')\varphi^\dagger(t, \mathbf{x}) \tag{3-85}$$

When $t' < t$ an antiparticle has been created by $\varphi(x')$, then absorbed by $\varphi^\dagger(x)$ at time t with a corresponding amplitude

$$\theta(t - t')\varphi^\dagger(t, \mathbf{x})\varphi(t', \mathbf{x}') \tag{3-86}$$

In both cases charge increases at x and decreases at x' independently of the causal propagation. In the case $t > t'$, instead of speaking of the creation of an antiparticle at x' subsequently absorbed at x, we may say that a hole has appeared at x' to be filled at the later time t. We thus recover Dirac's hole description of Chap. 2.

The sum of the two operators (3-85) and (3-86) is Dyson's time-ordered product

$$T\varphi(x')\varphi^\dagger(x) = \theta(t' - t)\varphi(x')\varphi^\dagger(x) + \theta(t - t')\varphi^\dagger(x)\varphi(x') \tag{3-87}$$

appropriately named since operators occur under the T symbol arranged from right to left with increasing times. Bosonic operators obviously commute under chronological ordering.

Let us act with the operator $\Box_{x'} + m^2$ on the time-ordered product (3-87). Care should be taken since the step functions are time dependent; therefore instead of finding zero we must obtain a distribution concentrated at equal times. Indeed,

$$\begin{aligned}\frac{\partial^2}{\partial t'^2} T\varphi(x')\varphi^\dagger(x) &= \frac{\partial}{\partial t'}\left\{T\frac{\partial}{\partial t'}\varphi(x')\varphi^\dagger(x) + \delta(t' - t)[\varphi(x'), \varphi^\dagger(x)]\right\}\\ &= T\frac{\partial^2}{\partial t'^2}\varphi(x')\varphi(x) + \delta(t' - t)\left[\frac{\partial}{\partial t'}\varphi(x'), \varphi^\dagger(x)\right]\end{aligned}$$

where we have used the fact that

$$\delta(t' - t)[\varphi(x'), \varphi^\dagger(x)] = i\delta(t' - t)\Delta(0, \mathbf{x}' - \mathbf{x}) = 0$$

Now, from (3-77) and (3-57),

$$\delta(t' - t)\left[\frac{\partial}{\partial t'}\varphi(x'), \varphi^\dagger(x)\right] = i\delta(t' - t)\frac{\partial}{\partial t'}\Delta(t' - t, \mathbf{x}' - \mathbf{x}) = -i\delta^4(x' - x)$$

and using $(\partial^2/\partial t'^2)\varphi(x') = (\Delta_{x'} - m^2)\varphi(x')$, we find

$$(\Box_{x'} + m^2)T\varphi(x')\varphi^\dagger(x) = -i\delta^4(x' - x) \tag{3-88}$$

If instead of fulfilling the homogeneous Klein-Gordon equation the field φ would satisfy $(\Box + m^2)\varphi(x) = j(x)$, Eq. (3-88) would be replaced by

$$(\Box_{x'} + m^2)T\varphi(x')\varphi^\dagger(x) = Tj(x')\varphi^\dagger(x) - i\delta^4(x' - x) \tag{3-89}$$

It follows that the vacuum expectation value

$$G_F(x' - x) = i\langle 0|\, T\varphi(x')\varphi^\dagger(x)\, |0\rangle \tag{3-90}$$

is one of the Green functions of the Klein-Gordon operator. A little calculation shows that it is indeed the Feynman scalar propagator encountered in Chap. 1:

$$G_F(x - y) = -\int \frac{d^4k}{(2\pi)^4} \frac{1}{k^2 - m^2 + i\varepsilon} e^{-ik\cdot(x-y)} \tag{3-91}$$

We conclude that the quantization of a free, relativistic scalar field has produced a satisfactory description of spinless noninteracting particles, obeying the Bose-Einstein statistics, with or without charge. In the absence of interactions the number of field quanta is conserved.

3-1-5 Thermodynamic Equilibrium

In a rest frame with total three-momentum zero it may happen that the correct description of the above quantized system, instead of being the vacuum state, is a thermodynamic equilibrium at temperature $T = 1/k\beta$ (k is Boltzmann's constant) and chemical potential μ. This is well defined as long as the total number N of quanta is conserved. We return to a fixed volume quantization with discrete momenta:

$$\mathbf{k} = \frac{2\pi}{V^{1/3}}\mathbf{n} \qquad n_1, n_2, n_3 \text{ integers}$$

whereupon integrals are replaced by sums

$$\int \frac{d^3k}{(2\pi)^3} \to \frac{1}{V}\Sigma_{\mathbf{k}}$$

We rescale the annihilation and creation operators by defining

$$a(k) = \sqrt{2\omega_k V}A_{\mathbf{k}}$$

in such a way that

$$[A_{\mathbf{k}}, A^\dagger_{\mathbf{k}'}] = \delta_{\mathbf{k},\mathbf{k}'}$$

with Kronecker symbols instead of delta functions. We have therefore

$$H = \int d\tilde{k}\, \omega_k a^\dagger(k)a(k) = \Sigma_{\mathbf{k}}\omega_k A^\dagger_{\mathbf{k}} A_{\mathbf{k}}$$

$$N = \int d\tilde{k}\, a^\dagger(k)a(k) = \Sigma_{\mathbf{k}} A^\dagger_{\mathbf{k}} A_{\mathbf{k}}$$

To simplify we have assumed the system to be neutral. The grand partition function $\mathscr{Z}$ is given by a trace in Fock space

$$\mathscr{Z} = \text{Tr}\, e^{-\beta(H-\mu N)} = \text{Tr}\exp\left[-\beta\, \Sigma_{\mathbf{k}}(\omega_k - \mu)A^\dagger_{\mathbf{k}}A_{\mathbf{k}}\right]$$

Each mode contributes a factorized term, and since Tr $e^{\lambda A^\dagger A} = (1 - e^{\lambda})^{-1}$,

$$\mathscr{Z} = \exp\left[-\Sigma_{\mathbf{k}} \ln\left(1 - e^{-\beta(\omega_k - \mu)}\right)\right]$$

In the large volume limit the thermodynamical potential is

$$\Omega = \log \mathscr{Z} = -V \int \frac{d^3k}{(2\pi)^3} \ln\left(1 - e^{-\beta(\omega_k - \mu)}\right)$$

From thermodynamics we know that if p is the pressure $\Omega = \beta p V$. Therefore

$$p = -\frac{1}{\beta} \int \frac{d^3k}{(2\pi)^3} \ln\left(1 - e^{-\beta(\omega_k - \mu)}\right) \tag{3-92}$$

while the mean values of the energy density $\bar{E}/V$ (without zero point oscillations) and particle density $\bar{N}/V$ are

$$\begin{aligned} \frac{\bar{E}}{V} &= -\frac{\partial \Omega}{V \partial \beta}\bigg|_{\beta\mu = \text{constant}} = \int \frac{d^3k}{(2\pi)^3} \frac{\omega_k}{e^{\beta(\omega_k - \mu)} - 1} \\ \frac{\bar{N}}{V} &= \frac{1}{\beta V} \frac{\partial \Omega}{\partial \mu}\bigg|_{\beta} = \int \frac{d^3k}{(2\pi)^3} \frac{1}{e^{\beta(\omega_k - \mu)} - 1} \end{aligned} \tag{3-93}$$

the familiar expressions corresponding to Bose's statistics. In the case $m = 0$, and when the number of particles is not conserved (that is, $\mu = 0$), we have a situation analogous with blackbody radiation except for polarization. The energy density becomes

$$\frac{\bar{E}}{V} = \int \frac{d^3k}{(2\pi)^3} \frac{|\mathbf{k}|}{e^{\beta|\mathbf{k}|} - 1} = \frac{4\pi}{(2\pi)^3} \int_0^\infty dk \, \frac{k^3}{\beta} \frac{d}{dk} \ln\left(1 - e^{-\beta k}\right)$$

Integrating by parts we obtain

$$\frac{\bar{E}}{V} = -\frac{3}{\beta} \int \frac{d^3k}{(2\pi)^3} \ln\left(1 - e^{-\beta|\mathbf{k}|}\right) = 3p \tag{3-94}$$

a classical relationship, familiar from blackbody radiation, stating that the average energy momentum tensor $\bar{\Theta}_{\mu\nu}$ which has only diagonal elements $\bar{\Theta}_{00} = \bar{E}/V$, $\bar{\Theta}_{ij} = p\delta_{ij}$ $(i, j = 1, 2, 3)$ is traceless:

$$\bar{\Theta}^{\mu}{}_{\mu} = \frac{\bar{E}}{V} - 3p = 0$$

Denoting by parentheses thermodynamic averages (we assume now $\mu = 0$ but keep the mass arbitrary)

$$(A) = \frac{\Sigma_\alpha \exp\{-\beta E_\alpha\} \langle \alpha | A | \alpha \rangle}{\Sigma_\alpha \exp\{-\beta E_\alpha\}}$$

let us study the propagation at finite temperature. Set

$$f(\omega) = \frac{1}{e^{\beta\omega} - 1}$$

Then

$$\begin{aligned} (A_{\mathbf{k}}^\dagger A_{\mathbf{k}'}) &= \delta_{\mathbf{k},\mathbf{k}'} f(\omega_k) \\ (A_{\mathbf{k}} A_{\mathbf{k}'}^\dagger) &= \delta_{\mathbf{k},\mathbf{k}'} [1 + f(\omega_k)] = -\delta_{\mathbf{k},\mathbf{k}'} f(-\omega_k) \\ (A_{\mathbf{k}} A_{\mathbf{k}'}) &= (A_{\mathbf{k}}^\dagger A_{\mathbf{k}'}^\dagger) = 0 \end{aligned}$$

In the infinite volume limit

$$(\varphi(x)\varphi(y)) = \int d\tilde{k} \, \{e^{-ik\cdot(x-y)}[1 + f(\omega_k)] + e^{ik\cdot(x-y)} f(\omega_k)\}$$

and the time-ordered product takes the form

$$(T\varphi(x)\varphi(y)) = i\int \frac{d^4k}{(2\pi)^4}\frac{e^{-ik\cdot(x-y)}}{k^2-m^2+i\varepsilon} + \int d\tilde{k}\,\frac{1}{e^{\beta\omega_k}-1}\,(e^{-ik\cdot(x-y)} + e^{ik\cdot(x-y)}) \tag{3-95}$$

This expression exhibits the propagator at finite temperature as the sum of the zero temperature Feynman propagator plus a temperature-dependent solution of the homogeneous Klein-Gordon equation which vanishes like $e^{-\beta m}$ when $\beta \to \infty (T \to 0)$. This propagator is interesting to study dynamical disturbances away from equilibrium.

The structure of the partition function suggests that we may encounter, in applications of perturbation theory to statistical systems, a different, "Euclidean," propagator. Starting from the basic dynamical field variables at fixed time, we define an evolution for imaginary times

$$x_0 = -i\hat{x}_0$$

where $\hat{x}_0$ is real, and at first restricted to the interval $[0, \beta]$ according to

$$\varphi(x) = e^{iHx_0}\,\varphi(0, \mathbf{x})\,e^{-iHx_0} = e^{H\hat{x}_0}\,\varphi(0, \mathbf{x})\,e^{-H\hat{x}_0}$$

Operator ordering with respect to $\hat{x}_0$ will be denoted by the symbol $\hat{T}$ and will satisfy

$$\left(-\frac{\partial^2}{\partial\hat{x}_0^2} - \Delta_x + m^2\right)[\hat{T}\varphi(x)\varphi(y)] = -i\delta(\hat{x}_0 - \hat{y}_0)\delta^3(\mathbf{x}-\mathbf{y})$$

It can be continued as a periodic function of $\hat{x}_0 - \hat{y}_0$ of period β. Indeed, assume at first $\hat{y}_0$ restricted to the interval $[0, \beta]$. Then from the cyclic character of the trace

$$\begin{aligned}
[\hat{T}\varphi(x)\varphi(y)]_{\hat{x}_0=0} &= (\operatorname{Tr} e^{-\beta H})^{-1}\operatorname{Tr}[e^{-\beta H}\,\varphi(y)\varphi(0,\mathbf{x})]\\
&= (\operatorname{Tr} e^{-\beta H})^{-1}\operatorname{Tr}[e^{-\beta H}\,e^{\beta H}\,\varphi(0,\mathbf{x})\,e^{-\beta H}\,\varphi(y)]\\
&= [\hat{T}\varphi(x)\varphi(y)]_{\hat{x}_0=\beta}
\end{aligned}$$

When this continuation is performed the $\delta(\hat{x}_0 - \hat{y}_0)$ function is to be understood as applied to periodic functions (functions on a circle) with $2\pi\delta(\alpha) = \Sigma_n e^{in\alpha}$. We can then write

$$\begin{aligned}
[\hat{T}\varphi(x)\varphi(y)] &= \frac{1}{\beta}\sum_{n=-\infty}^{+\infty}\exp\left[-2i\pi\frac{n}{\beta}(\hat{x}_0 - \hat{y}_0)\right]\int\frac{d^3k}{(2\pi)^3}\,e^{i\mathbf{k}\cdot(\mathbf{x}-\mathbf{y})}\,\frac{i}{k^2-m^2}\\
&= \frac{1}{\beta}\sum_{k^0=2i\pi n/\beta}\int\frac{d^3k}{(2\pi)^3}\,e^{-ik\cdot(x-y)}\,\frac{i}{k^2-m^2}
\end{aligned} \tag{3-96}$$

In the last expression use has been made of minkowskian notations with $x^0 = -i\hat{x}^0$ pure imaginary as well as k^0. If we denote by the symbol $\int^{(\beta)} d^4k/(2\pi)^4$ the combined discrete sum and integral

$$\frac{1}{\beta}\sum_{k^0=2\pi in/\beta}\int\frac{d^3k}{(2\pi)^3}$$

the structure of (3-96) is formally identical to the ordinary propagator. We observe, however, that the denominator

$$k^2 - m^2 = -\left[\left(\frac{2\pi n}{\beta}\right)^2 + \mathbf{k}^2 + m^2\right]$$

never vanishes. The reader should realize that (3-95) and (3-96) represent very different quantities relative to unrelated problems.

3-2 QUANTIZED RADIATION FIELD

The above thermodynamical example leads us at once to the radiation field. Indeed, Planck's blackbody spectrum was historically the first instance of field quantization. We have simply to generalize the preceding construction to the degrees of freedom described by the potential $A_\mu(x)$. Since we insist upon the local character of the theory, we cannot make use of the gauge-invariant tensor $F_{\mu\nu}$ as the fundamental dynamical variable. We have to face the gauge dependence of A_μ already discussed at length in Chap. 1. Various devices have been used to circumvent this difficulty. We shall not try to quote all of them here and present the Gupta-Bleuler indefinite metric quantization.

3-2-1 Indefinite Metric

We recall that the lagrangian

$$\mathscr{L} = -\tfrac{1}{4}F^2 \tag{3-97}$$

where F is expressed in terms of A, is unsuited for canonical quantization since the conjugate momentum of A_0 vanishes. If we were not to worry about manifest Lorentz covariance we could content ourselves with $\mathbf{A}$ by constraining A^0 and using a condition of the type $\operatorname{div} \mathbf{A} = 0$. This of course is physically reasonable, but does not fit our goals. We therefore use the procedure of modifying the equations of motion at the price of restoring Maxwell's theory by appropriate constraints on the physical states, since we shall introduce spurious degrees of freedom. The price will seem at first rather heavy, for not only will the Hilbert space of quantization be too large but instead of carrying a positive metric it shall allow for negative norm states! Thus the usual probabilistic interpretation of quantum mechanics will only emerge when we will have restricted ourselves to the physical quanta: the photons with only two polarizations. We also recall that to preserve gauge invariance, the principle which justifies the above procedure, we will have to make sure that the current remains conserved when introducing interactions.

We therefore modify (3-97) to read, with an arbitrary constant $\lambda \neq 0$,

$$\mathscr{L} = -\tfrac{1}{4}F^2 - \frac{\lambda}{2}(\partial \cdot A)^2 \tag{3-98}$$

so that Maxwell's equations are replaced by

$$\Box A_\mu - (1 - \lambda)\partial_\mu(\partial \cdot A) = 0 \tag{3-99}$$

and the conjugate momenta to the four components of A are

$$\pi^\rho = \frac{\partial \mathscr{L}}{\partial(\partial_0 A_\rho)} = F^{\rho 0} - \lambda g^{\rho 0}(\partial \cdot A) \tag{3-100}$$

In particular π^0 no longer vanishes.

Whether the right-hand side of Eq. (3-99) vanishes, as it does in this free case, or not if we replace it by a conserved current, we find that

$$\Box(\partial \cdot A) = 0 \tag{3-101}$$

so that $\partial \cdot A$ is a free scalar field. Classically we argued that appropriate conditions could be imposed on the boundary data in order that $\partial \cdot A$ vanishes everywhere as a consequence of Eq. (3-101), hence restoring Maxwell's theory in the Lorentz gauge, $\partial \cdot A = 0$. This cannot be achieved in the quantum case. Indeed, we shall assume the canonical commutation rules

$$[A_\rho(t, \mathbf{x}), \pi^\nu(t, \mathbf{y})] = ig_\rho{}^\nu \delta^3(\mathbf{x} - \mathbf{y}) \tag{3-102}$$

The operator equation $\partial \cdot A = 0$ is then inconsistent, since up to a constant factor this is π^0, and the latter has nonvanishing commutators with A^0 at equal times. We see at once that quantization according to (3-102) has overdone its job by adding unphysical features which will have to be eliminated. Even though all calculations can be pursued, keeping λ arbitrary, in the interest of simplicity we shall henceforth set $\lambda = 1$, in which case Eq. (3-99) reduces to

$$\Box A^\mu = 0 \qquad \lambda = 1 \tag{3-103}$$

In the quantum case the choice of λ is called by abuse of language a choice of gauge, and $\lambda = 1$ is referred to as the Feynman gauge.

The price for simplicity is that we will not directly verify that our results are λ independent. Within the framework of the free field we leave it as an exercise to the reader. He or she may as well discuss alternative forms of (3-98), where the added term might be taken as $(\lambda/2)[\partial \cdot A - \varphi(x)]^2$ with $\varphi(x)$ a c-number classical field or $(\lambda/2)[n(x) \cdot A]^2$ and $n(x)$ a classical vector field (perhaps constant).

The commutation rules (3-102) are completed by requiring that

$$[A_\rho(t, \mathbf{x}), A_\nu(t, \mathbf{y})] = [\pi_\rho(t, \mathbf{x}), \pi_\nu(t, \mathbf{y})] = 0 \tag{3-104}$$

From (3-104) it follows that the spatial derivatives of A_ρ commute at equal times so that (3-102) and (3-104) can be recast in the following form:

$$\begin{aligned} [A_\rho(t, \mathbf{x}), A_\nu(t, \mathbf{y})] &= [\dot{A}_\rho(t, \mathbf{x}), \dot{A}_\nu(t, \mathbf{y})] = 0 \\ [\dot{A}_\rho(t, \mathbf{x}), A_\nu(t, \mathbf{y})] &= ig_{\rho\nu}\, \delta^3(\mathbf{x} - \mathbf{y}) \end{aligned} \tag{3-105}$$

analogous to the scalar case repeated four times. Corresponding to the reality of the classical potential components we assume their quantized counterparts to be hermitian.

The dynamical system is therefore specified by (3-103) and (3-105). We note, however, a slight difference as compared to the scalar case, where the nontrivial commutator was

$$[\varphi(t, \mathbf{x}), \dot{\varphi}(t, \mathbf{y})] = i\delta^3(\mathbf{x} - \mathbf{y}) \tag{3-106}$$

Only three out of the four components of $A_\mu(t, \mathbf{x})$ behave as in Eq. (3-106). The commutator involving $A_0(t, \mathbf{x})$ has a reversed sign, as if the role of the field and its conjugate variable were interchanged. This is related with the four-vector character of A_μ and generates, as we shall soon realize, a very different Hilbert space when we "solve" (3-105) as we did in the preceding section.

In order to fulfill this program, we Fourier analyze the field in terms of plane wave solutions of Eq. (3-103) and set

$$A_\mu(x) = \int d\tilde{k} \sum_{\lambda=0}^{3} [a^{(\lambda)}(k)\varepsilon_\mu^{(\lambda)}(k)\, e^{-ik\cdot x} + a^{(\lambda)\dagger}(k)\varepsilon_\mu^{(\lambda)*}(k)\, e^{ik\cdot x}] \tag{3-107}$$

with, as in (3-35),

$$d\tilde{k} = \frac{d^3k}{2k_0(2\pi)^3} \qquad k_0 = |\mathbf{k}| \tag{3-108}$$

For each k on the positive light cone, $\varepsilon_\mu^{(\lambda)}(k)$ are a set of four linearly independent vectors which we may assume real. The symbol † refers to an hermitian conjugation with respect to a nondegenerate scalar product in Fock space. As indicated in (3-108) the field quanta are massless.

In order to exhibit the content of (3-107) we shall make a definite choice of polarization vectors $\varepsilon^{(\lambda)}(k)$ with respect to a fixed reference frame. Let n stand for the time axis $n^2 = 1$, $n^0 > 0$. We choose $\varepsilon^{(1)}$ and $\varepsilon^{(2)}$ in the plane orthogonal to k and n such that

$$\varepsilon^{(\lambda)}(k)\cdot\varepsilon^{(\lambda')}(k) = -\delta_{\lambda,\lambda'} \qquad \lambda, \lambda' = 1, 2$$

We then pick $\varepsilon^{(3)}$ in the plane (k, n) orthogonal to n and normalized

$$\varepsilon^{(3)}(k)\cdot n = 0 \qquad [\varepsilon^{(3)}(k)]^2 = -1$$

Finally we take $\varepsilon^{(0)}$ equal to n. With these conventions we call respectively $\varepsilon^{(1)}$ and $\varepsilon^{(2)}$ transverse, $\varepsilon^{(3)}$ longitudinal, and $\varepsilon^{(0)}$ scalar polarizations. In the frame where $n^0 = 1$ and $\mathbf{k}$ is along the third axis, we have simply

$$\varepsilon^{(0)} = \begin{pmatrix}1\\0\\0\\0\end{pmatrix} \quad \varepsilon^{(1)} = \begin{pmatrix}0\\1\\0\\0\end{pmatrix} \quad \varepsilon^{(2)} = \begin{pmatrix}0\\0\\1\\0\end{pmatrix} \quad \varepsilon^{(3)} = \begin{pmatrix}0\\0\\0\\1\end{pmatrix}$$

In any case we have

$$\sum_\lambda \frac{\varepsilon_\mu^{(\lambda)}(k)\varepsilon_\nu^{(\lambda)*}(k)}{\varepsilon^{(\lambda)}(k)\cdot\varepsilon^{(\lambda)*}(k)} = g_{\mu\nu} \qquad \varepsilon^{(\lambda)}(k)\cdot\varepsilon^{(\lambda')*}(k) = g^{\lambda\lambda'} \tag{3-109}$$

The denominator in the first expression is required because of the indefinite scalar product. From (3-107), A_μ automatically satisfies the field equations. The commutation relations (3-105) follow provided

$$[a^{(\lambda)}(k), a^{(\lambda')\dagger}(k')] = -g^{\lambda\lambda'}2k^0(2\pi)^3\delta^3(\mathbf{k}-\mathbf{k}') \tag{3-110}$$

Furthermore, we readily find at arbitrary times

$$[A_\mu(x), A_\nu(y)] = -ig_{\mu\nu}\Delta(x-y) \tag{3-111}$$

with Δ defined as in (3-56) for $m = 0$. We feel secure as we see covariant expressions such as (3-111) emerge naturally. We would then proceed by introducing a vacuum state annihilated by the various operators $a^{(\lambda)}(k)$:

$$a^{(\lambda)}(k)|0\rangle = 0 \qquad \lambda = 0, 1, 2, 3$$

But somehow we feel uneasy, as we have seemingly introduced twice as many polarization states as are usually attributed to photons. The trouble is explicit when we construct a one-particle state with scalar polarization:

$$|1\rangle = \int d\tilde{k}\, f(k) a^{(0)\dagger}(k)|0\rangle$$

Let us compute

$$\langle 1|1\rangle = \int\int d\tilde{k}\, d\tilde{k}'\, f(k)^* f(k') \langle 0|\, a^{(0)}(k) a^{(0)\dagger}(k')\,|0\rangle$$

According to (3-110) we find the paradoxical result that

$$\langle 1|1\rangle = -\langle 0|0\rangle \int d\tilde{k}\, |f(k)|^2$$

Fock space has an indefinite metric! The same calculation repeated with the three other types of polarization leads to states with positive metric. What about the probabilistic interpretation of quantum mechanics?

As long as we discuss the free case, the situation is not as catastrophic as it may seem, but interactions may excite the unwanted states. The crucial observation is that up to now we are not really dealing with Maxwell's theory since we have explicitly modified the lagrangian. To recover it, we would like to set $\partial \cdot A = 0$, but this is impossible as an operator equation. We may therefore at least try to select the states $|\psi\rangle$ such that the Lorentz condition holds in the mean

$$\langle \psi|\partial \cdot A|\psi\rangle = 0 \tag{3-112}$$

To preserve the linear structure of the physical Hilbert space $\mathscr{H}_1$, we require that the annihilation (positive frequency) part of $\partial \cdot A$ annihilates $\mathscr{H}_1$:

$$\partial^\mu A_\mu^{(+)}|\psi\rangle = 0 \tag{3-113}$$

Clearly (3-112) will follow from (3-113).

Let us now analyze the states belonging to $\mathscr{H}_1$. Since condition (3-113) is linear, we can consider basis states obtained by acting on the vacuum with (smeared) products of creation operators pertaining to the various polarizations. We therefore factorize these states as

$$|\psi\rangle = |\psi_T\rangle|\phi\rangle \tag{3-114}$$

where $|\psi_T\rangle$ corresponds to transverse photons and $|\phi\rangle$ is obtained by acting on the vacuum with longitudinal and scalar operators. This decomposition depends on our conventions for the polarization vectors. It is clear that it is sufficient to examine the consequence of condition (3-113) on such states as $|\phi\rangle$ since $\partial \cdot A^{(+)}$

only involves longitudinal and scalar polarizations:

$$i\partial \cdot A^{(+)} = \int d\tilde{k}\, e^{-ik\cdot x} \sum_{\lambda=0,3} a^{(\lambda)}(k)\varepsilon^{(\lambda)}(k)\cdot k \tag{3-115}$$

so that (3-113) reduces to

$$\sum_{\lambda=0,3} k\cdot\varepsilon^{(\lambda)}(k)a^{(\lambda)}(k)\,|\phi\rangle = 0 \tag{3-116}$$

It would be too much to expect that (3-116) could determine $|\phi\rangle$ entirely. Furthermore, we must allow for the arbitrariness in the choice of the transverse polarization vectors, to which we may add at will a term proportional to k. This arbitrariness must be reflected in the $|\phi\rangle$ part of the state so that we may at most expect that a given physical situation is represented by a class of equivalent vectors in $\mathscr{H}_1$. We want, of course, to verify that vectors in this subspace have positive square norm.

An equivalent form of (3-116) is

$$[a^{(0)}(k) - a^{(3)}(k)]\,|\phi\rangle = 0 \tag{3-117}$$

A state $|\phi\rangle$ may be a linear combination of states corresponding to 0, 1, 2, ... scalar or longitudinal "photons:"

$$\begin{aligned} |\phi\rangle &= C_0|\phi_0\rangle + C_1|\phi_1\rangle + \cdots + C_n|\phi_n\rangle + \cdots \\ |\phi_0\rangle &\equiv |0\rangle \end{aligned} \tag{3-118}$$

The number operator for these photons is, by virtue of (3-110) (note the minus sign),

$$N' = \int d\tilde{k}\,[a^{(3)\dagger}(k)a^{(3)}(k) - a^{(0)\dagger}(k)a^{(0)}(k)] \tag{3-119}$$

A state $|\phi_n\rangle$ such that (3-117) is satisfied will be such that

$$n\langle\phi_n|\phi_n\rangle = 0$$

Hence

$$\langle\phi_n|\phi_n\rangle = \delta_{n,0}$$

If $n \neq 0$ such a state therefore has a zero norm. A general solution is of the form (3-118) with

$$\langle\phi|\phi\rangle = |C_0|^2 \geq 0$$

the coefficients C_i remaining arbitrary and each of the $|\phi_n\rangle$ fulfilling (3-117). It remains to show that the arbitrariness in $|\phi\rangle$ does not affect the physical observables. Consider the energy, for instance. The hamiltonian reads

$$\begin{aligned} H &= \int d^3x :\pi^\mu \dot{A}_\mu - \mathscr{L}: = \frac{1}{2}\int d^3x : \sum_{i=1}^{3} [\dot{A}_i^2 + (\nabla A_i)^2] - \dot{A}_0^2 - (\nabla A_0)^2 : \\ &= \int d\tilde{k}\,\omega_k\left[\sum_1^3 a^{(\lambda)\dagger}(k)a^{(\lambda)}(k) - a^{(0)\dagger}(k)a^{(0)}(k)\right] \end{aligned} \tag{3-120}$$

When acting with H on a state in $\mathscr{H}_1$ we find in general a contribution from the unphysical part $|\phi\rangle$, but it vanishes when we take the mean value since

$$\frac{\langle\psi|H|\psi\rangle}{\langle\psi|\psi\rangle} = \frac{\langle\psi_T| \int d\tilde{k}\, \omega_k \sum_{\lambda=1,2} a^{(\lambda)\dagger}(k)a^{(\lambda)}(k)|\psi_T\rangle}{\langle\psi_T|\psi_T\rangle} \tag{3-121}$$

by virtue of (3-117). As far as momentum is concerned, the formula is analogous with ω_k replaced by $\mathbf{k}$. Insofar as only average values are observed, we see that not only have negative probabilities disappeared when restricting ourselves to $\mathscr{H}_1$ but also scalar and longitudinal photons do not contribute. Only the two physical transverse polarization states manifest themselves.

It is nice to realize that the arbitrariness on $|\phi\rangle$ is reflected by the fact that the mean value of A_μ in such a state is a pure gradient. Let us compute it as an exercise. We readily derive from (3-118) that

$$\langle\phi| A_\mu(x)|\phi\rangle = C_0^* C_1 \langle\phi_0| \int d\tilde{k}\, e^{-ik\cdot x}\, [\varepsilon_\mu^{(3)}(k)a^{(3)}(k) + \varepsilon_\mu^{(0)}(k)a^{(0)}(k)]\,|\phi_1\rangle + \text{complex conjugate (cc)}$$

All other components $|\phi_n\rangle$ do not contribute since A_μ changes n by a unit; $|\phi_0\rangle$ can be taken equal to $|0\rangle$, and $|\phi_1\rangle$ is necessarily of the form

$$|\phi_1\rangle = \int d\tilde{q}\, f(q)\,[a^{(3)\dagger}(q) - a^{(0)\dagger}(q)]\,|0\rangle$$

Choosing real polarization vectors we have

$$\langle\phi| A_\mu(x)|\phi\rangle = \int d\tilde{k}\, [\varepsilon_\mu^{(3)}(k) + \varepsilon_\mu^{(0)}(k)]\,[C_0^* C_1\, e^{-ik\cdot x} f(k) + \text{cc}]$$

We recall that $\varepsilon^{(0)}(k) = n$ and by its definition

$$\varepsilon^{(3)}(k) = \frac{k - n(k\cdot n)}{(k\cdot n)}$$

Therefore,

$$\varepsilon_\mu^{(3)}(k) + \varepsilon_\mu^{(0)}(k) = \frac{1}{(n\cdot k)}\, k_\mu$$

Hence, provided the integral is meaningful,

$$\langle\phi| A_\mu(x)|\phi\rangle = \partial_\mu \int \frac{d\tilde{k}}{(k\cdot n)}\, [iC_0^* C_1\, e^{-ik\cdot x} f(k) + \text{cc}] = \partial_\mu \Lambda(x) \tag{3-122}$$

where the scalar c-number function $\Lambda(x)$ is a solution of $\Box\Lambda(x) = 0$, and may be chosen at will by adjusting the vector $|\phi\rangle$. This is another proof that the arbitrariness of $|\phi\rangle$ reflects the gauge arbitrariness of A_μ without practical consequence.

We may select in the equivalence class of state vectors of the form $|\psi_T\rangle|\phi\rangle$ a convenient representative by imposing that $|\phi\rangle \equiv |0\rangle$, this again with reference to a choice of basis polarization vectors. Instead of taking the transverse ones, real and orthogonal, we frequently choose the complex combinations $(\varepsilon^{(1)} \pm i\varepsilon^{(2)})/\sqrt{2}$ to represent the two helicities of the photon.

We shall not elaborate in detail the proof of Lorentz invariance. It is clear that $\mathscr{H}_1$ is Lorentz invariant as well as the equivalence class representing a given physical state. The reader is invited to construct explicitly the generators $M^{\mu\nu}$ of Lorentz transformations. He or she will verify that photons are not only massless but carry an helicity ± 1. It is sometimes said that the photon has spin 1.

In spite of the fact that only vectors in $\mathscr{H}_1$ receive a physical interpretation it is worth emphasizing that the full indefinite norm Fock space is necessary to preserve the locality properties. They appear generally in complete sums over intermediate states.

As an exercise one can easily extend the discussion of Sec. 3-1-5 to recover the blackbody spectrum within the present formalism.

3-2-2 Propagator

We expect the propagator to be defined, as in the scalar case, by the vacuum expectation value of the time-ordered product. Up to now we have dealt with the Feynman gauge. Time ordering is performed as in Eq. (3-87) by multiplying products of fields by step functions

$$TA_\mu(x)A_\nu(y) = \theta(x^0 - y^0)A_\mu(x)A_\nu(y) + \theta(y^0 - x^0)A_\nu(y)A_\mu(x) \tag{3-123}$$

This, however, may sometimes be too naive. This expression is not a priori defined at equal times. Fortunately, we do not encounter this difficulty in the present case. Inserting the expansion (3-107) and using the commutators (3-110) we find

$$\langle 0|\, TA_\mu(x)A_\nu(y)\,|0\rangle = ig_{\mu\nu}G_F(x-y)\big|_{m=0} = -ig_{\mu\nu}\int \frac{d^4k}{(2\pi)^4}\frac{e^{-ik\cdot(x-y)}}{k^2+i\varepsilon} \tag{3-124}$$

Each component of the field behaves independently and we note a change in sign between the spatial and time components. Explicitly, when $m = 0$ the Feynman propagator in x space reads

$$G_F(x) = -\int \frac{d^4k}{(2\pi)^4}\frac{e^{-ik\cdot x}}{k^2+i\varepsilon} = \frac{1}{4i\pi^2}\frac{1}{x^2 - i\varepsilon} \tag{3-125}$$

The relation of these formulas with the measuring process of electromagnetic fields is discussed in the classical papers of Bohr and Rosenfeld.

Let us make a final comment concerning an arbitrary choice of the parameter λ in Eq. (3-98). From it we derive the field equations (3-99) and equal-time commutators with a more complex structure:

$$\begin{aligned}
&[A_\mu(t,\mathbf{x}), A_\nu(t,\mathbf{y})] = 0\\
&[\dot{A}_\mu(t,\mathbf{x}), A_\nu(t,\mathbf{y})] = ig_{\mu\nu}\left(1 + \frac{1-\lambda}{\lambda}g_{\mu 0}\right)\delta^3(\mathbf{x}-\mathbf{y})\\
&[\dot{A}_i(t,\mathbf{x}), \dot{A}_j(t,\mathbf{y})] = [\dot{A}_0(t,\mathbf{x}), \dot{A}_0(t,\mathbf{y})] = 0\\
&[\dot{A}_0(t,\mathbf{x}), \dot{A}_i(t,\mathbf{y})] = i\frac{\lambda-1}{\lambda}\frac{\partial}{\partial x^i}\delta^3(\mathbf{x}-\mathbf{y})
\end{aligned} \tag{3-126}$$

The different components are therefore coupled. If we write the classical solutions of (3-99) as

$$A_\mu(x) = \int \frac{d^4k}{(2\pi)^3} e^{-ik\cdot x} \tilde{A}_\mu(k)$$

we get the homogeneous matrix condition

$$[k^2 g^\mu{}_\nu - (1-\lambda)k^\mu k_\nu]\tilde{A}^\nu(k) = 0 \tag{3-127}$$

In order to find nontrivial solutions it is necessary that the determinant $\lambda(k^2)^4$ vanishes, which implies that $\tilde{A}_\mu(k)$ has its support on the cone $k^2 = 0$. It cannot, however, be of the simple form $a_\mu(k)\delta(k^2)$; this would imply that $A_\mu(x)$ satisfies the homogeneous Klein-Gordon equation, which is not the case. Necessarily, $\tilde{A}_\mu(x)$ involves $\delta(k^2)$ and $\delta'(k^2)$. An equivalent way of looking at this is to observe that for nonvanishing k^2 we have

$$[k^2 I - (1-\lambda)k\otimes k]^{-1} = \frac{I}{k^2} + \frac{1-\lambda}{\lambda}\frac{k\otimes k}{(k^2)^2} \tag{3-128}$$

We have left to the reader the task of going through the details of the construction of the indefinite metric space of states in this case. Let us, however, mention the construction of the propagator. We have with the naive definition

$$\begin{aligned}[\Box_x g^\mu{}_\rho - (1-\lambda)\partial^\mu\partial_\rho]\langle 0|\, TA^\rho(x)A^\nu(y)\,|0\rangle &= \langle 0|\, T[\Box_x g^\mu{}_\rho - (1-\lambda)\partial^\mu\partial_\rho]A^\rho(x)A^\nu(y)\,|0\rangle \\ &+ \delta(x^0-y^0)\langle 0|\,[\dot{A}^\mu(x), A^\nu(y)]\,|0\rangle \\ &- (1-\lambda)g^{\mu 0}\delta(x^0-y^0)\langle 0|\,[\partial_\rho A^\rho(x), A^\nu(y)]\,|0\rangle\end{aligned} \tag{3-129}$$

Using the field equations and equal-time commutators we observe that noncovariant terms conspire to disappear, and we are left with

$$[\Box_x g^\mu{}_\rho - (1-\lambda)\partial^\mu\partial_\rho]\langle 0|\, TA^\rho(x)A^\nu(y)\,|0\rangle = ig^{\mu\nu}\delta^4(x-y) \tag{3-130}$$

Therefore, using (3-128) and Feynman's $i\varepsilon$ prescription,

$$\langle 0|\, TA^\mu(x)A^\nu(y)\,|0\rangle = -i\int \frac{d^4k}{(2\pi)^4} e^{-ik\cdot(x-y)}\left[\frac{g^{\mu\nu}}{k^2+i\varepsilon} + \frac{1-\lambda}{\lambda}\frac{k^\mu k^\nu}{(k^2+i\varepsilon)^2}\right] \tag{3-131}$$

It is not obvious that the denominator in the second term reads $(k^2+i\varepsilon)^2$ and not $(k^2)^2 + i\varepsilon$, for instance. This could only be checked by a complete calculation, but it will soon become clear when we study the massive vector case.

When $\lambda = 1$ we recover the Feynman gauge. When $\lambda \to 0$ we see the type of singularity which arises. The limiting $\lambda \to \infty$ case is called the Landau gauge. Physical results should not be affected by the value of λ. If, for instance, we compute an integral of the type

$$\int d^4y\, \langle 0|\, TA^\mu(x)A^\nu(y)\,|0\rangle j_\nu(y)$$

with a smooth conserved current we can easily check that it is indeed independent of the value of λ.

3-2-3 Massive Vector Field

According to Maxwell's theory photons are massless or, equivalently, electromagnetic forces are of infinite range. This vanishing mass is at the origin of the infrared catastrophe, the emission of infinitely many soft photons whenever a charged particle is accelerated. Let us investigate how a small mass would affect this behavior. Kinematically, a new (longitudinal) polarization state would be present. Couplings should therefore be such that, in the limit where the mass μ goes to zero, the ratio of the couplings of longitudinal to transverse modes should vanish.

Classical spin 1 massive particles may be described by the Proca equations for the four-vector field $A_\mu(x)$:

$$\partial_\rho F^{\rho\nu} + \mu^2 A^\nu = 0 \qquad F^{\rho\nu} = \partial^\rho A^\nu - \partial^\nu A^\rho \tag{3-132}$$

Taking the divergence of this equation we find

$$\mu^2 \partial \cdot A = 0 \tag{3-133}$$

For $\mu^2 \neq 0$, A is divergenceless and Eq. (3-132) reduces to

$$(\square + \mu^2)A_\rho = 0 \qquad \partial \cdot A = 0 \tag{3-134}$$

The vanishing of $\partial \cdot A$ ensures that one of the four degrees of freedom of A_μ is eliminated in a covariant way, so that the field quanta will indeed carry spin 1. The lagrangian leading to (3-132) is easily recognized as

$$\mathscr{L} = -\tfrac{1}{4}F^2 + \tfrac{1}{2}\mu^2 A^2 \tag{3-135}$$

Note the plus sign in front of the mass term to be compared to the analogous $-(m^2/2)\varphi^2$ term in the scalar case (3-7). This is easily recalled if we consider that space-like components, contributing negatively to the invariant square, represent the physical degrees of freedom.

After quantization the field $A_\rho(x)$ will have the following expansion:

$$A_\rho(x) = \int d\tilde{k} \sum_{\lambda=1,2,3} \left[a^{(\lambda)}(k)\varepsilon_\rho^{(\lambda)}(k)\, e^{-ik\cdot x} + a^{(\lambda)\dagger}(k)\varepsilon_\rho^{(\lambda)*}(k)\, e^{ik\cdot x}\right]\Big|_{k^0=\sqrt{k^2+\mu^2}}$$

$$[a^{(\lambda)}(k), a^{(\lambda')\dagger}(k')] = \delta_{\lambda\lambda'} 2k^0(2\pi)^3\delta^3(\mathbf{k}-\mathbf{k}') \tag{3-136}$$

The three space-like orthonormalized vectors $\varepsilon_\rho^{(\lambda)}(k)$ are simultaneously orthogonal to the time-like vector k_ρ so that if we assume them real

$$\varepsilon^{(\lambda)}(k)\cdot\varepsilon^{(\lambda')}(k) = \delta_{\lambda\lambda'} \qquad \sum_\lambda \varepsilon_\rho^{(\lambda)}(k)\varepsilon_\nu^{(\lambda)}(k) = -\left(g_{\rho\nu} - \frac{k_\rho k_\nu}{\mu^2}\right)$$

The naive construction of a time-ordered product leads to a noncovariant propagator

$$\langle 0|TA_\rho(x)A_\nu(y)|0\rangle = -i\int \frac{d^4k}{(2\pi)^4}\, e^{-ik\cdot(x-y)}\, \frac{g_{\rho\nu} - k_\rho k_\nu/\mu^2}{k^2-\mu^2+i\varepsilon} - \frac{2i}{\mu^2}\,\delta_{\rho 0}\delta_{\nu 0}\delta^4(x-y) \tag{3-137}$$

while the commutator equals

$$[A_\rho(x), A_\nu(y)] = -\int \frac{d^4k}{(2\pi)^3}\,\varepsilon(k^0)\delta(k^2-\mu^2)\, e^{-ik\cdot(x-y)}\left(g_{\rho\nu} - \frac{k_\rho k_\nu}{\mu^2}\right) \tag{3-138}$$

The noncovariant contribution to the time-ordered product is, however, a localized distribution at coinciding points. We have already noticed that time ordering left such a distribution undefined. We can therefore define a covariant propagator by simply omitting these terms without affecting its properties. We shall give a more thorough discussion of such phenomena in Chap. 5.

If we were not interested in describing the limiting zero-mass case the above description of massive spin 1 particles would be perfectly sufficient. It uses a positive norm Fock space with only physical states. But we encounter singularities as we try to set $\mu^2 = 0$. This is reflected in the fact that the limiting lagrangian $-\frac{1}{4}F^2$ is unsuitable without a gauge prescription. We seek for an equivalent description admitting the correct zero-mass limit.

The trick is to introduce, very much as we did for the electromagnetic field, an indefinite metric Fock space with an auxiliary condition. We study therefore Stueckelberg's lagrangian

$$\mathscr{L} = -\tfrac{1}{4}F^2 + \tfrac{1}{2}\mu^2 A^2 - \tfrac{1}{2}\lambda(\partial \cdot A)^2 \tag{3-139}$$

in which for $\lambda \neq 0$ the limit $\mu^2 \to 0$ is not singular. To recover a gauge-invariant theory in the limiting case it is essential that the field A_μ be coupled to a conserved current.

From (3-139) the variational equations read

$$(\Box + \mu^2)A^\rho - (1 - \lambda)\partial^\rho(\partial \cdot A) = 0 \tag{3-140}$$

If we take the divergence of both sides we now get

$$\lambda\left[\Box + \frac{\mu^2}{\lambda}\right]\partial \cdot A = 0 \tag{3-141}$$

which remains valid if on the right-hand side of (3-140) we have a conserved current instead of zero.

For $\lambda \neq 0$, $\partial \cdot A$ is a scalar field satisfying a Klein-Gordon equation with square mass $m^2 = \mu^2/\lambda$ (m^2 could be negative; however we shall assume λ to be positive for this not to happen). The field

$$A_\rho^T = A_\rho + \frac{1}{m^2}\partial_\rho \partial \cdot A = A_\rho + \frac{\lambda}{\mu^2}\partial_\rho \partial \cdot A \tag{3-142}$$

is divergenceless as a consequence of (3-141)

$$\partial^\rho A_\rho^T = 0 \tag{3-143}$$

The corresponding splitting of A into a "transverse" (spin 1) and "scalar" part is

$$A_\rho = A_\rho^T - \frac{\lambda}{\mu^2}\partial_\rho(\partial \cdot A) \tag{3-144}$$

Canonical quantization introduces four creation and annihilation operators satisfying

$$
\begin{aligned}
[a^{(\lambda)}(k), a^{(\lambda')\dagger}(k')] &= \delta_{\lambda\lambda'}(2\pi)^3 2\sqrt{\mathbf{k}^2+\mu^2}\,\delta(\mathbf{k}-\mathbf{k}') \qquad 1 \le \lambda, \lambda' \le 3 \\
[a^{(0)}(k), a^{(0)\dagger}(k')] &= -(2\pi)^3 2\sqrt{\mathbf{k}^2+m^2}\,\delta(\mathbf{k}-\mathbf{k}')
\end{aligned} \tag{3-145}
$$

and all other commutators vanish. The minus sign in the last commutator (3-145) is a signal of the indefinite metric case. Two different masses occur in the vector and scalar case. The field itself has components on the two hyperboloids:

$$
\begin{aligned}
A_\rho(x) = &\int \frac{d^3k}{2(2\pi)^3\sqrt{\mathbf{k}^2+\mu^2}} \sum_{\lambda=1}^{3} [a^{(\lambda)}(k)\varepsilon_\rho^{(\lambda)}(k)\, e^{-ik\cdot x} + a^{(\lambda)\dagger}(k)\varepsilon_\rho^{(\lambda)*}(k)\, e^{ik\cdot x}] \\
&+ \int \frac{d^3k}{2(2\pi)^3\sqrt{\mathbf{k}^2+m^2}} \frac{k_\rho}{\mu} [a^{(0)}(k)\, e^{-ik\cdot x} + a^{(0)\dagger}(k)\, e^{ik\cdot x}]
\end{aligned} \tag{3-146}
$$

As before $\varepsilon^{(\lambda)}(k)$, for $\lambda = 1, 2, 3$ are three orthonormal space-like vectors orthogonal to $k(k^2 = \mu^2)$. We may easily verify that Eqs. (3-143) and (3-140) are satisfied by this expression while the corresponding covariant propagator is

$$
\begin{aligned}
\langle 0|\, TA_\rho(x)A_\nu(y)\,|0\rangle &= -i \int \frac{d^4k}{(2\pi)^4} e^{-ik\cdot(x-y)} \left(\frac{g_{\rho\nu} - k_\rho k_\nu/\mu^2}{k^2-\mu^2+i\varepsilon} + \frac{k_\rho k_\nu/\mu^2}{k^2-m^2+i\varepsilon} \right) \\
m^2 &= \frac{\mu^2}{\lambda}
\end{aligned} \tag{3-147}
$$

and the commutator reads

$$
\begin{aligned}
[A_\rho(x), A_\nu(y)] = -\int \frac{d^3k}{(2\pi)^3} \Bigg\{ &\left[\frac{1}{2k_0} (e^{-ik\cdot(x-y)} - e^{ik\cdot(x-y)}) \left(g_{\rho\nu} - \frac{k_\rho k_\nu}{\mu^2} \right) \right]_{k^0=\sqrt{\mathbf{k}^2+\mu^2}} \\
&+ \left[\frac{1}{2k_0} (e^{-ik\cdot(x-y)} - e^{ik\cdot(x-y)}) \frac{k_\rho k_\nu}{\mu^2} \right]_{k^0=\sqrt{\mathbf{k}^2+m^2}} \Bigg\}
\end{aligned} \tag{3-148}
$$

The reader is invited to study the limits $\mu \to 0$ and $\lambda \to 0$.

(*a*) When $\lambda \to 0$, $\mu \neq 0$, show that we recover Proca's theory.

(*b*) If $\lambda \neq 0$ and $\mu \to 0$, m goes to zero. Show that in this limit the Green functions tend to the values given in the preceding subsection.

In particular, since

$$
\frac{g_{\rho\nu} - k_\rho k_\nu/\mu^2}{k^2-\mu^2+i\varepsilon} + \frac{k_\rho k_\nu/\mu^2}{k^2-m^2+i\varepsilon} = \frac{g_{\rho\nu}}{k^2-\mu^2+i\varepsilon} + \frac{1-\lambda}{\lambda} \frac{k_\rho k_\nu}{(k^2-\mu^2/\lambda+i\varepsilon)(k^2-\mu^2+i\varepsilon)} \tag{3-149}
$$

We see that the double-pole term tends to $1/(k^2+i\varepsilon)^2$ as anticipated in Eq. (3-131). We shall return to this zero-mass limit in the following, in particular in Chap. 4. It should also be noticed that in k space the propagators in the Stueckelberg formalism behave as $1/k^2$ for large k, while this is not the case for the massive Proca propagator (3-137). As a supplementary exercise one may check the covariance of the theory and construct the generators P_μ and $M_{\mu\nu}$ of the Poincaré group. We have not offered here a very thorough description of the kinematical properties of photons. On this point the reader is referred to the works quoted in the notes. Rather, we had in mind to pave the way to recent developments in gauge theories, to be considered in Chap. 12.

We close this subsection by describing the principle of a method allowing us to set an upper limit to the photon mass using terrestrial measurements. This is the so-called constant field method of Schrödinger. Assume as a first approximation that the earth can be simulated by a perfect magnetic point dipole **M**. The corresponding localized conserved current is such that

$$\mathbf{M} = \int d^3x\, \mathbf{x} \times \mathbf{j}(\mathbf{x})$$

and can be written $\mathbf{j} = -\frac{1}{2}\mathbf{M} \times \nabla\delta^3(\mathbf{x})$. In the static case, let us derive from Proca's equations the corresponding vector potential such that $\operatorname{div} \mathbf{A} = 0$, a condition compatible with these equations. It follows that

$$(-\Delta + \mu^2)\mathbf{A} = \mathbf{j} = -\frac{1}{2}\mathbf{M} \times \nabla\delta^3(\mathbf{x}) = -\frac{i}{2}\int \frac{d^3k}{(2\pi)^3}\, e^{i\mathbf{k}\cdot\mathbf{x}}\,\mathbf{M} \times \mathbf{k}$$

a solution of which is

$$\mathbf{A}(\mathbf{x}) = -\frac{i}{2}\int \frac{d^3k}{(2\pi)^3}\,\mathbf{M} \times \mathbf{k}\,\frac{e^{i\mathbf{k}\cdot\mathbf{x}}}{\mathbf{k}^2 + \mu^2} = -\mathbf{M} \times \nabla\left(\frac{e^{-\mu r}}{8\pi r}\right)$$

The corresponding magnetic induction is

$$\mathbf{B} = \operatorname{curl} \mathbf{A} = -\nabla \times (\mathbf{M} \times \nabla)\frac{e^{-\mu r}}{8\pi r} = [(\mathbf{M}\cdot\nabla)\nabla - \mathbf{M}\Delta]\,\frac{e^{-\mu r}}{8\pi r}$$

Let us take the z axis along **M** with unit vector $\hat{z}$, so that **B** has the form

$$\mathbf{B} = \frac{e^{-\mu r}}{8\pi r^3}|\mathbf{M}|\{[\hat{r}(\hat{r}\cdot\hat{z}) - \tfrac{1}{3}\hat{z}](\mu^2 r^2 + 3\mu r + 3) - \tfrac{2}{3}\hat{z}\mu^2 r^2\}$$

The field has been split into two parts. In the first $[\hat{r}(\hat{r}\cdot\hat{z}) - \frac{1}{3}\hat{z}]$ has been factored out, corresponding to the angular distribution of the magnetic field in the limit $\mu \to 0$. At constant distance r there appears a uniform added term antiparallel to the earth's dipole. A careful scan of the angular distribution of the magnetic field allows us to set an upper limit to the mass μ of the order of 4×10^{-48} g (3×10^{-15} eV $\sim 10^{-10}$ cm^{-1}). More recent measurements using a different method have improved this result by an order of magnitude.

3-2-4 Vacuum Fluctuations

It is instructive to study simple effects arising from the field quantization before embarking on the coupled nonlinear problems. Especially interesting are those which do not seem to rely on the particle interpretation—in the present case, the photon aspect. We give a schematic description of two such situations which have to do with the observability of differences in vacuum fluctuations. We have encountered such phenomena when presenting Welton's interpretation of the Lamb shift in Chap. 2.

We can take into account simple macroscopic sources by modifying boundary conditions on the field which was considered up to now in free space. This procedure is to some extent unsatisfactory since it does not describe the microscopic mechanism responsible for these boundary conditions. But it is suitable for elementary calculations. The original observation of Casimir (1948) is that, in the vacuum, the electromagnetic field does not really vanish but rather fluctuates (compare Sec. 3-1-2). If we introduce macroscopic bodies—even uncharged—some work will be necessary to enforce appropriate boundary conditions. Intuition on the sign of this effect is lacking, so that work here is meant in some algebraic sense, meaning the difference in zero point energies between the two configurations. We had originally disregarded the (infinite) contribution $\Sigma \frac{1}{2}\hbar\omega_\alpha$ to the hamiltonian by arguing that it was unobservable. However, its variation can be measured.

Let us illustrate this point for the simple configuration of two large parallel perfectly conducting plates as considered first by Casimir. Of course, we can study different geometries and different materials with similar results (except perhaps for crucial signs). We idealize the plates by two large parallel squares of size L at a distance a (see Fig. 3-1) with $a \ll L$. Consider the energy per unit surface of the conductor with respect to the vacuum. Its derivative will be a force per unit surface with dimension $ML^{-1}T^{-2}$ (where M is mass, L length, and T time). The only quantities entering the problem are $\hbar$, c and the separation a (the boundary conditions $\mathbf{E}$ perpendicular and $\mathbf{B}$ parallel to the plate at the interface do not introduce any dimensional quantity). Of course, the effect is proportional to $\hbar$, as is the zero point energy. The force per unit surface is therefore proportional to $\hbar c/a^4$, the only quantity with the required dimension. We shall see that it is attractive.

Consider the modes inside the volume $L^2 a$, where $L \gg a$ and we ignore the edge contributions. As we know, only transverse modes contribute to the energy. If the component k_z perpendicular to the plates is different from zero it can only take discrete values $k_z = n\pi/a$ $(n = 1, 2, \ldots)$ to allow for the nodes on the plates and there are two polarization states. If, however, k_z vanishes only one mode survives so that the zero point energy of the configuration is

$$E = \sum \tfrac{1}{2}\hbar\omega_\alpha = \frac{\hbar c}{2}\sum_\alpha |\mathbf{k}_\alpha| = \frac{\hbar c}{2}\int L^2 \frac{d^2k_\parallel}{(2\pi)^2}\left[|\mathbf{k}_\parallel| + 2\sum_{n=1}^{\infty}\left(\mathbf{k}_\parallel^2 + \frac{n^2\pi^2}{a^2}\right)^{1/2}\right]$$

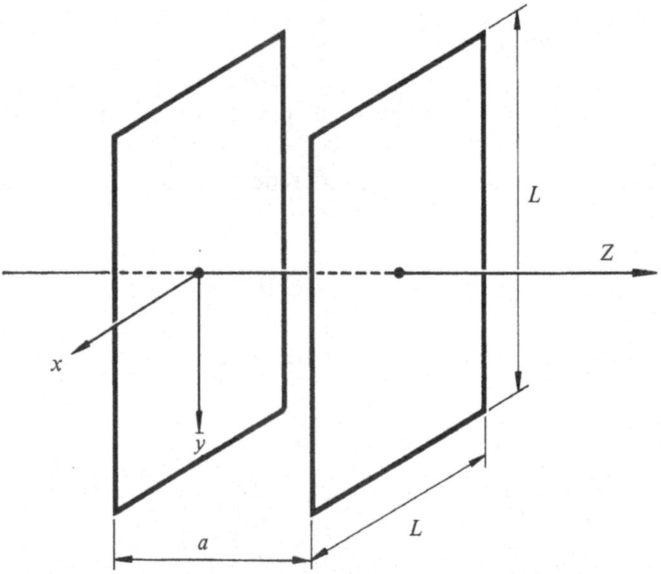

Figure 3-1 Casimir effect between two parallel plates.

As it stands this expression is, of course, meaningless, being infinite. But we must subtract the free value which contributes in this same volume a quantity

$$E_0 = \frac{\hbar c}{2} \int \frac{L^2 d^2 k_\parallel}{(2\pi)^2} \int_{-\infty}^{+\infty} \frac{a\,dk_z}{2\pi}\, 2\sqrt{\mathbf{k}_\parallel^2 + k_z^2}$$

$$= \frac{\hbar c}{2} \int \frac{L^2 d^2 k_\parallel}{(2\pi)^2} \int_0^\infty dn\, 2\sqrt{\mathbf{k}_\parallel^2 + n^2\pi^2/a^2}$$

Therefore the energy per unit surface is

$$\mathscr{E} = \frac{E - E_0}{L^2} = \frac{\hbar c}{2\pi} \int_0^\infty k\,dk \left(\frac{k}{2} + \sum_{n=1}^{\infty} \sqrt{k^2 + n^2\pi^2/a^2} - \int_0^\infty dn\, \sqrt{k^2 + n^2\pi^2/a^2} \right)$$

This quantity is apparently still not defined due to ultraviolet (large k) divergences. However, for wavelengths shorter than the atomic size it is unrealistic to use a perfect conductor approximation. Let us therefore introduce in the above integral a smooth cutoff function $f(k)$ equal to unity for $k \lesssim k_m$ and vanishing for $k \gg k_m$ where k_m is of the order of the inverse atomic size. Set $u = a^2k^2/\pi$; then

$$\mathscr{E} = \hbar c \frac{\pi^2}{4a^3} \int_0^\infty du \left[\frac{\sqrt{u}}{2} f\left(\frac{\pi}{a}\sqrt{u}\right) + \sum_1^\infty \sqrt{u + n^2}\, f\left(\frac{\pi}{a}\sqrt{u + n^2}\right) \right.$$
$$\left. - \int_0^\infty dn\, \sqrt{u + n^2}\, f\left(\frac{\pi}{a}\sqrt{u + n^2}\right) \right]$$

$$\mathscr{E} = \hbar c \frac{\pi}{4a^3} \left[\tfrac{1}{2}F(0) + F(1) + F(2) + \cdots - \int_0^\infty dn\, F(n) \right]$$

Here we have defined

$$F(n) = \int_0^\infty du\, \sqrt{u + n^2}\, f\left(\frac{\pi}{a}\sqrt{u + n^2}\right)$$

The interchange of sums and integrals was justified due to the absolute convergence in the presence of the cutoff function. As $n \to \infty$, $F(n) \to 0$. We can use the Euler-MacLaurin formula to compute the difference between the sum and integral occurring in the above bracket:

$$\tfrac{1}{2}F(0) + F(1) + F(2) + \cdots - \int_0^\infty dn\, F(n) = -\frac{1}{2!} B_2 F'(0) - \frac{1}{4!} B_4 F'''(0) + \cdots$$

The Bernoulli numbers B_ν are defined through the series

$$\frac{y}{e^y - 1} = \sum_{\nu=0}^{\infty} B_\nu \frac{y^\nu}{\nu!}$$

and $B_2 = \frac{1}{6}$, $B_4 = -\frac{1}{30}, \ldots$. We have

$$F(n) = \int_{n^2}^\infty du\, \sqrt{u}\, f\left(\frac{\pi\sqrt{u}}{a}\right) \qquad F'(n) = -2n^2 f\left(\frac{\pi n}{a}\right)$$

We assume that $f(0) = 1$, while all its derivatives vanish at the origin, so that $F'(0) = 0$, $F'''(0) = -4$, and higher derivatives of F are equal to zero. All reference to the cutoff has therefore disappeared from the final result

$$\mathscr{E} = \frac{\hbar c \pi^2}{a^3}\frac{B_4}{4!} = -\frac{\pi^2}{720}\frac{\hbar c}{a^3}$$

The force per unit area $\mathscr{F}$ reads

$$\mathscr{F} = -\frac{\pi^2}{240}\frac{\hbar c}{a^4} = -\frac{0.013}{(a_{\mu\mathrm{m}})^4}\,\mathrm{dyn/cm}^2 \tag{3-150}$$

and its sign corresponds to attraction.

This very tiny force has been demonstrated experimentally by Sparnay (1958), who was able to observe both its magnitude and dependence on the interplate distance!

The above derivation may be criticized on account of the fact that we have seemingly disregarded the effects outside the plates. In the present case they turn out to cancel exactly. The lesson is that vacuum fluctuations manifest themselves under very different circumstances than those encountered in particle creation or absorption. By considering various types of bodies influencing the vacuum configuration we may give an interesting interpretation of the forces acting on them. This is one line of thought that may be kept in mind and in which we may recognize some of the origins of Schwinger's source theory approach to quantum field phenomena.

As another example let us give a brief discussion of Van der Waals forces among neutral atoms or molecules. It can be presented by studying the field fluctuations in the presence of the two systems very much as above. After all, we should be able to give a microscopic description of the Casimir effect among macroscopic bodies using the residual forces among constituents. Rather than actually doing that we shall follow a presentation due to Feinberg and Sucher. A more detailed treatment is given in Chap. 7. The classical energy density of the electromagnetic field is $(\mathbf{E}^2 + \mathbf{B}^2)/2$, as we recall from Chap. 1. A neutral system may interact with an electromagnetic field. Let $\alpha_E(\alpha_B)$ be the static electric (magnetic) susceptibility. In a static electric field if ρ denotes the charge density, $\int d^3x\, \rho(x) = 0$ because of neutrality. The polarization is defined as $\mathbf{P} = \int d^3x\, \rho(x)\mathbf{x}$ and the corresponding contribution to the energy due to a change in the electric field $\delta\mathbf{E} = -\nabla\delta V$ is $\int d^3x\, \rho\delta V = -\int d^3x\, \rho(x)\mathbf{x}\cdot\delta\mathbf{E} = -\mathbf{P}\cdot\delta\mathbf{E}$ if $\delta\mathbf{E}$ is almost constant over the extent of the system. The susceptibility is such that $\mathbf{P} = \alpha_E\mathbf{E}$ for weak fields; hence under a small change $\delta\mathbf{E}$, the variation of the interaction energy is $-\alpha_E\mathbf{E}\cdot\delta\mathbf{E} = \delta[-\alpha_E(\mathbf{E}^2/2)]$. A similar reasoning yields the magnetic contribution, so that for the interaction energy we have

$$E_{\mathrm{int}} = -\tfrac{1}{2}(\alpha_E\mathbf{E}^2 + \alpha_B\mathbf{B}^2)$$

We attempt a relativistic phenomenological quantum generalization of this result. We want to write an interaction part of the lagrangian such that when integrated over three-space it reproduces in a static field and in the nonrelativistic limit, the negative of the above expression. The neutral system may be described by an hermitian scalar field φ, and the only Lorentz scalar quadratic in $F_{\mu\nu}$ and in φ having the required property is

$$-\mathscr{L}_{\mathrm{int}} = g_1\partial_\alpha\varphi\partial_\beta\varphi F^{\alpha\rho}F^{\beta}{}_{\rho} + g_2\varphi^2F^2$$

with g_1 and g_2 related to α_E and α_B as follows:

$$g_1 = \frac{1}{2m}(\alpha_E + \alpha_B) \qquad g_2 = -\frac{m}{4}\alpha_B$$

These relations and also the desire of having $\mathscr{L}_{\text{int}}$ quadratic in φ (instead of linear, for instance) stem from the fact that, in order to make the comparison with the nonrelativistic limit, we have to take averages in a one-particle state. In this nonrelativistic limit $\langle 1|\int d^3x\colon \varphi^2(x)\colon|1\rangle \simeq 1/m\langle 1|1\rangle$ and $\langle 1|\int d^3x\colon(\partial_0\varphi)^2\colon|1\rangle \simeq m\langle 1|1\rangle$. Since $-\mathscr{L}_{\text{int}} = \mathscr{H}_{\text{int}}$, the above expression enables us to obtain an asymptotic form of the static potential between two neutral polarizable systems. The reasoning is patterned after a similar derivation of the Coulomb potential between charged systems which goes as follows. The interaction energy of charge 1 with the field is written $\int d^3x\, j^{(1)}(x)\cdot A(x)$ but $A(x)$ is generated by charge 2 as $A(x) = \int d^4y\, G(x-y)j^{(2)}(y)$. Since $j^{(2)}$ is assumed to be time independent we may first integrate over y^0 using

$$G_F(x-y) = \frac{1}{4i\pi^2}\frac{1}{x_0^2 - \mathbf{x}^2 - i\varepsilon}$$

This leads to

$$A = \int d^3y \frac{1}{4\pi}\frac{1}{|\mathbf{x}-\mathbf{y}|} j^{(2)}(\mathbf{y})$$

and hence to the usual Coulomb interaction

$$E_{\text{int}} = \int d^3x\, d^3y\, j^{(1)}(\mathbf{x}) \frac{1}{4\pi|\mathbf{x}-\mathbf{y}|} j^{(2)}(\mathbf{y})$$

Let us return to our case. We are obviously interested in Green functions of quadratic operators such as F^2 (occurring as the coefficient of g_2), for instance. If $G_{\rho\sigma,\alpha\beta}(x-y)$ stands for $\langle 0|\, TF_{\rho\sigma}(x)F_{\alpha\beta}(y)\,|0\rangle$, a Green function for F^2 will be proportional to the square of the above expression, that is, $G_{\rho\sigma,\alpha\beta}(x-y)G^{\rho\sigma,\alpha\beta}(x-y)$. This distribution is typically of the form $1/(x^2 - i\varepsilon)^4$. The contribution arising from g_1 has a similar structure. Consequently, we expect a potential energy proportional to

$$V(|\mathbf{x}-\mathbf{y}|) \sim \int dy_0 \frac{1}{[(x_0-y_0)^2 - (\mathbf{x}-\mathbf{y})^2 - i\varepsilon]^4} \sim \frac{\text{constant}}{|\mathbf{x}-\mathbf{y}|^7}$$

This result originally derived by Casimir and Polder is at variance with the nonrelativistic treatment of Van der Waals forces, the outcome of which is a potential with a decrease of $|\mathbf{x}-\mathbf{y}|^{-6}$ at a large distance. We shall derive in Chap. 7 the full expression given by Feinberg and Sucher for the potential at large separations:

$$V_{ab}^{\text{asympt}}(|\mathbf{x}-\mathbf{y}|) = -\frac{1}{(4\pi)^3|\mathbf{x}-\mathbf{y}|^7}[23(\alpha_E^a\alpha_E^b + \alpha_B^a\alpha_B^b) - 7(\alpha_E^a\alpha_B^b + \alpha_E^b\alpha_B^a)]$$

and comment on its domain of validity, thereby explaining the discrepancy with the $|\mathbf{x}-\mathbf{y}|^{-6}$ nonrelativistic law.

3-3 DIRAC FIELD AND EXCLUSION PRINCIPLE

In the preceding chapter we have already introduced Dirac particles and presented the arguments requiring such a theory to be described as a many-body theory.

Canonical quantization has led up to now to an interpretation in terms of Bose-Einstein particles, but experiment shows undoubtedly that spin $\frac{1}{2}$ particles such as electrons and nucleons satisfy Pauli's exclusion principle and obey Fermi-Dirac statistics. We have thus to modify the rules of the game when dealing with half-integer spin particles.

3-3-1 Anticommutators

We have to consider the Dirac equation on a similar footing to Maxwell's equations. Both are classical field equations. We have therefore to introduce a lagrangian and conjugate momenta for the complex fields ψ and $\psi^\dagger$ (or $\overline{\psi} = \psi^\dagger \gamma^0$). In the free case the equations

$$(i\gamma \cdot \overrightarrow{\partial} - m)\psi = 0 \qquad \overline{\psi}(i\gamma \cdot \overleftarrow{\partial} + m) = 0 \tag{3-151}$$

follow from a variation of an action using as density the first-order lagrangian

$$\mathscr{L} = \frac{i}{2}[\overline{\psi}\gamma^\mu(\partial_\mu \psi) - (\partial_\mu \overline{\psi})\gamma^\mu \psi] - m\overline{\psi}\psi \tag{3-152}$$

We have, for instance,

$$\frac{\partial \mathscr{L}}{\partial \overline{\psi}} - \partial_\mu \frac{\partial \mathscr{L}}{\partial(\partial_\mu \overline{\psi})} = \frac{i}{2}\gamma^\mu \partial_\mu \psi - m\psi + \frac{i}{2}\partial_\mu(\gamma^\mu \psi) = 0$$

and similarly when varying with respect to ψ. When ψ and $\overline{\psi}$ satisfy the field equations the lagrangian (3-152) vanishes. Since canonical quantization does not work, we shall rather use the following procedure. Assume ψ and $\overline{\psi}$ to be quantized operators acting in some Hilbert space. Let us inquire under which conditions the Poincaré group generators, built according to Noether's theorem, transform the fields as required by the transformation laws studied in Chap. 2. Since these generators, as well as all observable quantities, are expressed in terms of the dynamical variables ψ and $\overline{\psi}$, this will ensure the covariance of the theory and we will then verify that these observables do commute at space-like separations.

The energy momentum tensor density is obtained from (3-152), using the equations of motion as

$$\Theta^{\mu\nu} = \partial^\nu \overline{\psi} \frac{\partial \mathscr{L}}{\partial(\partial_\mu \overline{\psi})} + \frac{\partial \mathscr{L}}{\partial(\partial_\mu \psi)} \partial^\nu \psi - g^{\mu\nu}\mathscr{L} = \frac{i}{2}\overline{\psi}\gamma^\mu \overleftrightarrow{\partial}^\nu \psi \tag{3-153}$$

For lack of any prescription we have arbitrarily ordered this expression by setting $\overline{\psi}$ to the left of ψ. It is clear that we proceed heuristically. The tensor $\Theta^{\mu\nu}$ does not seem symmetric either. Let us therefore study the generalized angular momentum density taking correctly into account internal variables (i.e., spinor indices). Consider an infinitesimal homogeneous Lorentz transformation under which we must have

$$\psi(x) \to \psi'(x) = S(\Lambda)\psi(\Lambda^{-1}x) = \psi(x) + \delta\psi(x)$$

$$\Lambda = I - \frac{i}{2}\,\delta\omega_{\alpha\beta}M^{\alpha\beta} \qquad M^{\alpha\beta}{}_{\mu\nu} = -M^{\beta\alpha}{}_{\mu\nu} = i(g^{\alpha}{}_{\mu}g^{\beta}{}_{\nu} - g^{\alpha}{}_{\nu}g^{\beta}{}_{\mu})$$

$$S(\Lambda) = I - \frac{i}{4}\,\delta\omega_{\alpha\beta}\sigma^{\alpha\beta} \tag{3-154}$$

$$\delta\psi(x) = -\frac{i}{2}\,\delta\omega_{\alpha\beta}\left[\frac{\sigma^{\alpha\beta}}{2} + i(x^{\alpha}\partial^{\beta} - x^{\beta}\partial^{\alpha})\right]\psi(x)$$

$$= -\frac{i}{2}\,\delta\omega_{\alpha\beta}L^{\alpha\beta}\psi(x)$$

The operator $L^{\alpha\beta} = \frac{1}{2}\sigma^{\alpha\beta} + i(x^{\alpha}\partial^{\beta} - x^{\beta}\partial^{\alpha})$ acts on the spinor indices as well as the configuration variables. Correspondingly,

$$\delta\bar{\psi}(x) = \delta\psi(x)^{\dagger}\gamma^{0} = \frac{i}{2}\,\bar{\psi}(x)\bar{L}^{\alpha\beta}\delta\omega_{\alpha\beta}$$

with $$\bar{L}^{\alpha\beta} = \gamma^{0}L^{\alpha\beta\dagger}\gamma^{0} = \frac{\sigma^{\alpha\beta}}{2} - i(x^{\alpha}\overleftarrow{\partial}^{\beta} - x^{\beta}\overleftarrow{\partial}^{\alpha})$$

If $\delta\omega^{\alpha\beta}$ is x independent we verify that the action is invariant. Thus when $\delta\omega^{\alpha\beta}$ varies with x we find as a variation around the stationary point

$$\delta I = \int d^4x\,\tfrac{1}{2}\bar{\psi}\left[\left\{\gamma^{\mu}, \frac{\sigma^{\alpha\beta}}{2}\right\} + i(x^{\alpha}\overleftrightarrow{\partial}^{\beta} - x^{\beta}\overleftrightarrow{\partial}^{\alpha})\gamma^{\mu}\right]\psi\,\frac{\partial_{\mu}\delta\omega_{\alpha\beta}}{2}$$

$$= -\int d^4x\,(\partial_{\mu}J^{\mu,\alpha\beta})\,\frac{\delta\omega_{\alpha\beta}}{2} = 0$$

A candidate for the generalized angular momentum tensor density is therefore

$$J^{\mu,\alpha\beta} = \tfrac{1}{2}\bar{\psi}\left[i(x^{\alpha}\overleftrightarrow{\partial}^{\beta} - x^{\beta}\overleftrightarrow{\partial}^{\alpha})\gamma^{\mu} + \left\{\gamma^{\mu}, \frac{\sigma^{\alpha\beta}}{2}\right\}\right]\psi$$

$$= x^{\alpha}\Theta^{\mu\beta} - x^{\beta}\Theta^{\mu\alpha} + \tfrac{1}{2}\bar{\psi}\left\{\gamma^{\mu}, \frac{\sigma^{\alpha\beta}}{2}\right\}\psi \tag{3-155}$$

We can verify directly that $\partial_{\mu}J^{\mu,\alpha\beta} = 0$. This tensor has an orbital and a spin part. The above construction gives a similar role to ψ and $\bar{\psi}$. The space integrals of $\Theta^{0\mu}$ and $J^{0,\alpha\beta}$ will yield the quantum generators of the group when we find a commutation prescription for the fields. We can also add that the invariance under phase transformations $\psi \to e^{i\alpha}\psi$, $\bar{\psi} \to e^{-i\alpha}\bar{\psi}$ leads to a conserved current

$$J^{\mu} = \bar{\psi}\gamma^{\mu}\psi \tag{3-156}$$

which was used in the one-particle theory of Chap. 2 to normalize the states when integrating J^0. It will be necessary to give a more general interpretation in the quantized picture.

To implement our program let us now expand the operators $\psi(x)$ and $\bar{\psi}(x)$ in terms of the c-number plane wave solutions of the Dirac equation, with operator-valued amplitudes b, $b^\dagger$, d, and $d^\dagger$:

$$
\begin{aligned}
\psi(x) &= \int \frac{d^3k}{(2\pi)^3} \frac{m}{k_0} \sum_{\alpha=1,2} [b_\alpha(k) u^{(\alpha)}(k)\, e^{-ik\cdot x} + d_\alpha^\dagger(k) v^{(\alpha)}(k)\, e^{ik\cdot x}] \\
\bar{\psi}(x) &= \int \frac{d^3k}{(2\pi)^3} \frac{m}{k_0} \sum_{\alpha=1,2} [b_\alpha^\dagger(k) \bar{u}^{(\alpha)}(k)\, e^{ik\cdot x} + d_\alpha(k) \bar{v}^{(\alpha)}(k)\, e^{-ik\cdot x}]
\end{aligned}
\tag{3-157}
$$

The spinors u and v were given in the previous chapter, Eqs. (2-37) and (2-38). The operators b and d must satisfy commutation rules such that

$$
\psi(x+a) = e^{iP\cdot a}\psi(x)\, e^{-iP\cdot a}
$$

or in differential form

$$
\partial_\mu \psi(x) = i[P_\mu, \psi(x)] \qquad \partial_\mu \bar{\psi}(x) = i[P_\mu, \bar{\psi}(x)] \tag{3-158}
$$

We express P_μ using the decomposition (3-157) with the result that

$$
P_\mu = \int d^3x\, \Theta^0{}_\mu = \int \frac{d^3k}{(2\pi)^3} \frac{m}{k^0} k_\mu \sum_\alpha [b_\alpha^\dagger(k) b_\alpha(k) - d_\alpha(k) d_\alpha^\dagger(k)] \tag{3-159}
$$

from which we have to subtract the vacuum contribution. We read from (3-159) that if the vacuum is defined in such a way that $b_\alpha(k)|0\rangle = d_\alpha(k)|0\rangle = 0$, and if we were to quantize according to commutators, b particles and d particles would contribute with opposite signs to the energy. The theory would not admit a stable ground state.

From (3-158) translational invariance is satisfied provided

$$
\begin{aligned}
[P_\mu, b_\alpha(k)] &= -k_\mu b_\alpha(k) \qquad & [P_\mu, d_\alpha(k)] &= -k_\mu d_\alpha(k) \\
[P_\mu, b_\alpha^\dagger(k)] &= k_\mu b_\alpha^\dagger(k) \qquad & [P_\mu, d_\alpha^\dagger(k)] &= k_\mu d_\alpha^\dagger(k)
\end{aligned}
\tag{3-160}
$$

Replacing P_μ by its expansion, this is equivalent to

$$
\left[\sum_\beta [b_\beta^\dagger(q) b_\beta(q) - d_\beta(q) d_\beta^\dagger(q)], b_\alpha(k)\right] = -(2\pi)^3 \frac{k^0}{m} \delta^3(\mathbf{k}-\mathbf{q}) b_\alpha(k)
$$

and three analogous relations. If we assume that

$$
[d_\beta^\dagger(q) d_\beta(q), b_\alpha(k)] = 0
$$

this condition can be written

$$
\sum_\beta [b_\beta^\dagger(q)\{b_\beta(q), b_\alpha(k)\} - \{b_\beta^\dagger(q), b_\alpha(k)\} b_\beta(q)] = -(2\pi)^3 \frac{k^0}{m} \delta^3(\mathbf{k}-\mathbf{q}) b_\alpha(k)
$$

Therefore, we can insure the correct interpretation of energy and momentum using, as an alternative, anticommutators (the above curly brackets) instead of commutators between the fundamental creation and annihilation operators $b^\dagger$, $d^\dagger$, b, and d. This solves at once the stability question mentioned above since, after vacuum energy subtraction, we are left with only positive contributions to the energy. We set, therefore,

$$\begin{aligned}\{b_\alpha(k), b_\beta^\dagger(q)\} &= (2\pi)^3 \frac{k^0}{m} \delta^3(\mathbf{k}-\mathbf{q})\delta_{\alpha\beta} \\ \{d_\alpha(k), d_\beta^\dagger(q)\} &= (2\pi)^3 \frac{k^0}{m} \delta^3(\mathbf{k}-\mathbf{q})\delta_{\alpha\beta}\end{aligned} \tag{3-161}$$

and all other anticommutators vanish. As a consequence,

$$\{\psi_\xi(t, \mathbf{x}), \psi_\eta^\dagger(t, \mathbf{y})\} = \delta^3(\mathbf{x}-\mathbf{y})\delta_{\xi\eta} \tag{3-162}$$

We may then easily verify that conditions (3-160) are fulfilled, together with the analogous ones stemming from the requirement of covariance under homogeneous transformations and involving the commutators of the generators $J^{\alpha\beta} = \int d^3x\, J^{0,\alpha\beta}$ with the field. The interpretation of (3-160) is that $b^\dagger(k)$ and $d^\dagger(k)$ create [$b(k)$ and $d(k)$ destroy] a four-momentum k^μ.

Wick products can be generalized to Fermi fields. When reordering the creation operators to the left of the annihilation ones, a sign corresponding to the parity of the permutation must be introduced. The correct definition of total energy momentum is therefore

$$\begin{aligned}P^\mu &= \int \frac{d^3k}{(2\pi)^3} \frac{m}{k^0} k^\mu \sum_{\alpha=1,2} :b_\alpha^\dagger(k) b_\alpha(k) - d_\alpha(k) d_\alpha^\dagger(k): \\ &= \int \frac{d^3k}{(2\pi)^3} \frac{m}{k^0} k^\mu \sum_{\alpha=1,2} [b_\alpha^\dagger(k) b_\alpha(k) + d_\alpha^\dagger(k) d_\alpha(k)]\end{aligned} \tag{3-163}$$

leading to a sum of positive contributions to the energy of a quantum state. Let us elaborate the structure of the corresponding Fock space.

3-3-2 Fock Space for Fermions

We focus first on one-particle states. The necessary smearing in momentum space is implicitly understood. For a given four-momentum we observe that we now have a fourfold degeneracy. Denote the corresponding states $|a\rangle$ $(a = 1, 2, 3, 4)$:

$$|1\rangle = b_1^\dagger(k)|0\rangle \qquad |2\rangle = b_2^\dagger(k)|0\rangle \qquad |3\rangle = d_1^\dagger(k)|0\rangle \qquad |4\rangle = d_2^\dagger(k)|0\rangle$$

They satisfy $P_\mu|a\rangle = k_\mu|a\rangle$. We look for observables commuting with P_μ in order to distinguish these states. One of these observables is the charge (normalized to ± 1 for a one-particle state)

$$\begin{aligned}Q &= \int d^3x\, j^0(x) = \int d^3x :\psi^\dagger(x)\psi(x): \\ &= \int d\tilde{k} \sum_{\alpha=1,2} [b_\alpha^\dagger(k) b_\alpha(k) - d_\alpha^\dagger(k) d_\alpha(k)]\end{aligned} \tag{3-164}$$

When dealing with Dirac particles the notation $d\tilde{k}$ will mean $[d^3k/(2\pi)^3](m/k^0)$ instead of $d^3k/(2\pi)^3 2k_0$ as it is for bosons. Since the vacuum has zero charge and

$$[Q, b_\alpha^\dagger(k)] = b_\alpha^\dagger(k) \qquad [Q, d_\alpha^\dagger(k)] = -d_\alpha^\dagger(k)$$

we have

$$Q\,|a\rangle = \begin{cases} +\,|a\rangle & a = 1, 2 \\ -\,|a\rangle & a = 3, 4 \end{cases}$$

In contradistinction to the "classical" case where we attempted to interpret (3-164) as a positive square norm it is seen that in the quantum case this is not so. The use of anticommutators has totally reversed the situation; energy is now positive and charge an indefinite quantity.

Dirac's quantized theory therefore describes particles of two types. In electrodynamics eQ will be the electric charge and ψ will annihilate electrons and create positrons. Since $[Q, P_\mu] = 0$, Q is time independent, and $[Q, \psi(x)] = -\psi(x)$ and $[Q, \bar{\psi}(x)] = \bar{\psi}(x)$.

To characterize the one-particle states completely it remains to introduce spin. In analogy with (3-158) we have for Lorentz transformations

$$[J^{\mu\nu}, \psi(x)] = -\left[i(x^\mu\partial^\nu - x^\nu\partial^\mu) + \frac{\sigma^{\mu\nu}}{2}\right]\psi(x) \tag{3-165}$$

Since

$$b_\alpha(k) = \int d^3x\, \bar{u}^{(\alpha)}(k)\, e^{ik\cdot x}\gamma^0\psi(x)$$

$$d_\alpha^\dagger(k) = \int d^3x\, \bar{v}^{(\alpha)}(k)\, e^{-ik\cdot x}\gamma^0\psi(x)$$

we find

$$\begin{aligned} [J^{\mu\nu}, b_\alpha^\dagger(k)] &= \int d^3x\, \bar{\psi}(x)\left[\frac{1}{i}(x^\mu \overleftarrow{\partial}^\nu - x^\nu \overleftarrow{\partial}^\mu) + \frac{\sigma^{\mu\nu}}{2}\right]\gamma^0 u^{(\alpha)}(k)\, e^{-ik\cdot x} \\ [J^{\mu\nu}, d_\alpha^\dagger(k)] &= -\int d^3x\, \bar{v}^{(\alpha)}(k)\, e^{-ik\cdot x}\gamma^0\left[i(x^\mu \overrightarrow{\partial}^\nu - x^\nu \overrightarrow{\partial}^\mu) + \frac{\sigma^{\mu\nu}}{2}\right]\psi(x) \end{aligned} \tag{3-166}$$

Consider now the action of the Pauli-Lubanski operator $W_\sigma = -\frac{1}{2}\varepsilon_{\sigma\mu\nu\rho}J^{\mu\nu}P^\rho$ on a state $|a\rangle$. The operator P^ρ is replaced by its eigenvalue k^ρ. Let $t \equiv (1, 0, 0, 0)$ be the time axis and

$$n = \left(t - k\frac{k\cdot t}{m^2}\right)\frac{m}{|\mathbf{k}|}$$

a normalized space-like four-vector orthogonal to k in the two-plane (t, k). Its space component is obviously along $\mathbf{k}$. Choose the third axis along $\mathbf{k}$. Then $(W\cdot n/m)\,|a\rangle$ reduces to

$$\frac{W\cdot n}{m}\,|a\rangle = J^{12}\,|a\rangle$$

When acting on the state $|a\rangle$ it is clear that the orbital contribution to J^{12} disappears, and since $J^{12}|0\rangle = 0$ we are left with

$$\frac{W\cdot n}{m} b_\alpha^\dagger(k)|0\rangle = \frac{m}{k^0}\sum_\beta u^{\dagger(\beta)}(k)\frac{\sigma^{12}}{2}u^{(\alpha)}(k)b_\beta^\dagger(k)|0\rangle$$

To proceed we have to make a choice for the spinor basis $u^{(\alpha)}$, $v^{(\alpha)}$. We set

$$u^{(\alpha)}(k) = \frac{\not k + m}{\sqrt{2m(k^0+m)}}\begin{pmatrix}\varphi^{(\alpha)}\\ 0\end{pmatrix}$$

with
$$\frac{\boldsymbol{\sigma}\cdot\mathbf{k}}{|\mathbf{k}|}\varphi^{(\alpha)} = \begin{cases}\varphi^{(\alpha)} & \alpha = 1\\ -\varphi^{(\alpha)} & \alpha = 2\end{cases}$$

The corresponding states are helicity states. Since the third axis was chosen along $\mathbf{k}$ we find

$$\frac{m}{k^0}u^{\dagger(\beta)}(k)\frac{\sigma_{12}}{2}u^{(\alpha)}(k) = \frac{\varepsilon_\alpha}{2}\delta^{\alpha\beta} \qquad \varepsilon_1 = 1,\ \varepsilon_2 = -1$$

with the result that

$$\frac{W\cdot n}{m}|a\rangle = \begin{cases}\frac{1}{2}|a\rangle & a = 1\\ -\frac{1}{2}|a\rangle & a = 2\end{cases} \tag{3-167}$$

Similarly,

$$\frac{W\cdot n}{m} d_\alpha^\dagger(k)|0\rangle = -\frac{m}{k^0}\sum_\beta v^{\dagger(\alpha)}(k)\frac{\sigma_{12}}{2}v^{(\beta)}(k)d_\beta^\dagger(k)|0\rangle$$

Defining

$$v^{(\alpha)}(k) = \frac{-\not k + m}{\sqrt{2m(k^0+m)}}\begin{pmatrix}0\\ \chi^{(\alpha)}\end{pmatrix}$$

with
$$\frac{\boldsymbol{\sigma}\cdot\mathbf{k}}{|\mathbf{k}|}\chi^{(\alpha)} = \begin{cases}\chi^{(\alpha)} & \alpha = 1\\ -\chi^{(\alpha)} & \alpha = 2\end{cases}$$

we obtain

$$\frac{m}{k^0}v^{\dagger(\alpha)}(k)\frac{\sigma_{12}}{2}v^{(\beta)}(k) = \frac{\varepsilon_\alpha}{2}\delta^{\alpha\beta} \qquad \varepsilon_1 = 1,\ \varepsilon_2 = -1$$

and
$$\frac{W\cdot n}{m}|a\rangle = \begin{cases}-\frac{1}{2}|a\rangle & a = 3\\ +\frac{1}{2}|a\rangle & a = 4\end{cases} \tag{3-168}$$

Formulas (3-167) and (3-168) complete the characterization of states: $|1\rangle$ and $|4\rangle$ have helicity $\frac{1}{2}$, $|2\rangle$ and $|3\rangle$ minus $\frac{1}{2}$.

We observe the typical inversion between the spinors u and v. It is the spinor $v^{(4)}$ with spin $\sigma_3 = -1$ which corresponds to positive helicity, in agreement with the hole interpretation.

Consider now multiparticle states. Denote the creation operators by the collective symbol $a^\dagger$, omitting therefore all indices of momentum, charge, or spin. A basis of Fock space is generated by the states

$$a^\dagger(1)\cdots a^\dagger(n)|0\rangle$$

From the anticommutation properties of the creation operators these states will be antisymmetric in the wave function arguments $1,\ldots,n$ and, in particular, will vanish if two of those coincide. We recover in this formalism Pauli's exclusion principle, valid for electrons and more generally for all identical fermions. Quantization using anticommutators has naturally generated Fermi-Dirac statistics.

The reader might enjoy discussing the free relativistic electron gas at thermal equilibrium. What happens if only the total charge is conserved and not electron and positron numbers separately?

3-3-3 Relation between Spin and Statistics—Propagator

From the relations (3-161) follows the anticommutator of two free fields at arbitrary separations. Clearly $\psi(x)$ and $\psi(x')$ anticommute; similarly for $\bar\psi(x)$, $\bar\psi(x')$, while

$$\{\psi_\xi(x), \bar\psi_{\xi'}(x')\} = \int d\tilde{k} \sum_\alpha [e^{-ik\cdot(x-y)}\, u_\xi^{(\alpha)}(k)\bar{u}_{\xi'}^{(\alpha)}(k) + e^{ik\cdot(x-y)}\, v_\xi^{(\alpha)}(k)\bar{v}_{\xi'}^{(\alpha)}(k)] \quad \text{(3-169)}$$

where ξ and ξ' are Dirac indices. From (2-40) and (2-41) we have

$$\sum_\alpha u_\xi^{(\alpha)}(k)\bar{u}_{\xi'}^{(\alpha)}(k') = \left(\frac{\not{k}+m}{2m}\right)_{\xi\xi'} \qquad \sum_\alpha v_\xi^{(\alpha)}(k)\bar{v}_{\xi'}^{(\alpha)}(k) = \left(\frac{\not{k}-m}{2m}\right)_{\xi\xi'}$$

It follows that

$$\{\psi_\xi(x), \bar\psi_{\xi'}(x')\} = (i\not\partial_x + m)_{\xi\xi'}\, i\Delta(x - x') \quad \text{(3-170)}$$

where $i\Delta(x - y)$ is the expression encountered in (3-77) and (3-56) in the analogous case for Bose fields. We conclude from (3-170) that when $(x - y)$ is a space-like interval the anticommutator vanishes. Moreover, if $x^0 = y^0 = t$,

$$\{\psi_\xi(t, \mathbf{x}), \bar\psi_{\xi'}(t, \mathbf{y})\} = -\gamma^0_{\xi\xi'}\, \partial_0\Delta(x^0 - y^0, \mathbf{x} - \mathbf{y})\big|_{x^0=y^0} = \gamma^0_{\xi\xi'}\delta^3(\mathbf{x} - \mathbf{y}) \quad \text{(3-171)}$$

in agreement with (3-162). In the sense of anticommutation $i\psi^\dagger$ appears as conjugate to ψ, as suggested by the lagrangian (3-152), even though we did not follow the procedure of canonical quantization.

We could wonder what would have followed from such canonical quantization. Spin $\frac{1}{2}$ particles would appear as bosons and energy would not be bounded from below. Furthermore, the commutator between two fields would be given by an expression similar to (3-169) with a relative minus sign between the two terms on the right-hand side. As a consequence, the result would read

$$(i\partial\!\!\!/_x + m) \int \frac{d^3k}{2k^0(2\pi)^3} (e^{-ik\cdot(x-y)} + e^{ik\cdot(x-y)})$$

This integral is an even solution of the homogeneous Klein-Gordon equation

$$\Delta_1(x-y) = \int \frac{d^3k}{2k^0(2\pi)^3} (e^{-ik\cdot(x-y)} + e^{ik\cdot(x-y)}) \tag{3-172}$$

which does not vanish for $(x-y)^2 < 0$.

Study the large separation behavior of $\Delta_1(x-y)$ in the space-like region.

In other words, local observables would not commute at equal times and locality would be violated. Similar violations would occur in a scalar theory quantized according to Fermi statistics.

These properties are instances of the relation between spin and statistics following from a local relativistic quantum theory. Half-integer spin fields have to be quantized according to Fermi-Dirac statistics, i.e., with anticommutators, and integer spin fields according to Bose-Einstein statistics, i.e., with commutators. A more general proof can be given in an axiomatic framework. It is to some extent fascinating to note that such a deep connection, essential to the stability of matter in circumstances apparently very remote from the relativistic domain, does require these concepts. We do not know any alternative basis for it.

The time-ordered product of two Dirac fields defined as

$$T\psi_\xi(x)\bar{\psi}_{\xi'}(y) = \theta(x^0 - y^0)\psi_\xi(x)\bar{\psi}_{\xi'}(y) - \theta(y^0 - x^0)\bar{\psi}_{\xi'}(y)\psi_\xi(x) \tag{3-173}$$

is such that

$$\begin{aligned} T\psi_\xi(x)\bar{\psi}_{\xi'}(y) &= \langle 0|\, T\psi_\xi(x)\bar{\psi}_{\xi'}(y)\,|0\rangle + {:}\psi_\xi(x)\bar{\psi}_{\xi'}(y){:} \\ \langle 0|\, T\psi_\xi(x)\bar{\psi}_{\xi'}(y)\,|0\rangle &= iS(x-y)_{\xi\xi'} \\ S(x-y) &= \int \frac{d^4k}{(2\pi)^4} e^{-ik\cdot(x-y)} \frac{k\!\!\!/ + m}{k^2 - m^2 + i\varepsilon} \end{aligned} \tag{3-174}$$

which agrees with (2-113). We note that

$$S(x) = -(i\partial\!\!\!/ + m)G_F(x) = (i\partial\!\!\!/ + m) \int \frac{d^4k}{(2\pi)^4} \frac{e^{-ik\cdot x}}{k^2 - m^2 + i\varepsilon} \tag{3-175}$$

Since $G_F(x)$ is even in x it follows that

$$S^T(-x) = -(-i\partial\!\!\!/^T + m)G_F(x)$$

With the help of the matrix C, introduced in (2-97), such that $C\gamma^{\mu T}C^{-1} = -\gamma^\mu$, we have

$$CS^T(-x)C^{-1} = S(x) \tag{3-176}$$

This symmetry property generalizes the one of $G_F(x)$. In the next subsection we shall investigate its meaning.

3-4 DISCRETE SYMMETRIES

Symmetries are, paradoxically, best understood when they are broken by external agents such as a measuring apparatus. Here we review briefly some discrete symmetries arising in the context of noninteracting quantum fields. Kinematic invariances have been discussed before and the subject of internal symmetries will be reconsidered in Chap. 11.

From the active point of view we understand by symmetry a one-to-one correspondence between states or, rather, density matrices of pure states $|\phi\rangle\langle\phi| \rightarrow |\phi^{\text{tr}}\rangle\langle\phi^{\text{tr}}|$ preserving transition probabilities

$$|\langle\phi_f^{\text{tr}}|\, e^{-iH(t_f-t_i)}\,|\phi_i^{\text{tr}}\rangle|^2 = |\langle\phi_f|\, e^{-iH(t_f-t_i)}\,|\phi_i\rangle|^2$$

This definition is somehow restrictive since it singles out the time variable, so that pure Lorentz transformations or time reversal have to be treated independently. This will be done below for the latter invariance. From Wigner's theorem, any one-to-one mapping of pure-state density matrices $\rho \rightarrow \rho'$, with the property that trace $(\rho_1\rho_2) =$ trace $(\rho'_1\rho'_2)$ can be extended to a unitary or antiunitary operator in Hilbert space.

Starting from an implementation of the symmetry on density matrices, the phases of the states can be adjusted in such a way as to produce a linear (or antilinear) operator on the Hilbert space. A symmetry group will be generally realized through a projective representation with $U(g_1)U(g_2) = e^{i\omega(g_1,g_2)}\,U(g_1, g_2)$ where $\omega(g_1, g_2)$ is a phase.

In the above restricted sense, every unitary operator U commuting with the hamiltonian generates a symmetry. While we shall mostly deal here with the Dirac field, the reader is invited to reconsider the cases of scalar or vector Bose fields.

3-4-1 Parity

We assume that it is possible, with appropriate modification of the experimental apparatus, to obtain the parity transformed state. From the field point of view we look for a unitary operator $\mathscr{P}$ satisfying

$$\mathscr{P}\psi(x)\mathscr{P}^\dagger = \eta_p\gamma^0\psi(\tilde{x}) \qquad \tilde{x}^\mu = x_\mu,\ |\eta_p| = 1 \tag{3-177}$$

since from the previous chapter we know that $\gamma^0\psi(\tilde{x})$ satisfies the parity transformed Dirac equation. We thus expect $\mathscr{P}$ to commute with H. Equation (3-177) entails

$$\begin{aligned} \mathscr{P}b_\alpha(p)\mathscr{P}^\dagger &= \eta_p b_\alpha(\tilde{p}) \qquad & \gamma^0 u^{(\alpha)}(\tilde{p}) &= u^{(\alpha)}(p) \\ \mathscr{P}d_\alpha(p)\mathscr{P}^\dagger &= -\eta_p^* d_\alpha(\tilde{p}) \qquad & \gamma^0 v^{(\alpha)}(\tilde{p}) &= -v^{(\alpha)}(p) \end{aligned} \tag{3-178}$$

Irrespective of the phase η_p, the relative parity of a fermion-antifermion system is minus one and $\mathscr{P}_\psi^2$ is a multiple of the identity. A unitary operator fulfilling the above requirements is

$$\mathscr{P} = \exp i \int d\tilde{k} \sum_{\alpha=1,2} \left\{ b_\alpha^\dagger(k)\left[\lambda b_\alpha(k) + \frac{\pi}{2} b_\alpha(\tilde{k})\right] - d_\alpha^\dagger(k)\left[(\lambda+\pi)d_\alpha(k) + \frac{\pi}{2} d_\alpha(\tilde{k})\right]\right\} \tag{3-179}$$

where λ is arbitrary and $\eta_p = e^{-i(\lambda+\pi/2)}$. In particular, $\eta_p = 1$ if $\lambda = -\pi/2$. A simple means of verifying that $\mathscr{P}$ is indeed a symmetry is to observe that under a time translation

$$e^{iHa}\, b_\alpha(k)\, e^{-iHa} = e^{-ik^0 a}\, b_\alpha(k) \qquad e^{iHa}\, d_\alpha(k)\, e^{-iHa} = e^{-ik^0 a}\, d_\alpha(k)$$

Hence,

$$e^{iHa}\, \mathscr{P}\, e^{-iHa} = \mathscr{P}$$

We may further verify that $\mathscr{P} P^\mu \mathscr{P}^\dagger = P_\mu$. In a coupled theory the complete parity operator will be the product of those pertaining to the different fields. Construct in particular $\mathscr{P}_A$ for the electromagnetic field such that

$$\mathscr{P}_A A^\mu(x) \mathscr{P}_A^\dagger = A_\mu(\tilde{x}) \tag{3-180}$$

Show that the bilinear Dirac current has a similar behavior:

$$\mathscr{P}_\psi :\bar{\psi}(x)\gamma^\mu\psi(x): \mathscr{P}_\psi^\dagger = :\bar{\psi}(\tilde{x})\gamma_\mu\psi(\tilde{x}): \tag{3-181}$$

from which follows, using $d^4x = d^4\tilde{x}$, that the interaction lagrangian

$$\int d^4x :\bar{\psi}(x)\gamma^\mu\psi(x): A_\mu(x)$$

is parity invariant.

3-4-2 Charge Conjugation

We have seen that quantized fields may describe particles of opposite charge with identical masses and spin. The corresponding charge may have a different interpretation according to the physical context. It can be electric, baryonic, leptonic, etc., whichever the case may be. Charge conjugation invariance therefore implies

1. The existence of antiparticles
2. The symmetric behavior of both kinds of quanta

We had a first example with the charged scalar field. It may, however, occur that particles and antiparticles are identical. Such is the case for photons, where the corresponding operator $\mathscr{C}$ just reverses the sign of the field

$$\mathscr{C} A_\mu(x) \mathscr{C}^\dagger = -A_\mu(x) \tag{3-182}$$

for reasons to become clear below.

From Chap. 2 the corresponding action on a Dirac field is

$$\mathscr{C}\psi(x)\mathscr{C}^\dagger = \eta_c C\bar{\psi}^T(x) \tag{3-183}$$

where transposition refers only to Dirac indices. In the standard representation of γ matrices

$$C = i\gamma^0\gamma^2$$

To be definite, we choose creation and annihilation operators for the helicity states $b(k, \pm)$, $d(k, \pm)$. Then

$$u(k, \pm 1) = \frac{\not{k} + m}{\sqrt{2m(k^0 + m)}}\begin{pmatrix}\varphi_\pm(\hat{k}) \\ 0\end{pmatrix} \tag{3-184a}$$

where $\qquad \boldsymbol{\sigma}\cdot\hat{k}\varphi_\pm(\hat{k}) = \pm\varphi_\pm(\hat{k}) \qquad \varphi_\varepsilon^\dagger(\hat{k})\varphi_{\varepsilon'}(\hat{k}) = \delta_{\varepsilon,\varepsilon'}$

and by definition

$$v(k, \pm) = C\bar{u}^T(k, \pm) = \frac{-\not{k} + m}{\sqrt{2m(m + k^0)}}\begin{pmatrix}0 \\ \chi_\pm(\hat{k})\end{pmatrix} \tag{3-184b}$$

In the usual representation $\chi_\pm(\hat{k}) = -i\sigma_2\varphi_\pm^*(\hat{k})$, and we can verify that $\boldsymbol{\sigma}\cdot\hat{k}\chi_\pm(\hat{k}) = \mp\chi_\pm(\hat{k})$. Furthermore,

$$C\bar{v}^T(k, \pm) = u(k, \pm) \tag{3-185}$$

With these choices we readily derive from (3-183) that

$$\begin{aligned}\mathscr{C}b(k, \pm)\mathscr{C}^\dagger &= \eta_c d(k, \pm) \\ \mathscr{C}d^\dagger(k, \pm)\mathscr{C}^\dagger &= \eta_c b^\dagger(k, \pm)\end{aligned} \tag{3-186}$$

This could have been imposed at first, with (3-183) following as a consequence. Up to a phase factor, $\mathscr{C}$ interchanges particles and antiparticles with the same momentum, energy, and helicity. The vacuum is left invariant. An explicit expression for $\mathscr{C}$ is

$$\mathscr{C}_\psi = \mathscr{C}_1\mathscr{C}_2 \tag{3-187}$$

$$\begin{aligned}\mathscr{C}_1 &= \exp -i\int d\tilde{k} \sum_{\varepsilon=\pm 1} \lambda[b^\dagger(k, \varepsilon)b(k, \varepsilon) - d^\dagger(k, \varepsilon)d(k, \varepsilon)] \\ \mathscr{C}_2 &= \exp i\frac{\pi}{2}\int d\tilde{k} \sum_{\varepsilon=\pm 1} [b^\dagger(k, \varepsilon) - d^\dagger(k, \varepsilon)][b(k, \varepsilon) - d(k, \varepsilon)]\end{aligned} \tag{3-188}$$

with $\eta_c = e^{i\lambda}$. The only effect of $\mathscr{C}_1$ is to carry this phase, and $\eta_c = 1$ corresponds to $\mathscr{C}_1 = 1$. The reader will also check that the current $:\bar{\psi}\gamma_\mu\psi:$ is odd under charge conjugation.

As an application let us classify, according to charge conjugation, the lowest bound states of a fermion-antifermion system, the prototype of which is positronium.

The latter is an (e^+e^-) system analogous to the hydrogen atom (pe^-) with the proton replaced by a positron. A nonrelativistic description is justified as a first approximation, due to the weakness of electromagnetic binding forces. The ground state is an s wave, $n = 1$, but hyperfine effects split a triplet orthopositronium state ($^3S_1$, if we use the notation $^{2S+1}L_J$) from a singlet ($^1S_0$) parapositronium state. Simplified wave functions correct from the quantum number point of view are written, using a fixed axis for spin quantization, as

$$|J = 1, M = 0, \text{ortho}\rangle = \int d^3\mathbf{q}\ \varphi_1(|\mathbf{q}|)[b_-^\dagger(\mathbf{q})d_+^\dagger(-\mathbf{q}) + b_+^\dagger(\mathbf{q})d_-^\dagger(-\mathbf{q})]\,|0\rangle$$

$$|J = 0, M = 0, \text{para}\rangle = \int d^3\mathbf{q}\ \varphi_0(|\mathbf{q}|)[b_-^\dagger(\mathbf{q})d_+^\dagger(-\mathbf{q}) - b_+^\dagger(\mathbf{q})d_-^\dagger(-\mathbf{q})]\,|0\rangle$$

The relative momentum wave functions φ_1 and φ_0 only depend on the magnitude of $\mathbf{q}$; $b_\pm^\dagger(\mathbf{q})$ [or $d_\pm^\dagger(\mathbf{q})$] denotes an electron (positron)-creation operator of momentum $\mathbf{q}$ with spin $\pm\frac{1}{2}$ along a fixed axis. Charge conjugation reads

$$\mathscr{C}b_\pm^\dagger(\mathbf{q})\mathscr{C}^\dagger = \eta_c^* d_\pm^\dagger(\mathbf{q}) \qquad \mathscr{C}d_\pm^\dagger(\mathbf{q})\mathscr{C}^\dagger = \eta_c b_\pm^\dagger(\mathbf{q})$$

The arbitrary phase η_c disappears when $\mathscr{C}$ acts on these states with the result that

$$\begin{aligned}\mathscr{C}\,|\text{ortho}\rangle &= -|\text{ortho}\rangle \\ \mathscr{C}\,|\text{para}\rangle &= |\text{para}\rangle\end{aligned} \tag{3-189}$$

The signs arise as follows. Charge conjugation interchanges electron and positron, as a result of which the relative momentum changes sign, leading to a factor of $(-1)^L = 1$ for s waves; the spin indices are interchanged, leading to a plus (minus) sign for a triplet (singlet) state. Finally, there is an additional minus sign arising from the anticommutation of $b^\dagger$ and $d^\dagger$ operators. This is an indirect and unexpected manifestation of Fermi-Dirac statistics.

The positronium states are unstable and have a slow decay by photon emission. From (3-182) the electromagnetic potential is odd under $\mathscr{C}$. This is, in fact, a condition for electromagnetic interactions to be invariant under $\mathscr{C}$. Hence an n-photon state behaves as

$$\mathscr{C}\,|n\rangle = (-1)^n\,|n\rangle$$

Correspondingly, orthopositronium must decay into an odd number of photons, and parapositronium into an even number. One photon decay is forbidden for the ortho state by energy momentum conservation. It must decay into at least three photons, while parapositronium can decay into two photons and has therefore a much shorter lifetime. The coupling constant being the fine structure constant α, for the lifetime τ we expect the ratio $\tau_{\text{singlet}}/\tau_{\text{triplet}} \sim 0(\alpha)$. We shall compute these quantities in Chap. 5. To lowest order in α,

$$\begin{aligned}\Gamma_s &= \tau_s^{-1} = \frac{m\alpha^5}{2} = 0.53 \times 10^{-5}\ \text{eV} & \tau_s &= 1.25 \times 10^{-10}\ \text{s} \\ \Gamma_t &= \tau_t^{-1} = 2\,\frac{\pi^2 - 9}{9\pi}\,m\alpha^6 = 4.75 \times 10^{-9}\ \text{eV} & \tau_t &= 1.410 \times 10^{-7}\ \text{s}\end{aligned} \tag{3-190}$$

A numerical accident makes τ_s smaller by an order of magnitude than $\alpha\tau_t$. The hyperfine splitting will be discussed in Chap. 10.

3-4-3 Time Reversal

Classically the meaning of time-reversal invariance is clear. By reversing the velocities in what used to be the final configuration, a system retraces its way back to some original configuration, if the fundamental dynamics enjoys this invariance. Hence initial and final states are interchanged with identical positions and opposite velocities. This interchange has the consequence that in quantum mechanics the corresponding operator $\mathscr{T}$ is antiunitary, that is,

$$\langle \mathscr{T}\varphi | \mathscr{T}\psi \rangle = \langle \psi | \varphi \rangle \qquad \mathscr{T}^\dagger \mathscr{T} = 1$$

The requirement of invariance under $\mathscr{T}$ means that

$$\mathscr{T} H \mathscr{T}^\dagger = H$$

if H is the hamiltonian. We assume time-translation invariance throughout. Consequently,

$$\mathscr{T} e^{-iH(t_2 - t_1)} \mathscr{T}^\dagger = e^{-iH(t_1 - t_2)}$$

The transition amplitude from the state $|\varphi_i\rangle$ at time t_i to the state $|\varphi_f\rangle$ at time t_f is equal to the corresponding amplitude from state $|\mathscr{T}\varphi_f\rangle$ at time t_i to $|\mathscr{T}\varphi_i\rangle$ at time t_f. Indeed,

$$\begin{aligned} \langle \varphi_f | e^{-iH(t_f - t_i)} | \varphi_i \rangle &= \langle \mathscr{T}\varphi_f | (\mathscr{T} e^{-iH(t_f - t_i)} \mathscr{T}^\dagger) | \mathscr{T}\varphi_i \rangle^* \\ &= \langle \mathscr{T}\varphi_f | e^{-iH(t_i - t_f)} | \mathscr{T}\varphi_i \rangle^* \\ &= \langle \mathscr{T}\varphi_i | e^{-iH(t_f - t_i)} | \mathscr{T}\varphi_f \rangle \end{aligned} \tag{3-191}$$

Since momenta, but not positions, are reversed, orbital angular momenta change sign under $\mathscr{T}$, and so must spins. In particular, helicities are unchanged.

For a scalar relativistic quantum field φ satisfying the Klein-Gordon equation, $\mathscr{T}\varphi(t, \mathbf{x})\mathscr{T}^\dagger$ equals $\varphi(-t, \mathbf{x})$, possibly up to a sign (since φ is hermitian). This leads to

$$\begin{aligned} \mathscr{T} a(k) \mathscr{T}^\dagger &= \eta a(\tilde{k}) \\ \mathscr{T}\varphi(t, \mathbf{x})\mathscr{T}^\dagger &= \eta\varphi(-t, \mathbf{x}) \end{aligned} \qquad \eta = \pm 1 \tag{3-192}$$

For the electromagnetic potential the corresponding transformation is

$$\mathscr{T} A^\mu(x) \mathscr{T}^\dagger = A_\mu(-\tilde{x}) \tag{3-193}$$

Consider finally a spinor field, and let us choose the helicity basis. The requirement is

$$\begin{aligned} \mathscr{T} b^\dagger(k, \varepsilon) \mathscr{T}^\dagger &= \eta_T^* b^\dagger(\tilde{k}, \varepsilon)\, e^{i\theta_b(k,\varepsilon)} \\ \mathscr{T} d^\dagger(k, \varepsilon) \mathscr{T}^\dagger &= \eta_T d^\dagger(\tilde{k}, \varepsilon)\, e^{-i\theta_d(k,\varepsilon)} \end{aligned} \tag{3-194}$$

with η_T a fixed phase. We want to choose $\theta_b(k, \varepsilon)$ and $\theta_d(k, \varepsilon)$ in such a way that $\mathscr{T}\psi(x)\mathscr{T}^\dagger$ satisfies the time-reversed Dirac equation. From the antilinear character of $\mathscr{T}$ we obtain

$$\mathscr{T}\psi(x)\mathscr{T}^\dagger = \eta_T \int d\tilde{k} \sum_\varepsilon [b(\tilde{k}, \varepsilon) u^*(k, \varepsilon) e^{-i\theta_b(k,\varepsilon)} e^{ik\cdot x} + d^\dagger(\tilde{k}, \varepsilon) v^*(k, \varepsilon) e^{-i\theta_d(k,\varepsilon)} e^{-ik\cdot x}]$$

The spinors u and v are defined in (3-184). We change k into $\tilde{k}$ in the integral. If there exists a fixed matrix A such that

$$\begin{aligned} Au(k, \varepsilon) &= e^{-i\theta_b(\tilde{k},\varepsilon)} u^*(\tilde{k}, \varepsilon) \\ Av(k, \varepsilon) &= e^{-i\theta_d(\tilde{k},\varepsilon)} v^*(\tilde{k}, \varepsilon) \end{aligned} \tag{3-195}$$

then we shall have

$$\mathscr{T}\psi(t, \mathbf{x})\mathscr{T}^\dagger = \eta_T A\psi(-t, \mathbf{x}) \tag{3-196}$$

From the fact that γ^0 is hermitian and γ antihermitian, it follows that

$$(\not{k}^T - m)u^*(\tilde{k}, \varepsilon) = 0 \qquad (\not{k}^T + m)v^*(\tilde{k}, \varepsilon) = 0$$

Multiplication by $\gamma^5 C$ yields

$$(\not{k} - m)\gamma^5 C u^*(\tilde{k}, \varepsilon) = 0 \qquad (\not{k} + m)\gamma^5 C v^*(\tilde{k}, \varepsilon) = 0$$

Consequently, we expect that A^{-1} is equal to $\gamma^5 C$ up to a phase provided that the two component spinors $\varphi_\pm(\hat{k})$ and $\varphi_\pm(-\hat{k})$ are correctly related. Indeed, in the standard representation of γ matrices

$$\gamma^5 C u^*(\tilde{k}, \varepsilon) = \gamma^5 C \frac{\tilde{\not{k}}^* + m}{\sqrt{2m(k^0 + m)}} \begin{pmatrix} \varphi_\varepsilon^*(-\hat{k}) \\ 0 \end{pmatrix}$$

$$= \frac{\not{k} + m}{\sqrt{2m(k^0 + m)}} \begin{pmatrix} -i\sigma_2 & 0 \\ 0 & -i\sigma_2 \end{pmatrix} \begin{pmatrix} \varphi_\varepsilon^*(-\hat{k}) \\ 0 \end{pmatrix}$$

Therefore all that is required is that $-i\sigma_2\varphi_\varepsilon^*(-\hat{k})$ be equal to $\varphi_\varepsilon(\hat{k})$ up to a phase. This is the case since $-i\sigma_2$ is unitary and

$$(\boldsymbol{\sigma}\cdot\hat{k})[-i\sigma_2\varphi_\varepsilon^*(-\hat{k})] = -i\sigma_2[-(\boldsymbol{\sigma}\cdot\hat{k})^*\varphi_\varepsilon^*(-\hat{k})] = \varepsilon[-i\sigma_2\varphi_\varepsilon^*(-\hat{k})]$$

Hence

$$-i\sigma_2\varphi_\varepsilon^*(-\hat{k}) = e^{i\theta_b(k,\varepsilon)}\varphi_\varepsilon(\hat{k})$$

and

$$\gamma^5 C u^*(\hat{k}, \varepsilon) = e^{i\theta_b(k,\varepsilon)} u(k, \varepsilon)$$

Now $(\gamma^5 C)^{-1} = -\gamma^5 C$; thus, up to a factor which can be absorbed in η_T we have indeed found the matrix A required in Eq. (3-195), at least for the u spinors. A similar calculation holds for v. The standard choice of phase is

$$A = i\gamma^1\gamma^3 = -i\gamma^5 C \tag{3-197}$$

and Eq. (3-196) is finally established.

3-4-4 Summary

For the Dirac field we can now summarize the various transformation properties of quadratic forms under discrete symmetries. Define them according to their tensor character:

$$\begin{aligned} S(x) &= :\bar{\psi}(x)\psi(x): \\ V^\mu(x) &= :\bar{\psi}(x)\gamma^\mu\psi(x): \\ T^{\mu\nu}(x) &= :\bar{\psi}(x)\sigma^{\mu\nu}\psi(x): \\ A^\mu(x) &= :\bar{\psi}(x)\gamma^5\gamma^\mu\psi(x): \\ P(x) &= i:\bar{\psi}(x)\gamma^5\psi(x): \end{aligned} \tag{3-198}$$

The last factor i is chosen for hermiticity. The symbol A for the pseudovector (or "axial" vector) current (see below) is conventional and should, of course, not be confused with the vector potential. Using the results of the previous subsections and keeping in mind that the field anticommute and that $\mathscr{T}$ is antiunitary, we have the following table:

	$S(x)$	$V^\mu(x)$	$T^{\mu\nu}(x)$	$A^\mu(x)$	$P(x)$
$\mathscr{P}$	$S(\tilde{x})$	$V_\mu(\tilde{x})$	$T_{\mu\nu}(\tilde{x})$	$-A_\mu(\tilde{x})$	$-P(\tilde{x})$
$\mathscr{C}$	$S(x)$	$-V^\mu(x)$	$-T^{\mu\nu}(x)$	$A^\mu(x)$	$P(x)$
$\mathscr{T}$	$S(-\tilde{x})$	$V_\mu(-\tilde{x})$	$-T_{\mu\nu}(-\tilde{x})$	$A_\mu(-\tilde{x})$	$-P(-\tilde{x})$
Θ	$S(-x)$	$-V^\mu(-x)$	$T^{\mu\nu}(-x)$	$-A^\mu(-x)$	$P(-x)$

(3-199)

In this table each entry represents the action of the operator on the corresponding density. For instance, $\mathscr{T}T^{\mu\nu}(x)\mathscr{T}^\dagger = -T_{\mu\nu}(-\tilde{x})$. We have added a line for the combined $\Theta = \mathscr{P}\mathscr{C}\mathscr{T}$ antiunitary operation. The corresponding transformation laws of the electromagnetic vector potential $A^\mu_{\text{em}}(x)$ are

$$\begin{aligned} \mathscr{P}A^\mu_{\text{em}}(x)\mathscr{P}^\dagger &= A_{\text{em},\mu}(\tilde{x}) & \mathscr{C}A^\mu_{\text{em}}(x)\mathscr{C}^\dagger &= -A^\mu_{\text{em}}(x) \\ \mathscr{T}A^\mu_{\text{em}}(x)\mathscr{T}^\dagger &= A_{\text{em},\mu}(-\tilde{x}) & \Theta A^\mu_{\text{em}}(x)\Theta^\dagger &= -A^\mu_{\text{em}}(-x) \end{aligned} \tag{3-200}$$

As a result the combination

$$:\bar{\psi}(x)\gamma_\mu\psi(x):A^\mu_{\text{em}}(x)$$

behaves as a scalar density $S(x)$. Finally, the behavior of $A^\mu(x)$ and $P(x)$ under parity justifies the denominations pseudovector current and pseudoscalar density respectively. All phase arbitrariness has disappeared from table (3-199).

In this section we have discussed the most frequent discrete symmetries and constructed the corresponding unitary (or antiunitary) operators in terms of the fundamental free fields. The necessity of such a construction is clear, for only then can we be sure that the transformed states do exist. It is a different matter to inquire whether the dynamics is left invariant when interactions are taken into account. This requires us to examine whether $UHU^\dagger = H$. When we consider such interacting fields, we shall try to define the symmetry operators acting on the states and the fields according to the above rules. The question will then be: Is the invariance implemented at the dynamical level?

A fundamental property of local quantum theory first discussed by Pauli, Zumino, and Schwinger states that, in any case, Θ remains an invariance of the theory. This is the famous PCT theorem. We sketch here a proof restricted to lagrangian field theory, referring the rigorous-minded reader to the works quoted in the notes for a more general treatment. Let a local quantum field theory be described by an action principle involving an hermitian Lorentz-invariant lagrangian. This is a combination of local scalar densities expressed in terms of the basic fields, eventually normal ordered. The quantization respects the spin statistics relation. The PCT theorem then states that even if P, C, and T are not separately invariances, their combination is. This implies, in particular, the existence of antiparticles for charged fields (with masses and spins equal to those of the corresponding particles) and the particle-antiparticle identification for neutral fields. In summary we want to show that the lagrangian $\mathscr{L}(x)$ behaves under Θ as a neutral scalar field

$$\Theta\mathscr{L}(x)\Theta^\dagger = \mathscr{L}(-x) \tag{3-201}$$

We should have added that for a scalar field in general

$$\Theta\varphi(x)\Theta^\dagger = \varphi^\dagger(-x) \tag{3-202}$$

If (3-201) is satisfied the action will remain invariant. We shall be satisfied here with a lagrangian depending on scalar (φ), vector (A_μ), and spinor (ψ) fields possibly carrying internal indices. The theorem remains true for higher spin fields. It is clear that $\mathscr{L}$ can be constructed using A_μ, ψ, $\bar{\psi}$ and only hermitian scalar fields combining if necessary the charged ones into $\varphi_1 = (\varphi + \varphi^\dagger)/\sqrt{2}$ and $\varphi_2 = (\varphi - \varphi^\dagger)/i\sqrt{2}$ ($\Theta\varphi_i(x)\Theta^\dagger = \varphi_i(-x)$). We can then summarize the action of Θ as follows:

1. It changes the arguments of fields from x to $-x$.
2. As a consequence derivatives change sign, $\partial_\mu \to -\partial_\mu$.
3. A vector field $A^\mu(x)$ gets an extra minus sign, i.e., behaves as the gradient of a scalar field.
4. Any quadratic form in $(\bar{\psi}, \psi)$ gets a sign $(-1)^P$ where P denotes the number of Lorentz indices carried by γ matrices or derivatives. This follows from table (3-199) and rule 2.
5. Finally, constants are complex conjugated.

Since $\mathscr{L}$ is a scalar, each term in a monomial expansion is obtained by contracting an even number of Lorentz indices. The minus signs coming from vector fields, derivatives, or quadratic forms in $(\bar{\psi}, \psi)$ are cancelled. The net effect is thus to change $\mathscr{L}(x)$ into $\mathscr{L}^\dagger(-x) = \mathscr{L}(-x)$, since the order of operators is irrelevant under normal ordering as vector and scalar fields commute and the anticommutation of the ψ fields has already been taken into account in (3-199). As a consequence, if the vacuum is invariant under Θ, so will be the dynamics.

It is interesting to check that equal-time commutation rules are invariant under Θ. This involves some subtlety since interactions may modify the expression of conjugate variables. Such is the case, for instance, for electromagnetic interactions of charged bosons and more generally when the interaction part of $\mathscr{L}$ involves derivatives. We may also verify that the hamiltonian density behaves as

$$\Theta\mathscr{H}(0, \mathbf{x})\Theta^\dagger = \mathscr{H}(0, -\mathbf{x})$$

and thus

$$\Theta H\Theta^\dagger = H$$

The above proof ignores the difficulties associated with the construction of an interacting field theory. The latter has to give a precise meaning to otherwise undefined products of operators. Nevertheless, the algebraic properties implied in the PCT theorem are preserved by this operation.

As an application we discuss form factors, a relativistic generalization of charge distributions. Consider first the matrix elements of the current $V^\mu(x)$ of a free Dirac theory between two one-particle states

$$\langle p', \beta| V^\mu(x) |p, \alpha\rangle = e^{-i(p-p')\cdot x} \langle p', \beta| V^\mu(0) |p, \alpha\rangle$$

The greek indices $\alpha, \beta, \ldots, \delta$ denote polarizations, $|p, \alpha\rangle = b_\alpha^\dagger(p)|0\rangle$, and

$$V^\mu(0) = :\bar{\psi}(0)\gamma^\mu\psi(0): = \iint d\tilde{k}_1\, d\tilde{k}_2 \sum_{\gamma,\delta} \bar{u}^{(\gamma)}(k_1)\gamma^\mu u^{(\delta)}(k_2) b_\gamma^\dagger(k_1) b_\delta(k_2) + \cdots$$

We have not explicitly given the terms with vanishing contribution to the matrix element. From

$$\langle p', \beta| b_\gamma^\dagger(k_1) b_\delta(k_2) |p, \alpha\rangle = (2\pi)^6 \frac{k_1^0 k_2^0}{m^2} \delta^3(k_1 - p')\delta^3(k_2 - p)\delta_{\beta\gamma}\delta_{\delta\alpha}$$

it follows that

$$\langle p', \beta| V^\mu(0) |p, \alpha\rangle = \bar{u}^{(\beta)}(p')\gamma^\mu u^{(\alpha)}(p)$$

This is in agreement with the definition of the charge

$$\begin{aligned}\langle p', \beta| Q |p, \alpha\rangle &= \int d^3x \langle p', \beta| V^0(0, \mathbf{x}) |p, \alpha\rangle \\ &= \int d^3x\, e^{i(\mathbf{p}'-\mathbf{p})\cdot \mathbf{x}}\, \bar{u}^{(\beta)}(p')\gamma^0 u^{(\alpha)}(p) \\ &= (2\pi)^3\delta^3(\mathbf{p} - \mathbf{p}')\bar{u}^{(\beta)}(p')\gamma^0 u^{(\alpha)}(p) = \langle p', \beta\,|\, p, \alpha\rangle\end{aligned}$$

Recall that $\bar{u}_\beta(p')\gamma^0 u_\alpha(p)$ is normalized to $\delta_{\alpha\beta}(p^0/m)$.

Assume now that $j^\mu(x)$ is an hermitian four-vector current density with the same transformation properties under discrete symmetries as $V^\mu(x)$, and let us write the general structure of its one-particle matrix element

$$\langle p', \beta| j^\mu(x) |p, \alpha\rangle = e^{-i(p-p')\cdot x}\, \bar{u}^{(\beta)}(p')O^\mu(p', p)u^{(\alpha)}(p)$$

In the sequel we always understand that the numerical matrix $O^\mu(p', p)$ is sandwiched between the two spinors $\bar{u}$ and u. From Lorentz covariance $O^\mu(p', p)$ must obey

$$S(\Lambda)O^\mu(p', p)S^{-1}(\Lambda) = (\Lambda^{-1})^\mu{}_\nu O^\nu(\Lambda p', \Lambda p)$$

and

$$\gamma^0 O^\mu(p', p)^\dagger \gamma^0 = O^\mu(p', p)$$

from hermiticity. We can take into account the fact that the spinors satisfy the free Dirac equation by replacing any $O^\mu(p', p)$ by

$$O^\mu(p', p) \to \frac{\not{p}' + m}{2m} O^\mu(p', p) \frac{\not{p} + m}{2m}$$

In particular this entails the Gordon identity

$$\gamma^\mu \to \frac{1}{2m}\left[(p' + p)^\mu + i\sigma^{\mu\nu}(p' - p)_\nu\right]$$

We also require the current $j^\mu(x)$ to be conserved

$$0 = \langle p', \beta | \partial_\mu j^\mu(x) | p, \alpha \rangle = i(p' - p)_\mu \langle p', \beta | j^\mu(x) | p, \alpha \rangle$$

Thus $(p' - p)_\mu O^\mu(p', p)$ vanishes. Let us note q, the space-like difference $p' - p$; since $p^2 = p'^2 = m^2$ the only scalar invariant is q^2. Allowing for the most general decomposition of O^μ on the basis of the 16 Dirac matrices, and using Lorentz invariance, conservation, and hermiticity, we find

$$\bar{u}^{(\beta)}(p')O^\mu(p', p)u^{(\alpha)}(p) = \bar{u}^{(\beta)}(p')\left[\gamma^\mu F_1(q^2) + i\frac{\sigma^{\mu\nu}q_\nu}{2m}F_2(q^2) + \gamma^5\frac{\sigma^{\mu\nu}q_\nu}{2m}F_3(q^2)\right]u^{(\alpha)}(p) \tag{3-203}$$

with $F_i(q^2)$ being real form factors—functions of the square momentum transfer q^2. Quantizing spin along a fixed axis we derive from Eqs. (3-178) and (3-199) that symmetry under parity implies

$$\langle p', \beta | j^\mu(x) | p, \alpha \rangle = \langle \tilde{p}', \beta | j_\mu(\tilde{x}) | \tilde{p}, \alpha \rangle$$

Since $u^{(\alpha)}(\tilde{p}) = \gamma^0 u^{(\alpha)}(p)$ this means $O^\mu(p', p) = \gamma^0 O_\mu(\tilde{p}', \tilde{p})\gamma^0$, the consequence of which is $F_3(q^2) = -F_3(q^2)$. Thus parity conservation alone yields $F_3 = 0$. However, it could happen that parity is violated by some interaction (in practice, weak interactions).

It was long believed that parity times charge conjugation remains a symmetry of all interactions. From the PCT theorem this meant that T was also a symmetry. This imposes $F_i^*(q^2) = \varepsilon_i F_i(q^2)$ with $\varepsilon_1 = \varepsilon_2 = -\varepsilon_3 = 1$. From hermiticity we know that the F functions are real. Therefore again $F_3 = 0$. It is known, however, that T is not a symmetry of all interactions, since the discovery of its violation by Fitch and Cronin in the neutral K meson decays.

As an example let us assume that $j^\mu(x)$ is the electromagnetic current of the interacting theory. By considering two normalized states close to rest, the reader will have no difficulty in showing that the following correspondence holds:

$$eF_1(0) = \text{charge } Q$$

$$\frac{e}{2m}[F_1(0) + F_2(0)] = \text{magnetic moment } \mu \quad (\text{nonrelativistic magnetic coupling } -\mu\boldsymbol{\sigma}\cdot\mathbf{B})$$

$$\frac{-e}{2m}F_3(0) = \text{electric dipole moment } d \quad (\text{nonrelativistic electric coupling } -d\boldsymbol{\sigma}\cdot\mathbf{E}).$$

Extensive measurements of nucleon-form factors have been performed using electron scattering off hydrogen and deuterium targets. For this case F_1 and F_2 are now known in a wide range of q^2. However, up to now F_3 has not shown up. In particular, bounds for the static electric dipole moment of the neutron have been reported (Ramsey) showing d neutron$/e \lesssim 10^{-23}$ cm. The magnitude of this electric dipole moment is a sensitive test of the mechanism of CP violation.

For a charged particle, the total gyromagnetic ratio is $2[F_1(0) + F_2(0)]/F_1(0)$. For the proton, for instance (with the convention $e = |e|$), $F_1(0) = 1$, $F_2(0) = 1.79$, and the gyromagnetic ratio is 5.58. In the literature we often find the Sachs combinations $G_E = F_1 + (q^2/4m^2)F_2$ and $G_M = F_1 + F_2$ which are more convenient for displaying the experimental data on scattering. These form factors extend analytically to positive (time-like) values of q^2 where they are related to the process photon $\rightarrow$ fermion + antifermion.

In the case of scalar (or pseudoscalar) particles the corresponding matrix element of the current has the form

$$\langle p' | j^\mu(x) | p \rangle = e^{iq\cdot x}\frac{(p' + p)^\mu}{2m}F(q^2) \tag{3-204}$$

We close this section with an algebraic exercise. It deals with quartic products of Dirac fields organized in covariant scalar densities as they occur for instance in the effective Fermi lagrangian for low-energy weak interactions. Its purpose is to examine the effect of rearranging the order in which these fields are coupled together to obtain a Lorentz scalar, the basic blocks being the quadratic quantities occurring in table (3-199). This is the Fierz theorem. Denote collectively by Γ^α the 16 Dirac matrices

$$\begin{array}{ccccc} \Gamma_S & \Gamma_V^\mu & \Gamma_T^{\mu\nu} & \Gamma_A^\mu & \Gamma_P \\ I & \gamma^\mu & \sigma^{\mu\nu} = \dfrac{i}{2}[\gamma^\mu, \gamma^\nu] & \gamma^5\gamma^\mu & i\gamma^5 \end{array} \tag{3-205}$$

Set $\Gamma_\alpha = (\Gamma^\alpha)^{-1}$ and observe that $\gamma^0(\Gamma^\alpha)^\dagger\gamma^0 = \Gamma^\alpha$ in such a way that the forms $\bar\psi\Gamma^\alpha\psi$ are hermitian. With these notations

$$\operatorname{Tr} \Gamma^\alpha\Gamma_\beta = 4\delta^\alpha{}_\beta \qquad 1 \le \alpha, \beta \le 16 \tag{3-206}$$

Any 4×4 matrix X has an expansion

$$X = x_\alpha\Gamma^\alpha = \tfrac{1}{4}\Gamma^\alpha \operatorname{Tr}(X\Gamma_\alpha) = \tfrac{1}{4}\Gamma_\alpha \operatorname{Tr}(X\Gamma^\alpha)$$

By identifying the coefficient of X_{ab} we find the following identity where Latin indices run from 1 to 4:

$$\delta_{a\bar a}\delta_{b\bar b} = \tfrac{1}{4}\Gamma_{\alpha,b\bar a}\Gamma^\alpha_{\bar a b} \tag{3-207}$$

Applying this to $\Gamma^\alpha\Gamma_\beta$ we have

$$\Gamma^\alpha\Gamma_\beta = \rho^\alpha{}_{\beta\gamma}\Gamma^\gamma \qquad \rho^\alpha{}_{\beta\gamma} \equiv \tfrac{1}{4}\operatorname{Tr}\Gamma^\alpha\Gamma_\beta\Gamma_\gamma \tag{3-208}$$

Writing $\Gamma_\alpha = \varepsilon_\alpha\Gamma^\alpha$ where $\varepsilon_\alpha = \pm 1$, it is readily seen that $\rho_{\alpha\beta\gamma} = \varepsilon_\alpha\rho^\alpha{}_{\beta\gamma}$ is invariant by cyclic permutation and that given the values of two of its indices it is nonvanishing for only one value of the third. The only nonvanishing terms are

$$\rho^\alpha{}_{\beta S} = \rho^\alpha{}_{S\beta} = \delta^\alpha{}_\beta$$

and

$$\begin{aligned}
\rho_{V_\mu T_{\rho\sigma} V_\nu} &= -\rho_{V_\mu V_\nu T_{\rho\sigma}} = i(g_{\rho\mu}g_{\nu\sigma} - g_{\mu\sigma}g_{\nu\rho}) \\
\rho_{V_\mu T_{\rho\sigma} A_\nu} &= \rho_{T_{\sigma\rho} V_\mu A_\nu} = \varepsilon_{\mu\nu\rho\sigma} \\
\rho_{V_\mu A_\nu P} &= -\rho_{A_\nu V_\mu P} = -ig_{\mu\nu} \\
\rho_{T_{\mu\nu} A_\sigma A_\rho} &= -\rho_{A_\sigma T_{\mu\nu} A_\rho} = i(g_{\mu\sigma}g_{\nu\rho} - g_{\mu\rho}g_{\nu\sigma}) \\
\rho_{T_{\mu\nu} P T_{\rho\sigma}} &= \rho_{P T_{\mu\nu} T_{\rho\sigma}} = \varepsilon_{\mu\nu\rho\sigma} \\
\rho_{T_{\mu\nu} T_{\rho\sigma} T_{\tau\varepsilon}} &= i[g_{\mu\rho}(g_{\nu\tau}g_{\varepsilon\sigma} - g_{\nu\varepsilon}g_{\tau\sigma}) + g_{\mu\sigma}(g_{\nu\varepsilon}g_{\rho\tau} - g_{\nu\tau}g_{\varepsilon\rho}) \\
&\quad + g_{\mu\tau}(g_{\nu\sigma}g_{\rho\varepsilon} - g_{\nu\rho}g_{\sigma\varepsilon}) + g_{\mu\varepsilon}(g_{\tau\sigma}g_{\nu\rho} - g_{\nu\sigma}g_{\rho\tau})]
\end{aligned} \tag{3-209}$$

Consider now the five Lorentz scalars

$$\begin{aligned}
s &= \bar u(4)u(2)\bar u(3)u(1) \\
v &= \bar u(4)\gamma^\mu u(2)\bar u(3)\gamma_\mu u(1) \\
t &= \tfrac{1}{2}\bar u(4)\sigma^{\mu\nu}u(2)\bar u(3)\sigma_{\mu\nu}u(1) \\
a &= \bar u(4)\gamma^5\gamma^\mu u(2)\bar u(3)\gamma_\mu\gamma^5 u(1) \\
p &= \bar u(4)\gamma^5 u(2)\bar u(3)\gamma^5 u(1)
\end{aligned} \tag{3-210}$$

Here $u(1)$ denotes $u(p_1, \alpha_1)$, the spinor with momentum p_1 and polarization α_1. We shall use the symbol $s(4, 2; 3, 1)$ to characterize the way in which the Dirac indices are contracted, and similarly for the other amplitudes. Any one of these quantities may be written as

$$b(4, 2; 3, 1) = \bar u_{a_3}(3)\bar u_{a_4}(4)\Gamma^\alpha_{a_4a_2}\Gamma_{\alpha,a_3a_1}u_{a_1}(1)u_{a_2}(2)$$

How are these quantities related to the corresponding ones where 4 is contracted with 1 and 3 with 2? The theorem states that there exists a numerical 5×5 matrix F relating the two sets of quantities. This matrix will then necessarily be equal to its inverse. Before proceeding, let us remark that we could ask a similar question in x space with operators of the type $:\bar\psi_4(x)\psi_2(x)\bar\psi_3(x)\psi_1(x):$. The relevant matrix would then be $-F$ due to Fermi field anticommutation. In any of the five quantities $b(4,2;3,1)$ we can use Eq. (3-207) to rewrite

$$u_{a_2}(2)u_{a_1}(1) = \delta_{a_2\bar{a}_2}\delta_{a_1\bar{a}_1}u_{\bar{a}_2}(2)u_{\bar{a}_1}(1) = \tfrac{1}{4}\Gamma_{\beta,a_2\bar{a}_1}\Gamma^{\beta}_{a_1\bar{a}_2}u_{\bar{a}_2}(2)u_{\bar{a}_1}(1)$$

Thus, with the help of (3-208),

$$\begin{aligned} b(4,2;3,1) &= \tfrac{1}{4}\bar{u}_{a_4}(4)\bar{u}_{a_3}(3)\Gamma^{\alpha}_{a_4a_2}\Gamma_{\beta,a_2\bar{a}_1}\Gamma_{\alpha,a_3a_1}\Gamma^{\beta}_{a_1\bar{a}_2}u_{\bar{a}_2}(2)u_{\bar{a}_1}(1) \\ &= \tfrac{1}{4}\rho^{\alpha}{}_{\beta\gamma}\rho_{\alpha}{}^{\beta\delta}\bar{u}(4)\Gamma^{\gamma}u(1)\bar{u}(3)\Gamma_{\delta}u(2) \end{aligned}$$

It remains to use (3-209) to obtain the desired result:

$$\begin{pmatrix} s \\ v \\ t \\ a \\ p \end{pmatrix}(4,2;3,1) = \frac{1}{4}\begin{bmatrix} 1 & 1 & 1 & 1 & 1 \\ 4 & -2 & 0 & 2 & -4 \\ 6 & 0 & -2 & 0 & 6 \\ 4 & 2 & 0 & -2 & -4 \\ 1 & -1 & 1 & -1 & 1 \end{bmatrix}\begin{pmatrix} s \\ v \\ t \\ a \\ p \end{pmatrix}(4,1;3,2)$$

The reader may check that $F^2 = 1$ and diagonalize the matrix. What is the behavior of the various amplitudes under the discrete symmetries?

NOTES

The physical meaning of the quantum commutation relations in electrodynamics is analyzed in N. Bohr and L. Rosenfeld, *Kgl. Danske Videnskab. Selsk. Mat.-Fys. Medd.*, vol. 12, p. 8, 1933, and *Phys. Rev.*, vol. 78, p. 794, 1950. For limits on the photon mass see A. S. Goldhaber and M. M. Nieto, *Rev. Mod. Phys.*, vol. 43, p. 277, 1971. Macroscopic effects of vacuum fluctuations were considered by H. B. G. Casimir, *Proc. Kon. Ned. Akad. Wetenschap.*, ser. B, vol. 51, p. 793, 1948. See also M. Fierz, *Helv. Phys. Acta*, vol. 33, p. 855, 1960; T. M. Boyer, *Annals of Physics (New York)*, vol. 56, p. 474, 1970; and R. Balian and B. Duplantier, *Annals of Physics (New York)*, vol. 112, p. 165, 1978. Experimental evidence is discussed by M. J. Sparnaay, *Physica*, vol. 24, p. 751, 1958. A study on Van der Waals forces is found in the work of G. Feinberg and J. Sucher, *Phys. Rev.*, ser. A, vol. 2, p. 2395, 1970. Quantum field theory at finite temperature is presented, for instance, in L. Dolan and R. Jackiw, *Phys. Rev.*, ser. D, vol. 9, p. 3320, 1974.

For a general axiomatic formulation of field theory and a derivation of some fundamental properties, see R. F. Streater and A. S. Wightman, "*PCT*, Spin and Statistics, and All That," Benjamin, New York, 1964, and R. Jost, "The General Theory of Quantized Fields," AMS, Providence, R.I., 1965.

CHAPTER

FOUR

INTERACTION WITH AN EXTERNAL FIELD

The interaction with an external field yields a simple example of a dynamical system. We introduce the important concepts of interaction representation and Wick's identities. Applications include the radiation of a classical source and the infrared catastrophe, the physical counterpart of which in the fermionic case is the process of pair creation by a *c*-number electromagnetic field.

4-1 QUANTIZED ELECTROMAGNETIC FIELD INTERACTING WITH A CLASSICAL SOURCE

4-1-1 Emission Probabilities

We shall first consider the interaction of the quantized electromagnetic field with an external given source. The physical problem is thus the emission or absorption of photons by a classical conserved current $j_\mu(x)$:

$$\partial_\mu j^\mu(x) = 0 \tag{4-1}$$

From the electromagnetic action

$$I = -\int d^4x \, (\tfrac{1}{4} F_{\mu\nu} F^{\mu\nu} + j_\mu A^\mu) \tag{4-2}$$

we derive the equation of motion

$$\partial_\mu F^{\mu\nu} = \Box A^\nu - \partial^\nu \partial \cdot A = j^\nu \tag{4-3}$$

As in the free case, we encounter a difficulty in the quantization of this field, which may be solved by the addition to the action of the term

$$-\tfrac{1}{2}\int d^4x\,(\partial\cdot A)^2$$

In this gauge, the equation of motion reduces to

$$\Box A^\mu = j^\mu \tag{4-4}$$

The operators $A_\mu(x)$ act in a space with an indefinite metric and satisfy the same commutation relations as in the free case

$$[A_\mu(t,\mathbf{x}), \dot{A}_\nu(t,\mathbf{y})] = -g_{\mu\nu}i\delta^3(\mathbf{x}-\mathbf{y}) \tag{4-5}$$

If we assume that the solution of (4-4) is defined on the same Fock space as the free field, the relation between the free and interacting fields amounts to a canonical transformation. In quantum mechanics with a finite number of degrees of freedom, this would be implemented by a unitary transformation. However, for an infinite number of degrees of freedom, there may be a difficulty. We illustrate this point with a simple model.

Let a finite quantum system be composed of N half-integer spins at the vertices of a three-dimensional cubic lattice. They may carry, for instance, a magnetic moment. The observables are the $3N$ operators $\sigma_1(n)$, $\sigma_2(n)$, $\sigma_3(n)$, n being the site index. The states are linear combinations of the eigenvectors of the σ_3, denoted by $|\pm\pm\cdots\pm\rangle$. They are generated by the action of the σ_- upon the state $|0\rangle \equiv |++\cdots+\rangle$. Consider now the rotated operators

$$\tau_1(n) = \sigma_1(n)\cos\theta - \sigma_3(n)\sin\theta$$

$$\tau_2(n) = \sigma_2(n)$$

$$\tau_3(n) = \sigma_1(n)\sin\theta + \sigma_3(n)\cos\theta$$

Clearly,

$$\tau_\alpha(n) = \exp\left[-i\frac{\theta}{2}\sum_{p=1}^{N}\sigma_2(p)\right]\sigma_\alpha(n)\exp\left[i\frac{\theta}{2}\sum_{p=1}^{N}\sigma_2(p)\right]$$

This means that the operators σ and τ which generate the same algebra are unitarily equivalent. Let $|\theta\rangle$ denote the new ground state of the τ:

$$|\theta\rangle = \exp\left[-i\frac{\theta}{2}\sum_{p=1}^{N}\sigma_2(p)\right]|0\rangle$$

It is easy to compute the scalar product $\langle 0|\theta\rangle = (\cos\theta/2)^N$. Assume that in the limit $N\to\infty$, the Hilbert space of states is constructed from the ground state $|0\rangle \equiv |++\cdots+\rangle$ by the action of a finite number of creation operators σ_- and Cauchy completion. We may again rotate the operators σ into τ and define the state $|\theta\rangle$ in analogy with $|0\rangle$. We may then seek a unitary transformation $U(\theta)$ which would implement this rotation:

$$|\theta\rangle = U(\theta)|0\rangle \qquad \tau_\alpha(n) = U(\theta)\sigma_\alpha(n)U^\dagger(\theta)$$

Such a unitary operator clearly does not exist when $N \to \infty$. Any scalar product of a rotated state with a nonrotated one vanishes in this limit: for instance, $\langle 0|\theta\rangle = (\cos\theta/2)^N \xrightarrow[N\to\infty]{} 0$. The moral is that, in the case of an infinite number of degrees of freedom, physically equivalent observables, i.e., realizing the same algebra of commutation relations, etc., are not necessarily unitarily equivalent. We have to keep this in mind when we observe that our field A_μ satisfies the free commutation relations (4-5).

Returning to the field equation (4-4), we may write a particular c-number solution

$$A_c^\mu(x) = \int d^4y\, G(x-y) j^\mu(y)$$

expressed in terms of some Green function $G(x-y)$ of the d'Alembertian operator $\Box_x$:

$$\Box_x G(x-y) = \delta^4(x-y)$$

The general solution of Eq. (4-4) is therefore

$$A_\mu(x) = A_\mu^{(0)}(x) + \int d^4y\, G(x-y) j_\mu(y) \tag{4-6}$$

where $A^{(0)}$ is a quantum free field. The precise form of Green function is specified by the boundary conditions. Assume that the current $j^\mu(y)$ has been switched on adiabatically on a finite time interval. Using the advanced and retarded Green functions of Eq. (1-170)

$$G_{\substack{\text{ret}\\ \text{adv}}}(x) = -\frac{1}{(2\pi)^4}\int d^4p \frac{e^{-ip\cdot x}}{(p_0 \pm i\varepsilon)^2 - \mathbf{p}^2} = \frac{1}{2\pi}\theta(\pm x_0)\delta(x^2)$$

we may write

$$\begin{aligned} A^\mu(x) &= A_{\text{in}}^\mu(x) + \int d^4y\, G_{\text{ret}}(x-y) j^\mu(y) \\ &= A_{\text{out}}^\mu(x) + \int d^4y\, G_{\text{adv}}(x-y) j^\mu(y) \end{aligned} \tag{4-7}$$

The free fields A_{in}^μ and A_{out}^μ describe the photon field before and after its interaction with the current j. In a formal sense, we have

$$\begin{aligned} \lim_{x_0\to-\infty} A^\mu(x) &= A_{\text{in}}^\mu(x) \\ \lim_{x_0\to+\infty} A^\mu(x) &= A_{\text{out}}^\mu(x) \end{aligned} \tag{4-8}$$

The precise mathematical meaning of these expressions depends on the source j. We may require at least that the matrix elements of some local average of the fields between normalized states satisfy these relations. This weak limit relies crucially on the assumption of an adiabatic vanishing of the source as $|t| \to \infty$.

This is often not actually realized and therefore Eq. (4-8) will have to be taken with a grain of salt.

The construction takes place in a given Hilbert space, the Fock space of incoming photons, for instance. The vacuum is then the state annihilated by the operators $a_{\text{in}}^{(\lambda)}(k)$, and we try to find the canonical transformation represented by a unitary operator S which connects the in- and out-fields, that is,

$$A_{\text{out}}^{\mu}(x) = S^{-1} A_{\text{in}}^{\mu}(x) S \tag{4-9}$$

and in- and out-states according to

$$|\text{out}\rangle = S^{-1}\,|\text{in}\rangle = S^{\dagger}\,|\text{in}\rangle$$

or

$$|\text{in}\rangle = S\,|\text{out}\rangle \tag{4-10}$$

Suppose that at time $t = -\infty$ the system is in a definite state, the vacuum for instance, i.e., contains no (physical) photon. The final state has some computable probability to contain zero, one, two, etc., emitted photons. For example, the probability amplitude to remain in the ground state is

$$\langle 0\ \text{out}|0\ \text{in}\rangle = \langle 0\ \text{in}|S|0\ \text{in}\rangle = \langle 0\ \text{out}|S|0\ \text{out}\rangle$$

For this interpretation to make sense, we should verify that the probability $p_0 = |\langle 0\ \text{out}|0\ \text{in}\rangle|^2$ is less than one. The probabilities $p_1, p_2, \ldots,$ the final polarizations, and the angular distributions are computable in an analogous way. Therefore, the operator S contains all the information about the final state.

We now turn to its determination. Equation (4-7) implies that

$$\begin{aligned} A_{\text{out}}^{\mu}(x) &= A_{\text{in}}^{\mu}(x) + \int d^4y\,[G_{\text{ret}}(x-y) - G_{\text{adv}}(x-y)]j^{\mu}(y) \\ &= A_{\text{in}}^{\mu}(x) + A_{\text{cl}}^{\mu}(x) \end{aligned} \tag{4-11}$$

The second term of the right-hand side of (4-11) is, of course, a solution of the homogeneous equation, and is nothing but the classical field A_{cl}^{μ} radiated by the current j (1-206). The combination $G^{(-)} \equiv G_{\text{ret}} - G_{\text{adv}}$ has already been encountered in Chap. 1 [see Eq. (1-173)]:

$$\begin{aligned} G^{(-)}(x) = G_{\text{ret}}(x) - G_{\text{adv}}(x) &= \frac{i}{(2\pi)^3}\int d^4p\, e^{-ip\cdot x}\,\varepsilon(p_0)\delta(p^2) \\ &= \frac{1}{2\pi}\,\varepsilon(x^0)\delta(x^2) = -\Delta(x) \end{aligned}$$

It coincides up to a sign with the commutator Δ of scalar massless free fields (3-56). This enables us to rewrite (4-11) as

$$A_{\text{out}}^{\mu}(x) = S^{-1} A_{\text{in}}^{\mu}(x) S = A_{\text{in}}^{\mu}(x) - i\int d^4y\,[A_{\text{in}}^{\mu}(x), A_{\text{in}}(y)\cdot j(y)] \tag{4-12}$$

This is reminiscent of the formula (2-81):

$$e^A B\, e^{-A} = B + [A, B] + \frac{1}{2!}[A,[A, B]] + \cdots$$

Indeed, only the first two terms of the right-hand side do not vanish, since A and B represent free fields, the commutator of which is a c number. Therefore, we may write

$$\begin{aligned} S &= \exp\left[-i\int d^4x\, A_{\text{in}}(x)\cdot j(x)\right] \\ &= \exp\left[-i\int d^4x\, A_{\text{out}}(x)\cdot j(x)\right] \end{aligned} \tag{4-13}$$

This form does satisfy all the conditions, including unitarity in the indefinite metric space. Only a c-number phase, depending possibly on j, is still arbitrary in S; the latter does not affect the physical quantities.

It is convenient to rewrite S in normal order. We decompose A^{μ}_{in} as a sum of annihilation $A^{\mu(+)}_{\text{in}}$ and creation $A^{\mu(-)}_{\text{in}}$ operators. We observe that $\partial_\mu A^{\mu(+)}_{\text{in}} = \partial_\mu A^{\mu(+)}_{\text{out}}$, and therefore S must leave the positive metric physical subspace invariant. The commutator of $A^{\mu(-)}$ and $A^{\nu(+)}$ is a c-number function:

$$[A^{\mu(-)}_{\text{in}}(x), A^{\nu(+)}_{\text{in}}(y)] = g^{\mu\nu}\int \frac{d^4k}{(2\pi)^3}\, e^{-ik\cdot(x-y)}\theta(k^0)\delta(k^2) \tag{4-14}$$

Therefore we may use the identity

$$e^A\, e^B = e^{A+B+[A,\, B]/2} \tag{4-15}$$

which is valid whenever $[A,[A, B]] = [B,[A, B]] = 0$, and write the normal form of S:

$$\begin{aligned} S &= \exp\left[-i\int d^4y\, A_{\text{in}}(y)\cdot j(y)\right] \\ &= \exp\left[-i\int d^4y\, A^{(-)}_{\text{in}}(y)\cdot j(y)\right]\exp\left[-i\int d^4y\, A^{(+)}_{\text{in}}(y)\cdot j(y)\right] \\ &\quad\times \exp\left\{\tfrac{1}{2}\iint d^4x\, d^4y\, [A^{(-)}_{\text{in}}(x)\cdot j(x), A^{(+)}_{\text{in}}(y)\cdot j(y)]\right\} \end{aligned} \tag{4-16}$$

We introduce the Fourier transform of the current $j(x)$:

$$J^\mu(k) = \int d^4x\, j^\mu(x)\, e^{-ik\cdot x} \tag{4-17}$$

The real character of j and its conservation law are expressed by

$$J^\mu(-k) = J^{\mu *}(k) \qquad \text{and} \qquad k_\mu J^\mu(k) = 0 \tag{4-18}$$

The exponent of the last term in Eq. (4-16) is

$$\frac{1}{2}\iint d^4x\, d^4y\, [A_{\text{in}}^{(-)}(x)\cdot j(x), A_{\text{in}}^{(+)}(y)\cdot j(y)] = \frac{1}{2}\int \frac{d^3k}{2k^0(2\pi)^3}\, J^*(k)\cdot J(k)\big|_{k^0=|\mathbf{k}|}$$

As expected, only the Fourier components of the source corresponding to light-like arguments do contribute. Moreover, for $k^2 = 0$, we may decompose

$$J^\mu(k) = k^\mu J_l(k) + J_{\text{tr}}^\mu(k) \tag{4-19}$$

where $J_l(k)$ is a number and $J_{\text{tr}}^\mu(k)$ is a space-like vector orthogonal to k. For instance, if $k = (k^0, \mathbf{k})$, we introduce the space-like four-vectors $\varepsilon_1 = (0, \mathbf{e}_1)$, $\varepsilon_2 = (0, \mathbf{e}_2)$ with $\mathbf{e}_1^2 = \mathbf{e}_2^2 = 1$, $\mathbf{e}_1 \cdot \mathbf{e}_2 = \mathbf{e}_1 \cdot \mathbf{k} = \mathbf{e}_2 \cdot \mathbf{k} = 0$. We may then choose $J_{\text{tr}}^\mu(k) = -\sum_{i=1,2} J_i(k)\varepsilon_i^\mu(k)$ with $J_i(k) = \varepsilon_i \cdot J(k)$. Using this decomposition, it is easy to see that

$$J^*(k)\cdot J(k) = J_{\text{tr}}^*(k)\cdot J_{\text{tr}}(k) = -[|J_1(k)|^2 + |J_2(k)|^2]$$

In other words, only transverse components contribute to the last term of (4-16):

$$\begin{aligned} S = {}& \exp\left[-i\int d^4x\, A_{\text{in}}^{(-)}(x)\cdot j(x)\right]\exp\left[-i\int d^4x\, A_{\text{in}}^{(+)}(x)\cdot j(x)\right] \\ & \times \exp\left\{-\tfrac{1}{2}\int d\tilde{k}\,[|J_1(k)|^2 + |J_2(k)|^2]\right\} \end{aligned} \tag{4-20}$$

The probabilities of emission are expressed in terms of these transverse components. Indeed, for p_0 we find

$$\begin{aligned} p_0 &= |\langle 0\,\text{out}|0\,\text{in}\rangle|^2 = |\langle 0\,\text{in}|S|0\,\text{in}\rangle|^2 \\ &= \exp\left\{-\int d\tilde{k}\,[|J_1(k)|^2 + |J_2(k)|^2]\right\} \end{aligned} \tag{4-21}$$

To compute the probability p_n corresponding to the emission of n photons when neither the momenta nor the polarizations are observed, we recall that an n-photon state is expressed by

$$|k_1\lambda_1, k_2\lambda_2, \ldots, k_n\lambda_n\rangle = a^{\dagger(\lambda_1)}(k_1)\cdots a^{\dagger(\lambda_n)}(k_n)\,|0\rangle$$

with the normalization

$$\begin{aligned} &\langle k_1\lambda_1, \ldots, k_n\lambda_n | k_1'\lambda_1', \ldots, k_n'\lambda_n'\rangle \\ &\quad = \sum \delta_{\lambda_1\lambda'_{P_1}} \delta_{\lambda_n\lambda'_{P_n}} (2\pi)^3\, 2k_1^0\, \delta^3(\mathbf{k}_1 - \mathbf{k}_{P_1})\cdots(2\pi)^3\, 2k_n^0\, \delta^3(\mathbf{k}_n - \mathbf{k}_{P_n}) \end{aligned}$$

where the sum runs over permutations.

Owing to Bose statistics, the projector over the n photon states reads

$$P_n = \frac{1}{n!}\int d\tilde{k}_1 \cdots d\tilde{k}_n \sum_{\lambda_i = 1,2} |k_1\lambda_1, \ldots, k_n\lambda_n\rangle\langle k_1\lambda_1, \ldots, k_n\lambda_n| \tag{4-22}$$

We have therefore to consider the matrix element

$$\begin{aligned}\langle k_1\lambda_1,\ldots,k_n\lambda_n,\text{out}|0\text{ in}\rangle &= \langle k_1\lambda_1,\ldots,k_n\lambda_n,\text{in}|S|0\text{ in}\rangle\\ &= \exp\left\{-\tfrac{1}{2}\int d\tilde{k}\,[|J_1(k)|^2+|J_2(k)|^2]\right\}\\ &\quad\times\langle k_1\lambda_1,\ldots,k_n\lambda_n,\text{in}|\exp\left[-i\int d^4x\,A_{\text{in}}^{(-)}(x)\cdot j(x)\right]|0\text{ in}\rangle\end{aligned} \tag{4-23}$$

Now

$$\begin{aligned}\int d^4x\,A_{\text{in}}^{(-)}(x)\cdot j(x) &= \int d\tilde{k}\sum_{\lambda=0}^{3} a^{(\lambda)\dagger}(k)\varepsilon^{(\lambda)}(k)\cdot J(k)\\ &= \int d\tilde{k}\sum_{\lambda=1,2} a^{(\lambda)\dagger}(k)J_\lambda(k)\end{aligned}$$

Current conservation has again limited the sum to the transverse degrees of freedom; if we include the "longitudinal" and "scalar" photons in the projector P_n, their contribution cancels automatically. In Eq. (4-23), the term with n creation operators gives the only contribution:

$$\begin{aligned}\frac{(-i)^n}{n!}\langle k_1\lambda_1,\ldots,k_n\lambda_n|\int d\tilde{q}_1\cdots d\tilde{q}_n\,a^{(\sigma_1)\dagger}(q_1)J_{\sigma_1}(q_1)\cdots a^{(\sigma_n)\dagger}(q_n)J_{\sigma_n}(q_n)|0\rangle\\ =(-i)^nJ_{\lambda_1}(k_1)\cdots J_{\lambda_n}(k_n)\end{aligned} \tag{4-24}$$

The factor $1/n!$ has disappeared, since there were $n!$ equal terms. The probability p_n reads

$$\begin{aligned}p_n &= \langle 0\text{ in}|P_n|0\text{ in}\rangle\\ &= \frac{1}{n!}\left\{\int d\tilde{q}\,[|J_1(q)|^2+|J_2(q)|^2]\right\}^n\exp\left\{-\int d\tilde{k}\,[|J_1(k)|^2+|J_2(k)|^2]\right\}\end{aligned} \tag{4-25}$$

Defining

$$\bar{n}\equiv\int d\tilde{k}\,[|J_1(k)|^2+|J_2(k)|^2] \tag{4-26}$$

we obtain the Poisson distribution

$$p_n = e^{-\bar{n}}\frac{\bar{n}^n}{n!} \tag{4-27}$$

The distribution is normalized:

$$\sum_0^\infty p_n = 1$$

and the average number of emitted photons is $\bar{n}$:

$$\sum_{0}^{\infty} np_n = \bar{n}$$

The Poisson distribution (4-27) reflects the statistical independence of the emission of successive photons, which may also be seen on the factorized structure of the matrix element (4-24). On the other hand, let us examine the nature of the final state. We start at time $-\infty$ from the vacuum state $|0 \text{ in}\rangle$ which is an eigenvector of the annihilation part $A_{\text{in}}^{\mu(+)}(x)$:

$$A_{\text{in}}^{\mu(+)}(x)\,|0 \text{ in}\rangle = 0$$

We may replace the time evolution of the state by that of the operator, and thus consider

$$A_{\text{out}}^{\mu(+)}(x)\,|0 \text{ in}\rangle = S^{-1}A_{\text{in}}^{\mu(+)}(x)S\,|0 \text{ in}\rangle = A_{\text{cl}}^{\mu(+)}(x)\,|0 \text{ in}\rangle \tag{4-28}$$

Here $A_{\text{cl}}^{\mu(+)}$ is the positive frequency part of the classical field involved in Eq. (4-11):

$$A_{\text{cl}}^{\mu(+)}(x) = \frac{i}{(2\pi)^3}\int d^3k\,\frac{1}{2k^0}\,e^{-ik\cdot x}J^{\mu}(k)\big|_{k^0=|\mathbf{k}|} \tag{4-29}$$

Hence we have

$$\langle 0 \text{ in}|\,A_{\text{out}}^{\mu}(x)\,|0 \text{ in}\rangle = A_{\text{cl}}^{\mu}(x) = \int d^4y\; G^{(-)}(x-y)j^{\mu}(y) \tag{4-30}$$

Equation (4-28) means that the final state is a coherent state (compare Chap. 3). This is not contradictory with the Poisson distribution of the emission. Indeed, there is a deep connection between this property of the state to be an eigenstate of the operator $A^{(+)}$ and the statistical independence of successive emissions. In a loose sense, neglecting quantum fluctuations, the field in the final state is $A_{\text{cl}}(x)$. This is what is expressed by Eq. (4-30).

4-1-2 Emitted Energy and the Infrared Catastrophe

To obtain an average value of the energy emitted in the process we consider

$$\begin{aligned}\bar{\mathscr{E}} &= \langle 0 \text{ in}|\,H(A_{\text{out}})\,|0 \text{ in}\rangle \\ &= \langle 0 \text{ in}|\,S^{-1}H(A_{\text{in}})S\,|0 \text{ in}\rangle \\ &= \langle 0 \text{ in}|\,S^{-1}\int d\tilde{k}\;k^0 \sum_{\lambda=1,2} a_{\text{in}}^{(\lambda)\dagger}(k)a_{\text{in}}^{(\lambda)}(k)S\,|0 \text{ in}\rangle\end{aligned} \tag{4-31}$$

and observe that the contributions of unphysical degrees of freedom cancel in the sum as expected. From Eq. (4-11), it follows that

$$a_{\text{out}}^{(\lambda)\dagger}(k) = S^{-1}a_{\text{in}}^{(\lambda)\dagger}(k)S = a_{\text{in}}^{(\lambda)\dagger}(k) - iJ_{\lambda}(k) \qquad \lambda = 1, 2$$

and thus

$$\bar{\mathscr{E}} = \int d\tilde{k}\, k^0[|J_1(k)|^2 + |J_2(k)|^2] = \frac{1}{2}\int \frac{d^3k}{(2\pi)^3}[|J_1(k)|^2 + |J_2(k)|^2] \tag{4-32}$$

This result agrees with the classical theory! Indeed, the radiated electromagnetic field $F^{\mu\nu}$ corresponding to the potential A^μ_{cl} is

$$F^{\mu\nu}_{\text{cl}}(x) = \partial^\mu A^\nu_{\text{cl}} - \partial^\nu A^\mu_{\text{cl}} = \int \frac{d^4k}{(2\pi)^3}\,\varepsilon(k^0)\delta(k^2)\,e^{-ik\cdot x}[k^\mu J^\nu(k) - k^\nu J^\mu(k)]$$

The energy density is

$$\varepsilon(x) = \frac{1}{2}\left\{\sum_{\alpha=1}^{3}[F^{0\alpha}_{\text{cl}}(x)]^2 + \sum_{1\le\alpha<\beta\le 3}[F^{\alpha\beta}_{\text{cl}}(x)]^2\right\}$$

and its integral, averaged over a long period of time, reads

$$\bar{\mathscr{E}}_{\text{cl}} = \overline{\int d^3x\,\varepsilon(x)} = \int d\tilde{k}\,\frac{1}{2k^0}\left[\sum_{\alpha=1}^{3}|k^0J^\alpha(k) - k^\alpha J^0(k)|^2 \right.$$
$$\left. + \sum_{1\le\alpha<\beta\le 3}|k^\alpha J^\beta(k) - k^\beta J^\alpha(k)|^2\right]$$

Using the decomposition (4-19) on the light cone, the longitudinal component of J does not contribute, and we get

$$\bar{\mathscr{E}}_{\text{cl}} = \int d\tilde{k}\,k^0[|J_1(k)|^2 + |J_2(k)|^2] \tag{4-33}$$

in agreement with (4-32). The interpretation of this result and of the one provided by Eq. (4-26) is that the number of photons emitted in a phase space element $d\tilde{k}$ is

$$d\bar{n} = \hbar^{-1}\,d\tilde{k}[|J_1(k)|^2 + |J_2(k)|^2] \tag{4-34}$$

and their energy is

$$d\bar{\mathscr{E}} = \hbar k^0\,d\bar{n} \tag{4-35}$$

In particular, if a finite amount of energy is emitted at low frequency, the number of soft photons $d\bar{n} = d\bar{\mathscr{E}}/\hbar k^0$ tends to blow up. Typically, it may happen for certain currents J that $\int d\bar{\mathscr{E}}$ is finite, whereas $\int d\bar{\mathscr{E}}/\hbar k^0$ diverges. This is the famous infrared catastrophe, already encountered in Chap. 1. We saw there that when a charge is suddenly accelerated from momentum p_i to momentum p_f, it creates a current

$$J^\mu(k) = ie\left(\frac{p_f{}^\mu}{p_f\cdot k} - \frac{p_i{}^\mu}{p_i\cdot k}\right)$$

As a consequence, it radiates a flow of photons, of finite total energy $\mathscr{E} \sim \int dk^0$. Unfortunately, the total number of particles diverges as $\bar{n} \sim \int dk^0/\hbar k^0$. Therefore it is impossible to abstract a charged particle from its radiation field. However, as stressed in Chap. 1, it is meaningless to count the number of emitted soft photons, and the only physically measurable quantity is the emitted energy.

Mathematically, the situation is really catastrophic. Indeed, when $\bar{n} \to \infty$, we learn from Eqs. (4-21) and (4-26) that

$$|\langle 0 \text{ out} | 0 \text{ in} \rangle| = e^{-\bar{n}/2} \to 0$$

Every matrix element between in- and out-states vanishes. Clearly, it is impossible to construct the "out" Fock space from the "in" space, nor to find a unitary operator S. This is exactly the same situation as in the simple model discussed at the beginning of this section. For a system with an infinite number of degrees of freedom, and under certain circumstances (here $\bar{n} = \infty$), inequivalent representations of the canonical commutation relations may exist. It is no wonder that we get into trouble here, since we try to describe the final states as superpositions of states of a finite number of photons, while we know that their actual number is infinite.

Physically, we may cut off a part of phase space, i.e., we decide to observe only final photons within a certain range of energy momentum. This indeed corresponds to an experimental limitation. Every photon detector has a given finite resolution, and photons of energy lower than this resolution are therefore unobservable. Let R be the unobserved region of phase and C_R its complement. The total probability of emission in R only, i.e., to detect no photon

$$p^R = \sum_{n=0}^{\infty} p_n^R = \sum_{n=0}^{\infty} e^{-\bar{n}} \frac{(\bar{n}_R)^n}{n!} = e^{-(\bar{n}-\bar{n}_R)} = \exp(-\bar{n}_{C_R})$$

is nonvanishing since

$$\bar{n}_{C_R} = \int_{C_R} d\tilde{k} \, [|J_1(k)|^2 + |J_2(k)|^2]$$

is finite, while each term of the sum vanishes as $\bar{n} \to \infty$. The observed radiated energy $\bar{\mathscr{E}}(C_R) = \bar{\mathscr{E}} - \bar{\mathscr{E}}(R)$ is as close to $\bar{\mathscr{E}}$ as desired, for small enough R, and the probability of detecting at least a photon with momentum outside R, $1 - \exp(-\bar{n}_{C_R})$, is finite.

There is an alternative possibility to avoid the slippery subtleties of the previous treatment of infrared divergences, at least of its mathematical difficulties. We give the photon a small mass μ and use the Stueckelberg gauge of Chap. 3 to quantize such a field. This will cut off the low-energy region since now $k^0 > \mu$ and therefore remove the infrared divergence. However, we have to verify that the extra degrees of freedom do not introduce any spurious effect, namely, that observable quantities are not affected in the limit $\mu \to 0$.

We thus consider a massive photon field coupled to a conserved current. In the massless case, we recall that the indefinite metric state has played no role; only transverse degrees of freedom have been excited. Similarly, here, in the Stueckelberg gauge, the conservation law of the current implies that only the transverse field

$$A_\mu^T = A_\mu + \frac{\lambda}{\mu^2} \partial_\mu(\partial \cdot A)$$

is coupled to the current

$$\int A_\mu^T(x) j^\mu(x)\, d^4x = \int A_\mu(x) j^\mu(x)\, d^4x$$

Therefore the negative metric states disappear. However, the longitudinal polarization state still contributes. We may compute $\bar{n}$, for instance, from Eqs. (4-16) and (3-148):

$$\bar{n} = -\int d^4x\, d^4y\, [A_{\text{in}}^{(-)}(x)\cdot j(x), A_{\text{in}}^{(+)}(y)\cdot j(y)]$$

$$= -\int \frac{d^3k}{2k^0(2\pi)^3} J(k)\cdot J^*(k)\Big|_{k^0=(\mathbf{k}^2+\mu^2)^{1/2}} \tag{4-36}$$

Typically for an accelerated charge, $J\cdot J^* \sim 1/k_0^2$, as we recalled above. Cutting off integrals at some $|\mathbf{k}| = k_{\text{max}}$, we get a potentially divergent $\bar{n}$:

$$\bar{n} \sim \int_0^{k_{\text{max}}} \frac{dk\, k^2}{(k^2+\mu^2)^{3/2}} \sim \ln\frac{k_{\text{max}}}{\mu}$$

while the emitted energy

$$\bar{\mathscr{E}} \sim \int_0^{k_{\text{max}}} \frac{dk\, k^2}{k^2+\mu^2} \sim k_{\text{max}} - \mu\frac{\pi}{2}$$

has a smooth behavior as $\mu \to 0$. We also find that the contribution $\bar{n}_L$ of longitudinal photons of given momentum $\mathbf{k}$ is vanishingly small with respect to that of transverse photons $\bar{n}_T$:

$$\frac{\bar{n}_L}{\bar{n}_T} \sim \frac{|J_l|^2 - |J_0|^2}{|\mathbf{J}_{\text{tr}}|^2} = \frac{\mu^2}{|\mathbf{k}|^2}\frac{|J_0|^2}{|\mathbf{J}_{\text{tr}}|^2} \tag{4-37}$$

where current conservation has been used to write $|J_l| = (k_0/|\mathbf{k}|)|J_0|$. The introduction of this small photon mass has provided a convenient regularization of the infrared divergence. However, the existence of a third state of polarization might modify the blackbody radiation spectrum by a factor 3/2. This difficulty is bypassed by the assumption that the equilibrium time for the third mode may be so large that the effect is totally unobservable.

It must be clear from this discussion that the main features of the infrared divergences of quantum electrodynamics are essentially classical, and depend on the nature of the external current and on the experimental resolution. However, quantum effects manifest themselves, e.g., in the fluctuation of the number of emitted photons. Assuming $\bar{n}$ to be finite, we compute

$$\overline{\Delta n^2} \equiv \overline{n^2} - \bar{n}^2 = \bar{n}$$

and therefore

$$\frac{\overline{\Delta n}}{\bar{n}} = \bar{n}^{-1/2} \tag{4-38}$$

These fluctuations are very small if $\bar{n}$ is large. Similarly, the energy fluctuations read

$$\overline{\Delta\mathscr{E}^2} = (\overline{\mathscr{E}^2} - \bar{\mathscr{E}}^2) = \int d\tilde{k}\, k_0^2\, [|J_1(k)|^2 + |J_2(k)|^2] \tag{4-39}$$

If the radiated frequencies are strongly peaked about the value $\langle k^0 \rangle$, we have

$$\overline{\Delta \mathscr{E}} \sim (\hbar \langle k^0 \rangle \bar{\mathscr{E}})^{1/2} \qquad \text{and} \qquad \bar{\mathscr{E}} \sim \bar{n}\hbar \langle k^0 \rangle$$

and thus, as expected,

$$\frac{\overline{\Delta \mathscr{E}}}{\bar{\mathscr{E}}} \sim \frac{1}{\sqrt{\bar{n}}} = \frac{\overline{\Delta n}}{\bar{n}} \tag{4-40}$$

4-1-3 Induced Absorption and Emission

We can use the explicit forms (4-13) and (4-16) obtained for the S matrix to discuss briefly some interesting situations. Namely, we may wonder how the presence of photons in the initial state affects the radiation of the source j.

We have seen in Eq. (4-28) that a classical current creates a coherent state from the vacuum. This coherent state may in turn be considered as the initial state for a second source j. In other words, the first source j_{inc} which generates the initial state

$$|j_{\text{inc}}\rangle = \exp\left[-i \int d^4x \, A_{\text{in}}(x) \cdot j_{\text{inc}}(x)\right] |0 \text{ in}\rangle \tag{4-41}$$

is assumed to be well separated from the second source $j(x)$. The projection of the final state on out-states reads

$$\begin{aligned}
\langle b \text{ out} | j_{\text{inc}}\rangle &= \langle b \text{ in}| \exp\left[-i \int d^4x \, A_{\text{in}}(x) \cdot j(x)\right] \exp\left[-i \int d^4y \, A_{\text{in}}(y) \cdot j_{\text{inc}}(y)\right] |0 \text{ in}\rangle \\
&= \exp\left[-\frac{i}{2} \int\int d^4x \, d^4y \, j_\mu(x) G^{(-)}(x-y) j^\mu_{\text{inc}}(y)\right] \\
&\quad \times \langle b \text{ in}| \exp\left\{-i \int d^4y \, A_{\text{in}}(y) \cdot [j_{\text{inc}}(y) + j(y)]\right\} |0 \text{ in}\rangle
\end{aligned} \tag{4-42}$$

The first factor in the last expression is a pure phase, independent of the state b, and thus unobservable. The second term tells us that the final state is produced by the sum $j_{\text{inc}} + j$. The total number $\bar{n}_{\text{tot}}$ of photons is obtained by substituting $J + J_{\text{inc}}$ for J in Eq. (4-26), and the (average) number of radiated photons, defined as the difference $\bar{n}_{\text{tot}} - \bar{n}_{\text{inc}}$, is

$$\bar{n}_{\text{rad}} = \bar{n}_{\text{tot}} - \bar{n}_{\text{inc}} = \int d\tilde{k} \, |\mathbf{J}_{\text{tr}}(k)|^2 + 2 \, \text{Re} \int d\tilde{k} \, \mathbf{J}_{\text{tr}}(k) \cdot \mathbf{J}^*_{\text{inc,tr}}(k) \tag{4-43}$$

The first term on the right-hand side of (4-43) is the number of photons emitted by the source j alone, while the second one is a typical interference term, which represents the stimulated absorption or emission. We observe that this term is linear in j_{inc} (or in $F^{\mu\nu}_{\text{inc}}$) and that terms corresponding to different frequencies are decoupled. This reflects again the independence of the various modes. The reader may compute the radiated energy in the presence of j_{inc} and observe again this interference phenomenon. On the one hand, we know that two classical sources interfere and energy interferences must imply interferences in photon number since $d\bar{n}(k) = d\bar{\mathscr{E}}(k)/\hbar k^0$. On the other hand, in view of the stochastic nature of the emission, it is rather surprising to find an interference in the number of emitted photons. This shows clearly the limitation of this stochastic interpretation. Emission and absorption are connected since an emitting source may also absorb photons. Then the equality $\bar{n}_{\text{tot}} = \bar{n}_{\text{inc}} + \bar{n}_j$ would mean the impossibility of absorption!

It is instructive to compare this situation with the case where the initial state has a definite number of photons. For simplicity, we assume all photons to be in the same mode:

$$|\bar{n}_{\rm inc}\ {\rm in}\rangle = \frac{1}{\sqrt{\bar{n}_{\rm inc}!}}\left[\int d\tilde{k} \sum_{\lambda=1,2} f_\lambda(k) a^{(\lambda)\dagger}(k)\right]^{\bar{n}_{\rm inc}} |0\ {\rm in}\rangle \tag{4-44}$$

The function f_λ is normalized:

$$\int d\tilde{k} \sum_{\lambda=1,2} |f_\lambda(k)|^2 = 1$$

and it is understood that $f(k)$ is peaked about a mean value. We shall only compute the average number of photons in the final state

$$\bar{n}_{\rm tot} = \langle \bar{n}_{\rm inc}\ {\rm in}|N_{\rm out}|\bar{n}_{\rm inc}\ {\rm in}\rangle$$

where

$$\begin{aligned} N_{\rm out} &= \int d\tilde{k} \sum_{\lambda=1,2} a^{(\lambda)\dagger}_{\rm out}(k) a^{(\lambda)}_{\rm out}(k) = S^\dagger \int d\tilde{k} \sum_\lambda a^{(\lambda)\dagger}_{\rm in}(k) a^{(\lambda)}_{\rm in}(k) S \\ &= \int d\tilde{k} \sum_\lambda [a^{(\lambda)\dagger}_{\rm in}(k) - iJ_\lambda(k)][a^{(\lambda)}_{\rm in}(k) + iJ^*_\lambda(k)] \end{aligned} \tag{4-45}$$

We only count the number of transverse photons, since longitudinal and scalar photons are absent in the initial state and are not emitted. Thus

$$\begin{aligned} \bar{n}_{\rm tot} &= \bar{n}_{\rm inc} + \int d\tilde{k}\,[|J_1(k)|^2 + |J_2(k)|^2] \\ &\quad + \langle \bar{n}_{\rm inc}| \int d\tilde{k} \left[\sum_\lambda i a^{(\lambda)\dagger}(k) J_\lambda(k) + \text{hermitian conjugate}\right] |\bar{n}_{\rm inc}\rangle \\ &= \bar{n}_{\rm inc} + \int d\tilde{k}\,[|J_1(k)|^2 + |J_2(k)|^2] \end{aligned} \tag{4-46}$$

The average number of radiated photons is the same as in the absence of photons in the initial state! On the average, there is no stimulated absorption or emission. But this does not mean that the emission probabilities are unchanged. Indeed, we may compute the probability of finding $\bar{m}$ photons in the final state. An easy calculation leads to

$$\begin{aligned} p_{\bar{n}_{\rm inc}\to\bar{m}} &= \frac{1}{\bar{n}_{\rm inc}!\,\bar{m}!} \frac{\partial^{\bar{n}_{\rm inc}}}{\partial z^{\bar{n}_{\rm inc}}} \frac{\partial^{\bar{n}_{\rm inc}}}{\partial \bar{z}^{\bar{n}_{\rm inc}}} \left(\exp - \left\{ \bar{n}_j + \int d\tilde{k} \left[z \sum_{\lambda=1,2} J^*_\lambda(k) f_\lambda(k) + \text{complex conjugate} \right] \right\} \right. \\ &\quad \times \left. \left[\int d\tilde{k} \sum_{\lambda=1,2} |J_\lambda(k) + z f_\lambda(k)|^2 \right]^{\bar{m}} \right) \Bigg|_{z=\bar{z}=0} \end{aligned} \tag{4-47}$$

For a weak source J, we may keep the lowest order in J. The only nonvanishing transition probabilities to this order are

$$\begin{aligned} p_{\bar{n}_{\rm inc}\to\bar{n}_{\rm inc}+1} &= \int d\tilde{k} \sum_{\lambda=1,2} |J_\lambda(k)|^2 + \bar{n}_{\rm inc} \left| \int d\tilde{k} \sum_{\lambda=1,2} f^*_\lambda(k) J_\lambda(k) \right|^2 \\ p_{\bar{n}_{\rm inc}\to\bar{n}_{\rm inc}} &= 1 - \int d\tilde{k} \sum_{\lambda=1,2} |J_\lambda(k)|^2 - 2\bar{n}_{\rm inc} \left| \int d\tilde{k} \sum_{\lambda=1,2} f^*_\lambda(k) J_\lambda(k) \right|^2 \\ p_{\bar{n}_{\rm inc}\to\bar{n}_{\rm inc}-1} &= \bar{n}_{\rm inc} \left| \int d\tilde{k} \sum_{\lambda=1,2} f^*_\lambda(k) J_\lambda(k) \right|^2 = 1 - p_{\bar{n}_{\rm inc}\to\bar{n}_{\rm inc}} - p_{\bar{n}_{\rm inc}\to\bar{n}_{\rm inc}+1} \end{aligned} \tag{4-48}$$

The first expression may be compared with

$$p_{0\to1} = \int d\tilde{k} \sum_{\lambda=1,2} |J_\lambda(k)|^2$$

The presence of photons in the initial state has raised the probability of emission. This is the basic result of stimulated emission. From Eqs. (4-48) it is also easy to check that the average number of radiated photons is still $\bar{n}_j$.

4-1-4 *S* Matrix and Evolution Operator

The derivation of the S matrix in Eq. (4-13) relied on the assumption that the source was not an independent dynamical variable, i.e., that its time evolution was given once and for all. This method has left an arbitrary, source-dependent phase in S. It is therefore appropriate to set up now the general formalism and to compare its results in this specific case. The idea is to construct the operator that realizes the time-dependent canonical transformation relating the interacting field A^μ to the incoming field A^μ_{in}:

$$A^\mu(t, \mathbf{x}) = U^{-1}(t)\, A^\mu_{\text{in}}(t, \mathbf{x})\, U(t) \tag{4-49}$$

As already discussed and expressed in Eq. (4-8), A^μ_{in} is some weak limit of A^μ as $t \to -\infty$, and, consequently, in that limit U reduces to the identity

$$\lim_{t \to -\infty} U(t) = I \tag{4-50}$$

In the canonical quantization, we deal with the operators $A(t)$ and $\pi(t)$ that satisfy

$$\frac{\partial}{\partial t} A(t, \mathbf{x}) = i[H(t), A(t, \mathbf{x})] \tag{4-51}$$

and an analogous equation for $\pi(t)$. Here $H(t) = H[A(t), \pi(t), j(t)]$ denotes the hamiltonian. Similarly, the in-field A_{in} satisfies

$$\frac{\partial}{\partial t} A_{\text{in}}(t, \mathbf{x}) = i[H_0^{\text{in}}, A_{\text{in}}(t, \mathbf{x})]$$

where H_0^{in} is the time-independent free hamiltonian [compare Eq. (3-120)] expressed in terms of the in-creation and annihilation operators. Let us derive the time evolution equation for the operator $U(t)$. Unitarity requires that

$$\left[\frac{d}{dt} U(t)\right] U^{-1}(t) + U(t) \frac{d}{dt} U^{-1}(t) = 0$$

Moreover, it follows from Eq. (4-49) that

$$U(t) H\big(A(t), \pi(t), j(t)\big) U^{-1}(t) = H\big(A_{\text{in}}(t), \pi_{\text{in}}(t), j(t)\big)$$

Now

$$\frac{\partial}{\partial t} A_{\text{in}}(t, \mathbf{x}) = \frac{\partial}{\partial t} [U(t) A(t, \mathbf{x}) U^{-1}(t)]$$

$$= \frac{dU(t)}{dt} U^{-1}(t) A_{\text{in}}(t, \mathbf{x}) + iU(t)\big[H\big(A(t), \pi(t), j(t)\big), A(t, \mathbf{x})\big] U^{-1}(t)$$

$$+ A_{\text{in}}(t, \mathbf{x}) U(t) \frac{d}{dt} U^{-1}(t)$$

$$= \left[\frac{dU(t)}{dt} U^{-1}(t) + iH\big(A_{\text{in}}(t), \pi_{\text{in}}(t), j(t)\big), A_{\text{in}}(t, \mathbf{x}) \right]$$

$$= i\left[H_0^{\text{in}}, A_{\text{in}}(t, \mathbf{x})\right]$$

with an analogous equation for π_{in}. We conclude that the operator

$$\frac{dU(t)}{dt} U^{-1}(t) + i\left[H\big(A_{\text{in}}(t), \pi_{\text{in}}(t), j(t)\big) - H_0^{\text{in}}\right]$$

commutes with every "in" operator and is thus a c number. This number will not contribute to normalized matrix elements of U (this point will be justified later), and we discard it hereafter. Then U satisfies the evolution equation

$$\begin{aligned} i \frac{dU}{dt} &= \left[H\big(A_{\text{in}}(t), \pi_{\text{in}}(t), j(t)\big) - H_0^{\text{in}}\right] U \\ &\equiv H_{\text{int}}\big(A_{\text{in}}(t), \pi_{\text{in}}(t), j(t)\big) U \equiv H_I(t) U \end{aligned} \tag{4-52}$$

In our case $H_0^{\text{in}} = H\big(A_{\text{in}}(t), \pi_{\text{in}}(t), j\big)\big|_{j=0}$, and the interaction hamiltonian $H_I(t)$ vanishes with j. The evolution dictated by the operator $U(t)$ will thus be referred to as the interaction representation. Notice that $H_I(t)$ depends on time both through the time dependence of $j(t)$ and of $A_{\text{in}}(t)$ and $\pi_{\text{in}}(t)$. Equation (4-52) may be solved by iteration of the corresponding integral equation:

$$U(t) = I - i \int_{-\infty}^{t} dt_1 \, H_I(t_1) U(t_1) \tag{4-53}$$

where the boundary condition (4-50) has been taken into account, with the result that

$$\begin{aligned} U(t) = I &- i \int_{-\infty}^{t} dt_1 \, H_I(t_1) + (-i)^2 \int_{-\infty}^{t} dt_1 \int_{-\infty}^{t_1} dt_2 \, H_I(t_1) H_I(t_2) + \cdots \\ &+ (-i)^n \int_{-\infty}^{t} dt_1 \int_{-\infty}^{t_1} dt_2 \cdots \int_{-\infty}^{t_{n-1}} dt_n \, H_I(t_1) \cdots H_I(t_n) + \cdots \end{aligned} \tag{4-54}$$

In these expressions, the ordering of the operators is important: it is related to the ordering of their time arguments. We are led to the general definitions of the time-ordered product of n operators (T product, for short):

$$T[A_1(t_1) \cdots A_n(t_n)] = \sum_P \theta(t_{P_1}, t_{P_2}, \ldots, t_{P_n}) \varepsilon_P A_{P_1}(t_{P_1}) \cdots A_{P_n}(t_{P_n}) \tag{4-55}$$

where the sum runs over all permutations P, the θ function enforces the condition

$$t_{P_1} \geq t_{P_2} \geq \cdots \geq t_{P_n}$$

and ε_P denotes the signature of the fermion operators permutation involved in this product. (The latter point is of no use now, but will serve us soon.) Here, the T product is symmetric and we may write

$$\begin{aligned} U(t) &= \sum_0^\infty (-i)^n \int_{-\infty}^{t} dt_1 \int_{-\infty}^{t_1} dt_2 \cdots \int_{-\infty}^{t_{n-1}} dt_n\, T[H_I(t_1)\cdots H_I(t_n)] \\ &= \sum_{n=0}^\infty \frac{(-i)^n}{n!} \int_{-\infty}^{t} dt_1 \int_{-\infty}^{t} dt_2 \cdots \int_{-\infty}^{t} dt_n\, T[H_I(t_1)\cdots H_I(t_n)] \\ &= T\exp\left[-i\int_{-\infty}^{t} dt'\, H_I(t')\right] \end{aligned} \tag{4-56}$$

The last expression is symbolic. Its meaning is given by the previous line. However, T products of exponentials enjoy an important property, which may be proved easily. For $t_1 \le t_2 \le t_3$,

$$T\exp\int_{t_1}^{t_3} dt\, 0(t) = T\exp\int_{t_2}^{t_3} dt\, 0(t)\, T\exp\int_{t_1}^{t_2} dt\, 0(t)$$

If $H_I(t) \to 0$ as $t \to -\infty$, the operator $U(t)$ of Eq. (4-56) does satisfy the requirement (4-50). This is what happens in our case if the current is adiabatically switched off in the remote past. The S matrix is defined as the limit

$$S = \lim_{t\to+\infty} U(t) = T\exp\left[-i\int_{-\infty}^{\infty} dt'\, H_I(t')\right] \tag{4-57}$$

In the present case, $H_I = \int j(x)\cdot A_{\text{in}}(x)\, d^3x$, and we get

$$S = T\exp\left[-i\int d^4x\, j_\mu(x)\, A^\mu_{\text{in}}(x)\right] \tag{4-58}$$

In this expression, the T symbol contains the only reference to a definite time coordinate. This looks noncovariant, but owing to local commutativity the density $\mathscr{H}_I(x) \equiv j^\mu(x)A_\mu(x)$ commutes with $\mathscr{H}_I(y)$ if $(x-y)^2 < 0$. Therefore a change of frame does not modify the expression of the T product. In every theory without derivative coupling we have

$$H_I(t) = \int d^3x\, \mathscr{H}_I(x) = -\int d^3x\, \mathscr{L}_I(x) \tag{4-59}$$

where it is understood that x^0 stands for t and that all operators are in-operators. Finally, the general "covariant" expression for the S matrix reads

$$S = T\exp\left[i\int d^4x\, \mathscr{L}_I(x)\right] \tag{4-60}$$

The quotation marks recall that our previous argument of locality does not rule out the occurrence of short-distance, a priori noncovariant singularities, when two arguments coincide. The treatment of these singularities is the subject of renormalization theory.

We want to compare the general expression [(4-58), (4-60)] with our previous result (4-13). For the two quantities to agree, the difference should be an unobservable phase. This is what we are going to check, using the property of free fields to have a c-number commutator. We give here a simple heuristic proof and let the reader elaborate it further. At any rate, this property is related to the important Wick theorem, which will soon be examined in detail. Let us write the S matrix as a limit

$$S = \lim_{\substack{t_i \to -\infty \\ t_f \to +\infty}} T \exp\left[-i \int_{t_i}^{t_f} dt\, H_I(t)\right]$$

divide the interval $t_f - t_i$ into N equal ones $\Delta t = (t_f - t_i)/N$, and denote by $t_k = t_i + [(2k-1)/2]\Delta t$ ($k = 1, \ldots, N$) the midpoint of each one. In the large N limit the T product is approximated by

$$T \exp\left[-i \int_{t_i}^{t_f} dt\, H_I(t)\right] = e^{-i\Delta t\, H_I(t_N)}\, e^{-i\Delta t\, H_I(t_{N-1})} \cdots e^{-i\Delta t\, H_I(t_1)}$$

Since the commutator $[H_I(t), H_I(t')]$ is a c number, we use the identity (4-15) to rewrite this product as

$$\exp\left\{-i\Delta t \sum_1^N H_I(t_k) - \tfrac{1}{2}\Delta t^2 \sum_{1 \le k < l \le N} [H_I(t_k), H_I(t_l)]\right\}$$

Letting N go to infinity and taking then the limits $t_i \to -\infty$, $t_f \to \infty$, we find

$$S = \exp\left[-i \int d^4x\, A_{\text{in}}(x)\cdot j(x)\right] \exp\left\{-\tfrac{1}{2} \iint d^4x\, d^4y\, \theta(x^0 - y^0) \times [A_{\text{in}}(x)\cdot j(x), A_{\text{in}}(y)\cdot j(y)]\right\} \tag{4-61}$$

The operators $A_{\text{in}}(x)\cdot j(x)$ and $A_{\text{in}}(y)\cdot j(y)$ are hermitian free fields; therefore their commutator is an imaginary c number. Indeed, we know the expression of the retarded commutator [(1-169) and (1-170)]:

$$\theta(x^0 - y^0)[A^\mu_{\text{in}}(x), A^\nu_{\text{in}}(y)] = ig^{\mu\nu}\theta(x^0 - y^0)G^{(-)}(x - y) = ig^{\mu\nu} G_{\text{ret}}(x - y)$$

where

$$G_{\text{ret}}(x - y) = -\int \frac{d^4k}{(2\pi)^4} \frac{e^{-ik\cdot(x-y)}}{(k_0 + i\varepsilon)^2 - \mathbf{k}^2} = \frac{1}{2\pi}\theta(x^0 - y^0)\delta[(x - y)^2]$$

Finally, we find

$$S = \exp\left[-i \int d^4x\, A_{\text{in}}(x)\cdot j(x)\right] \exp\left[-\frac{i}{2} \iint d^4x\, d^4y\, j^\mu(x) G_{\text{ret}}(x - y) j_\mu(y)\right] \tag{4-62}$$

As in Eq. (4-16), we may write the normal form of this expression:

$$S = :\exp\left[-i\int d^4x\, A_{\rm in}(x)\cdot j(x)\right]:\exp\left(\frac{1}{2}\int d^4x\, d^4y\, j_\mu(x)\{[A_{\rm in}^{(-)\mu}(x), A_{\rm in}^{(+)\nu}(y)] - \theta(x^0-y^0)[A_{\rm in}^\mu(x), A_{\rm in}^\nu(y)]\}\, j_\nu(y)\right)$$

The quantity within the brackets in the second exponential is a c number, and therefore is equal to its vacuum expectation value (in the state $|0\text{ in}\rangle$). We omit for simplicity the subscripts "in" and write

$$\begin{aligned}
\langle 0|\,[A^{(-)\mu}(x), A^{(+)\nu}(y)] - \theta(x^0-y^0)[A^\mu(x), A^\nu(y)]\,|0\rangle
&= -\langle 0|\,A^\nu(y)A^\mu(x) + \theta(x^0-y^0)[A^\mu(x), A^\nu(y)]\,|0\rangle \\
&= -\langle 0|\,\theta(x^0-y^0)A^\mu(x)A^\nu(y) + \theta(y^0-x^0)A^\nu(y)A^\mu(x)\,|0\rangle \\
&= -\langle 0|\,TA^\mu(x)A^\nu(y)\,|0\rangle \\
&= ig^{\mu\nu}\int \frac{d^4k}{(2\pi)^4}\frac{e^{-ik\cdot(x-y)}}{k^2+i\varepsilon}
\end{aligned}$$

To summarize, the normal form of the S matrix reads

$$S = :\exp\left[-i\int d^4x\, A_{\rm in}(x)\cdot j(x)\right]: \times \exp\left[-\frac{1}{2}\iint d^4x\, d^4y\,\langle 0\text{ in}|\,TA_{\rm in}(x)\cdot j(x)\,A_{\rm in}(y)\cdot j(y)\,|0\text{ in}\rangle\right] \quad (4\text{-}63)$$

We emphasize again that this result differs only by an unobservable phase from the expression given in Eq. (4-16).

4-2 WICK'S THEOREM

This section is devoted to a study of the algebraic manipulations that allow one to reorder a T product of operators, such as those appearing in the interaction representation, into a normal form, more convenient for the actual computation of matrix elements in the Fock space. In the sequel, all the fields are free fields.

4-2-1 Bose Fields

We have just derived, for a free massless vector field, the identity

$$T\exp\left[-i\int d^4x\, A(x)\cdot j(x)\right] = :\exp\left[-i\int d^4x\, A(x)\cdot j(x)\right]: \times \exp\left[-\frac{1}{2}\iint d^4x\, d^4y\langle 0|\,TA(x)\cdot j(x)A(y)\cdot j(y)\,|0\rangle\right] \quad (4\text{-}64)$$

This result depends neither on the vector nature nor on the masslessness of the field. The same formula holds for any boson field. Therefore, we shall not specify whether we deal with scalar, or vector, tensor, etc., fields, and omit hereafter all Lorentz indices. Equation (4-64) may be considered as the generating function of a set of identities, known as the Wick theorem. We expand the exponentials and identify the coefficient of $j(x_1)\cdots j(x_n)$, symmetrized over $x_1 \cdots x_n$ (since ultimately this product and its coefficient is integrated over $x_1 \cdots x_n$). We obtain

$$T[A(x_1)A(x_2)] = :A(x_1)A(x_2): + \langle 0|\, TA(x_1)A(x_2)\,|0\rangle$$

$$\begin{aligned}T[A(x_1)A(x_2)A(x_3)] = {}&:A(x_1)A(x_2)A(x_3): + :A(x_1):\langle 0|\, TA(x_2)A(x_3)\,|0\rangle \\ &+ :A(x_2):\langle 0|\, TA(x_1)A(x_3)\,|0\rangle \\ &+ :A(x_3):\langle 0|\, TA(x_1)A(x_2)\,|0\rangle\end{aligned}$$

$$\begin{aligned}T[A(x_1)\cdots A(x_n)] = {}&:A(x_1)\cdots A(x_n): \\ &+ \sum_{k<l} :A(x_1)\cdots \widehat{A(x_k)}\cdots \widehat{A(x_l)}\cdots A(x_n):\ \langle 0|\, TA(x_k)A(x_l)\,|0\rangle + \cdots \\ &+ \sum_{k_1<k_2<\cdots<k_{2p}} :A(x_1)\cdots \widehat{A(x_{k_1})}\cdots \widehat{A(x_{k_{2p}})}\cdots A(x_n): \\ &\times \sum_P \langle 0|\, T[A(x_{k_{P_1}})A(x_{k_{P_2}})]\,|0\rangle \cdots \langle 0|\, T[A(x_{k_{P_{2p-1}}})A(x_{k_{P_{2p}}})]\,|0\rangle + \cdots\end{aligned} \tag{4-65}$$

In these formulas, the caret above a term means that it is to be omitted from the product, and the sum $\sum_P$ runs over all permutations that lead to different expressions. In words, Eq. (4-65) expresses the T product as the sum of all possible normal products where some pairs of fields have been omitted and replaced by their contraction, i.e., the vacuum expectation value of their T product.

It is a good exercise to prove directly the identities (4-65). This is most easily done by induction. We also let readers convince themselves that they may be generalized to the T product of distinct fields $TA_1(x_1)\cdots A_n(x_n)$. They may be further extended to an expression of the form

$$T{:}[A(x_1)\cdots A(x_k)]{:}\cdots{:}[A(x_l)\cdots A(x_n)]{:}$$

with the restriction that only contractions between distinct normal ordered products occur. Identities similar to (4-65) may be written in order to express the ordinary product $A_1(x_1)\cdots A_n(x_n)$ in terms of normal products. The only modification is that now contractions represent the vacuum expectation value of the ordinary product

$$\begin{aligned}A_1(x_1)\cdots A_n(x_n) = {}&\sum_P :A_1(x_1)\cdots \widehat{A_{k_1}(x_{k_1})}\cdots \widehat{A_{k_{2p}}(x_{k_{2p}})}\cdots A(x_n): \\ &\times \{\langle 0|\, A_{k_1}(x_{k_1})A_{k_2}(x_{k_2})\,|0\rangle \cdots \langle 0|\, A_{k_{2p-1}}(x_{k_{2p-1}})\, A_{k_{2p}}(x_{k_{2p}})\,|0\rangle \\ &+ \text{permutations}\}\end{aligned}$$

Finally, analogous identities connect T products and ordinary products. In that case, the contraction stands for a retarded Green function; for instance,

$$TA(x)A(y) = A(x)A(y) + \theta(y^0 - x^0)[A(y), A(x)]$$

From Eq. (4-65), it follows that

$$\begin{aligned} \langle 0|\, T[A(x_1)\cdots A(x_{2p-1})]\,|0\rangle &= 0 \\ \langle 0|\, TA(x_1)\cdots A(x_{2p})\,|0\rangle &= \sum_P \langle 0|\, TA(x_{P_1})A(x_{P_2})\,|0\rangle \cdots \\ &\quad \times \langle 0|\, TA(x_{P_{2p-1}})A(x_{P_{2p}})\,|0\rangle \end{aligned} \tag{4-66}$$

The symmetric form of the right-hand side has been called "haffnian" by Caianiello. Using (4-66), Eq. (4-65) may be rewritten in the weaker form

$$\begin{aligned} TA(x_1)\cdots A(x_n) = \sum_{p=0}^{[n/2]} \sum_{k_1 < \cdots < k_{2p}} &: A(x_1)\cdots \widehat{A(x_{k_1})} \cdots \widehat{A(x_{k_{2p}})} \cdots A(x_n) : \\ &\times \langle 0|\, T[A(x_{k_1})\cdots A(x_{k_{2p}})]\,|0\rangle \end{aligned} \tag{4-67}$$

4-2-2 Fermi Fields

When Fermi fields are involved the only modifications to be made in Eq. (4-65) are the signs. As in the Bose case, we shall derive an identity relating the generating function of the T products to the one of normal products. The derivation for boson fields was based on the identity (4-15). In order to deal only with commutators, when fermion fields are present, we introduce anticommuting sources $\bar{\eta}$ and η for ψ and $\bar{\psi}$. These quantities anticommute among themselves as well as with ψ and $\bar{\psi}$. We insist on their purely mathematical role (they are elements of a Grassmann—i.e., anticommuting—algebra). They enable us to write the identity

$$[\bar{\eta}(x)\psi(x), \bar{\psi}(y)\eta(y)] = \bar{\eta}(x)\{\psi(x), \bar{\psi}(y)\}\eta(y) \tag{4-68}$$

All other commutators of $\bar{\eta}\psi$, $\bar{\psi}\eta$ vanish. We now introduce the fictitious interaction lagrangian

$$\mathscr{L}_I(x) = \bar{\eta}(x)\psi(x) + \bar{\psi}(x)\eta(x) \tag{4-69}$$

and the corresponding S matrix

$$S = T \exp i \int d^4x\, [\bar{\eta}(x)\psi(x) + \bar{\psi}(x)\eta(x)] \tag{4-70}$$

Since the commutator $[\mathscr{L}_I(x), \mathscr{L}_I(y)]$ commutes with $\mathscr{L}_I(z)$, we may use again our favorite identity (4-15) and get

$$\begin{aligned} T \exp i \int d^4x\, [\bar{\eta}(x)\psi(x) + \bar{\psi}(x)\eta(x)] &= \exp\left\{ i \int d^4x \big(\bar{\eta}(x)\psi(x) + \bar{\psi}(x)\eta(x)\big)\right\} \\ \times \exp\left\{ -\tfrac{1}{2} \iint d^4x\, d^4y\, \theta(x^0 - y^0) \right. & \left. [\bar{\eta}(x)\psi(x) + \bar{\psi}(x)\eta(x), \bar{\eta}(y)\psi(y) + \bar{\psi}(y)\eta(y)] \right\} \end{aligned} \tag{4-71}$$

In the same way, we may decompose $\bar{\eta}\psi + \bar{\psi}\eta$ into a sum of annihilation, positive frequency part (superscript +), and creation part (superscript −), and write

$$\exp i \int d^4x\, [\bar{\eta}(x)\psi(x) + \bar{\psi}(x)\eta(x)] = \exp i \int d^4x\, [\bar{\eta}(x)\psi^{(-)}(x) + \bar{\psi}^{(-)}(x)\eta(x)]$$
$$\times \exp i \int d^4x\, [\bar{\eta}(x)\psi^{(+)}(x) + \bar{\psi}^{(+)}(x)\eta(x)]$$
$$\times \exp \left\{ \tfrac{1}{2} \iint d^4x\, d^4y\, [\bar{\eta}(x)\psi^{(-)}(x) + \bar{\psi}^{(-)}(x)\eta(x), \bar{\eta}(y)\psi^{(+)}(y) + \bar{\psi}^{(+)}(y)\eta(y)] \right\} \tag{4-72}$$

As above, we replace the c number:

$$\int d^4x\, d^4y\, \{\theta(x^0 - y^0)[\bar{\eta}(x)\psi(x) + \bar{\psi}(x)\eta(x), \bar{\eta}(y)\psi(y) + \bar{\psi}(y)\eta(y)]$$
$$- [\bar{\eta}(x)\psi^{(-)}(x) + \bar{\psi}^{(-)}(x)\bar{\eta}(x), \bar{\eta}(y)\psi^{(+)}(y) + \bar{\psi}^{(+)}(y)\eta(y)]\}$$

by its vacuum expectation value, that is,

$$\langle 0| \iint d^4x\, d^4y\, \{\theta(x^0 - y^0)[\mathscr{L}_I(x), \mathscr{L}_I(y)] + \mathscr{L}_I(y)\,\mathscr{L}_I(x)\}\, |0\rangle$$
$$= \langle 0| \iint d^4x\, d^4y\; T[\mathscr{L}_I(x)\,\mathscr{L}_I(y)]\, |0\rangle$$

Owing to the anticommutation of the η, this is nothing but

$$\iint d^4x\, d^4y\, [\bar{\eta}_\alpha(x)\langle 0|\, \theta(x^0 - y^0)\psi_\alpha(x)\bar{\psi}_\beta(y) - \theta(y^0 - x^0)\bar{\psi}_\beta(y)\psi_\alpha(x)\,|0\rangle\, \eta_\beta(y)$$
$$+ \eta_\alpha(x)\langle 0|\, \theta(x^0 - y^0)\bar{\psi}_\alpha(x)\psi_\beta(y) - \theta(y^0 - x^0)\psi_\beta(y)\bar{\psi}_\alpha(x)\,|0\rangle\, \bar{\eta}_\beta(y)]$$

Notice that the other terms in $\psi\psi$ or $\bar{\psi}\bar{\psi}$ have disappeared in the integration. We recognize the definition of the T product of Fermi fields [see Eqs. (3-173) or (4-55)]:

$$\iint d^4x\, d^4y\, [\bar{\eta}_\alpha(x)\langle 0|\, T\psi_\alpha(x)\bar{\psi}_\beta(y)\,|0\rangle\, \eta_\beta(y) + \eta_\alpha(x)\langle 0|\, T\bar{\psi}_\alpha(x)\psi_\beta(y)\,|0\rangle\, \bar{\eta}_\beta(y)]$$

Both terms give the same result, after an interchange of the variables x and y under the integration sign. Finally, we obtain

$$T \exp i \int d^4x\, [\bar{\eta}(x)\psi(x) + \bar{\psi}(x)\eta(x)] = :\exp i \int d^4x\, [\bar{\eta}(x)\psi(x) + \bar{\psi}(x)\eta(x)]:$$
$$\times \exp\left[- \iint d^4x\, d^4y\, \bar{\eta}(x)\langle 0|\, T\psi(x)\,\bar{\psi}(y)\,|0\rangle\, \eta(y) \right] \tag{4-73}$$

The only difference with the corresponding formula for Bose fields (4-64) lies in the absence of the factor $\frac{1}{2}$ in front of the last exponent. This is due to the fermion charge. Had we started with a charged, nonhermitian Bose field A, we would have written $\mathscr{L}_I(x) = [A^*(x)j(x) + j^*(x)A(x)]$ and obtained two equal contributions in the exponent $\langle 0| \iint d^4x\, d^4y\, T[\mathscr{L}_I(x)\mathscr{L}_I(y)]\,|0\rangle$, and therefore no factor $\frac{1}{2}$ either. The explicit relations between time-ordered and normal products follow when expanding (4-73) and taking into account the anticommutation of the η, $\bar{\eta}$. For example, to second order in the fields, we get

$$\frac{1}{2}\iint d^4x\, d^4y\; T[\bar{\eta}(x)\psi(x) + \bar{\psi}(x)\eta(x)]\,[\bar{\eta}(y)\psi(y) + \bar{\psi}(y)\eta(y)]$$
$$= \frac{1}{2}\iint d^4x\, d^4y\; :[\bar{\eta}(x)\psi(x) + \bar{\psi}(x)\eta(x)]\,[\bar{\eta}(y)\psi(y) + \bar{\psi}(y)\eta(y)]:$$
$$+ \iint d^4x\, d^4y\; \bar{\eta}(x)\,\langle 0|\, T[\psi(x)\bar{\psi}(y)]\,|0\rangle\, \eta(y)$$

and after identification

$$\begin{aligned} T\psi(x)\psi(y) &= :\psi(x)\psi(y): \\ T\bar{\psi}(x)\bar{\psi}(y) &= :\bar{\psi}(x)\bar{\psi}(y): \\ T\psi(x)\bar{\psi}(y) &= :\psi(x)\bar{\psi}(y): + \langle 0|\, T\psi(x)\bar{\psi}(y)\,|0\rangle \\ T\bar{\psi}(x)\psi(y) &= :\bar{\psi}(x)\psi(y): + \langle 0|\, T\bar{\psi}(x)\psi(y)\,|0\rangle \end{aligned} \tag{4-74}$$

We note that the two last expressions are equivalent, since

$$T[\psi_\xi(x)\bar{\psi}_\eta(y)] = -T[\bar{\psi}_\eta(y)\psi_\xi(x)]$$

(ξ and η are spinor indices). For the sake of compactness, it is convenient to include in the argument of ψ the space-time coordinate, the Dirac index, and a discrete index which distinguishes $\psi_\alpha(x)$ from $\bar{\psi}_\alpha(x)$. With this convention, the general expression of Wick's theorem for Fermi fields reads

$$T[\psi(1)\cdots\psi(n)] = \sum_{p=0}^{[n/2]} \sum_{P} \sigma_P :\psi(1)\cdots\widehat{\psi(k_1)}\cdots\widehat{\psi(k_{2p})}\cdots\psi(n):$$
$$\times \langle 0|\, T\psi(k_{P_1})\psi(k_{P_2})\,|0\rangle \cdots \langle 0|\, T\psi(k_{P_{2p-1}})\psi(k_{P_{2p}})\,|0\rangle \tag{4-75}$$

One sums over all distinct contractions, and $\sigma_P = \pm 1$ is the signature of the permutation which transforms $\{1,\ldots,n\}$ into $\{1,\ldots,\widehat{k_1},\ldots,\widehat{k_{2p}},\ldots,n,\, k_{P_1},\ldots,k_{P_{2p}}\}$. Let us illustrate this rule on the T product of four fields:

$$\begin{aligned} T\{\psi(1)\psi(2)\psi(3)\psi(4)\} &= :\psi(1)\psi(2)\psi(3)\psi(4): + :\psi(1)\psi(2): \langle 0|\, T\psi(3)\psi(4)\,|0\rangle \\ &- :\psi(1)\psi(3): \langle 0|\, T\psi(2)\psi(4)\,|0\rangle + :\psi(1)\psi(4): \langle 0|\, T\psi(2)\psi(3)\,|0\rangle \\ &+ :\psi(2)\psi(3): \langle 0|\, T\psi(1)\psi(4)\,|0\rangle - :\psi(2)\psi(4): \langle 0|\, T\psi(1)\psi(3)\,|0\rangle \\ &+ :\psi(3)\psi(4): \langle 0|\, T\psi(1)\psi(2)\,|0\rangle + \langle 0|\, T\psi(1)\psi(2)\,|0\rangle \langle 0|\, T\psi(3)\psi(4)\,|0\rangle \\ &- \langle 0|\, T\psi(1)\psi(3)\,|0\rangle \langle 0|\, T\psi(2)\psi(4)\,|0\rangle + \langle 0|\, T\psi(1)\psi(4)\,|0\rangle \langle 0|\, T\psi(2)\psi(3)\,|0\rangle \end{aligned}$$

Equation (4-75) may be specialized to the vacuum expectation value:

$$\langle 0|\, T\psi(1)\cdots\psi(2n-1)\,|0\rangle = 0$$

$$\begin{aligned}\langle 0|\, T\psi(1)\cdots\psi(2n)\,|0\rangle &= \sum_{\text{distinct terms}} \varepsilon_P \langle 0|\, T\psi(P_1)\psi(P_2)\,|0\rangle \cdots \\ &\quad \times \langle 0|\, T\psi(P_{2n-1})\psi(P_{2n})\,|0\rangle \\ &= \sum_{\text{permutations } P} \frac{1}{2^n n!} \varepsilon_P \langle 0|\, T\psi(P_1)\psi(P_2)\,|0\rangle \cdots \\ &\quad \times \langle 0|\, T\psi(P_{2n-1})\psi(P_{2n})\,|0\rangle\end{aligned} \tag{4-76}$$

The expression on the right-hand side of (4-76) is a pfaffian, i.e., the square root of the determinant of the antisymmetric $2n \times 2n$ matrix with generic element $\langle 0|\, T\psi(i)\psi(j)\,|0\rangle$.

4-2-3 General Case

In the general case, where both fermion and boson fields are involved, the expression of the Wick theorem may be easily derived. Distinct fermion fields anticommute, distinct Bose fields commute, and a boson field is assumed to commute with a fermion field (the latter point is just a matter of convention). We then decide to use as above a compact notation: ψ denotes either a boson or a fermion field, with a discrete index to distinguish its nature. It must then be clear to the reader that the formula (4-75) is still valid, provided we understand that the sign σ_P is the signature of the permutation of the Fermi fields only. This result follows directly from the generating function (4-73), which has the same form whatever the fields are.

The Wick theorem discussed in this section has a close relationship to the computation of the integral of a polynomial of several variables, with a gaussian weight. This analogy is not fortuitous, as we shall see later in the discussion of functional integrals.

4-3 QUANTIZED DIRAC FIELD INTERACTING WITH A CLASSICAL POTENTIAL

4-3-1 General Formalism

After this algebraic interlude we shall now discuss another case of a quadratic lagrangian, namely, the electron-positron field in the presence of a classical electromagnetic field. Electron-positron pairs will possibly be created in this external field. This is the physical counterpart of the problem discussed in Sec. 4-1, whereas the creation or annihilation of single electrons by the anticommuting source of the last section was its mathematical counterpart.

Besides interesting phenomena such as pair creation—which has clearly no classical equivalent—this study should also give us some insight into the role of the Dirac equation for the quantized field ψ:

$$[i\partial\!\!\!/ - eA\!\!\!/(x) - m]\psi(x) = 0 \tag{4-77}$$

Here $A^\mu(x)$ is a given c number external field. This equation corresponds to the interaction lagrangian

$$\mathscr{L}_I(x) = -\mathscr{H}_I(x) = -e\bar{\psi}(x)\gamma^\mu\psi(x)A_\mu(x) \tag{4-78}$$

All the steps leading to the interaction representation and to the S matrix in Sec. 4-1-4 may be repeated here. This results in the expression

$$S = T\exp\left[-ie\int d^4x\, \bar{\psi}_{\rm in}(x)\gamma^\mu\psi_{\rm in}(x)A_\mu(x)\right] \tag{4-79}$$

But now, it is not trivial to put this expression into a normal form. This is because the interaction lagrangian is quadratic in the fields, rather than linear, and therefore the commutator of two currents $\bar{\psi}\gamma^\mu\psi$ is not a c number. Notice also that we do not know the general solution of the Dirac equation (4-77) in an arbitrary field A_μ.

In spite of these difficulties, we have all the machinery of the Wicks theorem at our disposal, and we can find a formal solution of the problem. Let us concentrate on the probability amplitude of emitting no pair:

$$\begin{aligned} S_0(A) &= \langle 0\ \text{in}|\, S\, |0\ \text{in}\rangle \\ &= \sum_{n=0}^{\infty}\frac{(-ie)^n}{n!}\int dx_1\cdots dx_n\,\langle 0|\, T[\bar{\psi}(x_1)A\!\!\!/(x_1)\psi(x_1)\cdots\bar{\psi}(x_n)A\!\!\!/(x_n)\psi(x_n)]\,|0\rangle \end{aligned} \tag{4-80}$$

where "in" subscripts have been omitted. According to the Wick theorem, each term in (4-80) is a sum of products of contractions of the form

$$\langle 0|\, TA\!\!\!/(x_k)\psi(x_k)\bar{\psi}(x_l)\,|0\rangle$$

For x_k and x_l given, let us introduce the 4×4 matrix

$$C(\alpha_k, x_k; \alpha_l, x_l) = -ie\sum_\alpha \langle 0|\, T[A\!\!\!/_{\alpha_k\alpha}(x_k)\psi_\alpha(x_k)\bar{\psi}_{\alpha_l}(x_l)]\,|0\rangle \tag{4-81}$$

In terms of C, $S_0(A)$ reads

$$S_0(A) = \sum_0^\infty \frac{(-1)^n}{n!}\int dx_1\cdots dx_n \sum_P \varepsilon_P \sum_{\alpha_1,\ldots,\alpha_n} C(\alpha_1, x_1; \alpha_{P_1}, x_{P_1})\cdots C(\alpha_n, x_n; \alpha_{P_n}, x_{P_n}) \tag{4-82}$$

It is convenient to consider the discrete indices α_i and the continuous variables x_i on the same footing and to group them in a bracket notation $|x, \alpha\rangle$. The space of these vectors $|x, \alpha\rangle$ is nothing but the space of classical spinors. We introduce the matrix Γ of generic element

$$\langle x, \alpha|\,\Gamma\,|y, \beta\rangle = C(x, \alpha; y, \beta) \tag{4-83}$$

Using this matrix notation, we may write

$$S_0(A) = \operatorname{Det}(I - \Gamma) = \exp[\operatorname{Tr}\ln(I - \Gamma)] \tag{4-84}$$

These expressions are indeed true for any finite-dimensional matrix Γ:

$$\begin{aligned}\det(I - \Gamma) &= \exp[\operatorname{tr}\ln(I - \Gamma)]\\ &= 1 - \sum_a \Gamma_{aa} + \frac{(-1)^2}{2}\sum_{a,b}(\Gamma_{aa}\Gamma_{bb} - \Gamma_{ab}\Gamma_{ba}) + \cdots\\ &\quad + \frac{(-1)^n}{n!}\sum_{a_1\cdots a_n}\ \sum_{\substack{P\\ \text{permutations}\\ \text{of }(1,\ldots,n)}} \varepsilon_P \Gamma_{a_1a_{P_1}}\cdots\Gamma_{a_na_{P_n}} + \cdots\end{aligned} \tag{4-85}$$

as a result of the Cayley-Hamilton formula for a determinant.

This expansion stops at the dth order for a finite d-dimensional matrix. The term of order d is, of course, $(-1)^d \det\Gamma$. This formula expresses the determinant as a sum of traces of antisymmetrized tensor products. There exists a similar formula for the inverse determinant

$$\begin{aligned}\det^{-1}(I - \Gamma) &= \exp[-\operatorname{tr}\ln(I - \Gamma)]\\ &= 1 + \sum_a \Gamma_{aa} + \tfrac{1}{2}\sum_{a,b}(\Gamma_{aa}\Gamma_{bb} + \Gamma_{ab}\Gamma_{ba}) + \cdots\\ &\quad + \frac{1}{n!}\sum_{a_1\cdots a_n}\ \sum_{\substack{P\\ \text{permutations}\\ \text{of }(1,\ldots,n)}} \Gamma_{a_1a_{P_1}}\cdots\Gamma_{a_na_{P_n}} + \cdots\end{aligned} \tag{4-86}$$

that is, as a sum of traces of symmetrized tensor products. This latter expression would be relevant for the problem of quantized boson fields coupled to an external field.

These expressions extend to the case of infinite matrices in the framework of the Fredholm theory, under proper assumptions on the behavior of Γ. We use capital letters for Det and Tr to recall that these operations imply integrations over continuous variables. The use of normal products in (4-78) would amount to the assumption that $\Gamma_{aa} \equiv 0$, and therefore to dropping the term $\sum \Gamma_{aa}$ in (4-85).

As in the last section of Chap. 2, it is convenient to introduce operators X_μ and P_μ, defined on the states $|x, \alpha\rangle$:

$$X_\mu|x, \alpha\rangle = x_\mu|x, \alpha\rangle$$

$$\langle x, \alpha|P_\mu|\varphi\rangle = i\frac{\partial}{\partial x^\mu}\langle x, \alpha|\varphi\rangle$$

They enjoy the canonical commutation relation

$$[X_\mu, P_\nu] = -ig_{\mu\nu} \tag{4-87}$$

and eigenvectors of P_μ, denoted $|p\rangle$, are such that

$$\langle p|x\rangle = \langle x|p\rangle^* = \frac{e^{ip\cdot x}}{(2\pi)^2}$$

Using these matrix notations, together with the expression (3-174) of the Dirac propagator, we find

$$\Gamma = e\not{A}(x)\,\frac{1}{\not{P} - m + i\varepsilon}$$

$$\begin{aligned} S_0(A) &= \operatorname{Det}\left[I - e\not{A}(x)\,\frac{1}{\not{P} - m + i\varepsilon}\right] \\ &= \operatorname{Det}\left\{[\not{P} - e\not{A}(x) - m + i\varepsilon]\,\frac{1}{\not{P} - m + i\varepsilon}\right\} \\ &= \exp\left\{-\operatorname{Tr}\ln\left[(\not{P} - m)\,\frac{1}{\not{P} - e\not{A}(x) - m + i\varepsilon}\right]\right\} \end{aligned} \tag{4-88}$$

The last expression exhibits the Feynman propagator in the external field $A_\mu(x)$.

It is a good exercise to check gauge invariance. Under a gauge transformation

$$A_\mu(x) \to A_\mu(x) + \partial_\mu \Phi(x)$$

$$\not{P} - e\not{A}(x) - m \to \not{P} - e\not{A}(x) - e\not{\partial}\Phi(x) - m$$

The relations (4-87) imply that

$$e^{-ie\Phi(x)}[\not{P} - e\not{A}(x) - m]\,e^{ie\Phi(x)} = \not{P} - e\not{A}(x) - e\not{\partial}\phi(x) - m$$

Therefore a gauge transformation amounts to a unitary transformation in the space of the classical spinors. It does not affect the determinants and leaves $S_0(A)$ invariant.

Let us examine the consequences of the unitarity of the S matrix. It is convenient to introduce a one-body scattering operator $\mathscr{T}(A)$ defined by

$$\mathscr{T}(A) = e\not{A}(x) + e\not{A}(x)\,\frac{1}{\not{P} - m + i\varepsilon}\,\mathscr{T}(A) \tag{4-89}$$

We assume the potential $A_\mu(x)$ to be real; for every operator B on the space of vectors $|x, \alpha\rangle$, we set

$$\bar{B} \equiv \gamma^0 B^\dagger \gamma^0$$

where the hermitian conjugation acts on the indices α and x. This operation changes the sign of $i\varepsilon$ in $\mathscr{T}(A)$:

$$\bar{\mathscr{T}}(A) = e\not{A}(x) + e\not{A}(x)\,\frac{1}{\not{P} - m - i\varepsilon}\,\bar{\mathscr{T}}(A) = e\not{A}(x) + \bar{\mathscr{T}}(A)\,\frac{1}{\not{P} - m - i\varepsilon}\,e\not{A}(x) \tag{4-90}$$

Thus

$$e\not{A}(x) = \bar{\mathscr{T}}(A) - \bar{\mathscr{T}}(A)\,\frac{1}{\not{P} - m - i\varepsilon}\,e\not{A}(x)$$

The insertion of this expression into (4-89) leads to

$$\mathscr{T}(A) = e\not{A}(x) + \left[\bar{\mathscr{T}}(A) - \bar{\mathscr{T}}(A)\,\frac{1}{\not{P} - m - i\varepsilon}\,e\not{A}(x)\right]\frac{1}{\not{P} - m + i\varepsilon}\,\mathscr{T}(A)$$

$$= e\not{A}(x) + \bar{\mathscr{T}}(A)\,\frac{1}{\not{P} - m + i\varepsilon}\,\mathscr{T}(A) - \bar{\mathscr{T}}(A)\,\frac{1}{\not{P} - m - i\varepsilon}\,[\mathscr{T}(A) - e\not{A}(x)]$$

$$= \bar{\mathscr{T}}(A)\left[\frac{1}{\not{P} - m + i\varepsilon} - \frac{1}{\not{P} - m - i\varepsilon}\right]\mathscr{T}(A) + e\not{A}(x) + \bar{\mathscr{T}}(A)\,\frac{1}{\not{P} - m - i\varepsilon}\,e\not{A}(x)$$

$$= \bar{\mathscr{T}}(A)\left[\frac{1}{\not{P} - m + i\varepsilon} - \frac{1}{\not{P} - m - i\varepsilon}\right]\mathscr{T}(A) + \bar{\mathscr{T}}(A)$$

Thus

$$\begin{aligned}
\mathscr{T}(A) - \bar{\mathscr{T}}(A) &= \bar{\mathscr{T}}(A)\left(\frac{1}{\not{P} - m + i\varepsilon} - \frac{1}{\not{P} - m - i\varepsilon}\right)\mathscr{T}(A) \\
&= \mathscr{T}(A)\left(\frac{1}{\not{P} - m + i\varepsilon} - \frac{1}{\not{P} - m - i\varepsilon}\right)\bar{\mathscr{T}}(A) \\
&= \bar{\mathscr{T}}(A)\left[\frac{2\pi}{i}(\not{P} + m)\delta(P^2 - m^2)\right]\mathscr{T}(A)
\end{aligned} \tag{4-91}$$

The operator $2\pi(\not{P} + m)\delta(P^2 - m^2)$ may be decomposed into a sum of projectors over positive and negative energy states

$$\begin{aligned}
2\pi(\not{P} + m)\delta(P^2 - m^2) &= \rho^{(+)} + \rho^{(-)} \\
\rho^{(\pm)}(P) &\equiv 2\pi(\not{P} + m)\theta(\pm P^0)\delta(P^2 - m^2)
\end{aligned} \tag{4-92}$$

These operators which project on the mass shell will emerge naturally in the computation of physical quantities, such as $|S_0(A)|^2$.

If we return to the original expression (4-82) of $S_0(A)$, every term in the right-hand side may be decomposed into cycles:

$$C(\alpha_1, x_1\,;\,\alpha_2, x_2)C(\alpha_2, x_2\,;\,\alpha_3, x_3)\cdots C(\alpha_k, x_k\,;\,\alpha_1, x_1)$$

Had we used the retarded propagator instead of the Feynman one, all these cyclic terms would vanish, since the retarded Green function has its support inside the future cone and thus a cyclic term has an empty support. As noted in Chap. 2, the retarded propagator has the form

$$\frac{\not{P} + m}{(P + i\varepsilon)^2 - m^2} = \frac{1}{\not{P} - m + i\not{\varepsilon}}$$

where ε_μ is an infinitesimal positive time-like four-vector. We conclude that

$$\operatorname{Det}\left[I - e\not{A}(x)\frac{1}{\not{P} - m + i\not{\varepsilon}}\right] = 1 \tag{4-93}$$

But

$$\frac{1}{\not{P} - m + i\not{\varepsilon}} = (\not{P} + m)\left[\mathrm{PP}\left(\frac{1}{P^2 - m^2}\right) - i\pi\,\varepsilon(P^0)\delta(P^2 - m^2)\right]$$

where PP means principal part. Therefore,

$$\frac{1}{\not{P}-m+i\varepsilon}-\frac{1}{\not{P}-m+i\not{\varepsilon}}=-i\rho^{(-)}$$

and

$$\begin{aligned}
1 &= \operatorname{Det}\left[I-e\not{A}(x)\frac{1}{\not{P}-m+i\not{\varepsilon}}\right]\\
&= \operatorname{Det}\left[I-e\not{A}(x)\frac{1}{\not{P}-m+i\varepsilon}-ie\not{A}(x)\rho^{(-)}\right]\\
&= \operatorname{Det}\left\{I-e\not{A}(x)\frac{1}{\not{P}-m+i\varepsilon}-i\left[I-e\not{A}(x)\frac{1}{\not{P}-m+i\varepsilon}\right]\mathscr{T}(A)\rho^{(-)}\right\}\\
&= \operatorname{Det}\left\{\left[I-e\not{A}(x)\frac{1}{\not{P}-m+i\varepsilon}\right]\left[I-i\mathscr{T}(A)\rho^{(-)}\right]\right\}\\
&= \operatorname{Det}\left[I-e\not{A}(x)\frac{1}{\not{P}-m+i\varepsilon}\right]\operatorname{Det}\left[I-i\mathscr{T}(A)\rho^{(-)}\right]
\end{aligned} \tag{4-94}$$

From these expressions and from (4-91), it follows that

$$\begin{aligned}
[S_0(A)]^{-1} &= \operatorname{Det}\left[I-i\mathscr{T}(A)\rho^{(-)}\right]\\
|S_0(A)|^{-2} &= \operatorname{Det}\{[I-i\mathscr{T}(A)\rho^{(-)}][I+i\bar{\mathscr{T}}(A)\rho^{(-)}]\}\\
&= \operatorname{Det}\{I+i[\bar{\mathscr{T}}(A)-\mathscr{T}(A)]\rho^{(-)}+\mathscr{T}(A)\rho^{(-)}\bar{\mathscr{T}}(A)\rho^{(-)}\}\\
&= \operatorname{Det}[I-\mathscr{T}(A)\rho^{(+)}\bar{\mathscr{T}}(A)\rho^{(-)}]\\
&= \exp\{\operatorname{Tr}\ln[I-\mathscr{T}(A)\rho^{(+)}\bar{\mathscr{T}}(A)\rho^{(-)}]\}
\end{aligned} \tag{4-95}$$

We write

$$|S_0(A)|^2=\exp\left[-\int d^4x\, w(x)\right] \tag{4-96}$$

with

$$w(x)=\operatorname{tr}\langle x|\ln[I-\mathscr{T}(A)\rho^{(+)}\bar{\mathscr{T}}(A)\rho^{(-)}]|x\rangle \tag{4-97}$$

(where the trace now refers only to the Dirac indices), and we interpret $w(x)$ as a probability density for pair creation. If we divide the four-dimensional space into small cells of volume Δx_i about the points x_i, we may approximate

$$|S_0(A)|^2\simeq\prod_i[1-\Delta x_i w(x_i)]$$

which confirms the previous interpretation. Each cell gives an independent contribution $w(x_i)$ to the emission probability. This may also be seen by an explicit computation of the probability of emission of one, two, etc., pairs. Because of the presence of the projectors $\rho^{(\pm)}$, the probability $w(x)\,d^4x$ is expressed in terms of on-shell matrix elements of $\mathscr{T}$ between a particle and a hole state. This is made clearer by the introduction of a reduced operator t in the following way. Using the normalized positive and negative energy solutions $u^{(a)}(p)$, $v^{(a)}(p)$, introduced in Chap. 2, we define the matrix t on the mass shell by

$$\langle \mathbf{p}a|\,t\,|\mathbf{p}'b\rangle = 2\pi \sum_{\alpha,\beta} \frac{m}{\omega_p^{1/2}\omega_{p'}^{1/2}}\, \bar{u}_\alpha^{(a)}(p)\langle p, \alpha|\mathscr{T}|p', \beta\rangle v_\beta^{(b)}(-p') \tag{4-98}$$

with $\omega_p \equiv (\mathbf{p}^2 + m^2)^{1/2}$. From the expressions of the projectors

$$\sum_{a=1,2} u^{(a)}(p)\bar{u}^{(a)}(p) = \frac{\not p + m}{2m} \qquad \text{for } p^0 > 0$$

$$\sum_{a=1,2} v^{(a)}(-p)\bar{v}^{(a)}(-p) = -\frac{\not p + m}{2m} \qquad \text{for } p^0 < 0$$

we then find

$$W_{\text{tot}} = \int d^4x\, w(x) = \int d^3p \operatorname{tr} \ln(1 + tt^\dagger) \tag{4-99}$$

This exhibits clearly the positivity of W_{tot}. We shall now compute explicitly this probability in two special instances.

4-3-2 Emission Rate to Lowest Order

It is possible to get an expansion of W in powers of $\alpha = e^2/4\pi$ by expanding both the logarithm and $\mathscr{T}(A)$ in Eq. (4-97). To lowest order, we find

$$W^{(1)} = -\operatorname{Tr}(e\not A\rho^{(+)} e\not A\rho^{(-)}) \tag{4-100}$$

We set

$$\langle p|\, A_\mu(X)\,|p'\rangle = \frac{1}{(2\pi)^2}\, a_\mu(p - p') = \frac{1}{(2\pi)^4} \int d^4x\, A_\mu(x)\, e^{i(p-p')\cdot x} \tag{4-101}$$

Thus

$$W^{(1)} = \frac{\alpha}{\pi} \int d^4q\, \delta^4(q - p_1 - p_2) \int \frac{d^3p_1}{2\omega_1} \frac{d^3p_2}{2\omega_2} \operatorname{tr}\left[(\not p_1 + m)\not a(q)(\not p_2 - m)\not a(-q)\right] \tag{4-102}$$

But if $q = p_1 + p_2$, we have

$$\tfrac{1}{4}\operatorname{tr}\left[(\not p_1 + m)\not a(q)(\not p_2 - m)\not a(-q)\right]$$
$$= \left[-\frac{q^2}{2}\, a(q)\cdot a(-q) + p_1\cdot a(q) p_2\cdot a(-q) + p_1\cdot a(-q) p_2\cdot a(q)\right]$$

We then compute the integrals $[q^2 = (p_1 + p_2)^2 \geq 4m^2]$:

$$I = \int \frac{d^3p_1}{\omega_1} \frac{d^3p_2}{\omega_2} \delta^4(q - p_1 - p_2) = 4\pi \int_{2m}^{\infty} p_1 \frac{d\omega_1}{\omega_1} \delta(\sqrt{q^2} - 2\omega_1)$$

$$= 2\pi \left(1 - \frac{4m^2}{q^2}\right)^{1/2} \theta\left(1 - \frac{4m^2}{q^2}\right) \tag{4-103a}$$

$$I^{\mu\nu} = \iint \frac{d^3p_1}{\omega_1} \frac{d^3p_2}{\omega_2} \delta^4(q - p_1 - p_2) p_1^\mu p_2^\nu$$

$$= \frac{I}{12}\left[\frac{q^\mu q^\nu}{q^2}(2q^2 + 4m^2) + g^{\mu\nu}(q^2 - 4m^2)\right] \tag{4-103b}$$

Indeed, $I_{\mu\nu}$ has the general form $I_{\mu\nu} = I_1 g_{\mu\nu} + I_2 q_\mu q_\nu$ and successive contractions with $q_\mu q_\nu$ and $g_{\mu\nu}$ lead to (4-103*b*). This yields for $W^{(1)}$:

$$W^{(1)} = \frac{\alpha}{3} \int \frac{d^4q}{q^2} \left(1 - \frac{4m^2}{q^2}\right)^{1/2} \theta(q^0)\theta\left(1 - \frac{4m^2}{q^2}\right)(2q^2 + 4m^2)$$
$$\times \{q \cdot a(q) q \cdot a(-q) - q^2 a(q) \cdot a(-q)\} \tag{4-104}$$

This result will be made more transparent if we introduce the Fourier transform of the electromagnetic field

$$F_{\mu\nu}(q) = -i[q_\mu a_\nu(q) - q_\nu a_\mu(q)]$$
$$\tfrac{1}{2} F_{\mu\nu}(q) F^{\mu\nu}(-q) = |\mathbf{B}(q)|^2 - |\mathbf{E}(q)|^2$$
$$= q^2 a(q) \cdot a(-q) - q \cdot a(q) q \cdot a(-q)$$

Finally, since the integrand in (4-104) is even, we may forget the restriction $q^0 > 2m$:

$$W^{(1)} = \frac{\alpha}{3} \int d^4q\, \theta(q^2 - 4m^2)[|\mathbf{E}(q)|^2 - |\mathbf{B}(q)|^2]\left(1 - \frac{4m^2}{q^2}\right)^{1/2}\left(1 + \frac{2m^2}{q^2}\right) \tag{4-105}$$

This is a nontrivial result of field theory, but it could have been obtained through a careful interpretation of the hole theory. Observe that for $q^2 > 0$, there exists a frame where $\mathbf{q} = 0$ and thus $\mathbf{B}(q) = -i\mathbf{q} \times \mathbf{a}(q)$ vanishes; this implies that $|\mathbf{E}(q)|^2 - |\mathbf{B}(q)|^2 \geq 0$, that is, pair creation is an electric effect. However, electromagnetic modes with $q^2 > 4m^2$ are not that easy to generate by the current technology!

We quote also the corresponding rate for the pair creation of spinless charged bosons, to lowest order

$$W^{(1)}_{\text{bosons}} = \frac{\alpha}{12} \int d^4q\, \theta(q^2 - 4m^2)[|\mathbf{E}(q)|^2 - |\mathbf{B}(q)|^2]\left(1 - \frac{4m^2}{q^2}\right)^{3/2} \tag{4-106}$$

4-3-3 Pair Creation in a Constant Uniform Electric Field

In the case of a constant uniform field, we are able to derive an exact, nonperturbative result. We return to Eq. (4-88) and write

$$\ln S_0(A) = \operatorname{Tr} \ln \left\{ [\not{P} - e\not{A}(X) - m + i\varepsilon] \frac{1}{\not{P} - m + i\varepsilon} \right\} \tag{4-107}$$

Since the trace of an operator is invariant under transposition and since the charge conjugation matrix C satisfies

$$C\gamma_\mu C^{-1} = -\gamma_\mu^T$$

we may write

$$\ln S_0(A) = \operatorname{Tr} \ln \left\{ [\not{P} - e\not{A}(X) + m - i\varepsilon] \frac{1}{\not{P} + m - i\varepsilon} \right\} \tag{4-108}$$

and the sum of (4-107) and (4-108) reads

$$2 \ln S_0(A) = \operatorname{Tr} \ln \left(\{ [\not{P} - e\not{A}(X)]^2 - m^2 + i\varepsilon \} \frac{1}{P^2 - m^2 + i\varepsilon} \right) \tag{4-109}$$

The useful identity

$$\ln \frac{a}{b} = \int_0^\infty \frac{ds}{s} \left(e^{is(b+i\varepsilon)} - e^{is(a+i\varepsilon)} \right) \tag{4-110}$$

enables us to write for the desired probability

$$\begin{aligned} w(x) &= \operatorname{Re} \int_0^\infty \frac{ds}{s} e^{-is(m^2 - i\varepsilon)} \operatorname{tr} \left(\langle x | e^{is(\not{P} - e\not{A}(x))^2} | x \rangle - \langle x | e^{isP^2} | x \rangle \right) \\ &= \operatorname{Re} \operatorname{tr} \int_0^\infty \frac{ds}{s} e^{-is(m^2 - i\varepsilon)} \\ &\quad \langle x | \left(\exp \left\{ is \left[(P - eA(x))^2 + \frac{e}{2} \sigma_{\mu\nu} F^{\mu\nu}(x) \right] \right\} - e^{isP^2} \right) | x \rangle \end{aligned} \tag{4-111}$$

For a constant field, this probability should not depend on x. Moreover, $\sigma_{\mu\nu}F^{\mu\nu}$ commutes with all the other operators, and we may compute its exponential. From now on, we assume that the electromagnetic field is purely electric (we have seen in the last section that the pair creation is an electric effect) and that the electric field is along the z axis. We also choose a gauge such that only $A^3(x) = -Et(t \equiv x^0)$ is nonvanishing. Then

$$\operatorname{tr} e^{ise\sigma_{\mu\nu}F^{\mu\nu}/2} = 4 \cosh(seE) \tag{4-112}$$

and using the commutation relation $[X_0, P_0] = -i$, we get

$$\begin{aligned} (P - eA)^2 - m^2 &= P_0^2 - \mathbf{P}_T^2 - (P^3 + eEX^0)^2 \\ &= e^{-iP^0P^3/eE} (P_0^2 - \mathbf{P}_T^2 - e^2E^2X^{02}) \, e^{iP^0P^3/eE} \end{aligned} \tag{4-113}$$

Therefore,

$$\operatorname{tr}\langle x|\, e^{is[(P-eA)^2+\sigma_{\mu\nu}F^{\mu\nu}/2]}\,|x\rangle$$

$$= 4\cosh(seE)\int \frac{d^3p}{(2\pi)^4}\, d\omega\, d\omega'\, e^{i(\omega'-\omega)(t+p^3/eE)-isp_T^2}\langle\omega|\, e^{is(P_0^2-e^2E^2X_0^2)}\,|\omega'\rangle$$

$$= \frac{2eE}{(2\pi)^2 is}\cosh(eEs)\int_{-\infty}^{\infty} d\omega\, \langle\omega|\, e^{is(P_0^2-e^2E^2X_0^2)}\,|\omega\rangle \tag{4-114}$$

The last integral may be considered as the trace of the evolution operator of an harmonic oscillator with a purely imaginary frequency. This results from the correspondence $P_0 \to P$, $-X_0 \to Q$, $2ieE \to \omega_0$, $\frac{1}{2} \to m_0$, $P_0^2 - e^2E^2X_0^2 \to P^2/2m_0 + \frac{1}{2}m_0\omega_0^2Q^2$. The energy levels of such a system are well known and it follows that

$$\operatorname{Tr}\exp\left[is\left(\frac{P^2}{2m_0}+\frac{m_0\omega_0^2}{2}Q^2\right)\right] = \sum_{n=0}^{\infty}\exp\left[is\left(n+\frac{1}{2}\right)\omega_0\right]$$

$$= \frac{i}{2\sin s\omega_0/2} \tag{4-115}$$

so that

$$\int_{-\infty}^{\infty} d\omega\, \langle\omega|\, e^{is(P_0^2-e^2E^2X_0^2)}\,|\omega\rangle = \frac{1}{2\sinh(seE)} \tag{4-116}$$

Collecting all the terms, we find

$$w = -\frac{1}{(2\pi)^2}\int_0^{\infty}\frac{ds}{s^2}\left[eE\coth(eEs) - \frac{1}{s}\right]\operatorname{Re}\left(ie^{-is(m^2-i\varepsilon)}\right) \tag{4-117}$$

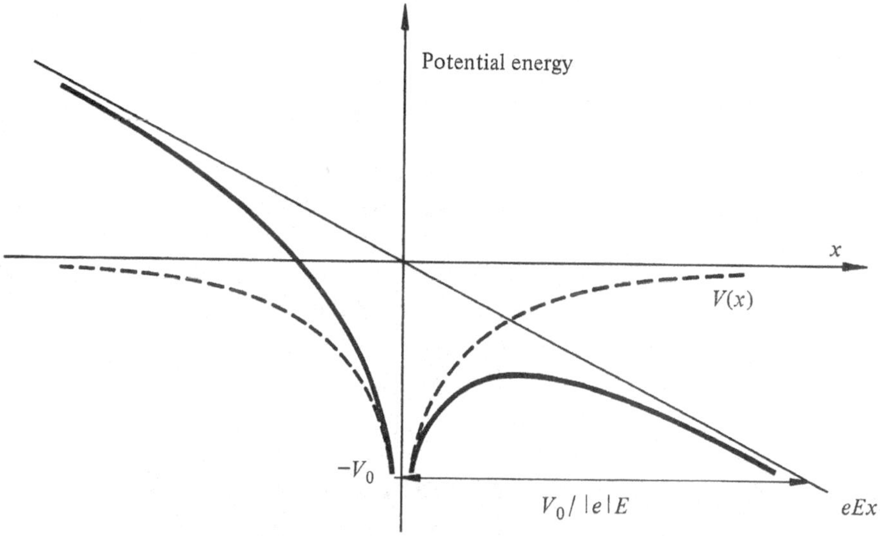

Figure 4-1 Potential energy (solid line) of an electron submitted to the binding potential $V(x)$ (dashed line) and to the electric potential eEx.

The term $1/s$ corresponds to the subtraction at $e = 0$ in (4-111). It must be noted that the integral converges at both ends: the infrared convergence at $s \to \infty$ is insured by the $m^2 - i\varepsilon$ prescription, while at $s \to 0$, $\operatorname{Re}(ie^{-ism^2}) = \sin sm^2 \simeq sm^2$, the bracket vanishes like s and the integrand is finite. The integral in (4-117) may be computed by the method of residues. It is first transformed into an integral over $(-\infty, +\infty)$; the integration contour is then deformed to encircle the negative imaginary axis and to pick up the contributions of the poles of the coth function. This leads to the final expression for the probability of creating pairs per unit volume and unit time:

$$w = \frac{\alpha E^2}{\pi^2} \sum_{n=1}^{\infty} \frac{1}{n^2} \exp\left(- \frac{n\pi m^2}{|eE|} \right) \tag{4-118}$$

Pair creation in a constant field has never been directly observed, because of the smallness of typical $|eE|$ as compared to m^2. We recall, for instance, that the electric field E_{at} on a Bohr orbit of an hydrogen atom is of order $eE_{\text{at}} \sim m^2\alpha^3 \sim 4 \times 10^{-7}\, m^2$.

The analogous expression for creation of spinless bosons reads

$$w_{\text{bosons}} = \frac{\alpha E^2}{2\pi^2} \sum_{n=1}^{\infty} \frac{(-1)^{n+1}}{n^2} \exp\left(- \frac{n\pi m^2}{|eE|} \right) \tag{4-119}$$

In these formulas, the essential factor is nonperturbative: $\exp(-\pi m^2/|eE|)$. This is reminiscent of the quantum phenomenon of tunneling through a potential barrier. Indeed, in atomic physics, let us consider an electron bound by the potential $V(x)$ (dashed line in Fig. 4-1) and submitted to an additional electric potential eEx. The resulting potential is depicted by the solid line in Fig. 4-1. For a binding energy $-V_0$ the ionization probability is proportional to

$$\exp - 2 \int_0^{V_0/|eE|} dx\, \sqrt{2m(V_0 - |eE|x)} = \exp\left(- \frac{4}{3} \sqrt{2mV_0}\, \frac{V_0}{|eE|} \right)$$

If we consider a negative energy electron as trapped in a potential well $|V_0| \sim 2m$, we get a factor $\exp(-\text{constant}\, m^2/|eE|)$ in qualitative agreement with the exact result.

4-3-4 The Euler-Heisenberg Effective Lagrangian

A classical electromagnetic field may produce pairs as a consequence of quantum effects. This means that the dynamics of a classical electromagnetic field contains quantum corrections. Until now, we have studied an absorption effect. Dispersive effects exist as well. We summarize these quantum corrections in an effective lagrangian

$$\mathscr{L}_{\text{eff}} = \mathscr{L}_0 + \delta\mathscr{L}$$

with

$$\mathscr{L}_0 = -\tfrac{1}{4} F_{\mu\nu} F^{\mu\nu} = \tfrac{1}{2}(\mathbf{E}^2 - \mathbf{B}^2) \tag{4-120}$$

$$\delta\mathscr{L} \equiv \delta\mathscr{L}[(\mathbf{E}^2 - \mathbf{B}^2), (\mathbf{E} \cdot \mathbf{B})^2]$$

where the correction $\delta\mathscr{L}$ is assumed to be local, at least for a slowly varying field, and has been written as a function of the invariants $\mathbf{E}^2 - \mathbf{B}^2$ and $(\mathbf{E}\cdot\mathbf{B})^2$. We know that the average properties of polarizability in matter are represented by the introduction of the electric induction **D** and the magnetic field **H**. Similarly, the role of $\delta\mathscr{L}$ is to take into account the polarizability of the vacuum, i.e., the interaction of the electromagnetic field with the vacuum fluctuations of the electron-positron field. The amplitude $S_0(A)$ will be present in any computation of an electromagnetic effect, since it represents the dynamics of the vacuum fluctuations. We expect, therefore, that $\delta\mathscr{L}$ is given by

$$\langle 0|S|0\rangle = S_0(A) = \exp\left[i\int d^4x\, \delta\mathscr{L}(x)\right] \tag{4-121}$$

The real part of $\delta\mathscr{L}$ describes dispersive effects and its imaginary part, the absorptive ones. The functional $\delta\mathscr{L}(A)$ may have a complicated form, but if we have only to deal with low frequencies it is sufficient to consider a slowly varying field, or even a constant field to lowest order. The method of computation of the last section will then be relevant. For a pure electric field, we find

$$\delta\mathscr{L}(E) = \frac{1}{8\pi^2}\int_0^\infty \frac{ds}{s^2}\left[eE\coth(eEs) - \frac{1}{s}\right]e^{-ism^2} \tag{4-122}$$

while in the general case

$$\delta\mathscr{L} = \frac{1}{8\pi^2}\int_0^\infty \frac{ds}{s}\, e^{-ism^2}\left[e^2ab\,\frac{\cosh(eas)\cos(ebs)}{\sinh(eas)\sin(ebs)} - \frac{1}{s^2}\right] \tag{4-123}$$

where $a^2 - b^2 \equiv \mathbf{E}^2 - \mathbf{B}^2$, $ab \equiv \mathbf{E}\cdot\mathbf{B}$. The reader may also derive (4-123) from the expression for the Dirac propagator given in Sec. 2-5-4 using the relation

$$\frac{\delta}{\delta A^\mu(x)}\int d^4x\, \delta\mathscr{L} = e\,\mathrm{tr}\,\langle x|\gamma_\mu \frac{i}{i\not\partial - e\not{A} - m}|x\rangle$$

In contrast with (4-117), the integral in (4-123) has a divergent part proportional to $a^2 - b^2$, which will be cured by a renormalization. The idea is to rewrite

$$\delta\mathscr{L} = \tfrac{1}{2}C(\mathbf{E}^2 - \mathbf{B}^2) + \delta\mathscr{L}_{\text{finite}}$$

where C is infinite, and to modify $\mathscr{L}_0$ in (4-120) into $(1 - C)\mathscr{L}_0$; this modification is undetectable, since only $\mathscr{L}_0 + \delta\mathscr{L}$ may be observed by assumption. The latter now reads

$$\mathscr{L}_{\text{eff}} = (1 - C)\frac{\mathbf{E}^2 - \mathbf{B}^2}{2} + \delta\mathscr{L} = \frac{\mathbf{E}^2 - \mathbf{B}^2}{2} + \delta\mathscr{L}_{\text{finite}} \tag{4-124}$$

$\delta\mathscr{L}_{\text{finite}}$ is obtained from $\delta\mathscr{L}$ by subtracting the divergent term; there is, of course, some arbitrariness in this subtraction. We shall discuss at length this renormalization operation and its arbitrariness in the following chapters.

The absorptive part of $\delta\mathscr{L}$ vanishes with all its derivatives at $e = 0$. Only dispersive effects may be computed by a perturbative calculation. For instance, we find, by expanding (4-123) to second order in α,

$$\delta \mathscr{L}^{(4)} = \frac{2\alpha^2}{45m^4} [(\mathbf{E}^2 - \mathbf{B}^2)^2 + 7(\mathbf{E} \cdot \mathbf{B})^2] \tag{4-125}$$

The Euler-Heisenberg effective lagrangian (4-123) and (4-124) may be used to discuss various nonlinear effects due to the quantum corrections.

NOTES

The physical interpretation of the infrared catastrophe dates back to the work of F. Bloch and A. Nordsieck, *Phys. Rev.*, vol. 52, p. 54, 1937. A thorough treatment within perturbation theory is given in D. R. Yennie, S. C. Frautschi, and H. Suura, *Ann. Phys.* (*New York*), vol. 13, p. 379, 1961. The use of coherent states to provide a bridge between quantum and classical electromagnetism has been pioneered by R. J. Glauber, *Phys. Rev.*, vol. 131, p. 2766, 1963. Further developments can be found, for instance, in "Quantum Optics and Electronics," edited by C. De Witt, A. Blandin, and C. Cohen-Tannoudji, Les Houches, 1964, Gordon and Breach, New York, 1965.

The interaction representation is due to F. J. Dyson, *Phys. Rev.*, vol. 75, p. 486, 1949, while the original work of G. C. Wick is in *Phys. Rev.*, vol. 80, p. 268, 1950.

The effective lagrangian for the electromagnetic field arises from the work of W. Heisenberg and H. Euler, *Z. Physik*, vol. 98, p. 714, 1936, and is worked out in detail by J. Schwinger, *Phys. Rev.*, vol. 82, p. 664, 1951, which is the source for Sec. 4-3. See also J. Schwinger, *Phys. Rev.*, vol. 93, p. 615, and vol. 94, p. 1362, 1954.

CHAPTER

FIVE

ELEMENTARY PROCESSES

We discuss the relation between measurements of cross sections and the dynamical content of field theory provided by Green functions through the reduction formalism of Lehmann, Symanzik, and Zimmermann. This is followed by elementary applications to electromagnetic processes to lowest order. General properties of unitarity and causality are presented and used to introduce partial wave expansions and dispersion relations.

5-1 *S* MATRIX AND ASYMPTOTIC THEORY

We have to complete our kinematics before embarking on the perturbative evaluation of transition amplitudes. We present in this chapter the general framework which will enable us to relate these calculations to the physical processes of interest.

In practice a major tool of investigation is particle scattering off various targets. On a macroscopic scale, interaction times are extremely small. It is therefore out of the question to follow in detail the time evolution during the elementary scattering events. We can only give the following picture. Long before the collision, well-separated wave packets evolve independently and freely. As already discussed in the special instances of Chap. 4, the set of these incoming states builds up a Fock space, the in-space with associated free fields. It is important to note that these in-states must represent exactly the individual characteristics of isolated particles, such as mass and charge. In other words, self-interaction effects must be absorbed in these measurable parameters. The collision process then follows, involving scattering, absorption, or creation of new particles, submitted to the fundamental conservation laws of energy, momentum, angular momentum, parity, charge conjugation, internal symmetries, etc. Long after the collision, free wave packets separate, representing the outgoing states. They are again described by free particle kinematics and by corresponding free fields. According to the postulates of quantum mechanics the amplitude

$$\langle b, \text{out} \,|\, a, \text{in} \rangle$$

enables us to obtain the probability that an incoming state $|a\rangle$ will evolve in time and be measured in the $|b\rangle$ state.

The out-states can represent the incoming states for a successive process. This is the case, for instance, when one prepares secondary beams. Therefore an isomorphism must exist among the in- and out-Fock space. Our goal is to relate the above transition amplitudes to the actual measurements. In other words, we want to establish the relativistic conventions of Fermi's golden rule and give the expression for cross sections.

5-1-1 Cross Sections

To take a simple case we first envisage the scattering of two distinct, spinless particles. The incoming state is written in terms of the incident wave packets in momentum space:

$$|i, \text{in}\rangle = \int \frac{d^3p_1}{2p_1^0(2\pi)^3} \frac{d^3p_2}{2p_2^0(2\pi)^3} f_1(p_1) f_2(p_2) \,|p_1, p_2, \text{in}\rangle \tag{5-1}$$

With these wave packets are associated positive energy solutions $\tilde{f}(x)$ of the Klein-Gordon equation with corresponding mass:

$$\tilde{f}(x) = \int \frac{d^3p}{2p^0(2\pi)^3} e^{-ip\cdot x} f(p) \tag{5-2}$$

and a flux given by

$$i \int d^3x\, \tilde{f}^*(x) \overleftrightarrow{\partial_0} \tilde{f}(x) = \int \frac{d^3p}{2p^0(2\pi)^3} |f(p)|^2 \tag{5-3}$$

The transition probability to a final state $|f, \text{out}\rangle$ is

$$W_{f\leftarrow i} = |\langle f, \text{out} \,|\, i, \text{in}\rangle|^2 \tag{5-4}$$

In the absence of external sources, translation invariance implies that the matrix elements vanish if energy and momentum are not conserved. Furthermore, as discussed in Chap. 4, the assumption of an isomorphism between out- and in-states implies the existence of a unitary operator S, commonly called the S matrix, such that

$$\langle f, \text{out} \,|\, i, \text{in}\rangle = \langle f, \text{in} |\, S \,| i, \text{in}\rangle \tag{5-5}$$

Unitarity of S is necessary to conserve probabilities. Strictly speaking, S applies the out-states on in-states according to (5-5). We shall loosely say that we take its matrix elements in the in-space and unless necessary we will drop the index "in." It is then decomposed as

$$S = I + iT \tag{5-6}$$

where T contains the information on the interactions. If we idealize the final state as a plane wave state we can extract the delta function of energy momentum conservation according to

$$\langle f | T | p_1, p_2 \rangle = (2\pi)^4 \delta^4(P_f - p_1 - p_2) \langle f | \mathscr{T} | p_1, p_2 \rangle \tag{5-7}$$

and the reduced operator $\mathscr{T}$ acts on the energy shell. When we substitute the decomposition (5-6) into the matrix element (5-5), the identity only contributes to forward scattering and represents a part of the incident wave packet unaffected by interactions. In most experiments we are only interested in the deflected part. This justifies keeping the sole contribution of $\mathscr{T}$ in the transition probability

$$\begin{aligned} W_{f \leftarrow i} = \int d\tilde{p}_1 \, d\tilde{p}_2 \, d\tilde{p}'_1 \, d\tilde{p}'_2 \, f_1^*(p_1) f_2^*(p_2) f_1(p'_1) f_2(p'_2) (2\pi)^4 \delta^4(p_1 + p_2 - p'_1 - p'_2) \\ \times (2\pi)^4 \delta^4(P_f - p_1 - p_2) \langle f | \mathscr{T} | p_1, p_2 \rangle^* \langle f | \mathscr{T} | p'_1, p'_2 \rangle \end{aligned} \tag{5-8}$$

If the final state is not a sharply defined eigenstate of momentum the above formula has to be slightly generalized in an obvious way. In most cases, one prepares the initial particles with almost well-defined momenta with a negligible width on the scale of the variation of the matrix elements of $\mathscr{T}$. In short, $f_i(p_i)$ is peaked around a mean value $\bar{p}_i$ with a width Δp_i so that

$$\langle f | \mathscr{T} | p'_1, p'_2 \rangle \simeq \langle f | \mathscr{T} | \bar{p}_1, \bar{p}_2 \rangle \simeq \langle f | \mathscr{T} | p_1, p_2 \rangle$$

Using this approximation and the integral representation

$$(2\pi)^4 \delta^4(p_1 + p_2 - p'_1 - p'_2) = \int d^4x \, e^{-ix \cdot (p'_1 + p'_2 - p_1 - p_2)}$$

we find

$$W_{f \leftarrow i} = \int d^4x \, | \tilde{f}_1(x) |^2 \, | \tilde{f}_2(x) |^2 (2\pi)^4 \delta^4(P_f - \bar{p}_1 - \bar{p}_2) | \langle f | \mathscr{T} | \bar{p}_1, \bar{p}_2 \rangle |^2 \tag{5-9}$$

which can be interpreted as a transition probability per unit time and unit volume:

$$\frac{dW_{f \leftarrow i}}{dV \, dt} = | \tilde{f}_1(x) |^2 \, | \tilde{f}_2(x) |^2 (2\pi)^4 \delta^4(P_f - \bar{p}_1 - \bar{p}_2) | \langle f | \mathscr{T} | \bar{p}_1, \bar{p}_2 \rangle |^2 \tag{5-10}$$

Now $\tilde{f}_i(x) = e^{-i\bar{p}_i \cdot x} F_i(x)$ with $F_i(x)$ a slowly varying function of x in such a way that

$$i\tilde{f}^*(x) \overleftrightarrow{\partial}_\mu \tilde{f}(x) \simeq 2\bar{p}_\mu | \tilde{f}(x) |^2 \tag{5-11}$$

To be concrete let us assume that particles of type 1 are the incident ones in the laboratory frame on particles of type 2 at rest. The number of particles in the target per unit volume is $dn_2/dV = 2\bar{p}_2^0|\tilde{f}_2(x)|^2$, $\bar{p}_2^0 = m_2$. The incident flux is given by velocity $|\bar{\mathbf{p}}_1|/\bar{p}_1^0$ times density $2\bar{p}_1^0|\tilde{f}_1(x)|^2$ equal to $2|\bar{\mathbf{p}}_1||\tilde{f}_1(x)|^2$.

This gives the following interpretation of Eq. (5-10): $dW_{f\leftarrow i}/dV\,dt$ = transition probability from state $|i\rangle$ to state $|f\rangle$ per unit time and unit volume = target density $[2m_2|\tilde{f}_2(x)|^2]$ × incident flux $[2|\bar{\mathbf{p}}_1|\,|\tilde{f}_1(x)|^2]$ × cross section $d\sigma$:

$$d\sigma = (2\pi)^4\delta^4(P_f - \bar{p}_1 - \bar{p}_2)\frac{1}{4m_2|\bar{\mathbf{p}}_1|}|\langle f|\mathscr{T}|\bar{p}_1, \bar{p}_2\rangle|^2 \tag{5-12}$$

The cross section is the transition probability per scatterer in the target and per unit incident flux. The idealization of a final state with well-defined momentum is corrected by integrating over an energy and momentum resolution Δ. For instance, if we consider a final state of n distinct spinless particles (Fig. 5-1), the cross section is given by

$$d\sigma_{n\leftarrow 2} = \frac{1}{4[(p_1\cdot p_2)^2 - m_1^2m_2^2]^{1/2}}\int_\Delta d\tilde{p}_3\cdots d\tilde{p}_{n+2}|\langle p_3,\ldots,p_{n+2}|\mathscr{T}|p_1,p_2\rangle|^2$$
$$\times (2\pi)^4\delta^4(p_1 + p_2 - p_3 - \cdots p_{n+2}) \tag{5-13}$$

We have expressed the factor $m_2|\mathbf{p}_1|$ in terms of relativistic invariants by remarking that in the laboratory frame

$$m_2|\mathbf{p}_1| = m_2[(p_1^0)^2 - m_1^2]^{1/2} = [(p_2\cdot p_1)^2 - m_1^2m_2^2]^{1/2}$$

In this way cross sections have been defined as Lorentz-invariant concepts. We remind the reader that $d\tilde{p}$ denotes the Lorentz-invariant measure $d\tilde{p} = d^3p/2p^0(2\pi)^3$ for bosons. These formulas assume, of course, that one-particle states are normalized according to

$$\langle p'|p\rangle = 2p^0(2\pi)^3\delta^3(\mathbf{p}' - \mathbf{p})$$

that is, are generated out of the vacuum state by canonical creation operators. This normalization gives the interpretation of the flux densities expressed in terms of the wave functions f.

When dealing with massive fermions the factor $2p^0$ will be replaced by p^0/m. For instance, if the two colliding particles are fermions the factor $\frac{1}{4}$ in front of (5-13) is replaced by m_1m_2, and if fermions occur among the final particles the phase space element $d\tilde{p}$ is to be understood as $(m/p^0)[d^3p/(2\pi)^3]$. However, if

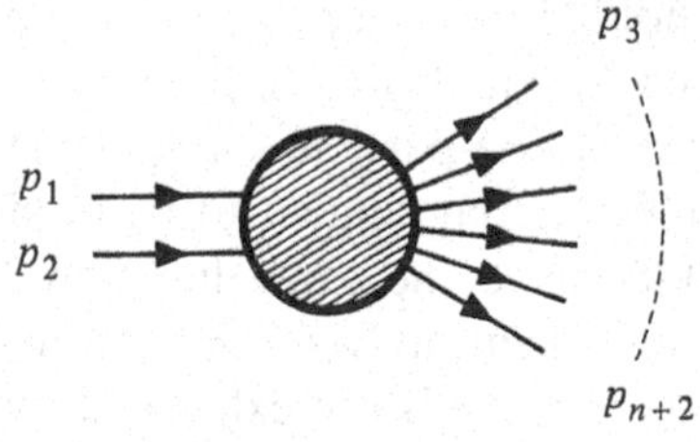

Figure 5-1 Kinematics of a general scattering process.

massless fermions are involved in the scattering process, we keep the boson phase space factors.

The reader is urged to study the modifications to (5-13) when some of the particles are identical and/or carry some spin. What happens if the initial state is not a pure state but has to be described using a polarization density matrix? We have presented here the connection between the invariant transition amplitude $\mathscr{T}$ and the cross section. Another physically important instance is the computation of the decay rate of an unstable composite system. This will be illustrated in the case of positronium in Sec. 5-2-3.

5-1-2 Asymptotic Theory

The central problem in the study of collisions is the calculation of S-matrix elements between on-shell states. Lagrangian field theory provides us with algorithms to compute these quantities, as we saw in the previous chapter where we investigated the behavior of a quantized field coupled to external sources. Local dynamics will enable us to scrutinize in close detail this S operator and to express it in terms of more elementary quantities, the fundamental Green functions. To keep the formulation as simple as possible we shall develop the argument in the framework of a self-interacting scalar field. The generalization to more elaborate cases, such as electrodynamics, only involves more kinematics without affecting the logic. It will be left to the sequel.

Let us summarize the situation. A Fock space of states is generated from a unique vacuum by a free field denoted by $\varphi_{\text{in}}(x)$. This is the stage on which the whole dynamical process takes place. In the best of all worlds, physical observables would be expressible in terms of this unique free field. In particular this ought to be the case for the interacting field $\varphi(x)$. Intuitively we imagine the relation between these two fields as follows. In the remote past, $\varphi_{\text{in}}(x)$ is a suitable limit of $\varphi(x)$. This refers, of course, to some definite process under consideration and only applies when the elementary participants in a collision are well separated from each other. To implement this idea we may assume that the coupling terms in the equations of motion are affected by some adiabatic cutoff function equal to one at finite times and vanishing smoothly when $|t| \to \infty$. All physical quantities have then to be understood in the limit when this adiabatic switching is removed. The adiabatic hypothesis asserts that under these conditions

$$x_0 \to -\infty \qquad \varphi(x) \to Z^{1/2}\varphi_{\text{in}}(x) \tag{5-14a}$$

Two questions immediately arise. What is the meaning of the factor $Z^{1/2}$? In what mathematical sense do we expect the limit to hold? First, the factor in (5-14a) is due to the fact that fields are naturally normalized by equal-time commutation relations, and φ_{in} acting on the vacuum only creates one-particle states while φ will also generate states with extra pairs (assuming the lagrangian to be even in φ). The amplitudes $\langle 1|\,\varphi_{\text{in}}(x)\,|0\rangle$ and $\langle 1|\,\varphi(x)\,|0\rangle$ have the same functional dependence on x dictated by the kinematics. The normalization factor $Z^{1/2}$ therefore takes into account that the content of the state $\varphi(x)\,|0\rangle$ is not exhausted by the matrix elements $\langle 1|\varphi(x)|0\rangle$, as would be true if φ was replaced by φ_{in}. Intuitively we thus expect $Z^{1/2}$ to be a number between zero and one.

Moreover, the limit involved in Eq. (5-14*a*) can only be a weak one, i.e., valid for each matrix element separately. If not, we would conclude that the commutator of two fields φ is equal, up to Z, to the corresponding c number commutator of free fields. In such circumstances canonical quantization would require $Z = 1$ and we would soon discover that φ is a free field! This shows that the adiabatic condition has to be treated with care. The constant Z can be related to a spectral representation of the commutator of two interacting fields, due to Lehmann and Källen, as follows. Consider the vacuum expectation value of this commutator. Assume the field hermitian for simplicity. Using translation invariance

$$\langle 0|\,[\varphi(x), \varphi(y)]\,|0\rangle = \sum_{\alpha} [\langle 0|\,\varphi(0)\,|\alpha\rangle\, e^{-ip_\alpha\cdot(x-y)}\langle\alpha|\,\varphi(0)\,|0\rangle - (x \leftrightarrow y)] \tag{5-15}$$

where the sum runs over a complete set of positive energy states α. To compare it with the commutator of two free fields of mass m,

$$i\Delta(x-y\,;m) = \int \frac{d^4q}{(2\pi)^3}\,\varepsilon(q^0)\delta(q^2-m^2)\,e^{-iq\cdot(x-y)} \tag{5-16}$$

we insert in (5-15) the identity

$$1 = \int d^4q\,\delta^4(q-p_\alpha)$$

and obtain

$$\langle 0|\,[\varphi(x), \varphi(y)]\,|0\rangle = \int \frac{d^4q}{(2\pi)^3}\,\rho(q)(e^{-iq\cdot(x-y)} - e^{iq\cdot(x-y)})$$

We have introduced the density

$$\rho(q) = (2\pi)^3 \sum_{\alpha} \delta^4(q-p_\alpha)|\,\langle 0|\,\varphi(0)\,|\alpha\rangle\,|^2$$

It is obviously positive and vanishes when q is not in the forward light cone; furthermore, it is invariant under a Lorentz transformation as required by the corresponding property of the field φ. It can therefore be written

$$\rho(q) = \sigma(q^2)\theta(q^0) \qquad \text{with } \sigma(q^2) = 0 \text{ if } q^2 < 0$$

In general this is a positive measure where δ-function singularities might possibly occur. We finally obtain a superposition of free commutator contributions with positive weights (Källen-Lehmann representation)

$$\langle 0|\,[\varphi(x), \varphi(y)]\,|0\rangle = i\int_0^\infty dm'^2\,\sigma(m'^2)\Delta(x-y\,;m') \tag{5-17}$$

Following the same line of reasoning it is easy to see that a similar result holds for the vacuum expectation value of the time-ordered product, with the same spectral function ρ.

We can explicitly separate in (5-17) the contribution of the one-particle state by using the asymptotic condition to write

$$\langle 0| [\varphi(x), \varphi(y)] |0\rangle = iZ\Delta(x-y; m) + i\int_{m_1^2}^{\infty} dm'^2\, \sigma(m'^2)\Delta(x-y; m') \tag{5-18}$$

Here m stands for the mass of the particles and $m_1 > m$ is the multiparticle threshold. If the interaction lagrangian does not involve field derivatives, $\dot{\varphi}$ will be conjugate to φ. By taking the time derivative of both sides of Eq. (5-18) and identifying the coefficient of $\delta^3(\mathbf{x}-\mathbf{y})$, we find that canonical quantization implies

$$1 = Z + \int_{m_1^2}^{\infty} dm'^2\, \sigma(m'^2) \tag{5-19}$$

Positivity of σ implies correctly that

$$0 \le Z < 1 \tag{5-20}$$

The value 1 is excluded if some matrix element of φ different from $\langle 1| \varphi(x) |0\rangle$ and $\langle 0| \varphi(x) |1\rangle$ does not vanish. Clearly $Z = 1$ requires $\varphi = \varphi_{\text{in}}$.

From the Källen-Lehmann representation it follows easily that the asymptotic limit cannot be understood in the strong sense. We can also check that as $|x^0 - y^0| \to \infty$ the dominant contribution arises from the one-particle contribution, the remaining terms being damped by strongly oscillating factors. By the same token (5-19) puts bounds on the density $\sigma(m^2)$. Many-particle states will give to $\sigma(m^2)$ a support extending to infinity and there will be a strong tendency to violate the positivity of Z.

In a manner analogous to the asymptotic limit (5-14a) in the remote past, it is expected that we also have

$$x_0 \to +\infty \qquad \varphi(x) \to Z^{1/2}\varphi_{\text{out}}(x) \tag{5-14b}$$

where φ_{out} is again a free field with the same mass m as φ_{in} and the same constant $Z^{1/2}$. Uniqueness of the vacuum implies $|0, \text{in}\rangle = |0, \text{out}\rangle = |0\rangle$ (a possible relative phase is conventionally set equal to one). Moreover, we assume that one-particle states are stable. Under these conditions $|1, \text{in}\rangle = |1, \text{out}\rangle$. Since $\langle 0| \varphi(x) |1\rangle$ has the same dependence on x as the corresponding matrix elements of φ_{in} or φ_{out}, the normalization of which is fixed by their free field character, we necessarily have

$$\langle 0| \varphi(x) |1\rangle = Z^{1/2}\langle 0| \varphi_{\text{in}}(x) |1\rangle = Z^{1/2}\langle 0| \varphi_{\text{out}}(x) |1\rangle \tag{5-21}$$

The S matrix induces the isomorphism between in- and out-states. From Eq. (5-5), it follows that

$$\begin{aligned} \varphi_{\text{in}}(x) &= S\varphi_{\text{out}}(x)S^{-1} \\ |i, \text{in}\rangle &= S\,|i, \text{out}\rangle \\ \langle f, \text{in}|\, S\,|i, \text{in}\rangle &= \langle f, \text{out}|\, S\,|i, \text{out}\rangle \end{aligned} \tag{5-22}$$

Moreover,

$$\begin{aligned} \langle 0|S|0\rangle &= \langle 0|0\rangle = 1 \\ \langle 1|S|1'\rangle &= \langle 1|1'\rangle \end{aligned} \tag{5-23}$$

where indices have been omitted from the one-particle states and we have only translated the stability of these states.

The unitary S matrix has to commute with Poincaré transformations to implement the covariance of the theory

$$U(a, \Lambda)SU^{-1}(a, \Lambda) = S \tag{5-24}$$

One can also express deeper locality properties of the S matrix if, following Bogoliubov and Shirkov, one allows for space-time dependence of the coupling constants in the lagrangian. We refer the reader to their textbook for a systematic use of this approach.

5-1-3 Reduction Formulas

Here we shall use the asymptotic conditions (5-14*a*, *b*) to relate S-matrix elements to the general Green functions of the interacting fields. Consider the transition amplitude $\langle p_1, \ldots, \text{out} \,|\, q_1, \ldots, \text{in}\rangle$, where for simplicity we have omitted the smearing test functions necessary for normalization. We shall reduce this element by extracting one by one the in- or out-creation operators which allow us to build the initial or final state. By definition

$$\begin{aligned} \langle p_1, \ldots, \text{out} \,|\, q_1, \ldots, \text{in}\rangle &= \langle p_1, \ldots, \text{out} \,|\, a^\dagger_{\text{in}}(q_1) \,|\, q_2, \ldots, \text{in}\rangle \\ &= \int_t d^3x \, e^{-iq_1 \cdot x} \frac{1}{i} \overleftrightarrow{\partial_0} \langle p_1, \ldots, \text{out} \,|\, \varphi_{\text{in}}(x) \,|\, q_2, \ldots, \text{in}\rangle \end{aligned}$$

The last integral is performed at an arbitrary time t. Let us therefore choose a large negative value to enable us to substitute $Z^{-1/2}\varphi$ for φ_{in}:

$$\langle p_1, \ldots, \text{out} \,|\, q_1, \ldots, \text{in}\rangle = \lim_{t \to -\infty} Z^{-1/2} \int_t d^3x \, e^{-iq_1 \cdot x} \frac{1}{i} \overleftrightarrow{\partial_0} \langle p_1, \ldots, \text{out} \,|\, \varphi(x) \,|\, q_2, \ldots, \text{in}\rangle$$

Now for an arbitrary integral

$$\left(\lim_{t \to +\infty} - \lim_{t \to -\infty} \right) \int d^3x \, \psi(\mathbf{x}, t) = \lim_{t_f \to +\infty, t_i \to -\infty} \int_{t_i}^{t_f} dt \frac{\partial}{\partial t} \int d^3x \, \psi(\mathbf{x}, t)$$

Using the asymptotic condition for arbitrarily large positive times we find

$$\begin{aligned} \langle p_1, \ldots, \text{out} \,|\, q_1, \ldots, \text{in}\rangle &= \langle p_1, \ldots, \text{out} \,|\, a^\dagger_{\text{out}}(q_1) \,|\, q_2, \ldots, \text{in}\rangle \\ &\quad + iZ^{-1/2} \int d^4x \, \partial_0 [e^{-iq_1 \cdot x} \overleftrightarrow{\partial_0} \langle p_1, \ldots, \text{out} \,|\, \varphi(x) \,|\, q_2, \ldots, \text{in}\rangle] \end{aligned} \tag{5-25}$$

where the integral extends over all space-time. The first term on the right-hand side of Eq. (5-25) is a sum of disconnected terms (corresponding to situations where at least one particle is not affected by the collision process):

$$\langle p_1,\ldots,\text{out}\,|\,a^\dagger_{\text{out}}(q_1)\,|\,q_2,\ldots,\text{in}\rangle = \sum_1^n 2p_k^0(2\pi)^3\delta^3(\mathbf{p}_k - \mathbf{q}_1)$$

$$\times\ \langle p_1,\ldots,\widehat{p_k},\ldots,\text{out}\,|\,q_2,\ldots,\text{in}\rangle$$

We have assumed here that all particles belong to the same species. The necessary changes when this is not the case are obvious. Observe that disconnected terms disappear when none of the initial and final momenta coincide.

Let us now take a closer look at the second term on the right-hand side of Eq. (5-25). Since this is understood as a kernel to be tested with wave packets, partial integrations over space variables are legitimate. This is, of course, not the case for time variables, otherwise all the calculations would collapse. With this in mind, denoting by m the mass of the particles and recalling that $q_1^2 = m^2$,

$$\int d^4x\ \partial_0[e^{-iq_1\cdot x}\overleftrightarrow{\partial_0}\langle\beta,\text{out}\,|\,\varphi(x)\,|\,\alpha,\text{in}\rangle]$$

$$= \int d^4x\ \{[(-\Delta + m^2)\,e^{-iq_1\cdot x}]\langle\beta,\text{out}\,|\,\varphi(x)\,|\,\alpha,\text{in}\rangle + e^{-iq_1\cdot x}\partial_0^2\langle\beta,\text{out}\,|\,\varphi(x)\,|\,\alpha,\text{in}\rangle\}$$

$$= \int d^4x\ e^{-iq_1\cdot x}(\Box + m^2)\langle\beta,\text{out}\,|\,\varphi(x)\,|\,\alpha,\text{in}\rangle$$

The first stage of the reduction formula takes the following form:

$$\langle p_1,\ldots,p_n,\text{out}\,|\,q_1,\ldots,q_l,\text{in}\rangle$$

$$= \sum_1^n 2p_k^0(2\pi)^3\delta^3(\mathbf{p}_k - \mathbf{q}_1)\langle p_1,\ldots,\widehat{p_k},\ldots,p_n,\text{out}\,|\,q_2,\ldots,q_l,\text{in}\rangle$$

$$+\ iZ^{-1/2}\int d^4x\ e^{-iq_1\cdot x}\,(\Box + m^2)\langle p_1,\ldots,p_n,\text{out}\,|\,\varphi(x)\,|\,q_2,\ldots,q_l,\text{in}\rangle \qquad (5\text{-}26)$$

The same steps can now be repeated again and again. To be specific, we shall do this for a particle in the final state. As far as the disconnected terms are concerned, nothing new emerges and we shall omit to write the corresponding contributions. This is, however, not true for the second term, for which we may write

$$\langle p_1,\ldots,\text{out}\,|\,\varphi(x_1)\,|\,q_2,\ldots,\text{in}\rangle = \langle p_2,\ldots,\text{out}\,|\,a_{\text{out}}(p_1)\varphi(x_1)\,|\,q_2,\ldots,\text{in}\rangle$$

$$= \lim_{y_1^0\to\infty} iZ^{-1/2}\int d^3y_1\, e^{ip_1\cdot y_1}\overleftrightarrow{\partial}_{y_1^0}$$

$$\times\ \langle p_2,\ldots,\text{out}\,|\,\varphi(y_1)\varphi(x_1)\,|\,q_2,\ldots,\text{in}\rangle$$

We would like to replace the last integral by a four-dimensional one as we did before. Clearly some trick is needed in order for the operator $a_{\text{in}}(p_1)$ coming from the lower bound of integration to operate directly on the state $|q_2,\ldots,\text{in}\rangle$. The time ordering of the fields

$$T\varphi(y_1)\varphi(x_1) = \begin{cases}\varphi(y_1)\varphi(x_1) & y_1^0 > x_1^0 \\ \varphi(x_1)\varphi(y_1) & x_1^0 > y_1^0\end{cases}$$

has just the right property of setting the operators in the convenient order relevant for the two limits. Without affecting the previous expression we can therefore substitute the T product and proceed as in the first step with the result that

$$\langle p_1,\ldots,\text{out}|\,\varphi(x_1)\,|q_2,\ldots,\text{in}\rangle = \langle p_2,\ldots,\text{out}|\,\varphi(x_1)a_{\text{in}}(p_1)\,|q_2,\ldots,\text{in}\rangle$$
$$+\, iZ^{-1/2}\int d^4y_1\, e^{ip_1\cdot y_1}(\Box_{y_1}+m^2)\langle p_2,\ldots,\text{out}|\,T\varphi(y_1)\varphi(x_1)\,|q_2,\ldots,\text{in}\rangle$$

On the right-hand side the first term will yield disconnected terms.

Once two particles have been reduced the answer looks like

$$\langle p_1,\ldots,\text{out}|q_1,\ldots,\text{in}\rangle = \langle p_1,\ldots,\text{in}|\,S\,|q_1,\ldots,\text{in}\rangle$$
$$= \text{disconnected terms}$$
$$+\,(iZ^{-1/2})^2\int d^4x_1\, d^4y_1\, e^{ip_1\cdot y_1 - iq_1\cdot x_1}(\Box_{y_1}+m^2)(\Box_{x_1}+m^2)$$
$$\times\,\langle p_2,\ldots,\text{out}|\,T\varphi(y_1)\varphi(x_1)\,|q_2,\ldots,\text{in}\rangle \tag{5-27}$$

Disconnected terms obtained so far involve one or two δ^3 functions. The same reasoning can now be carried further until all incoming and outgoing particles have been reduced

$$\langle p_1,\ldots,p_n,\text{out}|q_1,\ldots,q_l,\text{in}\rangle = \langle p_1,\ldots,p_n,\text{in}|\,S\,|q_1,\ldots,q_l,\text{out}\rangle$$
$$= \text{disconnected terms}$$
$$+\,(iZ^{-1/2})^{n+l}\int d^4y_1\ldots d^4x_l\,\exp\left(i\sum_1^n p_k\cdot y_k - \sum_1^l q_r\cdot x_r\right)$$
$$\times\,(\Box_{y_1}+m^2)\cdots(\Box_{x_l}+m^2)\,\langle 0|\,T\varphi(y_1)\cdots\varphi(x_l)\,|0\rangle \tag{5-28}$$

The remarkable feature of these expressions obtained by Lehmann, Symanzik, and Zimmermann is the relation they provide between the on-shell transition amplitudes and the general Green functions of the interacting theory. The latter are precisely the vacuum expectation values of time-ordered field products. Relation (5-28) implies that in momentum space the Green functions have poles in the variables p_i^2, where p_i is conjugate to x_i. Up to a normalization constant, the S-matrix element is nothing but the residue of this multiple pole.

We also notice a remarkable symmetry between incoming $(q_1,\ldots)$ and outgoing $(p_1,\ldots)$ momenta. It is convenient to replace the outgoing set $(p_1,\ldots)$ by an incoming one $(-p_1,\ldots)$ with negative energy component. In this way all momenta are on the same footing.

5-1-4 Generating Functional

It is possible to obtain a compact operator form for the set of relations (5-28). To simplify notations let us absorb a factor $Z^{-1/2}$ in φ in such a way that within this subsection the normalization of φ is taken as $\lim_{t\to\pm\infty} \varphi = \varphi_{\substack{\text{out}\\ \text{in}}}$. We shall also use a unique symbol for creation and annihilation operators, in agreement with a previous remark, by setting for an arbitrary four-vector k on the mass shell $k^2 = m^2$:

$$a^{\#}_{\substack{\text{in}\\ \text{out}}}(k) = \begin{cases} a_{\substack{\text{in}\\ \text{out}}}(k) & k^0 > 0 \\ a^{\dagger}_{\substack{\text{in}\\ \text{out}}}(-k) & k^0 < 0 \end{cases} \tag{5-29}$$

in such a way that

$$a^{\#}_{\substack{\text{in}\\ \text{out}}}(k) = i\varepsilon(k^0)\int d^3x\, e^{ik\cdot x}\overleftrightarrow{\partial_0}\varphi_{\substack{\text{in}\\ \text{out}}}(x) \tag{5-30}$$

Let $B_1, \ldots, B_n$ be local Bose fields. Equation (5-28) is equivalent in operator form to the relation

$$[ST[B_1(x_1)\cdots B_n(x_n)], a^{\#}_{\text{in}}(k)] = -i\varepsilon(k^0)S\int d^4y\, e^{ik\cdot y}(\square_y + m^2) \times T[\varphi(y)B(x_1)\cdots B_n(x_n)] \tag{5-31}$$

The proof follows easily by taking in-state matrix elements, using $a^{\#}_{\text{in}}S = Sa^{\#}_{\text{out}}$, and expressing $a^{\#}$ through (5-30) in terms of the field. The desired relation is then obtained as above by replacing boundary terms by an integral over the time derivative. Iteration yields the more general commutator

$$[\cdots[ST[B_1(x_1)\cdots B_n(x_n)], a^{\#}_{\text{in}}(k_1)], \ldots, a^{\#}_{\text{in}}(k_p)]$$
$$= (-i)^p\varepsilon(k_1^0)\cdots\varepsilon(k_p^0)\int d^4y_1\cdots d^4y_p \exp\left(i\sum_1^p k_r\cdot y_r\right)(\square_{y_1} + m^2)\cdots(\square_{y_p} + m^2)$$
$$\times ST[\varphi(y_1)\cdots\varphi(y_p)B_1(x_1)\cdots B_n(x_n)] \tag{5-32}$$

This could be applied to the case of several types of particles and fields.

To understand the use of (5-32) as a mean of generating S itself, let us remind ourselves of the construction of Fock space as a product of Hilbert spaces, each one corresponding to a definite mode (indexed here by the momentum k). For a given oscillator, with fundamental operators such that $[a, a^{\dagger}] = 1$ and a vacuum state $|0\rangle$, the normalized excited states are $|n\rangle = a^{\dagger n}/(n!)^{1/2}|0\rangle$. The projector $|0\rangle\langle 0|$ has the following representation:

$$|0\rangle\langle 0| = :e^{-a^{\dagger}a}: \tag{5-33}$$

This is indeed a hermitian operator leaving the vacuum state invariant and such that

$$[:e^{-a^{\dagger}a}:, a^{\dagger}] = \sum_1^{\infty}\frac{(-a^{\dagger})^n}{n!}[a^n, a^{\dagger}] = \sum_1^{\infty}\frac{(-a^{\dagger})^n a^{n-1}}{(n-1)!} = -a^{\dagger}:e^{-a^{\dagger}a}:$$

or

$$:e^{-a^{\dagger}a}:a^{\dagger} = 0$$

Consequently, $:e^{-a^\dagger a}:|n\rangle = \delta_{n0}|n\rangle$, proving (5-33).

Now let Q stand for an operator sufficiently regular to give a meaning to the subsequent operations. We expand it in terms of its matrix elements as

$$Q = \sum_{n,m=0}^{\infty} |m\rangle\langle m|\, Q\,|n\rangle\langle n| = \frac{a^{\dagger m}|0\rangle\langle 0|\,a^n}{m!\,n!}\,\langle 0|\,a^m Q a^{\dagger n}|0\rangle$$

$$= \sum_{m,n=0}^{\infty} \frac{a^{\dagger m}}{m!}\frac{\partial^m}{\partial\alpha^m} :e^{-a^\dagger a}: \frac{a^n}{n!}\frac{\partial^n}{\partial\bar{\alpha}^n}\langle 0|\,e^{\alpha a}\,Q\,e^{\bar{\alpha}a^\dagger}\,|0\rangle\Big|_{\alpha=\bar{\alpha}=0}$$

In this expression α and $\bar{\alpha}$ are taken as independent variables while the operators in front of the matrix element are in normal order. Therefore we obtain

$$Q = :e^{a^\dagger \partial/\partial\alpha + a\,\partial/\partial\bar{\alpha} - a^\dagger a}: \langle 0|\,e^{\alpha a}\,Q\,e^{\bar{\alpha}a^\dagger}\,|0\rangle\Big|_{\alpha=\bar{\alpha}=0}$$

This is not quite the desired result where we would like to find the matrix elements of Q between normalized coherent states, i.e., states obtained from the vacuum by the action of the unitary operator $e^{\bar{\alpha}a^\dagger - \alpha a}$. However, the relation

$$e^{\bar{\alpha}a^\dagger - \alpha a}|0\rangle = e^{-\alpha\bar{\alpha}/2}\,e^{\bar{\alpha}a^\dagger}\,|0\rangle$$

enables us to rewrite

$$Q = :e^{a^\dagger \partial/\partial\alpha + a\,\partial/\partial\bar{\alpha} - a^\dagger a}: e^{\alpha\bar{\alpha}}\,\langle 0|\,e^{\alpha a - \bar{\alpha}a^\dagger}\,Q\,e^{\bar{\alpha}a^\dagger - \alpha a}\,|0\rangle\Big|_{\alpha=\bar{\alpha}=0}$$

Moreover, we may commute $e^{\alpha\bar{\alpha}}$ with the derivatives acting to its left in order to be able to use the condition $\alpha = \bar{\alpha} = 0$ at the end. Under the normal product sign a and $a^\dagger$ commute and we note that $e^{a^\dagger \partial/\partial\alpha}$ translates α by an amount $a^\dagger$. These remarks lead to the final expression

$$Q = :e^{a^\dagger \partial/\partial\alpha + a\,\partial/\partial\bar{\alpha}}: \langle 0|\,e^{\alpha a - \bar{\alpha}a^\dagger}\,Q\,e^{\bar{\alpha}a^\dagger - \alpha a}\,|0\rangle\Big|_{\alpha=\bar{\alpha}=0} \tag{5-34}$$

It would, of course, be incorrect to conclude superficially from (5-34) that the sole knowledge of diagonal matrix elements of Q in the coherent state basis is sufficient to reconstruct the operator itself. As the formula indicates α and $\bar{\alpha}$ have to be taken as independent variables in order to carry out the required derivatives.

Returning now to the physical Fock space with infinitely many degrees of freedom we can write the unit operator as

$$I = \sum_{p=0}^{\infty} \frac{1}{p!} \int d\tilde{k}_1 \cdots d\tilde{k}_p\, |k_1, \ldots, k_p\rangle\langle k_1, \ldots, k_p| \tag{5-35}$$

where the initial term is understood as the vacuum projector $|0\rangle\langle 0|$, given by a generalization of Eq. (5-33), namely,

$$|0\rangle\langle 0| = :\exp\left[-\int d\tilde{k}\; a_{\text{in}}^\dagger(k) a_{\text{in}}(k)\right]: \tag{5-36}$$

so that the S matrix can be represented as

$$S = :\exp \int d\tilde{k} \left[a_{\text{in}}^\dagger(k)\,\frac{\delta}{\delta\alpha(k)} + a_{\text{in}}(k)\,\frac{\delta}{\delta\bar{\alpha}(k)}\right]:$$

$$\times \langle 0|\exp \int d^3k\; [a_{\text{in}}(k)\alpha(k) - a_{\text{in}}^\dagger(k)\bar{\alpha}(k)]S$$

$$\times \exp \int d^3k\,[a_{\text{in}}^\dagger(k)\bar{\alpha}(k) - a_{\text{in}}(k)\alpha(k)]\,|0\rangle\Big|_{\alpha=\bar{\alpha}=0} \tag{5-37}$$

in complete analogy with (5-34). Due to our normalization convention the measure d^3k and not $d\tilde{k}$ appears in the coherent state exponent. We shall now combine this result with Eq. (5-32) applied to the case where the B are set equal to unity. Introducing a source on the right-hand side of this equation it takes the simple form

$$\varepsilon(k_1^0)\cdots\varepsilon(k_p^0)[a_{\text{in}}^\#(k_1),\ldots,[a_{\text{in}}^\#(k_p),S]\cdots]$$
$$= S\prod_{r=1}^{p}\left[\int d^4y_r\, e^{ik_r\cdot y_r}(\Box_{y_r}+m^2)\frac{\delta}{\delta j(y_r)}\right]T\exp i\int d^4x\, j(x)\varphi(x)\bigg|_{j=0}$$

Since the vacuum is stable, that is, $S|0\rangle = |0\rangle$, it then follows that

$$\langle 0|\exp\int d^3k[a_{\text{in}}(k)\alpha(k) - a_{\text{in}}^\dagger(k)\bar{\alpha}(k)]S\exp\int d^3k\,[a_{\text{in}}^\dagger(k)\bar{\alpha}(k) - a_{\text{in}}(k)\alpha(k)]\,|0\rangle$$
$$= \exp\left\{\int d^4y\,\tilde{\alpha}(y)(\Box_y+m^2)\frac{\delta}{\delta j(y)}\right\}\langle 0|\,T\exp i\int d^4x\, j(x)\varphi(x)\,|0\rangle\bigg|_{j=0}$$

We have written $\tilde{\alpha}(y)$ for the solution of the homogeneous Klein-Gordon equation expressed as

$$\tilde{\alpha}(y) = \int d^3k\,[\alpha(k)\,e^{ik\cdot y} + \bar{\alpha}(k)\,e^{-ik\cdot y}]$$

Inserting this matrix element into Eq. (5-37) we see that

$$S = :\exp\left\{\int d^4y\,\varphi_{\text{in}}(y)(\Box_y+m^2)\frac{\delta}{\delta j(y)}\right\}:\,\langle 0|\,T\exp i\int d^4x\, j(x)\varphi(x)\,|0\rangle\bigg|_{j=0} \tag{5-38}$$

We have naturally been led to introduce the free in-field with its Fourier decomposition [reciprocal of (5-30)]

$$\varphi_{\text{in}}(y) = \int d\tilde{k}\,\,[a_{\text{in}}(k)\,e^{-ik\cdot y} + a_{\text{in}}^\dagger(k)\,e^{ik\cdot y}]$$

We should remind the reader that a proper normalization of the field φ requires its replacement by $Z^{-1/2}\varphi$ in the last T product. The outcome of this lengthy algebra is a compact relationship between the S matrix and a generating functional for the interacting field Green functions

$$Z(j) = \langle 0|\,T\exp i\int d^4x\,\varphi(x)j(x)\,|0\rangle \tag{5-39}$$

This bears some formal resemblance to the external source problems considered in Chap. 4. The difference is, of course, that we do not have any simple closed expression for $Z(j)$ in the present case.

The functional $Z(j)$ lies at the heart of field theory. We might be tempted to think that it contains far more information than is needed to compute the on-shell transition amplitudes. With present-day techniques, however, this is not true for several reasons. The full Green functions are needed to remedy in a consistent way the ultraviolet infinities occurring in the perturbative approach. Off-shell equations are also required to discuss bound states. Furthermore, short-distance behavior of interactions is ideally studied in the field theoretic framework. Nevertheless, notable achievements are to be credited to the S-matrix approach which abstracts from a relationship such as Eq. (5-38) only very general properties of scattering amplitudes and proceeds from there. Section 5-3 will give a simple-minded sketch of this line of thought which lies beyond the scope of this book.

5-1-5 Connected Parts

Some disconnected parts in the transition amplitudes (but not all) were not shown in Eq. (5-28) but can now be thoroughly studied with the help of Eq. (5-38). In simple terms some subsets of particles may interact independently from each other in a collision process. It is therefore natural to look for a definition of connectedness in such a way that an S-matrix element is the sum in all possible ways of products of connected ones involving subsets of interacting particles:

$$\langle p_I | S | q_J \rangle = \sum_{\cup_\alpha I_\alpha = I,\ \cup_\beta J_\beta = J} \prod_{\alpha,\beta} \langle p_{I_\alpha} | S^c | q_{J_\beta} \rangle \tag{5-40}$$

We use a shorthand notation for a collection of indices, $|p_I\rangle = \prod_{i \in I} a^\dagger(p_i)|0\rangle$, $\alpha(p_I) = \prod_{i \in I} \alpha(p_i)$. The connected matrix elements are defined recursively by the right-hand side of the equation where the sum runs over all partitions of the set of initial and final particles. If $|I|$ denotes the number of elements of I, we define

$$\begin{aligned} s^c(\alpha, \bar{\alpha}) &= \sum_{|I|,|J|} \frac{1}{|I|!\,|J|!} \int \prod_{i \in I} d^3 p_i \prod_{j \in J} d^3 q_j\, \alpha(p_I) \langle p_I | S^c | q_J \rangle \bar{\alpha}(q_J) \\ &= \langle 0 | \exp \int d^3 p\, \alpha(p) a(p) S^c \exp \int d^3 q\, \bar{\alpha}(q) a^\dagger(q) | 0 \rangle \end{aligned} \tag{5-41}$$

with $s^c(0, 0) = 0$. Similarly,

$$s(\alpha, \bar{\alpha}) = \langle 0 | \exp \int d^3 p\, \alpha(p) a(p) S \exp \int d^3 q\, \bar{\alpha}(q) a^\dagger(q) | 0 \rangle \tag{5-42}$$

such that $s(0, 0) = 1$. The recursion relations (5-40) are simply interpreted as

$$s(\alpha, \bar{\alpha}) = e^{s^c(\alpha, \bar{\alpha})} \tag{5-43}$$

If we use the definition (5-39) for the generating functional $Z(j)$ we may write

$$s(\alpha, \bar{\alpha}) = \exp \left\{ \int d^3 k\, 2k^0 (2\pi)^3 \alpha(k) \bar{\alpha}(k) + \int d^4 y\, \bar{\alpha}(y) (\Box_y + m^2) \frac{\delta}{\delta j(y)} \right\} Z(j) \bigg|_{j=0} \tag{5-44}$$

Comparing (5-43) and (5-44) it is natural to define the connected Green functions through

$$Z(j) = e^{G_c(j)} \tag{5-45}$$

Since $\exp\left[\int d^4 y\, \bar{\alpha}(y)(\Box_y + m^2)\, \delta/\delta j(y)\right]$ is a translation operator, we find by combining (5-43), (5-44), and (5-45) the relation

$$s^c(\alpha, \bar{\alpha}) = \int d^3 k\, 2k^0 (2\pi)^3 \alpha(k) \bar{\alpha}(k) + G_c[(\Box + m^2)\bar{\alpha}] \tag{5-46}$$

Let us make this formula more explicit. In the same way as we write‡

‡ For the 2-point function this differs by a factor i from the standard choice in the free-field case [compare for instance with Eq. (3-90)] but it is convenient in the general discussion. We hope that the reader will not find it difficult to convert from one convention to the other according to the context.

$$Z(j) = \langle 0| T \exp i \int d^4x\, \varphi(x)j(x)|0\rangle = \sum_0^\infty \frac{i^n}{n!} \int d^4x_1 \cdots d^4x_n\, j(x_1)\cdots j(x_n) G(x_1, \ldots, x_n) \tag{5-47}$$

with

$$G(x_1, \ldots, x_n) = \langle 0| T\varphi(x_1)\cdots\varphi(x_n)|0\rangle$$

let us define

$$G_c(j) = \sum_1^\infty \frac{i^n}{n!} \int d^4x_1 \cdots d^4x_n\, j(x_1)\cdots j(x_n) G_c(x_1, \ldots, x_n) \tag{5-48}$$

Then the meaning of Eq. (5-46) is that

$$s^c(\alpha, \bar{\alpha}) = \int d^3k\, 2k^0(2\pi)^3\alpha(k)\bar{\alpha}(k) + \sum_1^\infty \frac{i^n}{n!} \int d^4y_1 \cdots d^4y_n\, \tilde{\alpha}(y_1)\cdots\tilde{\alpha}(y_n)$$
$$\times (\Box_{y_1} + m^2)\cdots(\Box_{y_n} + m^2) G_c(y_1, \ldots, y_n) \tag{5-49}$$

If we insert the definition of $\tilde{\alpha}$ and the expansion (5-41) we see that for $n = |I| + |J| > 2$ we obtain the basic reduction formulas

$$\langle p_I | S^c | q_J \rangle = i^n \int d^4y_1 \cdots d^4y_n \exp\left(i \sum_1^{|I|} p_i \cdot y_i - \sum_1^{|J|} q_j \cdot y_{|I|+j} \right) \prod_1^n (\Box_{y_i} + m^2) G_c(y_1, \ldots, y_n) \tag{5-50}$$

expressing connected S-matrix elements in terms of connected Green functions. When compared to Eq. (5-28) we see that obviously disconnected terms have disappeared. Extra disconnected ones have also been suppressed as G_c has been substituted for G. Recall that the correct normalization of φ requires dividing the right-hand side by $Z^{n/2}$. The case $n = 2$ has been set aside. According to (5-40) we have

$$\langle p| S^c |q\rangle = \langle p| S |q\rangle = 2p^0(2\pi)^3\delta^3(\mathbf{p} - \mathbf{q}) \tag{5-51}$$

Moreover, we have to assume that $G(x) = 0$, meaning that the vacuum expectation value of the field vanishes. Using the Källen-Lehmann representation (5-17), the two-point Green function may be written

$$G_c(x, y) = G(x, y) = \langle 0| T\varphi(x)\varphi(y)|0\rangle$$
$$= i \int_0^\infty d\mu^2\, \sigma(\mu^2) \int \frac{d^4k}{(2\pi)^4} \frac{1}{k^2 - \mu^2 + i\varepsilon} e^{-ik\cdot(x-y)}$$

implying that $G_c(x, y)$ depends only on $x - y$ and has a single pole in the Fourier transform variable at $k^2 = m^2$. This means that

$$\int d^4y_1 \int d^4y_2\, e^{i(p\cdot y_1 - q\cdot y_2)} (\Box_{y_1} + m^2)(\Box_{y_2} + m^2) G_c(y_1, y_2) = 0$$

Hence, only the first term on the right-hand side of Eq. (5-49) contributes to $\langle p| S^c |q\rangle$ and we can readily verify that relation (5-51) follows. Even if $G(x) \neq 0$, translation invariance requires that it be x independent so that $G_c(x, y)$ differs from $G(x, y)$ by at most a constant and the preceding reasoning still applies. The reader will have no difficulty in proving that

$$G(x_1, \ldots, x_n) = \sum_{\cup I_\alpha = I} \prod_\alpha G_c(x_{I_\alpha}) \tag{5-52}$$

where $I = (1, \ldots, n)$. Thus a natural definition of connected S-matrix elements has led us to a definition of connected Green functions independent of any diagrammatic expansion. A more elaborate discussion will be given in Chap. 6. We note that sometimes the convention that $Z(j) = e^{iG_c(j)}$ is used. Let us conclude by noting that in an elastic two-body scattering process the separation into connected parts just corresponds to setting $S = I + iT$ and retaining only T in the transition amplitude used to construct the cross section.

5-1-6 Fermions

We generalize the preceding formalism to the case where spin $\frac{1}{2}$ fermions, described by the interpolating spinor field $\psi(x)$, are present in the initial or final states. The normalization constant affecting the field is traditionally called Z_2 in electromagnetism so that in the weak sense

$$\psi(x) \underset{t \to \pm\infty}{\longrightarrow} Z_2^{1/2} \psi_{\substack{\text{out} \\ \text{in}}}(x) \tag{5-53}$$

The vacuum expectation value of the anticommutator has the form

$$i\langle 0| \{\psi_\alpha(x), \overline{\psi}_\beta(y)\} |0\rangle = i \sum_n [\langle 0| \psi_\alpha(0) |n\rangle\langle n| \overline{\psi}_\beta(0) |0\rangle\, e^{-ip_n \cdot (x-y)}$$
$$+ \langle 0| \overline{\psi}_\beta(0) |n\rangle\langle n| \psi_\alpha(0) |0\rangle\, e^{ip_n \cdot (x-y)}]$$

The matrix spectral density

$$\rho_{\alpha\beta}(q) = (2\pi)^3 \sum_n \delta^4(p_n - q)\langle 0| \psi_\alpha(0) |n\rangle\langle n| \overline{\psi}_\beta(0) |0\rangle$$

can be expanded as a combination of the sixteen independent 4×4 matrices. From relativistic invariance it follows that

$$\rho(q) = \rho_1(q^2)\not{q} + \rho_2(q^2) + \rho_3(q^2)\not{q}\gamma^5 + \rho_4(q^2)\gamma^5$$

If we deal with a theory invariant under parity, as is the case for electromagnetism, a unitary operator $\mathscr{P}$ exists such that

$$\mathscr{P}|0\rangle = |0\rangle \qquad \mathscr{P}\psi(x)\mathscr{P}^{-1} = \gamma^0\psi(\tilde{x})$$

This leads to the condition

$$\rho(q) = \gamma^0\rho(\tilde{q})\gamma^0 \qquad \tilde{q} = (q^0, -\mathbf{q})$$

and requires the terms involving γ^5 to vanish, that is, $\rho_3 = \rho_4 = 0$. Consequently,

$$\rho(q) = \rho_1(q^2)\not{q} + \rho_2(q^2) \qquad \rho_i(q^2) = 0 \qquad \text{if } q^2 < 0$$

From PCT invariance the antiunitary operator Θ leaves the vacuum invariant and acts on the fields as

$$\Theta\psi_\alpha(x)\Theta^{-1} = i\gamma^5_{\alpha\beta}\psi^\dagger_\beta(-x) \qquad \Theta\overline{\psi}_\alpha(x)\Theta^{-1} = -i\psi_\beta(-x)(\gamma^5\gamma^0)_{\beta\alpha}$$

This enables us to relate the second term in the anticommutator to the first according to

$$\langle 0| \overline{\psi}_\beta(y)\psi_\alpha(x) |0\rangle = -\gamma^5_{\alpha\tau}\langle 0| \psi_\tau(-x)\overline{\psi}_{\tau'}(-y) |0\rangle\gamma^5_{\tau'\beta}$$

Hence

$$i\langle 0| \{\psi_\alpha(x), \overline{\psi}_\beta(y)\} |0\rangle = i \int \frac{d^4q}{(2\pi)^3}\, \theta(q^0)[\rho_1(q^2)i\not{\partial}_x + \rho_2(q^2)]_{\alpha\beta}$$
$$\times (e^{-iq \cdot (x-y)} - e^{iq \cdot (x-y)}) \tag{5-54}$$

Using the support properties of $\rho_i(q^2)$ on the positive real axis we find a superposition of free-field anticommutators:

$$i\langle 0|\{\psi^{\text{in}}(x), \overline{\psi}^{\text{in}}(y)\}|0\rangle = -(i\partial\!\!\!/_x + m)\Delta(x-y, m) = S(x-y, m)$$
$$\Delta(x-y, m) = i\int \frac{d^3k}{2k^0(2\pi)^3}(e^{-ik\cdot(x-y)} - e^{ik\cdot(x-y)}) \tag{5-55}$$

In the interacting case we find

$$i\langle 0|\{\psi_\alpha(x), \overline{\psi}_\beta(y)\}|0\rangle = -\int_0^\infty d\mu^2\,[\rho_1(\mu^2)i\partial\!\!\!/_x + \rho_2(\mu^2)]_{\alpha\beta}\Delta(x-y,\mu)$$
$$= \int_0^\infty d\mu^2\{\rho_1(\mu^2)S_{\alpha\beta}(x-y,\mu)$$
$$+ [\mu\rho_1(\mu^2) - \rho_2(\mu^2)]\delta_{\alpha\beta}\Delta(x-y,\mu)\} \tag{5-56}$$

It is understood that μ is the positive square root of μ^2. An analogous formula holds for the propagator with the same spectral densities ρ_1 and ρ_2:

$$\langle 0|T\psi(x)\overline{\psi}(y)|0\rangle = i\int \frac{d^4k}{(2\pi)^4}e^{-ik\cdot(x-y)}\int_0^\infty d\mu^2\,\frac{k\!\!\!/\rho_1(\mu^2) + \rho_2(\mu^2)}{k^2 - \mu^2 + i\varepsilon} \tag{5-57}$$

From the relation $\rho^\dagger(q) = \gamma^0\rho(q)\gamma^0$ it follows that ρ_1 and ρ_2 are real. Moreover, since the quantities $\text{tr}\,[\gamma^0\rho(q)]$ and $\text{tr}\,[\gamma^0(q\!\!\!/ - \mu)\rho(q)(q\!\!\!/ - \mu)]$ (with $q^2 = \mu^2$) are proportional respectively to the positive quantities $\sum_\alpha|\langle 0|\psi_\alpha(0)|n\rangle|^2$ and $\sum_\alpha|\langle 0|[(i\partial\!\!\!/ - \mu)\psi]_\alpha|n\rangle|^2$, we have the following positivity conditions:

$$\rho_1(\mu^2) \geq 0 \qquad \mu\rho_1(\mu^2) - \rho_2(\mu^2) \geq 0 \tag{5-58}$$

Finally, it is possible to extract the one-particle contribution. Using invariance under parity, we come to the conclusion that $\mu\rho_1(\mu^2) - \rho_2(\mu^2)$ has a support only on the continuum states, so that in (5-56) the first term only receives a contribution from the discrete one-particle state for $\mu = m$. This assumes, of course, that we can isolate a one-particle state from the continuum, and is not valid in the strict sense with massless particles such as photons present. The small fictitious mass of the photons will therefore be removed at the final stage of the calculations after coping with possible infrared divergences. Otherwise the single (charged)-particle pole becomes a gauge-dependent branch-point singularity, and the treatment gets slightly cumbersome.

Assuming, therefore, that ρ_1 contains a term of the form $Z_2\delta(\mu^2 - m^2)$ and that m_1 stands for the continuum threshold,

$$i\langle 0|\{\psi(x), \overline{\psi}(y)\}|0\rangle = Z_2S(x-y, m) - \int_{m_1^2}^\infty d\mu^2[\rho_1(\mu^2)i\partial\!\!\!/_x + \rho_2(\mu^2)]\Delta(x-y,\mu) \tag{5-59}$$

If we let $x^0 = y^0$ the left-hand side reduces to the canonical anticommutator $i\delta^3(\mathbf{x} - \mathbf{y})\gamma^0_{\alpha\beta}$. Using the identities

$$\Delta(x-y,\mu)\big|_{x^0=y^0}=0 \qquad \partial_0\Delta(x-y,\mu)\big|_{x^0=y^0}=-\delta^3(\mathbf{x}-\mathbf{y})$$

it therefore follows that

$$1=Z_2+\int_{m_1^2}^{\infty}d\mu^2\,\rho_1(\mu^2)$$

which, combined with positivity (5-58), leads to the same conclusion as in the scalar case, namely,

$$0\le Z_2<1 \tag{5-60}$$

From this point, we can proceed to the reduction formulas as in Sec. 5-1-3. Dirac operators replace Klein-Gordon ones. The free field has the Fourier decomposition

$$\psi_{\text{in}}(x)=\int\frac{d^3k}{(2\pi)^3}\frac{m}{k^0}\sum_{\varepsilon=\pm1}[b_{\text{in}}(k,\varepsilon)u(k,\varepsilon)\,e^{-ik\cdot x}+d_{\text{in}}^{\dagger}(k,\varepsilon)v(k,\varepsilon)\,e^{ik\cdot x}]$$

with fixed helicity spinors satisfying $v(k,\varepsilon)=C\bar{u}^T(k,\varepsilon)$. The creation and annihilation operators are expressed as

$$\begin{aligned}
b_{\text{in}}(k,\varepsilon)&=\int_t d^3x\,\bar{u}(k,\varepsilon)\,e^{ik\cdot x}\gamma^0\psi_{\text{in}}(x)\\
d_{\text{in}}^{\dagger}(k,\varepsilon)&=\int_t d^3x\,\bar{v}(k,\varepsilon)\,e^{-ik\cdot x}\gamma^0\psi_{\text{in}}(x)\\
b_{\text{in}}^{\dagger}(k,\varepsilon)&=\int_t d^3x\,\overline{\psi}_{\text{in}}(k,\varepsilon)\gamma^0\,e^{-ik\cdot x}u(k,\varepsilon)\\
d_{\text{in}}(k,\varepsilon)&=\int_t d^3x\,\overline{\psi}_{\text{in}}(k,\varepsilon)\gamma^0\,e^{ik\cdot x}\,v(k,\varepsilon)
\end{aligned}\tag{5-61}$$

A sign ambiguity affects one-particle states of given helicity and momentum. Ignoring finite size wave packets we can write for an electron, say, in the initial state [compare with (5-26)]

$$\begin{aligned}
\langle\text{out}|\,b_{\text{in}}^{\dagger}(k,\varepsilon)\,|\text{in}\rangle&=\lim_{t\to-\infty}Z_2^{-1/2}\int d^3x\,\langle\text{out}|\,\overline{\psi}(x)\gamma^0\,|\text{in}\rangle\,e^{-ik\cdot x}u(k,\varepsilon)\\
&=\langle\text{out}|\,b_{\text{out}}^{\dagger}(k,\varepsilon)\,|\text{in}\rangle-iZ^{-1/2}\int d^3x\Big[\langle\text{out}|\,\overline{\psi}(x)\,|\text{in}\rangle\\
&\quad\times\frac{1}{i}\overleftarrow{\partial}_0\gamma^0u(k,\varepsilon)\,e^{-ik\cdot x}+\langle\text{out}|\,\overline{\psi}(x)\,|\text{in}\rangle\frac{1}{i}\overrightarrow{\partial}_0\gamma^0u(k,\varepsilon)\,e^{-ik\cdot x}\Big]
\end{aligned}$$

The spinor $u(k,\varepsilon)\,e^{-ik\cdot x}$ satisfies the homogeneous Dirac equation, so that its time derivative can be replaced by space derivatives. Integration by parts leads to

$$\langle\text{out}|\,b_{\text{in}}^{\dagger}(k,\varepsilon)\,|\text{in}\rangle=\text{disc.}-iZ_2^{-1/2}\int d^4x\,\langle\text{out}|\,\overline{\psi}(x)\,|\text{in}\rangle(-i\overleftarrow{\not\partial}_x-m)u(k,\varepsilon)\,e^{-ik\cdot x} \tag{5-62a}$$

Here disc. stands for a disconnected term and (5-62a) expresses the result of reducing one particle in the initial state. We have corresponding expressions for the reduction of an antiparticle in the initial state, or for a particle or antiparticle in the final state. These read successively:

$$\langle\text{out}|\, d_{\text{in}}^{\dagger}(k,\varepsilon)\,|\text{in}\rangle = \text{disc.} + iZ_2^{-1/2}\int d^4x\, \bar{v}(k,\varepsilon)\, e^{-ik\cdot x}\,(i\not{\partial}_x - m)\langle\text{out}|\,\psi(x)\,|\text{in}\rangle \tag{5-62b}$$

$$\langle\text{out}|\, b_{\text{out}}(k,\varepsilon)\,|\text{in}\rangle = \text{disc.} - iZ_2^{-1/2}\int d^4x\, \bar{u}(k,\varepsilon)\, e^{ik\cdot x}(i\not{\partial}_x - m)\langle\text{out}|\,\psi(x)\,|\text{in}\rangle \tag{5-62c}$$

$$\langle\text{out}|\, d_{\text{out}}(k,\varepsilon)\,|\text{in}\rangle = \text{disc.} + iZ_2^{-1/2}\int d^4x\, \langle\text{out}|\,\bar{\psi}(x)\,|\text{in}\rangle(-i\overleftarrow{\not{\partial}}_x - m)v(k,\varepsilon)\, e^{ik\cdot x} \tag{5-62d}$$

When comparing (5-62a) and (5-62d), corresponding to an electron entering the interaction process or a positron leaving it, we observe the characteristic substitution

$$u(k,\varepsilon)\, e^{-ik\cdot x} \to -v(k,\varepsilon)\, e^{ik\cdot x}$$

discussed in the hole formalism (Chap. 2). The positron in the final state is equivalent to an electron with an opposite energy momentum in the initial state. The same remarks apply to the pair (5-62b) and (5-62c).

It is now easy to continue the reduction process, using the same trick as before, namely, a time-ordered product, except that we have to keep track of the anticommutation properties of fermion fields. To simplify the general expression we omit the explicit polarization dependence of spinors and operators.

Assume that particles labeled $(k_1, \ldots)$ and antiparticles $(k_1', \ldots)$ are the incoming ones, and particles $(q_1, \ldots)$ and antiparticles $(q_1', \ldots)$ the outgoing ones. The conjugate space-time variables are respectively denoted x, x', y, and y'. Finally, n and n' stand for the total numbers of particles and antiparticles. The scattering amplitude takes the form

$$\begin{aligned}
&\langle\text{out}|\cdots d_{\text{out}}(q_1')\cdots b_{\text{out}}(q_1) b_{\text{in}}^{\dagger}(k_1)\cdots d_{\text{in}}^{\dagger}(k_1')\cdots|\text{in}\rangle \\
&\quad = \text{disc.} + (-iZ_2^{-1/2})^n (iZ_2^{-1/2})^{n'} \int d^4x_1 \cdots d^4y_1' \cdots \\
&\quad \times \exp[-i\Sigma(k\cdot x + k'\cdot x' - q\cdot y - q'\cdot y')]\cdots \\
&\quad \times \bar{u}(q_1)(i\overrightarrow{\not{\partial}}_{y_1} - m)\cdots \bar{v}(k_1')(i\overrightarrow{\not{\partial}}_{x_1'} - m)\langle 0|\, T[\cdots\bar{\psi}(y_1')\cdots\psi(y_1)\bar{\psi}(x_1)\cdots\psi(x_1')\cdots]\,|0\rangle \\
&\quad \times (-i\overleftarrow{\not{\partial}}_{x_1} - m)u(k_1)\cdots(-i\overleftarrow{\not{\partial}}_{y_1'} - m)v(q_1')\cdots
\end{aligned} \tag{5-63}$$

By changing perhaps the sign, we can of course reorder the ψ and $\bar{\psi}$ fields under the T symbol.

The generalized Green functions are the vacuum expectation values of the time-ordered products of fields. The underlying hypothesis is that to each species of particles there corresponds not only an asymptotic free field but also an interacting (or interpolating) field. Some of these fields could be composite in terms of the elementary lagrangian variables if bound states can be generated. This is a nontrivial situation which requires a thorough investigation. Such bound states manifest themselves as poles of some Green functions in the corresponding energy momentum variable.

Define the connected Green functions for processes involving fermion fields and discuss the corresponding generating functionals.

5-1-7 Photons

We expect additional complications in the case of photons because of gauge invariance. Following the Stueckelberg approach we introduce a fictitious mass μ and an indefinite metric Hilbert space. Physical vector photons have the mass μ while the scalar ghost states have the mass $m(\mu^2/m^2 = \lambda)$. The interacting field satisfies the modified Maxwell equation

$$(\Box + \mu^2)A^\rho - (1 - \lambda)\partial^\rho \partial \cdot A = j^\rho \tag{5-64}$$

while the conserved current j vanishes in the case of free fields. The field conjugate to A_ρ is

$$\pi^\rho = \partial^\rho A^0 - \partial^0 A^\rho - \lambda g^{\rho 0} \partial \cdot A \tag{5-65}$$

Consider the vacuum expectation value of the field commutator. Inserting a complete set of positive energy intermediate states we obtain the general form

$$\langle 0|\,[A_\rho(x), A_\nu(y)]\,|0\rangle = -\int_0^\infty dM^2 \int \frac{d^3k}{2k^0(2\pi)^3}(e^{-ik\cdot(x-y)} - e^{ik\cdot(x-y)}) \times \left[\pi_1(M^2) g_{\rho\nu} - \pi_2(M^2)\frac{k_\rho k_\nu}{M^2}\right] \tag{5-66}$$

It is understood that k^0 stands for $(\mathbf{k}^2 + M^2)^{1/2}$ and the lower limit of the spectral measure support is larger or equal to inf (μ^2, m^2). As a consequence of current conservation the scalar component is a free field

$$(\Box + m^2)\partial \cdot A = 0 \tag{5-67}$$

We therefore write that

$$\langle 0|\,[(\Box + m^2)\partial \cdot A(x), A_\nu(y)]\,|0\rangle = 0$$

This yields the relation

$$(m^2 - M^2)[\pi_1(M^2) - \pi_2(M^2)] = 0$$

which means that π_1 and π_2 differ at most by a term proportional to $\delta(M^2 - m^2)$.

For convenience we set

$$\pi_1(M^2) = \pi(M^2)$$

$$\pi_2(M^2) = \pi(M^2) - \frac{Z_3 z}{\lambda}\,\delta(M^2 - m^2)$$

Hence

$$\langle 0|\,[A_\rho(x), A_\nu(y)]\,|0\rangle = -\int_0^\infty dM^2 \int \frac{d^3k}{2k^0(2\pi)^3}(e^{-ik\cdot(x-y)} - e^{ik\cdot(x-y)})$$
$$\times \left[\pi(M^2)\left(g_{\rho\nu} - \frac{k_\rho k_\nu}{M^2}\right) + \frac{Z_3 z}{\lambda}\frac{k_\rho k_\nu}{M^2}\,\delta(M^2 - m^2)\right] \tag{5-68}$$

which exhibits a transverse part weighted with $\pi(M^2)$ and a longitudinal one concentrated at $M^2 = m^2$.

We insist now that the equal-time commutator vanishes. On the right-hand side the corresponding contribution is proportional to a gradient of a δ function. Setting its coefficient equal to zero yields the sum rule

$$Z_3 z = \int_0^\infty dM^2\,\pi(M^2)\,\frac{\mu^2}{M^2} \tag{5-69}$$

Next we evaluate the equal-time commutator of the potential with its conjugate field. Using Eq. (5-65), a simple calculation yields

$$\langle 0|\,[A_\rho(x), \pi^\nu(y)]\,|0\rangle|_{x^0=y^0} = i\delta^3(\mathbf{x}-\mathbf{y})\left\{g_\rho^\nu \int_0^\infty dM^2\,\pi(M^2)\right.$$
$$\left. + g_\rho^0 g_0^\nu\left[Z_3 z - \int_0^\infty dM^2\,\pi(M^2)\right]\right\} \tag{5-70}$$

We would like this commutator to be canonical, which means equal to $ig_\rho^\nu\delta^3(\mathbf{x}-\mathbf{y})$. As far as space components are concerned this is achieved by setting

$$\int_0^\infty dM^2\,\pi(M^2) = 1 \tag{5-71}$$

In order that this result extend to the time components we would have to satisfy

$$Z_3 z = \int_0^\infty dM^2\,\pi(M^2)$$

This is, however, incompatible in general with Eq. (5-69). We are going to see that $\pi(M^2)$ is a positive measure. Therefore the validity of both relations leads to a measure of support $M^2 = \mu^2$ and can only hold for a free field. The best that can be achieved is to allow for different normalizations of the space and time canonical commutation rules, i.e., to assume that

$$
\begin{aligned}
[A_\rho(x), \pi^\nu(y)]\,|_{x^0=y^0} &= i\delta^3(\mathbf{x}-\mathbf{y})(g_\rho{}^\nu + a g_{\rho 0} g^{\nu 0}) \\
a = Z_3 z - 1 &= \int_0^\infty dM^2\, \pi(M^2)\, \frac{\mu^2 - M^2}{M^2}
\end{aligned}
\tag{5-72}
$$

Up to now, no reference has been made to the asymptotic condition. Since dynamics does not affect the various parts of the field in a similar fashion, it would be clearly unwise to assume that (even weakly) $A_\rho(x)$ tends for $x^0 \to -\infty$ to a free field up to normalization. It is more realistic to define a transverse component of this incoming field as

$$
A_\rho^{T,\mathrm{in}}(x) = A_\rho^{\mathrm{in}}(x) + \frac{1}{m^2}\,\partial_\rho \partial \cdot A^{\mathrm{in}}(x) \tag{5-73a}
$$

and to assume

$$
A_\rho(x) \underset{x^0 \to -\infty}{\longrightarrow} Z_3^{1/2}\left[A_\rho^{T,\mathrm{in}} - z^{1/2}\,\frac{1}{m^2}\,\partial_\rho \partial \cdot A^{\mathrm{in}}(x)\right] \tag{5-73b}
$$

This condition entails in particular that

$$
\langle 0|\, A_\rho(x)\,|1\rangle = Z_3^{1/2}\left[\langle 0|\, A_\rho^{T,\mathrm{in}}(x)\,|1\rangle - z^{1/2}\,\frac{1}{m^2}\,\partial_\rho \langle 0|\,\partial \cdot A^{\mathrm{in}}(x)\,|1\rangle\right] \tag{5-74}
$$

For the three transversely polarized states only the first term contributes, while only the second matters if we deal with a scalar state. The one-particle contribution to the commutator can then be worked out as

$$
\begin{aligned}
\langle 0|\,[A_\rho(x), A_\nu(y)]\,|0\rangle\big|_{1\,\mathrm{particle}} &= -Z_3 \int dM^2 \int \frac{d^3k}{2k^0(2\pi)^3}\,(e^{-ik\cdot(x-y)} - e^{ik\cdot(x-y)}) \\
&\quad \times \left[\delta(M^2-\mu^2)\left(g_{\rho\nu} - \frac{k_\rho k_\nu}{M^2}\right) + \frac{z}{\lambda}\,\frac{k_\rho k_\nu}{M^2}\,\delta(M^2 - m^2)\right]
\end{aligned}
\tag{5-75}
$$

If M_0^2 stands for the continuum threshold we obtain

$$
\begin{aligned}
&\langle 0|\,[A_\rho(x), A_\nu(y)]\,|0\rangle \\
&= -Z_3 \int \frac{d^3k}{2k^0(2\pi)^3}\,(e^{-ik\cdot(x-y)} - e^{ik\cdot(x-y)})\left(g_{\rho\nu} - \frac{k_\rho k_\nu}{\mu^2}\right)\bigg|_{k^0=(\mathbf{k}^2+\mu^2)^{1/2}} \\
&\quad - Z_3 z \int \frac{d^3k}{2k^0(2\pi)^3}\,(e^{-ik\cdot(x-y)} - e^{ik\cdot(x-y)})\,\frac{k_\rho k_\nu}{\mu^2}\bigg|_{k^0=(\mathbf{k}^2+m^2)^{1/2}} \\
&\quad - \int_{M_0^2}^\infty dM^2\, \pi(M^2) \int \frac{d^3k}{2k^0(2\pi)^3}\,(e^{-ik\cdot(x-y)} - e^{ik\cdot(x-y)})\left(g_{\rho\nu} - \frac{k_\rho k_\nu}{M^2}\right)\bigg|_{k^0=(\mathbf{k}^2+M^2)^{1/2}}
\end{aligned}
\tag{5-76}
$$

$$1 = Z_3 + \int_{M_0^2}^{\infty} dM^2\, \pi(M^2) \tag{5-77}$$

$$Z_3(z-1) = \int_{M_0^2}^{\infty} dM^2\, \pi(M^2)\, \frac{\mu^2}{M^2} \tag{5-78}$$

Accordingly, the deviation from canonical behavior is due only to the continuum contribution and vanishes for a free field since (5-72) reads

$$a = \int_{M_0^2}^{\infty} dM^2\, \pi(M^2) \frac{\mu^2 - M^2}{M^2} \tag{5-79}$$

The corresponding decomposition for the covariant propagator is

$$i\langle 0|\, T[A_\rho(x)A_\nu(y)]\,|0\rangle = \int \frac{d^4k}{(2\pi)^4}\, e^{-ik\cdot(x-y)} \Bigg[Z_3\left(\frac{g_{\rho\nu} - k_\rho k_\nu/\mu^2}{k^2 - \mu^2 + i\varepsilon} + z\, \frac{k_\rho k_\nu/\mu^2}{k^2 - m^2 + i\varepsilon}\right)$$
$$+ \int_{M_0^2}^{\infty} dM^2\, \pi(M^2)\, \frac{g_{\rho\nu} - k_\rho k_\nu/M^2}{k^2 - M^2 + i\varepsilon} \Bigg] \tag{5-80}$$

(see the discussion of the problem of covariance at the end of this subsection). Even though the longitudinal component is dynamically decoupled it turns out that the covariant expressions written above imply a nontrivial normalization factor $Z_3 z$ for this term; Eqs. (5-77) and (5-78) do not allow us to set $z = 1$ or $Z_3 z = 1$.

As soon as we have shown that the measure $\pi(M^2)$ is positive, in spite of the indefinite metric state, the physical interpretation of Z_3 will become clear, as it will play a role analogous to Z in the scalar field case.

To do so, we extract from Eqs. (5-64) and (5-76) the vacuum expectation value of the current commutator:

$$\langle 0|\, [j_\rho(x), j_\nu(y)]\,|0\rangle = -\int_{M_0^2}^{\infty} dM^2 \int \frac{d^3k}{2k^0(2\pi)^3}\, (e^{-ik\cdot(x-y)} - e^{ik\cdot(x-y)})$$
$$\times\, \pi(M^2)(\mu^2 - M^2)^2 \left(g_{\rho\nu} - \frac{k_\rho k_\nu}{M^2}\right) \tag{5-81}$$

This formula exhibits (1) the disappearance of one-particle contributions: $\langle 0|\, j\,|1 \text{ photon}\rangle = 0$ and (2) the current conservation. In a space with definite metric we would easily derive from (5-81) that the density $\pi(M^2)$ is positive. However, we have here

$$\sum_\alpha \delta(p_\alpha^2 - M^2)(-1)^{n_\alpha} \langle 0|\, j_\rho(x)\,|\alpha\rangle\langle\alpha|\, j_\nu(y)\,|0\rangle$$
$$= -\int \frac{d^3k}{2k^0(2\pi)^3}\, e^{-ik\cdot(x-y)}\, \pi(M^2)(\mu^2 - M^2)^2 \left(g_{\rho\nu} - \frac{k_\rho k_\nu}{M^2}\right) \tag{5-82}$$

where n_α denotes the number of scalar photons in the intermediate state $|\alpha\rangle$ and $\langle 0|\, j_\nu(0)\,|\alpha\rangle^* = \langle\alpha|\, j_\nu(0)\,|0\rangle$. In Chap. 4 we have encountered examples of cancellation between longitudinal and scalar contributions. Here we expect a different mechanism since the two kinds have different masses. We first note that states containing one scalar particle do not show up in (5-81). Furthermore, Eqs. (5-64) and (5-76) indicate that

$$\langle 0|\,[j_\rho(x), \partial\cdot A(y)]\,|0\rangle = 0 \tag{5-83}$$

This relation can readily be extended to other matrix elements, so that more generally we have the operator relation

$$[j_\rho(x), \partial\cdot A(y)] = 0 \tag{5-84}$$

which expresses the gauge invariance of the current and reminds us that $\partial\cdot A$ is a free field:

$$\partial\cdot A = (zZ_3)^{1/2}\partial\cdot A^{\text{in}} = (zZ_3)^{1/2}\partial\cdot A^{\text{out}} \tag{5-85}$$

Combining (5-84) and (5-85) it follows that the matrix elements $\langle 0|\,j_\rho(x)\,|\alpha\rangle$ such that $|\alpha\rangle$ includes scalar photons have to vanish. This shows that all terms on the right-hand side contribute with a positive sign and

$$\pi(M^2) > 0 \tag{5-86}$$

Thus the interpretation of the sum rule (5-77) is the same as in the scalar case. A closer study will reveal that Z_3 is in fact independent of the gauge parameter λ, as the preceding remarks suggest.

Deriving reduction formulas for states involving transverse photons is then straightforward. Let us consider, for instance, a process involving the emission of a photon with momentum k and polarization ε ($\varepsilon\cdot k = 0$):

$$\begin{aligned}\langle\beta, k, \varepsilon, \text{out}\,|\,\alpha, \text{in}\rangle &= \langle\beta, \text{out}\,|\,a^{(\varepsilon)}_{\text{out}}(k)\,|\,\alpha, \text{in}\rangle\\ &= -i\int_{t_f} d^3x\, e^{ik\cdot x}\overleftrightarrow{\partial_0}\,\langle\beta, \text{out}\,|\,\varepsilon\cdot A^{T,\text{out}}(x)\,|\,\alpha, \text{in}\rangle\end{aligned}$$

An asymptotic relation analogous to (5-73) with $t = t_f \to \infty$ and A^{in} replaced by A^{out} holds. The above matrix element can therefore be written

$$\lim_{t_f\to\infty} -iZ_3^{-1/2}\int_{t_f} d^3x\, e^{ik\cdot x}\overleftrightarrow{\partial_0}\,\langle\beta, \text{out}\,|\,\varepsilon\cdot A(x) + \frac{1}{m^2}\,\varepsilon\cdot\partial\,\partial\cdot A(x)\,|\,\alpha, \text{in}\rangle$$

$$= \text{disc.} - iZ_3^{-1/2}\int d^4x\, e^{ik\cdot x}(\Box_x + \mu^2)\langle\beta, \text{out}\,|\,\varepsilon\cdot A(x) + \frac{1}{m^2}\,\varepsilon\cdot\partial\,\partial\cdot A(x)\,|\,\alpha, \text{in}\rangle$$

Now

$$(\Box_x + \mu^2)\,\frac{1}{m^2}\,\partial_\rho\,\partial\cdot A(x) = \frac{\mu^2 - m^2}{m^2}\,\partial_\rho\,\partial\cdot A(x) = (\lambda - 1)\partial_\rho\partial\cdot A(x)$$

The combination occurring in the matrix element is therefore $\varepsilon\cdot j(x)$ and the final expression, with a possible disconnected term, reads

$$\langle\beta, k, \varepsilon, \text{out}\,|\,\alpha, \text{in}\rangle = \text{disc.} - iZ_3^{-1/2}\int d^4x\, e^{ik\cdot x}\langle\beta, \text{out}\,|\,\varepsilon\cdot j(x)\,|\,\alpha, \text{in}\rangle \tag{5-87}$$

This procedure can be carried out again and again. We quote the result pertaining to the case where an initial photon (k_i, ε_i) is scattered to a final state (k_f, ε_f). We get

$$\langle \beta, k_f, \varepsilon_f, \text{out} | \alpha, k_i, \varepsilon_i, \text{in} \rangle = \text{disc.} - Z_3^{-1} \int d^4x\, d^4y\, e^{i(k_f \cdot x - k_i \cdot y)} (\Box_y + \mu^2)$$
$$\times \langle \beta, \text{out} | T\varepsilon_f \cdot j(x) \left[\varepsilon_i \cdot A(y) + \frac{1}{m^2} \varepsilon_i \cdot \partial\, \partial \cdot A(y) \right] | \alpha, \text{in} \rangle$$

After the action of the Klein-Gordon operator we would like to replace the matrix element by $\langle \beta, \text{out} | T[\varepsilon_f \cdot j(x) \varepsilon_i \cdot j(y)] | \alpha, \text{in} \rangle$. The two quantities can at most differ by a distribution concentrated on the manifold $x = y$, due to the ambiguity arising in the definition of time ordering. We adapt our definition of the T symbol in such a way that the two expressions are identical:

$$\langle \beta, k_f, \varepsilon_f, \text{out} | \alpha, k_i, \varepsilon_i, \text{in} \rangle$$
$$= \text{disc.} - Z_3^{-1} \int d^4x\, d^4y\, e^{i(k_f \cdot x - k_i \cdot y)} \langle \beta, \text{out} | T\varepsilon_f \cdot j(x) \varepsilon_i \cdot j(x) | \alpha, \text{in} \rangle \tag{5-88}$$

This formula will be used for an elementary discussion of the Compton effect.

We return to the proper definition of covariant time ordering since we skipped any detail between the expressions (5-76) for the commutator and (5-80) for the covariant T product. Things are not so simple when dealing with tensor operators, and so deserve some comments. For the purpose of illustration let us study the product of two conserved currents. Denote by $p(M^2)$ the spectral function $(\mu^2 - M^2)^2 \pi(M^2)$ so that

$$W_{\rho\nu}(x - y) = \langle 0 | j_\rho(x) j_\nu(y) | 0 \rangle$$
$$= -\int_{M_0^2}^{\infty} dM^2\, p(M^2) \int \frac{d^4k}{(2\pi)^3} \theta(k^0) \delta(k^2 - M^2)\, e^{-ik \cdot (x-y)} \left(g_{\rho\nu} - \frac{k_\rho k_\nu}{M^2} \right) \tag{5-89}$$

Current conservation is explicit and the minus sign in front of the right-hand side follows from our Lorentz metric in agreement with a positive $p(M^2)$. Let

$$W_M(x - y) = \int \frac{d^4k}{(2\pi)^3} \delta(k^2 - M^2) \theta(k^0)\, e^{-ik \cdot (x-y)} = \langle 0 | \varphi(x) \varphi(y) | 0 \rangle \tag{5-90}$$

be the Wightman function of two scalar free fields of mass M. For the commutator we obviously have

$$\langle 0 | [j_\rho(x), j_\nu(y)] | 0 \rangle = W_{\rho\nu}(x - y) - W_{\nu\rho}(y - x) \tag{5-91}$$

We shall first attempt to define a naive time ordering (denoted $\tilde{T}$) as

$$\langle 0 | \tilde{T} j_\rho(x) j_\nu(y) | 0 \rangle = \theta(x^0 - y^0) \langle 0 | j_\rho(x) j_\nu(y) | 0 \rangle + \theta(y^0 - x^0) \langle 0 | j_\nu(y) j_\rho(x) | 0 \rangle \tag{5-92}$$

Unfortunately, using (5-89) we shall soon discover that $\tilde{T}$ does not lead to a covariant quantity. For a free field of mass M,

$$i\langle 0 | T\varphi(x)\varphi(y) | 0 \rangle = i\theta(x^0 - y^0) W_M(x - y) + i\theta(y^0 - x^0) W_M(y - x)$$
$$= -\int \frac{d^4k}{(2\pi)^4} e^{-ik \cdot (x-y)} \frac{1}{k^2 - M^2 + i\varepsilon} = G_F(y - x) \tag{5-93}$$

Consequently

$$i\langle 0|\,\tilde{T}j_\rho(x)j_\nu(y)\,|0\rangle = -\int_{M_0^2}^{\infty} dM^2\, \rho(M^2)\left[i\theta(x^0-y^0)\left(g_{\rho\nu} - \frac{\partial_\rho^x\partial_\nu^y}{M^2}\right)W_M(x-y) + (x\leftrightarrow y)\right] \tag{5-94}$$

We define a covariant T product as

$$i\langle 0|\,Tj_\rho(x)j_\nu(y)\,|0\rangle = iT_{\rho\nu}^{(1)}(x-y) + i\Delta T_{\rho\nu}(x,y) \tag{5-95}$$

with

$$\begin{aligned} iT_{\rho\nu}^{(1)}(x-y) &= -\int_{M_0^2}^{\infty} dM^2\, \rho(M^2)\left[i\left(g_{\rho\nu} - \frac{\partial_\rho^x\partial_\nu^y}{M^2}\right)\theta(x^0-y^0)W_M(x-y) + (x\leftrightarrow y)\right] \\ &= -\int_{M_0^2}^{\infty} dM^2\, \rho(M^2)\left(g_{\rho\nu} - \frac{\partial_\rho^x\partial_\nu^y}{M^2}\right)G_F(x-y) \\ &= \int_{M_0^2}^{\infty} dM^2\, \rho(M^2)\int\frac{d^4k}{(2\pi)^4}\, e^{-ik\cdot(x-y)}\,\frac{g_{\rho\nu} - k_\rho k_\nu/M^2}{k^2-M^2+i\varepsilon} \end{aligned} \tag{5-96}$$

Here $i\Delta T_{\rho\nu}(x, y)$ is a local (i.e., concentrated on $x = y$) covariant term which we leave undetermined for the time being. We shall see that it can be adjusted to take care of current conservation. Let us study the difference between the two time-ordered quantities. To do this we rewrite the term under square brackets in Eq. (5-96), using the fact that $W_M(x-y)$ depends only on $(x-y)^2$.

$$\begin{aligned} [\,] &= i\left(g_{\rho\nu} - \frac{\partial_\rho^x\partial_\nu^y}{M^2}\right)\theta(x^0-y^0)W_M(x-y) + (x\leftrightarrow y) \\ &= i\left[\theta(x^0-y^0)\left(g_{\rho\nu} - \frac{\partial_\rho^x\partial_\nu^y}{M^2}\right)W_M(x-y) + (x\leftrightarrow y)\right] \\ &\quad + \frac{i}{M^2}\, g_{\rho 0}\,\delta(x^0-y^0)\partial_\nu^x[W_M(x-y) - W_M(y-x)] \end{aligned}$$

From canonical quantization it is clear that

$$ig_{\rho 0}\delta(x^0-y^0)\partial_\nu^x[W_M(x-y) - W_M(y-x)] = g_{\rho 0}g_{\nu 0}\delta^4(y-x)$$

Therefore

$$iT_{\rho\nu}^{(1)}(x-y) - i\langle 0|\,\tilde{T}j_\rho(x)j_\nu(y)\,|0\rangle = -g_{\rho 0}g_{\nu 0}\delta^4(x-y)\int_{M_0^2}^{\infty} dM^2\,\frac{\rho(M^2)}{M^2} \tag{5-97}$$

The difference is a contact term concentrated on $x = y$ but not covariant.

Let us use this expression to study an interesting property of equal-time commutators, or rather of their vacuum expectation value in the present case. It follows from (5-96) that

$$\begin{aligned} \partial^\rho iT_{\rho\nu}^{(1)}(x-y) &= -i\int_{M_0^2}^{\infty} dM^2\, \rho(M^2)\int\frac{d^4k}{(2\pi)^4}\, e^{-ik\cdot(x-y)}\, k_\nu\,\frac{1-k^2/M^2}{k^2-M^2+i\varepsilon} \\ &= -\partial_\nu\delta^4(x-y)\int_{M_0^2}^{\infty} dM^2\,\frac{\rho(M^2)}{M^2} \end{aligned}$$

According to Eq. (5-97) this must be equal to

$$-g_{\nu 0}\partial_0\delta^4(x-y)\int_{M_0^2}^{\infty} dM^2\,\frac{\rho(M^2)}{M^2} + i\partial_x^\rho\langle 0|\,\tilde{T}j_\rho(x)j_\nu(y)\,|0\rangle$$

The current is conserved so that the divergence of the naive time-ordered symbol must equal the equal-time commutator; hence

$$-(\partial_\nu - g_{\nu 0}\partial_0)\delta^4(x-y)\int_{M_0^2}^{\infty} dM^2 \frac{\rho(M^2)}{M^2} = i\delta(x^0-y^0)\langle 0|\,[j_0(x), j_\nu(y)]\,|0\rangle$$

which means

$$\begin{aligned} &\langle 0|\,[j_0(x), j_0(y)]0\rangle|_{x^0=y^0} = 0 \\ &\langle 0|\,[j_0(x), j_k(y)]\,|0\rangle|_{x^0=y^0} = i\partial_k\delta^3(\mathbf{x}-\mathbf{y})\int_{M_0^2}^{\infty} dM^2\frac{\rho(M^2)}{M^2} \end{aligned} \tag{5-98}$$

The fact that a gradient of a delta function appears in the equal-time commutator of a time and space component of the current, as a consequence of covariance and positivity, was first noticed by Schwinger—hence the denomination of the c-number Schwinger term to recall that it even occurs in the vacuum expectation value. The vanishing of the equal-time commutator of two time components of the currents reflects the fact that integrated over space they generate the conserved electric charge. If the theory allows for a larger invariance group than $U(1)$, the currents would carry internal indices corresponding to the various generators of the symmetry, and the discussion of equal-time commutators would involve the Lie algebra of the group (see Chap. 11).

We can now return to the term $i\Delta T_{\rho\nu}(x, y)$ occurring in Eq. (5-95) and chosen to satisfy

$$i\partial_x^\rho\langle 0|\,Tj_\rho(x)j_\nu(y)\,|0\rangle = 0 \tag{5-99}$$

The choice

$$i\Delta T_{\rho\nu}(x,y) = g_{\rho\nu}\delta^4(x-y)\int_{M_0^2}^{\infty} dM^2\frac{\rho(M^2)}{M^2} \tag{5-100}$$

satisfies all the requirements, so that the complete expression of the covariant T product is

$$i\langle 0|\,Tj_\rho(x)j_\nu(y)\,|0\rangle = \int_{M_0}^{\infty} dM^2\frac{\rho(M^2)}{M^2}\int\frac{d^4k}{(2\pi)^4}\,e^{-ik\cdot(x-y)}\,\frac{k^2 g_{\rho\nu}-k_\rho k_\nu}{k^2-M^2-i\varepsilon} \tag{5-101}$$

5-2 APPLICATIONS

Before presenting in the following chapter the covariant perturbation theory, let us illustrate the methods developed so far on some simple examples involving photons and charged fermions.

5-2-1 Compton Effect

We computed in Chap. 1 the classical elastic Thomson scattering of light off a charged center. In 1923 A. H. Compton, in his study of X-rays, discovered the scattering process with the frequency shift which carries his name.

To be specific the target is taken as a free electron. We shall base our reasoning on formula (5-88) to compute the amplitude of the effect to lowest order in e, the electron charge. To this order Z_3 is equal to one and the current can be identified with the free electron current

$$j_\rho(x) = e :\overline{\psi}_{\text{in}}(x)\gamma_\rho\psi_{\text{in}}(x): \tag{5-102}$$

Henceforth we drop the index "in" and identify the in and out electronic states which means that the amplitude is evaluated to order e^2. The process itself is depicted on Fig. 5-2 where wiggly lines represent photons and solid ones electrons.

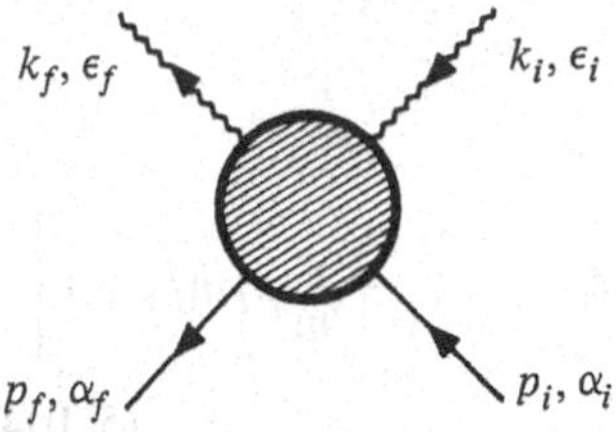

Figure 5-2 General kinematics of the Compton effect.

The connected S-matrix element (c stands for connected) is

$$S^c_{f\leftarrow i} = -e^2 \iint d^4x\, d^4y\, e^{i(k_f \cdot x - k_i \cdot y)} \langle p_f | T : \overline{\psi}(x) \not\epsilon_f \psi(x): \, : \overline{\psi}(y) \not\epsilon_i \psi(y): | p_i \rangle^c \quad (5\text{-}103)$$

The time-ordered operator is expressed by Wick's theorem in terms of normal products involving as coefficients the contractions

$$\psi_\xi(x)\overline{\psi}_{\xi'}(y) = -\overline{\psi}_{\xi'}(y)\psi_\xi(x) = iS^F_{\xi\xi'}(x-y) = i\int \frac{d^4q}{(2\pi)^4}\, e^{-iq\cdot(x-y)} \left(\frac{1}{\not q - m + i\varepsilon}\right)_{\xi\xi'}$$

$$\psi_\xi(x)\psi_{\xi'}(y) = \overline{\psi}_\xi(x)\overline{\psi}_{\xi'}(y) = 0$$

Let us apply Eq. (4-75); we omit contractions between operators referring to the same point because of the normal ordering of the current. Thus

$$T : \overline{\psi}(x)\gamma_\rho\psi(x): \, : \overline{\psi}(y)\gamma_\nu\psi(y):$$
$$= i : \overline{\psi}(x)\gamma_\rho S^F(x-y)\gamma_\nu\psi(y): + i : \overline{\psi}(y)\gamma_\nu S^F(y-x)\gamma_\rho\psi(x): + \cdots$$

The remaining terms do not contribute to the connected matrix element. We expand the free fields ψ and $\overline{\psi}$ in terms of creation and annihilation operators [Eq. (3-157)]. If α_i and α_f denote the initial and final electron polarizations

$$|p_i\rangle = b^\dagger(p_i, \alpha_i)|0\rangle \qquad |p_f\rangle = b^\dagger(p_f, \alpha_f)|0\rangle$$

we have typically to evaluate

$$\langle 0 | b(p_f, \alpha_f) : \overline{\psi}_\xi(x)\psi_\eta(y): b^\dagger(p_i, \alpha_i)|0\rangle = \iint \frac{d^3q_1}{(2\pi)^3}\frac{d^3q_2}{(2\pi)^3}\frac{m^2}{q_1^0 q_2^0}\, e^{i(q_1\cdot x - q_2 \cdot y)}$$
$$\times \sum_{\alpha_1, \alpha_2} \bar{u}_\xi(q_1, \alpha_1)u_\eta(q_2, \alpha_2)\langle 0 | b(p_f, \alpha_f) b^\dagger(q_1, \alpha_1) b(q_2, \alpha_2) b^\dagger(q_i, \alpha_i)|0\rangle$$
$$= e^{i(p_f \cdot x - p_i \cdot y)}\bar{u}_\xi(p_f, \alpha_f)u_\eta(p_i, \alpha_i)$$

This allows us to compute $S^c_{f\leftarrow i}$ as

$$S^c_{f\leftarrow i} = -ie^2 \iint d^4x\, d^4y\, e^{i(k_f\cdot x - k_i \cdot y)} \int \frac{d^4q}{(2\pi)^4}$$
$$\times \left[e^{i(p_f - q)\cdot x - i(p_i - q)\cdot y}\, \bar{u}(p_f, \alpha_f)\not\epsilon_f \frac{1}{\not q - m + i\varepsilon}\not\epsilon_i u(p_i, \alpha_i) \right.$$
$$\left. + e^{i(p_f - q)\cdot y - i(p_i - q)\cdot x}\, \bar{u}(p_f, \alpha_f)\not\epsilon_i \frac{1}{\not q - m + i\varepsilon}\not\epsilon_f u(p_i, \alpha_i) \right]$$

We perform the integrals over x, y, and q, so that

$$S^c_{f\leftarrow i} = -ie^2(2\pi)^4\delta^4(k_f + p_f - k_i - p_i)$$
$$\times\left[\bar{u}(p_f,\alpha_f)\not{\varepsilon}_f \frac{1}{\not{p}_i + \not{k}_i - m}\not{\varepsilon}_i u(p_i,\alpha_i) + \bar{u}(p_f,\alpha_f)\not{\varepsilon}_i \frac{1}{\not{p}_i - \not{k}_f - m}\not{\varepsilon}_f u(p_i,\alpha_i)\right]$$

(5-104)

Integrals over configuration space variables have generated the energy momentum conservation δ function. The $i\varepsilon$ prescription in fermion propagators can be dropped since by adding or subtracting a light-like vector (k) to an on-shell momentum we necessarily leave the mass shell.

From the discussion given in Sec. 5-1, the coefficient in brackets is the reduced transition matrix element

$$\mathscr{T}_{f\leftarrow i} = -e^2\bar{u}(p_f,\alpha_f)\left(\not{\varepsilon}_f \frac{1}{\not{p}_i + \not{k}_i - m}\not{\varepsilon}_i + \not{\varepsilon}_i \frac{1}{\not{p}_i - \not{k}_f - m}\not{\varepsilon}_f\right)u(p_i,\alpha_i) \tag{5-105}$$

The two terms occurring in $\mathscr{T}$ can be represented by Feynman diagrams (Fig. 5-3). External lines carry the polarization functions ε, u, or $\bar{u}$ and the corresponding momenta. Internal lines correspond to the propagators. Vertices which represent the interaction $e\gamma$ are such that energy momentum is conserved at each of them. This explains why the first propagator corresponds to $k_i + p_i = k_f + p_f$ and the second to $p_i - k_f = p_f - k_i$. Total energy momentum is therefore conserved throughout the diagram. The limit $\mu \to 0$ for photons is trivial and it is easily checked that if one of the polarizations is longitudinal (index 3) the contribution vanishes. We can therefore limit ourselves to the two transverse polarizations.

Using (5-105) let us compute the scattering of photons on unpolarized electrons when the final electron polarization is not detected. This amounts to averaging the probabilities on initial (electron) polarizations and summing over the final ones. Up to kinematical factors we therefore have to evaluate $\frac{1}{2}\sum_{\alpha_i,\alpha_f}|\mathscr{T}_{f\leftarrow i}|^2$. The scattering cross section is given by Eq. (5-13) with the appropriate modifications due to the presence of an initial electron [flux factor $(2/m)p_i\cdot k_i$] and of an electron in the final state [phase space factor $m\, d^3p_f/(2\pi)^3 p_f^0$], leading to the expression

$$d\sigma = \frac{1}{2}\sum_{\alpha_i,\alpha_f}(2\pi)^4\delta^{(4)}(p_f + k_f - p_i - k_i)|\mathscr{T}_{f\leftarrow i}|^2 \frac{m}{2p_i\cdot k_i}\frac{d^3k_f}{2k_f^0(2\pi)^3}\frac{m\,d^3p_f}{(2\pi)^3p_f^0}$$

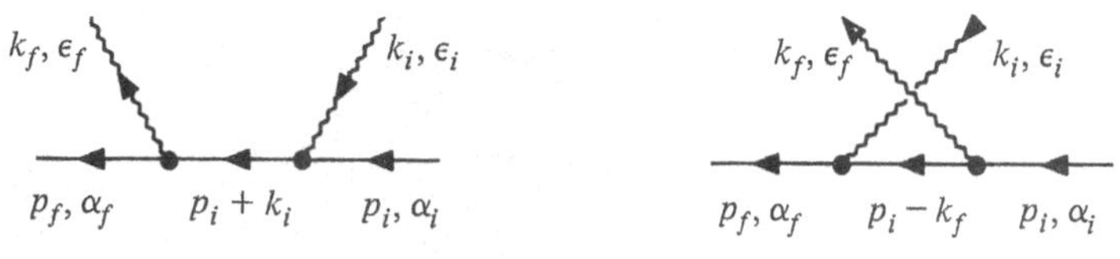

Figure 5-3 Lowest order Feynman diagrams for the Compton effect.

Let us use laboratory variables with

$$\int d^3p_f\, d^3k_f\, \delta^4(p_f + k_f - p_i - k_i) = (k_f^0)^2 \left| \frac{dk_f^0}{d(p_f^0 + k_f^0)} \right| d\Omega$$

where Ω is the emitted photon solid angle measured with respect to the incident momentum. We have

$$\frac{dp_f^0 + dk_f^0}{dk_f^0} = 1 + \frac{1}{2p_f^0}\frac{d(p_f^0)^2}{dk_f^0} = 1 + \frac{1}{2p_f^0}\frac{d}{dk_f^0}[m^2 + (\mathbf{k}_f - \mathbf{k}_i)^2] = \frac{p_f \cdot k_f}{p_f^0 k_f^0}$$

Hence

$$d\sigma = \alpha^2 \frac{1}{2} \sum_{\alpha_f, \alpha_i} |\bar{u} \cdots u|^2 \, \frac{(k_f^0 m)^2}{p_i \cdot k_i p_f \cdot k_f}\, d\Omega$$

where the kinematical factor is derived from momentum conservation

$$(p_i + k_i)^2 = (p_f + k_f)^2 \qquad p_i \cdot k_i = p_f \cdot k_f \overset{\text{lab}}{=} mk_i^0$$

For photons we use the notation $k = k^0 = |\mathbf{k}|$, and we are left with

$$d\sigma = \frac{\alpha^2}{2} \sum_{\alpha_f, \alpha_i} \Bigg| \bar{u}(p_f, \alpha_f) \not{\varepsilon}_f \frac{1}{\not{p}_i + \not{k}_i - m} \not{\varepsilon}_i u(p_i, \varepsilon_i) + \bar{u}(p_f, \alpha_f) \not{\varepsilon}_i \frac{1}{\not{p}_i - \not{k}_f - m} \not{\varepsilon}_f u(p_i, \alpha_i) \Bigg|^2 \left(\frac{k_f}{k_i}\right)^2 d\Omega \tag{5-106}$$

Rationalizing the denominators and using Dirac's equation the matrix element can be rewritten

$$\begin{aligned}
&\bar{u}(p_f, \alpha_f)\left(\not{\varepsilon}_f \frac{1}{\not{p}_i + \not{k}_i - m} \not{\varepsilon}_i + \not{\varepsilon}_i \frac{1}{\not{p}_i - \not{k}_f - m} \not{\varepsilon}_f\right) u(p_i, \alpha_i) \\
&\quad = \bar{u}(p_f, \alpha_f)\left(\not{\varepsilon}_f \frac{\not{p}_i + \not{k}_i + m}{2p_i \cdot k_i} \not{\varepsilon}_i - \not{\varepsilon}_i \frac{\not{p}_i - \not{k}_f + m}{2p_i \cdot k_f} \not{\varepsilon}_f\right) u(p_i, \alpha_i) \\
&\quad = \bar{u}(p_f, \alpha_f)\left[\not{\varepsilon}_f \frac{2(p_i + k_i)\cdot \varepsilon_i - \not{\varepsilon}_i \not{k}_i}{2p_i \cdot k_i} - \not{\varepsilon}_i \frac{2(p_i - k_f)\cdot \varepsilon_f + \not{\varepsilon}_f \not{k}_f}{2p_i \cdot k_f}\right] u(p_i, \alpha_i)
\end{aligned}$$

The transverse photon polarizations ε_i, ε_f can be chosen orthogonal to the "time" axis p_i and to k_i (for ε_i) or k_f (for ε_f). Hence the previous matrix element takes the form

$$-\bar{u}(p_f, \alpha_f)\left(\frac{\not{\varepsilon}_f \not{\varepsilon}_i \not{k}_i}{2p_i \cdot k_i} + \frac{\not{\varepsilon}_i \not{\varepsilon}_f \not{k}_f}{2p_i \cdot k_f}\right) u(p_i, \alpha_i)$$

We should have noted that the total contribution of the two Feynman diagrams is invariant under the exchange

$$(\varepsilon_i, k_i) \leftrightarrow (\varepsilon_f, -k_f) \tag{5-107}$$

permuting initial and final photon variables with a sign reversal of the momentum.

This is an example of a new symmetry of transition amplitudes, called crossing symmetry, which will be elaborated later (Sec. 5-3-2). Using

$$\sum_{\alpha} u_{\xi}(p,\alpha)\bar{u}_{\xi'}(p,\alpha) = \left(\frac{\not{p}+m}{2m}\right)_{\xi\xi'} \tag{5-108}$$

we can now sum over electron polarizations

$$X \equiv \sum_{\alpha_i,\alpha_f} |\bar{u}(p_f,\alpha_f)Ou(p_i,\alpha_i)|^2 = \operatorname{tr}\left(O\frac{\not{p}_i+m}{2m}\bar{O}\frac{\not{p}_f+m}{2m}\right) \tag{5-109}$$

with $\bar{O} = \gamma^0 O^\dagger \gamma^0$, and the trace is to be taken over Dirac indices. In the present case we find the cumbersome expression

$$X = \operatorname{tr}\left[\left(\frac{\not{\varepsilon}_f\not{\varepsilon}_i\not{k}_i}{2p_i\cdot k_i} + \frac{\not{\varepsilon}_i\not{\varepsilon}_f\not{k}_f}{2p_i\cdot k_f}\right)\frac{\not{p}_i+m}{2m}\left(\frac{\not{k}_i\not{\varepsilon}_i\not{\varepsilon}_f}{2p_i\cdot k_i} + \frac{\not{k}_f\not{\varepsilon}_f\not{\varepsilon}_i}{2p_i\cdot k_f}\right)\frac{\not{p}_f+m}{2m}\right] \tag{5-110}$$

The trace can be divided into four terms: two diagonal ones, each of which is obtained from the other by the substitution (5-107), and two crossed terms. The latter are equal since the trace of an odd number of γ matrices vanishes and

$$\operatorname{tr}(\gamma_1\gamma_2\cdots\gamma_{2n}) = \operatorname{tr}(\gamma_{2n}\cdots\gamma_2\gamma_1)$$

as follows by transposition and relating γ to γ^T through the charge conjugation matrix C. Two nontrivial traces remain to be computed. We try to make the best possible use of the identities $k_i^2 = k_f^2 = 0$, $\varepsilon_i^2 = \varepsilon_f^2 = -1$ by applying repeatedly the identity $\not{a}\not{b} = -\not{b}\not{a} + 2a\cdot b$ to move the corresponding momenta adjacent to each other

$$\begin{aligned} T_1 &= \operatorname{tr}[\not{\varepsilon}_f\not{\varepsilon}_i\not{k}_i(\not{p}_i+m)\not{k}_i\not{\varepsilon}_i\not{\varepsilon}_f(\not{p}_f+m)] = \operatorname{tr}[\not{\varepsilon}_f\not{\varepsilon}_i\not{k}_i\not{p}_i\not{k}_i\not{\varepsilon}_i\not{\varepsilon}_f\not{p}_f] \\ &= 2p_i\cdot k_i \operatorname{tr}(\not{\varepsilon}_f\not{\varepsilon}_i\not{k}_i\not{\varepsilon}_i\not{\varepsilon}_f\not{p}_f) = 2p_i\cdot k_i \operatorname{tr}(\not{\varepsilon}_f\not{k}_i\not{\varepsilon}_f\not{p}_f) \\ &= 8p_i\cdot k_i(2\varepsilon_f\cdot k_i\,\varepsilon_f\cdot p_f + k_i\cdot p_f) = 8p_i\cdot k_i[2(\varepsilon_f\cdot k_i)^2 + k_f\cdot p_i] \end{aligned}$$

where at the last stage we have used momentum conservation which implies $k_i\cdot p_f = k_f\cdot p_i$ and $\varepsilon_f\cdot p_f = \varepsilon_f\cdot(p_i+k_i-k_f) = \varepsilon_f\cdot k_i$. (We have chosen $\varepsilon_f\cdot p_i = \varepsilon_i\cdot p_i = 0$). Similarly,

$$\begin{aligned} T_2 &= \operatorname{tr}[\not{\varepsilon}_f\not{\varepsilon}_i\not{k}_i(\not{p}_i+m)\not{k}_f\not{\varepsilon}_f\not{\varepsilon}_i(\not{p}_f+m)] \\ &= \operatorname{tr}[\not{\varepsilon}_f\not{\varepsilon}_i\not{k}_i(\not{p}_i+m)\not{k}_f\not{\varepsilon}_f\not{\varepsilon}_i(\not{p}_i+\not{k}_i-\not{k}_f+m)] \\ &= \operatorname{tr}[\not{k}_i(\not{p}_i+m)\not{k}_f\not{\varepsilon}_f\not{\varepsilon}_i\not{\varepsilon}_f\not{\varepsilon}_i(\not{p}_i+m)] + 2k_f\cdot\varepsilon_i \operatorname{tr}(\not{\varepsilon}_i\not{k}_i\not{p}_i\not{k}_f) \\ &\quad -2k_i\cdot\varepsilon_f \operatorname{tr}(\not{k}_i\not{p}_i\not{k}_f\not{\varepsilon}_f) \\ &= 2k_i\cdot p_i \operatorname{tr}(\not{k}_f\not{\varepsilon}_f\not{\varepsilon}_i\not{\varepsilon}_f\not{\varepsilon}_i\not{p}_i) + 8(k_f\cdot\varepsilon_i)^2k_i\cdot p_i - 8(k_i\cdot\varepsilon_f)^2p_i\cdot k_f \\ &= 8k_i\cdot p_i\,k_f\cdot p_f[2(\varepsilon_f\cdot\varepsilon_i)^2-1] + 8(k_f\cdot\varepsilon_i)^2k_i\cdot p_i - 8(k_i\cdot\varepsilon_f)^2k_f\cdot p_i \end{aligned}$$

The second crossed term T_3 is equal to T_2; finally, T_4 is obtained from T_1 by the substitution (5-107). Adding these quantities with the appropriate factors occurring in Eq. (5-110) and using laboratory kinematics yields

$$X = \frac{1}{2m^2}\left[\frac{k_f}{k_i} + \frac{k_i}{k_f} + 4(\varepsilon_f \cdot \varepsilon_i)^2 - 2\right] \tag{5-111}$$

Inserting this in Eq. (5-106) we get finally the Klein-Nishina (1929) scattering cross section for Compton scattering off free electrons:

$$\frac{d\sigma}{d\Omega} = \frac{\alpha^2}{4m^2}\left(\frac{k_f}{k_i}\right)^2\left[\frac{k_f}{k_i} + \frac{k_i}{k_f} + 4(\varepsilon_f \cdot \varepsilon_i)^2 - 2\right] \tag{5-112}$$

The frequency shift is readily related to the scattering angle θ by squaring the four-vector equality $k_f - k_i = p_i - p_f$. We get $-2k_f^0 k_i^0(1 - \cos\theta) = 2m(k_f^0 - k_i^0)$. Thus

$$k_f = \frac{k_i}{1 + (k_i/m)(1 - \cos\theta)}$$

In the low-energy limit $k_i/m \to 0$, formula (5-112) reduces to the Thomson expression (see Sec. 1-3-2)

$$\frac{d\sigma}{d\Omega} = \frac{\alpha^2}{m^2}(\varepsilon_f \cdot \varepsilon_i)^2 \qquad \frac{\alpha}{m} = r_0 = 2.8 \times 10^{-13}\ \text{cm} \tag{5-113}$$

Near the forward direction, i.e., when $\theta \to 0$, the Klein-Nishina expression also reduces to the nonrelativistic result (5-113) whatever the incident energy may be.

If the incident photon beam is unpolarized and the final photon polarization is undetected it is necessary as in the case of electrons to average over ε_i and sum over ε_f to obtain the unpolarized cross section

$$\overline{\frac{d\sigma}{d\Omega}} = \frac{1}{2}\sum_{\varepsilon_i, \varepsilon_f}\frac{d\sigma}{d\Omega}$$

As in Chap. 1, $\sum_{\varepsilon_i, \varepsilon_f}(\varepsilon_i \cdot \varepsilon_f)^2 = 1 + \cos^2\theta$, leading to

$$\overline{\frac{d\sigma}{d\Omega}} = \frac{\alpha^2}{2m^2}\left(\frac{k_f}{k_i}\right)^2\left(\frac{k_f}{k_i} + \frac{k_i}{k_f} - \sin^2\theta\right) \tag{5-114}$$

A final integration over angles ($x = \cos\theta$) gives the total cross section

$$\bar{\sigma} = \frac{\pi\alpha^2}{m^2}\int_{-1}^{+1} dx \left\{\frac{1}{[1+(k_i/m)(1-x)]^3} + \frac{1}{1+(k_i/m)(1-x)} - \frac{1-x^2}{[1+(k_i/m)(1-x)]^2}\right\}$$

If σ_0 denotes the total Thomson cross section and ω the ratio of the initial energy to the electron rest energy

$$\sigma_0 = \frac{8\pi\alpha^2}{3m^2} \qquad \omega = \frac{k_i}{m} \tag{5-115}$$

we obtain

$$\bar{\sigma} = \sigma_0 \frac{3}{4}\left\{\frac{1+\omega}{\omega^3}\left[\frac{2\omega(1+\omega)}{1+2\omega} - \ln(1+2\omega)\right] + \frac{\ln(1+2\omega)}{2\omega} - \frac{1+3\omega}{(1+2\omega)^2}\right\} \tag{5-116}$$

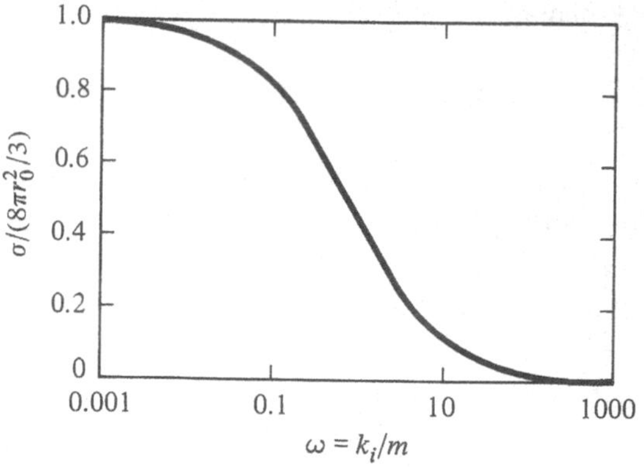

Figure 5-4 Ratio of the unpolarized Compton cross section to the Thomson cross section as a function of the incident energy $\omega = k_i/m$.

The low-energy ($\omega \ll 1$) and high-energy ($\omega \gg 1$) limits follow from Eq. (5-116):

$$\frac{\bar{\sigma}}{\sigma_0} = \begin{cases} 1 - 2\omega + O(\omega^2) & \omega \ll 1 \\ \dfrac{3}{8\omega}\left[\ln 2\omega + \frac{1}{2} + O\left(\dfrac{\ln \omega}{\omega}\right)\right] & \omega \gg 1 \end{cases} \tag{5-117}$$

The ratio $\bar{\sigma}/\sigma_0$ is plotted on Fig. 5-4.

We have verified the crossing symmetry of the Compton amplitude to lowest order in e. The reader can extend this property to all orders by examining the expression (5-88). As shown by the reduction formulas of Secs. 5-1-3 and 5-1-4 this can be generalized to more complex processes involving two charged conjugate particles occurring one in the initial state and the other in the final state. The amplitude is invariant by interchanging the incoming particle and outgoing antiparticle and reversing their four-momenta. The corresponding amplitude is, of course, unphysical since it involves negative energies, but the analytic continuation of amplitudes can be obtained through the reduction formulas themselves (see below in Sec. 5-3-2).

5-2-2 Pair Annihilation

The preceding calculations may be adapted to obtain the rate of a related process, namely, charged-pair annihilation into two photons, the opposite process from the one studied in Chap. 4. The starting point is an expression similar to (5-88) but corresponding to the reduction of two photons in the final state. However, Bose statistics implies that if Ω denotes the solid angle of one of the emitted photons, the total cross section will be obtained by integrating $\frac{1}{2}d\sigma/d\Omega$ over 4π, that is, $d\sigma/d\Omega$ over 2π. Of course, our expressions have to be symmetric in the interchange of the two photons.

The kinematics is recalled in Fig. 5-5a where an electron (p_1, α_1) and a positron (p_2, α_2) annihilate to emit photons (k_1, ε_1), (k_2, ε_2). The corresponding amplitude is

$$S_{f\leftarrow i} = -Z_3^{-1} \iint d^4x\, d^4y\, e^{i(k_1 \cdot x + k_2 \cdot y)} \langle 0 |\, T\{\varepsilon_1 \cdot j(x) \varepsilon_2 \cdot j(y)\}\, | p_1, \alpha_1, p_2, \alpha_2, \text{in} \rangle$$

Just as before, this is computed to lowest order taking into account the same

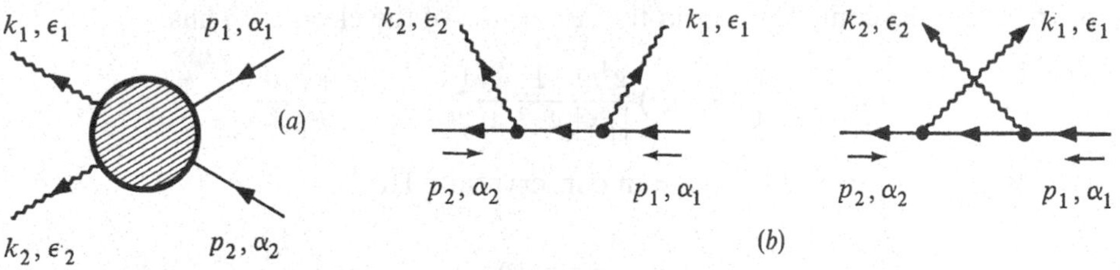

Figure 5-5 Pair annihilation: (*a*) general kinematics and (*b*) lowest order Feynman diagrams.

type of contributions. Straightforward calculations lead to

$$S_{f\leftarrow i} = -ie^2(2\pi)^4\delta^4(k_1 + k_2 - p_1 - p_2)$$
$$\times\, \bar{v}(p_2, \alpha_2)\left\{\not{\varepsilon}_2 \frac{1}{\not{p}_1 - \not{k}_1 - m}\not{\varepsilon}_1 + \not{\varepsilon}_1 \frac{1}{\not{p}_1 - \not{k}_2 - m}\not{\varepsilon}_2\right\}u(p_1, \alpha_1) \tag{5-118}$$

showing a remarkable analogy with the corresponding Compton amplitude. This is exhibited in Fig. 5-5*b*, where the fermionic lines are oriented according to the flux of charge *e*. This amounts to a reverse orientation as far as the positron momentum flow is concerned. A simple substitution allows us to derive Eq. (5-118) from Eq. (5-104) for the Compton effect. It reads

$$\begin{array}{lll} (k_i, \varepsilon_i) \to (-k_1, \varepsilon_1) & u(p_i, \alpha_i) \to u(p_1, \alpha_1) & p_i \to p_1 \\ (k_f, \varepsilon_f) \to (k_2, \varepsilon_2) & \bar{u}(p_f, \alpha_f) \to \bar{v}(p_2, \alpha_2) & p_f \to -p_2 \end{array} \tag{5-119}$$

Crossing symmetry is replaced by the Bose symmetry of the annihilation amplitude and the substitution (5-119) yields another instance of the crossing relations between processes.

From the matrix element (5-118) we derive the cross section for unpolarized electrons and positrons. The flux factor is $(1/m^2)[(p_1 \cdot p_2)^2 - m^4]^{1/2}$ and the statistical average yields a factor of $\frac{1}{4}$. Use is made of a relation analogous to Eq. (5-109) involving

$$-\sum_\alpha v(p, \alpha)\bar{v}(p, \alpha) = \frac{-\not{p} + m}{2m}$$

The cross section takes the form

$$d\sigma = \frac{-m^2e^4}{[(p_1 \cdot p_2)^2 - m^4]^{1/2}} \int \frac{1}{4} \operatorname{tr}\left\{\left(\frac{\not{\varepsilon}_2\not{k}_1\not{\varepsilon}_1}{2p_1 \cdot k_1} + \frac{\not{\varepsilon}_1\not{k}_2\not{\varepsilon}_2}{2p_1 \cdot k_2}\right)\frac{\not{p}_1 + m}{2m}\right.$$
$$\times \left.\left(\frac{\not{\varepsilon}_1\not{k}_1\not{\varepsilon}_2}{2p_1 \cdot k_1} + \frac{\not{\varepsilon}_2\not{k}_2\not{\varepsilon}_1}{2p_1 \cdot k_2}\right)\frac{-\not{p}_2 + m}{2m}\right\}(2\pi)^4\delta^4(k_1 + k_2 - p_1 - p_2)\frac{d^3k_1}{2k_1^0(2\pi)^3}\frac{d^3k_2}{2k_2^0(2\pi)^3}$$

The photon polarizations ε_1 and ε_2 have been taken orthogonal to p_1 (and k_1 or k_2 respectively). The trace is given by the substitution rule (5-119) applied to (5-111)

$$\operatorname{tr}(\cdots) = \frac{1}{2m^2}\left[-\left(\frac{k_2}{k_1} + \frac{k_1}{k_2}\right) + 4(\varepsilon_1 \cdot \varepsilon_2)^2 - 2\right]$$

We perform the computations in the rest frame of the electron. Thus

$$d\sigma = -\frac{1}{4}\int \operatorname{tr}(\cdots)\frac{e^4 m}{|\mathbf{p}_2|}\frac{1}{4\pi^2}\frac{1}{4k_1^0 k_2^0}\,d\Omega_1\,\frac{k_1^{0^2}\,dk_1^0}{d(k_1^0+k_2^0)}$$

with k_2^0 given by energy momentum conservation. Hence

$$\frac{d(k_1^0+k_2^0)}{dk_1^0} = 1 + \frac{1}{2k_2^0}\frac{d}{dk_1^0}(\mathbf{p}_2 - \mathbf{k}_1)^2 = \frac{k_1\cdot k_2}{k_1^0 k_2^0}$$

$$k_1\cdot k_2 = m(m+E_2) = k_1^0(m + E_2 - |\mathbf{p}_2|\cos\theta)$$

Finally,

$$\frac{d\sigma}{d\Omega} = \frac{\alpha^2}{8|\mathbf{p}_2|}\frac{m+E_2}{(m+E_2-|\mathbf{p}_2|\cos\theta)^2}\left[\frac{k_2}{k_1}+\frac{k_1}{k_2}-4(\varepsilon_1\cdot\varepsilon_2)+2\right] \tag{5-120}$$

yields the differential cross section for the emission of one of the photons in the direction defined by the solid angle Ω, corresponding to a positron $(E_2, \mathbf{p}_2)$ impinging on an electron at rest. In (5-120), k_1 and k_2 stand for the photon energies with

$$\frac{k_2}{k_1} = \frac{E_2 - |\mathbf{p}_2|\cos\theta}{m} \tag{5-121}$$

To obtain the total cross section we sum over photon polarizations:

$$\sum_{\varepsilon_1,\varepsilon_2}(\varepsilon_1\cdot\varepsilon_2)^2 = 1 + \left[1 - \left(\frac{m}{k_1}+\frac{m}{k_2}\right)\right]^2$$

and integrate over half the total 4π solid angle with the result

$$\sigma = \frac{\pi r_0^2}{1+\gamma}\left[\frac{\gamma^2+4\gamma+1}{\gamma^2-1}\ln\left(\gamma+\sqrt{\gamma^2-1}\right) - \frac{\gamma+3}{\sqrt{\gamma^2-1}}\right] \tag{5-122}$$

with

$$r_0 = \frac{\alpha}{m} \qquad \gamma = \frac{E_2}{m} \geq 1$$

In the two limiting cases $\gamma \to 1$ (threshold) and $\gamma \to \infty$ (high energy) we find

$$\sigma = \begin{cases} \dfrac{\pi r_0^2}{\sqrt{1-\gamma^{-2}}} = \dfrac{\pi r_0^2}{v_2} & \gamma \to 1,\ v_2 = \text{positron velocity} \\[2ex] \dfrac{\pi r_0^2}{\gamma}[\ln(2\gamma) - 1] & \gamma \to \infty \end{cases} \tag{5-123}$$

The Born approximation used so far is a priori rather poor at low energies where electron and positron interact through long-range Coulomb forces. If, however, (5-123) is used as a first guess we find that the annihilation probability per unit time of a slowly moving positron given by

$$w = \sigma \times \text{incident flux} \times \text{target density} \simeq \sigma v_2 n Z = nZ\pi r_0^2 \tag{5-124}$$

is independent of its velocity. Here n denotes the number of atoms per unit volume and Z their atomic number. A low-energy positron in a medium has therefore a lifetime approximately given by

$$\tau = w^{-1} \simeq (nZ\pi r_0^2)^{-1} \tag{5-125}$$

In lead, for instance, $\tau \sim 10^{-10}$ s. The general formula (5-122) was first obtained by Dirac in 1930.

5-2-3 Positronium Lifetime

We have just studied the free annihilation of an electron and a positron neglecting the Coulomb force. The latter can manifest itself through the binding of the two particles in the neutral positronium system, a lepton analog of the atomic bound states already discussed in Chap. 2. We may extend our previous results to obtain the decay rates of the para- and orthopositronium s-wave ground states. Since annihilation is a slow process the width of these unstable states is very small compared to the nonrelativistic binding energy. This justifies a two-step treatment, neglecting at first annihilation to compute the binding and then neglecting binding to compute annihilation. In a rigorous approach positronium would not appear as a stable asymptotic state and would only occur among unstable excitations. To zeroth order, the electron and positron are described by a nonrelativistic isotropic wave function

$$\begin{aligned}\psi(r) &= \frac{1}{(\pi a^3)^{1/2}}\, e^{-r/a} \\ &= \int \frac{d^3q}{(2\pi)^3}\, e^{i\mathbf{q}\cdot\mathbf{r}}\, \tilde{\psi}(\mathbf{q}) = \int \frac{d^3q}{(2\pi)^3}\, e^{i\mathbf{q}\cdot\mathbf{r}}\, \frac{8(\pi a^3)^{1/2}}{(1+a^2q^2)^2}\end{aligned} \tag{5-126}$$

where a is twice the Bohr radius of atomic hydrogen (since the reduced mass is half an electron mass)

$$a = 2a_0 = \frac{2}{m\alpha} \tag{5-127}$$

The meaningful electron or positron momenta are of order $1/a = \alpha m/2 \ll m$ and annihilation can be computed in a first-order approximation as if both particles were relatively at rest. The second step is therefore to compute the width of the singlet and triplet states for free particles at rest.

In Chap. 3 it was established that charge conjugation implied that the singlet state decays into an even number of photons, the lowest possible number therefore being two. We use our preceding result multiplied by 4, since instead of averaging over four polarization states we have here the decay of a given spin state.

Furthermore, we have to reconsider the reasoning of Sec. 5-1-1, giving the relation between cross section and transition probability per unit time, denoted here by τ_{sing}^{-1}. To obtain the latter we multiply 4σ by a normalization factor constructed as follows. We substitute for the flux factor

$$m^2[(p_1\cdot p_2)^2 - m^4]^{-1/2} = \frac{m}{|\mathbf{p}_2|} \underset{|\mathbf{p}_2|\to 0}{\simeq} v_2^{-1}$$

a weight

$$\left| \int \frac{d^3q}{(2\pi)^3} \tilde{\psi}(q) \right|^2 = |\psi(0)|^2 = \frac{1}{\pi a^3}$$

This arises by repeating the calculation leading from Eq. (5-4) to Eq. (5-13) in the case at hand. Thus

$$\tau_{\text{sing}}^{-1} = \frac{1}{\pi a^3} \lim_{\substack{E_2 \to m \\ v_2 \to 0}} (4v_2\sigma) = \frac{1}{\pi a^3}(4\pi r_0^2) = \frac{m\alpha^5}{2} \tag{5-128}$$

We note indeed that the ratio $\tau_{\text{sing}}^{-1}/B$, where B is the binding energy $B = m\alpha^2/4$, is of order $\alpha^3 \sim 10^{-6}$ justifying the above treatment.

The previous calculation may be interpreted as giving the decay probability per unit time in terms of the transition matrix element at threshold as

$$\tau^{-1} = \sum_f (2\pi)^4 \delta(P_f^0 - 2m)\delta^3(\mathbf{P}_f) \,|\langle f | \mathcal{T} | e^+e^- \rangle|^2 \,|\psi(0)|^2 \tag{5-129}$$

This formula enables us to discuss the orthopositronium case where the minimum number of emitted photons is three. From the reduction formula to lowest order

$$S_{f\leftarrow i} = (-ie)^3 \int d^4x_1\, d^4x_2\, d^4x_3 \exp\left(i \sum_1^3 k_j \cdot x_j\right)$$

$$\times \langle 0 | T[:\overline{\psi}(x_1)\not{\varepsilon}_1\psi(x_1):\ :\overline{\psi}(x_2)\not{\varepsilon}_2\psi(x_2):\ :\overline{\psi}(x_3)\not{\varepsilon}_3\psi(x_3):] \,|\, e^+e^-, \text{triplet}\rangle \tag{5-130}$$

Wick's theorem yields for the matrix element

$$\sum_{\text{permutations}} \langle 0| :\overline{\psi}(x_1)\not{\varepsilon}_1 iS(x_1 - x_2)\not{\varepsilon}_2 iS(x_2 - x_3)\not{\varepsilon}_3\psi(x_3): | e^+e^-, \text{triplet}\rangle$$

which inserted into (5-130) and after integration gives

$$S_{f\leftarrow i} = i(2\pi)^4\delta(P_f^0 - 2m)\delta^3(\mathbf{P}_f)\mathcal{T}_{fi}$$

$$= -ie^3(2\pi)^4\delta(P_f^0 - 2m)\delta^3(\mathbf{P}_f) \sum_{\text{permutations}} \bar{v}^{(\alpha)}\not{\varepsilon}_1 \frac{1}{\not{p} - \not{k}_2 - \not{k}_3 - m} \tag{5-131}$$

$$\times \not{\varepsilon}_2 \frac{1}{\not{p} - \not{k}_3 - m} \not{\varepsilon}_3 u^{(\beta)}$$

Here $u^{(\beta)}$ and $\bar{v}^{(\alpha)}$ represent the Dirac spinors of the electron and positron at rest, and a projection on the triplet state has to be performed. The common energy momentum of the fermions is $p \equiv (m, \mathbf{0})$ while (k_1, ε_1), (k_2, ε_2), and (k_3, ε_3) are the momenta and polarizations of the final photons with $P_f = \sum_1^3 k_j$. Of course, $\mathcal{T}$ is symmetric in the photon variables as required by Bose statistics. The integral over phase space will have to be divided by 3!. The result expressed in Eq. (5-131) may be represented by the sum of six Feynman diagrams, such as the one shown in Fig. 5-6*a*, obtained by permuting the photon variables. To

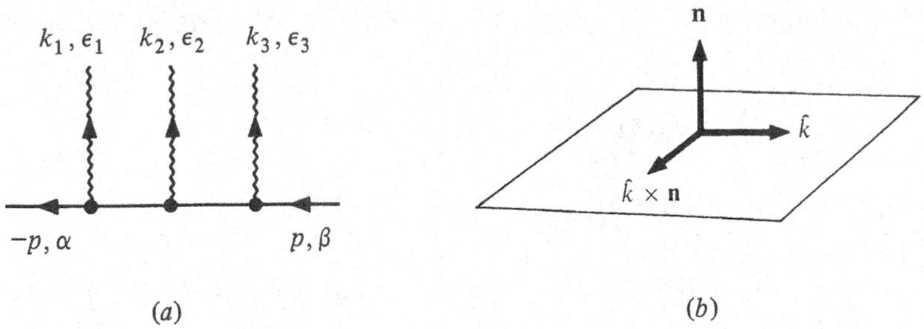

Figure 5-6 Three-photon decay of orthopositronium. (*a*) One of the six diagrams describing the process to lowest order. (*b*) Choice of axis to measure the photon polarizations; **n** is the normal to the reaction plane.

proceed with the calculation let us introduce the notation

$$\mathscr{T}_{f\leftarrow i} = -e^3 \chi_+^{c\dagger} \Sigma a_{123} \chi_- \qquad \chi_+^{c\dagger} = \chi_+^T i\sigma_2 \tag{5-132}$$

with $\chi_\pm$ standing for the two-component spinors describing the fermions. We shall illustrate a method different from the one which consists of the computation of traces when squaring $\mathscr{T}_{f\leftarrow i}$, namely, we shall explicitly obtain the 2×2 matrix a_{123}. With obvious notations the latter is derived from a 4×4 matrix as follows:

$$a_{123} = (0, 1)\not{\varepsilon}_1 \frac{1}{\not{p} - \not{k}_2 - \not{k}_3 - m} \not{\varepsilon}_2 \frac{1}{\not{p} - \not{k}_3 - m} \not{\varepsilon}_3 \begin{pmatrix} 1 \\ 0 \end{pmatrix}$$

Set $\omega_j = k_j^0$, $\hat{k}_j = \mathbf{k}_j/\omega_j$. Choose, furthermore, $\varepsilon_j \cdot p = 0$ for $j = 1, 2, 3$ and let us write $\boldsymbol{\delta}_j$ for $\hat{k}_j \times \varepsilon_j$. This is a unit vector since $\hat{k}_j \cdot \varepsilon_j = 0$. Then

$$\begin{aligned}(0, 1)\not{\varepsilon}_1 \frac{1}{\not{p} - \not{k}_2 - \not{k}_3 - m} &= (0, 1)\not{\varepsilon}_1 \frac{\not{k}_1 - \not{p} + m}{(k_1 - p)^2 - m^2} = -(0, 1)\frac{\not{\varepsilon}_1 \not{k}_1}{2m\omega_1} \\ &= -\frac{1}{2m\omega_1}(\boldsymbol{\sigma} \cdot \varepsilon_1, 0)\begin{pmatrix} \omega_1 & -\boldsymbol{\sigma} \cdot \mathbf{k}_1 \\ \boldsymbol{\sigma} \cdot \mathbf{k}_1 & -\omega_1 \end{pmatrix} \\ &= -\frac{1}{2m}(\boldsymbol{\sigma} \cdot \varepsilon_1, i\boldsymbol{\sigma} \cdot \boldsymbol{\delta}_1)\end{aligned}$$

Similarly,

$$\frac{1}{\not{p} - \not{k}_3 - m} \not{\varepsilon}_3 \begin{pmatrix} 1 \\ 0 \end{pmatrix} = -\frac{1}{2m}\begin{pmatrix} i\boldsymbol{\sigma} \cdot \boldsymbol{\delta}_3 \\ \boldsymbol{\sigma} \cdot \varepsilon_3 \end{pmatrix}$$

Therefore,

$$\begin{aligned}a_{123} &= \frac{1}{4m^2}(\boldsymbol{\sigma} \cdot \varepsilon_1, i\boldsymbol{\sigma} \cdot \boldsymbol{\delta}_1)\begin{pmatrix} 0 & -\boldsymbol{\sigma} \cdot \varepsilon_2 \\ \boldsymbol{\sigma} \cdot \varepsilon_2 & 0 \end{pmatrix}\begin{pmatrix} i\boldsymbol{\sigma} \cdot \boldsymbol{\delta}_3 \\ \boldsymbol{\sigma} \cdot \varepsilon_3 \end{pmatrix} \\ &= -\frac{1}{4m^2}(\boldsymbol{\sigma} \cdot \varepsilon_1\, \boldsymbol{\sigma} \cdot \varepsilon_2\, \boldsymbol{\sigma} \cdot \varepsilon_3 + \boldsymbol{\sigma} \cdot \boldsymbol{\delta}_1\, \boldsymbol{\sigma} \cdot \varepsilon_2\, \boldsymbol{\sigma} \cdot \boldsymbol{\delta}_3)\end{aligned}$$

The required combination $\sum_{\text{perm}} a_{123}$ is of the form $\boldsymbol{\sigma}\cdot\mathbf{V}$. Indeed its trace vanishes since it is proportional to

$$\sum_{\text{perm}} [(\varepsilon_1, \varepsilon_2, \varepsilon_3) + (\delta_1, \varepsilon_2, \delta_3)] = 0$$

Recall that tr $\sigma_i\sigma_j\sigma_k = 2i\varepsilon_{ijk}$. Thus $\sum_{\text{perm}} a_{123} = \boldsymbol{\sigma}\cdot\mathbf{V}$, where

$$\mathbf{V} = -\frac{1}{2m^2}\sum_{\text{cycl}} [\varepsilon_1(\varepsilon_2\cdot\varepsilon_3 - \delta_2\cdot\delta_3) + \delta_1(\varepsilon_3\cdot\delta_2 + \varepsilon_2\cdot\delta_3)] \tag{5-133}$$

Clearly the lifetime of orthopositronium is independent of its polarization state. However, the angular distribution when we observe the photon polarizations depends on the initial polarization state. We shall limit ourselves here to the total rate which can be computed for an unpolarized orthopositronium by averaging over the three spin states. Therefore,

$$\begin{aligned}\tau^{-1}_{\text{triplet}} &= \frac{1}{6}\int \frac{d^3k_1\,d^3k_2\,d^3k_3}{8\omega_1\omega_2\omega_3(2\pi)^9}(2\pi)^4\delta(\Sigma\,\omega_j - 2m)\delta^3(\Sigma\,\mathbf{k}_j)\\ &\quad\times\frac{e^6}{\pi a^3}\sum_{\substack{\text{photon}\\ \text{polarizations}}} \tfrac{1}{3}(|V_+|^2 + |V_0|^2 + |V_-|^2)\end{aligned} \tag{5-134}$$

In writing (5-134) we have taken into account the factor $i\sigma_2$ explicitly written in Eq. (5-132). For instance, the decay amplitude of the state $|J=1, J_z=1\rangle = |\frac{1}{2},\frac{1}{2}\rangle$ is proportional to the matrix element $\chi^T_{1/2}i\sigma_2\boldsymbol{\sigma}\cdot\mathbf{V}\chi_{1/2} = V_1 + iV_2 \equiv V_+$, while the corresponding amplitude for the state $|J=1, J_z=0\rangle = 1/\sqrt{2}|\frac{1}{2},-\frac{1}{2}\rangle + 1/\sqrt{2}|-\frac{1}{2},\frac{1}{2}\rangle$ involves

$$\frac{1}{\sqrt{2}}(\chi^T_{-1/2}i\sigma_2\boldsymbol{\sigma}\cdot\mathbf{V}\chi_{1/2} + \chi^T_{1/2}i\sigma_2\boldsymbol{\sigma}\cdot\mathbf{V}\chi_{-1/2}) = -\sqrt{2}V_3 \equiv V_0$$

The last amplitude for $|J=1, J_z=-1\rangle$ introduces $V_- = -V_1 + iV_2$. Therefore,

$$\tfrac{1}{3}(|V_+|^2 + |V_0|^2 + |V_-|^2) = \tfrac{2}{3}(V_1^2 + V_2^2 + V_3^2) = \tfrac{2}{3}\mathbf{V}^2$$

To complete the calculation it is convenient to use complex polarizations such that $\varepsilon^{\pm} = \varepsilon \pm i\delta$ satisfying

$$\begin{aligned}\varepsilon^{\pm 2} &= 0\\ \varepsilon^+\cdot\varepsilon^- &= 2\\ \varepsilon_1^{\pm}\cdot\varepsilon_2^{\pm} &= \varepsilon_1\cdot\varepsilon_2 - \delta_1\cdot\delta_2 \pm i(\varepsilon_1\cdot\delta_2 + \varepsilon_2\cdot\delta_1)\end{aligned}$$

Thus

$$\mathbf{V} = -\frac{1}{4m^2}\sum_{\text{cycl}} \varepsilon_1^+(\varepsilon_2^-\cdot\varepsilon_3^-) + \varepsilon_1^-(\varepsilon_2^+\cdot\varepsilon_3^+)$$

$$\begin{aligned}\tfrac{2}{3}\mathbf{V}^2 &= \frac{1}{12m^4}\,\text{Re}\sum_{\text{cycl}} \{2|\varepsilon_2^+\cdot\varepsilon_3^+|^2 + 2(\varepsilon_1^+\cdot\varepsilon_2^+)(\varepsilon_2^-\cdot\varepsilon_3^-)(\varepsilon_3^-\cdot\varepsilon_1^-)\\ &\quad + [(\varepsilon_1^+\cdot\varepsilon_2^-)(\varepsilon_2^-\cdot\varepsilon_3^-)(\varepsilon_3^+\cdot\varepsilon_1^+) + \text{cc}]\}\end{aligned}$$

We now sum over photon polarizations, choosing for each photon two orthogonal basis states (*a*) and (*b*) with $\boldsymbol{\varepsilon}$ along the normal to the reaction plane or along $\hat{k} \times \mathbf{n}$ (see Fig. 5-6*b*):

$$(a) \quad \boldsymbol{\varepsilon} = \mathbf{n} \qquad \boldsymbol{\delta} = \hat{k} \times \mathbf{n} \qquad \boldsymbol{\varepsilon}^+ = \boldsymbol{\alpha} = \mathbf{n} + i\hat{k} \times \mathbf{n} \qquad \boldsymbol{\varepsilon}^- = \boldsymbol{\alpha}^* = \mathbf{n} - i\hat{k} \times \mathbf{n}$$

$$(b) \quad \boldsymbol{\varepsilon} = \hat{k} \times \mathbf{n} \qquad \boldsymbol{\delta} = -\mathbf{n} \qquad \boldsymbol{\varepsilon}^+ = -i\boldsymbol{\alpha} \qquad \boldsymbol{\varepsilon}^- = i\boldsymbol{\alpha}^*$$

Thus

$$\sum_{\text{photon polarizations}} \tfrac{2}{3}\mathbf{V}^2 = \frac{4}{3m^4} \sum_{\text{cycl}} |\boldsymbol{\alpha}_2 \cdot \boldsymbol{\alpha}_3|^2$$

Only the first term in the expression for $\mathbf{V}$ has contributed to the sum over photon polarizations. Moreover,

$$\boldsymbol{\alpha}_2 \cdot \boldsymbol{\alpha}_3 = (\mathbf{n} + i\hat{k}_2 \times \mathbf{n}) \cdot (\mathbf{n} + i\hat{k}_3 \times \mathbf{n}) = 1 - \cos\theta_{23}$$

with $\cos\theta_{23} = \hat{k}_2 \cdot \hat{k}_3$. The final expression

$$\tfrac{1}{3}(|V_+|^2 + |V_0|^2 + |V_-|^2) = \tfrac{2}{3}\mathbf{V}^2 = \frac{4}{3m^4} \sum_{\text{cycl}} (1 - \cos\theta_{23})^2 \tag{5-135}$$

may now be inserted into (5-134) to yield

$$\tau_{\text{triplet}}^{-1} = \frac{\alpha^3 m^3}{8\pi} \frac{(4\pi)^3\alpha^3}{m^4} \frac{4}{3} \frac{1}{6} \frac{1}{(2\pi)^5} \int \frac{d^3k_1\, d^3k_2\, d^3k_3}{8\omega_1\omega_2\omega_3} \delta^3(\textstyle\sum \mathbf{k}_j)\delta(\Sigma\, \omega_j - 2m) \times \sum_{\text{cycl}} (1 - \cos\theta_{23})^2$$

Each term in the cyclic sum contributes equally. Therefore

$$\tau_{\text{triplet}}^{-1} = \frac{\alpha^6}{6\pi m} \iint_D d\omega_1\, d\omega_2\, (1 - \cos\theta_{12})^2$$
$$1 - \cos\theta_{12} = 2m\, \frac{\omega_1 + \omega_2 - m}{\omega_1\omega_2} \tag{5-136}$$

and the range of integration is the Mandelstam triangle $D: \omega_1 + \omega_2 \geq m$, $0 \leq \omega_1, \omega_2 \leq m$. Let us write

$$\tau_{\text{triplet}}^{-1} = \frac{2\alpha^6}{3\pi} mA \qquad A = \int_0^1 dx \int_{1-x}^1 dy\, \frac{(x + y - 1)^2}{x^2 y^2} \tag{5-137}$$

The integral A gives

$$A = \int_0^1 \frac{dx}{x^2} [x(1 - x) + x + 2(1 - x) \ln(1 - x)] = -3 + 2\sum_1^\infty \frac{1}{n^2}$$

The last sum $\sum_1^\infty 1/n^2$ equals $\pi^2/6$ so that $A = (\pi^2 - 9)/3$ and we get the final value quoted in Chap. 3:

$$\tau_{\text{triplet}}^{-1} = m\, \frac{2\alpha^6}{9\pi} (\pi^2 - 9) \tag{5-138}$$

From the above expressions one can easily derive the energy spectrum of the emitted photons.

5-2-4 Bremsstrahlung

In the first chapter we have given the classical radiation rate for the emission of photons when a charged particle suffers an abrupt change in velocity. We want to give here a more precise treatment, taking as an example Coulomb scattering off a fixed nucleus (charge $-Ze$) discussed in Chap. 2. The matrix element of interest is given by Eq. (5-87) if we deal with one photon emission. Assuming

$$\frac{Z\alpha}{v} \ll 1 \tag{5-139}$$

where v is the incident electron velocity, we shall perform an expansion in Ze as well as e and retain only the leading contribution. Proceeding heuristically, the external field is accounted for by replacing free spinors by solutions of the Dirac equation in the presence of the external field A^{ext} such that

$$A_0^{\text{ext}} = -\frac{Ze}{4\pi|\mathbf{x}|} \qquad \mathbf{A}^{\text{ext}} = 0$$

Let therefore ψ^{ext} denote the solutions of

$$(i\partial\!\!\!/ - eA\!\!\!/^{\text{ext}} - m)\psi^{\text{ext}}(x) = 0$$

subject to appropriate boundary conditions. We write the radiation amplitude

$$S_{f\leftarrow i} = -ie\int d^4x\, e^{ik\cdot x}\, \bar{\psi}^{\text{ext}}_{\text{out}}(x)\varepsilon\!\!/\psi^{\text{ext}}_{\text{in}}(x) \tag{5-140}$$

with $k^2 = 0$, $\varepsilon^2 = -1$, $\varepsilon\cdot k = 0$. The incident electron is specified asymptotically by its momentum p_i and polarization α, and the outgoing one by p_f and β. To lowest order in A^{ext} we have seen in Chap. 2 that

$$\begin{aligned}\psi^{\text{ext}}_{\text{in}}(x) &= \psi_{\text{in}}(x) + \int d^4y\, S^F(x-y)\, eA\!\!\!/^{\text{ext}}(y)\psi_{\text{in}}(y) + \cdots \\ \bar{\psi}^{\text{ext}}_{\text{out}}(x) &= \bar{\psi}_{\text{out}}(x) + \int d^4y\, \bar{\psi}_{\text{out}}(y)\, eA\!\!\!/^{\text{ext}}(y)S^F(y-x) + \cdots\end{aligned} \tag{5-141}$$

Here ψ_{in}, $\bar{\psi}_{\text{out}}$ stand for the corresponding solutions of the free Dirac equation

$$\begin{aligned}\psi_{\text{in}}(x) &= e^{-ip_i\cdot x}u(p_i, \alpha) \\ \bar{\psi}_{\text{out}}(x) &= e^{ip_f\cdot x}\bar{u}(p_f, \beta)\end{aligned} \tag{5-142}$$

where, of course, $p_i^2 = p_f^2 = m^2$. When substituting (5-141) in (5-140) we observe that the contribution proportional to $\int d^4x\, e^{ik\cdot x}\bar{\psi}_{\text{out}}(x)\varepsilon\!\!/\psi_{\text{in}}(x)$ vanishes for $k^2 = 0$. Therefore to lowest order

$$\begin{aligned}S_{f\leftarrow i} = -ie^2\int d^4x\, d^4y\, [&e^{ik\cdot x}\,\bar{\psi}_{\text{out}}(x)\varepsilon\!\!/ S^F(x-y)A\!\!\!/^{\text{ext}}(y)\psi_{\text{in}}(y) \\ &+ e^{ik\cdot y}\bar{\psi}_{\text{out}}(x)A\!\!\!/^{\text{ext}}(x)S^F(x-y)\varepsilon\!\!/\psi_{\text{in}}(y)]\end{aligned} \tag{5-143}$$

We note an analogy with the Compton amplitude where the external potential $A^{\text{ext}}(x)$ replaces the free photon wave function $e^{-ik\cdot x}\varepsilon(k)$. Using (5-142) together with the Fourier transform of the potential

$$A_0^{\text{ext}}(x) = -\frac{Ze}{4\pi|\mathbf{x}|} = -Ze\int\frac{d^4q}{(2\pi)^3}\,\delta(q^0)\,e^{-iq\cdot x}\,\frac{1}{|\mathbf{q}|^2} \tag{5-144}$$

we readily find

$$\begin{aligned} S_{f\leftarrow i} &= \frac{iZe^3}{|\mathbf{p}_f + \mathbf{k} - \mathbf{p}_i|^2}\,2\pi\delta(p_i^0 - p_f^0 - k_f^0) \\ &\quad\times \bar{u}(p_f,\beta)\left[\not{\varepsilon}\,\frac{1}{\not{p}_f + \not{k} - m}\,\gamma^0 + \gamma^0\,\frac{1}{\not{p}_i - \not{k} - m}\,\not{\varepsilon}\right]u(p_i,\alpha) \end{aligned} \tag{5-145}$$

Only energy conservation has survived since space-translation invariance is broken by the presence of the center of force.

Once more the derivation of the cross section has to be modified with the result that

$$d\sigma = \frac{mZ^2e^6}{p_i^0|\mathbf{v}_i|}\int 2\pi\delta(p_f^0 + k^0 - p_i^0)\left|\frac{[\;]}{|\mathbf{p}_f + \mathbf{k} - \mathbf{p}_i|^2}\right|^2\frac{md^3p_f\,d^3k}{2p_f^0k^0(2\pi)^6} \tag{5-146}$$

where $p_i^0|\mathbf{v}_i|/m = |\mathbf{p}_i|/m$ is the flux factor and the square bracket is the one occurring in Eq. (5-145).

To obtain the unpolarized cross section we sum over ε and β and average over α. To simplify the notations we write $\mathbf{q} = (\mathbf{p}_f + \mathbf{k} - \mathbf{p}_i)$ for the transferred momentum, let $\omega = k^0$ and Ω_γ, Ω_e be the directions of the outgoing photon and electron respectively. Thus

$$d\sigma = \frac{Z^2\alpha^3}{\pi^2}\,\frac{m^2|\mathbf{p}_f|}{|\mathbf{p}_i|\;|\mathbf{q}|^4}\,F\omega\,d\omega\,d\Omega_\gamma\,d\Omega_e \tag{5-147}$$

$$\begin{aligned} F &= \frac{1}{2}\sum_\varepsilon \operatorname{tr}\left[\left(\not{\varepsilon}\,\frac{\not{p}_f + \not{k} + m}{2p_f\cdot k}\,\gamma^0 - \gamma^0\,\frac{\not{p}_i - \not{k} + m}{2p_i\cdot k}\,\not{\varepsilon}\right)\left(\frac{\not{p}_i + m}{2m}\right)\right. \\ &\quad\left.\times\left(\gamma^0\,\frac{\not{p}_f + \not{k} + m}{2p_f\cdot k}\,\not{\varepsilon} - \not{\varepsilon}\,\frac{\not{p}_i - \not{k} + m}{2p_i\cdot k}\,\gamma^0\right)\left(\frac{\not{p}_f + m}{2m}\right)\right] \end{aligned} \tag{5-148}$$

The quantity F can be decomposed into three parts:

$$F = \frac{1}{2^5m^2}(F_1 + F_2 + F_3)$$

$$F_1 = \frac{1}{(p_f\cdot k)^2}\sum_\varepsilon \operatorname{tr}\left[\not{\varepsilon}(\not{p}_f + \not{k} + m)\gamma^0(\not{p}_f + \not{k} + m)\not{\varepsilon}(\not{p}_f + m)\right]$$

$$F_2 = F_1(p_i \leftrightarrow -p_f)$$

$$F_3 = -\frac{1}{(p_f\cdot k)(p_i\cdot k)}\operatorname{tr}\{[\gamma^0(\not{p}_i - \not{k} + m)\not{\varepsilon}(\not{p}_i + m)\gamma^0(\not{p}_f + \not{k} + m)\not{\varepsilon}(\not{p}_f + m)] + (p_i \leftrightarrow -p_f)\}$$

Choosing $\varepsilon^0 = 0$ we find

$$F_1 = \frac{8}{(p_f \cdot k)^2} \sum_{\varepsilon} [2(\varepsilon \cdot p_f)^2(m^2 + 2p_i^0 p_f^0 + 2p_i^0 \omega - p_i \cdot p_f - p_i \cdot k) + 2\varepsilon \cdot p_f \, \varepsilon \cdot p_i \, k \cdot p_f$$
$$+ 2p_i^0 \omega \, k \cdot p_f - p_i \cdot k \, p_j \cdot k]$$
$$F_2 = F_1(p_i \leftrightarrow -p_f)$$
$$F_3 = \frac{16}{(p_f \cdot k)(p_i \cdot k)} \sum_{\varepsilon} [\varepsilon \cdot p_i \, \varepsilon \cdot p_f(p_i \cdot k - p_f \cdot k + 2p_i \cdot p_f - 4p_i^0 p_f^0 - 2m^2)$$
$$+ (\varepsilon \cdot p_f)^2 k \cdot p_i - (\varepsilon \cdot p_i)^2 k \cdot p_f + p_i \cdot k \, p_f \cdot k - m^2 \omega^2$$
$$+ \omega(\omega \, p_i \cdot p_f - p_i^0 \, p_f \cdot k - p_f^0 \, p_i \cdot k)]$$

Let θ_f be the angle between the vectors $\mathbf{k}$ and $\mathbf{p}_f$, θ_i the angle between $\mathbf{k}$ and $\mathbf{p}_i$, and φ the angle between the planes $(\mathbf{k}, \mathbf{p}_f)$ and $(\mathbf{k}, \mathbf{p}_i)$, as shown on Fig. 5-7.

The sum over photon polarizations can be performed using

$$\sum_{\varepsilon} (\boldsymbol{\varepsilon} \cdot \mathbf{p}_f)^2 = |\mathbf{p}_f|^2 \sin^2 \theta_f \qquad \sum_{\varepsilon} (\boldsymbol{\varepsilon} \cdot \mathbf{p}_i)^2 = |\mathbf{p}_i|^2 \sin^2 \theta_i$$
$$\sum_{\varepsilon} (\boldsymbol{\varepsilon} \cdot \mathbf{p}_f)(\boldsymbol{\varepsilon} \cdot \mathbf{p}_i) = |\mathbf{p}_f| \, |\mathbf{p}_i| \sin \theta_f \sin \theta_i \cos \varphi$$

For the final formula we shall use the notations

$$p_f^0 = E_f \qquad p_i^0 = E_i \qquad |\mathbf{p}_f| = p_f \qquad |\mathbf{p}_i| = p_i \qquad |\mathbf{q}|^2 = q^2 \tag{5-149}$$

The differential cross section as found by Bethe and Heitler in 1934 can now be written as

$$d\sigma = \frac{Z^2 \alpha^3}{(2\pi)^2} \frac{p_f}{p_i q^4} \frac{d\omega}{\omega} d\Omega_\gamma \, d\Omega_e$$
$$\times \left[\frac{p_f^2 \sin^2 \theta_f}{(E_f - p_f \cos \theta_f)^2} (4E_i^2 - q^2) + \frac{p_i^2 \sin^2 \theta_i}{(E_i - p_i \cos \theta_i)^2} (4E_f^2 - q^2) \right. \tag{5-150}$$
$$+ 2\omega^2 \frac{p_i^2 \sin^2 \theta_i + p_f^2 \sin^2 \theta_f}{(E_f - p_f \cos \theta_f)(E_i - p_i \sin \theta_i)} - 2 \frac{p_f p_i \sin \theta_i \sin \theta_f \cos \varphi}{(E_f - p_f \cos \theta_f)(E_i - p_i \cos \theta_i)}$$
$$\left. \times (4E_i E_f - q^2 + 2\omega^2) \right]$$

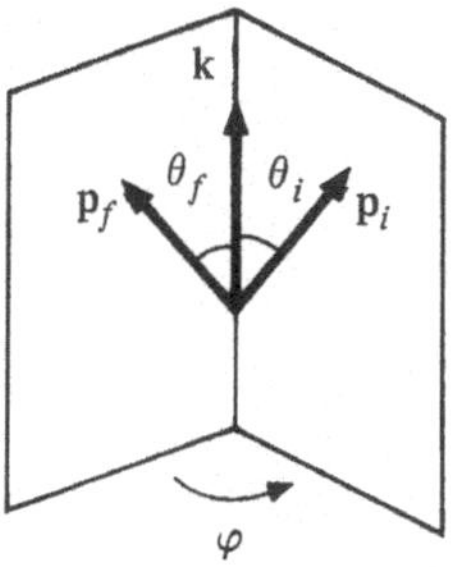

Figure 5-7 Kinematics of the bremsstrahlung process.

This result has a rather complicated structure exhibiting the by now familiar catastrophic (!) behavior when $\omega \to 0$.

In the limit $\omega \to 0$, Eqs. (5-147) and (5-148) may be written approximately

$$\frac{d\sigma}{d\Omega_e}\underset{k\to 0}{\simeq}\left(\frac{d\sigma}{d\Omega_e}\right)_{\text{elastic}} e^2 \frac{d^3k}{2\omega(2\pi)^3}\sum_{\varepsilon}\left(\frac{\varepsilon\cdot p_f}{k\cdot p_f}-\frac{\varepsilon\cdot p_i}{k\cdot p_i}\right)^2 \tag{5-151}$$

as claimed in Chap. 1, where the results on the integrated cross section were already given.

In this example one realizes that for k small, practical observation does not allow us to distinguish between the elastic cross section including correction to order e^2 and the inelastic one treated to the same order. We can therefore only expect that the sum of these two has a finite limit to this same order. This will be discussed further in Chap. 7.

5-3 UNITARITY AND CAUSALITY

The reduction formulas have related the measurable on-shell scattering amplitudes to the general Green functions. Most of the developments in the remainder of this book will be devoted to the computation of these functions in the framework of a local relativistic dynamics described in terms of a lagrangian. Nevertheless, it is important to stress and abstract the important physical requirements such as unitarity and locality common to the various dynamical systems and possibly more general than the present-day applications. One of the main outcomes of such an analysis is the proof of a specific analytic structure of amplitudes leading to dispersion relations in one or several variables, complemented by rules providing the boundary values along the various discontinuities. It is, however, not our purpose to investigate these developments in detail except to assert their importance and quote some significant results. In the next chapter we shall see how the analytic properties are reflected in the perturbative expressions.

5-3-1 Unitarity and Partial Wave Decomposition

Unitarity of the S matrix reflects the fundamental principle of probability conservation. Even though we have to introduce in certain instances the artificial device of an indefinite metric in Hilbert space, the physical quantities always refer to states with positive norm, preserved through the time evolution. Formal developments should not obscure this important fact, reflected in a number of relations, the prototype of which is the optical theorem of Bohr, Peierls, and Placzek.

Let us therefore look at the restrictions implied by the unitarity of the S matrix written as

$$S^\dagger S = I = I + i(T - T^\dagger) + T^\dagger T \tag{5-152}$$

in accordance with Eq. (5-6). Inserting momentum conservation

$$\langle f|\,T\,|i\rangle = (2\pi)^4\delta^4(P_f - P_i)\mathscr{T}_{fi} \tag{5-153}$$

we obtain the nonlinear relation among amplitudes:

$$(\mathscr{T}_{fi} - \mathscr{T}^*_{if}) = i \sum_n (2\pi)^4 \delta^4(P_n - P_i) \mathscr{T}^*_{nf} \mathscr{T}_{ni} \tag{5-154}$$

involving a sum over all possible intermediate states $|n\rangle$ coupled to $|i\rangle$ and $|f\rangle$. In practical cases the scattering process is initiated by a two-body state. Let us concentrate on two-body elastic scattering. Generalizing this definition we may assume that the particles exchange spin or internal quantum numbers, in agreement with the corresponding conservation laws, provided they remain within the same symmetry multiplet (see Chap. 11). For simplicity we ignore the case of identical particles. First choose $|f\rangle = |i\rangle$ in Eq. (5-154). This corresponds to forward scattering with spin and internal variables equal in the initial and final configuration. We have to assume here the absence of long-range forces. The left-hand side of Eq. (5-154) reduces then to $2i \operatorname{Im} \mathscr{T}_{ii}$. The right-hand side is related to the total cross section $\sigma_{\text{tot}}(i)$ up to a flux factor contributed by the initial state. To be specific, denote by (m_a, S_a) and (m_b, S_b) the masses and spins of the initial particles. For a given spin state i and total center of mass energy squared s we derive, from Eq. (5-13),

$$\sigma_{\text{tot}}(i) = \frac{1}{2\lambda^{1/2}(s, m_a^2, m_b^2)} \sum_n (2\pi)^4 \delta^4(P_n - P_i) \mathscr{T}^*_{ni} \mathscr{T}_{ni} \tag{5-155}$$

where

$$\lambda(x_1, x_2, x_3) \equiv (x_1^2 + x_2^2 + x_3^2) - 2x_1x_2 - 2x_2x_3 - 2x_3x_1 \tag{5-155a}$$

and we have used in (5-155) a normalization of states appropriate to an invariant phase space element $d^3p/2E(2\pi)^3$. The common center of mass three-momentum $|\mathbf{p}|$ of particles A and B is related to s through $4s\mathbf{p}^2 = \lambda(s, m_a^2, m_b^2)$. Using (5-155) we find the optical theorem in the form

$$\operatorname{Im} \mathscr{T}_{ii} = \lambda^{1/2}(s, m_a^2, m_b^2)\sigma_{\text{tot}}(i) \tag{5-156}$$

The amplitude $\mathscr{T}_{ii}$ enters in the expression of the forward elastic cross section. Assume the polarizations to be such that the initial state is invariant under rotations around the incident momentum. The elastic cross section can then be integrated over the azimuthal angle and expressed in terms of the momentum transfer t, instead of the cosine of the scattering angle. In the process $A + B \to A + B$ call (p_a, p_b) and (p'_a, p'_b) respectively the initial and final momenta satisfying energy momentum conservation $p_a + p_b = p'_a + p'_b$. The Mandelstam variables are defined as

$$\begin{aligned} s &= (p_a + p_b)^2 = (p'_a + p'_b)^2 \\ t &= (p_a - p'_a)^2 = (p_b - p'_b)^2 \qquad s + t + u = 2(m_a^2 + m_b^2) \\ u &= (p_a - p'_b)^2 = (p_b - p'_a)^2 \end{aligned} \tag{5-157}$$

The differential elastic cross section can be written as

$$\frac{d\sigma_{\text{el}}}{dt}(s, t) = \frac{|\mathscr{T}(s, t)|^2}{16\pi\lambda(s, m_a^2, m_b^2)} \tag{5-158}$$

Forward scattering is characterized by $t = 0$ and $\mathscr{T}(s, 0)$ is what we denoted as $\mathscr{T}_{ii}$ above. Consequently,

$$\frac{d\sigma_{el}}{dt}(s, 0) = \frac{1}{16\pi}\left[\frac{(\mathrm{Re}\,\mathscr{T}_{ii})^2}{\lambda(s, m_a^2, m_b^2)} + \sigma_{tot}^2(i)\right] > \frac{\sigma_{tot}^2(i)}{16\pi} \tag{5-159}$$

In very high-energy collisions it may happen that the imaginary part of the forward-scattering amplitude dominates over the real one. In this case Eq. (5-159) may serve to normalize the differential elastic cross section, assuming the total cross section to be known.

It is possible to extract further consequences from the unitarity condition if we succeed in diagonalizing the $\mathscr{T}$ operator, at least partially. For two-body scattering this is achieved by using as a basis the eigenstates of the total angular momentum.

The helicity formalism of Jacob and Wick is best suited to perform the corresponding projection. Call λ_a and λ_b (λ'_a, λ'_b) the helicities of the initial (final) particles in the center of mass frame where $\mathbf{p}_a + \mathbf{p}_b = \mathbf{p}'_a + \mathbf{p}'_b = 0$. For a two-body state in this frame let θ and φ be the polar angles of the relative three-momentum $\mathbf{p}$ with respect to a fixed system of axis. Consider $R_{\theta,\varphi}$, the product of a rotation of angle θ around the $\hat{y}$ axis followed by a rotation of angle φ around the $\hat{z}$ axis. This rotation transforms the unit vector $\hat{z}$ into the vector $\mathbf{p}/|\mathbf{p}|$. A state of total angular momentum J with projection M along $\hat{z}$ is obtained as

$$|J, M; \lambda_a, \lambda_b\rangle = \left(\frac{2J+1}{4\pi}\right)^{1/2} \int d\varphi \sin\theta\, d\theta\, \mathscr{D}^J_{\lambda_a - \lambda_b, M}(R^{-1}_{\theta,\varphi})\,|p_a, \lambda_a; p_b, \lambda_b\rangle \tag{5-160}$$

Helicity is defined as the component of spin along the three-momentum, thereby generalizing the case of Dirac particles, and the notion extends to massless particles.

We recall that the Wigner $\mathscr{D}$ functions are obtained from the representations of the $SU(2)$ group of special unitary 2×2 matrices as follows. Any rotation R acting on an arbitrary vector $\mathbf{p}$ can be represented by a pair U, $-U$ through $U\boldsymbol{\sigma}\cdot\mathbf{p}U^\dagger = \boldsymbol{\sigma}\cdot(R\mathbf{p})$. If we consider more generally an arbitrary 2×2 complex matrix $A = \begin{pmatrix} a & b \\ c & d \end{pmatrix}$ it may be identified as a linear operator on homogeneous polynomials in two variables $u_{1/2}, u_{-1/2}$. To obtain $\mathscr{D}^j_{m,m'}$, we extend this action on polynomials of degree $2j$, a basis of which is $u_{jm} = [(j+m)!(j-m)!]^{-1/2} u_{1/2}^{j+m} u_{-1/2}^{j-m}$. If

$$u'_{1/2} = au_{1/2} + bu_{-1/2}$$

$$u'_{-1/2} = cu_{1/2} + du_{-1/2}$$

$$u'_{j,m} = \mathscr{D}^j_{m,m'}(A) u_{j,m'}$$

then

$$\mathscr{D}^j_{m,m'}(A) = [(j+m)!(j-m)!(j+m')!(j-m')!]^{-1/2} \sum_{\substack{n_1+n_2=j+m,\, n_3+n_4=j-m \\ n_1+n_3=j+m',\, n_2+n_4=j-m'}} \frac{a^{n_1}}{n_1!}\frac{b^{n_2}}{n_2!}\frac{c^{n_3}}{n_3!}\frac{d^{n_4}}{n_4!} \tag{5-161a}$$

These matrices enjoy the properties

$$\mathscr{D}^j(A_1A_2) = \mathscr{D}^j(A_1)\mathscr{D}^j(A_2) \qquad \mathscr{D}^j(A^T) = \mathscr{D}^j(A)^T$$

$$\mathscr{D}^j(A^*) = \mathscr{D}^j(A)^* \qquad \mathscr{D}^j_{mm'}(\lambda I) = \lambda^{2j}\delta_{m,m'} \tag{5-161b}$$

$$\mathscr{D}^j_{mm'}(e^{i\sigma_3\theta/2}) = e^{im\theta}\,\delta_{m,m'}$$

In particular if A is unitary, so is $\mathscr{D}^j(A)$. Returning to the group $SU(2)$ the notation in Eq. (5-160) identifies the rotations $R_{\theta,\varphi}$ with its two-by-two representative in such a way that

$$\begin{gathered}\mathscr{D}^J_{\lambda_a-\lambda_b,M}(R^{-1}_{\theta,\varphi}) = \mathscr{D}^{J*}_{M,\lambda_a-\lambda_b}(\varphi,\theta,0) = e^{iM\varphi}\, d^J_{M,\lambda_a-\lambda_b}(\theta) \\ d^J_{m_1,m_2}(\theta) = (-1)^{m_1-m_2} d^J_{-m_1,-m_2}(\theta) = (-1)^{m_1-m_2} d^J_{m_2,m_1}(\theta)\end{gathered} \tag{5-161c}$$

Phase conventions may be chosen in such a way that if η_a and η_b stand for the intrinsic parities of particles A and B the transformation laws of these states under parity $\mathscr{P}$ and time reversal $\mathscr{T}$ read‡

$$\begin{aligned}\mathscr{P}\,|J,M;\lambda_a,\lambda_b\rangle &= \eta_a\eta_b(-1)^{J-s_a-s_b}\,|J,M;-\lambda_a,-\lambda_b\rangle \\ \mathscr{T}\,|J,M;\lambda_a,\lambda_b\rangle &= (-1)^{J-M}\,|J,-M;\lambda_a,\lambda_b\rangle\end{aligned} \tag{5-162}$$

These relations reflect the fact that helicity is odd under parity and even under time reversal.

Using the Wigner-Eckart theorem we may then write for the matrix element $\mathscr{T}_{fi}$ the following expansion taking rotational invariance into account

$$\begin{aligned}\mathscr{T}_{fi} &= \langle p'_a,\lambda'_a;p'_b,\lambda'_b|\,\mathscr{T}\,|p_a,\lambda_a;p_b,\lambda_b\rangle \\ &= 16\pi\sum_J (2J+1)\mathscr{T}^J_{\lambda'_a,\lambda'_b;\lambda_a,\lambda_b}(s)\mathscr{D}^{J*}_{\lambda_a-\lambda_b,\lambda'_a-\lambda'_b}(\varphi,\theta,0)\end{aligned} \tag{5-163}$$

A kinematical factor arising from the relation

$$\int \frac{d^3p_a\,d^3p_b}{(2\pi)^3(2E_a)(2\pi)^3(2E_b)}(2\pi)^4\delta^4(p_a+p_b-P) = \frac{\lambda^{1/2}(s,m_a^2,m_b^2)}{8s(2\pi)^2}\,d\Omega_p$$

will appear in the expression of unitarity given below. In Eq. (5-163) the $\hat{z}$ axis is chosen along the incident momentum $\mathbf{p}_a$ and θ, φ are the polar angles of $\mathbf{p}'_a$. Finally, the sum over J runs over integer or half-integer values according to whether an even or odd number of half-integer spins is present in the initial or final state. The Jacob and Wick expansion (5-163) is easily generalized to an arbitrary two-by-two scattering process.

If invariance under parity applies

$$\mathscr{T}^J_{\lambda'_a,\lambda'_b;\lambda_a,\lambda_b}(s) = \mathscr{T}^J_{-\lambda'_a,-\lambda'_b;-\lambda_a,-\lambda_b}(s) \tag{5-164}$$

Similarly, if time-reversal invariance holds

$$\mathscr{T}^J_{\lambda'_a,\lambda'_b;\lambda_a,\lambda_b}(s) = \mathscr{T}^J_{\lambda_a,\lambda_b;\lambda'_a,\lambda'_b}(s) \tag{5-165}$$

expressing the symmetry of the scattering matrix.

When the total energy $s^{1/2}$ is below the inelastic threshold only the initial two-body channel contributes to the sum over intermediate states in Eq. (5-154). Projecting on angular momentum J results in the relation

$$\mathscr{T}^J(s) - \mathscr{T}^{J\dagger}(s) = 2i\lambda^{1/2}(s,m_a^2,m_b^2)s^{-1}\mathscr{T}^{J\dagger}(s)\mathscr{T}^J(s) \tag{5-166}$$

‡The reader will not confuse the time-reversal operator $\mathscr{T}$ with the scattering amplitude used throughout this section.

where $\mathscr{T}^J(s)$ is considered as a $(2S_a + 1)(2S_b + 1) \times (2S_a + 1)(2S_b + 1)$ matrix in helicity space. Time-reversal invariance may be used to replace the left-hand side by $2i \operatorname{Im} \mathscr{T}^J(s)$.

If the particles are spinless the formulas simplify considerably. In particular, $\mathscr{D}^{J*}_{0,0}(\varphi, \theta, 0)$ is nothing but the Legendre polynomial $P_J(\cos\theta)$, while Eq. (5-166) is solved by introducing a phase shift $\delta_J(s)$ through

$$\frac{2\lambda^{1/2}}{s} \mathscr{T}^J(s) = -i(e^{2i\delta_J(s)} - 1) = 2e^{i\delta_J(s)} \sin \delta_J(s) \tag{5-167}$$

If the matrix $\mathscr{T}^J(s)$ is diagonal in some appropriate basis the same expression is true for each diagonal element.

Such is the case for instance for pion-nucleon scattering. The two independent amplitudes $\mathscr{T}^J_{1/2,1/2}$ and $\mathscr{T}^J_{-1/2,1/2}$ may be replaced by

$$\mathscr{T}^J_{\pm} = \mathscr{T}^J_{1/2,1/2} \pm \mathscr{T}^J_{-1/2,1/2}$$

corresponding to scattering in a state of given parity equal to $\pm(-1)^{J+1/2}$.

Let us collect the expressions valid for elastic spinless scattering. With q standing for the center of mass momentum, θ the scattering angle, and δ_l the lth partial wave phase shift (possibly complex),

$$\begin{aligned}
\frac{d\sigma}{dt} &= \frac{1}{64\pi s q^2} |\mathscr{T}|^2 \\
\mathscr{T} &= 16\pi \sum_0^{\infty} (2l + 1) \mathscr{T}_l P_l(\cos\theta) \\
\mathscr{T}_l &= \frac{s^{1/2}}{2q} e^{i\delta_l} \sin \delta_l \\
\sigma_{\text{tot}} &= \frac{1}{2qs^{1/2}} \operatorname{Im} \mathscr{T}(\cos\theta = 1) = \frac{8\pi}{qs^{1/2}} \sum_0^{\infty} (2l + 1) \operatorname{Im} \mathscr{T}_l \\
|\mathscr{T}_l|^2 &\le \frac{s^{1/2}}{2q} \operatorname{Im} \mathscr{T}_l
\end{aligned} \tag{5-168}$$

The last inequality turns into an equality below the inelastic threshold.

5-3-2 Causality and Analyticity

The macroscopic requirement that effects become manifest after the action of causes not only requires that causes be identifiable without ambiguity but also implies the concept of thermodynamic irreversibility, a phenomenon outside the realm of microscopic physics. Fortunately, the finite velocity of signals allows a weaker formulation which does not imply the choice of a privileged direction for the time arrow. We adhere to the view that space-like separated regions do not influence each other. Stated differently, local observables pertaining to such regions commute. We have seen that this can be further specialized to the commutation (or anticommutation) of the fundamental fields. For simplicity let us only discuss here the case of Bose fields. Mathematically the corresponding properties of field commutators translate into analytic properties of their Fourier transforms. This observation is the basis of the original treatment by Kramers and Kronig of the diffractive index of light in a medium relating dispersion and absorption. Hence the name "dispersion relations" for the analytic representations of the scattering amplitudes.

An example will illustrate these ideas. Consider the elastic scattering of particle A (mass m_a) on a target particle B (mass m_b). Both particles are assumed spinless to avoid cumbersome technical details without changing the essential conclusions, but may carry a charge. We therefore distinguish particle A from antiparticle $\bar{A}$ and use a complex field φ to describe both of them. If q_1 and q_2 denote the initial and final momenta of particle A and p_1 and p_2 the corresponding ones of particle B, the connected part of the scattering amplitude may be expressed as

$$S_{fi} = -\int d^4x\, d^4y\, e^{i(q_2 \cdot y - q_1 \cdot x)}(\Box_y + m_a^2)(\Box_x + m_a^2)\langle p_2|T\varphi^\dagger(y)\varphi(x)|p_1\rangle \qquad (5\text{-}169)$$

We have absorbed a factor $Z^{-1/2}$ into the definition of the field. With q_1 and q_2 inside the forward light cone, the time-ordered product in the right-hand side of Eq. (5-169) may be replaced by a retarded commutator

$$T\varphi^\dagger(y)\varphi(x) \to \theta(y^0 - x^0)[\varphi^\dagger(y), \varphi(x)]$$

without affecting the value of S_{fi}. This can be seen from the original derivation given in Sec. 5-1-3. Define the source $j(x)$ of the field $\varphi(x)$ through

$$(\Box + m_a^2)\varphi(x) = j(x) \qquad (5\text{-}170)$$

and assume for simplicity that φ and j commute at equal times. Taking translation invariance into account we may write

$$\begin{aligned} S_{fi} &= (2\pi)^4\delta^4(p_2 + q_2 - p_1 - q_1)i\mathcal{T} \\ \mathcal{T} &= i\int d^4z\, e^{iq \cdot z}\langle p_2|\,\theta(z^0)\left[j^\dagger\left(\frac{z}{2}\right), j\left(-\frac{z}{2}\right)\right]|p_1\rangle \\ q &= \tfrac{1}{2}(q_1 + q_2) \end{aligned} \qquad (5\text{-}171)$$

Lorentz invariance implies that $\mathcal{T}$ depends only on the scalar products among the momenta, i.e., on two out of the three Mandelstam variables for on-shell particles.

From locality, the retarded commutator $\langle p_2|\,\theta(z^0)[j^\dagger(z/2), j(-z/2)]\,|p_1\rangle$ vanishes unless $z^2 > 0$, $z_0 > 0$. Inspection of Eq. (5-171) reveals that $\mathcal{T}$ is an analytic function of the four-vector q in the so-called forward tube defined by the condition that Im q be a positive time-like vector. This follows from the assumption that the matrix elements of the fields are tempered (i.e., polynomially bounded) distributions. Indeed, if $q = q_R + iq_I$, the exponential in (5-171) provides a damping factor $e^{-z \cdot q_I}$ when both z and q_I are positive time-like vectors.

This example shows the direct relationship between the local properties of relativistic field theories and the analyticity of Green functions.

Before we analyze the mathematical consequences of this result let us recall the property of crossing symmetry. Instead of the process $A(q_1) + B(p_1) \to A(q_2) + B(p_2)$ suppose that we were studying the reaction $\bar{A}(\bar{q}_1) + B(p_1) \to \bar{A}(\bar{q}_2) + B(p_2)$ involving the scattering of antiparticle $\bar{A}$ on the same target B. The corresponding amplitude $\bar{\mathscr{T}}$ is given by

$$\bar{\mathscr{T}} = i \int d^4z\, e^{i\bar{q}\cdot z} \langle p_2 | \theta(z^0) \left[j\left(\frac{z}{2}\right), j^\dagger\left(-\frac{z}{2}\right) \right] | p_1 \rangle$$

$$\bar{q} = \tfrac{1}{2}(\bar{q}_1 + \bar{q}_2)$$

Changing the integration variable z into $-z$, this can also be written as

$$\bar{\mathscr{T}} = i \int d^4z\, e^{-i\bar{q}\cdot z} \langle p_2 | -\theta(-z_0) \left[j^\dagger\left(\frac{z}{2}\right), j\left(-\frac{z}{2}\right) \right] | p_1 \rangle \qquad (5\text{-}172)$$

This amplitude differs from $\mathscr{T}$ given by Eq. (5-171) in two respects. The momentum $q = \frac{1}{2}(q_1 + q_2)$ has been replaced by $-\bar{q} = \frac{1}{2}(-\bar{q}_2 - \bar{q}_1)$ and the retarded commutator by the advanced one

$$\langle p_2 | \theta(z^0) \left[j^\dagger\left(\frac{z}{2}\right), j\left(-\frac{z}{2}\right) \right] | p_1 \rangle \to \langle p_2 | -\theta(-z^0) \left[j^\dagger\left(\frac{z}{2}\right), j\left(-\frac{z}{2}\right) \right] | p_1 \rangle$$

We may study the quantities $\mathscr{T}$ and $\bar{\mathscr{T}}$ for arbitrary values of their arguments, instead of dealing with the actual physical processes where q and $\bar{q}$ lie in the forward light cone. Let us show that

$$\mathscr{T}(q) - \bar{\mathscr{T}}(\bar{q})\big|_{\bar{q}=-q} = iC(q)$$

vanishes in a certain domain. This is the Fourier transform of the commutator

$$\begin{aligned} C(q) &= \int d^4z\, e^{iz\cdot q} \langle p_2 | \left[j^\dagger\left(\frac{z}{2}\right), j\left(-\frac{z}{2}\right) \right] | p_1 \rangle \\ &= (2\pi)^4 \sum_n \langle p_2 | j^\dagger(0) | n \rangle \langle n | j(0) | p_1 \rangle \delta^4\left(q + \frac{p_1 + p_2}{2} - p_n\right) \qquad (5\text{-}173) \\ &\quad - (2\pi)^4 \sum_n \langle p_2 | j(0) | n \rangle \langle n | j^\dagger(0) | p_1 \rangle \delta^4\left(q - \frac{p_1 + p_2}{2} + p_n\right) \end{aligned}$$

If the vacuum is an isolated point in the spectrum, meaning the absence of massless particles, the states $|n\rangle$ contributing to each of the above sums will be such that there exists a lowest positive value p_n^2 in each term (a vacuum contribution is excluded by the connectedness hypothesis). Consequently, $C(q)$ will vanish outside a region bounded by the two hyperboloid sheets depicted in Fig. 5-8. One of them corresponds to a mass M_+, the other to M_-, and they are centered respectively at $-\frac{1}{2}(p_1 + p_2)$ and $\frac{1}{2}(p_1 + p_2)$. We conclude that the amplitudes $\mathscr{T}(p_2, q_2; p_1, q_1)$ and $\mathscr{T}(p_2, -q_1; p_1, -q_2)$ coincide at unphysical points corresponding to the unshaded region of Fig. 5-8. This is the property of crossing symmetry, which relates processes where a particle appearing on one side of a reaction is replaced by an antiparticle of opposite (therefore unphysical)

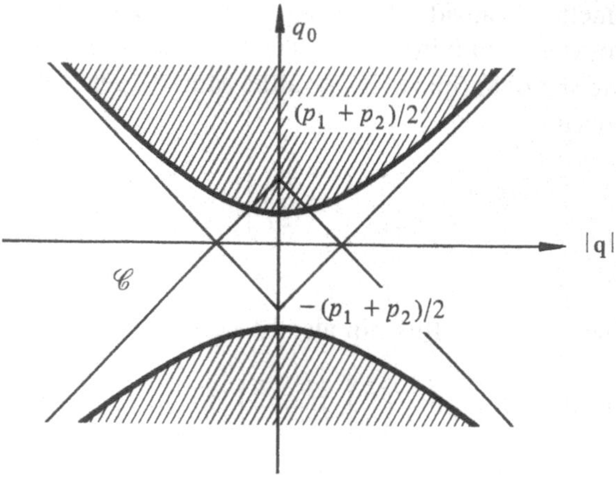

Figure 5-8 The two hyperboloid sheets limiting the support of the commutator in momentum space. The unshaded area is the coincidence region denoted $\mathscr{C}$.

four-momentum on the other side of the reaction. The statement of crossing symmetry will become meaningful if we succeed in showing that analytic continuation allows access to the coincidence region starting from physical values for the momenta. This property is a remarkable consequence of field theory.

We denote $\bar{\mathscr{T}}(q)$ the expression given in (5-172) with q substituted for $-\bar{q}$ and observe that $\bar{\mathscr{T}}(q)$ is analytic in the backward tube, i.e., for Im q a negative time-like vector.

Thus $\mathscr{T}(q)$ and $\bar{\mathscr{T}}(q)$ have a priori disjoint domains of analyticity and coincide on a real domain $\mathscr{C}$. The celebrated edge of the wedge theorem, due to Bremermann, Oehme, and Taylor, allows us to conclude that $\mathscr{T}$ and $\bar{\mathscr{T}}$ are analytic continuations of each other, and moreover that their common domain of analyticity is larger than the union of the forward and backward tubes and $\mathscr{C}$.

If we were dealing with functions of one complex variable only, the problem would be easily settled. Indeed, if $f^{\pm}(z)$ are analytic in the upper and lower complex z plane respectively and coincide on a segment of the real axis, a simple application of Cauchy's theorem proves that they are branches of the same analytic function. The coincidence points are in fact analyticity points. The present situation is obviously more complicated. Around each coincidence point there are holes corresponding to space-like imaginary directions where neither $\mathscr{T}$ nor $\bar{\mathscr{T}}$ are defined. Hence the pictorial name "edge of the wedge." Moreover, analysis in several complex variables uncovers new properties without equivalence in the case of one complex variable. Such is the notion of a holomorphy envelope. A function of several complex variables analytic in a domain $\mathscr{D}$ can at least be extended to a domain $\bar{\mathscr{D}} \supset \mathscr{D}$, with the property that through any of its boundary points we can find an analytic manifold, i.e., the set of zeros of an analytic function, lying entirely outside $\bar{\mathscr{D}}$ except for the boundary point in question. This property is referred to as pseudoconvexity (in analogy with ordinary convexity), with analytic manifolds replacing planes. In the case at hand, even though a purely geometric construction of $\bar{\mathscr{D}}$ is possible, an interesting representation generalizing the Källen-Lehmann representation for the vacuum expectation value of the commutator is available to solve the problem.

The reader may wonder whether the original analyticity domain, including the forward and backward tubes, was not enough in simple cases such as forward scattering with $p_1 = p_2 = p$, $q_1 = q_2 = q$. The following remark will dissipate doubts. Let a complex vector q lie on the mass shell and write $q = q_R + iq_I$. Then the condition $m_a^2 = q^2 = q_R^2 - q_I^2 + 2iq_R \cdot q_I$ states that q_R and q_I

are orthogonal and that it is impossible for q_I to be time-like. If it were so then from the orthogonality condition q_R would be space-like, $q_R^2 < 0$, and $q_R^2 - q_I^2$ would be negative. Thus only the q with negative or complex square mass belong to the tube. For such unphysical values a dispersion relation can be written at once, but we are left with the task of finding a suitable analytic continuation in q^2 in order to obtain some information about the physical process.

5-3-3 The Jost-Lehmann-Dyson Representation

This is a representation for the matrix element of the commutator $C(q)$ given in Eq. (5-173) consistent with the spectrum properties in momentum space and the support properties in configuration space. Recall the expression for the free-field commutator $\Delta(x, \mu^2)$ [Eqs. (3-55) and (3-56)] which vanishes outside the light cone and has an odd Fourier transform concentrated on the hyperboloid of mass μ^2. The Jost-Lehmann-Dyson representation gives $C(q)$ as a superposition in configuration space or a convolution in momentum space of such free-field commutators in the form

$$C(q) = \int_S d^4k\, d\mu^2\, \varepsilon(q^0 - k^0)\delta[(q-k)^2 - \mu^2]\rho(k, \mu^2) \tag{5-174}$$

The weight ρ is nonvanishing only for those values of k and μ^2 such that the hyperboloid $(q-k)^2 = \mu^2$ does not intersect the region $\mathscr{C}$, in which case it is called admissible. Such hyperboloids will have a center k lying in a bounded domain equal to the intersection of the future light cone centered at $-[(p_1 + p_2)/2]$ and of the past light cone centered at $(p_1 + p_2)/2$. The lowest limit on the μ integration is then

$$\mu_{\text{inf}} = \max\left\{0, M_+ - \sqrt{\left(\frac{p_1 + p_2}{2} + k\right)^2}, M_- - \sqrt{\left(\frac{p_1 + p_2}{2} - k\right)^2}\right\}$$

It is clear that $C(q)$ given by (5-174) has all the required properties. Dyson has shown that the converse is also true, namely, that $C(q)$ necessarily admits a representation of this form. We refer the interested reader to the articles quoted in the notes.

To derive the corresponding representation for the amplitude, we simply have to compare Eqs. (5-171) and (5-173). Up to a factor i, the difference lies in the presence of a step function multiplying the commutator in configuration space. We have to be careful about this multiplication of distributions which might leave some undefined contribution concentrated at the origin in z space. Ignoring this complication for the time being, we obtain in momentum space a convolution which yields

$$\mathscr{T}(q) = -\frac{1}{2\pi}\int_S \frac{d^4k\, d\mu^2\, \rho(k, \mu^2)}{(q-k)^2 - \mu^2} \tag{5-175}$$

The physical amplitude is recovered by taking the limit of real positive time-like values of q, with a vanishingly small imaginary part in the forward light cone.

It might happen that the integral over μ^2 diverges for large values, a reflection in momentum space of the poor definition of the retarded commutator. However, subtractions could be performed in the form

$$\mathscr{T}(q) = -\frac{1}{2\pi}\int_S d^4k\, d\mu^2 \frac{\rho(k,\mu^2)}{(q-k)^2-\mu^2}\left[\frac{(q-k)^2+\mu_0^2}{\mu^2+\mu_0^2}\right]^n + P(q) \quad (5\text{-}176)$$

with n an integer and $P(q)$ a polynomial in q, without altering the analytic properties following from Eq. (5-175). We shall therefore omit writing down these subtractions explicitly.

The representation (5-175) exhibits $\mathscr{T}(q)$ as an analytic function for all real and complex points which do not lie on admissible hyperboloids.

This analyticity domain $\bar{\mathscr{D}}$, derived from the Jost-Lehmann-Dyson representation, is hard to visualize. We borrow from the work of Bros, Froissart, Omnès, and Stora the following helpful geometrical construction. The trick is to introduce a fifth coordinate z and map the original complex four-dimensional q space on the quadratic surface $P: z = q^2$. In this way the denominator in Eq. (5-175) is replaced by the linear expression (in z and q) $z - 2q\cdot k + k^2 - \mu^2$ and the admissible hyperboloids by the corresponding planes with (k, μ^2) belonging to S. We then associate to each complex point (z, q) a complex straight line $L(z, q)$ described parametrically by $\operatorname{Re} z + \lambda \operatorname{Im} z$, $\operatorname{Re} q + \lambda \operatorname{Im} q$. If (z, q) belongs to P so does its conjugate, so that $L(z, q)$ intersects P at two points (corresponding to $\lambda = \pm i$). From the properties of quadratic surfaces it follows that the real points of $L(z, q)$ (obtained for λ real) cannot be on P, so that on a real picture $L(z, q)$ lies entirely above or entirely below P. [A special case would be that $L(z, q)$ is entirely contained in P.]

The real coincidence region is mapped into P in a region limited by the two planes $[(p_1 + p_2)/2]^2 \pm q\cdot(p_1 + p_2) + z < M_\pm^2$ and the admissible planes do not cross this coincidence region $\mathscr{C}$. From the fact that $\mu^2 > 0$ it follows that they are above a plane tangent to P. Since we are dealing with planes, this defines an excluded region, the convex hull of $\mathscr{C}$ (call it $\bar{\mathscr{C}}$), which cannot be intersected by an admissible plane (Fig. 5-9). We now have a description of the analyticity domain as follows. A complex point (z, q) belongs to it under one of the following circumstances:

(*a*) $L(z, q)$ lies entirely below P. This case corresponds to the original tubes.
(*b*) $L(z, q)$ intersects $\bar{\mathscr{C}}$. Indeed if (z, q) would belong to an admissible plane, the same would be true for its conjugate. From linearity $L(z, q)$ would then belong to the admissible plane and would not intersect $\bar{\mathscr{C}}$.

5-3-4 Forward Dispersion Relations

We are now in a position to examine the elastic forward amplitude when $q_1 = q_2 = q$, $p_1 = p_2 = p$, as a function of incoming energy $\nu = q\cdot p/m_b = (s - m_a^2 - m_b^2)/2m_b$. We wish to show that it is analytic in a cut plane, with the cuts running along parts of the real axis. This is a nontrivial result in view of the remark at the end of Sec. 5-3-2. Of course, we require that $q^2 = m_a^2$, the physical mass-shell condition.

This is obtained by intersecting the surface P, from the previous construction, by the plane $z = m_a^2$, leading to a hyperboloid. We have to study the corresponding intersection of $\bar{\mathscr{C}}$. For a complex q lying on the mass shell, the real line $L(m_a^2, q)$ does not intersect this hyperboloid with which it has already the two complex points (m_a^2, q) and (m_a^2, q^*) in common. We have noticed that for fixed p

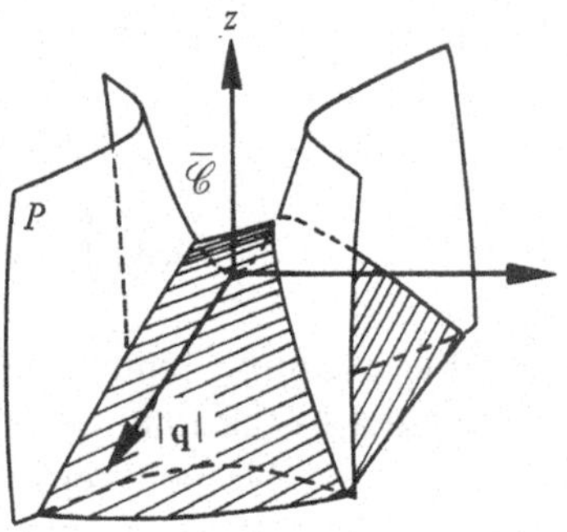

Figure 5-9 The region $\overline{\mathscr{C}}$ (shaded area), convex hull of the coincidence region in five-dimensional space.

Lorentz invariance implies that $\mathscr{T}$ is only a function of the component of q along p. Taking the latter along the time axis means that we can restrict our attention to vectors q with only two non-vanishing components, q_0 and q_1, say. The problem is in effect two dimensional with a third component z added for convenience. The complex point (m_a^2, q) on the hyperbola $q_0^2 - q_1^2 = m_a^2$ will have $(\text{Im } q)^2 < 0$ and will be a point of analyticity provided $L(m_a^2, q)$ intersects the convex hull $\overline{\mathscr{C}}$ in the (q_0, q_1) plane. The condition for this is that $\mathscr{C}$ meets both branches of the mass-shell hyperbola.

Now $\mathscr{C}$ is given by $(p \pm q)^2 < M_\pm^2$ or, equivalently, $m_b^2 \pm 2p \cdot q + z < M_\pm^2$. The condition that $\mathscr{C}$ meets the upper branch is $m_b^2 + 2m_a m_b + m_a^2 \leq M_+^2$ or $(m_a + m_b)^2 \leq M_+^2$. Similarly, the condition that it meets the lower branch is $(m_a + m_b)^2 \leq M_-^2$. These are therefore the criteria that eliminate the occurrence of complex singularities in the complex ν plane. On the other hand, we know that $M_\pm^2 \leq (m_a + m_b)^2$, corresponding to intermediate states $A + B$ or $\bar{A} + B$ in the sums of Eq. (5-173). From this we conclude that forward dispersion relations (without complex singularities) can at best be proved marginally using the above method.

It may happen that M_+ or M_-, or both, are strictly smaller than the elastic threshold. One case frequently encountered can be disposed of without too much trouble. This is when an intermediate isolated state [mass $M_0^2 < (m_a + m_b)^2$] occurs below the threshold. Its contribution is then a pole in the energy variable. Multiplying the amplitude by $(p \pm q)^2 - M_0^2$ does not modify the analytic properties but eliminates the singularity. In pion-nucleon scattering, for instance, the nucleon intermediate state contributes such a pole.

Let us assume that apart from such poles the condition $M_\pm^2 \geq (m_a + m_b)^2$ is satisfied. We then conclude that in the case of forward scattering the following properties hold:

1. The amplitude is analytic in the complex ν plane except for cuts extending along the real axis from m_a to $+\infty$ and from $-\infty$ to $-m_a$ and possible poles in between.
2. The amplitude is real in between the cuts so that its discontinuity across the cuts is purely imaginary. This follows from Eq. (5-173) applied to forward scattering.
3. The amplitude is bounded by a polynomial for large ν. This stems from the tempered character assumed for the fields and holds a fortiori for complex ν.
4. Finally, if $A \equiv \bar{A}$ crossing symmetry requires that $\mathscr{T}(-\nu^*) = \mathscr{T}(\nu)^*$.

Using this information and Cauchy's formula, a simple analytic representation is obtained. For simplicity assume that, as in property 4, $A \equiv \bar{A}$ and that $\mathscr{T}$

vanishes at infinity. Then we can write

$$\mathscr{T}(\nu) = \text{poles} + \frac{1}{\pi}\int_{(m_a+m_b)^2}^{\infty} d\nu' \,\text{Im}\, \mathscr{T}(\nu')\left(\frac{1}{\nu'-\nu} + \frac{1}{\nu'+\nu}\right) \tag{5-177}$$

If $\mathscr{T}(\nu)$ would grow like a power of ν a similar representation would hold for a function obtained by dividing ν by a real polynomial in ν.

We must emphasize that only physically measurable quantities enter the above representation. The absorptive part Im $\mathscr{T}(\nu)$ is related to the total cross section through the optical theorem, Eq. (5-156), while the possible pole terms have as residues products of coupling constants. Indeed, formulas of the type (5-177) enable us to measure such coupling constants.

As an illustration, consider the time-honored example of pion-nucleon scattering. For arbitrary charge we write the two independent amplitudes in the form

$$\mathscr{T} = \bar{u}(p_2)[A + \tfrac{1}{2}(\not{q}_1 + \not{q}_2)B]u(p_1) \tag{5-178}$$

with $\bar{u}(p_2)$, $u(p_1)$ the on-shell Dirac spinors describing the final (momentum p_2) and initial (momentum p_1) nucleons. The incoming pion has momentum q_1 and the final one q_2. The kinematical invariants are

$$\begin{aligned} s &= W^2 = (p_1 + q_1)^2 \\ t &= (q_1 - q_2)^2 = -2q^2(1 - \cos\theta) \\ u &= (p_1 - q_2)^2 = 2M^2 + 2\mu^2 - s - t \end{aligned} \tag{5-179}$$

Here M and μ are respectively the nucleon and pion mass, θ is the center of mass scattering angle, q is the pion three-momentum in this frame, and W is the total center of mass energy in such a way that

$$q^2 = \frac{[(W+M)^2 - \mu^2][(W-M)^2 - \mu^2]}{4W^2} \tag{5-180}$$

The differential cross section is, according to Sec. 5-1-1,

$$\frac{d\sigma}{d\Omega} = \left(\frac{M}{4\pi W}\right)^2 |\mathscr{T}|^2 \tag{5-181}$$

and in (5-178) the functions A and B only depend on the invariants. Anticipating the discussion of symmetries (Chap. 11) we know that strong interactions are isospin invariant and that the nucleon has isospin $\frac{1}{2}$ and the pion isospin 1. Therefore all the channels are described in terms of two reduced matrix elements corresponding to total isospin $\frac{1}{2}$ or $\frac{3}{2}$, that is, A and B are combinations of A^1, A^3 and B^1, B^3 respectively for any process with definite charges. This gives the following table:

$$\begin{aligned} &\left.\begin{matrix}\pi^+ p \to \pi^+ p \\ \pi^- n \to \pi^- n\end{matrix}\right\} \mathscr{T}^3 = \mathscr{T}^{(+)} - \mathscr{T}^{(-)} \\ &\left.\begin{matrix}\pi^- p \to \pi^- p \\ \pi^+ n \to \pi^+ n\end{matrix}\right\} \tfrac{1}{3}(\mathscr{T}^3 + 2\mathscr{T}^1) = \mathscr{T}^{(+)} + \mathscr{T}^{(-)} \\ &\left.\begin{matrix}\pi^- p \to \pi^0 n \\ \pi^+ n \to \pi^0 p\end{matrix}\right\} \frac{\sqrt{2}}{3}(\mathscr{T}^3 - \mathscr{T}^1) = -\sqrt{2}\mathscr{T}^{(-)} \end{aligned} \tag{5-182}$$

The last is called a charge exchange reaction. Since only two reduced amplitudes describe three measurable processes, this implies triangular inequalities among cross sections. It is convenient to introduce the combinations

$$\mathcal{T}^{(+)} = \frac{\mathcal{T}^1 + 2\mathcal{T}^3}{3} \qquad \mathcal{T}^{(\pm)}(q_1, q_2) = \pm\mathcal{T}^{(\pm)}(-q_2, -q_1) \qquad \mathcal{T}^{(-)} = \frac{\mathcal{T}^1 - \mathcal{T}^3}{3} \tag{5-183}$$

respectively even or odd under $s \leftrightarrow u$ exchange. From the explicit structure of Eq. (5-178) and using the variable

$$\nu = \frac{s-u}{4M} \tag{5-184}$$

odd under crossing (note that for $t = 0$ it coincides with the laboratory incoming pion energy), it follows that

$$A^{(\pm)}(-\nu, t) = \pm A^{(\pm)}(\nu, t) \qquad B^{(\pm)}(-\nu, t) = \mp B^{(\pm)}(\nu, t) \tag{5-185}$$

At $t = 0$ the amplitudes are analytic (in ν) in a plane cut along the real axis from the elastic threshold $[s = (M+\mu)^2, \nu = \mu]$ to infinity (the s cut) and from minus infinity to the u channel threshold $[u = (M+\mu)^2, \nu = -\mu]$.

Furthermore, isolated poles occur along the real axis corresponding to the nucleon intermediate state in the s and u channels. We may write the effective pseudoscalar pion-nucleon interaction

$$\mathscr{L}_{\pi N} = ig_{\pi NN}\bar{\psi}\pi_\alpha\tau_\alpha\gamma_5\psi \tag{5-186}$$

with τ_α being the nucleon isospin matrices and π_α ($\alpha = 1, 2, 3$) the three hermitian components of the pion field. Thus the pole contributions are

$$\mathcal{T}_{\text{pole}} = g_{\pi NN}^2\bar{u}(p_2)\tau_\beta\tau_\alpha\gamma_5\frac{\not{p}_1 + \not{q}_1 + M}{\mu^2 + 2p_1\cdot q_1}\gamma_5 u(p_1) + g_{\pi NN}^2\bar{u}(p_2)\tau_\alpha\tau_\beta\gamma_5\frac{\not{p}_1 - \not{q}_1 + M}{\mu^2 - 2p_1\cdot q_1}\gamma_5 u(p_1) \tag{5-187}$$

Now $2p_1 \cdot q_1 = 2M\nu - t/2$; we may use the Dirac equation satisfied by on-shell spinors to write

$$\mathcal{T}_{\text{pole}} = \frac{g_{\pi NN}^2}{2M}\bar{u}(p_2)\left[(\delta_{\alpha\beta} + \tfrac{1}{2}[\tau_\beta, \tau_\alpha])\frac{(\not{q}_1 + \not{q}_2)/2}{\nu_p - \nu} - (\delta_{\alpha\beta} - \tfrac{1}{2}[\tau_\beta, \tau_\alpha])\frac{(\not{q}_1 + \not{q}_2)/2}{\nu_p + \nu}\right]u(p_1) \qquad \nu_p = -\frac{\mu^2}{2M} + \frac{t}{4M} \tag{5-188}$$

We recognize that the poles only contribute to the B amplitudes with

$$B_p^{(+)} = \frac{g_{\pi NN}^2}{2M}\left(\frac{1}{\nu_p - \nu} - \frac{1}{\nu_p + \nu}\right) \qquad B_p^{(-)} = \frac{g_{\pi NN}^2}{2M}\left(\frac{1}{\nu_p - \nu} + \frac{1}{\nu_p + \nu}\right) \tag{5-189}$$

Note that the location of the poles is t dependent.

Inserting the value of ν_p at $t = 0$ and ignoring for the moment the question of subtractions, we find the forward dispersion relations in the form

$$A^{(\pm)}(\nu, 0) = \frac{1}{\pi}\int_\mu^\infty d\nu' \,\text{Im}\, A^{(\pm)}(\nu', 0)\left(\frac{1}{\nu' - \nu} \pm \frac{1}{\nu' + \nu}\right)$$
$$B^{(\pm)}(\nu, 0) = \frac{g_{\pi NN}^2}{2M}\left(\frac{1}{\nu_p - \nu} \mp \frac{1}{\nu_p + \nu}\right) + \frac{1}{\pi}\int_\mu^\infty d\nu' \,\text{Im}\, B^{(\pm)}(\nu', 0)\left(\frac{1}{\nu' - \nu} \mp \frac{1}{\nu' + \nu}\right) \tag{5-190}$$

In fact analyticity in s may be proved for fixed (negative) t in a finite range $-t_M \le t \le 0$, so that (5-190) may be generalized to

$$A^{(\pm)}(\nu, t) = \frac{1}{\pi}\int_{\mu+t/4M}^{\infty} d\nu' \operatorname{Im} A^{(\pm)}(\nu', t)\left(\frac{1}{\nu'-\nu} \pm \frac{1}{\nu'+\nu}\right)$$
$$B^{(\pm)}(\nu, t) = \frac{g_{\pi NN}^2}{2M}\left(\frac{1}{\nu_p - \nu} \mp \frac{1}{\nu_p + \nu}\right) + \frac{1}{\pi}\int_{\mu+t/4M}^{\infty} d\nu' \operatorname{Im} B^{(\pm)}(\nu', t)\left(\frac{1}{\nu'-\nu} \mp \frac{1}{\nu'+\nu}\right) \tag{5-191}$$

These relations may be used in conjunction with a partial wave analysis along the lines of Sec. 5-3-1. For this purpose, we rewrite the quantity $(M/4\pi W)\mathscr{T}$ appearing in the cross section (5-181) as

$$\frac{M}{4\pi W}\bar{u}(p_2, \lambda_2)[A + \tfrac{1}{2}(\not{q}_1 + \not{q}_2)B]u(p_1, \lambda_1) = \chi_2^\dagger(\lambda_2)\left[f_1 + \frac{(\boldsymbol{\sigma}\cdot\mathbf{q}_2)(\boldsymbol{\sigma}\cdot\mathbf{q}_1)}{\mathbf{q}^2} f_2\right]\chi_1(\lambda_1) \tag{5-192}$$

where λ_1 and λ_2 are the initial and final nucleon helicities and χ, $\chi^\dagger$ Pauli spinors in such a way that [compare with (2-37)]

$$u(p, \lambda) = \frac{\not{p} + M}{\sqrt{2M(M+E)}}\begin{pmatrix}\chi(\lambda)\\ 0\end{pmatrix}$$

With E denoting the nucleon center of mass energy this yields

$$f_1 = \frac{E+M}{8\pi W}[A + (W-M)B]$$
$$f_2 = \frac{E-M}{8\pi W}[-A + (W+M)B] \tag{5-193}$$

The 2×2 matrix appearing in (5-192) admits an angular decomposition of the following form:

$$f_{\lambda_2,\lambda_1} = \frac{1}{q}\sum_J (J + \tfrac{1}{2})\, f^J_{\lambda_2,\lambda_1}\mathscr{D}^{J*}_{\lambda_1,\lambda_2}(\varphi, \theta, 0) \tag{5-194}$$

where J runs over half-integer values.

Invariance under parity yields $f^J_{1/2,1/2} = f^J_{-1/2,-1/2}$ and $f^J_{1/2,-1/2} = f^J_{-1/2,1/2}$. Time reversal does not give any new information. Using the spinors of definite helicities we have, at $\varphi = 0$,

$$f_{1/2,1/2} = f_{-1/2,-1/2} = (f_1 + f_2)\chi_2^\dagger(\tfrac{1}{2})\chi_1(\tfrac{1}{2}) = (f_1 + f_2)\cos\frac{\theta}{2}$$
$$f_{1/2,-1/2} = -f_{-1/2,1/2} = (f_1 - f_2)\chi_2^\dagger(\tfrac{1}{2})\chi_1(-\tfrac{1}{2}) = (f_1 - f_2)\sin\frac{\theta}{2} \tag{5-195}$$

As was already mentioned, parity diagonalizes the matrix f_{λ_1,λ_2} in such a way that

$$f^J_{\pm} = f^J_{1/2,1/2} \pm f^J_{-1/2,1/2} = 2e^{i\delta_{J,\pm}}\sin\delta_{J,\pm} \tag{5-196}$$

The phase shifts $\delta_{J,\pm}$ are real below the inelastic threshold and correspond to the scattering in the channel of total angular momentum J and parity $\pm(-1)^{J+1/2}$. In nonrelativistic notation these states would be obtained from well-defined orbital angular momentum l with $J = l \pm \frac{1}{2}$ so that the corresponding phase shifts are also traditionally denoted $\delta_{l,+}$ and $\delta_{l+1,-}$ respectively. Let us now use the explicit formula for the Wigner $\mathscr{D}$ functions:

$$(J + \tfrac{1}{2})d^J_{1/2,1/2} = \cos\frac{\theta}{2}(P'_{J+1/2} - P'_{J-1/2})$$
$$(J + \tfrac{1}{2})d^J_{-1/2,1/2} = \sin\frac{\theta}{2}(P'_{J+1/2} + P'_{J-1/2}) \tag{5-197}$$

With the l notation ($\delta_{J,+} \equiv \delta_{l,+}$, $\delta_{J,-} \equiv \delta_{l+1,-}$) this yields the following expansions

$$
\begin{aligned}
f_1(\cos\theta) &= \sum_{l=0}^{\infty} P'_{l+1}(\cos\theta) f_{l,+} - P'_{l-1}(\cos\theta) f_{l,-} \\
f_2(\cos\theta) &= \sum_{l=1}^{\infty} P'_l(\cos\theta)(f_{l,+} - f_{l,-}) \\
f_{l,\pm} &= \frac{e^{i\delta_{l,\pm}} \sin\delta_{l,\pm}}{q}
\end{aligned}
\tag{5-198}
$$

Of course, for $l = 0$ the only surviving amplitude is $f_{0,+}$. The question of subtractions in the dispersion relations should now be investigated. We postpone some general statements until the next subsection except to say that in pion-nucleon scattering it can be shown that the only amplitude requiring a subtraction is $A^{(+)}$.

A classical application of Eqs. (5-190) is to the determination of the effective coupling constant. The principle of the method is to evaluate accurately from low-energy phase shift analysis the B amplitude for π^+-proton elastic scattering (it is, in fact, dominated by the $J = \frac{3}{2}$, $T = \frac{3}{2}$ resonance) and to compare its real part in the forward direction with the value given by the dispersion integral. The latter is also dominated by its low-energy contribution so that the approach is consistent. The value quoted by Hamilton and Woolcock is

$$
f^2 = \frac{1}{4\pi}\left(\frac{g_{\pi NN}\mu}{2M}\right)^2 = 0.081 \pm 0.002 \tag{5-199}
$$

Dispersion integrals play a leading role in disentangling the various phase shifts at intermediate energies. As an exercise we suggest extending the formalism to the t-channel reaction $\pi\pi \to N\bar{N}$.

5-3-5 Momentum Transfer Analyticity

In the sequel, we disregard the spin of the external particles. Even though it is not a straightforward matter to include it, this can be done without modifying the results. Instead of using as a starting point of our discussion a reduction formula involving a particle in both the initial and final states, as we did in Sec. 5-3-2, let us rather reduce two initial particles in an elastic process with momenta p_1 and q_1 and associated fields ψ and φ. Thus we would initially write the connected S-matrix element

$$
\langle p_2, q_2 | S | p_1, q_1 \rangle_c = -\int dx^4\, d^4y\, e^{-i(p_1 \cdot x + q_1 \cdot y)} (\square_x + m_b^2)(\square_y + m_a^2) \langle p_2, q_2 | T\psi(x)\varphi(y) | 0 \rangle_c \tag{5-200}
$$

Returning again to the derivation of this formula, we observe that we might as well have used a retarded commutator instead of a chronological product in the physical region of the elastic process. Dropping inessential contact terms, we may therefore consider the replacement (with j_ψ and j_φ the sources of the ψ and φ fields)

$$
(\square_x + m_b^2)(\square_y + m_a^2) \langle p_2, q_2 | T\psi(x)\varphi(y) | 0 \rangle_c \to \langle p_2, q_2 | \theta(x^0 - y^0)[j_\psi(x), j_\varphi(y)] | 0 \rangle_c \tag{5-201}
$$

Taking momentum conservation into account, we obtain

$$
\begin{aligned}
\langle p_2, q_2 | S | p_1, q_1 \rangle_c &= (2\pi)^4 \delta^4(p_2 + q_2 - p_1 - q_1) i\mathscr{T}_{fi} \\
\mathscr{T}_{fi} &= i \int d^4x\, e^{i(q_1 - p_1)\cdot x/2} \langle p_2, q_2 | \theta(x^0) \left[j_\psi\left(\frac{x}{2}\right), j_\varphi\left(-\frac{x}{2}\right) \right] | 0 \rangle_c
\end{aligned}
\tag{5-202}
$$

This representation shows that $\mathscr{T}$ is analytic in the variable $q = (q_1 - p_1)/2$, as follows from the Jost-Lehmann-Dyson formula, and we may repeat an analysis similar to the one given for forward scattering. We take the total incoming momentum $p = p_1 + q_1$ as a time axis. Thus we shall work in the center of mass frame and exploit (5-202) to obtain analytic properties in the cosine of the scattering angle. The coincidence region of advanced and retarded commutator, which could be used as well, is limited by the hyperboloids corresponding to the lowest masses M_1 and M_2 of intermediate states $|n\rangle$ contributing respectively to the matrix elements

$$\begin{array}{ll} \langle p_2, q_2 | j_\psi(0) | n \rangle \langle n | j_\varphi(0) | 0 \rangle & \langle p_2, q_2 | j_\varphi(0) | n \rangle \langle n | j_\psi(0) | 0 \rangle \\ \left(\frac{p}{2} + q\right)^2 < M_1^2 & \left(\frac{p}{2} - q\right)^2 < M_2^2 \end{array} \tag{5-203}$$

The stability of particles a (momentum q_1) and b (momentum p_1) imply that

$$M_1^2 > m_a^2 \qquad M_2^2 > m_b^2 \tag{5-204}$$

For on-shell scattering

$$\left(\frac{p}{2} + q\right)^2 = q_1^2 = m_a^2 \qquad \left(\frac{p}{2} - q\right)^2 = p_1^2 = m_b^2 \tag{5-205}$$

Seen in the five-dimensional space z, q (recall that we work on the surface P such that $z = q^2$) these two linear conditions determine a three-dimensional linear manifold which must clearly have a real intersection with P. Moreover, the planes defined by

$$\frac{p^2}{4} + 2p \cdot q + z = M_1^2 \qquad \frac{p^2}{4} - 2p \cdot q + z = M_2^2 \tag{5-206}$$

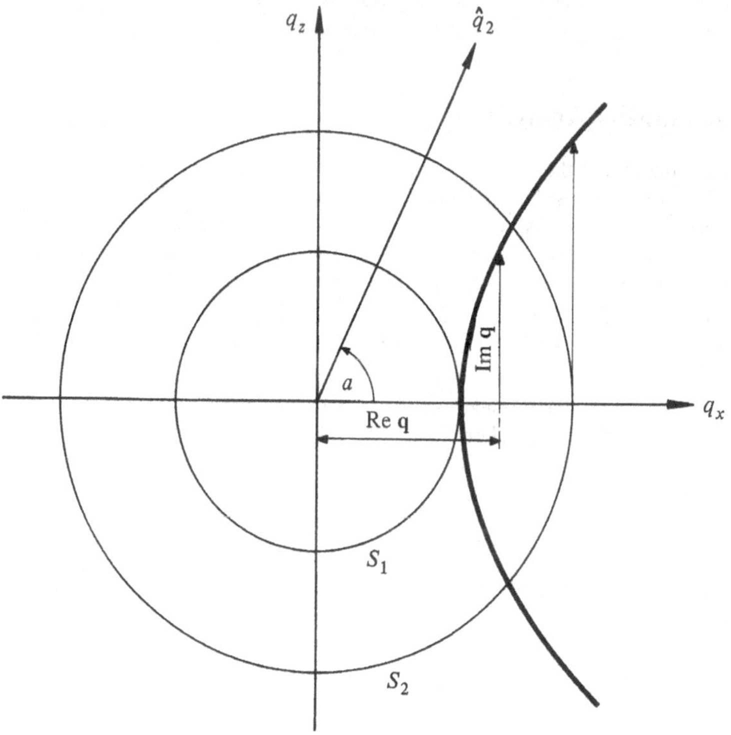

Figure 5-10 The analyticity domain in the relative three-momentum on the mass shell with real intersection S_1.

intersect P on two hyperboloids limiting $\mathscr{C}$. We denote by $\overline{\mathscr{C}}$ its convex hull. By virtue of the stability conditions the three-dimensional linear manifold representing the mass shell (5-205) intersects P along a sphere S_1 of constant relative three-momentum $\mathbf{q}^2 = \mathbf{p}_1^2 = \mathbf{q}_1^2$. The intersection with $\overline{\mathscr{C}}$ is also a sphere S_2 of radius R larger than $|\mathbf{p}_1|$ and is a function of the masses (Fig. 5-10). Thus the analyticity domain may be entirely described in terms of $\mathbf{q}$. For an arbitrary complex point on S_1,

$$\begin{aligned} \operatorname{Re}\mathbf{q}\cdot\operatorname{Im}\mathbf{q} &= 0 \\ \operatorname{Re}\mathbf{q}^2 - \operatorname{Im}\mathbf{q}^2 &= \mathbf{q}^2 \end{aligned} \tag{5-207}$$

We recall that $\mathbf{q}^2$ is to be thought of as the common square length of all three-momenta in the center of mass for elastic scattering. To be a point of analyticity the complex point $\operatorname{Re}\mathbf{q} + i\operatorname{Im}\mathbf{q}$ must be such that the real line $\operatorname{Re}\mathbf{q} + \lambda\operatorname{Im}\mathbf{q}$ intersects S_2. Since $\operatorname{Im}\mathbf{q}$ is orthogonal to $\operatorname{Re}\mathbf{q}$ this implies that $\operatorname{Re}\mathbf{q}$ lies inside S_2 while the first condition says that $\operatorname{Re}\mathbf{q}^2 > |\mathbf{p}_1|^2$. The boundary corresponds to $(\operatorname{Re}\mathbf{q})^2 = R^2$. Let us translate these results in terms of the scattering angle

$$x = \cos\theta = \frac{\mathbf{q}_1\cdot\mathbf{q}_2}{\sqrt{\mathbf{q}_1^2\mathbf{q}_2^2}}$$

The length square $\mathbf{q}_1^2$ is also fixed to $\mathbf{q}^2$. Thus $\sqrt{\mathbf{q}^2}x = \mathbf{q}\cdot\hat{q}_2$, where $\hat{q}_2$ is a unit fixed vector and the variable x varies inside a complex domain bounded by an ellipse, as follows from Eqs. (5-207),

$$\sqrt{\mathbf{q}_1^2}\operatorname{Re}x = R\cos a \qquad \sqrt{\mathbf{q}_1^2}\operatorname{Im}x = \sqrt{R^2 - \mathbf{q}_1^2}\sin a \tag{5-208}$$

This is called the small Lehmann ellipse

$$\frac{\operatorname{Re}^2 x}{R^2} + \frac{\operatorname{Im}^2 x}{R^2 - \mathbf{q}^2} = \frac{1}{\mathbf{q}^2} \tag{5-209}$$

The foci of this ellipse are the points $x = \pm 1$ and the semimajor axis is $R/\sqrt{\mathbf{q}^2}$.

More generally, the proof applies to any matrix element of the form $\langle p_{\text{out}}|p_1, q_1, \text{in}\rangle$ and gives sufficient analyticity properties to allow for a convergent spherical harmonic expansion in relative angles, expressing the rotational invariance of the S matrix.

An elliptic domain is the natural convergence region of a partial wave decomposition, a series in Legendre polynomials in the simplest case. To compute the actual value of R is not so easy. It turns out that

$$\frac{R^2}{\mathbf{q}^2} = 1 + \frac{(M_1^2 - m_a^2)(M_2^2 - m_b^2)}{\mathbf{q}^2[s - (M_1 - M_2)^2]} \tag{5-210}$$

The existence of an elliptic analyticity domain for the elastic amplitude implies that its absorptive part is analytic in a larger ellipse (the large Lehmann ellipse). This may be seen below the inelastic threshold as follows. Recall from Eq. (5-168) that

$$\begin{aligned} \mathscr{T} &= 16\pi\sum_0^\infty (2l+1)\mathscr{T}_l(s)P_l(\cos\theta) \\ \operatorname{Im}\mathscr{T}_l &= \frac{2q^0}{\sqrt{s}}|\mathscr{T}_l|^2 \end{aligned} \tag{5-211}$$

From an estimate of Legendre polynomials for large l it follows that a necessary and sufficient condition (generalizing Abel's criterion for Taylor series) for the expansion (5-211) to converge in an ellipse of semimajor axis $\cos\theta_0 > 1$ is that the least upper bound on $|\mathscr{T}_l|^{1/l}$ be

$$\frac{1}{\cos\theta_0 + \sqrt{\cos^2\theta_0 - 1}}$$

The absorptive (imaginary) part is defined through a similar series with $\mathscr{T}_l$ replaced by Im $\mathscr{T}_l$. From (5-211) it converges in an ellipse of semimajor axis $\cos\theta_1$ given by

$$\frac{1}{\cos\theta_1 + \sqrt{\cos^2\theta_1 - 1}} = \left(\frac{1}{\cos\theta_0 + \sqrt{\cos^2\theta_0 - 1}}\right)^2 = \frac{1}{2\cos^2\theta_0 - 1 + \sqrt{(2\cos^2\theta_0 - 1)^2 - 1}}$$

that is,

$$\cos\theta_1 = \cos 2\theta_0$$

These analytic properties in s and t obtained separately have been generalized to analytic properties in the joint variables through the work of Mandelstam, Lehmann, Bros, Epstein, and Glaser. Martin and his collaborators have furthermore made a profitable use of positivity properties of the absorptive part. We shall have the occasion in Chap. 6 to return to this subject in the context of the perturbative expansion.

To close this section let us mention two important results which put constraints on the large-energy behavior and are in close relation with analytic properties.

The first of these is the bound obtained by Froissart on the behavior of total cross sections for large s. The intuitive idea is that if strong interactions are of finite range the total cross sections should be bounded at infinity. Things are, in fact, not as simple, even from the experimental point of view. The reason lies in the fact that velocity-dependent forces are also expected when interactions are mediated by the exchange of particles with spin.

The assumptions needed in the derivation are the following:

1. The enlargement of the analyticity domain is such that for fixed $|t| < t_0$ dispersion relations in s may be written with at most n subtractions. A polynomial bound follows in the axiomatic framework and is also a term-by-term consequence of lagrangian field theory. In fact, t_0 may be proved to be equal to $4m_\pi^2$ for pion-nucleon scattering, for instance.
2. From the relation $t = -2q^2(1 - \cos\theta)$ and from the fact that the Legendre series for the absorptive part may be shown to converge within an ellipse of semimajor axis $\cos\theta_0 = 1 + t_0/2q^2$ (an improvement over the large Lehmann ellipse arising from positivity), it follows that

$$\text{Im}\,\mathscr{T}_l < \frac{s^n}{(2l+1)P_l(1 + t_0/2q^2)} \tag{5-212}$$

For fixed x larger than one, $P_l(x)$ is positive and increases exponentially with l.

In essence $P_l(x)$ behaves as $(x + \sqrt{x^2 - 1})^l$ so that the series for the absorptive part is effectively cut off at some $l_{\max}$ of the order

$$l_{\max} = n \ln s \frac{q}{\sqrt{t_0}} \tag{5-213}$$

for s large. From unitarity

$$|\mathscr{T}_l|^2 < \frac{\sqrt{s}}{2q}\,\text{Im}\,\mathscr{T}_l$$

the series for the amplitude in the physical region where $|P_l(\cos\theta)| \leq 1$ is bounded by $16\pi \sum_0^{l_{max}} (2l+1)|\mathscr{T}_l|$ and again from unitarity $|\mathscr{T}_l| < \sqrt{s}/2q \sim 1$. Thus

$$|\mathscr{T}(s, \cos\theta)| < 16\pi l_{max}^2 = \text{constant } s(\ln s)^2 \tag{5-214}$$

An accurate estimate of the constant may be obtained in terms of the maximum number of subtractions n. The latter can, in fact, be reduced to two using a similar bound as (5-214) in the u channel and the Phragmen-Lindelöf theorem to show that the bound $s(\ln s)^2$ holds in the complex directions; thus $n = 2$. The final estimates of the Froissart bounds are then

$$|\mathscr{T}(s, \cos\theta)| < 16\pi \frac{q^2}{t_0} (\ln s)^2 \tag{5-215}$$

and from the optical theorem

$$\sigma_{tot} < \frac{1}{t_0} (\ln s)^2 \tag{5-216}$$

Even though the scale of the logarithm is not ascertained, it is striking that the trend exhibited by the high-energy data obtained by the latest generation of accelerators is in qualitative agreement with the prediction of this bound. It is interesting to interpret the angular momentum cutoff (5-213) in terms of an effective opacity radius $\rho = l_{max}/q \sim \ln s/\sqrt{t_0}$. On the grounds of a potential analogy, ρ would be of the order $1/\sqrt{t_0}$ ($1/2m_\pi$ for pion-nucleon scattering). It only differs from this naive prediction by a logarithmic increase with energy.

Another property predicted on the basis of analytic properties is the asymptotic equality of particle and antiparticle total cross sections from a given target.

The theoretical observation is originally due to Pomeranchuk. The statement of his theorem needs some qualification in view of the fact that total cross sections may increase as $(\ln s)^2$ for large s. If σ and $\bar{\sigma}$ denote the two cross sections, then under mild assumptions we can show that $\sigma/\bar{\sigma} \to 1$ as s goes to infinity.

Much more could, and should, be said on the whole subject of the interplay of locality and unitarity as dictating the framework of scattering. The reader will undoubtedly find a more substantial treatment in the large literature devoted to this subject.

NOTES

For the Källen-Lehmann representation see G. Källen, "Quantum Electrodynamics," Springer-Verlag, Berlin, 1972. The asymptotic theory is covered in the work of H. Lehmann, K. Symanzik, and W. Zimmermann, *Nuovo Cimento*, vol. 1, p. 205, 1955. Additional terms in the current commutation relations were discussed by J. Schwinger, *Phys. Rev. Lett.*, vol. 3, p. 296, 1959.

The perturbative calculations are by now classroom exercises. A wealth of information on the Compton effect can be found in R. D. Evans, *Handbuch der Physik*, vol. XXXIV, p. 218, 1958.

On the foundations of dispersion relations the reader may consult some of the following references according to his or her taste: N. N. Bogoliubov and D. V. Shirkov, "Introduction to the Theory of Quantized Fields," Interscience, New York, 1959; Proceedings of the 1960 Les Houches Summer School, "Dispersion Relations and Elementary Particles," edited by C. de Witt and R. Omnès, Hermann, Paris, 1960, in particular the lectures of M. L. Goldberger, A. S. Wightman, and R. Omnès; 1960 Scottish Universities' Summer School, "Dispersion Relations," edited by G. R. Screaton, Oliver and Boyd, Edinburgh, 1961; and Proceedings of the International 1964 School of Physics Enrico Fermi, "Dispersion Relations and Their Connection with Causality," edited by E. P. Wigner, Academic Press, New York, 1964, in particular the lectures by M. Froissart.

On pion-nucleon scattering see, for instance, the lectures of J. Hamilton in "Strong Interactions and High Energy Physics," 1963 Scottish Universities' Summer School, edited by R. G. Moorhouse, Oliver and Boyd, Edinburgh, 1964.

For a summary of the work on analyticity and its physical consequences derived from first principles, including a discussion of various high-energy bounds, see the lectures of A. Martin in "Physique des Particules," Les Houches 1971 Summer School, edited by C. de Witt and C. Itzykson, Gordon and Breach, New York, 1973.

The foundations of S matrix theory are studied in "The S-Matrix," by D. Iagolnitzer, North Holland, Amsterdam, 1978.

CHAPTER

SIX

PERTURBATION THEORY

We develop the relativistic perturbation theory and present the technique of Feynman diagrams. Some attention is paid to the case where the interaction contains derivatives. We identify the expansion as a series in Planck's constant and study some elementary topological properties. The parametric representation of Feynman integrals is introduced and used to define the euclidean continuation. Analytic properties and discontinuity formulas are indicated.

6-1 INTERACTION REPRESENTATION AND FEYNMAN RULES

The basic problem is the calculation of Green functions, i.e., vacuum expectation values of time-ordered products of interacting fields

$$G(x_1, \ldots, x_n) = \langle 0|\, T[\varphi(x_1) \cdots \varphi(x_n)]\, |0\rangle \tag{6-1}$$

For brevity's sake we shall consider a generic scalar field (or collection of fields) φ, and shall focus our attention on quantum electrodynamics in the next section. It is convenient to collect all the Green functions in a generating functional

$$\begin{aligned} Z(j) &= \sum_{n=0}^{\infty} \frac{i^n}{n!} \int d^4x_1 \cdots d^4x_n\, j(x_1) \cdots j(x_n) G(x_1, \ldots, x_n) \\ &= \langle 0|\, T \exp\left[i \int d^4x\, j(x) \varphi(x) \right] |0\rangle \end{aligned} \tag{6-2}$$

The S matrix may be expressed in the compact form we saw in Eq. (5-38):

$$S = :\exp\left(\int d^4y \left\{\varphi_{\text{in}}(y)\left[(\Box_y + m^2)\frac{1}{Z^{1/2}}\frac{\delta}{\delta j(y)}\right]\right\}\right): Z(j)|_{j=0} \tag{6-3}$$

This relation is itself a generating functional for reduction formulas. The field $\varphi(x)$ satisfies dynamical equations derived from a lagrangian

$$\mathscr{L}(\varphi) = \mathscr{L}_0(\varphi) + \mathscr{L}_{\text{int}}(\varphi) \tag{6-4}$$

where $\mathscr{L}_0$ is a quadratic free lagrangian and $\mathscr{L}_{\text{int}}$ is an interaction lagrangian. For simplicity $\mathscr{L}_{\text{int}}$ is assumed at first to contain no field derivative. Since we are generally unable to find an exact expression of Green functions, we rely on a perturbative method, with $\mathscr{L}_{\text{int}}$ considered as a small perturbation to $\mathscr{L}_0$. Generally $\mathscr{L}_{\text{int}}$ depends on one or several coupling constants and the perturbative expansion will be a power series in this or these coupling constants. Even when the coupling constant is small, as in electrodynamics where $\alpha = e^2/4\pi\hbar c = 1/137$, we do not expect it to converge. As we shall see later, there are some indications that it is, rather, a divergent asymptotic series. In any case, the perturbative expansion may be considered as a formal mathematical series, from which much information may be extracted.

When the theory contains several coupling constants, the natural "small" parameter may be identified as $\hbar$. The corresponding series has also a topological characterization. It is a loopwise expansion, according to the increasing number of independent loops of the associated diagrams. Therefore, the expansion may simultaneously be regarded as a small coupling series, a semiclassical expansion about the free-field configuration, or a topologically well-defined procedure.

As in the previous chapters, we proceed to a heuristic construction and neglect at first various difficulties such as the infinite volume limit, the possible occurrence of ultraviolet divergences, or the proper definition of the lagrangian. These points will be reconsidered in the following.

6-1-1 Self-Interacting Scalar Field

As explained above, we first confine ourselves to a single scalar field φ. We have shown in Chap. 4 that there exists formally a unitary operator $U(t)$ which transforms the free field φ_{in} into the interacting one $\varphi(t)$:

$$\varphi(x) = U^{-1}(t)\varphi_{\text{in}}(x)U(t) \qquad \text{with } t = x^0 \tag{6-5}$$

where

$$U(t) = T\exp\left[-i\int_{-\infty}^{t} dt'\, H_{\text{int}}(t')\right] \tag{6-6}$$

Since the coupling terms contain no field derivative,

$$H_{\text{int}}(t) = -\int_{x^0=t} d^3x\, \mathscr{L}_{\text{int}}[\varphi_{\text{in}}(x)] \tag{6-7}$$

More generally, let $U(t_2, t_1)$ stand for

$$U(t_2, t_1) = T \exp\left[i \int_{t_1}^{t_2} dt' \int d^3x \, \mathscr{L}_{\text{int}}(\mathbf{x}, t') \right]$$
$$U(t, t) = I \tag{6-8}$$

where $\mathscr{L}_{\text{int}}$ is expressed as before in terms of φ_{in}. This operator enjoys the following properties:

$$U(t, -\infty) = U(t) \qquad U(t_1, t_2)U(t_2, t_3) = U(t_1, t_3)$$

We shall now derive the fundamental relation

$$\langle 0| T \exp\left[i \int d^4x \, j(x)\varphi(x) \right] |0\rangle$$
$$= \frac{\langle 0| T \exp\left[i \int d^4x \, \{\mathscr{L}_{\text{int}}[\varphi_{\text{in}}(x)] + j(x)\varphi_{\text{in}}(x)\} \right] |0\rangle}{\langle 0| T \exp\left\{ i \int d^4x \, \mathscr{L}_{\text{int}}[\varphi_{\text{in}}(x)] \right\} |0\rangle} \tag{6-9}$$

or, equivalently, after identification of the term of degree n in j:

$$G(x_1, \ldots, x_n) = \frac{\langle 0| T\varphi_{\text{in}}(x_1) \cdots \varphi_{\text{in}}(x_n) \exp\left\{ i \int d^4x \, \mathscr{L}_{\text{int}}[\varphi_{\text{in}}(x)] \right\} |0\rangle}{\langle 0| T \exp\left\{ i \int d^4x \, \mathscr{L}_{\text{int}}[\varphi_{\text{in}}(x)] \right\} |0\rangle} \tag{6-10}$$

In the right-hand side of Eqs. (6-9) and (6-10), the T symbol acts on the whole expression. This means that after an expansion of the exponential in powers of $\mathscr{L}_{\text{int}}$ all the fields φ_{in} have to be time ordered. The vacuum state $|0\rangle$ is to be understood as the "in" vacuum state $|0 \text{ in}\rangle$.

We assume that the interaction is adiabatically switched off in the remote past; therefore $\lim_{t \to -\infty} U(t) = I$. Let us consider a set of time-ordered points $x_1, \ldots, x_n$, that is, satisfying $x_1^0 > x_2^0 > \cdots > x_n^0$. We then have that

$$\begin{aligned} \langle 0| T\varphi(x_1) \cdots \varphi(x_n) |0\rangle &= \langle 0| \varphi(x_1) \cdots \varphi(x_n) |0\rangle \\ &= \langle 0| U^{-1}(t_1)\varphi_{\text{in}}(x_1) U(t_1, t_2) \varphi_{\text{in}}(x_2) \cdots \\ &\quad \times U(t_{n-1}, t_n) \varphi_{\text{in}}(x_n) U(t_n) |0\rangle \end{aligned}$$

Introducing a very large time t such that $t \gg t_1$, $-t \ll t_n$, we may write

$$U(t_n) = U(t_n, -t) U(-t)$$

and

$$U^{-1}(t_1) = U^{-1}(t) U(t, t_1)$$

Thus

$$\langle 0| T\varphi(x_1) \cdots \varphi(x_n) |0\rangle$$
$$= \langle 0| U^{-1}(t) T\varphi_{\text{in}}(x_1) \cdots \varphi_{\text{in}}(x_n) \exp\left[-i \int_{-t}^{t} dt' \, H_{\text{int}}(t') \right] U(-t) |0\rangle \tag{6-11}$$

This formula remains valid for an arbitrary order of times $x_1^0, \ldots, x_n^0$ within the interval $(-t, t)$. Letting t go to infinity, we get

$$\lim_{t \to \infty} U(-t)|0\rangle = |0\rangle \tag{6-12}$$

while in the same limit $\langle 0| U^{-1}(t)$ must be equal to $\langle 0|$ up to a phase since the vacuum is assumed to be stable. Therefore

$$\begin{aligned}\lim_{t\to\infty} \langle 0| U^{-1}(t) &= \lim_{t\to\infty} \langle 0| U^{-1}(t)|0\rangle\langle 0| \\ &= \lim_{t\to\infty} \frac{1}{\langle 0| U(t)|0\rangle} \langle 0|\end{aligned} \tag{6-13}$$

Inserting (6-12) and (6-13) into (6-11) leads to the desired Eq. (6-10). This derivation is nothing but heuristic, so we may consider Eqs. (6-9) and (6-10) as a mere definition.

It is gratifying—and this was not a trivial point in the early days of relativistic perturbation theory—that Eqs. (6-9) and (6-10) lead to explicitly covariant expressions. From Wick's theorem, the evaluation of T products reduces to combinations of normal products and covariant propagators. The bookkeeping of these combinations may be translated into a diagrammatic language, which we shall now present.

We expand the exponential in the numerator in Eq. (6-10). From now on, the index "in" will be omitted:

$$\begin{aligned}G(x_1, \ldots, x_n)\langle 0| T \exp\left[i \int d^4x\, \mathscr{L}_{\text{int}}(x)\right]|0\rangle \\ = \sum_{p=0}^{\infty} \frac{i^p}{p!} \langle 0| T\varphi(x_1)\cdots\varphi(x_n) \int d^4y_1 \cdots d^4y_p\, \mathscr{L}_{\text{int}}(y_1)\cdots\mathscr{L}_{\text{int}}(y_p)|0\rangle\end{aligned} \tag{6-14}$$

To be explicit, we shall consider the so-called φ^4 theory:

$$\mathscr{L}_{\text{int}}[\varphi(x)] = \frac{-\lambda}{4!} :\varphi^4(x): \tag{6-14a}$$

but all the following reasoning may be easily generalized to an arbitrary polynomial nonderivative interaction. The normalization $1/4!$ has been introduced for later convenience, as it will eliminate combinatorial factors. The minus sign in front of the coupling constant would be the one used in a classical field theory of total lagrangian $(\partial_\mu\varphi)^2/2 - m^2\varphi^2/2 - \lambda\varphi^4/4!$ to insure the stability of the solution $\varphi = 0$ of the equation of motion. After substitution in Eq. (6-14), we get

$$\begin{aligned}G(x_1, \ldots, x_n)\langle 0| T \exp\left[-\frac{i\lambda}{4!} \int d^4x :\varphi^4(x):\right]|0\rangle \\ = \sum_0^{\infty} \frac{(-i\lambda)^p}{p!} \langle 0| T\varphi(x_1)\cdots\varphi(x_n) \int d^4y_1 \cdots d^4y_p \frac{:\varphi^4(y_1):}{4!} \cdots \frac{:\varphi^4(y_p):}{4!}|0\rangle\end{aligned} \tag{6-15}$$

To evaluate the vacuum expectation value of the time-ordered product of free fields we use Wick's theorem [Eq. (4-65)] with the elementary contraction

$$\overline{\varphi(x)\varphi(y)} = \langle 0 | \, T\varphi(x)\varphi(y) \, | 0 \rangle = \int \frac{d^4k}{(2\pi)^4} \frac{i}{k^2 - m^2 + i\varepsilon} e^{-ik\cdot(x-y)} \tag{6-16}$$

We recall that two fields appearing under the same symbol : : must not be contracted. A given set of contractions is conveniently represented by a diagram, the contribution of which results from rules, which we first describe in configuration space. Since only Green functions with an even number of external fields are nonvanishing in the theory (6-14*a*), hereafter we replace n by $2n$. To the pth-order term

$$(-i\lambda)^p \langle 0 | \, T\varphi(x_1)\cdots\varphi(x_{2n}) \frac{:\varphi^4(y_1):}{4!} \cdots \frac{:\varphi^4(y_p):}{4!} \, | 0 \rangle \tag{6-17}$$

we associate a sum of contributions, each of which corresponds to a Feynman diagram. On this diagram, $x_1, \ldots, x_{2n}$ are the endpoints of the external lines and $y_1, \ldots, y_p$ are the vertices. Four lines originate from each vertex, and all the lines join two distinct points: x_a and x_b, or x_a and y_α, or y_α and $y_\beta (\alpha \neq \beta)$. A line joining z_i to z_j ($z = x_a$ or y_α) corresponds to the propagator $\overline{\varphi(z_i)\varphi(z_j)}$. To each vertex, we associate the weight $-i\lambda$. Finally, there is a symmetry factor resulting from the number of possible contractions leading to the same diagram. If we contract all the four fields originating from a monomial $:\varphi^4(y_\alpha):/4!$ with four fields of different arguments, we get a combinatorial factor $4!$ which will compensate the factor $1/4!$. On the other hand, if two lines originating from y_α end at the same point y_β, the factors $1/4!$ of vertices y_α and y_β are only partially suppressed. In the case depicted in Fig. 6-1*a*, the factor is $1/2!$. This results from $1/4! \times 1/4!$ (initial vertex factors) $\times\, 4 \times 3$ (number of possible contractions $z_1 - y_1$, $z_2 - y_1$) $\times\, 4 \times 3$ (contractions $z_3 - y_2$, $z_4 - y_2$) $\times\, 2$ (contractions between y_1 and y_2). Similarly, in the case of Fig. 6-1*b*, the factor is $1/3!$. More generally, the resulting symmetry factor is $1/S$, where S is the order of the symmetry group of the diagram, i.e., of the permutation group of the lines, when the vertices are fixed.

We summarize the Feynman rules for the T product (6-17):

1. Draw all distinct diagrams with $2n$ external points $x_1, \ldots, x_{2n}$ and p vertices $y_1, \ldots, y_p$, and sum all the contributions according to the following rules.

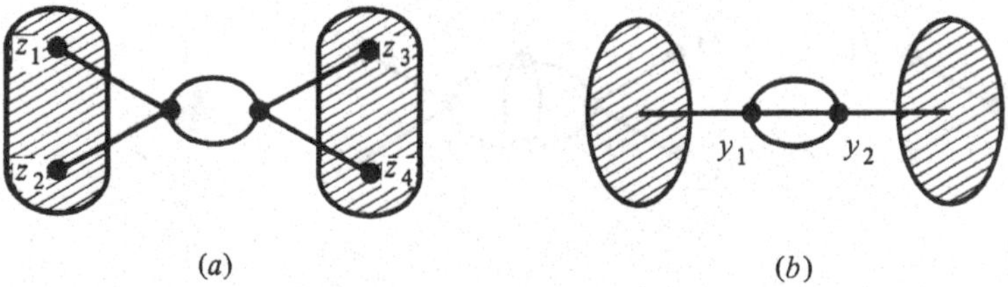

Figure 6-1 Examples of diagrams with a symmetry factor: in case (*a*), $S = 2!$; in case (*b*), $S = 3!$.

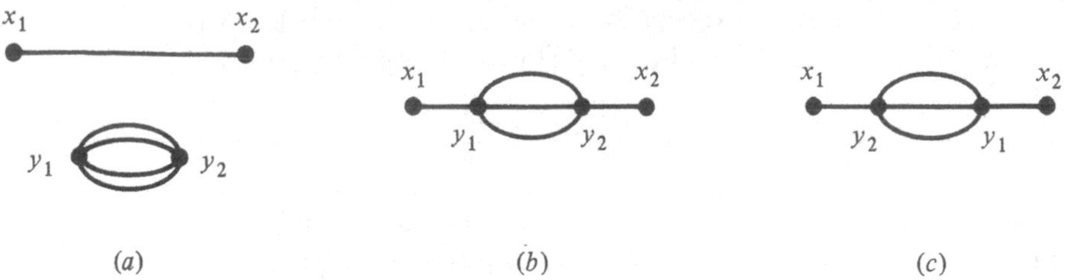

Figure 6-2 Low-order contributions to the two-point function in φ^4 theory.

2. To each vertex attach a factor $-i\lambda$.
3. To each line between z_i and z_j attach a factor $\overline{\varphi(z_i)\varphi(z_j)}$ given by (6-16).
4. Each diagram has to be divided by a symmetry factor S.

As an example, let us examine the two-point function $(-i\lambda)^p \langle 0 | \, T\varphi(x_1)\varphi(x_2)$ $:\varphi^4(y_1):/4! \cdots :\varphi^4(y_p):/4! \, |0\rangle$ to low orders. For $p = 0$, we get $\overline{\varphi(x_1)\varphi(x_2)}$, while there is no $p = 1$ contribution, since we forbid contractions $\overline{\varphi(y)\varphi(y)}$. For $p = 2$, there are three contributions, depicted in Fig. 6-2

(a) $$\frac{(-i\lambda)^2}{4!} \overline{\varphi(x_1)\varphi(x_2)} \, [\overline{\varphi(y_1)\varphi(y_2)}]^4$$

(b) $$\frac{(-i\lambda)^2}{3!} \overline{\varphi(x_1)\varphi(y_1)} \, [\overline{\varphi(y_1)\varphi(y_2)}]^3 \, \overline{\varphi(y_2)\varphi(x_2)}$$

(c) The same with $y_1 \leftrightarrow y_2$

Let us emphasize the importance and the meaning of the word "distinct" in rule 1 above. In configuration space, diagrams are considered as distinct when they are topologically different, all the points x_i and y_j being fixed. For instance, diagrams (*b*) and (*c*) of Fig. 6-2 are distinct, whereas diagrams (*a*) and (*b*) of Fig. 6-3 are not; in the latter case only one of them has to be taken into account.

Let us divide the diagrams into two classes. The first class contains no vacuum-vacuum subdiagram, i.e., subdiagrams (like the one of Fig. 6-2*a*) which are not connected to the external points; the second class includes such subdiagrams. To each diagram of the first class, we may associate a set of second-class diagrams, by addition of vacuum-vacuum subdiagrams. Denoting by the

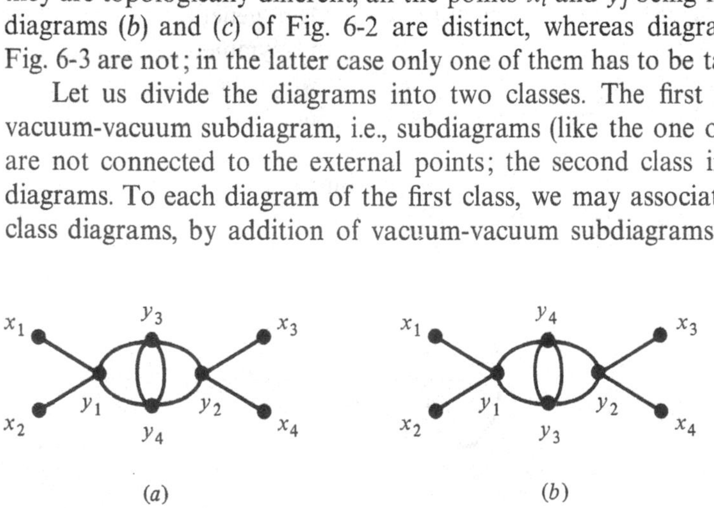

Figure 6-3 Example of indistinguishable diagrams; only one of them has to be taken into account.

superscript (1) the first-class contribution to the Green function, we may write

$$G(x_1,\ldots,x_{2n})\langle 0|\,T\exp\left\{i\int d^4x\,\mathscr{L}_{\text{int}}[\varphi(x)]\right\}|0\rangle$$

$$= \sum_{p=0}^{\infty}\frac{i^p}{p!}\int d^4y_1\cdots d^4y_p\sum_{k=0}^{p}\binom{p}{k}\langle 0|\,T\varphi(x_1)\cdots\varphi(x_{2n})\mathscr{L}_{\text{int}}(y_1)\cdots\mathscr{L}_{\text{int}}(y_k)|0\rangle^{(1)}$$

$$\times\langle 0|\,T\mathscr{L}_{\text{int}}(y_{k+1})\cdots\mathscr{L}_{\text{int}}(y_p)|0\rangle \qquad (6\text{-}18)$$

since the $(p-k)$th-order vacuum-vacuum subdiagram corresponds to the vacuum expectation value of the time-ordered product of $p-k$ interaction lagrangians. Moreover, the combinatorial factor

$$\binom{p}{k} = \frac{p!}{k!(p-k)!}$$

results from the fact that, after integration over $y_1,\ldots,y_p$, all possible permutations of the y give rise to equal contributions. It is easy to see that the vacuum amplitudes $\langle 0|\,T\exp[i\int d^4x\,\mathscr{L}_{\text{int}}(x)]\,|0\rangle$ factorizes on both sides of Eq. (6-18) and

$$G(x_1,\ldots,x_{2n})$$

$$= \sum_{p=0}^{\infty}\frac{i^p}{p!}\int d^4y_1\cdots d^4y_p\,\langle 0|\,T\varphi(x_1)\cdots\varphi(x_{2n})\,\mathscr{L}_{\text{int}}(y_1)\cdots\mathscr{L}_{\text{int}}(y_p)\,|0\rangle^{(1)} \qquad (6\text{-}19)$$

Only diagrams without vacuum subdiagrams contribute to the right-hand side of Eq. (6-19).

It may be simpler to give the Feynman rules in momentum space, i.e., to consider the Fourier transform $\tilde{G}$:

$$\begin{aligned} G(x_1,\ldots,x_{2n}) &= \int\frac{d^4p_1}{(2\pi)^4}\cdots\frac{d^4p_{2n}}{(2\pi)^4}\exp\left(i\sum_1^{2n}p_j\cdot x_j\right)\tilde{G}(p_1,\ldots,p_{2n}) \\ \tilde{G}(p_1,\ldots,p_{2n}) &= \int d^4x_1\cdots d^4x_{2n}\exp\left(-i\sum_{j=1}^{2n}p_j\cdot x_j\right)G(x_1,\ldots,x_{2n}) \end{aligned} \qquad (6\text{-}20)$$

With our conventions, all the p are incoming momenta. Translation invariance entails momentum conservation $\sum_1^{2n}p_j = 0$, which implies

$$\tilde{G}(p_1,\ldots,p_{2n}) = (2\pi)^4\delta^4(p_1+\cdots+p_{2n})G(p_1,\ldots,p_{2n}) \qquad (6\text{-}21)$$

For simplicity we use the same symbol in x or p space. We have to integrate over all space-time points x and y to get $\tilde{G}$ from (6-19) and (6-20). Using the Fourier representation (6-16) of the elementary contraction, the integration over x_i produces a contribution $i(p_i^2 - m^2 + i\varepsilon)^{-1}$ for each external line. We are left with the integrations over the y_j and the k_l, which are the momenta running through the internal lines. Let us concentrate on the contribution attached to a given configuration of lines and vertices for fixed internal momenta and given external configuration arguments $x_1,\ldots,x_{2n}$. Any permutation of the labels $y_1,\ldots,y_p$ of the internal vertices yields the same contribution. However, not all of the $p!$ permutations of the y need generate topologically distinct diagrams in the original configuration space. If this were the case we would get a factor $p!$ compensating the factor $(p!)^{-1}$ appearing in Eq. (6-19). It may happen for some diagrams that two (or more) vertices play identical roles (this is the case in the example of Fig. 6-3a where y_3 and y_4 are such vertices). If the p vertices are divided into groups of $v_1, v_2, \ldots, v_s$ vertices playing the same role ($v_1 + \cdots + v_s = p$), the integration over the y variables yields a degeneracy factor $p!/v_1! \cdots v_s!$. This compensates only partially the $(p!)^{-1}$ factor of Eq. (6-19) and leaves a vertex symmetry factor $1/v_1! \cdots v_s!$, which multiplies the line symmetry factor discussed previously. For instance, the diagram of Fig. 6-3a has, in momentum space, a global symmetry factor $\frac{1}{2} \times \frac{1}{2}$.

Finally, every y integral is of the form

$$\int d^4 y_j \, e^{-iy_j \cdot q_j} = (2\pi)^4 \delta^4(q_j)$$

where q_j denotes the sum of all external or internal momenta flowing into the vertex y_j. Momentum is conserved at each vertex, and is thus globally conserved.

Feynman rules in momentum space may now be written, for the Green function $\tilde{G}(p_1, \ldots, p_{2n})$, as follows:

1. Draw all topologically distinct diagrams with $2n$ external lines of incoming momenta $p_1, \ldots, p_{2n}$ and without vacuum subdiagrams. For each diagram, denote by $k_1, \ldots, k_I$ the momenta of internal lines. In a scalar theory without derivative coupling, the choice of an orientation of the internal lines is irrelevant.
2. To the jth external line, assign the factor $i/(p_j^2 - m^2 + i\varepsilon)$.
3. To the lth internal line, assign $[d^4 k_l/(2\pi)^4][i/(k_l^2 - m^2 + i\varepsilon)]$.
4. To each vertex, assign $(-i\lambda)(2\pi)^4 \delta^4(q_j)$ where q_j is the sum of all incoming momenta at vertex j.
5. Integrate over the k variables the product of all these contributions and divide by the symmetry factor of internal lines and vertices of the diagram.
6. Sum the contributions of all topologically distinct diagrams.

Generally, if V is the number of vertices (called p previously) and I the number of internal lines, there are $V - 1$ conservation rules among the k after separation of the global conservation (6-21), and thus at most $I - V + 1$ nontrivial integrations.

The Feynman rules must be slightly modified when some external lines are connected to no vertex at all. For instance, the single propagator of Fig. 6-4 gives in configuration space the contribution

$$\int \frac{d^4 p}{(2\pi)^4} \frac{i}{p^2 - m^2 + i\varepsilon} e^{-ip \cdot (x_k - x_l)}$$

and in momentum space

$$\frac{i}{p_k^2 - m^2 + i\varepsilon} (2\pi)^4 \delta^4(p_k + p_l)$$

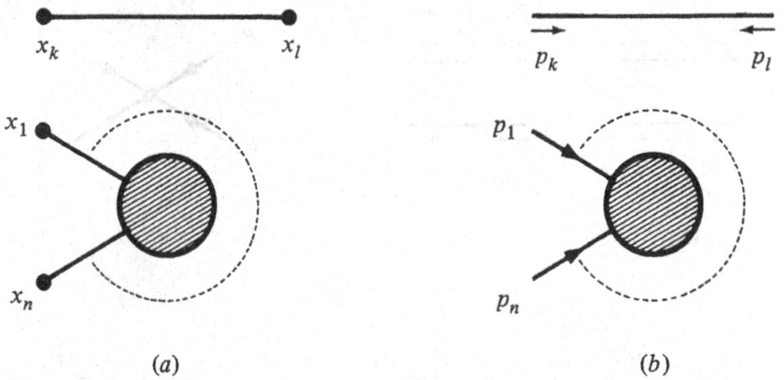

Figure 6-4 Diagrams with a short-circuited propagator, in configuration space (*a*) or in momentum space (*b*).

Let us illustrate these rules to lowest order for the two- and four-point Green functions. Factoring out the global momentum conservation δ function, the contributions of the diagrams of Fig. 6-5 to the two-point function $G_2(p, -p)$ read

$$(a) \qquad \frac{i}{p^2 - m^2 + i\varepsilon}$$

$$(b) \qquad \frac{(-i\lambda)^2}{3!}\left(\frac{i}{p^2 - m^2 + i\varepsilon}\right)^2 \int \frac{d^4k_1\, d^4k_2}{(2\pi)^4(2\pi)^4}$$

$$\times \frac{i^3}{(k_1^2 - m^2 + i\varepsilon)(k_2^2 - m^2 + i\varepsilon)[(p - k_1 - k_2)^2 - m^2 + i\varepsilon]}$$

Similarly, the contributions from the diagrams of Fig. 6-6 to $G_4(p_1, p_2, p_3, p_4)$ (with $p_1 + p_2 + p_3 + p_4 = 0$) are

$$(a) \qquad i^2(2\pi)^4\left[\frac{\delta^4(p_1 + p_2)}{(p_1^2 - m^2 + i\varepsilon)(p_3^2 - m^2 + i\varepsilon)} + \frac{\delta^4(p_1 + p_3)}{(p_1^2 - m^2 + i\varepsilon)(p_2^2 - m^2 + i\varepsilon)} + \frac{\delta^4(p_1 + p_4)}{(p_1^2 - m^2 + i\varepsilon)(p_2^2 - m^2 + i\varepsilon)}\right]$$

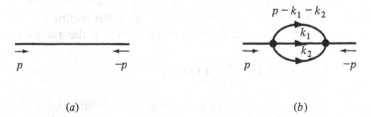

Figure 6-5 Low-order contributions to the two-point function in momentum space.

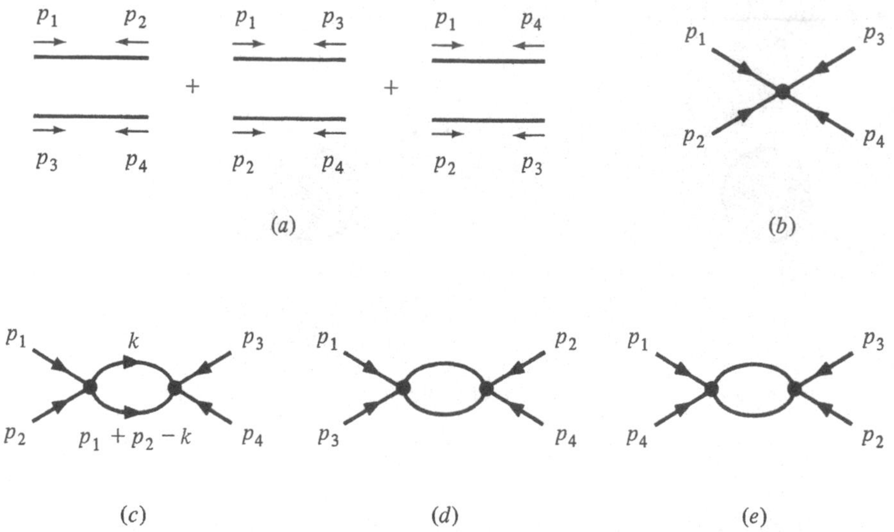

Figure 6-6 Low-order contributions to the four-point function.

(b) $$\frac{i^4(-i\lambda)}{(p_1^2-m^2+i\varepsilon)(p_2^2-m^2+i\varepsilon)(p_3^2-m^2+i\varepsilon)(p_4^2-m^2+i\varepsilon)}$$

(c) $$\frac{1}{2}\prod_{j=1}^{4}\frac{i}{p_j^2-m^2+i\varepsilon}(-i\lambda)^2 \times \int \frac{d^4k}{(2\pi)^4}\frac{i^2}{(k^2-m^2+i\varepsilon)[(p_1+p_2-k)^2-m^2+i\varepsilon]}$$

(d) or (e) Same as (c), with $p_2 \leftrightarrow p_3$ or $p_2 \leftrightarrow p_4$

In the computation of the corresponding S-matrix elements the external propagators $i(p_i^2-m^2+i\varepsilon)^{-1}$ are just compensated by the $i(\Box_{x_i}+m^2)$ operator of the reduction formulas (5-28). In the latter, the last integration over x_i identifies the incoming momentum of the Green function as defined in (6-20) with the physical on-shell momentum of the S-matrix element. Therefore, to get the contribution of a given Feynman diagram to the S matrix, we omit the external propagators and put the external momenta on the mass shell. As we shall see later, this rule must be supplemented by mass and wave-function renormalizations.

When a diagram is disconnected, the Green function G still contains δ functions expressing the conservation of a partial sum of external momenta. In Chap. 5, we have defined the connected Green functions G_c by the recursive expression

$$G(x_1, \ldots, x_n) = \sum_{\cup I_\alpha = I}\prod_\alpha G_c(x_{I_\alpha}) \tag{6-22}$$

where $I \equiv \{1, \ldots, n\}$, $x_I = \{x_1, \ldots, x_n\}$, together with the initial condition $G(x_1, x_2) = G_c(x_1, x_2)$ (for the φ^4 theory). The diagrams without the vacuum subdiagram

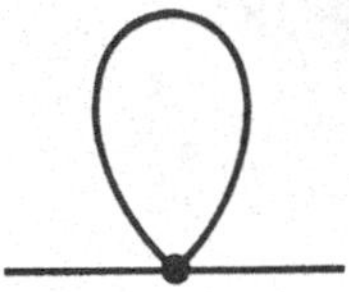

Figure 6-7 A "tadpole" diagram.

involved in the perturbative expansion of $G(x_1, x_2)$ are (topologically) connected. More generally, it will be shown by induction that only connected diagrams contribute to $G_c(x_1, \ldots, x_{2n})$. Let us assume that this is true up to the $(2n-2)$-point function, and let us rewrite (6-22) as

$$G(x_1, \ldots, x_{2n}) = G_c(x_1, \ldots, x_{2n}) + \sum_{\substack{\cup I_\alpha = I \\ |\alpha| > 1}} \prod_\alpha G_c(x_{I_\alpha}) \tag{6-23}$$

After substitution of the diagrammatic expansion for $G(x_1, \ldots, x_{2n})$ and for each G_c in the sum of the right-hand side, each contribution from a disconnected diagram to $G(x_1, \ldots, x_{2n})$ may be identified with a term of the sum, and vice versa. This shows that the algebraic definition of connectivity [Eq. (6-22)] may be identified with the topological one.

Let us return to the role of the normal ordering of the interaction lagrangian in (6-14*a*). Had we kept the ordinary product $\mathscr{L}_{\text{int}} = -\lambda/4!\,\varphi^4(x)$, new diagrams would have emerged, namely, those with contractions of fields pertaining to the same vertex. For instance, the "tadpole" diagram of Fig. 6-7 would appear in the two-point function. In certain circumstances, it may be more convenient to keep the ordinary product and to deal with tadpole diagrams. This is, for instance, what happens when we want to perform a local inhomogeneous transformation on the field (a translation, say), while keeping some symmetry properties. For the φ^4 theory, passing from one ordering to the other simply amounts to a change of the mass term. The latter change is actually infinite, but may be included in the mass renormalization that we shall study soon.

After this derivation of the Feynman rules for a self-interacting scalar field, we shall now do—more rapidly—the same job for more elaborate and physically more interesting theories, namely, for spinor and scalar quantum electrodynamics. Other cases involving fields with an internal symmetry will be also encountered in the following chapters. At any rate, the general method for deriving Feynman rules should be clear. In any doubtful situation, in particular when symmetry factors are dubious, it may be safer to return to the starting point, that is, Eq. (6-10) and Wick's theorem.

6-1-2 Feynman Rules for Spinor Electrodynamics

The total lagrangian describing the interacting system of photons, electrons, and positrons reads

$$\mathscr{L} = \mathscr{L}_0^{\gamma} + \mathscr{L}_0^{e^+,e^-} + \mathscr{L}_{\text{int}} \tag{6-24}$$

where

$$\mathscr{L}_0^{\gamma} = -\frac{1}{4}F^2 + \frac{\mu^2}{2}A^2 - \frac{\lambda}{2}(\partial \cdot A)^2$$

$$\mathscr{L}_0^{e^+,e^-} = \frac{i}{2}\bar{\psi}\gamma^{\mu}\overleftrightarrow{\partial}_{\mu}\psi - m\bar{\psi}\psi \tag{6-25}$$

$$\mathscr{L}_{\text{int}} = -e\bar{\psi}\gamma^{\mu}\psi A_{\mu}$$

The various terms have already been encountered in Chaps. 3 and 4. In the first one, we introduce a massive photon of mass μ much smaller than the electron mass m, and an indefinite metric is needed. The interaction term results from the minimal coupling prescription $i\partial_{\mu} \to i\partial_{\mu} - eA_{\mu}$ in the electron lagrangian.

We shall use distinct notations for the two types of propagators involved in the contractions. The electron-positron propagator will be denoted by a solid line, oriented in the direction of the charge (e) propagation:

$x, \alpha \quad y, \beta$

$$= \langle 0|\, T\psi_{\alpha}(x)\bar{\psi}_{\beta}(y)\,|0\rangle = \overbracket{\psi_{\alpha}(x)\bar{\psi}_{\beta}}(y)$$

$$= \int \frac{d^4k}{(2\pi)^4}\, e^{-ik\cdot(x-y)} \left(\frac{i}{\not{k} - m + i\varepsilon}\right)_{\alpha\beta} \tag{6-26}$$

As is well known, this propagator is not a symmetric function of x and y but satisfies Eq. (3-176). On the other hand, the photon propagator will be marked by a wavy line:

$x, \rho \quad y, \nu$

$$= \langle 0|\, TA_{\rho}(x)A_{\nu}(y)\,|0\rangle = \overbracket{A_{\rho}(x)A_{\nu}}(y)$$

$$= \int \frac{d^4k}{(2\pi)^4}\, e^{-ik\cdot(x-y)}(-i)\left(\frac{g_{\nu\rho} - k_{\nu}k_{\rho}/\mu^2}{k^2 - \mu^2 + i\varepsilon} + \frac{k_{\nu}k_{\rho}/\mu^2}{k^2 - M^2 + i\varepsilon}\right) \tag{6-27}$$

To avoid confusion with the electron mass m, we have set $M^2 \equiv \mu^2/\lambda$. Notice that λ and M^2 are not expected to appear in the physical results.

Because of charge conservation, Green functions have an equal number of ψ and $\bar{\psi}$ fields:

$$G(x_1, \ldots, x_n; x_{n+1}, \ldots, x_{2n}; y_1, \ldots, y_p)$$
$$= \langle 0|\, T\psi(x_1)\cdots\psi(x_n)\bar{\psi}(x_{n+1})\cdots\bar{\psi}(x_{2n})A_{\nu_1}(y_1)\cdots A_{\nu_p}(y_p)\,|0\rangle \tag{6-28}$$

For brevity's sake, we have omitted spinor and vector indices in the arguments of G. As in Eq. (6-14), we derive an expression for G in terms of in-fields:

$$G(x_1 \cdots x_{2n}; y_1 \cdots y_p)$$
$$= \frac{\langle 0|\, T\psi_{\text{in}}(x_1)\cdots\bar{\psi}_{\text{in}}(x_{2n})A^{\text{in}}_{\nu_1}(y_1)\cdots A^{\text{in}}_{\nu_p}(y_p)\exp i\int d^4z\, \mathscr{L}_{\text{int}}(z)\,|0\rangle}{\langle 0|\, T\exp i\int d^4z\, \mathscr{L}_{\text{int}}(z)\,|0\rangle} \tag{6-29}$$

where $\mathscr{L}_{\text{int}}(z) = \mathscr{L}_{\text{int}}[\psi_{\text{in}}(z), \bar{\psi}_{\text{in}}(z), A_{\text{in}}(z)]$ is supposed to be normal ordered. The "in" indices will be omitted in the sequel. Expanding (6-29) in powers of e and using Wick's theorem leads to Feynman diagrams built out of the propagators (6-26) and (6-27) and a vertex

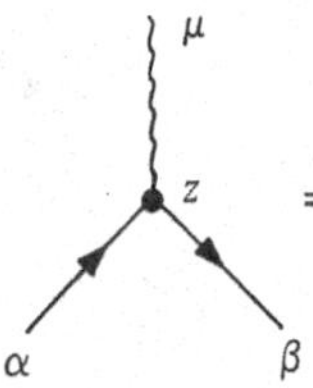

$$= -ie(\gamma_\mu)_{\beta\alpha} \tag{6-30}$$

This vertex possesses one vector and two spinor indices which are contracted with the corresponding indices of the photons and fermion propagators. As in the scalar case, the role of the denominator in Eq. (6-27) is to eliminate the vacuum-vacuum subdiagrams.

Let us collect the signs introduced by the Wick rules for fermions. From charge conservation, there are two kinds of fermion lines in the diagram, closed loops or open lines ending at a point $x_i (1 \le i \le n)$, and coming from a point $x_{k_i} (n + 1 \le k_i \le 2n)$. A closed loop is composed of a sequence of propagators

$$\psi(z_1)\bar{\psi}(z_k)\psi(z_k)\bar{\psi}(z_l) \cdots \psi(z_q)\bar{\psi}(z_1) \tag{6-31}$$

between fields appearing in interaction lagrangians. Since the latter commute under the T symbol, we may reorder the T product as

$$\langle 0| T \cdots [\bar{\psi}(z_1) A\!\!\!/(z_1)\psi(z_1)]\,[\bar{\psi}(z_k) A\!\!\!/(z_k)\psi(z_k)] \cdots [\bar{\psi}(z_q) A\!\!\!/(z_q)\psi(z_q)] \cdots |0\rangle$$

without introducing any sign. The product of contractions (6-31) is obtained after the permutation of $\bar{\psi}(z_1)$ with an odd number of fields. Therefore, a minus sign is to be associated to each fermion loop.

On the other hand, the open lines define a permutation of the points $x_i : x_1 x_{k_1} x_2 x_{k_2} \cdots x_n x_{k_n}$ where we recall that x_{k_p} is the origin of the line ending at x_p. There is then an extra sign, which is the signature of the permutation

$$(1, 2, \ldots, 2n) \to (1, k_1, 2, k_2, \ldots, n, k_n)$$

Diagrams which differ only by the orientation of a fermion loop contribute both only when they are topologically distinct. For instance, the two diagrams of Fig. 6-8 are clearly identical; only one of them contributes to $G(y_1, y_2)$. On the

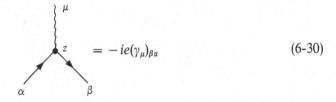

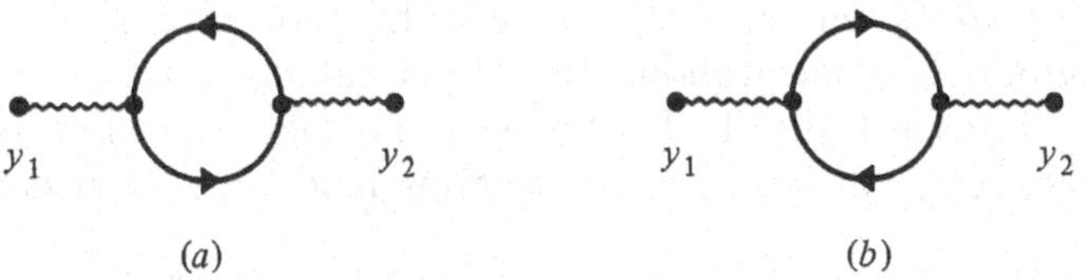

Figure 6-8 Example of identical diagrams in spinor electrodynamics; only one of them has to be retained.

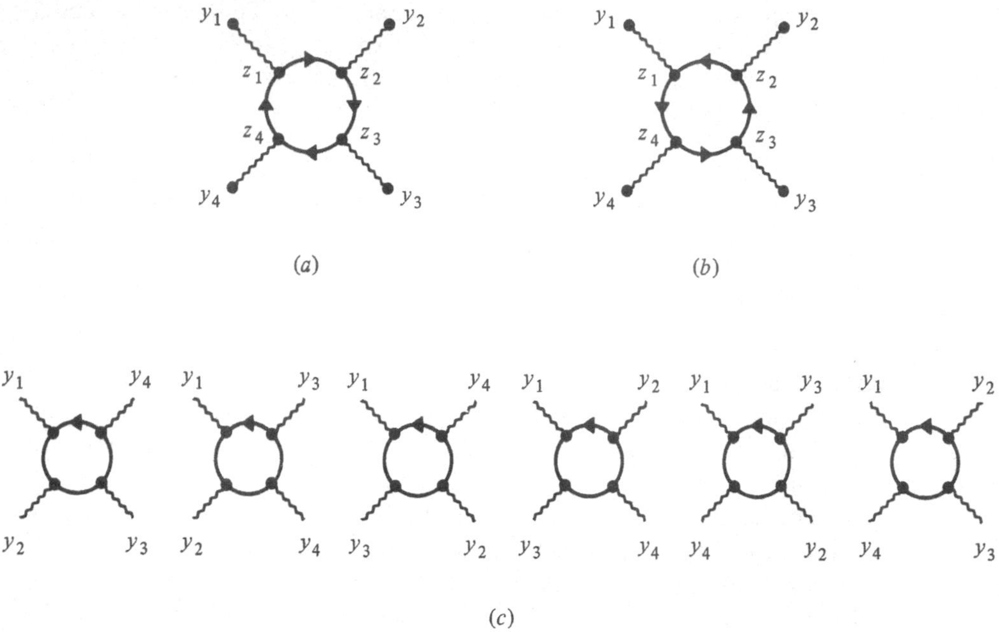

Figure 6-9 The four-photon amplitude to lowest order. The diagrams (*a*) and (*b*) are not identical in configuration space. After summation over the z variables, we are left with the six distinct diagrams depicted in (*c*).

contrary, the first two diagrams of Fig. 6-9 contribute both to the four-photon function (photon-photon scattering). They correspond to the two distinct sets of contractions:

(*a*) $$\bar{\psi}(z_1)A\!\!\!/(z_1)\psi(z_1)\bar{\psi}(z_2)A\!\!\!/(z_2)\psi(z_2)\bar{\psi}(z_3)A\!\!\!/(z_3)\psi(z_3)\bar{\psi}(z_4)A\!\!\!/(z_4)\psi(z_4)$$

and

(*b*) $$\bar{\psi}(z_1)A\!\!\!/(z_1)\psi(z_1)\bar{\psi}(z_2)A\!\!\!/(z_2)\psi(z_2)\bar{\psi}(z_3)A\!\!\!/(z_3)\psi(z_3)\bar{\psi}(z_4)A\!\!\!/(z_4)\psi(z_4)$$

where $A(z_1)$ is contracted with $A(y_1)$, etc. In configuration space, the four-point function $G(y_1, \ldots, y_4)$ receives contributions from $4! \times 3!$ distinct diagrams, obtained from those of Fig. 6-9*a* through permutations of z_1, z_2, z_3, z_4 and y_2, y_3, y_4; the diagram of Fig. 6-9*b*, viz, is derived from Fig. 6-9*a* by permuting $y_2 \leftrightarrow y_4$, $z_2 \leftrightarrow z_4$. When carrying out the z integrations, the $4!$ permutations over the z give a factor that compensates the $1/4!$ coming from the expansion of $\exp i \int dz\, \mathscr{L}_{\text{int}}(z)$. We are left with six distinct diagrams (and no factor) (see Fig. 6-9*c*).

The representation in momentum space is obtained by Fourier transformation. For a connected function, we set

$$G_c(x_1, \ldots, x_n; x_{n+1}, \ldots, x_{2n}; y_1, \ldots, y_p)$$
$$= \int \prod_1^{2n} \frac{d^4 p_i}{(2\pi)^4} \prod_1^{p} \frac{d^4 q_j}{(2\pi)^4} \exp\left(i \sum_1^{2n} p_i \cdot x_i + i \sum_1^{p} q_j \cdot y_j\right)$$
$$\times \tilde{G}_c(p_1, \ldots, p_n; p_{n+1}, \ldots, p_{2n}; q_1, \ldots, q_p) \tag{6-32}$$

where all momenta are incoming. Let us now summarize the Feynman rules for the computation of $\tilde{G}_c$. Draw all possible topologically distinct diagrams and associate the factors enumerated in Table 6-1. Then contract all spinor indices along the fermion lines (and thus compute a trace for each closed loop) and all vector indices along the photon lines. Finally, carry out all the integrations over internal momenta.

Table 6-1 Feynman rules for spinor electrodynamics

1. *External lines*

Incoming fermion

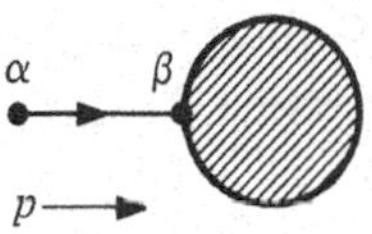

$$\left(\frac{i}{\not p - m + i\varepsilon}\right)_{\beta\alpha}$$

Outgoing fermion

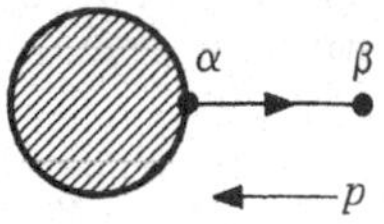

$$\left[\frac{i}{(-\not p) - m + i\varepsilon}\right]_{\beta\alpha}$$

Incident photon

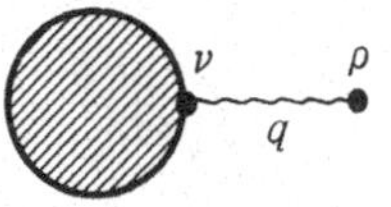

$$-i\left(\frac{g_{\rho\nu} - q_\rho q_\nu/\mu^2}{q^2 - \mu^2 + i\varepsilon} + \frac{q_\rho q_\nu/\mu^2}{q^2 - M^2 + i\varepsilon}\right)$$

2. *Vertices*

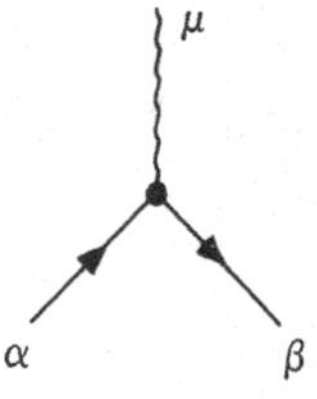

$$-ie(\gamma_\mu)_{\beta\alpha}(2\pi)^4 \delta^4(\textstyle\sum \text{incoming momenta})$$

3. *Propagators*

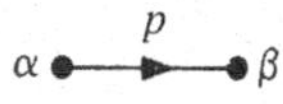

$$\frac{d^4 p}{(2\pi)^4}\left(\frac{i}{\not p - m + i\varepsilon}\right)_{\beta\alpha}$$

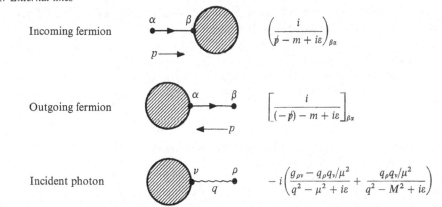

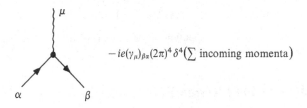

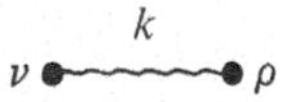

$$\frac{d^4 k}{(2\pi)^4}(-i)\left(\frac{g_{\nu\rho} - k_\nu k_\rho/\mu^2}{k^2 - \mu^2 + i\varepsilon} + \frac{k_\nu k_\rho/\mu^2}{k^2 - M^2 + i\varepsilon}\right)$$

4. A minus sign for every closed fermion loop and a global sign, depending on the configuration of external lines, computed as indicated above

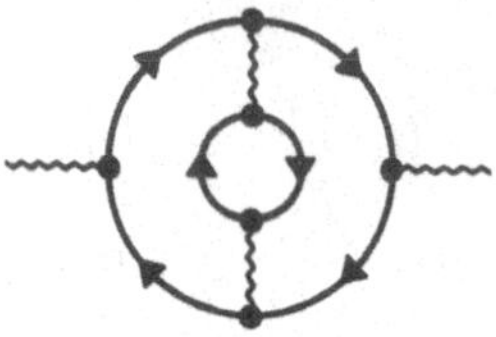

Figure 6-10 A diagram of spinor electrodynamics without any symmetry factor.

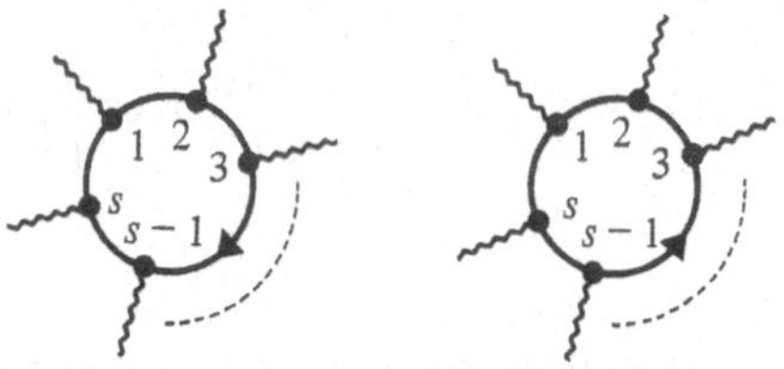

Figure 6-11 Two diagrams with opposite orientations of the spinor loop.

For a diagram of order r (diagram with r vertices), the integrations over the r variables z and their permutations cancel the factor $1/r!$ coming from the expansion of $\exp i \int d^4z\, \mathscr{L}_{\text{int}}(z)$. Therefore, with this set of rules, when only topologically distinct diagrams are taken into account, no symmetry factor occurs in fermion electrodynamics. The reader is invited to compare the diagram of Figs. 6-8*a* or 6-10 with the analogous scalar diagrams of Figs. 6-6*c* or 6-3*a*. The former have no symmetry factor when a single orientation is selected for each fermion loop; the latter have weights respectively of $1/2!$ and $(1/2!)^2$.

Furry's theorem All the diagrams containing a fermion loop with an odd number of vertices may be omitted in the computation of a Green function. Indeed, the two orientations of the loop (Fig. 6-11) lead to contributions of opposite sign. To show that, we write the contribution corresponding to the first orientation as

$$G_1 = \operatorname{tr}\left[\gamma_{\mu_1} S_F(z_1, z_s)\gamma_{\mu_s} S_F(z_s, z_{s-1})\gamma_{\mu_{s-1}} \cdots \gamma_{\mu_2} S_F(z_2, z_1)\right]$$

and we appeal to the existence of a matrix C satisfying [compare with (3-176)]

$$\begin{aligned} C S_F(x, y) C^{-1} &= S_F^T(y, x) \\ C\gamma_\mu C^{-1} &= -\gamma_\mu{}^T \end{aligned} \tag{6-33}$$

in order to insert CC^{-1} between propagators and rewrite

$$\begin{aligned} G_1 &= (-1)^s \operatorname{tr}\left[\gamma_{\mu_1}^T S_F^T(z_s, z_1)\gamma_{\mu_s}^T S_F^T(z_{s-1}, z_s)\gamma_{\mu_{s-1}}^T \cdots \gamma_{\mu_2}^T S_F^T(z_1, z_2)\right] \\ &= (-1)^s \operatorname{tr}\left[\gamma_{\mu_1} S_F(z_1, z_2)\gamma_{\mu_2} \cdots \gamma_{\mu_s} S_F(z_s, z_1)\right] \end{aligned} \tag{6-34}$$

Up to the sign $(-1)^s$, this is exactly the contribution of the second orientation. For odd s, the two contributions cancel, which proves the statement.

6-1-3 Electron-Electron and Electron-Positron Scattering

As an illustration of the diagrammatic machinery, let us compute the cross section for electron-electron scattering to lowest order. We start from the reduction formula derived in the previous chapter:

$$\begin{aligned} S_{fi}^c = \left(\frac{-i}{Z_2^{1/2}}\right)^4 &\int d^4x_1\, d^4x_2\, d^4x_1'\, d^4x_2'\, e^{i(p_1'\cdot x_1' + p_2'\cdot x_2' - p_1\cdot x_1 - p_2\cdot x_2)} \\ &\times \bar{u}(p_1', \varepsilon_1')(i\not{\partial}_{x_1'} - m)\bar{u}(p_2', \varepsilon_2')(i\not{\partial}_{x_2'} - m)\langle 0|\, T\psi(x_1')\psi(x_2')\bar{\psi}(x_1)\bar{\psi}(x_2)\,|0\rangle^c \\ &\times (-i\overleftarrow{\not{\partial}}_{x_1} - m)u(p_1, \varepsilon_1)(-i\overleftarrow{\not{\partial}}_{x_2} - m)u(p_2, \varepsilon_2) \end{aligned} \tag{6-35}$$

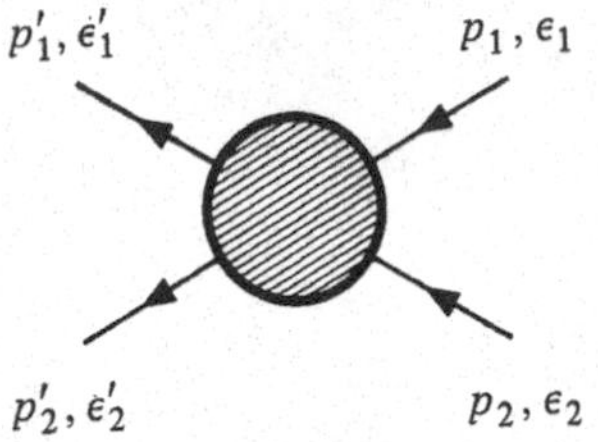

Figure 6-12 General form of the diagrams contributing to electron-electron scattering.

The notations are summarized on Fig. 6-12; notice that here p'_1, p'_2 are the outgoing electron momenta. To lowest order, the two diagrams of Fig. 6-13 contribute to the Green function, and $Z_2 = 1$. The Dirac operators and the integrations in (6-35) compensate the factors attached to the external lines and put the external momenta on the mass shell. Use of momentum conservation at every vertex implies in this case that no integration is left. The two diagrams have a relative minus sign and

$$S^c_{fi} = (2\pi)^4 \delta^4(p'_1 + p'_2 - p_1 - p_2)(-ie)^2$$
$$\times \Bigg[-\bar{u}(p'_1, \varepsilon'_1)\gamma^\nu u(p_1, \varepsilon_1)(-i)\left(\frac{g_{\nu\rho} - k^{(1)}_\nu k^{(1)}_\rho/\mu^2}{k^{(1)2} - \mu^2} + \frac{k^{(1)}_\nu k^{(1)}_\rho/\mu^2}{k^{(1)2} - M^2} \right) \bar{u}(p'_2, \varepsilon'_2)\gamma^\rho u(p_2, \varepsilon_2)$$
$$+ \bar{u}(p'_2, \varepsilon'_2)\gamma^\nu u(p_1, \varepsilon_1)(-i)\left(\frac{g_{\nu\rho} - k^{(2)}_\nu k^{(2)}_\rho/\mu^2}{k^{(2)2} - \mu^2} + \frac{k^{(2)}_\nu k^{(2)}_\rho/\mu^2}{k^{(2)2} - M^2} \right) \bar{u}(p'_1, \varepsilon'_1)\gamma^\rho u(p_2, \varepsilon_2) \Bigg] \tag{6-36}$$

where

$$k^{(1)} \equiv p_1 - p'_1 = p'_2 - p_2$$
$$k^{(2)} \equiv p_1 - p'_2 = p'_1 - p_2$$

Therefore,

$$k^{(1)}_\nu \bar{u}(p'_1, \varepsilon'_1)\gamma^\nu u(p_1, \varepsilon_1) = \bar{u}(p'_1, \varepsilon'_1)(\not{p}_1 - \not{p}'_1)u(p_1, \varepsilon_1) = 0$$

and

$$k^{(2)}_\nu \bar{u}(p'_2, \varepsilon'_2)\gamma^\nu u(p_1, \varepsilon_1) = \bar{u}(p'_2, \varepsilon'_2)(\not{p}_1 - \not{p}'_2)u(p_1, \varepsilon_1) = 0$$

The terms in $k_\nu k_\rho$ in the propagators give no contribution, and in the computation of the elastic cross section we may take the $\mu = 0$ limit without encountering any infrared divergence:

$$S^c_{fi} = -ie^2(2\pi)^4 \delta^4(p'_1 + p'_2 - p_1 - p_2)\mathscr{T}$$

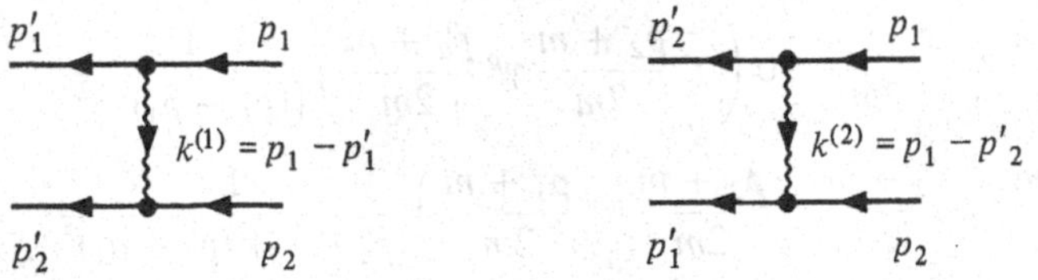

Figure 6-13 Lowest-order contributions to electron-electron scattering.

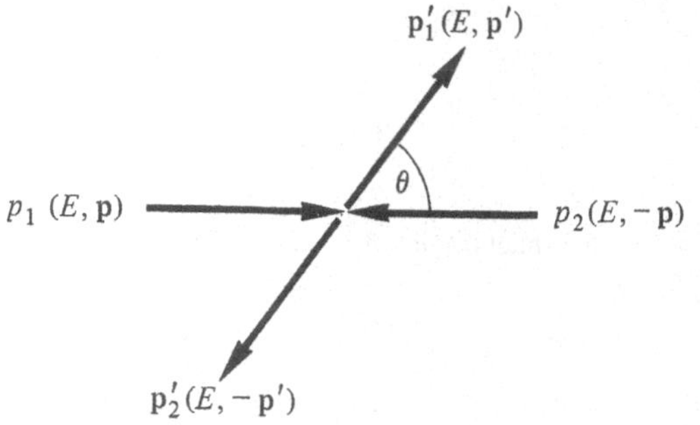

Figure 6-14 Kinematics of electron-electron scattering in the center of mass frame.

with

$$\mathscr{T} = \frac{\bar{u}(p_1', \varepsilon_1')\gamma^\nu u(p_1, \varepsilon_1)\bar{u}(p_2', \varepsilon_2')\gamma_\nu u(p_2, \varepsilon_2)}{(p_1 - p_1')^2} - \frac{\bar{u}(p_2', \varepsilon_2')\gamma^\nu u(p_1, \varepsilon_1)\bar{u}(p_1', \varepsilon_1')\gamma_\nu u(p_2, \varepsilon_2)}{(p_1 - p_2')^2} \tag{6-37}$$

The antisymmetry of the initial or final states is now manifest. There is no factor due to the identity of the particles, but of course the total cross section will be obtained by integrating over only half of the final phase space.

Let us compute the differential cross section for unpolarized initial beams, when the final polarizations are not observed. The kinematics of the reaction in the center of mass frame is represented in Fig. 6-14, where θ is the scattering angle in this frame, the energy E is conserved, and we denote $|\mathbf{p}| = |\mathbf{p}'| = p = \sqrt{E^2 - m^2}$. Using the general formula (5-13), with due care paid to the fermion normalization, we get

$$d\sigma = \frac{m^4}{[(p_1 \cdot p_2)^2 - m^4]^{1/2}} \int \frac{d^3p_1'}{(2\pi)^3 E_1'} \frac{d^3p_2'}{(2\pi)^3 E_2'} (2\pi)^4 \delta^4(p_1' + p_2' - p_1 - p_2) e^4 \overline{|\mathscr{T}|^2} \tag{6-38}$$

In the square modulus of the amplitude, an average and a sum over initial and final polarizations are understood. After integration of the energy momentum conservation δ function

$$\frac{d\sigma}{d\Omega} = \frac{m^2 e^4}{4E^2(2\pi)^2} \overline{|\mathscr{T}|^2} \tag{6-39}$$

with

$$\begin{aligned}
\overline{|\mathscr{T}|^2} &= \frac{1}{4} \sum_{\substack{\varepsilon_1\varepsilon_2 \\ \varepsilon_1'\varepsilon_2'}} |\mathscr{T}|^2 \\
&= \frac{1}{4}\left\{ \operatorname{tr}\left(\gamma_\nu \frac{\not{p}_1 + m}{2m} \gamma_\rho \frac{\not{p}_1' + m}{2m}\right) \operatorname{tr}\left(\gamma^\nu \frac{\not{p}_2 + m}{2m} \gamma^\rho \frac{\not{p}_2' + m}{2m}\right) \frac{1}{[(p_1' - p_1)^2]^2} \right. \\
&\quad - \operatorname{tr}\left(\gamma_\nu \frac{\not{p}_1 + m}{2m} \gamma_\rho \frac{\not{p}_2' + m}{2m} \gamma^\nu \frac{\not{p}_2 + m}{2m} \gamma^\rho \frac{\not{p}_1' + m}{2m}\right) \frac{1}{(p_1' - p_1)^2 (p_2' - p_1)^2} \\
&\quad \left. + (p_1' \leftrightarrow p_2') \right\}
\end{aligned} \tag{6-40}$$

The traces are readily evaluated

$$\operatorname{tr}\left[\gamma_\nu(\not p_1 + m)\gamma_\rho(\not p_1' + m)\right] = 4(p_{1\nu}p_{1\rho}' - g_{\nu\rho}p_1\cdot p_1' + p_{1\rho}p_{1\nu}' + m^2 g_{\nu\rho})$$

$$\operatorname{tr}\left[\gamma_\nu(\not p_1 + m)\gamma_\rho(\not p_1' + m)\right]\operatorname{tr}\left[\gamma^\nu(\not p_2 + m)\gamma^\rho(\not p_2' + m)\right]$$
$$= 32[(p_1\cdot p_2)^2 + (p_1\cdot p_2')^2 + 2m^2(p_1\cdot p_2' - p_1\cdot p_2)]$$

and

$$\gamma_\nu(\not p_1 + m)\gamma_\rho(\not p_2' + m)\gamma^\nu = -2\not p_2'\gamma_\rho\not p_1 + 4m(p_{2\rho}' + p_{1\rho}) - 2m^2\gamma_\rho$$

$$\operatorname{tr}\left[\gamma_\nu(\not p_1 + m)\gamma_\rho(\not p_2' + m)\gamma^\nu(\not p_2 + m)\gamma^\rho(\not p_1' + m)\right] = -32[(p_1\cdot p_2)^2 - 2m^2 p_1\cdot p_2]$$

Therefore,

$$\overline{|\mathscr{T}|^2} = \frac{1}{2m^4}\left\{\frac{(p_1\cdot p_2)^2 + (p_1\cdot p_2')^2 + 2m^2(p_1\cdot p_2' - p_1\cdot p_2)}{[(p_1' - p_1)^2]^2}\right.$$
$$+ \frac{(p_1\cdot p_2)^2 + (p_1\cdot p_1')^2 + 2m^2(p_1\cdot p_1' - p_1\cdot p_2)}{[(p_2' - p_1)^2]^2}$$
$$\left. + 2\frac{(p_1\cdot p_2)^2 - 2m^2 p_1\cdot p_2}{(p_1' - p_1)^2(p_2' - p_1)^2}\right\} \tag{6-41}$$

We may express all invariants in terms of the energy E and scattering angle θ:

$$p_1\cdot p_2 = 2E^2 - m^2$$
$$p_1\cdot p_1' = E^2(1 - \cos\theta) + m^2\cos\theta$$
$$p_1\cdot p_2' = E^2(1 + \cos\theta) - m^2\cos\theta$$

We finally obtain the Møller formula (1932)

$$\frac{d\sigma}{d\Omega} = \frac{\alpha^2(2E^2 - m^2)^2}{4E^2(E^2 - m^2)^2}\left[\frac{4}{\sin^4\theta} - \frac{3}{\sin^2\theta} + \frac{(E^2 - m^2)^2}{(2E^2 - m^2)^2}\left(1 + \frac{4}{\sin^2\theta}\right)\right] \tag{6-42}$$

In the ultrarelativistic limit, where $m/E \to 0$,

$$\left.\frac{d\sigma}{d\Omega}\right|_{\text{ur}} = \frac{\alpha^2}{E^2}\left(\frac{4}{\sin^4\theta} - \frac{2}{\sin^2\theta} + \frac{1}{4}\right) = \frac{\alpha^2}{4E^2}\left(\frac{1}{\sin^4\theta/2} + \frac{1}{\cos^4\theta/2} + 1\right) \tag{6-43}$$

and in the nonrelativistic limit, $E^2 \simeq m^2$, $v^2 = (E^2 - m^2)/E^2$,

$$\left.\frac{d\sigma}{d\Omega}\right|_{\text{nr}} = \left(\frac{\alpha}{m}\right)^2\frac{1}{4v^4}\left(\frac{4}{\sin^4\theta} - \frac{3}{\sin^2\theta}\right)$$
$$= \left(\frac{\alpha}{m}\right)^2\frac{1}{16v^4}\left(\frac{1}{\sin^4\theta/2} + \frac{1}{\cos^4\theta/2} - \frac{1}{\sin^2\theta/2\cos^2\theta/2}\right) \tag{6-44}$$

a result obtained first by Mott (1930). We let the reader reexpress these cross sections in the laboratory frame. It is also instructive to compare (6-44) with the nonrelativistic limit of the Rutherford formula for Coulomb scattering, where we set $Z = 1$ with a reduced mass equal to $m/2$.

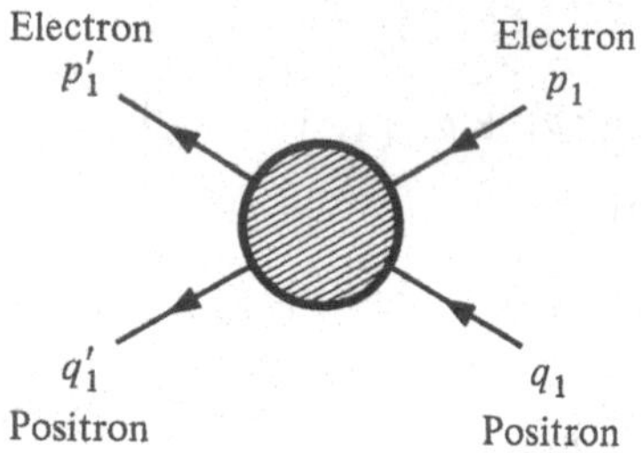

Figure 6-15 General form of the diagrams contributing to electron-positron scattering.

Let us now consider electron-positron scattering. The kinematics and lowest-order diagrams are depicted in Figs. 6-15 and 6-16. Polarization indices are omitted and in Fig. 6-16 four-momenta are oriented according to the charge flow. The scattering amplitude may then be obtained from (6-37) by substituting

$$\begin{aligned} u(p_1) &\to u(p_1) \qquad & p_1 &\to p_1 \\ u(p'_1) &\to u(p'_1) \qquad & p'_1 &\to p'_1 \\ u(p_2) &\to v(q'_1) \qquad & p_2 &\to -q'_1 \\ u(p'_2) &\to v(q_1) \qquad & p'_2 &\to -q_1 \end{aligned} \tag{6-45}$$

and by changing the sign of the amplitude. The center of mass cross section is readily computed:

$$\frac{d\sigma}{d\Omega} = \frac{m^4 e^4}{4E^2(2\pi)^2} \overline{|\mathcal{T}|^2}$$

with

$$\begin{aligned} \overline{|\mathcal{T}|^2} = \frac{1}{2m^4} \bigg\{ & \frac{(p_1 \cdot q'_1)^2 + (p_1 \cdot q_1)^2 - 2m^2(p_1 \cdot q_1 - p_1 \cdot q'_1)}{[(p'_1 - p_1)^2]^2} \\ & + \frac{(p_1 \cdot q'_1)^2 + (p_1 \cdot p'_1)^2 + 2m^2(p_1 \cdot p'_1 + p_1 \cdot q'_1)}{[(p_1 + q_1)^2]^2} \\ & + 2\, \frac{(p_1 \cdot q'_1)^2 + 2m^2 p_1 \cdot q'_1}{(p_1 - p'_1)^2 (p_1 + q_1)^2} \bigg\} \end{aligned} \tag{6-46}$$

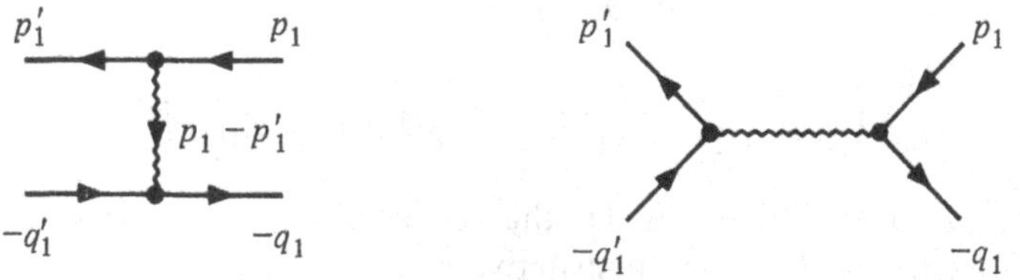

Figure 6-16 Lowest-order contributions to electron-positron scattering.

It is then straightforward to write the expression of the cross section:

$$\frac{d\sigma^{e^-e^+}}{d\Omega} = \frac{\alpha}{2E^2}\left[\frac{5}{4} - \frac{8E^4 - m^4}{E^2(E^2 - m^2)(1 - \cos\theta)} + \frac{(2E^2 - m^2)^2}{2(E^2 - m^2)^2(1 - \cos\theta)^2} + \frac{2E^4(-1 + 2\cos\theta + \cos^2\theta) + 4E^2m^2(1 - \cos\theta)(2 + \cos\theta) + 2m^4\cos^2\theta}{16E^4}\right] \tag{6-47}$$

The ultrarelativistic and nonrelativistic limits are respectively

$$\frac{d\sigma^{e^-e^+}}{d\Omega_{\text{ur}}} = \frac{\alpha^2}{8E^2}\left[\frac{1 + \cos^4\theta/2}{\sin^4\theta/2} + \frac{1}{2}(1 + \cos^2\theta) - 2\frac{\cos^4\theta/2}{\sin^2\theta/2}\right] \tag{6-48}$$

$$\frac{d\sigma^{e^-e^+}}{d\Omega_{\text{nr}}} = \left(\frac{\alpha}{m}\right)^2 \frac{1}{16v^4\sin^4\theta/2} \tag{6-49}$$

Notice that the annihilation diagram does not contribute to the latter case. These expressions are due to Bhabha (1936).

The results of Eqs. (6-42) and (6-47) may be compared with experimental data. At low energies we show in Fig. 6-17 some data of Ashkin, Page, and Woodward (1954) for electron-electron scattering at 90 degrees. Møller's formula (6-42) gives a good agreement, in contrast with the corresponding expression [to be derived below, Eq. (6-65)] which would apply to spinless particles. Results for

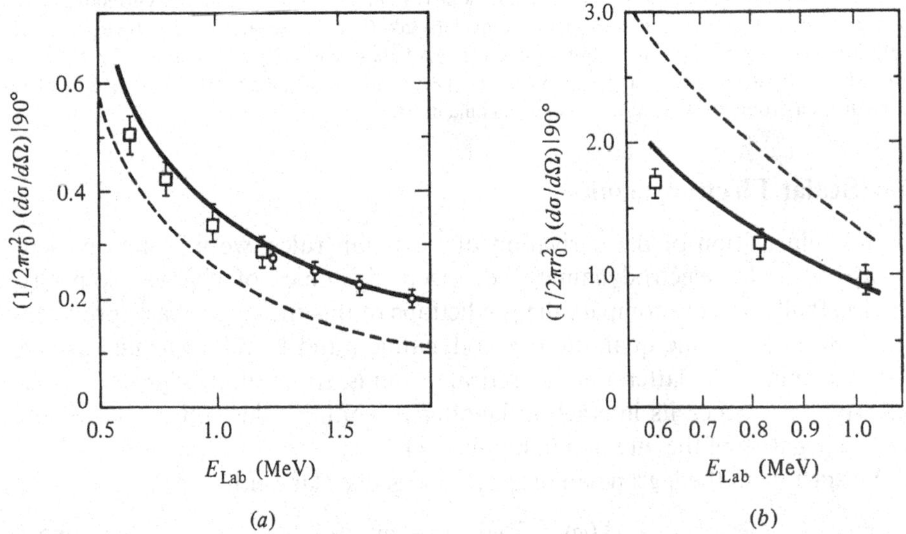

Figure 6-17 Experimental data for electron-electron and electron-positron scattering at $\theta = 90°$ as a function of the incident electron energy in the laboratory frame. (*a*) Electron-electron scattering. The solid line represents the Møller formula, the broken one the Møller formula when the spin terms are omitted. (*b*) Electron-positron scattering. The solid line is the Bhabha formula, the broken one the prediction when annihilation terms are deleted. (*From A. Ashkin, L. A. Page, and W. M. Woodward, Phys. Rev., vol. 94, p. 357, 1974.*)

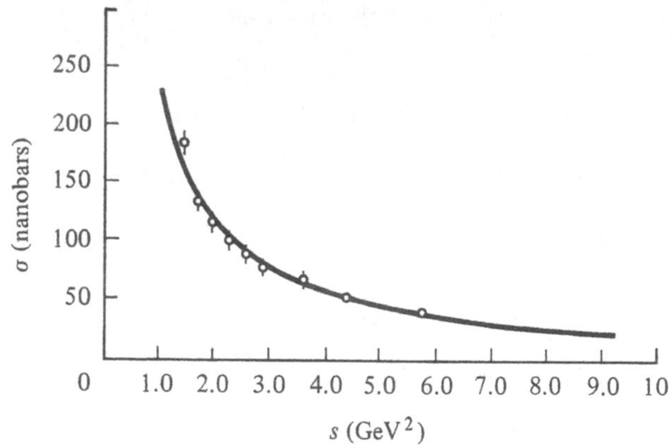

Figure 6-18 The cross section for Bhabha scattering at high energy, for scattering angle $45° < \theta < 135°$ as a function of total energy momentum square s. The solid line is calculated from quantum electrodynamics with first-order radiative corrections. (*From K. Strauch in "Proceedings of the Sixth Symposium on Electron and Photon Interactions at High Energies," edited by H. Rollnik and W. Pfeil, North-Holland, Amsterdam, 1974.*)

electron-positron scattering are also fitted quite well by Bhabha's cross section. Omission of the annihilation term would spoil this agreement. The energy of the incident particle in the laboratory frame plotted on the abscissa is chosen in the intermediate range where neither the nonrelativistic nor the ultrarelativistic approximation is valid. The numerical values show a significant departure from the ratio 2:1 between e^-e^- and e^-e^+ cross sections, expected on the basis of a naive argument of indiscernability of the two electrons.

The predictions of these lowest-order calculations may be amended by radiative corrections (see Chap. 7) to allow a comparison with more recent data taken at the extreme relativistic energies of the electron-positron storage rings. For square center of mass energies in the range 2 to 9 GeV^2 the agreement between theory and experiment is spectacular, as demonstrated in Fig. 6-18. The data are from a group working at the Adone machine in Frascati.

6-1-4 Scalar Electrodynamics

As a last illustration of the derivation of Feynman rules, we will now present the case of scalar electrodynamics, i.e., electrodynamics of charged spin-zero particles. It allows us to compare the predictions of this theory for basic processes with those of fermionic quantum electrodynamics, and therefore to understand the role of spin in the latter. On theoretical grounds, scalar electrodynamics is an interesting case, since its interaction lagrangian contains derivatives and differs from the negative of the interaction hamiltonian.

We start from the lagrangian of a free charged scalar field

$$\mathscr{L}_0(\varphi) = \partial_\mu \varphi^\dagger \partial^\mu \varphi - m^2 \varphi^\dagger \varphi \tag{6-50}$$

(caution, no factor $\frac{1}{2}$!) and perform the minimal substitution $\partial_\mu \varphi \to (\partial_\mu + ieA_\mu)\varphi$. Adding the electromagnetic lagrangian $\mathscr{L}_0^\gamma(A)$ of Eq. (6-25) yields

$$\mathscr{L} = \mathscr{L}_0(\varphi) + \mathscr{L}_0^\gamma(A) + \mathscr{L}_{\text{int}} \tag{6-51a}$$

$$\mathscr{L}_{\text{int}} = -ieA^\mu(\varphi^\dagger \overleftrightarrow{\partial}_\mu \varphi) + e^2 A^2 \varphi^\dagger \varphi \tag{6-51b}$$

Since the interaction lagrangian contains derivatives, the canonical momenta are modified:

$$\begin{aligned} \pi^\dagger &= \frac{\partial \mathscr{L}}{\partial(\partial_0 \varphi^\dagger)} = \partial^0 \varphi + ieA^0 \varphi \\ \pi &= \frac{\partial \mathscr{L}}{\partial(\partial_0 \varphi)} = \partial^0 \varphi^\dagger - ieA^0 \varphi^\dagger \end{aligned} \tag{6-52}$$

The hamiltonian reads

$$\begin{aligned} H &= H_0^\gamma + H_0(\varphi, \pi) + H_{\text{int}} \\ H_0(\varphi, \pi) + H_{\text{int}} &= \int d^3x \, (\pi\dot{\varphi} + \pi^\dagger \dot{\varphi}^\dagger - \mathscr{L}_0 - \mathscr{L}_{\text{int}}) \end{aligned} \tag{6-53}$$

Here H_0 is the free hamiltonian of Eq. (3-76):

$$H_0(\varphi, \pi) = \int d^3x \, (\pi^\dagger \pi + \partial_k \varphi^\dagger \, \partial_k \varphi + m^2 \, \varphi^\dagger \varphi) \tag{6-54}$$

and H_{int} takes the form (the spatial index k takes the values 1, 2, 3)

$$H_{\text{int}} = \int d^3x \, \mathscr{H}_{\text{int}} = \int d^3x \left[ieA^0(\pi^\dagger \varphi^\dagger - \pi\varphi) + ieA^k(\varphi^\dagger \overleftrightarrow{\partial_k} \varphi) - e^2 A^k A_k \varphi^\dagger \varphi\right] \tag{6-55}$$

If we substitute (6-52) into (6-55), we observe that $\mathscr{H}_{\text{int}}(\varphi) = -\mathscr{L}_{\text{int}}(A, \varphi) - e^2 A^{02} \varphi^\dagger \varphi$. On the other hand, we may perform a canonical transformation to the interaction representation; φ, π, A are changed into φ_{in}, π_{in}, A_{in}, and $\pi_{\text{in}} = \partial_0 \varphi_{\text{in}}^\dagger$. Thus, in this representation

$$\begin{aligned} \mathscr{H}_{\text{int}} &= ieA_{\text{in}}^0(\pi_{\text{in}}^\dagger \varphi_{\text{in}}^\dagger - \pi_{\text{in}} \varphi_{\text{in}}) + ieA_{\text{in}}^k \varphi_{\text{in}}^\dagger \overleftrightarrow{\partial_k} \varphi_{\text{in}} - e^2 A_{\text{in},k} A_{\text{in}}^k \varphi_{\text{in}}^\dagger \varphi_{\text{in}} \\ &= ieA_{\text{in}}^\mu \varphi_{\text{in}}^\dagger \overleftrightarrow{\partial_\mu} \varphi_{\text{in}} - e^2 A_{\text{in}}^2 \varphi_{\text{in}}^\dagger \varphi_{\text{in}} + e^2 A_{\text{in}}^{02} \varphi_{\text{in}}^\dagger \varphi_{\text{in}} \\ &= -\mathscr{L}_{\text{int}}(A_{\text{in}}, \varphi_{\text{in}}) + e^2 A_{\text{in}}^{02} \varphi_{\text{in}}^\dagger \varphi_{\text{in}} \end{aligned} \tag{6-56}$$

The hamiltonian and minus the lagrangian differ by a noncovariant term (notice the change of sign in front of this term) which seems to jeopardize the covariance of the theory.

Consider, however, a typical Green function of the theory

$$\begin{aligned} &G(x_1, \ldots, x_n; x_{n+1}, \ldots, x_{2n}; y_1, \ldots, y_p) \\ &= \langle 0 | \, T\varphi(x_1) \cdots \varphi(x_n) \varphi^\dagger(x_{n+1}) \cdots \varphi^\dagger(x_{2n}) A(y_1) \cdots A(y_p) \, | 0 \rangle \\ &= \frac{\langle 0 | \, T\varphi_{\text{in}}(x_1) \cdots A_{\text{in}}(y_p) \exp\left[-i \int d^4z \, \mathscr{H}_{\text{int}}(z)\right] | 0 \rangle}{\langle 0 | \, T \exp\left[-i \int d^4z \, \mathscr{H}_{\text{int}}(z)\right] | 0 \rangle} \end{aligned} \tag{6-57}$$

There is a second source of noncovariant terms in the chronological products of derivative terms. To end up with covariant results, we expect these two types of noncovariant terms to cancel exactly. This is indeed what happens.

We give a formal proof of this property, since we sidestep the problems connected with the ultraviolet divergences of the perturbation expansion.

Let us look more closely at the noncovariant contractions induced by the derivative terms. In the following, the "in" indices will be dropped. We list the relevant propagators

$$\langle 0|\, T\varphi(x)\varphi^\dagger(y)\,|0\rangle = \langle 0|\, T\varphi(y)\varphi^\dagger(x)\,|0\rangle = i\int \frac{d^4k}{(2\pi)^4}\frac{e^{-ik\cdot(x-y)}}{k^2-m^2+i\varepsilon} \tag{6-58a}$$

$$\langle 0|\, T\partial_\mu\varphi(x)\varphi^\dagger(y)\,|0\rangle = \partial^x_\mu\langle 0|\, T\varphi(x)\varphi^\dagger(y)\,|0\rangle \tag{6-58b}$$

$$\langle 0|\, T\partial_\mu\varphi(x)\partial_\nu\varphi^\dagger(y)\,|0\rangle = \partial^x_\mu\partial^y_\nu\langle 0|\, T\varphi(x)\varphi^\dagger(y)\,|0\rangle - ig_{\mu 0}g_{\nu 0}\delta^4(x-y) \tag{6-58c}$$

Since the φ propagators are the only ones with noncovariant terms, we may consider the field A as classical in the following argument and handle all Green functions of φ and $\varphi^\dagger$ by introducing a generating functional

$$Z(j) \equiv \langle 0|\, T\exp -i\int d^4z\,[\mathscr{H}_{\text{int}}(y) + j^*(y)\varphi(y) + j(y)\varphi^\dagger(y)]\,|0\rangle$$

$$= \langle 0|\, T\sum_r \frac{e^r}{r!}\int d^4z_1\cdots d^4z_r\; A^{\mu_1}(z_1)\varphi^\dagger(z_1)\overleftrightarrow{\partial}_{\mu_1}\varphi(z_1)\cdots A^{\mu_r}(z_r)\varphi^\dagger(z_r)\overleftrightarrow{\partial}_{\mu_r}\varphi(z_r)$$

$$\times\exp -i\int d^4z\,(-e^2A^2\varphi^\dagger\varphi + e^2A^{02}\varphi^\dagger\varphi + j^*\varphi + j\varphi^\dagger)\,|0\rangle \tag{6-59}$$

Only the derivative part of the interaction hamiltonian has been expanded in powers of e, since it gives rise to the noncovariant contraction (6-58c). We introduce a further notation: the covariant chronological product $\hat{T}$ coincides with the T product of Eqs. (6-58), but for the last term of (6-58c):

$$\langle 0|\, \hat{T}\partial_\mu\varphi(x)\partial_\nu\varphi^\dagger(y)\,|0\rangle \equiv \partial^x_\mu\partial^y_\nu\langle 0|\, T\varphi(x)\varphi^\dagger(y)\,|0\rangle \tag{6-60}$$

Let us examine the contribution of s noncovariant contractions $-ig_{\mu 0}g_{\nu 0}\delta^4(x-y)$. There are $\binom{r}{2s} = r!/(2s)!\,(r-2s)!$ choices of $(2s)$ vertices among r, and $(2s)!/2^s s!$ ways of contracting them into s pairs. Each noncovariant contraction of two vertices gives

$$eA^\mu(x)\varphi(x)[\langle 0|\, T\partial_\mu\varphi^\dagger(x)\partial_\nu\varphi(y)\,|0\rangle - \langle 0|\, \hat{T}\partial_\mu\varphi^\dagger(x)\partial_\nu\varphi(y)\,|0\rangle](-e)A^\nu(y)\varphi^\dagger(y) + (\varphi\leftrightarrow\varphi^\dagger)$$
$$= 2ie^2A_0^2\varphi^\dagger(x)\varphi(x)\delta^4(x-y)$$

where we have implicitly assumed Wick ordering. Therefore, we obtain

$$Z(j) = \langle 0|\, \hat{T}\sum_{r=0}^{\infty}\frac{e^r}{r!}\left[\int d^4z'\, A^\mu(z')\varphi^\dagger(z')\overleftrightarrow{\partial}_\mu\varphi(z')\right]^r \sum_{s=0}^{\infty}\frac{(ie^2)^s}{s!}\left[\int d^4z''\, A_0^2(z'')\varphi^\dagger(z'')\varphi(z'')\right]^s$$

$$\times\exp -i\int d^4z\,(-e^2A^2\varphi^\dagger\varphi + e^2A_0^2\varphi^\dagger\varphi + j\varphi^\dagger + j^*\varphi)\,|0\rangle \tag{6-61}$$

We see that the noncovariant contractions may be exponentiated and their contribution compensates the noncovariant vertex $e^2A_0^2\varphi^\dagger\varphi$.

We may finally write, with a covariant $\hat{T}$ product (6-60),

Table 6-2 Feynman rules for scalar electrodynamics

Photon	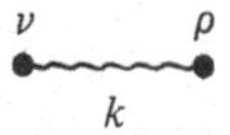	$-i\left(\frac{g_{\nu\rho}-k_\nu k_\rho/\mu^2}{k^2-\mu^2+i\varepsilon}+\frac{k_\nu k_\rho/\mu^2}{k^2-M^2+i\varepsilon}\right)\frac{d^4k}{(2\pi)^4}$
Boson	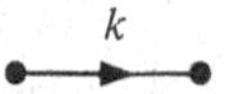	$\frac{i}{k^2-m^2+i\varepsilon}\frac{d^4k}{(2\pi)^4}$
Vertices	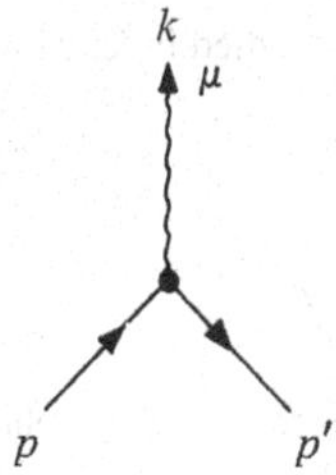	$-ie(p_\mu+p'_\mu)(2\pi)^4\delta^4(p-p'-k)$
	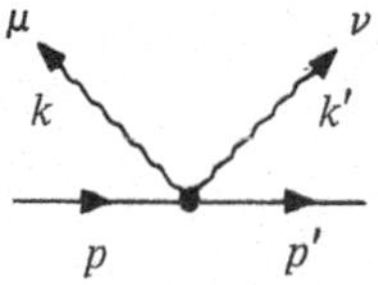	$2ie^2g_{\mu\nu}(2\pi)^4\delta^4(p-p'-k-k')$

$$G(x_1,\ldots,y_p)=\frac{\langle 0|\,\hat{T}\varphi_{\text{in}}(x_1)\cdots A_{\text{in}}(y_p)\exp i\int d^4z\,\mathscr{L}_{\text{int}}(z)\,|0\rangle}{\langle 0|\,\hat{T}\exp i\int d^4z\,\mathscr{L}_{\text{int}}(z)\,|0\rangle} \tag{6-62}$$

and a covariant lagrangian

$$\mathscr{L}_{\text{int}}=-ieA^\mu_{\text{in}}\varphi^\dagger_{\text{in}}\overleftrightarrow{\partial}_\mu\varphi_{\text{in}}+e^2A^2_{\text{in}}\varphi^\dagger_{\text{in}}\varphi_{\text{in}} \tag{6-63}$$

The resulting Feynman rules are summarized in Table 6-2. The factor two in the second type of vertex ("seagull term") comes from the fact that the quadratic term in A is A^2 and not $A^2/2!$. Also remember that in scalar electrodynamics a symmetry factor $\frac{1}{2}$ arises whenever a pair of photons is contracted between a pair of seagull vertices (Fig. 6-19). On the contrary, closed scalar loops being oriented give no symmetry factor, as long as only topologically distinct loops are taken into account.

Figure 6-19 A diagram of scalar electrodynamics affected with a symmetry factor $1/S=\frac{1}{2}$.

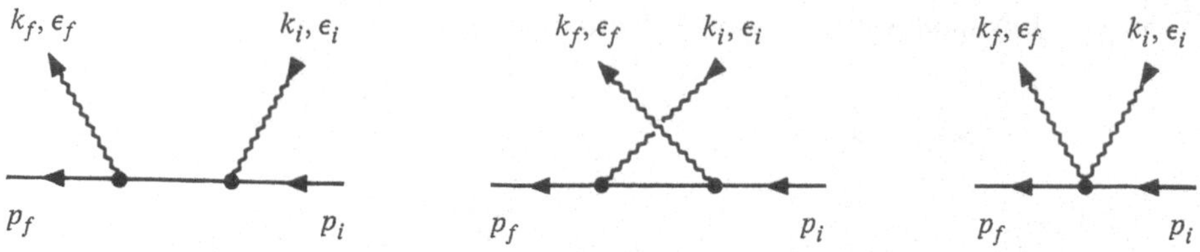

Figure 6-20 Lowest-order diagrams for Compton scattering in scalar electrodynamics.

We shall now review briefly some basic processes, already studied in fermionic electrodynamics.

(*a*) *Compton scattering*
With the same notations as in Sec. 5-2-1, we have

$$\frac{d\sigma}{d\Omega} = \frac{\alpha^2}{4m^2}\left(\frac{k_f}{k_i}\right)^2 |\mathscr{T}|^2$$

To lowest order, the three diagrams of Fig. 6-20 contribute, but if we choose polarization vectors satisfying $\varepsilon_i \cdot p_i = \varepsilon_f \cdot p_i = 0$, only the third one is nonvanishing:

$$|\mathscr{T}|^2 = 4(\varepsilon_f \cdot \varepsilon_i)^2$$

and

$$\frac{d\sigma}{d\Omega} = \frac{\alpha^2}{m^2}\left(\frac{k_f}{k_i}\right)^2 (\varepsilon_f \cdot \varepsilon_i)^2 \tag{6-64}$$

to be compared with the Klein-Nishina formula (5-112).

(*b*) *Scattering of two identical charges*
We use the same notations and diagrams as in Sec. 6-1-3 and Figs. 6-12 and 6-13. In the center of mass frame

$$\frac{d\sigma}{d\Omega} = \frac{\alpha^2}{16E^2}|\mathscr{T}|^2$$

with

$$|\mathscr{T}|^2 = 4\left[\frac{p_1 \cdot p_2 + p_1 \cdot p_2'}{(p_1 - p_1')^2} + \frac{p_1 \cdot p_2 + p_1 \cdot p_1'}{(p_1 - p_2')^2}\right]$$

and

$$\frac{d\sigma}{d\Omega} = \frac{\alpha^2}{4E^2}\left(\frac{2E^2 - m^2}{E^2 - m^2}\right)^2 \left[\frac{2}{\sin^2\theta} - \frac{E^2 - m^2}{2E^2 - m^2}\right]^2 \tag{6-65}$$

to be compared with Eq. (6-42).

(*c*) *Pair annihilation into two photons*
We use the same notations as in Sec. 5-2-2 and the same diagrams as in Fig. 6-20:

$$\frac{d\sigma}{d\Omega} = \frac{\alpha^2}{4}\frac{m + E_2}{(m + E_2 - p_2\cos\theta)^2 p_2}|\mathscr{T}|^2 \tag{6-66}$$

with $|\mathscr{T}|^2 = 4(\varepsilon_f \cdot \varepsilon_i)^2$. If final polarizations are not observed

$$\sigma_{\text{tot}} = \frac{\pi r_0^2}{1 + \gamma}\left[\frac{2(3 + \gamma)}{\sqrt{\gamma^2 - 1}} - \frac{4\gamma}{\gamma^2 - 1}\ln\left(\gamma + \sqrt{\gamma^2 - 1}\right)\right] \tag{6-67}$$

with $\gamma = E_2/m$ and $r_0 = \alpha/m$. In the nonrelativistic limit, $\gamma \to 1$ and $\sigma_{\text{tot}} \sim 2\pi r_0^2/v_2$, that is, twice as large as the cross section $e^+e^- \to \gamma\gamma$.

(*d*) *Bremsstrahlung*
The same notations as in Sec. 5-2-4 are used; the diagrams are depicted in Fig. 6-21. We pick a polarization vector ε such that $\varepsilon \cdot k = 0$, $\varepsilon^0 = 0$. Then the third diagram does not contribute:

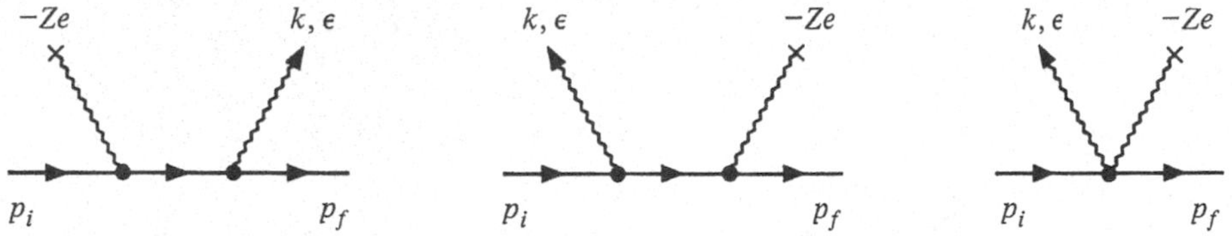

Figure 6-21 Lowest-order diagrams for bremsstrahlung in scalar electrodynamics.

$$S_{fi} = \frac{-ie^2(-Ze)}{(\mathbf{p}_f + \mathbf{k} - \mathbf{p}_i)^2} 2\pi\delta(p_f^0 + k^0 - p_i^0) \left[\frac{2p_f \cdot \varepsilon(p_i^0 + p_f^0 + k^0)}{(p_f + k)^2 - m^2} + \frac{2p_i \cdot \varepsilon(p_i^0 + p_f^0 - k^0)}{(p_i - k)^2 - m^2} \right]$$

$$d\sigma = \frac{Z^2\alpha^3}{\pi^2} \frac{p_f}{p_i} \frac{\omega \, d\omega \, d\Omega_\gamma \, d\Omega}{q^4} \left(\frac{p_f \cdot \varepsilon}{p_f \cdot k} p_i^0 + \frac{p_i \cdot \varepsilon}{p_i \cdot k} p_f^0 \right)^2$$

Summing over the polarizations

$$d\sigma_{\text{unpol}} = \frac{Z^2\alpha^3}{(2\pi)^2} \frac{p_f}{p_i} \frac{1}{q^4} \frac{d\omega}{\omega} \, d\Omega_\gamma \, d\Omega$$
$$\times \left[\frac{p_f^2 \sin^2\theta_f}{(E_f - p_f \cos\theta_f)^2} 4E_i^2 + \frac{p_i^2 \sin^2\theta_i}{(E_i - p_i \cos\theta_i)^2} 4E_f^2 + \frac{2p_i p_f \sin\theta_i \sin\theta_f \cos\varphi}{(E_i - p_i \cos\theta_i)(E_f - p_f \cos\theta_f)} 4E_i E_f \right] \tag{6-68}$$

6-2 DIAGRAMMATICS

The Born diagrams—or tree diagrams—encountered so far were very simple, since they implied no integration over internal momenta. In the general case, however, we have to deal with more complicated diagrams. This section will be devoted to a survey of the general properties and terminology concerning Feynman diagrams. For the sake of simplicity, most of this material will be presented in the case of a scalar theory of a single neutral field, except when explicitly stated.

6-2-1 Loopwise Expansion

The loopwise perturbative expansion, i.e., the expansion according to the increasing number of independent loops of connected Feynman diagrams, may be identified with an expansion in powers of $\hbar$. By definition, the number of independent loops is nothing but the number of independent internal four-momenta in the diagrams, when conservation laws at each vertex have been taken into account. For a connected diagram, with I internal lines and V vertices, we have V δ^4 functions expressing this conservation, and after extracting the conservation of the incoming momenta, we are left with $V - 1$ constraints. Therefore, the number of independent momenta or loops is

$$L = I - (V - 1) \tag{6-69}$$

Notice that L is not the number of faces or closed circuits that may be drawn

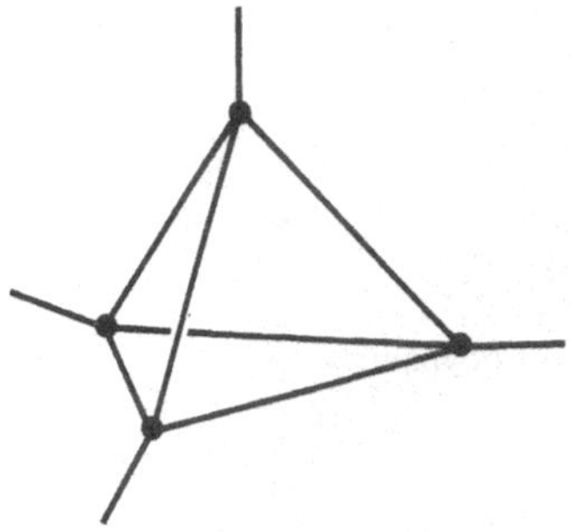

Figure 6-22 The tetrahedron diagram has only three independent loops.

on the internal lines of the diagram. For instance, the tetrahedron diagram of Fig. 6-22 has four closed circuits but only three independent loops.

To find the connection between L and the power of $\hbar$, we collect all factors $\hbar$. We leave aside the factor $\hbar$ that gives the mass term a correct dimension. In other words, the Klein-Gordon equation should read $[\partial_x^2 + (mc/\hbar)^2]\varphi = 0$, indicating that the mass term is of quantum origin. This phenomenon is disregarded in the sequel. There are thus two origins of such factors. First the commutation (or anticommutation) relations imply a factor $\hbar$, for example, $[\varphi(\mathbf{x}), \pi(\mathbf{y})] = i\hbar\delta^3(\mathbf{x} - \mathbf{y})$, which leads to a factor $\hbar$ in each propagator

$$\langle T\varphi(x)\varphi(y)\rangle = \int \frac{d^4k}{(2\pi)^4} e^{ik\cdot(x-y)} \frac{i\hbar}{k^2 - m^2 + i\varepsilon}$$

Second, the evolution operator $e^{-iHt/\hbar}$ contains $\hbar$ explicitly, and so does $\exp\left[(i/\hbar)\int \mathscr{L}_{\text{int}}(\varphi_{\text{in}})\, d^4z\right]$. Thus, we have a factor $\hbar$ for each propagator and $\hbar^{-1}$ for each vertex. The total power for a diagram with E external lines is $\hbar^{I+E-V} = \hbar^{E-1+L}$. For a fixed number of external lines, i.e., for a given Green function, the announced result follows.

In units where $c = 1$, a scalar field φ has dimension $[\varphi] \sim (\text{energy/length})^{1/2}$, the coupling constant λ of $\mathscr{L}_{\text{int}} = \lambda\varphi^4$ has dimension $[\hbar]^{-1} \sim (\text{energy} \times \text{length})^{-1}$, and a spinor field has dimension $[\psi] \sim (\text{energy})^{1/2}/\text{length}$. This gives the right dimension $[\hbar]$ to the actions

$$I = \int d^4z \left[\frac{1}{2}(\partial\varphi)^2 - \frac{1}{2}\left(\frac{m}{\hbar}\right)^2 \varphi^2 + \lambda\varphi^4\right] \quad \text{or} \quad I = \int d^4z\, \bar{\psi}\left(i\partial\!\!\!/ - \frac{M}{\hbar}\right)\psi$$

Of course, the lowest-order diagrams in $\hbar$ are the Born diagrams without any loop, such as those computed in the preceding section.

The reader may then wonder why this topological loopwise expansion coincides with the expansion according to powers of the coupling constant, in a theory with a single coupling constant. This is because in such a theory there exist auxiliary relations between V, the number of vertices (the power of λ), and L. Take, for example, the φ^4 theory; counting the total number of incident lines at each vertex tells us that

$$4V = E + 2I$$

for a diagram with E external lines. Eliminating I with the help of (6-69), we get $L - 1 + E/2 = V$(E is even).

6-2-2 Truncated and Proper Diagrams

We introduce some terminology that will prove useful in the sequel.

The truncated functions are defined through the multiplication of Green functions in momentum space (without the δ^4 function of total energy momentum) by the inverse two-point functions pertaining to each external line:

$$G^{(n)}_{\text{trunc}}(p_1, \ldots, p_n) \equiv \prod_{k=1}^{n} [G^{(2)}(p_k, -p_k)]^{-1} G^{(n)}(p_1, \ldots, p_n) \qquad n > 2 \qquad (6\text{-}70)$$

The two-point function $G^{(2)}$ is referred to as the complete propagator. For $p^2 \sim m^2$, we have $[G^{(2)}(p, -p)]^{-1} \sim (iZ)^{-1}(p^2 - m^2)$ where Z is the wave function renormalization introduced in Chap. 5. Hence, up to powers of Z, the on-shell values of these truncated functions are the quantities entering the reduction formulas. For instance, the connected part of the matrix element of Eq. (5-28) reads

$$\langle p_1, \ldots, p_n \text{ out} | q_1, \ldots, q_m \text{ in} \rangle_c = Z^{(n+m)/2} G^{(n+m)}_{\text{trunc}}(-p_1, \ldots, -p_n, q_1, \ldots, q_m) \Big|_{p_i^2 = q_j^2 = m^2}$$
$$\times (2\pi)^4 \delta^4(\Sigma p_i - \Sigma q_j)$$

The perturbative expansion of truncated functions is expressed in terms of truncated diagrams, i.e., that have no self-energy part on their external lines. Moreover, in the Feynman rules, no factor or propagator is attributed to the external lines (see Fig. 6-23 for illustration). Finally, if we restore the factors $\hbar$, as indicated in the last subsection, L-loop truncated diagrams get a factor $\hbar^{L-1}$.

We finally define proper or one-particle irreducible diagrams. Those are truncated connected diagrams which remain connected when an arbitrary internal line is cut (see Fig. 6-23).

The proper functions, defined by their perturbative expansion in terms of proper diagrams, are the building blocks of perturbation theory, since the integrations over internal momenta may be carried out independently in each proper subdiagram of a given diagram. For the same reason, they play a central role

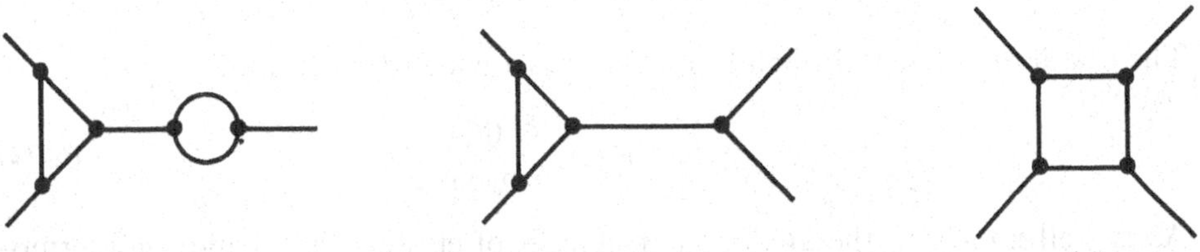

(a) (b) (c)

Figure 6-23 Examples of a nontruncated diagram (a), of a truncated but not proper diagram (b), and of a proper diagram (c). In cases (b) and (c), no factor is ascribed to the external lines.

in the renormalization program (see Chap. 8), since it is necessary and sufficient to make them finite to get rid of all ultraviolet divergences.

Beside this topological definition, the proper functions may be defined algebraically. As shown by Jona-Lasinio, the generating functional of proper vertices is the Legendre transform of the generating functional of connected diagrams. The latter, denoted $G_c(j)$, has been defined in Chap. 5 as the logarithm of the generating functional of all Green's functions

$$Z(j) = \langle 0 | T \exp \left[i \int d^4x \, \varphi(x) j(x) \right] | 0 \rangle = e^{G_c(j)} \tag{6-71a}$$

We proved in Sec. 6-1 that it corresponds indeed to connected Feynman diagrams with

$$G_c(j) = \sum_{n=1}^{\infty} \frac{i^n}{n!} \int d^4x_1 \cdots d^4x_n \, G_c(x_1, \ldots, x_n) j(x_1) \cdots j(x_n) \tag{6-71b}$$

We now construct the Legendre transform of G_c as follows. Let $\varphi_c(x, j)$ be the functional of j defined through

$$\varphi_c(x, j) = \frac{\delta}{i\delta j(x)} G_c(j) \tag{6-72}$$

and let us assume that the relation $\varphi_c(x) = \varphi_c(x, j)$ may be inverted to yield $j(x) = j_c(x, \varphi_c)$. This is possible at least as a formal series provided $(\delta G_c/\delta j)|_{j=0} = 0$, which means that the one-point function vanishes and $[\delta^2 G_c/\delta j(x)\delta j(y)]|_{j=0} \neq 0$. This is what we assume in the following. The subscript c in φ_c aims at reminding us that φ_c is an ordinary c-number function, not to be confused with the quantum field φ.

The functional $\Gamma(\varphi_c)$ is by definition

$$i\Gamma(\varphi_c) = \left[G_c(j) - i \int d^4x \, j(x)\varphi_c(x) \right] \Bigg|_{j(x) = j_c(x, \varphi_c)} \tag{6-73}$$

The factor i has been introduced for later convenience. From (6-73), it follows by differentiation with respect to $\varphi_c(x)$ that

$$i \frac{\delta}{\partial \varphi_c(x)} \Gamma(\varphi_c) = \int d^4y \left\{ \left[\frac{\delta G_c(j)}{\delta j(y)} - i\varphi_c(y) \right]_{j = j_c(\varphi_c)} \frac{\delta j_c(y, \varphi_c)}{\delta \varphi_c(x)} \right\} - ij_c(x, \varphi_c)$$

The first term of the right-hand side vanishes, owing to (6-72), and

$$j_c(x, \varphi_c) = -\frac{\delta \Gamma(\varphi_c)}{\delta \varphi_c(x)} \tag{6-74}$$

As is well known in the analogous instances of classical mechanics or thermodynamics, the Legendre transformation is involutive. We also notice that $i\Gamma(\varphi_c)$ may be regarded as the value of $G_c(j) - i \int d^4x \, j(x)\varphi_c(x)$ (j and φ_c independent) at its stationary point in j.

We want to show that $\Gamma(\varphi_c)$ is the generating functional of proper functions $\Gamma(x_1,\ldots,x_n)$:

$$\Gamma(\varphi_c) = \sum \frac{1}{n!} \int d^4x_1 \cdots d^4x_n \, \Gamma^{(n)}(x_1,\ldots,x_n)\varphi_c(x_1)\cdots\varphi_c(x_n) \tag{6-75}$$

We differentiate Eq. (6-72) with respect to $\varphi_c(y)$ and get

$$\begin{aligned} \delta^4(x-y) &= \frac{\delta}{\delta\varphi_c(y)}\left[\frac{\delta}{i\delta j(x)} G_c(j)\big|_{j=j_c(\varphi_c)}\right] \\ &= -i \int d^4z \, \frac{\delta j_c(z,\varphi_c)}{\delta\varphi_c(y)} \frac{\delta^2 G_c(j)}{\delta j(z)\delta j(x)}\bigg|_{j=j_c(\varphi_c)} \\ &= i \int d^4z \, \frac{\delta^2\Gamma(\varphi_c)}{\delta\varphi_c(z)\delta\varphi_c(y)} \frac{\delta^2 G_c(j)}{\delta j(z)\delta j(x)}\bigg|_{j=j_c(\varphi_c)} \end{aligned} \tag{6-76}$$

where in the last equation use has been made of Eq. (6-74). Hence, the kernel $[\delta^2\Gamma/\delta\varphi_c(y)\delta\varphi_c(z)]$ is the inverse of $i\,[\delta^2 G_c/\delta j(z)\delta j(x)]|_{j=j_c}$. We may now set $\varphi_c = 0$, according to the assumption that $(\delta G_c/\delta j)|_{j=0} = 0$. The previous identity tells us that the connected two-point function $G_c^{(2)}(z-x) = -[\delta^2 G_c/\delta j(z)\delta j(x)]|_{j=0}$ is the inverse (for convolution) of the function

$$-i\Gamma^{(2)}(y-z) = -i[\delta^2\Gamma/\delta\varphi_c(y)\delta\varphi_c(z)]_{\varphi_c=0}$$

From translational invariance $\Gamma^{(2)}$ depends only on the difference $y-z$ and

$$\int d^4z \, \Gamma^{(2)}(y-z)G_c^{(2)}(z-x) = i\delta^4(x-y) \tag{6-77}$$

In momentum space, this reads

$$G^{(2)}(p,-p)\Gamma^{(2)}(p,-p) = i$$

where the Fourier transforms of the Γ are defined as in Eqs. (6-20) and (6-21). Henceforth, we shall use shorter notations for the two-point function $G^{(2)}(p) \equiv G^{(2)}(p,-p)$, $\Gamma^{(2)}(p) \equiv \Gamma^{(2)}(p,-p)$. If we write the former as

$$G^{(2)}(p) = \frac{i}{p^2 - m^2 - \Sigma(p)} \tag{6-78}$$

where $\Sigma(p)$ is called the self-energy, we have

$$\Gamma^{(2)}(p) = i[G^{(2)}(p)]^{-1} = p^2 - m^2 - \Sigma(p) \tag{6-79}$$

We may now expand (6-78) as (see Fig. 6-24)

$$\begin{aligned} G^{(2)}(p) &= \frac{i}{p^2-m^2} + \frac{i}{p^2-m^2}\frac{1}{i}\Sigma(p)\frac{i}{p^2-m^2} \\ &\quad + \frac{i}{p^2-m^2}\frac{1}{i}\Sigma(p)\frac{i}{p^2-m^2}\frac{1}{i}\Sigma(p)\frac{i}{p^2-m^2} + \cdots \end{aligned}$$

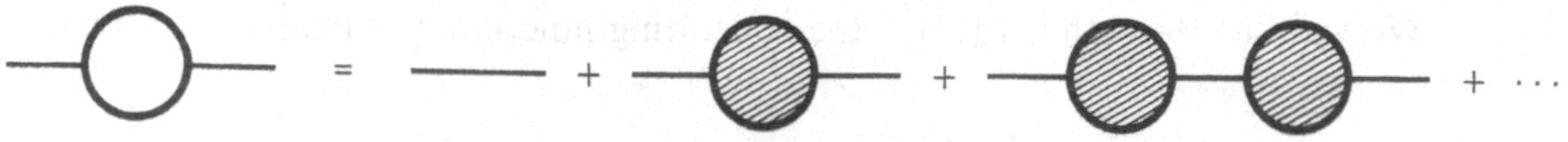

$$\frac{i}{p^2-m^2}+\frac{i}{p^2-m^2}\,\frac{\Sigma}{i}\,\frac{i}{p^2-m^2}+\frac{i}{p^2-m^2}\,\frac{\Sigma}{i}\,\frac{i}{p^2-m^2}\,\frac{\Sigma}{i}\,\frac{i}{p^2-m^2}+\cdots$$

Figure 6-24 Graphical representation of Eq. (6-79). The white blob represents the complete propagator $G^{(2)}(p)$, the shaded one the proper self-energy $-i\Sigma$.

which means that $-i\Sigma$ is the one-particle irreducible perturbative contribution to the two-point function. We obtain $i\Gamma^{(2)}$ by addition of a zeroth-order contribution $i(p^2 - m^2)$.

The interpretation of higher functions $\Gamma^{(n)}$ as one-particle irreducible (proper) functions may be derived by successive differentiations at $\varphi_c = 0$ of the master identity (6-76). For instance, after one more differentiation with respect to $\varphi_c(u)$, we get (see Fig. 6-25)

$$0 = i\int d^4z\, \Gamma^{(3)}(z, y, u)G^{(2)}(z, x) + \int d^4z\, d^4v\, \Gamma^{(2)}(z, y)G^{(3)}(x, z, v)\Gamma^{(2)}(v, u)$$

and, using (6-77),

$$\begin{aligned}\Gamma^{(3)}(x, y, z) &= \int d^4x'\, d^4y'\, d^4z'\, G^{(3)}(x', y', z')\Gamma^{(2)}(x', x)\Gamma^{(2)}(y', y)\Gamma^{(2)}(z', z)\\ &= -iG^{(3)}_{\text{trunc}}(x, y, z)\end{aligned} \tag{6-80}$$

Thus $i\Gamma^{(3)}$ is the truncated function (or one-particle irreducible; there is no distinction for the three-point function). The reader may pursue this process and be convinced that the higher $i\Gamma^{(n)}$ are indeed the one-particle irreducible functions.

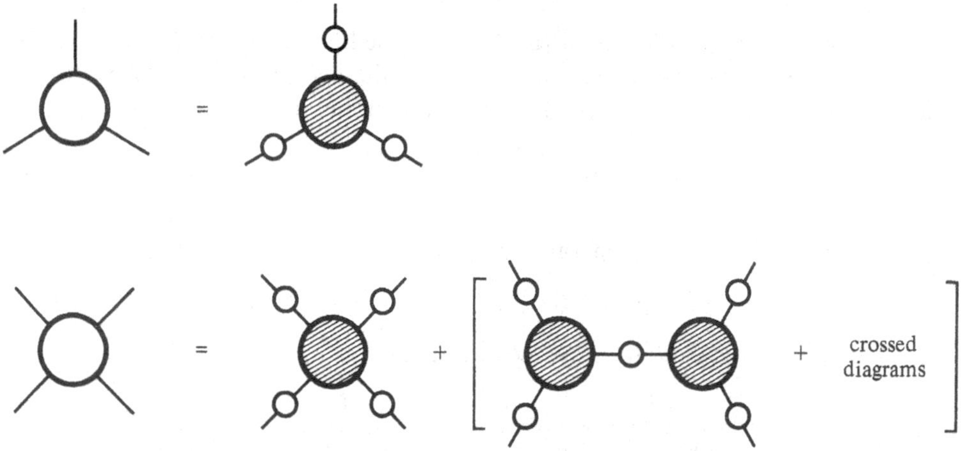

Figure 6-25 Graphical representation of Eq. (6-80) and of a similar identity for the four-point function. As in Fig. 6-24, the white blobs represent connected functions $G^{(n)}$, the shaded ones proper functions $i\Gamma^{(n)}$.

It may be simpler to rewrite $\Gamma(\varphi_c) = \frac{1}{2}\int d^4x\, d^4y\, \Gamma^{(2)}(x-y)\varphi_c(x)\varphi_c(y) + \hat{\Gamma}(\varphi_c)$ and to solve Eq. (6-74), together with (6-72) as

$$-\int d^4x\, \Gamma^{(2)}(y-x)\,\frac{\delta G_c}{i\delta j(x)} = j(y) + \frac{\delta}{\delta\varphi_c(y)}\,\hat{\Gamma}\left(\varphi_c = \frac{\delta G_c}{i\delta j}\right)$$

Using Eq. (6-77), we get

$$\frac{\delta G_c(j)}{i\delta j(x)} = i\int d^4y\, G_c^{(2)}(x-y)\left[j(y) + \frac{\delta\hat{\Gamma}}{\delta\varphi_c(y)}\left(\frac{\delta G_c(j)}{i\delta j}\right)\right] \tag{6-81}$$

This identity appears most clearly in the pictorial representation of Fig. 6-26. Any connected diagram either contributes to the two-point function $G_c^{(2)}$ or has a tree structure made of the generalized vertices $\Gamma^{(n)}(n \geq 3)$ connected with complete propagators $G_c^{(2)}$. On the other hand, we know that any connected diagram may be decomposed into one-particle irreducible subdiagrams. To identify the latter with the $i\Gamma^{(n)}$, it remains to see that all diagrams appear with a correct counting factor. This is just a matter of inspection.

The interpretation of these remarkable identities will be pursued in Chap. 9, where we shall develop the method of functional integration. We also mention that further Legendre transformations might be used to define two-particle irreducible kernels, etc.

From this discussion, it must be clear that, to lowest order (tree diagrams), $\Gamma(\varphi_c)$ reduces to

$$\Gamma_{\text{tree}}(\varphi_c) = \frac{1}{\hbar}\int d^4x\, \mathscr{L}[\varphi_c(x)] \tag{6-82}$$

The $\hbar$ factor has been restored according to the rule derived above. Indeed, the proper two-point function is given to lowest order by

$$\Gamma^{(2)}(x, y) = \frac{1}{\hbar}\int \frac{d^4p_1}{(2\pi)^4}\frac{d^4p_2}{(2\pi)^4}\, e^{i(p_1\cdot x + p_2\cdot y)}(2\pi)^4\delta^4(p_1+p_2)(p_1^2 - m^2)$$

and thus contributes

$$\tfrac{1}{2}\int d^4x\, d^4y\, \Gamma^{(2)}(x, y)\varphi_c(x)\varphi_c(y) = \tfrac{1}{2}\int d^4x\, [(\partial\varphi_c)^2 - m^2\varphi_c^2]$$

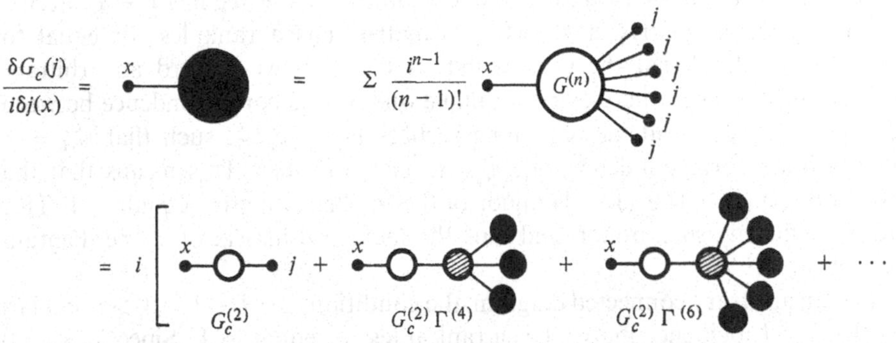

Figure 6-26 Diagrammatic representation of the identity (6-81). The conventions are the same as on Figs. 6-24 and 6-25; the black blob represents the generating functional $\delta G_c(j)/i\delta j(x)$.

to $\Gamma(\varphi_c)$, whereas the higher $\Gamma^{(n)}$, to the same order $\hbar^{-1}$, reduce to the original vertices of the lagrangian.

The importance of the functional $\Gamma(\varphi)$ stems from the fact that it generalizes to higher orders the classical action; $\Gamma(\varphi)$ is often referred to as the effective action. It plays an important role in the discussion of the stable fundamental field configurations, for instance in the discussion of spontaneous symmetry breaking, as will be seen in Chap. 11.

The previous considerations may be extended without difficulty to several species of scalar fields, or to fields with a nonzero spin. In the case of half-integer spin, we have to use anticommuting sources (as in Sec. 4-3), and the "classical" field ψ_c—the analog of φ_c—also belongs to such an anticommuting algebra. Some care is needed in the manipulation of these anticommuting functions, but the Legendre transformation still yields the generating functional of proper vertices.

6-2-3 Parametric Representation

We shall now consider an arbitrary proper diagram G in a scalar theory and compute the corresponding contribution $I(G)$ as given by Feynman rules. We assume that G has no tadpole, i.e., has no internal line starting and ending at the same vertex. As before, I and V denote the number of internal lines and vertices, respectively. It is convenient to give an arbitrary orientation to each internal line. Define the incidence matrix $\{\varepsilon_{vl}\}$, with indices running over vertices and internal lines respectively, as

$$\varepsilon_{vl} = \begin{cases} 1 & \text{if the vertex } v \text{ is the starting point of the line } l \\ -1 & \text{if the vertex } v \text{ is the endpoint of the line } l \\ 0 & \text{if } l \text{ is not incident on } v \end{cases}$$

This $V \times I$ matrix characterizes the topology of the diagram. We introduce the following definitions. A subdiagram G' of G is a subset of vertices and lines of G, such that no vertex is isolated; we do not assume that, given two vertices of G', all lines connecting them in G belong to G'. A tree on G will be a subdiagram containing all vertices of G but no loop. G may be reconstructed from one of its trees by addition of lines. From Eq. (6-69), we know that a tree has $V - 1$ internal lines. Its incidence matrix, a $V \times (V - 1)$ matrix, has a rank less or equal to $V - 1$. Let us show that the rank is indeed $V - 1$. If we discard an arbitrary vertex of the tree, then there exists a unique one-to-one correspondence between the lines $l_1, \ldots, l_{V-1}$ and the remaining vertices $v_1, \ldots, v_{V-1}$, such that $\varepsilon_{v_k l_k} \neq 0$. For any other correspondence, one $\varepsilon_{v_k l_k}$ at least vanishes. This means that the corresponding $(V - 1) \times (V - 1)$ minor of the incidence matrix equals ± 1. This holds true for any such minor, and thus the incidence matrix of a tree diagram has rank $V - 1$.

For an arbitrary connected diagram, the condition $L = I + 1 - V > 0$ implies that the $V \times I$ incidence matrix has a rank ρ less or equal to V. Since $\sum_v \varepsilon_{vl} = 0$ for any $l = 1, \ldots, I$, $\rho \leq V - 1$, and since ε is obtained by the addition of further columns to the rank $V - 1$ matrix of a tree subdiagram, $\rho = V - 1$.

Let us now consider G as a Feynman diagram contributing to some proper Green function $\Gamma(p_1, \ldots, p_n)$. The p are incoming momenta, $\sum_1^n p_i = 0$, and we denote by $P_v = \sum p_{k_v}$ the sum of incoming momenta at the vertex v; clearly, $\sum_{v=1}^V P_v = 0$. The contribution $\tilde{I}_G(P)$ of G to $i\tilde{\Gamma}(p_i) = (2\pi)^4 \delta^4(\sum p_i) i\Gamma(p_i)$ depends only on the P, provided of course that we consider a theory without derivative couplings. This quantity has the form

$$\tilde{I}_G(P) = \frac{C(G)}{S(G)} (2\pi)^4 \delta^4(\textstyle\sum P) I_G(P)$$

$$= \frac{C(G)}{S(G)} \int \prod_{l=1}^{I} \frac{d^4 k_l}{(2\pi)^4} \left(\frac{i}{k_l^2 - m_l^2 + i\varepsilon} \right) \prod_{v=1}^{V} (2\pi)^4 \delta^4 \left(P_v - \sum_{l=1}^{I} \varepsilon_{vl} k_l \right) \tag{6-83}$$

Here $S(G)$ is the symmetry factor of the diagram and $C(G)$ stands for all factors pertaining to the vertices; for instance, $C(G) = (-i\lambda)^V$ in the $\lambda(\varphi^4/4!)$ theory. The four-momentum k_l has been oriented according to the orientation chosen along the lth internal line. The theory under study might involve various species of scalar fields—hence the subscripts on the masses m_l. In a scalar theory with derivative couplings or if nonzero spin fields are introduced, polynomials in k would appear in the numerator of the integrand of (6-83). This is only a minor complication and will be illustrated in practical instances in the subsequent chapters. Next we use the following integral representations of the free scalar propagator and of the δ^4 function:

$$\frac{i}{p^2 - m^2 + i\varepsilon} = \int_0^\infty d\alpha \, e^{i\alpha(p^2 - m^2 + i\varepsilon)} \tag{6-84a}$$

$$(2\pi)^4 \delta^4 \left(P_v - \sum_l \varepsilon_{vl} k_l \right) = \int d^4 y_v \exp \left[-i y_v \cdot \left(P_v - \sum_l \varepsilon_{vl} k_l \right) \right] \tag{6-84b}$$

In (6-84a), the integral converges at the upper limit owing to the presence of $i\varepsilon$. This $i\varepsilon$ will be omitted in the sequel (m^2 may be regarded as having a small imaginary part).

We insert the representations (6-84) into (6-83) and boldly interchange the order of integrations. This is illegitimate if the integrals are not absolutely convergent, which occurs frequently, but will be justified in Chap. 8 by the processes of regularization and renormalization. The integrations over each four-momentum k_l may be carried out easily

$$\int \frac{d^4 k_l}{(2\pi)^4} \exp \left[i\alpha_l \left(k_l^2 + \frac{1}{\alpha_l} \sum_v y_v \cdot \varepsilon_{vl} k_l \right) \right]$$

$$= \int \frac{d^4 k_l}{(2\pi)^4} \exp \left\{ i\alpha_l \left[\left(k_l + \frac{1}{2\alpha_l} \sum_v y_v \varepsilon_{vl} \right)^2 - \frac{1}{4\alpha_l^2} \left(\sum_v y_v \varepsilon_{vl} \right)^2 \right] \right\}$$

$$= A A^{*3} \exp \left[-\frac{i}{4\alpha_l} \left(\sum_v y_v \varepsilon_{vl} \right)^2 \right]$$

where A stands for the Fresnel integral

$$A = \int \frac{dk}{2\pi} e^{i\alpha k^2} = \frac{e^{i\pi/4}}{\sqrt{4\pi\alpha_l}}$$

$$AA^{*3} = \frac{-i}{(4\pi\alpha_l)^2}$$

Therefore

$$(2\pi)^4 \delta^4(\sum P) I_G(P)$$

$$= \int \prod_v d^4 y_v \int_0^\infty \prod_l \left(d\alpha_l \frac{-i \exp\left\{-i\left[\alpha_l m_l^2 + \left(\sum_v y_v \varepsilon_{vl}\right)^2 /4\alpha_l\right]\right\}}{(4\pi\alpha_l)^2} \right) e^{-i\sum_v y_v \cdot P_v} \tag{6-85}$$

The integration variables y are then reshuffled by the change of variables

$$y_1 = z_1 + z_V$$

$$y_2 = z_2 + z_V$$

$$\cdots\cdots\cdots\cdots\cdots\cdots$$

$$y_{V-1} = z_{V-1} + z_V$$

$$y_V = z_V$$

the jacobian of which is one. Since $\sum_v \varepsilon_{vl} = 0$ (the line l connects two and only two vertices), z_V appears only in the last exponent of (6-85). Therefore, integration over z_V yields the δ^4 function of energy momentum conservation, leading to

$$I_G(P) =$$

$$\int \prod_{v=1}^{V-1} d^4 z_v \int_0^\infty \prod_l \left(d\alpha_l \frac{-i \exp\left\{-i\left[\alpha_l m_l^2 + \left(\sum_{v=1}^{V-1} z_v \varepsilon_{vl}\right)^2 /4\alpha_l\right]\right\}}{(4\pi\alpha_l)^2} \right) e^{-i\sum_{v=1}^{V-1} z_v \cdot P_v}$$

The $(V-1) \times (V-1)$ matrix $d_G(\alpha)$ defined as

$$[d_G(\alpha)]_{v_1 v_2} = \sum_l \varepsilon_{v_1 l} \frac{1}{\alpha_l} \varepsilon_{v_2 l}$$

may be shown to be nonsingular, and its determinant reads

$$\Delta_G(\alpha) \equiv \det[d_G(\alpha)] = \sum_{\mathcal{T}} \prod_{l \in \mathcal{T}} \frac{1}{\alpha_l}$$

where the sum runs over all trees of G. Since every tree of G has $V-1$ lines, Δ_G is clearly a homogeneous polynomial in the α_l^{-1}. For $\alpha > 0$, Δ is positive. The remaining $(V-1)$ four-vectors z may then be integrated over and the result is

$$I_G(P) = \int_0^\infty \prod_l (d\alpha_l\, e^{-im_l^2\alpha_l}) \frac{\exp i\left\{\sum_{v_1,v_2=1}^{V-1} P_{v_1}\cdot[d_G^{-1}(\alpha)]_{v_1v_2}P_{v_2}\right\}}{[i(4\pi)^2]^L[\alpha_1\cdots\alpha_I\Delta_G(\alpha)]^2}$$

This formula expresses $I_G(P)$ as a function of the invariant scalar products of external momenta $P_{v_1}\cdot P_{v_2}$. The functions of α that appear in the exponent or in the denominator may be given a form independent of the choice of the vertex V. On the one hand, the denominator

$$\mathscr{P}_G(\alpha) \equiv \alpha_1\cdots\alpha_I\Delta_G(\alpha) = \sum_{\mathscr{T}} \prod_{l\notin\mathscr{T}} \alpha_l \tag{6-86}$$

is a homogeneous polynomial of degree L clearly independent of V. On the other hand, the quadratic form

$$Q_G(P, \alpha) \equiv \sum_{v_1,v_2=1}^{V-1} P_{v_1}\cdot[d_G^{-1}(\alpha)]_{v_1v_2}P_{v_2}$$

may be expressed as

$$Q_G(P, \alpha) = \frac{1}{\mathscr{P}_G(\alpha)} \sum_C s_C \prod_{l\in C} \alpha_l \tag{6-87}$$

where the sum runs over all possible "cuts" C of $L+1$ lines that divide the diagram into two and exactly two connected parts $G_1(C)$, $G_2(C)$. Such partitions are obtained from the trees of G by cutting an $(L+1)$th line. Then s_C denotes the square of the momenta P entering $G_1(C)$ or $G_2(C)$:

$$s_C = \left(\sum_{v\in G_1(C)} P_v\right)^2 = \left(\sum_{v\in G_2(C)} P_v\right)^2 \tag{6-88}$$

Therefore Q_G is a ratio of two homogeneous polynomials of degrees $L+1$ and L respectively.

The final parametric form reads

$$I_G(P) = \int_0^\infty \prod_1^I (d\alpha_l\, e^{-i\alpha_l m_l^2}) \frac{e^{iQ_G(P,\alpha)}}{[i(4\pi)^2]^L[\mathscr{P}_G(\alpha)]^2} \tag{6-89}$$

with $\mathscr{P}_G$ and Q_G given in Eqs. (6-86) to (6-88) and we recall that $L = I + 1 - V$. The proof of the results (6-86) to (6-88) and detailed properties of the functions $\mathscr{P}_G$, Q_G skipped in this elementary presentation may be found in the references.

Let us illustrate the rules (6-86) to (6-88) on an explicit example. The diagram of Fig. 6-27 contributes to the four-point function in the φ^4 theory. We recall that the symmetry factor is $S(G) = 2$ and $C(G) = (-i\lambda)^3$. The trees $\mathscr{T}$ are $(l_1 l_2)$, $(l_1 l_3)$, $(l_1 l_4)$, $(l_2 l_3)$, $(l_2 l_4)$. Hence the polynomial $\mathscr{P}_G$ is equal to

$$\mathscr{P}_G = \alpha_3\alpha_4 + \alpha_2\alpha_4 + \alpha_2\alpha_3 + \alpha_1\alpha_4 + \alpha_1\alpha_3$$

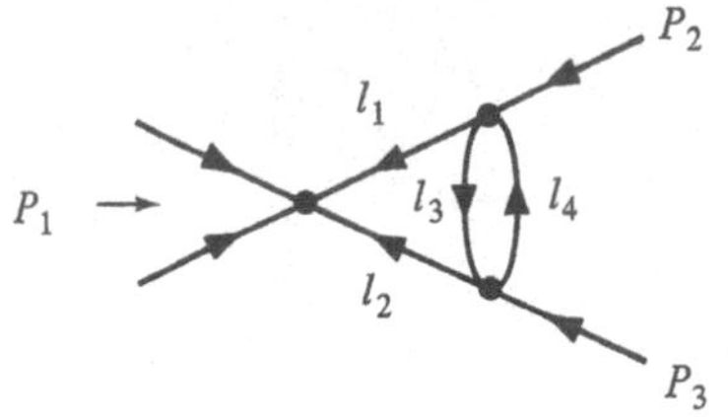

Figure 6-27 Sample diagram contributing to the four-point function.

We now list the cuts C and the corresponding s_C:

$$\begin{array}{ll} (l_1 l_2 l_3) & s_C = P_1^2 = (P_2 + P_3)^2 \\ (l_1 l_2 l_4) & s_C = P_1^2 = (P_2 + P_3)^2 \\ (l_1 l_3 l_4) & s_C = P_2^2 = (P_1 + P_3)^2 \\ (l_2 l_3 l_4) & s_C = P_3^2 = (P_1 + P_2)^2 \end{array}$$

Choosing P_1 and P_2 as independent variables, we have

$$Q_G(P, \alpha) = \mathscr{P}_G^{-1}[P_1^2 \alpha_1 \alpha_2(\alpha_3 + \alpha_4) + P_2^2 \alpha_1 \alpha_3 \alpha_4 + (P_1 + P_2)^2 \alpha_2 \alpha_3 \alpha_4]$$

We may use the homogeneity properties of $\mathscr{P}_G$ and Q_G:

$$Q_G(P, \lambda\alpha) = \lambda Q_G(P, \alpha)$$

$$\mathscr{P}(\lambda\alpha) = \lambda^L \mathscr{P}_G(\alpha)$$

to further simplify Eq. (6-89). We insert in the integrand the identity

$$1 = \int_0^\infty d\lambda\, \delta\left(\lambda - \sum_i^I \alpha_l\right)$$

and rescale $\alpha_l \to \lambda \alpha_l$. Then

$$I_G(P) = \frac{1}{[i(4\pi)^2]^L} \int_0^1 \prod_1^I d\alpha_l \frac{\delta(1 - \sum \alpha_l)}{[\mathscr{P}_G(\alpha)]^2} \int_0^\infty \frac{d\lambda}{\lambda} \lambda^{I-2L} \exp\left\{i\lambda\left[Q_G(P, \alpha) - \sum_l m_l^2 \alpha_l\right]\right\} \tag{6-90}$$

The integration over the α has been reduced to the interval (0, 1) because of the δ function. In the λ integration, the convergence at the upper limit does not give any trouble, owing to the $i\varepsilon$ implicit in m_l^2. The integral converges at the lower end provided $2V - I - 2 > 0$. In terms of Euler's gamma function,

$$I_G(P) = \frac{i^{3(V-1)-2I}}{(4\pi)^{2L}} \Gamma(2V - I - 2) \int_0^1 \prod_1^I d\alpha_l \frac{\delta\left(1 - \sum_1^I \alpha_l\right)}{[\mathscr{P}_G(\alpha)]^2 \left[Q_G(P, \alpha) - \sum_1^I \alpha_l m_l^2\right]^{2V-I-2}} \tag{6-91}$$

The convergence condition, rewritten as $2L - 4V + 4 < 0$, just expresses that $I_G(P)$ has a negative dimension in an energy scale. This may be read on Eq. (6-83) or on

Eq. (6-91), since Q_G has dimension $[E]^2$. When this dimension is nonnegative, the integral diverges at the lower limit, $\lambda \sim 0$, which is of course the translation in this parametric representation of the ultraviolet divergence at large k of the integral (6-83). This relation between dimension and occurrence of divergences foreshadows the power counting that will be our touchstone in the renormalization procedure.

The parametric representations (6-89) to (6-91) are interesting and useful in many respects. They provide an explicit form which may (or may not!) be convenient in practical computations. They may also be used to study analyticity properties of Green functions or to carry out the renormalization program. Finally, they exhibit fruitful analogies with the theory of electric circuits.

6-2-4 Euclidean Green Functions

The Green functions of a (scalar) theory are functions of the invariant scalar products of their external momenta, which are real Lorentz four-vectors. As we have seen at the end of the preceding chapter, and as we shall see in the next section, these functions enjoy analyticity properties and may be continued to unphysical regions. Here, we dwell on the continuation to the euclidean region where field theory presents interesting analogies with statistical mechanical problems. We shall not formulate this euclidean theory *ab initio*, but rather derive its perturbative rules. Again, for simplicity, we deal with a scalar theory.

Consider a proper function $\Gamma^{(n)}(p_1, \ldots, p_n)$ and assume that its arguments satisfy the condition

$$\left(\sum_{k=1}^{n} u_k p_k \right)^2 \leq 0 \qquad \text{(euclidean region)} \tag{6-92}$$

for any real (u_k). The manifold satisfying this condition is a linear space-like subspace (a hyperspace) of momentum space. If we consider $\Gamma^{(n)}$ as a function of the invariants $p_i \cdot p_j$ $(i, j = 1, \ldots, n-1)$ let us compute the dimension of the total manifold in the absence of this constraint, and the one of the submanifold (6-92). We must, of course, remember that in a four-dimensional space, any set of more than four vectors is linearly dependent. Hence for $n \geq 6$, the p are not independent. Consequently, the invariants $p_i \cdot p_j$ belong to a manifold of dimension $4n - 10$ for $n \geq 4$ (1 and 3 for $n = 2$, 3 respectively). If condition (6-92) is taken into account this restricts them to a submanifold of dimension $3n - 6$ for $n \geq 4$ (1 and 3 for $n = 2$, 3). For $n = 4$, both have dimension 6 (for instance, s, t, u, $p_1^2, \ldots, p_4^2$, with $\sum p_i^2 = s + t + u$), whereas for $n \geq 5$, the euclidean submanifold (6-92) has a dimension less than that of the whole space.

In the euclidean domain (6-92), all the p are orthogonal to some time-like vector n. After a Lorentz transformation, we may choose $n = (1, 0, 0, 0)$ and thus $p_i^0 = 0$ for all i. This enables us to associate to each p_i a vector $\hat{p}_i$ of a euclidean R^4 space: in that frame $\hat{p}_i^0 = p_i^0 = 0$ and $\hat{\mathbf{p}}_i = \mathbf{p}_i$. Then

$$\hat{p}_i \cdot \hat{p}_j = \hat{p}_i^0 \hat{p}_j^0 + \hat{p}_i^1 \hat{p}_j^1 + \hat{p}_i^2 \hat{p}_j^2 + \hat{p}_i^3 \hat{p}_j^3 = -p_i \cdot p_j$$

If we now pick a diagram contributing to $\Gamma^{(n)}$ and express its contribution according to Eq. (6-89), we may define P_v, the sum of the p_i entering the vertex v, and to each "cut" C of the diagram we associate [compare (6-87) and (6-88)]

$$\hat{s}_C \equiv \left(\sum_{v \in G_1(C)} \hat{P}_v \right)^2 = -\left(\sum_{v \in G_1(C)} P_v \right)^2 = -s_C$$

$$Q_G(\hat{P}, \alpha) = \frac{1}{\mathscr{P}_G(\alpha)} \sum_C \hat{s}_C \prod_{l \in C} \alpha_l = -Q_G(P, \alpha)$$

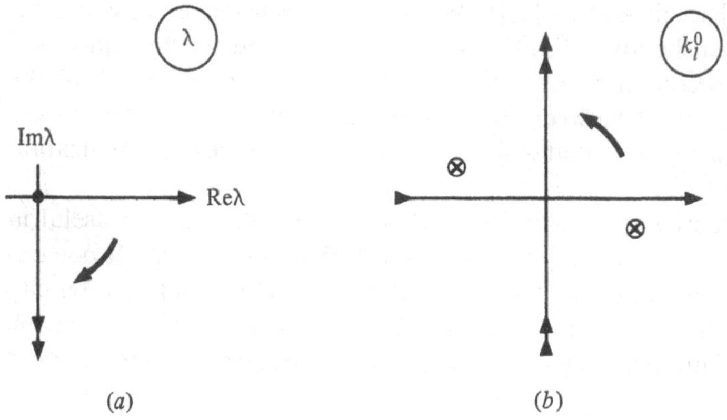

Figure 6-28 The Wick rotation in parametric space (*a*) or in momentum space (*b*). The initial contour of integration is indicated by a single arrow, the final one by a double arrow. On (*b*), the crosses ⊗ stand for the position of the poles $k_l^0 = \pm[(\mathbf{k}_l + \Sigma \mathbf{p})^2 + m_l^2 - i\varepsilon]^{1/2}$.

Therefore, on the manifold (6-92), we may rewrite (6-90) as

$$I_G(P) = \int_0^1 \left(\prod_1^I d\alpha_l\right) \delta(1 - \sum \alpha_l) \int_0^\infty \frac{d\lambda}{\lambda} \lambda^{I-2L} \frac{\exp\{-i\lambda[Q_G(\hat{P}, \alpha) + \sum \alpha_l m_l^2]\}}{[i(4\pi)^2]^L [\mathscr{P}_G(\alpha)]^2}$$

We recall that the m_l^2 have implicitly a negative imaginary part. This, together with the positivity properties of the bracket of the exponent (for $m_l^2 \geq 0$), allows us to rotate by $-\pi/2$ (i.e., clockwise) the integration contour in the complex λ plane (Fig. 6-28). This is, of course, equivalent to a simultaneous $(-\pi/2)$ rotation of all the variables α in Eq. (6-89). It must be stressed that the $i\varepsilon$ prescription has played a crucial role in telling us which rotation was allowed and that the Wick rotation, as it is usually called, is illegitimate if some m_l^2 happen to be negative. In momentum space [compare with (6-83)], the Wick rotation amounts to a $\pi/2$ rotation (counterclockwise) of all k_l^0 in the frame where all $p_i^0 = 0$. This is in agreement with the position of the poles of the propagator, located at

$$k_l^0 = \pm[(\mathbf{k}_l + \sum \mathbf{p})^2 + m_l^2 - i\varepsilon)]^{1/2}$$

After this rotation, we get in the euclidean region

$$I_G(P) \equiv (-i)^{V-1} \hat{I}_G(\hat{P}) = (-i)^{V-1} \int_0^\infty \prod_1^I d\alpha_l \frac{\exp -[Q_G(\hat{P}, \alpha) + \sum \alpha_l m_l^2]}{(4\pi)^{2L}[\mathscr{P}_G(\alpha)]^2} \tag{6-93}$$

The factor $C(G)$ in front of $I_G(P)$ [compare with (6-83)] contains i^V coming from the expansion of $\exp[i \int d^4z\, \mathscr{L}_{\text{int}}(z)]$ to the Vth order. We set $C(G) = i^V \hat{C}(G)$ and rewrite (6-93) as

$$\frac{C(G)}{S(G)} I_G(P) = i \frac{\hat{C}(G)}{S(G)} \hat{I}_G(\hat{P})$$

For instance, in the $\lambda\varphi^4$ theory, $\hat{C}(G) = (-\lambda)^V$. We recall from the analysis of Sec. 6-2-2 that the proper functions have been identified with the $i\Gamma^{(n)}$ defined through the Legendre transformation (6-73) and (6-75). Therefore, thanks to this last factor i, these functions $\Gamma^{(n)}$ coincide with the real, euclidean Green functions $\hat{\Gamma}^{(n)}$:

$$\Gamma^{(n)}(p_1, \ldots, p_n) = \hat{\Gamma}^{(n)}(\hat{p}_1, \ldots, \hat{p}_n)$$

defined as sums of the contribution $[\hat{C}(G)/S(G)]\, \hat{I}_G(\hat{P})$ of each diagram. It is easy to see, by repeating backward the same computation as in Sec. 6-2-3, that

$$(2\pi)^4\delta^4(\sum \hat{p})\frac{\hat{C}(G)}{S(G)}\hat{I}_G(\hat{P}) = \frac{\hat{C}(G)}{S(G)}\int \prod_l \frac{d^4\hat{k}_l}{(2\pi)^4}\frac{1}{\hat{k}_l^2 + m_l^2}\prod_v (2\pi)^4\delta^4\left(\hat{P}_{vl} - \sum_l \varepsilon_{vl}\hat{k}_l\right) \tag{6-94}$$

where now $\hat{k}_l$ are euclidean four-momenta, $\hat{k}_l^2 = \sum_{i=0}^3 (\hat{k}_l^i)^2$. Here, as in Eq. (6-93), the integrand enjoys positivity properties. Of course, Eq. (6-94) means that $\hat{\Gamma}$ may be computed by euclidean Feynman rules:

$$\begin{array}{lll} \text{Vertex} & -\lambda & \text{instead of } -i\lambda \\ \text{Propagator} & \dfrac{1}{\hat{k}^2 + m^2} & \text{instead of } \dfrac{i}{k^2 - m^2 + i\varepsilon} \end{array} \tag{6-95}$$

From these rules, we learn that connected Green functions (with external propagators) are related by

$$G_c^{(n)}(p_1, \ldots, p_n) = i(-i)^n \hat{G}_c^{(n)}(\hat{p}_1, \ldots, \hat{p}_n) \tag{6-96}$$

in the coincidence region (6-92).

In configuration space, these functions read

$$\hat{G}_c(x_1, \ldots, x_n) = \int \prod_1^n \frac{d^4 q_k}{(2\pi)^4} e^{-i\hat{q}_k \cdot x_k}(2\pi)^4\delta^4(\sum q_i)\hat{G}_c(\hat{q}_1, \cdots, \hat{q}_n)$$

with $x \cdot \hat{q} = x^0 q^0 + x^1 q^1 + x^2 q^2 + x^3 q^3$. We then introduce the generating functional

$$\hat{G}_c(j) = \sum \frac{1}{n!}\int dx_1 \cdots dx_n j(x_1) \cdots j(x_n)\hat{G}_c(x_1, \cdots, x_n)$$

and a similar expression for $\hat{\Gamma}(\varphi_c)$. Collecting all factors i in Eqs. (6-71*b*), (6-72), (6-73), and (6-96), it is then straightforward to check that $\hat{\Gamma}(\varphi_c)$ is the Legendre transform of $\hat{G}_c$:

$$\hat{\Gamma}(\varphi_c) = \left[\hat{G}_c(j) - \int d^4x\, j(x)\varphi_c(x)\right] \tag{6-97}$$

where φ_c and j are related by

$$\varphi_c(x) = \frac{\delta\hat{G}_c(j)}{\delta j(x)} \tag{6-98}$$

The lowest-order euclidean functional in the φ^4 theory is

$$\hat{\Gamma}_{\text{tree}}(\varphi_c) = -\int d^4x\, \mathscr{L}_E[\varphi_c(x)] \equiv -\int d^4x \left(\frac{1}{2}\sum_{k=0}^{3}\partial_k\varphi_c\partial_k\varphi_c + \frac{m^2}{2}\varphi_c^2 + \frac{\lambda}{4!}\varphi_c^4\right)$$

Notice the change in sign of the $(\partial_0\varphi)^2$ term as compared to (6-82).

The euclidean theory may be considered as a statistical mechanical theory, the weight of each field configuration being $\exp\left[-\int d^4z\, \mathscr{L}_E(\varphi)\right]$. This will appear more transparent after the introduction of functional integrals (Chap. 9).

As an exercise, the reader may discuss the meaning of the Wick rotation when spin $\frac{1}{2}$ fields are present, namely, what happens to the γ matrices and what is the relation between ψ and $\overline{\psi}$. He or she may then write the Feynman rules for euclidean quantum electrodynamics.

6-3 ANALYTICITY PROPERTIES

In the last section of Chap. 5, analyticity properties of amplitudes were derived from the general principle of local causality. As a typical example, the two-body forward elastic scattering amplitude has been proved to be analytic in the energy variable, under suitable conditions on the values of the masses of the particles. The analyticity domain is a cut plane with two branch points at $s = M_+^2$ and $u = 2(m_a^2 + m_b^2) - s = M_-^2$ (at $t = 0$), corresponding to the lowest intermediate states in the direct (s) and crossed (u) channels. The discontinuities across the cuts were related through the optical theorem to the total cross section of the channel. Therefore, we expected that the amplitude will have singularities along the real axis at all values of the energy corresponding to the opening of a new threshold, i.e., the onset of a new possible final state. For instance, the $\pi N \to \pi N$ forward scattering amplitude is expected to have singularities at s (or u) $= (m_N + m_\pi)^2$, $(m_N + 2m_\pi)^2$, etc.

In the present section, we shall present a sketchy discussion of analyticity properties of Feynman integrals. The interest of such a study is threefold. First, it is important to test in the framework of perturbation theory the general analyticity properties derived rigorously. Whenever we are able to prove from general axioms the existence of an analyticity domain, the contribution of an arbitrary diagram must go through the successive steps of the proof and must therefore enjoy these analyticity properties, or rather their equivalent perturbative expression. However, when the masses are such that the general proof fails, the study of individual diagrams may prove useful. Second, such a study allows to search for complex singularities or to study analyticity in several variables. Even though we suspect that the perturbation series does not converge and therefore that the exact amplitudes may have different properties from those of individual diagrams, this investigation can be suggestive of fruitful conjectures. Finally, dispersion relations in one or several variables may be a useful device for the computation of Feynman amplitudes.

We shall frequently refer to the physical sheet of a scattering amplitude. We have in mind the region reached by analytic continuation, starting from threshold and using the Feynman $i\varepsilon$ prescription. In this process we do not consider the possibility of crossing the various cuts arising from the singularity structure of the amplitude.

6-3-1 Landau Equations

Two types of singularities arise in a function defined by an integral

$$F(z) = \int_C dx\, f(x, z)$$

along a contour C, for instance, an interval (a, b) of the real axis. The function $f(x, z)$ is assumed to be analytic in the variables x, z, but for singularities located at $x = x_k(z)$. Clearly $F(z)$ is analytic for any value of z such that there exists an open neighborhood of C free of singularity. If some singularity $x_k(z)$ approaches the contour C, and if the contour may be deformed to avoid it, then F remains analytic. Therefore, singularities of F as a function of z are expected in two cases where the contour C may no longer be deformed:

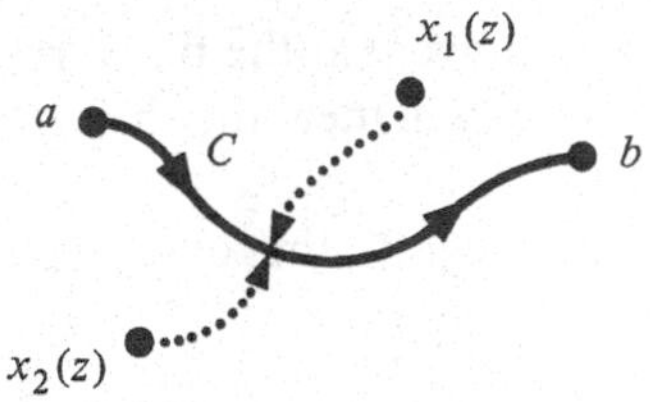

Figure 6-29 Singularity arising from the pinching of the contour between two singularities $x_1(z)$ and $x_2(z)$ of the integrand.

1. One of the singularities $x_r(z)$ approaches one of the endpoints of the contour: as $z \to z_0$, $x_r(z) \to a$ or b; z_0 is called an endpoint singularity.
2. The contour C is pinched between two singularities $x_1(z)$ and $x_2(z)$ (Fig. 6-29): as $z \to z_0$, $x_1(z)$ and $x_2(z)$ approach the contour from below and from above; z_0 is a pinch singularity.

Elementary examples are provided by the following integrals:

$$F(z) = \int_{-1}^{1} \frac{dx}{x^2 + z} = \frac{2}{\sqrt{z}} \arctan \frac{1}{\sqrt{z}} \tag{6-99a}$$

$$F(z) = \int_{0}^{1} \frac{dx}{(x-2)(x-z)} = \frac{1}{z-2} \ln \left[\frac{2(z-1)}{z}\right] \tag{6-99b}$$

$$F(z) = \int_{0}^{1} \frac{dx}{1 - x^2 z} = \frac{1}{2\sqrt{z}} \ln \frac{1+\sqrt{z}}{1-\sqrt{z}} \tag{6-99c}$$

In the first example, $z = 0$ is a pinch singularity. As $z \to 0_+$, the contour is trapped between the two poles $x = i\sqrt{z}$ and $x = -i\sqrt{z}$. In the second case, $z = 0$ and $z = 1$ are endpoint singularities. The pinch singularity at $z = 2$ appears in all Riemann sheets of the logarithm, except for the first one (corresponding to the principal determination). Indeed in the first sheet, the contour of integration does not cross the point $x = 2$. However, going to further sheets obliges us to deform the contour of integration. Pinching may now happen. The same phenomenon occurs in the third example (6-99c). Here the singularity at $z = 1$ is an endpoint singularity while the appearance of the singularity $z = 0$ in every sheet except the first one comes from a pinching at infinity. This is clear if we change the integration variable x into $u = 1/x$.

This discussion must be extended to functions of several complex variables x_i and z_j, $F(z_j) = \int_H (\prod_i dx_i) f(x_i, z_j)$. The boundaries of the integration domain, the hypercontour H, are specified by a set of analytic relations $\tilde{\mathscr{S}}_r(x, z) = 0$. The singularities of the integrand $f(x_i, z_j)$ take place on analytic manifolds $\mathscr{S}_s(x, z) = 0$. Singularities occur when the hypercontour H is pinched between two or more surfaces of singularities or when a surface of singularity meets a boundary surface. More precisely, it may be shown that a necessary condition of singularity is that there exists a set of complex parameters λ_s, $\tilde{\lambda}_r$ not all equal to zero such that at $x_i = x_i^0$, $z_i = z_j^0$:

$$\begin{aligned} \lambda_s \mathscr{S}_s(x^0, z^0) &= 0 \qquad \text{for all } s \\ \tilde{\lambda}_r \tilde{\mathscr{S}}_r(x^0, z^0) &= 0 \qquad \text{for all } r \\ \frac{\partial}{\partial x_i}\left[\sum_r \tilde{\lambda}_r \tilde{\mathscr{S}}_r(x, z) + \sum_s \lambda_s \mathscr{S}_s(x, z)\right]\Bigg|_{x^0, z^0} &= 0 \qquad \text{for all } i \end{aligned} \tag{6-100}$$

The last condition expresses that the hypersurfaces are tangent at the pinching point. This is only a necessary condition. Determining whether the hypercontour is really pinched requires a detailed study.

We apply these general results to the case of Feynman integrals, first expressed in minkovskian momentum space. We consider

$$I_G(P) = \int \prod_{l=1}^{L} \frac{d^4 q_l}{(2\pi)^4} \prod_{i=1}^{I} \frac{i}{k_i^2 - m_i^2 + i\varepsilon} \tag{6-101}$$

The notations are the same as in Eq. (6-83), but all internal momenta k_i have been expressed in terms of a set of loop variables q_l and of the external momenta P. The boundaries of the integration domain are at infinity. In this elementary discussion, we shall disregard the possible occurrence of endpoint singularities at infinity and introduce no term $\tilde{\mathscr{S}}_r$. The singularities of the integrand are given by the equations $\mathscr{S}_i \equiv k_i^2 - m_i^2 = 0$. The Landau equations are nothing but the expression of the necessary conditions (6-100) in this case:

$$\begin{cases} \lambda_i(k_i^2 - m_i^2) = 0 & \text{for all } i = 1, \ldots, I & \quad (6\text{-}102a) \\ \displaystyle\sum_i \lambda_i k_i \cdot \frac{\partial k_i}{\partial q_l} = 0 & l = 1, \ldots, L & \quad (6\text{-}102b) \end{cases}$$

The second equation may be rewritten

$$\sum_{i \in \mathscr{L}_l} (\pm)\lambda_i k_i = 0 \tag{6-103}$$

using the fact that those k that depend linearly (and with a coefficient ± 1) on the loop variable q_l lie along a single loop, denoted by $\mathscr{L}_l$. The interpretation of the first equation (6-102a) is that singularities occur only when, for every internal line, either the four-momentum is on its mass shell $k_i^2 = m_i^2$ or the parameter λ_i vanishes. In the latter case, the ith line never appears in the singularity equations. The singularity is the one of a reduced diagram where the ith line has been contracted to a point.

It is instructive to see how the Landau equations are derived in other integral representations. Before considering the parametric form studied in Sec. 6-2-3, we introduce a mixed representation

$$\frac{1}{(I-1)!} I_G(P) = i^I \int \prod_{l=1}^{L} \frac{d^4 q_l}{(2\pi)^4} \int_0^1 \prod_{i=1}^{I} d\alpha_i \frac{\delta(1 - \sum \alpha_i)}{\left[\sum_{i=1}^{I} \alpha_i(k_i^2 - m_i^2)\right]^I} \tag{6-104}$$

obtained from Eq. (6-101) by means of the identity

$$\frac{1}{A_1 \cdots A_I} = (I-1)! \int_0^1 \frac{d\alpha_1 \cdots d\alpha_I \, \delta(1 - \sum \alpha_i)}{\left(\sum_{i=1}^{I} \alpha_i A_i\right)^I} \tag{6-105}$$

Now the singularities of $I_G(P)$ arise from the zeros of the denominator $\mathscr{S} = \sum_1^I \alpha_i(k_i^2 - m_i^2)$ pinching the hypercontour in (α, q) space or meeting the boundaries $\tilde{\mathscr{S}}_i = \alpha_i = 0$. The singularity equations read

$$\lambda'\mathscr{S} = 0 \tag{6-106a}$$

$$\lambda'_i \alpha_i = 0 \qquad i = 1, \ldots, I \tag{6-106b}$$

$$\frac{\partial}{\partial q_l}\left(\lambda'\mathscr{S} + \sum_i \lambda'_i \alpha_i\right) = 0 \qquad l = 1, \ldots, L \tag{6-106c}$$

$$\frac{\partial}{\partial \alpha_j}\left(\lambda'\mathscr{S} + \sum_i \lambda'_i \alpha_i\right) = 0 \qquad j = 1, \ldots, I \tag{6-106d}$$

The case $\lambda' = 0$ leads to $\lambda'_i = 0$ for all i. We discard this trivial solution and assume $\mathscr{S} = 0$. Then Eq. (6-106c) gives

$$\frac{\partial \mathscr{S}}{\partial q_l} = 2 \sum_i \alpha_i k_i \cdot \frac{\partial k_i}{\partial q_l} = 0$$

while Eq. (6-106b, d) lead to

$$\alpha_j(k_j^2 - m_j^2) = 0 \qquad j = 1, \ldots, I$$

We recover the previous equations (6-102) with the λ replaced by the α.

The parametric representation (6-91), valid when the diagram is superficially ultraviolet finite [that is, $2L - 4(V - 1) < 0$], is obtained from Eq. (6-104) by integration over the loop momenta q. Disregarding possible singularities coming from the zeros of $\mathscr{P}_G(\alpha)$, we take $\mathscr{S} = Q_G(P, \alpha) - \sum_{i=1}^{I} \alpha_i m_i^2$, $\tilde{\mathscr{S}}_i = \alpha_i$, and write singularity conditions as

$$\begin{aligned} &\lambda'\mathscr{S} = 0 \\ &\lambda'_i \alpha_i = 0 \qquad i = 1, \ldots, I \\ &\frac{\partial}{\partial \alpha_j}\left(\lambda'\mathscr{S} + \sum_i \lambda'_i \alpha_i\right) = 0 \qquad j = 1, \ldots, I \end{aligned} \tag{6-107}$$

Again $\lambda' = 0$ is a trivial solution. The conditions may be rewritten as $\alpha_j \partial\mathscr{S}/\partial\alpha_j = 0$ for each j, which implies $\mathscr{S} = \sum_j \alpha_j \partial\mathscr{S}/\partial\alpha_j = 0$, owing to the homogeneity of $\mathscr{S}$. To show the equivalence with the Landau equations (6-102) we need to reintroduce a momentum variable k_i for each internal line, defined in terms of the external momenta and of the α parameters. They are chosen to satisfy an equation of the type (6-103) where λ_i is replaced by α_i. With this definition the condition $\alpha_j \partial\mathscr{S}/\partial\alpha_j = 0$ boils down to Eq. (6-102a).

In the search for solutions to the systems (6-106) or (6-107), the condition $\sum \alpha_i = 1$ may be omitted, because of the homogeneity property of the function $\mathscr{S}$.

In either representation, the solution corresponding to λ_i (or α_i) $\neq 0$ for all i is referred to as the leading singularity, while a solution where λ_i (or α_i) $= 0$ for $i \in \mathscr{I}$ is called a nonleading singularity. In the associated reduced diagram $R = G/\mathscr{I}$, where all lines $i \in \mathscr{I}$ have been contracted to a point, this singularity is a leading singularity.

The interest of the Landau equations is to cast the problem of determining the location of singularities in an algebraic form. However, finding the general solution of these equations, even for simple diagrams, is an arduous task. We shall pursue this program only in the case of real singularities.

6-3-2 Real Singularities

Real singularities are those occurring for real values of the invariants $s_{ij} = P_i \cdot P_j$ on the physical sheet. Notice that these real values of the invariants do not necessarily correspond to a physically possible kinematical configuration. For instance, the region $s = (p_a + p_b)^2 < (m_a + m_b)^2$ is a real but nonphysical interval for elastic scattering of two on-shell particles of masses m_a, m_b.

For real values of the invariants s_{ij} and a nonvanishing imaginary part $(-i\varepsilon)$ given to the internal masses, no singularity appears along the real contour of integration in parametric space. Therefore, singularities appear only in the limit $\varepsilon \to 0$.

It may be shown that any real solution of the Landau equations corresponds to a pinching of the integration contour as $\varepsilon \to 0$ and thus gives rise to a singularity of the integral.

To see this we use the parametric form (6-91) of the integral. Consider first a leading singularity, occurring at some real value $s = s^{(0)}$ of the invariants. The Landau equations $\partial\mathscr{S}/\partial\alpha_j = 0$ have a real solution $\alpha_j = \alpha_j^{(0)}$ $(j = 1, \ldots, I)$. In the vicinity of this point,

$$\Delta\mathscr{S} = \frac{1}{2}\sum_{i,j=1}^{I-1} \left.\frac{\partial^2\mathscr{S}}{\partial\alpha_i\partial\alpha_j}\right|_{\alpha^{(0)},s^{(0)}} \Delta\alpha_i\Delta\alpha_j + O(\Delta\alpha^3)$$

After diagonalization of this quadratic form

$$\Delta\mathscr{S} = \tfrac{1}{2}\sum \sigma_i \Delta\beta_i^2$$

with real eigenvalues σ_i, we see that in any variable β_i $(i = 1, \ldots, I-1)$, the solutions of $\Delta\mathscr{S} - i\varepsilon = 0$ approach the real axis from both sides, except maybe for exceptional configurations where $\sigma_i = 0$. Therefore, there is a pinch singularity in every successive β_i integration. The case of a nonleading singularity is easily disposed of. Suppose that n parameters α_i vanish at the solution of the Landau equations: $\alpha_i^{(0)} = 0$, $i \in \mathscr{I}$. After integration over those parameters in a neighborhood of the origin, we are back to the first case.

The same method may be used to study the type of singularity. Suppose that we are near the singularity, $s \simeq s^{(0)}$, and consider the contribution to the integral from a neighborhood of $\alpha = \alpha^{(0)}$. We write

$$\Delta\mathscr{S} = \sum \frac{\partial\mathscr{S}}{\partial s_a}\Delta s_a + \sum_{\substack{i\in\mathscr{I}\\ i=1,\ldots,n}} \frac{\partial\mathscr{S}}{\partial\alpha_i}\Delta\alpha_i + \frac{1}{2}\sum_{\substack{i,j\in R\\ i,j=n+1,\ldots,I-1}} \frac{\partial^2\mathscr{S}}{\partial\alpha_i\partial\alpha_j}\Delta\alpha_i\Delta\alpha_j$$

where the first term comes from the variation $\Delta s_a = s_a - s_a^{(0)}$ of the invariants, the second one from the n parameters $(i \in \mathscr{I})$ near zero, and the third from the parameters of the reduced diagram R. As the α_i, $i \in \mathscr{I}$, vanish, the denominator $\mathscr{P}_G(\alpha)$ in Eq. (6-91) vanishes as a power $L(\mathscr{I})$, where $L(\mathscr{I})$ is the number of independent loops of the subdiagram associated to $\mathscr{I}$. A simple counting argument yields the following behavior for the integral $I_G(P)$ as a function of the number of loops and internal lines of the reduced diagram

$$I_G(P) \sim (\Delta s)^k$$

where
$$k = (2L - I) + [n - 2L(\mathscr{I})] + \tfrac{1}{2}(I - 1 - n) = \tfrac{1}{2}[4L(R) - I(R) - 1] \tag{6-108}$$

Whenever k is a nonnegative integer, this behavior may be modified by logarithms

$$I_G(P)_{s\sim s^{(0)}} \sim (\Delta s)^k \ln^m|\Delta s| \qquad k = 0, 1, \ldots \tag{6-109}$$

This result is actually independent of the singularity being real. We conclude that, typically, Feynman integrals have logarithmic or square root branch points as functions of the invariant scalar products $s_{ij} = P_i \cdot P_j$.

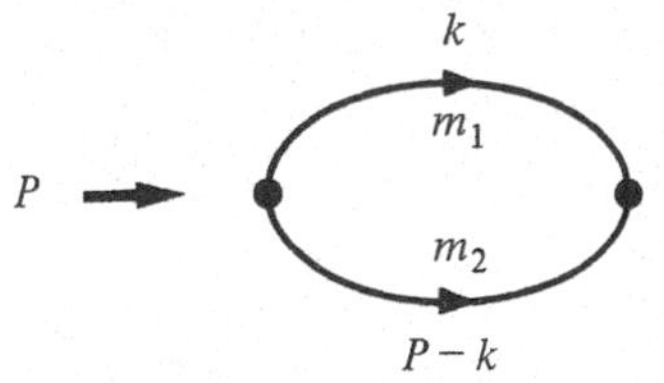

Figure 6-30 The bubble diagram.

Interesting properties may be proved concerning real singularities. In parametric space, at the solution of the Landau equations, the derivatives $\partial\mathscr{S}/\partial\alpha_j$ vanish for all nonzero $\alpha_j^{(0)}$. Therefore, such a solution corresponds either to a local extremum or to a saddle-point of $\mathscr{S}$ on the compact set $\Delta = \{0 \leq \alpha_i \leq 1, \sum \alpha_i = 1\}$. On the one hand, it may be shown that it cannot be a minimum. For instance, the self-energy diagram of Fig. 6-30 is singular at $P^2 = (m_1 + m_2)^2$. For this diagram the function reads

$$\mathscr{S} = \frac{\alpha_1\alpha_2 P^2}{\alpha_1 + \alpha_2} - \alpha_1 m_1^2 - \alpha_2 m_2^2$$

and for $P^2 = (m_1 + m_2)^2$ has a maximum zero on Δ at $\alpha_1 = m_2/(m_1 + m_2)$, $\alpha_2 = m_1/(m_1 + m_2)$, the solution to the Landau equations. On the other hand, saddle-point configurations do not seem to appear for Feynman integrals (although no proof of this point is known to the authors). Consequently, we restrict ourselves to solutions corresponding to local maxima of $\mathscr{S}$, called thresholds, where both real and imaginary parts of the amplitude are singular. The imaginary part gets a new additive contribution above the threshold, as we shall see, while the higher-order derivatives of the real part are divergent.

A further classification of thresholds appears useful. Normal thresholds are those singularities whose occurrence is expected from unitarity as recalled in the introduction of this section. Let us introduce the following definition. A set of l internal lines to G is said to be an intermediate state of G if cutting $l - 1$ lines among them leaves the diagram connected, while a further cut disconnects it into two parts G_1, G_2, each having incident external momenta. Notice that, in contrast with the concept of cut C introduced in Eq. (6-87), G_1 and G_2 may still contain loops. As in Eq. (6-88), we define (see Fig. 6-31)

$$s = \left(\sum_{v \in G_1} P_v\right)^2 = \left(\sum_{v \in G_2} P_v\right)^2$$

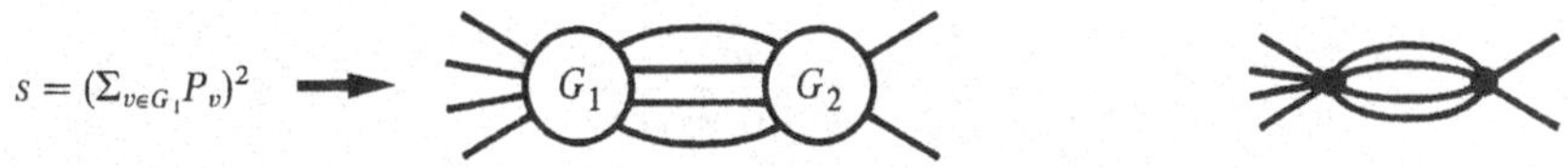

Figure 6-31 A diagram with an intermediate state of masses $m_1, \ldots, m_l$ and the corresponding reduced diagram.

Then

$$s = s^{(0)} = \Big(\sum_{\substack{\text{intermediate}\\ \text{state}}} m_i\Big)^2 \tag{6-110}$$

is a normal threshold of the amplitude. Physically, $s^{(0)}$ is the smallest value of the invariant s such that the physical state of masses $m_1, \ldots, m_l$ can be created.

Let us show that such an intermediate state does correspond to a solution of Landau's equations. We equate to zero all α pertaining to lines that do not belong to the intermediate state, say $\alpha_{l+1}, \ldots, \alpha_I$. Then for the reduced diagram

$$\mathscr{S}(\alpha_1, \ldots, \alpha_l, 0, \ldots, 0) = \frac{s}{\sum_{i=1}^{l} \alpha_i^{-1}} - \sum_{i=1}^{l} \alpha_i m_i^2$$

The equations $\partial\mathscr{S}/\partial\alpha_i = 0$ have a solution at $\sqrt{s} = \sum m_i$, $\alpha_i = m_i^{-1}/(\sum m_i^{-1})$, and this is easily seen to be a zero and a maximum of $\mathscr{S}$.

Contrary to naive expectations, there exist other real singularities, called anomalous thresholds. These singularities are somehow troublesome since their contribution to the absorptive part of the amplitude cannot be directly related through unitarity to physical processes. The condition for the existence of anomalous thresholds in a scattering amplitude corresponds to a situation where the axiomatic derivation of dispersion relations is no longer valid. It is therefore important to control the possible occurrence of these new singularities. This is a difficult problem and we shall only illustrate this phenomenon on simple examples in the next subsection.

In momentum space, the distinction between normal and anomalous thresholds may be interpreted in terms of the dimension of the space spanned by the internal momenta k_i of the reduced diagram, at the solution of the Landau equations. For a normal threshold, these momenta form a one-dimensional space. This is obvious from the set of equations (6-102) for the reduced diagram depicted in Fig. 6-30. These equations express that all the momenta k_i $(i = 1, \ldots, l)$ are collinear, and hence collinear with the external momentum $P = \sum_{v \in G_1}^{i} P_v$. On the contrary, the dimension is larger than one for an anomalous threshold.

6-3-3 Real Singularities of Simple Diagrams

The general considerations will be illustrated on one-loop diagrams. First we consider the bubble diagram of Fig. 6-30. We are interested in analyticity properties in the variable $s = P^2$, where P is the total energy momentum entering one or the other vertex. The diagram may represent a self-energy contribution in a φ^3 theory, or a scattering amplitude in a φ^4 theory, etc. At any rate, the solution of the Landau equations is trivial. We write the integral in momentum space as

$$T_G(P^2) = \frac{-i}{(2\pi)^4} \int \frac{d^4k}{(k^2 - m_1^2)[(P-k)^2 - m_2^2]} \tag{6-111}$$

Here instead of $I_G(P)$ defined in (6-83) and (6-101) we consider the quantity $T_G(P) = (-i)^{V+1} I_G(P)$ that contributes additively to the scattering amplitude of Eq. (5-171). The Landau equations express the existence of two real numbers λ_1 and λ_2 such that

$$\begin{cases} \lambda_1(k^2 - m_1^2) = 0 \\ \lambda_2[(P-k)^2 - m_2^2] = 0 \\ \lambda_1 k_\mu - \lambda_2 (P-k)_\mu = 0 \end{cases} \tag{6-112}$$

This means that P and k are collinear, and it follows that $m_1^2\lambda_1^2 = m_2^2\lambda_2^2$, $s = P^2 = (m_1 \pm m_2)^2$. The value $(m_1 + m_2)^2$ is the expected normal threshold, while the other one, $(m_1 - m_2)^2$, will soon be shown not to occur on the physical sheet. It seems that we are cheating, since the integral (6-111) is logarithmically ultraviolet divergent and requires a renormalization. Only the value of $T_G(P^2)$ subtracted at some point is meaningful, but this subtraction does not modify either the Landau equations or the singularity structure. In parametric space, the subtracted integral reads

$$T_G(s) - T_G(s_1) = -\frac{1}{(4\pi)^2}\int_0^1 \frac{d\alpha_1\, d\alpha_2\, \delta(1-\alpha_1-\alpha_2)}{(\alpha_1+\alpha_2)^2} \ln\left[\frac{\alpha_1\alpha_2 s/(\alpha_1+\alpha_2) - \alpha_1 m_1^2 - \alpha_2 m_2^2}{\alpha_1\alpha_2 s_1/(\alpha_1+\alpha_2) - \alpha_1 m_1^2 - \alpha_2 m_2^2}\right]$$

as is readily seen from Eq. (6-90) if the integration over λ is carried out after subtraction of the integrand at an arbitrary value s_1. The singularities come from possible zeros of $\mathscr{S} = \alpha_1\alpha_2 s/(\alpha_1 + \alpha_2) - \alpha_1 m_1^2 - \alpha_2 m_2^2$ and the equations $\partial\mathscr{S}/\partial\alpha_i = 0$ $(i = 1, 2)$ lead to

$$m_1^2 = \frac{\alpha_2^2 s}{(\alpha_1+\alpha_2)^2} \qquad m_2^2 = \frac{\alpha_1^2 s}{(\alpha_1+\alpha_2)^2}$$

Hence $s = (m_1 + m_2)^2$ since $\alpha_1, \alpha_2 > 0$. The integral may be computed explicitly:

$$4\pi^2[T_G(s) - T_G(s_1)] = \left[\frac{\lambda^{1/2}}{2s}\ln\left(\frac{m_1^2+m_2^2-s+\lambda^{1/2}}{m_1^2+m_2^2-s-\lambda^{1/2}}\right) - \frac{m_1^2-m_2^2}{2s}\ln\frac{m_1^2}{m_2^2}\right] - [s = s_1] \tag{6-113}$$

In this expression the last term reminds us that the amplitude is subtracted at $s = s_1$, and λ stands for the function introduced in (5-155a):

$$\lambda = \lambda(s, m_1^2, m_2^2) = s^2 + m_1^4 + m_2^4 - 2(m_1^2 + m_2^2)s - 2m_1^2 m_2^2 = [s - (m_1+m_2)^2][s - (m_1-m_2)^2]$$

The argument of the logarithm can be rewritten as

$$\frac{m_1^2+m_2^2-s+\lambda^{1/2}}{m_1^2+m_2^2-s-\lambda^{1/2}} = \left[\frac{\sqrt{(m_1+m_2)^2-s}+\sqrt{(m_1-m_2)^2-s}}{\sqrt{(m_1+m_2)^2-s}-\sqrt{(m_1-m_2)^2-s}}\right]^2$$

and in this form it is clear that in the physical sheet, corresponding to the first determination of the logarithm, $T_G(s)$ is free from singularity at $s = (m_1 - m_2)^2$. On the other hand, singularities at $s = 0$ and $s = (m_1 - m_2)^2$ occur in the unphysical sheets.

Even though this is not the main purpose of this subsection, we may elaborate a little on these complex singularities. The singularity at $s = (m_1 - m_2)^2$, also called a pseudothreshold, is seen to correspond to a solution of the Landau equations in parametric space at a negative value of α_1 or α_2, thus outside the initial integration interval. This means that going to unphysical sheets has forced us to deform the contour of integration from $0 \le \alpha \le 1$ to a complex path. This is the analog of the phenomenon discussed in the example (6-99b). Similarly, the singularity at $s = 0$ is the analog of the case (6-99c). This singularity, which did not appear as a solution of the first set of Landau equations (6-112) is an illustration of the so-called second-type singularities that we have discarded by assuming that the boundaries at infinity of the integration domain in momentum space were irrelevant. They are not, actually, because a pinching at infinity may occur which leads to the singularity at $s = 0$, as seen in the form (6-111) of the integral. The Landau equations express that the hyperboloids $k^2 = m_1^2$ and $(P - k)^2 = m_2^2$ are tangent, which occurs when $P^2 = (m_1 \pm m_2)^2$; but the contact may also take place at infinity, when the centers coincide or are separated by a zero length vector, that is, $s = P^2 = 0$.

The appearance of a singularity at $s = 0$ must be a warning. The analysis of complex singularities is indeed . . . complex!

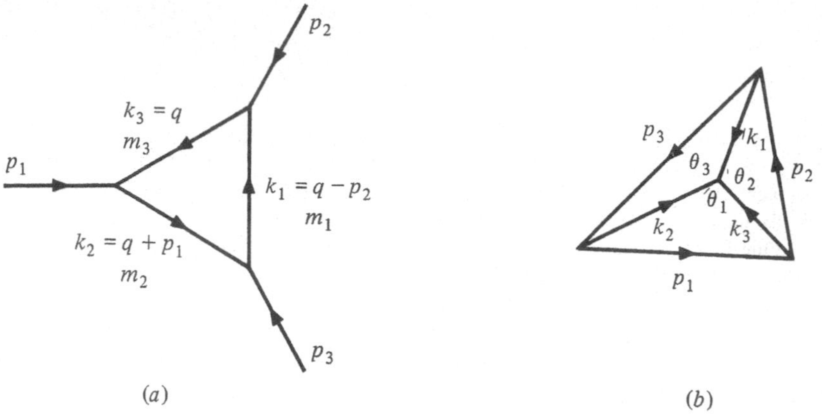

Figure 6-32 The triangle vertex diagram, and its dual diagram.

In the physical sheet, the imaginary discontinuity of $T_G(s)$ across the cut running from $(m_1 + m_2)^2$ to $+\infty$ is

$$T_G(s + i\varepsilon) - T_G(s - i\varepsilon) = 2i\Delta T_G(s) = 2i\,\frac{\lambda^{1/2}(s, m_1^2, m_2^2)}{4\pi s}\,\theta[s - (m_1 + m_2)^2] \tag{6-114}$$

From this absorptive part, the amplitude $T_G(s) - T_G(s_1)$ may be reconstructed by a once-subtracted dispersion relation, i.e., by the Cauchy formula along a contour encircling the cut:

$$T_G(s) - T_G(s_1) = \frac{s - s_1}{\pi}\int_{(m_1+m_2)^2}^{\infty} \frac{ds'\,\mathrm{Im}\,T_G(s')}{(s' - s_1)(s' - s)} \tag{6-115}$$

We now turn to the vertex diagram of Fig. 6-32*a*:

$$T_G(p_1^2, p_2^2, p_3^2) = -i\int \frac{d^4q}{(2\pi)^4}\,\frac{1}{(k_1^2 - m_1^2)(k_2^2 - m_2^2)(k_3^2 - m_3^3)} \tag{6-116}$$

with $k_1 = q - p_2$, $k_2 = q + p_1$, and $k_3 = q$. The Landau equations consist of the system

$$\lambda_i(k_i^2 - m_i^2) = 0 \qquad i = 1, 2, 3$$
$$\sum_i \lambda_i k_i = 0$$

For the leading singularity, none of the λ vanish; therefore the determinant of scalar products $(k_i \cdot k_j)$ must vanish. This is more conveniently rewritten as

$$\begin{vmatrix} 1 & y_{12} & y_{13} \\ y_{12} & 1 & y_{23} \\ y_{13} & y_{23} & 1 \end{vmatrix} = 0 \tag{6-117}$$

where $y_{ij} = k_i \cdot k_j / m_i m_j = -(p_k^2 - m_i^2 - m_j^2)/2m_i m_j$, (i, j, k) being a permutation of (1, 2, 3).

Similarly, for the nonleading singularity occurring at $\lambda_3 = 0$, say, we must have

$$\begin{vmatrix} 1 & y_{12} \\ y_{12} & 1 \end{vmatrix} = 0$$

that is,

$$y_{12} = \pm 1 \qquad p_3^2 = (m_1 \pm m_2)^2$$

We recover the normal threshold at $s = p_3^2 = (m_1 + m_2)^2$ while, as in the case of the bubble diagram, the singularity at $p_3^2 = (m_1 - m_2)^2$ does not show up in the physical sheet. This may be seen by the analysis of the Landau equations in parametric space. The discontinuity across the cut between $(m_1 + m_2)^2$ and $+\infty$ is (with $s = p_3^2$)

$$T_G(s + i\varepsilon) - T_G(s - i\varepsilon) = \frac{2i}{4\pi\lambda^{1/2}(m_1^2, m_2^2, s)} \ln \frac{a + b}{a - b} \tag{6-118}$$

where
$$a = s^2 - s(p_1^2 + p_2^2 + m_1^2 + m_2^2 - 2m_3^2) - (m_1^2 - m_2^2)(p_1^2 - p_2^2)$$

and
$$b = \lambda^{1/2}(m_1^2, m_2^2, s)\lambda^{1/2}(p_1^2, p_2^2, s)$$

Of course, similar normal thresholds exist in the channels p_1^2 and p_2^2.

To analyze the anomalous threshold, we introduce the dual diagram of Fig. 6-32*b*. It is a tetrahedron in three-dimensional euclidean space with edges of square lengths p_i^2 and k_i^2. This is possible since $p_i^2 > 0$ and the k_i are on-shell, $k_i^2 = m_i^2$. The stability conditions of internal and external particles insure that the angles θ_i $(0 \le \theta_i \le \pi)$ defined through

$$\cos\theta_3 = -\frac{p_3^2 - m_1^2 - m_2^2}{2m_1m_2}$$

are real. Equation (6-117) implies that the dual diagram lies in a plane. Finally, the reality condition requires the central point of Fig. 6-32*b* to lie inside the triangle. This means $\theta_1 + \theta_2 > \pi$ or, equivalently, $\cos\theta_1 + \cos\theta_2 < 0$, that is,

$$m_1(p_1^2 - m_2^2 - m_3^2) + m_2(p_2^2 - m_1^2 - m_3^2) > 0 \tag{6-119}$$

Let us derive this result using the parametric representation, in the special case where $p_1^2 = p_2^2 \equiv p^2$, $m_1^2 = m_2^2 \equiv m^2$. The function $\mathscr{S}$ reads

$$\mathscr{S} = \frac{\alpha_1\alpha_2 p_3^2 + \alpha_2\alpha_3 p_1^2 + \alpha_3\alpha_1 p_2^2}{\alpha_1 + \alpha_2 + \alpha_3} - \alpha_1 m_1^2 - \alpha_2 m_2^2 - \alpha_3 m_3^2$$

and the Landau equations, after elimination of $\alpha_1 = 1 - \alpha_2 - \alpha_3$, yield, for $\alpha = \alpha_2 = \alpha_3$ $(0 < \alpha < \frac{1}{2}$ since $0 < \alpha_1 < 1)$,

$$\begin{cases} \alpha^2(4p^2 - p_3^2) = m_3^2 \\ (4p^2 - p_3^2)(\alpha^2 - \alpha) = m^2 - p^2 \end{cases}$$

The solution α lies between 0 and $\frac{1}{2}$ provided

$$p^2 > m^2 + m_3^2 \tag{6-120}$$

This expresses condition (6-119) in that special case. The anomalous threshold takes place at

$$p_3^2 = -\frac{1}{m_3^2}\lambda(p^2, m^2, m_3^2) = 4m^2 - \frac{1}{m_3^2}(p^2 - m^2 - m_3^2)^2$$

If we consider a physical instance such as the electromagnetic form factor of the nucleon, it seems that we are in trouble since an anomalous threshold might occur below the normal one. Indeed, the lowest state coupled to the photon is the two-pion system, $m = m_1 = m_2 \ge m_\pi$, and the lowest one coupled to the on-shell nucleon $(p^2 = p_1^2 = p_2^2 = M_N^2)$ is the pion-nucleon state; hence $m + m_3 \ge m_\pi + M_N$. If we take $m_3 = m = (m_\pi + M_N)/2$, all the previous conditions are fulfilled; Eq. (6-120) reduces to

$$M_N^2 \ge \tfrac{1}{2}(M_N + m_\pi)^2 \qquad \text{that is, } (\sqrt{2} - 1)M_N \ge m_\pi$$

and is obviously satisfied by the experimental masses. Fortunately, this reasoning has not taken into account the various conservation laws, e.g., the conservation of baryonic charge which tells us that any state coupled to the nucleon must contain at least one baryon, of mass larger than M_N. Thus the condition (6-120) cannot be fulfilled and we are safe. Similarly, anomalous thresholds are not present in the form factors of pions or kaons, but they appear for hyperons or deuterons.

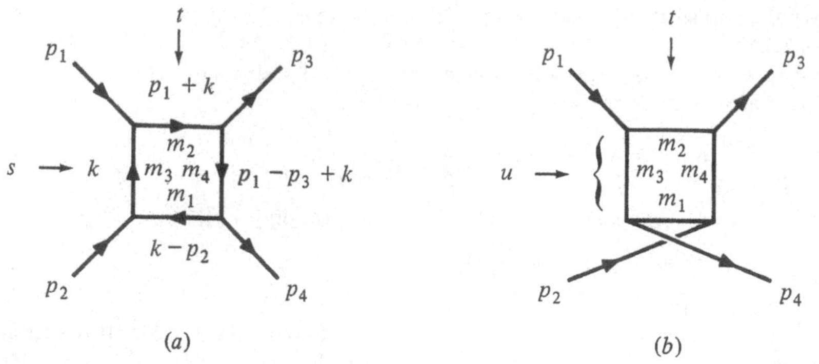

Figure 6-33 The box diagrams in the $s - t$ channel (*a*) and in the crossed $u - t$ channel (*b*).

The analyticity properties of the scattering amplitude depicted in Fig. 6-33*a*, the box diagram, may be studied along the same lines:

$$T_G = -i \int \frac{d^4k}{(2\pi)^4} \frac{1}{[(p_1 + k)^2 - m_2^2][(p_1 - p_3 + k)^2 - m_4^2][(p_2 - k)^2 - m_1^2](k^2 - m_3^2)} \tag{6-121}$$

In the s channel, we find a normal threshold at $s = (m_1 + m_2)^2$ [in the t channel at $t = (m_3 + m_4)^2$]. The discontinuity across the corresponding cut may be evaluated:

$$T_G(s + i\varepsilon, t) - T_G(s - i\varepsilon, t) = 2i\Delta_s T_G(s, t)$$
$$= 2i \frac{2s\lambda^{1/2}(s, m_1^2, m_2^2)}{4\pi D^{1/2}} \ln \left[\frac{(ac - bd \cos\theta) + D^{1/2}}{(ac - bd\cos\theta) - D^{1/2}}\right] \theta[s - (m_1 + m_2)^2] \tag{6-122}$$

where θ stands for the center of mass scattering angle

$$\cos\theta = \frac{s}{\lambda^{1/2}(s, p_1^2, p_2^2)\lambda^{1/2}(s, p_3^2, p_4^2)}\left[t - u + \frac{(p_1^2 - p_2^2)(p_3^2 - p_4^2)}{s}\right]$$

The expressions of a and b have been given in (6-118), and c and d are obtained from a and b by changing p_1 into p_3, p_2 into p_4, m_3 into m_4. Finally,

$$D = (ac - bd\cos\theta)^2 - (a^2 - b^2)(c^2 - d^2)$$

These expressions simplify in the case of equal masses $m_1^2 = m_2^2 = m^2$, $m_3^2 = m_4^2 = m'^2$, $p_1^2 = p_2^2 = p_3^2 = p_4^2 = p^2$ to

$$\frac{2s\lambda^{1/2}(s, m_1^2, m_2^2)}{D^{1/2}} = \{st[(s - 4m^2)(t - 4m'^2) - 4(p^2 - m^2 - m'^2)^2]\}^{-1/2} \tag{6-123}$$

The discontinuity $\Delta_s T_G$ considered as a function of t has itself a discontinuity in the t channel, called the double spectral function ρ_{st}:

$$\rho_{st}(s, t) = \frac{1}{2i}[\Delta_s T_G(s, t + i\varepsilon) - \Delta_s T_G(s, t - i\varepsilon)]$$
$$= \frac{2s\lambda^{1/2}(s, m_1^2, m_2^2)}{4D^{1/2}} \theta[s - (m_1 + m_2)^2]\theta[t - (m_3 + m_4)^2]\theta(D) \tag{6-124}$$

Therefore, $T_G(s, t)$ satisfies fixed s or fixed t dispersion relations, or even an analytic representation in the set of variables s and t:

$$T_G(s, t) = \frac{1}{\pi} \int_{(m_1+m_2)^2}^{\infty} ds' \frac{\Delta_s T_G(s', t)}{s' - s}$$
$$= \frac{1}{\pi^2} \iint ds'\, dt' \frac{\rho_{st}(s', t')}{(s' - s)(t' - t)} \tag{6-125a}$$

The second expression in (6-125*a*) involving the double discontinuity ρ_{st} is a special case of a Mandelstam representation. The box diagram considered here has vanishing double spectral functions ρ_{su} or ρ_{tu}. This is not the case of the crossed diagram of Fig. 6-33*b*, for which ρ_{tu} only is nonvanishing, and which therefore satisfies the fixed *s* dispersion relation

$$T_G(t, u) = \frac{1}{\pi} \int_{(m_3+m_4)^2}^{\infty} dt' \frac{\Delta_t T_G(t', u' = \sum p_i^2 - t' - s)}{t' - t} + \frac{1}{\pi} \int_{(m_1+m_2)^2}^{\infty} du' \frac{\Delta_u T_G(t' = \sum p_i^2 - s - u', u')}{u' - u} \tag{6-125b}$$

or the Mandelstam representation

$$T_G(t, u) = \frac{1}{\pi^2} \iint dt'\, du' \frac{\rho_{tu}(t', u')}{(t' - t)(u' - u)} \tag{6-125c}$$

All this is not completely correct. In the foregoing, we have assumed that all external particles are stable against decay: $p_1^2 < (m_2 + m_3)^2$, etc., and that no anomalous threshold occurs. The precise conditions for this may be studied in the same way as in the vertex case.

It has been conjectured that, more generally, in favorable cases such as pion-nucleon scattering, the physical scattering amplitude satisfies a general Mandelstam representation of the form

$$\begin{aligned}
T_G(s, t, u) = T_G(s_1, t_1, u_1) &+ \frac{(s - s_1)}{\pi} \int_{M_s^2}^{\infty} \frac{\rho_1(s')}{(s' - s)(s' - s_1)} ds' \\
&+ \frac{(t - t_1)}{\pi} \int_{M_t^2}^{\infty} \frac{\rho_2(t')}{(t' - t)(t' - t_1)} dt' + \frac{(u - u_1)}{\pi} \int_{M_u^2}^{\infty} \frac{\rho_3(u')}{(u' - u)(u' - u_1)} du' \\
&+ \frac{(s - s_1)(u - u_1)}{\pi^2} \iint ds'\, du' \frac{\rho_{su}(s', u')}{(s' - s)(s' - s_1)(u' - u)(u' - u_1)} \\
&+ \frac{(t - t_1)(u - u_1)}{\pi^2} \iint dt'\, du' \frac{\rho_{tu}(t', u')}{(t' - t)(t' - t_1)(u' - u)(u' - u_1)} \\
&+ \frac{(s - s_1)(t - t_1)}{\pi^2} \iint ds'\, dt' \frac{\rho_{st}(s', t')}{(s' - s)(s - s_1)(t' - t)(t' - t_1)}
\end{aligned} \tag{6-126}$$

where the double spectral functions ρ_{st}, ρ_{tu}, and ρ_{us} are nonvanishing in definite regions (the shaded areas of Fig. 6-34). In Eq. (6-126), we have assumed that subtractions were necessary.

Beyond its illustrative purpose, the study of these simple diagrams may be useful as a preliminary step before tackling more complicated cases. For instance, through the use of a majorization technique, analyticity properties of higher-order diagrams may be deduced from those of simpler "primitive diagrams." For this subject, we refer to more specialized references.

6-3-4 Physical-Region Singularities. Cutkosky Rules

Physical-region singularities correspond to real external four-momenta. Such is the case of normal thresholds, while anomalous thresholds do not appear as a leading singularity in the physical region if the stability condition of external particles holds, i.e., if for each vertex the square of the incident four-momentum P_v^2 is smaller than the lowest normal threshold of that channel.

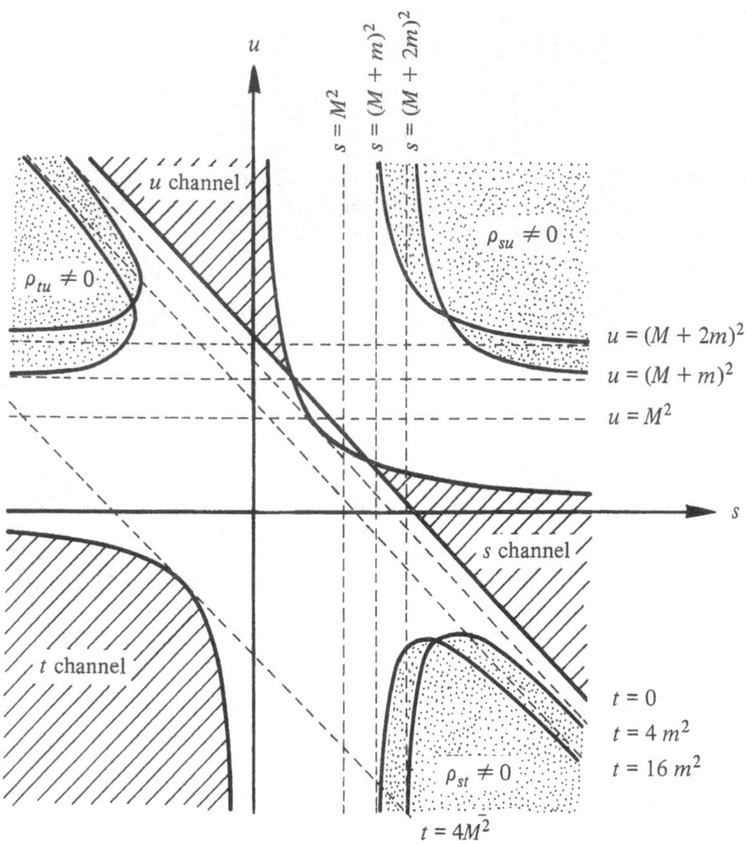

Figure 6-34 The *stu* plane for pion-nucleon scattering. The areas shaded by diagonal lines are the physical regions of the s channel $\pi N \to \pi N$, u channel $\pi N \to \pi N$, and t channel $\pi\pi \to N\bar{N}$; m and M stand for the pion and nucleon masses respectively. The double spectral regions have been shaded by dots.

Coleman and Norton have found the following simple interpretation. The leading singularity of a diagram G occurs in the physical region if and only if the vertices of G may be regarded as real space-time points and their internal lines as the trajectories of real on-shell relativistic particles.

To prove this statement, it may be more convenient to use the mixed representation of Eq. (6-104). For physical-region singularities, the integration variables α_i and q_l must take real values at the solution of the Landau equations. The latter read [compare with Eqs. (6-102), (6-103), and (6-106)]

$$k_i^2 = m_i^2 \tag{6-127a}$$

$$\sum_{i \in \mathscr{L}_l} (\pm)\alpha_i k_i = 0 \tag{6-127b}$$

The reality of the external momenta P and of the integration variables q implies the reality of the k. These four-momenta are on their mass shell, owing to (6-127a).

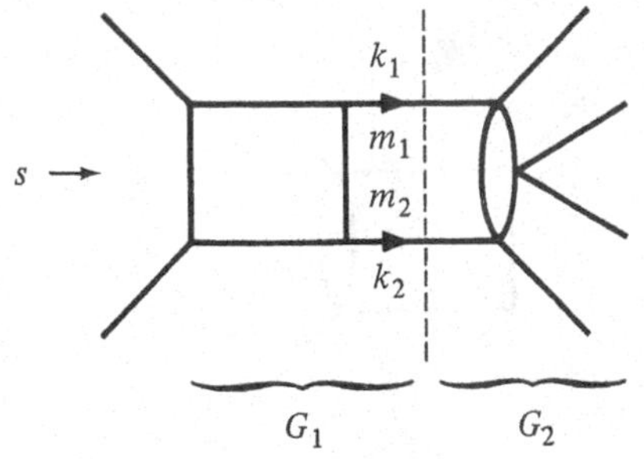

Figure 6-35 Illustration of the Cutkosky rule: the cut represented by a broken line contributes to the singularity at $s = (m_1 + m_2)^2$. The momenta k_1 and k_2 are on their mass shell: $k_1^2 = m_1^2$, $k_2^2 = m_2^2$.

As for Eq. (6-127*b*), it means that if particle i propagates with momentum $(\pm k_i)$ during the proper time interval $\alpha_i m_i$ $(\alpha_i > 0)$, the total space-time displacement along a closed loop is zero. In other words, the diagram may be considered as a genuine physical process.

The Cutkosky rules give a compact expression for the discontinuity across the cut arising from a physical-region singularity. Let $I_G(P)$ be the Feynman amplitude of Eq. (6-101) and $T_G(P) = (-i)^{V+1} I_G(P)$ the corresponding contribution to the scattering amplitude. The discontinuity $2i\Delta T_G(P)$ in a given channel associated with a specific cut is obtained as follows. Let $(I - I_1 - I_2)$ internal momenta be on-shell, $k_j^2 = m_j^2$, and $I_1 + I_2$ momenta off-shell. These may be subdivided into two groups pertaining to two subdiagrams G_1 and G_2, each with I_i internal lines and V_i vertices (Fig. 6-35). The following formula holds:

$$\Delta T_G(P) = \frac{1}{2}(-i)^{V_1 - V_2} \int \prod_1^L \frac{d^4 q_l}{(2\pi)^4} \prod_{l_1=1}^{I_1} \frac{i}{k_{l_1}^2 - m_{l_1}^2 + i\varepsilon}$$
$$\times \prod_{l_2=1}^{I_2} \frac{-i}{k_{l_2}^2 - m_{l_2}^2 - i\varepsilon} \prod_{j=I_1+I_2+1}^{I} (2\pi)\theta(k_j^0)\delta(k_j^2 - m_j^2) \qquad (6\text{-}128)$$

The role of the δ functions is to put the particles corresponding to the intermediate state on their positive energy mass shell. In the physical region, Eq. (6-128) follows from the fact that individual diagrams satisfy the unitarity condition.

The following elegant proof is borrowed from Nakanishi. We start from an arbitrary diagram, assumed for simplicity to arise in a scalar theory. We observe that we may always construct a hermitian lagrangian such that the lowest-order scattering amplitude for a definite process is the diagram under consideration. To this end, we attach to each internal line l of the diagram a different species of field, φ_l, of mass m_l (the mass of the corresponding propagator). Moreover, if an external line (or set of lines) of momentum P_v enters the vertex v, we attach to this vertex a field ϕ_v, of squared mass $M_v^2 = P_v^2$. Provided $P_v^2 > 0$, the initial diagram may now be regarded as the lowest-order contribution to the on-shell scattering amplitude of the particles described by $\phi_1, \phi_2, \ldots$, derived from the lagrangian

$$\mathscr{L}_1 = \sum_{l=1}^{I} \tfrac{1}{2}[(\partial\varphi_l)^2 - m_l^2\varphi_l^2] + \sum_v \tfrac{1}{2}[(\partial\phi_v)^2 - M_v^2\phi_v^2] + \sum_{v=1}^{V} \phi_v \left(\prod_{l:\,\varepsilon_{vl}\neq 0} \varphi_l \right) \qquad (6\text{-}129)$$

In the last sum, ϕ_v must be replaced by one if no external line is incident to v, and the product runs over all internal lines incident to v. This construction is illustrated in Fig. 6-36. If i denotes the

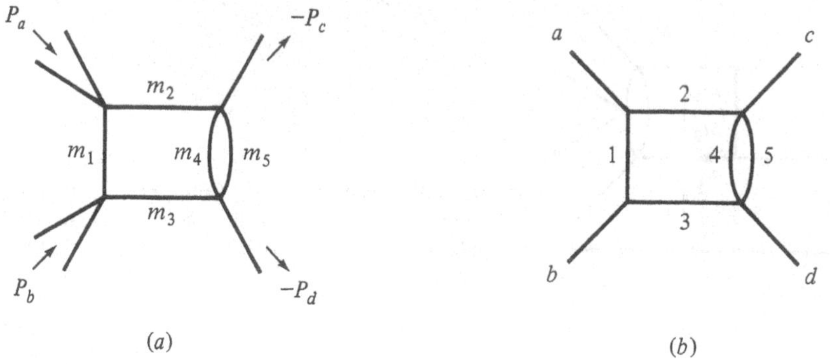

Figure 6-36 The diagram (a) has the same analyticity properties as the diagram (b), which may be regarded as the lowest-order contribution to the on-shell amplitude $a + b \to c + d$ in the theory described by

$$\mathscr{L}_1 = \sum_{l=1}^{5} \tfrac{1}{2}[(\partial\varphi_l)^2 - m_l^2\varphi_l^2] + \sum_{a,b,c,d} \tfrac{1}{2}[(\partial\phi_a)^2 - M_a^2\phi_a^2] + \phi_a\varphi_1\varphi_2 + \phi_b\varphi_1\varphi_3 + \phi_c\varphi_2\varphi_4\varphi_5 + \phi_d\varphi_3\varphi_4\varphi_5$$

initial state (characterized by $P_v^0 > 0$) and f the final one, it is straightforward to check that

$$\mathscr{T}_{fi} = T_G(P)$$

where $\mathscr{T}_{fi}$ is computed from $\mathscr{L}_1$, while T_G is simply the amplitude associated with the initial Feynman diagram G. Since unitarity holds for on-shell processes [Eq. (5-154)], we have

$$\mathscr{T}_{fi} - \mathscr{T}_{if}^* = i \sum_n (2\pi)^4 \delta^4(P_n - P_i) \mathscr{T}_{nf}^* \mathscr{T}_{ni} = 2i\Delta T_G(P) \tag{6-130}$$

In the language of Feynman diagrams, the sum runs over all possible intermediate physical states n, that is, such that the intermediate momenta are on their mass shell

$$k_l^2 = m_l^2 \qquad k_l^0 > 0 \qquad l \in \mathscr{I}$$

Unitarity thus yields (6-131)

$$2\Delta T_G(P) = \sum T_{G_1}(P) T_{G_2}^*(P)$$

where the sum runs over all partitions of G into two subparts G_1, G_2; G_1 (respectively G_2) is connected to the initial state i (respectively final f) and G is the union $G_1 \cup G_2 \cup \{\mathscr{I}\}$. It also includes an integral over the intermediate-state phase space. We thus recover the Cutkosky rules.

The reader will verify this rule in the case of one-loop diagrams.

We may also try to extend it beyond the physical region, thereby generalizing unitarity. This is, for instance, what is needed, if we want to compute the double spectral functions defined in the last subsection. However, proving these general Cutkosky rules requires nontrivial methods. In particular, much care is necessary, since the meaning of the $\delta^{(+)}$ distribution for complex k becomes hazardous.

NOTES

Covariant perturbation theory, born in the mid-1940s, is due to S. Tomonoga, *Prog. Theor. Phys.*, vol. I, p. 27, 1946; J. Schwinger, *Phys. Rev.*, vol. 74, p. 1439,

1948, vol. 75, p. 651, 1949, vol. 76, p. 790, 1949; R. P. Feynman, *Phys. Rev.*, vol. 76, p. 769, 1949; and F. J. Dyson, *Phys. Rev.*, vol. 75, pp. 486 and 1736, 1949. The definition of a covariant propagator was anticipated in the work of E. C. G. Stueckelberg; see his paper with D. Rivier, *Phys. Rev.*, vol. 74, p. 218, 1948.

Scalar electrodynamics was considered by P. T. Matthews, *Phys. Rev.*, vol. 80, p. 292, 1950. The method used in the text is from F. Rohrlich, *Phys. Rev.*, vol. 80, p. 666, 1950.

The early calculations on electron-electron and electron-positron scattering are recorded in the book by N. F. Mott and H. S. W. Massey, "Theory of Atomic Collisions," Oxford, 1956; see especially chap. XXII. Polarization effects are discussed in W. H. McMaster, *Rev. Mod. Phys.*, vol. 33, p. 8, 1961.

The Legendre transformation was introduced in field theory by G. Jona-Lasinio, *Nuovo Cimento*, vol. 34, p. 1790, 1964. For a general survey of diagrammatics see G. 't Hooft and M. Veltman, "Diagrammar," Cern report 73-9, Geneva, 1973.

Topological properties and parametric representations of Feynman diagrams are studied in the book by N. Nakanishi, "Graph Theory and Feynman Integrals," Gordon and Breach, New York, 1970, which also contains an introduction to the analyticity properties.

Among the many contributors to this study let us quote L. D. Landau, *Nucl. Phys.*, vol. 13, p. 181, 1959; S. Mandelstam, *Phys. Rev.*, vol. 112, p. 1344, 1958, vol. 115, p. 1741, 1959; and R. E. Cutkosky, *J. Math. Phys.*, vol. 1, p. 429, 1960. An interpretation of physical-region singularities is found in S. Coleman and R. E. Norton, *Nuovo Cimento*, vol. 38, p. 438, 1965.

More details and references can be found in "The Analytic *S*-Matrix" by R. J. Eden, P. V. Landshoff, D. I. Olive, and J. C. Polkinghorne, Cambridge University Press, 1966; "Analytic Properties of Feynman Diagrams in Quantum Field Theory" by I. T. Todorov, Pergamon Press, Oxford, 1971; and in the textbook of J. D. Bjorken and S. D. Drell, "Relativistic Quantum Fields," McGraw-Hill, New York, 1965.

CHAPTER

SEVEN

RADIATIVE CORRECTIONS

The renormalization program of quantum field theory is presented and carried out to the order of one-loop diagrams in electrodynamics. It is then applied to the calculation of the magnetic moment anomaly, radiative corrections to Coulomb scattering (involving an analysis of infrared divergences), the atomic Lamb shift, and photon-photon scattering. We also include a discussion of induced electromagnetic long-range forces between neutral particles in the relativistic regime.

7-1 ONE-LOOP RENORMALIZATION

We undertake in this chapter the study of higher orders of perturbation theory. What appears, at first sight, as a straightforward exercise, requiring perhaps analytical skills, turns out to be a highly nontrivial problem due to the presence of ultraviolet divergences. The general presentation of the renormalization theory is postponed to a later chapter. In order to get some familiarity with the subject we concentrate here on the computation of radiative corrections to lowest order in quantum electrodynamics. This enables us to see how we extract sensible results from apparently ill-defined expressions, to compare them with experimental values, and to progressively introduce the concepts of renormalization. A serious drawback of this approach is that electrodynamics is in that respect a rather involved theory. We have to cope with gauge invariance and to disentangle infrared from ultraviolet divergences. Nevertheless, its amazing successes certainly make this effort worth while and justify inverting the logical order.

The parameters such as masses and coupling constants which appear in the lagrangian are not directly measurable quantities. In the classical point-particle theory, for instance, we must add to the bare mass an electromagnetic contribution to obtain the physical inertial mass. The latter is, of course, finite while the former may well be infinite. We shall therefore give an operational definition to the fundamental parameters (finite in number). Renormalization theory will then show that the perturbative expressions for Green functions are finite when expressed in terms of these physical parameters. Masses will generally be defined as isolated poles of two-point functions. The corresponding residues, which appear as multiplicative constants of scattering amplitudes, will be absorbed into the definition of renormalized fields. Finally, coupling constants will be chosen by fixing the value of certain amplitudes at appropriate points in momentum space.

In order to carry out this program it is better to deal first with well-defined finite quantities. The origin of divergences lies in the singular character of Green functions at short relative distances. Equivalently, in momentum space the Fourier transforms do not vanish fast enough at infinity. In an intermediate step we are then led to regularize the theory, i.e., to replace the original expressions by smoother ones such that the integrals become finite. We shall thus proceed in three steps: (1) regularize, (2) renormalize, and (3) eliminate the regularizing parameters. Renormalization will be successful if finite quantities are obtained as a result of this process.

7-1-1 Vacuum Polarization

Let us first take a look at the photon propagator in momentum space. To the free contribution

$$G^{[0]}_{\rho\nu}(k) = -i\left(\frac{g_{\rho\nu} - k_\rho k_\nu/\mu^2}{k^2 - \mu^2 + i\varepsilon} + \frac{k_\rho k_\nu/\mu^2}{k^2 - M^2 + i\varepsilon}\right) \qquad M^2 = \frac{\mu^2}{\lambda} \tag{7-1}$$

we should add a correction which, according to the rules of Chap. 6, is given to lowest order by (see Fig. 7-1)

$$\begin{aligned} G^{[1]}_{\rho\nu}(k) &= G^{[0]}_{\rho\rho'}(k)\bar{\omega}^{\rho'\nu'}(k)G^{[0]}_{\nu'\nu}(k) \\ \bar{\omega}^{\rho\nu}(k) &= -(-ie)^2 \int \frac{d^4p}{(2\pi)^4} \operatorname{tr}\left(\gamma^\rho \frac{i}{\not{p} - m + i\varepsilon}\gamma^\nu \frac{i}{\not{p} - \not{k} - m + i\varepsilon}\right) \end{aligned} \tag{7-2}$$

The additional minus sign arises from the fermion loop. The integral seems quadratically divergent for large internal momentum p. To give it a meaning we use the Pauli-Villars regularization. This amounts to minimally coupling the

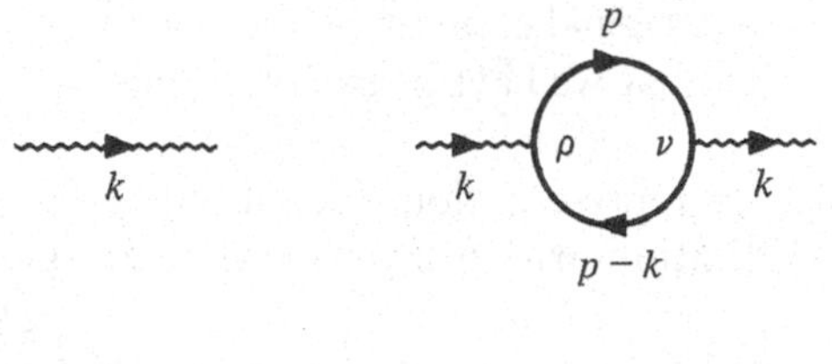

Figure 7-1 Photon propagator to lowest order.

photons to additional spinor fields with a very large mass $\lambda_s m$. These fields might correspond to indefinite metric sectors of the Hilbert space. As far as the vacuum polarization tensor $\bar{\omega}^{\rho\nu}(k)$ is concerned, this prescription implies the replacement

$$\bar{\omega}^{\rho\nu}(k, m) \to \bar{\omega}^{\rho\nu}(k, m) + \sum_{s=1}^{S} C_s \bar{\omega}^{\rho\nu}(k, \lambda_s m) \tag{7-3}$$

where the substitution is understood under the integral sign of Eq. (7-2). Were it not for the divergence of the integral we would recover the original value in the limit $\lambda_s \to \infty$. The constants C_s will be chosen in order to remove this divergence. The minimal coupling of the additional fields implies that gauge invariance is preserved in this regularization procedure.

Denote collectively by the symbol Λ the large masses $\lambda_s m$ and let $\bar{\omega}^{\rho\nu}(k, m, \Lambda)$ stand for the right-hand side of (7-3). We have

$$\begin{aligned}\bar{\omega}^{\rho\nu}(k, m, \Lambda) &= -e^2 \int \frac{d^4p}{(2\pi)^4} \left\{ \frac{\operatorname{tr} \gamma^\rho(\not{p} + m)\gamma^\nu(\not{p} - \not{k} + m)}{(p^2 - m^2 + i\varepsilon)[(p - k)^2 - m^2 + i\varepsilon]} \right. \\ &\quad \left. + \sum_{s=1}^{S} C_s(m \to \lambda_s m) \right\} \\ &= -4e^2 \int \frac{d^4p}{(2\pi)^4} \left\{ \frac{p^\rho(p-k)^\nu + p^\nu(p-k)^\rho - g^{\rho\nu}(p^2 - p\cdot k - m^2)}{(p^2 - m^2 + i\varepsilon)[(p-k)^2 - m^2 + i\varepsilon]} \right. \\ &\quad \left. + \sum_{s=1}^{S} C_s(m \to \lambda_s m) \right\}\end{aligned} \tag{7-4}$$

The function $\bar{\omega}^{\rho\nu}(k)$ is the Fourier transform of the current-current Green function. To any order, current conservation leads to the condition

$$k_\rho \bar{\omega}^{\rho\nu}(k) = 0 \tag{7-5}$$

which is formally satisfied, since

$$\begin{aligned}k_\rho \bar{\omega}^{\rho\nu}(k) &= -e^2 \int \frac{d^4p}{(2\pi)^4} \operatorname{tr} \left(\not{k} \frac{1}{\not{p} - m + i\varepsilon} \gamma^\nu \frac{1}{\not{p} - \not{k} - m + i\varepsilon} \right) \\ &= -e^2 \int \frac{d^4p}{(2\pi)^4} \operatorname{tr} \left(\gamma^\nu \frac{1}{\not{p} - \not{k} - m + i\varepsilon} - \gamma^\nu \frac{1}{\not{p} - m + i\varepsilon} \right)\end{aligned}$$

where we have replaced $\not{k}$ by $(\not{p} - m + i\varepsilon) - (\not{p} - \not{k} - m + i\varepsilon)$ and used the cyclic invariance of the trace. If the integral were convergent a translation of integration variable $p \to p + k$ in the first term would insure that condition (7-5) is fulfilled. While this manipulation is meaningless for the original expression, the same steps are justified for the regularized tensor, showing that we have a sensible regularization.

To evaluate (7-4) we use the parametric representation introduced in the previous chapter with the additional complication arising from the numerators in the fermion propagators. We therefore write

$$\bar{\omega}_{\rho\nu}(k, m, \Lambda) = -4e^2 \int \frac{d^4p}{(2\pi)^4} \left[\left[\frac{\partial}{\partial z_1^\rho}\frac{\partial}{\partial z_2^\nu} + \frac{\partial}{\partial z_1^\nu}\frac{\partial}{\partial z_2^\rho} - g_{\rho\nu}\left(\frac{\partial}{\partial z_1}\cdot\frac{\partial}{\partial z_2} + m^2\right)\right]\right.$$
$$\times \int_0^\infty d\alpha_1 \int_0^\infty d\alpha_2 \exp(i\{\alpha_1(p^2 - m^2) + \alpha_2[(p-k)^2 - m^2]$$
$$\left. + z_1\cdot p + z_2\cdot(p-k)\}) + \sum_{s=1}^{S} C_s(m \to \lambda_s m)\right]_{z_1=z_2=0}$$

Auxiliary four-vectors z_1 and z_2 have been introduced to generate the integrand in (7-4) by differentiation. Integrating over p and performing the required derivatives we obtain

$$\bar{\omega}_{\rho\nu}(k, m, \Lambda) = \frac{i\alpha}{\pi}\int_0^\infty\int_0^\infty \frac{d\alpha_1\, d\alpha_2}{(\alpha_1+\alpha_2)^2}\left(\left\{\frac{2\alpha_1\alpha_2}{(\alpha_1+\alpha_2)^2}k_\rho k_\nu - g_{\rho\nu}\left[\frac{\alpha_1\alpha_2 k^2}{(\alpha_1+\alpha_2)^2}\right.\right.\right.$$
$$\left.\left. - \frac{i}{\alpha_1+\alpha_2} + m^2\right]\right\}\exp\left\{i\left[-m^2(\alpha_1+\alpha_2) + \frac{\alpha_1\alpha_2}{\alpha_1+\alpha_2}k^2\right]\right\}$$
$$\left. + \sum_{s=1}^{S} C_s(m \to \lambda_s m)\right)$$

The polynomial in k appearing in the integral can be rearranged to read

$$2(k_\rho k_\nu - g_{\rho\nu}k^2)\frac{\alpha_1\alpha_2}{(\alpha_1+\alpha_2)^2} - g_{\rho\nu}\left[m^2 - \frac{\alpha_1\alpha_2}{(\alpha_1+\alpha_2)^2}k^2 - \frac{i}{(\alpha_1+\alpha_2)}\right]$$

Our calculation has remained covariant but $\bar{\omega}_{\rho\nu}$ does not explicitly exhibit current conservation, which requires it to be proportional to the combination $k_\rho k_\nu - g_{\rho\nu}k^2$. The second term in the above expression does not have this structure and its contribution reads $g_{\rho\nu}\Delta\bar{\omega}$, with

$$\Delta\bar{\omega} = -\frac{i\alpha}{\pi}\int_0^\infty\int_0^\infty \frac{d\alpha_1\, d\alpha_2}{(\alpha_1+\alpha_2)^2}\sum_{s=0}^{S} C_s\left[-\frac{\alpha_1\alpha_2 k^2}{(\alpha_1+\alpha_2)^2} - \frac{i}{\alpha_1+\alpha_2} + m_s^2\right]$$
$$\times \exp\left\{i\left[-m_s^2(\alpha_1+\alpha_2) + \frac{\alpha_1\alpha_2}{\alpha_1+\alpha_2}k^2\right]\right\}$$

where by convention $C_0 = 1$, $m_0^2 = m^2$, $m_s^2 = \lambda_s^2 m^2$. This can also be written

$$\Delta\bar{\omega} = -\frac{i\alpha}{\pi}\int_0^\infty\int_0^\infty \frac{d\alpha_1\, d\alpha_2}{(\alpha_1+\alpha_2)^3}\, i\rho\frac{\partial}{\partial\rho}\frac{1}{\rho}\sum_{s=0}^{S} C_s \exp\left\{i\rho\left[-m_s^2(\alpha_1+\alpha_2)\right.\right.$$
$$\left.\left.\left. + \frac{\alpha_1\alpha_2}{\alpha_1+\alpha_2}k^2\right]\right\}\right|_{\rho=1}$$

For fixed m_s the constants C_s are chosen in such a way that the integrals converge in the vicinity of $\alpha_1, \alpha_2 \to 0$. The convergence for large α is insured by the $i\varepsilon$ prescription implicit in the masses. We may then interchange derivatives on ρ with integration. The expression

$$\int_0^\infty \int_0^\infty \frac{d\alpha_1\, d\alpha_2}{(\alpha_1+\alpha_2)^3 \rho} \sum_{s=0}^{\infty} C_s \exp\left\{ i\rho \left[-m_s^2(\alpha_1+\alpha_2) + \frac{\alpha_1\alpha_2}{\alpha_1+\alpha_2} k^2 \right] \right\}$$

is then found to be ρ independent as shown by a change of variable $\alpha_i \to \rho^{-1}\alpha_i$. Therefore $\Delta\bar{\omega}$ vanishes. The remaining term of the vacuum polarization tensor reads

$$\bar{\omega}_{\rho\nu} = -i(g_{\rho\nu}k^2 - k_\rho k_\nu)\bar{\omega}$$

$$\bar{\omega}(k^2, m, \Lambda) = \frac{2\alpha}{\pi} \int_0^\infty \int_0^\infty d\alpha_1\, d\alpha_2 \frac{\alpha_1\alpha_2}{(\alpha_1+\alpha_2)^4} \times \sum_{s=0}^{S} C_s \exp\left\{ i \left[-m_s^2(\alpha_1+\alpha_2) + \frac{\alpha_1\alpha_2}{\alpha_1+\alpha_2} k^2 \right] \right\} \tag{7-6}$$

Using the homogeneity properties we introduce a factor $1 = \int_0^\infty d\rho\, \delta(\rho - \alpha_1 - \alpha_2)$ under the integral sign and change the variables according to $\alpha_i \to \rho\alpha_i$. We obtain

$$\bar{\omega}(k^2, m, \Lambda) = \frac{2\alpha}{\pi} \int_0^1 \int_0^1 d\alpha_1\, d\alpha_2\, \delta(1 - \alpha_1 - \alpha_2)\alpha_1\alpha_2 \int_0^\infty \frac{d\rho}{\rho} \sum_{s=0}^{S} C_s\, e^{i\rho[-m_s^2 + \alpha_1\alpha_2 k^2]}$$

The integral over ρ looks logarithmically divergent at the point $\rho = 0$. Having factorized two powers of momentum in passing from $\bar{\omega}^{\rho\nu}$ to $\bar{\omega}$ we have decreased the degree of divergence from two to zero. This is an example of the close relationship between the degree of divergence and the dimensionality of integrals.

If we choose the coefficients C_s in such a way that

$$\sum_{s=0}^{S} C_s = 1 + \sum_{s=1}^{S} C_s = 0 \tag{7-7}$$

the regularized integral over ρ will be convergent. Furthermore, let us pick k such that $k^2 < 4m^2$, which corresponds to the threshold of pair creation. Since α_1, α_2 are between zero and one and satisfy $\alpha_1 + \alpha_2 = 1$, it follows that $\alpha_1\alpha_2 \leq \frac{1}{4}$. Then $(m_s^2 - \alpha_1\alpha_2 k^2)$ is positive and the integration contour in the complex ρ plane can be rotated by $-\pi/2$ in such a way that the integral reads

$$\lim_{\eta\to 0} \int_\eta^\infty \frac{d\rho}{\rho} \sum_{s=0}^{S} C_s\, e^{-\rho(m_s^2 - \alpha_1\alpha_2 k^2)}$$

$$= \lim_{\eta\to 0} \sum_{s=0}^{S} C_s \left(-e^{-\rho} \ln \rho \big|_{\eta(m_s^2 - \alpha_1\alpha_2 k^2)} + \int_0^\infty d\rho\, e^{-\rho} \ln \rho \right)$$

Condition (7-7) eliminates the dangerous $\ln \eta$ term in the integrated expression. For fixed $k^2 < 4m^2$ and $m_s^2 = \lambda_s^2 m^2 \to \infty$ the result takes the form

$$-\left[\ln (m^2 - \alpha_1\alpha_2 k^2) + \sum_{s=1}^{\infty} C_s \ln m_s^2 \right] = -\left[\ln\left(1 - \frac{\alpha_1\alpha_2 k^2}{m^2}\right) + \sum_{s=1}^{\infty} C_s \ln \lambda_s^2 \right]$$

where we have neglected k^2 as compared to m_s^2. Let us define Λ such that

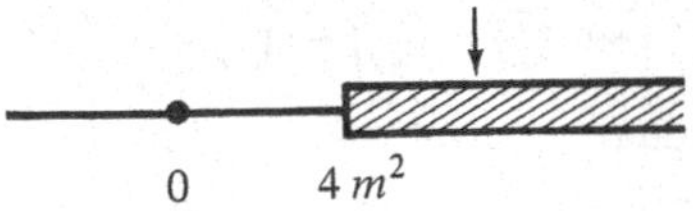

Figure 7-2 The complex k^2 plane for the vacuum polarization. The arrow indicates how the physical region is reached above threshold.

$$\sum_{s=1}^{\infty} C_s \ln \lambda_s^2 = -\ln \frac{\Lambda^2}{m^2} \tag{7-8}$$

We end up with the following expression for the regularized vacuum polarization:

$$\bar{\omega}(k^2, m, \Lambda) = -\frac{2\alpha}{\pi} \int_0^1 d\beta\, \beta(1-\beta) \left\{ -\ln \frac{\Lambda^2}{m^2} + \ln \left[1 - \beta(1-\beta) \frac{k^2}{m^2} \right] \right\}$$

The remaining integral can readily be performed, leading to the analytic expression

$$\bar{\omega}(k^2, m, \Lambda) = -\frac{\alpha}{3\pi} \left\{ -\ln \frac{\Lambda^2}{m^2} + \frac{1}{3} + 2\left(1 + \frac{2m^2}{k^2}\right) \left[\left(\frac{4m^2}{k^2} - 1 \right)^{1/2} \times \operatorname{arccot} \left(\frac{4m^2}{k^2} - 1 \right)^{1/2} - 1 \right] \right\} \tag{7-9}$$

The calculation has been carried out under the assumption $k^2 < 4m^2$. The corresponding function may be continued in the complex k^2 plane. The values for $k^2 > 4m^2$ are obtained by taking a limiting value from above the cut, starting at the point $k^2 = 4m^2$ (Fig. 7-2). The discontinuity across the cut is $\bar{\omega}(k^2 + i\varepsilon) - \bar{\omega}(k^2 - i\varepsilon) = 2i \operatorname{Im} \bar{\omega}(k^2 + i\varepsilon)$. This can also be derived using the last integral representation given above by setting $\beta = (1-u)/2$, integrating by parts, and changing the variable to $u = (1 - 4m^2/k'^2)^{1/2}$. We find that

$$\bar{\omega}(k^2, m, \Lambda) = \frac{\alpha}{3\pi} \ln \frac{\Lambda^2}{m^2} + \frac{\alpha}{3\pi} k^2 \int_{4m^2}^{\infty} \frac{dk'^2}{k'^2} \frac{1}{k'^2 - k^2} \left(1 - \frac{4m^2}{k'^2}\right)^{1/2} \left(1 + \frac{2m^2}{k'^2}\right) \tag{7-10}$$

a once-subtracted dispersion relation exhibiting the above analytic properties, with

$$\operatorname{Im} \bar{\omega}(k^2, m) = \frac{\alpha}{3} \left(1 - \frac{4m^2}{k^2}\right)^{1/2} \left(1 + \frac{2m^2}{k^2}\right) \tag{7-11}$$

This absorptive part is independent of the regularizing cutoff Λ and hence cannot be affected by renormalization. It coincides with our earlier calculation of pair production by an external field, Eq. (4-105), expressed as the square of the corresponding amplitude (Fig. 7-3) and given to this order by

$$W^{(1)} = \operatorname{Re} \int d^4k\, a_\rho(k) \bar{\omega}^{\rho\nu}(k) a_\nu(-k) = \int d^4k \operatorname{Im} \bar{\omega}(k) [|\mathbf{E}(k)|^2 - |\mathbf{B}(k)|^2]$$

The only effect of regularization has been to provide the constant $(\alpha/3\pi) \ln (\Lambda^2/m^2)$, which is of course divergent if we let Λ go to infinity.

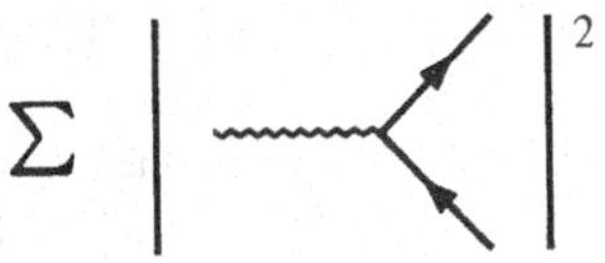

Figure 7-3 The probability for pair production in an external field giving the discontinuity of the vacuum polarization tensor.

Figure 7-4 Photon propagator in terms of the vacuum polarization.

Before applying renormalization, let us return to the original expressions (7-1) and (7-2). We do not really intend to compute the photon propagator like that! The contribution $G^{[1]}$ added to $G^{[0]}$ would produce a double pole besides the simple one. When we were discussing the ordering of perturbation theory according to the number of loops we had in mind the one-particle irreducible Green functions, the inverse propagator for the case at hand. The zero-loop contribution to this quantity is $G^{[0]-1}_{\rho\nu}$ and its one-loop contribution is $-\bar{\omega}^{[1]}_{\rho\nu}$, where we have added the suffix 1 to indicate that it was evaluated to order one:

$$G^{[0]-1}_{\rho\nu} - \bar{\omega}^{[1]}_{\rho\nu} = i[(k^2 - \mu^2)g_{\rho\nu} - (1 - \lambda)k_\rho k_\nu + (g_{\rho\nu}k^2 - k_\rho k_\nu)\bar{\omega}^{[1]}] \tag{7-12}$$

Inverting this expression we obtain the propagator represented diagrammatically by a sum, each term of which is a string of bubbles (Fig. 7-4). If more generally we succeed in showing to all orders that the vacuum polarization tensor has the form of a scalar function multiplying the combination $(g_{\rho\nu}k^2 - k_\rho k_\nu)$, then Eq. (7-12) with the index one omitted will give the general relation between the propagator and $\bar{\omega} = \bar{\omega}^{[1]} + \bar{\omega}^{[2]} + \cdots$ as

$$iG_{\rho\nu} = \frac{g_{\rho\nu} - [1 + \bar{\omega}(k^2)]k_\rho k_\nu/\mu^2}{k^2[1 + \bar{\omega}(k^2)] - \mu^2} + \frac{k_\rho k_\nu}{\mu^2}\frac{1}{k^2 - \mu^2/\lambda} \tag{7-13}$$

Equation (7-13) can be compared with formula (5-80). At the time we were ignoring mass renormalization as well as divergences. To be more precise, assume that the denominator in (7-13) admits an integral representation of the form

$$\frac{1}{k^2[1 + \bar{\omega}(k^2)] - \mu^2} = \int_0^\infty dk'^2 \frac{\rho(k'^2)}{k^2 - k'^2} \tag{7-14}$$

If $\bar{\omega}$ continues to satisfy a dispersion relation of the type (7-10) (with a threshold at the origin as $\mu^2 \to 0$, instead of $4m^2$ as it is to lowest order), we shall have the relation

$$\rho(k^2) = \frac{k^2}{\pi}\frac{\operatorname{Im}\bar{\omega}(k^2)}{\{k^2[1 + \operatorname{Re}\bar{\omega}(k^2)] - \mu^2\}^2 + [k^2 \operatorname{Im}\bar{\omega}(k^2)]^2} + \text{pole contributions}$$

If the integrals that we write remain meaningful we find

$$\begin{aligned}\frac{1 + \bar{\omega}(k^2)}{k^2[1 + \bar{\omega}(k^2)] - \mu^2} &= \frac{1}{k^2}\left\{1 + \frac{\mu^2}{k^2[1 + \bar{\omega}(k^2)] - \mu^2}\right\} \\ &= \frac{1}{k^2} + \frac{\mu^2}{k^2}\int_0^\infty dk'^2 \frac{\rho(k'^2)}{k^2 - k'^2} \\ &= \mu^2 \int_0^\infty dk'^2 \frac{\rho(k'^2)}{k'^2(k^2 - k'^2)} + \frac{1}{k^2}\left[1 - \mu^2 \int_0^\infty \frac{dk'^2}{k'^2}\rho(k'^2)\right]\end{aligned} \tag{7-15}$$

According to (7-14) taken for $k = 0$ and assuming $\bar{\omega}(0)$ finite, the last term in brackets in the above formula vanishes. Consequently, Eq. (7-13) reads

$$iG_{\rho\nu}(k) = \int_0^\infty dk'^2\, \rho(k'^2)\, \frac{g_{\rho\nu} - k_\rho k_\nu/k'^2}{k^2 - k'^2} + \frac{k_\rho k_\nu}{\mu^2}\, \frac{1}{k^2 - \mu^2/\lambda} \tag{7-16}$$

This is indeed comparable to Eq. (5-80).

The resummation implied by Eqs. (7-12) and (7-13) has generated a new phenomenon. Indeed, the isolated pole at $k^2 = \mu^2$ of the free theory has at best been moved, and corresponds now to the zero of the expressions $k^2[1 + \bar{\omega}(k^2)] - \mu^2$. As far as the photon propagator is concerned this is not too drastic. Indeed, μ^2 has only been introduced for convenience, in order to cut off the infrared divergences in intermediate calculations. We shall always have to compute combinations such as $j_1^\rho(k) G_{\rho\nu}(k) j_2^\nu(-k)$ where j_1 and j_2 are conserved currents $[k \cdot j(k) = 0]$ in order to extract physical information. This combination eliminates the terms in $k_\rho k_\nu$ and enables us to consider the limit $\mu^2 \to 0$. If instead we would have set $\mu^2 = 0$ then G would have taken the form

$$iG_{\rho\nu}(k) = \frac{g_{\rho\nu}}{k^2[1 + \bar{\omega}(k^2)]} + \frac{k_\rho k_\nu}{k^4}\, \frac{1 + \bar{\omega}(k^2) - \lambda}{\lambda[1 + \bar{\omega}(k^2)]} \qquad \mu^2 = 0 \tag{7-17}$$

Unless $\bar{\omega}(0) = \infty$, the quantity $k^2[1 + \bar{\omega}(k^2)]$ still vanishes at $k^2 = 0$.

In a loose way we may characterize the property $k^2[1 + \bar{\omega}(k^2)]\,|_{k^2=0} = 0$ by saying that there is no genuine mass renormalization for photons. But wave function renormalization is there, since the residue of the pole at $k^2 = 0$ is now equal to $[1 + \bar{\omega}(0)]^{-1}$, instead of being one. One-loop calculations reveal a remarkable circumstance. Namely, under regularization we find that the only potentially divergent and therefore unknown term in $\bar{\omega}$ is a constant which may be taken as its value at $k^2 = 0$. If only $j \cdot G \cdot j$ is measurable we may use two widely separated identical sources to define what we mean by the charge squared. We define it to be the coefficient of the $1/4\pi r$ Coulomb static potential at large distances. This appears here as $e^2 = e_0^2/[1 + \bar{\omega}(0)] = Z_3 e_0^2$ where we have introduced an index zero to characterize the bare coupling constant used up to now in the perturbative series. The above definition of Z_3 agrees with the one given in Chap. 5 in the limit $\mu^2 = 0$. We conclude therefore that

$$\begin{aligned} e^2 &= Z_3 e_0^2 \\ Z_3 &= \frac{1}{1 + \bar{\omega}(0)} = 1 - \frac{\alpha}{3\pi} \ln \frac{\Lambda^2}{m^2} + \cdots \end{aligned} \tag{7-18}$$

To be consistent we have to express all quantities as series in a parameter counting the number of loops L in the bare Feynman diagrams. As we have seen in Sec. 6-2-1, it is natural to multiply such diagrams with a factor $\hbar^L$. For this reason if we concentrate on terms of order $\hbar^1$ we may substitute α for α_0 in the finite remainder $\bar{\omega}^{[1]}(k^2) - \bar{\omega}^{[1]}(0)$, since this is already of order $\hbar$ as is $\alpha - \alpha_0$. This also justifies the use of α in the right-hand side of (7-18).

According to this definition of charge the renormalized propagator reads

$$iG^R_{\rho\nu}(k) = \frac{g_{\rho\nu}}{k^2[1 + \bar{\omega}(k^2) - \bar{\omega}(0)]} + \text{terms in } k_\rho k_\nu \tag{7-19}$$

It will have residue 1 at the photon pole $k^2 = 0$ and will be expressed in terms of the physical charge.

We see the renormalization program emerging: undetermined and potentially divergent quantities disappear when Green functions are expressed in terms of physical renormalized quantities.

When we set $\sum_1^S C_s \ln \lambda_s^2 = -\ln(\Lambda^2/m^2)$ the sign was chosen in agreement with condition (7-7). The intuitive content of (7-18) is that the bare charge has been screened at large distances by a factor $Z_3 = 1 - (\alpha/3\pi)\log(\Lambda^2/m^2)$ smaller than one and positive as long as Λ^2/m^2 is not overwhelmingly large. Vacuum polarization corresponding to the creation of virtual electron-positron pairs has reduced the charge of a test particle as seen by a distant one.

We may take another point of view in deriving equations such as (7-19). It amounts to saying that the original lagrangian is only a bookkeeping device enabling us to define the perturbative amplitudes. It was postulated from some intuitive correspondence principle between classical and quantum mechanics. As such it may be amended by quantum corrections. Let us therefore assume that to lowest order $\mathscr{L}$ is expressed in terms of physical parameters. It will then be necessary to construct perturbative corrections of the form $\delta\mathscr{L}$, called counterterms (the coefficients of which will become infinite as the cutoff goes to infinity), in order to maintain the proper definition of these physical parameters:

$$\begin{aligned} \mathscr{L}_{\text{tot}} &= \mathscr{L}(e, m, \ldots) + \delta\mathscr{L} \\ \delta\mathscr{L} &= \delta\mathscr{L}^{[1]} + \delta\mathscr{L}^{[2]} + \cdots \end{aligned} \tag{7-20}$$

The various pieces of $\delta\mathscr{L}$ are classified according to the corresponding powers of $\hbar$ counting the number of loops in the original diagrams. The fact that this number of loops may be expressed as powers of $\hbar$ ties in nicely with the previous remarks.

Our calculation may then be interpreted as identifying a contribution in $\delta\mathscr{L}^{[1]}$ of the form

$$\delta\mathscr{L}^{[1]}_{F^2} = -\tfrac{1}{4}(Z_3 - 1)^{[1]} F_{\rho\nu}F^{\rho\nu} \tag{7-21}$$

Adding such a term to $\mathscr{L}$ will produce an additional diagram to order $\hbar$ (Fig. 7-5) and will modify the original (regularized) tensor according to

$$\bar{\omega}_{\rho\nu}(k) \to \bar{\omega}_{\rho\nu}(k) - (Z_3 - 1)^{[1]} i(g_{\rho\nu}k^2 - k_\rho k_\nu)$$

Figure 7-5 The two diagrams of order $\hbar$ for vacuum polarization in the counterterm approach.

Therefore

$$\bar{\omega}(k^2, \Lambda) \to \bar{\omega}(k^2, \Lambda) + (Z_3 - 1)^{[1]} = \frac{\alpha}{3\pi} \ln \frac{\Lambda^2}{m^2} + (Z_3 - 1)^{[1]} + [\bar{\omega}(k^2) - \bar{\omega}(0)]$$

The new rule implies that we use α everywhere and not α_0. Also, in equation (7-9) the term independent of $\ln(\Lambda^2/m^2)$ vanishes for $k^2 = 0$, justifying our introduction of a renormalized quantity $\bar{\omega}^R(k^2) = \bar{\omega}(k^2) - \bar{\omega}(0)$. From Eq. (7-18) it follows that $(Z_3 - 1)^{[1]} + (\alpha/3\pi)\ln(\Lambda^2/m^2) = 0$ and therefore

$$\bar{\omega}^R(k^2) = \frac{\alpha}{3\pi} k^2 \int_{4m^2}^{\infty} \frac{dk'^2}{k'^2} \frac{1}{k'^2 - k^2} \left(1 - \frac{4m^2}{k'^2}\right)^{1/2} \left(1 + \frac{2m^2}{k'^2}\right)$$

$$= -\frac{\alpha}{3\pi} \left\{ \frac{1}{3} + 2\left(1 + \frac{2m^2}{k^2}\right)\left[\left(\frac{4m^2}{k^2} - 1\right)^{1/2} \operatorname{arccot}\left(\frac{4m^2}{k^2} - 1\right)^{1/2} - 1\right]\right\} \tag{7-22}$$

This expression is meaningful for real k^2 smaller than $4m^2$. It has to be analytically continued above the cut for $k^2 > 4m^2$. Note the following interesting limits:

$$\bar{\omega}^R(k^2) = \begin{cases} \dfrac{\alpha}{15\pi} \dfrac{k^2}{m^2} + O\left(\dfrac{k^4}{m^4}\right) & k^2 \to 0 \\ -\dfrac{\alpha}{3\pi} \ln\left(-\dfrac{k^2}{m^2}\right) + \dfrac{5\alpha}{3\pi} + O\left[\alpha \dfrac{m^2}{k^2} \ln\left(-\dfrac{k^2}{m^2}\right)\right] & k^2 \to -\infty \end{cases}$$

The behavior at large negative k^2 has a striking similarity (including the coefficient) with the original divergent term involving the cutoff. This is, of course, not an accident, and will be better understood in due time.

The screening of charge implied by the vacuum polarization has physical consequences. Using the renormalized expressions and concentrating on static situations where $k^2 = -\mathbf{k}^2$, we see that the Coulomb law is modified by the replacement $e^2 \to e^2/[1 + \bar{\omega}(-\mathbf{k}^2)]$ very much as in ordinary dielectric materials where we would have $e^2/\varepsilon(\mathbf{k})$ with ε as the dielectric constant. For small enough $\mathbf{k}^2$ (as compared to m^2 which sets the scale) we can approximate the Coulomb interaction as

$$\frac{e^2}{\mathbf{k}^2} \to \frac{e^2}{\mathbf{k}^2[1 + \bar{\omega}^R(-\mathbf{k}^2)]} \simeq \frac{e^2}{\mathbf{k}^2}\left(1 + \frac{\alpha}{15\pi} \frac{\mathbf{k}^2}{m^2}\right)$$

a result originally due to Uehling.

In configuration space for an infinitely heavy nucleus of charge $-Ze$ located at the origin this means a correction

$$V(r) = -\frac{Ze^2}{4\pi r} \to \left(1 - \frac{\alpha}{15\pi} \frac{\Delta}{m^2}\right) \frac{-Ze^2}{4\pi r} = \frac{-Ze^2}{4\pi r} - \frac{\alpha}{15\pi} \frac{Ze^2}{m^2} \delta^3(\mathbf{r})$$

The effect of the added term has to be computed to first order, otherwise it would be inconsistent to keep only terms of order α in $\bar{\omega}^R$. Expanding around zero to leading order in $(\mathbf{k}^2/m^2)$ has produced a singularity in configuration space. However, we only need the mean value of the potential in unperturbed states so that it is not crucial that the actual correction, strongly peaked for small r, is replaced by a δ function. Observe that this enhancement at small r is in agreement with the idea that for smaller distances we recover the unshielded coupling to the "bare" charge.

Applied to an hydrogen-like atom the above perturbation leads to a displacement of s-wave levels according to

$$\delta E_{n,l} = -\frac{Z\alpha e^2}{15\pi m^2}\int d^3r\, \psi^*_{n,l}(\mathbf{r})\delta^3(r)\psi_{n,l}(\mathbf{r})$$

$$= -\frac{Z\alpha e^2}{15\pi m^2}\delta_{l,0}|\psi_{n,0}(0)|^2 = -\frac{4}{15\pi}\frac{Z^4\alpha^5}{n^3}m\delta_{l,0} \tag{7-23}$$

where we have used the value of the Coulomb wave function at the origin $\psi_{n,0}(0) = (\pi n^3 a^3)^{-1/2}$, $a = (Zm\alpha)^{-1}$. In the Dirac theory the two $n = 2, j = \frac{1}{2}$ levels of opposite parity $2S_{1/2}$ and $2P_{1/2}$ are degenerate. The vacuum polarization correction has the effect of lowering the s state by an amount $\Delta E/h = (1/h)(E_{2S_{1/2}} - E_{2P_{1/2}}) \simeq -27$ MHz (Fig. 7-6). In his 1947 measurement of this level splitting, Lamb obtained a value of the order of $+1000$ MHz, proving that other effects largely overcompensate vacuum polarization. In Sec. 6-3-2 the full theory will be shown to agree with experiment, providing indirect evidence therefore for this screening of the charge.

It is possible, of course, to avoid the use of the small $\mathbf{k}^2$ approximation. A static charge located at the origin will induce a modified Coulomb potential to order α given by

$$\frac{e}{4\pi r} \rightarrow V(r) = \frac{e}{4\pi r}Q(r)$$

$$Q(r) = 1 + \frac{2\alpha}{3\pi}\int_1^\infty du\, e^{-2mru}\left(1 + \frac{1}{2u^2}\right)\frac{(u^2-1)^{1/2}}{u^2} \tag{7-24}$$

$$Q(r) = \begin{cases} 1 + \dfrac{\alpha}{3\pi}\left[\ln\dfrac{1}{(mr)^2} - 2\gamma - \dfrac{5}{3} + \cdots\right] & mr \ll 1 \\[2ex] 1 + \dfrac{\alpha}{4\pi^{1/2}(mr)^{3/2}}e^{-2mr} + \cdots & mr \gg 1 \end{cases}$$

with γ equal to Euler's constant $-\int_0^\infty du \ln u\, e^{-u} = 0.5772\ldots$. According to the definition of charge $Q(\infty) = 1$. We see again that as r decreases $Q(r)$ increases, even becoming infinite as r tends to zero. The approximation used in Eq. (7-23) is only valid in the mean.

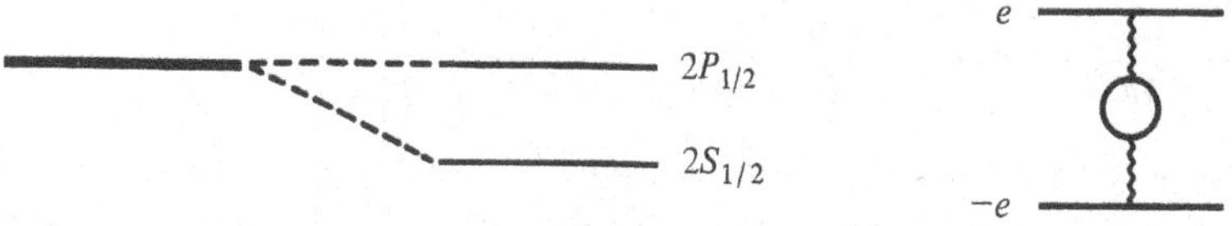

Figure 7-6 The vacuum polarization contribution to the $2S_{1/2} - 2P_{1/2}$ hydrogen splitting.

Finally, we remark that the above conventions amount to the use of a photon propagator with residue equal to unity at the photon pole (when $\mu^2 \to 0$). The factor $Z_3^{-1/2}$ attached to photon lines in scattering amplitudes has therefore been replaced by one.

7-1-2 Electron Propagator

The first nontrivial contribution to the one-particle irreducible Green function with two external electron lines (also called electron self-energy) is shown diagrammatically on Fig. 7-7 and reads

$$-i\Sigma(p) = (-ie)^2 \int \frac{d^4k}{(2\pi)^4} \frac{1}{i} \left[\frac{g_{\rho\sigma}}{k^2 - \mu^2 + i\varepsilon} + \frac{(1-\lambda)k_\rho k_\sigma}{(k^2 - \mu^2 + i\varepsilon)(\lambda k^2 - \mu^2 + i\varepsilon)} \right] \times \gamma^\rho \frac{i}{\not{p} - \not{k} - m + i\varepsilon} \gamma^\sigma \tag{7-25}$$

This expression, a 4×4 matrix function of p, unfortunately suffers from all possible diseases. It has an apparent linear ultraviolet divergence (meaning that the integrand looks like $\int d^4k/|k|^3$ for large k). We shall see that this can be reduced to a logarithmic divergence. Infrared divergences might also creep in for small k and special values of the external momentum; we therefore keep $\mu^2 \neq 0$ to be on the safe side. Also gauge dependence is expected, as would be the case in nonrelativistic physics for a wave function of a charged particle.

To deal with meaningful quantities it is necessary, as before, to use a regularization of the free propagators. To avoid cumbersome notation we shall assume that a cutoff Λ^2 has been introduced in an appropriate manner, without being more explicit. We write

$$\Sigma(p) = \Sigma^a(p) + \frac{1-\lambda}{\lambda} \Sigma^b(p) \tag{7-26}$$

in such a way that $\Sigma^a(p)$ corresponds to the choice of Feynman gauge ($\lambda = 1$), and use the parametric representation of propagators to express $\Sigma^{a,b}$ after integration over the loop momentum k as

$$\Sigma^a(p) = \frac{\alpha}{2\pi} \int_0^\infty \int_0^\infty \frac{d\alpha_1 \, d\alpha_2}{(\alpha_1 + \alpha_2)^2} \left(2m - \frac{\alpha_1}{\alpha_1 + \alpha_2} \not{p} \right) \times \exp\left[i \left(\frac{\alpha_1 \alpha_2}{\alpha_1 + \alpha_2} p^2 - \alpha_1 \mu^2 - \alpha_2 m^2 \right) \right] \tag{7-27a}$$

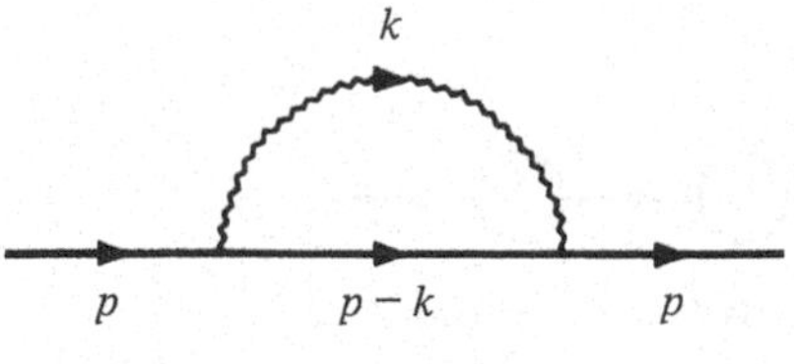

Figure 7-7 Electron self-energy to lowest order.

$$\Sigma^b(p) = -\frac{i\alpha}{4\pi}\int_0^\infty\int_0^\infty\int_0^\infty \frac{d\alpha_1\,d\alpha_2\,d\alpha_3}{(\alpha_1+\alpha_2+\alpha_3)^3}\left\{m\left(2i+\frac{\alpha_2^2p^2}{\alpha_1+\alpha_2+\alpha_3}\right)\right.$$
$$\left.+\not p\left[-i-\frac{3i\alpha_2}{\alpha_1+\alpha_2+\alpha_3}+\frac{\alpha_2^2(\alpha_1+\alpha_3)p^2}{(\alpha_1+\alpha_2+\alpha_3)^2}\right]\right\}$$
$$\times\exp\left\{i\left[\frac{\alpha_2(\alpha_1+\alpha_3)}{\alpha_1+\alpha_2+\alpha_3}p^2-\alpha_1\mu^2-\alpha_3\frac{\mu^2}{\lambda}-\alpha_2m^2\right]\right\} \tag{7-27b}$$

The masses are understood as having a vanishing negative imaginary part, ensuring convergence for the large α.

As in the case of photons, the self-energy parts are classified according to the number of loops, and the propagator itself is obtained by summing the series (Fig. 7-8):

$$\frac{i}{\not p-m}+\frac{i}{\not p-m}\left[-i\Sigma(p)\right]\frac{i}{\not p-m}$$
$$+\frac{i}{\not p-m}\left[-i\Sigma(p)\right]\frac{i}{\not p-m}\left[-i\Sigma(p)\right]\frac{i}{\not p-m}+\cdots=\frac{i}{\not p-m-\Sigma(p)} \tag{7-28}$$

We proceed by computing $\Sigma^a(p,\Lambda)$, where the cutoff Λ will soon make its appearance when we try to integrate over a common dilatation factor of the α parameters:

$$\Sigma^a(p,\Lambda) = \frac{\alpha}{2\pi}\int_0^\infty\int_0^\infty d\alpha_1\,d\alpha_2\,\delta(1-\alpha_1-\alpha_2)$$
$$\times\int_0^\infty\frac{d\rho}{\rho}(2m-\alpha_1\not p)\,e^{i\rho(\alpha_1\alpha_2p^2-\alpha_1\mu^2-\alpha_2m^2)} \tag{7-29}$$

The integral over ρ is only logarithmically divergent due to the extraction of the kinematical factors m and $\not p$, in contradistinction to the classical linear divergence of self-mass (Chap. 1). Were m so large that $|p| \ll \Lambda \ll m$, corresponding to a static classical limit, then we would recover this linear divergence in Λ.

In order to have a relativistic invariant regularization we may subtract from Σ^a the contribution of a fictitious massive photon with μ replaced by Λ. The present value of Λ is therefore unrelated to the one used in the vacuum polarization. Should it be necessary, for large but finite cutoff, to study the Λ-dependent parts, it would be wise to introduce a uniform regularization procedure. We therefore use the substitution

$$\int_0^\infty\frac{d\rho}{\rho}e^{i\rho(\alpha_1\alpha_2p^2-\alpha_1\mu^2-\alpha_2m^2)}\to\int_0^\infty\frac{d\rho}{\rho}\left(e^{i\rho(\alpha_1\alpha_2p^2-\alpha_1\mu^2-\alpha_2m^2)}-e^{-i\rho\alpha_1\Lambda^2}\right)$$

Figure 7-8 The electron propagator expressed as a power series in the self-energy Σ.

where all other massive parameters have been neglected as compared to Λ^2 when the latter tends to infinity. The identity

$$\int_0^\infty \frac{d\rho}{\rho}\,(e^{i\rho(z_1+i\varepsilon)} - e^{i\rho(z_2+i\varepsilon)}) = \ln\frac{z_2}{z_1}$$

allows us to write

$$\Sigma^a(p, \Lambda) = \frac{\alpha}{2\pi}\int_0^1 d\beta\,(2m - \beta\not{p})\ln\left[\frac{\beta\Lambda^2}{m^2(1-\beta) + \beta\mu^2 - \beta(1-\beta)p^2 - i\varepsilon}\right] \tag{7-30}$$

The Λ dependence affects the coefficients of a linear term in $\not{p}$, the remaining expression being finite.

The situation is slightly complicated by an infrared problem. As long as μ^2 remains finite the quadratic form appearing in the logarithm is positive for $p^2 < (m+\mu)^2$. Indeed, for $0 \le \beta \le 1$ and p^2 below threshold we have $0 \le \beta(1-\beta) \le \frac{1}{4}$ and

$$\begin{aligned}(1-\beta)m^2 + \beta\mu^2 - \beta(1-\beta)p^2 &\ge (1-\beta)m^2 + \beta\mu^2 - \beta(1-\beta)(m+\mu)^2\\ &= [(1-\beta)m - \beta\mu]^2\end{aligned}$$

In this region Σ^a will be "real," that is, $\overline{\Sigma^a} \equiv \gamma^0\Sigma^{a\dagger}\gamma^0 = \Sigma^a$. Beyond this point Σ^a will acquire an "imaginary" part corresponding to the process

Virtual electron $\to$ real electron + photon

This is analogous to the photon case beyond the threshold for pair creation. The bothersome point here is that for $\mu^2 \to 0$ the threshold coincides with the physical mass shell $p^2 = m^2$. In the real world it will not be possible to isolate in the propagator the electron pole from a cut starting at the same location. This explains our insistence on keeping $\mu^2 > 0$. When $\mu^2 = 0$, one of the zeros of the quadratic form reaches the limit $\beta = 1$ of the integration interval, and expansions around the singular point $p^2 = m^2$ are no longer possible.

A similar but distinct phenomenon occurs also in the photon case at much higher orders (four or more loops) corresponding to intermediate states with three, five, . . . , photons. It leads to power-type corrections to the large-distance behavior of the photon propagator with extremely small coefficients. Due to its smallness this effect has not yet been measured.

We may, however, consider the limit $\mu^2 \to 0$ for values $p^2 < m^2$ without encountering singularities. This has the virtue of greatly simplifying the integral in Eq. (7-30), leading to the result

$$\begin{aligned}\Sigma^a(\not{p}, \Lambda, \mu = 0) = \frac{\alpha}{2\pi}\Bigg\{&\ln\frac{\Lambda^2}{m^2}\left(2m - \frac{1}{2}\not{p}\right) + 2m\left[1 + \frac{m^2 - p^2}{p^2}\ln\left(1 - \frac{p^2}{m^2}\right)\right]\\ &- \frac{1}{2}\not{p}\left[\frac{3}{2} + \frac{m^4 - (p^2)^2}{(p^2)^2}\ln\left(1 - \frac{p^2}{m^2}\right) + \frac{m^2}{p^2}\right]\Bigg\}\end{aligned} \tag{7-31}$$

As p^2 tends to m^2 this expression remains finite but its derivatives become infinite.

Insertion of such a self-energy in formula (7-28) obviously shifts the position of the propagator pole and modifies its residue. Two equivalent attitudes are again possible. Either we call m_0 (and e_0) the parameters occurring in the original lagrangian, define m to be the modified position of the pole, and reexpress the Green functions in terms of physical mass and coupling constant. Or we avoid mentioning bare parameters and introduce counterterms in the lagrangian in such a way as to keep the position and residue of this pole unchanged.

We may do that by expanding $\sum$ considered as a function of $\not{p}(\not{p}^2 = p^2)$ around $\not{p} = m$:

$$\begin{aligned}\sum(\not{p}, \Lambda) &= \delta m(\Lambda) - [Z_2^{-1}(\Lambda) - 1](\not{p} - m) + Z_2^{-1}(\Lambda) \sum_R(\not{p}) \\ \delta m &= \sum(\not{p}, \Lambda)\big|_{\not{p}=m} \\ -[Z_2^{-1}(\Lambda) - 1] &= \frac{\partial \sum}{\partial \not{p}}(\not{p}, \Lambda)\big|_{\not{p}=m}\end{aligned} \tag{7-32}$$

To order $\hbar$ the factor $Z_2^{-1}(\Lambda)$ in front of $\sum_R(\not{p})$ can be ignored and the first equation appears as a Taylor expansion since

$$\sum_R(\not{p})\big|_{\not{p}=m} = 0 \qquad \text{and} \qquad \frac{\partial}{\partial \not{p}} \sum_R(\not{p})\big|_{\not{p}=m} = 0$$

However, it is more transparent to interpret this relation as

$$\not{p} - m - \sum_R(\not{p}) = Z_2(\Lambda)[\not{p} - m_0 - \sum(\not{p})]$$

In the Feynman gauge we use (7-30) to extract δm and Z_2^{-1}. In fact, for δm we can ignore the infrared infinities and apply Eq. (7-31):

$$\delta m^a(\Lambda) = m \frac{3\alpha}{4\pi}\left(\ln \frac{\Lambda^2}{m^2} + \frac{1}{2}\right) \tag{7-33}$$

To extract $Z_2^{-1}(\Lambda) - 1$ we have to return to (7-30) and keep the contributions singular or finite when $\mu^2 \to 0$. This yields

$$\begin{aligned}[Z_2^{-1}(\Lambda) - 1]^a &= -\frac{\partial}{\partial \not{p}} \sum^a(\not{p}, \Lambda, \mu)\big|_{\not{p}=m} \\ &= \frac{\alpha}{2\pi}\int_0^1 d\beta\, \beta \left\{\ln\left[\frac{\beta\Lambda^2}{(1-\beta)^2 m^2}\right] - \frac{2(1-\beta)(2-\beta)}{(1-\beta)^2 + \beta\mu^2/m^2}\right\} \\ &= \frac{\alpha}{2\pi}\left[\frac{1}{2}\ln\frac{\Lambda^2}{m^2} + \ln\frac{\mu^2}{m^2} + \frac{9}{4} + O\left(\frac{\mu}{m}\right)\right]\end{aligned} \tag{7-34}$$

The renormalized value of $\sum$ is obtained by subtracting $\delta m - (Z_2^{-1} - 1)(\not{p} - m)$. In the range $|p^2 - m^2| \gg \mu^2$ we use (7-31) with the result

$$\textstyle\sum_R^a(p) = \frac{\alpha}{2\pi}\left\{2m\left[\frac{5}{8} + \frac{m^2 - p^2}{p^2}\ln\left(1 - \frac{p^2}{m^2}\right)\right] - \frac{\not{p}}{2}\left[\frac{3}{2} + \frac{m^2}{p^2} + \frac{m^4 - (p^2)^2}{(p^2)^2}\ln\left(1 - \frac{p^2}{m^2}\right)\right] + (\not{p} - m)\left(\frac{9}{4} + \ln\frac{\mu^2}{m^2}\right)\right\} \tag{7-35}$$

For $p^2 - m^2 \gg \mu^2$ the logarithm acquires an imaginary part obtained by analytic continuation above the cut starting at $(m + \mu)^2 : \ln(1 - p^2/m^2) \to \ln(p^2/m^2 - 1) - i\pi$. The corresponding absorptive part of $\sum$ in the limit $\mu^2 \to 0$ is

$$\text{abs}\textstyle\sum_R^a(p) = \alpha\left(1 - \frac{m^2}{p^2}\right)\left[m - \frac{\not{p}}{4}\left(1 + \frac{m^2}{p^2}\right)\right]\theta(p^2 - m^2) \tag{7-36}$$

A separate study is required to investigate $\sum_R$ in the vicinity of $p^2 = m^2$. By insisting that the renormalization be carried on-shell, additional infrared singularities originally absent from Eq. (7-31) have made their appearance.

Let us see how the previous results valid in the Feynman gauge are modified in an arbitrary gauge. In other words, we want to compute $\sum^b$, which we rewrite in momentum space as follows:

$$\textstyle\sum^b(p) = -\frac{ie^2}{(2\pi)^4}\int d^4k\, \frac{1}{(k^2 - \mu^2 + i\varepsilon)(k^2 - \mu^2/\lambda + i\varepsilon)} \times\left[\frac{\not{k}(p^2 - m^2)}{(p - k)^2 - m^2 + i\varepsilon} - \frac{k^2(\not{p} - m)}{(p - k)^2 - m^2 + i\varepsilon} - \not{k}\right] \tag{7-37}$$

Let I_1, I_2, I_3 stand for the contributions of the three successive terms in the integrand. Provided we use a covariant regularization, I_3 will vanish, being odd in k (it can only be proportional to $\not{p}$ but the corresponding integral does not depend on p). After regularization we see that $\sum^b$ vanishes for $\not{p} = m$ and hence does not contribute to δm. In fact, I_1 is ultraviolet divergent. However, its derivative at $\not{p} = m$ is infrared singular and μ^2 will have to be kept finite in order to extract its contribution to wave function renormalization, i.e., to $(Z_2^{-1} - 1)^b$. Finally, I_2 has a logarithmic ultraviolet divergence but is infrared convergent. For $|p^2 - m^2| \gg \mu^2$ we find

$$\frac{1-\lambda}{\lambda}\textstyle\sum^b(p, \Lambda) = \frac{1-\lambda}{\lambda}\frac{\alpha}{4\pi}\left\{\not{p}\,\frac{p^2 - m^2}{p^2}\left[1 + \frac{m^2}{p^2}\ln\left(1 - \frac{p^2}{m^2}\right)\right] - (\not{p} - m)\left[\ln\frac{\Lambda^2}{m^2} + 1 - \frac{p^2 - m^2}{p^2}\ln\left(1 - \frac{p^2}{m^2}\right)\right]\right\} \tag{7-38}$$

In the vicinity of $p^2 = m^2$ the first bracket $[1 + (m^2/p^2)\ln(1 - p^2/m^2)]$ is to be replaced by

$$\frac{1}{2}\left[1 + \frac{\ln(\mu^2/\lambda m^2) - \lambda\ln(\mu^2/m^2)}{1 - \lambda}\right]$$

Consequently, we can split $[(1-\lambda)/\lambda]\sum^b$ as the sum of a renormalized part plus a term vanishing linearly with $\not{p} - m$ in the following form valid for $|p^2 - m^2| \gg \mu^2$:

$$\begin{aligned}\frac{1-\lambda}{\lambda}\sum{}^b(p,\Lambda) &= -(\not{p}-m)\frac{\alpha}{4\pi}\left[\frac{1-\lambda}{\lambda}\ln\frac{\Lambda^2}{m^2} + \ln\frac{\mu^2}{m^2} - \frac{1}{\lambda}\ln\frac{\mu^2}{\lambda m^2}\right] \\ &\quad + \frac{\alpha}{4\pi}\frac{1-\lambda}{\lambda}\left\{\not{p}\frac{p^2-m^2}{p^2}\left[1 + \frac{m^2}{p^2}\ln\left(1-\frac{p^2}{m^2}\right)\right]\right. \\ &\quad - (\not{p}-m)\left[1 - \frac{p^2-m^2}{p^2}\ln\left(1-\frac{p^2}{m^2}\right)\right. \\ &\quad \left.\left. + \frac{\ln(\mu^2/\lambda m^2) - \lambda\ln(\mu^2/m^2)}{1-\lambda}\right]\right\}\end{aligned} \tag{7-39}$$

Therefore

$$[Z_2(\Lambda)^{-1} - 1]^b = \frac{\alpha}{4\pi}\left(\frac{1-\lambda}{\lambda}\ln\frac{\Lambda^2}{m^2} + \ln\frac{\mu^2}{m^2} - \frac{1}{\lambda}\ln\frac{\mu^2}{\lambda m^2}\right) \tag{7-40}$$

In formulas (7-39) and (7-40) the meaning of the cutoff Λ is the same as the one used for the computation of $\sum^a$.

Considering the combination $(Z_2^{-1} - 1)^{a+b}$ we observe that the infrared singular term $\ln(\mu^2/m^2)$ has a coefficient proportional to $3 - 1/\lambda$. The choice $\lambda = \frac{1}{3}$ (Yennie and Fried) eliminates the infrared divergences to this order. In the limit $\mu^2 \to 0$ it corresponds to a photon propagator $-i(g_{\rho\sigma}/k^2 + 2\,k_\rho k_\sigma/k^4)$. As a consequence, $\sum_R^{a+b}$ is also free of infrared singularities.

We have now to compensate the ultraviolet divergences by adding counterterms to the lagrangian in order to maintain the proper definition of physical parameters. The original lagrangian contains a term $-m\overline{\psi}\psi$. This is replaced by $-m_0\overline{\psi}\psi = -m\overline{\psi}\psi + \delta m\overline{\psi}\psi$. To order one we therefore introduce

$$\delta\mathscr{L}^{[1]}_{\overline{\psi}\psi} = \delta m\overline{\psi}\psi \tag{7-41}$$

with

$$\delta m = \delta m^a = \frac{3\alpha}{4\pi}m\left(\ln\frac{\Lambda^2}{m^2} + \frac{1}{2}\right) \tag{7-42}$$

Note that δm is gauge independent, a physically reasonable result, and only logarithmically divergent. The added piece to the lagrangian will contribute to the proper two-point function a term $i\delta m$ (Fig. 7-9). According to the definition (7-25) this will subtract δm from $\sum$ and insure that m is indeed the physical mass. In the sense of formal power series, we have to order one $(Z_2 - 1)^{[1]} = -(Z_2^{-1} - 1)^{[1]}$. Thus the counterterm

Figure 7-9 Diagrammatic representation of the one-loop counterterms $\delta m\overline{\psi}\psi + (Z_2 - 1)[(i/2)\overline{\psi}\overleftrightarrow{\not{\partial}}\psi - m\overline{\psi}\psi]$ to the electron two-point function.

$$\delta\mathscr{L}^{[1]}_{(i/2)\bar{\psi}\overleftrightarrow{\not\partial}\psi - m\bar{\psi}\psi} = (Z_2 - 1)\left(\frac{i}{2}\bar{\psi}\overleftrightarrow{\not\partial}\psi - m\bar{\psi}\psi\right) \tag{7-43}$$

with
$$(Z_2 - 1) = -\frac{\alpha}{4\pi}\left(\frac{1}{\lambda}\ln\frac{\Lambda^2}{m^2} + 3\ln\frac{\mu^2}{m} - \frac{1}{\lambda}\ln\frac{\mu^2}{\lambda m^2} + \frac{9}{4}\right) \tag{7-44}$$

will compensate to order one the divergent term in $\not p - m$. The remaining finite propagator $[\not p - m - \sum_R^{a+b}(p)]^{-1}$ will have a pole at $\not p = m$ with unit residue.

Remarks

(*a*) To order $\hbar$ we cannot distinguish between counterterms of the form $\delta m\bar{\psi}\psi$ and $Z_2\delta m\bar{\psi}\psi$ since $(Z_2\delta m)^{[1]} = \delta m^{[1]}$.

(*b*) In the Landau gauge $\lambda \to \infty$ the ultraviolet divergences cancel in Z_2^{-1} to order one. Unfortunately, no unique choice of gauge eliminates all ultraviolet divergences perturbatively.

(*c*) All counterterms introduced up to now have a structure similar to the terms in the original lagrangian, pointing toward the success of the renormalization program.

The original observation that the electron self-mass is only logarithmically divergent is due to Weisskopf (1939), and the first complete calculation to Karplus and Kroll (1950).

7-1-3 Vertex Function

After studying the two-point functions we now face the three-point vertex function $(e\bar{e}\gamma)$. The unique one-particle irreducible diagram pertaining to this process is shown in Fig. 7-10 and is given by the expression

$$-ie\Gamma_\mu^{[1]}(p', p) = (-ie)^3\int\frac{d^4k}{(2\pi)^4}\frac{1}{i}\left[\frac{g_{\rho\sigma}}{k^2 - \mu^2 + i\varepsilon} + \frac{(1-\lambda)k_\rho k_\sigma}{(k^2 - \mu^2 + i\varepsilon)(\lambda k^2 - \mu^2 + i\varepsilon)}\right]$$
$$\times\left(\gamma^\sigma\frac{i}{\not p' - \not k - m + i\varepsilon}\gamma_\mu\frac{i}{\not p - \not k - m + i\varepsilon}\gamma^\rho\right) \tag{7-45}$$

Using the same notation, the zeroth-order value is γ^μ. Accordingly we shall write the complete one-particle irreducible three-point function

$$\Lambda_\mu(p', p) = \gamma_\mu + \Gamma_\mu(p', p) \tag{7-46}$$

with Eq. (7-45) giving the first nontrivial contribution to Γ_μ.

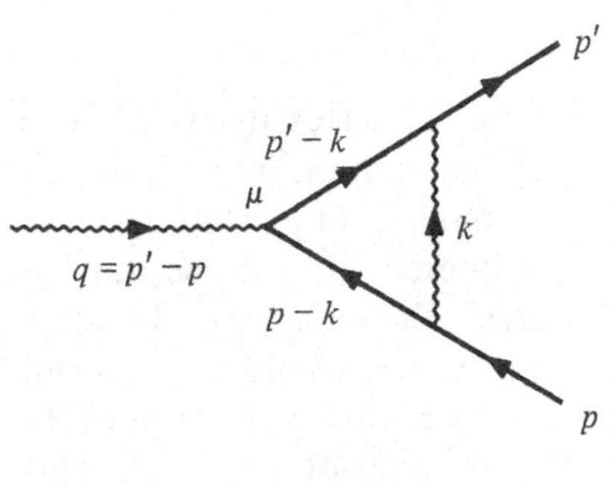

Figure 7-10 Vertex diagram to the one-loop order.

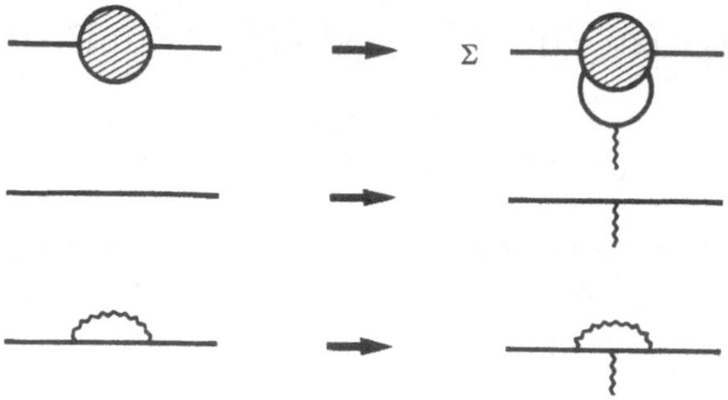

Figure 7-11 Insertion of an external photon line in an electron self-energy diagram leading to the Ward identity.

If we recall the introduction of the electromagnetic coupling through the minimal substitution $\not{p} \to \not{p} - e\not{A}$ we may expect a close relationship between the electron propagator and the vertex function. This is indeed the case and is embodied in an identity due to Ward.

Consider the insertion of an external photon line of zero momentum in an electron self-energy diagram, in all possible manners, on the internal charged particle propagators (Fig. 7-11). The term of zeroth order yields $i(\not{p} - m) \to -ie\not{A}$, while the one-loop contribution as depicted on Fig. 7-11 leads to $-ie\sum^{[1]}(p) \to -ieA^\mu \Gamma_\mu^{[1]}(p, p)$, and so on. The graphical correspondence may be interpreted as follows. In each internal charged particle propagator of a self-energy diagram we substitute $\not{p} - e\not{A}$ for the running momentum $\not{p}$ with A_μ a fixed constant four-vector. We then expand in powers of eA and extract the coefficient of the linear term. This is $\Gamma_\mu(p, p)$.

For instance, if we apply the procedure to the one-loop contribution to $\sum(p)$ we find

$$\sum(p) \to \frac{\partial}{\partial eA^\mu}(-ie)^2 i \int \frac{d^4k}{(2\pi)^4}\frac{1}{i}\left[\frac{g_{\rho\sigma}}{k^2 - \mu^2 + i\varepsilon} + \frac{(1-\lambda)k_\rho k_\sigma}{(k^2 - \mu^2 + i\varepsilon)(\lambda k^2 - \mu^2 + i\varepsilon)}\right]$$

$$\times \gamma^\rho \frac{i}{\not{p} - \not{k} - e\not{A} - m + i\varepsilon}\gamma^\sigma \bigg|_{A=0}$$

Since

$$\frac{1}{\not{p} - \not{k} - e\not{A} - m + i\varepsilon} = \frac{1}{\not{p} - \not{k} - m + i\varepsilon} + \frac{1}{\not{p} - \not{k} - m + i\varepsilon} e\not{A} \frac{1}{\not{p} - \not{k} - e\not{A} - m + i\varepsilon}$$

after differentiation, and setting A equal to zero, we recover an expression for $\Gamma^\mu(p, p)$ consistent with Eq. (7-45).

We could ask whether the above operation yields all vertex diagrams. The answer is obviously "no." Consider in a self-energy diagram an internal closed fermion loop. According to Furry's theorem an even number of photon lines are attached to this loop (if we agree to cut through each internal photon line starting and ending on this loop). Again according to Furry's theorem, it is therefore impossible from the above procedure to find a vertex contribution where an internal fermion loop would be attached to the rest of the diagram by an odd number of photon lines, such as depicted on the example of Fig. 7-12. This

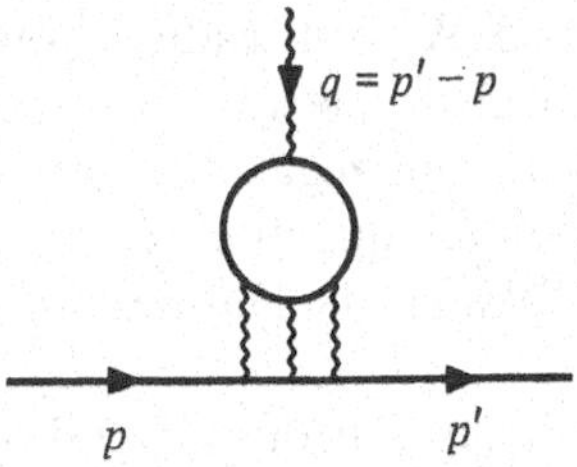

Figure 7-12 Example of a diagram that cannot be obtained from $\Sigma(p)$ by the procedure described in the text.

means that we obtain only those diagrams where the external photon is attached to the fermionic line that carries the charge flow originating in the external line. This may at first seem troublesome if we are to find a relation between $\sum$ and Γ. Fortunately, the sum of all diagrams where the external line is attached in all possible ways to an internal electron loop vanishes when evaluated at zero momentum transfer q. This we can see after a suitable Pauli-Villars regularization of the loop by isolating a factor of the form

$$\sum \int d^4p \operatorname{tr} \left(\frac{1}{\not{p} - m + i\varepsilon} \gamma_\mu \frac{1}{\not{p} - m + i\varepsilon} \gamma_2 \frac{1}{\not{p} + \not{q}_2 - m + i\varepsilon} \times \gamma_3 \frac{1}{\not{p} + \not{q}_2 + \not{q}_3 - m + i\varepsilon} \cdots \gamma_{2p} \right)$$

where the sum of all momenta entering the loop $q_2 + q_3 + \cdots + q_{2p}$ vanishes. As before,

$$\frac{1}{\not{p} - m + i\varepsilon} \gamma_\mu \frac{1}{\not{p} - m + i\varepsilon} = - \frac{\partial}{\partial p^\mu} \frac{1}{\not{p} - m + i\varepsilon}$$

Summing over all possible insertions of the external photon for fixed values of the internal photon momenta we get a total derivative as an integrand. The integral vanishes provided it is regularized in a gauge invariant way. We conclude that the prescription of attaching an external photon may be restricted to the distinguished set of electron propagators carrying the external charge flow. The same set may be seen to carry the electron momentum flow and the prescription is then equivalent to a derivative with respect to this momentum. This yields the Ward identity as

$$\Gamma_\mu(p, p) = - \frac{\partial}{\partial p^\mu} \sum(p) \tag{7-47a}$$

If we include the zeroth-order contribution this reads

$$\Lambda_\mu(p, p) = \gamma_\mu + \Gamma_\mu(p, p) = \frac{\partial}{\partial p^\mu} [\not{p} - m - \sum(p)] \tag{7-47b}$$

In the next chapter this identity will be generalized for nonvanishing momentum transfer in the form due to Takahashi:

$$(p' - p)^\mu \Lambda_\mu(p', p) = [\not{p}' - m - \sum(p')] - [\not{p} - m - \sum(p)] \tag{7-48}$$

and its relation with current conservation will be discussed. We can readily show that (7-47*a*) follows from (7-48). The sketch of a proof given above obviously requires some care. In particular, a consistent gauge-invariant regularization is needed. In fact one of the goals of the renormalization of quantum electrodynamics is precisely to transform the formal Ward-Takahashi identities into relations among renormalized finite quantities.

We return to the one-loop calculation and drop the index 1. From Eq. (7-45), $\Gamma(p', p) - \Gamma(p, p)$ is finite, and $\Gamma(p, p)$ is known from the Ward identity in terms of $\sum(p)$. This means that

$$\begin{aligned}\Gamma_\mu(p', p) &= -\frac{\partial}{\partial p^\mu}\sum(p) + \Gamma_\mu(p', p) - \Gamma_\mu(p, p)\\ &= \gamma_\mu(Z_2^{-1} - 1) + \Gamma_\mu(p', p) - \Gamma_\mu(p, p) - Z_2^{-1}\frac{\partial}{\partial p^\mu}\sum{}^R(p) \qquad (7\text{-}49a)\end{aligned}$$

We define the (infinite) vertex renormalization constant Z_1 through

$$\Gamma_\mu(p', p) = \gamma_\mu(Z_1^{-1} - 1) + Z_1^{-1}\Gamma_\mu^R(p', p) \qquad (7\text{-}49b)$$

Again the factor Z_1^{-1} in front of the renormalized vertex correction Γ_μ^R can be replaced by unity, when computing to first order, and $\Gamma_\mu^R(p', p)$ is normalized by requiring that on-shell ($p'^2 = p^2 = m^2$) and for zero momentum transfer, its matrix element between Dirac spinors vanishes:

$$\bar{u}(p')\Gamma_\mu^R(p', p)u(p)\Big|_{\substack{p^2 = p'^2 = m^2\\ p = p'}} = 0 \qquad (7\text{-}50)$$

Clearly $\Gamma_\mu(p', p) - \Gamma_\mu(p, p)$ vanishes when $p = p'$ and so does the matrix element of $(\partial/\partial p^\mu)\sum^R(p)$, since this operator is proportional to $(\not{p} - m)$. Comparing (7-49*a*) and (7-49*b*), we reach the conclusion that

$$Z_2 = Z_1 \qquad (7\text{-}51a)$$

and

$$\Gamma_\mu^R(p', p) = Z_1[\Gamma_\mu(p', p) - \Gamma_\mu(p, p)] - \frac{\partial}{\partial p^\mu}\sum{}^R(p) \qquad (7\text{-}51b)$$

The Ward identity remains valid in a theory with massive photons. We leave it to the reader to be convinced that a one-loop counterterm of the form

$$\delta\mathscr{L}^{[1]}{}_{\bar{\psi}A\psi} = -e(Z_1 - 1)\bar{\psi}\not{A}\psi \qquad (7\text{-}52)$$

will indeed cancel the vertex logarithmic divergence to this order.

What remains to be done is to compute $\Gamma_\mu^R(p', p)$. We shall satisfy ourselves with the case where p and p' are on-shell and Γ_ρ is to be considered as sandwiched between Dirac spinors as $\bar{u}(p')\Gamma_\rho(p', p)u(p)$. We shall, however, not write the spinors explicitly but we freely use the Gordon relation

$$\bar{u}(p')\gamma_\rho u(p) = \frac{1}{2m}\bar{u}(p')[(p + p')_\rho + i\sigma_{\rho\nu}(p' - p)^\nu]u(p) \qquad (7\text{-}53)$$

The quantity that we intend to calculate has a gauge-independent finite part. Indeed, in Eq. (7-45) the term proportional to $(1-\lambda)$ involves an integral over

$$\not{k}\frac{1}{\not{p}'-\not{k}-m+i\varepsilon}\gamma_\rho\frac{1}{\not{p}-\not{k}-m+i\varepsilon}\not{k}$$

We can replace $\not{k}$ to the right by $\not{k}-\not{p}+m$ and by $\not{k}-\not{p}'+m$ to the left, and thus find a quantity independent of p and p' which is thus absorbed in $(Z_1^{-1}-1)$ by virtue of the normalization condition. As far as $(Z_1^{-1}-1)$ is concerned, we know that it depends on the choice of gauge. We therefore set $\lambda=1$ to write

$$\Gamma_\rho(p',p)=i(-ie)^2\int\frac{d^4k}{(2\pi)^4}\frac{1}{k^2-\mu^2+i\varepsilon}\gamma_\sigma\frac{\not{p}'-\not{k}+m}{(p'-k)^2-m^2+i\varepsilon}\gamma_\rho$$
$$\times\frac{\not{p}-\not{k}+m}{(p-k)^2-m^2+i\varepsilon}\gamma^\sigma$$

The numerator in this expression may be written using the mass shell condition as

$$4\left\{\gamma_\rho\left[(p'-k)\cdot(p-k)-\frac{k^2}{2}\right]+(p'+p-k)_\rho\not{k}-mk_\rho\right\}$$

An auxiliary integral is

$$\int\frac{d^4k}{(2\pi)^4}e^{ik\cdot z}\frac{1}{(k^2-\mu^2+i\varepsilon)(k^2-2p'\cdot k+i\varepsilon)(k^2-2p\cdot k+i\varepsilon)}$$
$$=-\frac{1}{(4\pi)^2}\int_0^\infty\frac{d\alpha_1\,d\alpha_2\,d\alpha_3}{(\alpha_1+\alpha_2+\alpha_3)^2}\exp\left\{-i\left[\alpha_1\mu^2+\frac{(z/2-\alpha_2p'-\alpha_3p)^2}{\alpha_1+\alpha_2+\alpha_3}\right]\right\}$$

The introduction of a factor $e^{ik\cdot z}$ in the integrand allows us to obtain the required expression in the numerator of Γ_ρ by differentiation. After symmetrization in α_2 and α_3 we obtain the following representation:

$$\Gamma_\rho(p',p)=\frac{\alpha}{i\pi}\int_0^\infty\frac{d\alpha_1\,d\alpha_2\,d\alpha_3}{(\alpha_1+\alpha_2+\alpha_3)^3}\left(\gamma_\rho\left\{(\alpha_1+\alpha_2+\alpha_3)p\cdot p'-(\alpha_2+\alpha_3)\frac{(p'+p)^2}{2}+\frac{i}{2}\right.\right.$$
$$\left.\left.+\frac{1}{2(\alpha_1+\alpha_2+\alpha_3)}[m^2(\alpha_2+\alpha_3)^2-\alpha_2\alpha_3q^2]\right\}+\frac{m}{2}(p'+p)_\rho\frac{\alpha_1(\alpha_2+\alpha_3)}{\alpha_1+\alpha_2+\alpha_3}\right)$$
$$\times\exp\left\{\frac{i}{\alpha_1+\alpha_2+\alpha_3}[\alpha_2\alpha_3q^2-(\alpha_2+\alpha_3)^2m^2-\alpha_1(\alpha_1+\alpha_2+\alpha_3)\mu^2]\right\}$$

where we have set $q=p'-p$. When we attempt to perform the integral over the homogeneity scale of the α_j parameters, we encounter the expected logarithmic singularity which we subtract taking into account the normalization condition (7-50). We observe that $q^2\le 0$ if p and p' lie on the mass shell. After a Wick rotation $\alpha_j \to \alpha_j/i$ we obtain

$$\Gamma_\rho^R(p', p) = -\frac{\alpha}{\pi}\int_0^\infty d\alpha_1\, d\alpha_2\, d\alpha_3\, \delta\left(1 - \sum_1^3 \alpha_j\right)$$

$$\times \left\{\frac{\gamma_\rho\{m^2 - q^2/2 - (\alpha_2+\alpha_3)(2m^2 - q^2/2) + \frac{1}{2}[(\alpha_2+\alpha_3)^2 m^2 - \alpha_2\alpha_3 q^2]\}}{m^2(\alpha_2+\alpha_3)^2 + \alpha_1\mu^2 - \alpha_2\alpha_3 q^2}\right.$$

$$+ (m/2)\frac{(p'+p)_\rho \alpha_1(\alpha_2+\alpha_3)}{m^2(\alpha_2+\alpha_3)^2 + \alpha_1\mu^2 - \alpha_2\alpha_3 q^2}$$

$$\left. + \frac{1}{2}\gamma_\rho \ln\left[m^2(\alpha_2+\alpha_3)^2 + \alpha_1\mu^2 - \alpha_2\alpha_3 q^2\right]\right\} - (p' = p, q = 0)$$

since from (7-51*b*) the renormalization of the on-shell vertex function to order one amounts to a subtraction at $q = 0$. We isolate the form factors $F_1(q^2)$ and $F_2(q^2)$ defined as

$$\Gamma_\rho^R(p', p) = \gamma_\rho F_1(q^2) + \frac{i}{2m}\sigma_{\rho\nu} q^\nu F_2(q^2) \tag{7-54}$$

We are, of course, only computing the contributions of order α to these form factors. The term of order zero adds one to F_1. We find

$$F_1(q^2) = -\frac{\alpha}{\pi}\int_0^\infty d\alpha_1\, d\alpha_2\, d\alpha_3\, \delta\left(1 - \sum_1^3 \alpha_j\right)$$

$$\times \left\{\frac{m^2 - q^2/2 - (\alpha_2+\alpha_3)(2m^2 - q^2/2) + m^2[(\alpha_2+\alpha_3) - \frac{1}{2}(\alpha_2+\alpha_3)^2] - \alpha_2\alpha_3(q^2/2)}{m^2(\alpha_2+\alpha_3)^2 + \alpha_1\mu^2 - \alpha_2\alpha_3 q^2}\right.$$

$$\left. + \frac{1}{2}\ln\left[m^2(\alpha_2+\alpha_3)^2 + \alpha_1\mu^2 - \alpha_2\alpha_3 q^2\right] - (q^2 = 0)\right\} \tag{7-55}$$

$$F_2(q^2) = \frac{\alpha}{\pi}\int_0^\infty d\alpha_1\, d\alpha_2\, d\alpha_3\, \delta\left(1 - \sum_1^3 \alpha_j\right)\frac{m^2\alpha_1(\alpha_2+\alpha_3)}{m^2(\alpha_2+\alpha_3)^2 + \alpha_1\mu^2 - \alpha_2\alpha_3 q^2} \tag{7-56}$$

The momentum transfer can be parametrized in terms of an hyperbolic angle θ such that

$$p \cdot p' = m^2 \cosh\theta \qquad q^2 = -4m^2 \sinh^2\frac{\theta}{2} \tag{7-57}$$

The easiest calculation is that of the magnetic form factor F_2. It does not involve any infrared divergence to this order so we can set $\mu^2 = 0$. Integration over α_1 yields

$$F_2(q^2) = \frac{\alpha}{\pi}\int_0^1 d\alpha_2\, d\alpha_3\, \theta(1 - \alpha_2 - \alpha_3)\frac{\alpha_2 + \alpha_3 - (\alpha_2+\alpha_3)^2}{\alpha_2^2 + \alpha_3^2 + 2\alpha_2\alpha_3\cosh\theta}$$

Introducing a new homogeneity parameter, this can be further reduced to

$$F_2(q^2) = \frac{\alpha}{\pi}\int_0^1 d\alpha_2\, d\alpha_3\, \delta(1-\alpha_2-\alpha_3)\, \frac{1}{\alpha_2^2+\alpha_3^2+2\alpha_2\alpha_3\cosh\theta}\int_0^1 dx\,(1-x)$$

$$= \frac{\alpha}{2\pi}\int_0^1 d\beta\, \frac{1}{\beta^2+(1-\beta)^2+2\beta(1-\beta)\cosh\theta}$$

$$= \frac{\alpha}{2\pi}\int_0^1 \frac{d\beta}{\beta^2}\, \frac{1}{\{1+[(1-\beta)/\beta]\,e^{\theta}\}\{1+[(1-\beta)/\beta]\,e^{-\theta}\}}$$

The last integral can easily be performed using $(1-\beta)/\beta$ as an integration variable and the result is

$$F_2(q^2) = \frac{\alpha}{2\pi}\,\frac{\theta}{\sinh\theta} \tag{7-58}$$

In particular, when $q^2 \to 0$, $\theta \to 0$ and

$$F_2(0) = \frac{\alpha}{2\pi} \tag{7-59}$$

We turn to the calculation of F_1, paying attention to infrared divergences. After integration over α_1 we find that F_1 may be expressed as the sum of four terms:

$$F_1(q^2) = \sum_j^4 \mathscr{T}_j + \text{constant}$$

and the constant is adjusted so that $F_1(0) = 0$. The various contributions in the limit of small μ^2 are as follows:

$$\mathscr{T}_1 = -\frac{\alpha}{\pi}\int_0^1 d\alpha_2 \int_0^{1-\alpha_2} d\alpha_3\, \frac{\cosh\theta}{\alpha_2^2+\alpha_3^2+2\alpha_2\alpha_3\cosh\theta+(\mu^2/m^2)(1-\alpha_2-\alpha_3)}$$

$$= -\frac{\alpha}{\pi}\cosh\theta \int_0^1 d\beta \int_0^1 dx$$

$$\times\, \frac{x}{x^2[\beta^2+(1-\beta)^2+2\beta(1-\beta)\cosh\theta]+(\mu^2/m^2)(1-x)}$$

$$= -\frac{\alpha}{2\pi}\cosh\theta \int_0^1 d\beta\, \frac{\ln[\beta^2+(1-\beta)^2+2\beta(1-\beta)\cosh\theta]-\ln(\mu^2/m^2)}{\beta^2+(1-\beta)^2+2\beta(1-\beta)\cosh\theta}$$

Changing variables according to

$$1-2\beta = \frac{\tanh\varphi}{\tanh(\theta/2)}$$

reduces the integral to

$$\mathscr{T}_1 = \frac{\alpha}{\pi}\,\theta\coth\theta\,\ln\frac{\mu}{m} - \frac{2\alpha}{\pi}\coth\theta\int_0^{\theta/2} d\varphi\,\varphi\tanh\varphi$$

The other pieces of the puzzle are

$$\mathscr{T}_2 = \frac{\alpha}{\pi} \cosh\theta \int_0^1 d\alpha_2 \int_0^{1-\alpha_2} d\alpha_3 \frac{\alpha_2 + \alpha_3}{\alpha_2^2 + \alpha_3^2 + 2\alpha_2\alpha_3 \cosh\theta} = \frac{\alpha}{\pi}\theta \coth\theta$$

$$\mathscr{T}_3 = \frac{2\alpha}{\pi}(1 - \cosh\theta) \int_0^1 d\alpha_2 \int_0^{1-\alpha_2} d\alpha_3 \frac{\alpha_2\alpha_3}{\alpha_2^2 + \alpha_3^2 + 2\alpha_2\alpha_3 \cosh\theta} = \frac{\alpha}{2\pi}\left(\frac{\theta}{\sinh\theta} - 1\right)$$

$$\mathscr{T}_4 = -\frac{\alpha}{2\pi}\int_0^1 d\alpha_2 \int_0^{1-\alpha_2} d\alpha_3 \ln\left[(\alpha_2+\alpha_3)^2 + 4\alpha_2\alpha_3 \sinh^2\frac{\theta}{2}\right]$$

$$= -\frac{\alpha}{2\pi}\frac{\theta/2}{\tanh(\theta/2)} + \text{constant}$$

Finally, adjusting the normalization, we obtain the complete expression

$$F_1(q^2) = \frac{\alpha}{\pi}\left[\left(\ln\frac{\mu}{m} + 1\right)(\theta\coth\theta - 1) - 2\coth\theta \int_0^{\theta/2} d\varphi\, \varphi \tanh\varphi - \frac{\theta}{4}\tanh\frac{\theta}{2}\right] \tag{7-60}$$

In the vicinity of $q^2 = 0$ we have $q^2/m^2 \simeq -\theta^2$ so that, to leading order in q^2,

$$F_1(q^2)_{q^2\to 0} \simeq \frac{\alpha q^2}{3\pi m^2}\left(\ln\frac{m}{\mu} - \frac{3}{8}\right) \tag{7-61}$$

This means that the complete expression of the renormalized vertex function evaluated between spinors takes the form

$$\Lambda_\rho^R = \gamma_\rho\left[1 + \frac{\alpha q^2}{3\pi m^2}\left(\ln\frac{m}{\mu} - \frac{3}{8}\right)\right] + \frac{i}{2m}\sigma_{\rho\nu}q^\nu \frac{\alpha}{2\pi} + \cdots$$

$$= \frac{(p+p')_\rho}{2m}\left[1 + \frac{\alpha q^2}{3\pi m^2}\left(\ln\frac{m}{\mu} - \frac{3}{8}\right)\right] + \frac{i}{2m}\sigma_{\rho\nu}q^\nu\left(1 + \frac{\alpha}{2\pi}\right) + \cdots \tag{7-62}$$

By analytic continuation in the complex variable θ we obtain Γ_ρ^R for positive values of q^2. Setting, for instance, $\theta = i\psi$ we reach the expression valid for q^2 between 0 and $4m^2$, where F_1 and F_2 continue to be real to this order:

$$q^2 = 4m^2 \sin^2\frac{\psi}{2} \qquad 0 \le \psi \le \pi \tag{7-63}$$

where

$$F_1 = \frac{\alpha}{\pi}\left[\left(\ln\frac{\mu}{m} + 1\right)(\psi\cot\psi - 1) + 2\cot\psi \int_0^{\psi/2} d\varphi\, \varphi\tan\varphi + \frac{\psi}{4}\tan\frac{\psi}{2}\right]$$

$$F_2 = \frac{\alpha}{2\pi}\frac{\psi}{\sin\psi} \tag{7-64}$$

A useful exercise is to continue these expressions beyond $4m^2$ and give an interpretation of the imaginary part of the form factors. What is the behavior for large $|q^2|$?

7-1-4 Summary

Let us present a summary of what has been achieved by eliminating ultraviolet divergences. Up to now we have defined three divergent constants:

1. $Z_1 = Z_2$, gauge dependent as well as infrared divergent, but equal by virtue of the Ward identity.
2. Z_3 and δm both gauge independent and infrared finite. We have computed $Z_1 - 1 = Z_2 - 1$, $Z_3 - 1$, and δm to order $\hbar$ in such a way that the lagrangian

$$\mathscr{L} = -\frac{1}{4}F^2 + \frac{\hat{\mu}^2}{2}A^2 - \frac{\lambda}{2}(\partial\cdot A)^2 + \frac{i}{2}\bar{\psi}\overleftrightarrow{\not\partial}\psi - m\bar{\psi}\psi - e\bar{\psi}\not{A}\psi$$
$$-\frac{1}{4}(Z_3 - 1)F^2 + (Z_2 - 1)\left(\frac{i}{2}\bar{\psi}\overleftrightarrow{\not\partial}\psi - m\bar{\psi}\psi\right) + Z_2\,\delta m\bar{\psi}\psi$$
$$-e(Z_1 - 1)\bar{\psi}\not{A}\psi \tag{7-65}$$

will produce finite renormalized two- and three-point functions, up to the same order. We shall see that the same holds for higher Green functions.

The photon "mass term" has been denoted $(\hat{\mu}^2/2)A^2$. As we noted, the true position of the pole in the massive theory is displaced by radiative corrections by an amount related to the wave function renormalization constant Z_3, since the vacuum polarization tensor has only one overall divergence. It is simpler to define the corresponding dimensional parameter as (minus) the coefficient of the term $g_{\rho\sigma}$ in the inverse propagator at zero momentum:

$$\hat{\mu}^2 = \frac{\mu_0^2}{1 + \bar{\omega}(0)} = Z_3\mu_0^2 \tag{7-66}$$

Here μ_0 stands for the (infinite) bare mass. Therefore, $\hat{\mu}^2 A^2 = Z_3\mu_0^2 A^2$. In a massive vector boson theory the real mass μ would be related to $\hat{\mu}$ by a finite renormalization factor.

It is now necessary to convince ourselves that higher Green functions are finite, since we have now exhausted our freedom of choice by selecting the measurable mass, coupling constant, and pole residues. This analysis is of interest because there still exists a potentially dangerous amplitude for photon-photon scattering. If we were dealing with spin-zero charged bosons, a new physical constant, apparently absent at the level of Born diagrams, would emerge from the needs of renormalization. In this case we have to give the normalization of the charged particle elastic scattering amplitude.

The mechanism responsible for ultraviolet divergences in one-loop integrals may be characterized as follows. If k denotes the momentum flowing through the loop we are interested in the large-k behavior of the integrand in the Feynman integral. Each fermion propagator contributes a factor $|k|^{-1}$ and each boson propagator a factor $|k|^{-2}$. Vertices with a γ_μ coupling are k independent. Proper functions are classified according to the number of fermion external lines zero, two, four, etc. Typical diagrams are shown in Fig. 7-13.

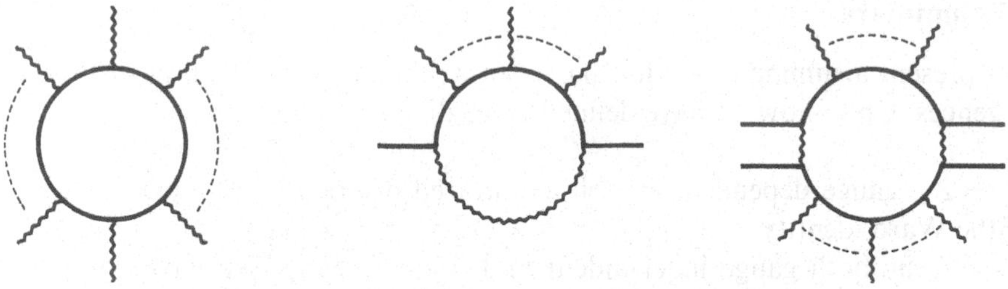

Figure 7-13 Examples of one-loop diagrams with zero, two, four, . . ., external fermion lines.

If Λ stands for some ultraviolet cutoff, the behavior of the integral will be

$$\Lambda^{4-I_F-2I_B} \qquad 4 - I_F - 2I_B > 0$$

$$\ln \Lambda \qquad 4 - I_F - 2I_B = 0$$

with $I_F(I_B)$ denoting the number of fermion (boson) lines in the loop. The integral will be superficially divergent if

$$\omega(G) = 4 - I_F - 2I_B \geq 0 \tag{7-67}$$

where $\omega(G)$ is called the superficial degree of divergence. In electrodynamics the kinematics somewhat complicates the matter. We have already encountered examples such as in the case of vacuum polarization ($I_B = 0$, $I_F = 2$) where the above counting seemed to indicate a quadratic divergence $\omega = 2$, but current conservation allowed the extraction of a factor $(k^2 g^{\rho\sigma} - k^\rho k^\sigma)$ leaving only a logarithmic divergence. Similarly for the electron self-energy $\omega = 1$, but again the fact that $\sum$ was found proportional to the dimensional quantities m and $\not{p}$ reduced the degree of divergence by one unit. For the vertex function the counting is correct and $\omega = 0$. But even in this case only one form factor was found divergent. We may express $\omega(G)$ in terms of the (even) number of external fermionic lines E_F and of the number E_B of bosonic external lines by using the fact that two fermion lines and one boson line meet at each vertex. Along the loop the number of vertices V is equal to the number of internal propagators

$$V = I_F + I_B$$

and according to the previous argument

$$2V = 2I_F + E_F \qquad V = 2I_B + E_B$$

since each propagator meets two vertices. Eliminating V, I_F, and I_B we find

$$\omega = 4 - \tfrac{3}{2}E_F - E_B \tag{7-68}$$

Therefore we derive the following table:

E_F	E_B		Superficial degree of divergence	Effective degree of divergence
0	2	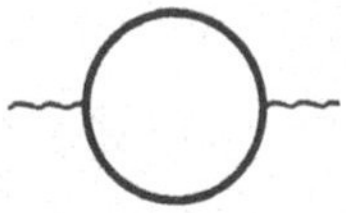	2	0
0	3	Excluded by charge conjugation invariance		
0	4	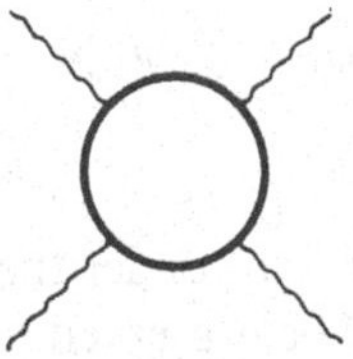	0	
2	0	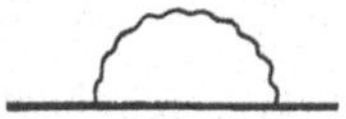	1	0
2	1	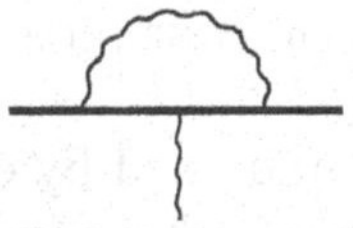	0	0

(7-69)

As the number of external lines is increased the integral becomes more convergent and table (7-69) shows that we have exhausted all possibilities except for photon-photon scattering. However, the latter is in fact convergent, again due to gauge invariance, which dictates some kinematical structure leaving a well-behaved integral. This follows from our earlier discussion of the Euler-Heisenberg effective lagrangian (Sec. 3-3-4) and will be discussed in detail below (Sec. 7-3-1).

To order $\hbar$ we now have a finite predictive theory satisfying perturbatively the general principles of unitarity and causality, since the counterterms amount to subtractions in dispersion relations, i.e., they do not modify the analyticity properties nor the discontinuity rules. Another confirmation stems from the fact that we are using a local hermitian lagrangian.

Let us now return to Eq. (7-65) and reorganize the various terms as

$$\mathscr{L} = -\frac{1}{4}Z_3F^2 + \frac{1}{2}\mu_0^2 Z_3 A^2 - \frac{\lambda}{2}(\partial \cdot A)^2 + Z_2\left[\frac{i}{2}\overline{\psi}\overleftrightarrow{\not\partial}\psi - (m - \delta m)\overline{\psi}\psi\right] - Z_1 e\overline{\psi}\not{A}\psi \tag{7-70}$$

This expression leads to a natural interpretation. The fields ψ, $\overline{\psi}$, and A occurring here are renormalized fields corresponding to propagators with residue equal to unity at the one-particle pole. Define unrenormalized fields ψ_0, $\overline{\psi}_0$, and A_0 and cutoff-dependent bare parameters μ_0^2, m_0, e_0 according to

$$\begin{aligned} \psi_0 &= Z_2^{1/2}(\Lambda)\psi & m_0 &= m - \delta m(\Lambda) \\ \overline{\psi}_0 &= Z_2^{1/2}(\Lambda)\overline{\psi} & \mu_0^2 &= Z_3^{-1}(\Lambda)\hat{\mu}^2 \\ A_0 &= Z_3^{1/2}(\Lambda)A & e_0 &= Z_1(\Lambda)Z_2^{-1}(\Lambda)Z_3^{-1/2}(\Lambda)e = Z_3^{-1/2}(\Lambda)e \end{aligned} \tag{7-71}$$

Then the lagrangian (7-70) expressed in terms of bare quantities is similar to the original lagrangian

$$\mathscr{L} = -\frac{1}{4}F_0^2 + \frac{\mu_0^2}{2}A_0^2 - \frac{\lambda}{2}Z_3^{-1}(\Lambda)(\partial \cdot A_0)^2 + \frac{i}{2}\overline{\psi}_0 \overleftrightarrow{\not{\partial}} \psi_0 - m_0\overline{\psi}_0\psi_0 - e_0\overline{\psi}_0 \not{A}_0 \psi_0 \tag{7-72}$$

Not only do finitely many types of counterterms suffice to render the perturbative series term-by-term convergent (a property which will be preserved to all orders in $\hbar$, as we shall see) but the very structure of the theory is preserved by renormalization.

It is worth emphasizing the role of the Ward identity, a result of which is

$$e_0 A_0 = eA \tag{7-73}$$

giving a meaning to the scale of the electromagnetic interaction and preserving the meaning of minimal coupling.

Note also that the A^2 and $(\partial \cdot A)^2$ terms were not affected by counterterms, a special property of electrodynamics. In the same way that we introduced a bare "mass" μ_0 we can define a bare parameter λ_0 through

$$\lambda_0 = Z_3^{-1}(\Lambda)\lambda$$

The lagrangian (7-72) gives rise to renormalized Green functions defined as follows:

$$\begin{aligned} &G_0(p_1, \ldots, p_{2n}, k_1, \ldots, k_l; \mu_0, m_0, e_0, \lambda_0, \Lambda) \\ &\qquad = Z_2^n(\Lambda)Z_3^{l/2}(\Lambda)G_R(p_1, \ldots, p_{2n}, k_1, \ldots, k_l; \mu, m, e, \lambda) \end{aligned} \tag{7-74}$$

Here $p_1, \ldots, p_{2n}$ $(k_1, \ldots, k_l)$ stand for the external fermion (photon) momenta. Again this holds perturbatively in $\hbar$, and up to now we may only assert this up to order $\hbar^1$. The relations (7-71) and (7-74) justify the denomination of multiplicative renormalization.

The wave function renormalization constants $Z_2^{1/2}$ and $Z_3^{1/2}$ appear to be infinite when expanded perturbatively. This shows, at least in this sense, that one of the working hypotheses of this approach, namely, the canonical equivalence of the free and interacting picture, is not mathematically well founded. We have succeeded, however, in overcoming this difficulty, and we shall discover later on that there is some interesting physical meaning in these apparently infinite quantities.

The extension of this program to all orders will be the subject of Chap. 8.

7-2 RADIATIVE CORRECTIONS TO THE INTERACTION WITH AN EXTERNAL FIELD

The remainder of this chapter deals with examples of radiative corrections. One of the issues will be to make sense out of the infrared divergences arising from the long-range electromagnetic forces. This survey is only indicative and without any pretense to completeness.

7-2-1 Effective Interaction and Anomalous Magnetic Moment

In Chaps. 2 and 4 we have already presented various aspects of the interaction of charged particles with c-number external fields. Here we want to discuss quantum corrections generated when we substitute to the quantum field A_q the sum $A_q + A_c$, where A_c is the classical field.

To first order in A_c we derive an effective interaction represented by the first three diagrams of Fig. 7-14 including one-loop quantum corrections. This means that the elementary interaction $e\gamma_\rho A_c^\rho$ has been replaced by

$$eA_c^\rho\gamma_\rho \to eA_c^\rho(\gamma_\rho + \Gamma_\rho + \bar{\omega}_{\rho\nu}G^{\nu\sigma}\gamma_\sigma) \tag{7-75}$$

where Γ and $\bar{\omega}$ have been computed in the preceding section and G is the free photon propagator. We have assumed the electrons on mass shell, disregarding therefore the self-energy insertions on the external lines which are assumed to be absorbed in the correct definition of the mass and proper normalization of the states.

At low momentum transfer $q = p' - p$, keeping only dominant terms as $\mu \to 0$, we obtain

$$eA_c^\rho\left\{\gamma_\rho\left[1 + \frac{\alpha q^2}{3\pi m^2}\left(\ln\frac{m}{\mu} - \frac{3}{8} - \frac{1}{5}\right)\right] + \frac{i}{2m}\frac{\alpha}{2\pi}\sigma_{\rho\nu}q^\nu\right\} \tag{7-76}$$

When deriving (7-76) the gauge-dependent terms of $G^{\nu\sigma}$ proportional to $q^\nu q^\sigma$ have not contributed when acting on the vacuum polarization. Also a term in $\not{q} = \not{p}' - \not{p}$ has disappeared when evaluated on the mass shell. Consequently, the remaining expression remains gauge invariant in the sense that if A_c^ρ is replaced by $A_c^\rho + q^\rho f(q)$ the added term vanishes on the mass shell.

The quasistatic limit $q \to 0$ allows a simple interpretation. In configuration space the momentum transfer q can be replaced by the derivative operator $i\partial$

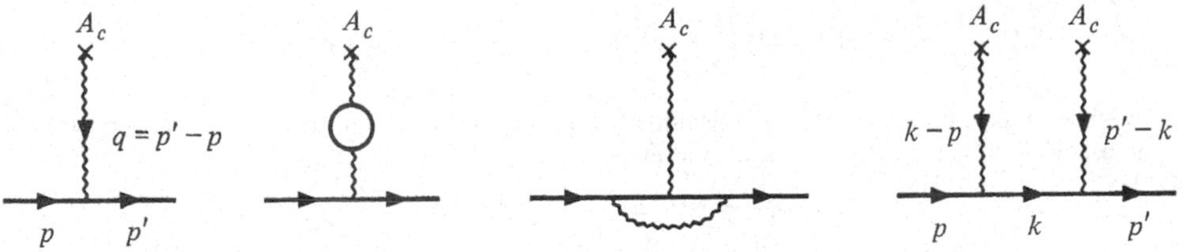

Figure 7-14 Lowest-order contributions to the interaction with an external field.

acting on A_c. Using Gordon's identity we obtain the two equivalent expressions of the interaction hamiltonian of a slowly varying (in space and time) external field with a charged spinning particle:

$$\Delta H \simeq e \int d^3x \left\{ \bar{\psi}(x)\gamma_\rho\psi(x)\left[1 - \frac{\alpha}{3\pi m^2}\left(\ln\frac{m}{\mu} - \frac{3}{8} - \frac{1}{5}\right)\Box\right]A_c^\rho(x) \right.$$
$$\left. + \frac{\alpha}{2\pi}\frac{1}{2m}\bar{\psi}(x)\sigma_{\nu\rho}\psi(x)\partial^\nu A_c^\rho(x)\right\}$$
$$\simeq e \int d^3x \left\{ \frac{i}{2m}\bar{\psi}(x)\overleftrightarrow{\partial}_\rho\psi(x)\left[1 - \frac{\alpha}{3\pi m^2}\left(\ln\frac{m}{\mu} - \frac{3}{8} - \frac{1}{5}\right)\Box\right]A_c^\rho(x) \right.$$
$$\left. + \left(1 + \frac{\alpha}{2\pi}\right)\frac{1}{4m}\bar{\psi}(x)\sigma_{\nu\rho}\psi(x)F_c^{\nu\rho}(x)\right\} \tag{7-77}$$

In the second form the first term introduces a convective current analogous to the one pertaining to scalar particles, and its interaction with A_c is affected by infrared divergences. The second term has a simple interpretation. In the limit of a constant F_c it reduces to an effective magnetic dipole energy. Let us assume that F represents a constant magnetic field $F^{12} = -B^3$, $F^{23} = -B^1$, $F^{31} = -B^2$, $\sigma_{12} = \sigma_3$, etc. It then reads

$$-\mathbf{B}\cdot\boldsymbol{\mu} = -\mathbf{B}\cdot\left[\frac{e}{2m}\left(1 + \frac{\alpha}{2\pi}\right)2\int d^3x\, \bar{\psi}(x)\frac{\boldsymbol{\sigma}}{2}\psi(x)\right] \tag{7-78}$$

When discussing the Dirac equation we found a gyromagnetic ratio 2 leading to a magnetic dipole moment

$$\frac{e}{2m}\boldsymbol{\sigma} = 2\frac{e}{2m}\frac{\boldsymbol{\sigma}}{2}$$

We see that to order $\hbar$ quantum corrections modify this gyromagnetic ratio written in the form $g = 2(1 + a)$, where a is called the anomaly, by $a = \alpha/2\pi$. This result first derived by Schwinger in 1948 has been confirmed by numerous accurate experiments, which enable us to test higher-order corrections.

The electron anomaly is mass independent to lowest order. Therefore, the same result applies as well to any elementary spin $\frac{1}{2}$ particle such as the muon. A complete theory to higher orders requires the consideration of all charged particles interacting with the electromagnetic field. This leads to a difference in these anomalies arising from mass differences. At the present time, calculations have been pushed to third order (three loops). They read

$$a_e^{\text{th}} = \frac{1}{2}\frac{\alpha}{\pi} - 0.328\,478\,445\left(\frac{\alpha}{\pi}\right)^2 + 1.183(11)\left(\frac{\alpha}{\pi}\right)^3 = 1\,159\,652\,359(282) \times 10^{-12}$$

(Numbers in parentheses indicate uncertainties.) The latest measurements are of a similar accuracy and are likely to be improved in the near future:

$$a_e^{\text{exp}} = 1\,159\,652\,410\,(200) \times 10^{-12}$$

To cope with such an improvement the theoretical program involves obtaining the $(\alpha/\pi)^3$ term with greater accuracy and evaluating the next order [since $(\alpha/\pi)^4 \sim 29 \times 10^{-12}$] represented by as many as 891 Feynman diagrams! Minor contributions involve the insertions of a muon-loop vacuum polarization correction in the virtual photon propagator ($\sim 2 \times 10^{-12}$) together with hadronic ($\sim 1.4 \times 10^{-12}$) or weak ($\sim 0.05 \times 10^{-12}$) contributions. The measurement of the electron anomaly is likely to become in the near future one of the most precise means of obtaining an accurate value of the fine structure constant α, by a unique combination of theoretical and experimental skills.

In the case of the muon anomaly the uncertainties on hadronic corrections set a limit on the accuracy of theoretical predictions. The most recent measurement gives

$$a_\mu^{\text{exp}} = 1\,165\,922\,(9) \times 10^{-9}$$

The pure electrodynamic contributions are

$$\begin{aligned} a_\mu^{\text{qed}} &= \frac{1}{2}\frac{\alpha}{\pi} + 0.765\,782\left(\frac{\alpha}{\pi}\right)^2 + 24.45\,(0.06)\left(\frac{\alpha}{\pi}\right)^3 + 128.3\,(71.4)\left(\frac{\alpha}{\pi}\right)^4 \\ &= 1\,165\,851.8\,(2.4) \times 10^{-9} \end{aligned}$$

where the α^4 term is an estimate using the large ratio m_μ/m_e, and the hadronic correction is of order

$$a_\mu^{\text{hadr}} = 66.7\,(9.4) \times 10^{-9}$$

The theory is in reasonable agreement with experiment with a prediction

$$a_\mu^{\text{th}} = 1\,165\,919\,(10) \times 10^{-9}$$

Weak interaction effects are expected to contribute at the level of $a_\mu^{\text{weak}} \sim 2 \times 10^{-9}$.

The first term in the effective hamiltonian (7-77) appears troublesome in view of its infrared singularity. This arose from our insistence on having an isolated pole in the corresponding Green function. However, such a state can never be separated from those including an arbitrary number of soft photons. The latter are always excited as soon as a charged particle suffers a change in velocity, no matter how small. Therefore we have to include processes involving real emission or absorption of soft photons in the external field as soon as we discuss the virtual electromagnetic effects embodied in Eq. (7-77). The fictitious photon mass disappears from the final answer to a physically correct question, being replaced by the experimental resolution.

7-2-2 Radiative Corrections to Coulomb Scattering

As an example let us consider Coulomb scattering on a nucleus of small Z so that $Z\alpha$ may still be considered as a small parameter. Atomic corrections (a screening effect in a first approximation) will be neglected. Using the effective interaction (7-75) let us compute all corrections up to order α^3 to the Mott cross section [Eq. (2-130)] for unpolarized electrons. If E and p stand for the energy and magnitude of the electron three-momentum and β denotes its velocity $\beta = p/E$, we recall that the scattering cross section at angle θ was given by

$$\left(\frac{d\sigma}{d\Omega}\right)_{\text{Mott}} = \frac{Z^2\alpha^2}{4p^2\beta^2 \sin^4(\theta/2)}\left(1 - \beta^2 \sin^2\frac{\theta}{2}\right) \tag{7-79}$$

Since the first corrections will be of order $Z^2\alpha^3$ it will be mandatory to include the second Born approximation whose interference with the leading term will produce a contribution of order $Z^3\alpha^3$. This is comparable with the effect of vertex and vacuum polarization renormalization provided Z is small enough. The interest of pushing calculations that far is that it also enables us to discuss another aspect of infrared singularities related to the long-range character of the Coulomb forces. The latter induces an infinite phase shift on the scattered plane waves. To prevent it we may introduce a screening factor which in a consistent theory would be related to the fictitious photon mass μ. In order to show that this type of infrared divergence is cancelled in the cross section we shall distinguish it by introducing an independent screening length σ and by studying the limit $\sigma \to 0$ separately.

The diagrams contributing to the scattering amplitude are shown in Fig. 7-14 with the external potential A_c equal to the screened Coulomb interaction

$$\mathbf{A}_c = 0 \qquad A_c^0(x) = -\frac{Ze\, e^{-\sigma r}}{4\pi r} = Ze \int \frac{d^4q}{(2\pi)^3} \frac{\delta(q_0)}{q^2 - \sigma^2}\, e^{iq\cdot x}$$

The first three diagrams (1), (2), and (3) of Fig. 7-14 yield

$$S_{123} = -iZe^2(2\pi)\delta(p_0 - p'_0)\frac{1}{q^2 - \sigma^2}\bar{u}(p')\gamma^0\left\{1 + \frac{\alpha}{\pi}\left[\left(1 + \ln\frac{\mu}{m}\right)(2\varphi \coth 2\varphi - 1)\right.\right.$$
$$-2\coth 2\varphi \int_0^\varphi d\varphi'\, \varphi' \tanh\varphi' - \frac{1}{2}\varphi\tanh\varphi$$
$$\left.\left. + \left(1 - \frac{\coth^2\varphi}{3}\right)(\varphi\coth\varphi - 1) + \frac{1}{9}\right] - \frac{\not{q}}{2m}\frac{\alpha}{\pi}\frac{\varphi}{\sinh 2\varphi}\right\}u(p) \tag{7-80}$$

with $p\cdot p' = m^2\cosh 2\varphi$ and $q = p' - p$. The fourth diagram adds the quantity

$$S_4 = (-iZe)^2\int\frac{d^4k}{(2\pi)^4}\bar{u}(p')\left[\frac{2\pi\delta(p'_0 - k_0)}{(p' - k)^2 - \sigma^2}e\gamma^0\frac{i}{\not{k} - m + i\varepsilon}e\gamma^0\frac{2\pi\delta(k^0 - p^0)}{(k - p)^2 - \sigma^2}\right]u(p)$$

From the kinematical constraints, $p^0 = p'^0 = E$ and $|\mathbf{p}| = |\mathbf{p}'|$. Set

$$I_1 = \int d^3k\,\frac{1}{[(\mathbf{p}' - \mathbf{k})^2 + \sigma^2][(\mathbf{p} - \mathbf{k})^2 + \sigma^2](\mathbf{p}^2 - \mathbf{k}^2 + i\varepsilon)}$$

$$\frac{1}{2}(\mathbf{p} + \mathbf{p}')I_2 = \int d^3k\,\frac{\mathbf{k}}{[(\mathbf{p}' - \mathbf{k})^2 + \sigma^2][(\mathbf{p} - \mathbf{k})^2 + \sigma^2](\mathbf{p}^2 - \mathbf{k}^2 + i\varepsilon)}$$

Then S_4 takes the form

$$S_4 = -2i\frac{Z^2\alpha^2}{\pi}2\pi\delta(p'_0 - p_0)\bar{u}(p')[m(I_1 - I_2) + \gamma^0 E(I_1 + I_2)]u(p) \tag{7-81}$$

We are, of course, only interested in the small σ limits of I_1 and I_2. Feynman's identity

$$\frac{1}{(a + \lambda)(b + \lambda)} = -\frac{\partial}{\partial\lambda}\int_0^1 d\beta\,\frac{1}{[\beta a + (1 - \beta)b + \lambda]}$$

enables us to write

$$I_1 = -\int_0^1 d\beta \frac{\partial}{2M\partial M} \int d^3k \frac{1}{(\mathbf{p}^2 - \mathbf{k}^2 + i\varepsilon)[(\mathbf{k} - \mathbf{P})^2 + M^2]}$$

where $M^2 = \sigma^2 + 4\beta(1-\beta)\mathbf{p}^2 \sin^2(\theta/2)$ and $\mathbf{P} = \beta\mathbf{p} + (1-\beta)\mathbf{p}'$. Similarly,

$$\frac{1}{2}(p + p')_a I_2 = \int_0^1 d\beta \left(\frac{\partial}{2\partial P_a} - \frac{P_a}{2M} \frac{\partial}{\partial M} \right) \int d^3k \frac{1}{(\mathbf{p}^2 - \mathbf{k}^2 + i\varepsilon)[(\mathbf{k} - \mathbf{P})^2 + M^2]}$$

The intermediate integral is readily performed with the result that

$$\int d^3k \frac{1}{(\mathbf{p}^2 - \mathbf{k}^2 + i\varepsilon)[(\mathbf{k} - \mathbf{P})^2 + M^2]} = \frac{i\pi^2}{P} \ln \frac{p - P + iM}{p + P + iM}$$

with an adequate choice of branch cuts to define the logarithm. The corresponding expressions for I_1 and I_2 in the limit $\sigma \to 0$ are

$$I_1 = \frac{\pi^2}{2ip^3 \sin^2(\theta/2)} \ln \frac{2p \sin(\theta/2)}{\sigma} \tag{7-82}$$

$$I_2 = \frac{\pi^2}{2p^3 \cos^2(\theta/2)} \left\{ \frac{\pi}{2} \left[1 - \frac{1}{\sin(\theta/2)} \right] - i \left[\frac{1}{\sin^2(\theta/2)} \ln \frac{2p \sin(\theta/2)}{\sigma} + \ln \frac{\sigma}{2p} \right] \right\} \tag{7-83}$$

Inserting these expressions in Eq. (7-81) we obtain the last matrix element. The unpolarized cross section is therefore given by

$$\frac{d\sigma}{d\Omega} = \frac{4Z^2\alpha^2 m^2}{|\mathbf{q}|^4} \frac{1}{2} \sum_{\text{pol}} |\bar{u}(p') T u(p)|^2 \tag{7-84}$$

with T standing for the coefficient of $(iZe^2/|\mathbf{q}|^2)\, 2\pi\delta(p^0 - p'^0)$ in the sum $S_{123} + S_4$, that is,

$$T = \gamma^0(1 + A) + \gamma^0 \frac{\not{q}}{2m} B + C$$

$$A = \frac{\alpha}{\pi} \left[\left(1 + \ln \frac{\mu}{m} \right) (2\varphi \coth 2\varphi - 1) - 2 \coth 2\varphi \int_0^\varphi d\varphi' \, \varphi' \tanh \varphi' - \frac{\varphi}{2} \tanh \varphi \right.$$
$$\left. + \left(1 - \frac{1}{3} \coth^2 \varphi \right) (\varphi \coth \varphi - 1) + \frac{1}{9} \right] - \frac{Z\alpha}{2\pi^2} |\mathbf{q}|^2 E(I_1 + I_2) \tag{7-85}$$

$$B = -\frac{\alpha}{\pi} \frac{\varphi}{\sinh 2\varphi}$$

$$C = -\frac{Z\alpha}{2\pi^2} m |\mathbf{q}|^2 (I_1 - I_2)$$

To be consistent, in Eq. (7-84) we keep only terms up to order α^3. Note that the only complex quantities are I_1 and I_2. The sum over polarizations yields

$$m^2 \sum_{\text{pol}} |\bar{u}(p')Tu(p)|^2 = m^2 \operatorname{tr}\left(T \frac{\not{p}+m}{2m} \bar{T} \frac{\not{p}'+m}{2m}\right)$$
$$= (1 + 2 \operatorname{Re} A)(2E^2)[1 - \beta^2 \sin^2(\theta/2)]$$
$$+ 2B(2E^2)\beta^2 \sin^2(\theta/2) + 2 \operatorname{Re} C\,(2mE) + O(\alpha^2)$$

All the dependence on the screening factor σ which was occurring in the imaginary parts of A and C has disappeared as expected, and the elastic cross section including the first nontrivial radiative corrections reads

$$\left(\frac{d\sigma}{d\Omega}\right)_{\text{elastic}} = \left(\frac{d\sigma}{d\Omega}\right)_{\text{Mott}} \left\{1 + \frac{2\alpha}{\pi}\left[\left(1 + \ln\frac{\mu}{m}\right)(2\varphi \coth 2\varphi - 1)\right.\right.$$
$$- 2\coth 2\varphi \int_0^{\varphi} d\varphi'\, \varphi' \tanh\varphi' - \frac{\varphi}{2}\tanh\varphi$$
$$\left. + \left(1 - \frac{\coth^2\varphi}{3}\right)(\varphi\coth\varphi - 1) + \frac{1}{9} - \frac{\varphi}{\sinh 2\varphi}\frac{\beta^2\sin^2(\theta/2)}{1-\beta^2\sin^2(\theta/2)}\right]$$
$$\left. + Z\alpha\pi\frac{\beta\sin(\theta/2)[1-\sin(\theta/2)]}{1-\beta^2\sin^2(\theta/2)}\right\} \tag{7-86}$$

We recall that E and p are the electron energy and momentum, $\beta = p/E$ its velocity, θ the scattering angle, and $\cosh 2\varphi = p \cdot p'/m^2$, $\sinh\varphi = p/m \sin(\theta/2)$, and $|\mathbf{q}|^2 = 4m^2 \sinh^2\varphi = 4p^2 \sin^2(\theta/2)$.

In Eq. (7-86) we are left with the genuine infrared divergence. To obtain a sensible result we will have to compute and add the cross section for emitting soft photons.

7-2-3 Soft Bremsstrahlung

In Sec. 5-2-4 we have already encountered electron bremsstrahlung in a Coulomb field. We obtained the Bethe-Heitler formula exhibiting a typical $d\omega/\omega$ spectrum, where ω is the photon energy or, what amounts to the same, the electron energy loss. If we consider a typical energy resolution ΔE we would expect by integration a probability of radiation of the order of $Z^2\alpha^3 \ln(\Delta E/\mu)$, which might well compensate the analogous term in the elastic cross section. Our present objective is hence to integrate the Bethe-Heitler expression on photon variables in a range which we define as $\omega \leq \Delta E$. We shall assume that $\mu \ll \Delta E \ll E$. An objection may be raised to the effect that in Chap. 5 we practically assumed that $\mu \approx 0$. In fact, the only dangerous term in the limit $\mu/E \ll \Delta E/E \to 0$ arises from the propagators. We may therefore use the formula (5-151) valid for vanishing photon momentum k. If this contribution is denoted $[(d\sigma/d\Omega)(\Delta E)]_{\text{inelastic}}$, we have

$$\left[\frac{d\sigma}{d\Omega}(\Delta E)\right]_{\text{inelastic}} = \left(\frac{d\sigma}{d\Omega}\right)_{\text{Mott}} \int_{\omega\leq\Delta E} \frac{d^3k}{2\omega(2\pi)^3} e^2 \left[\frac{2p\cdot p'}{k\cdot p\, k\cdot p'} - \frac{m^2}{(k\cdot p)^2} - \frac{m^2}{(k\cdot p')^2}\right] \tag{7-87}$$

The resolution could vary with the kinematical conditions, which would require a more careful evaluation using the complete Bethe-Heitler expression. Here we only want to demonstrate the basic compensation mechanism. In formula (7-87) we have in principle $E = E' + \omega$, but we are only looking for the dominant contributions in ΔE. Therefore, whenever allowed we shall take the limit $E' \to E$, $|\mathbf{p}| \to |\mathbf{p}'|$. Of course, $\omega = (\mathbf{k}^2 + \mu^2)^{1/2}$. Let B denote the coefficient of $(d\sigma/d\Omega)_{\text{Mott}}$. Using Feynman's technique we write

$$B = \frac{\alpha}{4\pi^2}\int_{-1}^{+1} dz \int_{\omega \le \Delta E} \frac{d^3k}{2\omega}\left[\frac{2p\cdot p'}{(k\cdot P)^2} - \frac{m^2}{(k\cdot p)^2} - \frac{m^2}{(k\cdot p')^2}\right]$$

where $P = \frac{1}{2}[(p + p') + z(p' - p)]$. Taking into account that $|\mathbf{k}| < [(\Delta E)^2 - \mu^2]^{1/2} \sim \Delta E$ we find

$$\int_{\omega \le \Delta E} \frac{d^3k}{\omega(k\cdot P)^2} = 4\pi \int_0^{\Delta E} \frac{d|\mathbf{k}|}{(\mathbf{k}^2 + \mu^2)^{1/2}} \frac{|\mathbf{k}|^2}{[\mathbf{k}^2(P_0^2 - \mathbf{P}^2) + \mu^2 P_0^2]}$$

$$= 4\pi \int_0^{\arctan(\Delta E/\mu)} d\psi \, \frac{\sin\psi \tan\psi}{P_0^2 - \mathbf{P}^2 \sin^2\psi}$$

$$\simeq \frac{4\pi}{P_0^2 - \mathbf{P}^2}\left(\ln\frac{2\Delta E}{\mu} - \frac{P_0}{2|\mathbf{P}|}\ln\frac{P_0 + |\mathbf{P}|}{P_0 - |\mathbf{P}|}\right)$$

Inserting this in the expression of B with the notation $p = |\mathbf{p}|$, $p' = |\mathbf{p}'|$, and $P = |\mathbf{P}|$, which we hope will not confuse the reader, we are left with one-dimensional integrals over z:

$$B = \frac{\alpha}{\pi}\left\{2\ln\frac{2\Delta E}{\mu}\left[-1 + \frac{p\cdot p'}{2}\int_{-1}^{+1} dz \frac{1}{P_0(z)^2 - P(z)^2}\right] + \frac{E}{2p}\ln\frac{E+p}{E-p} + \frac{E'}{2p'}\ln\frac{E'+p'}{E'-p'}\right.$$
$$\left. - \frac{p\cdot p'}{2}\int_{-1}^{+1} dz \frac{1}{P_0(z)^2 - P(z)^2}\frac{P_0(z)}{P(z)}\ln\frac{P_0(z)+P(z)}{P_0(z)-P(z)}\right\}$$

We observe that $P_0(z)^2 - P(z)^2 = m^2(\cosh^2\varphi - z^2\sinh^2\varphi)$, using our previous notations. Consequently,

$$\frac{p\cdot p'}{2}\int_{-1}^{+1} dz \frac{1}{P_0(z)^2 - P(z)^2} = \cosh 2\varphi \int_0^1 dz \frac{1}{\cosh^2\varphi - z^2\sinh^2\varphi} = 2\varphi\coth 2\varphi$$

For the last integral we introduce the variable $\xi = P(z)/\beta P_0(z)$ and take the limit $E' \to E$, $p' \to p$ up to corrections of order $\Delta E/E$. We see that ξ is also equal to $[\cos^2(\theta/2) + z^2\sin^2(\theta/2)]^{1/2}$ with

$$\frac{p\cdot p'}{2}\int_{-1}^{+1} dz \frac{1}{P_0^2(z) - P^2(z)}\frac{P_0(z)}{P(z)}\ln\frac{P_0(z)+P(z)}{P_0(z)-P(z)}$$
$$= \cosh 2\varphi \frac{1-\beta^2}{\beta\sin(\theta/2)}\int_{\cos(\theta/2)}^1 d\xi \frac{1}{(1-\beta^2\xi^2)[\xi^2 - \cos^2(\theta/2)]^{1/2}}\ln\frac{1+\beta\xi}{1-\beta\xi}$$

Putting everything together, the soft photon emission cross section reads

$$\left[\frac{d\sigma}{d\Omega}(\Delta E)\right]_{\text{inelastic}} = \left(\frac{d\sigma}{d\Omega}\right)_{\text{Mott}} \frac{2\alpha}{\pi}\left\{(2\varphi \coth 2\varphi - 1)\ln\frac{2\Delta E}{\mu} + \frac{1}{2\beta}\ln\frac{1+\beta}{1-\beta}\right.$$
$$\left. - \frac{1}{2}\cosh 2\varphi \frac{1-\beta^2}{\beta \sin(\theta/2)}\int_{\cos(\theta/2)}^{1} d\xi \frac{1}{(1-\beta^2\xi^2)[\xi^2 - \cos^2(\theta/2)]^{1/2}}\ln\frac{1+\beta\xi}{1-\beta\xi}\right\} \tag{7-88}$$

7-2-4 Finite Inclusive Cross Section

The physically measurable quantity, which we denote by $(d\sigma/d\Omega)(\Delta E)$ is the sum of elastic and inelastic contributions

$$\frac{d\sigma}{d\Omega}(\Delta E) = \left(\frac{d\sigma}{d\Omega}\right)_{\text{elastic}} + \left[\frac{d\sigma}{d\Omega}(\Delta E)\right]_{\text{inelastic}} \tag{7-89}$$

Both terms are evaluated to the same order α^3 and are given respectively by Eqs. (7-86) and (7-88). Schwinger, who first computed these corrections, has introduced the notations δ_R and δ_B for the relative corrections arising from virtual or real photon emission and second-order Born approximation. In this way

$$\frac{d\sigma}{d\Omega}(\Delta E) = \left(\frac{d\sigma}{d\Omega}\right)_{\text{Mott}} (1 - \delta_R + \delta_B) \tag{7-90}$$

When adding the results of Eqs. (7-86) and (7-88) the infrared cutoff drops out altogether, as expected. The fictitious mass μ has only been useful to give a meaning to the intermediate steps of the calculation and the $\mu \to 0$ limit is now perfectly legitimate when care has been taken to allow for the experimental resolution ΔE. The final result is, of course, sensitive to ΔE.

When adding the two contributions we may use the following identity which can be obtained by comparing the changes of variables performed when integrating $(d\sigma/d\Omega)_{\text{elastic}}$ and $[(d\sigma/d\Omega)(\Delta E)]_{\text{inelastic}}$. Recalling that $\sinh\varphi = [\beta/(1-\beta^2)^{1/2}] \times \sin(\theta/2)$, we have

$$\int_0^{\varphi} d\psi\, \psi \tanh\psi = -\frac{\varphi}{2}\ln(1-\beta^2) + \frac{(1-\beta^2)}{4\sin(\theta/2)}\sinh 2\varphi$$
$$\times \int_{\cos(\theta/2)}^{1} d\xi \frac{\xi \ln(1-\beta^2\xi^2)}{(1-\beta^2\xi^2)[\xi^2 - \cos^2(\theta/2)]^{1/2}}$$

Hence we find

$$\delta_R = \frac{2\alpha}{\pi}\left\{(1 - 2\varphi\coth 2\varphi)\left(1 + \ln\frac{2\Delta E}{m}\right) + \varphi\tanh\varphi + (1 - \varphi\tanh\varphi)\right.$$
$$\times \left(1 - \frac{\coth^2\varphi}{3}\right) - \frac{1}{9} + \frac{1}{2\beta}\ln\frac{1-\beta}{1+\beta} - \varphi\coth 2\varphi \ln(1-\beta^2)$$

$$+\frac{\beta^2\sin^2(\theta/2)}{1-\beta^2\sin^2(\theta/2)}\frac{\varphi}{\sinh 2\varphi}+\frac{1}{2}\cosh 2\varphi\,\frac{1-\beta^2}{\beta\sin(\theta/2)}$$

$$\times\int_{\cos(\theta/2)}^{1}d\xi\,\frac{1}{[\xi^2-\cos^2(\theta/2)]^{1/2}}\left[\frac{\ln(1+\beta\xi)}{1-\beta\xi}-\frac{\ln(1-\beta\xi)}{1+\beta\xi}\right]\Bigg\} \tag{7-91}$$

$$\delta_B = Z\alpha\pi\,\frac{\beta\sin(\theta/2)[1-\sin(\theta/2)]}{1-\beta^2\sin^2(\theta/2)} \tag{7-92}$$

Numerical investigation reveals that these corrections are far from negligible as they increase with energy. When $\Delta E/m \to 0$ the previous formulas become invalid. A criterion for validity is

$$\frac{\alpha}{\pi}\ln\frac{m}{\Delta E} < 1$$

To go beyond this result, it is necessary to include higher-order terms involving the emission of several or even an infinite number of photons. The contributions of soft photons is then known to factorize in exponential form, showing that the process has zero probability in the limit $\Delta E/m \to 0$.

7-3 NEW EFFECTS

Higher-order terms in the perturbation series induce new effects, some of which will now be briefly presented.

7-3-1 Photon-Photon Scattering

Four photon interactions have no classical counterpart and arise through quantum fluctuations of virtual charged particle pairs. From a theoretical point of view it is of interest to demonstrate how gauge invariance and current conservation cure the remaining potentially divergent amplitude, as was anticipated in the discussion of Sec. 7-1.

Let us first sketch some dimensional arguments. The basic amplitude (Fig. 7-15) is of order α^2 so that the cross section takes the form

$$\sigma \sim \int \frac{d^3k_3}{\omega_3}\frac{d^3k_4}{\omega_4}\frac{\alpha^4|M|^2}{k_1\cdot k_2}\,\delta^4(k_1+k_2-k_3-k_4) \tag{7-93}$$

with obvious notations for the momenta and energies of the photons. The dimen-

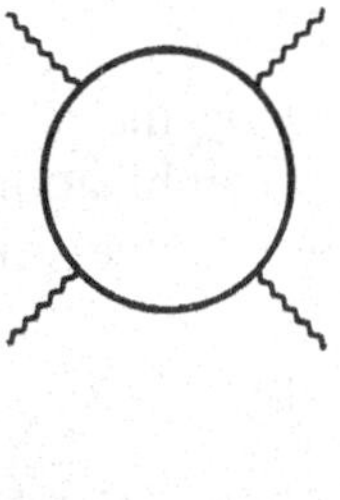

Figure 7-15 Basic diagram for photon-photon scattering.

sional parameters are, for instance, the common energy ω in the center of mass frame and the electron mass m. Here M stands for a dimensionless matrix element so that σ behaves correctly as a surface.

We expect gauge invariance to allow for the extraction of four powers of the momenta from M. This will be subsequently confirmed. Consequently, $|M|^2$ must behave at least as $(\omega/m)^8$ for small ω. In this limit $k_1 \cdot k_2$ is proportional to ω^2. We therefore predict that at low energies

$$\sigma \sim a\alpha^4 \left(\frac{\omega}{m}\right)^6 \frac{1}{m^2} \qquad \frac{\omega}{m} \ll 1 \tag{7-94}$$

where a is a numerical constant. Since $\alpha^4/m^2 \sim 10$ μbarns, this cross section is exceedingly small up to the hundred-keV region. On the other hand, for $\omega/m \gg 1$, due to the convergence of the process we do not encounter mass singularities so that simple dimensional counting indicates that

$$\sigma \sim b\alpha^4 \frac{1}{\omega^2} \qquad \frac{\omega}{m} \gg 1 \tag{7-95}$$

with b a second numerical constant. In other words, it is likely that σ peaks for typical values of ω/m of order one.

We shall not present a fully fledged calculation, but instead will show that the results of the last section in Chap. 4 allow us to compute the numerical constant a of the low-energy behavior. For that matter we observe that the Euler-Heisenberg lagrangian [Eq. (4-123)] involves a summation of all one-loop diagrams, each one integrated over the same constant electromagnetic field as many times as there are external lines. Note that **E** and **B** are linear in the external momenta. Furthermore, at a given order in e the sum of all diagrams (corresponding to permutations of the external line indices) is gauge invariant. When contracted with a polarization vector $\varepsilon_\rho(k)$ for an on-shell process it should be invariant in the substitution $\varepsilon_\rho(k) \to \varepsilon_\rho(k) + \lambda k_\rho$. This means that this contracted quantity can only depend on $\varepsilon(k)$ through the combination $\varepsilon_\rho(k)k_\sigma - \varepsilon_\sigma(k)k_\rho$. Up to a factor i this is the Fourier transform $f_{\rho\sigma}(k)$ of the corresponding electromagnetic field. The term proportional to e^4 will then involve at least four powers of the momenta. In the low-energy limit it will then be sufficient to study the coefficient of the combination $\prod_1^4 f^{(i)}_{\rho_i\sigma_i}$ at zero frequency. If the external electromagnetic fields coincide, the corresponding quantity is the fourth-order term in the effective lagrangian computed in a constant field, which was given by Eq. (4-125) as

$$i\delta\mathscr{L}_4 = i\frac{2\alpha^2}{45m^4}\left[(\mathbf{E}^2 - \mathbf{B}^2)^2 + 7(\mathbf{E}\cdot\mathbf{B})^2\right] = \frac{i2\alpha^2}{45m^4}\left[\left(\frac{f^2}{2}\right)^2 + 7\left(\frac{f\cdot\tilde{f}}{4}\right)^2\right] \tag{7-96}$$

To recover the scattering amplitude we have simply to replace f by the sum $f^{(1)} + f^{(2)} + f^{(3)} + f^{(4)}$ and divide the coefficient of the multilinear combination $\prod_1^4 f^{(i)}$ by 4! With the substitution $f_{\rho\sigma} = i(k_\rho\varepsilon_\sigma - k_\sigma\varepsilon_\rho)$ the required value will be obtained in the limit $\omega/m \to 0$.

As a shorthand notation we identify $f_\rho{}^\sigma$ with a matrix, and the trace symbol runs over the four values of the Lorentz index. It is then easy to see that

$$(f^2)^2 = (\mathrm{tr}\, f^2)^2 \qquad (f \cdot \tilde{f})^2 = 4\, \mathrm{tr}\, f^4 - 2(\mathrm{tr}\, f^2)^2$$

From the above instructions the amplitude M is found to be

$$\begin{aligned} M = -\frac{i\alpha^2}{45m^4}\frac{1}{12}\, [5(&\mathrm{tr}\, f^{(1)}f^{(2)}\mathrm{tr}\, f^{(3)}f^{(4)} + \mathrm{tr}\, f^{(1)}f^{(3)}\, \mathrm{tr}\, f^{(2)}f^{(4)} \\ &+ \mathrm{tr}\, f^{(1)}f^{(4)}\mathrm{tr}\, f^{(2)}f^{(3)}) - 7\, \mathrm{tr}\,(f^{(1)}f^{(2)}f^{(3)}f^{(4)} + f^{(2)}f^{(1)}f^{(3)}f^{(4)} \\ &+ f^{(3)}f^{(1)}f^{(2)}f^{(4)} + f^{(2)}f^{(3)}f^{(1)}f^{(4)} + f^{(3)}f^{(2)}f^{(1)}f^{(4)} \\ &+ f^{(1)}f^{(3)}f^{(2)}f^{(4)})] \end{aligned} \tag{7-97}$$

The unpolarized cross section involves the average of the absolute square of M:

$$\overline{|M|^2} = \tfrac{1}{4} \sum_{\varepsilon^{(i)}} |M|^2$$

In the center of mass frame it reads

$$\begin{aligned} d\sigma &= \frac{1}{4(k_1 \cdot k_2)} \int \frac{d^3k_3}{2\omega_3(2\pi)^3}\frac{d^3k_4}{2\omega_4(2\pi)^3}(2\pi)^4\delta^4(k_3 + k_4 - k_1 - k_2)\overline{|M|^2} \\ &= \frac{1}{64\omega^2(2\pi)^2}\overline{|M|^2}\, d\Omega \end{aligned} \tag{7-98}$$

To perform the sum over polarizations we note that

$$\sum_{\varepsilon} f_\rho^{*\sigma} f_{\rho'}{}^{\sigma'} = -(g_{\rho\rho'}k^\sigma k^{\sigma'} + k_\rho k_{\rho'} g^{\sigma\sigma'} - g_\rho{}^{\sigma'}k^\sigma k_{\rho'} - g_{\rho'}{}^\sigma k_\rho k^{\sigma'})$$

A tedious calculation yields

$$\begin{aligned} \frac{d\sigma}{d\Omega} &= \frac{1}{(2\pi)^2}\frac{1}{2\omega^2}\frac{\alpha^4}{(90)^2m^8} \\ &\quad \times 139\, [(k_1 \cdot k_2)^2(k_3 \cdot k_4)^2 + (k_1 \cdot k_3)^2(k_2 \cdot k_4)^2 + (k_1 \cdot k_4)^2(k_2 \cdot k_3)^2] \end{aligned} \tag{7-99}$$

The second invariant combination which could have entered this expression, that is, $\sum_{\text{distinct permutations}} (k_1 \cdot k_2)(k_2 \cdot k_3)(k_3 \cdot k_4)(k_4 \cdot k_1)$ is equal to half the quantity under brackets in Eq. (7-99). If θ is the center of mass scattering angle we have

$$k_1 \cdot k_2 = k_3 \cdot k_4 = 2\omega^2$$

$$k_1 \cdot k_3 = k_2 \cdot k_4 = \omega^2(1 - \cos\theta)$$

$$k_1 \cdot k_4 = k_2 \cdot k_3 = \omega^2(1 + \cos\theta)$$

so that the unpolarized differential cross section is equal to

$$\frac{d\sigma}{d\Omega} = \frac{1}{(2\pi)^2}\frac{139}{(90)^2}\alpha^4\left(\frac{\omega}{m}\right)^6\frac{1}{m^2}(3 + \cos^2\theta)^2 \qquad \omega/m \ll 1 \tag{7-100}$$

From Bose statistics this expression is symmetrical in the exchange $\theta \to \pi - \theta$. To obtain the total elastic cross section it is therefore required to integrate only over half the unit sphere in solid angle, with the result

$$\sigma = \frac{1}{2\pi} \frac{139}{(90)^2} \left(\frac{56}{11}\right) \alpha^4 \left(\frac{\omega}{m}\right)^6 \frac{1}{m^2} \qquad \omega/m \ll 1 \tag{7-101}$$

This result agrees with our previous estimate therefore producing the required coefficient a.

As an exercise the reader may compute the coefficient b entering the high-energy behavior (7-95) of the cross section. The photon-photon induced interaction implies the possibility of a coherent scattering on charged targets when two of the lines in the diagram of Fig. 7-15 refer to Coulomb interactions with a nucleus. This can be related to pair creation in the nuclear field and may be compared with the Compton process. An alternative process worth investigating is the photon "splitting" amplitude when a photon hitting a target $(-Ze)$ yields two photons.

7-3-2 Lamb Shift

The 1947 measurement by Lamb and Retherford of the shift between the hydrogen states $2S_{1/2}$ and $2P_{1/2}$ predicted to be degenerate according to the Dirac theory has been a memorable event stimulating the development of quantum field theory. Here we shall content ourselves with a simple account of the effect, leaving aside the more basic questions related to the formulations of a relativistic bound-state theory. In fact what Lamb and Retherford measured was the $2P_{3/2} - 2S_{1/2}$ transitions as a function of an applied magnetic field. In the limit of zero field the observed value was approximately 1000 MHz lower than what could be expected from the fine structure interval $m\alpha^4/32 \sim 10{,}960$ MHz.

To analyze this discrepancy we shall use, following the original work of Bethe, a combination of Dirac's theory and radiative corrections, as discussed earlier in this chapter. We have seen that the vacuum polarization, by modifying the effective potential, had the effect of increasing the binding of the $2S_{1/2}$ level by 27 MHz relative to the $2P_{1/2}$ state. This cannot therefore be the main effect since the experimental observation requires the $2S_{1/2}$ level to lie well above the $2P_{1/2}$ one. Of course, a full theory requires to take into account all effects of the same order and to include nuclear corrections (magnetic moment, recoil, form factors, polarizability, etc.). Moreover, all excited levels are metastable, meaning that each one has a natural line width.

In Chap. 2 we recalled Welton's argument implying the interaction of the bound electron with the fluctuating vacuum electric field and leading to an estimated shift with the correct sign and order of magnitude:

$$\delta E_{n,l} \simeq \frac{4}{3\pi} m \frac{Z^4\alpha^5}{n^3} \ln \frac{1}{Z\alpha} \delta_{l,0} \tag{7-102}$$

We now want to give a more quantitative evaluation.

Our starting point will be the effective interaction (7-77) for electrons interacting with an external field. This leads to a modified Dirac equation of the form

$$\left[i\partial\!\!\!/ - m - \left\{\gamma_\rho\left[1 - \frac{\alpha}{3\pi m^2}\left(\ln\frac{m}{\mu} - \frac{3}{8} - \frac{1}{5}\right)\Box\right] + \frac{\alpha}{2\pi}\frac{\sigma_{\nu\rho}}{2m}\partial^\nu\right\}eA^\rho(x)\right]\psi(x) = 0 \tag{7-103}$$

including vacuum polarization and vertex corrections evaluated for a slowly variable external field. The differential operators inside the curly brackets operate on the Coulomb potential $A^\rho(x)$ only.

The quantity ψ is to be interpreted as a matrix element of the field operator between the vacuum and a one-electron state in the presence of the external potential. According to the perturbative rules, the added term can only be taken into account to first order. Otherwise our calculation would be logically inconsistent. The unperturbed states are solutions of the Dirac equation

$$(i\partial\!\!\!/ - m - eA\!\!\!/)\psi(x) = 0 \qquad A^\rho(x) = -\frac{Ze}{4\pi|\mathbf{x}|}g^{\rho 0} \tag{7-104}$$

which have been discussed in Sec. 2-3. Consequently, the level shift is given by the expression

$$\delta E = e\alpha\int d^3x\,\psi^\dagger(\mathbf{x})\left\{\frac{1}{3\pi m^2}\left(\ln\frac{m}{\mu} - \frac{3}{8} - \frac{1}{5}\right)\Delta[A^0(\mathbf{x})] + \frac{i}{4\pi m}\,\boldsymbol{\gamma}\cdot\mathbf{E}(\mathbf{x})\right\}\psi(\mathbf{x}) \tag{7-105}$$

where ψ stands for a solution pertaining to the quantum numbers (n, l, j) of Eq. (7-104).

As a first approximation we may even use the nonrelativistic value for ψ in Eq. (7-105) in such a way that

$$\Delta A^0(\mathbf{x}) = Ze\delta^3(\mathbf{x}) \qquad |\psi_n^{\mathrm{nr}}(0)|^2 = \frac{1}{\pi a_n^3} = \frac{(Z\alpha m)^3}{\pi n^3}$$

This yields

$$e\int d^3\mathbf{x}\,\psi_{n,l,j}^\dagger(\mathbf{x})[\Delta A^0(\mathbf{x})]\psi_{n,l,j}(\mathbf{x}) = \delta_{l,0}\frac{4m^3}{n^3}(Z\alpha)^4$$

Define now the separate contributions of the two terms in (7-105) as

$$\delta E = \delta E^{(1)} + \delta E^{(2)} \tag{7-106}$$

It follows that

$$\begin{aligned}\delta E_{n,l,j}^{(1)} &= \frac{e\alpha}{3\pi m^2}\int d^3\mathbf{x}\,\psi_{n,l,j}^\dagger(\mathbf{x})\left(\ln\frac{m}{\mu} - \frac{3}{8} - \frac{1}{5}\right)[\Delta A^0(x)]\,\psi_{n,l,j}(\mathbf{x}) \\ &= \delta_{l,0}\frac{4m\alpha}{3\pi n^3}(Z\alpha)^4\left(\ln\frac{m}{\mu} - \frac{3}{8} - \frac{1}{5}\right)\end{aligned} \tag{7-107}$$

The relativistic wave functions are singular at the origin. A more rigorous treatment would require avoiding an expansion of the effective interaction in powers of q^2/m^2. Since only mean values of the potential enter the expression (7-105), the above result remains accurate to order $\alpha(Z\alpha)^4$.

We turn now to the second term $\delta E^{(2)}$, which we intend to compute in the same spirit. Some limited use of the Dirac equation is necessary since γ matrices relate small and large components. To this end we express $\delta E^{(2)}_{n,l,j}$ as

$$\delta E^{(2)}_{n,l,j} = \frac{ie\alpha}{4\pi m} \int d^3x\, \psi^\dagger_{n,l,j}(\mathbf{x}) \boldsymbol{\gamma} \cdot \mathbf{E}(\mathbf{x}) \psi_{n,l,j}(\mathbf{x}) \tag{7-108}$$

which represents the effect of the anomalous magnetic moment inducing an electric dipole moment for a moving electron. The large (φ) and small (χ) components of ψ are approximately related by

$$\chi(x) = -i\frac{\boldsymbol{\sigma} \cdot \boldsymbol{\nabla}}{2m}\varphi(x)$$

ignoring the effects of the Coulomb potential. Since $e\mathbf{E}(\mathbf{x}) = -Z\alpha(\mathbf{x}/|\mathbf{x}|^3)$ and

$$\psi^\dagger \frac{\boldsymbol{\gamma} \cdot \mathbf{x}}{|\mathbf{x}|^3} \psi = \varphi^\dagger \frac{\boldsymbol{\sigma} \cdot \mathbf{x}}{|\mathbf{x}|^3} \chi - \chi^\dagger \frac{\boldsymbol{\sigma} \cdot \mathbf{x}}{|\mathbf{x}|^3} \varphi \simeq \frac{1}{2im}\left(\varphi^\dagger \frac{\boldsymbol{\sigma} \cdot \mathbf{x}\, \boldsymbol{\sigma} \cdot \boldsymbol{\nabla}}{|\mathbf{x}|^3} \varphi + \text{hc}\right)$$

we find after an integration by parts that

$$\int d^3x\, \psi^\dagger(\mathbf{x}) \frac{\boldsymbol{\gamma} \cdot \mathbf{x}}{|\mathbf{x}|^3} \psi(\mathbf{x}) \simeq \frac{i}{2m} \int d^3x\, \varphi^\dagger \left[\boldsymbol{\sigma} \cdot \boldsymbol{\nabla}, \frac{\boldsymbol{\sigma} \cdot \mathbf{x}}{|\mathbf{x}|^3}\right] \varphi$$

As in the case of the hyperfine interaction, some care is required when handling such a singular quantity as $\mathbf{x}/|\mathbf{x}|^3$. We recall that $\mathbf{x}/|\mathbf{x}|^3 = -\boldsymbol{\nabla} 1/|\mathbf{x}|$ and $\Delta 1/|\mathbf{x}| = -4\pi\delta^3(\mathbf{x})$. We can take for φ a Pauli spinor and use the expression of the nonrelativistic angular momentum operator $\mathbf{L} = (1/i)\mathbf{x} \times \boldsymbol{\nabla}$ together with the identity

$$[\sigma_a A_a, \sigma_b B_b] = \delta_{ab}(A_a B_b - B_b A_a) + i\varepsilon_{abc}\sigma_c(A_a B_b + B_b A_a)$$

to obtain

$$\delta E^{(2)}_{n,l,j} = \frac{Z\alpha^2}{8\pi m^2} \int d^3\mathbf{x}\, \varphi^\dagger_{n,l,j}(\mathbf{x}) \left[4\pi\delta^3(\mathbf{x}) + 4\frac{\mathbf{L}\cdot\mathbf{S}}{|\mathbf{x}|^3}\right] \varphi_{n,l,j}(\mathbf{x})$$

We know the value of the wave function at the origin. The matrix element of $\mathbf{L}\cdot\mathbf{S}$ is equal to $(1 - \delta_{l,0})[j(j+1) - l(l+1) - \frac{3}{4}]/2$. Finally, the mean value of $1/|\mathbf{x}|^3$ for states of angular momenta equal or larger to one is

$$\left\langle \frac{1}{|\mathbf{x}|^3} \right\rangle_{n,l,j} = \frac{2}{l(l+1)(2l+1)n^3}(Zm\alpha)^3$$

Putting everything together we find

$$\delta E^{(2)}_{n,l,j} = \frac{\alpha(Z\alpha)^4}{2\pi n^3} m \frac{1}{2l+1} C_{j,l}$$

$$C_{j,l} = \delta_{l,0} + (1 - \delta_{l,0}) \frac{j(j+1) - l(l+1) - \frac{3}{4}}{l(l+1)}$$

$$= \begin{cases} \dfrac{1}{l+1} & \text{if } j = l + \frac{1}{2} \\ -\dfrac{1}{l} & \text{if } j = l - \frac{1}{2}, \quad l \geq 1 \end{cases} \tag{7-109}$$

For states with nonvanishing orbital momentum, $\delta E^{(1)}$ vanishes, while $\delta E^{(2)}$ gives a small contribution of order $\alpha(Z\alpha)^4$ originating from the electron anomaly.

For s states our calculation of δE still contains the fictitious mass μ and diverges as $\mu \to 0$, a signal that we have omitted some essential contribution. As the qualitative argument indicated, an effective infrared cutoff was expected of the order of the Bohr radius $(Zm\alpha)^{-1}$ instead of the arbitrary value μ^{-1}. The mistake can be traced to the use of the effective interaction (7-77) for arbitrary large wavelengths. For wavelengths comparable to or larger than the Bohr radius the radiative corrections should in turn take into account the Coulomb interaction, i.e., the fact that they apply not to a free but to a bound electron.

In order to correct for this effect we use the observation that the ratio of the binding energy to the rest mass energy is very small, of order $\frac{1}{2}Z^2\alpha^2$. Let us introduce a cutoff K on the three-momentum of virtual photons such that $mZ^2\alpha^2 \ll K \ll m$. For $k < K$ it will be possible to treat the electrons in a nonrelativistic approximation and we shall be able to take the nuclear potential into account. For $k > K$ we will neglect the effects of binding and use the previous results, except for the fact that we need an accurate relation between K and μ.

To do this we return to Sec. 7-2-3, where we computed the rate of soft bremsstrahlung by integrating the quantity

$$B = \frac{\alpha}{4\pi^2} \int_{\omega \leq \Delta E} \frac{d^3k}{2\omega} \left[\frac{2p \cdot p'}{k \cdot pk \cdot p'} - \frac{m^2}{(k \cdot p)^2} - \frac{m^2}{(k \cdot p')^2} \right]$$

The latter was evaluated for $\mu \ll \Delta E$ but $\mu \neq 0$. Let us repeat the calculation setting $\mu = 0$ and using instead the lower cutoff $K (K \ll \Delta E)$ on photon three-momenta. Denote such a quantity as B_1. The comparison of B with B_1 will offer a means of relating K and μ. Keeping the definition of the hyperbolic angle φ such that $p \cdot p' = m^2 \cosh 2\varphi$ we find

$$B_1 = \frac{2\alpha}{\pi}(2\varphi \coth 2\varphi - 1) \ln \frac{\Delta E}{K}$$

Our previous result for B was [Eq. (7-88)]

$$B = \frac{2\alpha}{\pi} \left\{ (2\varphi \coth 2\varphi - 1) \ln \frac{2\Delta E}{\mu} + \frac{1}{2\beta} \ln \frac{1+\beta}{1-\beta} \right.$$
$$\left. - \frac{1}{2} \cosh 2\varphi \frac{1-\beta^2}{\beta \sin(\theta/2)} \int_{\cos(\theta/2)}^{1} d\xi \frac{1}{(1-\beta^2\xi^2)[\xi^2 - \cos^2(\theta/2)]^{1/2}} \ln \frac{1+\beta\xi}{1-\beta\xi} \right\}$$

For bound electrons the velocity is small ($\beta \sim Z\alpha$). It is therefore appropriate to match B_1 and B in the limiting case where $\beta \to 0$, $\varphi \sim \beta \sin(\theta/2)$. Under these circumstances the integral appearing in the expression for B is approximately equal to

$$2\beta \sin\frac{\theta}{2}\left[1 + \frac{4}{9}\beta^2\left(1 + 2\cos^2\frac{\theta}{2}\right) + \cdots\right]$$

We may then identify B and B_1 provided that

$$\ln\frac{\mu}{2K} = -\frac{5}{6} \tag{7-110}$$

The Lamb-shift calculation only involves one virtual photon line. Furthermore, the nucleus dictates a privileged frame for which the splitting $k \lessgtr K$ is legitimate. We decompose accordingly $\delta E^{(1)}$ into two parts, $\delta E^{>}$ and $\delta E^{<}$. For $\delta E^{>}$ we use the result (7-107) with the substitution implied by (7-110). The remaining piece, $\delta E^{<}$, requires a new calculation. Our expression is more correctly

$$\begin{aligned} \delta E^{(1)} &= \delta E^{<} + \delta E^{>} \\ \delta E^{>} &= \frac{4}{3\pi} m\alpha \frac{(Z\alpha)^4}{n^3}\left(\ln\frac{m}{2K} + \frac{5}{6} - \frac{3}{8} - \frac{1}{5}\right)\delta_{l,0} \end{aligned} \tag{7-111}$$

This will be meaningful if the dependence on K cancels between $\delta E^{<}$ and $\delta E^{>}$.

To obtain $\delta E^{<}$ we return to the original procedure of radiative corrections with the proviso that the momenta occurring in the virtual radiation field are limited by K and hence can be treated by second-order nonrelativistic perturbation theory starting from a Schrödinger equation of the form

$$(E + \delta E^{<})\psi = \left[\frac{1}{2m}\left(\frac{1}{i}\nabla - e\mathbf{A}_q\right)^2 + e(A^0 + A_q^0)\right]\psi \tag{7-112}$$

Here ψ is a wave function for both the electron and the radiation field (A_q^0, $\mathbf{A}_q$) which describes in the unperturbed state a bound electron and the electromagnetic vacuum. The quantum part effectively reduces to

$$\begin{aligned} \mathbf{A}_q(\mathbf{x}) &= \int_{|\mathbf{k}|<K} \frac{d^3k}{2|\mathbf{k}|(2\pi)^3}\left[\sum_{\lambda=1,2} \varepsilon_\lambda(\mathbf{k}) a^{(\lambda)}(k)\, e^{i\mathbf{k}\cdot\mathbf{x}} + \text{hc}\right] \\ A_q^0(\mathbf{x}) &= 0 \end{aligned} \tag{7-113}$$

We treat the interaction with the radiation field to lowest nonvanishing order and find two contributions. The first one is a "seagull" contribution arising from the quadratic term in $\mathbf{A}_q$. It may be reabsorbed in E as a contribution to mass renormalization, since it affects indiscriminately all levels. The second one has the form

$$\delta E_n^< = \sum_{n'} \int_{k<K} \frac{d^3k}{2k(2\pi)^3} \sum_\lambda \frac{e^2}{(2m)^2}$$
$$\times \int d^3x \frac{|\psi_n^*(\mathbf{x})[(1/i)\nabla\cdot\varepsilon_\lambda(\mathbf{k})\, e^{i\mathbf{k}\cdot\mathbf{x}} + e^{i\mathbf{k}\cdot\mathbf{x}}\varepsilon_\lambda(\mathbf{k})\cdot(1/i)\nabla]\psi_{n'}(\mathbf{x})|^2}{E_n - E_{n'} - k}, \tag{7-114}$$

where E_n and $E_{n'}$ stand for the unperturbed levels, $E_n \simeq m - mZ^2\alpha^2/2n^2$. The range of the $\mathbf{x}$ integration is effectively limited by the Bohr radius. Accordingly, $K|\mathbf{x}| \lesssim K/mZ\alpha$. From our hypothesis on K this may be considered as very small, justifying the dipole approximation where $e^{i\mathbf{k}\cdot\mathbf{x}}$ is replaced by unity. Define $\mathbf{v}_{\text{op}} = (1/im)\nabla$ and note that

$$\sum_{\lambda=1,2} |\langle n|\mathbf{v}_{\text{op}}\cdot\varepsilon_\lambda(\mathbf{k})|n'\rangle|^2 = \left(\delta^{ab} - \frac{k^a k^b}{k^2}\right)\langle n|v^a_{\text{op}}|n'\rangle\langle n'|v^b_{\text{op}}|n\rangle$$

and
$$\int \frac{d\Omega}{4\pi}\left(\delta^{ab} - \frac{k^a k^b}{k^2}\right) = \frac{2}{3}\,\delta_{ab}$$

We therefore conclude that

$$\delta E_n^< = \frac{2\alpha}{3\pi}\sum_{n'}\int_0^K \frac{dk\, k}{E_n - E_{n'} - k}|\langle n|\mathbf{v}_{\text{op}}|n'\rangle|^2 \tag{7-115}$$

Formula (7-115) is not yet the correct one since it does not take into account mass renormalization in the sense that we are using in $\mathbf{v}_{\text{op}}$ the expression of the physical mass. The contribution of very soft photons ($k < K$) to the self-mass has yet to be subtracted from it. In other words, a counterterm of the form $-(\delta m/2m^2)\mathbf{p}_{\text{op}}{}^2 = -(\delta m/2)\mathbf{v}_{\text{op}}{}^2$ had to be inserted in the hamiltonian with δm adjusted in such a way that $\delta E^<$ vanishes for a free electron. Since $\langle n|\mathbf{v}_{\text{op}}{}^2|n\rangle = \sum_{n'}|\langle n|\mathbf{v}_{\text{op}}|n'\rangle|^2$ and, owing to the fact that $K \gg |E_n - E_{n'}|$, it is clear that the correct expression for $\delta E^<$ is

$$\begin{aligned}\delta E_n^< &= \frac{2\alpha}{3\pi}\sum_{n'}\int_0^K dk\, k\left(\frac{1}{E_n - E_{n'} - k} + \frac{1}{k}\right)|\langle n|\mathbf{v}_{\text{op}}|n'\rangle|^2 \\ &= \frac{2\alpha}{3\pi}\sum_{n'}(E_{n'} - E_n)\ln\left|\frac{K}{E_n - E_{n'}}\right||\langle n|\mathbf{v}_{\text{op}}|n'\rangle|^2\end{aligned} \tag{7-116}$$

At that stage we can only do a numerical evaluation. Bethe, who was the first to determine this nonrelativistic contribution, has introduced the following logarithmic average. For a state $|n\rangle$ pertaining to an s wave define $\langle E_n\rangle$ through

$$\ln\langle E_n\rangle = \frac{\sum_{n'}|\langle n|\mathbf{v}_{\text{op}}|n'\rangle|^2(E_{n'} - E_n)\ln|E_{n'} - E_n|}{\sum_{n'}|\langle n|\mathbf{v}_{\text{op}}|n'\rangle|^2(E_{n'} - E_n)} \qquad l = 0 \tag{7-117}$$

using an arbitrary, but fixed, energy scale. This definition becomes meaningless for higher waves ($l \neq 0$) where the denominator vanishes, as we shall soon realize. Therefore, in this case $\delta E^<$ becomes independent of K, a fortunate circumstance.

For this case we define, nevertheless, a quantity $\langle E_n \rangle$ through

$$\delta E_n^< = \frac{2\alpha}{3\pi} \sum_{n'} |\langle n | \mathbf{v}_{\text{op}} | n' \rangle|^2 (E_{n'} - E_n) \ln \frac{1}{|E_n - E_{n'}|}$$

$$= \frac{4\alpha}{3\pi} \frac{(Z\alpha)^4}{n^3} m \ln \frac{Z^2 \text{ Ryd}}{\langle E_n \rangle} \qquad l \neq 0 \tag{7-118}$$

$$\text{Ryd} = \frac{m\alpha^2}{2}$$

For s waves, using Eq. (7-117) we have

$$\delta E_n^< = \frac{2\alpha}{3\pi} \sum_{n'} |\langle n | \mathbf{v}_{\text{op}} | n' \rangle|^2 (E_{n'} - E_n) \ln \frac{K}{\langle E_n \rangle} \qquad l = 0$$

Let us transform this expression. If H_0 is the Schrödinger hamiltonian including the Coulomb potential we may write

$$\sum_{n'} (E_{n'} - E_n) |\langle n | \mathbf{v}_{\text{op}} | n' \rangle|^2 = \langle n | [\mathbf{v}_{\text{op}}, H_0] \cdot \mathbf{v}_{\text{op}} | n \rangle = \langle n | \mathbf{v}_{\text{op}} \cdot [H_0, \mathbf{v}_{\text{op}}] | n \rangle$$

$$= -\frac{1}{2m^2} \langle n | \left\{ \left[\mathbf{p}_{\text{op}}, \frac{Z\alpha}{|\mathbf{x}|} \right] \cdot \mathbf{p}_{\text{op}} + \mathbf{p}_{\text{op}} \cdot \left[\frac{Z\alpha}{|\mathbf{x}|}, \mathbf{p}_{\text{op}} \right] \right\} | n \rangle$$

$$= \frac{1}{2m^2} \int d^3x \, \nabla \left(\frac{Z\alpha}{|\mathbf{x}|} \right) \cdot \nabla [\psi^*(\mathbf{x}) \psi(\mathbf{x})] = \frac{Z\alpha}{2m^2} 4\pi |\psi(0)|^2$$

$$= \frac{2m(Z\alpha)^4}{n^3} \delta_{l,0}$$

It follows then that for s waves

$$\delta E_n^< = \frac{4\alpha}{3\pi} \frac{(Z\alpha)^4}{n^3} m \ln \frac{K}{\langle E_n \rangle} \qquad l = 0 \tag{7-119}$$

We now combine all the pieces: $\delta E^<$ given by (7-118) and (7-119), $\delta E^>$ given by (7-111), and $\delta E^{(2)}$ by (7-109) (which also contributes for $l = 0$). The arbitrary quantities μ and K disappear and the Lamb shift is found to be

$$\delta E_{n,l,j} = \frac{4\alpha}{3\pi} \frac{(Z\alpha)^4}{n^3} m \begin{cases} \ln \dfrac{m}{2\langle E_{n,0} \rangle} + \dfrac{19}{30} & \text{for } l = 0 \\[2ex] \ln \dfrac{Z^2 \text{ Ryd}}{\langle E_{n,l} \rangle} + \dfrac{3}{8} \dfrac{C_{l,j}}{2l+1} & \text{for } l \neq 0 \end{cases} \tag{7-120}$$

with $C_{l,j}$ given in (7-109).

The main contribution was computed by Bethe while the $\alpha(Z\alpha)^4$ term was obtained by Kroll and Lamb, and by French and Weisskopf.

For atomic hydrogen using the numerical values

$$\langle E_{2S} \rangle = 16.640 \text{ Ryd} \qquad \langle E_{2P} \rangle = 0.9704 \text{ Ryd} \tag{7-121}$$

the value for the $2S_{1/2} - 2P_{1/2}$ splitting is predicted by formula (7-120) to be

$$E_{2S_{1/2}} - E_{2P_{1/2}} = \frac{m\alpha^5}{6\pi}\left(\ln \frac{m\langle E_{2P}\rangle}{2\text{ Ryd }\langle E_{2S}\rangle} + \frac{91}{120}\right) = 1\,052.1 \text{ MHz} \tag{7-122}$$

which compares favorably with the 1953 measurement of Triebwasser, Dayhoff, and Lamb of 1 057.8 ± 0.1 MHz.

To improve the calculation it is necessary to develop a more elaborate theoretical framework. Further corrections involve a refinement on the electron propagation in the Coulomb field (higher powers and logarithms in $Z\alpha$) and require a treatment of higher-order radiative corrections. Terms implying the nuclear recoil [in $(m/M)(Z\alpha)^4$ and $(m/M)(Z\alpha)^5$] and effects of a finite nuclear radius R_N [in $(R_N m)^2(Z\alpha)^4$] have also to be included to obtain the latest theoretical values of Erickson (1971):

$$1\,057.916 \pm 0.010 \text{ MHz}$$

or Mohr (1975):

$$1\,057.864 \pm 0.014 \text{ MHz}$$

where the main uncertainty arises from higher orders in the binding ($Z\alpha$) in the second-order (α^2) electron self-energy. The latest experimental values are due to Lundeen and Pipkin (1975):

$$1\,057.893 \pm 0.020 \text{ MHz}$$

and Andrews and Newton (1976):

$$1\,057.862 \pm 0.020 \text{ MHz}$$

7-3-3 Van der Waals Forces at Large Distances

We return to the question of the large distance potential between polarizable neutral systems introduced in Chap. 3. Our derivation is modeled after the work of Feinberg and Sucher.

We have seen that a phenomenological hamiltonian density could account for the interaction energy of a polarizable neutral particle described by the scalar hermitian field φ with a slowly varying field. It reads

$$\begin{gathered}\mathscr{H}_{\text{int}} = g_1 \partial_\alpha \varphi \partial_\beta \varphi F^{\alpha\gamma} F^\beta{}_\gamma + g_2 \varphi^2 F^2 \\ g_1 = \frac{\alpha_E + \alpha_B}{2m} \\ g_2 = -\frac{m}{4}\alpha_B\end{gathered} \tag{7-123}$$

The couplings g_1 and g_2 are related to the electric (α_E) and magnetic (α_B) susceptibilities in such a way that a particle at rest would contribute an interaction energy of the form

$$\mathscr{E}_{\text{int}} = -\alpha_E \frac{\mathbf{E}^2}{2} - \alpha_B \frac{\mathbf{B}^2}{2} \tag{7-124}$$

The type of system we have in mind may be, for instance, an atom whose internal structure is disregarded except for the parameters α_E and α_B. Note that these quantities have the dimension of a volume. To define a static interaction potential between such systems we proceed as follows. Assume α_E and α_B small enough to identify the Born term of a scattering amplitude in the low-energy regime with the Fourier transform of the potential according to Fermi's golden rule

$$d\sigma_{ab}{}^{\text{nr}} = \frac{1}{v_{ab}} |-i\tilde{V}(\mathbf{q})|^2 \frac{d^3 p_f}{(2\pi)^3} 2\pi\delta(E_f - E_i) \tag{7-125}$$

Here v_{ab} stands for the relative velocity of particles a and b, $\mathbf{q}$ is the transfer momentum, and $\tilde{V}(\mathbf{q})$

is such that

$$V(\mathbf{r}) = \int \frac{d^3q}{(2\pi)^3} \tilde{V}(\mathbf{q})\, e^{i\mathbf{q}\cdot\mathbf{r}} \tag{7-126}$$

Let $\mathscr{F}$ be the fully relativistic scattering amplitude arising to lowest order from the interaction (7-123) (see Fig. 7-16). Let us study the threshold behavior of the relativistic cross section, i.e., the limit $(p_a + p_b)^2 \to (m_a + m_b)^2$:

$$d\sigma_{ab}{}^R = \frac{1}{4[(p_a \cdot p_b)^2 - m_a^2 m_b^2]^{1/2}} |\mathscr{F}|^2 \frac{d^3p'_a\, d^3p'_b}{(2\pi)^3(2E'_a)(2\pi)^3(2E'_b)} (2\pi)^4 \delta^4(p'_a + p'_b - p_a - p_b)$$

In this limit $p_a \cdot p_b \simeq m_a m_b + (p^2/2)[(m_a + m_b)^2/m_a m_b]$ with $p = |\mathbf{p}|$ equal to the magnitude of the common three-momentum of particles a and b in the center of mass frame. Therefore,

$$[(p_a \cdot p_b)^2 - m_a^2 m_b^2]^{1/2} \simeq (m_a + m_b)p = m_a m_b \left| \frac{\mathbf{p}}{m_a} - \frac{-\mathbf{p}}{m_b} \right| = m_a m_b v_{ab}$$

and

$$d\sigma^R \to \frac{1}{v_{ab}} \left| \frac{\mathscr{F}}{4m_a m_b} \right|^2 \frac{d^3p_f}{(2\pi)^3} 2\pi\delta(E_f - E_i) \tag{7-127}$$

This means that close to threshold

$$-i\tilde{V}(\mathbf{q}) = \frac{\mathscr{F}}{4m_a m_b} \tag{7-128}$$

where $-\mathbf{q}^2$ can be identified with the square relativistic momentum transfer $q^2 = (p_a - p'_a)^2$, and the center of mass energy square $s = (p_a + p_b)^2$ is taken equal to $(m_a + m_b)^2$.

From (7-123) we can compute the amplitude $\mathscr{F}$ to order g^2 as represented in Fig. 7-16. This is again a one-loop diagram with a strong ultraviolet divergence. A theory based on (7-123) is not renormalizable due to the dimensionality of the coupling constants. However, we agreed to use it only as a phenomenological description and as such to any order we may perform a finite number of ultraviolet subtractions without modifying the long-range behavior of V which we intend to find. Therefore for our purpose we are safe. The elementary contraction needed to apply perturbation theory is

$$\begin{gathered}\langle 0|\, TF_{\mu\nu}(x)F_{\rho\sigma}(y)\,|0\rangle = -i \int \frac{d^4k}{(2\pi)^4} \frac{e^{-ik\cdot(x-y)}}{k^2 + i\varepsilon} K_{\mu\nu,\rho\sigma}(k) \\ K_{\mu\nu,\rho\sigma}(k) = k_\mu k_\rho g_{\nu\sigma} - k_\nu k_\rho g_{\mu\sigma} - k_\mu k_\sigma g_{\nu\rho} + k_\nu k_\sigma g_{\mu\rho}\end{gathered} \tag{7-129}$$

Before any subtraction the Feynman rules provide us with an amplitude $\mathscr{F}$ of the form

$$\begin{aligned}\mathscr{F} &= \int \frac{d^4k\, d^4k'\, \delta^4(k + k' - q)}{(2\pi)^4(k^2 + i\varepsilon)(k'^2 + i\varepsilon)} d(k, k') \\ d(k, k') &= 8g_2^a g_2^b K_{\mu\nu,\rho\sigma}(k) K^{\mu\nu,\rho\sigma}(k') \\ &\quad + [4g_1^a g_2^b (p_\alpha^a p'^a_\beta + p_\beta^a p'^a_\alpha) K^{\alpha\nu,\rho\sigma}(k) K^\beta{}_{\nu,\rho\sigma}(k') + (a \leftrightarrow b)] \\ &\quad + g_1^a g_1^b [(p_\alpha^a p'^a_\beta + p_\beta^a p'^a_\alpha) K^{\alpha\nu,\alpha'\nu'}(k) K^\beta{}_\nu{}^{,\beta'}{}_{\nu'}(k')(p_{\alpha'}^b p'^b_{\beta'} + p_{\beta'}^b p'^b_{\alpha'}) + (k \leftrightarrow k')]\end{aligned} \tag{7-130}$$

A convenient way to perform the ultraviolet subtractions without modifying the long-range behavior follows from the observation that for fixed $s = (p_a + p_b)^2$, $\mathscr{F}$ is analytic (and purely imaginary) along the semiaxis $q^2 < 0$. Its cut along the real axis arises from its real part and we are only interested in the vicinity of $q^2 = 0$. We therefore write

$$\mathscr{F} = \frac{1}{i\pi} \int_0^{M^2} dm^2 \frac{\mathscr{F}_R(m^2)}{m^2 - q^2 - i\varepsilon} + \Delta\mathscr{F}(q^2) \tag{7-131}$$

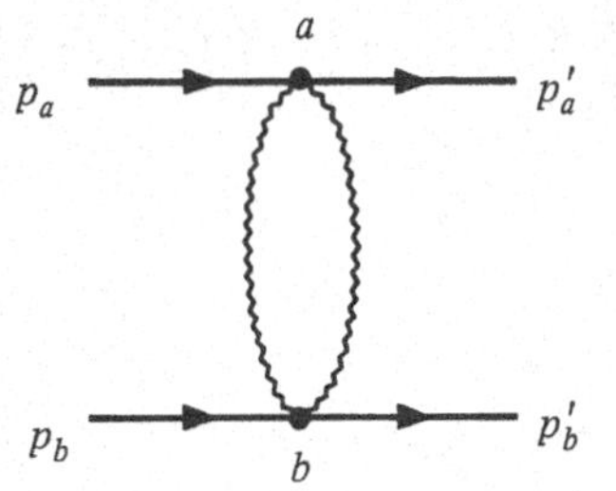

Figure 7-16 Interaction between neutral systems arising from electric and magnetic susceptibilities.

with $\Delta\mathscr{F}$ regular in the vicinity of $q^2 = 0$ and M^2 positive and arbitrary. Consequently, from (7-126) and (7-128) the asymptotic part of the potential will be given by

$$V_{\text{as}}(r) = \frac{1}{16\pi^2 m_a m_b r} \int_0^{M^2} dm^2\, \mathscr{F}_R(m^2)\, e^{-mr}$$
$$= \frac{1}{16\pi^2 m_a m_b r^3} \int_0^{M^2 r^2} dy^2\, \mathscr{F}_R\left(\frac{y^2}{r^2}\right) e^{-y} \tag{7-132}$$

We shall soon see that $\mathscr{F}_R(q^2)$ behaves as $-A(q^2)^2$ for $q^2 \to 0$ so that the dominant part of V will be obtained by taking the limit $Mr \to \infty$ in (7-132), that is,

$$V_{\text{as}}(r) = \frac{-A}{16\pi^2 m_a m_b r^7} \int_0^{\infty} dx^2\, x^4\, e^{-x} = -\frac{5!A}{8\pi^2 m_a m_b r^7} \tag{7-133}$$

It remains to compute A. To do so we observe that each denominator in (7-130) may be split into two terms according to $1/(k^2 + i\varepsilon) = \mathrm{PP}/k^2 - i\pi\delta(k^2)$ where PP means principal part. This we abbreviate as $P - i\pi\delta$. Up to the polynomial d the integrand in (7-130) takes the form

$$(P - i\pi\delta)(P' - i\pi\delta') = PP' - \pi^2\delta\delta' - i\pi(\delta P' + \delta' P)$$

This is, of course, nothing but the Fourier transform of $[G_F(x-y)]^2$ in configuration space with G_F the Feynman propagator. The effect of d is then to act as a polynomial of derivatives on this quantity. Now if instead of using G_F^2 we were to use the product $G_{\text{adv}}(x-y)G_{\text{ret}}(x-y)$ with at most a support at $x = y$ we would obtain by a Fourier transformation an ill-defined polynomial of the momenta without contribution to the real part of $\mathscr{F}$. Consequently, as far as we are concerned, the replacement

$$\frac{1}{k^2 + i\varepsilon}\,\frac{1}{k'^2 + i\varepsilon} \to \frac{1}{k^2 + i(k^0/|k^0|)\varepsilon}\,\frac{1}{k'^2 - i(k'^0/|k'^0|)\varepsilon}$$

would lead to zero. This means that we may subtract to the previous contribution the combination

$$(P - i\pi\varepsilon\delta)(P' + i\pi\varepsilon'\delta') = PP' + \pi^2\varepsilon\varepsilon'\delta\delta' + i\pi(P\varepsilon'\delta' - P'\varepsilon\delta)$$

with $\varepsilon\delta$ standing for $(k^0/|k^0|)\,\delta(k^2)$. The result is the quantity

$$(PP' - \pi^2\delta\delta') - (PP' + \pi^2\varepsilon\varepsilon'\delta\delta') = -\pi^2(1 + \varepsilon\varepsilon')\delta\delta'$$

Note that $1 + \varepsilon\varepsilon' = 2$ if k^0 and k'^0 are of the same sign and zero otherwise, and that the δ functions project the photons on-shell. For $q^2 > 0$ the conservation of energy momentum $k + k' = q$ implies that we may keep only one of the two possibilities $k^0 > 0$, $k'^0 > 0$, for instance. We have now

$$\mathscr{F}_R = -\frac{1}{2} \int \frac{d^3k}{2k^0(2\pi)^3}\,\frac{d^3k'}{2k'^0(2\pi)^3}\,(2\pi)^4\delta^4(k + k' - q)\, d(k, k')\big|_{k^2 = k'^2 = 0} \tag{7-134}$$

This lengthy derivation is nothing but an application of the Cutkosky rules (Chap. 6) to this special case. Of course, $\mathscr{F}_R$ as given by Eq. (7-134) is a convergent quantity and we simply want its leading behavior near $q^2 = 0$. We first evaluate $d(k, k')$ given by (7-130). The explicit expression is slightly cumbersome:

$$\begin{aligned} d(k, k')\big|_{k^2=k'^2=0} &= 64 g_2^a g_2^b (k\cdot k')^2 + 16 g_1^a g_2^b (p^a\cdot p'^a)(k\cdot k')^2 \\ &+ 16 g_1^b g_2^a (p^b\cdot p'^b)(k\cdot k')^2 + 4 g_1^a g_1^b [(p^a\cdot p^b)(p'^a\cdot p'^b)(k\cdot k')^2 \\ &- (p^a\cdot p^b)(p'^a\cdot k)(p'^b\cdot k')(k\cdot k') - (p^a\cdot p^b)(p'^a\cdot k')(p'^b\cdot k)(k\cdot k') \\ &- (p'^a\cdot p'^b)(p^a\cdot k')(p^b\cdot k)(k\cdot k') - (p'^a\cdot p'^b)(p^a\cdot k)(p^b\cdot k')(k\cdot k') \\ &+ (p'^a\cdot p^b)(p^a\cdot p'^b)(k\cdot k')^2 - (p'^a\cdot p^b)(p^a\cdot k)(p'^b\cdot k')(k\cdot k') \\ &- (p'^a\cdot p^b)(p^a\cdot k')(p'^b\cdot k)(k\cdot k') - (p^a\cdot p'^b)(p'^a\cdot k')(p^b\cdot k)(k\cdot k') \\ &- (p^a\cdot p'^b)(p'^a\cdot k)(p^b\cdot k')(k\cdot k') + (p^a\cdot p'^a)(p^b\cdot k)(p'^b\cdot k')(k\cdot k') \\ &+ (p^a\cdot p'^a)(p^b\cdot k')(p'^b\cdot k)(k\cdot k') + (p^b\cdot p'^b)(p^a\cdot k)(p'^a\cdot k')(k\cdot k') \\ &+ (p^b\cdot p'^b)(p^a\cdot k')(p'^a\cdot k)(k\cdot k') + 2(p^a\cdot k')(p'^a\cdot k')(p^b\cdot k)(p'^b\cdot k) \\ &+ 2(p^a\cdot k)(p'^a\cdot k)(p^b\cdot k')(p'^b\cdot k')] \end{aligned} \tag{7-135}$$

We have $2k\cdot k' = q^2$ and we look for the coefficient of $(q^2)^2$ after integration. Therefore the scalar products multiplying $(q^2)^2$ of the type $p^a\cdot p^b$ or $p^a\cdot p'^a, \ldots,$ may be replaced by $m_a m_b, m_a^2, \ldots$. We set

$$\mathscr{F}_R = -\frac{1}{2}\int \frac{d^3k}{2k^0(2\pi)^3}\,\frac{d^3k'}{2k'^0(2\pi)^3}\,(2\pi)^4\delta^4(k+k'-q)\,d = -\frac{\theta(q^2)}{16\pi}\,\langle d\rangle \tag{7-136}$$

where $\langle d\rangle$ is a function of q^2 alone close to the threshold and reads

$$\langle d\rangle = \frac{\displaystyle\int (d^3k/k^0)(d^3k'/k'^0)\,\delta^4(k+k'-q)d}{\displaystyle\int (d^3k/k^0)(d^3k'/k'^0)\,\delta^4(k+k'-q)} \tag{7-137}$$

We need

$$\langle k_\mu k'_\nu\rangle = \tfrac{1}{6}q_\mu q_\nu + \tfrac{1}{12}g_{\mu\nu}q^2 \tag{7-138}$$

obtained by remarking that the trace is $q^2/2$, and

$$\begin{aligned} \langle k'_\mu k'_\nu k_\alpha k_\beta\rangle &= a q_\mu q_\nu q_\alpha q_\beta + b(q_\mu q_\nu g_{\alpha\beta} + q_\alpha q_\beta g_{\mu\nu})q^2 \\ &+ cq^2(q_\mu q_\alpha g_{\nu\beta} + q_\nu q_\alpha g_{\mu\beta} + q_\mu q_\beta g_{\nu\alpha} + q_\nu q_\beta g_{\mu\alpha}) \\ &+ d(q^2)^2 g_{\mu\nu}g_{\alpha\beta} + e(q^2)^2(g_{\mu\alpha}g_{\nu\beta} + g_{\nu\alpha}g_{\mu\beta}) \end{aligned} \tag{7-139}$$

The form (7-139) is obtained by neglecting all terms behaving as higher powers of q^2 and using the obvious symmetries in the exchanges $(\mu\leftrightarrow\nu)$, $(\alpha\leftrightarrow\beta)$, $(\mu, \nu\leftrightarrow\alpha, \beta)$. To get the five coefficients a, b, c, d, and e we use the following five conditions:

(*a*) Taking the partial trace over (μ, ν) we must find zero (mass shell condition) implying

$$a + 4b + 4c = 0 \qquad b + 4d + 2e = 0$$

(*b*) Taking the partial trace over (ν, α) we obtain $q^2/2\,\langle k'_\mu k_\beta\rangle$ already computed; therefore,

$$a + 2b + 6c = \tfrac{1}{12} \qquad c + d + 5e = \tfrac{1}{24}$$

(*c*) Finally, contracting with $q^\mu q^\nu q^\alpha q^\beta$ yields $(q^2/2)^4$; thus

$$a + 2b + 4c + d + 2e = \tfrac{1}{16}$$

Solving this system we obtain the coefficients needed in the sequel

$$d = e = \frac{1}{15 \times 16} \tag{7-140}$$

At threshold we have

$$\begin{aligned}\langle d \rangle = 64 g_2^a g_2^b \left(\frac{q^2}{2}\right)^2 &+ 16 g_1^a g_2^b \left(\frac{q^2}{2}\right)^2 m_a^2 + 16 g_1^b g_2^a \left(\frac{q^2}{2}\right)^2 m_b^2 + 4 g_1^a g_1^b \left(2 m_a^2 m_b^2 \left(\frac{q^2}{2}\right)^2\right. \\
&- \frac{q^2}{12} \{(p^a \cdot p^b)[2(q \cdot p'^a)(q \cdot p'^b) + q^2 (p'^a \cdot p'^b)] + (p'^a \cdot p'^b)[2(q \cdot p^a)(q \cdot p^b) + q^2 (p^a \cdot p^b)] \\
&+ (p'^a \cdot p^b)[2(q \cdot p^a)(q \cdot p'^b) + q^2 (p^a \cdot p'^b)] + (p^a \cdot p'^b)[2(q \cdot p'^a)(q \cdot p'^b) + q^2 (p'^a \cdot p^b)]\} \\
&+ \frac{q^2}{12} \{(p^a \cdot p'^a)[2(q \cdot p^b)(q \cdot p'^b) + q^2 (p^b \cdot p'^b)] + (p^b \cdot p'^b)[2(q \cdot p^a)(q \cdot p'^a) + q^2 (p^a \cdot p'^a)]\} \\
&\left. + 4 \langle (p^a \cdot k')(p'^a \cdot k')(p^b \cdot k)(p'^b \cdot k) \rangle \right)\end{aligned} \tag{7-141}$$

But $q \cdot p^a$, $q \cdot p'^a$, $q \cdot p^b$, $q \cdot p'^b$ are all proportional to q^2. For instance,

$$q \cdot p^a = (p'^a - p^a)\left(\frac{p^a + p'^a}{2} + \frac{p^a - p'^a}{2}\right) = -\frac{q^2}{2}$$

Therefore when extracting the coefficient of $(q^2)^2$ all terms involving as coefficient scalar products of q with an external momentum are negligible. In particular, the contribution of the last mean value reduces to

$$\langle (p^a \cdot k')(p'^a \cdot k')(p^b \cdot k)(p'^b \cdot k) \rangle \simeq m_a^2 m_b^2 (q^2)^2 (d + 2e) = \frac{m_a^2 m_b^2 (q^2)^2}{80}$$

The coefficient A defined as $\mathscr{F}_R = -A(q^2)^2 + \cdots$ is thus equal to

$$\begin{aligned}A &= \frac{1}{16\pi}\left[16 g_2^a g_2^b + 4(m_a^2 g_1^a g_2^b + m_b^2 g_1^b g_2^a) + \frac{23}{15} m_a^2 m_b^2 g_1^a g_1^b\right] \\
&= \frac{m_a m_b}{16\pi}\left[\alpha_B^a \alpha_B^b - \frac{1}{2}(\alpha_E^a + \alpha_B^a)\alpha_B^b - \frac{1}{2}(\alpha_E^b + \alpha_B^b)\alpha_B^a + \frac{23}{60}(\alpha_E^a + \alpha_B^a)(\alpha_E^b + \alpha_B^b)\right]\end{aligned} \tag{7-142}$$

Returning to Eq. (7-133) we obtain the asymptotic behavior of the Van der Waals force as

$$V_{\text{as}}(r) = -\frac{[23(\alpha_E^a \alpha_E^b + \alpha_B^a \alpha_B^b) - 7(\alpha_E^a \alpha_B^b + \alpha_E^b \alpha_B^a)]}{(4\pi)^3 r^7} \tag{7-143}$$

Let us close this section with a brief comment on the relation between the relativistic potential in $1/r^7$ and the usual nonrelativistic $1/r^6$ law, disregarding the effects of retardation. In a nonrelativistic calculation we only include the instantaneous Coulomb forces. To simplify, let us consider two hydrogen atoms at distance $|\mathbf{r}_a - \mathbf{r}_b|$ with electrons located at $\mathbf{r}_a + \mathbf{r}_1$ and $\mathbf{r}_b + \mathbf{r}_2$. Neglecting altogether spin and the Pauli principle, the interaction potential may be written as

$$V = \frac{e^2}{4\pi}\left(\frac{1}{|\mathbf{r}_a - \mathbf{r}_b|} + \frac{1}{|\mathbf{r}_a + \mathbf{r}_1 - \mathbf{r}_b - \mathbf{r}_2|} - \frac{1}{|\mathbf{r}_a + \mathbf{r}_1 - \mathbf{r}_b|} - \frac{1}{|\mathbf{r}_b + \mathbf{r}_2 - \mathbf{r}_a|}\right)$$

At large separations, neglecting angular dependences, this is qualitatively of the form $(e^2/4\pi r_{ab}^3) r_1 r_2$. However, due to this angular dependence the first effect of this perturbation vanishes (we are using an adiabatic approximation). To second order we obtain a negative contribution of the type

$$V^{\text{nr}} \sim -\frac{1}{(r_{ab})^6} \langle e^2 r_1^2 \rangle_a \langle e^2 r_2^2 \rangle_b \frac{1}{\Delta E}$$

where ΔE is a characteristic atomic excitation energy. Similarly, the (electric) susceptibility α for each atom is obtained by treating a perturbation $e\mathbf{E}\cdot\mathbf{r}$ to second order, with the result that

$$\alpha \sim \frac{\langle e^2 r^2 \rangle}{\Delta E}$$

Altogether we find

$$V^{\mathrm{nr}} \sim -\frac{\Delta E \alpha_a \alpha_b}{(r_{ab})^6}$$

We therefore expect that an interpolating formula between the r^{-6} and r^{-7} laws involves a characteristic retardation time τ proportional to $\hbar/\Delta E$. The transition between the two regimes would occur at distances of order $c\tau = \hbar c/\Delta E$, where ΔE could be taken as a fraction of the ionization energy. For hydrogen this estimate would be of order 10^2 times the Bohr radius. In the practical theory of liquids this implies that we are always justified to work in the framework of the nonrelativistic approximation.

NOTES

Radiative corrections are treated in all textbooks and many of the original articles are reproduced in Schwinger's "Quantum Electrodynamics," Dover, New York, 1958, already quoted. Let us give a short list.

Subjects covered are as follows: gauge-invariant regularization in W. Pauli and F. Villars, *Rev. Mod. Phys.*, vol. 21, p. 434, 1949; vacuum polarization, limit $k^2 \to 0$, in E. A. Uehling, *Phys. Rev.*, vol. 48, p. 55, 1935; complete treatment in J. Schwinger, *Phys. Rev.*, vol. 75, p. 651, 1949; Ward identities in J. C. Ward, *Phys. Rev.*, vol. 78, p. 182, 1950, and Y. Takahashi, *Nuovo Cim.*, vol. 6, p. 371, 1957; electron anomaly to order α in J. Schwinger, *Phys. Rev.*, vol. 73, p. 416, 1948; and electron form factors to order α in R. P. Feynman, *Phys. Rev.*, vol. 76, p. 769, 1949. Work in the 1960s on higher-order corrections is reviewed in the report of B. Lautrup, A. Peterman, and E. de Rafael, *Physics Reports*, vol. 3C, p. 193, 1972. The most recent theoretical value quoted in the text comes from the work of P. Cvitanovic and T. Kinoshita, *Phys. Rev.*, ser. D, vol. 10, p. 4007, 1974. A large number of theorists have also contributed to this effort. The astonishingly accurate experimental value is due to R. S. Van Dyck, P. B. Schwinberg, and H. G. Dehmelt, *Phys. Rev. Lett.*, vol. 38, p. 310, 1977. The experimental value of the muon anomaly has been taken from the work of J. Bailey, K. Borer, F. Combley, H. Drumm, F. Farlcy, J. Field, W. Flegel, P. Hattersley, F. Krienen, F. Lange, E. Picasso, and W. Von Rüden, *Phys. Lett.*, vol. 68B, p. 191, 1977, and the theoretical prediction from the review of J. Calmet, S. Narison, M. Perrottet, and E. de Rafael, *Rev. Mod. Phys.*, vol. 49, p. 21, 1977. The original treatment of radiative corrections to Coulomb scattering is due to J. Schwinger, *Phys. Rev.*, vol. 75, p. 1912, 1949. We follow in part the presentation in the book by A. I. Akhiezer and V. B. Berestetskii, "Quantum Electrodynamics," Interscience Publishers, New York, 1965, which is extremely rich in detailed calculations. For infrared divergences we remind the reader of the work of D. Yennie, S. Frautschi, and H. Suura quoted in Chap. 4. As for the Lamb shift the general aspects of hydrogenic atomic systems are covered in H. A. Bethe and E. Salpeter, "Quantum Mechanics of One and Two Electron Atoms," Springer-Verlag, Berlin, 1957. The crucial experimental evidence comes from the experiment of W. E. Lamb and R. C. Retherford, *Phys. Rev.*, vol. 72, p. 241, 1947. A more refined measurement in the 1950s is due to S. Triebwasser, E. S. Dayhoff, and W. E. Lamb, *Phys. Rev.*, vol. 89, p. 98, 1953.

The recent values quoted in the text are from S. R. Lundeen and F. M. Pipkin, *Phys. Rev. Lett.*, vol. 34, p. 1368, 1975, and D. A. Andrews and G. Newton, *Phys. Rev. Lett.*, vol. 37, p. 1254, 1976. The early theoretical contributions are those of H. A. Bethe, *Phys. Rev.*, vol. 72, p. 339, 1947, for the nonrelativistic part, and of N. M. Kroll and W. E. Lamb, *Phys. Rev.*, vol. 75, p. 388, 1949, and J. B. French and V. F. Weisskopf, *Phys. Rev.*, vol. 75, p. 1240, 1949, for the relativistic one. For a consistent theory of the effect to all orders see, for instance, G. W. Erickson and D. R. Yennie, *Annals of Physics*, vol. 35, pp. 271 and 447, 1965. The theoretical prediction has been taken from the work of G. W. Erickson, *Phys. Rev. Lett.*, vol. 27, p. 780, 1971, and P. J. Mohr, *Phys. Rev. Lett.*, vol. 34, p. 1050, 1975.

For a recent estimate of the accuracy achieved in these higher-order computations one may consult the review by T. Kinoshita in "Perspective of Quantum Physics," published by Iwanami Shoten, Tokyo.

Needless to say, the previous short list of references does not do any justice to the enormous and beautiful experimental and theoretical work in quantum electrodynamics.

CHAPTER

EIGHT

RENORMALIZATION

Renormalization to all orders is at the heart of field theory. After an introduction devoted to a study of various regularization methods we present the Bogoliubov-Zimmermann subtraction scheme. We give indications on the convergence proof of renormalized integrals, and study ultraviolet behavior (Weinberg's theorem) and massless theories. Renormalization of composite operators is briefly discussed. The interplay between gauge invariance and renormalization is dealt with in the last part of the chapter.

8-1 REGULARIZATION AND POWER COUNTING

8-1-1 Introduction

This chapter is devoted to a systematic study of the renormalization procedure. The underlying philosophy has already been presented in the preceding chapter, in the case of quantum electrodynamics. We have seen that divergences may be absorbed in a redefinition of the various parameters, mass, coupling constant, etc., of the theory. It is convenient to get rid of the difficulties specific to electrodynamics, namely, gauge invariance, and to study first the renormalization of a scalar theory. The new problems arising from the existence of symmetries will be considered at the end of this chapter for quantum electrodynamics, and in Chaps. 11 and 12 for other internal symmetries.

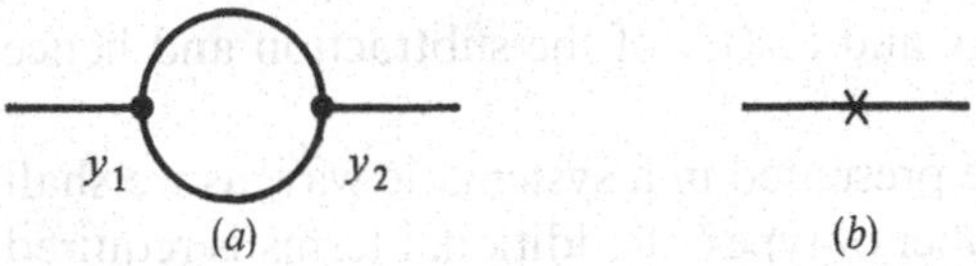

Figure 8-1 A divergent diagram and the associated counterterm.

The nature and properties of renormalization were first formulated and studied by the founding fathers of quantum field theory, Tomonaga, Feynman, Dyson, Schwinger, etc. Important contributions were then made by Salam, Weinberg, Bogoliubov and Parasiuk, and Hepp. More recently, Zimmermann and his followers have further clarified the systematics of the renormalization operation, while Epstein and Glaser presented an axiomatic approach.

The mathematical nature of the problem is clear. Divergences occur in perturbative computations because of lack of care in the multiplication of distributions. For instance, the amputated self-energy diagram of Fig. 8-1*a* is the ill-defined expression $[G_F(y_1 - y_2)]^2$. As it stands, this expression does not make sense; in momentum space, its Fourier transform is logarithmically divergent. As we saw in Chap. 7, such divergent expressions must be subtracted. Since we demand that the subtractions be local in configuration space, this operation amounts to a redefinition of the parameters of the lagrangian by an infinite amount. Therefore, we abandon the idea of using or observing the parameters of the initial lagrangian, the so-called "bare" quantities, and reexpress everything in terms of the finite "renormalized" and observable parameters.

In the previous instance, we replace $G_F^2(y_1 - y_2)$ by

$$\Pi(y_1 - y_2) = [G_F(y_1 - y_2)]^2 - S(y_1 - y_2)$$

where S is a distribution concentrated at the origin, chosen in such a way as to make the whole expression Π meaningful. Here a term proportional to a δ function suffices. In Fourier representation

$$[G_F(y_1 - y_2)]^2 = \int \frac{d^4p}{(2\pi)^4} e^{ip\cdot(y_1 - y_2)} \int \frac{d^4k}{(2\pi)^4} \frac{1}{(k^2 - m^2 + i\varepsilon)[(p-k)^2 - m^2 + i\varepsilon]}$$

is replaced, for instance, by the well-defined expression

$$\Pi(y_1 - y_2) = \int \frac{d^4p}{(2\pi)^4} e^{ip\cdot(y_1 - y_2)} \int \frac{d^4k}{(2\pi)^4} \times \left[\frac{1}{(k^2 - m^2 + i\varepsilon)[(p-k)^2 - m^2 + i\varepsilon]} - \frac{1}{(k^2 - m^2 + i\varepsilon)^2}\right]$$

and we may write formally

$$\Pi(y_1 - y_2) = [G_F(y_1 - y_2)]^2 + A\delta(y_1 - y_2) \tag{8-1}$$

Strictly speaking, the constant A is infinite, but formally the second term in (8-1) may be regarded as coming from the zeroth-order counterterm depicted in Fig. 8-1*b*. By imposing a definite normalization condition on Π, we may fix unambiguously the finite part of A. Let us emphasize that this renormalization procedure works because of the locality and reality of the subtraction and hence of the counterterm.

This subtraction operation may be presented in a systematic way, as we shall see in the following. If only a finite number of types of additional terms is required in the lagrangian to eliminate the divergences of Green functions, the renormalized theory will depend only on a finite number of parameters. Such a theory is called renormalizable or super-renormalizable. It remains to prove that the renormalized integrals are indeed finite and satisfy the constraints of locality and unitarity. We shall evoke briefly these points.

A slightly different approach, proposed by Dyson and Schwinger, relies on a set of integral equations between proper Green functions (Sec. 10-1). It also unfortunately involves infinite multiplicative renormalization constants at intermediate stages.

Finally, the most orthodox procedure of Epstein and Glaser relies directly on the axioms of local field theory in configuration space. It is free of mathematically undefined quantities but hides the multiplicative structure of renormalization. The latter will lead under proper interpretation to the study of the renormalization group (Chap. 13).

8-1-2 Regularization

In order to give a meaning to otherwise formal and divergent expressions, it is important as a preliminary step to regularize the perturbative expansion. After performing the renormalization subtractions, we shall be free to remove this regularization. Several procedures have been devised. The final results are finite and independent of this choice.

The most naive prescription would be to perform a Wick rotation and to cut off the large values of the (euclidean) four-momenta of each loop. Every such integration would be restricted to a compact sphere $(k^2)^{1/2} < \Lambda$. This certainly makes any Feynman amplitude finite, but it ruins our beloved Poincaré (or rather euclidean) invariance. Therefore it is used only exceptionally, in heuristic arguments. A variant consists in discretizing space-time, i.e., in assuming that configuration variables x_μ take only discrete values corresponding, say, to the vertices of a regular lattice. Clearly, cutting off the small distances is equivalent to cutting off the large momenta. Here, too, rotational invariance is lost.

A covariant regularization is obtained by substituting to the Feynman propagator $G_F(x-y, m)$ an expression of the form

$$G_F^{\text{reg}}(x-y) = G_F(x-y, m) + \sum_k C_k G_F(x-y, M_k) \tag{8-2}$$

where the coefficients C_k are judiciously chosen, as functions of m and the M, so as to remove some of the singularities of G_F. Which degree of singularity is acceptable in order that every Feynman diagram of a given theory be finite, and therefore how many terms are needed in the sum (8-2), will be clear as soon as we have given a criterion of convergence. Here it suffices to say that any given diagram may be made finite by such a substitution. This is the case of the self-energy contribution considered in the introduction, for instance, if we substitute

$$G_F(x-y, m) \to G_F(x-y, m) - G_F(x-y, M)$$

or in momentum space

$$\frac{1}{k^2-m^2} \to \frac{1}{k^2-m^2} - \frac{1}{k^2-M^2}$$

In the general case, if we denote collectively all the auxiliary masses M_k by Λ, the initial propagator is recovered by letting the cutoff Λ go to infinity.

Modifying the large-momentum behavior of the propagator may also be performed on its parametric representation. It then amounts to modify the small α integration region in the form

$$\int_0^\infty d\alpha\, e^{i\alpha(k^2-m^2+i\varepsilon)} \to \int_0^\infty d\alpha\, \rho_\Lambda(\alpha)\, e^{i\alpha(k^2-m^2+i\varepsilon)}$$

where the function $\rho_\Lambda(\alpha)$ vanishes with some of its derivatives at $\alpha=0$. We demand also that as the cutoff is removed, that is, $\Lambda\to\infty$, $\rho_\Lambda(\alpha)\to 1$ for all $\alpha>0$. For example, we may take $\rho_\Lambda(\alpha)=\theta(\alpha-1/\Lambda^2)$ or $\rho(\alpha)=\alpha^{\lambda-1}$ where $\lambda\to 1$ as $\Lambda\to\infty$; the latter case corresponds to a propagator $(k^2-m^2+i\varepsilon)^{-\lambda}$ in momentum space. This type of regularization has been extensively studied by Speer.

In some instances, it may be important to use a more subtle procedure, in order to maintain some invariance properties such as gauge invariance. The Pauli-Villars regularization, already encountered in the previous chapter, has this property. Every photon propagator is replaced by a sum of the form (8-2). On the other hand, only fermion propagators pertaining to internal closed fermion loops are modified. More precisely, to a closed loop with $2n$ vertices, we associate

$$\operatorname{tr}\left[\prod_{p=1}^{2n} \gamma_{\mu_p} S_F(z_p - z_{p-1}, m)\right] - \sum_{s=1}^{S} C_s \operatorname{tr}\left[\prod_{p=1}^{2n} \gamma_{\mu_p} S_F(z_p - z_{p-1}, M_s)\right] \tag{8-3}$$

where $z_0 \equiv z_{2n}$. We shall reexamine this regularization in greater detail in Sec. 8-4-2.

Divergent Feynman integrals would be ultraviolet convergent in a smaller space-time dimension. In the dimensional regularization of 't Hooft and Veltman, the Feynman integrals are computed for an arbitrary integer space-time dimension d. The result of the integration may then be analytically continued to arbitrary real or even complex values of d. In this regularization, ultraviolet divergences manifest themselves as poles at rational or integer values of d. We are eventually interested in letting d go to four, the physical space-time dimension, and therefore we focus on the simple or multiple poles at $d=4$. This continuation may also be defined in the presence of Dirac γ matrices, except for the case of the γ_5 matrix which is quite particular to $d=4$ (or in general to even space-time dimensions). The virtue of this method is that it automatically preserves internal symmetries that do not involve the γ_5 matrix. Technically, all the manipulations that we perform to check Ward identities (see Chap. 7 or Sec. 8-4) such as shifts of integration variables, contractions of Lorentz indices, etc., are consistent with the regularization.

The analytic continuation to d-dimensional space-time is most easily performed, after a Wick rotation to the euclidean region, on the parametric form derived in Chap. 6. Indeed, we may compute the amplitude of Eq. (6-94) in an arbitrary integer dimension d (we drop henceforth the caret notation for euclidean momenta):

$$(2\pi)^d \delta^d(\textstyle\sum p) I_G(P) = \displaystyle\int \prod_{l=1}^{I} \frac{d^d k_l}{(2\pi)^d} \frac{1}{k_l^2 + m_l^2} \prod_v (2\pi)^d \delta^d(P_v - \sum_l \varepsilon_{vl} k_l) \tag{8-4}$$

The method of Sec. 6-2-3 is easily seen to go through, and the result reads

$$I_G(P) = \int_0^\infty \prod_{l=1}^{I} d\alpha_l \frac{\exp - [Q_G(P, \alpha) + \sum \alpha_l m_l^2]}{[(4\pi)^L \mathscr{P}_G(\alpha)]^{d/2}} \tag{8-5}$$

with the same functions Q_G and $\mathscr{P}_G$ as those defined in Eqs. (6-86) and (6-87). The integration over the homogeneity parameter λ of the α is convergent at $\lambda = 0$ if $I - dL/2 = L(1 - d/2) + V - 1 > 0$:

$$\begin{aligned} I_G(P) &= \int_0^\infty \frac{d\lambda}{\lambda} \lambda^{I - dL/2} \int_0^1 \prod_{l=1}^{I} d\alpha_l \frac{\exp - \lambda[Q_G(P, \alpha) + \sum \alpha_l m_l^2]}{[(4\pi)^L \mathscr{P}_G(\alpha)]^{d/2}} \delta(1 - \textstyle\sum \alpha) \\ &= \Gamma\left(I - \frac{dL}{2}\right) \int_0^1 \prod_{l=1}^{I} d\alpha_l\, \delta(1 - \textstyle\sum \alpha) \dfrac{[Q_G(P, \alpha) + \sum \alpha_l m_l^2]^{dL/2 - I}}{[(4\pi)^L \mathscr{P}_G(\alpha)]^{d/2}} \end{aligned} \tag{8-6}$$

For instance, for $p, n \geq 1$,

$$\begin{aligned} I_G(p) &= \int \frac{d^d k}{(2\pi)^d} \frac{1}{[(p-k)^2 + m_1^2]^n (k^2 + m_2^2)^p} \\ &= \frac{1}{(4\pi)^{d/2}} \frac{\Gamma(n + p - d/2)}{\Gamma(n)\Gamma(p)} \int_0^1 d\alpha\, \alpha^{n-1} (1-\alpha)^{p-1} [\alpha(1-\alpha)p^2 + \alpha m_1^2 + (1-\alpha) m_2^2]^{d/2 - n - p} \end{aligned} \tag{8-7}$$

The whole dependence on d may be explicited by performing the α integration. For low enough d, the result is finite. When $d \to 4$, ultraviolet divergences may arise from the Euler function in front of (8-6), or from the α integration, or from both. However, for one-loop diagrams such as (8-7), only $\Gamma(I - dL/2)$ diverges. If $I - 2L$ is a nonpositive integer, then

$$\Gamma\left(I - \frac{dL}{2}\right) \underset{d \to 4}{\sim} \frac{(-1)^{2L - I}}{(2L - I)!} \frac{2}{(4 - d)L} \tag{8-8}$$

and $I_G(P)$ has a simple pole at $d = 4$. When internal divergences coming from the α integration are also present, this pole may become a multiple one (compare with the examples below).

For completeness, we have to say how to deal with integrals involving Lorentz vectors and/or spinors. The former case does not offer any difficulty. Four-vectors are transformed into d-vectors, the integrations are performed, and the result continued to arbitrary dimensions. For instance, again in euclidean space,

$$\begin{aligned} \int \frac{d^d k}{(2\pi)^d} \frac{k_\mu}{[(p-k)^2 + m_1^2]^n (k^2 + m_2^2)^p} &= \frac{p_\mu}{(4\pi)^{d/2}} \frac{\Gamma(n + p - d/2)}{\Gamma(n)\Gamma(p)} \\ &\quad \times \int_0^1 d\alpha\, \alpha^n (1-\alpha)^{p-1} [\alpha(1-\alpha)p^2 + \alpha m_1^2 + (1-\alpha) m_2^2]^{d/2 - n - p} \end{aligned} \tag{8-9a}$$

$$\int \frac{d^dk}{(2\pi)^d} \frac{k_\mu k_\nu}{[(p-k)^2 + m_1^2]^n (k^2 + m_2^2)^p} = \frac{1}{(4\pi)^{d/2}\Gamma(n)\Gamma(p)}$$

$$\times \left\{ \int_0^1 d\alpha\, \alpha^{n-1}(1-\alpha)^{p-1}[\alpha(1-\alpha)p^2 + \alpha m_1^2 + (1-\alpha)m_2^2]^{d/2+1-n-p} \frac{\delta_{\mu\nu}}{2} \Gamma\left(n+p-\frac{d}{2}-1\right)\right.$$

$$\left. + \int_0^1 d\alpha\, \alpha^{n+1}(1-\alpha)^{p-1}[\alpha(1-\alpha)p^2 + \alpha m_1^2 + (1-\alpha)m_2^2]^{d/2-n-p} p_\mu p_\nu \Gamma\left(n+p-\frac{d}{2}\right)\right\} \tag{8-9b}$$

In this expression, $\delta_{\mu\nu}$ is the unit tensor in d-dimensional space, satisfying

$$\sum_\mu \delta_{\mu\mu} = d \tag{8-10}$$

This condition is maintained in the continuation to noninteger dimensions. Such a prescription ensures the consistency of this continuation with algebraic manipulations such as contractions or shifts of integration variables. For instance, it is easy to check that

$$\int \frac{d^dk}{(2\pi)^d} \frac{k_\mu k_\nu \delta^{\mu\nu} - m_2^2}{[(p-k)^2 - m_1^2]^n (k^2 - m_2^2)^p} = \int \frac{d^dk}{(2\pi)^d} \frac{1}{[(p-k)^2 - m_1^2]^n (k^2 - m_2^2)^{p-1}}$$

Spinors require more care. First, in the original four-dimensional Feynman integrand, we distinguish γ matrices of fermion loops from those pertaining to fermion lines connected with external lines. The latter are reduced by means of projection operators in four dimensions. We therefore have to consider for each diagram a collection of form factors involving only γ matrices pertaining to loops. They are supposed to satisfy the rules

$$\{\gamma_\mu, \gamma_\nu\} = -2\delta_{\mu\nu} \tag{8-11a}$$

(remember that we have performed a Wick rotation and hence our γ matrices are now antihermitian)

$$\text{tr (odd number of } \gamma \text{ matrices)} = 0 \tag{8-11b}$$

$$\text{tr } I = f(d) \tag{8-11c}$$

where $f(d)$ is an arbitrary smooth function satisfying $f(4) = 4$, for instance, $f(d) = 4$ or $f(d) = d$. The form of f is irrelevant for the procedure of regularization. (This would not be irrelevant if we really wanted to define the theory in d dimensions.) From the rules (8-11), we may reconstruct the whole set of identities for contractions and traces of products of γ matrices that we listed in the Appendix for $d = 4$ in the Minkowskian case. For instance,

$$\text{tr } \gamma_\mu\gamma_\nu = -f(d)\delta_{\mu\nu}$$

$$\gamma_\mu\gamma_\mu = -dI$$

$$\gamma_\mu\gamma_\nu\gamma_\mu = (d-2)\gamma_\nu$$

We have not defined the d-dimensional analog of γ_5. This is because the usual definition of γ_5 appeals to the existence of $\varepsilon_{\mu\nu\rho\sigma}$, the completely antisymmetric tensor which is specific to $d = 4$. We conclude that we cannot continue fermion loops containing an odd number of γ_5 matrices. This innocent-looking limitation is the manifestation, in the framework of this dimensional regularization, of a serious problem, namely, the possible appearance of chiral anomalies (Chaps. 11 and 12).

This set of prescriptions may look like a rule of thumb to a skeptical reader. Its consistency, though quite likely and checked in everyday computations, has to the best of our knowledge never been completely proved. Especially embarrassing is the case of massless theories. For instance, we encounter integrals of the form

$$\int \frac{d^dk}{(2\pi)^d} \frac{k_{\mu_1} \cdots k_{\mu_p}}{(k^2)^n}$$

that do not depend on any scale. The analytic continuation of such an integral is ill defined since there is no dimension d where it is meaningful. It is either infrared or ultraviolet divergent according to whether $d + p - 2n \leq 0$ or ≥ 0. We shall disregard these problems and adhere, whenever necessary, to the prescription that such massless tadpole integrals vanish in the dimensional regularization.

In computations with this regularization, we must remember that in d dimensions some coupling constants may acquire a dimension. For instance, the electric charge, the coefficient of

$$\int d^dx\, A_\mu(x)\,\bar{\psi}(x)\gamma^\mu\psi(x)$$

in the dimensionless action ($\hbar = 1$) has a dimension in a mass scale

$$[e] = d - (d-1) - \frac{d-2}{2} = \frac{4-d}{2}$$

since $[\psi] = (d-1)/2$ and $[A] = (d-2)/2$. We therefore replace e by $\mu^{(4-d)/2}e'$, where e' is dimensionless and μ is either one of the masses of the problem or an arbitrary energy scale if all particles are massless. When expanding the result near $d = 4$, we will therefore generate logarithms of this scale.

Whether in the neighborhood of $d = 4$ we also expand the factors $(4\pi)^{-Ld/2}$ in Eq. (8-6) is a matter of taste, as is the definition of $f(d)$ in (8-11c). The important point is that in computing counterterms in a gauge-invariant way, or in comparing diagrams to check Ward identities, we always consider classes of diagrams with the same total number of loops L, and hence the same power of 4π, and the same number of fermion loops, and hence the same power of $f(d)$. A change of prescription will not affect the validity of Ward identities and will only modify counterterms by a finite amount.

As an illustration consider the vacuum polarization in scalar electrodynamics. Feynman rules for this theory have been given in Chap. 6. After a Wick rotation (for an euclidean external momentum p) the two diagrams of Fig. 8-2 give the contributions

$$\Gamma^{(a)}_{\mu\nu} = -2e^2 \int \frac{d^dk}{(2\pi)^d} \frac{\delta_{\mu\nu}}{k^2 + m^2} = -\frac{2e^2\delta_{\mu\nu}}{(4\pi)^{d/2}}\Gamma\left(1 - \frac{d}{2}\right)(m^2)^{d/2-1}$$

and

$$\Gamma^{(b)}_{\mu\nu} = e^2 \int \frac{d^dk}{(2\pi)^d} \frac{(2k+p)_\mu(2k+p)_\nu}{[(p+k)^2 + m^2](k^2 + m^2)}$$

$$= \frac{e^2}{(4\pi)^{d/2}}\left\{2\delta_{\mu\nu}\Gamma\left(1 - \frac{d}{2}\right)\int_0^1 d\alpha\,[\alpha(1-\alpha)p^2 + m^2]^{d/2-1}\right.$$

$$\left. + p_\mu p_\nu \Gamma\left(2 - \frac{d}{2}\right)\int_0^1 d\alpha\,(1-2\alpha)^2[\alpha(1-\alpha)p^2 + m^2]^{d/2-2}\right\}$$

According to Eq. (8-8), let us isolate the pole at $d = 4$ in the sum $\Gamma^{(a)} + \Gamma^{(b)}$:

$$\Gamma^{(a)}_{\mu\nu} + \Gamma^{(b)}_{\mu\nu} = \frac{e^2}{(4\pi)^2}\frac{2}{4-d}\frac{1}{3}(p_\mu p_\nu - p^2\delta_{\mu\nu}) + \text{regular terms} \tag{8-12}$$

We verify that (a) the divergent terms proportional to m^2 have cancelled in the sum, and (b) the tensor structure of the divergence, and hence of the counterterm, is transverse to p. Both results are in agreement with what was expected from gauge invariance. Computation of the finite part is left to the reader.

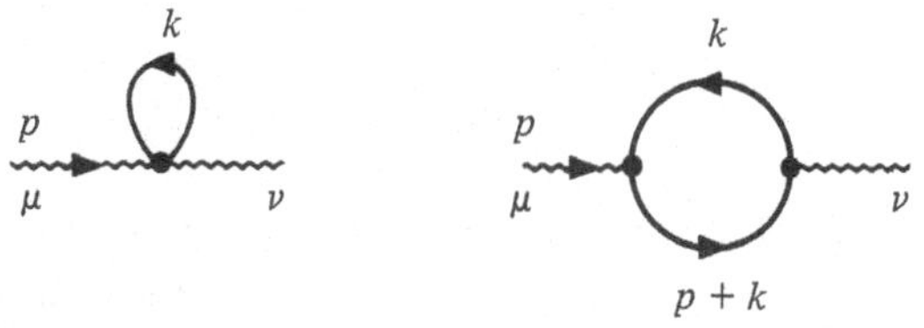

Figure 8-2 One-loop vacuum polarization in scalar electrodynamics.

Other regularization schemes may be invented. The important point is that the final renormalized result does not depend on the choice of regularization. Strictly speaking, finding a regularization and showing that it makes all diagrams finite is not sufficient. We should also prove that the structure of divergences is such that they may be removed by acceptable counterterms, i.e., by local and hermitian polynomials in the fields. Divergences of the form $\log p^2 \log \Lambda^2$, for instance, or with a complex coefficient would be catastrophes. At the very end we shall, however, proceed in a different way. We shall prove that sufficiently subtracted Feynman integrands lead to a finite theory. Since these subtractions do correspond to the introduction of acceptable counterterms, the result will follow a posteriori. In other words, no regularization will be involved in the forthcoming proof of finiteness. However, its outcome and meaning would be obscured without recourse to an implicit regularization, which is therefore a convenient and very common tool.

8-1-3 Power Counting

We have already used the concept of superficial degree of divergence whenever we have based arguments about the convergence of Feynman integrals on dimensional considerations. Here we develop this concept in a systematic way.

In this section and the following, we shall only face the problem of ultraviolet divergences, and postpone to a further study the possible infrared troubles arising from the masslessness of some particles. To be specific, we assume for the time being that all fields are massive.

A naive way to estimate whether a given Feynman diagram G is convergent is to dilate simultaneously its internal momenta by a common factor λ, $k_l \to \lambda k_l$ and to look for its behavior $I_G \sim \lambda^\omega$ as λ goes to infinity. In the parametric representation, this amounts to investigating the integrand when all the α tend to zero at the same rate $\alpha_l \to \lambda^{-2}\alpha_l$. We expect, and will soon prove, that if the overall power of λ, called the superficial degree of divergence and denoted ω, is nonnegative, $\omega \geq 0$, the integral is generally divergent. If it is negative, some subintegrations may still be divergent, and the integral is said to be superficially convergent.

We consider a theory involving spin 0 or 1 boson fields and spin $\frac{1}{2}$ fermion fields. The fermion propagators behave for large momentum λk as λ^{-1} and the boson ones as λ^{-2}. Here we assume that massive vector fields have a propagator in the Stueckelberg gauge [Eq. (3-147)]. We shall reexamine in Chap. 12 what happens to massive fields coupled to nonconserved currents. The case of higher-spin fields will not be considered here.

We use the notations of Chap. 6. A vertex v of the diagram G carries a power λ^{δ_v} if the corresponding term in the interaction lagrangian involves field derivatives. Each integration d^4q over the loop momenta contributes a power λ^4. If L denotes the number of independent loops, I_B and I_F the number of internal boson and fermion lines respectively, and V the total number of vertices, the superficial degree of divergence ω of G reads

$$\omega(G) = 4L + \left(\sum_{\text{vertices}} \delta_v \right) - I_F - 2I_B \tag{8-13a}$$

or

$$\omega(G) - 4 = 3I_F + 2I_B + \sum_{\text{vertices}} (\delta_v - 4) \tag{8-13b}$$

owing the relation (6-69):

$$L = I_B + I_F + 1 - V \tag{8-14}$$

The quantity δ_v counts the number of derivatives acting on the fields at vertex v contracted to give internal propagators. If f_v (respectively b_v) denotes the number of internal fermion (respectively boson) lines incident to the vertex v, we obviously have

$$I_F = \tfrac{1}{2} \sum_v f_v \qquad \text{and} \qquad I_B = \tfrac{1}{2} \sum_v b_v \tag{8-15}$$

since every internal line is counted twice in the sum over the vertices. Thus Eq. (8-13b) may be rewritten

$$\omega(G) - 4 = \sum_v (\hat{\omega}_v - 4) \tag{8-13c}$$

where

$$\hat{\omega}_v \equiv \delta_v + \tfrac{3}{2} f_v + b_v \tag{8-16}$$

The interpretation of this index attached to the vertex v is clear if we remember the dimensional considerations of Sec. 6-2-1. A spin $\frac{1}{2}$ fermion field is given the dimension $\frac{3}{2}$ in mass scale and a boson field the dimension 1. Therefore ω_v is the contribution to the dimension of the interaction monomial of the f_v internal fermions, b_v internal bosons, and δ_v derivatives on internal fields. Alternatively, if ω_v is the dimension of the monomial of $\mathscr{L}_{\text{int}}$ attached to the vertex v, including external and internal fields, that is,

$$\omega_v = \text{total number of boson fields} + \tfrac{3}{2} \text{ times the total number of fermion fields} + \text{total number of field derivatives} \tag{8-17}$$

and if E_F and E_B stand for the number of external fermion and boson lines of the diagram, it is clear from (8-13c) that

$$\omega(G) - 4 = \sum_{\text{vertices}} (\omega_v - 4) - \tfrac{3}{2} E_F - E_B - \delta \tag{8-18}$$

where δ is the total power of external momenta factorized from the Feynman integral. Of course, the dimension ω_v of the vertex does not include the contribution of the dimensioned coupling constant pertaining to this vertex, g_v. As already noticed in Chap. 6,

$$\omega_v + [g_v] = 4 \tag{8-19}$$

It is not surprising to see that the convergence of Feynman integrals has to do with the dimension of the coupling constants. Indeed, if $\omega_v < 4$ for all vertices, the coupling constant g_v has a positive dimension. Going to higher and higher orders in perturbation theory yields more and more powers of g and forces the Feynman integrand to vanish faster and faster at large momenta in order to maintain the total dimension fixed. Conversely, if all $\omega_v > 4$, we expect the integrals to be more and more divergent. All these concepts apply to any subdiagram as well. In general, a diagram with $\omega(G) \geq 0$ is called superficially divergent.

We are now led to a classification of field theories into three classes:

1. Nonrenormalizable theories are those containing at least an interaction monomial of degree $\omega_v > 4$. For a given Green function Eq. (8-18) shows that the superficial degree of divergence grows with the number of vertices, i.e., with the order of perturbation theory. Any function becomes divergent at a high enough order, as will be demonstrated below.
2. Renormalizable theories are the most interesting. All their interaction monomials have $\omega_v \leq 4$, and at least one of them has $\omega_v = 4$. If all monomials have $\omega_v = 4$, we see from (8-18) that all diagrams contributing to a given function have the same degree of divergence. Only a finite number of Green's functions gives rise to overall divergences.
3. Super-renormalizable theories have only vertices with $\omega_v < 4$. The degree of divergence decreases with the order of perturbation. Such theories have only a finite number of divergent diagrams.

The word "nonrenormalizable" may be misleading. It does not mean that such theories cannot be made finite but rather that the proliferation of their divergences, and hence of counterterms, make them unrealistic in the framework of perturbation theory. After renormalization they will depend on an infinite set of arbitrary parameters barring any deeper principle allowing to relate them. We shall not consider them any more. On the other hand, super-renormalizable theories form a too restricted class and are often pathological.

From the rule (8-17), it is easy to list by simple inspection all the possible renormalizable or super-renormalizable theories. We have to construct all possible interaction monomials that are Lorentz scalars, hermitian, and have a degree ω_v less or equal to 4, from derivatives, scalar fields φ, Dirac fields ψ, and vector fields A_μ (supposed to be endowed with a Stueckelberg propagator). The monomials $\overline{\psi}\psi\varphi$ or $\overline{\psi}\gamma_5\psi\varphi$, φ^4 or $(A^2)^2$, $\overline{\psi}A\!\!\!/\psi$ (or perhaps $\overline{\psi}A\!\!\!/\gamma_5\psi$), $\varphi^\dagger\partial_\mu\varphi A^\mu$, $\varphi^\dagger\varphi A^2$, as well as monomials usually appearing in the kinetic lagrangian, $\overline{\psi}\partial\!\!\!/\psi$, $(\partial\varphi)^2$, $(\partial_\mu A_\nu)^2$, $(\partial_\mu A^\mu)^2$, have $\omega_v = 4$ and exhaust the list of renormalizable monomials, up to the introduction of several species of each field and possible internal symmetries. The terms φ^3, $\overline{\psi}\psi$, $\overline{\psi}\gamma_5\psi$, φ^2, A^2 have $\omega_v = 3$ or 2, and thus lead to super-renormalizable lagrangians. Among nonrenormalizable theories, let us quote the pseudovector coupling $\overline{\psi}\gamma_\rho\gamma_5\psi\partial^\rho\varphi$, the Fermi coupling $\overline{\psi}\gamma_\rho(1-\gamma_5)\psi\overline{\psi}\gamma^\rho(1-\gamma_5)\psi$, or higher-degree monomials in φ: φ^5, φ^6, etc.

The previous analysis and list have been presented for a four-dimensional space-time. For purposes of generalization, or for the needs of statistical mechanics, it may be necessary to extend the analysis to different dimensions. The reader will find no difficulty in deriving the analog of Eqs. (8-13) to (8-18) or in determining in which dimension the Fermi theory or the scalar theories φ^3, φ^6, $(\partial_\mu \varphi)^2 P(\varphi)$, where P is an arbitrary polynomial, are renormalizable.

8-1-4 Convergence Theorem

Let us analyze more closely the relation between the convergence of Feynman integrals and power counting. We first note that as long as the propagators contain an imaginary part $i\varepsilon$ it is equivalent to study the convergence in minkowskian or in euclidean space; only the latter case will be considered here.

The limit $\varepsilon \to +0$ has been studied by Bogoliubov and Parasiuk and by Hepp. They have shown the equivalence between the absolute convergence in the euclidean region defining an analytic function of the external momenta and the convergence of the corresponding Feynman integral toward a tempered (i.e., polynomially bounded) distribution in Minkowski space in the limit $\varepsilon \to +0$.

We use here a restrictive definition of a subdiagram g of a diagram G as a subset of vertices of G and of all the internal lines joining them in G. To each connected proper (i.e., one-particle irreducible) diagram, we associate the family $\mathscr{F}$ of all its connected proper subdiagrams. Of course, $\mathscr{F}$ contains G itself.

Theorem If $\omega(g) < 0$ for all $g \in \mathscr{F}$, then the Feynman integral corresponding to G is absolutely convergent (in the euclidean region).

To simplify the expressions, we only present the proof in the case of a scalar theory without derivative couplings, using the parametric representation. In the euclidean region, it reads

$$I_G(P) = \int_0^\infty d\alpha_1 \cdots d\alpha_I \, \frac{\exp - \left[\sum \alpha_l m_l^2 + Q(P, \alpha)\right]}{(4\pi)^{2L} \mathscr{P}(\alpha)^2} \tag{8-20}$$

We assume that $m_l^2 > 0$ for all l and recall that Q is a quadratic, positive definite form in the external momenta and a homogeneous rational fraction of degree 1 in the α. Hence, only the convergence at $\alpha = 0$ is questionable. The exponent being bounded at the origin plays no role. The polynomial $\mathscr{P}$ is a sum of monomials of degree L [compare with (6-86)]:

$$\mathscr{P}(\alpha) = \sum_{\text{trees } \mathscr{T}} \prod_{l \notin \mathscr{T}} \alpha_l \tag{8-21}$$

Following Hepp, we divide the integration domain into sectors

$$0 \le \alpha_{\pi_1} \le \alpha_{\pi_2} \le \cdots \le \alpha_{\pi_I}$$

where π is a permutation of $(1, 2, \ldots, I)$. We shall prove the convergence of the integral sector by sector. To each sector corresponds a family of nested subsets γ_l of lines of G (not necessarily subdiagrams):

$$\gamma_1 \subset \gamma_2 \subset \cdots \subset \gamma_I \equiv G$$

where γ_l contains the lines pertaining to $(\alpha_{\pi_1}, \ldots, \alpha_{\pi_l})$. For notational simplicity we present the reasoning for the first sector corresponding to the identical permutation. In this sector, we perform a change of variables

$$\begin{array}{llll} \alpha_1 = \beta_1^2 \beta_2^2 & \cdots & \beta_I^2 \\ \alpha_2 = \quad \beta_2^2 & \cdots & \beta_I^2 \\ \cdots\cdots\cdots\cdots\cdots\cdots & & \\ \alpha_{I-1} = & \beta_{I-1}^2 & \beta_I^2 \\ \alpha_I = & & \beta_I^2 \end{array} \tag{8-22}$$

the jacobian of which is

$$\frac{D(\alpha_1, \ldots, \alpha_I)}{D(\beta_1, \ldots, \beta_I)} = 2^I \beta_1 \beta_2^3 \cdots \beta_I^{2I-1}$$

In these β variables the integration domain Δ reads

$$0 \leq \beta_I \leq \infty \qquad \text{and} \qquad 0 \leq \beta_l \leq 1 \qquad \text{for } 1 \leq l \leq I-1$$

We may define the superficial degree of divergence of each γ_l, even though γ_l is not necessarily connected:

$$\omega_l \equiv \omega(\gamma_l) = 2L_l - 4I_l \tag{8-23}$$

where $I_l = l$ is the number of internal lines of γ_l. The number of independent loops L_l reads

$$L_l = I_l + C_l - V_l \tag{8-24}$$

in terms of the number of connected parts C_l and vertices V_l. This generalization of formula (6-69) may be easily proved by induction. Of course, we have $L_1 = 0$, $L_I = L$. As a function of the β, $\mathscr{P}$ is a polynomial, the distinct monomials of which have coefficients equal to unity. Let us prove that it is of the form

$$\mathscr{P} = \beta_1^{2L_1} \beta_2^{2L_2} \cdots \beta_I^{2L_I} [1 + O(\beta)] \tag{8-25}$$

First, when all the α are simultaneously dilated by a factor ρ,

$$\mathscr{P}(\rho\alpha_1, \ldots, \rho\alpha_I) = \rho^L \mathscr{P}(\alpha_1, \ldots, \alpha_I)$$

Let us now investigate the behavior of $\mathscr{P}$ when only the α belonging to γ_l are dilated by a factor ρ:

$$\mathscr{P}(\rho\alpha_1, \ldots, \rho\alpha_l, \alpha_{l+1}, \ldots, \alpha_I) \tag{8-26}$$

Equation (8-21) expresses $\mathscr{P}$ in terms of trees on G. Each tree $\mathscr{T}$ of G projects on γ_l along the union of $C_l' \geq C_l$ connected trees $\mathscr{T}_1, \ldots, \mathscr{T}_{C_l'}$. This will contribute in (8-21) a term behaving under the dilatation (8-26) as a power of ρ equal to the number of lines of γ_l that do not belong to

$$\mathscr{U} = \mathscr{T}_1 \cup \mathscr{T}_2 \cup \cdots \cup \mathscr{T}_{C_l'}$$

Since the set $\mathscr{U}$ satisfies an equation of the type (8-24) with a vanishing number of loops (it is a union of trees) and joins all vertices of γ_l, its total number of lines is

$$I_{\mathscr{U}} = V_l - C_l'$$

Hence the power of ρ in the corresponding monomial of $\mathscr{P}$ is $I_l - I_{\mathscr{U}} = l - V_l + C_l' \geq l - V_l + C_l = L_l$. The lowest power of ρ is reached by the trees $\mathscr{T}$ of G such that $C_l' = C_l$, that is, by these trees that project according to

$$\mathscr{T}\Big|_{\text{proj}, \gamma_l} = \bigcup_1^{C_l} \mathscr{T}_l$$

All these trees $\mathscr{T}$ may be generated by constructing independent connected trees in each connected part of γ_l and completing their union into a tree of G. Therefore, their total contribution to $\mathscr{P}$ factorizes into $\mathscr{P}_{\gamma_l}$, that is, the polynomial $\mathscr{P}$ attached to γ_l by the rule (8-21), times $\mathscr{P}_{G/\gamma_l}$, where the reduced diagram G/γ_l is obtained by contracting all the lines and vertices of each connected part of γ_l into a unique vertex. We thus conclude that

$$\mathscr{P}(\rho\alpha_1, \ldots, \rho\alpha_l, \alpha_{l+1}, \ldots, \alpha_I) = \rho^{L_l}[\mathscr{P}_{\gamma_l}(\alpha_1, \ldots, \alpha_l)\mathscr{P}_{G/\gamma_l}(\alpha_{l+1}, \ldots, \alpha_I) + O(\rho)] \tag{8-27}$$

We now return to the variables β. Obviously β_I^2 is the homogeneity factor of the parameters pertaining to $\gamma_I = G$, β_{I-1}^2 that of $\gamma_{I-1}, \ldots$, etc. Since the γ are nested, Eq. (8-27) may be used repeatedly:

$$\begin{aligned}
&\mathscr{P}(\beta_1^2\beta_2^2\cdots\beta_I^2, \beta_2^2\cdots\beta_I^2, \ldots, \beta_{I-1}^2\beta_I^2, \beta_I^2)\\
&\quad = \beta_I^{2L_I}\mathscr{P}(\beta_1^2\cdots\beta_{I-1}^2, \beta_2^2\cdots\beta_{I-1}^2, \ldots, \beta_{I-1}^2, 1)\\
&\quad = \beta_I^{2L_I}\beta_{I-1}^{2L_{I-1}}[\mathscr{P}_{\gamma_{I-1}}(\beta_1^2\cdots\beta_{I-2}^2, \beta_2^2\cdots\beta_{I-2}^2, \ldots, \beta_{I-2}^2, 1)\mathscr{P}_{G/\gamma_{I-1}}(1) + O(\beta_{I-1}^2)]\\
&\quad = \cdots
\end{aligned} \tag{8-28}$$

with the result (8-25). The coefficient in front of the leading term is 1, since a unique monomial of the initial polynomial $\mathscr{P}$ may have this combination of powers of the β. We return to the integral (8-20) in the sector Δ. Under the assumption of the theorem, namely, that $\omega(g) < 0$ for any proper diagram g, it is easy to show that $\omega(\gamma) < 0$ for any subset γ, proper or not. Therefore the integrand is majorized up to a factor by

$$\frac{\beta_1^1\beta_2^3\cdots\beta_I^{2I-1}}{\beta_1^{4L_1}\beta_2^{4L_2}\cdots\beta_I^{4L_I}} = \prod_{l=1}^{I} \beta_l^{-4L_l+2l-1} = \prod_{l=1}^{I} \beta_l^{-\omega_l-1}$$

and since $\omega_l < 0$, the integral $\int_0 \Pi\, d\beta_l\, \beta_l^{-\omega_l-1}$ is absolutely convergent at the origin. We conclude that the integral (8-20) is absolutely convergent in every sector. QED

For a scalar theory without derivative coupling, the foregoing proof has also shown that as soon as a proper subdiagram has a nonnegative superficial degree of divergence the Feynman integral is divergent. Indeed, there is a divergence in at least one sector and, the integral being positive definite, no cancellation may occur. On the other hand, we have encountered examples in electrodynamics of cancellations between different terms appearing in the numerator of the Feynman integrand in momentum space. For instance, we have shown that the vacuum polarization is only logarithmically rather than quadratically divergent and that light-by-light scattering is convergent.

Later, we shall have to show that after subtraction of all subdiagrams such that $\omega(g) \geq 0$ the integral is absolutely convergent. Notice that the previous considerations on absolute convergence justify a posteriori the manipulations—interchange of orders of integrations or changes of integration variables—performed in the derivation of (8-20). As a bonus we also find how many terms have to be subtracted to the conventional propagator (8-2) in order to regularize the theory. For instance, in the φ^4 theory, the replacement $(k^2 - m^2)^{-1} \to (k^2 - m^2)^{-1} - (k^2 - \Lambda^2)^{-1}$ makes every diagram finite but the one-loop tadpole, since the superficial degree of divergence now reads

$$\omega(G) = 4L - 4I = 4(1 - V)$$

which is negative for $V > 1$. The one-loop tadpole may be either regularized separately or discarded by the use of a Wick ordering prescription. In this way, we end up with a finite regularized theory. Such an analysis may be carried out in the case of quantum electrodynamics (see Sec. 8-4-2).

A useful corollary of the previous theorem is the following. If a diagram G has no superficially divergent subdiagram, $\omega(g) < 0$ for all subdiagrams $g \neq G$, but is itself superficially divergent, $\omega(G) \geq 0$, then the divergent part of its amplitude is a polynomial of degree less or equal to $\omega(G)$ in the external momenta P and in the internal masses. Indeed, since $\omega(G)$ measures the degree of homogeneity of $I_G(P)$ in the momenta and masses, the $[\omega(G) + 1]$-th derivatives with respect to the P_i and the m_l have a degree minus one, and hence are superficially convergent. By virtue of the theorem, the derivatives $[\partial^{\omega+1}/(\partial P)^{\omega+1}]I_G(P)$ [or $(\partial^{\omega+1}/\partial m^{\omega+1})I_G$, or mixed derivatives of this order] are finite.

That our proof of the theorem is still valid is an easy exercise left to the reader.

Finally, we get

$$I_{G,\mathrm{reg}}(P, m, \Lambda) = I_{\mathrm{fin}}(P) + D(P, m, \Lambda) \tag{8-29}$$

where $I_{\mathrm{fin}}(P)$ is finite as the cutoff is removed and D is a polynomial in the P and the m of degree less or equal to $\omega(G)$.

8-2 RENORMALIZATION

8-2-1 Normalization Conditions and Structure of the Counterterms

We shall now come to grips with the renormalization operation itself. Let us once again repeat the general ideas and emphasize the outcome of this operation.

Our aim is to express the proper Green functions in terms of renormalized Feynman integrals associated with the initial diagrams. This may be achieved by means of three equivalent procedures. In the first approach, the one presented in Chap. 7, we add to the initial lagrangian a formal series (in $\hbar$) of counterterms. This in turn amounts to an order-by-order redefinition of the parameters of the theory: the "bare" parameters appearing in the lagrangian are implicit functions of the renormalized ones. The former are unobservable and divergent as the regularization is removed, while the latter are the real finite parameters of the theory, mass, coupling constants, etc. Finally, we will see in the next subsections that these two procedures are equivalent to an algorithm of subtraction of the integrand. This operation, due to Bogoliubov, has the merit of providing, diagram by diagram, a finite result without any intermediate recourse to a regularization.

We will try to juggle as skilfully as possible with these three equivalent approaches and to use the most appropriate one for each problem. For instance, we shall first discuss the recursive construction of the counterterms and then use the relation between bare and renormalized theories to stress the multiplicative character of the renormalization process. Finally, Bogoliubov's subtraction operation will offer a convenient framework for an (heuristic) proof of the fundamental convergence theorem.

The construction of counterterms proceeds by induction. We assume that the theory has been made finite up to a given order $\hbar^{L-1}$ (i.e., a number $L-1$ of loops), through the introduction of judicious counterterms. According to power counting [Eqs. (8-13) to (8-18)], to the next order $\hbar^L$, only a finite number of proper functions have a nonnegative superficial degree of divergence. Except for possible cancellations arising from the existence of symmetries, we associate to each of them a local monomial in the fields and their derivatives. It is, of course, a Lorentz scalar; its structure reflects the nature of the Green function since it must contribute itself to the process under study and its extra contribution to this function to that order must cancel its divergence. There remains, of course, a finite arbitrariness. After introduction of a regularization, the coefficients of the counterterms are completely determined by normalization conditions, imposed on the superficially divergent Green functions. If these normalization conditions are satisfied to lowest order (the tree approximation as implied by the initial Lagrangian), demanding that the Green functions satisfy them order by order fixes unambiguously not only the infinite but also the finite part of the counterterms.

For instance, in the renormalizable φ^4 theory of a scalar field of mass m, the two-point function $\Gamma^{(2)}$ is quadratically divergent $[\omega(\Gamma^{(2)}) = 2]$. We demand that the renormalized $\Gamma_R^{(2)}$ satisfies

$$\Gamma_R^{(2)}(m^2) = 0 \qquad \frac{d}{dp^2}\Gamma_R^{(2)}(m^2) = 1 \tag{8-30}$$

This is a natural condition for the physical mass, since the interpretation of the theory in terms of particles requires that the complete propagator $G_R^{(2)}(p^2) = i[\Gamma_R^{(2)}(p^2)]^{-1}$ has a pole of residue i at $p^2 = m^2$. To order L, the regularized function $\Gamma_{\text{reg}}^{(2)}$ already renormalized up to order $L-1$ reads

$$\Gamma_{\text{reg}}^{(2)}(p^2) = \Gamma_{\text{fin}}^{(2)}(p^2, m^2, \Lambda^2) + p^2\Delta_1\Gamma^{(2)}(m^2, \Lambda^2) + \Delta_2\Gamma^{(2)}(m^2, \Lambda^2) \tag{8-31}$$

where $\Gamma_{\text{fin}}^{(2)}(p^2, m^2, \Lambda^2)$ is finite as $\Lambda^2 \to \infty$, whereas $\Delta_1\Gamma^{(2)}(m^2, \Lambda^2)$ behaves at worst as a power of $\ln(\Lambda^2/m^2)$, and $\Delta_2\Gamma^{(2)}$ as Λ^2 [times $\ln^p(\Lambda^2/m^2)$]. The counterterm to order L reads

$$\Delta\mathscr{L}^{[L]} = a\,\frac{(\partial\varphi)^2}{2} - bm^2\,\frac{\varphi^2}{2} \tag{8-32}$$

hence giving to $\Gamma^{(2)}$ an extra contribution of the form

$$\Delta\Gamma^{(2)} = ap^2 - bm^2 \tag{8-33}$$

Since the conditions (8-30) are already satisfied to lowest order, they yield to order L:

$$\left\{\begin{aligned} &\Gamma_R^{(2)[L]}(p^2 = m^2) = \Gamma_{\text{reg}}^{(2)}(p^2 = m^2) + (a-b)m^2 = 0 \\ &\frac{\partial}{\partial p^2}\Gamma_R^{(2)[L]}(p^2)\big|_{p^2=m^2} = \frac{\partial\Gamma_{\text{reg}}^{(2)}}{\partial p^2}\bigg|_{p^2=m^2} + a = 0 \end{aligned}\right. \tag{8-34}$$

and therefore determine a and b. A similar normalization condition is imposed on the other superficially divergent function, namely, the four-point function of the φ^4 theory. A physically sensible and symmetric condition consists in requiring that this function takes the value $-\lambda$ (the renormalized coupling constant) at the on-shell (but unphysical) point S_m:

$$S_m \begin{cases} p_1^2 = p_2^2 = p_3^2 = p_4^2 = m^2 & \Gamma_R^{(4)}|_{S_m} = -\lambda \\ s = t = u = \dfrac{4m^2}{3} \end{cases} \tag{8-35}$$

This is clearly in agreement with the lowest-order value $\Gamma^{(4)[0]} = -\lambda$. Analogous normalization conditions can and must be introduced in any renormalizable theory.

Of course, the foregoing is by no means a proof that divergences may be disposed of by counterterms. In particular, we have not completely proved that $\omega(\Gamma^{(2)}) = 2$ implies that $\Gamma_{\text{reg}}^{(2)}$ has the behavior (8-31). Indeed, we are not really in the conditions of the corollary (8-29), since all subdiagrams are not necessarily superficially convergent, but may have been renormalized by lower-order counterterms. That our reasoning is nevertheless correct will actually follow a posteriori, when we will have shown that this procedure does lead to a finite renormalized theory. Here we only want to recall the logic of the method and stress the necessity of normalization conditions. At this point, it may be worth recalling that the difference between renormalizable and nonrenormalizable theories lies in the number of these conditions. While in the former a finite number of conditions suffice to define the theory in terms of a finite number of renormalized parameters, the renormalization of the latter requires an infinite set of such conditions; ultimately, a nonrenormalizable theory will depend on an infinite number of parameters.

There is a great amount of arbitrariness in the choice of the normalization conditions. The only proviso is that they must be satisfied to lowest order, so as to fix unambiguously the subtractions to higher orders. We shall reexamine this point in Sec. 8-2-5 below. Owing to this arbitrariness, it may be more convenient to use a less physical, intermediate renormalization. When all the fields have a nonvanishing mass, it is safe to choose normalization conditions at the origin in momentum space. In the above example of the φ^4 theory, the conditions will read

$$\begin{aligned} &\Gamma_R^{(2)}(p^2)|_{p^2=0} = -m^2 \qquad \frac{d}{dp^2}\Gamma_R^{(2)}|_{p^2=0} = 1 \\ &\Gamma_R^{(4)}(0,0,0,0) = -\lambda \end{aligned} \tag{8-36}$$

The quantity m^2 as defined by (8-36) is no longer the physical mass, even though it is related to it.

When dealing with particles of nonzero spin, the tensor structure of Green's functions must be taken into account. It may happen that only the form factors of some tensors are divergent (for instance, in the study of the vertex function in the last chapter, F_2 was found finite and F_1 divergent). Only the latter requires subtractions and normalization conditions.

A consequence of the previous recursive construction of the counterterms concerns their structure. In a renormalizable theory, the counterterms satisfy the criterion of renormalizability. This is what we found in spinor electrodynamics to the one-loop approximation, where the only counterterms had the form

$$:(\partial_\mu A_\nu - \partial_\nu A_\mu)^2: \qquad :\overline{\psi}\psi: \qquad \frac{i}{2}:\overline{\psi}\overleftrightarrow{\partial\!\!\!/}\psi: \qquad \overline{\psi}A\!\!\!/\psi$$

Similarly, in the φ^4 theory, the counterterms are $:(\partial\varphi)^2:$, $:\varphi^2:$, $:\varphi^4:$ of degree less or equal to four. In scalar electrodynamics, the situation is slightly different, since starting from the monomials generated by minimal coupling we find counterterms of the same structure, plus a new term of the type $:(\varphi^\dagger\varphi)^2:$. Therefore, we end up with a mixture of electrodynamics and φ^4 self-coupling, but the theory remains renormalizable. This also means that scalar electrodynamics depends on a supplementary unexpected parameter, namely, the value of the four-point function at a given point.

More generally, if a proper diagram G of a renormalizable theory is superficially divergent,

$$0 \leq \omega(G) \leq 4 - \tfrac{3}{2}E_F - E_B - \delta$$

[we use Eq. (8-18), where $\omega_v \leq 4$], the corresponding counterterm has E_F fermion fields, E_B boson fields, and a number δ of derivatives. Therefore, its dimension, in the sense of Eq. (8-17) is

$$\omega_{v_c} = \tfrac{3}{2}E_F + E_B + \delta \leq 4 - \omega(G) \tag{8-37}$$

The counterterm thus generated has a dimension less or equal to four; hence it is also renormalizable.

Whenever the counterterms have the same structure as monomials of the initial lagrangian, they may be considered as redefining the parameters of the theory. The quantities appearing in the lagrangian, resulting from the addition of counterterms to the initial monomials, will be referred to as bare parameters. The bare parameters are determined order by order in perturbation theory as functions of the renormalized quantities, so that the renormalization conditions are satisfied. We discussed this construction in the previous chapter for electrodynamics (see also Sec. 8-4 below). In the case of the φ^4 theory, we write the lagrangian plus its counterterms as

$$\begin{aligned}\mathscr{L} + \Delta\mathscr{L} = \mathscr{L}_R &= \frac{1}{2}(\partial_\mu\varphi)^2 - \frac{m^2}{2}\varphi^2 - \frac{\lambda}{4!}\varphi^4 \\ &\quad + \frac{1}{2}(Z-1)(\partial_\mu\varphi)^2 - \frac{1}{2}(Zm_0^2 - m^2)\varphi^2 - \frac{1}{4!}(Z^2\lambda_0 - \lambda)\varphi^4 \\ &= \frac{1}{2}Z(\partial_\mu\varphi)^2 - \frac{Zm_0^2}{2}\varphi^2 - \frac{Z^2\lambda_0}{4!}\varphi^4\end{aligned}$$

$$= \frac{1}{2}(\partial_\mu \varphi_0)^2 - \frac{m_0^2}{2}\varphi_0^2 - \frac{\lambda_0}{4!}\varphi_0^4 \tag{8-38}$$

where $$\varphi_0 \equiv Z^{1/2}\varphi$$

This is sometimes referred to as the renormalized lagrangian, an unfortunate denomination, since its coefficients are infinite. The lagrangian $\mathscr{L}_R$ has the same expression as the initial one, up to the replacements $\varphi \to \varphi_0$, $\lambda \to \lambda_0$, $m \to m_0$. We recall that in perturbation theory

$$\begin{aligned} \lambda_0(\lambda, \Lambda/m) &= \lambda[1 + O(\lambda\hbar)] \\ \frac{m_0(\lambda, \Lambda/m)}{m} &= 1 + O(\lambda\hbar) \\ Z(\lambda, \Lambda/m) &= 1 + O(\lambda^2\hbar^2) \end{aligned} \tag{8-39}$$

The last relation is particular to the φ^4 interaction. A priori we would expect $Z = 1 + O(\hbar)$.

The fact that renormalization boils down to a redefinition of the parameters implies that unrenormalized (or bare) and renormalized Green functions are related through

$$\begin{aligned} G_R^{(n)}(p_1, \ldots, p_n, m, \lambda) &= Z^{-n/2} G_{\mathrm{reg}}^{(n)}(p_1, \ldots, p_n, m_0, \lambda_0, \Lambda) \\ \Gamma_R^{(n)}(p_1, \ldots, p_n, m, \lambda) &= Z^{n/2} \Gamma_{\mathrm{reg}}^{(n)}(p_1, \ldots, p_n, m_0, \lambda_0, \Lambda) \end{aligned} \tag{8-40}$$

This relation holds for connected and proper functions respectively. Indeed, the connected renormalized functions are computed from the lagrangian $\mathscr{L}_R = \mathscr{L} + \Delta\mathscr{L}$ by adding a source term $j\varphi$, while the unrenormalized ones may be computed from $\mathscr{L}_R$ (in the presence of a regularization) by coupling a source to φ_0, that is, $j\varphi_0$. Taking n derivatives with respect to j leads to the announced relation. The limit $\Lambda \to \infty$ is understood on the right-hand side of Eqs. (8-40). Finally, proper functions are obtained after elimination of one-particle reducible diagrams, an operation which commutes with renormalization, and amputation by the renormalized or unrenormalized propagator respectively—whence the change in the power of Z.

8-2-2 Bogoliubov's Recursion Formula

In the preceding subsection, counterterms have been associated to proper functions, i.e., to a sum of Feynman diagrams computed to a given order. The existence of normalization conditions enables us to associate a counterterm to each superficially divergent diagram, computed using the requirement that it satisfies the normalization condition. A subtlety is involved when owing to cancellations caused by symmetries (Bose or Fermi symmetry, internal symmetry, etc.) individual diagrams may be more divergent than their sum to a given order. In that case, it is safer to consider only sets of diagrams—gauge-invariant sets in quantum electrodynamics—that exhibit these cancellations. We shall avoid this unessential complication in the sequel by restricting ourselves to a scalar nonderivative theory.

Let us now investigate the effect of lower-order counterterms on a given proper Feynman diagram G. Indeed, G may contain superficially divergent proper subdiagrams $\gamma : \omega(\gamma) \geq 0$. To each of these subdiagrams we may associate a counterterm of order $\hbar^{L_\gamma}$ (L_γ being the number of loops of γ). Let $\mathscr{I}_G$ denote the integrand of the Feynman diagram G in momentum space, $\overline{\mathscr{R}}_G$ the one when all lower-order counterterms are taken into account, and $\mathscr{R}_G$ the renormalized integrand that leads to a finite integral. If G is superficially convergent,

$$\mathscr{R}_G = \overline{\mathscr{R}}_G \qquad \text{for } \omega(G) < 0 \tag{8-41}$$

However, if $\omega(G) \geq 0$, $\mathscr{R}_G$ differs from $\overline{\mathscr{R}}_G$ by the contribution of the counterterm attached to G itself. Alternatively, $\overline{\mathscr{R}}_G$ has to be subtracted to yield $\mathscr{R}_G$ that leads to a finite integral satisfying the normalization condition. After integration over the internal momenta $\int (\mathscr{R}_G - \overline{\mathscr{R}}_G)$ must be a polynomial of degree less or equal to $\omega(G)$ in the independent external momenta of G. In the framework of the intermediate renormalization (subtraction at zero momentum), $\overline{\mathscr{R}}_G - \mathscr{R}_G$ will be nothing but the Taylor expansion $T_G \overline{\mathscr{R}}_G$ of $\overline{\mathscr{R}}_G$ in the external momenta at the origin, up to the order $\omega(G)$ included. We shall write

$$\mathscr{R}_G = (1 - T_G)\overline{\mathscr{R}}_G \qquad \text{for } \omega(G) \geq 0 \tag{8-42}$$

We are left with the problem of relating $\overline{\mathscr{R}}_G$ to the initial integrand $\mathscr{I}_G$. The difference between $\overline{\mathscr{R}}_G$ and $\mathscr{I}_G$ comes from the counterterms associated with the renormalization parts γ of G. Following Zimmermann, this refers to proper superficially divergent subdiagrams, hence containing all the lines of G that join two of their vertices, according to our convention of Sec. 8-1-4. The contribution of the counterterm associated with γ is $-T_\gamma \overline{\mathscr{R}}_\gamma$, and when inserted in the diagram G in place of γ this gives

$$\mathscr{I}_{G/\gamma}(-T_\gamma \overline{\mathscr{R}}_\gamma) \tag{8-43}$$

Here $T_\gamma \overline{\mathscr{R}}_\gamma$ denotes the Taylor expansion of the modified integrand $\overline{\mathscr{R}}_\gamma$, in the independent external momenta of γ, up to the order $\omega(\gamma)$ (included).

This Taylor expansion is not as well defined as in the case of G itself, since the distinction between internal and external independent momenta requires more care. We shall admit here that a definition of these external variables is always possible and refer the scrupulous reader to the references quoted in the notes.

On the other hand, $\mathscr{I}_{G/\gamma}$ stands for the contribution to the integrand of the lines and vertices external to γ in the initial diagram. It contains in particular the propagators pertaining to the lines that join γ to the rest of G. (See Fig. 8-3 for an illustration.) Equation (8-43) contains the contribution of the counterterm relative to γ (and possibly of those contributing to $\overline{\mathscr{R}}_\gamma$). It is now easy to write the contribution of two counterterms relative to two disjoint renormalization

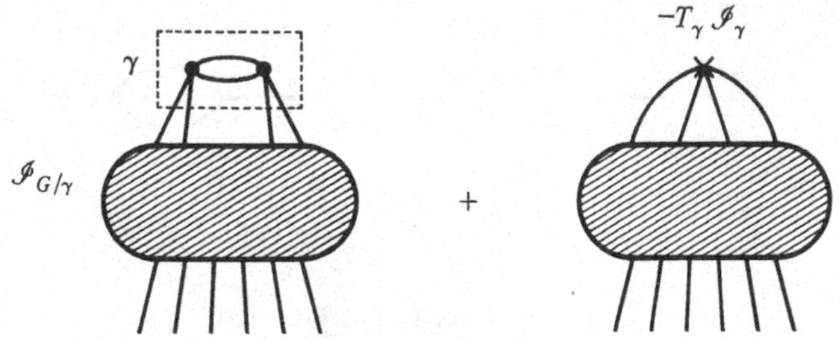

Figure 8-3 Subtracting an internal renormalization part γ in the integrand $\mathscr{I}_G$.

parts γ_1 and γ_2, where disjoint means that they have no common line or vertex: $\gamma_1 \cap \gamma_2 = \varnothing$. This contribution reads

$$\mathscr{I}_{G/\{\gamma_1,\gamma_2\}}(-T_{\gamma_1}\overline{\mathscr{R}}_{\gamma_1})(-T_{\gamma_2}\overline{\mathscr{R}}_{\gamma_2}) \tag{8-44}$$

where $\mathscr{I}_{G/\{\gamma_1,\gamma_2\}}$ refers to the contribution to the initial integrand of all lines and vertices of G except for those belonging to γ_1 or γ_2. The reason why we only consider disjoint renormalization parts is that otherwise both could not be replaced by their corresponding counterterm. This enumeration of the possible contributions to $\overline{\mathscr{R}}_G$ may be pursued. The resulting recursion formula, due to Bogoliubov, reads

$$\overline{\mathscr{R}}_G = \mathscr{I}_G + \sum_{\substack{\{\gamma_1,\ldots,\gamma_s\}\\ \gamma_a \cap \gamma_b = \varnothing}} \mathscr{I}_{G/\{\gamma_1,\ldots,\gamma_s\}} \prod_{a=1}^{s} (-T_{\gamma_a}\overline{\mathscr{R}}_{\gamma_a}) \tag{8-45}$$

The first term on the right-hand side is the initial integrand, and the sum runs over all families of disjoint renormalization parts. Equation (8-45) after iteration gives, together with Eqs. (8-41) and (8-42), the expression of the renormalized integrand. If G contains no superficially divergent proper subdiagram, then $\overline{\mathscr{R}}_G = \mathscr{I}_G$. This was the case of all diagrams studied in Chap. 7. For one-loop diagrams, either $\omega(G) < 0$ and $\mathscr{R}_G = \mathscr{I}_G$, or $\omega(G) \geq 0$ and $\mathscr{R}_G = (1 - T_G)\mathscr{I}_G$.

Let us now present simple examples involving more loops. For the sake of simplicity, we consider the theory of a scalar field φ endowed with the interaction $\mathscr{L}_{\text{int}} = -\lambda\varphi^3/3!$ in a six-dimensional space-time. Though such a theory suffers from serious pathologies—its hamiltonian is not bounded from below—it does make sense to develop its perturbative expansion. It is a renormalizable theory (in six dimensions). The power counting of Eq. (8-18) is replaced by $\omega(G) = 6L - 2I = 6 - 2E$. Consider the diagram of Fig. 8-4. The subdiagram γ inside the dotted box is the only renormalization part,

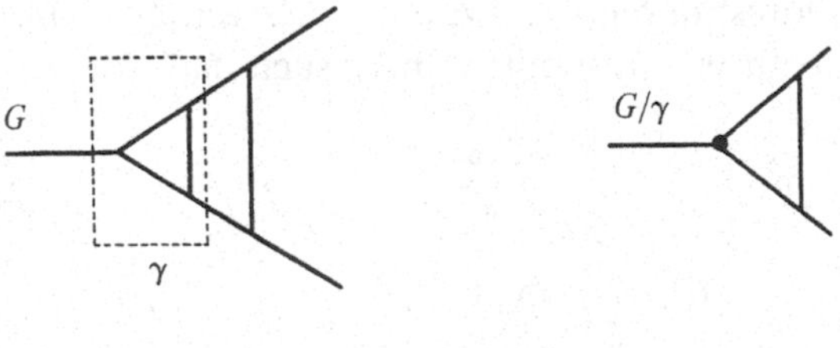

Figure 8-4 Nested divergences.

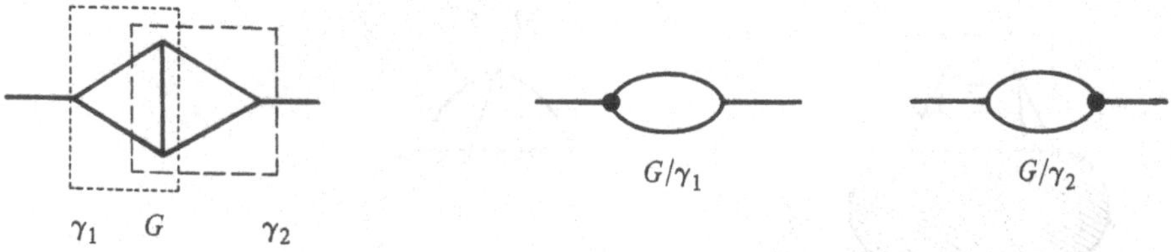

Figure 8-5 Overlapping divergences.

besides G itself. According to (8-45),

$$\overline{\mathscr{R}}_G = \mathscr{I}_G + \mathscr{I}_{G/\gamma}(-T_\gamma \overline{\mathscr{R}}_\gamma)$$
$$\overline{\mathscr{R}}_\gamma = \mathscr{I}_\gamma$$

whence we deduce

$$\overline{\mathscr{R}}_G = \mathscr{I}_G + \mathscr{I}_{G/\gamma}(-T_\gamma \mathscr{I}_\gamma)$$
$$\equiv (1 - T_\gamma)\mathscr{I}_G$$

with a slight abuse of notations, and, following (8-42),

$$\mathscr{R}_G = (1 - T_G)\overline{\mathscr{R}}_G = (1 - T_G)(1 - T_\gamma)\mathscr{I}_G \tag{8-46}$$

We see that to the nested diagrams $\gamma \subset G$ have been associated two factors $(1 - T)$. This would still be true if we were to add more (parallel) rungs to this diagram. It is a good exercise to verify that the renormalized integrand leads to a finite integral.

We turn to the diagram of Fig. 8-5. It is superficially divergent $[\omega(G) = 2]$, as well as its two renormalization parts γ_1 and γ_2 $[\omega(\gamma_1) = \omega(\gamma_2) = 0]$. Here a new phenomenon arises; γ_1 and γ_2 are neither nested nor disjoint but overlapping. They share one line and two vertices, but none is included in the other. Consequently, Eq. (8-45) leads to

$$\overline{\mathscr{R}}_G = \mathscr{I}_G + \mathscr{I}_{G/\gamma_1}(-T_{\gamma_1}\overline{\mathscr{R}}_{\gamma_1}) + \mathscr{I}_{G/\gamma_2}(-T_{\gamma_2}\overline{\mathscr{R}}_{\gamma_2})$$
$$\overline{\mathscr{R}}_{\gamma_1} = \mathscr{I}_{\gamma_1}$$
$$\overline{\mathscr{R}}_{\gamma_2} = \mathscr{I}_{\gamma_2}$$

If we write $T_\gamma \mathscr{I}_G \equiv \mathscr{I}_{G/\gamma} T_\gamma \mathscr{I}_\gamma$ we finally have

$$\begin{aligned} \overline{\mathscr{R}}_G &= (1 - T_{\gamma_1} - T_{\gamma_2})\mathscr{I}_G \\ \mathscr{R}_G &= (1 - T_G)\overline{\mathscr{R}}_G = (1 - T_G)(1 - T_{\gamma_1} - T_{\gamma_2})\mathscr{I}_G \end{aligned} \tag{8-47}$$

We observe that $\mathscr{R}_G$ is not equal to $(1 - T_G)(1 - T_{\gamma_1})(1 - T_{\gamma_2})\mathscr{I}_G$. The supplementary terms, namely, $(1 - T_G)(T_{\gamma_1} T_{\gamma_2})$, just correspond to overlapping subdiagrams.

This diagram of Fig. 8-5 will be analyzed further in the case of four-dimensional quantum electrodynamics in Sec. 8-4-4.

8-2-3 Zimmermann's Explicit Solution

The preceding examples suggest the general solution of the recursion equation (8-45). Following Zimmermann, a forest of renormalization parts will be defined as a family U of proper superficially divergent subdiagrams γ such that

$$\text{if } \gamma_1 \text{ and } \gamma_2 \in U \quad \begin{cases} \text{either } \gamma_1 \subset \gamma_2 \\ \text{or} \quad \gamma_2 \subset \gamma_1 \\ \text{or} \quad \gamma_1 \cap \gamma_2 = \varnothing \end{cases} \tag{8-48}$$

We recall that $\gamma_1 \cap \gamma_2 = \varnothing$ means that these subdiagrams have no common vertex nor line. A forest may be empty. Moreover, we denote by V the forests of G such that, if G is itself superficially divergent, G does not belong to V. Of course, if G is not superficially divergent, the two sets of forests $\{U\}$ and $\{V\}$ are identical.

If a consistent set of internal momenta has been chosen to make the operations T_γ meaningful, two such operations pertaining to disjoint γ_1, γ_2 commute, whereas if $\gamma_1 \subset \gamma_2$, we understand that T_{γ_1} will stand to the right of T_{γ_2} [compare the examples of Eqs. (8-46) and (8-47)]. Under these circumstances, let us show that

$$\begin{aligned} \overline{\mathscr{R}}_G &= \sum_V \prod_{\gamma \in V} (-T_\gamma) \mathscr{I}_G \\ \mathscr{R}_G &= \sum_U \prod_{\gamma \in U} (-T_\gamma) \mathscr{I}_G \end{aligned} \tag{8-49}$$

is the solution of Eq. (8-45). First, we remark that to the empty forest corresponds the term $\mathscr{I}_G$. Second, if G is a renormalization part, we may associate to every V two forests of the U type, $U_1 = V$ and $U_2 = \{G\} \cup V$. Therefore Eq. (8-49) leads in this case to $\mathscr{R}_G = (1 - T_G)\overline{\mathscr{R}}_G$ whereas if $\omega(G) < 0$, $\mathscr{R}_G = \overline{\mathscr{R}}_G$ by definition of the V. In either case, this is in agreement with Eqs. (8-41) and (8-42). To prove that Eq. (8-49) is indeed the solution to Eq. (8-45), we proceed by induction, assuming that it has been proved for arbitrary diagrams up to a given number of loops $L - 1$. If G has L loops, let us write Eq. (8-45) and insert Eq. (8-49) for each $\overline{\mathscr{R}}_\gamma$. We get

$$\overline{\mathscr{R}}_G = \mathscr{I}_G + \sum_{\substack{\{\gamma_1, \ldots, \gamma_s\} \\ \gamma_a \cap \gamma_b = \varnothing}} \mathscr{I}_{G/\{\gamma_1, \ldots, \gamma_s\}} \prod_{a=1}^{s} - T_{\gamma_a} \left[\sum_{V_{\gamma_a}} \prod_{\gamma' \in V_{\gamma_a}} (-T_{\gamma'}) \mathscr{I}_{\gamma_a} \right] \tag{8-50}$$

The notations are cumbersome but obvious. It is then a matter of a simple inspection to check that Eq. (8-50) generates all the terms of (8-49), corresponding to forests of extremal elements $\gamma_1, \ldots, \gamma_s$. By definition, an extremal element of a forest is an element γ not included in any other one of the forest. The summation over the disjoint sets $\{\gamma_1, \ldots, \gamma_s\}$ generates all the forests of G once and only once.

We have been discrete about the choice of normalization conditions. Since the nature of the Taylor subtraction depends on this choice, the most convenient one is the intermediate renormalization defined above, i.e., the subtraction at zero momentum. When a different point is chosen, for instance, in massless theories where subtracting at zero momentum is forbidden (see Sec. 8-3-1), some care is required to maintain Lorentz invariance.

Supplementary subtractions, i.e., beyond the degree $\omega(G)$ prescribed by power counting, may be performed consistently in an analogous way. We shall not develop this point here, and refer the reader to the references (see also some remarks at the end of Sec. 8-2-6).

The construction of $\mathscr{R}$ has involved only disjoint renormalization parts, i.e., without any common vertex or line. As a point of consistency, let us show that the renormalized integrand for an articulate (or one-vertex reducible) diagram may be factorized (Fig. 8-6). As an example, we consider the diagram of Fig. 8-6*b* of the four-dimensional φ^4 theory. The forests U are

$$\varnothing, \{\gamma_1\}, \{\gamma_2\}, \{\gamma_1, G\}, \{\gamma_2, G\}, \{G\}$$

Hence

$$\mathscr{R}_G = (1 - T_G - T_{\gamma_1} - T_{\gamma_2} + T_G T_{\gamma_1} + T_G T_{\gamma_2}) \mathscr{I}_{\gamma_1} \mathscr{I}_{\gamma_2}$$

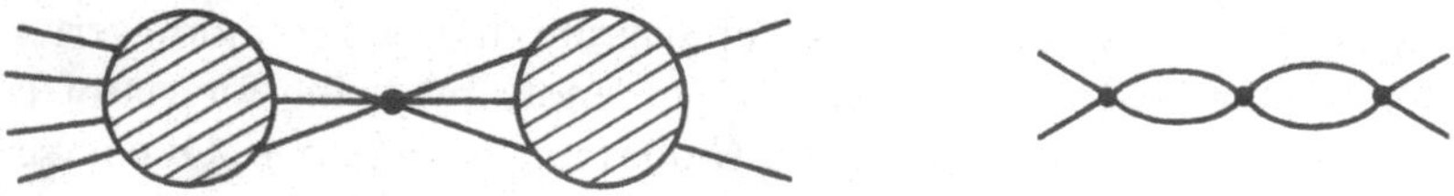

(a) (b)

Figure 8-6 Articulate (one-vertex reducible) diagrams.

For such a diagram, $T_G = T_{\gamma_1} T_{\gamma_2}$. Since G, γ_1, and γ_2 have the same degree of divergence, and since a Taylor expansion of degree ω does not affect a polynomial of degree ω, we have

$$T_G = T_G T_{\gamma_1} = T_G T_{\gamma_2} = T_{\gamma_1} T_{\gamma_2}$$

Thus, $\mathscr{R}_G$ also reads

$$\mathscr{R}_G = (1 - T_{\gamma_1})\mathscr{I}_{\gamma_1}(1 - T_{\gamma_2})\mathscr{I}_{\gamma_2} = \mathscr{R}_{\gamma_1}\mathscr{R}_{\gamma_2}$$

This proof is readily generalized.

It is straightforward to verify that in the case where all the renormalization parts of G are nested, Eq. (8-49) reduces to the product of operators $(1 - T_\gamma)$ over all the γ. The general case consists in expanding the same product and omitting all terms corresponding to overlapping subdiagrams. These two statements are a mere generalization of the properties found in the particular examples at the end of Sec. 8-2-2.

The result expressed in Eq. (8-49) is a major achievement but is not completely satisfactory until we prove that it leads to a convergent integral. Especially embarrassing is the case of overlapping divergences. Is the prescription contained in Eq. (8-49) really sufficient to provide a finite expression in such instances? As we shall see, the answer is "yes." The proof of this result will render the use of an intermediate regularization unnecessary since the subtraction of the integrand leads to a convergent Feynman integral. However, it is often more convenient to deal with regularized amplitudes rather than with the cumbersome expression (8-49). Before giving a sketchy proof of convergence, we will first discuss the form of the subtractions in the parametric representation.

8-2-4 Renormalization in Parametric Space

It is possible to reexpress the subtraction prescriptions in the parametric representation. This formulation allows a simpler proof of the convergence theorem. We shall outline the major steps.

As a preliminary example, consider again the one-loop self-energy diagram (Fig. 8-7) of the six-dimensional φ^3 theory. Its value in the euclidean region is

$$I(P^2) = \int_0^\infty d\alpha_1\, d\alpha_2\, \frac{e^{-[m^2(\alpha_1+\alpha_2)+Q(P,\alpha)]}}{(4\pi)^3 \mathscr{P}(\alpha)^3}$$

with $\mathscr{P}(\alpha) = \alpha_1 + \alpha_2$, $Q(P, \alpha) = [\alpha_1\alpha_2/(\alpha_1 + \alpha_2)]P^2$. We find a quadratic divergence due to the non-

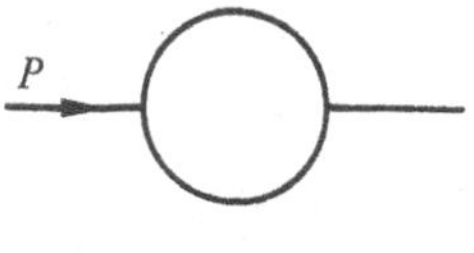

Figure 8-7 The bubble diagram.

integrability at the origin of the parametric space. If we use the intermediate renormalization, the renormalized form will be obtained after subtraction of the truncated Taylor expansion in P^2 at the origin

$$I_R(P^2) = I(P^2) - I(0) - P^2 \left.\frac{dI}{dP^2}\right|_0$$
$$= \int_0^\infty d\alpha_1\, d\alpha_2 \frac{e^{-m^2(\alpha_1+\alpha_2)}}{(4\pi)^3 \mathscr{P}(\alpha)^3}\left(e^{-Q(P,\alpha)} - e^{-Q(0,\alpha)} - P^2 \frac{d}{dP^2} e^{-Q(P,\alpha)}\Big|_0\right) \tag{8-51}$$

Under a simultaneous dilatation of the parameters α_1, α_2 by a factor λ, the subtracted integrand will behave as λ^{-1}. Owing to the extra factor λ arising from the measure $d\alpha_1\, d\alpha_2$, the integral will be convergent at the origin. Now the subtraction in P^2 may be transmuted into a subtraction over the parameter λ, if we recall the homogeneity properties of Q and $\mathscr{P}$ and observe that

$$Q(\rho P, \alpha) = Q(P, \rho^2 \alpha)$$

and

$$[\mathscr{P}(\alpha)]^{-3}\left(e^{-Q(P,\alpha)} - e^{-Q(0,\alpha)} - P^2 \frac{d}{dP^2} e^{-Q(P,\alpha)}\Big|_0\right)$$
$$= [\mathscr{P}(\alpha)]^{-3} e^{-Q(P,\alpha)} - \sum_0^2 \frac{1}{n!}\frac{\partial^n}{\partial\rho^n}\{[\mathscr{P}(\alpha)]^{-3} e^{-Q(\rho P,\alpha)}\}\big|_{\rho=0}$$
$$= [\mathscr{P}(\alpha)]^{-3} e^{-Q(P,\alpha)} - \sum_0^2 \frac{1}{n!}\frac{\partial^n}{\partial\rho^n}\{\rho^6 e^{-Q(P,\rho^2\alpha)}[\mathscr{P}(\rho^2\alpha)]^{-3}\}\big|_{\rho=0} \tag{8-52}$$

This explicit example leads us to the following definition. Let $f(\rho)$ be a function of ρ such that $\rho^p f(\rho)$ is differentiable at the origin for some integer p. We define the operator $\mathscr{T}^k$ as

$$\mathscr{T}^k f(\rho) = \rho^{-p_1} \sum_{s=0}^{k+p_1} \frac{\rho^s}{s!}\frac{d^s}{d\rho^s}[\rho^{p_1} f(\rho)]_{\rho=0} \qquad \forall p_1 \geq p \tag{8-53}$$

with k integer. It is straightforward to check that this definition does not depend on $p_1 \geq p$ and that it could be generalized to p noninteger. The expression (8-53) makes sense provided $k + p \geq 0$, which allows k to be negative. By convention, we set

$$\mathscr{T}^k \equiv 0 \qquad \text{if } k + p < 0$$

The essential property of this generalized Taylor expansion is that

$$(1 - \mathscr{T}^k) f(\rho) \sim O(\rho^{k+1}) \tag{8-54}$$

In the previous example, the renormalized integrand reads

$$(1 - \mathscr{T}^{-4})\{[\mathscr{P}(\rho^2\alpha)]^{-3} e^{-Q(P,\rho^2\alpha)}\}\big|_{\rho=1}$$

where the notation implies that the subtractions are performed at $\rho = 0$ and the function is then evaluated at $\rho = 1$.

These operations are generalized to arbitrary situations. Returning to a four-dimensional theory, the translation of formulas (8-49) yields the following renormalized amplitude:

$$I_R(P) = \frac{1}{(4\pi)^{2L}} \int_0^\infty \prod_1^I (d\alpha_s\, e^{-\alpha_s m_s^2}) \mathscr{R}(P, \alpha)$$
$$\mathscr{R}(P, \alpha) = \sum_U \prod_{\gamma \in U} (-\mathscr{T}_{\rho_\gamma}^{-2I_\gamma})\{[\mathscr{P}(\alpha)]^{-2} e^{-Q(P,\alpha)}\}\big|_{\rho_\gamma=1} \tag{8-55}$$

The operator $\mathscr{T}_{\rho_\gamma}$ acts on the parameters α_l pertaining to the renormalization part γ, after a rescaling $\alpha_l \to \rho_\gamma^2 \alpha_l$; I_γ denotes the number of internal lines of γ. The subtractions are performed at $\rho_\gamma = 0$ and

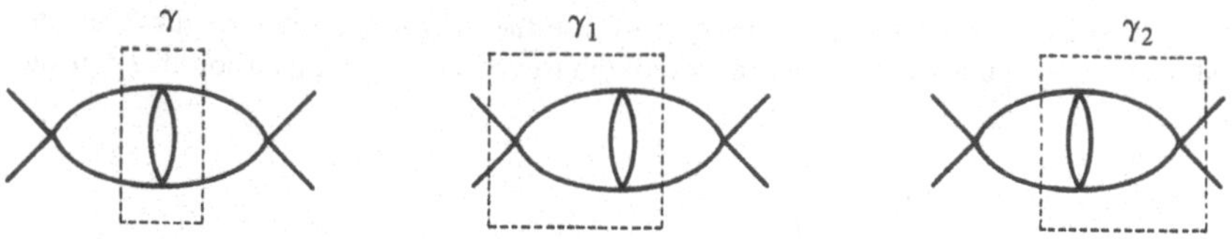

Figure 8-8 Overlapping divergences to be studied in parametric space.

the result evaluated at $\rho_\gamma = 1$. Finally, these operations are performed on all the γ belonging to a forest U, and a summation over the U is carried out ($U = \varnothing$ corresponds to the identity with no subtraction). In Eq. (8-55), we have taken for simplicity a scalar theory without derivative couplings, considered in the euclidean region and renormalized at the origin in momentum space. The generalization to nonzero spin, derivative couplings, higher dimensions, etc., offers no difficulty—only for notational ones.

This formulation of subtractions in parametric space enjoys remarkable algebraic properties. It may be shown that the complete expression (8-55) is independent of the ordering of the Taylor subtractions, although two particular $\mathscr{T}$ pertaining to overlapping subdiagrams do not commute in general. Moreover, for a given diagram, or a finite set of diagrams, there exists an upper bound on the powers p. Consequently, the operators $\mathscr{T}_{\rho_\gamma}$ depend only on the number of internal lines I_γ, and no longer on the particular topology of the diagram.

Even the last reference to the topology in (8-55), namely, the enumeration of forests and renormalization parts, may be disposed of. First, we prove that

$$\mathscr{R}(P, \alpha) = \Pi\,(1 - \mathscr{T}_{\rho_\gamma}^{-2I_\gamma})\{[\mathscr{P}(\alpha)]^{-2}\, e^{-Q(P,\alpha)}\}\,|_{\rho_\gamma = 1} \tag{8-56}$$

where the product runs over all the renormalization parts of the diagram. In other words, the subtractions pertaining to overlapping subdiagrams drop out. As an illustration let us verify this property on the example of Fig. 8-8. In four-dimensional space, the diagram G together with the subdiagrams γ_1, γ_2, and γ are renormalization parts. Showing the equivalence between Eqs. (8-55) and (8-56) amounts to proving the identity

$$0 = (1 - \mathscr{T}_\lambda^{-2I_G})(-\mathscr{T}_{\rho_1}^{-2I_{\gamma_1}})(1 - \mathscr{T}_\rho^{-2I_\gamma})(-\mathscr{T}_{\rho_2}^{-2I_{\gamma_2}})[\mathscr{P}^{-2}(\alpha)\, e^{-Q(P,\alpha)}]\,|_{\rho_1 = \rho_2 = \rho = \lambda = 1}$$

where the dilatation parameter for G has been denoted λ. If f stands for the function between brackets, it satisfies

$$f \equiv f(\lambda\rho_1\rho_2\rho, \lambda\rho_1, \lambda\rho_2)$$

since only those α belonging to γ are simultaneously dilated by ρ_1^2 and ρ_2^2. Therefore, the action of $\mathscr{T}_{\rho_2}$ yields

$$-\mathscr{T}_{\rho_2}^{-2I_{\gamma_2}} f = \sum_{k \le -2I_{\gamma_2}} (\lambda\rho_2)^k f_k(\rho_1\rho, \lambda\rho_1)$$

(The lower bound on k depends on the theory; in a scalar theory it is $-4L_{\gamma_2}$ in terms of the number of loops of γ_2.) Then

$$(1 - \mathscr{T}_\rho^{-2I_\gamma})(-\mathscr{T}_{\rho_2}^{-2I_{\gamma_2}}) f = \sum_{\substack{k \le -2I_{\gamma_2} \\ m > -2I_\gamma}} (\lambda\rho_2)^k (\rho_1\rho)^m f_{k,m}(\lambda\rho_1)$$

and

$$(-\mathscr{T}_{\rho_1}^{-2I_{\gamma_1}})(1 - \mathscr{T}_\rho^{-2I_\gamma})(-\mathscr{T}_{\rho_2}^{-2I_{\gamma_2}}) f = \sum_{\substack{k \le -2I_{\gamma_2} \\ m > -2I_\gamma \\ n+m \le -2I_{\gamma_1}}} (\lambda\rho_2)^k (\rho_1\rho)^m (\lambda\rho_1)^n f_{k,m,n}$$

Each individual term in the sum is homogeneous of degree $k + n$ in λ. But

$$k + n \le -2(I_{\gamma_1} + I_{\gamma_2}) - m \le -2(I_{\gamma_1} + I_{\gamma_2} - I_\gamma) \equiv -2I_G$$

Therefore the action of the last subtraction $(1 - \mathscr{T}_\lambda^{-2I_G})$ which retains only the terms of degree larger than $-2I_G$ gives zero. It is a mere exercise of bookkeeping to generalize this proof to arbitrary situations.

Finally, it is easy to show that if we perform subtractions on a homogeneity scale of subsets of parameters that do not correspond to superficially divergent proper subdiagrams, their effect drops out in the complete expression. Therefore, the final result of this analysis is

$$\mathscr{R}(P, \alpha) = \prod_g (1 - \mathscr{T}_{\rho_g}^{-2I_g})[\mathscr{P}^{-2}(\alpha)\, e^{-Q(P,\alpha)}]\big|_{\rho_g = 1} \tag{8-57}$$

where the product runs over the $(2^{I_G} - 1)$ nonempty sets of α parameters. Again, the result does not depend on the ordering. The merit of this last expression is twofold. First, it is independent of the topology of the diagram. Second, it enables us to understand, at least qualitatively, the arguments for the proof of convergence. Indeed, for any family g of parameters α, we may bring the operator $(1 - \mathscr{T}_{\rho_g}^{-2I_g})$ to the left of the product in (8-57), and it follows from (8-54) that the integrand behaves as $\rho_g^{-2I_g+1}$ and hence is integrable in ρ_g since the measure still contains a factor $\rho_g^{2I_g-1}\, d\rho_g$. That the possible singularity of an arbitrary subset of the α is integrable is unfortunately insufficient to insure the convergence of the integral. For instance, in the integral

$$\int_0^1 \int_0^1 d\alpha_1\, d\alpha_2 \frac{2\alpha_1}{(\alpha_1^2 + \alpha_2)^2}$$

the singularities corresponding to $\alpha_1 \to 0, \alpha_2 \neq 0$, or to $\alpha_2 \to 0, \alpha_1 \neq 0$, or to $\alpha_1 \sim \alpha_2 \to 0$ are integrable, but integration over α_1 leads to the divergent integral

$$\int_0^1 \frac{d\alpha_2}{\alpha_2(1 + \alpha_2)}$$

The proof that such phenomena do not occur for Feynman integrals is tedious. We will not reproduce it here. As in Sec. 8-1-4, we have to divide the integration domain into sectors and appeal to the homogeneity properties of the parametric functions.

We conclude this long and technical analysis by stating once again the important result of the Bogoliubov-Parasiuk-Hepp theorem. The subtraction operation described in Eqs. (8-45), (8-49), or (8-57) yields an absolutely convergent integral, and defines an analytic function of the momenta in the euclidean region and a tempered distribution in the Minkowski domain.

8-2-5 Finite Renormalizations

We have considered so far the subtractions of infinities. But the previous developments on the construction of counterterms, on the multiplicative character of the renormalization, or on the algebra of subtractions apply as well to finite renormalizations. This term refers to the operations required to modify the normalization conditions, thereby changing by a finite amount the (renormalized) parameters of the theory. Such is the case for instance, if we want to modify the normalization conditions (8-30) and (8-35) into (8-36). More generally, sticking to our φ^4 theory, let us consider the following set of renormalization conditions, depending on an arbitrary mass scale μ:

$$\begin{gathered} \Gamma_R^{(2)}(p^2)\big|_{p^2=\mu^2} = \mu^2 - m^2 \qquad \frac{\partial}{\partial p^2}\Gamma_R^{(2)}(p^2)\big|_{p^2=\mu^2} = 1 \\ \Gamma_R^{(4)}\big|_{S_\mu} = -\lambda \end{gathered} \tag{8-58}$$

where S_μ is defined as in Eq. (8-35) but with m replaced by μ.

This is a perfectly sensible choice since it is satisfied to lowest order, and it interpolates between conditions (8-30) and (8-35), and (8-36). It is, however, safer to choose μ such that the renormalization points $p^2 = \mu^2$ and S_μ lie inside the analyticity regions of the two- and four-point functions respectively. Otherwise, the above condition should be understood to hold for the real part of the amplitude only.

The theory depends now on two mass scales: m is the mass entering the propagator in Feynman diagrams and μ specifies the renormalization point. As for the physical mass, defined as the pole of the complete propagator, it is some function of m, μ, and λ, and may be computed order by order in perturbation theory. Also, the residue at the pole is no longer one, and the computation of S matrix elements should take it into account.

How are two renormalized theories corresponding to two different choices of μ related? Clearly each one may be reconstructed from the other through the introduction of finite counterterms, determined order by order to implement the new conditions. As in the case of the infinite renormalization, this is in turn equivalent to a redefinition of the parameters m and λ of the theory, provided we also allow for a finite wave-function renormalization of the field. These parameters being equal to the value of the Green functions at a given point μ, changing μ into μ' amounts to changing the parameters m, λ (and 1) into m', λ', and z. Hence

$$\Gamma_R^{(n)}(p_1, \ldots, p_n; m, \lambda, \mu) = z^{n/2}\Gamma_R^{(n)}(p_1, \ldots, p_n; m', \lambda', \mu') \tag{8-59}$$

where m', λ', and z are functions of m, λ, μ, and μ', computable order by order in perturbation theory.

This will be illustrated on the two-point functions of the φ^3 theory in a six-dimensional space. This renormalizable theory has the merit of having a nontrivial wave-function renormalization to the one-loop approximation. If $\mathscr{L}_{\text{int}} = -\lambda/3! :\varphi^3:$, the self-energy of Fig. 8-7 reads

$$i\Gamma^{(2)[1]}(P^2) = -\lambda^2 \int \frac{d^6k}{(2\pi)^6} \frac{1}{[(P-k)^2 - m^2](k^2 - m^2)}$$

where some regularization is understood and the superscript 1 refers to the one-loop correction. A straightforward computation yields

$$\Gamma^{(2)[1]}(P^2) = \frac{\lambda^2}{(4\pi)^3} \int_{1/\Lambda^2}^{\infty} \frac{d\rho}{\rho^2} \int_0^1 d\alpha\, e^{-\rho[m^2 - \alpha(1-\alpha)P^2]}$$

and

$$\frac{\partial^2 \Gamma^{(2)[1]}}{(\partial P^2)^2} = \frac{\lambda^2}{(4\pi)^3} \int_0^1 \frac{d\alpha\, \alpha^2(1-\alpha)^2}{m^2 - \alpha(1-\alpha)P^2}$$

The renormalized function $\Gamma_R^{(2)[1]}$ satisfying (8-58) reads

$$\Gamma_R^{(2)[1]}(P^2) = \frac{\lambda^2}{(4\pi)^3} \int_0^1 d\alpha \left\{[m^2 - \alpha(1-\alpha)P^2] \ln \frac{m^2 - \alpha(1-\alpha)P^2}{m^2 - \alpha(1-\alpha)\mu^2} + \alpha(1-\alpha)(P^2 - \mu^2)\right\} \tag{8-60}$$

where we do not bother to perform the α integration explicitly. If we now change μ into μ', it is easy to see that $\Gamma_R^{(2)} = P^2 - m^2 + \Gamma_R^{(2)[1]}$ satisfies Eq. (8-59) with

$$\begin{aligned} m'^2 &= m^2 + \frac{\lambda^2}{(4\pi)^3}\int_0^1 d\alpha \left[(1-\alpha+\alpha^2)m^2 \ln \frac{m^2-\alpha(1-\alpha)\mu^2}{m^2-\alpha(1-\alpha)\mu'^2} + \alpha(1-\alpha)(\mu^2-\mu'^2)\right] \\ z &= 1 + \frac{\lambda^2}{(4\pi)^3}\int_0^1 d\alpha\, \alpha(1-\alpha) \ln \frac{m^2-\alpha(1-\alpha)\mu^2}{m^2-\alpha(1-\alpha)\mu'^2} \end{aligned} \tag{8-61}$$

The function λ' would be determined by the calculation of the one-loop three-point function. This will be left as an exercise to the reader.

This may be generalized to arbitrary conditions, in any renormalizable theory, with a conclusion similar to Eq. (8-59). The information conveyed in Eq. (8-59), namely, the equivalence of changing renormalization points or renormalized parameters, is referred to as renormalization group invariance. The implications of this equation or of its infinitesimal form—the so-called renormalization group equation—will be studied in Chap. 13. We shall see that the innocent-looking freedom in the choice of normalization conditions has nontrivial and important consequences.

8-2-6 Composite Operators

The Green functions considered so far involved only elementary fields, i.e., dynamical variables entering the lagrangian. The renormalization procedure performed either through Bogoliubov's subtraction R operation or through the introduction of counterterms extends to a wider class of functions involving composite operators. By this we mean local monomials of field operators and their derivatives. A prototype of such an operator is the electromagnetic current $\overline{\psi}\gamma_\mu\psi$. Composite operators play an important role in many developments.

For the sake of simplicity, we shall again discuss only the scalar φ^4 theory. Composite operators are of the form φ^2, $(\partial\varphi)^2$, $\varphi\Box\varphi$, φ^4, φ^6, etc., but also $\varphi\partial_\mu\varphi$, $\partial_\mu\varphi\,\partial_\nu\varphi, \ldots$, if we construct vector- or tensor-like operators. For power counting, these operators have clearly the dimensions $\omega_i = 2, 4, 4, 4, 6, \ldots, 3, 4, \ldots$ respectively. To deal with Green functions containing insertions of these operators $O_i(x)$, it is convenient to add sources χ_i coupled to them in the action. Consequently,

$$Z(j, \chi) = \langle 0|\, T \exp\left\{i \int d^4x\, [j(x)\varphi(x) + \chi_i(x)O_i(x)]\right\}|0\rangle \tag{8-62}$$

will generate these new functions. Connected Green functions will be obtained from the logarithm of Z, as in Eq. (6-71), while the Legendre transformation of Chap. 6, performed on the source j only, will generate the proper functions, i.e., one-particle irreducible but with an arbitrary number of O_i insertions. When restricting our attention to a finite number of those, we shall perform a finite number of derivatives with respect to the χ_i and then set $\chi_i = 0$.

If we consider a diagram with N insertions of operators of dimensions ω_i, a mere application of Eq. (8-13c) reveals that the new superficial degree of divergence ω' differs from the one in the absence of insertions by an amount

$$\omega' - \omega = \sum_{1}^{N} (\omega_i - 4) \tag{8-63}$$

Insertions of operators of degree less or equal to four in a superficially convergent diagram preserve the convergence, whereas insertions of degree larger than four deteriorate the power counting. However, whatever the (finite) number of insertions of composite operators, there exists a subtraction prescription or, equivalently, counterterms that make proper Green's functions finite. For instance, assume that the two-point (two external φ) function

$$\langle T O_1(x_1) O_2(x_2) \cdots O_N(x_N) \varphi(y) \varphi(z) \rangle_{\text{proper}} \tag{8-64}$$

is superficially divergent with degree ω'. There exists a local counterterm, quadratic in φ and proportional to $\chi_1(x) \cdots \chi_N(x)$, with a polynomial of derivatives of degree less or equal to ω'. Clearly this counterterm will contribute only to the function (8-64). This will now be exemplified in simple instances.

1. *Insertion of the φ^2 operator.* This operator has a dimension $\omega(\varphi^2) = 2$ and therefore its insertion improves power counting. There are two superficially divergent proper functions with φ^2 insertions. They are depicted in Fig. 8-9. Both have $\omega = 0$. To the first, we associated a counterterm quadratic in χ but φ independent, while the second requires a counterterm of the form $\frac{1}{2}\chi\varphi^2$. The original term $\frac{1}{2}\chi\varphi^2$ in the lagrangian is thus changed into

$$\tfrac{1}{2}(1 + \delta_1)\chi\varphi^2 + \delta_2 \chi^2$$

with δ_1, δ_2 two divergent scalar quantities. If we rewrite

$$1 + \delta_1 = Z Z_{\varphi^2}$$

as a definition of Z_{φ^2} in terms of the wave function renormalization Z of φ, we see that n-point functions with a single φ^2 insertion satisfy

$$\Gamma^{(n)}_{\varphi^2, R}(q; p_1, \ldots, p_n, \lambda, m) = Z_{\varphi^2} Z^{n/2} \Gamma^{(n)}_{\varphi^2, \text{reg}}(q; p_1, \ldots, p_n, \lambda_0, m_0) \tag{8-65}$$

Here q stands for the momentum entering the diagram at the φ^2 vertex. We see that φ^2 is multiplicatively renormalized, as φ, with a wave-function renormalization constant Z_{φ^2}. The insertion of several operators is a straightforward generalization

$$\Gamma^{(n)}_{\varphi^2\varphi^2, R}(q_1, q_2; p_1, \ldots, p_n, \lambda, m) =$$
$$Z^2_{\varphi^2} Z^{n/2} \Gamma^{(n)}_{\varphi^2\varphi^2, \text{reg}}(q_1, q_2; p_1, \ldots, p_n, \lambda_0, m_0) + 2\delta_2 \delta_{n,0} \tag{8-66}$$

Figure 8-9 Divergent diagrams with φ^2 insertions.

where the last term takes account of the vacuum diagrams ($n = 0$) of Fig. 8-9a.

Of course, these new subtractions require new normalization conditions to be fully specified. For instance, we may impose conditions at zero momentum

$$\Gamma^{(2)}_{\varphi^2,R}(0;0) = 1 \qquad \Gamma^{(0)}_{\varphi^2\varphi^2,R}(0,0) = 0 \tag{8-67}$$

in agreement with lowest order.

The renormalization of this φ^2 operator is not independent of mass renormalization. Since the addition of a term $\frac{1}{2}\chi(x)\varphi^2(x)$ in the lagrangian may be regarded as an x-dependent variation of the mass $m^2 \to m^2 - \chi(x)$, it follows that

$$\Gamma^{(2)}_{\varphi^2,R}(0;p;\lambda,m) = -Z_{\varphi^2}\frac{\partial}{\partial m_0^2}\Gamma^{(2)}_R(p;\lambda,m)\big|_{\lambda_0}$$

where the derivative is taken at fixed λ_0. Therefore, if subtractions are performed at zero momentum as in Eqs. (8-36) and (8-67) we conclude that

$$Z_{\varphi^2} = \left.\frac{\partial m_0^2}{\partial m^2}\right|_{\lambda_0} \tag{8-68}$$

2. *Insertion of the φ^4 operator*. The insertion of operators of dimension four does not affect power counting. The counterterms linear in the source χ_{φ^4}, that is, counterterms relevant for a single insertion, will be combinations of $\{O_i, i = 1,\ldots,4\} \equiv \{\varphi^2(x), \varphi^4(x), (\partial\varphi)^2(x), \varphi\Box\varphi(x)\}$ (the last two are equivalent when integrated over x, but differ in general). These operators are in fact mixed by renormalization. Consequently, φ^4 may not be considered independently from those. To be more specific, the generalization of Eq. (8-65) reads

$$\Gamma^{(n)}_{O_i,R}(q;p_1,\ldots,p_n;\lambda,m) = \sum_j Z^{n/2}Z_{ij}\Gamma^{(n)}_{O_j,\text{reg}}(q;p_1,\ldots,p_n;\lambda_0,m_0) \tag{8-69}$$

This amounts to saying that the renormalized insertion of φ^4 requires the following source term in the lagrangian:

$$\chi_{\varphi^4}\left[Z_{21}Z\frac{\varphi^2}{2} + Z_{22}Z^2\frac{\varphi^4}{4!} + Z_{23}Z\frac{(\partial\varphi)^2}{2} + Z_{24}Z\frac{\varphi\Box\varphi}{2}\right]$$

where the Z_{ij} are determined after introduction of suitable normalization conditions. Finally, from the previous study, we know that φ^2 is multiplicatively renormalizable, which means that $Z_{1j} = 0$ for $j \neq 1$ and $Z_{11} = Z_{\varphi^2}$.

More generally, in a renormalizable theory, a complete set of composite operators O_i of dimension less or equal to a given number D, and with the same quantum numbers, is multiplicatively renormalizable in the previous matrix form, at least as long as we deal with a single insertion. Moreover, $Z_{ij} = 0$ if dim $O_i <$ dim O_j (the matrix Z_{ij} is not symmetric). Both results are obvious consequences of the power counting of Eq. (8-63).

As an exercise, the reader may investigate the relation between the Z_{ij} for dimension four operators and the bare constants λ_0, m_0, Z. He or she may also carry out the analysis of the renormalization of dimension six operators, or the one of the tensor operators, such as the energy momentum tensor.

In some instances, it may be interesting to assign to an operator a dimension greater than the one coming from dimensional counting, and to renormalize it accordingly. For instance, if instead of considering φ^2 as an operator of dimension two we assign to it the dimension four, more subtractions (and normalization conditions) will be necessary. The new (or "hard") operator, denoted $N_4(\varphi^2)$ to distinguish it from the old (or "soft") one $N_2(\varphi^2)$, will be on the same footing as φ^4, $(\partial\varphi)^2$, and $\varphi \Box \varphi$. In particular, Z_{1j} will no longer vanish. This generalization, useful in some applications, has been introduced by Zimmermann.

In the case of several insertions, with the rule (8-63) as Ariadne's clue, we find that counterterms (multilinear in the sources) keep dimension four as long as the initial operators O_i have dimension four, but that their dimension increases as $\dim(O_i) > 4$. For instance, a double insertion of φ^4 and $\varphi \Box \varphi$ requires again φ^2, φ^4, $\varphi \Box \varphi$, and $(\partial\varphi)^2$ counterterms, while the insertion of φ^4 and φ^6 also involves all operators of dimension six and a double insertion of φ^6 leads to counterterms of dimension eight!

8-3 ZERO-MASS LIMIT, ASYMPTOTIC BEHAVIOR, AND WEINBERG'S THEOREM

Until now, we have sidestepped all problems related to massless particles. The object of this section is to present an heuristic discussion on some aspects of these questions and their relation with Weinberg's theorem. The latter deals with the behavior of Feynman diagrams as external momenta become very large.

8-3-1 Massless Theories

When dealing with massless particles in the internal propagators, we face the risk of encountering new divergences. We are going to show that nothing of this sort happens as long as we keep away from some particular values of the external momenta. In the euclidean (i.e., Wick-rotated) version of the theory, when all vertices of the lagrangian have a dimension four (or higher) the proper functions are finite at any nonzero and nonexceptional value of the external momenta. Nonexceptional momenta are configurations such that no partial sum of the incoming momenta p_i vanishes. The integrations remain finite for small values of the internal momenta because the external momenta provide a lower cutoff.

This result breaks down in theories containing super-renormalizable couplings. For instance, the diagram of Fig. 8-10a with one φ^2 vertex (or, equivalently, a φ^2 insertion of zero momentum) is

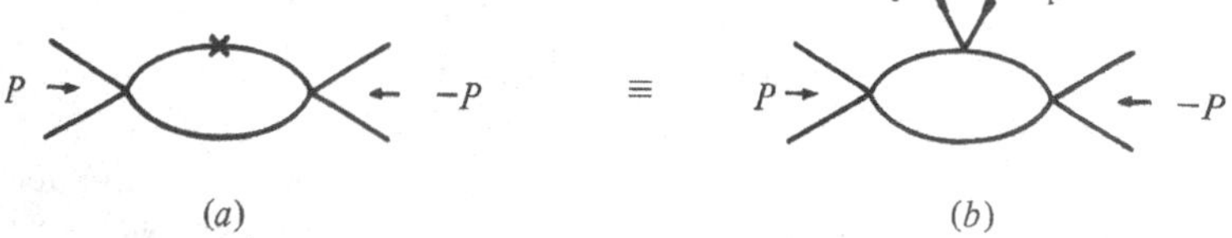

Figure 8-10 Infrared-divergent diagram.

infrared divergent:

$$I_G(P) \sim \int \frac{d^4k}{(k^2)^2(P-k)^2}$$

This amplitude could also be considered as a six-point function in a φ^4 theory (Fig. 8-10*b*), but then it is evaluated for an exceptional configuration.

A rigorous proof requires careful slicings of the integration domain, very much as for the theorem of Sec. 8-1-4. Here we shall just give a simple argument based on power counting, discarding possible ultraviolet problems.

Infrared power counting amounts to finding out how many internal momenta may become soft in the diagram while preserving momentum conservation at each vertex. Let us consider an N-point function with nonvanishing external momenta, and assume that a definite flow through the diagram has been chosen for these hard external momenta. Since the latter are not exceptional, any internal line irrigated by them cannot have a vanishing momentum, as all loop momenta go to zero. Moreover, the same property shows that these hard internal lines form a connected pattern on the diagram (the heavy lines in the example of Fig. 8-11). Consequently, as far as infrared behavior is concerned, we can contract them into a single vertex. To this vertex are attached N hard external lines and i internal lines, with $i \geq 2$ since the diagram is one-particle irreducible. Let I, L, V_3, and V_4 denote the number of internal lines, loops, three- and four-point vertices respectively in the contracted diagram. We have the usual topological relations

$$L = I + 1 - (V_3 + V_4 + 1)$$

$$N + 2I = 3V_3 + 4V_4 + (N + i)$$

Since, by assumption, all vertices of the lagrangian have degree four, one power of momentum is attached to each three-point vertex. Therefore, the degree of homogeneity of the contracted diagram, which gives the superficial degree of infrared divergence when all internal loop momenta are simultaneously small, is

$$\begin{aligned}\omega_{\text{ir}} &= 4L - 2I + V_3 \\ &= i \geq 2\end{aligned}$$

At least heuristically, this ensures infrared convergence.

We might wonder whether relaxing the hypothesis that all internal loop momenta after contraction are small would not upset the previous power counting. For instance, let us assume that a number $I_H \leq I$ of internal momenta are kept hard and that they form a pattern with L_H loops and V_{3H} and V_{4H} vertices of each species. It would seem that ω_{ir} is changed into $\omega'_{\text{ir}} = \omega_{\text{ir}} - \Delta\omega_{\text{ir}}$:

$$\Delta\omega_{\text{ir}} = 4L_H + V_{3H} - 2I_H$$

and that we might be in trouble. Fortunately, this is not the case. The hard pattern may be considered as a diagram (possibly disconnected), the external lines of which are all soft by construction.

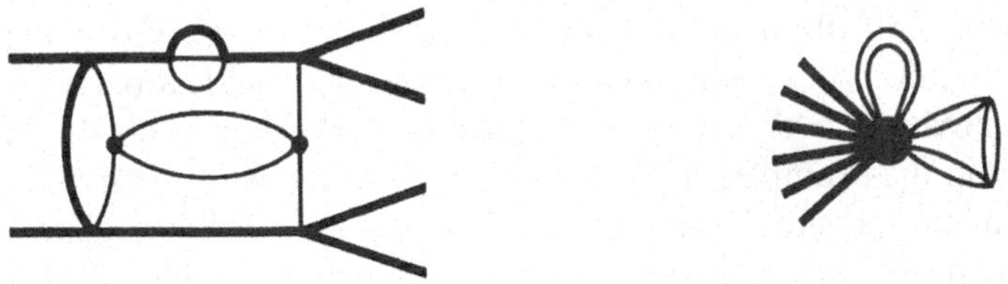

Figure 8-11 Flow of hard momenta through a diagram, and the corresponding contracted diagram.

Thus it behaves as a power $\Delta\omega_{\text{ir}}$ in these momenta. Therefore ω_{ir} is not affected. This is not surprising since ω_{ir} is the homogeneity degree of the diagram.

We may show even more. When a single momentum p vanishes, the others being nonzero and nonexceptional, Green functions remain finite.

The reader will find no difficulty in extending the previous simple argument to that case and in showing that ω_{ir} is at most reduced by one unit.

This crude argument has neglected possible ultraviolet problems. In order that renormalization does not upset the result it is mandatory to choose judicious normalization conditions. Subtractions at zero momentum must be avoided, since Green's functions are generally divergent at that point. In a zero-mass theory, it is safe to choose renormalization points at euclidean values of the momenta, for example, $p^2 = -\mu^2 < 0$, instead of Eq. (8-34) or (8-36). The necessity of introducing a non-zero normalization point implies that a theory where the physical mass parameters vanish involves nevertheless a mass scale μ. The independence of the physical quantities with respect to this arbitrary choice leads to renormalization group constraints, to be discussed later.

The previous considerations apply as well to theories involving both massless and massive particles, such as quantum electrodynamics. Green's functions are finite at any nonzero and nonexceptional euclidean values of the momenta. When more than one external momentum vanishes a case-by-case analysis is required.

In summary, if we let all (or some of) the internal masses of a Feynman diagram go to zero, we do not encounter any singularity, provided:

1. All the vertices have degree four.
2. The external momenta are not exceptional.
3. There is at most one soft external momentum.
4. Renormalization has been carried out at some fixed euclidean point.

What happens when external momenta are continued from euclidean to physical on-shell values is a harder question. A corollary of the previous theorem is of interest. Consider a proper two-point function and assume that analytic continuation to the minkowskian region may be performed, avoiding threshold singularities. The Green function remains finite, and so does its absorptive part. Owing to the Cutkosky rules (see Chap. 6), this means that any total decay rate to final states involving massless particles is finite. This result due to Kinoshita is to be compared with a theorem proved by Lee and Nauenberg. According to this theorem, any transition probability in a theory involving massless particles is finite, provided summation over degenerate states is performed. This is, of course, what we found in the examples of Chaps. 4 and 7, where we were mainly concerned with soft emission. It should be understood that additional divergences may occur when energetic, collinear, massless particles are produced. These cases are also included in the previous discussion.

8-3-2 Ultraviolet Behavior and Weinberg's Theorem

We look for a precise relation between the superficial degree of ultraviolet divergence $\omega(G)$ and the behavior of Feynman integrals when all the external momenta are large. Rescaling all these momenta by a common large factor λ is equivalent, on dimensional grounds, to dividing all internal masses by the same factor λ. Therefore, the problem is closely related to the massless limit considered in the previous subsection.

For simplicity, consider once again a scalar theory without derivative couplings. We select a Green function evaluated at euclidean external momenta and restrict ourselves, for the time being, to the ultraviolet convergent case. After integration over a global homogeneity variable, Eq. (8-20) becomes

$$I_G(P) = \frac{\Gamma[-\omega(G)/2]}{(4\pi)^{2L}} \int_0^1 \prod_l d\alpha_l\, \delta(1 - \sum \alpha_l)\, \frac{[\sum \alpha_l m_l^2 + Q(P, \alpha)]^{\omega(G)/2}}{[\mathscr{P}(\alpha)]^2}$$

If the P are dilated, $P \to \lambda P$, the integral behaves as $\lambda^{\omega(G)}$ [remember that $\omega(G) < 0$ by assumption] provided that

$$\int \prod d\alpha_l\, \delta(1 - \sum \alpha_l)\, \frac{[Q(P, \alpha)]^{\omega(G)/2}}{[\mathscr{P}(\alpha)]^2}$$

converges. This in turn depends on the existence of a zero-mass limit of $I_G(P)$. We see in particular that the large-λ behavior is related to the configuration of the P. From the preceding subsection, we know that if the euclidean momenta are nonvanishing and nonexceptional, such a limit will exist. In this case, the asymptotic behavior is given by the ultraviolet power counting

$$I_G(\lambda P) \underset{\lambda \to \infty}{\sim} \lambda^{\omega(G)} \tag{8-70}$$

This result extends to the case of superficially divergent, but renormalized, diagrams. However, the power behavior may be modified by powers of logarithms

$$I_G(\lambda P) \sim \lambda^{\omega(G)} \sum_0^{q_{\max}} C_q \ln^q \lambda\, [1 + O(\lambda^{-1})] \tag{8-71}$$

If the zero-mass limit is ill defined, e.g., if the diagram contains vertices of degree less than four, and/or exceptional external momenta, we expect some departure from this behavior, namely, higher powers than $\lambda^{\omega(G)}$ and possible logarithmic corrections.

To illustrate this point, consider the one-loop diagram of Fig. 8-12 with $n + 2$ lines and vertices: n of these vertices represent mass insertions. The superficial degree is $\omega(G) = -2n$. The corresponding (euclidean) integral reads

$$I_G(P) = \int \frac{d^4k}{(2\pi)^4} \frac{1}{(k^2 + m^2)^{n+1}[(P + k)^2 + m^2]}$$

$$= \frac{1}{(4\pi)^2 n} \int_0^1 d\alpha\, \alpha^n [m^2 + \alpha(1 - \alpha)P^2]^{-n}$$

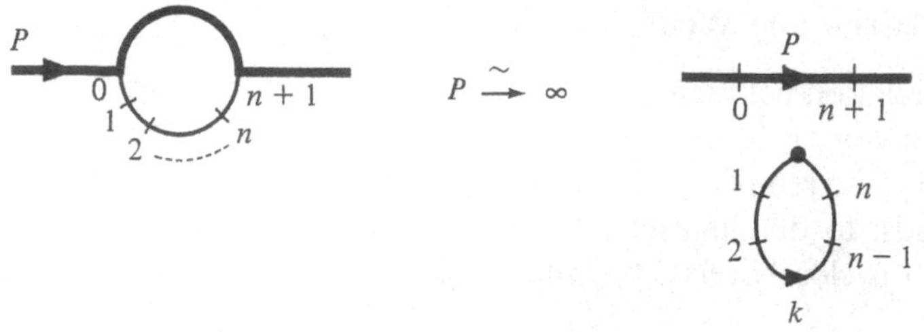

Figure 8-12 A self-energy diagram with n φ^2 insertions.

As $P^2 \to \infty$, $I_G(P)$ behaves as

$$I_G(P) \underset{P^2\to\infty}{\sim} \begin{cases} \dfrac{1}{P^2}\ln\dfrac{P^2}{m^2} & \text{for } n = 1 \\[2ex] \dfrac{m^2}{P^2}\dfrac{1}{(m^2)^n} & \text{for } n > 1 \end{cases}$$

A superficial estimate would have been $(P^2)^{-n}$. This may be interpreted by saying that the large momentum flows through a single propagator (the upper one in Fig. 8-12), giving a power $(P^2)^{-1}$. The coefficient arising from the lower chain of propagators, $\int^{|P|} d^4k/(k^2+m^2)^{n+1}$, behaves as $\ln P^2$ for $n = 1$, or as a constant for $n > 1$.

In the general case studied by Weinberg, the large-λ behavior is related to the minimal number of propagators irrigated by the large momentum flow. The following result holds in a renormalizable (or super-renormalizable) theory with no massless particles. When the euclidean momenta (possibly exceptional) are scaled by a large factor λ, a Feynman integral I_G is majorized by

$$I_G(\lambda P) \sim \lambda^{\Omega+\varepsilon} \tag{8-72}$$

where ε is a symbolic notation to express the fact that the power behavior λ^Ω is corrected by an integral power of logarithms and Ω stands for

$$\Omega = \sup_g \omega(g) \tag{8-73}$$

Here g runs over all subsets of G such that (1) the reduced diagram G/g has a vanishing total external momentum entering each of its vertices and (2) each connected part of G/g is attached to some external lines. In short, this means that the (large) external momenta may flow through g without irrigating G/g (see Fig. 8-13). Finally, $\omega(g)$ is the ultraviolet superficial degree of divergence of g.

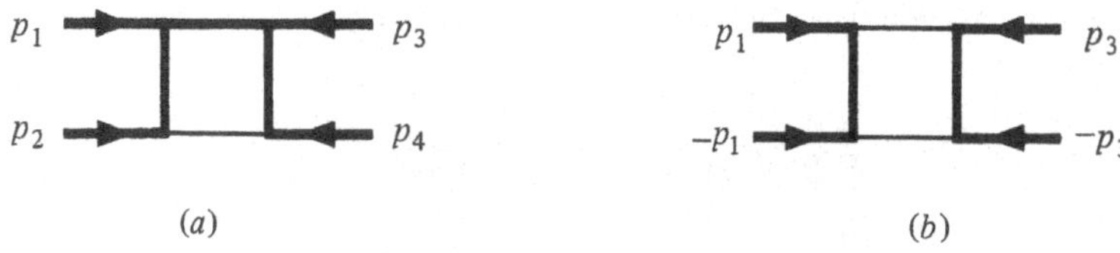

Figure 8-13 Possible flows of the large external momenta (heavy lines) through the box diagram, in a nonexceptional (*a*) and in an exceptional (*b*) configuration.

When the theory is strictly renormalizable (no super-renormalizable coupling) and when momenta are nonexceptional, it may be shown that the upper bound Ω is reached by $\omega(G)$ in agreement with our previous considerations. The reader may apply these general rules to the previous example of Fig. 8-12.

These results may be extended to minkowskian momenta or to configurations where only subsets of momenta become large. Bounds on the power of the logarithm of λ can also be obtained.

8-4 THE CASE OF QUANTUM ELECTRODYNAMICS

This section is devoted to the specific problems of quantum electrodynamics, namely, those related to gauge invariance and Ward identities. All these matters have already been encountered and treated within the one-loop approximation in Sec. 7-1-4 and our purpose here is to carry out the analysis to all orders, in a way consistent with renormalization. We shall first derive the Ward identities in a more systematic and algebraic way than in Chap. 7.

8-4-1 Formal Derivation of the Ward-Takahashi Identities

We start from the lagrangian (6-25) with a photon mass μ. Through Noether's theorem the conservation of the current

$$j_\rho(x) = e:\overline{\psi}(x)\gamma_\rho\psi(x): \qquad \partial^\rho j_\rho(x) = 0 \tag{8-74}$$

was a consequence of the invariance of the lagrangian under the global phase transformations

$$\psi(x) \to e^{i\chi}\psi(x) \qquad \chi \text{ constant}$$

This property implies relations between Green functions involving a single current operator and an arbitrary number of fields ψ, $\overline{\psi}$, A.

Our program is to derive first a set of covariant identities as a consequence of current conservation and the fact that Green's functions may be expressed as time-ordered products in Minkowski space. We then try to maintain these relations through renormalization. The expressions we obtain in this way have a Lorentz covariant form and it is understood that we work with covariant T products (see Sec. 5-1-7).

If the caret above a term means its omission, we have

$$\begin{aligned}
&\partial_x^\rho \langle 0|\, T j_\rho(x)\psi(x_1)\overline{\psi}(y_1)\cdots\overline{\psi}(y_n)A_{\rho_1}(z_1)\cdots A_{\rho_p}(z_p)\,|0\rangle \\
&\quad = \sum_{i=1}^{n} \langle 0|\, T\{[j_0(x),\psi(x_i)]\delta(x^0 - x_i^0)\overline{\psi}(y_i) + \psi(x_i)[j_0(x),\overline{\psi}(y_i)]\delta(x^0 - y_i^0)\} \\
&\qquad \times \psi(x_1)\overline{\psi}(y_1)\cdots\widehat{\psi(x_i)\overline{\psi}(y_i)}\cdots A_{\rho_p}(z_p)\,|0\rangle \\
&\quad + \sum_{j=1}^{p} \langle 0|\, T\psi(x_1)\cdots\overline{\psi}(y_n)A_{\rho_1}(z_1)\cdots[j_0(x),A_{\rho_j}(z_j)]\delta(x^0 - z_j^0)\cdots A_{\rho_p}(z_p)\,|0\rangle
\end{aligned} \tag{8-75}$$

The term containing $\partial_\rho j^\rho$ has dropped out, while the remaining terms just come from the x^0 dependence implicit in the T product. We now appeal to the canonical commutation rules:

$$[j_0(x), \psi(x')]\delta(x^0 - x'^0) = -e\psi(x)\delta^4(x - x')$$
$$[j_0(x), \bar{\psi}(x')]\delta(x^0 - x'^0) = e\bar{\psi}(x)\delta^4(x - x') \tag{8-76}$$
$$[j_0(x), A_\rho(x')]\delta(x^0 - x'^0) = 0$$

which express that ψ, $\bar{\psi}$, and A create quanta of electric charge $Q = \int j_0(\mathbf{x}, t)\, d^3x$ equal to $-e$, e, and zero respectively. In contrast with the explicit form of the current, which follows from the minimal coupling prescription, the previous commutation rules, or at least their integrated versions, are crucial for the conservation of the charge. The Ward-Takahashi identity (8-75) reads

$$\partial_x^\rho \langle 0| Tj_\rho(x)\psi(x_1)\bar{\psi}(y_1)\cdots A_{\rho_p}(z_p)|0\rangle$$
$$= e\langle 0| T\psi(x_1)\bar{\psi}(y_1)\cdots A_{\rho_p}(z_p)|0\rangle \sum_{i=1}^{n} [\delta^4(x - y_i) - \delta^4(x - x_i)] \tag{8-77}$$

We shall investigate further the following cases:

$n = 1$	$p = 0$	self-energy and vertex
$n = 0$	$p = 1$	vacuum polarization
$n = 0$	$p = 3$	photon-photon scattering

We remark that (8-77) holds also for the connected part of Green functions. The forthcoming discussion is formal, as we sidestep ultraviolet divergences. It will be justified in the next subsection, where we exhibit a regularization that preserves the identities.

1. Let $G_{\rho\sigma}$ be the complete photon propagator and $G^{[0]}_{\rho\sigma}$ the free one (Fig. 8-14):

$$G_{\rho\sigma}(x) = G^{[0]}_{\rho\sigma}(x) - i\int d^4x'\, G^{[0]}_{\rho\sigma'}(x - x')\langle 0| Tj_{\sigma'}(x')A_\sigma(x)|0\rangle \tag{8-78}$$

It follows from Eq. (8-77) that

$$\partial_x^\rho G_{\rho\sigma}(x) = \partial_x^\rho G^{[0]}_{\rho\sigma}(x) \tag{8-79}$$

Interpreted in momentum space, this is seen to imply the transversity of the

Figure 8-14 The vacuum polarization in quantum electrodynamics.

vacuum polarization

$$
\begin{aligned}
k^\rho \bar{\omega}_{\rho\sigma}(k) &= 0 \\
\bar{\omega}_{\rho\sigma}(k) &= -i(g_{\rho\sigma}k^2 - k_\rho k_\sigma)\bar{\omega}(k^2)
\end{aligned} \tag{8-80}
$$

which generalizes the result derived in Eq. (7-6).

Indeed, in momentum space, Eq. (8-79) reads

$$
k^\rho G_{\rho\sigma}(k) = k^\rho G^{[0]}_{\rho\sigma}(k) = -i\frac{M^2}{\mu^2(k^2 - M^2)}
$$

where $M^2 = \mu^2/\lambda$ or, after multiplication by $G^{-1}_{\sigma\rho'} = -i\Gamma_{\sigma\rho'}$,

$$
k_\rho = -\frac{M^2}{\mu^2(k^2 - M^2)} k^\sigma \Gamma_{\sigma\rho}(k) \tag{8-81}
$$

If we parametrize $\Gamma_{\sigma\rho}$ in the form

$$
\begin{aligned}
\Gamma_{\sigma\rho}(k) &= -[g_{\rho\sigma}k^2 - k_\rho k_\sigma(1-\lambda)] + g_{\rho\sigma}\mu^2 + \bar{\omega}_{\rho\sigma}(k) \\
&= A(k^2)(g_{\rho\sigma}k^2 - k_\rho k_\sigma) + B(k^2)k_\rho k_\sigma
\end{aligned} \tag{8-82}
$$

the identity (8-81) tells us that

$$
B(k^2) = \lambda\frac{k^2 - M^2}{k^2}
$$

In other words, $B(k^2)$ is not affected by radiative correction and $\bar{\omega}_{\rho\sigma}$ is transverse.

2. The relation between the electron self-energy and the vertex functions is obtained by considering the complete (not necessarily proper but certainly connected) vertex function $\mathscr{V}_\rho(p', p)$ defined as (Fig. 8-15)

$$
\begin{aligned}
&-ie(2\pi)^4\delta^4(p' - p - q)\mathscr{V}_\rho(p', p) \\
&\quad = \int d^4x\, d^4x_1\, d^4y_1\, e^{i(p'\cdot x_1 - p\cdot y_1 - q\cdot x)}\langle 0|\, TA_\rho(x)\psi(x_1)\bar{\psi}(y_1)\,|0\rangle \\
&\quad = -iG^{[0]}_{\rho\sigma}(q)\int d^4x\, d^4x_1\, d^4y_1\, e^{i(p'\cdot x_1 - p\cdot y_1 - q\cdot x)}\langle 0|\, Tj^\sigma(x)\psi(x_1)\bar{\psi}(y_1)\,|0\rangle
\end{aligned} \tag{8-83}
$$

$-ie(2\pi)^4\delta^4(p' - p - q)\mathscr{V}_\rho$ =

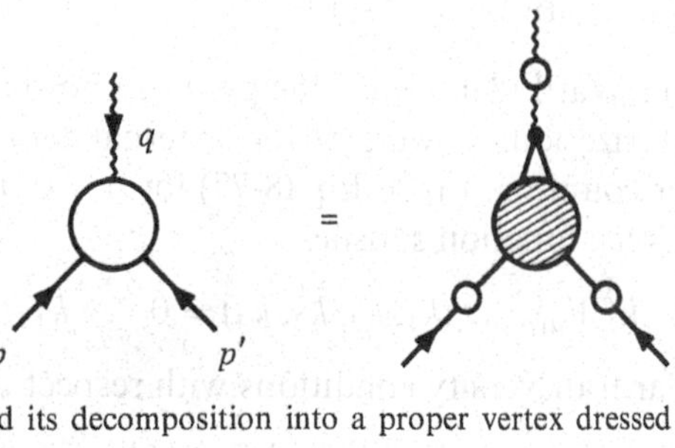

Figure 8-15 The vertex function and its decomposition into a proper vertex dressed with complete propagators.

In terms of the vertex function $\Lambda_\rho(p', p)$ already encountered in Eq. (7-46) $[\Lambda_\rho^{[0]}(p', p) = \gamma_\rho]$ and of the complete electron propagator $iS(p)[S^{[0]}(p) = (\not{p} - m)^{-1}]$, we have (compare Fig. 8-15)

$$\mathscr{V}_\rho(p', p) = G_{\rho\rho'}(q)[iS(p')\Lambda^{\rho'}(p', p)iS(p)] \tag{8-84}$$

Contracting Eq. (8-83) with k_ρ and using Eqs. (8-84) and (8-79), we get

$$e(2\pi)^4\delta^4(p' - p - q)q^\rho G_{\rho\sigma}^{[0]}(q)S(p')\Lambda^\sigma(p', p)S(p)$$
$$= -q^\rho G_{\rho\sigma}^{[0]}(q)\int d^4x\, d^4x_1\, d^4y_1\, e^{i(p'\cdot x_1 - p\cdot y_1 - q\cdot x)}\langle 0|\, Tj^\sigma(x)\psi(x_1)\bar{\psi}(y_1)\,|0\rangle \tag{8-85}$$

Since $q^\rho G_{\rho\sigma}^{[0]}(q)$ is proportional to q^σ, it follows that

$$e(2\pi)^4\delta^4(p' - p - q)S(p')q^\rho\Lambda_\rho(p', p)S(p)$$
$$= i\int d^4x\, d^4x_1\, d^4y_1 e^{i(p'\cdot x_1 - p\cdot y_1 - q\cdot x)}\,\partial_x^\rho\langle 0|\, Tj_\rho(x)\psi(x_1)\bar{\psi}(y_1)\,|0\rangle$$

We may now use the general identity (8-77) for $n = 1, p = 0$:

$$e(2\pi)^4\delta^4(p' - p - q)S(p')q^\rho\Lambda_\rho(p', p)S(p)$$
$$= ie\int d^4x\, d^4x_1\, d^4y_1\, e^{i(p'\cdot x_1 - p\cdot y_1 - q\cdot x)}\langle 0|\, T\psi(x_1)\bar{\psi}(y_1)\,|0\rangle$$
$$\times [\delta^4(x - y_1) - \delta^4(x - x_1)]$$

Therefore,

$$S(p')q^\rho\Lambda_\rho(p', p)S(p) = S(p) - S(p') \tag{8-86}$$

$$q^\rho\Lambda_\rho(p', p) = S^{-1}(p') - S^{-1}(p) = [\not{p}' - m - \textstyle\sum(p')] - [\not{p} - m - \textstyle\sum(p)] \tag{8-87}$$

Differentiating with respect to p'^ρ at $q = 0$ yields

$$\Lambda_\rho(p, p) = \frac{\partial}{\partial p^\rho}S^{-1}(p) \tag{8-88}$$

in agreement with Eq. (7-47*b*).

3. Finally, the Ward identity for the photon-photon scattering amplitude enables us to factorize four powers of the external momenta and hence to improve the power counting. From Eq. (8-77) for $n = 0$, $p = 3$, we learn that the four-photon Green function satisfies

$$k_1^{\rho_1}\Gamma_{\rho_1\rho_2\rho_3\rho_4}(k_1, k_2, k_3, k_4) = 0 \qquad k_1 + k_2 + k_3 + k_4 = 0$$

and similar transversity conditions with respect to k_2, k_3, k_4. It follows that

$$\Gamma_{\rho_1\rho_2\rho_3\rho_4}(k_1, k_2, k_3, k_4) = k_1^\sigma\Gamma_{1,\sigma\rho_1\rho_2\rho_3\rho_4}(p_1, p_2, p_3, p_4)$$

where Γ_1 is antisymmetric in the first pair of indices σ, ρ_1. This factorization of photon momenta may be pursued for k_2, k_3, k_4, without introducing singularities. We end up with a four-point function whose effective superficial degree of divergence is minus four instead of zero.

It remains to show that these identities are preserved by the regularization and renormalization operations.

8-4-2 Pauli-Villars Regularization to All Orders

We choose to work here with the traditional Pauli-Villars regularization, rather than the dimensional regularization. The latter will, however, be illustrated in the two-loop computation of Sec. 8-4-4. We recall that the Pauli-Villars method amounts to regularizing the fermion loops and photon propagators independently. The photon propagator is replaced in a standard fashion by a superposition of free propagators. Fermions loops, however, are treated as a whole, each one standing now for a sum of contributions corresponding to fermions of different masses, minimally coupled to the electromagnetic current (Fig. 8-16). Explicitly,

$$I = \sum_{s=0}^{S} C_s \int \frac{d^4p}{(2\pi)^4} \operatorname{tr} \left(\frac{1}{\not p - M_s + i\varepsilon} \gamma_{\mu_1} \frac{1}{\not p + \not q_1 - M_s + i\varepsilon} \cdots \gamma_{\mu_{2n}} \right) \tag{8-89}$$

with the convention that $C_0 \equiv 1$, $M_0 = m$, and $q_1 \cdots q_{2n}$ are the momenta entering the loop. We rationalize the denominators, compute the trace, and expand both the numerators and the denominators in powers of M_s^2. We find

$$\begin{aligned} I &= \sum_{s=0}^{S} C_s \int \frac{d^4p}{(2\pi)^4} \frac{P_n(p^2, p\cdot q_i, q_i^2) + M_s^2 P_{n-1}(p^2, p\cdot q_i, q_i^2) + \cdots}{Q_{2n}(p^2, p\cdot q_i, q_i^2) + M_s^2 Q_{2n-1}(p^2, p\cdot q_i, q_i^2) + \cdots} \\ &= \sum_{s=0}^{S} C_s \int \frac{d^4p}{(2\pi)^4} \left[\frac{P_n}{Q_{2n}} + M_s^2 \left(\frac{P_{n-1}}{Q_{2n}} - \frac{P_n Q_{2n-1}}{Q_{2n}^2} \right) + \cdots \right] \end{aligned} \tag{8-90}$$

where P_k and Q_k are polynomials in p of degree less or equal to $2k$. For large $|p|$, the coefficient of M_s^{2k} behaves as $|p|^{-2n-2k}$. Therefore, if we impose two conditions

$$\sum_{s=0}^{S} C_s = 0 \qquad \sum_{s=0}^{S} C_s M_s^2 = 0 \tag{8-91}$$

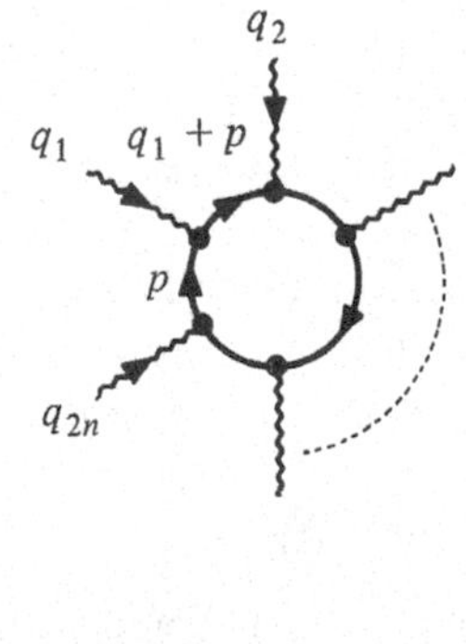

Figure 8-16 A fermion loop.

the integrand of any fermion loop behaves as $|p|^{-6}$ (we discard vacuum diagrams; hence $n \geq 1$) and is thus superficially convergent. Such conditions may be realized through the introduction of only two auxiliary masses

$$M_1^2 = m^2 + 2\Lambda^2 \qquad M_2^2 = m^2 + \Lambda^2 \qquad (\text{and } M_0^2 \equiv m^2) \tag{8-92}$$

where Λ^2 is the cutoff (which will ultimately go to infinity). This choice is such that

$$C_1 = \frac{M_2^2 - m^2}{M_1^2 - M_2^2} = 1 \qquad C_2 = \frac{M_1^2 - m^2}{M_2^2 - M_1^2} = -2 \tag{8-93}$$

The reason for insisting that C_1 and C_2 be integers will soon become clear. As for the photon propagator, a single subtraction

$$G_{\rho\sigma,\text{reg}}(k) = \left[-i \left(\frac{g_{\rho\sigma} - k_\rho k_\sigma/\mu^2}{k^2 - \mu^2} - \frac{k_\rho k_\sigma/\mu^2}{k^2 - \mu^2/\lambda} \right) \right] - [\mu^2 \to \mu_1^2] \tag{8-94}$$

will make it smooth enough to render all diagrams convergent. It is understood that $\mu_1^2 \to \infty$ as $\Lambda^2 \to \infty$. All the previous prescriptions follow from a regularized lagrangian

$$\mathscr{L}_{\text{reg}} = \frac{1}{\mu_1^2 - \mu^2} \left[-\frac{1}{4} (\partial_\rho A_\sigma - \partial_\sigma A_\rho)(\Box + \mu^2 + \mu_1^2)(\partial^\rho A^\sigma - \partial^\sigma A^\rho) + \frac{\mu^2 \mu_1^2}{2} A^2 \right.$$
$$\left. - \frac{\lambda}{2} \partial \cdot A (\lambda \Box + \mu^2 + \mu_1^2) \partial \cdot A \right] + \sum_{s=0}^{3} \bar{\psi}_s (i\not{\partial} - e\not{A} - M_s) \psi_s \tag{8-95}$$

After integration by parts, the quadratic part in A reads

$$\mathscr{L}_A = \frac{1}{2(\mu_1^2 - \mu^2)} A_\rho K^{\rho\sigma}(\mu^2) K_{\sigma\nu}(\mu_1^2) A^\nu$$
$$= \tfrac{1}{2} A_\tau K^{\tau\rho}(\mu^2) [K(\mu_1^2) - K(\mu^2)]^{-1}{}_{\rho\sigma} K^{\sigma\nu}(\mu_1^2) A_\nu$$

where the differential operator $K_{\rho\sigma}(\mu^2)$ is defined as

$$K_{\rho\sigma}(\mu^2) = g_{\rho\sigma}(\Box + \mu^2) - \partial_\rho \partial_\sigma (1 - \lambda)$$

It is then clear that the A propagator, i.e., the inverse of the quadratic form appearing in $\mathscr{L}_A$, is $i[K^{-1}(\mu^2) - K^{-1}(\mu_1^2)]_{\rho\sigma}$. On the other hand, we have introduced in (8-95) three auxiliary fields ($\psi_0 \equiv \psi$), minimally coupled to the electromagnetic field and endowed with the masses M_1, $M_2 \equiv M_3$ satisfying (8-92). Moreover, ψ_1 is considered as an ordinary Fermi field, whereas ψ_2 and ψ_3 are quantized according to the Bose statistics! The effect of these strange rules is, of course, to reproduce the prescription (8-93). Because of the degeneracy between ψ_2 and ψ_3 and of the absence of the minus sign in their closed loops, $C_2 = -2$.

We have reached our aim. The theory has been regularized in a satisfactory way, since $\mathscr{L}_{\text{reg}}$ of Eq. (8-95) is gauge invariant (up to the photon mass term and the gauge-fixing transverse terms in $\partial \cdot A$) and the Ward identities of the previous paragraph are clearly satisfied.

8-4-3 Renormalization

We now have to show that renormalization may be carried out while preserving the Ward identities provided we choose suitable normalization conditions. We require that

$$\left\{\begin{array}{ll}\Gamma_R^{(2)}(\not p)\big|_{\not p=m} \equiv S_R^{-1}(\not p)\big|_{\not p=m} = 0 & \text{(8-96}a\text{)}\\ \left.\dfrac{\partial\Gamma_R^{(2)}(\not p)}{\partial\not p}\right|_{\not p=m} = 1 & \text{(8-96}b\text{)}\\ \Lambda_R^\rho(p,p)\big|_{\not p=m} = \gamma^\rho & \text{(8-96}c\text{)}\\ \qquad\Gamma_R^{\rho\sigma}(k) = -\{(k^2g^{\rho\sigma} - k^\rho k^\sigma)[1+\bar\omega_R(k^2)] - g^{\rho\sigma}\mu^2 + \lambda k^\rho k^\sigma\} & \text{(8-96}d\text{)}\\ \qquad\bar\omega_R(0) = 0 & \text{(8-96}e\text{)}\end{array}\right.$$

These conditions are obviously satisfied to lowest order. Equations (8-96*b*, *c*) are in agreement with the identity (8-87), while Eqs. (8-96*d*, *e*) incorporate the information derived from (8-79) and (8-82). As for the conditions (8-96*a*, *b*), they define the physical mass of the electron, since they guarantee that the complete propagator S has a pole of residue one at $\not p = m$. Similarly, condition (8-96*c*) will lead to a physical definition of the charge as the coupling between an on-shell fermion and a photon of vanishing momentum. However, as already noticed in Chap. 7, the conditions (8-96*b*, *c*, *e*) may no longer be maintained as the photon mass μ goes to zero. Indeed, the derivative of the self-energy on the mass shell $(\partial/\partial\not p)\sum(\not p)\big|_{\not p=m}$, and consequently the vertex function, are plagued with infrared divergences as μ^2 approaches zero. Therefore, if we insist on taking the limit $\mu\to 0$, other normalization conditions must be chosen. To remain on the safe side we keep here μ^2 small but finite.

The proof that Ward identities are preserved by renormalization proceeds by induction. We assume that they hold up to a given order $\hbar^L$. In other words, we have determined to this order the bare quantities Z_1, Z_2, Z_3, m_0, μ_0^2, and λ_0 in such a way that the lagrangian

$$\begin{aligned}\mathscr{L}^{[L]} &= -\frac{Z_3}{4}(\partial^\mu A^\nu - \partial^\nu A^\mu)(\partial_\mu A_\nu - \partial_\nu A_\mu) - \frac{\lambda_0 Z_3}{2}(\partial^\mu A_\mu)^2\\ &\quad + \frac{\mu_0^2 Z_3}{2}A^2 + Z_2\bar\psi(i\not\partial - m_0)\psi + eZ_1\bar\psi\not A\psi\\ &= -\frac{1}{4}(\partial^\mu A_0^\nu - \partial^\nu A_0^\mu)(\partial_\mu A_{0\nu} - \partial_\nu A_{0\mu}) - \frac{\lambda_0}{2}(\partial\cdot A_0)^2 + \frac{\mu_0^2}{2}A_0^2\\ &\quad + \bar\psi_0(i\not\partial - m_0 - e_0\not A_0)\psi_0\end{aligned} \qquad \text{(8-97)}$$

leads to renormalized Green functions satisfying (8-96). We remind the reader that

$$A_0 = Z_3^{1/2}A \qquad \psi_0 = Z_2^{1/2}\psi \qquad \text{and} \qquad e_0 = \frac{Z_1 e}{Z_2 Z_3^{1/2}} \qquad \text{(8-98)}$$

Moreover, we assume that, to this order, the Ward identities imply that

$$Z_3\mu_0^2 = \mu^2 \qquad Z_3\lambda_0 = \lambda \qquad \text{and} \qquad Z_1 = Z_2 \tag{8-99}$$

To the next order $\hbar^{L+1}$, the Green functions are still divergent. We introduce a gauge-invariant regularization such as the one exhibited in Sec. 8-4-2 to regularize the theory and compute the Green functions using $\mathscr{L}^{[L]}$, that is, taking account of all lower-order counterterms. Because of the structure of $\mathscr{L}^{[L]}$, we observe that these regularized functions $\Gamma_{\text{reg}}^{[L+1]}$ satisfy the identities derived in Sec. 8-4-1. Using these identities and the normalization conditions (8-96), it is straightforward to verify that the new counterterms needed to order $L+1$ have the same structure as in lower orders [namely, that no $(A^2)^2$ counterterm is necessary] and that they still enjoy (8-99).

For instance

$$Z_2^{[L+1]} = \frac{\partial}{\partial \not p}\left[\Gamma_R^{(2)[L+1]}(\not p) - \Gamma_{\text{reg}}^{(2)[L+1]}(\not p)\right]_{\not p = m} = -\frac{\partial}{\partial \not p}\Gamma_{\text{reg}}^{(2)[L+1]}(\not p)\Big|_{\not p = m}$$

and

$$Z_1^{[L+1]}\gamma_\rho = \left[\Lambda_{\rho R}^{[L+1]}(p, p) - \Lambda_{\rho,\text{reg}}^{[L+1]}(p, p)\right]_{\not p = m} = -\Lambda_{\rho,\text{reg}}^{[L+1]}(p, p)\Big|_{\not p = m}$$

Finally,

$$Z_2^{[L+1]} = Z_1^{[L+1]}$$

follows from Eq. (8-88).

In the renormalized theory we end up with functions related to the bare regularized ones through

$$\Gamma_R^{(2n,p)}(p_1', p_1, \ldots, p_n', p_n, q_1, \ldots, q_p, m, \mu, e, \lambda)$$
$$= \lim_{\Lambda\to\infty} Z_3^{p/2} Z_2^n \Gamma_{\text{reg}}^{(2n,p)}(p_1', \ldots, q_p, m_0, \mu_0, e_0, \lambda_0, \Lambda) \tag{8-100}$$

These renormalized Green functions satisfy the Ward identities as a trivial consequence of this multiplicative character of renormalization.

It is important to emphasize the role of the identity $Z_1 = Z_2$ in the charge renormalization $e_0 = eZ_3^{-1/2}$. If several species of charged particles (electron, muon, etc.) are coupled to the electromagnetic field, the identities $Z_1^{(e)} = Z_2^{(e)}$, $Z_1^{(\mu)} = Z_2^{(\mu)}, \ldots$, guarantee that renormalization is universal. The concept of charge universality only makes sense owing to this identity.

It would be more appropriate to say that the ratio of renormalized to bare charge is independent of the type of charged particle, since in this restricted framework there is no natural explanation for charge quantization. This is not the case for certain unified models of weak and electromagnetic interactions, where the electromagnetic gauge invariance corresponds to a subgroup of a larger simple group of invariance (see Chap. 11).

In summary the method described in this section is to express the consequences of a symmetry in terms of Ward identities and to verify their consistency with renormalization. It will again be encountered in different contexts in the following: chiral symmetry, nonabelian gauge symmetries, etc.

8-4-4 Two-Loop Vacuum Polarization

The computation of the two-loop vacuum polarization will be presented in massless euclidean quantum electrodynamics ($\mu = m = 0$), using dimensional regularization.

It is rather instructive in several respects:

(*a*) It provides an example of features specific to higher-order corrections, i.e., not visible at the one-loop order.
(*b*) It shows how dimensional regularization may be applied in a case involving spinors.
(*c*) It is convincing evidence that the renormalization program does work, even when overlapping divergences are present.
(*d*) It illustrates the statements concerning massless theories and asymptotic behavior. The zero-mass computation may also be regarded as yielding the asymptotic (large-k) behavior of the vacuum polarization $\bar{\omega}(k^2)$ in massive quantum electrodynamics.
(*e*) It serves as a test of the general results derived in the previous subsections, namely, the transversity of the vacuum polarization tensor.
(*f*) Finally, it exhibits an interesting property of the vacuum polarization, namely, unexpected cancellations at large momentum. We shall elaborate on this point at the end of the calculation.

Since we are working in the euclidean version of the theory, with a dimensional regularization, let us first list some useful formulas. As in Eq. (8-11), the antihermitian matrices satisfy

$$\{\gamma_\mu, \gamma_\nu\} = -2\delta_{\mu\nu} \tag{8-101}$$

and, by convention, we choose [compare with (8-11*c*)]

$$\mathrm{tr}\, I = 4 \tag{8-102}$$

All standard identities for contractions and traces of γ matrices then follow (d is the euclidean space dimension):

$$\begin{aligned}
\gamma_\rho\gamma_\rho &= -d \\
\not{a}^2 &= -a^2 \\
\gamma_\rho\gamma_\mu\gamma_\rho &= (d-2)\gamma_\mu \\
\gamma_\rho\gamma_\mu\gamma_\nu\gamma_\rho &= -(d-4)\gamma_\mu\gamma_\nu + 4\delta_{\mu\nu} \\
\gamma_\rho\gamma_\mu\gamma_\nu\gamma_\sigma\gamma_\rho &= (6-d)\gamma_\sigma\gamma_\nu\gamma_\mu - 2(d-4)(\delta_{\mu\nu}\gamma_\sigma - \delta_{\mu\sigma}\gamma_\nu + \delta_{\nu\sigma}\gamma_\mu) \\
\mathrm{tr}\,\gamma_\mu\gamma_\nu &= -4\delta_{\mu\nu} \\
\mathrm{tr}\,\gamma_\mu\gamma_\nu\gamma_\rho\gamma_\sigma &= 4(\delta_{\mu\nu}\delta_{\rho\sigma} - \delta_{\mu\rho}\delta_{\nu\sigma} + \delta_{\mu\sigma}\delta_{\nu\rho}) \\
\mathrm{tr}\,\gamma_\mu\gamma_\nu\gamma_\rho\gamma_\sigma\gamma_\tau\gamma_\upsilon &= -4(\delta_{\mu\nu}\delta_{\rho\sigma}\delta_{\tau\upsilon} - \text{perm})
\end{aligned} \tag{8-103}$$

In euclidean space, the action of massless quantum electrodynamics reads

$$I_E = \int d^dx \left[\bar{\psi}(i\not{\partial} - e\not{A})\psi + \frac{1}{4}(\partial_\mu A_\nu - \partial_\nu A_\mu)^2 + \frac{\lambda}{2}(\partial_\mu A_\mu)^2\right] \tag{8-104}$$

In the Feynman gauge, $\lambda = 1$, the perturbative rules derived from e^{-I_E} read

Fermion propagator 

$$\left(\frac{1}{\not{p}}\right)_{\beta\alpha} = -\frac{(\not{p})_{\beta\alpha}}{p^2}$$

Photon propagator 

$$\frac{\delta_{\mu\nu}}{k^2}$$

Vertex

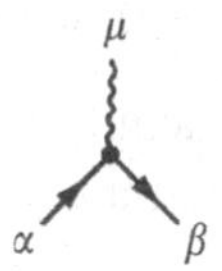

$e(\gamma_\mu)_{\beta\alpha}$

and, of course, a minus sign for each fermion loop. When working in dimension $d \neq 4$, the charge e acquires a dimension $(4-d)/2$, as already noticed in Sec. 8-1-2. If μ is an arbitrary mass scale, we write

$$e = \mu^{(4-d)/2} e' \tag{8-105}$$

where e' is dimensionless. We shall use this expression whenever we expand our results near $d = 4$, in order to recover correct homogeneity properties. In this way the unavoidable mass scale creeps into the massless theory.

As a warming up, we first recalculate the one-loop vacuum polarization, fermion self-energy, and vertex in this formalism.

The one-loop self-energy of Fig. 7-5 is

$$\begin{aligned}\Gamma_{\rho\sigma}(k) &= -e^2 \int \frac{d^d p}{(2\pi)^d} \frac{\operatorname{tr}[\not p \gamma_\sigma(\not p + \not k)\gamma_\rho]}{p^2(p+k)^2} \\ &= -e^2 \operatorname{tr}(\gamma_\mu\gamma_\rho\gamma_\nu\gamma_\sigma) \frac{1}{(4\pi)^{d/2}} B\left(\frac{d}{2}, \frac{d}{2}\right)(k^2)^{d/2-2} \\ &\quad \times \left[-k_\mu k_\nu \Gamma\left(2 - \frac{d}{2}\right) + \frac{\delta_{\mu\nu}}{2} k^2 \Gamma\left(1 - \frac{d}{2}\right)\right]\end{aligned}$$

The integral was performed using Eq. (8-9), where $m_1^2 = m_2^2 = 0$, in terms of Euler's B function

$$B(x, y) = \int_0^1 d\alpha\, \alpha^{x-1}(1-\alpha)^{y-1} = \frac{\Gamma(x)\Gamma(y)}{\Gamma(x+y)}$$

Defining $\varepsilon = 4 - d$ we are led to

$$\begin{aligned}\Gamma_{\rho\sigma}(k) &= -\frac{8e'^2}{(4\pi)^{d/2}} B\left(\frac{d}{2}, \frac{d}{2}\right)\Gamma\left(2 - \frac{d}{2}\right)\left(\frac{k^2}{\mu^2}\right)^{d/2-2}(\delta_{\rho\sigma}k^2 - k_\rho k_\sigma) \\ &\underset{d\simeq 4}{\simeq} -\frac{\alpha}{3\pi}\left[\frac{2}{\varepsilon} - \ln\frac{k^2}{\mu^2} + \text{constant} + O(\varepsilon)\right](\delta_{\mu\nu}k^2 - k_\mu k_\nu)\end{aligned} \tag{8-106}$$

in agreement with Eq. (7-9), if we identify $2/\varepsilon$ with $\ln(\Lambda^2/m^2)$. With the conventions of Chap. 6, the euclidean proper two-point function is the opposite of the inverse propagator. Therefore, after introduction of the counterterm

$$(Z_3 - 1)^{[1]} = -\frac{\alpha}{3\pi}\frac{2}{\varepsilon}$$

the renormalized vacuum polarization reads, in four dimensions,

$$\begin{aligned}\Gamma^{[1]}_{\rho\sigma} &= -(\delta_{\rho\sigma}k^2 - k_\rho k_\sigma)\bar\omega^{[1]}(k^2) \\ \bar\omega^{[1]}(k^2) &= -\frac{\alpha}{3\pi}\left(\ln\frac{k^2}{\mu^2} + \text{constant}\right)\end{aligned} \tag{8-107}$$

The constant is an uninteresting combination of π, γ (Euler constant), etc. Notice that we have used a new type of normalization conditions. Instead of fixing the value of $\bar\omega(k^2)$ at some point, we have decided to subtract from (8-106) the pole term only. We shall adhere to this convenient prescription, referred to as the minimal renormalization. Such a procedure satisfies automatically the Ward identities. The other two diagrams of Fig. 7-7 and 7-10 are computed in the same way. The fermion self-energy is

$$\begin{aligned}\Gamma^{(2)}(p) &= -e^2 \int \frac{d^dk}{(2\pi)^d} \frac{\gamma_\rho \not{k} \gamma_\rho}{(p-k)^2 k^2} \\ &= -e^2(d-2)\gamma_\mu \int \frac{d^dk}{(2\pi)^d} \frac{k_\mu}{(p-k)^2 k^2} \\ &= -\frac{e^2(d-2)(p^2)^{d/2-2}}{(4\pi)^{d/2}} \not{p}\,\Gamma\left(2-\frac{d}{2}\right) B\left(\frac{d}{2}-1, \frac{d}{2}\right) \\ &\underset{d\simeq 4}{\simeq} -\frac{e^2}{(4\pi)^2} \not{p}\left(\frac{2}{\varepsilon} + \text{finite terms}\right)\end{aligned} \tag{8-108}$$

Hence it requires a wave-function renormalization equal to

$$(Z_2 - 1)^{[1]} = -\frac{e^2}{(4\pi)^2}\frac{2}{\varepsilon} \tag{8-109}$$

As for the vertex function it reads

$$\begin{aligned}\Gamma_\mu(p, q) &= e^3 \int \frac{d^dk}{(2\pi)^d} \frac{\gamma_\rho(\not{k}+\not{q})\gamma_\mu \not{k}\gamma_\rho}{(p-k)^2(k+q)^2k^2} \\ &= e^3[(6-d)\gamma_\beta\gamma_\mu\gamma_\alpha - 2(d-4)(\delta_{\alpha\mu}\gamma_\beta - \delta_{\alpha\beta}\gamma_\mu + \delta_{\mu\beta}\gamma_\alpha)] \int \frac{d^dk}{(2\pi)^d} \frac{(k+q)_\alpha k_\beta}{(p-k)^2(k+q)^2k^2} \\ &\underset{d\simeq 4}{\simeq} \frac{e^3}{(4\pi)^2}\frac{2}{\varepsilon}\gamma_\mu + \text{finite terms}\end{aligned}$$

which leads to the counterterm

$$(Z_1 - 1)^{[1]} = -\frac{e^2}{(4\pi)^2}\frac{2}{\varepsilon} \tag{8-110}$$

We verify that the identity $Z_1 = Z_2$ is satisfied to this order, and that Z_1, Z_2, Z_3 are the same as in the massive minkowskian theory with the identification $2/\varepsilon = \ln(\Lambda^2/m^2)$. This computation was clearly unnecessary!

We now turn to the two-loop diagrams of Fig. 8-17a_1, a_2, b. The insertion of the counterterms

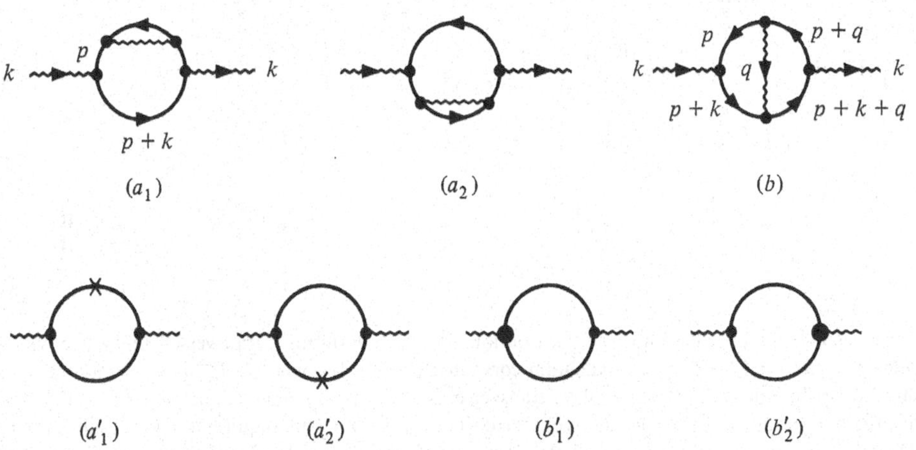

Figure 8-17 Two-loop contributions to the vacuum polarization. The contributions of order $\hbar$ counterterms are also depicted.

of order $\hbar$ is also depicted. The two diagrams a_1 and a_2 obviously give equal contributions. We first compute $\Gamma^{(a)}_{\rho\sigma} = \Gamma^{(a_1)}_{\rho\sigma} + \Gamma^{(a_2)}_{\rho\sigma}$ in terms of the fermion self-energy $\Gamma^{(2)}(p)$ of Eq. (8-108):

$$\Gamma^{(a)}_{\rho\sigma}(k) = -2e^2 \int \frac{d^d p}{(2\pi)^d} \operatorname{tr}\left[\frac{1}{\not p + \not k}\gamma_\rho \frac{1}{\not p}\Gamma^{(2)}(p)\frac{1}{\not p}\gamma_\sigma\right]$$
$$= \frac{2e^4(d-2)}{(4\pi)^{d/2}}\Gamma\left(2-\frac{d}{2}\right)B\left(\frac{d}{2}-1,\frac{d}{2}\right)\int \frac{d^d p}{(2\pi)^d}\frac{\operatorname{tr}[(\not p + \not k)\gamma_\rho \not p \gamma_\sigma]}{(p+k)^2(p^2)^{3-d/2}}$$

The integral over p is performed using the identity

$$\frac{1}{(p^2)^{3-d/2}} = \frac{1}{\Gamma(3-d/2)}\int_0^\infty dx\, x^{2-d/2}\, e^{-xp^2} \tag{8-111}$$

and the parametric machinery. This yields

$$\int \frac{d^d p}{(2\pi)^d}\frac{(p+k)_\mu p_\nu}{(p+k)^2(p^2)^{3-d/2}} = \frac{1}{(4\pi)^{d/2}}\frac{1}{\Gamma(3-d/2)}B\left(\frac{d}{2}, d-2\right)(k^2)^{d-4}$$
$$\times\left[-k_\mu k_\nu \Gamma(4-d) + \tfrac{1}{2}\delta_{\mu\nu}k^2\Gamma(3-d)\right]$$

and after some algebra

$$\Gamma^{(a)}_{\rho\sigma}(k) = \frac{8e^4(d-2)}{(4\pi)^d}\frac{B(d/2-1, d/2)B(d/2, d-2)(k^2)^{d-4}}{2-d/2}$$
$$\times\left[-2k_\mu k_\nu\Gamma(4-d) + \delta_{\mu\nu}k^2\Gamma(3-d)(1-d/2) + k^2\delta_{\mu\nu}\Gamma(4-d)\right] \tag{8-112}$$

Let us expand this expression near $d = 4$, $\Gamma(4-d) = 1/\varepsilon - \gamma + O(\varepsilon)$, $\Gamma(3-d) = -1/\varepsilon + (\gamma - 1)$, etc. We find

$$\Gamma^{(a)}_{\rho\sigma}(k) = \frac{\alpha^2}{3\pi^2}\left\{k_\rho k_\sigma\left[-\frac{1}{\varepsilon^2} + \frac{1}{\varepsilon}\left(\gamma + \ln\frac{k^2}{\mu^2} - \frac{7}{4}\right) - \frac{1}{2}\ln^2\frac{k^2}{\mu^2} + \ln\frac{k^2}{\mu^2}\left(-\gamma + \frac{7}{4}\right) + \cdots\right]\right.$$
$$\left. - \delta_{\rho\sigma}k^2\left[-\frac{1}{\varepsilon^2} + \frac{1}{\varepsilon}\left(\gamma + \ln\frac{k^2}{\mu^2} - 2\right) - \frac{1}{2}\ln^2\frac{k^2}{\mu^2} + \ln\frac{k^2}{\mu^2}(-\gamma + 2) + \cdots\right]\right\} \tag{8-113}$$

where constant terms have been omitted. The contribution $\Gamma^{(a')}_{\rho\sigma}$ of the counterterms (Fig. 8-17a_1', a_2') is also readily computed

$$\Gamma^{(a')}_{\rho\sigma}(k) = -2e^2\int\frac{d^d p}{(2\pi)^d}\operatorname{tr}\left\{\frac{1}{(\not p + \not k)}\gamma_\rho\frac{1}{\not p}\left[\frac{e^2}{(4\pi)^2}\not p\frac{2}{\varepsilon}\right]\frac{1}{\not p}\gamma_\sigma\right\}$$
$$= -\frac{\alpha^2}{\pi^2}\frac{2}{\varepsilon}B\left(\frac{d}{2},\frac{d}{2}\right)(k^2)^{d/2-2}(k^2\delta_{\mu\nu} - k_\mu k_\nu)\Gamma\left(2-\frac{d}{2}\right)$$
$$\simeq -\frac{\alpha^2}{3\pi^2}(k^2\delta_{\mu\nu} - k_\mu k_\nu)\left\{\frac{2}{\varepsilon^2} + \frac{1}{\varepsilon}\left(-\ln\frac{k^2}{\mu^2} + \frac{5}{3} - \gamma\right) + \left[\frac{1}{4}\ln^2\frac{p^2}{\mu^2} + \left(\frac{\gamma}{2} - \frac{5}{6}\right)\ln\frac{p^2}{\mu^2}\right]\right\}$$
$$\tag{8-114}$$

Three remarks are in order. First, the contribution of the counterterms is transverse—this was obvious since it is essentially the one-loop diagram computed previously—whereas $\Gamma^{(a)}_{\rho\sigma}$ is not. Only the sum of all contributions of Fig. 8-17 will be transverse. Second, we see that in the sum $\Gamma^{(a)} + \Gamma^{(a')}$ the divergent terms do not cancel. This was expected since the diagram still requires an overall subtraction. At first sight, it seems surprising that the dominant terms $1/\varepsilon^2$ (or $\ln^2 \Lambda^2$) do not cancel, since the internal divergences have been subtracted by the counterterms. But the renormalized fermion self-energy behaves as $\ln p^2$ at large p^2; hence its insertion in a superficially divergent diagram gives rise to a $\ln^2 \Lambda^2$ divergence

$$\int^{\Lambda} \frac{d^4p}{p^2(p+k)^2} \ln p^2 \sim \ln^2 \Lambda^2$$

The coefficient of the $\ln(k^2/\mu^2)$ term depends on the normalization condition for the self-energy diagram. As is seen from Eq. (8-114), changing the counterterm $(Z_2 - 1)^{[1]}$ by a finite amount results in a modification of the $\ln(k^2/\mu^2)$ term (and of the neglected constant term). However, this dependence will disappear in the complete expression of $\Gamma_{\sigma\rho}$. This illustrates the fact that the renormalization group equation (8-59) is only satisfied by a Green function to a given order, but not by individual diagrams.

We proceed to the much more cumbersome computation of Fig. 8-17b. With the notations of Fig. 8-17, the amplitude reads

$$\Gamma^{(b)}_{\rho\sigma}(k) = -e^4 \int \frac{d^dq}{(2\pi)^d q^2} \operatorname{tr}(\gamma_\nu\gamma_\sigma\gamma_\mu\gamma_\lambda\gamma_\alpha\gamma_\rho\gamma_\beta\gamma_\lambda) I_{\alpha\beta\mu\nu}(k, q) \tag{8-115}$$

where
$$I_{\alpha\beta\mu\nu} = \int \frac{d^dp}{(2\pi)^d} \frac{p_\beta(p+k)_\alpha(p+k+q)_\mu(p+q)_\nu}{p^2(p+k)^2(p+q+k)^2(p+q)^2}$$

The p integration will be first carried out. We introduce Feynman parameters $\alpha_1, \ldots, \alpha_4$ for the lines of momentum p, $p+q$, $p+q+k$, $p+k$ respectively. We get

$$I_{\alpha\beta\mu\nu} = \int_0^\infty d\alpha_1\, d\alpha_2\, d\alpha_3\, d\alpha_4\, J_{\alpha\beta\mu\nu}$$

$$J_{\alpha\beta\mu\nu} = \int \frac{d^dp}{(2\pi)^d} p_\beta(p+k)_\alpha(p+k+q)_\mu(p+q)_\nu \tag{8-116}$$
$$\times \exp - (\alpha_{23}q^2 + \alpha_{34}k^2 + 2\alpha_3 k\cdot q + \textstyle\sum p^2 + 2\alpha_{23} q\cdot p + 2\alpha_{34} k\cdot p)$$

where we have used the shorthand notations $\alpha_{23} = \alpha_2 + \alpha_3$, etc., $\sum = \alpha_1 + \alpha_2 + \alpha_3 + \alpha_4$. It is convenient to shift the integration variable

$$p' = p + \frac{\alpha_{23}}{\sum} q + \frac{\alpha_{34}}{\sum} k \equiv p + Q$$

and to rewrite the numerator of (8-116) as

$$p_\beta(p+k)_\alpha(p+k+q)_\mu(p+q)_\nu = (p'-Q)_\beta(p'+k-Q)_\alpha(p'+k+q-Q)_\mu(p'+q-Q)_\nu$$

We then have to deal with integrals of the form

$$\int \frac{d^dp'}{(2\pi)^d} \exp(-\textstyle\sum p'^2) = \frac{1}{(4\pi\sum)^{d/2}}$$

$$\int \frac{d^dp'}{(2\pi)^d} p'_\mu p'_\nu \exp(-\textstyle\sum p'^2) = \frac{1}{2\sum(4\pi\sum)^{d/2}}\delta_{\mu\nu} \tag{8-117}$$

$$\int \frac{d^dp'}{(2\pi)^d} p'_\mu p'_\nu p'_\alpha p'_\beta \exp(-\textstyle\sum p'^2) = \frac{1}{4\sum^2(4\pi\sum)^{d/2}}(\delta_{\mu\nu}\delta_{\alpha\beta} + \delta_{\mu\alpha}\delta_{\nu\beta} + \delta_{\mu\beta}\delta_{\nu\alpha})$$

while the integrals involving odd powers of p' vanish. We also need the trace of γ matrices. From (8-103), we find

$$\begin{aligned}
&\operatorname{tr}(\gamma_\nu\gamma_\sigma\gamma_\mu\gamma_\lambda\gamma_\alpha\gamma_\rho\gamma_\beta\gamma_\lambda) \\
&= -4(d-2)[\delta_{\alpha\rho}(\delta_{\nu\sigma}\delta_{\mu\beta} - \delta_{\nu\mu}\delta_{\sigma\beta} + \delta_{\nu\beta}\delta_{\sigma\mu}) + \delta_{\beta\rho}(\delta_{\nu\sigma}\delta_{\mu\alpha} - \delta_{\nu\mu}\delta_{\sigma\alpha} + \delta_{\nu\alpha}\delta_{\sigma\mu}) \\
&\quad - \delta_{\alpha\beta}(\delta_{\nu\sigma}\delta_{\mu\rho} - \delta_{\nu\mu}\delta_{\rho\sigma} + \delta_{\nu\rho}\delta_{\sigma\mu})] \\
&\quad -4(6-d)[\delta_{\nu\beta}(\delta_{\sigma\alpha}\delta_{\mu\rho} - \delta_{\rho\sigma}\delta_{\mu\alpha}) + \delta_{\nu\alpha}(\delta_{\rho\sigma}\delta_{\mu\beta} - \delta_{\sigma\beta}\delta_{\mu\rho}) + \delta_{\nu\rho}(\delta_{\sigma\beta}\delta_{\mu\alpha} - \delta_{\sigma\alpha}\delta_{\mu\beta})]
\end{aligned} \tag{8-118}$$

The first bracket is symmetric under the interchange of α and β, whereas the second one is antisymmetric. A tedious but straightforward algebra yields

$$(4\pi\textstyle\sum)^{d/2}\exp\left[\dfrac{\alpha_{23}\alpha_{14}q^2+\alpha_{12}\alpha_{34}k^2+2(\alpha_1\alpha_3-\alpha_2\alpha_4)q\cdot k}{\sum}\right]J_{\alpha\beta\mu\nu}\operatorname{tr}(\gamma_\nu\gamma_\sigma\gamma_\mu\gamma_\lambda\gamma_\alpha\gamma_\rho\gamma_\beta\gamma_\lambda)$$

$$\begin{aligned}
&=-4(d-2)\{Q_\rho Q_\sigma(k^2-2q^2)+Q_\rho q_\sigma[-k\cdot(k+2q)+4q\cdot Q]+Q_\rho k_\sigma[(k+2q)\cdot q-2k\cdot Q]\\
&\quad+k_\rho q_\sigma Q\cdot(k-2q)-k_\rho k_\sigma Q\cdot(q-Q)+2q_\rho q_\sigma Q\cdot(k-Q)-\delta_{\rho\sigma}(k+q-Q)\cdot(q-Q)Q\cdot(k-Q)\}\\
&\quad+4(6-d)\{(\delta_{\rho\sigma}k^2-k_\rho k_\sigma)Q\cdot(Q-q)-(\delta_{\rho\lambda}k\cdot Q-Q_\rho k_\lambda)[\delta_{\lambda\sigma}k\cdot(Q-q)-k_\lambda(Q-q)_\sigma]\}\\
&\quad-\frac{4(d-2)}{2\sum}\{(2-d)(2q_\rho q_\sigma-k_\rho k_\sigma)+(d-2)\delta_{\rho\sigma}[(k+q-Q)\cdot(q-Q)-Q\cdot(k-Q)]\\
&\quad+2\delta_{\rho\sigma}(k-2Q)\cdot(k+2q-2Q)\}\\
&\quad+\frac{4(6-d)}{2\sum}\{[k^2\delta_{\rho\sigma}-k_\rho k_\sigma](d-2)\}-\frac{4(d-2)}{4\sum^2}[\delta_{\rho\sigma}d(d+2)]\\
&\equiv-4(d-2)A_1+4(6-d)A_2-\frac{4(d-2)}{2\sum}B_1+\frac{4(6-d)}{2\sum}B_2-\frac{4(d-2)}{4\sum^2}C_1
\end{aligned}\tag{8-119}$$

where the origin and the denomination of each term is clear and we have used the symmetry between ρ and σ.

The q integration must now be carried out. We write

$$\frac{1}{q^2}=\frac{\alpha_{23}\alpha_{14}}{\sum}\int_0^\infty dx\exp\left(-\frac{x\alpha_{23}\alpha_{14}q^2}{\sum}\right)\tag{8-120}$$

As the integrand involves

$$\exp\left[-\frac{(x+1)\alpha_{23}\alpha_{14}q^2+\alpha_{12}\alpha_{34}k^2+2(\alpha_1\alpha_3-\alpha_2\alpha_4)q\cdot k}{\sum}\right]$$

times powers of q, we perform a new shift of integration variable

$$q=q'-\alpha k\tag{8-121}$$

where

$$\alpha=\frac{\alpha_1\alpha_3-\alpha_2\alpha_4}{(x+1)\alpha_{14}\alpha_{23}}$$

Accordingly,

$$Q=\hat{\alpha}_{23}q'+zk\tag{8-122}$$

with

$$z=\frac{\alpha_{34}}{\sum}-\frac{\alpha_1\alpha_3-\alpha_2\alpha_4}{\sum(x+1)\alpha_{14}}\qquad\hat{\alpha}_{23}=\frac{\alpha_{23}}{\sum}$$

Also, we call y the coefficient of q^2 in the exponential

$$y\equiv(x+1)\frac{\alpha_{23}\alpha_{14}}{\sum}\equiv\frac{1}{x'}\frac{\alpha_{23}\alpha_{14}}{\sum}\tag{8-123}$$

Therefore, the desired amplitude $\Gamma^{(b)}_{\rho\sigma}(k)$ reads

$$\begin{aligned}\Gamma^{(b)}_{\rho\sigma}(k)=&-e^4\int_0^\infty d\alpha_1\cdots d\alpha_4\int_0^1\frac{dx'}{x'^2}\frac{1}{(4\pi\sum)^{d/2}}\frac{\alpha_{23}\alpha_{14}}{\sum}\\
&\times\exp\left[-\frac{\alpha_{12}\alpha_{23}\alpha_{34}\alpha_{41}-x'(\alpha_1\alpha_3-\alpha_2\alpha_4)^2}{\sum\alpha_{23}\alpha_{14}}k^2\right]\end{aligned}$$

$$\times \int \frac{d^d q'}{(2\pi)^d} e^{-yq'^2} \left[-4(d-2)A_1 + \cdots - \frac{4(d-2)}{4\sum^2} C_2 \right] \tag{8-124}$$

It is then a pure matter of patience to compute

$$(4\pi y)^{d/2} \int \frac{d^d q'}{(2\pi)^d} e^{-yq'^2} A_1 = k_\rho k_\sigma \frac{\alpha_{23}\alpha_{14} + 2(\alpha_1 - \alpha_3)(\alpha_4 - \alpha_2)}{2y\sum^2}(d-2)$$

$$+ \delta_{\rho\sigma}\Bigg[(k^2)^2 z(1-z)(\alpha+z)(1-\alpha-z)$$

$$+ \frac{k^2}{2y}\left\{ 2\hat{\alpha}_{23}(1-\hat{\alpha}_{23})[\alpha + 2z - 2z(\alpha+z)] + \frac{2(\alpha_1-\alpha_3)(\alpha_4-\alpha_2)}{\sum^2} \right\}$$

$$- \frac{k^2 d}{2y}[(\alpha+z)(1-\alpha-z)\hat{\alpha}_{23}^2 + (1-\hat{\alpha}_{23})^2 z(1-z)]$$

$$+ \frac{d(d+2)}{4y^2}\hat{\alpha}_{23}^2(1-\hat{\alpha}_{23})^2 \Bigg]$$

$$(4\pi y)^{d/2} \int \frac{d^d q'}{(2\pi)^d} e^{-yq'^2} B_1 = (2-d)(2\alpha^2 - 1)k_\rho k_\sigma$$

$$-(d-2)\delta_{\rho\sigma}\left\{ \frac{1}{y} + k^2[(\alpha+z)(1-\alpha-z) + z(1-z)] - \frac{d}{2y}[\hat{\alpha}_{23}^2 + (1-\hat{\alpha}_{23})^2] \right\}$$

$$+ 2\delta_{\rho\sigma}\left[k^2(1-2z)(1-2z-2\alpha) - 2\hat{\alpha}_{23}(1-\hat{\alpha}_{23})\frac{d}{y} \right]$$

$$(4\pi y)^{d/2} \int \frac{d^d q'}{(2\pi)^d} e^{-yq'^2} A_2 = \frac{d-2}{2y}\hat{\alpha}_{23}(1-\hat{\alpha}_{23})(k_\rho k_\sigma - k^2 \delta_{\rho\sigma})$$

Consequently,

$$\Gamma_{\rho\sigma}^{(A_2+B_2)} = -\frac{e^4}{(4\pi)^d} 4(6-d)(d-2)(k^2\delta_{\rho\sigma} - k_\rho k_\sigma) \int_0^1 \frac{dx'}{x'^2} \int_0^\infty \frac{d\alpha_1 \cdots d\alpha_4 \alpha_{23}\alpha_{14}(1-x')}{2\sum^{d/2+2} y^{d/2}}$$

$$\times \exp\left[-\frac{\alpha_{12}\alpha_{23}\alpha_{34}\alpha_{41} - x'(\alpha_1\alpha_3 - \alpha_2\alpha_4)^2}{\sum \alpha_{23}\alpha_{14}} k^2 \right]$$

$$= -\frac{e^4}{(4\pi)^d} 2(6-d)(d-2)(k^2\delta_{\rho\sigma} - k_\rho k_\sigma)(k^2)^{d-4}\Gamma(4-d)$$

$$\times \int_0^1 dx' \, x'^{d/2-2}(1-x') \int_0^1 \frac{d\alpha_1 \cdots d\alpha_4 \delta(1-\alpha_1-\alpha_2-\alpha_3-\alpha_4)}{(\alpha_{14}\alpha_{23})^{3d/2-5}}$$

$$\times [\alpha_{12}\alpha_{23}\alpha_{34}\alpha_{41} - x'(\alpha_1\alpha_3 - \alpha_2\alpha_4)^2]^{d/2-4}$$

After the change of variables,

$$\alpha_1 = \beta u \qquad \alpha_2 = (1-\beta)v \qquad \alpha_3 = (1-\beta)(1-v) \qquad \alpha_4 = \beta(1-u)$$

$$\alpha_{14} = \beta \qquad \alpha_{23} = 1-\beta$$

$$[\alpha_{12}\alpha_{23}\alpha_{34}\alpha_{41} - x'(\alpha_1\alpha_3 - \alpha_2\alpha_4)^2] = \beta(1-\beta)\{[\beta(1-u) + (1-\beta)(1-v)]$$

$$\times [\beta u + (1-\beta)v] - x'\beta(1-\beta)(u-v)^2\}$$

$$\int_0^1 d\alpha_1 \cdots d\alpha_4 \delta(1-\alpha_1-\alpha_2-\alpha_3-\alpha_4)F(\alpha_i) = \int_0^1 d\beta\, \beta(1-\beta) \int_0^1 du \int_0^1 dv\, F(\alpha_i)$$

it is easy to check that this integral is convergent at $d = 4$. Therefore, neglecting as before constant terms in $\bar{\omega}(k^2)$:

$$\Gamma^{(A_2+B_2)}_{\rho\sigma} \underset{d=4-\varepsilon}{\simeq} \frac{\alpha^2}{4\pi^2}\left(\frac{1}{\varepsilon} - \ln\frac{k^2}{\mu^2} + \cdots\right)(k_\rho k_\sigma - \delta_{\rho\sigma}k^2) \tag{8-125}$$

Similarly, in the C_1 contribution

$$\Gamma^{(C_1)}_{\rho\sigma} = \frac{e^4(d-2)d(d+2)\delta_{\rho\sigma}}{(4\pi)^d}(k^2)^{d-3}\Gamma(3-d)$$
$$\times \int_0^1 dx'\, x'^{d/2-2}\int_0^1 \frac{d\alpha_1\cdots d\alpha_4\,\delta(1-\alpha_1-\alpha_2-\alpha_3-\alpha_4)}{(\alpha_{14}\alpha_{23})^{3d/2-4}}$$
$$\times\left[\alpha_{12}\alpha_{23}\alpha_{34}\alpha_{41} - x'(\alpha_1\alpha_3-\alpha_2\alpha_4)^2\right]^{d-3}$$

the β integration is also convergent at $d = 4$. Hence

$$\Gamma^{(C_1)}_{\rho\sigma} \simeq -\frac{3\alpha^2}{\pi^2}\delta_{\rho\sigma}k^2\left(\frac{1}{\varepsilon} - \ln\frac{k^2}{\mu^2} + \cdots\right)\int_0^1 dx'\int_0^1 d\beta\int_0^1 du\int_0^1 dv$$
$$\times\{[\beta(1-u)+(1-\beta)(1-v)][\beta u+(1-\beta)v] - x'\beta(1-\beta)(u-v)^2\}$$
$$\simeq -\frac{\alpha^2}{\pi^2}\frac{13}{24}\delta_{\rho\sigma}k^2\left(\frac{1}{\varepsilon} - \ln\frac{k^2}{\mu^2} + \cdots\right) \tag{8-126}$$

On the contrary, the contributions from A_1 and B_1 possess internal divergences at $\beta = 0$ or 1. For instance, the terms proportional to $k_\rho k_\sigma$ read

$$\Gamma^{(A_1+B_1)}_{\rho\sigma} = \frac{4e^4}{(4\pi)^d}(d-2)^2\Gamma(4-d)(k^2)^{d-4}k_\rho k_\sigma\int_0^1 dx'\, x'^{d/2-2}\int_0^1 du\int_0^1 dv$$
$$\times\int_0^1 d\beta\,[\beta(1-\beta)]^{2-d/2}\{[\beta(1-u)+(1-\beta)(1-v)]$$
$$\times[\beta u+(1-\beta)v] - x'\beta(1-\beta)(u-v)^2\}^{d-4}$$
$$\times\left\{x'\frac{\beta^2u(1-u)+(1-\beta)^2v(1-v)}{\beta(1-\beta)} - x'[uv+(1-u)(1-v)]\right.$$
$$\left.+\frac{1+x'}{2} - x'^2(u-v)^2\right\} + \delta^{\rho\sigma}\cdots$$

But the required expansion at $d \simeq 4$ is easily obtained if we observe that, for any function $F(\beta)$ regular at 0 and 1,

$$\int_0^1 d\beta\,[\beta(1-\beta)]^{-1+\varepsilon/2}F(\beta) = \int_0^1 d\beta\,[\beta^{-1+\varepsilon/2} + (1-\beta)^{-1+\varepsilon/2}]F(\beta) + O(\varepsilon)$$

Ultimately, we obtain

$$\Gamma^{(A_1+B_1)}_{\rho\sigma} \simeq \frac{\alpha^2}{3\pi^2}k_\rho k_\sigma\left[\frac{1}{\varepsilon^2} - \frac{1}{\varepsilon}\left(\ln\frac{k^2}{\mu^2} + \gamma - \frac{7}{4}\right) + \frac{1}{2}\ln^2\frac{k^2}{\mu^2} + \ln\frac{k^2}{\mu^2}\left(\gamma - \frac{7}{4}\right) + \cdots\right]$$
$$-\frac{\alpha^2}{3\pi^2}\delta_{\rho\sigma}k^2\left[\frac{1}{\varepsilon^2} + \frac{1}{\varepsilon}\left(\frac{3}{8} - \gamma - \ln\frac{k^2}{\mu^2}\right) + \frac{1}{2}\ln^2\frac{k^2}{\mu^2} + \ln\frac{k^2}{\mu^2}\left(\gamma - \frac{3}{8}\right) + \cdots\right] \tag{8-127}$$

The total contribution of Fig. 8-17*b* is therefore

$$\Gamma^{(b)}_{\rho\sigma} = \frac{\alpha^2}{3\pi^2}\left\{k_\rho k_\sigma\left[\frac{1}{\varepsilon^2} + \frac{1}{\varepsilon}\left(\frac{5}{2} - \gamma - \ln\frac{k^2}{\mu^2}\right) + \frac{1}{2}\ln^2\frac{k^2}{\mu^2} + \ln\frac{k^2}{\mu^2}\left(\gamma - \frac{5}{2}\right) + \cdots\right]\right.$$
$$\left.-\delta_{\rho\sigma}k^2\left[\frac{1}{\varepsilon^2} + \frac{1}{\varepsilon}\left(\frac{11}{4} - \gamma - \ln\frac{k^2}{\mu^2}\right) + \frac{1}{2}\ln^2\frac{k^2}{\mu^2} + \ln\frac{k^2}{\mu^2}\left(\gamma - \frac{11}{4}\right) + \cdots\right]\right\} \tag{8-128}$$

We should still compute the contribution of the vertex counterterms, as depicted in Fig. 8-17b'_1, b'_2. But owing to the Ward identity, it is easy to see that they cancel exactly the contribution of the self-energy counterterms (a'_1, a'_2). Indeed, the latter are proportional to $(Z_2^{-1} - 1)^{[1]} = -(Z_2 - 1)^{[1]}$, while the former are proportional to $(Z_1 - 1)^{[1]}$. This is, of course, particular to quantum electrodynamics.

Adding the two contributions (8-113) and (8-128), we finally get the result

$$\Gamma^{[2]}_{\rho\sigma} = (k_\rho k_\sigma - \delta_{\rho\sigma}k^2)\bar{\omega}^{[2]}(k^2)$$
$$\bar{\omega}^{[2]}(k^2) = -\frac{\alpha^2}{\pi^2}\left(-\frac{1}{4\varepsilon} + \frac{1}{4}\ln\frac{k^2}{\mu^2} + \cdots\right) \tag{8-129}$$

We observe all the desired features:

1. The divergent terms of the form $(1/\varepsilon)\ln(k^2/\mu^2)$ which could not be eliminated by a local counterterm have cancelled.
2. The vacuum polarization tensor is transverse, as expected from the Ward identity. This holds for both the divergent and the finite parts. The former may therefore be renormalized by a transverse counterterm of order $\hbar^2$. For large euclidean k^2, and up to order α^2, the renormalized function $\bar{\omega}_R(k^2)$ is then given by

$$\bar{\omega}_R(k^2) = -\frac{\alpha}{3\pi}\ln\frac{k^2}{\mu^2} - \frac{\alpha^2}{4\pi^2}\ln\frac{k^2}{\mu^2} + O(\alpha^3) \tag{8-130}$$

   The foolhardy reader may check that the finite terms that we have neglected are also transverse, or (even better) that the complete expression of $\bar{\omega}(k^2)$ is independent of the gauge parameter λ.
3. Fortunately, the result (8-130) coincides with the one obtained by other authors! We conclude that the dependence on the normalization condition of the self-energy has dropped out in the sum $\Gamma^{(a)} + \Gamma^{(b)}$. This is obvious since the two counterterms actually cancel.
4. Unexpectedly, we find that the divergent terms in $1/\varepsilon^2$ [which in a conventional regularization would read $\ln^2(\Lambda^2/\mu^2)$] and, accordingly, the $\ln^2(k^2/\mu^2)$ have disappeared. This is a general property of the vacuum polarization, valid to all orders when we restrict ourselves to diagrams with a single fermion loop. Indeed, it is possible to fix order by order the gauge parameter λ so as to make $Z_1 = Z_2$ equal to one. Consequently, the selected subset of diagrams does not exhibit any internal divergence and its contribution to the gauge-invariant quantity $\bar{\omega}(k^2)$ behaves as a single power of $\ln(\Lambda^2/k^2)$.

This has led to interesting speculations by Johnson, Willey, and Baker and by Adler. Could it possibly be that the coefficient $f(\alpha)$ of the logarithmic behavior vanishes for a nonzero value of α? Since the same function multiplies $\ln\Lambda^2$ (or $1/\varepsilon$), this would seem to indicate that a reordering of the perturbation series might eliminate ultraviolet divergences for some appropriate value of the bare coupling.

NOTES

A complete treatment of renormalization for electrodynamics was first given by F. J. Dyson, *Phys. Rev.*, vol. 75, p. 1736, 1949. One can follow the early evolution of the subject in Schwinger's book, "Quantum Electrodynamics," Dover, New York, 1958. The important contributions made by N. N. Bogoliubov and O. S. Parasiuk, *Acta Math.*, vol. 97, p. 227, 1957, appear also in the textbook of the first author and D. V. Shirkov, "Introduction to the Theory of Quantized Fields," Interscience, New York, 1959, and the work of K. Hepp, *Comm. Math. Phys.*, vol. 2, p. 301, 1966, on the convergence of renormalized Feynman integrals is developed in detail in his "Théorie de la Renormalisation," Springer-Verlag, Berlin, 1969. In the early 1970s, W. Zimmermann gave a comprehensive account of the renormalization, including his method of subtracting the integrands, in "Lectures on Elementary Particles and Quantum Field Theory," Brandeis Summer Institute 1970, edited by S. Deser, M. Grisaru, and H. Pendleton, MIT Press, 1970, and in *Ann. of Phys.* (*N.Y.*), vol. 77, p. 536, 1973. The elegant but somehow abstract work of H. Epstein and V. Glaser, *Annales de l'Institut Poincaré*, vol. XIX, p. 211, 1973, definitely settles the matter in showing that the procedure preserves all properties of local field theory.

Each of the numerous regularization procedures has its own merits. The Pauli-Villars scheme has been quoted above (Chap. 7). Dimensional regularization is discussed by G. 't Hooft and M. Veltman, *Nucl. Phys.*, ser. B, vol. 44, p. 189, 1972. See also E. R. Speer in "Renormalization Theory," Erice Summer School 1975, edited by G. Velo and A. S. Wightman, D. Reidel Publishing Company, Dordrecht, Holland, and Boston, Mass., 1976.

The convergence of renormalized integrals was studied by S. Weinberg, *Phys. Rev.*, vol. 118, p. 838, 1960, and by K. Hepp, *Comm. Math. Phys.*, vol. 2, p. 301, 1966. For some aspects of this question in parametric space, see T. Appelquist, *Ann. of Phys.* (*N.Y.*), vol. 54, p. 27, 1969, M. Bergère and J. B. Zuber, *Comm. Math. Phys.*, vol. 35, p. 113, 1974, and M. Bergère and Y. M. P. Lam, *Comm. Math. Phys.*, vol. 39, p. 1, 1974.

Massless theories are considered in K. Symanzik, *Comm. Math. Phys.*, vol. 34, p. 7, 1973. For infrared cancellations see the work of T. Kinoshita, *J. Math. Phys.*, vol. 3, p. 650, 1962, T. D. Lee and M. Nauenberg, *Phys. Rev.*, ser. B, vol. 133, p. 1549, 1964, and T. Kinoshita and A. Ukawa, *Phys. Rev.*, ser. D, vol. 13, p. 1573, 1976. Our heuristic presentation has been borrowed from E. C. Poggio and H. R. Quinn, *Phys. Rev.*, ser. D, vol. 14, p. 578, 1976.

The computation of vacuum polarization to order α^2 was carried out by R. Jost and J. M. Luttinger, *Helv. Phys. Acta*, vol. 23, p. 201, 1950. See also the textbook of J. D. Bjorken and S. D. Drell, "Relativistic Quantum Fields," McGraw-Hill, New York, 1965. The possibility of a finite quantum electrodynamics has been studied by K. Johnson, R. Willey, and M. Baker, *Phys. Rev.*, vol. 163, p. 1699, 1967, and S. L. Adler, *Phys. Rev.*, ser. D, vol. 5, p. 3021, 1972.

CHAPTER

NINE

FUNCTIONAL METHODS

The path integral formalism of Feynman and Kac provides a unified view of quantum mechanics, field theory, and statistical models. Starting from the case of finitely many degrees of freedom it is generalized to include fermionic systems and then extended to infinite systems. The steepest-descent method of integration exhibits the close relationship with classical mechanics and allows us to recover ordinary perturbation theory. Among various applications, we deal here with the concept of effective action, quantization of constrained systems, and evaluation of high orders in perturbation theory.

9-1 PATH INTEGRALS

The original suggestion of an alternative presentation of quantum mechanical amplitudes in terms of path integrals stems from the work of Dirac (1933) and was brilliantly elaborated by Feynman in the 1940s. Schwinger developed an equivalent approach based on functional differentiation. This work was first regarded with some suspicion due to the difficult mathematics required to give it a decent status. In the 1970s it has, however, proved to be the most flexible tool in suggesting new developments in field theory and therefore deserves a thorough presentation.

9-1-1 The Role of the Classical Action in Quantum Mechanics

Let us return to quantum mechanics to inquire about the role of the lagrangian formalism as opposed to the hamiltonian one. For simplicity let us discuss a system with one degree of freedom described by the pair of conjugate operators Q and P satisfying

$$[Q, P] = i \tag{9-1}$$

We use capital letters for operators to distinguish them from their classical c-number counterparts. Let the hamiltonian be

$$H(P, Q) = \frac{P^2}{2m} + V(Q) \tag{9-2}$$

and denote by $|a\rangle$, $|b\rangle, \ldots$ the states of the system. We are going to seek an expression for the transition amplitude

$$\langle b(t') | a(t) \rangle = \langle b | \, e^{-iH(t'-t)} \, | a \rangle \tag{9-3}$$

With the usual representation of the commutation rules (9-1) we could introduce the square integrable wave functions

$$a(q) = \langle q | a \rangle$$

$$\tilde{a}(p) = \langle p | a \rangle$$

and attempt to solve the Schrödinger partial differential equation arising from (9-2). Here the improper states $|q\rangle$ and $|p\rangle$ are such that

$$Q|q\rangle = q|q\rangle \qquad P|p\rangle = p|p\rangle$$

$$\langle q' | q \rangle = \delta(q' - q) \qquad \langle p' | p \rangle = \delta(p' - p) \qquad \langle q | p \rangle = \langle p | q \rangle^* = \frac{1}{\sqrt{2\pi}} e^{ipq} \tag{9-4}$$

$$\langle q | P | p \rangle = p \langle q | p \rangle = \frac{1}{i} \frac{\partial}{\partial q} \langle q | p \rangle$$

Returning to the transition amplitude, we note that we may use the superposition principle to insert a complete set of states at intermediate times. This is analogous to the use of the Huyghens' principle in optics. We split the time evolution in infinitesimal steps $t \to t + \Delta t$ and first evaluate

$$\langle q_2(t + \Delta t) | q_1(t) \rangle = \langle q_2 | \, e^{-i\Delta t H} \, | q_1 \rangle$$

The boundary condition requires that when Δt goes to zero the above amplitude reduces to $\delta(q_2 - q_1)$. For small Δt we expect that the matrix element is negligible when q_2 differs appreciably from q_1, whether its modulus decreases or its phase oscillates very rapidly. This suggests that we may substitute for the potential operator $V(Q)$ its value $V(q_1)$ or $V(q_2)$, and amounts to an approximation of the type

$$e^{-i\Delta t H} \sim e^{-i\Delta t (P^2/2m)} \, e^{-i\Delta t V(Q)}$$

The terms we have neglected involve the commutators $[P^2/2m, V(Q)]$ multiplying higher powers of Δt. They will be negligible if V has a slow variation in the neighborhood of q_1 and q_2. This means that in the short interval Δt we neglect the transfer of potential into kinetic energy. For the matrix element we obtain the estimate:

$$\langle q_2(t+\Delta t)|q_1(t)\rangle \simeq \langle q_2|\, e^{-i\Delta t(P^2/2m)}\, e^{-i\Delta t V(Q)}\,|q_1\rangle$$
$$= \left(\frac{m\, e^{-i\pi/2}}{2\pi\Delta t}\right)^{1/2} \exp i\left[\frac{m}{2}\frac{(q_2-q_1)^2}{\Delta t} - \Delta t\, V(q_1)\right] \tag{9-5}$$

We observe the consistency of this procedure. For $|q_2 - q_1| \gg (\hbar\Delta t/m)^{1/2}$ the strong oscillations damp the amplitude so that the correction terms are vanishingly small provided $|V'/V|(\hbar\Delta t/m)^{1/2} \ll 1$, where V' is the derivative of the potential. This may be used to obtain a more symmetric form by replacing $V(q_1)$ by $\frac{1}{2}[V(q_1) + V(q_2)]$, for instance.

Expression (9-5) is suggestive. Indeed, $(q_2 - q_1)/\Delta t$ is the analog of the velocity $\dot{q}$, and the exponent reduces to $i\Delta t L(q, \dot{q})$ where $L(q, \dot{q})$ is the Lagrange function

$$L(q, \dot{q}) = \tfrac{1}{2}m\dot{q}^2 - V(q) \tag{9-6}$$

More precisely, let $q(t')$ be the trajectory from q_1 to q_2 during the time interval $(t, t+\Delta t)$ according to classical mechanics, i.e., obeying the principle of least action. As we have just seen, in the limit $\Delta t \to 0$, only values of the kernel for q_2 in the vicinity of q_1 matter, in a domain of order $(\Delta t/m)^{1/2}$. The phase of the transition amplitude is then equivalent to the action

$$I(2, 1) = \int_{q_1(t)}^{q_2(t+\Delta t)} dt'\, L(q, \dot{q}) \tag{9-7}$$

evaluated along the classical trajectory from (q_1, t) to $(q_2, t + \Delta t)$. The latter is not too different from a linear path

$$q(t') = \left(1 + \frac{t-t'}{\Delta t}\right)q_1 + \frac{t'-t}{\Delta t}\, q_2$$

Hence

$$I(2, 1) \simeq \frac{m}{2}\frac{(q_2-q_1)^2}{\Delta t} - \int_t^{t+\Delta t} dt'\, V[q(t')] \simeq \frac{m}{2}\frac{(q_2-q_1)^2}{\Delta t} - \Delta t V(q_1) \tag{9-8}$$

and

$$\langle q_2(t+\Delta t)|q_1(t)\rangle \simeq \left(\frac{m\, e^{-i\pi/2}}{2\pi\Delta t}\right)^{1/2} e^{iI[q_2(t+\Delta t), q_1(t)]} \tag{9-9}$$

In the limit $\Delta t \to 0$ we check that this reduces to $\delta(q_2 - q_1)$.

For a finite time interval the superposition principle allows us to write

$$\langle q_f, t_f|q_i, t_i\rangle = \lim_{n\to\infty} \int \prod_{1}^{n-1} dq_p \prod_{0}^{n-1} \langle q_{p+1}, t_{p+1}|q_p, t_p\rangle$$
$$t_p = t_i + \frac{p}{n}(t_f - t_i) \qquad t_0 \equiv t_i \qquad t_n \equiv t_f \tag{9-10}$$

Hence

$$\langle q_f, t_f | q_i, t_i \rangle = \lim_{n\to\infty} \int \prod_1^{n-1} dq_p \left[\frac{nm\, e^{-i\pi/2}}{2\pi(t_f - t_i)} \right]^{n/2} \exp\left[i \int_{t_i}^{t_f} dt\, L(q, \dot{q}) \right] \tag{9-11}$$

As a matter of fact the original phase was really the sum

$$I = I(n, n-1) + I(n-1, n-2) + \cdots + I(1, 0)$$

that is, the action computed along a broken path as shown on Fig. 9-1. When the interval $\Delta t = (t_f - t_i)/n$ shrinks to zero, it is conceivable that we would obtain the action along an arbitrary path. To define precisely the set of functions $q(t)$ which contribute essentially in this limit is a nontrivial mathematical matter. For smooth enough $V(q)$ they are the so-called lipschitzian functions of order $\frac{1}{2}$ in the present case, which means that $|q(t') - q(t)|$ is bounded by a constant times $|t' - t|^{1/2}$. The careful reader is referred to the literature on this point. A physicist will, however, proceed without fear and succeed in extracting a surprising amount of information from formulas such as (9-11) without the need to investigate in more detail the implications of the limit $n \to \infty$.

We shall use the shorthand notation

$$\langle q_f, t_f | q_i, t_i \rangle = \int \mathscr{D}(q) \exp\left[i \int_{t_i}^{t_f} dt\, L(q, \dot{q}) \right] \tag{9-12}$$

The "measure" on the functional space of trajectories $q(t)$ is denoted $\mathscr{D}(q)$ by including the product of normalizing factors $(m\, e^{-i\pi/2}/2\pi\Delta t)^{1/2}$, and $q(t)$ is restricted by the boundary conditions $q(t_i) = q_i$, $q(t_f) = q_f$.

A crucial property of the path integral reflects the superposition principle. It implies that if t lies in the interval t_i, t_f we must have

$$\int \mathscr{D}(q)\, e^{iI(f,i)} = \int dq(t) \int \mathscr{D}(q)\, e^{iI(f,t)} \int \mathscr{D}(q)\, e^{iI(t,i)} \tag{9-13}$$

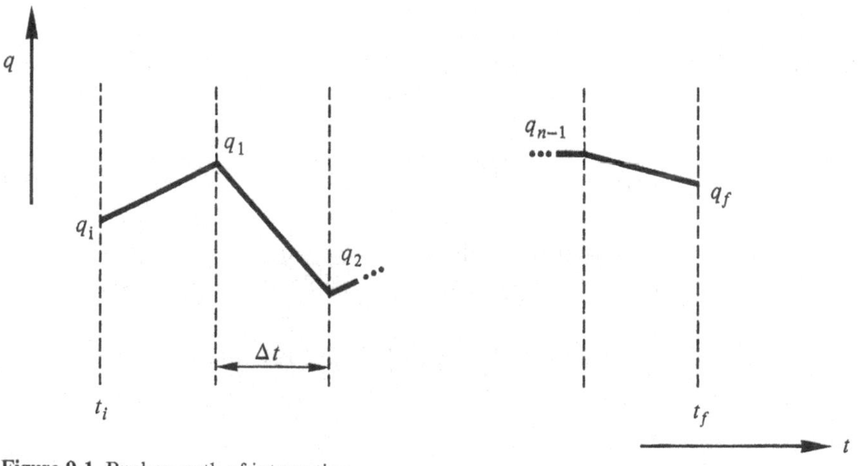

Figure 9-1 Broken path of integration.

If $\hbar$ is reinstated in these expressions the integrand becomes $e^{iI/\hbar}$. The classical limit $\hbar \to 0$ involves naturally the evaluation of the path integral by the stationary phase method. In this way the classical trajectories correspond to the extrema of the action. This presentation of quantum mechanics leads therefore in the most straightforward way to the dynamical principle of classical mechanics as a limiting case. If the classical trajectory from q_i to q_f is unique, we expect that up to a normalization factor, as $\hbar \to 0$,

$$\langle f \mid i \rangle \underset{\hbar \to 0}{\to} e^{(i/\hbar)I(f,i)}$$

Quantum mechanics can then be interpreted as implying fluctuations around this classical trajectory.

Mathematical subtleties associated with this formalism stem from the fact that the "measure" $\mathscr{D}(q)$ is complex and the integrand an oscillatory function. This suggests that an euclidean theory obtained after a Wick rotation to imaginary time might afford a more manageable mathematical entity. It now involves the matrix elements of the operator e^{-Ht} and corresponds to a shift from the Schrödinger equation to the heat, or diffusion, equation. Indeed, this was the situation considered by Wiener, who first discussed path integrals in a mathematical context. To summarize we have now answered the question raised in the title of this section, namely, we have recognized that the classical action appears in the quantum mechanical amplitudes as an exponent in the weight of trajectories in path integrals.

The formalism may be extended to matrix elements of operators. In this way we shall recover Schwinger's quantum mechanical action principle. Suppose, for instance, that we want to compute the matrix element of an operator $\mathcal{O}$ at an intermediate time t between t_i and t_f:

$$\begin{aligned}\langle q_f, t_f | \mathcal{O}(t) | q_i, t_i \rangle &= \langle q_f | e^{-iH(t_f - t)} \mathcal{O} e^{-iH(t - t_i)} | q_i \rangle \\ &= \int dq' \, dq'' \int \mathscr{D}(q) \, e^{iI(q_f, t_f; q'', t)} \langle q'' | \mathcal{O} | q' \rangle \int \mathscr{D}(q) \, e^{iI(q', t; q_i, t_i)}\end{aligned}$$

For simplicity let us assume that $\mathcal{O}$ is diagonal in the q representation

$$\langle q'' | \mathcal{O} | q' \rangle = \mathcal{O}(q') \delta(q'' - q')$$

The above expression reduces symbolically to

$$\langle q_f, t_f | \mathcal{O}(t) | q_i, t_i \rangle = \int \mathscr{D}(q) \, e^{iI(f,i)} \, \mathcal{O}[q(t)] \tag{9-14}$$

This may be further generalized to a time-ordered product of operators

$$\mathcal{O}_1(t_1)\mathcal{O}_2(t_2)\cdots \qquad t_1 \geq t_2 \geq \cdots$$

If they are all diagonal in the q representation we have

$$\langle q_f, t_f | \mathcal{O}_1(t_1)\mathcal{O}_2(t_2)\cdots | q_i, t_i \rangle = \int \mathscr{D}(q) \, e^{iI(f,i)} \, \mathcal{O}_1[q(t_1)]\mathcal{O}_2[q(t_2)]\cdots \tag{9-15}$$

Using these results we may consider the effect of an infinitesimal change in the dynamics between t_i and t_f (for instance, a slight variation of the potential or of the boundary conditions). The corresponding variation of the transition amplitude may be expressed as

$$\delta\langle q_f, t_f \,|\, q_i, t_i\rangle = i \int \mathscr{D}(q)\, e^{iI(f,i)}\; \delta I(f, i)$$

This has the form of (9-14) and (9-15) with an operator which we denote δI, depending, if the case may be, on the intermediate times, so that we can write in general

$$\delta\langle q_f, t_f \,|\, q_i, t_i\rangle = i\langle q_f, t_f |\, \delta I \,| q_i, t_i\rangle \tag{9-16}$$

This expression may be used as an infinitesimal substitute to functional integrals. For instance, if the variation is a result of an increase in the final time from t_f to $t_f + \delta t_f$ for which $\delta I = -H\delta t$, formula (9-16) reproduces Schrödinger's equation in the form

$$\delta\langle q_f, t_f \,|\, q_i, t_i\rangle = -i\langle q_f, t_f |\, H(t_f) \,| q_i, t_i\rangle\, \delta t_f$$

Let us check Eq. (9-11) for a simple harmonic oscillator

$$V(Q) = \frac{m\omega^2}{2}\, Q^2 \tag{9-17}$$

of frequency ω and energy levels $E_n = (n + \frac{1}{2})\omega$.

We apply (9-11) as it stands. On any short interval $(t, t + \Delta t)$ we approximate an arbitrary path by a linear one in such a way that the corresponding increment in the action is

$$\Delta I(2, 1) = \frac{m}{2}\left[\frac{(q_2 - q_1)^2}{\Delta t} - \omega^2 \Delta t\, \frac{q_2^2 + q_2 q_1 + q_1^2}{3}\right]$$

Set $t = t_f - t_i$. We find

$$\langle f \,|\, i\rangle = \lim_{n\to\infty} \int \prod_1^{n-1} dq_p \left(\frac{nm\, e^{-i\pi/2}}{2\pi t}\right)^{n/2} \exp\left\{i\frac{m}{2}\sum_1^n \left[\frac{n}{t}(q_p - q_{p-1})^2 - \frac{\omega^2 t}{3n}(q_p^2 + q_p q_{p-1} + q_{p-1}^2)\right]\right\} \tag{9-18}$$

We have to integrate the exponential of a quadratic form $e^{i(m/2)qMq}$ where q stands for the set $q_0 \equiv q_i, q_1, \ldots, q_{n-1}, q_n \equiv q_f$ and M is such that

$$M_{00} = M_{nn} = \frac{n}{t} - \frac{\omega^2}{3n}t$$

$$M_{k,k} = 2\left(\frac{n}{t} - \frac{\omega^2 t}{3n}\right)$$

$$M_{k,k+1} = M_{k-1,k} = -\left(\frac{n}{t} + \frac{\omega^2 t}{6n}\right) \qquad 1 \le k \le n-1$$

All other matrix elements are zero. The integration variables are $q_1, \ldots, q_{n-1}$. If N is the matrix obtained by deleting from M the rows and columns with indices 0 and n, we have

$$qMq = M_{00}(q_0^2 + q_n^2) + 2M_{01}(q_0 q_1 + q_{n-1}q_n) + \sum_{j,l=1}^{n-1} q_j N_{jl} q_l$$

Using the classical method to compute gaussian integrals, it follows that

$$\int d^{n-1}q\, e^{i(m/2)qNq} = \left(\frac{2\pi}{m} e^{i\pi/2}\right)^{(n-1)/2} (\det N)^{-1/2}$$

Linear terms in the exponential are taken into account by a translation, with the result that

$$\langle f \mid i \rangle = \lim_{n\to\infty} \left(\frac{nm\, e^{-i\pi/2}}{2\pi t}\right)^{n/2} \left(\frac{2\pi\, e^{i\pi/2}}{m}\right)^{(n-1)/2} (\det N)^{-1/2}$$

$$\times \exp\left\{\frac{im}{2}\left[M_{00}(q_0^2 + q_n^2) - M_{01}^2(N_{11}^{-1}q_0^2 + N_{n-1,n-1}^{-1}q_n^2 + 2N_{1,n-1}^{-1}q_0 q_n)\right]\right\}$$

Defining a to be equal to

$$a = \frac{1}{2}\frac{1 + \omega^2 t^2/6n^2}{1 - \omega^2 t^2/3n^2}$$

the determinant of N is given by

$$\det N = 2^{n-1}\left(\frac{n}{t} - \frac{\omega^2 t}{3n}\right)^{n-1} \det_{n-1} \begin{vmatrix} 1 & -a & 0 & \dots & \\ -a & 1 & -a & \dots & \\ 0 & -a & 1 & -a & \dots \\ \vdots & \vdots & \vdots & & \end{vmatrix}$$

The remaining determinant will be denoted $I_{n-1}(a)$. For fixed a it satisfies the recursion relation

$$I_p(a) = I_{p-1}(a) - a^2 I_{p-2}(a) \qquad I_0(a) = I_1(a) = 1$$

This is solved in the form

$$\begin{pmatrix} I_p(a) \\ I_{p-1}(a) \end{pmatrix} = \begin{pmatrix} 1 & -a^2 \\ 1 & 0 \end{pmatrix}^{p-1} \begin{pmatrix} 1 \\ 1 \end{pmatrix}$$

After diagonalization of the 2×2 matrix we find

$$I_{n-1}(a) = \frac{\lambda_+^n(a) - \lambda_-^n(a)}{\lambda_+(a) - \lambda_-(a)} \qquad \lambda_\pm(a) = \frac{1 \pm i\sqrt{4a^2 - 1}}{2} \underset{n\to\infty}{\sim} \tfrac{1}{2}e^{\pm i\omega t/n}$$

For large n we thus have

$$I_{n-1}(a) \sim \frac{n}{2^{n-1}} \frac{\sin \omega t}{\omega t}$$

Consequently,

$$\langle f \mid i \rangle = \left(\frac{m\omega\, e^{-i\pi/2}}{2\pi \sin \omega t}\right)^{1/2} \lim_{n\to\infty} \exp\left\{\frac{im}{2}\left[(q_i^2 + q_f^2)(M_{00} - M_{01}^2 N_{11}^{-1}) - 2q_i q_f M_{01}^2 N_{1,n-1}^{-1}\right]\right\}$$

When $\sin \omega t < 0$ a careful calculation allows one to choose the correct phase. Let us leave this point aside by assuming $\omega t < \pi$. To conclude, we have to evaluate

$$N_{1,1}^{-1} = \frac{t}{2n(1 - \omega^2 t^2/3n^2)} \frac{I_{n-2}(a)}{I_{n-1}(a)} \sim \frac{t}{n}\left(1 - \frac{\omega t}{n} \cot \omega t\right) + O(n^{-3})$$

$$N_{1,n-1}^{-1} = \frac{t}{2n(1 - \omega^2 t^2/3n^2)} \frac{a^{n-2}}{I_{n-1}(a)} \sim \frac{\omega t^2}{n^2 \sin \omega t} + O(n^{-3})$$

The final expression

$$\langle f|i\rangle = \left(\frac{m\omega\, e^{-i\pi/2}}{2\pi \sin \omega t}\right)^{1/2} \exp\left\{\frac{im\omega}{2}\left[(q_f^2 + q_i^2)\cot \omega t - \frac{2q_f q_i}{\sin \omega t}\right]\right\} \tag{9-19}$$

coincides, of course, with the result obtained through more traditional means. It is interesting to note that in this special case the coefficient of the exponential is equal to the value of the action for the actual classical trajectory over the finite time interval. This is a particular feature of quadratic hamiltonians.

As a second exercise the reader may apply the same method to a time-dependent hamiltonian describing the one-dimensional motion of a particle submitted to an external force:

$$H = \frac{P^2}{2m} - QF(t) \tag{9-20}$$

The corresponding amplitude is

$$\langle f|i\rangle_F = \left[\frac{m\, e^{-i\pi/2}}{2\pi(t_f - t_i)}\right]^{1/2} e^{iI(f,i)} \tag{9-21}$$

where again $I(f, i)$ stands for the action evaluated along the classical trajectory

$$I(f, i) = \frac{m}{2}\frac{(q_f - q_i)^2}{t_f - t_i} + \int_{t_i}^{t_f} dt\, F(t)\left(q_f\frac{t - t_i}{t_f - t_i} + q_i\frac{t_f - t}{t_f - t_i}\right)$$
$$+ \frac{1}{2m}\int_{t_i}^{t_f}\int_{t_i}^{t_f} dt'\, dt''\, F(t')\left[\frac{(t' - t_i)(t'' - t_i)}{t_f - t_i} - \operatorname{Inf}(t' - t_i, t'' - t_i)\right]F(t'') \tag{9-22}$$

The expression $G(t', t'') = t't''/T - \operatorname{Inf}(t', t'')$ is the symmetric Green function of the classical problem $\ddot{q} = F(t)/m$ already encountered in Chap. 1, satisfying the boundary conditions $G(0, t'') = G(T, t'') = 0$.

The analog of Eq. (9-16) yields

$$\frac{\delta}{i\delta F(t)}\langle f|i\rangle_F = \langle f|Q(t)|i\rangle_F \tag{9-23}$$

This result enables us to give an algebraic definition of the transition amplitude for an arbitrary hamiltonian of the form (9-2). In this case we may write

$$\int \mathscr{D}(q) \exp\left\{i\int_{t_i}^{t_f} dt\left[\frac{m\dot{q}^2}{2} - V(q)\right]\right\} = \exp\left\{-i\int_{t_i}^{t_f} dt\, V\left[\frac{\delta}{i\delta F(t)}\right]\right\}\langle f|i\rangle_F\bigg|_{F=0} \tag{9-24}$$

This formal expression becomes operational if we expand the exponential operator into a power series and generate the perturbation theory.

All the previous considerations can be extended to finitely many degrees of freedom without difficulty.

Let us now mention an apparently more general derivation of the evolution kernel (9-12). The idea is to use both the $|q\rangle$ and $|p\rangle$ bases of the Hilbert space. Let us indeed assume that the mixed matrix element of the hamiltonian may be given the form

$$\langle p|H|q\rangle = h(p, q)\frac{1}{\sqrt{2\pi}}e^{-ipq} \tag{9-25}$$

The "classical" value $h(p, q)$ will be equal to the quantum operator H ordered with the P to the left of the Q after substitution of c numbers to operators. This will not present any difficulty for hamiltonians of the form (9-2). For an infinitesimal time interval Δt we obtain from (9-25) the regular approximation

$$\langle p|e^{-i\Delta t H}|q\rangle \simeq \frac{1}{\sqrt{2\pi}}e^{-ipq - i\Delta t\, h(p,q)}$$

Therefore if we write

$$\langle q_f, t_f \mid q_i, t_i \rangle = \lim_{n\to\infty} \int \prod_1^n dp_s \prod_1^{n-1} dq_r \langle q_n \mid p_n \rangle \langle p_n \mid e^{-itH/n} \mid q_{n-1} \rangle \cdots \langle q_1 \mid p_1 \rangle \langle p_1 \mid e^{-itH/n} \mid q_0 \rangle$$

where as usual $q_0 \equiv q_i$, $q_n \equiv q_f$, $t = t_f - t_i$, we obtain

$$\langle q_f, t_f \mid q_i, t_i \rangle = \lim_{n\to\infty} \int \frac{\prod_1^n dp_s \prod_1^{n-1} dq_r}{(2\pi)^n}$$

$$\times \exp\left\{ i\left[p_n(q_n - q_{n-1}) + \cdots + p_1(q_1 - q_0) - \frac{t}{n} h(p_n, q_{n-1}) - \cdots - \frac{t}{n} h(p_1, q_0) \right] \right\}$$

$$= \int \mathscr{D}(p, q) \exp\left\{ i \int_{t_i}^{t_f} dt \, [p\dot{q} - h(p, q)] \right\} \tag{9-26}$$

The last expression is of course symbolic. Observe that we integrate over one more variable p in such a way that the boundary conditions only involve q_i and q_f. We recognize in the integrand the classical action

$$I(f, i) = \int_{t_i}^{t_f} dt \, [p\dot{q} - h(p, q)] \tag{9-27}$$

expressed in terms of the canonical variables p and q. If $h(p, q)$ depends on p quadratically, as is usual, the integral over p reduces to a gaussian. If we perform the latter, Eq. (9-26) reproduces (9-12).

The same formalism enables us to study scattering problems. For simplicity let us restrict ourselves to a one-dimensional short-range potential. Accordingly, the evolution reduces at large times to the free one dictated by H_0, and the S matrix is obtained as the limit

$$S = \lim_{\substack{t_f \to \infty \\ t_i \to -\infty}} e^{it_f H_0} e^{-i(t_f - t_i)H} e^{-it_i H_0} \tag{9-28}$$

Its matrix elements between states ψ_f and ψ_i are given by

$$\langle \psi_f \mid S \mid \psi_i \rangle = \lim \int dq_f \, dq_i \, \psi_f^*(q_f, t_f) \langle q_f, t_f \mid q_i, t_i \rangle \psi_i(q_i, t_i) \tag{9-29}$$

The notations are such that

$$\psi(q) = \int \frac{dp}{\sqrt{2\pi}} e^{ipq} \tilde{\psi}(p)$$

and $\psi(q, t)$ is defined as the solution of the free Schrödinger equation such that $\psi(q, 0) = \psi(q)$, that is,

$$\psi(q, t) = \int \frac{dp}{\sqrt{2\pi}} e^{i(pq - p^2 t/2m)} \tilde{\psi}(p) \tag{9-30}$$

As $|t| \to \infty$, $\psi(q, t)$ may be estimated by the stationary phase method as

$$\psi(q, t) \sim \left(\frac{m}{|t|} \right)^{1/2} e^{-i\pi/4 \operatorname{sign}(t) + imq^2/2t} \tilde{\psi}\left(\frac{mq}{t} \right) \tag{9-31}$$

This may be introduced in Eq. (9-29) where we change variables according to $q_f = p_f t_f/m$, $q_i = p_i t_i/m$, with the result

$$\langle \psi_f | S | \psi_i \rangle = \lim \int dp_f \, dp_i \left(\frac{e^{-i\pi} |t_f t_i|}{m^2} \right)^{1/2}$$
$$\times \tilde{\psi}_f^*(p_f) \, e^{-ip_f^2 t_f/2m} \left\langle \frac{p_f}{m} t_f, t_f \middle| \frac{p_i}{m} t_i, t_i \right\rangle e^{ip_i^2 t_i/2m} \, \tilde{\psi}_i(p_i) \qquad \text{(9-32)}$$

which may be checked in the trivial case $V = 0$, $S = 1$. When the path integral is used to express the matrix element of the time evolution operator, Eq. (9-32) admits an interesting interpretation. Indeed, it follows from this equation that the integral involves paths behaving asymptotically as free trajectories:

$$t \to \infty \qquad q(t) \to \frac{p_f t}{m}$$

$$t \to -\infty \qquad q(t) \to \frac{p_i t}{m}$$

In one dimension conservation of energy implies, of course, that $p_f = \pm p_i$ but the formalism is easily generalized to more interesting many-dimensional cases.

As an application let us present the eikonal approximation for three-dimensional scattering on a finite range potential $V(\mathbf{r})$ at large energy and small transfer.

We start from the representation of the evolution operator

$$\langle \mathbf{r}_2, t_2 | \mathbf{r}_1, t_1 \rangle = \int \mathscr{D}[\mathbf{r}(t)] \exp \left\{ i \int_{t_1}^{t_2} dt \left[\tfrac{1}{2} m \dot{\mathbf{r}}^2 - V[\mathbf{r}(t)] \right] \right\} \qquad \text{(9-33)}$$

In the free case ($V = 0$) this reduces to

$$\left[\frac{m \, e^{-i\pi/2}}{2\pi(t_2 - t_1)} \right]^{3/2} e^{im(\mathbf{r}_2 - \mathbf{r}_1)^2/2(t_2 - t_1)}$$

Under the physical conditions of interest, the path integral is dominated by the classical extremal trajectories, which for large impact parameters can be approximated by straight lines:

$$\mathbf{r}(t) = \mathbf{r}_1 + \frac{t - t_1}{t_2 - t_1} (\mathbf{r}_2 - \mathbf{r}_1) \qquad \text{(9-34)}$$

After factorization of energy conservation and a Fourier transformation to momentum space we readily derive the following expression for the matrix elements of the transition operator $T = (S - 1)/i$:

$$\left\langle \mathbf{p} - \frac{\mathbf{q}}{2} \middle| T \middle| \mathbf{p} + \frac{\mathbf{q}}{2} \right\rangle \simeq \frac{i|\mathbf{p}|}{m} \int \frac{d^2 b}{(2\pi)^3} e^{i\mathbf{q}\cdot\mathbf{b}} \left\{ \exp\left[-i \int_{-\infty}^{+\infty} dt \, V\left(\mathbf{b} + \frac{\mathbf{p}t}{m} \right) \right] - 1 \right\}$$

$$\mathbf{p}^2 \to \infty \qquad \frac{\mathbf{q}^2}{\mathbf{p}^2} \to 0 \qquad \text{(9-35)}$$

where $\mathbf{b}$ ($|\mathbf{b}|$ is the impact parameter) and $\mathbf{q}$ are two-dimensional vectors in the plane perpendicular to the average momentum.

This approximation has been generalized to the relativistic case. Consider the scattering

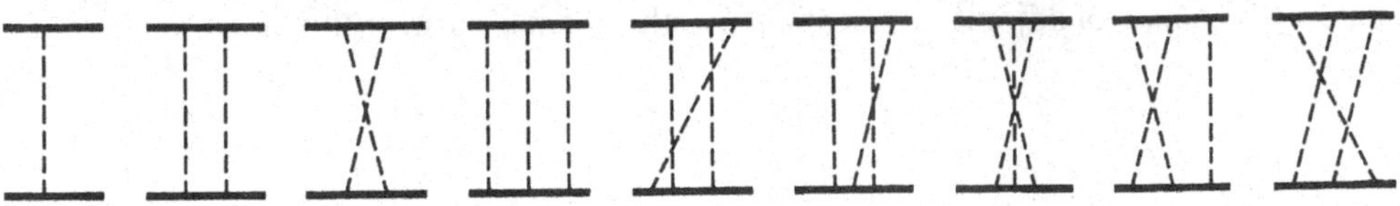

Figure 9-2 Crossed ladder diagrams.

amplitude corresponding to the set of crossed ladder diagrams, with the exchange of scalar bosons of mass μ and coupling constant g (Fig. 9-2). It can be represented by a path integral analogous to (9-32) and (9-33). The resulting amplitude for large s (square of the center of mass energy) and small transfer $t = -\mathbf{q}^2$ is given by

$$T(s,t) \simeq -2s \int d^2\mathbf{b}\, e^{i\mathbf{q}\cdot\mathbf{b}} \left[\exp\left(\frac{ig^2}{4\pi s}\int \frac{d^2\mathbf{k}}{2\pi}\frac{e^{i\mathbf{k}\cdot\mathbf{b}}}{\mathbf{k}^2+\mu^2}\right) - 1\right] \tag{9-36}$$

The reader may determine the behavior of (9-36) as a function of $s \to \infty$ and generalize further to electrodynamics where the corresponding amplitudes exhibit bound-state poles with the correct nonrelativistic limit [see Sec. 2-3-2].

9-1-2 Trajectories in the Bargmann-Fock Space

Up to now we have encountered integrals over trajectories in configuration space, Eq. (9-12), or phase space, Eq. (9-26). In both cases the boundary conditions involved the q or p at each limit. We want to introduce a new type of trajectories suited to a generalization to field theory. It is strongly inspired by the harmonic oscillator case since a field theory may be considered as an assembly of interacting oscillators. This will provide an easy parallel with the treatment of fermionic systems.

We make use of the coherent states of Bargmann and Fock discussed in Chap. 3. This gives a representation of destruction and creation operators

$$a = \frac{Q+iP}{\sqrt{2}} \qquad a^\dagger = \frac{Q-iP}{\sqrt{2}} \qquad [a, a^\dagger] = 1 \tag{9-37}$$

in a space of analytic functions of a complex variable denoted with a complex conjugation bar $\bar{\alpha}$ or $\bar{z}$. The reason for this choice will appear in the sequel. If the original oscillator problem involved a frequency ω we could first perform a canonical transformation $Q \to Q\omega^{-1/2}$, $P \to P\omega^{1/2}$ before introducing a and $a^\dagger$ according to (9-37). The analytic functions under consideration generate a Hilbert space with scalar product

$$\langle g | f \rangle = \int \frac{dz\, d\bar{z}}{2i\pi}\, e^{-z\bar{z}}\, \bar{g}(z) f(\bar{z}) \tag{9-38}$$

and we have the correspondence

$$a \to \frac{\partial}{\partial \bar{z}} \qquad a^\dagger \to \bar{z} \tag{9-39}$$

leading to a pair of adjoint operators. An orthonormal basis in this space is

$$f_n(\bar{z}) = \frac{\bar{z}^n}{\sqrt{n!}} \tag{9-40}$$

A unitary transformation maps this description on the more conventional set of square integrable functions of the configuration variable q. To the f_n correspond the well-known Hermite wave functions of the oscillator. Of course, the f_n are eigenfunctions of the operator $H = a^\dagger a = \bar{z}\partial/\partial\bar{z}$ with eigenvalue n.

In the conventional approach we characterize an operator A by its matrix elements $\langle q'|A|q\rangle$ in such a way that the action of A on a state vector ψ yields the wave function

$$[A\psi](q) = \int dq' \, \langle q|A|q'\rangle \psi(q')$$

Let us proceed similarly in the Bargmann-Fock space. If $|n\rangle$ stands for the state corresponding to the function f_n in Eq. (9-40) we write

$$A = \sum_{n,m} |n\rangle A_{n,m} \langle m| \qquad A_{n,m} = \langle n|A|m\rangle \tag{9-41}$$

Accordingly, for any state f,

$$\begin{aligned}\langle \bar{z}|Af\rangle = [Af](\bar{z}) &= \sum_{n,m} \frac{\bar{z}^n}{\sqrt{n!}} A_{n,m} \langle m|f\rangle \\ &= \sum_{n,m} \int \frac{d\xi\, d\bar{\xi}}{2\pi i} e^{-\bar{\xi}\xi} \frac{\bar{z}^n}{\sqrt{n!}} A_{n,m} \frac{\xi^m}{\sqrt{m!}} f(\bar{\xi})\end{aligned} \tag{9-42}$$

The kernel naturally associated to A, and denoted $A(\bar{z}, \xi)$, such that

$$[Af](\bar{z}) = \int \frac{d\xi\, d\bar{\xi}}{2\pi i} e^{-\bar{\xi}\xi} A(\bar{z}, \xi) f(\bar{\xi})$$

is therefore

$$A \to A(\bar{z}, \xi) = \sum_{n,m} \frac{\bar{z}^n}{\sqrt{n!}} A_{n,m} \frac{\xi^m}{\sqrt{m!}} \tag{9-43}$$

For sufficiently regular A, the function $A(\bar{z}, \xi)$ is an analytic function of the two complex variables ξ and $\bar{z}$. Incidentally, it is the desire to write the operators in this form which justified the choice of the complex conjugation bar over the argument of the analytic functions.

The representation (9-43) enjoys the following fundamental superposition property as a consequence of the orthogonality of the f_n:

$$A_1 A_2(\bar{z}, \xi) = \int \frac{d\eta\, d\bar{\eta}}{2\pi i} e^{-\bar{\eta}\eta} A_1(\bar{z}, \eta) A_2(\bar{\eta}, \xi) \tag{9-44}$$

Returning to Eq. (9-41) we rewrite it as

$$A = \sum_{n,m} \frac{a^{\dagger n}}{\sqrt{n!}} |0\rangle A_{n,m} \langle 0| \frac{a^m}{\sqrt{m!}}$$

We recall that the projector on the ground state $|0\rangle\langle 0|$ may be expressed in terms of a normal product as

$$|0\rangle\langle 0| = :e^{-a^\dagger a}: \tag{9-45}$$

Consequently,

$$A = \sum_{n,m} A_{n,m} :a^{\dagger n} e^{-a^\dagger a} a^m: = \sum_{n,m} A^N_{n,m} \frac{a^{\dagger n} a^m}{\sqrt{n!\,m!}} \tag{9-46}$$

This normal form suggests the definition of a normal kernel to represent the operator A. We denote it $A^N(\bar{z}, z)$ to distinguish it from our previous definition with $\bar{z}$ and z considered as independent variables

$$A^N(\bar{z}, z) = \sum_{n,m} \frac{\bar{z}^n}{\sqrt{n!}} A^N_{n,m} \frac{z^m}{\sqrt{m!}} \qquad A = :A^N(a^\dagger, a): \tag{9-47}$$

To obtain the relation between $A^N(\bar{z}, z)$ and $A(\bar{z}, z)$ we may either use Eq. (9-46) or observe that the Hilbert space of entire functions is endowed with a reproducing kernel analogous to the Dirac delta function in the form

$$\begin{aligned} f(\bar{z}) &= \sum_n \frac{\bar{z}^n}{\sqrt{n!}} \langle n | f \rangle = \sum_n \int \frac{d\xi\, d\bar{\xi}}{2\pi i} e^{-\bar{\xi}\xi} \frac{\bar{z}^n \xi^n}{n!} f(\bar{\xi}) \\ &= \int \frac{d\xi\, d\bar{\xi}}{2\pi i} e^{-\bar{\xi}\xi + \bar{z}\xi} f(\bar{\xi}) \end{aligned} \tag{9-48}$$

For the operator $a^{\dagger n} a^m$,

$$\begin{aligned} [a^{\dagger n} a^m f](\bar{z}) &= \bar{z}^n \frac{d^m}{d\bar{z}^m} f(\bar{z}) = \int \frac{d\xi\, d\bar{\xi}}{2\pi i} e^{-\bar{\xi}\xi + \bar{z}\xi} \bar{z}^n \frac{d^m}{d\bar{\xi}^m} f(\bar{\xi}) \\ &= \int \frac{d\xi\, d\bar{\xi}}{2\pi i} e^{-\bar{\xi}\xi + \bar{z}\xi} \bar{z}^n \xi^m f(\bar{\xi}) \end{aligned}$$

we find

$$A(\bar{z}, \xi) = e^{\bar{z}\xi} A^N(\bar{z}, \xi) \tag{9-49}$$

and this relation extends by linearity to an arbitrary operator.

Starting from Eqs. (9-44) and (9-49) we may compute the kernel corresponding to the evolution operator for a quantum mechanical problem. We assume the hamiltonian to be given in normal form in terms of the operators $a^\dagger$ and a. Its normal kernel is simply obtained by substituting complex numbers for the creation and annihilation operators. We denote this function by $h(\bar{z}, \xi)$. For an infinitesimal time interval Δt we have approximately

$$U(\bar{z}, \xi, \Delta t) \simeq e^{\bar{z}\xi - i\Delta t\, h(\bar{z}, \xi)} \tag{9-50}$$

This reduces correctly to the kernel of the identity when $\Delta t \to 0$. Repeated application of Eq. (9-44) over a finite interval leads to the path integral

$$U(\bar{z}_f, t_f; z_i, t_i) = \lim_{n\to\infty} \int \prod_1^{n-1} \frac{dz_k\, d\bar{z}_k}{2\pi i}$$

$$\times \exp\left[\bar{z}_n z_{n-1} - \sum_1^{n-1} \bar{z}_k z_k + \bar{z}_1 z_0 - i\,\frac{t_f - t_i}{n} \sum_0^{n-1} h(\bar{z}_{k+1}, z_k)\right] \tag{9-51}$$

We use for the limit the symbolic notation

$$U(\bar{z}_f, t_f; z_i, t_i) = \int \mathscr{D}(z, \bar{z}) \exp\left\{\frac{\bar{z}_f z_f + \bar{z}_i z_i}{2} + i \int_{t_i}^{t_f} dt \left[\frac{\dot{\bar{z}} z - \bar{z}\dot{z}}{2i} - h(\bar{z}, z)\right]\right\} \tag{9-52}$$

In this expression the integration variables $z(t_f)$ and $\bar{z}(t_i)$ remain independent from $\bar{z}(t_f)$ and $z(t_i)$, which are fixed by the boundary conditions.

We recognize once again the classical action in the exponent of (9-52). Indeed the form $p\, dq$ or, rather, $\frac{1}{2}(p\, dq - q\, dp)$ may be written according to (9-37) as

$$\tfrac{1}{2}(p\, dq - q\, dp) = \frac{1}{2i}(z\, d\bar{z} - \bar{z}\, dz) \tag{9-53}$$

which requires to treat z and $\bar{z}$ as independent variables. Clearly, all that is said above immediately generalizes to time-dependent hamiltonians and to several degrees of freedom.

Let us compute the evolution operator for an harmonic oscillator driven by an external time-dependent force in such a way that

$$H = \omega a^\dagger a - f(t) a^\dagger - \bar{f}(t) a \tag{9-54}$$

Here $\bar{f}$ is the complex conjugate of f. As illustrated by Eqs. (9-19) and (9-21) the result should be proportional to $e^{iI(f,i)}$ where $I(f, i)$ is the action computed on the extremal trajectory satisfying the classical equations of motion

$$\dot{z} + i[\omega z - f(t)] = 0 \qquad z(t_i) = z_i$$

$$\dot{\bar{z}} - i[\omega \bar{z} - \bar{f}(t)] = 0 \qquad \bar{z}(t_f) = \bar{z}_f$$

The solution is

$$z(t) = z_i\, e^{-i\omega(t - t_i)} + i \int_{t_i}^{t} dt'\, e^{-i\omega(t - t')} f(t')$$

$$\bar{z}(t) = \bar{z}_f\, e^{i\omega(t - t_f)} + i \int_{t}^{t_f} dt'\, e^{i\omega(t - t')} \bar{f}(t')$$

These quantities are obviously not conjugate due to the dissymmetry of boundary conditions. The corresponding exponent of the path integral along this trajectory is

$$iI = \tfrac{1}{2}(\bar{z}_f z_f + \bar{z}_i z_i) + \frac{1}{2}\int_{t_i}^{t_f} dt\, [\dot{\bar{z}} z - \bar{z}\dot{z} - 2ih(\bar{z}, z, t)]$$

$$= \bar{z}_f\, e^{-i\omega(t_f - t_i)} z_i + i \int_{t_i}^{t_f} dt\, [\bar{z}_f\, e^{-i\omega(t_f - t)} f(t) + \bar{f}(t)\, e^{-i\omega(t - t_i)} z_i] \tag{9-55a}$$

$$- \int_{t_i}^{t_f}\int_{t_i}^{t_f} dt\, dt'\, \bar{f}(t)\, e^{-i\omega(t - t')} f(t') \theta(t - t')$$

Explicit calculation of the path integral gives simply

$$U(\bar{z}_f, t_f; z_i, t_i) = e^{iI} \tag{9-55b}$$

Thus the case of the driven oscillator is very simple in this formalism and the kernel is regular everywhere.

The gaussian integration formula used repeatedly in these evaluations is worth quoting here as it occurs as a cornerstone of the application of path integrals. If A stands for the matrix of a nonsingular quadratic form the hermitian part of which is positive and z and u stand for column vectors of complex numbers,

$$\int \prod_1^N \frac{dz_k\, d\bar{z}_k}{2\pi i}\, e^{-\bar{z}Az + \bar{u}z + u\bar{z}} = (\det A)^{-1}\, e^{\bar{u}A^{-1}u} \tag{9-56}$$

Note that the exponent on the right-hand side is the saddle-point value of the exponent of the integrand.

9-1-3 Fermion Systems

Since path integrals exhibit the close relationship between classical and quantum mechanics it would seem a priori that we would encounter some difficulties when extending the treatment to fermions. Fortunately the relevant construction in terms of an anticommuting algebra has been devised by Berezin and we have already made some use of it in Chap. 4.

Let us start from a two-level system with the two operators a and $a^\dagger$ fulfilling

$$\{a, a^\dagger\} = 1 \qquad a^2 = a^{\dagger 2} = 0 \tag{9-57}$$

We shall try to represent them as acting on a Hilbert space of "analytic functions." The analogy with the infinite series in z and $\bar{z}$ used previously for bosons suggests the following. Let us consider series with complex coefficients in two anticommuting variables η and $\bar{\eta}$, that is, such that

$$\eta\bar{\eta} + \bar{\eta}\eta = 0 \qquad \eta^2 = \bar{\eta}^2 = 0 \tag{9-58}$$

These series reduce to polynomials of the form

$$P(\bar{\eta}, \eta) = p_0 + p_1\bar{\eta} + \tilde{p}_1\eta + p_{12}\eta\bar{\eta} \tag{9-59}$$

The set of these polynomials of dimension $2^2 = 4$ may be identified with the exterior algebra on a two-dimensional vector space (generated by homogeneous polynomials of degree one). The associative multiplication is defined in agreement with the rules (9-58). We also introduce the linear derivation

$$\partial P = \tilde{p}_1 + p_{12}\bar{\eta} \qquad \bar{\partial} P = p_1 - p_{12}\eta \tag{9-60}$$

On each monomial they act by suppressing the corresponding η (for ∂) or $\bar{\eta}$ (for $\bar{\partial}$) once the latter has been brought to the left; otherwise they give zero.

Define the subset of analytic functions by the condition

$$\partial f = 0 \tag{9-61}$$

which implies that f depends only on $\bar{\eta}$.

Note that

$$\partial^2 P = \bar{\partial}^2 P = 0 \qquad \text{and} \qquad \bar{\partial}\partial P = -\partial\bar{\partial}P = p_{12}$$

which means that polynomials in the derivative operators have the same structure as the original anticommuting algebra. The reader will observe that $\partial(P_1P_2)$ is not equal to $\partial P_1 P_2 + P_1 \partial P_2$ and will easily find the correct version of this identity. The construction may easily be generalized to several degrees of freedom. With $2n$ degrees of freedom, the exterior algebra will be of dimension 2^{2n} and the space of analytic functions of dimension 2^n.

Returning to the case $n = 1$ we write an analytic function as

$$f = f_0 + f_1\bar{\eta}$$

and define a scalar product such that

$$(g, f) = \bar{g}_0 f_0 + \bar{g}_1 f_1 \tag{9-62}$$

Here a bar on scalars means complex conjugation. Is it possible to represent this scalar product as an integral as in the boson case? The answer is "yes," provided we identify derivation and integration as follows. The integral symbol is defined by linearity starting from the requirements

$$\int d\bar{\eta}\,\bar{\eta} = 1 \qquad \int d\eta\,\eta = 1 \qquad \int d\bar{\eta}\,1 = \int d\eta\,1 = 0 \tag{9-63}$$

If we also agree that $d\eta$ and $d\bar{\eta}$ anticommute and that the rules (9-63) apply when $d\bar{\eta}$ and $\bar{\eta}$ or $d\eta$ and η are brought next to each other, we indeed see that integrals and derivatives are identical. Consequently,

$$\int d\eta\,P = \partial P \qquad \int d\bar{\eta}\,P = \bar{\partial}P \qquad \int d\bar{\eta}\,d\eta\,P = \bar{\partial}\partial P \tag{9-64}$$

As a consequence the integral of a derivative vanishes ($\partial^2 = \bar{\partial}^2 = 0$) and the procedure is easily extended to several degrees of freedom.

We have the possibility to make changes of integration variables under the integral sign. If we limit ourselves for the time being to linear transformations which automatically respect the structure (9-58), i.e., of the form

$$\begin{pmatrix}\eta\\ \bar{\eta}\end{pmatrix} = A\begin{pmatrix}\xi\\ \bar{\xi}\end{pmatrix}$$

where A is a nonsingular c-number matrix, a substitution in any polynomial P yields

$$P(\eta, \bar{\eta}) = Q(\xi, \bar{\xi})$$

and in particular

$$p_{12}\eta\bar{\eta} = q_{12}\xi\bar{\xi} = (p_{12} \det A)\xi\bar{\xi}$$

As a result

$$\int d\bar{\eta}\, d\eta\, P(\eta, \bar{\eta}) = \int d\bar{\xi}\, d\xi (\det A)^{-1} Q(\xi, \bar{\xi}) \tag{9-65}$$

This implies a rule differing from the usual one in the sense that the standard jacobian appears inverted since

$$\det A = J\left(\frac{\eta, \bar{\eta}}{\xi, \bar{\xi}}\right)$$

A basis being chosen to allow for a definition of analytic functions, let us define complex conjugation as

$$\bar{f}(\eta) = \bar{f}_0 + \bar{f}_1 \eta \tag{9-66}$$

We then verify that

$$(g, f) = \int d\bar{\eta}\, d\eta\, e^{-\bar{\eta}\eta}\, \bar{g}(\eta) f(\bar{\eta}) \tag{9-67}$$

analogous to formula (9-38) for bosons. We have now the following representation of a and $a^\dagger$ in terms of a pair of adjoint operators:

$$a \to \bar{\partial} \qquad a^\dagger \to \bar{\eta} \tag{9-68}$$

Obviously $a^2 = a^{\dagger 2} = 0$. Furthermore,

$$a(a^\dagger f) = f_0 \qquad a^\dagger(af) = f_1 \bar{\eta}$$

As a result

$$aa^\dagger + a^\dagger a = 1$$

and $\qquad (g, af) = \bar{g}_0 f_1 \qquad (f, a^\dagger g) = \bar{f}_1 g_0 = \overline{(g, af)}$

More generally, we can assign an integral kernel to a linear operator on the space $\mathscr{H}$ of analytic functions. Consider the orthonormal states $|0\rangle$ and $|1\rangle$ such that $a|0\rangle = 0$, $a^\dagger|0\rangle = |1\rangle$, corresponding to the functions 1 and $\bar{\eta}$. Let us write

$$A = \sum_{n,m} |n\rangle A_{n,m} \langle m|$$

$$(Af)(\bar{\eta}) = \int d\bar{\xi}\, d\xi\, e^{-\bar{\xi}\xi} A(\bar{\eta}, \xi) f(\bar{\xi}) \tag{9-69}$$

$$A(\bar{\eta}, \xi) = \sum_{n,m} \bar{\eta}^n A_{n,m} \xi^m$$

As in the case of bosons we represent the projector on the ground state as

$$|0\rangle\langle 0| = :e^{-a^\dagger a}: = 1 - a^\dagger a \tag{9-70}$$

and rewrite A in normal form:

$$A = \sum_{n,m} A_{n,m} : a^{\dagger n} e^{-a^\dagger a} a^m : = \sum_{n,m} A^N_{n,m} a^{\dagger n} a^m \tag{9-71}$$

The associated normal kernel

$$A^N(\bar{\eta}, \eta) = \sum_{n,m} A^N_{n,m} \bar{\eta}^n \eta^m \tag{9-72}$$

is related to the previous one through

$$A(\bar{\eta}, \eta) = e^{\bar{\eta}\eta} A^N(\bar{\eta}, \eta) \tag{9-73}$$

For instance, the integral kernel of the identity is $e^{\bar{\eta}\eta}$ so that

$$f(\bar{\eta}) = \int d\bar{\xi}\, d\xi\, e^{-\bar{\xi}\xi + \bar{\eta}\xi} f(\bar{\xi}) \tag{9-74}$$

while the product of operators is given by

$$A_1 A_2(\bar{\eta}, \eta) = \int d\bar{\xi}\, d\xi\, e^{-\bar{\xi}\xi} A_1(\bar{\eta}, \xi) A_2(\bar{\xi}, \eta) \tag{9-75}$$

The analog of formula (9-56) for gaussian integrals is

$$\int \prod_1^n d\bar{\eta}_k\, d\eta_k \exp\left[-\sum_{k,l} \bar{\eta}_k A_{kl} \eta_l + \sum_k (\bar{\eta}_k \xi_k + \bar{\xi}_k \eta_k)\right] = \det A \quad \exp\left[\sum_{k,l} \bar{\xi}_k (A^{-1})_{kl} \xi_l\right] \tag{9-76}$$

The close parallelism with the boson case allows an immediate transcription to obtain a path integral for transition amplitudes. Let $H(a^\dagger, a, t)$ be the normal ordered hamiltonian of a fermionic system. The corresponding normal kernel is obtained by substituting $\bar{\eta}$, η for $a^\dagger$, a in this order. Consequently, the kernel of the evolution operator is given by

$$U(\bar{\eta}_f, t_f; \eta_i, t_i) = \int \mathscr{D}(\bar{\eta}, \eta) \exp\left\{\frac{\bar{\eta}_f \eta_f + \bar{\eta}_i \eta_i}{2} + i \int_{t_i}^{t_f} dt \left[\frac{\bar{\eta}\dot{\eta} - \dot{\bar{\eta}}\eta}{2i} - h(\bar{\eta}, \eta, t)\right]\right\} \tag{9-77}$$

As an exercise consider the motion of a quantum spin $\frac{1}{2}$ submitted to the action of a constant field B along the z axis. The ground state is defined as corresponding to the value $S_z = \frac{1}{2}$ of the spin. The hamiltonian reads $H = \mu B(2a^\dagger a - 1)$ with μ the gyromagnetic ratio. Equation (9-77) shows that

$$U(\bar{\eta}_f, t_f; \eta_i, t_i) = \exp\left[i\mu B(t_f - t_i) + \bar{\eta}_f \eta_i e^{-2i\mu B(t_f - t_i)}\right] \tag{9-78}$$

Note the similarity with the harmonic oscillator. The treatment can be extended to a time-dependent field, including in particular a transverse field rotating at frequency ω.

What is striking when looking at Eq. (9-76) is that the integral of a quadratic form is also given, up to a factor, by the value on the "extremal trajectory," i.e., giving a stationary value to its exponent.

9-2 RELATIVISTIC FORMULATION

We generalize the previous approach to the infinite systems of interacting fields.

9-2-1 S Matrix and Green's Functions in Terms of Path Integrals

We start with an examination of the familiar case of a neutral scalar field coupled to an external real c-number source $j(x)$. The classical action is

$$I_0(\varphi, j) = \int d^4x \left[\tfrac{1}{2}(\partial\varphi^2) - \frac{m^2}{2}\varphi^2 + j\varphi \right] \tag{9-79}$$

and the quantum hamiltonian

$$H = \int d^3x \left[\tfrac{1}{2}\pi_{\text{op}}^2 + \tfrac{1}{2}(\nabla\varphi_{\text{op}})^2 + \frac{m^2}{2}\varphi_{\text{op}}^2 - j\varphi_{\text{op}} \right] \tag{9-80}$$

It describes an assembly of quantum oscillators coupled to varying external forces (Chaps. 3 and 4). At a given time the Fourier decomposition of the field is

$$\begin{aligned} \varphi_{\text{op}}(\mathbf{x}) &= \int d\tilde{k}\, [a(k)\, e^{i\mathbf{k}\cdot\mathbf{x}} + a^\dagger(k)\, e^{-i\mathbf{k}\cdot\mathbf{x}}] \\ \pi_{\text{op}}(\mathbf{x}) &= -i\int d\tilde{k}\, \omega(k)[a(k)\, e^{i\mathbf{k}\cdot\mathbf{x}} - a^\dagger(k)\, e^{-i\mathbf{k}\cdot\mathbf{x}}] \end{aligned} \tag{9-81}$$

in terms of which

$$\begin{aligned} H &= \int d\tilde{k}\, [\omega(k)a^\dagger(k)a(k) - f(t, \mathbf{k})a^\dagger(k) - \bar{f}(t, \mathbf{k})a(k)] \\ f(t, \mathbf{k}) &= \int d^3x\, e^{-i\mathbf{k}\cdot\mathbf{x}}\, j(\mathbf{x}, t) \end{aligned} \tag{9-82}$$

In this form the hamiltonian is diagonal and we can apply formula (9-55) which gives the integral kernel of the evolution operator

$$\begin{aligned} U(\bar{z}_f, t_f; z_i, t_i) = \exp\Bigg(\int d\tilde{k} \Bigg\{ &\bar{z}_f(k)\, e^{-i\omega(k)(t_f - t_i)}\, z_i(k) \\ &+ i\int_{t_i}^{t_f} dt[\bar{z}_f(k)\, e^{-i\omega(k)(t_f - t)}\, f(t, \mathbf{k}) + \bar{f}(t, \mathbf{k})\, e^{-i\omega(k)(t - t_i)}\, z_i(k)] \\ &- \frac{1}{2}\int_{t_i}^{t_f}\int_{t_i}^{t_f} dt\, dt'\, \bar{f}(t, \mathbf{k})\, e^{-i\omega(k)|t - t'|}\, f(t', \mathbf{k}) \Bigg\} \Bigg) \end{aligned} \tag{9-83}$$

With the source switched off at $|t| \to \infty$, the S matrix is defined as the limit of the operator $e^{it_f H_0}\, U(t_f, t_i)\, e^{-it_i H_0}$, where H_0 is obtained from H by setting $j = 0$. For coherent states, such as those used here, the action of e^{-itH_0} amounts simply to the shift $z \to z\, e^{-i\omega t}$, where ω is the frequency associated to the oscillator described by z. Consequently, the integral kernel of the S matrix reads

$$\mathscr{S}(\bar{z}_f, z_i) = \lim_{-t_i, t_f \to \infty} \exp \left\{ \int d\tilde{k} \left[\bar{z}_f(k) z_i(k) \right.\right.$$
$$+ i \int_{t_i}^{t_f} dt \left[\bar{z}_f(k)\, e^{i\omega(k)t} f(t, \mathbf{k}) + \bar{f}(t, \mathbf{k})\, e^{-i\omega(k)t} z_i(k) \right]$$
$$\left.\left. - \frac{1}{2} \int_{t_i}^{t_f} \int_{t_i}^{t_f} dt\, dt'\, \bar{f}(t, \mathbf{k})\, e^{-i\omega(k)|t-t'|} f(t', \mathbf{k}) \right] \right\} \tag{9-84}$$

From Eq. (9-49) the normal kernel will simply follow if we drop the first factor in the exponential. The remaining part is interpreted in terms of the classical asymptotic field

$$\varphi_{\text{as}}(\mathbf{x}, t) = \int d\tilde{k} \left[z_i(k)\, e^{-ik \cdot x} + \bar{z}_f(k)\, e^{ik \cdot x} \right] \tag{9-85}$$

solution of the homogeneous Klein-Gordon equation. Since $\bar{z}_f(k)$ is not the complex conjugate of $z_i(k)$, φ_{as} is given in terms of boundary conditions on positive frequencies for $t \to -\infty$ and negative ones for $t \to +\infty$. We recognize Feynman's mixed boundary condition. With these notations

$$\int d\tilde{k} \int_{-\infty}^{+\infty} dt \left[\bar{z}_f(k)\, e^{i\omega(k)t} f(t, \mathbf{k}) + \bar{f}(t, \mathbf{k})\, e^{-i\omega(k)t} z_i(k) \right]$$
$$= \int d^4x \int d\tilde{k}\, j(x) \left[\bar{z}_f(k)\, e^{i\omega(k)t - i\mathbf{k} \cdot \mathbf{x}} + z_i(k)\, e^{-i\omega(k)t + i\mathbf{k} \cdot \mathbf{x}} \right]$$
$$= \int d^4x\, j(x) \varphi_{\text{as}}(x)$$

Furthermore,

$$\int d\tilde{k} \iint dt\, dt'\, \bar{f}(t, \mathbf{k})\, e^{-i\omega(k)|t-t'|} f(t', \mathbf{k})$$
$$= \iint d^4x\, d^4x'\, j(x) j(x') \int d\tilde{k}\, e^{-i\omega(k)|t-t'| + i\mathbf{k} \cdot (\mathbf{x} - \mathbf{x}')}$$

The integral over k is the Feynman propagator

$$\int d\tilde{k}\, e^{-i\omega(k)|t-t'| + i\mathbf{k} \cdot (\mathbf{x} - \mathbf{x}')} = i \int \frac{d^4k}{(2\pi)^4} \frac{e^{-ik \cdot (x - x')}}{k^2 - m^2 + i\varepsilon}$$
$$= -iG_F(x - x')$$
$$= \langle 0 |\, T\varphi_{\text{op}}(x) \varphi_{\text{op}}(x')\, | 0 \rangle$$

where φ_{op} is the quantized scalar free field. Finally,

$$\mathscr{S}^N(\bar{z}_f, z_i)\big|_j = \exp\left[i\int d^4x\, j(x)\varphi_{\rm as}(x) + \frac{i}{2}\iint d^4x\, d^4x'\, j(x)G_F(x-x')j(x')\right] \tag{9-86}$$

This was indeed our result (4-63) and was used as a starting point in the discussion of Wick's theorem. The relativistic covariance as well as Feynman's $i\varepsilon$ prescription for propagators have followed naturally from the path integral formalism.

To obtain the S operator which will be denoted $S_0(j)$, we substitute $\varphi_{\rm op}$ to $\varphi_{\rm as}$ and normal order:

$$\begin{aligned} S_0(j) &= :\exp\left[i\int d^4x\, j(x)\varphi_{\rm op}(x)\right]: Z_0(j) \\ Z_0(j) &= \exp\left[\frac{i}{2}\iint d^4x\, d^4x'\, j(x)G_F(x-x')j(x')\right] \end{aligned} \tag{9-87}$$

Since

$$(\Box + m^2)\frac{1}{i}\frac{\delta}{\delta j(x)} Z_0(j) = j(x)Z_0(j)$$

formula (9-87) may be reinterpreted as

$$S_0(j) = :\exp\left\{\int d^4x \left[\varphi_{\rm op}(x)(\Box + m^2)\frac{\delta}{\delta j(x)}\right]\right\}: Z_0(j) \tag{9-88}$$

where $(\Box + m^2)(\delta/\delta j)$ acts only on $Z_0(j)$.

We consider now more complex interactions. Let us introduce, for instance, self-coupling through a potential $V(\varphi)$ in such a way that the action takes the form

$$I = \int d^4x\left[\tfrac{1}{2}(\partial\varphi)^2 - \frac{m^2}{2}\varphi^2 - V(\varphi)\right] \qquad I(\varphi, j) = I + \int d^4x\, j\varphi \tag{9-89}$$

As indicated by the notation, we take for simplicity $V(\varphi)$ to depend only on φ and not on its gradient. For instance, $V(\varphi)$ may be $\lambda\varphi^4/4!$. To derive the S matrix we use the same type of reasoning leading to Eqs. (9-23) and (9-24). In other words, we have the following relation between normal kernels:

$$\mathscr{S}^N = \exp\left[-i\int d^4x\, V\left(\frac{1}{i}\frac{\delta}{\delta j(x)}\right)\right]\mathscr{S}^N(j)\big|_{j=0} \tag{9-90}$$

with $\mathscr{S}^N(j)$ given by the expression (9-86).

The perturbative series follows from the expansion of the exponential operator in (9-90). The S matrix itself may be written

$$\begin{aligned} S &= :\exp\left[\int d^4x\, \varphi_{\rm op}(x)(\Box + m^2)\frac{\delta}{\delta j(x)}\right]: Z(j)\big|_{j=0} \\ Z(j) &= \exp\left[-i\int d^4x\, V\left(\frac{1}{i}\frac{\delta}{\delta j(x)}\right)\right]Z_0(j) \end{aligned} \tag{9-91}$$

We recognize in $Z(j)$ the generating functional of Green functions and the above formula is identical with Eq. (5-38). What is new here is that we have derived functional representations for the normal kernel of the S matrix and for $Z(j)$. Indeed, the operator $\exp\left[-i\int d^4x\, V(\delta/i\delta j(x))\right]$ generates the self-interaction leading to

$$\mathscr{S}(\bar{z}_f, z_i) = \lim \int \mathscr{D}(z, \bar{z}) \exp\left\{\int d\tilde{k}\left[\frac{\bar{z}_f(k)z_f(k) + \bar{z}_i(k)z_i(k)}{2} + iI(f, i)\right]\right\} \tag{9-92}$$

with

$$I(f, i) = \int_{t_i}^{t_f} dt \int d\tilde{k}\left[\frac{\dot{\bar{z}}(t, k)z(t, k) - \bar{z}(t, k)\dot{z}(t, k)}{2i} - \omega(k)\bar{z}(t, k)z(t, k) - h_{\text{int}}(\bar{z}, z)\right] \tag{9-93}$$

In the present case

$$h_{\text{int}} = \int d^3x\, V[\varphi(x)] \tag{9-94}$$

where φ is a functional of $z(t, k)$ and $\bar{z}(t, k)$ obtained by replacing in (9-81) the operators $a(t, k)$ and $a^\dagger(t, k)$ by these c numbers:

$$\begin{aligned}\varphi(\mathbf{x}, t) &= \int d\tilde{k}\left[z(t, k)\, e^{i\mathbf{k}\cdot\mathbf{x}} + \bar{z}(t, k)\, e^{-i\mathbf{k}\cdot\mathbf{x}}\right] \\ \pi(\mathbf{x}, t) &= -i\int d\tilde{k}\, \omega(k)\left[z(t, k)\, e^{i\mathbf{k}\cdot\mathbf{x}} - \bar{z}(t, k)\, e^{-i\mathbf{k}\cdot\mathbf{x}}\right]\end{aligned} \tag{9-95}$$

Before this substitution H_{int} has to be written in normal form. Finally, to generate the S matrix from the evolution operator the limits $t_f, -t_i \to \infty$ have to be supplemented by the free asymptotic conditions

$$\begin{aligned}\lim_{t\to\infty} \bar{z}(t, k) &= \bar{z}_f(k)\, e^{i\omega(k)t} \\ \lim_{t\to-\infty} z(t, k) &= z_i(k)\, e^{-i\omega(k)t}\end{aligned} \tag{9-96}$$

The normalization of the path integral follows from the previous case when V is identified with $j\varphi$. For shortness the measure will be abbreviated as $\mathscr{D}(\varphi, \pi)$ or even $\mathscr{D}(\varphi)$.

Similarly, the functionals $Z_0(j)$ and $Z(j)$ are expressed as integrals

$$Z_0(j) = \int \mathscr{D}(\varphi) \exp\left[iI_0(\varphi) + i\int d^4x\, \varphi(x)j(x)\right] \tag{9-97}$$

$$Z(j) = \int \mathscr{D}(\varphi) \exp\left[iI(\varphi) + i\int d^4x\, \varphi(x)j(x)\right] \tag{9-98}$$

$$I_0(\varphi) = \int d^4x\left[\tfrac{1}{2}(\partial\varphi)^2 - \frac{m^2}{2}\varphi^2(x)\right]$$

$$I(\varphi) = \int d^4x \left[\tfrac{1}{2}(\partial\varphi)^2 - \frac{m^2}{2}\varphi^2(x) - V(\varphi) \right]$$

It is understood that the measure involves a normalization factor such that $Z(0) = 1$.

From the expression for $Z_0(j)$ we obtain the integral

$$\int \mathscr{D}(\varphi)\varphi(x_1)\cdots\varphi(x_{2n})\, e^{iI_0(\varphi)} = \frac{(-i)^n}{2^n n!} \sum_{\text{permutations}} G_F(x_{P_1} - x_{P_2})\cdots G_F(x_{P_{2n-1}} - x_{P_{2n}}) \tag{9-99}$$

that is, the explicit form of Wick's theorem.

It is, of course, legitimate to consider the path integral (9-98) to be defined as a formal power series using Eq. (9-91) or (9-99). This allows us to check that a number of manipulations familiar when dealing with integrals—such as changing integration variables, integrations by parts, etc.—are justified. As an example let us investigate the effect of an infinitesimal change of variable of the form

$$\varphi(x) = \chi(x) + \varepsilon F(\chi, x) \tag{9-100}$$

where F is an arbitrary functional of χ subjected to the restriction that it admits an expansion in powers of χ. This change is canonical in the sense that the relation $\varphi \to \chi$ may be inverted (as a formal series). Taking into account the jacobian arising in the change of variable we obtain, to first order in ε,

$$Z(j) = \int \mathscr{D}(\chi)\left[1 + \varepsilon \int d^4x \frac{\delta F(\chi, x)}{\delta\chi(x)}\right]\left\{1 + i\varepsilon \int d^4x \left[\frac{\delta I(\chi)}{\delta\chi(x)} + j(x)\right] F(\chi, x)\right\}$$
$$\times \exp i\left[I(\chi) + \int d^4x\, j(x)\chi(x)\right] \tag{9-101}$$

Collecting the terms proportional to ε and extracting the factors in front of the integral by replacing χ by $\delta/i\delta j(x)$, we find

$$\int d^4x \left\{ i\left[\frac{\delta I}{\delta\chi(x)}\left(\frac{1}{i}\frac{\delta}{\delta j}\right) + j(x)\right] F\left(\frac{1}{i}\frac{\delta}{\delta j}, x\right) + \frac{\delta F}{\delta\chi(x)}\left(\frac{1}{i}\frac{\delta}{\delta j}, x\right)\right\} Z(j) = 0 \tag{9-102}$$

This may also be written

$$\int d^4x\, F\left(\frac{1}{i}\frac{\delta}{\delta j}, x\right)\left[\frac{\delta I}{\delta\chi(x)}\left(\frac{1}{i}\frac{\delta}{\delta j}\right) + j(x)\right] Z(j) = 0 \tag{9-103}$$

and can be checked directly from Eq. (9-98).

Let us specialize this general identity to the case where $F(\chi, x) = f(x)$, that is, the change of variable is simply a local translation on the field. Equation (9-103) reduces to

$$\left[\frac{\delta I}{\delta\chi(x)}\left(\frac{\delta}{i\delta j}\right) + j(x)\right] Z(j) = 0 \tag{9-104a}$$

In the scalar field case this reads explicitly

$$\left[(\Box + m^2)\frac{\delta}{i\delta j(x)} + V'\left(\frac{1}{i}\frac{\delta}{\delta j(x)}\right) - j(x)\right] Z(j) = 0 \tag{9-104b}$$

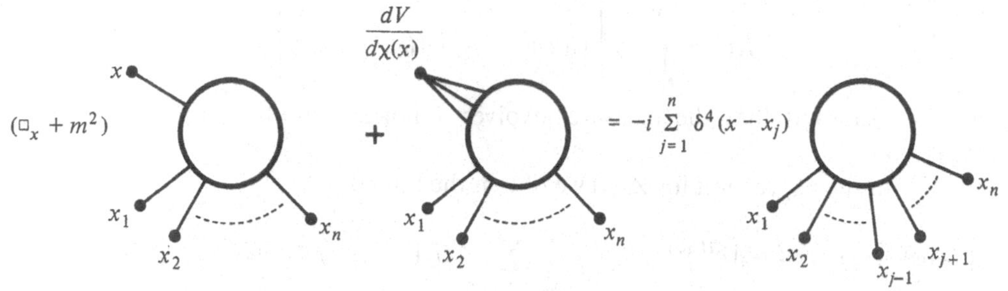

Figure 9-3 Graphical representation of the identity (9-104*a, b*).

This identity—a direct consequence of the equations of motion—relates functions with $n+1$, n, and $n-1$ points. The first involves a Klein-Gordon operator and the second an insertion of V'. It is depicted on Fig. 9-3.

Exercises

(*a*) Derive the corresponding identity satisfied by $G_c(j)$, the generating functional of connected Green functions.

(*b*) Observe that if the functional $F(\chi, x)$ is local, i.e., only involves χ at the point x, Eq. (9-102) introduces $\delta F/\delta\chi(x) = F'[\chi(x)]\delta(0)$. How can this term be interpreted? Show that this term is of higher order in $\hbar$ as compared to the other ones in (9-102) and that it is compensated by the Wick contraction between $F(\chi)$ and the piece $(\Box + m^2)\chi$ of $\delta I/\delta\chi$.

(*c*) Use the equation of motion to show that the insertion of the operator $\int d^4x\, \varphi(x)[\delta I/\delta\varphi(x)]$ counts the number of external lines of Green functions.

(*d*) Prove the equivalence theorem. The latter states that while an infinitesimal invertible change of field variable modifies the Green functions, it does not affect the S matrix. In other words, if $F(\varphi) = O(\varphi^2)$ the S matrix obtained by applying Eq. (9-91) to $Z'(j)$ given by

$$Z'(j) = \int \mathscr{D}(\varphi) \exp i\left\{I(\varphi) + \int d^4x\, j(x)[\varphi(x) + F(\varphi, x)]\right\} \tag{9-105}$$

coincides with the one computed with the help of $Z(j)$. Indeed, when amputating external lines as indicated by the operator $(\Box + m^2)$ in (9-91) the effect of $F(\varphi)$ only amounts to a wave-function renormalization.

(*e*) Show that even in the case where the interaction lagrangian contains derivative terms the perturbation expansion of the path integral leads to covariant Feynman rules.

(*f*) Finally, extend the formalism to Fermi fields. Write the generating functional $Z(j)$ for the Yukawa coupling of fermions to scalar bosons and for the case of electrodynamics. In this last case show that a change of variables of the form of an infinitesimal gauge transformation leads to the system of Ward identities discussed in Chap. 8.

These functional techniques to derive identities between Green functions will be used extensively in the following chapters.

9-2-2 Effective Action and Steepest-Descent Method

We have already defined the generating functional of connected Green functions $G_c(j)$ according to

$$Z(j) = \exp[G_c(j)] \tag{9-106}$$

and the effective action through the Legendre transformation

$$\varphi(x, j) = \frac{\delta}{i\delta j(x)} G_c(j)$$
$$i\Gamma(\varphi) = G_c(j) - i \int d^4x\, j(x)\varphi(x) \tag{9-107}$$

To evaluate $Z(j)$ from the path integral we may retain in the exponent the quadratic part of the action, expand the rest in a series, and apply Wick's theorem (9-99). The underlying assumption is that the coupling constant is small. A slightly more general reasoning is suggested by the path integral representation itself. Since the only integrals that can be performed in closed form are the gaussian ones, the idea is to use the steepest descent or stationary phase method (in the minkowskian case) to select the best point around which to expand in series. The small parameter is here Planck's constant $\hbar$—as is clear if the dimensions are explicited and the action I replaced by $I/\hbar$. Therefore, the above method corresponds naturally to the semiclassical approximation in quantum mechanics and leads to a series organized according to the number of loops.

Our first goal is therefore to find the extremal values of the exponent in the integral (9-98), that is, the fields φ_0 satisfying the classical equation

$$(\Box + m^2)\varphi_0 + V'(\varphi_0) = j \tag{9-108}$$

We shall assume that the solution reduces to the trivial one $\varphi_0 = 0$ for $j = 0$ so that, in the sense of formal series at least, the solution is then unique. The Green function to be used in the solution of (9-108) involves Feynman's $i\varepsilon$ prescription.

The assumption that only the trivial solution remains in the absence of source turns out to be invalid in a number of physically interesting cases. Such nontrivial extrema of the classical action require a case-by-case analysis for a proper interpretation.

In the vicinity of this extremal trajectory we shift the integration variable as $\varphi \to \varphi_0 + \varphi$, keep the quadratic part in the field in the exponential, and expand perturbatively the higher-order terms. By virtue of the stationarity condition (9-108) linear terms are absent, and as usual the normalization requires $Z(0) = 1$. Consequently, we find

$$Z(j) = e^{(i/\hbar)I(\varphi_0, j)} \int \mathscr{D}(\varphi) \exp\left(\frac{i}{\hbar}\int d^4x \left\{\tfrac{1}{2}(\partial\varphi)^2 - \frac{\varphi^2}{2}[m^2 + V''(\varphi_0)] - \sum_{p\geq 3} \frac{\varphi^p}{p!} V^{(p)}(\varphi_0)\right\}\right) \tag{9-109}$$

The new quadratic part is

$$\int d^4x \left\{\tfrac{1}{2}(\partial\varphi)^2 - \frac{\varphi^2}{2}[m^2 + V''(\varphi_0)]\right\} = -\int d^4x\, \tfrac{1}{2}\varphi[\Box + m^2 + V''(\varphi_0)]\varphi \tag{9-110}$$

and leads to a nontrivial propagator through its dependence on φ_0. To obtain the $\hbar$ expansion let us rescale the field as $\varphi \to \hbar^{1/2}\varphi$ so that

$$Z(j) = e^{G_c(j)/\hbar}$$

$$= e^{(i/\hbar)I(\varphi_0, j)} \int \mathscr{D}(\varphi) \exp\left(i \int d^4x \left\{\tfrac{1}{2}(\partial\varphi)^2 - \frac{\varphi^2}{2}\left[m^2 + V''(\varphi_0)\right] - \sum_{p \geq 3} \hbar^{p/2-1} \frac{\varphi^p}{p!} V^{(p)}(\varphi_0)\right\}\right) \tag{9-111}$$

Wick's theorem applied to (9-111) yields nonvanishing contributions only for even polynomials in φ. Only integer powers of $\hbar$ will therefore occur in the loop expansion. From Eq. (9-111) we read that the leading term (order $\hbar^0$) to $G_c(j)$ is $I(\varphi_0, j)$. Let us compute the next term. The integral over the quadratic part yields

$$\int \mathscr{D}(\varphi) \exp\left\{-i \int d^4x\, \varphi[\Box + m^2 + V''(\varphi_0)]\varphi\right\} = (\mathrm{Det}\, K_0^{-1} K_V)^{-1/2}$$

$$K_V = \{\Box_x + m^2 + V''[\varphi_0(x)]\}\delta^4(x-y) \tag{9-112}$$

$$K_0 = (\Box_x + m^2)\delta^4(x-y)$$

As in Chap. 4, we use capital letters for determinants and traces of operators of infinite dimension. The inverse of K_0, introduced by the normalization, has to be chosen as $G_F(x-y)$. Hence

$$K_0^{-1}K_V = \delta^4(x-y) + G_F(x-y)V''[\varphi_0(y)]$$

Since a determinant may be written

$$\mathrm{Det}\, A = e^{\mathrm{Tr}\,(\ln A)} \tag{9-113}$$

we find

$$-iG_c(j) = I(\varphi_0, j) + \frac{i}{2}\hbar\, \mathrm{Tr}\ln\left[1 + G_F V''(\varphi_0)\right] + O(\hbar^2) \tag{9-114}$$

To obtain $\Gamma(\varphi)$ we have to invert the relation

$$\varphi(x, j) = \frac{\delta G_c(j)}{i\delta j(x)}$$

According to (9-114) and (9-108) φ is given to leading order by φ_0 up to corrections of order $\hbar$. Moreover, since I is stationary at φ_0 we have $I(\varphi, j) - I(\varphi_0, j) = O(\hbar^2)$. Finally, we have to subtract $\int d^4x\, \varphi(x)j(x)$ from $-iG_c(j)$ to obtain Γ, given therefore by

$$\Gamma(\varphi) = I(\varphi) + \frac{i}{2}\hbar\, \mathrm{Tr}\ln\left[1 + G_F V''(\varphi)\right] + O(\hbar^2) \tag{9-115}$$

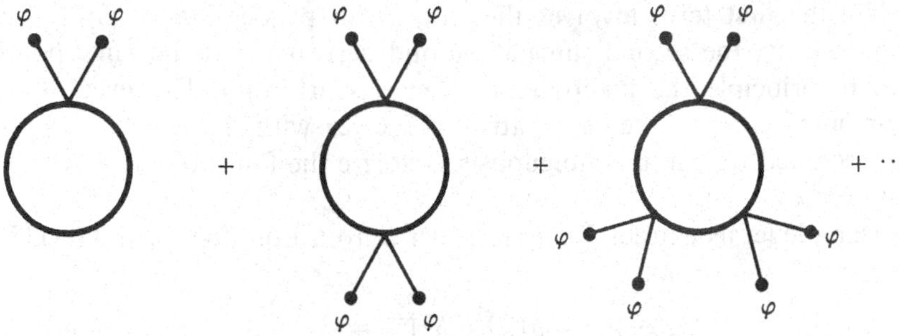

Figure 9-4 The effective potential to the one-loop order.

The perturbative interpretation of the second term is clear if we expand it as

$$\frac{i\hbar}{2}\operatorname{Tr}\ln\left[1+G_F V''(\varphi)\right] = i\hbar\sum_{n=1}^{\infty}\frac{(-1)^{n-1}}{2n}\operatorname{Tr}\{[G_F V''(\varphi)]^n\}$$

$$= i\hbar\sum_{n=1}^{\infty}\frac{(-1)^{n-1}}{2n}\int d^4z_1\cdots d^4z_n\, G_F(z_1-z_2)V''[\varphi(z_2)]G_F(z_2-z_3)\cdots$$

$$\cdots V''[\varphi(z_n)]G_F(z_n-z_1)V''[\varphi(z_1)]$$

This is the sum of the contributions of one-loop diagrams made of n propagators $-iG_F(x-y)$ and n vertices $-iV''(\varphi)$. It is depicted on Fig. 9-4 in the case of $V(\varphi)=\lambda\varphi^4/4!$. Notice that the factor $1/2n$ in front of each term of the sum is the symmetry factor of the corresponding diagram (n stands for the rotations, 2 for the reflection); similarly in the case of φ^4 theory, the factor $\frac{1}{2}$ in $V''(\varphi)=\lambda\varphi^2/2$ takes into account the symmetry between the two external legs attached to each vertex.

This expansion may be carried out to all orders. The successive terms of G_c are represented by connected Feynman diagrams generated by the interaction term $-\sum_{p\geq 3}\hbar^{p/2-1}(\varphi^p/p!)V^{(p)}(\varphi_0)$ with propagators obtained by the inversion of the kernel $\square + m^2 + V''(\varphi_0)$.

As far as $\Gamma(\varphi)$ is concerned, the result of the Legendre transformation is to select among the previous diagrams only the one-particle irreducible ones, and to replace φ_0 everywhere by the arbitrary argument φ. Of course, any actual calculation has to face ultraviolet subtractions.

In short, the steepest-descent or stationary phase method leads elegantly to the semiclassical expansion according to the number of loops. To go beyond perturbation theory requires either to expand around nontrivial extrema or to approximate the path integral in some utterly different way.

We return to $\Gamma(\varphi)$. We know that its expansion in φ generates the one-particle irreducible Green functions. As far as particle physics is concerned it is generally this aspect which is relevant. We may also insist on the role of $\Gamma(\varphi)$ as an effective action. Taking into account translation invariance we can find an expansion involving higher and higher derivatives in the field φ in the form

$$\Gamma(\varphi)=\int d^4x\left[-V_{\text{eff}}(\varphi)+\tfrac{1}{2}Z_{\text{eff}}(\varphi)(\partial\varphi)^2+\cdots\right] \tag{9-116}$$

In (9-116) the first term involves the sum of all proper functions at zero external momentum, the second sums all second derivatives at the same point, and so on. In principle the function $\varphi(x)$ remains arbitrary. However, if we wish to compute V_{eff} only, we may satisfy ourselves with a calculation for a constant φ provided we can unambiguously factorize the four-volume divergent integration over x.

As an example let us extract V_{eff} up to order $\hbar$ from Eqs. (9-114) and (9-115). In general,

$$V_{\text{eff}} = V_{\text{eff}}^{(0)} + \hbar V_{\text{eff}}^{(1)} + \hbar^2 V_{\text{eff}}^{(2)} + \cdots \tag{9-117}$$

Using (9-114) we have first

$$V_{\text{eff}}^{(0)} = \frac{m^2}{2}\varphi^2 + V(\varphi) \tag{9-118}$$

In the determinant occurring in (9-112) φ is now a constant and the propagator $[\Box + m^2 + V''(\varphi)]^{-1}$ is thus diagonal in momentum space:

$$\operatorname{Tr}\ln\{[\Box + m^2 + V''(\varphi)](\Box + m^2)^{-1}\}$$

$$= \int d^4x \int \frac{d^4p}{(2\pi)^4} \ln\left[1 - V''(\varphi)\frac{1}{p^2 - m^2 + i\varepsilon}\right] \tag{9-119}$$

$$V_{\text{eff}}^{(1)} = \frac{-i}{2}\int \frac{d^4p}{(2\pi)^4} \ln\left[1 - V''(\varphi)\frac{1}{p^2 - m^2 + i\varepsilon}\right]$$

This expression is, of course, meaningless before ultraviolet subtractions. To be concrete let us pick the potential

$$V(\varphi) = \frac{\lambda\varphi^4}{4!} \qquad V''(\varphi) = \frac{\lambda\varphi^2}{2}$$

To comply with the prescription of normal ordering we should also have added a term in φ^2. Otherwise we have to include tadpole diagrams, corresponding to a contraction of two fields at the same vertex.

This is precisely the only divergent one-loop diagram of the two-point function $\Gamma^{(2)}$ (Fig. 9-5*a*), and it contributes a quadratically divergent term proportional to φ^2 in the expansion of $V_{\text{eff}}^{(1)}$:

$$\frac{\lambda\varphi^2}{4}\int \frac{d^4p}{(2\pi)^4}\frac{i}{p^2 - m^2 + i\varepsilon}$$

To this order we also have the logarithmic divergence of the four-point function $\Gamma^{(4)}$ associated with the diagram of Fig. 9-5*b*:

$$\frac{i}{4}\left(\frac{\lambda\varphi^2}{2}\right)^2 \int \frac{d^4p}{(2\pi)^4}\frac{1}{(p^2 - m^2 + i\varepsilon)^2}$$

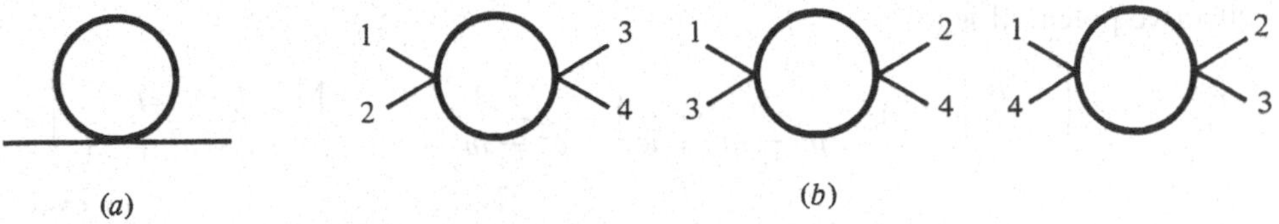

Figure 9-5 Divergent diagrams to the one-loop order.

Higher powers of φ lead to convergent integrals. Renormalization introduces counterterms needed to insure that

$$\Gamma^{(2)}(m_{\text{phys}}^2) = 0 \qquad \frac{d\Gamma^{(2)}}{dp^2}(m_{\text{phys}}^2) = 1 \tag{9-120}$$

and a constraint on the four-point function giving the meaning of the renormalized coupling constant, for instance, its value at the on-shell symmetric point S:

$$\begin{aligned} p_i^2 &= m_{\text{phys}}^2 \\ (p_i + p_j)^2 &= \tfrac{4}{3} m_{\text{phys}}^2 \qquad i \neq j \\ \Gamma^{(4)}(p_i)\big|_S &= -\lambda_{\text{phys}} \end{aligned} \tag{9-121}$$

These conditions are useful when dealing with some actual application to a scattering problem. They are, however, cumbersome to compute quantities in the effective action expanded at zero external momenta.

Up to a finite renormalization, we can replace (9-120) and (9-121) by

$$\Gamma^{(2)}(0) = -m^2 \qquad \frac{d\Gamma^{(2)}}{dp^2}(0) = 1 \qquad \Gamma^{(4)}(0) = -\lambda \tag{9-122}$$

To emphasize the distinction we have introduced the label "physical" to the mass and coupling occurring in the previous normalizations.

The virtue of Eqs. (9-122) is that they can be translated as conditions on the effective renormalized potential and function $Z_{\text{eff}}(\varphi)$ as

$$\begin{aligned} \frac{d^2}{d\varphi^2} V_{\text{eff}}(0) &= m^2 \\ \frac{d^4}{d\varphi^4} V_{\text{eff}}(0) &= \lambda \\ Z_{\text{eff}}(0) &= 1 \end{aligned} \tag{9-123}$$

These requirements are obviously fulfilled to order $\hbar^0$ as shown by (9-118). At this level we also have

$$Z_{\text{eff}}^{(0)} = 1 \tag{9-124}$$

The counterterms in φ^2 and φ^4 cancel the corresponding terms in the higher orders of V_{eff} and Z_{eff}. Consequently, the correct one-loop contribution to the effective potential is

$$V_{\text{eff}}^{(1)} = \frac{-i}{2}\int \frac{d^4p}{(2\pi)^4}\left[\ln\left(1 - \frac{\lambda\varphi^2/2}{p^2 - m^2 + i\varepsilon}\right) + \frac{\lambda\varphi^2/2}{p^2 - m^2 + i\varepsilon} + \frac{1}{2}\frac{(\lambda\varphi^2/2)^2}{(p^2 - m^2 + i\varepsilon)^2}\right] \tag{9-125}$$

It is useful at this stage to perform the Wick rotation $p_0 \to ip_0$. We denote by k the corresponding euclidean four-momentum so that

$$V_{\text{eff}}^{(1)} = \frac{1}{2}\int \frac{d^4k}{(2\pi)^4}\left[\ln\left(1 + \frac{\lambda\varphi^2/2}{k^2 + m^2}\right) - \frac{\lambda\varphi^2/2}{k^2 + m^2} + \frac{1}{2}\left(\frac{\lambda\varphi^2/2}{k^2 + m^2}\right)^2\right] \tag{9-126}$$

Performing the integral and adding this expression to the zeroth-order term given by (9-118) yields

$$V_{\text{eff}} = \frac{m^2\varphi^2}{2} + \frac{\lambda\varphi^4}{4!} + \frac{1}{(8\pi)^2}\left[\left(\frac{\lambda\varphi^2}{2} + m^2\right)^2 \ln\left(1 + \frac{\lambda\varphi^2}{2m^2}\right) - \frac{\lambda\varphi^2}{2}\left(\frac{3}{2}\frac{\lambda\varphi^2}{2} + m^2\right)\right] + \cdots \tag{9-127}$$

We observe that the large-φ behavior of $V(\varphi)$ is modified by the quantum corrections. To exhibit it more transparently, it is useful to introduce yet another set of normalization conventions to be able to set $m = 0$. We may define a new coupling constant λ_M such that

$$\lambda_M = \frac{d^4}{d\varphi^4} V_{\text{eff}}(\varphi)\bigg|_{\varphi = M} \tag{9-128}$$

From (9-127) it follows that

$$\lambda_M = \lambda + \frac{\lambda^2}{(8\pi)^2}\left[6 \ln \frac{\lambda M^2}{2m^2} + 16 + O\left(\frac{m^2}{M^2}\right)\right]$$

We use this definition in (9-127) and consider the limiting massless theory for which

$$V_{\text{eff}}(\varphi) = \lambda_M \frac{\varphi^4}{4!} + \frac{\lambda_M^2\varphi^4}{(16\pi)^2}\left(\ln\frac{\varphi^2}{M^2} - \frac{25}{6}\right) + \cdots \tag{9-129}$$

a result due to Coleman and Weinberg. It is not possible to directly set $m = 0$ in Eq. (9-127). This arises from the structure of the ultraviolet subtractions in Eq. (9-126), the second of which, designed to enforce the condition $(d^4 V_{\text{eff}}/d\varphi^4)(0) = 0$, introduces an infrared divergence in the limit $m = 0$. We have to choose an arbitrary but nonvanishing subtraction point $\varphi = M$ to define the massless theory. Stated differently, the Γ functions at zero momentum generated by the expansion of (9-127) are singular as m goes to zero. The behavior when $\varphi \to \infty$ is clear from the fact that the dimension of V is four so that dominant terms must be proportional to φ^4 up to logarithms in φ/M. Note that the arbitrariness in the point M implies that a related change in M and λ_M must leave V_{eff} invariant. We encounter here a manifestation of the renormalization group to be discussed at length later.

To first order in the loop expansion there is no wave function renormalization in the φ^4 model. Nevertheless, the function $Z_{\text{eff}}(\varphi)$ is nontrivial to this order even though it contains no logarithms. Using the effective action (9-115) it may be shown to be equal to

$$Z_{\text{eff}}(\varphi) = 1 + \frac{\lambda^2}{6(4\pi)^2}\frac{\varphi^2}{(2m^2+\lambda\varphi^2)} + \cdots \tag{9-130}$$

The massless limit may be obtained as in (9-129). Calculations have been pursued to higher orders. Let us quote here the results up to second order with the corresponding diagrams. We use the shorter notations

$$x = \frac{\lambda\varphi^2}{2m^2} \qquad \alpha = \frac{\hbar\lambda}{(4\pi)^2} \qquad V_{\text{eff}}(\varphi) = \frac{m^4}{\lambda}\,v(x,\alpha) \qquad Z_{\text{eff}}(\varphi) = z(x,\alpha) \tag{9-131}$$

and we obtain:

Diagram	v	z
	x	1
	$\frac{x^2}{6}$	0
	$\frac{\alpha}{4}\left[(1+x)^2\ln(1+x) - \left(x + \frac{3x^2}{2}\right)\right]$	$\frac{\alpha}{6}\frac{x}{1+x}$
	$\frac{\alpha^2}{8}[(1+x)\ln(1+x) - x]^2$	$\frac{\alpha^2}{12}\frac{x}{1+x}\left[\ln(1+x) + \frac{x}{1+x}\right]$
	$\frac{\alpha^2}{2}x\left[\frac{1+x}{2}\ln^2(1+x) - 2(1+x) \times \ln(1+x) + 2x\right]$	$\frac{\alpha^2}{6}\left\{-\frac{1}{2}\ln(1+x) + \frac{x}{1+x} + \frac{2x}{1+x}\left[\ln(1+x) + \frac{4A}{3} - 1\right] - \frac{x^2}{(1+x)^2}[\ln(1+x) + \frac{4}{3}(A-1)] + \frac{4}{3}\frac{x^2}{(1+x)^2} + B\right\}$ (9-132)

Here B is a constant depending on the normalization conventions and A is given by

$$A = -\int_0^1 du\,\frac{\ln u}{1-u+u^2} = \frac{3}{2}\sum_0^\infty\left[\frac{1}{(1+3p)^2} - \frac{1}{(2+3p)^2}\right] = 1.17195 \tag{9-133}$$

The effective action discussed here is similar to the Euler-Heisenberg lagrangian of electromagnetism (Sec. 4-3-4). We leave it to the reader to spell out the details of this relationship.

The S matrix has an expansion in powers of $\hbar$ similar to the one of Green's functions. Show that the two first terms for its normal kernel are given by

$$\mathscr{S}(\bar{z}_f, z_i) = \exp\left\{\frac{i}{\hbar}I(\varphi_{\text{class}}) - \tfrac{1}{2}\operatorname{tr}\ln[1 + G_F V''(\varphi_{\text{class}})] + O(\hbar)\right\} \tag{9-134}$$

where φ_{class} is the solution of

$$(\Box + m^2)\varphi_{\text{class}} + V'(\varphi_{\text{class}}) = 0 \tag{9-135}$$

with Feynman boundary conditions; φ_{class} is also given by the integral equation

$$\varphi_{\text{class}}(x) = \varphi_{\text{as}}(x) - \int d^4y\, G_F(x-y)\, V'[\varphi_{\text{class}}(y)] \tag{9-136}$$

with φ_{as} given by Eq. (9-85).

The effective action may be given a more physical definition. In particular, $V_{\text{eff}}(\varphi)$ may be interpreted as the ground-state energy density under the constraint that the mean value of the field is equal to φ uniformly. It enables us to explore possible instabilities of the system (Chap. 11).

9-3 CONSTRAINED SYSTEMS

A number of dynamical systems may be described by observables submitted to constraints at fixed time. As a familiar instance we may quote electrodynamics. The quantization of such systems is not a straightforward matter and we shall soon encounter even more severe difficulties in the case of nonabelian gauge theories. Path integrals provide an ideal framework to handle such systems because they maintain a close relationship with the classical case where a simple treatment may be given. The quantization method requires the elimination of as many pairs of canonically conjugate variables as there are constraints. The latter have to satisfy suitable compatibility conditions to be clarified below.

This technical development can be skipped in a first reading and reconsidered when turning to the concrete application to gauge fields (Chap. 12), where use will be made of the result (9-159).

9-3-1 General Discussion

Let a classical system with $n (n > 1)$ degrees of freedom be submitted first to a unique constraint

$$f(p, q) = 0 \tag{9-137}$$

Call C the $(2n - 1)$-dimensional manifold in phase space characterized by (9-137). Our considerations will apply locally, i.e., in a neighborhood of a point belonging to C. Furthermore, we should not distinguish between two functions f and F which both vanish on C, that is, such that $F(p, q) = \alpha(p, q) f(p, q)$. Let $\mathscr{A}$ be the ring of (differentiable) functions which vanish on C. We use the notation $F \sim 0$ to mean $F \in \mathscr{A}$.

We may incorporate the constraint in the action using a time-dependent Lagrange multiplier $\lambda(t)$. Equations of motion follow from the stationarity conditions of

$$I = \int dt\,[p\dot{q} - h(p,q) - \lambda(t) f(p,q)] \tag{9-138}$$

These include Eq. (9-137) obtained by varying λ, together with

$$\dot{q}_i = \frac{\partial h}{\partial p_i} + \lambda \frac{\partial f}{\partial p_i} \qquad \dot{p}_i = -\frac{\partial h}{\partial q_i} - \lambda \frac{\partial f}{\partial q_i} \qquad 1 \leq i \leq n \tag{9-139}$$

Of course, on C the variables (p_i, q_i) are too numerous. A natural compatibility condition is that the evolution (9-139) leaves the manifold C invariant. This reads in terms of Poisson brackets

$$\{h, f\} \sim 0 \tag{9-140}$$

More generally it follows that any $F \sim 0$ will have a Poisson bracket with h belonging to $\mathscr{A}$:

$$F \sim 0 \Rightarrow \{h, F\} \sim 0 \tag{9-141a}$$

$\mathscr{A}$ is stable under the Poisson bracket since it has a unique generator

$$F \sim 0 \Rightarrow \{f, F\} \sim 0 \tag{9-141b}$$

and Eqs. (9-141a, b) imply that this ring of functions (but not necessarily an individual member) is stable under the evolution (9-139). Let F be an arbitrary fixed element in $\mathscr{A}$. We define an equivalence relation E on C as follows. Consider the flow generated by F in phase space. In infinitesimal form it is described by the equations

$$\frac{dq_i}{du}(u) = \frac{\partial F}{\partial p_i} \qquad \frac{dp_i}{du}(u) = -\frac{\partial F}{\partial q_i} \qquad F \sim 0 \tag{9-142}$$

Two points on C will be equivalent if and only if they belong to the same trajectory of the flow (9-142). This equivalence relation E is (1) independent of the choice of F in $\mathscr{A}$ and (2) invariant under the time evolution.

Indeed, we note that if a line of flow passes through a point belonging to C it is entirely contained in C. Furthermore, if F' is another element in $\mathscr{A}$, since $F = aF'$, the tangent vectors to the two flows are proportional on C, which proves the first point. From (9-141) h is constant on the lines of flow in C. Finally, let G be an arbitrary function constant along the lines of flow (9-142) lying in C, that is, such that

$$F \sim 0 \Rightarrow \{G, F\} \sim 0 \tag{9-143}$$

Its time evolution during an infinitesimal time interval δt is given by $G \to G + \dot{G}\,\delta t$ with $\dot{G} = \{h + \lambda f, G\} \sim \{h, G\}$ and $\{\{h, G\}, F\} = \{h, \{G, F\}\} - \{G, \{h, F\}\} \sim 0$ from Jacobi's identity and conditions (9-141a) and (9-143). Therefore $G + \delta G$ is again constant along the lines of flow and this proves the second point.

The manifold C is thus split in a time-independent manner in a set of equivalence classes according to E. The factor space C/E may in turn be considered as a $2n - 2$ phase space provided some regularity conditions on the function f are fulfilled. All the flows of the type of Eq. (9-142) are equivalent on C and physical observables (such as the energy) are constant under these flows. It is thus sufficient to choose a representative in each class, again in a regular manner. To this end, we intersect the manifold C by another one described by the auxiliary condition

$$g(p, q) = 0 \tag{9-144}$$

in such a way that each line of flow admits a unique intersection with this transverse surface. This is ensured if g varies monotonically along each line of flow, which will follow if

$$\{f, g\} \neq 0 \tag{9-145}$$

With such a choice we can explicitly parametrize C/E by performing the canonical transformation

$$\begin{aligned} &(p_i, q_i) \to (P_j, Q_j) \qquad 1 \leq i, j \leq n \\ &P_n = g(p, q) \end{aligned} \tag{9-146}$$

In the new variables the bracket between f and g will be

$$\{f, g\} = -\frac{\partial f}{\partial Q_n}$$

and the condition (9-145) will enable us to solve the equation

$$f(Q, P) = 0 \qquad \frac{\partial f}{\partial Q_n} \neq 0 \tag{9-147}$$

for Q_n as a function of $Q_1, \ldots, Q_{n-1}, P_1, \ldots, P_n$:

$$Q_n \equiv Q_n(Q_1, \ldots, Q_{n-1}, P_1, \ldots, P_n) \tag{9-147a}$$

The definition of the factor space C/E is completed by condition (9-144) which takes the form

$$P_n = 0 \tag{9-148}$$

Finally, the quantity

$$H(Q_1, \ldots, Q_{n-1}, P_1, \ldots, P_{n-1}) = h(q_i, p_i)\Big|_{\substack{Q_n \equiv Q_n(Q_1, \ldots, P_{n-1}, 0) \\ P_n = 0}} \tag{9-149}$$

is the effective hamiltonian on the remaining space, as is readily checked.

The arbitrariness involved in the choice of the function g required to fulfil Eq. (9-145) may create some difficulty in a global definition throughout phase space.

It is easy to generalize this construction to the case of $m < n$ independent constraints

$$f_1(p, q) = 0, \ldots, f_m(p, q) = 0 \tag{9-150}$$

The ring $\mathscr{A}$ contains the smooth functions vanishing on the manifold C defined by the above equations, i.e., of the form

$$F(p, q) = \sum_{1}^{m} \alpha_k(p, q) f_k(p, q) \tag{9-151}$$

and we require that $\{h, F\}$ belongs to $\mathscr{A}$ if F does:

$$F \sim 0 \Rightarrow \{h, F\} \sim 0 \tag{9-152}$$

In order to be able to define on C an equivalence relation eliminating m other coordinates, we also require that $\mathscr{A}$ be invariant under the Poisson bracket

$$F_1, F_2 \sim 0 \Rightarrow \{F_1, F_2\} \sim 0 \tag{9-153}$$

a condition which was automatically fulfilled when $\mathscr{A}$ had a unique generator.

Proceeding as before, we define a fibration on C by considering through any of its points the set of trajectories generated by any element $F \sim 0$. The tangents to these trajectories build up a m-dimensional linear manifold in the $(2n - m)$-dimensional tangent space to C. The time-independent equivalence relation E identifies points belonging to the same m-dimensional manifold generated by $\mathscr{A}$, and C/E has a natural phase space structure. The latter may be explicited by the introduction of m auxiliary conditions

$$g_1(p, q) = 0, \ldots, g_m(p, q) = 0 \tag{9-154}$$

intersecting the m-dimensional fibers of C at a unique point. For this property to hold, a sufficient condition is that

$$\det \{g_k, f_l\} \neq 0 \qquad 1 \leq k, l \leq m \tag{9-155}$$

A canonical transformation $(q, p) \to (Q, P)$ is defined in such a way that

$$P_{n-m+1} = g_1, \ldots, P_n = g_m \tag{9-156}$$

Conditions (9-150) and (9-155) enable us to solve on C for $Q_{n-m+1}, \ldots, Q_n$ in terms of $Q_1, \ldots, Q_{n-m}, P_1, \ldots, P_n$, and C/E is finally characterized by

$$P_{n-m+1} = 0, \ldots, P_n = 0 \tag{9-157}$$

On the remaining space the dynamics is generated by a hamiltonian H obtained from the original $h(p, q)$ by the canonical transformation taking into account the above restriction procedure.

Show that this construction for m constraints follows recursively from the one given in the case of only one constraint.

To quantize such systems in terms of the independent canonical variables P and Q, the hamiltonian H, and the corresponding observables, we simply write transition amplitudes as

$$\langle f \mid i \rangle = \int \mathscr{D}(P, Q) \exp \left\{ i \int dt [P\dot{Q} - H(P, Q)] \right\} \tag{9-158}$$

In actual cases it will be generally unpractical to perform the elimination leading to the canonical parametrization of C/E. We therefore look for an expression of (9-158) in terms of the original constrained variables (p, q). To do this we rewrite the measure at each time in the path integral as

$$\prod_1^{n-m} dP_k\, dQ_k = \prod_1^n dP_k\, dQ_k \prod_{n-m+1}^{n} \delta(P_s)\delta[Q_s - Q_s(Q_1,\ldots,Q_{n-m}, P_1,\ldots,P_n)]$$

Since $\prod_1^n dP_k\, dQ_k$ is canonically invariant it is equal to $\prod_1^n dp_k\, dq_k$. Furthermore, $\prod_{n-m+1}^n \delta(P_s) = \prod_1^m \delta(g_k)$, and the familiar rule on δ-functions yields

$$\prod_{n-m+1}^{n} \delta[Q_s - Q_s(Q_1,\ldots,Q_{n-m}, P_1,\ldots,P_n)] = \prod_1^m \delta[f_k(p,q)]\, \frac{D(f_1,\ldots,f_m)}{D(Q_{n-m+1},\ldots,Q_n)}$$

From (9-156) it follows that the jacobian is nothing but det $\{g_k, f_l\}$ which we write for short as det $\{g, f\}$.

Using the integral representation

$$\prod_1^m \delta(f_k) = \int \prod_1^m \frac{d\lambda_k}{2\pi} \exp\left(-i\sum_1^m \lambda_k f_k\right)$$

and reassembling all the pieces we find

$$\langle f \mid i\rangle = \int \mathscr{D}(p,q,\lambda) \prod_t \left[\delta(g) \det\{g,f\}\right] \exp\left[i\int dt(p\dot{q} - h - \lambda f)\right] \quad (9\text{-}159)$$

The constraints are clearly exhibited and we recognize the action given in (9-138). The variables λ occur without conjugate quantities.

For this construction to be meaningful it is mandatory that (9-159) be unaffected by a different choice of auxiliary conditions $g_k = 0$. Let us verify this point in an infinitesimal form. Consider the effect of a small change

$$g_k + \delta g_k = 0 \qquad 1 \le k \le m \quad (9\text{-}160)$$

The linear system

$$\delta g_k = \sum_{s=1}^m \delta v_s \{f_s, g_k\} \quad (9\text{-}161)$$

admits a unique solution by virtue of (9-155). This means that

$$\delta g_k \sim \{\delta F, g_k\} \qquad \delta F = \sum_s \delta v_s f_s \quad (9\text{-}162)$$

The corresponding δF generates a canonical transformation

$$p \to p + \delta p \qquad q \to q + \delta q \qquad \delta p = \{\delta F, p\} \qquad \delta q = \{\delta F, q\} \quad (9\text{-}163)$$

leaving the measure $\Pi\, dp\, dq$ invariant. Remembering Eq. (9-153) it follows that δF also generates a nonsingular linear transformation on the constraints

$$f \to (1 + \delta A) f \quad (9\text{-}164)$$

with the matrix δA depending in general on the point (p, q). Finally, the action $\int dt(p\dot{q} - h)$ is at most modified by boundary terms required to take the new boundary conditions into account.

If we rewrite (9-159) after integration on λ:

$$\langle f|i\rangle = \int \mathscr{D}(p,q) \prod_t \left[\delta(f)\delta(g) \det\{g,f\}\right] \exp\left[i\int dt(p\dot{q}-h)\right]$$

we see that all quantities are defined modulo a function of $\mathscr{A}$, owing to the presence of $\delta(f)$. We then apply the canonical change (9-163); using

$$\prod_k \delta(f_k) \to \det(1+\delta A)^{-1} \prod_k \delta(f_k)$$

we see that

$$\prod_k \left[\delta(f_k)\delta(g_k)\right] \det\{g,f\} \to \det(1+\delta A)^{-1} \prod_k \left[\delta(f_k)\delta(g_k+\delta g_k)\right] \det\{g+\delta g, f+\delta f\}$$

since the differences $\delta g - \{\delta F, g\}$ and their Poisson brackets with the f and $f + \delta f$ vanish on C. Finally,

$$\det\{g+\delta g, f+\delta f\} = \det(1+\delta A)\det\{g+\delta g, f\}$$

In summary,

$$\prod_k \left[\delta(f_k)\delta(g_k)\right] \det\{g,f\} \to \prod_k \left[\delta(f_k)\delta(g_k+\delta g_k)\right] \det\{g+\delta g, f\} \tag{9-165}$$

Therefore we have proved that the path integral (9-159) is indeed independent of an infinitesimal variation in the auxiliary conditions up to boundary terms in the phase. Can the reader explain the precise role of these boundary terms equal to $\exp\{i[p\,\partial\delta F/\partial p]_i^f\}$? Can one generalize the arguments to time-dependent auxiliary conditions?

9-3-2 The Electromagnetic Field as an Example

To get acquainted with this quantization method let us return to the electromagnetic field coupled to a c-number external conserved current, with its action

$$I = \int d^4x \left(\frac{\mathbf{E}^2 - \mathbf{B}^2}{2} - \rho A^0 + \mathbf{j}\cdot\mathbf{A}\right)$$

Instead of using only the potential as a dynamical variable we choose the so-called first-order formalism, with fields and potential as primitive entities. We rewrite I as

$$I = \int d^4x \left[-\mathbf{E}\cdot(\nabla A^0 + \dot{\mathbf{A}}) - \mathbf{B}\cdot\operatorname{curl}\mathbf{A} + \frac{\mathbf{B}^2 - \mathbf{E}^2}{2} - \rho A^0 + \mathbf{j}\cdot\mathbf{A}\right] \tag{9-166}$$

This would reduce to the previous expression if we were to replace **E** and **B** in terms of A^0 and **A**. Varying the action with respect to the fields we recover the relations between field and potential

$$\mathbf{E} = -(\nabla A^0 + \dot{\mathbf{A}}) \qquad \mathbf{B} = \operatorname{curl}\mathbf{A} \tag{9-167a}$$

which imply the first set of homogeneous Maxwell equations

$$\operatorname{curl}\mathbf{E} + \dot{\mathbf{B}} = 0 \qquad \operatorname{div}\mathbf{B} = 0 \tag{9-167b}$$

The variation with respect to A yields the second pair of equations

$$\operatorname{div}\mathbf{E} = \rho \qquad \operatorname{curl}\mathbf{B} - \dot{\mathbf{E}} = \mathbf{j} \tag{9-168}$$

Note that

$$\frac{\delta}{\delta \mathbf{A}(\mathbf{x})} \int d^3y\, \mathbf{B}(\mathbf{y}) \cdot \operatorname{curl} \mathbf{A}(\mathbf{y}) = \operatorname{curl} \mathbf{B}(\mathbf{x})$$

Among these equations $\mathbf{B} = \operatorname{curl} \mathbf{A}$ and $\operatorname{div} \mathbf{E} = \rho$ appear as constraints. The first one may be simply solved by replacing $\mathbf{B}$ everywhere by curl $\mathbf{A}$ which amounts after a partial integration to rewriting the action as

$$I = \int d^4x \left\{ -\mathbf{E} \cdot \dot{\mathbf{A}} - \left[\frac{\mathbf{E}^2 + (\operatorname{curl} \mathbf{A})^2}{2} + \mathbf{j} \cdot \mathbf{A} \right] + A^0 (\operatorname{div} \mathbf{E} - \rho) \right\} \tag{9-169}$$

If this is compared to (9-138) we realize that A^0 plays, as did λ, the role of a Lagrange multiplier without any conjugate variable. We are thus led to identify the canonical variables (p_i, q_i) with $\mathbf{A}(\mathbf{x})$ and $\mathbf{E}(\mathbf{x})$ respectively. Poisson brackets are defined through

$$\{A_i(\mathbf{x}), E_j(\mathbf{y})\} = \delta_{ij} \delta^3(\mathbf{x} - \mathbf{y}) \tag{9-170}$$

It is necessary to generalize the compatibility equations (9-152) to cope with the case where ρ depends on time. Those are now written

$$F \sim 0 \Rightarrow \frac{\partial F}{\partial t} + \{h, F\} \sim 0 \tag{9-171}$$

where the time derivative operates on the explicit time dependence. In the present case, $F = \operatorname{div} \mathbf{E} - \rho$ and

$$\frac{\partial}{\partial t} (\operatorname{div} \mathbf{E} - \rho) = -\frac{\partial \rho}{\partial t}$$

The condition (9-171) reduces to

$$-\frac{\partial \rho}{\partial t} + \left\{ \int d^3y \left[\frac{\mathbf{E}^2 + (\operatorname{curl} \mathbf{A})^2}{2} + \mathbf{j} \cdot \mathbf{A} \right], \operatorname{div} \mathbf{E} - \rho \right\} = -\frac{\partial \rho}{\partial t} - \operatorname{div} \mathbf{j} = 0 \tag{9-172}$$

an identity by virtue of current conservation. Thus, apart from this slight generalization, the case at hand can be cast into the framework discussed in the preceding subsection.

It remains to choose auxiliary conditions $g(\mathbf{E}, \mathbf{A}) = 0$. This is, of course, arbitrary to a large extent. A serious simplification will follow if g is linear in the dynamical variables as is the constraint $\operatorname{div} \mathbf{E} - \rho = 0$. Under such circumstances $\det \{g, f\}$ will be independent of the variables and hence may be absorbed in the normalization of the path integral. A possible condition is

$$\partial_\mu A^\mu \equiv \dot{A}^0 + \operatorname{div} \mathbf{A} = c(x, t) \tag{9-173}$$

involving an arbitrary function c.

We return to covariant notations and write the transition amplitude as

$$\int \mathscr{D}(F, A) \prod_x \delta(\partial \cdot A - c) \exp \left\{ i \int d^4x \left[\tfrac{1}{4} F_{\mu\nu} F^{\mu\nu} - \tfrac{1}{2} F_{\mu\nu} (\partial^\mu A^\nu - \partial^\nu A^\mu) - j \cdot A \right] \right\} \tag{9-174}$$

The gaussian integral over **B** realizes automatically the substitution $\mathbf{B} \to \text{curl } \mathbf{A}$. Similarly, the integral over **E** substitutes for the electric field its value $-(\nabla A^0 + \dot{\mathbf{A}})$. When this is done we have for the amplitude

$$\int \mathscr{D}(A) \prod_x \delta(\partial \cdot A - c) \exp\left\{-i \int d^4x \left[\tfrac{1}{4}(\partial_\mu A_\nu - \partial_\nu A_\mu)(\partial^\mu A^\nu - \partial^\nu A^\mu) + j \cdot A\right]\right\} \tag{9-175}$$

This is not quite the expression used in the previous chapters. However, since (9-175) is independent of the arbitrary function $c(\mathbf{x}, t)$ we may complete this identification by integrating on these functions c with a weight $\exp\left[-(i\lambda/2) \int d^4x \, c^2\right]$. If we denote by $F_{\mu\nu}$ the quantity $\partial_\mu A_\nu - \partial_\nu A_\mu$, the final version of the path integral is the familiar one

$$\int \mathscr{D}(A) \exp\left\{-i \int d^4x \left[\frac{F^2}{4} + \frac{\lambda}{2}(\partial \cdot A)^2 + j \cdot A\right]\right\} \tag{9-176}$$

This presentation illustrates the arbitrariness implied by gauge invariance. It is also worth comparing it to the operator formalism.

9-4 LARGE ORDERS IN PERTURBATION THEORY

9-4-1 Introduction

Functional integrals provide us with new tools to investigate numerous aspects of field theory. As an illustration we close this chapter with a discussion of the behavior of perturbation theory at large orders. One of the goals of this study is of course an attempt at a better understanding of the amazing accuracy of the successive approximations in quantum electrodynamics and the other promising weak coupling models. Another motivation lies in the hope of overcoming the very limitations of the perturbative series and of coping with strong coupling phenomena. Even though present-day knowledge is far from being satisfactory, interesting developments have occurred which justify this endeavor. This will also afford us the occasion to introduce some useful techniques in dealing with path integrals.

The nature of the perturbative series is related to the analytic properties of Green functions in terms of the coupling constant in the vicinity of zero. Such a study is possible but extremely difficult.

Fortunately there exists a less rigorous, but manageable approach which leads to similar conclusions, namely, that the perturbative series, in all cases of interest, is strongly divergent. In spite of this fact it may be very useful, as will be explained below.

Thus the problem may be divided into two distinct parts. First, we look for an estimate of large orders in the expansion of Green functions as given by a well-defined set of Feynman rules and renormalization prescriptions. Some aspects of this program have been completed. It is noteworthy that this is independent of the question whether or not the series defines a unique mathematical quantity. For certain practical purposes this first step might well be sufficient. For instance, it could happen that quantum electrodynamics, as it now stands, is not completely consistent, but that power series in α are nevertheless asymptotic within the context of a deeper theory.

The second aspect of the problem is indeed to try to reconstruct from these expansions a unique theory according to some definite prescriptions. This is clearly a formidable task which requires some independent information. The latter has to be provided by a different construction. Nevertheless, we might discover some hints in the structure of the series which suggest a sensible way of summing it.

The typical nature of the divergence may easily be understood on a simple example. For this purpose we replace path integrals by ordinary ones such as

$$Z(g) = \frac{1}{\sqrt{2\pi}} \int_{-\infty}^{+\infty} d\varphi \, e^{-(\varphi^2/2 + g\varphi^4)} \tag{9-177}$$

Of course, this case is trivial enough that we might obtain $Z(g)$ in closed form. The point $g = 0$ is an essential singularity. When $\operatorname{Re} g$ becomes negative (and $|\operatorname{Arg} g| < \pi$) we may still rotate the integration contour in (9-177). This is no longer possible when g approaches the negative real axis and the integral blows up for large φ. We may think of φ as the value of a field at a point with $\varphi^2/2 + g\varphi^4$ as a caricature of the action. Negative values of g correspond to an unstable situation where the "potential" is not bounded from below. This is reflected in the perturbative expansion if we write

$$Z(g) = \sum_0^\infty g^k Z_k$$

$$Z_k = \frac{(-1)^k}{\sqrt{2\pi}} \int_{-\infty}^{+\infty} d\varphi \, \frac{\varphi^{4k}}{k!} \, e^{-\varphi^2/2} = (-1)^k 4^k \, \frac{\Gamma(2k + \frac{1}{2})}{\sqrt{\pi} k!} \tag{9-178}$$

Using Stirling's formula

$$k! \sim \sqrt{2\pi k} \; e^{k \ln k - k}$$

for large k, we find that Z_k behaves as

$$Z_k \sim \frac{(-16)^k}{\sqrt{\pi}} \, e^{(k-1/2)\ln k - k} \tag{9-179}$$

Nevertheless, the power series in g is asymptotic in the complex g plane cut along the negative real axis since

$$\left| Z(g) - \sum_0^n Z_k g^k \right| < \frac{4^{n+1}\Gamma(2n + \frac{3}{2})}{\sqrt{\pi}(n+1)!} \, \frac{|g|^{n+1}}{[\cos(\frac{1}{2}\operatorname{Arg} g)]^{2n+3/2}} \tag{9-180}$$

meaning that for fixed n and g small enough the right-hand side may be made arbitrarily small.

The asymptotic behavior of Z_k was obtained by applying Stirling's formula to the exact expression. The latter will not be available in more realistic cases. However, this suggests to apply the method of steepest descent for large k to

$$Z_k = \frac{(-1)^k}{k!} \sqrt{\frac{2}{\pi}} \int_0^\infty d\varphi \, e^{-\varphi^2/2 + 4k \ln \varphi} \tag{9-181}$$

The position of the saddle point is

$$\varphi_c^2 = 4k \tag{9-182}$$

and integration over quadratic deviations from φ_c yields

$$Z_k \sim \frac{(-1)^k}{k!} \sqrt{2} \, e^{2k \ln 4k - 2k} = \frac{(-16)^k}{\sqrt{\pi}} \, e^{(k-1/2)\ln k - k} \tag{9-183}$$

as before.

What can we do from such a divergent series besides using it in an asymptotic sense to evaluate the function for small g? In some fortunate circumstances, such as the one discussed here, the power series (9-178) contains in fact enough information to reconstruct the function unambiguously. Of course, it might be argued that we could add any function such as $\exp(-1/\sqrt{g})$ with vanishing derivatives at the origin along the real positive axis. However, within a well-defined class of functions excluding these pathologies [and to which $Z(g)$ belongs] a Borel transformation enables us to recover $Z(g)$ from its divergent perturbative series.

Introduce

$$B(t) = \sum_0^\infty \frac{Z_k t^k}{\Gamma(k + \frac{3}{2})} \tag{9-184}$$

where the choice of $\Gamma(k + \frac{3}{2})$ is justified by the growth given in Eq. (9-179) and could be slightly modified in analogous cases. The series on the right-hand side in (9-184) will converge within a circle of finite radius in the complex t plane. If $B(t)$ may be continued along the entire real positive axis and does not grow too fast at infinity, then $Z(g)$ will be given by

$$Z(g) = \int_0^\infty dt \, t^{1/2} B(gt) \, e^{-t} \tag{9-185}$$

Knowledge of the perturbative series is insufficient to prove Borel summability. It is, however, sufficient to disprove it in the case where $B(t)$ would be found to have a singularity on the positive real axis. This happens, for instance, if asymptotically the Z_k have equal phases.

For our simple example we find

$$\begin{aligned} B(t) &= \frac{2}{\sqrt{\pi}} \sqrt{1-u} \\ u &= \frac{\sqrt{1+16t} - 1}{\sqrt{1+16t} + 1} \end{aligned} \tag{9-186}$$

and $B(t)$ is analytic in a plane cut from $-\frac{1}{16}$ to $-\infty$.

To obtain a convergent expansion for $Z(g)$, let us map the cut t plane onto a circle, keeping the origin fixed (this is given here by the choice of variable u), and derive the convergent Taylor series for $\tilde{B}(u)$:

$$B(t) = \tilde{B}(u) = \sum_0^\infty b_k[u(t)]^k \tag{9-187}$$

from the knowledge of its expansion in t. Then

$$Z(g) = \sum_0^\infty b_k \int_0^\infty dt\, t^{1/2}[u(gt)]^k e^{-t} \tag{9-188}$$

Since b_k decreases here as $k^{-3/2}$ it is easy to see that this new series will converge as $e^{-3(k/3g^{1/2})^{2/3}}$.

The behavior exhibited in this example bears a close relationship with some divergences encountered in field theory since the expansion of $Z(g)$ in powers of g has coefficients equal to the number of vacuum diagrams in a φ^4 model. This follows readily from the fact that Wick's theorem applies to integrals of monomials over a gaussian weight.

This remark applies to other field theories as well. In electrodynamics, for instance, consider the integral

$$\begin{aligned}\tilde{Z}(j, \bar{\eta}, \eta) &= \frac{1}{\pi}\int dA\, d\bar{\psi}\, d\psi\, e^{-A^2/2 - \bar{\psi}(1-eA)\psi + \bar{\psi}\eta + \bar{\eta}\psi + jA} \\ &= \int \frac{dA}{1-eA} e^{-A^2/2 + \bar{\eta}\eta/(1-eA) + jA}\end{aligned} \tag{9-189}$$

where ψ and $\bar{\psi}$ are considered as complex conjugate c-number variables. The expansion of $\ln \tilde{Z}$ in powers of e generates the number of diagrams of the connected functions, except for the cancellations implied by Furry's theorem. These are implemented by symmetrizing the generating function of charged loops with respect to e, $\ln(1 - eA) \to \frac{1}{2}\ln(1 - e^2A^2)$, that is, by replacing $\tilde{Z}$ by

$$Z(j, \bar{\eta}, \eta) = \int \frac{dA}{\sqrt{1-e^2A^2}} e^{-A^2/2 + \bar{\eta}\eta/(1-eA) + jA} \tag{9-190}$$

The integral is meaningful for negative e^2. For the photon and electron propagators G and S, related to the vacuum polarization $\bar{\omega}$ and self-energy Σ, we find respectively

$$\begin{aligned}G &= (1-\bar{\omega})^{-1} = \langle A^2 \rangle \\ S &= (1-\Sigma)^{-1} = \left\langle \frac{1}{1-e^2A^2} \right\rangle\end{aligned} \tag{9-191}$$

where the average is over the measure $(dA/\sqrt{1-e^2A^2})\, e^{-A^2/2}$. Surprisingly these expressions coincide:

$$G = S = -2z\left[1 + \frac{d}{dz}\ln K_0(z)\right] \qquad z = -\frac{1}{4e^2} \tag{9-192}$$

with $K_0(z)$ the modified Bessel function

$$K_0(z) = \int_0^\infty d\theta\, e^{-z\cosh\theta}$$

Expansion of (9-192) for large z yields

$$\begin{aligned} G = S &= 1 + e^2 + 4e^4 + 25e^6 + 208e^8 + 2\,146e^{10} + 26\,368e^{12} + \cdots \\ \bar{\omega} = \Sigma &= e^2 + 3e^4 + 18e^6 + 153e^8 + 1\,638e^{10} + 20\,898e^{12} + \cdots \end{aligned} \tag{9-193}$$

to be compared with the number of diagrams for the vacuum polarization with one charged loop only:

$$\bar{\omega}_1 = \sum (2n-1)!!\, e^{2n} = e^2 + 3e^4 + 15e^6 + 105e^8 + 945e^{10} + 10\,395e^{12} + \cdots \tag{9-194}$$

Similarly, the generating function for vertex diagrams Γ is equal to

$$\Gamma = 4z(1-S)S^{-2}G^{-1} = 1 + e^2 + 7e^4 + 72e^6 + 891e^8 + 12\,672e^{10} + \cdots \tag{9-195}$$

Extremely courageous people are undertaking the computation of the 891 diagrams of the electron anomaly to eighth order, but will anybody ever dream of considering the 12 672 ones to tenth order!

9-4-2 Anharmonic Oscillator

Let us apply the previous ideas to the study of a quantum mechanical system. Although the method works for any polynomial potential we shall for definiteness consider the ground-state energy of an anharmonic oscillator with hamiltonian

$$H = \tfrac{1}{2}(p^2 + \varphi^2) + g\varphi^4 \tag{9-196}$$

The configuration variable is denoted φ and its conjugate momentum p to emphasize the formal analogy with higher-dimensional field theories. The problem of the expansion in powers of g may be considered in the framework of the Schrödinger equation. We expect an instability for infinitesimal negative g, which can be investigated by means of the WKB (Wentzel, Kramers, Brillouin) approximation. Since this method will not be available as such in higher dimensions it is instructive to use an alternative approach tailored after the previous example.

The matrix elements of the evolution operator e^{-itH} can be expressed as path integrals. So does the trace of $e^{-\beta H}$, which may be interpreted as the partition function of a canonical ensemble of oscillators. Here β^{-1} is equal to the absolute temperature multiplied by Boltzmann's constant and

$$F = -\frac{1}{\beta}\ln(\operatorname{Tr} e^{-\beta H}) \tag{9-197}$$

is the free energy. When the temperature goes to zero, or β to infinity, F reduces to the ground-state energy.

Thus we may represent the partition function $Z(g)$ as a Feynman-Kac path integral over the exponential of an euclidean action (with a change in the relative sign of kinetic and potential contributions as compared to the usual expression). The configurations $\varphi(t)$ are periodic in "time": $\varphi(0) = \varphi(\beta)$, to comply with the fact that we are computing a trace

$$\operatorname{Tr} e^{-\beta H} = Z(g) = \int \mathscr{D}\varphi \exp\left[-\int_0^\beta dt\left(\frac{\dot{\varphi}^2 + \varphi^2}{2} + g\varphi^4\right)\right] \tag{9-198}$$

The ground state energy is given by the formula

$$E(g) - E(0) = \lim_{\beta\to\infty} \left[-\frac{1}{\beta} \ln \frac{Z(g)}{Z(0)} \right] \tag{9-199}$$

with $Z(g)$ expressed through (9-198).

The expansion of $Z(g)/Z(0)$ in powers of g reads

$$\begin{aligned} \frac{Z(g)}{Z(0)} &= \sum_0^\infty Z_k g^k \\ Z_k &= \frac{(-1)^k}{k!} \frac{1}{Z(0)} \int \mathscr{D}\varphi \exp\left[-\int_0^\beta dt \, \frac{\dot{\varphi}^2 + \varphi^2}{2} + k \ln\left(\int_0^\beta dt \, \varphi^4 \right) \right] \end{aligned} \tag{9-200}$$

For large k we evaluate Z_k using the steepest-descent method. We look for a saddle point $\varphi_c(t)$ such that $\varphi_c(0) = \varphi_c(\beta)$ and minimizing the effective action

$$I = \int_0^\beta dt \, \frac{\dot{\varphi}^2 + \varphi^2}{2} - k \ln\left(\int_0^\beta dt \, \varphi^4 \right) \tag{9-201}$$

that is,

$$\ddot{\varphi}_c - \varphi_c + \frac{4k\varphi_c^3}{\displaystyle\int_0^\beta dt' \, \varphi_c^4(t')} = 0 \tag{9-202}$$

Rescaling φ_c according to

$$\varphi_c(t) = \left[\frac{4k}{\displaystyle\int_0^\beta dt' \, \tilde{\varphi}^4(t')} \right]^{1/2} \tilde{\varphi}(t) \tag{9-203}$$

we obtain the equation

$$\ddot{\tilde{\varphi}} - \tilde{\varphi} + \tilde{\varphi}^3 = 0 \tag{9-204}$$

expressing the fact that $\tilde{\varphi}$ is of order unity while φ_c grows like $k^{1/2}$. As compared to the usual equation of motion, Eq. (9-204) exhibits two changes of sign beyond the rescaling which has eliminated the magnitude of the coupling. First, the potential, and hence the force, has reversed its sign due to the rotation to euclidean time (another way to put it is to replace t by it so that the acceleration is reversed). Second, the effective coupling is negative as shown by the relative sign between the harmonic ($\tilde{\varphi}$) and anharmonic ($\tilde{\varphi}^3$) contributions. The effective potential is depicted in Fig. 9-6. The equation is translationally invariant in time and admits the symmetry $\tilde{\varphi} \to -\tilde{\varphi}$.

By relabeling t we may use a symmetric interval in time $[-\beta/2, \beta/2]$ instead of $[0, \beta]$. In the limit $\beta \to \infty$ the interval is infinite and the solution which minimizes the action is found to be

$$\tilde{\varphi}_\tau(t) = \frac{\sqrt{2}}{\cosh(t - \tau)} \tag{9-205}$$

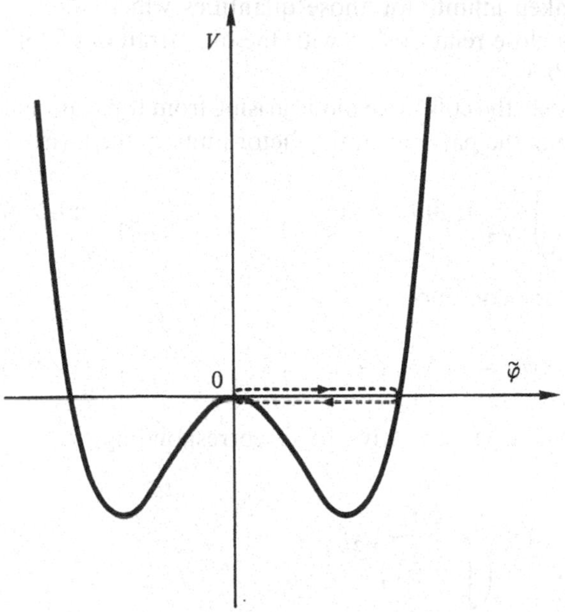

Figure 9-6 Effective potential in euclidean space $V(\tilde{\varphi}) = -\tilde{\varphi}^2/2 + \tilde{\varphi}^4/4$. The limiting motion is shown in dotted lines.

up to a global sign and with an arbitrary origin in time τ. Clearly this expression fulfils the periodic boundary conditions which are here supplemented by the condition that the action be finite. We thus have an infinity of degenerate saddle points satisfying

$$\int_{-\infty}^{+\infty} dt \, \frac{\tilde{\varphi}^2}{2} = 2 \qquad \int_{-\infty}^{+\infty} dt \, \frac{\tilde{\varphi}^4}{4} = \frac{4}{3} \qquad \int_{-\infty}^{+\infty} dt \, \frac{\dot{\tilde{\varphi}}^2}{2} = \frac{2}{3} \tag{9-206}$$

and

$$I_c = \lim_{\beta \to \infty} \left[\int_0^\beta dt \, \frac{\dot{\varphi}_c^2 + \varphi_c^2}{2} - k \ln \left(\int_0^\beta dt \, \varphi_c^4 \right) \right] = -k \ln (3k^2) + 2k \tag{9-207}$$

In this limit the relation between φ_c and $\tilde{\varphi}$ reads

$$\varphi_c = \sqrt{\frac{3k}{4}} \, \tilde{\varphi} \tag{9-208}$$

Due to the degeneracy of the saddle points which follow from the symmetries of the problem, the integration over quadratic deviations from the minimum requires some care. It is proper to call it a quantization problem since we look for the quantum frequencies of oscillations around a classical extremum of the action. The question will arise whenever a classical saddle point will be the starting point of an approximation to a path integral and degeneracies arise due to a continuous invariance. One or several modes will be of zero frequency. Here they correspond to time translations and should first be isolated to yield a factor β after integration over τ. It is, of course, crucial to keep β finite, albeit large, even though it may be taken infinite for those quantities which admit a finite limit. The problem bears a close relationship with the quantization of constrained systems studied in Sec. 9-3.

Thus we want to treat separately the collective mode arising from translational motion. To do this we introduce in the path integral a factor unity in the form

$$1 = \int_{-\beta/2}^{\beta/2} d\tau\, \delta(\theta_\varphi - \tau) \tag{9-209}$$

where θ_φ is defined implicitly by the condition

$$\int dt\, \psi(t - \theta_\varphi)\varphi(t) = 0 \tag{9-210}$$

and the function ψ is the normalized derivative of $\tilde{\varphi}$ corresponding, say, to $\tau = 0$:

$$\psi(t) = \frac{\dot{\tilde{\varphi}}(t)}{\left(\int dt\, \dot{\tilde{\varphi}}^2\right)^{1/2}} \tag{9-211}$$

The reason for this choice will soon appear clear.

If φ is translated by an amount τ through $\varphi(t) \to \varphi_\tau(t) = \varphi(t - \tau)$, this leaves the limiting action as well as the path integral measure invariant but θ_φ is changed into $\theta_\varphi + \tau$. Thus the integral over τ can be performed explicitly and yields, as expected, a factor β. Then

$$Z_k = \beta \frac{(-1)^k}{k!} e^{-I_c} \frac{1}{Z(0)} \int \mathscr{D}\varphi\, \delta(\theta_\varphi)\, e^{-(I - I_c)} \tag{9-212}$$

The δ function will select a unique saddle point except for a sign ambiguity. Expanding around one of the two solutions (the positive one, for instance) is allowed provided we multiply by a factor two, since from symmetry the two contributions are equal. We set

$$\varphi = \varphi_c + \chi \tag{9-213}$$

and keep the dominant terms in the expansion in χ. It then follows that

$$\delta(\theta_\varphi) = \left| \int dt\, \dot{\psi}(\varphi_c + \chi) \right| \delta\left(\int dt\, \psi\chi \right) \tag{9-214}$$

To leading order we drop $\int dt\, \dot{\psi}\chi$ and observe that

$$-\int dt\, \dot{\psi}\varphi_c = \sqrt{\frac{3k}{4}} \left(\int dt\, \dot{\tilde{\varphi}}^2 \right)^{1/2} = \sqrt{k} \tag{9-215}$$

while $I - I_c$ to second order in χ reads

$$I - I_c = \frac{1}{2}\left[\int dt \left(\dot{\chi}^2 + \chi^2 - \frac{6\chi^2}{\cosh^2 t}\right) + \left(\int dt\, u\chi\right)^2\right]$$
$$u = \frac{\sqrt{3}}{2}\tilde{\varphi}^3 \tag{9-216}$$

With Dirac's bracket notations we find

$$Z_k = \beta \frac{(-1)^k}{k!} e^{-I_c(2k^{1/2})} \frac{\int \mathscr{D}\chi\delta(\langle\psi|\chi\rangle) \exp\left[-\frac{1}{2}\langle\chi|(K + 1 + |u\rangle\langle u|)|\chi\rangle\right]}{\int \mathscr{D}\chi \exp\left[-\frac{1}{2}\langle\chi|(K_0 + 1)|\chi\rangle\right]} \tag{9-217}$$

where the Schrödinger operator K is

$$K = -\frac{d^2}{dt^2} - \frac{6}{\cosh^2 t} \tag{9-218}$$

and K_0 corresponds to the free part $K_0 = -d^2/dt^2$. The instructions are to integrate the normalized version of $\dot{\tilde{\varphi}}$ in the subspace orthogonal to ψ. The spectrum of $K + 1$ contains a discrete zero mode with eigenfunction given by ψ. Indeed, by differentiating Eq. (9-204) with respect to time and inserting the solution (9-205), we have

$$\left(-\frac{d^2}{dt^2} + 1 - \frac{6}{\cosh^2 t}\right)\dot{\tilde{\varphi}} = 0 \tag{9-219}$$

This does not affect the integration due to the δ function. Moreover, since ψ has one node there exists a unique normalized χ corresponding to a negative eigenvalue of $K + 1$. The remaining part of the spectrum is a positive continuum. It may be noted from the definition of u that $\langle\psi|u\rangle = 0$.

The operator K corresponds to a new Schrödinger problem with a so-called Bargmann potential which admits explicit solutions. We want to compute the nonvanishing determinant of $K + 1 + |u\rangle\langle u|$ in the space orthogonal to $|\psi\rangle$.

The final result for Z_k will then be

$$Z_k = \beta \frac{(-1)^k}{k!} e^{-I_c(2k^{1/2})} N$$
$$N^{-2} = 2\pi \frac{\det(K + 1 + |u\rangle\langle u|)_\perp}{\det(K_0 + 1)_\perp} \tag{9-220}$$

The factor 2π comes from the gaussian integration over the ψ mode omitted in the denominator when we consider $(K_0 + 1)_\perp$. This factor would have been modified had we not chosen to normalize ψ. The extra projector $|u\rangle\langle u|$ is easily disposed of by noticing that

$$(K + 1)\tilde{\varphi} = -2\tilde{\varphi}^3 = -\frac{4}{\sqrt{3}} u \tag{9-221}$$

As noted above, since $|u\rangle$ belongs to the subspace orthogonal to ψ, this means that Eq. (9-221) may be inverted in this space to yield

$$\begin{aligned}\det(K+1+|u\rangle\langle u|)_\perp &= \det(K+1)_\perp\left(1+\langle u|\frac{1}{K+1}|u\rangle\right)\\ &= \det(K+1)_\perp\left(1-\frac{\sqrt{3}}{4}\langle\tilde{\varphi}|u\rangle\right)\\ &= -\det(K+1)_\perp \end{aligned}\tag{9-222}$$

The negative sign is welcome because of the remaining unique negative eigenvalue of $(K+1)_\perp$. In practical terms, to separate the contribution of the transverse modes we may use a limiting procedure by replacing $K+1$ by $K+z$, and extracting the coefficient of $(z-1)$ as $z\to 1$. Thus

$$N^{-2} = -2\pi\frac{\det(K+1)_\perp}{\det(K_0+1)_\perp} = -2\pi\lim_{z\to 1}\frac{1}{z-1}\frac{\det(K+z)}{\det(K_0+z)}\tag{9-223}$$

Several methods may now be used to obtain the explicit expression of this Fredholm determinant. The most instructive, which may be generalized to an arbitrary one-dimensional potential, would be to relate the evaluation of fluctuations around a saddle point in the path integral and the ordinary WKB approximation for wave functions. We can also note in this special case that the fluctuation problem reduces to a soluble equation. Both methods have been used in the literature to obtain

$$\frac{\det(K+z)}{\det(K_0+z)} = \frac{\Gamma(1+\sqrt{z})\Gamma(\sqrt{z})}{\Gamma(3+\sqrt{z})\Gamma(\sqrt{z}-2)}\tag{9-224}$$

from which we derive

$$N^{-2} = \frac{\pi}{6}\tag{9-225}$$

Putting all factors together results in

$$Z_k = \beta\frac{(-1)^k}{k!}2k^{1/2}\sqrt{\frac{6}{\pi}}\,e^{k\ln(3k^2)-2k}\tag{9-226}$$

As expected this yields a $k!$ increase as

$$Z_k = \beta\Gamma(k+\tfrac{1}{2})(-3)^k\left(\frac{6}{\pi^3}\right)^{1/2}\tag{9-227}$$

Due to this wild behavior it is easy to translate this result in terms of the expansion for the energy. Indeed, the leading contribution of the coefficient of g^k in the series for $\log(1+\sum_1^\infty c_p g^p)$ is $c_k[1+O(1/k)]$ when the c_k grow like $k!$. Thus, applying formula (9-199) we find that the factor β drops out and we are left with

$$E(g) = \tfrac{1}{2} + \sum_{1}^{\infty} g^k E_k$$
$$E_k = (-1)^{k+1}\Gamma(k + \tfrac{1}{2})3^k\left(\frac{6}{\pi^3}\right)^{1/2}\left[1 + O\left(\frac{1}{k}\right)\right] \tag{9-228}$$

a result first derived by Bender and Wu.

In the case of the anharmonic oscillator, Graffi, Grecchi, and Simon have been able to prove that the function $E(g)$ is Borel summable, i.e., may be represented in terms of its Borel transform by a relation of the type (9-185).

We could then proceed to systematic corrections in powers of $1/k$ using the standard perturbation techniques around the saddle point. The method may also be extended to excited states by keeping terms of order $e^{-\beta}$ in the saddle-point action and expanding e^{-I_c} in powers of $e^{-\beta}$.

As a final remark we observe that when looking for periodic trajectories of period β we have other solutions with fractional periods $\beta/2$, $\beta/3, \ldots$, which would also lead to saddle points. These have a classical action two, three,..., times larger than the lowest one; thus they contribute only exponentially small corrections to the results.

The extension of these methods to field theory is in principle straightforward, at least insofar as renormalization is left aside. We look in a similar way to instabilities occurring for small coupling. These are responsible for singularities in the Borel transformed plane. As long as they do not reach the positive real axis (assumed to correspond to the physical situations) the theory is sound. This extra information enables us to use in the most effective way the first few terms of the perturbation series for an accurate determination of physical quantities. This program has encountered a great success in some applications to statistical mechanics of the bosonic φ^4 model in three dimensions. It can be generalized to include fermionic fields. Singularities on the positive real axis may, however, be also encountered which originate from actual instabilities even at the semiclassical level (such is the case in the theory of gauge fields presented in Chap. 12). By this we mean that to develop a perturbation series we have to select one among several degenerate minima of the classical energy. The divergence of the series then reflects the quantum tunneling between these ground states. Classical euclidean solutions to the field equations with finite action I interpolate between these states and contribute a transition amplitude $e^{-I/\hbar}$. To construct a meaningful theory it is then necessary to remove this degeneracy by introducing additional quantum numbers.

It is also possible that extra singularities occur as a result of renormalization. These seem to raise difficult problems about the consistency of renormalizable field theories.

NOTES

The idea of introducing path integrals via the superposition principle in quantum mechanics originates in the work of P. A. M. Dirac, *Physikalische Zeitschrift der Sowjetunion*, vol. 3, p. 64, 1933, and was developed by R. P. Feynman, *Rev. Mod. Phys.*, vol. 20, p. 367, 1948. An expanded version is found in the book by R. P. Feynman and A. R. Hibbs, “Quantum Mechanics and Path Integrals,”

McGraw-Hill, New York, 1965. The treatment of fermions was given by F. A. Berezin in "The Method of Second Quantization," Academic Press, New York, 1966. Among the numerous recent presentations we may quote the lectures of J. Zinn-Justin at the 1974 Bonn Summer Institute, edited by H. Rollnik and K. Dietz, "Trends in Elementary Particle Theory," Lecture Notes in Physics No. 37, Springer-Verlag, Berlin, 1975, and those of L. D. Faddeev, Les Houches Summer School 1975, "Methods in Field Theory," edited by R. Balian and J. Zinn-Justin, North-Holland, Amsterdam, 1976. The approach through functional derivatives has been developed by J. Schwinger, *Phys. Rev.*, vol. 82, p. 914, 1951.

For a discussion of the eikonal approximation in field theory see, for instance, H. D. I. Abarbanel in "Cargèse Lectures in Physics," vol. 5, edited by D. Bessis, Gordon and Breach, New York, 1970.

The effective potential is discussed in the paper by S. Coleman and E. Weinberg, *Phys. Rev.*, ser. D, vol. 7, p. 1888, 1973, and in the Erice 1973 lectures of S. Coleman entitled "Secret Symmetry" in "Laws of Hadronic Matter," Proceedings of the Eleventh Course, International School in Physics "Ettore Majorana," edited by A. Zichichi, Academic Press, New York, 1975. The calculations quoted in the text are described in J. Iliopoulos, C. Itzykson, and A. Martin, *Rev. Mod. Phys.*, vol. 47, p. 165, 1975.

P. A. M. Dirac gave the first general discussion of the quantization of constrained systems, in *Proc. Roy. Soc.*, ser. A, vol. 246, p. 326, 1958. It was elaborated in the context of path integrals by L. D. Faddeev, *Theoretical and Mathematical Physics*, vol. 1, p. 1, 1969.

Vacuum instabilities and their relation to the divergence of the perturbative expansion were suggested by F. J. Dyson, *Phys. Rev.*, vol. 85, p. 631, 1952. An early study of Borel summability was made by A. M. Jaffe, *Comm. Math. Phys.*, vol. 1, p. 127, 1965. It has been proved for the anharmonic oscillator by S. Graffi, V. Grecchi, and B. Simon, *Phys. Lett.*, ser. B, vol. 32, p. 631, 1970. Formulas for the anharmonic oscillator were obtained by C. M. Bender and T. T. Wu, *Phys. Rev.*, vol. 184, p. 1231, 1969; *Phys. Rev. Lett.*, vol. 27, p. 461, 1971; and *Phys. Rev.*, ser. D, vol. 7, p. 1620, 1973. Quantization around a classical extremum is discussed in R. F. Dashen, B. Hasslacher, and A. Neveu, *Phys. Rev.*, ser. D, vol. 10, p. 4114, 1974.

Large-orders estimates in field theory are studied by L. N. Lipatov, *JETP*, vol. 45, p. 216, 1977, and E. Brézin, J. C. Le Guillou, and J. Zinn-Justin, *Phys. Rev.*, ser. D, vol. 15, pp. 1544 and 1558, 1977.

CHAPTER

TEN

INTEGRAL EQUATIONS AND BOUND-STATE PROBLEMS

The purpose of establishing integral equations among amplitudes is to study non-perturbative properties. In particular we wish to find a suitable framework for the investigation of relativistic bound states. We shall concentrate our attention on this aspect leaving aside other interesting applications. The formalism is described in some detail and is illustrated on the example of the hyperfine splitting of positronium.

10-1 THE DYSON-SCHWINGER EQUATIONS

Using different techniques, Schwinger and Dyson have derived independently integral equations for Green functions as a consequence of the field equations, i.e., the specific structure of the dynamical lagrangian. This was already encountered in Chap. 9 for a self-coupled scalar field.

When properly renormalized these equations may be used as an alternative approach to perturbation theory. A closer look reveals that the system involves an infinite hierarchy of equations. Consequently, apart from the discussion of general properties their usefulness is limited by the approximations needed to bring them in manageable form. This unfortunate circumstance may be traced to the fact that the natural mathematical tool in this context is the functional calculus, a not too familiar instrument.

10-1-1 Field Equations

For definiteness we shall deal with electrodynamics, but the technique is general. We introduce as usual a generating functional with a source $J_\mu(x)$ for the electromagnetic potential and anticommuting $\eta(x)$ and $\bar{\eta}(x)$ sources for the electron-positron field. A path integral expression for this functional is

$$
\begin{aligned}
Z(J, \eta, \bar{\eta}) = e^{G(J,\eta,\bar{\eta})} &= \int \mathscr{D}(A, \psi, \bar{\psi})\, e^{i\sigma} \\
\sigma = I(A, \psi, \bar{\psi}) &+ \int d^4x\, [J_\mu(x)A^\mu(x) + \bar{\eta}(x)\psi(x) + \bar{\psi}(x)\eta(x)]
\end{aligned}
\tag{10-1}
$$

Here I is the action as a function of A, ψ, $\bar{\psi}$ given in (6-25), and a normalization factor is understood to insure that the connected functional, where we omit the index c for shortness, satisfies $G(0) = 0$.

To cope with renormalization, the action may be regularized and counterterms included. For the sake of simplicity these will not be explicitly spelled out until the end of the derivation.

Field equations will follow from the observation that the integral of a derivative vanishes. We have, for instance,

$$
\begin{aligned}
0 &= \int \mathscr{D}(A, \psi, \bar{\psi}) \left[\frac{\delta I}{\delta A^\mu(x)} + J_\mu(x) \right] e^{i\sigma} \\
&= \left[\frac{\delta I}{\delta A^\mu(x)} \left(\frac{1}{i\delta J}, \frac{\delta}{i\delta\bar{\eta}}, -\frac{\delta}{i\delta\eta} \right) + J_\mu(x) \right] Z(J, \eta, \bar{\eta})
\end{aligned}
\tag{10-2}
$$

Write

$$
\frac{\delta I}{\delta A^\mu(x)} = [\Box g_{\mu\nu} - (1 - \lambda)\partial_\mu\partial_\nu] A^\nu - e\bar{\psi}\gamma_\mu\psi \tag{10-3}
$$

Then Eq. (10-2) reads

$$
J_\mu(x) + [\Box g_{\mu\nu} - (1 - \lambda)\partial_\mu\partial_\nu] \frac{\delta G}{i\delta J_\nu(x)} - e\frac{\delta G}{\delta\eta}\gamma_\mu\frac{\delta G}{\delta\bar{\eta}} - e\frac{\delta}{\delta\eta}\left(\gamma_\mu\frac{\delta G}{\delta\bar{\eta}}\right) = 0 \tag{10-4}
$$

It is advantageous to perform the Legendre transformation to irreducible functions

$$
G(J, \eta, \bar{\eta}) = i\Gamma(A, \psi, \bar{\psi}) + i\int d^4x\, (J \cdot A + \bar{\psi}\eta + \bar{\eta}\psi) \tag{10-5}
$$

satisfying

$$
\begin{aligned}
A_\mu(x) &= \frac{1}{i}\frac{\delta G}{\delta J^\mu(x)} & \psi(x) &= \frac{1}{i}\frac{\delta G}{\delta\bar{\eta}(x)} & \bar{\psi}(x) &= -\frac{1}{i}\frac{\delta G}{\delta\eta(x)} \\
J_\mu(x) &= -\frac{\delta\Gamma}{\delta A^\mu(x)} & \eta(x) &= -\frac{\delta\Gamma}{\delta\bar{\psi}(x)} & \bar{\eta}(x) &= \frac{\delta\Gamma}{\delta\psi(x)}
\end{aligned}
\tag{10-6}
$$

Inserting these definitions in (10-4) and recognizing that

$$-\delta_{\alpha\beta}\delta^4(x-y) = \frac{1}{i}\int d^4z \, \frac{\delta^2 G}{\delta\eta_\alpha(x)\delta\bar{\eta}_\gamma(z)} \, \frac{\delta^2\Gamma}{\delta\psi_\gamma(z)\delta\bar{\psi}_\beta(y)}\bigg|_{\substack{\eta=\bar{\eta}=0\\ \psi=\bar{\psi}=0}} \tag{10-7}$$

we may reexpress it when the fermion sources vanish as

$$\frac{\delta\Gamma}{\delta A^\mu(x)}\bigg|_{\psi=\bar{\psi}=0} = [\Box g_{\mu\nu} - (1-\lambda)\partial_\mu\partial_\nu]A^\nu(x) - ie \operatorname{tr}\left[\gamma_\mu\left(\frac{\delta^2\Gamma}{\delta\bar{\psi}\delta\psi}\right)^{-1}(x, x)\right] \tag{10-8}$$

The inverse of $\delta^2\Gamma/\delta\bar{\psi}(x)\delta\psi(y)$ is the electron propagator including radiative corrections in the presence of an external field. We take one derivative with respect to A and set it equal to zero. In this process we encounter the irreducible vertex function (7-46)

$$\frac{\delta}{\delta A^\mu(x)} \, \frac{\delta^2\Gamma}{\delta\bar{\psi}(y)\delta\psi(z)}\bigg|_{A,\psi,\bar{\psi}=0} = e\Lambda_\mu(x; y, z) \tag{10-9}$$

the vacuum polarization amplitude

$$\frac{\delta^2\Gamma}{\delta A^\mu(x)\delta A^\nu(y)}\bigg|_{A,\psi,\bar{\psi}=0} = [\Box g_{\mu\nu} - (1-\lambda)\partial_\mu\partial_\nu]\delta^4(x-y) + (\Box g_{\mu\nu} - \partial_\mu\partial_\nu)\bar{\omega}(x-y) \tag{10-10}$$

and the complete electron propagator $iS(x, y)$, satisfying the relation (Fig. 10-1)

$$(g_{\mu\nu}\Box - \partial_\mu\partial_\nu)\bar{\omega}(x, y) = ie^2 \int d^4z_1 \, d^4z_2 \operatorname{tr}[\gamma_\mu S(x, z_1)\Lambda_\nu(y; z_1, z_2)S(z_2, x)] \tag{10-11}$$

The normalizations are such that to lowest order

$$\begin{aligned} \Lambda_\nu(y; z_1, z_2) &= \gamma_\nu\delta^4(y-z_1)\delta^4(y-z_2) + \cdots \\ S(x, y) &= \int \frac{d^4p}{(2\pi)^4} \, \frac{e^{-ip\cdot(x-y)}}{\not{p} - m - \sum(p)} \end{aligned} \tag{10-12}$$

In other words, Eq. (10-4) is a sophisticated form of Maxwell's equations of which (10-11) is only a small part. Higher functional derivatives would lead to further relations. The graphical interpretation given in Fig. 10-1 could, of course, be used for a direct derivation. Similarly, Dirac's equation will follow from the

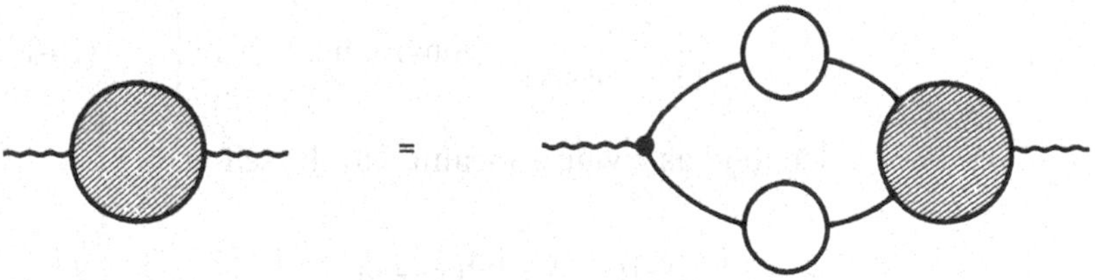

Figure 10-1 Integral equation for the vacuum polarization.

identity

$$0 = \int \mathscr{D}(A, \bar{\psi}, \psi) \left[\frac{\delta I}{\delta \bar{\psi}(x)} + \eta(x) \right] e^{i\sigma}$$
$$= \left[\frac{\delta I}{\delta \bar{\psi}(x)} \left(\frac{\delta}{i\delta J}, \frac{\delta}{i\delta \bar{\eta}}, -\frac{\delta}{i\delta \eta} \right) + \eta(x) \right] Z(J, \eta, \bar{\eta}) \tag{10-13}$$

It is convenient here to work with complete Green functions without disconnected photon amplitudes, obtained as derivatives of $Z(J, \eta, \bar{\eta})/Z(J, 0, 0)$. For shortness we write $Z(J, 0, 0) \equiv Z(J)$ and obtain

$$\left[\eta(x) + \left(i\partial\!\!\!/ - m - e\gamma^\mu \frac{\delta}{i\delta J^\mu(x)} \right) \frac{\delta}{i\delta \bar{\eta}(x)} \right] Z(J, \eta, \bar{\eta}) = 0 \tag{10-14}$$

We take a derivative with respect to η and set $\eta = \bar{\eta} = 0$

$$\delta^4(x - y) Z(J) - \left[i\partial\!\!\!/ - m - e\gamma^\mu \frac{\delta}{i\delta J^\mu(x)} \right] Z(J) S(x, y; J) = 0$$

where $S(x, y; J)$ describes the propagation in the presence of the source J. If we denote

$$A_\mu(x; J) = Z^{-1}(J) \frac{\delta}{i\delta J^\mu(x)} Z(J) = \frac{1}{i} \frac{\delta}{\delta J^\mu(x)} G(J, 0, 0)$$

this equation may be rewritten

$$\delta^4(x - y) - \left[i\partial\!\!\!/ - m - eA\!\!\!/(x; J) - e\gamma^\mu \frac{\delta}{i\delta J^\mu(x)} \right] S(x, y; J) = 0 \tag{10-15}$$

with a Kronecker δ function for spinor indices omitted. In (10-15) the differentiation with respect to J can be performed, the source set equal to zero, and as a consequence $A(x; J)|_{J=0} = 0$. This yields

$$(i\partial\!\!\!/ - m) S(x, y) - ie^2 \int d^4z \, d^4x' \, d^4y' \, \gamma_\mu G^{\mu\nu}(x, z) S(x, x')$$
$$\times \Lambda_\nu(z; x', y') S(y', y) = \delta^4(x - y) \tag{10-16}$$

since the complete three-point function involves the irreducible vertex convoluted with propagators for the two fermions and for the photon

$$G^{\mu\nu}(x, y) = \int d^4k \, e^{-ik \cdot (x-y)} \left\{ -\frac{g_{\mu\nu}}{k^2[1 + \bar{\omega}(k^2)]} + \text{longitudinal terms} \right\} \tag{10-17}$$

Equation (10-16) is best understood as giving a meaning to the self-mass operator in the form

$$[(i\partial\!\!\!/ - m - \textstyle\sum) S](x, y) = \delta^4(x - y) \tag{10-18}$$

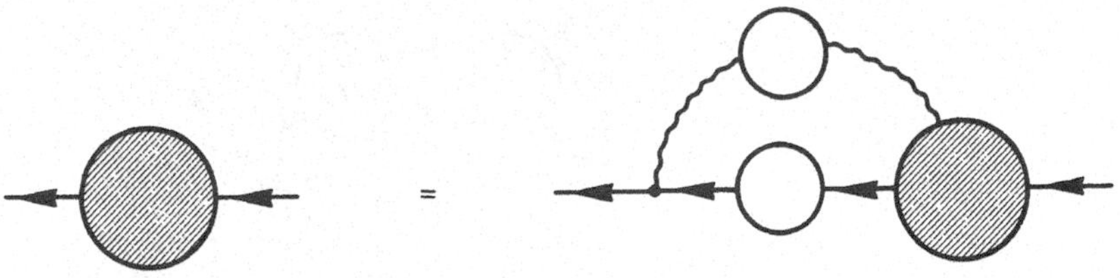

Figure 10-2 Representation of the self-mass operator.

with

$$\sum(x, y) = ie^2 \int d^4z \, d^4x' \, \gamma_\mu G^{\mu\nu}(x, z) S(x, x') \Lambda_\nu(z; x', y) \tag{10-19}$$

This formula is illustrated in Fig. 10-2.

If we set $\eta = \bar{\eta} = 0$, we may rewrite Eq. (10-4) in the form

$$[\Box g_{\mu\nu} - (1 - \lambda)\partial_\mu \partial_\nu] A^\nu(x; J) = -J_\mu - ie \, \mathrm{tr} \, [\gamma_\mu S(x, x; J)] \tag{10-20}$$

In conjunction with (10-15) this yields a closed functional system where the argument is the source J. Alternatively, we may write an equivalent system with an external potential A as argument. For this purpose let $G_{\nu\rho}(x, y; A)$ be the photon propagator. Differentiating Eq. (10-20) with respect to J we obtain

$$[\Box_x g_\mu{}^\nu - (1 - \lambda)\partial_\mu \partial^\nu] G_{\nu\rho}(x, y; A)$$

$$+ ie \int d^4z \, \mathrm{tr} \left[\gamma_\mu \frac{\delta S(x, x; A)}{\delta A^\nu(z)} G_{\nu\rho}(z, y; A) \right] = g_{\mu\rho} \delta^4(x - y) \tag{10-21}$$

while (10-15) may be rewritten as

$$[i\partial\!\!\!/_x - m - eA\!\!\!/(x)] S(x, y; A) - ie \int d^4z \, \gamma_\mu G^{\mu\rho}(x, z; A) \frac{\delta S(x, y; A)}{\delta A^\rho(z)} = \delta^4(x - y) \tag{10-22}$$

Again this is a complete set to determine S and G. These remarks are unfortunately of little help due to our inexperience in handling such expressions.

Returning to Eq. (10-14) we can obtain further information if we take higher derivatives with respect to spinor sources. Of great interest is the Green function involving the propagation of two charged particles: $S(x_1, x_2; y_1, y_2; J)$. Proceeding as before we find

$$\left[i\partial\!\!\!/_{x_1} - m - eA\!\!\!/(x_1; J) - e\gamma^\mu \frac{\delta}{i\delta J^\mu(x_1)} \right] S(x_1, x_2; y_1, y_2; J)$$

$$= \delta^4(x_1 - y_1) S(x_2, y_2; J) - \delta^4(x_1 - y_2) S(x_2, y_1; J) \tag{10-23}$$

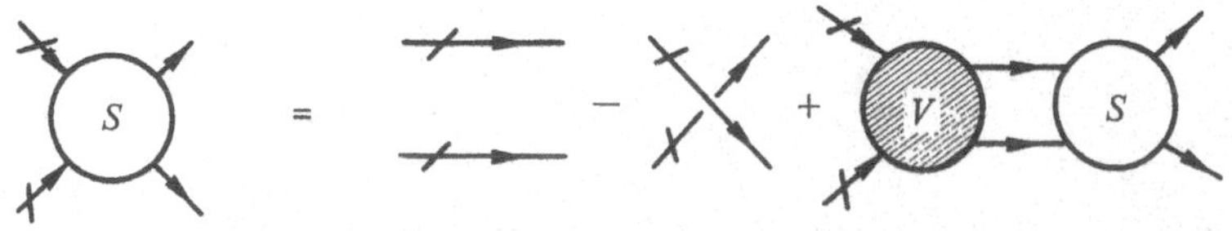

Figure 10-3 The kernel V of the Bethe-Salpeter equation. The slashes indicate truncation of the corresponding propagator lines.

The antisymmetric combination on the right-hand side is a reflection of the Pauli principle. Taking into account the relation (10-18) satisfied by the two-point function, we may act on the variable x_2 with the operator $i\partial\!\!\!/ - m - \sum$. (For different applications we might be interested to act on one of the two remaining variables y_1 or y_2.) At any rate we find, when setting $J = 0$,

$$\begin{aligned}
&(i\partial\!\!\!/_{x_2} - m - \textstyle\sum)(i\partial\!\!\!/_{x_1} - m - \textstyle\sum)S(x_1, x_2; y_1, y_2) \\
&\quad = \delta^4(x_1 - y_1)\delta^4(x_2 - y_2) - \delta^4(x_1 - y_2)\delta^4(x_2 - y_1) \\
&\quad + (i\partial\!\!\!/_{x_2} - m - \textstyle\sum)\left[e\gamma^\mu \frac{\delta}{\delta J^\mu(x_1)} - \textstyle\sum\right] S(x_1, y_1; x_2, y_2; J)\big|_{J=0}
\end{aligned} \tag{10-24}$$

The first term on the right-hand side is readily traced out as the contribution of the disconnected amplitude. The second term may best be understood from a graphical analysis. Written as a convolution integral

$$\int d^4z_1\, d^4z_2\, V(x_1, x_2; z_1, z_2)S(z_1, z_2; y_1, y_2)$$

it involves a kernel $V(x_1, x_2; y_1, y_2)$ described in terms of truncated four-fermion diagrams which cannot be disconnected by cutting two fermion lines (Fig. 10-3). To lowest order, and adding superscripts to γ matrices to recall on which indices they act,

$$V(x_1, x_2; y_1, y_2) = ie^2\gamma_\mu^{(1)}G^{\mu\nu}(x_1 - x_2)\gamma_\nu^{(2)}\delta^4(x_1 - y_1)\delta^4(x_2 - y_2) \tag{10-25}$$

With this notation we write the resulting Bethe-Salpeter equation in the form

$$\begin{aligned}
&(i\partial\!\!\!/_{x_2} - m - \textstyle\sum)(i\partial\!\!\!/_{x_1} - m - \textstyle\sum)S(x_1, x_2; y_1, y_2) \\
&\quad = \delta^4(x_1 - y_1)\delta^4(x_2 - y_2) - \delta^4(x_1 - y_2)\delta^4(x_2 - y_1) \\
&\quad + \int d^4z_1\, d^4z_2\, V(x_1, x_2; z_1, z_2)S(z_1, z_2; y_1, y_2)
\end{aligned} \tag{10-26}$$

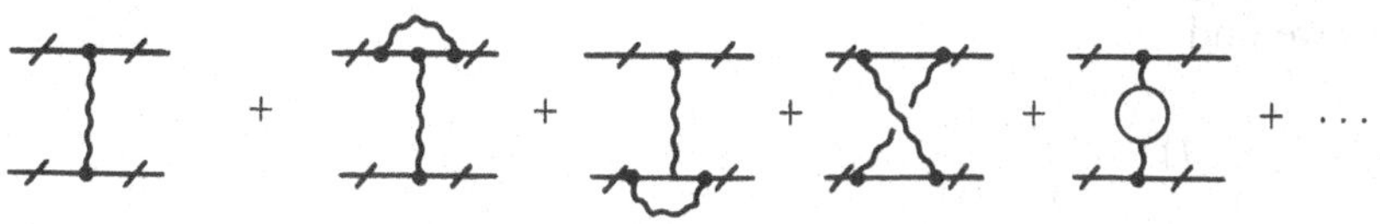

Figure 10-4 Bethe-Salpeter equation.

It is represented graphically in Fig. 10-4 where the slashes indicate amputations on the corresponding propagator lines.

As it stands this equation has been written in the doubly charged electron-electron channel. It can, of course, be established in the crossed electron-positron channel where it is suited for the investigation of positronium bound states.

10-1-2 Renormalization

The field equations obviously need renormalization. To avoid cumbersome notations we did not introduce the counterterms in the action. They have now to be reinstated. As this is not our main concern here we shall not elaborate these points in great detail. We only sketch the modifications brought upon the equations by taking the multiplicative renormalization of Green functions into account when expressing them in terms of the physical mass and coupling constant.

Consider, first, Eq. (10-15) for the electron propagator. In terms of electron wave function (Z_2) and vertex renormalization (Z_1) this should properly be written

$$\delta^4(x-y) - \left\{ Z_2(i\not\partial - m) - eZ_1 \left[\not A(x, J) + \gamma^\mu \frac{\delta}{i\delta J^\mu(x)} \right] \right\} S(x, y; J) = 0 \qquad (10\text{-}27)$$

The Ward identity requires $Z_1 = Z_2$, but both are infinite quantities. Similarly, the photon propagator equation (10-21) should read

$$Z_3[\Box_x g_\mu{}^\nu - \partial_\mu \partial^\nu] G_{\nu\rho}(x, y) + \lambda \partial_\mu \partial^\nu G_{\nu\rho}(x, y)$$
$$+ iZ_1 e \int d^4z \, \mathrm{tr} \left[\gamma_\mu \frac{\delta S(x, x)}{\delta A^\nu(z)} G_{\nu\rho}(z, y) \right] = g_{\mu\rho} \delta^4(x-y) \qquad (10\text{-}28)$$

implying again the infinite renormalization constants necessary to obtain finite results when iterating these equations for the finite Green functions G and S.

This is a rather unpleasant situation, except if we return to the perturbative expansion.

Exercises

(*a*) Derive the renormalized forms of Eqs. (10-11) and (10-19).

(*b*) Relate the vertex function to the Bethe-Salpeter-kernel and examine the renormalization properties of Eq. (10-26).

(*c*) Obtain the equivalent equations in the φ^4 model and discuss their renormalization properties.

10-2 RELATIVISTIC BOUND STATES

In the relativistic approach, bound states and resonances are identified by the occurrence of poles in Green functions. A simple extension of the Schrödinger equation is unfortunately not available, except in limiting cases such as those of static external sources discussed in the framework of Dirac's equation.

The general difficulty is in fact twofold. At first we have to face retardation effects which introduce an extra relative time variable in the problem. An alternative description uses the mediation of a field. But the quantum aspects of the latter cannot be ignored. Thus it appears that the mere concept of a two-body bound state, for instance, is only an oversimplification of the actual situation. In spite of various artefacts—which may be of great practical importance—it remains true that we have to return to the general field theoretic framework whenever we need a fundamental description. This is also the case if we want to introduce further radiative corrections.

A great amount of work has been devoted to the subject of which only a small, but hopefully representative, part will be surveyed here. We recall that we have already discussed aspects of hydrogen-like bound states in Chap. 2 and computed the Lamb-shift corrections to lowest order in Chap. 7.

10-2-1 Homogeneous Bethe-Salpeter Equation

Instead of dealing with the full complexity of the spinor problem, we shall examine for the time being a simpler model of scalar particles interacting through the exchange of another type of scalar particles. This is, of course, a theoretical exercise in order to exhibit some of the features of the real problem. The kernel V has also to be truncated. Moreover, we shall ignore the effects of statistics by assuming that the two "charged" particles are of a different kind.

In symbolic notation Eq. (10-26) may be rewritten

$$S^{(12)} = S^{(1)}S^{(2)} + S^{(1)}S^{(2)}VS^{(12)} \tag{10-29}$$

where $S^{(1)}$ stands now for the complete propagator of particle 1, soon to be approximated by a free propagator involving the physical mass, and renormalization effects have been discarded. Defining $-D$ as the inverse kernel of the product $S^{(1)}S^{(2)}$ this equation could formally be solved in the form

$$S^{(12)} = -(D + V)^{-1} \tag{10-30}$$

showing that poles may occur whenever $D + V$ has a zero eigenvalue. We are thus led to a homogeneous equation describing the properties of bound states. To be precise we define

$$\begin{aligned} S^{(12)} &= \langle 0|\, T\varphi_1(x_1)\varphi_2(x_2)\varphi_1^\dagger(y_1)\varphi_2^\dagger(y_2)\,|0\rangle \\ S^{(1)} &= \langle 0|\, T\varphi_1(x_1)\varphi_1^\dagger(y_1)\,|0\rangle \end{aligned} \tag{10-31}$$

A bound-state contribution of mass M to $S^{(12)}$ (assumed nondegenerate for simplicity) will be of the form

$$\int \frac{d^3P}{(2\pi)^3 2\sqrt{P^2 + M^2}}\, \chi_P(x_1, x_2)\bar{\chi}_P(y_1, y_2) \tag{10-32}$$

with

$$\begin{aligned} \chi_P(x_1, x_2) &= \langle 0|\, T\varphi_1(x_1)\varphi_2(x_2)\,|P\rangle \\ \bar{\chi}_P(y_1, y_2) &= \langle P|\, T\varphi_1^\dagger(y_1)\varphi_2^\dagger(y_2)\,|0\rangle = \langle 0|\, \bar{T}\varphi_1(y_1)\varphi_2(y_2)\,|P\rangle^* \end{aligned} \tag{10-33}$$

Here $\bar{T}$ denotes antichronological ordering. The generalization to a set of degenerate bound states is straightforward. Approximating $S^{(1)}$ and $S^{(2)}$ by free propagators

$$S^{(i)} = \int \frac{d^4k}{(2\pi)^4} \frac{i}{k^2 - m_i^2 + i\varepsilon} e^{-ik \cdot (x_1 - y_1)} \tag{10-34}$$

we find

$$(\Box_{x_1} + m_1^2)(\Box_{x_2} + m_2^2)\chi_P(x_1, x_2) + \int d^4z_1\, d^4z_2\, V(x_1, x_2; z_1, z_2)\chi_P(z_1, z_2) = 0 \tag{10-35}$$

$$\bar{\chi}_P(y_1, y_2)(\overleftarrow{\Box}_{y_1} + m_1^2)(\overleftarrow{\Box}_{y_2} + m_2^2) + \int d^4z_1\, d^4z_2\, \bar{\chi}_P(z_1, z_2)V(z_1, z_2; y_1, y_2) = 0$$

Equations (10-35) will be sufficient to illustrate the type of reasoning when dealing with Bethe-Salpeter equations.

In spite of some similarity with the nonrelativistic Schrödinger case, this is a very different type of equation. It is reflected in the larger number of configuration variables, in the fact that we deal with fourth-order integro-differential equations, in the presence of a kernel V which has to be specified perturbatively, and in the way the energy momentum of the bound state enters the equation.

From translational invariance, it follows that

$$\chi_P(x_1 + a, x_2 + a) = e^{-iP \cdot a}\chi_P(x_1, x_2) \tag{10-36}$$

It is natural to introduce the relative space-time coordinate $x = x_1 - x_2$. However, the overall configuration variable is a priori arbitrary. We may choose two positive quantities η_1 and η_2 such that

$$\begin{aligned} \eta_1 + \eta_2 &= 1 \qquad & x_1 &= X + \eta_2 x \\ X &= \eta_1 x_1 + \eta_2 x_2 \qquad & x_2 &= X - \eta_1 x \end{aligned} \tag{10-37}$$

a transformation with unit jacobian, and write the reduced Bethe-Salpeter amplitudes as

$$\begin{aligned} \chi_P(x_1, x_2) &= e^{-iP \cdot X}\chi_P(x) \\ \bar{\chi}_P(x_1, x_2) &= e^{iP \cdot X}\bar{\chi}_P(x) \end{aligned} \tag{10-38}$$

According to (10-33) χ and $\bar{\chi}$ are not to be confused with wave functions but stand rather for generalized form factors. The normalization conditions are therefore not straightforward, since they involve the relative time variable x_0. This innocent-looking question has been the subject of a long elaboration. The role of normalization is, of course, to provide the correct relation between the χ function and the four-point Green function. Furthermore, it is essential in selecting the proper solutions to Eq. (10-35). We return to the inhomogeneous Eq. (10-29). Introducing the overall momentum of the pair (1, 2) through

$$S^{(12)}(x_1, x_2; y_1, y_2 | P) = \int d^4a\, e^{iP \cdot a} S^{(12)}(x_1 + a, x_2 + a; y_1, y_2)$$

the bound state and its CPT transform contribute a pole in the variable P^2:

$$S^{(12)}(x_1, x_2; y_1, y_2 | P) = \frac{i\chi_P(x_1, x_2)\bar{\chi}_P(y_1, y_2)}{P^2 - M^2 + i\varepsilon} + R \tag{10-39}$$

with R regular in the vicinity of $P^2 = M^2$. The factorization property of the pole residue is crucial for the interpretation in terms of bound state and has to be suitably generalized in the case of degeneracies.

We iterate the equation $(D + V)S^{(12)} = -1$ in the form $S^{(12)}(D + V)S^{(12)} = -S^{(12)}$, insert (10-39), and use the fact that $(D + V)\chi = \bar{\chi}(D + V) = 0$ to compare the residues of both sides at $P^2 = M^2$. The result is symbolically

$$\lim_{P^2 \to M^2} \frac{\bar{\chi}(D + V)\chi}{P^2 - M^2} = i$$

or, equivalently, the covariant expression

$$\bar{\chi}\left[\frac{\partial}{\partial P^\mu}(D + V)\right]\chi = 2iP_\mu \tag{10-40}$$

where the left-hand side involves an integral over relative variables. In general the normalization condition depends on the "potential" V as opposed to the nonrelativistic case.

It is useful to have these equations also written in momentum space. Let p denote the conjugate variable to x. According to (10-37) we have

$$P = p_1 + p_2 \qquad p = \eta_2 p_1 - \eta_1 p_2 \tag{10-41}$$

where $p_1 = \eta_1 P + p$ and $p_2 = \eta_2 P - p$ refer to momenta carried by the fields φ_1 and φ_2 (Fig. 10-5). We have here no natural definition of the relative momentum as given by the separation of variables in a nonrelativistic motion, which results from the specific choice $\eta_{1,2} = m_{1,2}/(m_1 + m_2)$. In the relativistic case this choice may, however, be a good candidate for the purpose of comparison. Using the same symbols for the Fourier transformed quantities and taking translational invariance into account, we have

$$\chi_P(p) = \int e^{ip\cdot x}\chi_P(x)\, d^4x$$

$$\begin{aligned} &[(\eta_1 P + p)^2 - m_1^2][(\eta_2 P - p)^2 - m_2^2]\chi_P(p) + \int \frac{d^4p'}{(2\pi)^4} V(p, p'; P)\chi_P(p') = 0 \\ &\bar{\chi}_P(p)[(\eta_1 P + p)^2 - m_1^2][(\eta_2 P - p)^2 - m_2^2] + \int \frac{d^4p'}{(2\pi)^4} \bar{\chi}_P(p')V(p', p; P) = 0 \end{aligned} \tag{10-42}$$

The exchange of a scalar particle of mass μ and coupling constants g_1 and g_2 to particles 1 and 2, that is, such that the corresponding interaction lagrangian

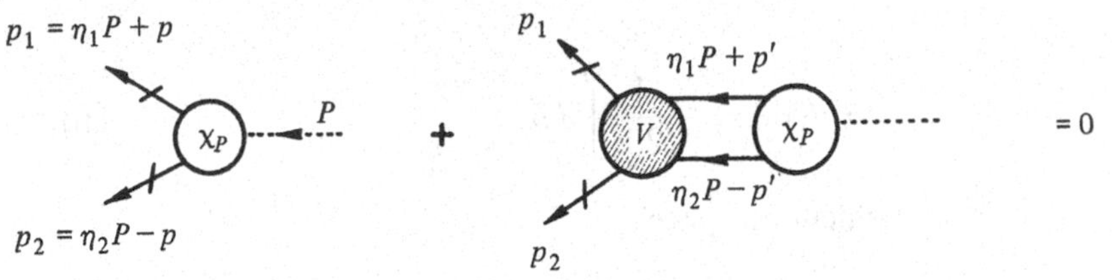

Figure 10-5 Homogeneous equation for the bound-state amplitude χ.

is $\mathscr{L}_{\rm int} = (g_1 \varphi_1^\dagger \varphi_1 + g_2 \varphi_2^\dagger \varphi_2)\phi$, leads in the Born approximation to

$$V_1(p, p'; P) = -\frac{ig_1 g_2}{(p-p')^2 - \mu^2 + i\varepsilon}$$

independent of P. Explicitly, the normalization condition reads

$$\int \frac{d^4p}{(2\pi)^4}\frac{d^4p'}{(2\pi)^4}\,\bar\chi_P(p')\,\frac{\partial}{\partial P^\mu}\,\{[\eta_1 P + p)^2 - m_1^2][(\eta_2 P - p)^2 - m_2^2](2\pi)^4\delta^4(p-p')$$
$$+ V(p', p; P)\}\chi_P(p) = 2iP_\mu \qquad (10\text{-}43)$$

With Eqs. (10-35) and (10-40), or equivalently (10-42) and (10-43), we have now the basic ingredients to study some definite models.

10-2-2 The Wick Rotation

Bound-state equations are derived in Minkowski space. In his early study, Wick was led to an analytic continuation to euclidean variables which was at the origin of the Wick rotation. It is easy to justify this procedure perturbatively, i.e., precisely, ignoring the possibility of new singularities such as those investigated here. Some care is required in order to extend it to the present situation. The essential physical point is to insure that stability criteria are met. Of course, the desirability of using this trick is to bring the equation in a form convenient for a simpler analysis.

Let us proceed in a straightforward manner by writing, for $X = 0$,

$$\begin{aligned}\chi_P(x) &= \theta(x^0) f(x, P) + \theta(-x^0) g(x, P)\\ \bar\chi_P(x) &= \theta(x^0) g^*(x, P) + \theta(-x^0) f^*(x, P)\end{aligned} \qquad (10\text{-}44)$$

with

$$\begin{aligned} f(x, P) &= \langle 0|\varphi_1(\eta_2 x)\varphi_2(-\eta_1 x)|P\rangle = \int \frac{d^4p}{(2\pi)^4}\, e^{-ip\cdot x} f(p; P)\\ g(x, P) &= \langle 0|\varphi_2(-\eta_1 x)\varphi_1(\eta_2 x)|P\rangle = \int \frac{d^4p}{(2\pi)^4}\, e^{-ip\cdot x} g(p; P)\end{aligned} \qquad (10\text{-}45)$$

Using

$$\theta(x_0) = -\frac{1}{2i\pi}\int d\omega\,\frac{e^{-i\omega x_0}}{\omega + i\varepsilon} \tag{10-46}$$

we obtain χ and $\bar{\chi}$ in momentum space

$$\begin{aligned}\chi_P(p) &= -\frac{1}{2i\pi}\int dq_0\,\frac{f(q_0, \mathbf{p}, P)}{p^0 - q^0 + i\varepsilon} + \frac{1}{2i\pi}\int dq_0\,\frac{g(q_0, \mathbf{p}, \mathrm{P})}{p^0 - q^0 - i\varepsilon}\\ \bar{\chi}_P(p) &= -\frac{1}{2i\pi}\int dq_0\,\frac{f^*(q_0, \mathbf{p}, P)}{p^0 - q^0 + i\varepsilon} + \frac{1}{2i\pi}\int dq_0\,\frac{g^*(q_0, \mathbf{p}, P)}{p^0 - q^0 - i\varepsilon}\end{aligned} \tag{10-47}$$

The relation between χ and $\bar{\chi}$ is such that their discontinuities are conjugate. A rotation to the imaginary axis will be allowed provided f and g have a suitable support in the variable q_0. This follows from their definition by inserting intermediate states

$$\begin{aligned}f(q, P) &= \int d^4x\, e^{iq\cdot x}\sum_n \langle 0|\varphi_1(0)|n\rangle\langle n|\varphi_2(0)|P\rangle\, e^{-i(p_n - \eta_1 P)\cdot x}\\ &= \sum_n (2\pi)^4\delta^4(q - p_n + \eta_1 P)\langle 0|\varphi_1(0)|n\rangle\langle n|\varphi_2(0)|P\rangle \\ g(p, P) &= \sum_n (2\pi)^4\delta^4(q + p_n - \eta_2 P)\langle 0|\varphi_2(0)|n\rangle\langle n|\varphi_1(0)|P\rangle\end{aligned} \tag{10-48}$$

The stability of particle 1 requires in the first expression $p_n^2 \geq m_1^2$, $p_n^0 > 0$. This means that $f(q, P)$ vanishes unless $(\eta_1 P + q)^2 \geq m_1^2$, $\eta_1 P_0 + q_0 > 0$. Likewise in the second expression $p_n^2 \geq m_2^2$, $p_n^0 > 0$, hence $g(q, P)$ vanishes unless $(\eta_2 P - q)^2 \geq m_2^2$, $\eta_2 P_0 - q_0 > 0$. In the representation (10-47), the integral of f (or f^*) extends from ω_+ to $+\infty$, the one of g (or g^*) from $-\infty$ to ω_-, with

$$\begin{aligned}\omega_+ &= \sqrt{m_1^2 + (\eta_1\mathbf{P} + \mathbf{p})^2} - \eta_1 P_0\\ \omega_- &= -\sqrt{m_2^2 + (\eta_2\mathbf{P} - \mathbf{p})^2} + \eta_2 P_0\end{aligned} \tag{10-49}$$

A rotation to the imaginary axis without encircling singularities will be possible if $\omega_- < \omega_+$. For a stable bound state $P_0 < \sqrt{(m_1 + m_2)^2 + \mathbf{P}^2}$. If we choose the center of mass frame $\mathbf{P} = 0$ and $\eta_{1,2} = m_{1,2}/(m_1 + m_2)$, then

$$\omega_+ = \sqrt{m_1^2 + \mathbf{p}^2} - \frac{m_1 P_0}{m_1 + m_2} > 0 \qquad \text{and} \qquad \omega_- = \frac{m_2 P_0}{m_1 + m_2} - \sqrt{m_2^2 + \mathbf{p}^2} < 0$$

Consequently, we have a gap between the two cuts in the integral representations (10-47).

We may now return to the Bethe-Salpeter equation and perform the Wick rotation using these results. For simplicity we retain the center of mass frame and the previous assignments for $\eta_{1,2}$. Furthermore, we limit ourselves to the so-called ladder approximation, i.e., the kernel V is evaluated in the Born approximation

$$V = -\frac{ig_1g_2}{(p-p')^2 - \mu^2 + i\varepsilon} \tag{10-50}$$

The equation takes the form

$$\left[\left(\frac{m_1}{m_1+m_2}P_0 + p_0\right)^2 - \mathbf{p}^2 - m_1^2\right]\left[\left(\frac{m_2}{m_1+m_2}P_0 - p_0\right)^2 - \mathbf{p}^2 - m_2^2\right]\chi_P(p)$$
$$- \frac{ig_1g_2}{(2\pi)^4}\int d^4p' \frac{\chi_P(p')}{(p-p')^2 - \mu^2 + i\varepsilon} = 0 \tag{10-51}$$

Consider the analytic continuation $p^0 \to p^0 e^{i\theta}$ with θ varying from 0 to $\pi/2$. For the first term in (10-51) we use the representation (10-47) and examine separately the cases $p^0 > 0$ and $p^0 < 0$. Using the properties $\omega_- < 0 < \omega_+$ we see that no singularity is encountered in the process of analytic continuation. For the second term we proceed to a simultaneous rotation $p^0 \to p^0 e^{i\theta}$, $p'^0 \to p'^0 e^{i\theta}$, and for the same reason we do not encounter any singularity from $\chi_P(p')$ nor from the denominator. Using abusively for $\theta = \pi/2$ the same notations for the functions of the euclidean argument the final result assumes the form

$$\left[\left(p_0 - \frac{im_1}{m_1+m_2}P_0\right)^2 + \mathbf{p}^2 + m_1^2\right]\left[\left(p_0 + \frac{im_2}{m_1+m_2}P_0\right)^2 + \mathbf{p}^2 + m_2^2\right]\chi_P(p)$$
$$- \frac{g_1g_2}{(2\pi)^4}\int d^4p' \frac{\chi_P(p')}{(p-p')^2 + \mu^2} = 0 \tag{10-52}$$

The metric used in (10-52) is the euclidean one $(p-p')^2 = \sum_{i=0}^{3}(p_i - p'_i)^2$.

The extension of the preceding method to higher kernels requires a more elaborate analysis in several variables. A similar study may also be carried out for the scattering case where part of the integration contour remains pinched between the two cuts. We leave it as an exercise to the reader to examine these cases.

10-2-3 Scalar Massless Exchange in the Ladder Approximation

We present the solution obtained by Wick and Cutkosky for the equation (10-52) when $\mu^2 = 0$ as an illustration of the methods and problems encountered in studying the relativistic bound-state equations. This choice is not completely arbitrary from the physical point of view since in spite of the oversimplifications resulting in part from the neglect of spin (particularly of the exchanged field) the situation has some analogy with the realistic cases such as positronium. Furthermore, it allows an almost analytic solution and exhibits peculiarities which may arise in more complex models.

The set of ladder diagrams looks at first as a natural generalization of nonrelativistic potential theory. However, such an approximation overlooks essential features of a relativistic quantum theory. By neglecting the crossed-ladder diagrams, it violates (s–u) crossing symmetry. It implies a correspondence between the proper time ordering of the relativistic interactions, thus mistreating the relativistic static limit when one of the masses gets large. This is a severe limitation when dealing with realistic problems. For instance, in electrodynamics, unless we find some compelling reasons to use it in a particular gauge (the noncovariant Coulomb gauge for instance), this approximation will not be gauge invariant; nor will it satisfy the reasonable criterion of reducing to the Klein-Gordon (or Dirac in the spin $\frac{1}{2}$ case) equation when one of the particles becomes very massive.

To fulfill this criterion it is necessary to include at least the set of crossed-ladder diagrams leading to an infinite kernel V. Find the corresponding approximation in the functional language of Eqs. (10-23) and (10-24).

With these limitations in mind we return to Eq. (10-52), where we set $\mu^2 = 0$. Wick and Cutkosky, observing the analogy with the hydrogen problem in momentum space, suggested the use of the stereographic map on a unit sphere in five-dimensional space. This method introduced by Fock in the nonrelativistic case enables us to use conformal transformations in a transparent manner and to exhibit the dynamical symmetry of the system. It is convenient to introduce the notation λ for the dimensionless quantity

$$\lambda = \frac{g_1 g_2}{16 m_1 m_2 \pi} \tag{10-53}$$

This choice reminds us that the g are dimensioned φ^3 coupling constants. It allows a comparison with the electromagnetic case, λ being the analog of the fine-structure constant. We shall first content ourselves with the equal-mass case $m_1 = m_2 = m$. Taking m as the energy unit, we find

$$\left[\left(p - i\frac{P}{2}\right)^2 + 1\right]\left[\left(p + i\frac{P}{2}\right)^2 + 1\right]\chi_P(p) = \frac{\lambda}{\pi^3}\int d^4p' \, \frac{\chi_P(p')}{(p - p')^2} \tag{10-54}$$

where P stands for $(P_0, \mathbf{0})$. This may be regarded as an eigenvalue equation for the coupling λ and appeal to Fredholm's theory will show that it admits a discrete spectrum. The stereographic projection to be considered in more detail in Chap. 13 amounts to an association of a point p in R^4 to a vector z in five-dimensional space on a sphere of radius $\frac{1}{2}\sqrt{1 - P^2/4}$ with four-dimensional pro-

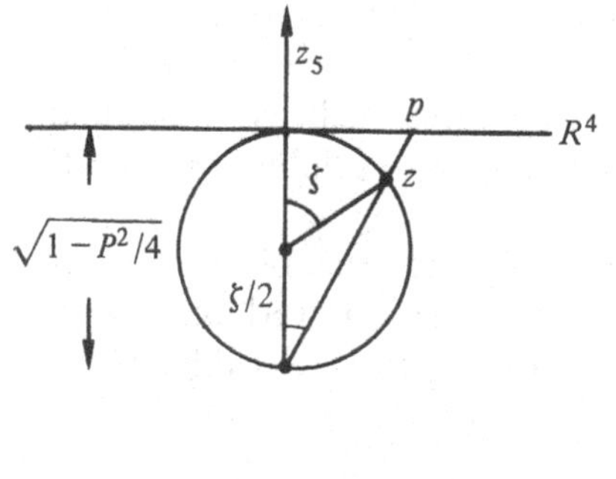

Figure 10-6 Projection from the four-dimensional p space onto a sphere of diameter $\sqrt{1 - P^2/4}$.

jection along p and polar angle ζ such that $\tan \zeta/2 = |p|/\sqrt{1 - P^2/4}$ (Fig. 10-6). We use (β, θ, φ) as the additional polar angles of p; hence $p \cdot P = |p||P| \cos \beta$, and

$$\left[\left(p - i\frac{P}{2}\right)^2 + 1\right]\left[\left(p + i\frac{P}{2}\right)^2 + 1\right] = \frac{1}{\cos^4 \zeta/2}\left(1 - \frac{P^2}{4}\right) \times \left(1 - \frac{P^2}{4} + \frac{P^2}{4} \sin^2 \zeta \cos^2 \beta\right) \tag{10-55}$$

If $d\Omega_4$ denotes the solid angle element on the sphere normalized according to $\int d\Omega_4 = 8\pi^2/3$ then

$$\frac{d^4p'}{(p - p')^2} = \frac{(1 - P^2/4)}{8} \frac{\cos^2 \zeta/2}{\cos^6 \zeta'/2} \frac{d\Omega'_4}{1 - \cos \alpha} \tag{10-56}$$

where α is the angle between the corresponding vectors z and z', namely $z \cdot z' = [(1 - P^2/4)/4] \cos \alpha$. Finally, we define a new function $K(z)$ through

$$K(z) = \frac{1}{\cos^6 \zeta/2} \chi(p) \tag{10-57}$$

It follows that

$$\left(1 - \frac{P^2}{4} + \frac{P^2}{4} \sin^2 \zeta \cos^2 \beta\right) K(z) = \frac{\lambda}{8\pi^3} \int \frac{d\Omega'_4}{1 - \cos \alpha} K(z') \tag{10-58}$$

The limiting case $P^2 = 0$ admits an O(5) invariance with K proportional to a five-dimensional spherical harmonic. Let $N - 1$ stand for the degree of this spherical harmonic, which means that multiplication by $|z|^{N-1}$ yields an harmonic function in five-dimensional space. To compute the eigenvalues [with degeneracy $N(N + 1)(2N + 1)/6$, $N = 1, 2, \ldots$] it is sufficient to apply (10-58) to special spherical harmonics depending only on the angle with the fifth axis. Those are orthogonal polynomials (generalizing the Legendre polynomials) obtained by expanding in powers of $z_</z_>$ the elementary Green function

$$\frac{1}{|z - z'|^3} = \frac{1}{|z_>|^3[1 + z_<^2/z_>^2 - 2(z_</z_>) \cos \alpha]^{3/2}}$$
$$\frac{1}{[1 + t^2 - 2t \cos \alpha]^{3/2}} = \sum_{N=1}^{\infty} C_{N-1}(\cos \alpha) t^{N-1} \tag{10-59}$$

Thus, choosing z along the fifth axis, λ_N is given by

$$C_{N-1}(1) = \frac{\lambda_N}{8\pi^3} \int d\Omega'_4 \frac{C_{N-1}(\zeta')}{1 - \cos \zeta'} = \frac{\lambda_N}{4\pi} \int_0^{\pi} \frac{d\zeta \sin^3 \zeta\, C_{N-1}(\zeta)}{1 - \cos \zeta}$$

or, equivalently for $|t| < 1$,

$$\sum_1^{\infty} \frac{t^{N-1}}{\lambda_N} C_{N-1}(1) = \frac{1}{4\pi} \int_0^{\pi} \frac{d\zeta \sin^3 \zeta}{1 - \cos \zeta} \frac{1}{(1 + t^2 - 2t \cos \zeta)^{3/2}} = \frac{1}{2\pi(1 - t)}$$

From its definition (10-59), $C_{N-1}(1) = N(N+1)/2$; thus

$$\frac{\lambda_N}{\pi} = N(N+1) \tag{10-60}$$

In the general case $0 < P^2 < 4$, Eq. (10-58) exhibits an O(4) invariance very much as in the hydrogenic case. This follows from the fact that only the combination $\sin\zeta \cos\beta$, proportional to the projection of z on the zeroth axis (i.e., along P), enters the equation, apart from O(5)-invariant expressions. Thus the O(4) invariance pertains to rotations leaving the zeroth axis invariant.

The unit sphere in R^5 can then be projected back on the euclidean space R^4 but now from a point along the zeroth axis. If γ is such that $\cos\gamma = \sin\zeta\cos\beta$, let q be the projected four-vector from the unit sphere with $|q| = \tan(\gamma/2)$ and $\tilde{\chi}(q) = \cos^6(\gamma/2)\,K(z)$. Performing similar transformations as in (10-55) and (10-56) we find a manifestly O(4)-invariant equation

$$\left[q^4 + 2q^2\left(1 - \frac{P^2}{2}\right) + 1\right]\tilde{\chi}(q) = \frac{\lambda}{\pi^3}\int \frac{d^4q'}{(q-q')^2}\,\tilde{\chi}(q') \tag{10-61}$$

The transformation from p to q and χ to $\tilde{\chi}$ can, of course, be performed directly. If n is a unit vector along the zeroth axis we have

$$p = \sqrt{1 - \frac{P^2}{4}}\left(n + 2\frac{q-n}{q^2 - 2q\cdot n + 1}\right)$$

and

$$\chi(p) = \left(1 - \frac{P^2}{4}\right)^3 (q^2 - 2q\cdot n + 1)\tilde{\chi}(q) \tag{10-62}$$

The invariance of Eq. (10-61) may be used to perform a partial wave analysis in R^4 using the O(4)-orthonormalized spherical harmonics $\mathscr{Y}_{nlm}(\hat{q})$. The positive integer $n \geq 1$ indexes the representation of O(4) of dimension n^2, analogous to the principal quantum number of atomic physics. The ordinary orbital momentum l takes values between 0 and $n-1$.

If O(4) is identified with $SU(2) \times SU(2)/Z_2$ then the $\mathscr{Y}_{nlm}$ form a basis for the representation (j,j) with $n = 2j+1$. The O(3) subgroup unmodified by the combined transformation (10-62) is the diagonal subgroup of $SU(2) \times SU(2)/Z_2$.

In R^4 we have a decomposition analogous to (10-59), that is,

$$\frac{1}{(q-q')^2} = \frac{2\pi^2}{q_>^2}\sum_{n=1}^{\infty}\left(\frac{q_<}{q_>}\right)^{n-1}\frac{1}{n}\sum_{l,m}\mathscr{Y}_{nlm}(\hat{q})\mathscr{Y}^*_{nlm}(\hat{q}') \tag{10-63}$$

Define a radial amplitude $R_n(q^2)$ through

$$\tilde{\chi}(q) = \frac{R_n(q^2)}{|q|^{n+1}[q^4 + 2q^2(1 - P^2/2) + 1]}\mathscr{Y}_{nlm}(\hat{q}) \tag{10-64}$$

Then (10-63) and (10-64) when inserted in (10-61) reduce it to a one-dimensional equation in the variable $x = q^2$:

$$R_n(x) = \frac{\lambda}{n\pi}\int_0^\infty dx'\left[\theta(x-x') + \left(\frac{x}{x'}\right)^n\theta(x'-x)\right]\frac{R_n(x')}{x'^2 + 2x'(1-P^2/2)+1} \tag{10-65}$$

equivalent to the differential equation

$$x\frac{d^2R_n}{dx^2} - (n-1)\frac{dR_n}{dx} + \frac{\lambda}{\pi}\frac{R_n}{x^2+2x(1-P^2/2)+1} = 0 \tag{10-66}$$

supplemented by the conditions that R_n vanishes as x^n for $x \to 0$ and R_n/x^n goes to zero as $x \to \infty$.

Wick and Cutkosky use the variables

$$z = \frac{x-1}{x+1} \tag{10-67}$$

ranging from -1 to $+1$ and the function g_n such that

$$R_n(x) = \frac{1}{(1-z)^n}g_n(z) \tag{10-68}$$

to write the equivalent forms of (10-65) and (10-66):

$$g_n(z) = \frac{\lambda}{2\pi n}\int_{-1}^{+1} dz'\left[\theta(z'-z)\left(\frac{1+z}{1+z'}\right)^n + \theta(z-z')\left(\frac{1-z}{1-z'}\right)^n\right]\frac{g_n(z')}{1-(P^2/4)(1-z'^2)} \tag{10-69}$$

and

$$\left[(1-z^2)\frac{d^2}{dz^2} + 2(n-1)z\frac{d}{dz} - n(n-1) + \frac{\lambda}{\pi}\frac{1}{1-P^2/4+(P^2/4)z^2}\right]g_n(z) = 0 \tag{10-70}$$

with $g_n(\pm 1) = 0$.

Standard analysis applied to either (10-66) or (10-70) shows that λ_n has a discrete spectrum for $0 \le P^2 \le 4$ which may be numerically computed. For fixed n (and P^2) we have a sequence of solutions indexed by an integer K ($K = 0, 1, \ldots$) which is equal to the number of zeros of the radial function (excluding boundary points).

The unequal-mass case may be given a similar treatment, exhibiting again an O(4) symmetry and mapping it, in effect, on the equal-mass case. We keep the assignment $\eta_{1,2} = m_{1,2}/(m_1+m_2)$ and define the following quantities:

$$\begin{aligned}
\Delta &= \frac{m_1 - m_2}{m_1 + m_2}\\
\varepsilon^2 &= (1-\Delta^2)\left[1 - \frac{P^2}{(m_1+m_2)^2}\right]\\
\operatorname{tg}\theta &= i\frac{\sqrt{(1-\Delta^2-\varepsilon^2)(1-\Delta^2)}}{\varepsilon\Delta}\\
p &= \frac{m_1+m_2}{2}\frac{\varepsilon}{D}\{\sin\theta\,(1+q^2)n + 2q - 2n[(1+\cos\theta)q\cdot n + \sin\theta]\}\\
D &= 1 + \cos\theta + (1-\cos\theta)q^2 - 2\sin\theta\; q\cdot n
\end{aligned} \tag{10-71}$$

The unit vector n is along P and these formulas define a complex conformal transformation. We deform the contour of integration in such a way that the final equation satisfied by

$$\tilde{\chi}(q) = D^{-3}\chi(P) \tag{10-72}$$

is real and reads

$$\left\{q^4 + 2q^2\left[\frac{2\varepsilon^2}{(1-\Delta^2)^2} - 1\right] + 1\right\}\tilde{\chi}(q) = \frac{\lambda}{\pi^3}\int d^4q' \frac{\tilde{\chi}(q')}{(q'-q)^2} \tag{10-73}$$

This is similar to (10-61) to which it reduces for $m_1 = m_2$ since

$$\frac{2\varepsilon^2}{(1-\Delta^2)^2} - 1 = \frac{2}{1-(m_1-m_2)^2/(m_1+m_2)^2}\left[1 - \frac{P^2}{(m_1+m_2)^2}\right] - 1$$

if we recall that in Eq. (10-61) the common mass m was taken as a unit of energy.

Thus by an appropriate change of variables we can translate the results of the equal-mass case into those pertaining to the unequal-mass case. We return, therefore, to the former and to Eqs. (10-66) or (10-70).

Of special interest is the behavior close to the threshold $P^2 = 4$ where we can approximate the kernel

$$\frac{1}{\pi[x^2 + 2x(1 - P^2/2) + 1]} \equiv \frac{1}{\pi[(x-1)^2 + x(4-P^2)]} \underset{P^2\to 4}{\simeq} \frac{\delta(x-1)}{\sqrt{4-P^2}}$$

Thus, from (10-65),

$$R_n(x) \simeq \frac{\lambda}{n\sqrt{4-P^2}} R_n(1)[\theta(x-1) + x^n\theta(1-x)] \tag{10-74}$$

Demanding consistency, i.e., that both sides agree for $x = 1$ [we set formally $\theta(0) = \frac{1}{2}$], requires

$$\sqrt{4-P^2} = \frac{\lambda}{n} \tag{10-75}$$

which is the familiar hydrogenic spectrum in terms of B, the binding energy $P^2 = (2 - B/m)^2$:

$$B = \frac{m}{2}\frac{\lambda^2}{2n^2} + \cdots \tag{10-76}$$

where $m/2$ is the reduced mass and λ is identified with $e^2/4\pi$. Unfortunately this is not the end of the story.

The expression (10-74) yields a function without nodes ($K = 0$) and therefore does not give the limit of the eigenvalues corresponding to $K \geq 1$. Those correspond to abnormal solutions for which λ does not vanish in the limit of zero binding and there is therefore no nonrelativistic analog! Wick and Cutkosky have shown that the corresponding eigenvalues for a function with K nodes tend in the limit $P^2 \to 4$ to an n-independent limit

$$\frac{\lambda_K}{\pi} \underset{P^2\to 4}{\simeq} \frac{1}{4} + \frac{\pi^2}{[\ln(1-P^2/4)]^2} \qquad K \geq 1 \tag{10-77}$$

This is not the only disease of this model. It is possible to study corrections to the sensible set of solutions ($K = 0$) which reproduces to lowest order the nonrelativistic result, Eqs. (10-75) and (10-76). The result of this analysis shows that $2n\varepsilon/\lambda$, with $\varepsilon = \sqrt{1 - P^2/4}$, has an expansion of the form

$$\frac{2n\varepsilon}{\lambda} = 1 + \varepsilon(a_{11} \ln \varepsilon + a_{12}) + \varepsilon^2[a_{21}(\ln \varepsilon)^2 + a_{22} \ln \varepsilon + a_{23}] + \cdots \tag{10-78}$$

with ε of order λ. These logarithmic terms are not present in a more physical approximation than the ladder series.

If this were not enough, Nakanishi has shown that certain solutions have a negative norm! They are called "ghosts," and it is unclear whether they result from the inadequacy of the approximation or from a deeper inconsistency of the theory.

The previous study may be generalized to the inhomogeneous equation for the amplitude, in particular to high-energy behavior in the crossed channel (corresponding to the exchanged particles). It is possible to find a Regge behavior $s^{\alpha(t)}$ with a computable trajectory function $\alpha(t)$.

Various attempts have been made to introduce an effective potential (or quasipotential) approximation to reduce the number of degrees of freedom of the relativistic bound-state problem, in particular the troublesome relative time. Even though they lead to interesting practical results, they mutilate one way or another the exact theory and generally introduce spurious singularities.

10-3 HYPERFINE SPLITTING IN POSITRONIUM

It should not be concluded that relativistic weak binding corrections cannot be obtained for two-body systems that agree with experiment. On the contrary, the positronium states give an example of a successful agreement. This will serve to illustrate the theory. To be completely fair, we should admit that accurate predictions require some artistic gifts from the practitioner. As yet no systematic method has been devised to obtain the corrections in a completely satisfactory way.

We quote here some of the significant results and refer to Secs. 2-3 and 5-2 for preliminary investigations. Even though in the study of positronium we restrict ourselves to an almost pure electromagnetic system, some of the methods are useful in other instances such as models of quark bound states of hadrons.

The energy difference between the higher triplet (ortho) and lower singlet (para) ground states of positronium, denoted respectively 1^3S_1 and 1^1S_0 (in the spectroscopic notation $n^{2S+1}L_J$), has now been measured with great accuracy. The values quoted for this hyperfine splitting ΔE_{ts} are

$$\begin{aligned} \Delta E_{\text{ts}} &= 2.033870\,(16) \times 10^5 \text{ MHz} \\ &\quad \text{(Mills and Bearman)} \\ \Delta E_{\text{ts}} &= 2.033849\,(12) \times 10^5 \text{ MHz} \\ &\quad \text{(Egan, Frieze, Hughes, and Yam)} \end{aligned} \tag{10-79}$$

The interval is sometimes also called the fine structure of positronium. Recently Mills, Berko, and Canter have also measured the spacing between $n = 2$ triplet excited levels

$$E(2^3S_1) - E(2^3P_2) = 8624 \pm 2.8 \text{ MHz} \tag{10-80}$$

We recall that all these states are unstable. The ground-state radiative width has already been discussed to lowest order (Sec. 5-2).

We may understand the magnitude and sign of the singlet-triplet splitting in positronium by noticing that it corresponds to the sum of two effects. The magnetic interaction is given by the Fermi estimate discussed in the context of the hydrogen atom (Sec. 2-3-2). In terms of the electron-positron parameters with gyromagnetic factors equal to 2, it reads

$$\Delta E_b = \frac{e^2}{6m^2} |\varphi_0|^2 \langle \boldsymbol{\sigma}_1 \cdot \boldsymbol{\sigma}_2 \rangle$$

where $|\varphi_0|^2$ is the square of the nonrelativistic wave function at the origin for a system of two equal-mass particles $|\varphi_0|^2 = (m\alpha)^3/8\pi$.

We have also to deal with a new effect corresponding to the annihilation channel, as was mentioned in the early discussion of electron-positron scattering (Sec. 6-1-3). If we restrict ourselves to the lowest-order effect it corresponds to the one-photon channel contributing an s-wave interaction energy of order α to the triplet state only because of its odd charge conjugation. The desired energy shift may be computed from an effective potential identified with the corresponding T-matrix element at the threshold ($S = I + iT$) multiplied by $|\varphi_0|^2$. From Chap. 6, the scattering amplitude at the threshold is

$$-e^2 \, \frac{\bar{v}(2)\gamma^\nu u(1)\bar{u}(1)\gamma_\nu v(2)}{(2m)^2}$$

We have been careful about the signs and $u(v)$ denotes the electron (positron) spinor at the threshold

$$u(1) = \begin{pmatrix} \chi_1 \\ 0 \end{pmatrix} \qquad \bar{u}(1) = (\chi_1^\dagger, 0) \qquad v(2) = C\bar{u}^T(2) = \begin{pmatrix} 0 \\ -i\sigma_2 \chi_2^* \end{pmatrix}$$

The matrix element reads explicitly

$$-\frac{e^2}{4m^2} \chi_2^T(-i\sigma_2)\boldsymbol{\sigma}\chi_1 \cdot \chi_1^\dagger \boldsymbol{\sigma}(-i\sigma_2)(\chi_2^\dagger)^T = \frac{e^2}{4m^2}\left(\frac{3}{2}\chi_2^\dagger\chi_2\chi_1^\dagger\chi_1 + \frac{1}{2}\chi_2^\dagger\boldsymbol{\sigma}\chi_2 \cdot \chi_1^\dagger\boldsymbol{\sigma}\chi_1\right)$$

where we have used some Fierz reshuffling. Therefore we find

$$\Delta E_A = \frac{e^2}{4m^2} |\varphi_0|^2 \left\langle \frac{3 + \boldsymbol{\sigma}_1 \cdot \boldsymbol{\sigma}_2}{2} \right\rangle$$

The term $\langle (3 + \boldsymbol{\sigma}_1 \cdot \boldsymbol{\sigma}_2)/2 \rangle$ is nothing but $\mathbf{S}^2$ where $\mathbf{S} = (\boldsymbol{\sigma}_1 + \boldsymbol{\sigma}_2)/2$ is the total spin; thus it contributes a positive amount to the triplet energy only. To lowest order, the total ground-state spin dependence to the energy displacement is thus expected to be

$$\Delta(E_b + E_A) = \frac{e^2}{2m^2} |\varphi_0|^2 \left\langle \frac{\boldsymbol{\sigma}_1 \cdot \boldsymbol{\sigma}_2}{3} + \frac{\mathbf{S}^2}{2} \right\rangle$$

To obtain the hyperfine splitting we subtract the values corresponding to $S = 1$ and 0. Thus

$$\Delta E_{ts} = \frac{7}{6}\alpha^2 \text{ Ryd} \tag{10-81}$$

where Ryd is the Rydberg constant

$$\text{Ryd} = \frac{m\alpha^2}{2} = 3.28984 \times 10^{15}\ \text{Hz} \tag{10-82}$$

Eq. (10-81) predicts $\Delta E = 2.04 \times 10^5$ MHz, which compares favorably with the experimental results up to order α. This is fortunately two orders of magnitude greater than the two-photon decay rate of the singlet state $\Gamma_{2\gamma} = \alpha^3$ Ryd. Hence the computation of the hyperfine splitting will be an accurate test both of quantum electrodynamics and of the application of the Bethe-Salpeter equation.

10-3-1 General Setting

Our presentation will be modeled after the original work of Karplus and Klein. The starting point is the analog of Eq. (10-26) written in the electron-positron channel. The bound-state amplitude $\chi = \langle 0 | \psi(x)\bar{\psi}(y) | P \rangle$ stands for a 4×4 matrix in the space of Dirac indices. It should be emphasized that χ is a gauge-covariant quantity. Consequently, it is not indifferent to choose a particular gauge if χ satisfies an approximately soluble equation. In practice, since relative velocities are small, we expect the nonrelativistic approximation to the binding energy to be a reasonable starting point. This is

$$E_0 = 2m - \frac{m\alpha^2}{4n^2} \tag{10-83}$$

and χ is given by a Pauli-Schrödinger wave function.

A noncovariant radiation gauge would seem the most appropriate to derive this result. This is, however, a dangerous choice since higher corrections will require renormalization. On the other hand, we know that to implement gauge covariance in a relativistic manner it is necessary to include in the kernel V an infinite series of crossed diagrams. We are facing a dilemma which will be resolved in practice in the following way. We will use a covariant gauge, to be specific the Feynman one, and will have to separate an instantaneous interaction from a retarded one. This is because we are unable to solve exactly the equivalent of the Wick-Cutkosky relativistic equation. The retarded interaction will then be treated as a perturbation on the same footing as higher-order terms in V. If we are led to smaller and smaller corrections which do agree with experiment, this procedure may at least be considered as a useful method in spite of its poor theoretical foundations. This will be true of the corrections of order α^3 Ryd. The development of the subject shows, however, that even the experts have some trouble with higher orders required for a comparison with the very accurate experimental resolution.

Several improvements have been suggested such as expanding χ in terms of Lorentz invariant scalar amplitudes multiplying covariant quantities, carrying out a complete angular analysis on the Wick rotated amplitude or attempting to find an equivalent but soluble form of the relativistic ladder approximation.

Ignoring at first radiative corrections, we write the equation satisfied by χ

in momentum space as

$$\left(\frac{P\!\!\!/}{2}+p\!\!\!/-m\right)\chi\left(\frac{P\!\!\!/}{2}-p\!\!\!/+m\right)=(V_B+V_a)\chi$$

$$V_B\chi = i\frac{\alpha}{4\pi^3}\int\frac{d^4p'}{(p-p')^2}\gamma_\mu\chi(p')\gamma^\mu \tag{10-84}$$

$$V_a\chi = -i\frac{\alpha}{4\pi^3}\frac{1}{P^2}\gamma^\mu\int d^4p' \operatorname{tr}[\gamma_\mu\chi(p')]$$

where V_B represents one-photon exchange in the crossed t channel and V_a is the one-photon annihilation interaction. Even at this early stage the complexity is somehow frightening.

It will be convenient to work with the quantity K obtained by acting with the antisymmetric charge conjugation matrix C on the second index of χ:

$$K_{\alpha\beta} = C_{\beta\beta'}\chi_{\alpha\beta'} \tag{10-85}$$

This amplitude has the same transformation properties as one corresponding to the particle-particle channel except that the latter has no electromagnetic bound state. It satisfies

$$\left(\frac{P\!\!\!/}{2}+p\!\!\!/-m\right)_1\left(\frac{P\!\!\!/}{2}-p\!\!\!/-m\right)_2 K=(V_B+V_a)K(p)$$

$$V_BK = i\frac{\alpha}{4\pi^3}\int\frac{d^4p'}{(p-p')^2}\gamma_{1\mu}\gamma_2^\mu K(p') \tag{10-86}$$

$$V_aK = -i\frac{\alpha}{4\pi^3}\frac{1}{P^2}\gamma_\mu C\int d^4p' \operatorname{tr}[\gamma^\mu K(p')C]$$

We have appended subscripts to the product of γ matrices in order to distinguish the electron from the positron variables.

The structure $\gamma_{1\mu}\gamma_2^\mu/k^2$, with $k=(p-p')$, arises from the use of the Feynman gauge. We can, however, split this interaction into an instantaneous and a retarded part in the total rest frame, $P=(E,\mathbf{0})$, according to

$$\frac{\gamma_1^0\gamma_2^0-\boldsymbol{\gamma}_1\cdot\boldsymbol{\gamma}_2}{k^2} = -\frac{\gamma_1^0\gamma_2^0}{\mathbf{k}^2}+\left(\frac{\gamma_1^0\gamma_2^0k_0^2}{k^2\mathbf{k}^2}-\frac{\boldsymbol{\gamma}_1\cdot\boldsymbol{\gamma}_2}{k^2}\right) \tag{10-87}$$

The second term on the right-hand side contains both retardation and magnetic interaction. It will be treated perturbatively (we wish we could do better); so will be the annihilation contribution.

The zeroth-order approximation for K is obtained by separating in V_B a Coulomb part V_0 from a remaining part noted V_b, according to (10-87). We write therefore in the instantaneous approximation

$$\left(\frac{P\!\!\!/}{2}+p\!\!\!/-m\right)_1\left(\frac{P\!\!\!/}{2}-p\!\!\!/-m\right)_2 K(p) = -i\frac{\alpha}{4\pi^3}\int\frac{d^4p'}{(\mathbf{p}-\mathbf{p}')^2}\gamma_1^0\gamma_2^0K(p') \tag{10-88}$$

This suggests the derivation of an equation for

$$\varphi(\mathbf{p}) = \int dp^0\, K(p^0, \mathbf{p}) \tag{10-89}$$

by dividing both sides of (10-88) by the wave operators and integrating over p^0. Thus

$$\varphi(\mathbf{p}) = -i\frac{\alpha}{4\pi^3}\int dp^0 \frac{(\not{P}/2 + \not{p} + m)_1}{(P/2+p)^2 - m^2 + i\varepsilon}\frac{(\not{P}/2 - \not{p} + m)_2}{(P/2-p)^2 - m^2 + i\varepsilon}$$

$$\times \int \frac{d^3p'}{(\mathbf{p}-\mathbf{p}'^2)}\gamma_1^0\gamma_2^0\varphi(\mathbf{p}') \tag{10-90}$$

The $i\varepsilon$ prescription in the denominators follows from the derivation of the equation. The p^0 integral converges and may be computed by closing the contour in the upper half plane encountering poles at $p_0 = \mp E/2 + \omega - i\varepsilon$ where $\omega = \sqrt{\mathbf{p}^2 + m^2}$. If we use the Dirac notations $\beta = \gamma^0$ and $\boldsymbol{\alpha} = \beta\boldsymbol{\gamma}$ and the definitions

$$H(\mathbf{p}) = \boldsymbol{\alpha}\cdot\mathbf{p} + \beta m$$

$$\Lambda^{\pm}(\mathbf{p}) = \frac{\omega \pm H(\mathbf{p})}{2\omega} \tag{10-91}$$

$$\Lambda^{\pm\pm} = \Lambda_1^{\pm}(\mathbf{p})\Lambda_2^{\pm}(-\mathbf{p})$$

the result of the integration over p^0 takes the form

$$\varphi(\mathbf{p}) = \frac{1}{2}\left(\frac{\Lambda^{++}}{\omega - E/2} + \frac{\Lambda^{--}}{\omega + E/2}\right)\frac{\alpha}{2\pi^2}\int\frac{d^3p'\,\varphi(\mathbf{p}')}{(\mathbf{p}-\mathbf{p}')^2} \tag{10-92}$$

Note the equivalence between the two formulations of Feynman's propagator

$$S = \frac{\not{p} + m}{p^2 - m^2 + i\varepsilon} = \left[\frac{\Lambda^+(\mathbf{p})}{p_0 - \omega + i\varepsilon} + \frac{\Lambda^-(\mathbf{p})}{p_0 + \omega - i\varepsilon}\right]\gamma^0$$

Since

$$H(\mathbf{p})\Lambda^{\pm}(\mathbf{p}) = \pm\omega\Lambda^{\pm}(\mathbf{p})$$

Eq. (10-92) can also be written as an effective one-body equation

$$[H_1(\mathbf{p}) + H_2(-\mathbf{p}) - E]\varphi(\mathbf{p}) = (\Lambda^{++} - \Lambda^{--})\frac{\alpha}{2\pi^2}\int\frac{d^3p'}{(\mathbf{p}-\mathbf{p}')^2}\varphi(\mathbf{p}') \tag{10-93}$$

Denoting $\varphi^{\pm\pm} = \Lambda^{\pm\pm}\varphi$, Eq. (10-93) is equivalent to a set of coupled equations

$$(2\omega - E)\varphi^{++}(\mathbf{p}) = \Lambda^{++}\frac{\alpha}{2\pi^2}\int\frac{d^3p'}{(\mathbf{p}-\mathbf{p}')^2}[\varphi^{++}(\mathbf{p}') + \varphi^{--}(\mathbf{p}')]$$

$$(2\omega + E)\varphi^{--}(\mathbf{p}) = \Lambda^{--}\frac{\alpha}{2\pi^2}\int\frac{d^3p'}{(\mathbf{p}-\mathbf{p}')^2}[\varphi^{++}(\mathbf{p}') + \varphi^{--}(\mathbf{p}')] \tag{10-94}$$

$$\varphi^{+-}(\mathbf{p}) = \varphi^{-+}(\mathbf{p}) = 0$$

Neither (10-88) nor (10-92) is sufficient for our purposes since it omits the retarded part V_b and the annihilation kernel crucial for the hyperfine splitting. When both of these will be included, we obtain the correct version of an equation originally discussed by Breit for such problems.

To discuss this effect, we will satisfy ourselves here with the form of perturbation theory originally derived by Salpeter. It is designed to circumvent the difficulty arising from the presence of the energy E, quadratically in the differential operator of the equation (and parametrically in the general kernel) and is suited to an instantaneous unperturbed interaction.

Returning to Eq. (10-88) we multiply both sides by $\gamma_1^0\gamma_2^0$ and obtain

$$\left[\frac{E}{2} + p_0 - H_1(\mathbf{p})\right]\left[\frac{E}{2} - p_0 - H_2(-\mathbf{p})\right]K = -\frac{i}{2\pi}\, v\varphi(\mathbf{p})$$
$$v\varphi(\mathbf{p}) = \frac{\alpha}{2\pi^2}\int \frac{d^3p'}{(\mathbf{p}-\mathbf{p}')^2}\,\varphi(\mathbf{p}') \tag{10-95}$$

Assuming φ to be known and using the same $i\varepsilon$ prescription to invert the operator D acting on K we have

$$K = D^{-1}\frac{1}{2i\pi}\, v\varphi$$
$$D^{-1} = \sum_{\pm\pm} \frac{\Lambda^{\pm\pm}}{[E/2 + p_0 \mp (\omega - i\varepsilon)][E/2 - p_0 \mp (\omega - i\varepsilon)]} \tag{10-96}$$

Integration over p_0 would lead back to Eq. (10-92). For shortness let us define η in such a way that

$$(H - E)\eta = v\varphi \tag{10-97}$$

where we have set

$$H(\mathbf{p}) = H_1(\mathbf{p}) + H_2(-\mathbf{p})$$

From Eq. (10-93) it follows that

$$\varphi = (\Lambda^{++} - \Lambda^{--})\eta \tag{10-98}$$

and K is given by

$$K = (H - E)\frac{1}{2i\pi}\, D^{-1}\eta \tag{10-99}$$

For a fixed η corresponding to a given solution of the equation we define the amplitude Q through an expression similar to (10-99) but with the energy E replaced by E'. Recall that D also depends on E so that

$$Q = (H - E')\frac{1}{2i\pi}\, D^{-1}(E')\eta \tag{10-100}$$

Consider the combination

$$\left[D(E') - \frac{1}{2i\pi}\int dp^0\, v\right]Q = (H - E')\frac{1}{2i\pi}\eta - \frac{1}{2i\pi}\int dp^0\, v(H - E')\frac{1}{2i\pi}D^{-1}(E')\eta$$

On the left-hand side when E' reduces to E and Q to K we find zero. On the right-hand side the only dependence on the variable p^0 is in the propagator $D^{-1}(E')$. The integral is the same as above, leading to

$$\left[D(E') - \frac{1}{2i\pi}\int dp^0\, v\right]Q = (H - E')\frac{1}{2i\pi}\eta - \frac{1}{2i\pi}v(\Lambda^{++} - \Lambda^{--})\eta$$

$$= -\frac{1}{2i\pi}(E' - E)\eta \tag{10-101}$$

The last equality follows from Eqs. (10-97) and (10-98). Similarly, if

$$\tilde{Q} = \eta^\dagger(H - E')\frac{1}{2i\pi}D^{-1}(E') \tag{10-102}$$

the hermiticity of H and v leads to

$$\tilde{Q}D(E') - \frac{1}{2i\pi}\int dp^0\, \tilde{Q}v = -\frac{1}{2i\pi}\eta^\dagger(E' - E) \tag{10-103}$$

We are now in a position to discuss the effect of a perturbation

$$\delta V = \gamma_1^0\gamma_2^0(V_b + V_a) \tag{10-104}$$

to the instantaneous Coulomb interaction. The new wave function K' and energy E' will satisfy

$$\left[D(E') - \frac{1}{2i\pi}\int dp^0\, v - \delta V\right]K' = 0 \tag{10-105}$$

We multiply this equation to the left by $\tilde{Q}$, defined as above relative to E', and integrate over the four-momentum p. From Eq. (10-103) we obtain

$$(E' - E)\int d^3p\, \eta^\dagger(\mathbf{p})\varphi(\mathbf{p}) = -2i\pi\int d^4p\, \tilde{Q}(p)(\delta V K')(p)$$

This is an exact result. To make it useful as a first-order result we simply replace φ' by φ and K' and Q by K. Thus

$$\Delta E^{(1)} = -\frac{i}{(2\pi)^2}\int d^4p\, \tilde{K}(p)(\delta V K)(p) \tag{10-106}$$

provided we use the normalization

$$\int\frac{d^3p}{(2\pi)^3}\eta^*(\mathbf{p})\varphi(\mathbf{p}) \equiv \int\frac{d^3p}{(2\pi)^3}[|\varphi^{++}(\mathbf{p})|^2 - |\varphi^{--}(\mathbf{p})|^2] = 1 \tag{10-107}$$

For future purpose we shall also need the second-order energy displacement.

It involves in principle the inverse of the complete propagator $[D(E) - 1/2i\pi \times \int dp^0\, v]^{-1}$. For the present applications we can, however, satisfy ourselves with the free two-particle propagator so that

$$\Delta E^{(2)} = -\frac{i}{(2\pi)^2}\int d^4p\, (\tilde{K}\delta V)(p) D^{-1}(E;p)(\delta V K)(p) \tag{10-108}$$

Investigate the normalization condition from the point of view developed in Sec. 10-2 and obtain its relation with this perturbative expansion of the bound-state energy.

10-3-2 Calculation to Order α^5

Focusing on the hyperfine problem means that we deal with the spherically symmetric $n = 1$ states. With an accurate enough solution to the unperturbed Eqs. (10-92) or (10-93) we may now compute the shift. To lowest order it will be made of the two contributions

$$\Delta E_b^{(1)} = \frac{\alpha}{\pi(2\pi)^4}\int d^4p\, d^4p'\, \bar{K}(p')\frac{1}{(p'-p)^2}\left[\gamma_1^0\gamma_2^0\frac{(p_0 - p_0')^2}{(\mathbf{p}-\mathbf{p}')^2} - \gamma_1\cdot\gamma_2\right]K(p)$$

$$\Delta E_a^{(1)} = -\frac{\alpha}{\pi(2\pi)^4}\frac{1}{4m^2}\int d^4p\, d^4p'\, \mathrm{tr}\,[\bar{K}(p')\gamma_\mu C]\,\mathrm{tr}\,[C\gamma^\mu K(p)] \tag{10-109}$$

with $\bar{K}$ standing for $\tilde{K}\gamma_1^0\gamma_2^0$.

With the inclusion of radiative corrections we intend to obtain ΔE to order α^5 while the leading term is of order α^4. It is therefore desirable to find a suitable approximation for the instantaneous wave-function φ solution of Eqs. (10-92) or (10-93). We recall that the corresponding Schrödinger wave function satisfying

$$\varphi_S(\mathbf{p}) = \frac{m}{\mathbf{p}^2 + m^2\alpha^2/4}\frac{\alpha}{2\pi^2}\int\frac{d^3p'}{(\mathbf{p}-\mathbf{p}')^2}\varphi_S(\mathbf{p}') \tag{10-110}$$

that is, with E given by the lowest-order approximation (10-83) and ω expanded to order p^2, is

$$\varphi_S(\mathbf{p}) = \frac{4\pi\alpha m}{(\mathbf{p}^2 + m^2\alpha^2/4)^2}\varphi_0 \tag{10-111}$$

where φ_0 is the wave function at the origin in configuration space

$$|\varphi_0|^2 = \frac{(m\alpha)^3}{8\pi} \tag{10-112}$$

If this expression [multiplied by the product of two spinors at rest, i.e., of the form $\binom{\chi_1}{0}\otimes\binom{\chi_2}{0}$ which will not be written explicitly] is inserted into (10-96), it yields an approximation which is sufficient for our purposes:

$$K(p) = \frac{2\alpha}{i} \sum \frac{\Lambda^{\pm\pm}}{[E_0/2 + p_0 \mp (\omega - i\varepsilon)][E_0/2 - p_0 \mp (\omega - i\varepsilon)]} \frac{\varphi_0}{(\mathbf{p}^2 + m^2\alpha^2/4)} \tag{10-113}$$

Here E_0 is given by Eq. (10-83). Therefore to the required accuracy

$$\Delta E_b^{(1)} = -\frac{2\alpha^3}{\pi^2} \frac{|\varphi_0|^2}{(2\pi)^3} \int \frac{d^4p\, d^4p'}{(p - p')^2 + i\varepsilon} \frac{1}{(\mathbf{p}^2 + m^2\alpha^2/4)(\mathbf{p}'^2 + m^2\alpha^2/4)}$$
$$\times \left\langle \sum \frac{\Lambda^{\pm\pm}}{[E_0/2 + p'_0 \mp (\omega' - i\varepsilon)][E_0/2 - p'_0 \mp (\omega' - i\varepsilon)]} \left[\frac{(p_0 - p'_0)^2}{(\mathbf{p} - \mathbf{p}')^2} - \boldsymbol{\alpha}_1 \cdot \boldsymbol{\alpha}_2 \right] \right.$$
$$\left. \sum \frac{\Lambda^{\pm\pm}}{[E_0/2 + p_0 \mp (\omega - i\varepsilon)][E_0/2 - p_0 \mp (\omega - i\varepsilon)]} \right\rangle \tag{10-114}$$

The bracket denotes the expectation value between the product of spinors at rest. We observe in front of (10-114) a factor α^6. Therefore the terms that we intend to compute can only arise from the contributions of order $1/\alpha^2$ or $1/\alpha$ to the integral. From the presence of the denominators $[\mathbf{p}^2 + m^2\alpha^2/4][\mathbf{p}'^2 + m^2\alpha^2/4]$ it is clear that the small $|\mathbf{p}|$, $|\mathbf{p}'|$ region yields the dominant term and that any factor $|\mathbf{p}|$ or $|\mathbf{p}'|$ will contribute an extra factor α. In the expectation value we find the retarded Coulomb interaction and the Breit interaction $\boldsymbol{\alpha}_1 \cdot \boldsymbol{\alpha}_2$. In the former, with a relative error of order α^2, we may replace $(p_0 - p'_0)^2$ by $(p - p')^2$, thereby suppressing the photon denominator. The p_0 and p'_0 integration can then be performed and the spin-dependent effect is seen to involve four powers of the three-momenta in the numerator. It is therefore negligible to the order α^5.

We come next to the Breit term in $\boldsymbol{\alpha}_1 \cdot \boldsymbol{\alpha}_2$. It is convenient to introduce the relative time variable by Fourier transforming the wave functions. We rewrite the bracket term:

$$\int \frac{dp_0\, dp'_0}{(p - p')^2 + i\varepsilon} \langle\,\rangle = \int \frac{2\pi\, dt\, dk_0}{k_0^2 - \mathbf{k}^2 + i\varepsilon} e^{itk_0} \langle D^{-1}(-t)(-\boldsymbol{\alpha}_1 \cdot \boldsymbol{\alpha}_2) D^{-1}(t) \rangle \tag{10-115}$$

where $\mathbf{k}$ stands for $\mathbf{p} - \mathbf{p}'$ and

$$D^{-1}(t) = \sum_{\varepsilon_1, \varepsilon_2} \Lambda^{\varepsilon_1\varepsilon_2} \int \frac{dp_0}{2\pi} e^{-itp_0} \frac{1}{[E_0/2 + p_0 - \varepsilon_1(\omega - i\varepsilon)][E_0/2 - p_0 - \varepsilon_2(\omega - i\varepsilon)]}$$
$$= -i \sum_{\varepsilon_1, \varepsilon_2} \frac{\Lambda^{\varepsilon_1\varepsilon_2}}{E_0 - (\varepsilon_1 + \varepsilon_2)\omega} \{[\theta(\varepsilon_1)\theta(t) - \theta(-\varepsilon_1)\theta(-t)]\, e^{-it(\varepsilon_1\omega - E_0/2)}$$
$$+ [\theta(\varepsilon_2)\theta(-t) - \theta(-\varepsilon_2)\theta(t)]\, e^{it(\varepsilon_2\omega - E_0/2)}\} \tag{10-116}$$

In this sum the only piece with a dangerous denominator corresponds to Λ^{++}, since $E_0 - 2\omega \simeq -(1/m)(\mathbf{p}^2 + m^2\alpha^2/4)$ and the small $\mathbf{p}^2$ region is predominant in the integration to come. If we were to look for the dominant α^4 contribution it would be sufficient to keep the term multiplying the projector over positive energies. However, to reach the next correction we are forced to use the complete expression. Contributions of order $\mathbf{p}^2$ may fortunately be dropped. This is so because they will generate a correction of relative order α^2 provided $\mathbf{p}^2$ is associated to a factor insuring a sufficient ultraviolet convergence. Replacing E_0 whenever possible by $2m$ a satisfactory approximation for $D^{-1}(t)$ is

$$D^{-1}(t) \simeq \frac{im}{2\omega(\mathbf{p}^2 + m^2\alpha^2/4)}\left(1 + \frac{\boldsymbol{\alpha}_1 \cdot \mathbf{p}}{2m}\right)\left(1 - \frac{\boldsymbol{\alpha}_2 \cdot \mathbf{p}}{2m}\right)$$
$$\times \left[(\omega + m)\, e^{-i|t|(\omega - m)} - (\omega - m)\, e^{-i|t|(\omega + m)}\right] \qquad (10\text{-}117)$$

Using the same principle as before and keeping in the matrix element only the term responsible for the hyperfine splitting we find

$$\left\langle\left(1 + \frac{\boldsymbol{\alpha}_1 \cdot \mathbf{p}'}{2m}\right)\left(1 - \frac{\boldsymbol{\alpha}_2 \cdot \mathbf{p}'}{2m}\right)(-\boldsymbol{\alpha}_1 \cdot \boldsymbol{\alpha}_2)\left(1 + \frac{\boldsymbol{\alpha}_1 \cdot \mathbf{p}}{2m}\right)\left(1 - \frac{\boldsymbol{\alpha}_2 \cdot \mathbf{p}}{2m}\right)\right\rangle$$
$$\simeq -\frac{\mathbf{k}^2}{4m^2}\langle(\boldsymbol{\sigma}_1 - \hat{k}\boldsymbol{\sigma}_1 \cdot \hat{k}) \cdot (\boldsymbol{\sigma}_2 - \hat{k}\boldsymbol{\sigma}_2 \cdot \hat{k})\rangle$$

The remaining part of the integral being rotationally invariant the right-hand side may be further replaced by $-(\mathbf{k}^2/4m^2)\frac{2}{3}\langle\boldsymbol{\sigma}_1 \cdot \boldsymbol{\sigma}_2\rangle$. Inserting those expressions into (10-115) and performing the t integral yields

$$\Delta E_b^{(1)} = -\frac{2\alpha^3}{3\pi}\frac{|\varphi_0|^2}{(2\pi)^3}\langle\boldsymbol{\sigma}_1 \cdot \boldsymbol{\sigma}_2\rangle \int \frac{d^3p\, d^3p'\, dk_0}{(\mathbf{p}^2 + \alpha^2m^2/4)^2(\mathbf{p}'^2 + \alpha^2m^2/4)^2(k_0^2 - \mathbf{k}^2 + i\varepsilon)}$$
$$\times \left(\frac{i}{k_0 - \omega - \omega' + 2m + i\varepsilon} - \frac{i}{k_0 + \omega + \omega' - 2m - i\varepsilon}\right)$$

The integral over k_0 is easily obtained by closing the contour in the upper half plane leaving

$$\Delta E_b^{(1)} = \frac{4\alpha^3}{3(2\pi)^3}|\varphi_0|^2\langle\boldsymbol{\sigma}_1 \cdot \boldsymbol{\sigma}_2\rangle \int \frac{d^3p\, d^3p'}{(\mathbf{p}^2 + \alpha^2m^2/4)^2(\mathbf{p}'^2 + \alpha^2m^2/4)^2}$$
$$\times \frac{|\mathbf{k}|}{4\omega\omega'}\left[\frac{(\omega + m)(\omega' + m)}{\omega + \omega' + |\mathbf{k}| - 2m} - 2\frac{(\omega\omega' - m^2)}{\omega + \omega' + |\mathbf{k}|} + \frac{(\omega - m)(\omega' - m)}{\omega + \omega' + |\mathbf{k}| + 2m}\right]$$
$$(10\text{-}118)$$

The last term can be omitted since the integral converges when we set α^2 equal to zero in both denominators. We use the identity $(\omega + m)(\omega' - m) + (\omega - m)(\omega' + m) = 2(\omega\omega' - m^2)$ and the symmetry of the integral to drop in the middle term one of the α^2 in the denominators, therefore obtaining

$$-2\int \frac{d^3p\, d^3p'}{\mathbf{p}^2(\mathbf{p}'^2 + m^2\alpha^2/4)^2}\frac{(\omega' + m)|\mathbf{k}|}{4(\omega + m)\omega\omega'(\omega + \omega' + |\mathbf{k}|)}$$

If we rescale $\mathbf{p}' \to (m\alpha/2)\mathbf{p}'$ and $\mathbf{p} \to m\mathbf{p}$ and keep the dominant α^{-1} contribution this yields a simple integral

$$-\frac{2}{m^2\alpha}(4\pi)^2\int_0^\infty \frac{dp}{\sqrt{1+p^2}}\,\frac{p}{(1+\sqrt{1+p^2})(1+p+\sqrt{1+p^2})}\int_0^\infty \frac{dp'\,p'^2}{(1+p'^2)^2} = -\frac{(2\pi)^3}{m^2\alpha}(1-\ln 2)$$

The first integral in (10-118) contains the dominant effect. We write it as

$$\int \frac{d^3p\,d^3p'}{(\mathbf{p}^2+m^2\alpha^2/4)^2(\mathbf{p}'^2+m^2\alpha^2/4)^2}\,\frac{(\omega+m)(\omega'+m)}{4\omega\omega'}\,\frac{|\mathbf{k}|}{\omega+\omega'+|\mathbf{k}|-2m}$$

$$=\frac{\pi}{2}\frac{(2\pi)^3}{m^2\alpha^2}+\int \frac{d^3p\,d^3p'}{(\mathbf{p}^2+m^2\alpha^2/4)^2(\mathbf{p}'^2+m^2\alpha^2/4)^2}\left[\frac{(\omega+m)(\omega'+m)}{4\omega\omega'}\,\frac{|\mathbf{k}|}{\omega+\omega'+|\mathbf{k}|-2m}-1\right]$$

We cannot apply here the same method as above in a straightforward way. If we were to do so we would generate an extra infrared divergence from the energy denominator $(\omega+\omega'+|\mathbf{k}|-2m)$. To overcome this difficulty we introduce artificially a photon mass μ, thereby replacing $|\mathbf{k}|$ by $\sqrt{\mathbf{k}^2+\mu^2}$ in the denominator. If μ is such that $m\alpha/2 \ll \mu \ll m$ we can proceed as before. The remaining integral is then easily evaluated as

$$-\frac{(2\pi)^3}{m^2\alpha}\left[1+\ln\left(\frac{m}{\mu}\right)\right]$$

Collecting our results we have

$$\Delta E_b^{(1)} = \frac{2\pi\alpha}{3m^2}|\varphi_0|^2\langle\boldsymbol{\sigma}_1\cdot\boldsymbol{\sigma}_2\rangle\left[1-\frac{4\alpha}{\pi}\left(1+\frac{1}{2}\ln\frac{m}{2\mu}\right)\right] \tag{10-119}$$

The α^4 contribution agrees, of course, with the original estimate. The next correction contains a spurious infrared divergence due to our way of estimating integrals (it is, of course, correct with μ chosen as explained above). When all terms of the same magnitude will be collected this $\ln(m/\mu)$ should (and will) disappear.

Instead of proceeding directly to the evaluation of the annihilation contribution we shall turn to the second-order effect to exhibit the announced cancellation. It would, however, be foolish to compute the second-order effect of the V_b potential without introducing the correction to the Bethe-Salpeter kernel due to the crossed exchange of two photons (Fig. 10-7b) to maintain gauge covariance. We denote the sum of these contributions by $\Delta E^{(2)}_{bb+x}$. In applying Eq. (10-108) we are going to make a further simplification. We observe that in this relation we have replaced the intermediate propagation of the electron-positron pair by the free propagation. This amounts in Fig. 10-7a to neglect the instantaneous interaction between the two V_b interactions. In the same spirit we may substitute

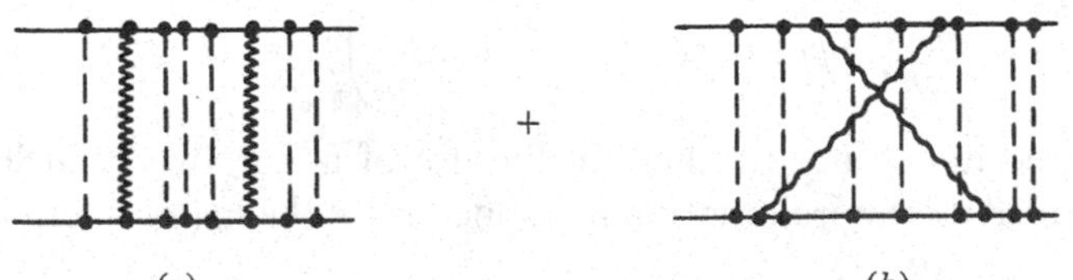

Figure 10-7 Second-order exchange: (a) the retarded interaction V_b to second order and (b) crossed photon exchange. Broken lines represent the instantaneous exchange, notched lines the retarded one, and wavy lines the covariant photon propagator.

for K (and $\tilde{K}$) its zeroth-order approximation, i.e., the wave function at the origin in configuration space

$$K(p) \to \int d^4p\, K(p) = (2\pi)^3 \varphi_0$$

In short,

$$\Delta E^{(2)}_{bb+x} = \frac{4i\alpha^2}{(2\pi)^2} |\varphi_0|^2 \int \frac{d^4k}{(k^2+i\varepsilon)^2} \Big\langle \Big\{ \Big(\frac{\gamma_1^0\gamma_2^0 k_0^2}{\mathbf{k}^2} - \gamma_1 \cdot \gamma_2 \Big)$$

$$\times \frac{(\not P/2 + \not k + m)_1 (\not P/2 - \not k + m)_2}{[(k_0+m^2) - \omega^2 + i\varepsilon][(k_0 - m)^2 - \omega^2 - i\varepsilon]} \Big(\frac{\gamma_1^0\gamma_2^0 k_0^2}{\mathbf{k}^2} - \gamma_1 \cdot \gamma_2 \Big) \Big\}$$

$$+ \frac{\gamma_1^\mu (\not P/2 + \not k + m)_1 \gamma_{1\nu} \gamma_2{}^\nu (\not P/2 + \not k + m)_2 \gamma_{2\mu}}{[(k_0 + m)^2 - \omega^2 + i\varepsilon]^2} \Big\rangle \qquad (10\text{-}120)$$

The second term is, of course, the crossed photon exchange. The notation ω is for $\sqrt{\mathbf{k}^2 + m^2}$ and P has been taken here equal to $(2m, \mathbf{0})$. We proceed to the matrix algebra by taking into account the fact that a spherical average over $\mathbf{k}$ may be performed. Thus the part contributing to the splitting is

$$\Big\langle \Big(\gamma_1^0\gamma_2^0 \frac{k_0^2}{\mathbf{k}^2} - \gamma_1 \cdot \gamma_2 \Big) \Big(\frac{\not P}{2} + \not k + m \Big)_1 \Big(\frac{\not P}{2} - \not k + m \Big)_2 \Big(\gamma_1^0\gamma_2^0 \frac{k_0^2}{\mathbf{k}^2} - \gamma_1 \cdot \gamma_2 \Big) \Big\rangle$$

$$\underset{\text{av}}{\to} \tfrac{2}{3} \langle \boldsymbol{\sigma}_1 \cdot \boldsymbol{\sigma}_2 \rangle k_0^2$$

$$\Big\langle \gamma_1^\mu \Big(\frac{\not P}{2} + \not k + m \Big)_1 \gamma_{1\nu} \gamma_2^\nu \Big(\frac{\not P}{2} + \not k + m \Big)_2 \gamma_{2\mu} \Big\rangle$$

$$\underset{\text{av}}{\to} \tfrac{2}{3} \langle \boldsymbol{\sigma}_1 \cdot \boldsymbol{\sigma}_2 \rangle (3k_0^2 - 2\mathbf{k}^2)$$

and we are left with

$$\Delta E^{(2)}_{bb+x} = \frac{8i\alpha^2}{3\pi} |\varphi_0|^2 \langle \boldsymbol{\sigma}_1 \cdot \boldsymbol{\sigma}_2 \rangle \int_0^\infty k^2\, dk \int_{-\infty}^{\infty} \frac{dk_0}{(k_0^2 - k^2 + i\varepsilon)^2}$$

$$\times \Big\{ \frac{k_0^2}{[(k_0+m)^2 - \omega^2 + i\varepsilon][(k_0 - m)^2 - \omega^2 + i\varepsilon]} + \frac{3k_0^2 - 2\mathbf{k}^2}{[(k_0 - m)^2 - \omega^2 + i\varepsilon]^2} \Big\}$$

We reinstate a photon mass μ and, using similar techniques as above, we find

$$\Delta E^{(2)}_{bb+x} = \frac{4\alpha^2}{3m^2} |\varphi_0|^2 \langle \boldsymbol{\sigma}_1 \cdot \boldsymbol{\sigma}_2 \rangle \Big(\frac{5}{4} + \ln \frac{m}{2\mu} \Big) \qquad (10\text{-}121)$$

The origin of the infrared divergence lies here in the use of a free two-particle propagator, but as expected this approximation is justified since the $\ln(m/\mu)$ terms cancel between (10-119) and (10-121).

Second-order radiative corrections $\Delta E^{(2)}_{bR}$ modify the vertices and the V_b potential through vacuum polarization. The latter correction does not affect the singlet-triplet splitting to order α^5 since it is a short-range effect while the former may be taken into account by including the anomalous magnetic moments of the electron and positron, i.e., by multiplying the dominant term in (10-119) by $(1 + \alpha/2\pi)^2$. Thus to the required order

$$\Delta E_{bR}^{(2)} = \frac{2\alpha^2}{3m^2} |\varphi_0|^2 \langle \boldsymbol{\sigma}_1 \cdot \boldsymbol{\sigma}_2 \rangle \tag{10-122}$$

We turn now to the annihilation part $\Delta E_a^{(1)}$ in Eq. (10-109), where the replacement of P^2 by $4m^2$ in the denominator was justified for the present calculation. A new difficulty arises here since in this contribution we implicitly encounter part of the vertex charge renormalization by including the Coulomb wave function. This is clearly indicated in Fig. 10-8. Care must be paid to the way in which the subtraction is carried out, because of the noncovariant splitting of the one-photon exchange potential. To restore the covariance of the procedure it is necessary to include the second-order terms $V_b D^{-1} V_a + V_a D^{-1} V_b$. This has the effect of completing the photon propagator to its covariant form. As a matter of fact, it follows from the approximation of a free two-particle propagator in the second-order energy displacement that the V_b potential acts immediately before or after the annihilation vertex. Of course, the leading α^4 contribution is insensitive to this effect. Thus we shall directly combine $\Delta E_a^{(1)} + \Delta E_{ab+ba}^{(2)}$ into

$$\Delta E_a^{(1)} + \Delta E_{ab+ba}^{(2)} = -\frac{\alpha}{4\pi(2\pi)^4 m^2} Z_1^2 \int d^4p' \,\mathrm{tr}\,\{[\bar{K}(p') + \Delta\bar{K}'(p')]\gamma_\mu C\}$$

$$\times \int d^4p \,\mathrm{tr}\,\{C\gamma^\mu[K(p) + \Delta K(p)]\}$$

$$\int d^4p \,\mathrm{tr}\,[C\gamma^\mu K(p)] = 4\pi\alpha m^2 \int \frac{d^3p}{\omega(\mathbf{p}^2 + m^2\alpha^2/2)^2} \,\mathrm{tr}\left\{C\gamma^\mu\left[\left(1 + \frac{\boldsymbol{\alpha}_1 \cdot \mathbf{p}}{2m}\right)\right.\right.$$

$$\left.\left.\times \left(1 - \frac{\boldsymbol{\alpha}_2 \cdot \mathbf{p}}{2m}\right) + \frac{\mathbf{p}^2}{4m^2}\right]\varphi_0\right\}$$

$$\int d^4p \,\mathrm{tr}\,[C\gamma^\mu \Delta K(p)] = 2i\alpha \int d^4p \left\{\frac{\mathrm{tr}\,[C\gamma^\mu D^{-1}(p, P)(p_0^2/\mathbf{p}^2 - \boldsymbol{\alpha}_1 \cdot \boldsymbol{\alpha}_2)\varphi_0]}{p^2 - \mu^2 + i\varepsilon}\right.$$

$$\left. - \frac{\mathrm{tr}\,[C\gamma^\mu D^{-1}(p, 0)\gamma_1{}^\nu \gamma_{2\nu}\varphi_0]}{p^2 - \Lambda^2 + i\varepsilon}\right\} \tag{10-123}$$

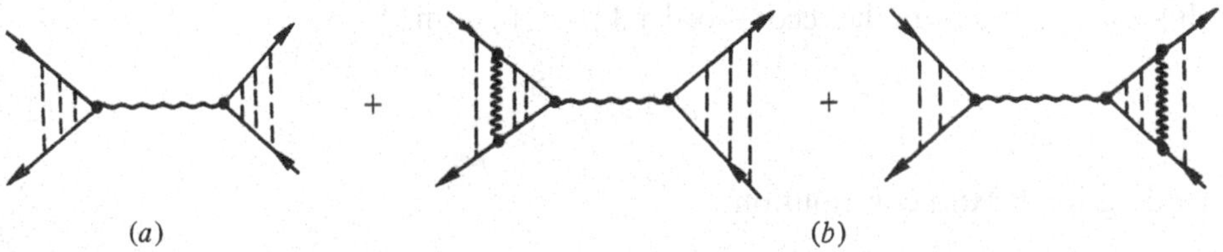

(*a*) (*b*)

Figure 10-8 Annihilation diagrams: (*a*) lowest-order term and (*b*) second-order contribution of the crossed term $V_a D^{-1} V_b + V_b D^{-1} V_a$.

Note that this time in $K(p)$ we have kept the complete two-particle free propagator, while φ_0 is still to be understood to involve the product $\chi_1 \otimes \chi_2$ of two-component spinors describing the polarization of the state and hence appears as a matrix under the trace sign. In $\Delta K(p)$ we have approximated the Coulomb wave function by its value at the origin in momentum space and used an infrared cutoff μ. The vertex renormalization is obtained as in Chap. 7 by subtracting the similar expression at a large photon mass Λ and using the renormalization constant $Z_1 = Z_2$ in the Feynman gauge computed by the same method as in Eq. (7-34):

$$Z_1 = 1 - \frac{\alpha}{2\pi}\left(\ln\frac{\Lambda}{m} + 2\ln\frac{\mu}{m} + \frac{9}{4}\right) \tag{10-124}$$

It was already noted in the previous subsection that only space-like values of virtual annihilation photon polarization (index μ) contribute to the splitting. Moreover, the charge conjugation matrix

$$C = i\gamma^2\gamma^0 = \begin{pmatrix} 0 & -i\sigma_2 \\ -i\sigma_2 & 0 \end{pmatrix}$$

is odd in the representation we are using, meaning that

$$\operatorname{tr}\left[C\gamma(\ )_1(\ )_2\varphi_0\right]$$

is to be interpreted as

$$(0, \chi_2^T(-i\sigma_2))\left[C^{-1}(\ )_2^T C\gamma(\ )_1\right]\begin{pmatrix}\chi_1 \\ 0\end{pmatrix}\varphi_0$$

Using this remark, a straightforward (but long and painful) calculation yields

$$\begin{aligned} \Delta E_a^{(1)} + \Delta E_{ab+ba}^{(2)} &= \frac{\pi\alpha}{m^2}\left(1 - \frac{4\alpha}{\pi}\right)\operatorname{tr}(\varphi_0^\dagger \gamma C)\cdot\operatorname{tr}(C\gamma\varphi_0) \\ &= -\frac{\pi\alpha}{m^2}\left(1 - \frac{4\alpha}{\pi}\right)[\chi_1^\dagger\sigma(-i\sigma_2)\chi_2^*]\cdot[\chi_2^T(-i\sigma_2)\sigma\chi_1]|\varphi_0|^2 \\ &= \frac{\pi\alpha}{m^2}\left(1 - \frac{4\alpha}{\pi}\right)|\varphi_0|^2\langle \mathbf{S}^2\rangle \end{aligned} \tag{10-125}$$

The above expression has still to be multiplied by $[1 - \bar{\omega}(4m^2)]$ to include the radiative correction due to the vacuum polarization to order α. In this formalism this effect arises from the second-order $V_a D^{-1} V_a$ term

$$\bar{\omega}(4m^2) = \frac{8\alpha}{9\pi} \tag{10-126}$$

leading to an extra contribution

$$\Delta E_{aa}^{(2)} = \frac{\pi\alpha}{m^2}\left(-\frac{8\alpha}{9\pi}\right)|\varphi_0|^2\langle \mathbf{S}^2\rangle \tag{10-127}$$

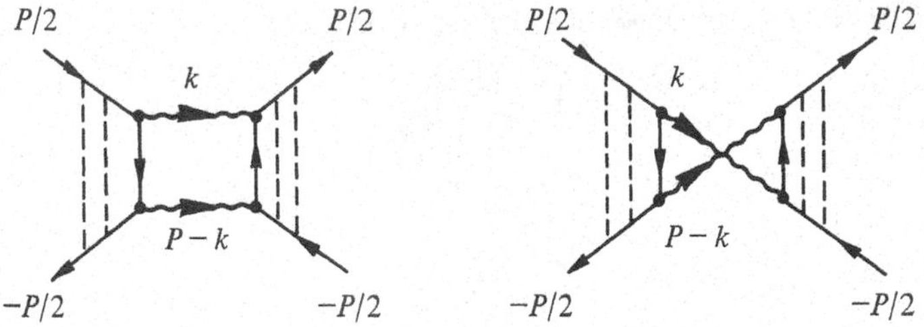

Figure 10-9 Two-photon annihilation diagrams.

The last piece of the puzzle is the two-photon annihilation term with (minus) its imaginary part equal to half the singlet level width (Fig. 10-9). Its expression is found by adding a new term in the Bethe-Salpeter kernel. It is best to return to Eq. (10-84) and to approximate the wave function by its nonrelativistic value at the origin, thereby obtaining the leading α^5 effect in the even-charge conjugation channel. To keep track of various coefficients it is simpler to use the relativistic notation with $u(1)$, $v(2)$ being spinors at rest:

$$\Delta E_{2\gamma}^{(2)} - i\frac{\Gamma_{2\gamma}}{2} = \frac{i\alpha^2}{\pi^2}|\varphi_0|^2 \int \frac{d^4k}{[k^2 + i\varepsilon][(P-k)^2 + i\varepsilon]} \bar{v}(2)\gamma^\nu \frac{1}{\not{P}/2 - \not{k} - m}\gamma^\mu u(1)$$
$$\times \bar{u}(1)\left(\gamma_\mu \frac{1}{\not{P}/2 - \not{k} - m}\gamma_\nu + \gamma_\nu \frac{1}{-\not{P}/2 + \not{k} - m}\gamma_\mu\right)v(2) \tag{10-128}$$

Evaluation of the integral and of the spin matrix element leaves us with a coefficient proportional to the projector on the singlet state:

$$\Delta E_{2\gamma}^{(2)} - \frac{i\Gamma_{2\gamma}}{2} = -\frac{\alpha^2}{m^2}|\varphi_0|^2\langle 2 - \mathbf{S}^2\rangle(2 - 2\ln 2 + i\pi) \tag{10-129}$$

with the imaginary part giving $\Gamma_{2\gamma} = m\alpha^5/2$ as found in (5-128).

We have now the complete value of the hyperfine splitting in positronium to order α^5. Taking Eqs. (10-119), (10-121), and (10-122) for the exchange channel and (10-125), (10-127), and (10-129) for the annihilation one, and subtracting the values in the $S = 1$ and $S = 0$ states, we find the Karplus and Klein result

$$\Delta E_{\text{ts}} = \frac{1}{2}\alpha^2 \,\text{Ryd}\left[\frac{7}{3} - \left(\frac{32}{9} + 2\ln 2\right)\frac{\alpha}{\pi}\right] \tag{10-130}$$

This value is not yet sufficient for a comparison with the experimental result. The reader will appreciate the effort needed to extract the α^6 terms. The difficulty lies partly in a good control of the recoil effects, i.e., of a correct treatment of the relativistic Coulomb wave functions, which leads in fact to terms of order $\alpha^6 \ln(1/\alpha)$. Other effects such as vacuum polarization in the exchange channel also contribute to this term, so that the correction is of the type $(\alpha^2/2)$ Ryd $[B\alpha^2 \ln(1/\alpha) + C]$. The latest value, reported by Lepage in 1977, is $B = -\frac{1}{6}$. When added to (10-130) this would yield the theoretical prediction 2.033774×10^5 MHz, to be compared with (10-7.9).

The Bethe-Salpeter formalism may also be applied to obtain radiative corrections to the decay widths, as well as to other systems such as the hydrogen atom. As an exercise the reader may derive, within the present formalism, the excited spectrum of positronium to order α^4:

$$E^S_{n,LJ} = 2m + \text{Ryd}\left(-\frac{1}{2n^2} + \frac{11\alpha^2}{32n^4} + \frac{\alpha^2}{n^3}\varepsilon^S_{LJ}\right)$$

$$\varepsilon^0_{LJ} = -\delta_{L,J}\frac{1}{2L+1}$$

$$\varepsilon^1_{L,J} = -\frac{1}{2L+1} + \frac{7}{6}\delta_{L,0} + \frac{1-2\delta_{L0}}{2(2L+1)} \times \begin{cases} \dfrac{3L+4}{(L+1)(2L+3)} & J = L+1 \\ -\dfrac{1}{L(L+1)} & J = L \\ -\dfrac{(3L-1)}{L(2L-1)} & J = L-1 \end{cases} \tag{10-131}$$

How does this compare to the triplet splitting of Eq. (10-80)?

NOTES

The field equations appear in the work of F. J. Dyson, *Phys. Rev.*, vol. 75, p. 1736, 1949, and J. Schwinger, *Proc. Nat. Acad. Sc.*, vol. 37, pp. 452 and 455, 1951.

Bound-state equations have a long history which can be traced from the book of H. A. Bethe and E. E. Salpeter, "Quantum Mechanics of One and Two Electron Atoms," Springer-Verlag, Berlin, 1957. Modern developments were prompted by the work of these authors, *Phys. Rev.*, vol. 82, p. 309, 1951, vol. 84, p. 1232, 1951, by J. Schwinger in the article just quoted, and by M. Gell-Mann and F. Low, *Phys. Rev.*, vol. 84, p. 350, 1951. The ladder exchange was treated by Y. Nambu, *Progr. Theor. Phys.*, vol. 5, p. 614, 1950, by G. C. Wick, *Phys. Rev.*, vol. 96, p. 1124, 1954, and by R. E. Cutkosky, *Phys. Rev.*, vol. 96, p. 1135, 1954. The review of N. Nakanishi, *Suppl. Prog. Theor. Phys.*, vol. 43, p. 1, 1969, covers his own work together with the many contributions up to the late 1960s.

The perturbation theory of the hyperfine splitting was developed by E. E. Salpeter, *Phys. Rev.*, vol. 87, p. 328, 1952, and by R. Karplus and A. Klein, *Phys. Rev.*, vol. 87, p. 848, 1952, in the positronium case. The recent accurate measurements are due to A. P. Mills and G. H. Bearman, *Phys. Rev. Lett.*, vol. 34, p. 246, 1975, and P. O. Egan, W. E. Frieze, V. W. Hughes, and M. H. Yam, *Phys. Lett.*, ser. A, vol. 54, p. 412, 1975. The excited triplet splitting reported in the text is from A. P. Mills, S. Berko, and K. F. Canter, *Phys. Rev. Lett.*, vol. 34, p. 1541, 1975. A. Stroscio has reviewed the subject of positronium in *Phys. Rep.*, ser. C, vol. 22, p. 215, 1975. Recent contributions are those by G. P. Lepage in *Phys. Rev.*, ser. A, vol. 16, p. 863, 1977, and by G. T. Bodwin and D. R. Yennie in *Phys. Rep.*, ser. C, vol. 43, p. 267, 1978, including reference to earlier work.

The positronium spectrum is discussed in J. Schwinger, "Particles, Sources and Fields," vol. 2, Addison-Wesley, Reading, 1973.

CHAPTER

ELEVEN

SYMMETRIES

The importance of symmetries in quantum systems and particle physics has led to a number of specific investigations. Model building and understanding of the fundamental interactions require the identification of the various invariances and the way in which they happen to be realized, approximated, or even violated. This is supported by the spectacular results of unitary symmetry, current algebra, and chiral invariance, and by the developing interest in the quark substructure of hadrons. On the other hand, similar considerations are at the heart of the theory of order-disorder transitions in macroscopic media, emphasizing once again the deep relationship between field theory and statistical mechanics. We present here some of the models, the study of which has brought to light specific field theoretic phenomena such as almost degenerate multiplets, spontaneous symmetry breaking, and quantum anomalies.

11-1 QUANTUM IMPLEMENTATION OF SYMMETRIES

We have already encountered several instances of symmetries. Some were discrete such as parity, time reversal, and charge conjugation. Others were continuous, as is the kinematical Lorentz invariance, or even space-time dependent, as in the case of gauge invariance. We shall now concentrate our attention on continuous internal symmetries and discuss the realization of quantum symmetries within lagrangian field theory, even though some of the statements and results do not require such a specific framework.

11-1-1 Statement of the Problem

Consider a theory invariant under a continuous group of transformations at the classical level. More precisely this means that the lagrangian (or hamiltonian) and hence the equations of motion are invariant under these transformations, as discussed in Chap. 1. The question is now to clarify the way in which this property is reflected in the quantum system, and in particular to study its consequences on the spectrum of states.

In the quantum framework the discussion of symmetries implies the existence of a group of transformations (local automorphisms) acting on the physical observables and therefore on the dynamical field variables. In infinitesimal form we are thus led to the construction of associated four-vector current densities and integrated charges. The point is then to know whether we can implement these transformations by unitary transformations on the Hilbert space of states. Locally we want to study the current conservation, while globally we ask whether the charges generate unitary operators commuting with the S matrix.

In Chap. 1 we have already described within the classical theory how we can associate to a family of infinitesimal transformations a corresponding set of Noether currents. Given the field transformations

$$\begin{aligned} \phi(x) &\to \phi(x) + \delta\phi(x) \\ \delta\phi(x) &= \mathscr{F}_a(x)\delta\alpha^a \end{aligned} \tag{11-1}$$

depending on the infinitesimal parameters $\delta\alpha^a$, the currents are defined through

$$j_\mu^a(x) = \frac{\partial \mathscr{L}}{\partial(\partial^\mu\phi)} \frac{\delta\phi(x)}{\delta\alpha^a} \tag{11-2}$$

where $\mathscr{L}$ is the lagrangian. This holds whether or not the action is invariant under these transformations, provided the lagrangian is a function of the fields and their first derivatives only. Using the Euler-Lagrange equations of motion the variation of $\mathscr{L}$ reduces to

$$\delta\mathscr{L} = \partial^\mu j_\mu^a \delta\alpha^a \tag{11-3}$$

The charges

$$Q^a(t) = \int d^3x \, j_0^a(\mathbf{x}, t) \tag{11-4}$$

are then the generators of these transformations through the Poisson bracket operation

$$\{Q^a(t), \phi(\mathbf{x}, t)\} = \frac{\delta\phi}{\delta\alpha^a}(\mathbf{x}, t) \tag{11-5}$$

Equation (11-5) assumes that the Poisson bracket of ϕ with $\delta\phi$ vanishes. This is the case if $\delta\phi$ does not depend on the conjugate momentum to ϕ. In particular, the quantities $Q^a(t)$ satisfy the commutation relations dictated by the Lie algebra of the transformation group with structure constants $C^{ab}{}_c$ [see Eq. (1-146)]:

$$\{Q^a(t), Q^b(t)\} = -C^{ab}{}_c Q^c(t) \tag{11-6}$$

According to (11-3) if $\mathscr{L}$ is invariant the currents are classically conserved:

$$\partial^\mu j^a_\mu = 0 \tag{11-7}$$

and the charges are time independent. This can be seen using Stokes theorem and neglecting spatial surface terms.

In the quantum theory we would like to reproduce the previous steps and in particular construct the conserved currents. However, since the latter are in general local polynomials in the fields it is not unlikely to encounter some difficulties. A preliminary regularization may sometimes be useful. It might, for instance, involve a slight relative shift of the arguments of the fields, the subtraction of divergent terms, etc. As a result, the validity of the conservation law may no longer appear trivial. In practice we do not always master the commutation rules of fields, and we may rather try to check whether it is consistent to assume that the conservation laws hold, order by order in perturbation theory, after an appropriate renormalization. This program is carried out by showing the equivalence with a set of identities to be fulfilled by the Green functions. These are similar to the Ward identities of electrodynamics and can be tested perturbatively. In the example of the chiral symmetry of the σ model of Gell-Mann and Lévy we shall discover that anomalies due to ultraviolet divergences may invalidate the classical conservation laws. This kind of study is therefore not academic.

It is also necessary to consider the global aspect of the problem, namely, the unitary implementation of symmetries and its effect on the spectrum. Two cases arise. The first is the most familiar one and has been studied in detail since the early works of Weyl and Wigner. It corresponds to a unitary representation $g \to U(g)$ of the symmetry group in the Hilbert space of states. Its action on the observables is given by

$$A \to {}^gA = U(g)AU^\dagger(g) \tag{11-8}$$

Such was the case of the Poincaré invariance. This involves the discussion of phase factors leading to representations of the simply connected covering group (Wigner's theorem). The states are classified in multiplets corresponding to irreducible representations. In particular, the vacuum or ground state corresponds to the identity representation. Symmetries sometimes rely on approximations, such as neglecting weaker interactions. This is the case of the isotopic invariance which is often said to be weakly broken by electromagnetic forces. We shall briefly review this situation below. It may, however, occur that the symmetry is only reflected at the dynamical level and that the vacuum is not invariant. We call it the Goldstone realization. The unitary operators $U(g)$ do not exist, and the symmetry appears spontaneously broken. States are no longer classified in multiplets and under certain conditions massless particles appear. This situation is not infrequent in physics, an example being ferromagnetism and spin waves. In particle physics the dynamical behavior of pions seems to approximate this scheme.

11-1-2 Behavior of the Ground State

The symmetry properties of a system may be characterized by the behavior of its ground state. As we shall see, if the vacuum is annihilated by a set of charges, the corresponding currents are conserved and the symmetry group unitarily implemented. Conversely, assuming conserved currents, if the vacuum is not invariant the symmetry is spontaneously broken.

Were the vacuum not unique, we could define an orthonormal basis in this lowest-energy subspace, diagonalizing all commuting hermitian observables. To simplify, assume a discrete set of such states $|n\rangle$. Consider any couple of observables $A(x)$, $B(y)$. When their arguments are infinitely space-like separated, only the intermediate translationally invariant ground states should contribute to the matrix elements of their product, according to a generalization of the Riemann-Lebesgue lemma

$$\lim_{|\mathbf{x}|\to\infty} \langle n| A(\mathbf{x})B(0)|m\rangle = \sum_p \langle n| A(0)|p\rangle\langle p| B(0)|m\rangle$$

Here $|p\rangle$ runs over the ground states. According to causality, the matrix element of the commutator $\langle n|[A(\mathbf{x}), B(0)]|m\rangle$ vanishes in the limit $|\mathbf{x}|\to\infty$. The matrices $\langle n| A(0)|m\rangle$ and $\langle n| B(0)|m\rangle$ commute and can be simultaneously diagonalized. Within a given sector the vacuum matrix element of a product of local operators factorizes when their space-like separations become large. This is called the cluster property.

An apparent degeneracy of the vacuum can occur as a result of a poor approximation. The real ground state is the unique combination which minimizes the energy. An example is provided by a quantum mechanical one-dimensional system with a potential $V(x) = (x^2 - 1)^2$ with two equal minima. Quantum mechanics allows for barrier tunneling and leads to a unique symmetric ground state.

We may approach the ground state in a translationally invariant way by requiring it to minimize the effective potential $V_{\text{eff}}(\phi)$. The latter plays the role of density of potential energy in a state with a given mean value of the field.

Let us quote an important property which follows from locality. Assume that a conserved current operator $j^\mu(x)$ has been constructed, so that $\partial_\mu j^\mu(x) = 0$. Define the integral of $j^0(\mathbf{x}, t)$ over a bounded region V of space, $Q_V(t)$, as

$$Q_V(t) = \int_V d^3x\, j^0(\mathbf{x}, t) \tag{11-9}$$

This operator clearly stands a better chance of being well defined than its limit $Q(t)$, the integral over all space. The commutator of $Q_V(t)$ with a local observable A is independent of time for V large enough:

$$\lim_{V\to\infty} \frac{d}{dt}[Q_V(t), A] = 0 \tag{11-10}$$

Indeed, from current conservation we have

$$0 = \int_V d^3x\,[\partial_\mu j^\mu(\mathbf{x}, t), A] = \frac{d}{dt}\int_V d^3x\,[j^0(\mathbf{x}, t), A] + \int_S d\mathbf{S}\cdot[\mathbf{j}(\mathbf{x}, t), A]$$

When V becomes large enough, the surface integral vanishes since the commutator involves local operators separated by a very large space-like interval.

The validity of this statement extends to more general situations. In particular, A may be replaced by any multilocal operator. The proof could even be extended to a nonrelativistic system provided that only short-range forces are present.

What seems an utter triviality—either the vacuum is invariant or it is not—covers, however, a wealth of situations with very different physical content. What if the vacuum is indeed invariant? A theorem of Coleman shows that the associated currents are then conserved. Assume that $Q(t)$, the space integral of j^0, is well defined (at least on a dense subset of the Hilbert space including the ground state) and annihilates the vacuum

$$Q(t)|0\rangle = \int d^3x\, j_0(\mathbf{x}, t)|0\rangle = 0 \tag{11-11}$$

If the theory admits an energy gap, any state with zero momentum will be such that

$$\langle n, \mathbf{P}_n = 0|\, j_0(\mathbf{x}, t)\,|0\rangle = 0$$

from translational invariance, since the above quantity is $\mathbf{x}$ independent with a vanishing integral. Here we cheat slightly since strictly speaking such a state is not normalizable. The necessary refinement may, however, be worked out. Consequently,

$$\langle n, \mathbf{P}_n = 0|\,\partial_0 j^0(\mathbf{x}, t)\,|0\rangle = \langle n, \mathbf{P}_n = 0|\,\partial_\mu j^\mu(\mathbf{x}, t)\,|0\rangle = 0 \tag{11-12}$$

The last equality follows from the fact that $\nabla \cdot \mathbf{j}(\mathbf{x}, t) = i[\mathbf{P}, \mathbf{j}(\mathbf{x}, t)]$. Assuming the matrix element of $\mathbf{j}(\mathbf{x}, t)$ to be meaningful the matrix element of its three-divergence must vanish between zero momentum states. Finally $\partial_\mu j^\mu(x)$ being a Lorentz scalar, Eq. (11-12) generalizes to arbitrary matrix elements between the vacuum and any other state. Hence $\partial_\mu j^\mu$ annihilates the vacuum. One may show that this is only possible for a local operator if it vanishes.

We conclude that

$$\partial_\mu j^\mu = 0 \tag{11-13}$$

The symmetry is exact and unitarily implemented by operators of the form $U = e^{i\omega Q}$.

11-2 MASS SPECTRUM, MULTIPLETS, AND GOLDSTONE BOSONS

11-2-1 The Octet Model of Gell-Mann and Ne'eman

When the internal symmetries form a compact Lie group, we may choose the phases of the representative operators to obtain a unitary representation of its covering group. This representation splits into irreducible components, acting on subspaces of the original Hilbert space. The latter is therefore generated by multiplets of states.

Here we shall briefly describe the $SU(3)$ symmetry or octet model valid with some degree of approximation at the level of strong interactions. It generalizes the familiar isospin invariance of nuclear forces introduced by Heisenberg in the 1930s.

Larger flavor groups, as they are called nowadays, manifest themselves at higher energies. There is already some evidence for an $SU(4)$ symmetry through the study of the new multi-GeV narrow resonances, and it may well be that this is not the end of the story.

The members of the elementary isodoublet, proton and neutron, are classified according to the eigenvalues of the third component of isospin T_3 with eigenvalues $\pm\frac{1}{2}$ in such a way that

$$Q = \frac{N}{2} + T_3 \tag{11-14}$$

Here N is the baryonic charge. In the lagrangian framework, the corresponding fields will form an isospinor ψ and, neglecting mass differences, the kinetic term

$$\mathscr{L}_0 = \overline{\psi}(i\partial\!\!\!/ + M)\psi$$

is invariant under isospin rotations

$$\psi \to U\psi \qquad \overline{\psi} \to \overline{\psi}U^\dagger \tag{11-15}$$

with U a 2×2 unimodular unitary matrix. In the simplest case we couple this field to the pion isovector $\boldsymbol{\pi}$ in an invariant way (see Chap. 5). For the pion, a relation analogous to (11-14) holds with $N = 0$ and T_3 taking the values $+1$, 0, or -1. Using the pseudoscalar character of $\boldsymbol{\pi}$ the only (renormalizable) invariant interaction is

$$\mathscr{L}_{\text{int}} = ig_{\pi NN}\overline{\psi}\gamma^5\boldsymbol{\tau}\cdot\boldsymbol{\pi}\psi \tag{11-16}$$

The boldface characters used for the Pauli matrices or the pion field refer to isospace. Equation (11-16) expresses in a compact form a set of relations on the various couplings

$$g_{\pi^+ pn} = \sqrt{2}g_{\pi^\circ pp} = -\sqrt{2}g_{\pi^\circ nn} \tag{11-17}$$

asserting the dynamical content of the symmetry. These relations were obtained by noting that the usual fields $\pi^{(+)}$, $\pi^{(-)} = (\pi^{(+)})^\dagger$, and $\pi^{(0)}$ are related to the cartesian coordinates through

$$\begin{aligned} \tfrac{1}{2}(\pi_1^2 + \pi_2^2 + \pi_3^2) &= \pi^{(-)}\pi^{(+)} + \tfrac{1}{2}\pi^{(0)2} \\ \pi^{(0)} &= \pi_3 \\ \pi^{(\pm)} &= \frac{1}{\sqrt{2}}(\pi_1 \mp i\pi_2) \end{aligned} \tag{11-18}$$

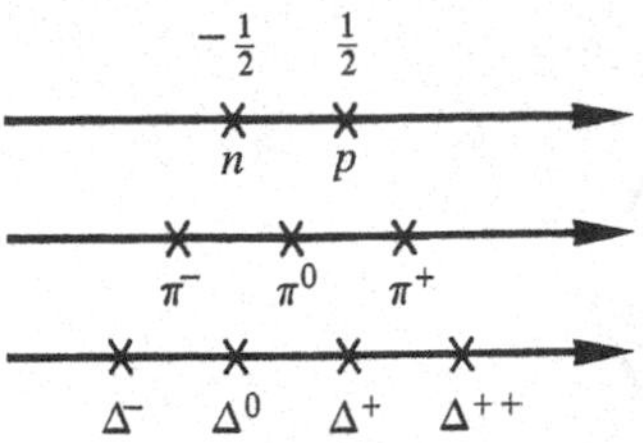

Figure 11-1 Weight diagrams for $SU(2)$.

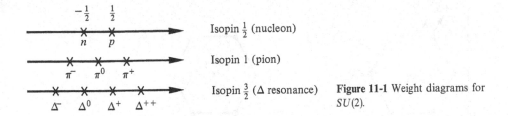

Note that $\pi^{(-)}$ creates a positive pion and annihilates a negative one. When discussing pion-nucleon scattering we mentioned some dynamical consequences of this symmetry such as triangular inequalities on cross sections. In the early 1960s, Gell-Mann and Ne'eman succeeded in obtaining a generalization of isospin symmetry to a larger group $SU(3)$. More precisely, all known multiplets correspond to representations of the factor group $SU(3)/Z_3$ where Z_3 is the abelian center of $SU(3)$ generated by the cubic roots of the identity. The latter therefore acts as the identity on all hadronic states which are called of triality zero.

We recall that the representations are conveniently described in terms of the Lie algebra of the infinitesimal hermitian generators of the group. The procedure is familiar in quantum mechanics. We diagonalize a maximal set of commuting generators (Cartan subalgebra). Basis states are represented by weight vectors with components equal to the eigenvalues of these operators in a space of dimension equal to the rank of the Lie algebra, i.e., the dimension of its Cartan subalgebra. For the familiar example of $SU(2)$, of rank one, weight diagrams are one-dimensional, the abscissa being the eigenvalue of T_3 (Fig. 11-1). A particular role is played by the adjoint representation acting on the Lie algebra itself through the linear action of commutation. In the case of $SU(2)$ it corresponds to isospin 1 with

$$\begin{aligned} T_\pm &= T_1 \pm iT_2 \\ [T_3, T_\pm] &= \pm T_\pm \end{aligned} \tag{11-19}$$

This extends to $SU(3)$ with eight (3×3 traceless hermitian) generators and a Lie algebra of rank two. The two diagonal generators are linear combinations of iospin T_3 and hypercharge Y, equal to the sum of baryonic number and strangeness. The Gell-Mann and Nishijima formula gives the relation with the charge generalizing Eq. (11-14):

$$Q = \frac{Y}{2} + T_3 \qquad Y = N + S \tag{11-20}$$

The lowest-dimensional octets of baryons and pseudoscalar mesons (corresponding to the adjoint representation of $SU(3)$] as well as the resonance decuplet which led to the discovery of the Ω^- particle are shown in Fig. 11-2.

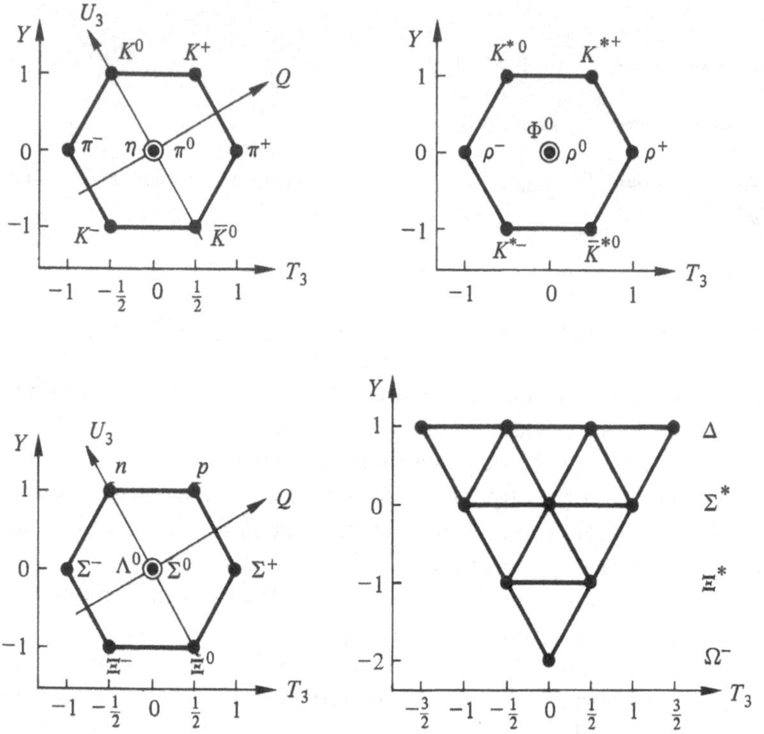

Figure 11-2 Meson and baryon octets and resonance decuplet in the Gell-Mann and Ne'eman model.

A basis of the Lie algebra generalizing the Pauli matrices is given by the Gell-Mann matrices

$$\lambda_1 = \begin{pmatrix} 0 & 1 & 0 \\ 1 & 0 & 0 \\ 0 & 0 & 0 \end{pmatrix} \quad \lambda_2 = \begin{pmatrix} 0 & -i & 0 \\ i & 0 & 0 \\ 0 & 0 & 0 \end{pmatrix} \quad \lambda_3 = \begin{pmatrix} 1 & 0 & 0 \\ 0 & -1 & 0 \\ 0 & 0 & 0 \end{pmatrix}$$

$$\lambda_4 = \begin{pmatrix} 0 & 0 & 1 \\ 0 & 0 & 0 \\ 1 & 0 & 0 \end{pmatrix} \quad \lambda_5 = \begin{pmatrix} 0 & 0 & -i \\ 0 & 0 & 0 \\ i & 0 & 0 \end{pmatrix} \quad \lambda_6 = \begin{pmatrix} 0 & 0 & 0 \\ 0 & 0 & 1 \\ 0 & 1 & 0 \end{pmatrix} \tag{11-21}$$

$$\lambda_7 = \begin{pmatrix} 0 & 0 & 0 \\ 0 & 0 & -i \\ 0 & i & 0 \end{pmatrix} \quad \lambda_8 = \begin{pmatrix} 1/\sqrt{3} & 0 & 0 \\ 0 & 1/\sqrt{3} & 0 \\ 0 & 0 & -2/\sqrt{3} \end{pmatrix}$$

These matrices are normalized according to

$$\operatorname{tr}(\lambda_k \lambda_l) = 2\delta_{lk}$$

and fulfil the commutation and anticommutation relations

$$[\lambda_k, \lambda_l] = 2if_{klm}\lambda_m \qquad \{\lambda_k, \lambda_l\} = \tfrac{4}{3}\delta_{kl} + 2d_{klm}\lambda_m \tag{11-22}$$

with f_{klm} totally antisymmetric and d_{klm} totally symmetric, with nonvanishing elements given by the following table:

$k\ l\ m$	$2f_{klm}$	$k\ l\ m$	$2d_{klm}$	$k\ l\ m$	$2d_{klm}$
1 2 3	2	1 1 8	$2/\sqrt{3}$	3 6 6	-1
1 4 7	1	1 4 6	1	3 7 7	-1
1 5 6	-1	1 5 7	1	4 4 8	$-1/\sqrt{3}$
2 4 6	1	2 2 8	$2/\sqrt{3}$	5 5 8	$-1/\sqrt{3}$
2 5 7	1	2 4 7	-1	6 6 8	$-1/\sqrt{3}$
3 4 5	1	2 5 6	1	7 7 8	$-1/\sqrt{3}$
3 6 7	-1	3 3 8	$2/\sqrt{3}$	8 8 8	$-2/\sqrt{3}$
4 5 8	$\sqrt{3}$	3 4 4	1		
6 7 8	$\sqrt{3}$	3 5 5	1		

(11-23)

From the time of its introduction the unitary symmetry has remained puzzling by the lack of multiplets with nonzero triality. The simplest of those would correspond to a triplet of hypothetical quarks, introduced by Gell-Mann and by Zweig. These particles [an isodoublet (u, d) and an isosinglet (s)] would have fractional charges and baryonic number (Fig. 11-3). If the unitary group turns out to be larger than $SU(3)$, more quarks carrying new quantum numbers would be required such as the charmed quark (c), an isosinglet of charge 2/3 and hypercharge 1/3. The corresponding new symmetry group $SU(4)$ is of rank three. The weight diagrams for its (four-dimensional) fundamental representation and (fifteen-dimensional) adjoint representation are shown in Fig. 11-4.

Many unsuccessful attempts have been made to detect free quarks, mostly on the basis of their fractional charge. It appears as if all known hadrons share the properties of quark-bound states. The idea that the latter might be permanently bound has led to the term "confinement" and several mechanisms have been

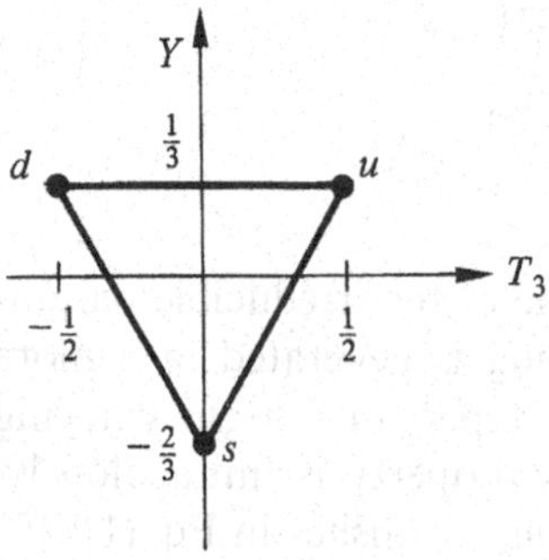

Figure 11-3 The quark triplet.

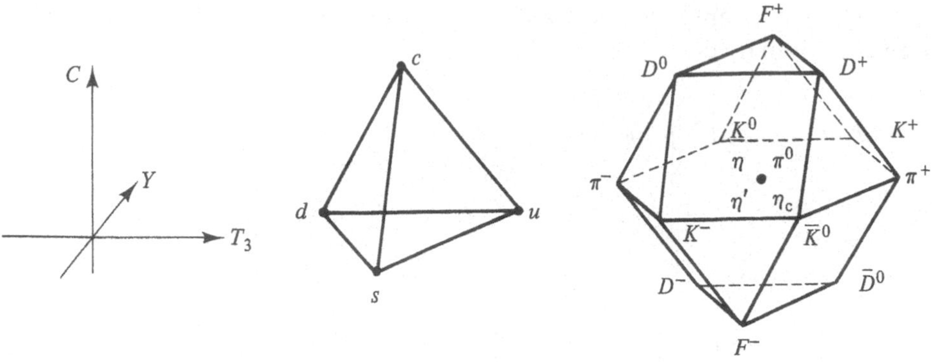

Figure 11-4 Fundamental and adjoint representations of $SU(4)$. We have displayed the new charmed mesons.

suggested. One of the most promising assumes an unbroken color symmetry according to a local invariance, to be discussed in the next chapters.

The octet symmetry is only approximately obeyed by strong interactions. This is even more true of any larger invariance group. Nevertheless, the pattern of symmetry breaking looks very regular, obeying the so-called octet dominance rule. To quote but one example, consider the mass differences between members of the same multiplet, for definiteness the baryon octet of nucleons, Λ, Σ, and Ξ particles. In an effective lagrangian, assuming only isospin invariance, the mass terms would have as coefficients m_N, m_Λ, m_Σ, m_Ξ. Each bilinear field product such as $\bar{p}p + \bar{n}n + \cdots$ transforms under $SU(3)$ as the direct product of two octet representations which can be split into irreducible parts according to the Clebsch-Gordan series

$$8 \otimes 8 = 1 \oplus 8' \oplus 8'' \oplus 10 \oplus \overline{10} \oplus 27 \tag{11-24}$$

Neutrality means that the non-self-conjugate representations 10 and $\overline{10}$ are absent in the mass matrix. This implies that the four physical masses may be reexpressed as linear combinations of four parameters multiplying the combinations having definite transformation properties according to

$$\begin{aligned} m_N &= (m_1 - 2m_8' + m_8'' - 3m_{27}) \\ m_\Lambda &= (m_1 - m_8' - m_8'' + 9m_{27}) \\ m_\Sigma &= (m_1 + m_8' + m_8'' + m_{27}) \\ m_\Xi &= (m_1 + m_8' - 2m_8'' - 3m_{27}) \end{aligned} \tag{11-25}$$

This reshuffling is, of course, of no interest unless some of the irreducible parameters vanish. Suppose, however, that the mass splitting is generated in a more fundamental lagrangian by a quark-mass term of the type $\bar{\psi}(a + b\lambda_8)\psi$ carrying only singlet and octet representations and that this property is (miraculously) preserved by the interactions. This would mean that m_{27} vanishes in Eq. (11-25), leading to the Gell-Mann and Okubo relation

$$\tfrac{1}{2}(m_N + m_\Xi) = \tfrac{1}{4}(3m_\Lambda + m_\Sigma) \tag{11-26}$$

experimentally well satisfied with

$$\tfrac{1}{2}(m_N + m_\Xi) = 1.128 \text{ GeV} \qquad \tfrac{1}{4}(3m_\Lambda + m_\Sigma) = 1.134 \text{ GeV}$$

For the pseudoscalar octet (K, π, η) the analogous formula involves empirically square masses

$$4m_K^2 = 3m_\eta^2 + m_\pi^2 \tag{11-27}$$

We should allow for some mixing with a ninth η' meson to complete a nonet. Using Eq. (11-27) at face value, it predicts $m_\eta^2 = 0.320 \text{ GeV}^2$, the experimental value being $m_\eta^2 = 0.301 \text{ GeV}^2$.

Octet dominance applied to the resonance decuplet leads to equal-mass spacing, in good agreement with the observed data and considered as a major triumph in its prediction of the complete set of quantum numbers of the Ω^- particle:

$$m = m_0 + m_8 Y \qquad \begin{aligned} m_{\Sigma^*} - m_\Delta &= 152 \text{ MeV} \\ m_{\Xi^*} - m_{\Sigma^*} &= 149 \text{ MeV} \\ m_\Omega - m_{\Xi^*} &= 139 \text{ MeV} \end{aligned} \tag{11-28}$$

Unitary symmetry leads to numerous relations among scattering amplitudes which are in satisfactory agreement with experiment. We shall have more to say about it when we discuss the applications of current algebra.

11-2-2 Spontaneous Symmetry Breaking

Spontaneous symmetry breaking occurs when the ground state is not invariant under the transformation group. The latter acts in a larger nonseparable Hilbert space, as was exemplified by the behavior under rotations of the infinite ferromagnet (Chap. 4). We can convince ourselves of this fact by the following heuristic remark. Let us try to compute the norm of the state $Q|0\rangle$, where Q is the would-be total charge

$$\langle 0| Q^2 |0\rangle = \int d^3x \, \langle 0| j_0(\mathbf{x}) Q |0\rangle \tag{11-29}$$

From translational invariance this is also equal to

$$\int d^3x \, \langle 0| j_0(0) Q |0\rangle$$

and is clearly infinite when $Q|0\rangle \neq 0$.

The most striking consequence analyzed by Goldstone is the appearance of massless particles when the broken symmetry is a continuous one. These states are generated by operators which would rotate the vacuum by an infinitesimal amount to a degenerate vacuum, since on physical grounds such a transformation does not cost any energy.

To prove Goldstone's theorem we assume the existence of a conserved current and consider any operator A such that

$$\delta a(t) \equiv \lim_{V\to\infty} \langle 0|[Q_V(t), A]|0\rangle \neq 0 \tag{11-30}$$

That there exists such an observable just expresses the noninvariance of the vacuum. Insert a complete set of intermediate states of definite four-momentum in this relation:

$$\begin{aligned}\delta a(t) &= \lim_{V\to\infty} \sum_n \int_V d^3x\, [\langle 0|j_0(0)|n\rangle\langle n|A|0\rangle e^{-iP_n\cdot x} - \langle 0|A|n\rangle\langle n|j_0(0)|0\rangle e^{iP_n\cdot x}] \\ &= \sum_n (2\pi)^3\delta^3(\mathbf{P}_n)[\langle 0|j_0(0)|n\rangle\langle n|A|0\rangle e^{-iE_nt} - \langle 0|A|n\rangle\langle n|j_0(0)|0\rangle e^{iE_nt}] \\ &\neq 0\end{aligned} \tag{11-31}$$

We have already shown in Eq. (11-10) that current conservation implies

$$\frac{d}{dt}\,\delta a(t) = 0 \tag{11-32}$$

Therefore

$$0 = \sum_n (2\pi)^3\delta^3(\mathbf{P}_n)E_n[\langle 0|j_0(0)|n\rangle\langle n|A|0\rangle e^{-iE_nt} + \langle 0|A|n\rangle\langle n|j_0(0)|0\rangle e^{iE_nt}] \tag{11-33}$$

It then follows from Eqs. (11-31) and (11-33) that a state $|n\rangle$ must exist such that $\langle 0|A|n\rangle\langle n|j_0(0)|0\rangle \neq 0$ for which $E_n\delta^3(\mathbf{P}_n)$ vanishes. This is a massless state with the same quantum numbers as j_0 (and A) since it is generated by this operator from the vacuum.

These massless states are called Goldstone bosons associated to the symmetry breaking. They are indeed bosons if j_0 is a boson-type operator. In a more general setting such as the recent supersymmetric theories j_μ could in fact carry half-integer spin and the associated massless particle would then be a fermion. In other words, the spin of these states bears a relation with the Lorentz transformation properties of $j_\mu(x)$.

A subtle point is that the massless states need not necessarily be observable. This remark applies to theories with an unphysical sector of unobservable states (such as quantum electrodynamics in the Gupta-Bleuler gauge) and may be relevant when we want to avoid the conclusion of the theorem.

The mechanism of spontaneous symmetry breaking is characterized by its aesthetic appeal, as opposed to the ad hoc breaking prescriptions such as those encountered when discussing the octet model. It is economical in the sense that it does not introduce any new parameter. It is also theoretically advantageous since it preserves renormalizability properties. As a counterpart the appearance of massless particles might be an unpleasant feature in certain applications. We shall encounter in the next chapter an elegant way to dispose of them using Higgs' mechanism.

As an extreme example of spontaneous symmetry breaking consider a free massless scalar field with a lagrangian

$$\mathscr{L} = \tfrac{1}{2}(\partial\phi)^2 \tag{11-34}$$

invariant under the field translations

$$\phi(x) \to \phi(x) + \lambda \tag{11-35}$$

The associated conserved current is

$$j^\mu(x) = \partial^\mu \phi(x) \tag{11-36}$$

The vacuum is, of course, not invariant under these transformations. If we substitute $\phi(x)$ for A in Eq. (11-30) we find

$$\delta\langle 0|\,\phi(x)\,|0\rangle = \lambda\langle 0\,|\,0\rangle \neq 0 \tag{11-37}$$

Alternatively, the total "charge"

$$\lim_{V\to\infty} Q_V(t) = \lim_{V\to\infty}\int_V d^3x\; \partial_0\phi(\mathbf{x}, t) \tag{11-38}$$

is not well defined. The Goldstone boson in this case is the quantum of the field ϕ itself. An interesting point about this example is that we may look at the state obtained from the vacuum through the action of $e^{i\lambda Q_V}$. From the Fourier decomposition of the field

$$\phi(x) = \int \frac{d^3k}{2k^0(2\pi)^3}\,[a(k)\,e^{-ik\cdot x} + a^\dagger(k)\,e^{ik\cdot x}] \tag{11-39}$$

it follows that in the infinite-volume limit the state

$$|\lambda\rangle = \lim_{V\to\infty} e^{i\lambda Q_V(0)}\,|0\rangle \tag{11-40}$$

takes the form

$$\begin{aligned}|\lambda\rangle &= \lim_{V\to\infty}\exp\left\{\lambda\int_V d^3x \int\frac{d^3k}{2(2\pi)^3}\,[a(k)\,e^{i\mathbf{k}\cdot\mathbf{x}} - a^\dagger(k)\,e^{-i\mathbf{k}\cdot\mathbf{x}}]\right\}|0\rangle \\ &= e^{(\lambda/2)[a(0)-a^\dagger(0)]}\,|0\rangle\end{aligned} \tag{11-41}$$

It appears therefore as a coherent superposition of zero-energy momentum states. To show that $\langle\lambda\,|\,0\rangle$ vanishes we have to be more careful in the handling of the infinite-volume limit. It is better to introduce a smooth spatial cutoff, for instance,

$$Q_V = \int d^3x\; \partial_0\phi(\mathbf{x}, 0)\, e^{-\mathbf{x}^2/V^{2/3}} \tag{11-42}$$

The reader can easily verify that

$$\langle 0\,|\,\lambda\rangle = e^{-\pi(\lambda V^{1/3})^2/64} \underset{V\to\infty}{\to} 0 \tag{11-43}$$

A physically meaningful model exhibiting spontaneous symmetry breaking is based on a set of n coupled scalar fields with a lagrangian

$$\mathscr{L}(\phi) = \tfrac{1}{2}(\partial\phi)^2 - \frac{\mu^2}{2}\phi^2 - \frac{\lambda}{4}(\phi^2)^2 \tag{11-44}$$

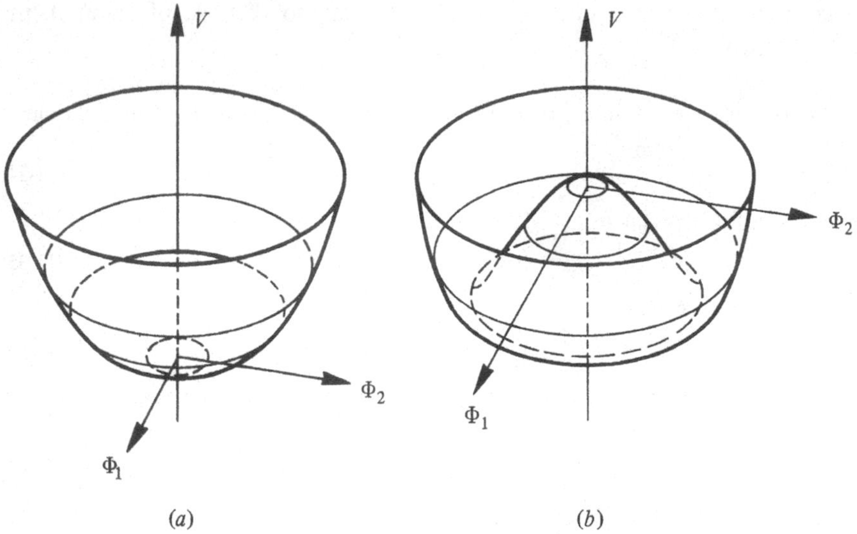

Figure 11-5 The potential $V(\phi) = (\mu^2/2)\phi^2 + (\lambda/4)(\phi^2)^2$: (a) $\mu^2 > 0$, (b) $\mu^2 < 0$.

invariant under an internal $O(n)$ symmetry group. The notation is such that ϕ stands for the column vector of these fields and symmetry breaking will occur if the bare-mass parameter μ^2 is negative. At the classical level the ground state is not described by a zero mean field value due to the fact that the potential

$$V(\phi) = \frac{\mu^2}{2}\phi^2 + \frac{\lambda}{4}(\phi^2)^2 \tag{11-45}$$

has a minimum for a nonvanishing value of the field (Fig. 11-5)

$$|\phi| = v = \left(-\frac{\mu^2}{\lambda}\right)^{1/2} \tag{11-46}$$

To insure the stability of the theory we assume, of course, that large values of ϕ are damped, which means $\lambda > 0$. It appears as if the vacuum is degenerate: every state with $|\phi| = v$ is a priori a candidate. However, any choice leads to an inequivalent Hilbert space. Note that such a choice of a ground-state value ϕ_v will leave an $O(n-1)$ subgroup unbroken, corresponding to those transformations which act as the identity on ϕ_v.

To be specific, let us pick a coordinate system in internal space such that $\phi_v = \langle 0|\phi|0\rangle$ has only its nth component nonvanishing, and parametrize the field in terms of a radial displacement ρ and $(n-1)$ orthogonal rotations as

$$\phi = e^{\mathbf{t}\cdot\xi/v}\begin{pmatrix} 0 \\ 0 \\ \vdots \\ v+\rho \end{pmatrix} \tag{11-47}$$

The t_a with $1 \leq a \leq n-1$ stand for $n-1$ generators of the $O(n)$ group acting effectively on the vector ϕ_v. In other words, the Lie algebra of $O(n)$ contains in addition to the $O(n-1)$ generators the $n-1$ t_a's. When expressed in terms of the new dynamical variables ρ and ξ the lagrangian reads

$$\mathscr{L} = \tfrac{1}{2}(\partial\rho)^2 + \tfrac{1}{2}(\partial\xi)^2 - \frac{\mu^2}{2}(v+\rho)^2 - \frac{\lambda}{4}(v+\rho)^4 \tag{11-48}$$

From this we see that the ξ correspond to $(n-1)$ decoupled massless fields, as predicted by Goldstone's theorem.

In the previous example, to be studied in more detail in Sec. 11-4, the ground-state degeneracy is apparent at the classical level. We have learnt, however, in Chap. 9 that quantum corrections can alter the potential, leading to the possibility of vacuum degeneracy as a result of radiative corrections. This appears to be the case for the following model for scalar electrodynamics, as pointed out by Coleman and Weinberg. A complex massless boson field is coupled minimally to an electromagnetic field. The lagrangian is

$$\mathscr{L} = \mathscr{L}_{\text{em}} + \tfrac{1}{2}(\partial\phi_1 - eA\phi_2)^2 + \tfrac{1}{2}(\partial\phi_2 + eA\phi_1)^2 - \frac{\lambda}{4!}(\phi_1^2 + \phi_2^2)^2 \tag{11-49}$$

The complex field ϕ is written in terms of its two real components ϕ_1 and ϕ_2. Moreover, the anharmonic coupling constant is denoted $\lambda/4!$ to compare the one-loop computation of the effective potential with the expression given by Eq. (9-129) for the lagrangian

$$\mathscr{L} = \tfrac{1}{2}(\partial\phi)^2 - \frac{\lambda}{4!}\phi^4 \tag{11-50}$$

which led to the result

$$V = \frac{\lambda}{4!}\phi^4 + \frac{\lambda^2\phi^4}{(16\pi)^2}\left(\ln\frac{\phi^2}{M^2} - \frac{25}{6}\right) \tag{11-51}$$

In Chap. 9 renormalization was performed at a value of the field equal to M to insure that

$$\lambda = \frac{d^4}{d\phi^4}V(\phi)\big|_{\phi=M} \tag{11-52}$$

Returning to scalar electrodynamics, we can repeat a similar calculation using the symmetry factors shown in Fig. 11-6 for the contribution of ϕ_1^4 to V. The result

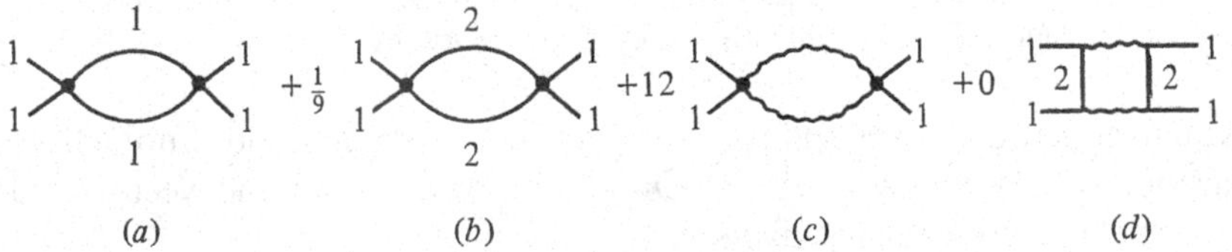

Figure 11-6 Relative weights of the one-loop diagrams for the Green function with four ϕ_1 fields.

reads

$$V = \frac{\lambda}{4!}(\phi_1^2 + \phi_2^2)^2 + \left(\frac{5}{1152\pi^2}\lambda^2 + \frac{3e^4}{64\pi^2}\right)(\phi_1^2 + \phi_2^2)^2\left[\ln\left(\frac{\phi_1^2 + \phi_2^2}{M^2}\right) - \frac{25}{6}\right] \tag{11-53}$$

The calculation was carried out in the Landau gauge. The last diagram, Fig. 11-6*d*, gives a vanishing contribution at zero momentum. A factor three in diagram (*c*) originates in the trace of $(g^\mu{}_\nu - k^\mu k_\nu/k^2)$. The remaining weights arise from the vertices.

It is consistent to assume that λ and e^4 are comparable. Indeed, the λ coupling is generated by the extra divergence occurring at order e^4. Therefore we may neglect λ^2 as compared to e^4. In this way the potential V of Eq. (11-53) has a minimum for a nonvanishing value $\phi_v^2 \equiv \phi_{v,1}^2 + \phi_{v,2}^2$ given by

$$\left[\frac{\lambda}{6} - 11\left(\frac{e^2}{4\pi}\right)^2\right] + 3\left(\frac{e^2}{4\pi}\right)^2 \ln\frac{\phi_v^2}{M^2} = 0 \tag{11-54}$$

If we choose the scale of ϕ such that $M^2 = \phi_v^2$, that is,

$$\lambda = \left.\frac{d^4V}{d\phi^4}\right|_{\phi^2 = \phi_v^2} \tag{11-55}$$

we find the relation

$$\lambda = 66\left(\frac{e^2}{4\pi}\right)^2 \tag{11-56}$$

giving indeed a value for λ of order e^4. The parameters of the theory are now e and ϕ_v.

A similar reasoning would not apply to the pure ϕ^4 theory of Eq. (11-50). It is true that the potential V has a minimum for a value ϕ_v such that

$$\frac{\lambda}{6} - \frac{11\lambda^2}{192\pi^2} + \frac{\lambda^2}{64\pi^2}\ln\frac{\phi_v^2}{M^2} = 0 \tag{11-57}$$

that is,

$$\lambda \ln\frac{\phi_v^2}{M^2} = -\frac{32}{3}\pi^2 + O(\lambda) \tag{11-58}$$

However, the quantity $\lambda \ln(\phi_v^2/M^2)$ is now sizable and we expect higher quantum corrections to be nonnegligible. Stated differently, it is not legitimate in this case to set $\phi_v^2 = M^2$, since this would imply a large value of λ for which the perturbative series cannot be trusted.

For scalar electrodynamics spontaneous symmetry breaking does not imply the appearance of any massless boson. On the contrary, both the vector and the scalar particle acquire masses given by

$$m_\phi^2 = 6\left(\frac{e^2}{4\pi}\right)^2 \phi_v^2$$
$$m_A^2 = e^2\phi_v^2 \tag{11-59}$$

with the striking consequence that

$$\frac{m_\phi^2}{m_A^2} = O(e^2) \tag{11-60}$$

The reason for such a behavior will be studied in more detail in the following chapter.

We conclude this section with a discussion of the role of the dimensionality of space-time. A theorem due to Mermin and Wagner states that a continuous symmetry can only be spontaneously broken in a dimension larger than two. For a discrete symmetry this lower critical dimensionality is one. This is, in fact, well known since in quantum mechanics with finitely many degrees of freedom (corresponding to one-dimensional field theory) tunneling between degenerate classical minima allows for a unique symmetric ground state. Alternatively, we may consider the discrete analog of a field theory, the simplest example being the Ising model in statistical mechanics. Path integrals are replaced by sums of terms of the form $e^{-E/kT}$ where E is the energy of a configuration. For the Ising model $E = -J\sum_{(ij)} \sigma_i\sigma_j$ where the sum runs over neighboring sites on a lattice and the discrete "spin" σ_i takes values of ± 1. Such a model admits a discrete symmetry corresponding to reversing all the spins $\sigma_i \to -\sigma_i$. This symmetry is spontaneously broken in the low-temperature phase below a critical point in dimension two or higher, but no transition occurs in dimension one (Peierls, 1938).

A similar model called the classical Heisenberg model replaces the variables σ_i by unit vectors $\mathbf{S}_i$ on a sphere. In this case a continuous $O(n)$ group operates if $\mathbf{S}_i$ has n components, and no spontaneous magnetization occurs below dimension three.

The Mermin-Wagner theorem has been restated by Coleman in the framework of field theory. We may establish this property by showing that the spontaneous breakdown of a continuous symmetry would lead to a Goldstone boson. But in a two-dimensional space-time it is not possible to construct a massless scalar field operator. Indeed, the corresponding two-point Wightman function

$$\langle 0|\,\phi(x)\phi(0)\,|0\rangle = \int \frac{d^2k}{2\pi}\,\theta(k^0)\delta(k^2)\,e^{ik\cdot x}$$
$$= \int_0^\infty \frac{dk^1}{2\pi k^1}\cos(k^1x^1)\,e^{ik^1x^0} \tag{11-61}$$

is a meaningless infrared divergent integral. No subtraction procedure may be devised to circumvent this difficulty without spoiling the fundamental properties of field theory, for instance, positivity of the Hilbert space metric. A massless scalar field theory is undefined in a two-dimensional world due to severe infrared divergences. In the statistical language, fluctuations overcome energy in destroying long-range order in this dimension.

A simple argument reveals the nature of this phenomenon. Let us use a discrete classical Heisenberg model on a lattice. Compare the two configurations shown in Fig. 11-7 where the orientation of the "spin" is allowed to vary along the direction of the first axis, say. The action is replaced by the energy proportional respectively to $E_a = -L^d$ and to $E_b = -L^{d-1}\sum_1^L \cos(\theta/L)$. The relative weight of these configurations is given by the Boltzmann factor

$$e^{\beta(E_b - E_a)} \sim e^{-\beta\theta^2 L^{d-2}/2}$$

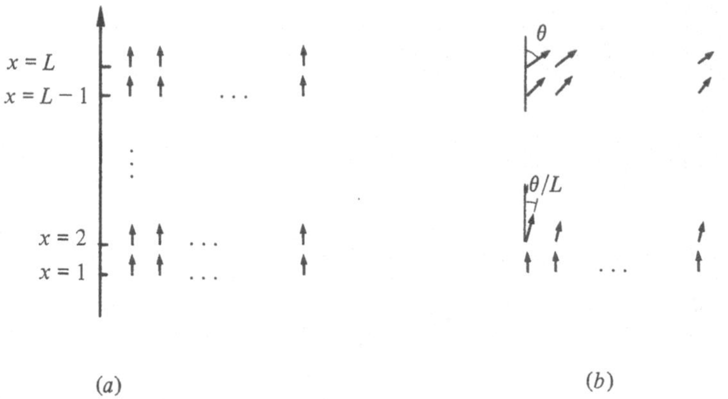

Figure 11-7 Two configurations for the classical Heisenberg model on a d-dimensional lattice. One of the directions has been singled out to show the effect of a continuous rotation of the average spin.

For d larger than 2 we see that the (b) configuration has a negligible weight in the thermodynamic limit and for sufficiently low temperature, meaning that order is favored. For $d = 2$ averaging over fluctuations will destroy this order. Similar arguments may be presented for a discrete symmetry showing that the lower critical dimension is then equal to one.

11-3 CURRENT ALGEBRA

11-3-1 Current Commutators

Weak interactions have led to consider in detail the structure and properties of hadronic currents. Phenomenologically these interactions are well described by an effective current-current lagrangian of the form

$$\mathscr{L} = -\frac{G}{\sqrt{2}} J^{\mu}(x) J^{\dagger}_{\mu}(x) \tag{11-62}$$

where G is Fermi's constant equal to

$$G = (1.026 \pm 0.001) \times 10^{-5} m_p^{-2} \tag{11-63}$$

The total current $J_\mu(x)$ is the sum of leptonic (l_μ) and hadronic (h_μ) contributions

$$J_\mu(x) = l_\mu(x) + h_\mu(x) \tag{11-64}$$

Disregarding the contribution of the newly discovered massive lepton, the leptonic current involves only negative helicities for the electron e^-, muon μ^-, and neutrinos ν_e, ν_μ:

$$l_\rho(x) = \overline{\psi}_e(x)\gamma_\rho(1 - \gamma_5)\psi_{\nu_e}(x) + \overline{\psi}_\mu(x)\gamma_\rho(1 - \gamma_5)\psi_{\nu_\mu}(x) \tag{11-65}$$

The hadronic current itself combines strangeness conserving $h_\rho^{(\Delta S=0)}$ and strangeness changing $h_\rho^{(\Delta S=1)}$ parts as

$$h_\rho = \cos\theta_c\, h_\rho^{(\Delta S=0)} + \sin\theta_c\, h_\rho^{(\Delta S=1)} \tag{11-66}$$

where θ_c is the Cabibbo angle equal approximately to

$$\theta_c \simeq 0.25 \tag{11-67}$$

Such a decomposition implies that we can relate the scale of the various components associated to transitions with different quantum numbers. Such a scale will be provided by the nonlinear algebra of current commutators.

Each of these currents is a superposition $V - A$ of a vector and an axial part, as was the leptonic current. In the framework of unitary symmetry, they belong to an octet of currents denoted V^a, A^a $(a = 1, \ldots, 8)$ with

$$\begin{aligned} h_\mu^{(\Delta S=0)} &= (V_\mu{}^1 - iV_\mu{}^2) - (A_\mu{}^1 - iA_\mu{}^2) \\ h_\mu^{(\Delta S=1)} &= (V_\mu{}^4 - iV_\mu{}^5) - (A_\mu{}^4 - iA_\mu{}^5) \end{aligned} \tag{11-68}$$

The reader should not be confused by the notation A_μ used here for an axial current and its meaning as a potential in electrodynamics. Both are traditional.

The current-current weak interaction, supplemented by these hypotheses, leads to remarkable universality properties. Consider, for instance, the matrix element measured in neutron β decay:

$$G\langle p|\, h_\mu^{(\Delta S=0)}\,|n\rangle = \bar{u}_p\gamma_\mu(G_V - G_A\gamma_5)u_n \tag{11-69}$$

The form factors G_V and G_A on the right-hand side are evaluated at zero momentum transfer due to the smallness of the neutron-proton mass difference. Within 1 percent the observed value of G_V coincides with the corresponding quantity measured in muon decay. The value quoted in (11-63) is the one extracted from muon decay taking radiative corrections into account. The most precise measurements involving strongly interacting particles are usually performed in allowed β transitions $(0^+ \to 0^+)$ among nuclear states. Care is also taken to correct for various effects, such as radiative corrections, Cabibbo angle, etc. Some recent values are

$$^{14}\text{O} \xrightarrow{\beta^+} {}^{14}\text{N} \qquad G_V/G = 1.006$$

$$^{26}\text{Al} \to {}^{26}\text{Mg} \qquad G_V/G = 1.011$$

The fact that $G_V(0)$ is not renormalized by strong interactions finds a natural interpretation if we assume with Gell-Mann and Feynman that the vector current V_μ is conserved and in fact generates the hadronic unitary symmetry. In other words, the $V_\mu{}^a$ $(a = 1, 2, 3)$ are the components of the isospin current and $V_\mu^{(\Delta S=1)}$ is approximately conserved if we neglect $SU(3)$ breaking. The hadronic electromagnetic current is therefore

$$j_\mu^{\text{em}} = V_\mu^3 + \frac{1}{\sqrt{3}} V_\mu{}^8$$

where for the purpose of comparison we omit the factor e in the definition of j^{em}. This hypothesis is a generalization of the electromagnetic case, where current conservation implies a universal charge renormalization. We recall from Chaps. 7 and 8 that the Ward identity expressing the conservation law implied that the charge renormalization was entirely due to vacuum polarization. All fermion interactions resulted only in the wave function renormalization constant $Z_2 = Z_1$.

Therefore some components of the vector hadronic weak current and the isovector part of the electromagnetic current are members of the same multiplet and simultaneously conserved. This conserved vector current (CVC) hypothesis of Gell-Mann and Feynman may be checked by comparing the weak and electromagnetic decay rates of members of the same isospin multiplet. In particle physics we find a prediction for the $\pi^+ \to \pi^0 + e^+ + \nu$ and $\pi^- \to \pi^0 + e^- + \bar{\nu}$ transitions. Due to the smallness of the momentum transfer, the amplitudes are directly normalized, through an isospin rotation, by the corresponding matrix elements of the electric charge. This leads to the rates

$$\Gamma_{\pi^\pm \to \pi^0 + (e\nu)} = \frac{G^2(m_{\pi^+} - m_{\pi^0})^5}{30\pi^3} = 0.45\ \mathrm{s}^{-1}$$

where terms of order $m_e^2/(m_{\pi^+} - m_{\pi^0})^2 \sim 10^{-2}$ have been neglected. This is to be compared to the experimental value

$$\Gamma_{\pi^\pm \to \pi^0 + (e\nu)} = (0.39 \pm 0.03)\ \mathrm{s}^{-1}$$

Accurate tests can also be performed in nuclear physics.

In contradistinction, the axial current does not appear to be conserved, even in the limit of an exact $SU(3)$ invariance. Consequently, the axial coupling constant as measured in β decay is not equal to the vector one:

$$\frac{G_A}{G_V} = 1.22 \pm 0.02 \tag{11-70}$$

It is, however, conceivable to study a limit where axial currents would be conserved, at least the strangeness conserving ones. This would correspond to an additional chiral $SU(2) \times SU(2)$ invariance group, generated by the charges Q^a and $Q_5{}^a$ $(a = 1, 2, 3)$.

If the fundamental dynamical variables suitable for a description of hadrons include the quark fields, an explicit expression may be derived for the currents

$$V_\mu{}^a(x) = \bar{q}(x)\gamma_\mu \frac{\lambda^a}{2} q(x) \qquad A_\mu{}^a(x) = \bar{q}(x)\gamma_\mu\gamma_5 \frac{\lambda^a}{2} q(x) \tag{11-71}$$

We may then obtain equal-time commutation relations for the currents as well as for the corresponding charges:

$$\begin{aligned} Q^a(t) &= \int d^3x\ V_0{}^a(\mathbf{x}, t) \\ Q_5{}^a(t) &= \int d^3x\ A_0{}^a(\mathbf{x}, t) \end{aligned} \tag{11-72}$$

Gell-Mann has postulated that these commutation relations derived from the quark model remain valid independently of this assumption on the hadronic substructure. If $SU(3)$ is not exactly valid some of the charges may depend on time, while the equal-time algebra remains time independent:

$$\begin{aligned}[][Q^a, Q^b] &= if_{abc}Q^c \\ [Q^a, Q_5{}^b] &= if_{abc}Q_5{}^c \\ [Q_5{}^a, Q_5{}^b] &= if_{abc}Q^c\end{aligned} \tag{11-73}$$

We recognize that the commutations rules are those of the Lie algebra of the group $SU(3) \times SU(3)$. This is readily verified by constructing the left- and right-handed combinations

$$Q_{\mp}{}^a = \tfrac{1}{2}(Q^a \mp Q_5{}^a) \tag{11-74}$$

which fulfil

$$\begin{aligned}[][Q_{\pm}{}^a, Q_{\pm}{}^b] &= if_{abc}Q_{\pm}{}^c \\ [Q_{+}{}^a, Q_{-}{}^b] &= 0\end{aligned} \tag{11-75}$$

The ordinary unitary group is the diagonal subgroup of $SU(3) \times SU(3)$, and parity exchanges the two sets of charges

$$\mathscr{P}Q_{+}\mathscr{P}^{-1} = Q_{-} \tag{11-76}$$

The charge algebra gives its meaning to the universality concept since it is basically nonlinear. For instance, from

$$[Q^{1+i2}, Q^{1-i2}] = 2Q^3 \tag{11-77}$$

it follows that the matrix element of the left-hand side entering a weak amplitude is universally normalized with reference to the isospin.

The relations (11-73) may be generalized in two steps. First we may write commutation relations between charges and currents expressing the behavior of $V_\mu{}^a$ and $A_\mu{}^a$ under the $SU(3) \times SU(3)$ transformations. This abstracts from the quark model the fact that currents belong to a representation $(1, 8) \oplus (8, 1)$:

$$\begin{aligned}[][Q^a(t), V_\mu{}^b(\mathbf{x}, t)] &= if_{abc}V_\mu{}^c(\mathbf{x}, t) \\ [Q^a(t), A_\mu{}^b(\mathbf{x}, t)] &= if_{abc}A_\mu{}^c(\mathbf{x}, t) \\ [Q_5{}^a(t), V_\mu{}^b(\mathbf{x}, t)] &= -if_{abc}A_\mu{}^c(\mathbf{x}, t) \\ [Q_5{}^a(t), A_\mu{}^b(\mathbf{x}, t)] &= if_{abc}V_\mu{}^c(\mathbf{x}, t)\end{aligned} \tag{11-78}$$

From Eqs. (11-78) an integral over space allows us to recover the charge commutation relations.

In a second step the quark model suggests writing equal-time commutators for the time components:

$$\begin{aligned}[][V_0{}^a(\mathbf{x}, t), V_0{}^b(\mathbf{y}, t)] &= if_{abc}V_0{}^c(\mathbf{x}, t)\delta^3(\mathbf{x} - \mathbf{y}) \\ [V_0{}^a(\mathbf{x}, t), A_0{}^b(\mathbf{y}, t)] &= if_{abc}A_0{}^c(\mathbf{x}, t)\delta^3(\mathbf{x} - \mathbf{y}) \\ [A_0{}^a(\mathbf{x}, t), A_0{}^b(\mathbf{y}, t)] &= if_{abc}V_0{}^c(\mathbf{x}, t)\delta^3(\mathbf{x} - \mathbf{y})\end{aligned} \tag{11-79}$$

These relations between time components seem safe. This is presumably not the case if we attempt a further generalization to include all the components. Terms involving derivatives of the δ function are likely to appear. We would have, for instance,

$$[V_0{}^a(\mathbf{x}, t), V_i^b(\mathbf{y}, t)] = if_{abc} V_i^c(\mathbf{x}, t)\delta^3(\mathbf{x} - \mathbf{y}) + S_{ab;ij}\partial_j\delta^3(\mathbf{x} - \mathbf{y}) \qquad (11\text{-}80)$$

Such additional terms were originally introduced by Schwinger in his discussion of the conserved electromagnetic current [$U(1)$ symmetry]. In this case let us assume that a relation of the type

$$[j_0(\mathbf{x}, t), j_i(\mathbf{y}, t)] = 0 \qquad (11\text{-}81)$$

holds. No δ-function contribution appears on the right-hand side due to the vanishing of the structure constants in this abelian case. As a consequence

$$[j_0(\mathbf{x}, t), \nabla \cdot \mathbf{j}(\mathbf{y}, t)] = 0 \qquad (11\text{-}82)$$

Through current conservation this also means

$$[j_0(\mathbf{x}, t), \partial_0 j_0(\mathbf{y}, t)] = 0 \qquad (11\text{-}83)$$

Taking the vacuum expectation value and inserting a complete set of eigenstates of the energy leads in the limit $\mathbf{x} \to \mathbf{y}$ to

$$\sum_n E_n |\langle 0| j_0(0) |n\rangle|^2 = 0 \qquad (11\text{-}84)$$

From the positivity of energy we would conclude that j_0 vanishes! In this case at least, Schwinger terms are unavoidable.

These extra contributions do not appear in the quantities $[Q, j_\mu]$ obtained by integration. The situation is even more intricate when dealing with equal-time commutators of space components which are strongly model dependent, and will not be used in the following.

We have already encountered Schwinger terms in Sec. 5-1-7 when we tried to write spectral representations for the vacuum expectation value of the time-ordered product or the commutator of electromagnetic currents. We observed a relation between the Schwinger term and the local noncovariant difference between the naive time-ordered product ($\tilde{T}$) and its covariant conserved version (T). We found that

$$\langle 0| Tj_\mu(x)j_\nu(y) |0\rangle = \langle 0| \tilde{T}j_\mu(x)j_\nu(y) |0\rangle - i(g_{\mu\nu} - g_{\mu 0}g_{\nu 0})\delta^4(x - y)\xi \qquad (11\text{-}85)$$

with ξ an integral over the spectral function

$$\xi = \int dM^2 \frac{\rho(M^2)}{M^2} \qquad (11\text{-}86)$$

On the other hand, it was also shown that

$$\langle 0| [j_0(\mathbf{x}, t), j_k(\mathbf{y}, t)] |0\rangle = i\partial_k\delta^3(\mathbf{x} - \mathbf{y})\xi \qquad (11\text{-}87)$$

If we assume such relations between Schwinger terms and noncovariant parts of the naive time-ordered product (seagull terms) for arbitrary currents and states, these undesirable contributions cancel in the applications of current algebra. Indeed, a typical result follows from a Ward identity satisfied by the covariant time-ordered product

$$\frac{\partial}{\partial x^\mu}\langle A|\,Tj^\mu(x)j^\nu(y)\,|B\rangle = \langle A|\,T\partial_\mu j^\mu(x)j^\nu(y)\,|B\rangle + \delta(x^0-y^0)\langle A|\,[j^0(x), j^\nu(y)]\,|B\rangle + \frac{\partial}{\partial x^\mu}(\text{seagull}) \tag{11-88}$$

If the last term is compensated by the Schwinger term arising from the equal-time commutator, both may be dropped and the identity reduces to the naive one:

$$\frac{\partial}{\partial x^\mu}\langle A|\,Tj^\mu(x)j^\nu(y)\,|B\rangle = \langle A|\,T\partial_\mu j^\mu(x)j^\nu(y)\,|B\rangle + \delta(x^0-y^0)\langle A|\,[j^0(x), j^\nu(y)]\,|B\rangle\,\Big|_{\text{naive}} \tag{11-89}$$

This will be taken for granted. The Ward identity will be used to obtain low-energy theorems and sum rules whenever some information on the current divergence is available. For axial currents we may postulate a partial conservation law (PCAC), the precise form of which will be stated below. Let us first, however, show how local commutation rules may be tested. This will enable us to introduce a new device, the so-called infinite momentum frame of Dirac, Fubini, and Furlan.

Choose nucleon states of equal momentum p for A and B and average over their polarization, thereby defining

$$W_{\mu\nu}(p,q) = \frac{1}{2\pi}\int d^4x\, e^{iq\cdot x}\,\frac{1}{2}\sum_{\text{pol}}\langle p|\,[j^a_\mu(x), j^b_\nu(0)]\,|p\rangle \tag{11-90}$$

An integration over q^0 generates the equal-time commutator

$$\int dq^0\, W_{00}(p,q) = \int d^3x\, e^{-i\mathbf{q}\cdot\mathbf{x}}\,\frac{1}{2}\sum_{\text{pol}}\langle p|\,[j^a_0(\mathbf{x},0), j^b_0(0)]\,|p\rangle \tag{11-91}$$

In typical applications the currents would be chosen as vector densities $j_0^{a,b} = V_0^{1\pm i2}$ or $V_0^{4\pm i5}$, with commutators equal to combinations of V_0^3 and V_0^8. For instance,

$$\begin{aligned}\int d^3x\, e^{-i\mathbf{q}\cdot\mathbf{x}}\,\frac{1}{2}\sum_{\text{pol}}\langle p|\,[V_0^{1+i2}(\mathbf{x},0), V_0^{1-i2}(0)]\,|p\rangle &= 4\int d^3x\,\delta^3(\mathbf{x})\,\frac{1}{2}\sum_{\text{pol}}\langle p|\,V_0^3(0)\,|p\rangle \\ &= 4\frac{p_0}{m}\,T^3\end{aligned} \tag{11-92}$$

where T^3 is the third component of the nucleon isospin and states have been normalized according to $\langle p|p'\rangle = (p_0/m)(2\pi)^3\delta^3(\mathbf{p}-\mathbf{p}')$.

The covariant amplitude $W_{\mu\nu}$ is decomposed according to

$$W_{\mu\nu} = -W_1 g_{\mu\nu} + W_2 \frac{p_\mu p_\nu}{m^2} - iW_3 \varepsilon_{\mu\nu\rho\sigma} \frac{p_\rho q_\sigma}{2m^2}$$

$$+ W_4 \frac{q_\mu q_\nu}{m^2} + W_5 \frac{p_\mu q_\nu + p_\nu q_\mu}{2m^2} + iW_6 \frac{p_\mu q_\nu - p_\nu q_\mu}{2m^2} \tag{11-93}$$

Here $W_1, \ldots, W_6$ are functions of the Lorentz invariants q^2 and $\nu = q \cdot p$. Consequently,

$$\int dq^0\, W_{00}(p, q) = p_0 \int d\nu \left(\frac{-W_1}{p_0^2} + \frac{W_2}{m^2} + \frac{q_0^2}{p_0^2 m^2} W_4 + \frac{q_0}{p_0 m^2} W_5 \right)_{\text{fixed } \mathbf{q}} \tag{11-94}$$

If we select a frame where $\mathbf{p} \cdot \mathbf{q} = 0$, and therefore such that $q_0 = \nu/p_0$ and $q^2 = (\nu/p_0)^2 - \mathbf{q}^2$, we may rewrite (11-94) as

$$\int dq^0\, W_{00}(p, q) = p_0 \int d\nu \left(\frac{-W_1}{p_0^2} + \frac{W_2}{m^2} + \frac{\nu^2 W_4}{p_0^4 m^2} + \frac{\nu W_5}{p_0^2 m^2} \right)_{\text{fixed } \mathbf{q}} \tag{11-95}$$

The interpretation of the sum rule is still complicated because of its dependence on p_0. Instead of choosing the nucleon rest frame, where $p_0 = m$, it is more clever to use a limiting one obtained by letting p_0 go to infinity with q^2 held fixed:

$$\begin{array}{cc} p_0, |\mathbf{p}| \to \infty & \mathbf{p} \cdot \mathbf{q} = 0 \\ q^2 \to -\mathbf{q}^2 & \text{fixed} \end{array} \tag{11-96}$$

If we are allowed to interchange limit and integration we find that

$$\frac{1}{m^2} \int d\nu\, W_2(\nu, q^2 = -\mathbf{q}^2) = \lim_{p_0 \to \infty} \frac{1}{p_0} \int dq^0\, W_{00} \tag{11-97}$$

This is a severe constraint in view of (11-91) and (11-92), since in particular it implies that the left-hand side is independent of q^2. It tests the local character of current algebra.

There are several applications of this type of sum rule. We quote only a few of them. The Cabibbo-Radicati sum rule is obtained when applying Eq. (11-97) to the isospin vector currents V^{1+i2}, V^{1-i2}. It reads

$$2 \frac{dF_1^V}{dq^2}(0) = \left[\frac{F_2^V(0)}{2m} \right]^2 + \frac{1}{2\pi^2 \alpha} \int_{\nu_{\text{thresh}}}^{\infty} \frac{d\nu}{\nu} (2\sigma_{1/2}^V - \sigma_{3/2}^V) \tag{11-98}$$

The notation $F_1^V(q^2)$, $F_2^V(q^2)$ stands for the isovector nucleon form factors, i.e., such that the matrix element of the electromagnetic current reads

$$\langle p_2 | j_\mu^{\text{em}}(0) | p_1 \rangle = \bar{u}_2 \left[\gamma_\mu \left(\frac{F_1^S + F_1^V \tau_3}{2} \right) + \frac{i\sigma_{\mu\nu}}{2m} q^\nu \left(\frac{F_2^S + F_2^V \tau_3}{2} \right) \right] u_1 \tag{11-99}$$

and $\sigma_{1/2}^V$, $\sigma_{3/2}^V$ are the total hadronic cross sections for the scattering of "isovector" photons off nucleons in the channels of total isospin $\frac{1}{2}$ or $\frac{3}{2}$ respectively. The

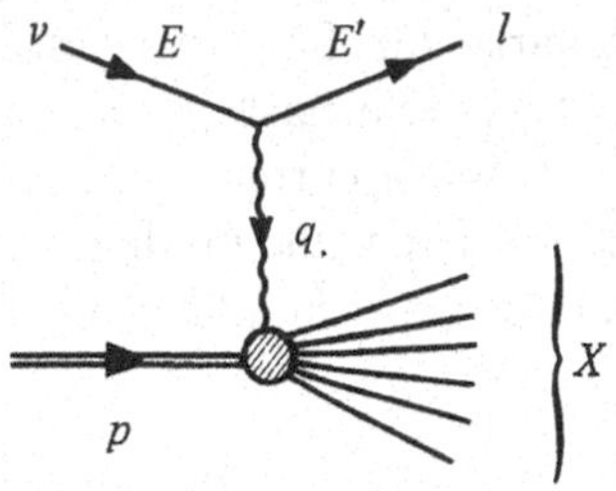

Figure 11-8 Inclusive process $\nu + N \to l + X$.

experimental values for the two sides of the sum rule are in good agreement. In the unit of the inverse square pion mass one finds

$$2\frac{dF_1^V}{dq^2}(0) = 0.132m_\pi^{-2} \qquad \left[\frac{F_2^V(0)}{2m}\right]^2 + \frac{1}{2\pi^2\alpha}\int_{\nu_{\text{thresh}}}^{\infty}\frac{d\nu}{\nu}(2\sigma_{1/2}^V - \sigma_{3/2}^V) = 0.126m_\pi^{-2}. \tag{11-100}$$

The inclusive cross section for the process neutrino (ν) or antineutrino ($\bar{\nu}$) + nucleon (N) → lepton (l) + X, where X is an unidentified hadronic state (Fig. 11-8), is expressed in terms of the structure functions W_1, W_2, W_3 appearing in the expression (11-93) as

$$\frac{d\sigma^{(\nu,\bar{\nu})}}{d|q^2|\,d\nu} = \frac{G^2}{2\pi m}\frac{E'}{E}\left(2W_1^{(\nu,\bar{\nu})}\sin^2\frac{\theta}{2} + W_2^{(\nu,\bar{\nu})}\cos^2\frac{\theta}{2} \mp W_3^{(\nu,\bar{\nu})}\frac{E+E'}{m}\sin^2\frac{\theta}{2}\right) \tag{11-101}$$

The kinematical variables ν and q^2 are related to the incident neutrino energy E, final lepton energy E', and laboratory scattering angle θ through

$$\begin{aligned} \nu &= p \cdot q = m(E - E') \\ q^2 &= -4EE'\sin^2\frac{\theta}{2} \end{aligned} \tag{11-102}$$

The lepton mass has been neglected for high-energy experiments. This explains the disappearance of W_4, W_5, W_6 in (11-101) and the simple expression for the momentum transfer. Such processes will be studied in more detail in Chap. 13.

For fixed q^2 and ν, when E (hence E') grows, θ tends to zero and the cross section may be approximated by the W_2 contribution alone. Therefore

$$\lim_{E\to\infty}\frac{d\sigma^{(\nu,\bar{\nu})}}{d|q^2|} = \frac{G^2}{2\pi m}\int_{-q^2/2}^{\infty} d\nu\, W_2^{(\nu,\bar{\nu})}(\nu, q^2) \tag{11-103}$$

Crossing symmetry implies

$$W_i^{(\nu)}(\nu, q^2) = -W_i^{(\bar{\nu})}(-\nu, q^2) \tag{11-104}$$

Current algebra then allows us to write the Adler sum rule

$$\lim_{E_1\to\infty}\left(\frac{d\sigma^{(\nu)}}{d|q^2|} - \frac{d\sigma^{(\bar{\nu})}}{d|q^2|}\right) = \frac{G^2}{\pi}\left[2T_3\cos^2\theta_c + \left(T_3 + \frac{3Y}{2}\right)\sin^2\theta_c\right] \tag{11-105}$$

with T_3 the third component of isospin, Y the hypercharge on the target, and θ_c the Cabibbo angle. An analogous result applies to the inclusive inelastic electron scattering $e + N \to e + X$ using the vector current V^{1+i2}. Since current algebra does not apply to the isoscalar part of the electromagnetic current the prediction obtained by Bjorken is only an inequality for the sum over neutron and proton targets:

$$\lim_{E_1 \to \infty} \left(\frac{d\sigma^{en}}{d|q^2|} + \frac{d\sigma^{ep}}{d|q^2|} \right) \geq \frac{2\pi\alpha^2}{(q^2)^2} \tag{11-106}$$

11-3-2 Approximate Conservation of the Axial Current and Chiral Symmetry

An $SU(2) \times SU(2)$ Lie algebra is generated by the commutation rules of vector and axial charges. Assuming that axial currents are approximately conserved, the resulting symmetry is called chiral symmetry. It is implemented in the Goldstone mode with the pions as massless particles. This hypothesis enables us to deduce new consequences from current algebra in the form of sum rules and low-energy theorems. The generalization to $SU(3) \times SU(3)$ including strangeness changing currents appears more questionable.

Consider the axial current matrix element between a pion state and the vacuum

$$\langle 0 | A^j_\mu(x) | \pi^k(p) \rangle = ip_\mu \delta^{jk} f_\pi e^{-ip \cdot x} \tag{11-107}$$

This amplitude determines the rate of the decay $\pi \to \mu^- \nu$ as

$$\Gamma_{\pi \to \mu\nu} = \frac{G^2 m_\mu^2 f_\pi^2 (m_\pi^2 - m_\mu^2)^2}{4\pi m_\pi^3} \cos^2 \theta_c \tag{11-108}$$

Experimentally, f_π is measured as

$$f_\pi \simeq 93 \text{ MeV} \tag{11-109}$$

From (11-107) the matrix element of $\partial^\mu A_\mu(x)$ is

$$\langle 0 | \partial^\mu A^j_\mu(x) | \pi^k(p) \rangle = m_\pi^2 \delta^{jk} f_\pi e^{-ip \cdot x} \tag{11-110}$$

Therefore current conservation implies $f_\pi m_\pi^2 = 0$. We have the choice between two possibilities. Either f_π or m_π^2 is zero, in contradiction to experimental facts. Nevertheless, let us boldly imagine a world with massless pions as a tentative approximation. If this $SU(2) \times SU(2)$ group is implemented with an invariant vacuum we are led to an unrealistic world with parity degenerate multiplets and corresponding selection rules. On the other hand, a Goldstone realization is in agreement with the fact that pions have a mass much smaller than other mesons.

Let us therefore explore the consequences of setting $\partial^\mu A_\mu = 0$ and $m_\pi^2 = 0$. Compute the axial current matrix element between nucleon states

$$\langle N(p_2) | A^j_\mu(0) | N(p_1) \rangle = \bar{u}(p_2) \frac{\tau^j}{2} [\gamma_\mu \gamma_5 g_A(q^2) + q_\mu \gamma_5 h_A(q^2)] u(p_1) \tag{11-111}$$

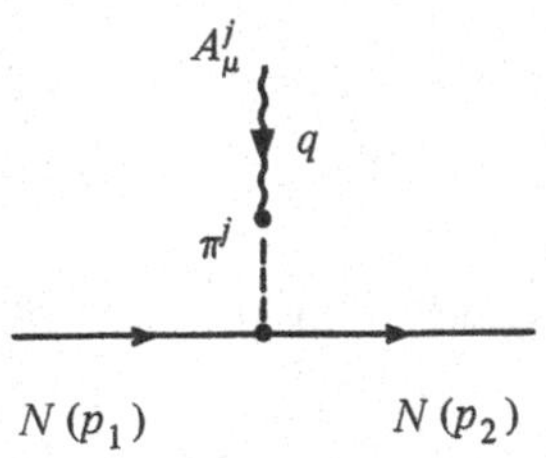

Figure 11-9 Pion contribution to the axial current matrix element between nucleon states.

with $q = p_2 - p_1$. Comparing with (11-69) we find

$$g_A(0) = G_A/G_V \simeq 1.22 \tag{11-112}$$

Current conservation implies

$$2mg_A(q^2) + q^2 h_A(q^2) = 0 \tag{11-113}$$

from which it is not wise to conclude that $g_A(0) = 0$. Indeed, the form factor $h(q^2)$ has a pole at $q^2 = 0$ corresponding to pion exchange (Fig. 11-9). The corresponding contribution reads

$$if_\pi \frac{q_\mu}{q^2} ig_{\pi NN}\bar{u}(p_2)\gamma_5\tau^j u(p_1) \tag{11-114}$$

with $g_{\pi NN}$ the pion-nucleon effective coupling constant (Sec. 5-3-4) approximately given by

$$\frac{g^2_{\pi NN}}{4\pi} = 14.6 \tag{11-115}$$

From (11-113) and (11-114) we obtain at zero transfer the Goldberger-Treiman relation

$$g_A(0) = \frac{G_A}{G_V} = g_{\pi NN}\frac{f_\pi}{m} \tag{11-116}$$

in agreement with the experimental data, within a 10 percent error [the right-hand side of (11-116) is 1.34 to be compared with 1.22]. This is a quite remarkable result since it relates quantities from strong ($g_{\pi NN}$) and weak (f_π, G_A/G_V) interactions. In the real world with massive pions we would like to have some control over the extrapolation from the value $q^2 = 0$ to $q^2 = m_\pi^2$. This is achieved by assuming a partial conservation law (PCAC)

$$\partial^\mu A^j_\mu(x) = m^2_\pi f_\pi \pi^j(x) \tag{11-117}$$

which identifies the divergence of the current with a smooth interpolating pion field. We know already that $\partial^\mu A_\mu$ has the correct quantum numbers. From Eq. (11-110) it may indeed be used to generate the asymptotic pion states if $m^2_\pi f_\pi \neq 0$. Equation (11-117) is supplemented by an hypothesis of a smooth extrapolation of form factors beyond the mass shell. This means that matrix elements of $\partial^\mu A_\mu$ are dominated by the pion pole for small values of the transfer momentum $|q^2| \lesssim m_\pi^2$:

$$D(q^2) = \langle A|\partial^\mu A_\mu|B\rangle = \frac{C}{q^2 - m_\pi^2} + \cdots \qquad (11\text{-}118)$$

where C is a residue to be determined. This is what is meant in practice by partial conservation of the axial current.

11-3-3 Low-Energy Theorems and Sum Rules

With current algebra and PCAC, we are in a position to derive low-energy theorems. The situation is analogous to the one prevailing in electrodynamics. We shall therefore begin with a study of low-energy Compton scattering.

To simplify we write the amplitude for photon scattering on a spinless target of unit charge (Fig. 11-10):

$$(2\pi)^4\delta^4(p_1 + k_1 - p_2 - k_2)\mathscr{T} = ie^2 \int d^4x\, d^4y\, \varepsilon_2^\mu \varepsilon_1^\nu\, e^{i(k_2\cdot x - k_1\cdot y)} \langle p_2|\, Tj_\mu(x)j_\nu(y)\,|p_1\rangle$$

$$\begin{aligned}\mathscr{T} &= ie^2 \int d^4x\, \varepsilon_2^\mu \varepsilon_1^\nu\, e^{ik_2\cdot x} \langle p_2|\, Tj_\mu(x)j_\nu(0)\,|p_1\rangle \qquad (11\text{-}119)\\ &= e^2 \varepsilon_2^\mu \varepsilon_1^\nu T_{\mu\nu}\end{aligned}$$

Choosing for convenience the polarization vectors such that $\varepsilon_1 \cdot k_1 = \varepsilon_2 \cdot k_2$ and taking Lorentz and time-reversal invariances into account, we write the general form of $T_{\mu\nu}$ as

$$T_{\mu\nu} = Ag_{\mu\nu} + BP_\mu P_\nu + C(P_\nu k_{1\mu} + P_\mu k_{2\nu}) + Dk_{1\mu}k_{2\nu} \qquad (11\text{-}120)$$

Here P stands for the average target momentum, $P = (p_1 + p_2)/2$, and the scalar amplitudes A, B, C, and D are free of kinematical singularities. From current conservation it follows that $k_2^\mu T_{\mu\nu} = 0$. Therefore

$$\begin{aligned} A + P\cdot k_2 C + k_1\cdot k_2 D &= 0 \\ P\cdot k_2 B + k_1 \cdot k_2 C &= 0 \end{aligned} \qquad (11\text{-}121)$$

In the low-energy limit k_1, $k_2 \to 0$, the amplitudes B, C, and D are dominated by their dynamical pole terms given by Born terms with renormalized residues.

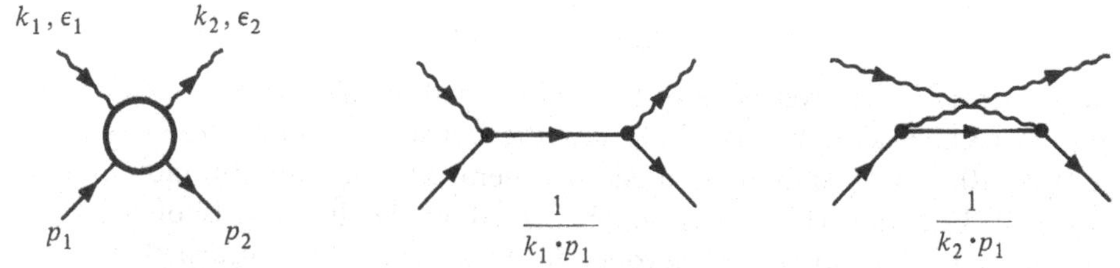

Figure 11-10 Compton amplitude and Born terms.

An elementary computation leads to the value of the amplitude A at threshold

$$\lim_{k_1,k_2\to 0} A = 2 \tag{11-122}$$

In this limit, but without any perturbative approximation, we therefore find for the cross section

$$\lim_{k_1,k_2\to 0} \frac{d\sigma}{d\Omega} = \frac{\alpha^2}{m^2}(\varepsilon_1 \cdot \varepsilon_2)^2 \tag{11-123}$$

in agreement with the classical evaluation of Chap. 1 and the lowest-order calculation (Sec. 5-2-1). Low, Gell-Mann, and Goldberger who first developed these arguments obtained from the same hypothesis the next term linear in k_1 and k_2 for spin $\frac{1}{2}$ targets. The amplitude $\mathscr{T}$ for forward scattering is written (with ω the common laboratory energy $\omega = p \cdot k/m$)

$$\mathscr{T} = f_1(\omega^2)\varepsilon_1 \cdot \varepsilon_2 + i\omega f_2(\omega^2)(\varepsilon_2 \times \varepsilon_1) \cdot \boldsymbol{\sigma} \tag{11-124}$$

The polarizations are now taken to have vanishing time components. The low-energy theorem then states

$$\begin{aligned} f_1(0) &= -\frac{e^2}{m} \\ f_2(0) &= -\frac{e^2}{8m^2}(g-2)^2 \end{aligned} \tag{11-125}$$

The amplitude f_2 involves the anomalous part of the magnetic moment of the target ($g - 2 = 3.58$ for the proton). The first relation is of course equivalent to (11-122). This prediction is difficult to test directly. It is better to transform it into a sum rule as suggested by Drell and Hearn. We use an unsubtracted dispersion relation for $f_2(\omega^2)$ [the corresponding one for $f_1(\omega^2)$ would require at least one subtraction] to write

$$f_2(\omega^2) = \frac{1}{\pi}\int_0^\infty d\omega'^2 \, \frac{\operatorname{Im} f_2(\omega'^2)}{\omega'^2 - \omega^2} = \frac{1}{2\pi}\int d\omega'^2 \, \frac{\sigma_-(\omega'^2) - \sigma_+(\omega'^2)}{\omega'^2 - \omega^2}$$

Here $\sigma_\pm(\omega^2)$ refer to the total cross section for a circularly polarized photon with its spin parallel or antiparallel to the spin of the target. The t-channel exchange implied by the helicity flip gives credence to the fact that the integral over the difference of cross sections stands a good chance of converging, hence justifying our assumption of an unsubtracted dispersion relation. From this relation we derive that

$$\frac{\alpha\pi^2(g-2)^2}{2m^2} = \int \frac{d\omega}{\omega}\,[\sigma_+(\omega) - \sigma_-(\omega)] \tag{11-126}$$

For protons the left-hand side is equal to 205 μb while the right-hand side is between 200 and 270 μb according to the treatment of high-energy data.

Let us apply similar techniques to amplitudes involving axial currents. This will give us some information on low-energy pion-nucleon scattering. We introduce the axial vector matrix element between initial nucleon and final pion-nucleon states:

$$(2\pi)^4\delta^4(p_1 + q_1 - p_2 - q_2)T_\mu^{jk} = \int d^4x\, e^{-iq_1\cdot x}\langle\pi^j(q_2)N(p_2)|\,A_\mu^k(x)\,|N(p_1)\rangle \tag{11-127}$$

According to (11-117) this is related to the pion-nucleon amplitude $\mathscr{T}_{\pi N}$. To see this we contract both sides of (11-127) with q_1 and get

$$\begin{aligned}(2\pi)^4\delta^4(p_1 + q_1 - p_2 - q_2)q_1^\mu T_\mu^{jk} &= -i\int d^4x\, e^{-iq_1\cdot x}\langle\pi^j(q_2)N(p_2)|\,\partial^\mu A_\mu^k(x)\,|N(p_1)\rangle \\ &= -i\frac{m_\pi^2}{m_\pi^2 - q_1^2} f_\pi \int d^4x\, e^{-iq_1\cdot x} \\ &\quad\times(\Box_x + m_\pi^2)\langle\pi^j(q_2)N(p_2)|\,\pi^k(x)\,|N(p_1)\rangle\end{aligned}$$

Hence

$$q_1^\mu T_\mu^{jk} = -if_\pi\frac{m_\pi^2}{m_\pi^2 - q_1^2}\mathscr{T}_{\pi N}^{jk}(p_2, q_2; p_1, q_1) \tag{11-128}$$

When q_1 goes to zero the left-hand side vanishes unless T_μ^{jk} has a singularity. The only one arises from the nucleon pole (Fig. 11-11). But contrary to what happened in the Compton case this singularity is compensated by the vanishing of the numerator and $q_1^\mu T_\mu^{jk} \sim q_1^2/q_1\cdot p_1 \to 0$ when $q_1 \to 0$.

This leads to the Adler compatibility condition for the extrapolated amplitude

$$\lim_{q_1\to 0}\mathscr{T}_{\pi N}^{jk}(p_2, q_2; p_1, q_1) = 0 \tag{11-129}$$

This soft pion limit is of course unphysical. An alternative derivation assumes directly $m_\pi = 0$ and $\partial_\mu A^\mu = 0$. From $q_\mu^1 T^\mu = 0$ we find (11-129) by isolating the pion pole from other singularities.

If we stick to this slightly unrealistic world with zero-mass pions, we can write a Ward identity similar to (11-89) for the amplitude

$$T_{\mu\nu}^{jk} = \int d^4x\, e^{iq\cdot x}\langle H(p_2)|\,TA_\mu^j(x)A_\nu^k(0)\,|H(p_1)\rangle \tag{11-130}$$

with H an arbitrary hadronic state.

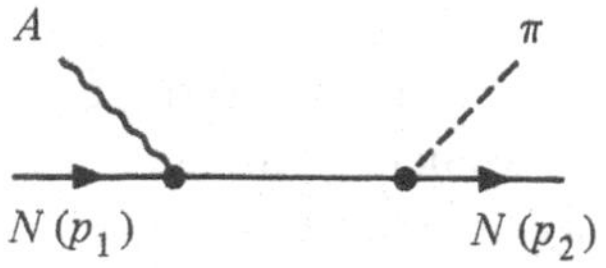

Figure 11-11 Nucleon pole contribution to the matrix element $\langle\pi N|A|N\rangle$.

It follows that

$$q^{\mu}T_{\mu\nu}^{jk} = i\int d^4x\, e^{iq\cdot x}\langle H(p_2)|\,\delta(x^0)[A_0^j(x), A_\nu^k(0)]\,|H(p_1)\rangle$$

$$= -\varepsilon^{jkl}\int d^4x\, e^{iq\cdot x}\,\delta^4(x)\langle H(p_2)|\,V_\nu^l(0)\,|H(p_1)\rangle$$

$$= -\varepsilon^{jkl}(p_1^\nu + p_2^\nu)T_H^l \tag{11-131}$$

We have used current algebra and the fact that V_μ^l is the isospin current, so that T_H^l stands for the isospin of the hadron. When $p_1 = p_2$, $T^{\mu\nu}$ has a double pole at $q^2 = 0$, with a residue proportional to the πH scattering amplitude at threshold

$$T_{\mu\nu}^{jk} = i\,\frac{q_\mu q_\nu}{(q^2)^2}\,f_\pi^2\mathscr{T}_{\pi H}^{jk} + \cdots \tag{11-132}$$

Consequently, the threshold value of $\mathscr{T}_{\pi H}$ (in the real world when $p\cdot q = m_\pi m_H$) is given by

$$\mathscr{T}_{\pi H}\Big|_{\text{threshold}} \simeq -\frac{m_\pi m_H}{f_\pi^2}\langle 2\mathbf{T}_H\cdot\mathbf{T}_\pi\rangle \tag{11-133}$$

If T stands for the total isospin in the s channel we have the relation

$$\langle 2\mathbf{T}_H\cdot\mathbf{T}_\pi\rangle = T(T+1) - T_H(T_H+1) - 2 \tag{11-134}$$

This result may be reexpressed in terms of the s-wave scattering lengths in the various isospin channels defined as

$$\mathscr{T}_{\pi H}^{T}\Big|_{\text{threshold}} = 8\pi(m_H + m_\pi)a_T \tag{11-135}$$

For pion-nucleon scattering this yields the values

$$a_{1/2} = 0.166m_\pi^{-1} \qquad a_{3/2} = -0.083m_\pi^{-1} \qquad a_{1/2} + 2a_{3/2} = 0 \tag{11-136}$$

in surprising agreement with the measured values

$$a_{1/2}^{\text{exp}} = (0.171 \pm 0.005)m_\pi^{-1} \qquad a_{3/2}^{\text{exp}} = -(0.088 \pm 0.004)m_\pi^{-1} \tag{11-137}$$

Similar successful calculations may be performed for pion-pion or pion-kaon scattering.

We can also reexpress these low-energy theorems in the form of sum rules using dispersion relations. We split the pion-nucleon amplitude in even or odd parts under crossing (Sec. 5-3-4):

$$\mathscr{T}_{\pi N}^{jk} = \mathscr{T}^{+}\delta^{jk} + i\varepsilon^{jkl}\tau^l\mathscr{T}^{-} \tag{11-138}$$

and use the variables $\nu = p\cdot q$ and transfer t. Phenomenological considerations predict the behavior of these amplitudes for large ν at $t = 0$:

$$\mathscr{T}^{+} \underset{\nu\to\infty}{\sim} \nu(\ln\nu)^{\alpha} \qquad \mathscr{T}^{-} \underset{\nu\to\infty}{\sim} \nu^{0.5}$$

Therefore it is likely that $\mathcal{T}^-(\nu, 0)/\nu$ satisfies an unsubtracted dispersion relation. The sum rule derived below will then provide a consistency check of this assumption. With the pole contribution of the nucleon intermediate state at $2\nu + m_\pi^2 = 0$, the forward dispersion relation reads

$$\frac{\mathcal{T}^-(\nu, 0)}{\nu} = \frac{m_\pi^2 g_{\pi NN}^2}{\nu^2 - m_\pi^4/4} + \frac{2}{\pi}\int_{m_\pi m_N}^{\infty} d\nu' \frac{\operatorname{Im}\mathcal{T}^-(\nu')}{\nu'^2 - \nu^2} \tag{11-139}$$

From (11-133) we obtain the value at threshold (neglecting m_π^2 as compared to m_N^2 in the Born term denominator):

$$\frac{1}{f_\pi^2} = \frac{g_{\pi NN}^2}{m_N^2} + \frac{2}{\pi}\int_{m_\pi m_N}^{\infty} \frac{d\nu}{\nu^2} \operatorname{Im}\mathcal{T}^-(\nu) + O\left(\frac{m_\pi^2}{m_N^2}\right) \tag{11-140}$$

Finally, we express $\operatorname{Im}\mathcal{T}$ in terms of the total cross sections with the help of the optical theorem and use the Goldberger-Treiman relation to substitute G_A/G_V for f_π^2. This leads to the Adler-Weisberger sum rule

$$1 - \frac{G_V^2}{G_A^2} = \frac{2m_N^2}{\pi g_{\pi NN}^2}\int_{m_N m_\pi}^{\infty} \frac{d\nu}{\nu} [\sigma_{\text{tot}}^{\pi^+ p}(\nu) - \sigma_{\text{tot}}^{\pi^- p}(\nu)] \tag{11-141}$$

The experimental agreement is again very convincing. A numerical evaluation leads to a value of G_A/G_V between 1.16 and 1.24 to be compared to the β-decay value of 1.22 ± 0.02.

There is an impressive list of applications of current algebra and soft pions techniques to weak semileptonic or nonleptonic decays for which the reader is referred to the literature.

In conclusion, chiral symmetry is a good approximation in hadronic physics. An extension to a $SU(3) \times SU(3)$ group is not reliable as shown by the large K and η masses. Gell-Mann, Oakes, and Renner have given a phenomenological description of the $SU(3) \times SU(3)$ breaking by writing an effective hamiltonian for strong interactions in the form

$$H = H_0 + H_1 \tag{11-142}$$

The term H_1 responsible for $SU(3) \times SU(3)$ breaking would be of the form

$$H_1 = \varepsilon_0 u_0 + \varepsilon_8 u_8 \tag{11-143}$$

with u_0 transforming as a scalar under $SU(3)$ and u_8 as the eighth component of an octet. A phenomenological analysis shows that $\varepsilon_8/\varepsilon_0$ is close to $-\sqrt{2}$ rather than zero. This indicates that $SU(2) \times SU(2)$ rather than $SU(3)$ would be a better approximate symmetry.

11-4 THE σ MODEL

We shall now describe a field theoretic model originally introduced by Gell-Mann and Lévy (1960) as an example realizing chiral symmetry and partial conservation of the axial current. The name σ model originates in one of their notations. We take this opportunity to study the interplay of renormalization and symmetries. Following the ideas of Lee and Symanzik this enables us to introduce the machinery of Ward identities which will be useful in the study of gauge fields.

11-4-1 Description of the Model

The σ model involves a fermionic isodoublet field ψ of zero bare mass, a triplet of pseudoscalar pions, and a scalar field σ. The corresponding lagrangian is written

$$\begin{aligned}\mathscr{L} &= \mathscr{L}_S + c\sigma \\ \mathscr{L}_S &= \bar{\psi}[i\partial\!\!\!/ + g(\sigma + i\boldsymbol{\pi}\cdot\boldsymbol{\tau}\gamma_5)]\psi + \tfrac{1}{2}[(\partial\boldsymbol{\pi})^2 + (\partial\sigma)^2] \\ &\quad - \frac{\mu^2}{2}(\sigma^2 + \boldsymbol{\pi}^2) - \frac{\lambda}{4}(\sigma^2 + \boldsymbol{\pi}^2)^2\end{aligned} \tag{11-144}$$

It is usually referred to as the linear model for a reason to be discussed later on.

The part $\mathscr{L}_S$ (the index S is for symmetric) is invariant under an $SU(2) \times SU(2)$ chiral group acting as follows. The right and left combinations $\psi_R = \frac{1}{2}(1 + \gamma_5)\psi$, $\psi_L = \frac{1}{2}(1 - \gamma_5)\psi$ transform respectively according to the representations $(\frac{1}{2}, 0)$ and $(0, \frac{1}{2})$ while the set $(\sigma, \boldsymbol{\pi})$ belongs to the $(\frac{1}{2}, \frac{1}{2})$ representation.

To see this, we rewrite the coupling term

$$\bar{\psi}(\sigma + i\boldsymbol{\pi}\cdot\boldsymbol{\tau}\gamma_5)\psi = \bar{\psi}_L(\sigma + i\boldsymbol{\pi}\cdot\boldsymbol{\tau})\psi_R + \bar{\psi}_R(\sigma - i\boldsymbol{\pi}\cdot\boldsymbol{\tau})\psi_L$$

If (U, V) stands for an element of $SU(2) \times SU(2)$ with U and V varying independently, the transformation $(\sigma + i\boldsymbol{\pi}\cdot\boldsymbol{\tau}) \to V(\sigma + i\boldsymbol{\pi}\cdot\boldsymbol{\tau})U^{-1}$ is an allowed one, leading to real σ' and $\boldsymbol{\pi}'$ fields. Therefore if we perform simultaneously the isospinor rotations $\psi_R \to U\psi_R$, $\psi_L \to V\psi_L$ the interaction term is obviously invariant. So is the quantity $\sigma^2 + \boldsymbol{\pi}^2$ proportional to the determinant of $(\sigma + i\boldsymbol{\pi}\cdot\boldsymbol{\tau})$. Finally, the kinetic term $\bar{\psi}i\partial\!\!\!/\psi$ is equal to $\bar{\psi}_R i\partial\!\!\!/\psi_R + \bar{\psi}_L i\partial\!\!\!/\psi_L$, showing the invariance of the full lagrangian.

The corresponding infinitesimal variations are generated by the chiral charges $Q^\alpha_{R,L}$ with α running from one to three:

$$\begin{aligned}
&[Q^\alpha_R, \psi_R] = -\tfrac{1}{2}\tau^\alpha\psi_R && [Q^\alpha_L, \psi_L] = -\tfrac{1}{2}\tau^\alpha\psi_L \\
&[Q^\alpha_R, \psi_L] = 0 && [Q^\alpha_L, \psi_R] = 0 \\
&[Q^\alpha_R, \sigma] = \frac{i}{2}\pi^\alpha && [Q^\alpha_L, \sigma] = -\frac{i}{2}\pi^\alpha \\
&[Q^\alpha_R, \pi^\beta] = -\frac{i}{2}\delta^{\alpha\beta}\sigma + \frac{i}{2}\varepsilon^{\alpha\beta\gamma}\pi^\gamma && [Q^\alpha_L, \pi^\beta] = \frac{i}{2}\delta^{\alpha\beta}\sigma + \frac{i}{2}\varepsilon^{\alpha\beta\gamma}\pi^\gamma
\end{aligned} \tag{11-145}$$

A compact notation for the $(\sigma, \boldsymbol{\pi})$ transformations is

$$\begin{aligned}
[Q^\alpha_R, (\sigma + i\boldsymbol{\pi}\cdot\boldsymbol{\tau})] &= (\sigma + i\boldsymbol{\pi}\cdot\boldsymbol{\tau})\frac{\tau^\alpha}{2} \\
[Q^\alpha_L, (\sigma + i\boldsymbol{\pi}\cdot\boldsymbol{\tau})] &= -\frac{\tau^\alpha}{2}(\sigma + i\boldsymbol{\pi}\cdot\boldsymbol{\tau})
\end{aligned} \tag{11-146}$$

a form which makes the invariance properties of the lagrangian easy to check. The Lie algebra of $SU(2) \times SU(2)$ is isomorphic to the one of its factor group $O(4) = SU(2) \times SU(2)/Z_2$ and $(\sigma, \boldsymbol{\pi})$ transforms as a vector under $O(4)$.

In the absence of the breaking term $c\sigma$, the vector and axial currents are conserved. They read

$$V_\mu^\alpha = \overline{\psi}\gamma_\mu \frac{\tau^\alpha}{2}\psi + \varepsilon^{\alpha\beta\gamma}\pi^\beta\partial_\mu\pi^\gamma$$
$$A_\mu^\alpha = \overline{\psi}\gamma_\mu\gamma_5 \frac{\tau^\alpha}{2}\psi + (\sigma\partial_\mu\pi^\alpha - \pi^\alpha\partial_\mu\sigma) \tag{11-147}$$

The breaking term $c\sigma$ leaves the diagonal $SU(2)$ group unbroken. The vector current remains conserved and the axial current, the expression of which is not modified, acquires a nonvanishing divergence

$$\partial^\mu A_\mu^\alpha = -c\pi^\alpha \tag{11-148}$$

This model therefore incorporates all the desirable features and the divergence of the axial current appears naturally as proportional to the pion field.

The linear breaking term implies that the quantum σ field has a nonvanishing vacuum expectation value $\langle 0|\sigma|0\rangle = v$, meaning that a perturbation expansion must be performed taking into account the fluctuations of this field around the value v instead of zero. This is achieved by shifting the field as

$$\sigma' = \sigma - v$$

and requiring that σ' has zero vacuum expectation value. Reexpressed in terms of σ' the complete lagrangian reads

$$\begin{aligned}\mathscr{L} = {} & \overline{\psi}[i\partial\!\!\!/ + gv + g(\sigma' + i\boldsymbol{\pi}\cdot\boldsymbol{\tau}\gamma_5)]\psi + \tfrac{1}{2}[(\partial\boldsymbol{\pi})^2 + (\partial\sigma')^2] \\ & - \tfrac{1}{2}(\mu^2 + 3\lambda v^2)\sigma'^2 - \tfrac{1}{2}(\mu^2 + \lambda v^2)\boldsymbol{\pi}^2 - \lambda v\sigma'(\sigma'^2 + \boldsymbol{\pi}^2) \\ & - \frac{\lambda}{4}(\boldsymbol{\pi}^2 + \sigma'^2)^2 + \sigma'(c - \mu^2 v - \lambda v^3)\end{aligned} \tag{11-149}$$

The effect of the translation has been threefold. The mass degeneracy of meson fields has disappeared. On inspection these masses now read

$$m_\pi^2 = \mu^2 + \lambda v^2$$
$$m_\sigma^2 = \mu^2 + 3\lambda v^2 \tag{11-150}$$

Furthermore, the fermion has acquired a mass equal to

$$m_N = -gv \tag{11-151}$$

Finally we find a new trilinear coupling $\sigma'\pi\pi$.

The vacuum expectation value v will be constrained to satisfy a complicated condition $\langle\sigma'\rangle = 0$. The best which can be done is to implement it perturbatively by requiring that tadpole diagrams for the transition $\sigma' \to$ vacuum vanish (Fig. 11-12). The Born approximation to this condition is, from (11-149),

$$c - \mu^2 v - \lambda v^3 = 0$$

Figure 11-12 Tadpole amplitude.

If we return to Eq. (11-117) we are led to the following identification:

$$f_\pi m_\pi^2 = -c = -v(\mu^2 + \lambda v^2) = -vm_\pi^2$$

Therefore,

$$\begin{aligned} f_\pi &= -v \\ m_N &= gf_\pi \end{aligned} \tag{11-152}$$

which is the Goldberger-Treiman relation to this order where $G_A/G_V = 1$.

When c goes to zero two distinct situations may arise. One possibility is that v also goes to zero, in which case we find the normal mode of symmetry described above with massless nucleons. An alternative possibility arises when $\mu^2 < 0$, and the limit corresponds to

$$v^2 = -\frac{\mu^2}{\lambda} \tag{11-153}$$

This is the Goldstone phenomenon studied in Sec. 11-2-2, with a vanishing pion mass

$$m_\pi^2 = \mu^2 + \lambda v^2 = 0$$

In this phase the σ model may be used to derive the low-energy theorems for pion-pion or pion-nucleon scattering.

11-4-2 Renormalization

The lagrangian $\mathscr{L}(\sigma', \pi, \psi)$ obtained above after translation of the σ field, or its version in the limit $c = 0$ corresponding to the Goldstone mode, is renormalizable in the sense of power counting. All the monomials in the interaction lagrangian are of dimension smaller or equal to four; the same is true of the possible counterterms. It remains to show that the lagrangian plus its counterterms has a similar form as in (11-149), a remnant of the original structure (11-144). In particular, what will be the fate of the PCAC relation (11-148)? We shall show that these properties will be preserved by performing the renormalization in the symmetric normal phase and proving that this is sufficient to treat the cases of explicit ($c \neq 0$) or spontaneous ($c = 0$, $\mu^2 < 0$) symmetry breaking.

To simplify matters we omit the fermion fields. A complete treatment does not reveal any new difficulty. We also use compact notations with ϕ a multiplet of n fields transforming according to the vector representation of a symmetry group $O(n)$. In the previous instance n was equal to four. We write the lagrangian

$$\mathscr{L} = \mathscr{L}_S + \mathbf{c}\cdot\boldsymbol{\phi}$$
$$\mathscr{L}_S = \tfrac{1}{2}(\partial\boldsymbol{\phi})^2 - \frac{\mu^2}{2}\boldsymbol{\phi}^2 - \frac{\lambda}{4}(\boldsymbol{\phi}^2)^2 \tag{11-154}$$

$\mathscr{L}_S$ is therefore invariant under the transformations

$$\boldsymbol{\phi} \to \boldsymbol{\phi} + \delta\boldsymbol{\phi} \qquad \delta\phi_j = \delta\omega_\alpha T^\alpha_{jk}\phi_k \tag{11-155}$$

where the T^α_{jk} are the representatives of the infinitesimal generators of the group, in the present case n by n antisymmetric real matrices $T^\alpha_{jk} + T^\alpha_{kj} = 0$.

It is convenient first to regularize the theory in an invariant manner. This may be done, for instance, by modifying the kinetic term into the form

$$\int d^4x\,(\partial\boldsymbol{\phi})^2 = \int d^4x\,\boldsymbol{\phi}(-\Box)\boldsymbol{\phi} \to -\int d^4x\,\boldsymbol{\phi}\Box\left(1 + a\frac{\Box}{\Lambda^2} + b\frac{\Box^2}{\Lambda^4} + \cdots\right)\boldsymbol{\phi}$$

leading to a propagator with a behavior smooth enough at large momentum to insure the convergence of all Feynman integrals. This regularization will be understood in the sequel without being explicitly written out.

Consider now the generating functional for connected Green functions in the symmetric theory

$$e^{G_S(\mathbf{j})} = \int \mathscr{D}(\boldsymbol{\phi}) \exp\left\{i\int d^4x\,[\mathscr{L}_S(\boldsymbol{\phi}) + \mathbf{j}\cdot\boldsymbol{\phi}]\right\} \tag{11-156}$$

As a consequence of the invariance of the (regularized) lagrangian under the transformations (11-155), $G_S(\mathbf{j})$ satisfies

$$G_S(\mathbf{j}) = G_S[\mathbf{j} + \delta\omega(T\mathbf{j})] \tag{11-157a}$$

or, equivalently,

$$\int d^4x\, j_k(x)T^\alpha_{kl}\frac{\delta G_S(\mathbf{j})}{\delta j_l(x)} = 0 \tag{11-157b}$$

In order to exhibit the structure of the divergences we need a similar identity for irreducible Green functions obtained after the Legendre transformation

$$i\Gamma_S(\boldsymbol{\phi}) + i\int d^4x\,\mathbf{j}\cdot\boldsymbol{\phi} = G_S(\mathbf{j})$$
$$\phi_k(x) = \frac{\delta G_S(\mathbf{j})}{i\delta j_k(x)} \tag{11-158}$$

Since conversely

$$j_k(x) = -\frac{\delta\Gamma_S(\boldsymbol{\phi})}{\delta\phi_k(x)}$$

we derive that $\Gamma_S(\boldsymbol{\phi})$ enjoys the same invariance properties under the transformations (11-155):

$$\int d^4x\, \phi_k(x) T^{\alpha}_{kl} \frac{\delta \Gamma_S(\phi)}{\delta \phi_l(x)} = 0 \tag{11-159}$$

Consequently all counterterms required by renormalization have to be symmetric, provided that the normalization conditions are. Indeed, assume the property true up to the L-loop order. Since $\mathscr{L}_S$ was the most general symmetric polynomial of degree four, the effect of counterterms could only renormalize the mass μ and the coupling constant λ, and affect multiplicatively (all components of) the field by the wave-function renormalization factor Z. Since $\mathscr{L}_S + \Delta\mathscr{L}_S^{(L)}$ enjoys the same invariance as $\mathscr{L}_S$ did, it follows that Γ computed with it up to order $(L+1)$ fulfils the same constraint (11-159). Its divergent part of order $(L+1)$ is the local symmetric polynomial generating the counterterms of order $(L+1)$. This inductive proof appears pedantic here. In more elaborate cases the parametrization of the transformation may be modified order by order by renormalization, but the framework described above remains useful.

Consider now the complete lagrangian $\mathscr{L}$ including the linear breaking term $\mathbf{c} \cdot \boldsymbol{\phi}$. A translation on $\boldsymbol{\phi}$:

$$\boldsymbol{\phi} = \boldsymbol{\phi}' + \mathbf{v} \qquad \langle \boldsymbol{\phi} \rangle = \mathbf{v} \tag{11-160}$$

is necessary in order to generate the correct perturbation theory. Omitting the prime, we have therefore

$$e^{G(\mathbf{j})} = \frac{\int \mathscr{D}(\phi) \exp\{i \int d^4x\, [\mathscr{L}_S(\boldsymbol{\phi} + \mathbf{v}) + \boldsymbol{\phi} \cdot (\mathbf{j} + \mathbf{c})]\}}{\int \mathscr{D}(\phi) \exp\{i \int d^4x\, [\mathscr{L}_S(\boldsymbol{\phi} + \mathbf{v}) + \boldsymbol{\phi} \cdot \mathbf{c}]\}} \tag{11-161}$$

where the denominator insures that $G(0) = 0$. Since $\boldsymbol{\phi}$ is a dummy variable we may undo the translation

$$e^{G(\mathbf{j})} = \frac{\int \mathscr{D}(\phi) \exp\{i \int d^4x\, [\mathscr{L}_S(\boldsymbol{\phi}) + (\boldsymbol{\phi} - \mathbf{v}) \cdot (\mathbf{j} + \mathbf{c})]\}}{\int \mathscr{D}(\phi) \exp\{i \int d^4x\, [\mathscr{L}_S(\boldsymbol{\phi}) + (\boldsymbol{\phi} - \mathbf{v}) \cdot \mathbf{c}]\}} \tag{11-162}$$

and express $G(\mathbf{j})$ in terms of the symmetric functional as

$$G(\mathbf{j}) = G_S(\mathbf{j} + \mathbf{c}) - G_S(\mathbf{c}) - i \int d^4x\, \mathbf{j} \cdot \mathbf{v} \tag{11-163}$$

where $\mathbf{v}$ is determined by the condition that the derivative of $G(\mathbf{j})$ with respect to $\mathbf{j}$ vanishes at $\mathbf{j} = 0$:

$$v_k = \left. \frac{\delta G_S(\mathbf{j})}{i \delta j_k(x)} \right|_{\mathbf{j}(x) = \mathbf{c}} \tag{11-164}$$

Similarly,

$$\begin{aligned} i\Gamma(\boldsymbol{\phi}) &= G(\mathbf{j}) - i \int d^4x\, \mathbf{j} \cdot \boldsymbol{\phi} \\ &= G_S(\mathbf{j} + \mathbf{c}) - G_S(\mathbf{c}) - i \int d^4x\, \mathbf{j} \cdot (\boldsymbol{\phi} + \mathbf{v}) \end{aligned} \tag{11-165}$$

with
$$\phi_k(x) = \frac{\delta G(\mathbf{j})}{i\delta j_k(x)} = \frac{\delta G_S(\mathbf{j}+\mathbf{c})}{i\delta j_k(x)} - v_k \qquad (11\text{-}166)$$

Thus

$$\Gamma(\boldsymbol{\phi}) = \Gamma_S(\boldsymbol{\phi}+\mathbf{v}) - \Gamma_S(\mathbf{v}) + \int d^4x\, \mathbf{c}\cdot\boldsymbol{\phi}(x)$$
$$c_k = -\frac{\delta\Gamma_S(\mathbf{v})}{\delta v_k} \qquad (11\text{-}167)$$

The identity (11-159) expressed in terms of $\Gamma(\boldsymbol{\phi})$ states that

$$\int d^4x\, [\phi_k(x) + v_k]\, T^\alpha_{kl}\left[\frac{\delta\Gamma(\boldsymbol{\phi})}{\delta\phi_l(x)} - c_l\right] = 0 \qquad (11\text{-}168)$$

Let us look in more detail at the content of (11-168). For instance, take one derivative with respect to $\boldsymbol{\phi}$ and set $\boldsymbol{\phi} = 0$:

$$T^\alpha_{kl}\left[\frac{\delta\Gamma(\boldsymbol{\phi})}{\delta\phi_l(x)} - c_l\right]\Bigg|_{\phi=0} + v_m T^\alpha_{ml}\int d^4y\, \frac{\delta^2\Gamma(\boldsymbol{\phi})}{\delta\phi_k(x)\delta\phi_l(y)}\Bigg|_{\phi=0} = 0 \qquad (11\text{-}169)$$

Taking into account that $\delta\Gamma(\boldsymbol{\phi})/\delta\phi_l(x)|_{\phi=0} = 0$, which follows from the definition of $\mathbf{v}$, and the antisymmetry of the matrices T, this means the following relation for the inverse propagator $\Gamma^{(2)}_{mk}(p^2)$ at zero momentum:

$$T^\alpha_{ml}[c_l\delta_{mk} + v_l\Gamma^{(2)}_{mk}(0)] = 0 \qquad (11\text{-}170)$$

This shows that $\mathbf{c}$ and $\mathbf{v}$ are collinear, a fact already clear from Eq. (11-167). If we think of $\mathbf{c}$ as an external magnetic field then the magnetization $\mathbf{v}$ is along $\mathbf{c}$. A mass parameter for the transverse states (with respect to $\mathbf{c}$) is defined according to

$$\Gamma^{(2)}_T(0) = -m_T^2 \qquad (11\text{-}171)$$

and
$$\mathbf{c} = m_T^2\mathbf{v} \qquad (11\text{-}172)$$

The quantity m_T plays the role of a pion mass in the σ model and generalizes a result obtained in Sec. 11-4-1. From (11-167), it follows that once the symmetric theory is renormalized so is the broken one. Explicitly, if

$$\Gamma_{SR}(\boldsymbol{\phi}, \mu^2, \lambda) = \Gamma_{S,\text{reg}}(Z^{1/2}\boldsymbol{\phi}, \mu^2 + \delta\mu^2, \lambda_0) \qquad (11\text{-}173)$$

the corresponding renormalized functional for the broken symmetry case is

$$\Gamma_R(\boldsymbol{\phi}, \mu^2, \lambda, \mathbf{c}) = \Gamma_{\text{reg}}(Z^{1/2}\boldsymbol{\phi}, \mu^2 + \delta\mu^2, \lambda_0, Z^{-1/2}\mathbf{c}) \qquad (11\text{-}174)$$

A rescaling of the symmetry breaking parameter $\mathbf{c}$ has been necessary:

$$\mathbf{c}_0 = Z^{-1/2}\mathbf{c} \qquad (11\text{-}175)$$

in such a way that $\mathbf{c}\cdot\boldsymbol{\phi}$ remains invariant, $\mathbf{c}\cdot\boldsymbol{\phi} = \mathbf{c}_0\cdot\boldsymbol{\phi}_0$, and insures that

$$\mathbf{v}_0 = Z^{1/2}\mathbf{v}$$
$$c_k = -\frac{\delta\Gamma_{SR}(\mathbf{v})}{\delta v_k} \tag{11-176}$$
$$c_{0k} = -\frac{\delta\Gamma_{S,\text{reg}}(\mathbf{v}_0)}{\delta v_{0k}}$$

Equation (11-167) implies that the amplitudes for the case with broken symmetry are obtained by resumming tadpole insertions in the symmetric theory. For instance, the p-point Green functions ($p > 1$) may be written as (Fig. 11-13)

$$\Gamma^{(p)}_{k_1,\ldots,k_p}(x_1,\ldots,x_p) = \sum_{n=0}^{\infty}\frac{v^n}{n!}\int d^4y_1\cdots d^4y_n\,\Gamma^{(n+p)}_{S,k_1,\ldots,k_p,1,\ldots,1}(x_1,\ldots,x_p,y_1,\ldots,y_n) \tag{11-177}$$

For compactness we have assumed $\mathbf{v}$ along the first axis in isotopic space.

This discussion of renormalization may also be carried out when the breaking terms are more complex and involve higher-dimension operators. The lesson to be drawn from the work of Symanzik is that breaking terms of dimension $\omega < 4$ only require counterterms of dimension lower or equal to ω. For instance, a mass-breaking term ($\omega = 2$) will not affect the counterterms of degree three or four which will remain symmetric. A soft breaking ($\omega < 4$) leaves a remnant of the initial symmetry. A hard breaking ($\omega = 4$) will a priori destroy completely the symmetry.

These properties on the ultraviolet divergences have their counterpart on the asymptotic behavior at large momenta of renormalized Green functions, at least in the euclidean region. This is another aspect of Weinberg's theorem (Sec. 8-3-2). Thus a soft breaking will not affect the asymptotic regime which will remain symmetric.

The reader may wonder what is the link between this discussion and Ward identities

$$\begin{aligned}\frac{\partial}{\partial x_\mu}&\langle Tj^\alpha_\mu(x)A_1(x_1)\cdots A_n(x_n)\rangle\\ &= \langle T\partial^\mu j^\alpha_\mu(x)A_1(x_1)\cdots A_n(x_n)\rangle + \sum_{p=1}^{n}\langle TA_1(x_1)\cdots\delta(x^0-x^0_p)[j^\alpha_0(x),A_p(x_p)]\cdots A_n(x_n)\rangle\end{aligned} \tag{11-178}$$

encountered in quantum electrodynamics or in the applications of current algebra. It is easy to convince oneself that the former are relations of the kind of (11-178) integrated over x. The fields $A_p(x_p)$ are identified with $\phi_{k_p}(x_p)$ and j is the Noether current $j^\alpha_\mu(x) = [\partial_\mu\phi_k(x)]T^\alpha_{kl}[\phi_l(x) + v_l]$. Upon integration the left-hand side vanishes as a total derivative without boundary contribution in the absence of massless states. On the right-hand side $\partial^\mu j^\alpha_\mu(x)$ is equal to $c_l T^\alpha_{lm}[\phi_m(x) + v_m]$ and the last term involves the variation of the field

$$\delta(x^0 - y^0)[j^\alpha_0(x), \phi_k(y)] = \delta^4(x-y)T^\alpha_{kl}[\phi_l(y) + v_l]$$

We therefore obtain the identity (11-168).

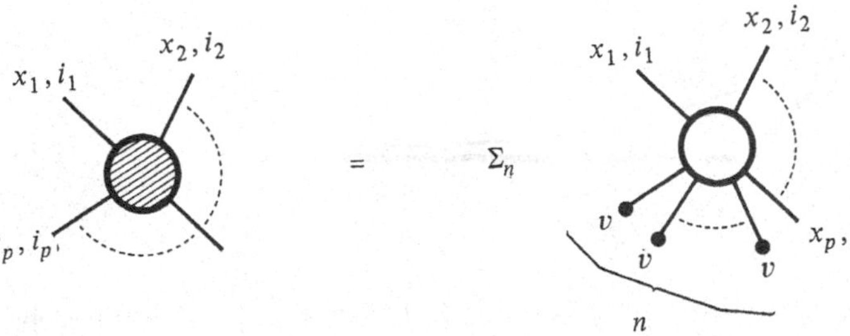

Figure 11-13 Summing tadpole diagrams.

We let the reader derive (11-178) from the functional integral (11-162) by observing that it is not affected by a change of variable $\phi \to \phi + \delta\phi$, $\delta\phi = (T^{\alpha}\phi)\delta\omega_{\alpha}(x)$ with an x-dependent $\delta\omega_{\alpha}(x)$. This dependence is responsible for a variation $\delta\mathscr{L}_S(\phi) = j\cdot\partial(\delta\omega)$ according to the definition of the current. A derivative with respect to $\delta\omega$ will yield Eq. (11-178). The renormalization of this identity may first be studied in the symmetric theory to show that the currents are not renormalized or, in other words, that $Z_j = 1$. Moreover, this property is not affected by a soft breaking term so that the multiplicative renormalization constant for the current remains equal to one. Are there new problems when we try to derive identities with several currents?

We are now in a position to discuss the interesting case with spontaneous symmetry breaking. At first sight it would seem reasonable to start with the renormalized symmetric theory with $\mu^2 > 0$ and then continue to the region $\mu^2 < 0$. The trouble is that this procedure involves a transition through a singular point. Physical quantities such that $m_T^2(\mu^2, c)$ will not be analytic in μ^2 for $c = 0$. To cope with this difficulty we may introduce a small linear breaking term ($c \neq 0$) to turn around the singularity. In Fig. 11-14 we show how the Goldstone phase may be reached along the path $\alpha\beta\gamma\delta$ by successively varying c and μ^2. From the previous analysis it follows that counterterms pertaining to the symmetric theory (point α)

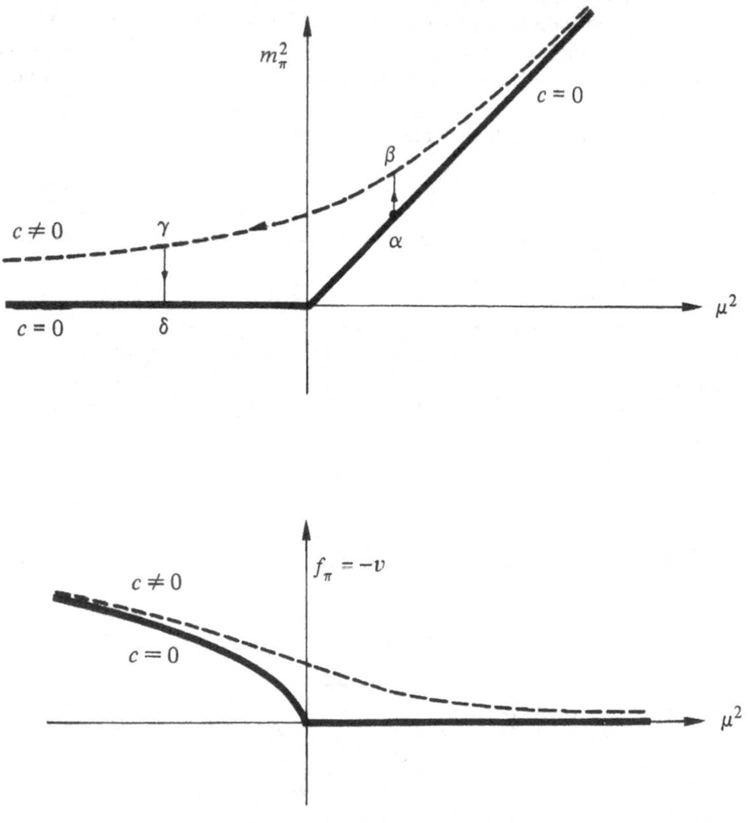

Figure 11-14 The behavior of m_π^2 and $f_\pi = -v$ as functions of μ^2 in the absence (solid line) and in the presence (broken line) of an external field c.

will insure the finiteness at the broken symmetry point β. Symmetric mass counterterms will lead from β to γ corresponding to the variation in μ^2. Finally, in the limit of vanishing c (point δ) the renormalized functions will satisfy the identity

$$\int d^4x \, [\phi_k(x) + v_k] \, T^\alpha_{kl} \frac{\delta\Gamma(\phi)}{\delta\phi_l(x)} = 0 \tag{11-179}$$

and the limit of Eq. (11-170) will read

$$v_l T^\alpha_{lm} \Gamma^{(2)}_{mk}(0) = 0 \tag{11-180}$$

This means that the $(n - 1)$ transverse bosons are massless, being the Goldstone bosons of the spontaneously broken symmetry.

Equation (11-179) implicitly contains all the relations among Green's functions in the Goldstone phase and implies a series of low-energy theorems.

The fact that the linear σ model and its Goldstone limit verify all the constraints of current algebra and PCAC even at the level of Born terms suggests the use of it as a phenomenological lagrangian, in the same spirit as the Fermi theory for weak interactions. We may in fact find a number of phenomenological interactions with this property. The spirit of this approach is to ignore renormalization problems and to make use of the relations arising among processes to lowest order.

An example is provided by the nonlinear σ model. The name originates from the nonlinear realization of the chiral group on the manifold

$$\sigma^2(x) + \boldsymbol{\pi}^2(x) = f_\pi^2 \tag{11-181}$$

Expressing σ in terms of the pion field the resulting lagrangian is written

$$\mathscr{L} = \tfrac{1}{2}(\partial\boldsymbol{\pi})^2 + \frac{1}{2} \frac{(\boldsymbol{\pi} \cdot \partial\boldsymbol{\pi})^2}{\sqrt{f_\pi^2 - \boldsymbol{\pi}^2}} \tag{11-182}$$

At the classical level we see that the chiral symmetry is realized in the Goldstone mode. The composite field $\sigma = \sqrt{f_\pi^2 - \boldsymbol{\pi}^2}$ has a nonvanishing value and the pion is massless.

Even though this theory is not renormalizable according to usual criteria in four-dimensional space-time, it has aroused great interest for its applications in statistical mechanics where it describes the continuous version of the classical Heisenberg model. Moreover, a theory such as (11-182) is renormalizable in dimension two where a scalar field is dimensionless and the Ward identities may be shown to constrain the structure of counterterms. It then enjoys a number of features in common with the recent field theories of strong interactions.

11-5 ANOMALIES

Up to now we have only encountered examples where renormalization did not conflict with the symmetries apparent at the classical level. In this section we discuss situations of great interest where the opposite is true, and new features are uncovered by renormalization. In Chap. 13 similar phenomena will appear when dealing with scale invariance. We introduce this subject with an apparent paradox in the application of current algebra.

11-5-1 The $\pi^0 \to 2\gamma$ Decay and Current Algebra

When reviewing the successes of current algebra in Sec. 11-3 we carefully omitted what seems at first a failure in its application to the π^0 decay in two photons (Fig. 11-15). Let us now study this process. The amplitude may be written

$$\mathscr{T} = \mathscr{T}(q^2)\big|_{q^2=m_\pi^2} = \lim_{q^2\to m_\pi^2} \varepsilon_1^\mu \varepsilon_2^\nu T_{\mu\nu}(q) \tag{11-183}$$

$$\mathscr{T}(q^2) = (q^2 - m_\pi^2)e^2\varepsilon_1^\mu\varepsilon_2^\nu \int d^4x\, d^4y\, \langle 0|\, Tj_\mu(x)j_\nu(y)\pi(0)\,|0\rangle\, e^{i(k_1\cdot x+k_2\cdot y)} \tag{11-184}$$

We have exhibited the time-ordered product of two electromagnetic currents coupled to the two photons (k_1, ε_1 and k_2, ε_2) and of the pion field $\pi(x)$. Here $q = k_1 + k_2$ and the photons are on-shell $k_1^2 = k_2^2 = 0$. In the spirit of PCAC (11-117) we are led to replace the neutral pion field by the divergence of the axial current, omitting the isospin index 3. Therefore,

$$\mathscr{T}(q^2) = \frac{m_\pi^2 - q^2}{f_\pi m_\pi^2}\, \varepsilon_1^\mu\varepsilon_2^\nu q^\rho T_{\mu\nu\rho} \tag{11-185}$$

with
$$T_{\mu\nu\rho} = -ie^2 \int d^4x\, d^4y\, e^{i(k_1\cdot x+k_2\cdot y)} \langle 0|\, Tj_\mu(x)j_\nu(y)A_\rho(0)\,|0\rangle \tag{11-186}$$

To extract the derivative acting on the axial current from the time-ordering symbol we have used the fact that the commutator $\delta(x^0 - y^0)[j_\mu(x), A_0(y)]$ vanishes for the quantum numbers entering (11-186). Let us now write the most general decomposition of the tensor $T_{\mu\nu\rho}(k_1, k_2)$. This involves taking into account a negative parity from the axial current, a symmetry in the combined exchange $(k_1, \mu) \leftrightarrow (k_2, \nu)$ from the Bose statistics of photons, and the transversity to k_1^μ and k_2^ν as a result of the electromagnetic current conservation

$$k_1^\mu T_{\mu\nu\rho} = k_2^\nu T_{\mu\nu\rho} = 0 \tag{11-187}$$

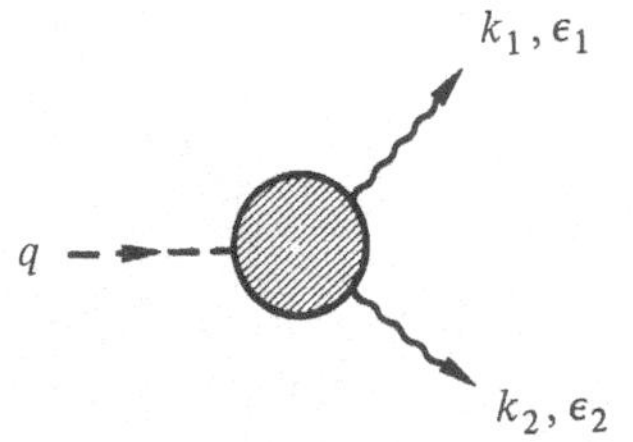

Figure 11-15 $\pi^0 \to 2\gamma$ decay.

With this information we may write the following expression when $k_1^2 = k_2^2 = 0$:

$$T_{\mu\nu\rho}(k_1, k_2) = \varepsilon_{\mu\nu\sigma\tau}k_1^\sigma k_2^\tau q_\rho T_1(q^2) + (\varepsilon_{\mu\rho\sigma\tau}k_{2\nu} - \varepsilon_{\nu\rho\sigma\tau}k_{1\mu})k_1^\sigma k_2^\tau T_2(q^2)$$
$$+ [(\varepsilon_{\mu\rho\sigma\tau}k_{1\nu} - \varepsilon_{\nu\rho\sigma\tau}k_{2\mu})k_1^\sigma k_2^\tau - \varepsilon_{\mu\nu\rho\sigma}(k_1^\sigma - k_2^\sigma)k_1 \cdot k_2]T_3(q^2) \quad (11\text{-}188)$$

Consequently,

$$q^\rho T_{\mu\nu\rho} = \varepsilon_{\mu\nu\sigma\tau}k_1^\sigma k_2^\tau q^2[T_1(q^2) + T_3(q^2)] \quad (11\text{-}189)$$

We evaluate these amplitudes to leading order in electromagnetism. Since no strongly interacting zero-mass particle is around (m_π^2 is considered to be different from zero) it follows from (11-189) that

$$\mathscr{T}(0) = 0 \quad (11\text{-}190)$$

For the soft pion theory to be valid, this should be interpreted as meaning that $\mathscr{T}(m_\pi^2)$ is suppressed:

$$\mathscr{T}(m_\pi^2) \simeq 0 \quad (11\text{-}191)$$

The dominant part of the $\pi^0 \to 2\gamma$ is forbidden according to this observation of Sutherland and of Veltman. Fortunately this conclusion turns out to be incorrect due to the effects of renormalization. We stress that this does not question the validity of the extrapolation from (11-190) to (11-191) but really means that the statements (11-189) or (11-190) have to be modified. A method of approaching the problem is to perform a perturbative calculation within a given model. The interpretation of the result will be that we have to modify the PCAC relation (11-117) in the presence of electromagnetic interactions.

11-5-2 Axial Anomaly in the σ Model

To be specific we use the σ model with fermions. For simplicity we keep only one Fermi field of charge $+1$ (the proton) and two mesons π^0 and σ. From Sec. 11-4-1 the lagrangian is

$$\mathscr{L} = \bar{\psi}[i\partial\!\!\!/ - m + g(\sigma + i\pi\gamma_5)]\psi + \tfrac{1}{2}(\partial\pi)^2 + \tfrac{1}{2}(\partial\sigma)^2 - \frac{m_\pi^2}{2}\pi^2 - \frac{m_\sigma^2}{2}\sigma^2$$
$$- \lambda v\sigma(\sigma^2 + \pi^2) - \frac{\lambda}{4}(\pi^2 + \sigma^2)^2 \quad (11\text{-}192)$$

To lowest order

$$m = -gv \qquad m_\sigma^2 - m_\pi^2 = 2\lambda v^2 \quad (11\text{-}193)$$

The proton is the only charged particle to contribute to the current; therefore

$$j_\mu = \bar{\psi}\gamma_\mu\psi$$

The axial current

$$A_\mu = \tfrac{1}{2}\bar{\psi}\gamma_\mu\gamma_5\psi + (\sigma\partial_\mu\pi - \pi\partial_\mu\sigma) + v\partial_\mu\pi \quad (11\text{-}194)$$

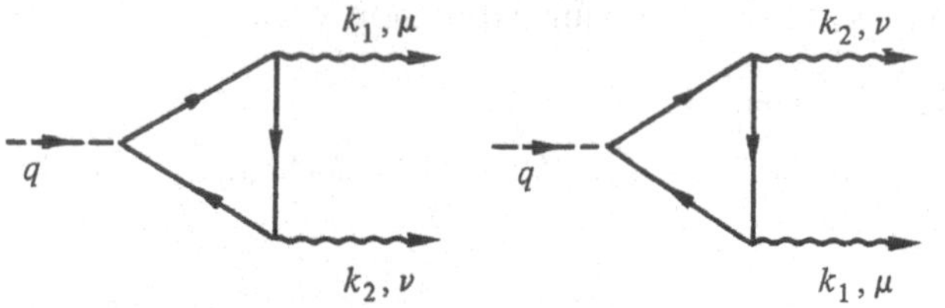

Figure 11-16 One-loop diagrams for π^0 decay.

has a divergence formally given by

$$\partial^\mu A_\mu = m_\pi^2 f_\pi \pi \qquad f_\pi = -v \tag{11-195}$$

The π^0-decay amplitude computed to the lowest one-loop order is (Fig. 11-16)

$$T_{\mu\nu} = T_{\mu\nu}^{(1)}(k_1, k_2) + T_{\nu\mu}^{(1)}(k_2, k_1) \tag{11-196}$$

$$T_{\mu\nu}^{(1)}(k_1, k_2) = ge^2 \int \frac{d^4p}{(2\pi)^4} \frac{\text{tr}\,[(\not{q} + \not{p} - \not{k}_1 + m)\gamma_\mu(\not{q} + \not{p} + m)\gamma_5(\not{p} + m)\gamma_\nu]}{[(q + p - k_1)^2 - m^2][(q + p)^2 - m^2](p^2 - m^2)}$$

The trace in the numerator is equal to $4im\varepsilon_{\mu\nu\rho\sigma}k_1^\rho k_2^\sigma$ (with $\varepsilon_{0123} = -1$) and the remaining integral is convergent:

$$\begin{aligned} T_{\mu\nu}^{(1)}(k_1, k_2) &= 4ige^2 m\varepsilon_{\mu\nu\rho\sigma}k_1^\rho k_2^\sigma \\ &\quad \times \int \frac{d^4p}{(2\pi)^4} \frac{1}{[(q + p - k_1)^2 - m^2][(q + p)^2 - m^2](p^2 - m^2)} \\ &= 4ge^2 m\varepsilon_{\mu\nu\rho\sigma}k_1^\rho k_2^\sigma \\ &\quad \times \frac{1}{(4\pi)^2} \int_0^1 dx \int_0^{1-x} dy \, \frac{1}{m^2 - xk_1^2 - yk_2^2 + (xk_1 - yk_2)^2} \end{aligned} \tag{11-197}$$

On-shell we have $k_1^2 = k_2^2 = 0$, $2k_1 \cdot k_2 = q^2$ and we find for the contribution of the two diagrams

$$\mathscr{T}(q^2) = \frac{e^2 gm}{2\pi^2} \varepsilon_{\mu\nu\rho\sigma}\varepsilon_1^\mu \varepsilon_2^\nu k_1^\rho k_2^\sigma \int_0^1 dx \int_0^{1-x} dy \, \frac{1}{m^2 - xyq^2} \tag{11-198}$$

This is in contradiction to Eq. (11-190) since $\mathscr{T}(0)$ is not zero, given in fact by

$$\mathscr{T}(0) = \frac{e^2 g}{4\pi^2 m} \varepsilon_{\mu\nu\rho\sigma}\varepsilon_1^\mu \varepsilon_2^\nu k_1^\rho k_2^\sigma \tag{11-199}$$

Note that (11-198) implies a smooth extrapolation in the neighborhood of $q^2 = 0$ with

$$\mathscr{T}(q^2) = \left(1 + \frac{q^2}{12m^2} + \cdots\right)\mathscr{T}(0) \tag{11-200}$$

Using $g/m = 1/f_\pi$, $f_\pi \sim 93$ MeV, the numerical rate obtained from the leading contribution $\mathscr{T}(0)$ is

$$\Gamma = \frac{1}{2m_\pi} \sum_{\varepsilon_1, \varepsilon_2} \int \frac{d^3k_1\, d^3k_2}{4(2\pi)^6 \omega_1 \omega_2} |\mathscr{T}|^2 (2\pi)^4 \delta(q - k_1 - k_2)$$

$$= \frac{\alpha^2 m_\pi^3}{64\pi^3 f_\pi^2} = 7.63 \text{ eV} \tag{11-201}$$

where care has been taken to integrate on a 2π solid angle to comply with Bose statistics. The experimental rate

$$\Gamma^{\text{exp}} = (7.37 \pm 1.5) \text{ eV} \tag{11-202}$$

compares favorably with the above estimate. In fact, as early as 1949, ignoring all subtleties of current algebra, Steinberger had derived a similar estimate.

In the quark model the electromagnetic current is coupled to fractionally charged fermions (charges Q_i). The axial curfent reads

$$A_\mu^3 = \sum_i \overline{\psi}_i \gamma_\mu \gamma_5 \frac{\tau_3^i}{2} \psi_i + \text{meson contributions}$$

The amplitude $\mathscr{T}$ would be multiplied by a factor $x = \sum_i \tau_3^i Q_i^2$. For a quark triplet (u, d, s), the charges are $Q_i = (\frac{2}{3}, \frac{1}{3}, -\frac{1}{3})$ and the τ_3^i are $(1, -1, 0)$; thus $x = \frac{1}{3}$. The agreement with experiment is lost unless quarks are three times degenerate according to a hidden color quantum number. This is in fact one of the most direct evidences of the necessity of such a degeneracy, together with the question of statistics in baryonic states.

We have now to understand what invalidated the Veltman-Sutherland result. For this purpose we compute the amplitude $T_{\mu\nu\rho}$ defined by Eq. (11-186) within the same model. From (11-194) the axial current is coupled either to the pion with a factor $iq_\rho v = -iq_\rho f_\pi$ or directly to the proton. We have therefore, to lowest order,

$$T_{\mu\nu\rho} = \hat{T}_{\mu\nu\rho} + f_\pi q_\rho \frac{1}{m_\pi^2 - q^2} T_{\mu\nu} \tag{11-203}$$

The first two diagrams of Fig. 11-17 contribute to $\hat{T}_{\mu\nu\rho}$; the last two involve the pion amplitude $T_{\mu\nu}$. The questionable identity written in (11-185) takes the form

$$\frac{m_\pi^2 - q^2}{f_\pi m_\pi^2} q^\rho T_{\mu\nu\rho} = T_{\mu\nu} \tag{11-204}$$

or, alternatively,

$$q^\rho \hat{T}_{\mu\nu\rho} = f_\pi T_{\mu\nu} = \frac{m}{g} T_{\mu\nu} \tag{11-205}$$

Of course we want to insure that renormalization preserves the electromagnetic current conservation (11-187). Let us compute the first two diagrams of Fig. 11-17.

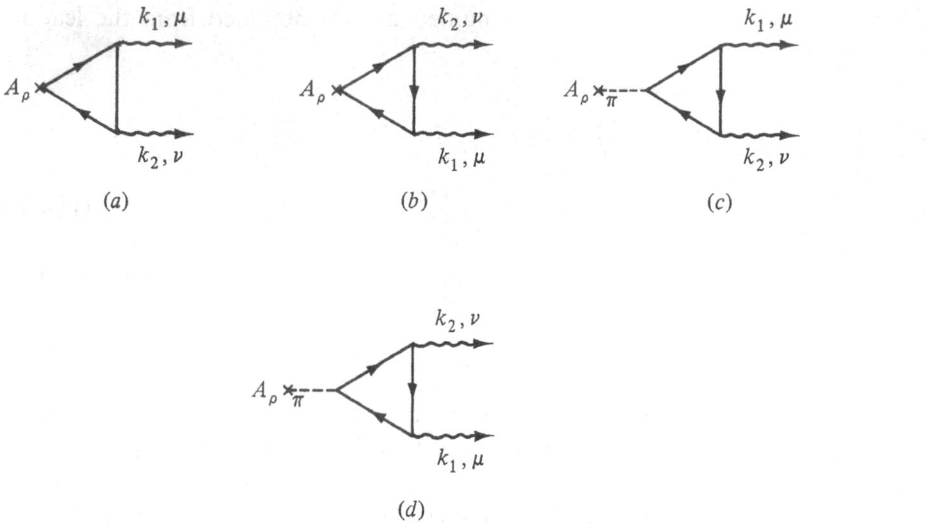

Figure 11-17 Lowest-order diagrams for the amplitude $T_{\mu\nu\rho}$.

With a factor $\frac{1}{2}$ arising from our definition (11-194), we find

$$\hat{T}^{(1)}_{\mu\nu\rho}(k_1, k_2) = \hat{T}^{(2)}_{\nu\mu\rho}(k_2, k_1)$$

$$= i\frac{e^2}{2}\int\frac{d^4p}{(2\pi)^4}\,\mathrm{tr}\left(\frac{i}{\not{p}-m}\gamma_\mu\frac{i}{\not{p}+\not{k}_1-m}\gamma_\rho\gamma_5\frac{i}{\not{p}-\not{k}_2-m}\gamma_\nu\right) \tag{11-206}$$

Contracting with q^ρ yields

$$q^\rho\hat{T}^{(1)}_{\mu\nu\rho}(k_1, k_2) = \frac{e^2}{2}\int\frac{d^4p}{(2\pi)^4}\frac{\mathrm{tr}\,[(\not{p}+m)\gamma_\mu(\not{p}+\not{k}_1+m)(\not{k}_1+\not{k}_2)\gamma_5(\not{p}-\not{k}_2+m)\gamma_\nu]}{(p^2-m^2)[(p+k_1)^2-m^2][(p-k_2)^2-m^2]} \tag{11-207}$$

The central term in the trace may be arranged in the form

$$\begin{aligned}(\not{p}+\not{k}_1+m)&(\not{k}_1+\not{k}_2)\gamma_5(\not{p}-\not{k}_2+m)\\ &= (\not{p}+\not{k}_1+m)(\not{p}+\not{k}_1-m+\not{k}_2+m-\not{p})\gamma_5(\not{p}-\not{k}_2+m)\\ &= 2m(\not{p}+\not{k}_1+m)\gamma_5(\not{p}-\not{k}_2+m)+[(p+k_1)^2-m^2]\gamma_5(\not{p}-\not{k}_2+m)\\ &\quad+(\not{p}+\not{k}_1+m)\gamma_5[(p-k_2)^2-m^2]\end{aligned}$$

Thus

$$\begin{aligned}q^\rho\hat{T}^{(1)}_{\mu\nu\rho}(k_1, k_2) &= \frac{m}{g}\,T^{(1)}_{\mu\nu}(k_1, k_2)\\ &+\frac{e^2}{2}\int\frac{d^4p}{(2\pi)^4}\,\mathrm{tr}\left\{\frac{(\not{p}+m)\gamma_\mu\gamma_5(\not{p}-\not{k}_2+m)\gamma_\nu}{(p^2-m^2)[(p-k_2)^2-m^2]}-\frac{(\not{p}+\not{k}_1+m)\gamma_\nu\gamma_5(\not{p}+m)\gamma_\mu}{[(p+k_1)^2-m^2](p^2-m^2)}\right\}\end{aligned} \tag{11-208}$$

The first term has been recognized as the amplitude $T^{(1)}_{\mu\nu}(k_1, k_2)$ occurring in π^0 decay [Eq. (11-196)]. Hence the validity of (11-205) depends on the remaining contribution vanishing. At first sight we would think that each term contributing to this integral is zero, as it is a Lorentz pseudotensor depending on one four-vector argument. Moreover, we might also argue that a translation of integration variable $p \to p + k_2$ would show a cancellation between the first term in the integrand of $q^\rho \hat{T}^{(1)}_{\mu\nu\rho}$ and the second in $q^\rho \hat{T}^{(2)}_{\mu\nu\rho}$. Both arguments are wrong because of the linear divergence of these integrals. A safe method is to use a Pauli-Villars gauge-invariant regularization. Dimensional regularization is not suited here, because of the presence of γ_5 or $\varepsilon_{\mu\nu\sigma\rho}$ particular to four-dimensional space-time.

The introduction of a large mass (M) regulator field transforms (11-208) into

$$q^\rho[\hat{T}^{(1)}_{\mu\nu\rho}(m) - \hat{T}^{(1)}_{\mu\nu\rho}(M)] = \left[\frac{m}{g}\, T^{(1)}_{\mu\nu}(m) - \frac{M}{g}\, T^{(1)}_{\mu\nu}(M)\right] \tag{11-209}$$

since the previous arguments are valid for the regulated finite integral appearing in (11-208). From the calculation given above we can easily find

$$\lim_{M\to\infty} \frac{M}{g}\, T^{(1)}_{\mu\nu}(M) = \frac{\alpha}{2\pi}\, \varepsilon_{\mu\nu\rho\sigma} k_1^\rho k_2^\sigma \tag{11-210}$$

The result of regularization is to uncover this finite term which represents the effect of the second integral in (11-208). Finally, by adding the contribution of diagrams (*a*) and (*b*) in Fig. 11-17 we find

$$q^\rho \hat{T}_{\mu\nu\rho} = \frac{m}{g}\, T_{\mu\nu} - \frac{\alpha}{\pi}\, \varepsilon_{\mu\nu\rho\sigma} k_1^\rho k_2^\sigma \tag{11-211}$$

with an anomalous term on the right-hand side correcting (11-205). Note that the identities (11-187) are still preserved.

We might ask whether it is not possible to eliminate this new term in (11-211) using a different subtraction scheme. After all, even though $q^\rho T_{\mu\nu\rho}$ is finite this is not the case for $\hat{T}_{\mu\nu\rho}$, which has a linear divergence. The unknown subtraction term $\Delta\hat{T}_{\mu\nu\rho}$ must be a linear polynomial in k_1, k_2, and transform as a Lorentz pseudotensor symmetric in the exchange $(k_1, \mu) \leftrightarrow (k_2, \nu)$. The only candidate has the form

$$\Delta\hat{T}_{\mu\nu\rho} = \text{constant } \varepsilon_{\mu\nu\rho\sigma}(k_1 - k_2)^\sigma \tag{11-212}$$

but does not satisfy (11-187). It is possible to find a Ward identity without anomalies for the axial current [i.e., of the form (11-205)], but the price is to lose ordinary gauge invariance.

The anomaly which has appeared in (11-211) is welcome since it leads to a satisfactory π^0 lifetime. Rather than accumulating all evils to preserve the PCAC relation, it is more fruitful to correct it to first order in α to read

$$\partial^\mu A^3_\mu = f_\pi m_\pi^2 \pi^0 - \frac{\alpha}{8\pi}\, \varepsilon_{\mu\nu\rho\sigma} F^{\mu\nu} F^{\rho\sigma} \tag{11-213}$$

where $F_{\mu\nu}$ is the electromagnetic field strength. No result obtained previously is affected by this correction. The impossibility of maintaining for the regularized or renormalized theory all the Ward identities implied by the classical approximation justifies the name given to these anomalous Ward identities. Let us study some of their properties and consequences.

11-5-3 General Properties

The phenomenon analyzed in the framework of the σ model is common to all theories with fermions coupled to axial currents when we try to satisfy simultaneously axial and vector Ward identities. It was discovered as early as 1951 by Schwinger and was explicitly studied by Adler for electrodynamics and Bell and Jackiw for the σ model.

In electrodynamics we want to verify the relation

$$\partial_\mu j_5^\mu = \partial_\mu(\overline{\psi}\gamma^\mu\gamma_5\psi) = 2mi\overline{\psi}\gamma_5\psi = 2mj_5 \tag{11-214}$$

which states that the failure of axial current conservation arises from the fermionic mass term. The axial current of Eq. (11-214) differs by a factor 2 from the one discussed above. A consequence of (11-214) would be a relation among Green functions, analogous to (11-205), of the form (Fig. 11-18)

$$\begin{aligned} q^\rho R_{\mu\nu\rho} &= ie^2 q^\rho \int d^4x\, d^4y\, e^{i(k_1\cdot x + k_2\cdot y)} \langle T j_\mu(x) j_\nu(y) j_{5\rho}(0)\rangle \\ &= 2me^2 \int d^4x\, d^4y\, e^{i(k_1\cdot x + k_2\cdot y)} \langle T j_\mu(x) j_\nu(y) j_5(0)\rangle \\ &= 2mR_{\mu\nu} \end{aligned} \tag{11-215}$$

Let us introduce the proper functions for the axial current and the pseudoscalar density:

$$\begin{aligned} \Gamma_{5\mu}(p', p) &= \int d^4x\, d^4y\, e^{i(p'\cdot y - p\cdot x)} \langle T j_{5\mu}(0)\psi(y)\overline{\psi}(x)\rangle_T \\ \Gamma_5(p', p) &= \int d^4x\, d^4y\, e^{i(p'\cdot y - p\cdot x)} \langle T j_5(0)\psi(y)\overline{\psi}(x)\rangle_T \end{aligned} \tag{11-216}$$

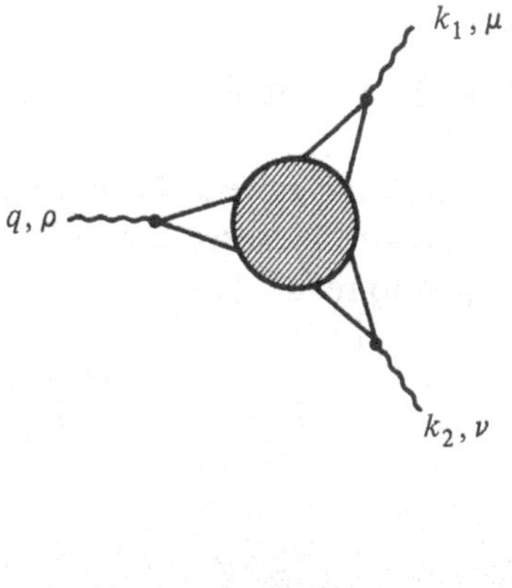

Figure 11-18 Green function with two vector and one axial currents.

where the subscript T means truncation of the fermion propagators. From (11-214)

$$(p'_\mu - p_\mu)\Gamma_5^\mu(p',p) = -2im\Gamma_5(p',p) + \gamma_5 S^{-1}(p) + S^{-1}(p')\gamma_5 \qquad (11\text{-}217)$$

According to our usual notations $iS(p)$ is the complete fermion propagator such that to lowest order $S^{-1}(p) = \not{p} - m$. Equation (11-217) is the analog of the Ward identity for the vertex function (8-87).

$$(p'_\mu - p_\mu)\Lambda^\mu(p',p) = S^{-1}(p') - S^{-1}(p) \qquad (11\text{-}218)$$

Equation (11-217) would imply a common multiplicative renormalization of mj_5 and j_5^μ with renormalization constant $Z_5 = 1$.

In fact, the identity (11-214) and its consequences, (11-215) and (11-217), are not verified perturbatively. They are modified by anomalies arising from the triangular diagram, as was the case for the σ model. The computation given in the previous section can readily be extended to the present case with the result

$$q^\rho R_{\mu\nu\rho} = 2mR_{\mu\nu} + \frac{2}{\pi}\alpha\varepsilon_{\mu\nu\rho\sigma}k_1^\rho k_2^\sigma \qquad (11\text{-}219)$$

as the corrected version of (11-215) to one-loop order. In operator form the anomaly reads

$$\partial_\mu j_5^\mu(x) = 2mj_5(x) - \frac{\alpha}{4\pi}\varepsilon_{\mu\nu\rho\sigma}F^{\mu\nu}F^{\rho\sigma} \qquad (11\text{-}220)$$

Similarly, (11-217) is modified as

$$(p'_\mu - p_\mu)\Gamma_5^\mu(p',p) = -2im\Gamma_5(p',p) + \gamma_5 S^{-1}(p) + S^{-1}(p')\gamma_5 + i\frac{\alpha}{4\pi}F(p',p)$$

$$F(p',p) = \int d^4x\, d^4x'\, e^{i(p'\cdot y - p\cdot x)}\langle T\psi(y)\overline{\psi}(x)\varepsilon_{\mu\nu\rho\sigma}F^{\mu\nu}F^{\rho\sigma}(0)\rangle_T \qquad (11\text{-}221)$$

As a consequence of (11-221) evaluated at zero momentum, $2mj_5$ has still a renormalization constant equal to unity. The same is no longer true for j_5^μ. The breaking of chiral invariance now involves a hard component $\varepsilon_{\mu\nu\rho\sigma}F^{\mu\nu}F^{\rho\sigma}$.

The structure of anomalies in a renormalizable theory with fermion fields endowed with an internal symmetry coupled to vector, axial, scalar, and pseudoscalar fields may be analyzed along the same lines. The general conclusion is that only diagrams containing fermion loops coupled to vector and axial currents lead to anomalies. Axial currents must occur in odd numbers along the loop. Moreover, any loop containing a scalar or pseudoscalar coupling may be eliminated by an adequate subtraction. If we insist on preserving the normal Ward identities for vector currents, the diagrams of Fig. 11-19 lead to anomalies for the axial ones. As in the triangular case, the manipulation performed when trying to verify the identity $\partial_\mu j_5^\mu = 2mj_5$ involve divergent integrals up to the pentagon diagram.

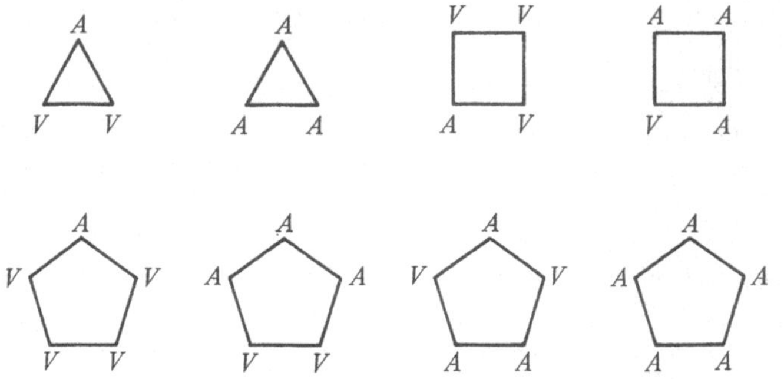

Figure 11-19 One-fermion-loop diagrams leading to anomalies in the axial Ward identities.

Let $\mathscr{V}_\mu$ and $\mathscr{A}_\mu$ stand for the vector and axial fields considered as matrices acting on the internal symmetry indices of fermions in such a way that the interaction lagrangian reads

$$\mathscr{L}_{\text{int}} = \bar{\psi}\gamma^\mu(\mathscr{V}_\mu + \gamma_5\mathscr{A}_\mu)\psi \tag{11-222}$$

The corresponding currents

$$J^\mu = \frac{\partial\mathscr{L}}{\partial\mathscr{V}_\mu^T} \qquad J_5^\mu = \frac{\partial\mathscr{L}}{\partial\mathscr{A}_\mu^T} \tag{11-223}$$

are also matrices in the internal symmetry indices. The vector Ward identities are by hypothesis the normal ones:

$$\partial_\mu J^\mu(x) = j(x) \tag{11-224}$$

while the axial current anomalies take the form

$$\partial_\mu J_5^\mu(x) = j_5(x) - \frac{1}{4\pi^2}\varepsilon^{\mu\nu\rho\sigma}(\tfrac{1}{4}F^V_{\mu\nu}F^V_{\rho\sigma} + \tfrac{1}{12}F^A_{\mu\nu}F^A_{\rho\sigma}$$
$$+ \tfrac{2}{3}i\mathscr{A}_\mu\mathscr{A}_\nu F^V_{\rho\sigma} + \tfrac{2}{3}iF^V_{\mu\nu}\mathscr{A}_\rho\mathscr{A}_\sigma + \tfrac{2}{3}i\mathscr{A}_\mu F^V_{\nu\rho}\mathscr{A}_\sigma - \tfrac{8}{3}\mathscr{A}_\mu\mathscr{A}_\nu\mathscr{A}_\rho\mathscr{A}_\sigma) \tag{11-225}$$

In these expressions j and j_5 denote the naive divergences of the currents and

$$\begin{aligned} F^V_{\mu\nu} &= \partial_\mu\mathscr{V}_\nu - \partial_\nu\mathscr{V}_\mu - i[\mathscr{V}_\mu, \mathscr{V}_\nu] - i[\mathscr{A}_\mu, \mathscr{A}_\nu] \\ F^A_{\mu\nu} &= \partial_\mu\mathscr{A}_\nu - \partial_\nu\mathscr{A}_\mu - i[\mathscr{A}_\mu, \mathscr{V}_\nu] - i[\mathscr{V}_\mu, \mathscr{A}_\nu] \end{aligned} \tag{11-226}$$

Neither the structure nor the coefficient appearing in expressions such as (11-213), (11-221), or (11-225) are modified by higher-order corrections provided a suitable definition is given to the renormalized operators. To see this, let us consider the case of electrodynamics, for instance. The argument relies on the existence of a regularization of the photon propagator, in the form $-\frac{1}{4}F_{\mu\nu}(1 + \Box^2/\Lambda^4)F^{\mu\nu}$, for instance, such that all diagrams of interest except those with one loop become superficially convergent. Indeed, the new power counting gives for a diagram with E_F external fermion lines, E_A photon lines or current insertions and L loops: $\omega = 8 - \frac{7}{2}E_F - E_A - 4L$. Therefore amplitudes with $L \geq 2$ will superficially converge. It is, of course, understood that the standard renormalization of quantum electrodynamics has disposed of all internal divergences. This regularization is gauge invariant and does not modify the structure of the axial current. Consequently, the only anomalies are those arising from one-loop subdiagrams which we have just analyzed.

We note that it is possible to define a different axial current with a normal Ward identity but violating gauge invariance. In electrodynamics, for instance, such would be the case for

$$\hat{j}_{\mu 5} = j_{\mu 5} + \frac{\alpha}{2\pi} \varepsilon_{\mu\nu\rho\sigma} A^\nu F^{\rho\sigma} \tag{11-227}$$

which satisfies

$$\partial_\mu \hat{j}_5^\mu = 2mj_5 \tag{11-228}$$

These different possibilities can be interpreted in a different language. We could try to construct directly the operator j_5^μ in terms of the fields ψ and $\overline{\psi}$ in the presence of an external potential. Since combinations such as $\psi(x)\overline{\psi}(y)$ are singular in the limit $x \to y$ we separate the arguments by an infinitesimal space-like interval ε, therefore defining

$$j_5^{(1)\mu}(x, \varepsilon) = \overline{\psi}(x + \varepsilon)\gamma^\mu\gamma_5\psi(x) \tag{11-229}$$

This new operator is not gauge invariant. Following Schwinger's suggestion, this is corrected as follows. We multiply (11-229) by a phase factor involving the integral of the vector potential along a space-like path from x to $x + \varepsilon$ along which the various components commute:

$$j_5^\mu(x, \varepsilon) = \overline{\psi}(x + \varepsilon)\gamma^\mu\gamma_5\psi(x) \exp\left[-ie\int_x^{x+\varepsilon} dz^\rho A_\rho(z)\right] \tag{11-230}$$

If we heuristically use the equation of motion, the divergence of this operator can be written as

$$\partial_\mu j_5^\mu(x, \varepsilon) = 2mi\overline{\psi}(x + \varepsilon)\gamma_5\psi(x) \exp\left[-ie\int_x^{x+\varepsilon} dz^\rho A_\rho(z)\right] - iej_5^\mu(x, \varepsilon)F_{\mu\nu}(x)\varepsilon^\nu[1 + O(\varepsilon)] \tag{11-231}$$

The second term on the right-hand side is singular in the limit $\varepsilon \to 0$. The vacuum matrix element of j_5^μ in the presence of an external field has a $1/\varepsilon$ behavior from which we recover in the limit

$$\langle 0|\partial_\mu j_5^\mu(x)|0\rangle = \langle 0|2mi\overline{\psi}(x)\gamma_5\psi(x)|0\rangle - \frac{\alpha}{4\pi} \varepsilon_{\mu\nu\rho\sigma} F^{\mu\nu}F^{\rho\sigma}(x) \tag{11-232}$$

The axial anomaly modifies the PCAC expression in the presence of electromagnetic interactions [Eq. (11-213)]. The added term is a hard one, changing the renormalization properties of the axial current. We may return to π^0 decay and express the amplitude as

$$\varepsilon_1^\mu \varepsilon_2^\nu T_{\mu\nu}(q^2) = (m_\pi^2 - q^2)\langle\gamma(k_1, \varepsilon_1), \gamma(k_2, \varepsilon_2)|\pi(0)|0\rangle \tag{11-233}$$

which, according to (11-213), may also be written

$$\varepsilon_1^\mu \varepsilon_2^\nu T_{\mu\nu}(q^2) = \frac{m_\pi^2 - q^2}{m_\pi^2 f_\pi}\Big[\langle\gamma(k_1, \varepsilon_1), \gamma(k_2, \varepsilon_2)|\partial^\rho A_\rho^3(0)|0\rangle + \frac{\alpha}{8\pi}\varepsilon_{\rho_1\sigma_1\rho_2\sigma_2}\langle\gamma(k_1, \varepsilon_1), \gamma(k_2, \varepsilon_2)|F^{\rho_1\sigma_1}F^{\rho_2\sigma_2}|0\rangle\Big] \tag{11-234}$$

To lowest order in α the second matrix element is equal to $(\alpha/\pi)\varepsilon_1^\mu \varepsilon_2^\nu \varepsilon_{\mu\nu\rho\sigma} k_1^\rho k_2^\sigma$, while the first one vanishes in the limit $q^2 = 0$. We therefore reach the conclusion that

$$T_{\mu\nu}(0) = \frac{\alpha}{\pi f_\pi} \varepsilon_{\mu\nu\rho\sigma} k_1^\rho k_2^\sigma \tag{11-235}$$

in agreement with (11-199). This low-energy theorem is in fact valid to all orders in α due to the nonrenormalization property of the anomaly. On the other hand, the extrapolation to $q_\pi^2 = m^2$ will depend on the order of the approximation.

We conclude that anomalies are not an artefact but a true product of renormalization, involving a deeper aspect of field theory. Each time a soluble model is available we can check them explicitly as in the two-dimensional Schwinger model of electrodynamics with massless fermions which exhibits a computable anomaly of the form

$$\partial_\mu j_5^\mu = c\varepsilon_{\mu\nu} F^{\mu\nu} \tag{11-236}$$

The appearance of anomalous dimensions in the asymptotic behavior (Chap. 13) will offer a new insight into these phenomena.

NOTES

Unitary symmetry is described by its fathers M. Gell-Mann and Y. Ne'eman in "The Eightfold Way," Benjamin, New York, 1964. The quark model was independently proposed by M. Gell-Mann, *Phys. Lett.*, vol. 8, p. 214, 1963, and by G. Zweig (unpublished CERN report, 1963).

The spectrum associated to symmetry breaking was discussed by J. Goldstone, *Nuov. Cim.*, vol. 19, p. 154, 1961; Y. Nambu and G. Jona-Lasinio, *Phys. Rev.*, vol. 122, p. 345, 1961; and J. Goldstone, A. Salam, and S. Weinberg, *Phys. Rev.*, vol. 127, p. 965, 1962. The impossibility of continuous symmetry breaking in dimension two was shown by N. D. Mermin and H. Wagner, *Phys. Rev. Lett.*, vol. 17, p. 1133, 1966, in the context of statistical mechanics, and was extended to field theory by S. Coleman, *Comm. Math. Phys.*, vol. 31, p. 259, 1973. The proof of phase transitions in lattice systems with a discrete symmetry is due to R. E. Peierls, *Phys. Rev.*, vol. 54, p. 918, 1938. The ground-state invariance was analyzed by S. Coleman, *J. Math. Phys.*, vol. 7, p. 787, 1966. A general survey of symmetry problems is given by G. S. Guralnik, C. R. Hagen, and T. W. Kibble in "Advances in Particle Physics," edited by R. L. Cool and R. E. Marshak, Interscience, New York, 1968; and by S. Coleman in "Laws of Hadronic Matter," edited by A. Zichichi, Academic Press, New York, 1975.

The abundant work on current algebra may be traced in the books of S. Adler and R. Dashen, "Current Algebras," Benjamin, New York, 1968; V. de Alfaro, S. Fubini, G. Furlan, and C. Rossetti, "Currents in Hadron Physics," North-Holland, Amsterdam, 1973; and in the lectures by S. Weinberg in "Lectures on Elementary Particles and Quantum Field Theory (Brandeis, 1970)," edited by S. Deser, M. Grisaru, and H. Pendleton, MIT Press, Cambridge, Mass., 1970; and S. B. Treiman in "Lectures on Current Algebra and Its Applications," Princeton University Press, 1972.

Phenomenological lagrangians are discussed by S. Gasiorowicz and D. A. Geffen, *Rev. Mod. Phys.*, vol. 41, p. 531, 1969.

References to specific points discussed in the text are as follows.

Sum rules

N. Cabibbo and L. A. Radicati, *Phys. Lett.*, vol. 19, p. 697, 1966.
S. Adler, *Phys. Rev.*, vol. 143, p. 1144, 1966.
J. D. Bjorken, *Phys. Rev. Lett.*, vol. 16, p. 408, 1966.

Consequences of PCAC

M. L. Goldberger and S. B. Treiman, *Phys. Rev.*, vol. 110, p. 1178, 1958.

Low-energy theorems

F. E. Low, *Phys. Rev.*, vol. 96, p. 1428, 1954, and vol. 110, p. 974, 1958.
M. Gell-Mann and M. L. Goldberger, *Phys. Rev.*, vol. 96, p. 1433, 1954.
S. D. Drell and A. C. Hearn, *Phys. Rev. Lett.*, vol. 16, p. 908, 1966.
S. L. Adler, *Phys. Rev.*, vol. 140, ser. B, p. 736, 1965.
W. I. Weisberger, *Phys. Rev.*, vol. 143, p. 1302, 1966.

Soft pion theorems

S. Adler, *Phys. Rev.*, vol. 139, ser. B, p. 1638, 1965.
S. Weinberg, *Phys. Rev. Lett.*, vol. 17, p. 616, 1966.

$SU(3) \times SU(3)$ breaking

M. Gell-Mann, R. J. Oakes, and B. Renner, *Phys. Rev.*, vol. 175, p. 2195, 1968.

The σ model was proposed by M. Gell-Mann and M. Lévy, *Nuov. Cim.*, vol. 16, p. 705, 1960, and is reviewed by B. W. Lee in "Chiral Dynamics," Gordon and Breach, 1972.

The neutral pion decay was considered by J. Steinberger, *Phys. Rev.*, vol. 76, p. 1180, 1949, and in the context of current algebra by D. G. Sutherland, *Nucl. Phys.*, ser. B, vol. 2, p. 433, and by M. Veltman, *Proc. Roy. Soc.*, ser. A, vol. 301, p. 107, 1967.

Anomalies appear in J. Schwinger's work, *Phys. Rev.*, vol. 82, p. 664, 1951, and were analyzed by S. Adler, *Phys. Rev.*, vol. 177, p. 2426, 1969, and by J. S. Bell and R. Jackiw, *Nuov. Cim.*, vol. 60, ser. A, p. 47, 1969. The general structure of anomalies is discussed by W. A. Bardeen, *Phys. Rev.*, vol. 184, p. 1848, 1969. See also the Brandeis 1970 Lectures by S. Adler, edited by S. Deser, M. Grisaru, and H. Pendleton, MIT Press, Cambridge, Mass., 1970, and R. Jackiw in "Lectures on Current Algebra and Its Applications," Princeton University Press, 1972. Massless two-dimensional quantum electrodynamics was solved by J. Schwinger, *Phys. Rev.*, vol. 128, p. 2425, 1962.

CHAPTER

TWELVE

NONABELIAN GAUGE FIELDS

After reviewing their geometrical background we present in detail the quantum theory of Yang-Mills fields. The emphasis is on the quantization and the renormalization procedures, both in the symmetric and spontaneously broken cases. Vacuum degeneracy and classical solutions are briefly evoked. Applications to the unified theory of weak and electromagnetic interactions are studied in the framework of the Weinberg-Salam model.

12-1 CLASSICAL THEORY

The generalization of gauge invariance to nonabelian groups was a very appealing and natural idea, first proposed by Yang and Mills in 1954. In spite of the large amount of work they inspired, these nonabelian gauge theories had a rather slow development until the end of the 1960s. The important problems of quantization, renormalization, and mass generation were then brilliantly solved. It is believed that a theory of this sort can provide a unified description of weak and electromagnetic interactions, and possibly also of strong interactions.

These models offer all possible intricacies. Their quantization and renormalization are difficult, they exhibit remarkable mechanisms of symmetry breaking, and they have a unique behavior at short distances, while most aspects of their long-range behavior are not yet fully elucidated. It has recently been realized that even the classical theory has a fascinating complexity.

It is almost impossible to present such a wealth of information within a few dozens of pages. We shall restrict ourselves to the basic topics and only mention some of the possible developments and problems.

12-1-1 The Gauge Field A_μ and the Tensor $F_{\mu\nu}$

The model considered by Yang and Mills was based on the isotopic symmetry, with the group $SU(2)$ of global invariance. Was it possible to transform this invariance into a local one, so that the reference frame used to define the isospins could vary from point to point? If so, the information that a particle produced at a given space-time point was a proton, say, was meaningless for a different observer, unless there existed a way to compare their two frames. This is the role of the gauge field, very much as in electrodynamics relative phases of charged fields at different points make sense only when compared via the electromagnetic potential.

To be more explicit, let us consider N fields (Lorentz scalars for simplicity) transforming according to an irreducible representation U of some compact Lie group G:

$$\phi(x) \to ({}^g\phi)(x) = U(g)\phi(x) \tag{12-1}$$

where $U(g)$ is an $N \times N$ unitary (or orthogonal) matrix. We assume the lagrangian invariant under the transformation (12-1). We want to build a more complex theory, still invariant when g depends on the space-time point x. Let $g(x)$ be such a G-valued function. The derivatives $\partial_\mu \phi$ have no longer any intrinsic meaning as we compare fields $\phi(x + \delta x)$ and $\phi(x)$ that transform independently under (12-1). It is necessary to introduce a new object to allow this comparison.

Consider an infinitesimal transformation

$$g = e + \delta\alpha_a t^a \tag{12-2}$$

where e is the identity of G and the t^a are elements of the Lie algebra satisfying the commutation relations

$$[t^a, t^b] = C^{ab}{}_c t^c \tag{12-3}$$

In the representation associated to the fields ϕ, antihermitian matrices T^a represent the elements t^a of the Lie algebra, so that under the transformation g of Eq. (12-2),

$$\begin{aligned} \phi(x) &\to ({}^g\phi)(x) = \phi(x) + \delta\phi(x) \\ \delta\phi(x) &= \delta\alpha_a T^a \phi(x) \equiv \delta\alpha\phi \end{aligned} \tag{12-4a}$$

If the infinitesimal parameters $\delta\alpha_a(x)$ depend on x, these transformations will be called (local) gauge transformations, and the transformation group—the gauge group—will formally be the infinite product $\prod_x G_x$:

$$\delta\phi(x) = \delta\alpha(x)\phi(x) = \delta\alpha_a(x) T^a \phi(x) \tag{12-4b}$$

We introduce the gauge fields $x \to A^\mu{}_a(x)$. They are vector fields and carry an index of the adjoint representation of G. We denote A^μ the corresponding element of the Lie algebra:

$$A^\mu(x) = A^\mu{}_a(x) t^a \tag{12-5}$$

and we will use the same notation for any of its representations, $A^\mu{}_a(x) T^a$. What is understood should be clear from the context.

To an infinitesimal path $(x, x + dx)$, the gauge field associates an element in the group

$$g(x + dx, x; A) = e + dx^\mu A_\mu(x) \tag{12-6}$$

providing the means to compare two neighboring frames. This generalizes to a finite path C going from x_1 to x_2. If $s \to x(s)$, $0 \le s \le 1$ is a parametrization of this path: $s(0) = x_1$, $s(1) = x_2$, the element of G associated to C is

$$\begin{aligned} g(C; A) &= P \exp\left(\int_0^1 ds \frac{dx}{ds} \cdot A\right) \\ &= \sum_0^\infty \int_0^1 ds_n \int_0^{s_n} ds_{n-1} \cdots \int_0^{s_2} ds_1 \frac{dx}{ds_n} \cdot A[x(s_n)] \cdots \frac{dx}{ds_1} \cdot A[x(s_1)] \end{aligned} \tag{12-7}$$

The path-ordering operation P is the analog of the familiar time ordering T.

In the language of differential geometry, the A^μ form a connection and define a parallel displacement of geometrical objects belonging to representation spaces of the group. A parallel displacement of ϕ from x to $x + dx$ is defined through

$$\phi_t(x) = {}^{g(x+dx,x)}\phi(x) = \phi(x) + dx \cdot A(x)\phi(x) \tag{12-8}$$

The definition of the covariant derivative follows naturally:

$$\begin{aligned} dx^\mu D_\mu \phi(x) &\equiv \phi(x + dx) - \phi_t(x) = dx^\mu[\partial_\mu - A_\mu(x)]\phi(x) \\ D_\mu &\equiv \partial_\mu - A_\mu(x) \end{aligned} \tag{12-9a}$$

or, in components,

$$(D_\mu)_A{}^B = \partial_\mu \delta_A{}^B - (T^c)_A{}^B A_{\mu c}(x) \tag{12-9b}$$

In particular, in the adjoint representation $(T^c)_a{}^b = C^{cb}{}_a = -C^{bc}{}_a$,

$$(D_\mu)_a{}^b = \partial_\mu \delta_a{}^b + C^{bc}{}_a A_{\mu c}(x) \tag{12-9c}$$

If G is the group $U(1)$, this reduces to the familiar concept in electrodynamics. The gauge field is endowed with a transformation law under gauge transformations such that $\phi_t(x)$ transforms as $\phi(x + dx)$. To this end, it is sufficient that

$$g(x + dx, x; A + \delta A) = g(x + dx)g(x + dx, x; A)g^{-1}(x) \tag{12-10}$$

Hence

$$\begin{aligned} \phi_t(x) &\to U[g(x + dx, x; A + \delta A)]U[g(x)]\phi(x) \\ &= U[g(x + dx)]\phi_t(x) \end{aligned} \tag{12-11}$$

as desired. In infinitesimal form, this yields

$$\delta D_\mu \phi(x) = \delta\alpha(x) D_\mu \phi(x) \tag{12-12}$$

which means that $D_\mu \phi(x)$ transforms as $\phi(x)$. Expanding (12-10) to first order in $\delta\alpha$ results in

$$\delta A_\mu = \partial_\mu \delta\alpha(x) + [\delta\alpha, A_\mu] = D_\mu \delta\alpha(x) \tag{12-13a}$$

where $\delta\alpha$ is considered as transforming under the adjoint representation. More explicitly,

$$\delta A_{\mu a}(x) = \partial_\mu \delta\alpha_a(x) + C^{bc}{}_a \delta\alpha_b(x) A_{\mu c}(x) \tag{12-13b}$$

For a finite gauge transformation, Eq. (12-10) gives

$$A_\mu(x) \to {}^g A_\mu(x) = [\partial_\mu g(x)]g^{-1}(x) + g(x)A_\mu(x)g^{-1}(x) \tag{12-14}$$

As compared to the abelian case, a new feature has emerged. For a constant g (or $\delta\alpha$), A_μ transforms as a charged field belonging to the adjoint representation, as indicated by the second term on the right-hand sides of Eqs. (12-13) and (12-14).

In classical electrodynamics, two potentials that only differ, locally, by a nonsingular gauge transformation are physically equivalent, and are then characterized by the same field strength tensor $F_{\mu\nu}$. We want to construct the analogous curvature tensor $F_{\mu\nu}$ in the nonabelian case. To this end, we consider the parallel displacement of some representative ϕ along an infinitesimal closed path C. After returning to the starting point, ϕ has rotated by

$$g(C; A) = P \exp\left[\int_C dx \cdot A(x)\right]$$

If l is the dimension of C, we expand this element up to order l^2:

$$g(C; A) = e + \int_C dx \cdot A + \int_C \int_C_{x_2 > x_1} dx_2 \cdot A(x_2)\, dx_1 \cdot A(x_1) + O(l^3)$$

where, as before, the ordering implies that the curve has been parametrized with $x_0 = x(0) = x(1)$. We also expand $A_\mu[x(s)]$:

$$A_\mu[x(s)] = A_\mu(x_0) + [x^\nu(s) - x_0^\nu]\partial_\nu A_\mu(x_0) + \cdots$$

and get

$$g(C; A) = e + \frac{1}{2}\iint_{\mathscr{A}} dx^\mu \wedge dx^\nu(\partial_\mu A_\nu - \partial_\nu A_\mu - [A_\mu, A_\nu]) + O(l^3) \tag{12-15}$$

where $\mathscr{A}$ is the infinitesimal area bounded by C. We set

$$\begin{aligned} F_{\mu\nu} &\equiv \partial_\mu A_\nu - \partial_\nu A_\mu - [A_\mu, A_\nu] \\ &= (\partial_\mu A_{\nu a} - \partial_\nu A_{\mu a} - C^{bc}{}_a A_{\mu b} A_{\nu c})t^a \end{aligned} \tag{12-16}$$

thereby generalizing the definition of electric and magnetic field. This tensor is often referred to as the strength tensor. A useful identity follows from the previous derivation:

$$[D_\mu, D_\nu] = -F_{\mu\nu} \tag{12-17}$$

In a gauge transformation, $g(C; A)$ transforms as

$$g(C; A) \to g(x_0)g(C; A)g^{-1}(x_0)$$

[cf. Eq. (12-10)]. Hence, F transforms as a charged field belonging to the adjoint representation

$$F(x) \to g(x)F(x)g^{-1}(x) \tag{12-18}$$

or in infinitesimal form

$$\begin{aligned} F &\to F + \delta F \\ \delta F &= [\delta\alpha(x), F] \\ \delta F_{\mu\nu a}(x) &= C^{bc}{}_a \delta\alpha_b(x) F_{\mu\nu c}(x) \end{aligned} \tag{12-19}$$

In particular, the covariant derivative of F reads

$$D_{\mu a}{}^b F_{\nu\rho b} t^a = [D_\mu, F_{\nu\rho}] = \partial_\mu F_{\nu\rho} - [A_\mu, F_{\nu\rho}]$$

In electromagnetism, the exterior derivative of the differential form $A_\mu(x)dx^\mu$, viz $\frac{1}{2}F_{\mu\nu}dx^\mu \wedge dx^\nu$ is a closed form of degree two:

$$\partial_\rho F_{\mu\nu} + \text{cyclic permutations} = 0 \tag{12-20}$$

This property is equivalent to the homogeneous Maxwell equations $\operatorname{div}\mathbf{B} = 0$, $\operatorname{curl}\mathbf{E} + \partial\mathbf{B}/\partial t = 0$. Reciprocally, given such a closed form, Poincaré's lemma entails the local existence of a potential A from which F is derived (Sec. 1-1-2). The situation is slightly different in the nonabelian case. There does exist an analogous identity

$$[D_\mu, F_{\nu\rho}] + [D_\nu, F_{\rho\mu}] + [D_\rho, F_{\mu\nu}] = 0 \tag{12-21}$$

as a consequence of Jacobi's identity and Eq. (12-17). Notice, however, that Eq. (12-21) assumes the existence of A, as it appears in the covariant derivative. Moreover, one can show that if $F_{\nu\rho}$ and A_μ satisfy (12-21), F is not necessarily the strength tensor associated with A. Accordingly, in contrast with the abelian case, the field strength tensor F does not determine uniquely all gauge-invariant quantities.

If $F_{\mu\nu}$ vanishes in the neighborhood of a point, A_μ is a pure gauge

$$F_{\mu\nu} = 0 \Leftrightarrow \exists g(x): A_\mu(x) = [\partial_\mu g(x)]g^{-1}(x) \tag{12-22}$$

Indeed, if $F = 0$, the integral of A_μ along a path C from the origin to x does not depend on the curve

$$g(x) = g(x, 0; A) = P\exp\left[\int_C dx \cdot A(x)\right]$$

by definition of F. This element $g(x)$ satisfies (12-22). Conversely, a pure gauge A_μ has obviously a vanishing F.

Owing to the gauge arbitrariness, we may sometimes demand that the potential locally satisfy a definite condition. This is referred to as a choice of gauge. For instance, let n^μ be a fixed four-vector. There exists a gauge transformation, $A \to A'$, such that

$$n \cdot A'(x) = 0 \tag{12-23}$$

This is the so-called axial gauge.

To show that this is possible, let us introduce a four-vector N^μ such that $N \cdot n \neq 0$ (for instance, $N = n$, if $n^2 \neq 0$). Every point x may be written in a unique way:

$$x = \lambda(x)n + x_\perp \qquad x_\perp \cdot N = 0$$

with

$$\lambda(x) = \frac{N \cdot x}{N \cdot n} \qquad x_\perp = x - n\frac{N \cdot x}{N \cdot n}$$

Consider the segment C, $x(s) = s\lambda(x)n + x_\perp (0 \leq s \leq 1)$, and the integral

$$g(x) = P \exp\left\{\int_0^1 ds \frac{dx}{ds} \cdot A[x(s)]\right\} = P \exp\left[\int_0^{\lambda(x)} d\lambda\, n \cdot A(\lambda n + x_\perp)\right]$$

Introducing

$$A'_\mu = (\partial_\mu g^{-1})g + g^{-1}A_\mu g$$

we observe that

$$n \cdot A' = (n \cdot \partial g^{-1})g + g^{-1} n \cdot Ag = g^{-1}(n \cdot A - n \cdot \partial)g$$

From the definition of $g(x)$, we have

$$\frac{\partial}{\partial \lambda(x)} g(x) = n \cdot \partial g(x) = n \cdot A(x)g(x)$$

which implies $n \cdot A' = 0$.

Similarly, we may show that through a gauge transformation we can always fulfil locally the Lorentz gauge condition

$$\partial \cdot A = 0 \tag{12-24}$$

or any condition obtained from (12-23) or (12-24) by replacing the right-hand side by a function taking its value in the Lie algebra.

12-1-2 Classical Dynamics

Our aim is now to define a gauge-invariant action. As far as the coupling to the various multiplets of charged fields (matter fields) is concerned, it suffices to use the minimal coupling prescription. We substitute everywhere the covariant derivative for the ordinary one:

$$\partial_\mu \to D_\mu = \partial_\mu - A_\mu \tag{12-25}$$

where $A_\mu = A_{\mu a}T^a$ is understood to act in the representation of the matter field. The part of the action depending on A only must be a Lorentz scalar, gauge-invariant quantity—at most quadratic in the derivatives. The only candidate is the trace of $F_{\mu\nu}F^{\mu\nu}$ in some irreducible representation. For a simple Lie group traces in different irreducible representations are proportional since there exists a single quadratic invariant in the Lie algebra. We pick, for instance, the fundamental representation (of smallest dimension) and normalize its generators according to

$$\operatorname{tr}(t^a t^b) = -\tfrac{1}{2}\delta^{ab} \tag{12-26}$$

where the minus sign is related to our convention that the t are antihermitian. For instance, in $SU(2)$ and $SU(3)$, we take respectively

$$t^a = \frac{i\sigma^a}{2} \qquad a = 1, 2, 3$$

and

$$t^a = \frac{i\lambda^a}{2} \qquad a = 1, \ldots, 8$$

The action pertaining to A reads

$$I = \frac{1}{2g^2} \int d^4x \operatorname{tr} (F_{\mu\nu} F^{\mu\nu}) = -\frac{1}{4g^2} \int d^4x \sum_a F_{\mu\nu a} F^{\mu\nu}{}_a \tag{12-27}$$

The dimensionless parameter g, not to be confused with a group element, plays the role of a coupling constant. This is evident after rescaling A into

$$A \to gA$$

$$I \to \frac{1}{2} \int d^4x \operatorname{tr} (F_{\mu\nu} F^{\mu\nu})$$

where

$$F_{\mu\nu} = \partial_\mu A_\nu - \partial_\nu A_\mu - g[A_\mu, A_\nu] = -\frac{1}{g}[D_\mu, D_\nu] \tag{12-28}$$

$$D_\mu = \partial_\mu - gA_\mu$$

For the sake of brevity, however, we will not perform this rescaling until we derive the Feynman rules.

For a general invariance group the Lie algebra is the direct sum of simple Lie algebras, plus abelian factors. To each of these terms a quadratic invariant and an independent coupling constant may be associated. An example is provided by the Weinberg-Salam model for the weak and electromagnetic interactions, based on the group $SU(2) \times U(1)$, with two coupling constants related to the Fermi coupling and the electric charge (see Sec. 12-6 below).

The classical equations of motion are readily derived from the stationarity condition of the action (12-27):

$$0 = \delta I = \frac{2}{g^2} \int d^4x \operatorname{tr} [\delta A^\nu (\partial^\mu F_{\mu\nu} - [A^\mu, F_{\mu\nu}])]$$

Hence

$$[D^\mu, F_{\mu\nu}] = \partial^\mu F_{\mu\nu} - [A^\mu, F_{\mu\nu}] = 0$$

or

$$(D^\mu)_a{}^b F_{\mu\nu b} = 0 \tag{12-29}$$

which provides a nonabelian generalization of the Maxwell equations. The nonlinear character of the equations (12-29) makes their resolution nontrivial.

These equations possess the desired covariance property. If A_μ is a solution, so are its gauge transforms. Also, it must be clear that the system (12-29) is a compatible system. In particular, the contraction with ∂_ν gives zero:

$$\begin{aligned}\partial^\nu \partial^\mu F_{\mu\nu} &= \partial^\nu [A^\mu, F_{\mu\nu}] \\ &= [\partial^\nu A^\mu, F_{\mu\nu}] + [A^\mu, \partial^\nu F_{\mu\nu}] \\ &= [\partial^\nu A^\mu, F_{\mu\nu}] + [A^\mu, [A^\nu, F_{\mu\nu}]] \\ &= \tfrac{1}{2}[\partial^\nu A^\mu - \partial^\mu A^\nu - [A^\nu, A^\mu], F_{\mu\nu}] = 0\end{aligned}$$

Setting $g = 1$ the canonical energy momentum tensor is [compare with Eq. (1-105)]

$$\tilde{\Theta}^{\mu\nu} = 2 \operatorname{tr} (F^{\mu\rho}\partial^\nu A_\rho - \tfrac{1}{4}g^{\mu\nu}F^{\rho\sigma}F_{\rho\sigma})$$

but is not gauge invariant. This may be cured by subtraction of a total derivative $\Delta\tilde{\Theta}^{\mu\nu}$:

$$\Delta\tilde{\Theta}^{\mu\nu} = 2 \operatorname{tr} [F^{\mu\rho}(\partial_\rho A^\nu + [A^\nu, A_\rho])] = 2\partial_\rho \operatorname{tr} (F^{\mu\rho}A^\nu)$$

where use has been made of the equation of motion (12-29):

$$\begin{aligned}\Theta^{\mu\nu} = \tilde{\Theta}^{\mu\nu} - \Delta\tilde{\Theta}^{\mu\nu} &= 2 \operatorname{tr} (F^{\mu\rho}F^\nu{}_\rho - \tfrac{1}{4}g^{\mu\nu}F^{\rho\sigma}F_{\rho\sigma}) \\ &= \sum_a (F^{\mu\rho}_a F_{\rho a}^{\ \nu} - \tfrac{1}{4}g^{\mu\nu}F^{\rho\sigma}_a F_{\sigma\rho a})\end{aligned} \tag{12-30}$$

We introduce the analogs of the electric and magnetic fields

$$E^i_a = F^{i0}{}_a \qquad B^i{}_a = -\tfrac{1}{2}\varepsilon_{ijk}F^{jk}{}_a \qquad i, j, k = 1, 2, 3 \tag{12-31}$$

where, as throughout this chapter, the indices i, j, k are space indices ($i, j, k = 1, 2, 3$), while a, b, c are indices of the Lie algebra. In terms of $\mathbf{E}$ and $\mathbf{B}$,

$$\begin{aligned}\Theta^{00} &= \frac{1}{2}\sum_a (\mathbf{E}_a \cdot \mathbf{E}_a + \mathbf{B}_a \cdot \mathbf{B}_a) = -\operatorname{tr} (\mathbf{E}^2 + \mathbf{B}^2) \\ \Theta^{0i} &= \sum_a (\mathbf{E}_a \times \mathbf{B}_a)^i = -2 \operatorname{tr} (\mathbf{E} \times \mathbf{B})^i\end{aligned} \tag{12-32}$$

12-1-3 Euclidean Solutions to the Classical Equations of Motion

The search for classical solutions is motivated by the belief that a semiclassical approach may shed some light on the underlying quantum world and that classical configurations of fields that make the action stationary play an important role.

Especially interesting are nondissipative configurations with a finite energy, i.e., such that their energy remains localized in a finite spatial region and is not radiated to infinity. Such objects are candidates to describe extended systems at the quantum level. These are coherent states of the fundamental fields, provided they are stable against decay. Stability may follow from some conservation law, perhaps of topological nature. These systems have received the name of solitons or energy lumps. As they arise from an expansion about a nontrivial stationary point of the action, these lumps and their quantum excitations exhibit features that could not have been suspected from the ordinary perturbative expansion. We will examine in Sec. 12-5-3 an example of a four-dimensional gauge theory involving scalar fields, which possesses finite energy solutions. On the other hand, it is possible to show that nontrivial finite-energy nondissipative solutions do not exist in nonabelian gauge theories with gauge fields only. In other words, any solution of this sort is equivalent to $A^\mu \equiv 0$.

Consider the Lagrange function for a time-independent solution, i.e., the space integral of the lagrangian

$$L = L_1 - L_2 = \frac{1}{2}\int d^3x\,(\mathbf{E}_a \cdot \mathbf{E}_a - \mathbf{B}_a \cdot \mathbf{B}_a)$$

with the notation of Eq. (12-31). The total energy H is the sum $L_1 + L_2$; hence its finiteness implies the finiteness of L_1, L_2, and L. A solution if it exists is unstable under the scale transformations

$$A_a^0(\mathbf{x}) \to \rho\lambda A_a^0(\lambda\mathbf{x})$$

$$A_a^i(\mathbf{x}) \to \lambda A_a^i(\lambda\mathbf{x})$$

The Lagrange function is transformed into

$$L \to \rho^2\lambda L_1 - \lambda L_2$$

but it should be stationary at $\rho = \lambda = 1$. This implies

$$L_1 = L_2 = 0$$

Hence

$$F_{\mu\nu} = 0$$

and, by a global extension of the local statement of Eq. (12-22), this means that the gauge field is a pure gauge everywhere. This scaling argument (due to Coleman) may be shown to prevent such time-independent solutions in any space dimension different from four. It has been extended to more general nondissipative configurations.

There exist, however, nontrivial four-dimensional euclidean solutions to the classical equations. Before explaining the nature and role of these euclidean solutions, let us analyze further the structure of the ground state in nonabelian gauge theories.

To carry out this analysis it is convenient to impose the gauge condition $A_0 = 0$. Classically the ground state must correspond to time-independent field configurations of vanishing energy density. We have therefore $F_{\mu\nu} \equiv 0$, which means that the field $\mathbf{A}$ is a pure gauge

$$\mathbf{A}(\mathbf{x}) = \partial[g(\mathbf{x})]g^{-1}(\mathbf{x}) \tag{12-33}$$

Furthermore, we assume that we may restrict ourselves to transformation functions $g(\mathbf{x})$ that have the same limit in all spatial directions. We may take this limit to be the identity in the group

$$g(\mathbf{x}) \xrightarrow[|\mathbf{x}|\to\infty]{} e \tag{12-34}$$

(there is actually no very convincing argument to justify this restriction). Under these circumstances, all the field configurations of the form (12-33) and (12-34) may be regarded as describing a ground state. We may wonder whether all these copies of the vacuum are equivalent, i.e., whether there exists a continuous gauge transformation vanishing at spatial infinity that connects any two of them. The surprising answer is generally "no." Suppose for definiteness that the gauge group is $SU(2)$. Any matrix of $SU(2)$ may be parametrized in the form

$$g(\mathbf{x}) = u_0 + i\mathbf{u} \cdot \boldsymbol{\sigma} \tag{12-35}$$

in terms of the Pauli matrices, and with u real satisfying $u_0^2 + \mathbf{u}^2 = 1$. Hence, $SU(2)$ is isomorphic to the three-dimensional sphere $S_3 : u_0^2 + u_1^2 + u_2^2 + u_3^2 = 1$. On the other hand, the whole three-dimensional space with all points at infinity identified is also topologically equivalent to S_3. Therefore, the gauge transformation $g(\mathbf{x})$ associated with each vacuum is a mapping from S_3 onto S_3. According to homotopy theory, such mappings fall into equivalence classes. Two mappings $\mathbf{x} \to g_1(\mathbf{x})$ and $\mathbf{x} \to g_2(\mathbf{x})$ belong to the same class if there exists a continuous deformation from $g_1(\mathbf{x})$ to $g_2(\mathbf{x})$. In the case at hand, the classes are labeled by a positive or negative integer called the winding number or Pontryagin index of the class. This integer characterizes the number of times S_3 is mapped onto itself. It is equal to

$$n = \frac{1}{2\pi^2} \int d^3x \det \{A_a^k(\mathbf{x})\} \tag{12-36}$$

with $iA_a^k(\mathbf{x})\sigma^a$ given by Eq. (12-33). Examples of representatives of the class $n = 1$ are

$$g_1(\mathbf{x}) = -\exp\left(i\pi \frac{\boldsymbol{\sigma}\cdot\mathbf{x}}{\sqrt{\mathbf{x}^2+1}}\right) \qquad \text{or} \qquad g_1(\mathbf{x}) = \frac{\mathbf{x}^2-1}{\mathbf{x}^2+1} + 2i\frac{\boldsymbol{\sigma}\cdot\mathbf{x}}{\mathbf{x}^2+1}$$

and $g_n(\mathbf{x}) = g_1{}^n(\mathbf{x})$ belongs to the nth class. In the case of other simple groups, the same conclusion holds, namely, that there exists a discrete set of inequivalent vacua $|n\rangle$ labeled by an integer n.

As explained in Chap. 11, such a degeneracy of the ground state is intolerable and is actually cured by quantum tunneling. The true vacuum is a linear superposition of the degenerate approximate vacua $|n\rangle$. Since the above gauge transformation shifts the winding number n by one unit, and since the true vacuum must be invariant—up to a phase—under any gauge transformation, it has the form

$$|\theta\rangle = \sum_{n=-\infty}^{\infty} e^{in\theta}|n\rangle \tag{12-37}$$

where θ is a new arbitrary (and unexpected!) parameter in the theory.

We may then try to understand better how tunneling takes place between the degenerate vacua $|n\rangle$. To this end, we recall the Feynman-Kac formula [Eq. (9-198)]

$$\langle n_2|e^{-TH}|n_1\rangle = \int \mathcal{D}(A)\, e^{-I} \tag{12-38}$$

where H is the hamiltonian of the system and I is the euclidean action between times 0 and T:

$$\begin{aligned} I &= \frac{1}{4g^2}\int_0^T d\tau \int d^3x\, F^{\mu\nu}{}_a F^{\mu\nu}{}_a \\ &= \frac{1}{2g^2}\int_0^T d\tau \int d^3x\, (\mathbf{E}_a\cdot\mathbf{E}_a + \mathbf{B}_a\cdot\mathbf{B}_a) \end{aligned} \tag{12-39}$$

We have extended the definition of the strength tensor in euclidean variables. The functional integral in (12-38) is restricted by the boundary conditions on the winding number $n(A)$:

$$\begin{aligned} n[A(\mathbf{x}, \tau = 0)] &= n_1 \\ n[A(\mathbf{x}, \tau = T)] &= n_2 \end{aligned} \tag{12-40}$$

We are slightly cheating here, since the measure $\mathcal{D}(A)$ has not yet been properly defined. Its precise definition will be clarified in the quantization procedure of Sec. 12-2. We see now an alternate explanation of the classification of ground states according to homotopy classes of $S_3 \to S_3$. For $T \to \infty$, we look at those configurations the neighborhood of which yields a finite contribution to the Feynman-Kac formula. Since their euclidean action is finite, $F_{\mu\nu}$ must vanish at infinity in all euclidean directions, which amounts to saying that A_μ is a pure gauge field and yields a mapping from S_3, the surface at infinity in a four-dimensional euclidean space, onto the group, for example, $SU(2) \sim S_3$. Moreover, it may be shown that the winding number n associated with this mapping may be written as the euclidean integral

$$n = \frac{1}{64\pi^2}\int d^4x\, \varepsilon^{\mu\nu\rho\sigma} F^{\mu\nu}{}_a F^{\rho\sigma}{}_a = \frac{1}{32\pi^2}\int d^4x\, F^{\mu\nu}{}_a \tilde{F}^{\mu\nu}{}_a \tag{12-41}$$

where the dual tensor is $\tilde{F}_a^{\mu\nu} \equiv \frac{1}{2}\varepsilon^{\mu\nu\rho\sigma}F^{\rho\sigma}{}_a$. We observe that

$$\begin{aligned} F^{\mu\nu}{}_a \tilde{F}^{\mu\nu}{}_a &= \partial^\mu K^\mu \\ K^\mu &= 2\varepsilon^{\mu\nu\rho\sigma} A^\nu{}_a(\partial^\rho A^\sigma{}_a - \tfrac{1}{3}\varepsilon_{abc} A^\rho{}_b A^\sigma_c) \end{aligned} \tag{12-42}$$

For very large T, we expect the integral (12-38) to be dominated by the neighborhood of stationary configurations which are solutions to the euclidean classical equation of motion

$$D^{\mu}{}_{ab}F^{\mu\nu}{}_b = 0 \tag{12-43}$$

satisfying the boundary condition (12-40) or, equivalently, such that

$$n \equiv n_2 - n_1 = \frac{1}{32\pi^2}\int d^4x\, F^{\mu\nu}{}_a\tilde{F}^{\mu\nu}{}_a \tag{12-44}$$

Such a solution has an action bounded from below in terms of n. This comes from the positivity of

$$0 \leq \int d^4x\,(F^{\mu\nu}{}_a \pm \tilde{F}^{\mu\nu}{}_a)^2 = 2\int d^4x\,[(F^{\mu\nu}{}_a)^2 \pm F^{\mu\nu}{}_a\tilde{F}^{\mu\nu}{}_a] \tag{12-45}$$

whence

$$I = \frac{1}{4g^2}\int d^4x\,(F^{\mu\nu}{}_a)^2 \geq \frac{1}{4g^2}\left|\int d^4x\, F^{\mu\nu}{}_a\tilde{F}^{\mu\nu}{}_a\right| = \frac{8\pi^2|n|}{g^2} \tag{12-46}$$

The inequality is saturated by the self-dual or anti-self-dual configurations $F = \pm\tilde{F}$, since the equations of motion are then automatically satisfied as a consequence of the identity (12-21).

Explicit solutions of arbitrary winding number have been recently discovered. The solution proposed by Belavin, Polyakov, Schwartz, and Tyupkin for $n = \pm 1$ reads

$$A_\mu = \frac{x^2}{x^2+\lambda^2}\,[\partial_\mu g(x)]g^{-1}(x) \tag{12-47}$$

with

$$g(x) = \frac{x_0 \pm i\boldsymbol{\sigma}\cdot\mathbf{x}}{(x^2)^{1/2}}$$

Clearly, at infinity A_μ reduces to a pure gauge, arising from the identity mapping from S_3 onto S_3. We thus expect and may check by a direct computation that this solution has $n = \pm 1$. More generally, if we use the parametrization

$$\begin{aligned} A^i{}_a &= (\varepsilon_{aik}\partial_k \mp \delta_{ai}\partial_0)\ln f \qquad i, k = 1, 2, 3 \\ A^0{}_a &= \pm\partial_a \ln f \end{aligned} \tag{12-48}$$

the equation $F = \pm\tilde{F}$ reduces to

$$\frac{1}{f}\Box_E f = 0 \qquad \Box_E \equiv \partial_0^2 + \boldsymbol{\nabla}^2 \tag{12-49}$$

The case $n = \pm 1$ corresponds to $f^{(1)} = 1 + \lambda^2/x^2$ but other solutions corresponding to a winding number $\pm n$ may be devised:

$$f^{(n)}(x) = \sum_{i=1}^{n+1}\frac{\lambda_i{}^2}{(x - x_i)^2} \tag{12-50}$$

They depend on $n + 1$ arbitrary scales λ_i and positions x_i. Such solutions have been called pseudoparticles (by reference to the imaginary time coordinate) or instantons (due to their local character in time, as compared to the three-dimensional solitons). It is known that further solutions must exist. For $SU(2)$, for instance, the general solution depends on $8n - 3$ parameters up to gauge transformation. We will not dwell on this rapidly evolving problem, nor on many other related topics such as the modifications to be brought to this picture in the presence of massless fermions.

Although fascinating, these global properties will be omitted in the following. This is legitimate because we shall concentrate on the perturbative expansion, which is insensitive to the choice of the vacuum we expand about.

12-1-4 Gauge Invariance and Constraints

The equations of motion are insufficient to determine the fields $A_\mu(x)$, given a set of Cauchy conditions at a time t_0. Two solutions obtained from each other through a gauge transformation $g(x)$ such that $g(x) = e$ for $t \le t_0$ satisfy the same Cauchy conditions but may differ for $t > t_0$. Therefore, the gauge arbitrariness has to be restricted through the introduction of auxiliary conditions which do not affect gauge-invariant physical observables. The following discussion, due to Faddeev, will lead us to the functional quantization outlined in Chap. 9.

We restrict ourselves to a simple compact Lie group. The action, as in the abelian case, is first rewritten in terms of independent variables F and A:

$$I = \frac{1}{g^2} \int d^4x \operatorname{tr} \left[(\partial_\mu A_\nu - \partial_\nu A_\mu - [A_\mu, A_\nu])F^{\mu\nu} - \tfrac{1}{2} F_{\mu\nu} F^{\mu\nu}\right] \tag{12-51}$$

By variation with respect to F and A, we obtain respectively (12-16) and (12-29). We introduce the notations $\mathbf{E}$ and $\mathbf{B}$ of Eq. (12-31) and make use of the time-independent relation between $\mathbf{A}$ and $\mathbf{B}$ to eliminate the latter. In the sequel, $\mathbf{B}$ will stand for $\mathbf{B}(\mathbf{A})$. We have

$$\begin{aligned} I &= \frac{2}{g^2} \int d^4x \operatorname{tr} \left[(\partial^0 \mathbf{A} + \boldsymbol{\nabla} A^0 - [A^0, \mathbf{A}]) \cdot \mathbf{E} + \tfrac{1}{2}(\mathbf{E}^2 + \mathbf{B}^2)\right] \\ &= \frac{2}{g^2} \int d^4x \operatorname{tr} \left[\partial^0 \mathbf{A} \cdot \mathbf{E} + \tfrac{1}{2}(\mathbf{E}^2 + \mathbf{B}^2) - A^0(\boldsymbol{\nabla} \cdot \mathbf{E} + [\mathbf{A}, \mathbf{E}])\right] \end{aligned} \tag{12-52}$$

after an integration by parts. Since $\frac{1}{2}(\mathbf{E}^2 + \mathbf{B}^2)$ is the energy density, we face a typical problem of constrained systems. The canonical variables p and q are here $\mathbf{A}/g$ and $\mathbf{E}/g$. The variables A^0 play the role of Lagrange multipliers for the constraints

$$\Gamma(x) \equiv \boldsymbol{\nabla} \cdot \mathbf{E} + [\mathbf{A}, \mathbf{E}] = 0 \tag{12-53}$$

which are the $\nu = 0$ components of the equations of motion (12-29). Owing to the diagonal metric (12-26) (which enables us to have completely antisymmetric structure constants C_{abc}), we write the equal-time Poisson brackets as

$$\{A^i{}_a(x), E^j{}_b(y)\}_{x_0 = y_0} = g^2 \delta^{ij} \delta_{ab} \delta^3(\mathbf{x} - \mathbf{y}) \tag{12-54}$$

We also need the brackets $\{H, \Gamma\}$ and $\{\Gamma, \Gamma\}$ where

$$H = \frac{1}{2g^2} \int d^3x \sum_a (\mathbf{E}_a^2 + \mathbf{B}_a^2) \tag{12-55}$$

Therefore

$$\begin{aligned} \{\mathbf{E}_a(x), \Gamma(y)\}_{x^0 = y^0} &= g^2 [\mathbf{E}(x), t^a] \delta^3(\mathbf{x} - \mathbf{y}) \\ \{\mathbf{A}_a(x), \Gamma(y)\}_{x^0 = y^0} &= g^2 [\mathbf{D}_y, t^a \delta^3(\mathbf{x} - \mathbf{y})] \end{aligned} \tag{12-56}$$

where $\mathbf{D}_y$ stands for $\boldsymbol{\nabla}_y + \mathbf{A}(y)$, and the commutation is understood for the derivatives and for the Lie algebra as well. On the other hand, in a time-independent infinitesimal gauge transformation, **E** and **A** transform as

$$\begin{aligned}\delta \mathbf{E}_a(x) &= C_{abc}\delta\alpha_b(x)\mathbf{E}_c(x)\\ &= \frac{1}{g^2}\int_{y_0=x_0} d^3y \left\{\mathbf{E}_a(x), \sum_b \delta\alpha_b(y)\Gamma_b(y)\right\}\\ \delta \mathbf{A}_a(x) &= -\nabla\delta\alpha_a(x) + C_{abc}\delta\alpha_b(x)\mathbf{A}_c(x)\\ &= \frac{1}{g^2}\int_{y_0=x_0} d^3y \left\{\mathbf{A}_a(x), \sum_b \delta\alpha_b(y)\Gamma_b(y)\right\}\end{aligned}$$

Hence, the Γ are the infinitesimal generators of time-independent gauge transformations in this hamiltonian formalism. Without any further calculation, we conclude that

$$\begin{aligned}\{\Gamma_a(x), \Gamma_b(y)\}_{x_0=y_0} &= g^2 C_{abc}\Gamma_c(y)\delta^3(\mathbf{x}-\mathbf{y})\\ \{H, \Gamma_a(x)\} &= 0\end{aligned} \tag{12-57}$$

The equations of motion read

$$\begin{aligned}\partial_0 E^i &= (\text{curl } \mathbf{B})^i + \varepsilon_{ijk}[A^j, B^k] + [A^0, E^i]\\ \partial_0 \mathbf{A} &= -\mathbf{E} - \nabla A^0 + [A^0, \mathbf{A}]\end{aligned} \tag{12-58}$$

and the constraints

$$\begin{aligned}\nabla\cdot\mathbf{E} + [\mathbf{A}, \mathbf{E}] &= 0\\ B^i &= (\text{curl } \mathbf{A})^i + \tfrac{1}{2}\varepsilon_{ijk}[A^j, A^k]\end{aligned} \tag{12-59}$$

The observables are functionals of **E** and **A**, restricted to the manifold (12-59), such that their Poisson brackets with the latter vanish on that manifold. As a consequence they are invariant under time-independent gauge transformations. Such is the case of the hamiltonian density, for instance.

12-2 QUANTIZATION OF GAUGE FIELDS

The next step is to quantize the theory and to derive the perturbative rules for the computation of Green functions. The physical interpretation, which was a valuable guide in the case of electrodynamics, is missing here. In particular, we do not know whether asymptotic massless states, similar to photons, will emerge from the quantization.

12-2-1 Constrained Quantization

The method of Sec. 9-3 is basically hamiltonian, and dynamical variables have been chosen in a noncovariant way. Nevertheless, we hope to end up with covariant expressions.

The r constraint equations

$$\Gamma = \nabla \cdot \mathbf{E} + [\mathbf{A}, \mathbf{E}] \equiv \mathbf{D} \cdot \mathbf{E} = 0 \tag{12-60}$$

(r is the order of the Lie group) have vanishing Poisson brackets with H or among themselves, on the manifold (12-60), as prescribed in (9-152) and (9-153). Two pairs $(\mathbf{A}, \mathbf{E})$ satisfying (12-60) are equivalent if they lie on the same trajectory of the flow (9-142):

$$\frac{d\mathbf{A}}{du} = \{\Gamma, \mathbf{A}\} \qquad \frac{d\mathbf{E}}{du} = \{\Gamma, \mathbf{E}\}$$

that is, if they are obtained from each other through a time-independent gauge transformation. We then have to introduce an auxiliary condition which selects a unique representative in each equivalence class.

It is always possible to perform a gauge transformation such that everywhere

$$A^3 = 0 \tag{12-61}$$

[see Eq. (12-23)]. These r auxiliary constraints of the axial gauge define a unique element on each trajectory, if we assume that the fields vanish fast enough at spatial infinity. Since gauge transformations which preserve the condition $A^3 = 0$ are independent of the variable x^3 and must reduce to the identity at infinity, they do so throughout space. This condition also prevents from performing any global transformation.

We have next to compute the Poisson brackets between the Γ and the auxiliary conditions. The advantage of the axial gauge (12-61) is that these brackets do not depend on the dynamical variables $\mathbf{E}$ and $\mathbf{A}$. Indeed, $\{\Gamma, A^3\}$ is an infinitesimal gauge transformation of A^3, but restricted to $A^3 = 0$, and does not depend on $\mathbf{A}$ or $\mathbf{E}$. The determinant of this bracket that appears in the path integral (9-159) may be absorbed in the normalization, and will not be written in what follows. The generating functional of the Green functions reads

$$e^{G_A(J)} = \int \mathscr{D}(\mathbf{E}, \mathbf{A}, A^0) \prod_x \delta(A^3)$$
$$\times \exp\left\{\frac{2i}{g^2}\int d^4x \operatorname{tr}\left[\partial^0\mathbf{A}\cdot\mathbf{E} + \tfrac{1}{2}(\mathbf{E}^2 + \mathbf{B}^2) - A^0\mathbf{D}\cdot\mathbf{E} - gA\cdot J\right]\right\} \tag{12-62}$$

We have added a source term for the four components of A^μ, although A^3 vanishes and A^0 may be integrated explicitly.

The introduction of a source term enables us to question the system. Some answers such as the Green functions may depend on the choice of gauge and are just an intermediary step to get physical information. It would be tempting, of course, to consider directly gauge-invariant quantities such as the S matrix. Unfortunately, the asymptotic states are unknown and any computation of the would-be S-matrix elements is plagued with severe—though interesting—infrared divergences. Moreover, the theory requires renormalization, and an algorithm for eliminating ultraviolet divergences which does not rely on the Green functions remains to be found.

Since $\mathbf{E}$ appears only quadratically in the exponent of (12-62), the gaussian integration may be carried out:

$$e^{G_A(J)} = \int \mathscr{D}(A^\mu) \prod_x \delta(A^3) \exp\left\{\frac{i}{g^2}\int d^4x\, [\mathscr{L} - 2g \operatorname{tr}(A \cdot J)]\right\} \tag{12-63}$$

$$\mathscr{L} = \tfrac{1}{2} \operatorname{tr}(F^{\mu\nu}F_{\mu\nu})$$

The subscript A of $G_A(J)$ stresses the fact that Green functions depend on the choice of gauge. In spite of the simplicity of this axial gauge, the Feynman rules that we may derive are not Lorentz covariant. It is therefore natural to consider more general conditions. We shall proceed in several steps.

We first study another noncovariant gauge, the Coulomb gauge, defined by the auxiliary condition

$$\operatorname{div} \mathbf{A} = 0 \tag{12-64}$$

As stated above, it is always possible to find a local gauge transformation so as to satisfy (12-64). This condition was generally considered as determining uniquely the gauge transformation g. Phrased differently, if $\operatorname{div} \mathbf{A} = 0$, it was traditionally stated that the solution in g of $\operatorname{div}({}^g\mathbf{A}) = 0$ reduces to the identity under suitable conditions at spatial infinity. This is what happens in the abelian case. If $\operatorname{div} \mathbf{A} = 0$, $\operatorname{div} \mathbf{A}' = 0$ with $\mathbf{A}' = \mathbf{A} + \nabla\Lambda$, the harmonic function Λ vanishes at infinity and hence everywhere. However, as pointed out recently by Gribov, this is no longer true in the nonabelian case. The equation for g involves $\mathbf{A}$ and admits solutions for $\mathbf{A}$ large enough. Given $\mathbf{A}$ such that $\operatorname{div} \mathbf{A} = 0$ let us look for a time-independent infinitesimal gauge transformation $A'^i = A^i + [\delta\alpha, A^i] + \partial^i\delta\alpha$ such that $\operatorname{div} \mathbf{A}' = 0$:

$$\partial_i(\partial^i\delta\alpha + [\delta\alpha, A^i]) = 0 \tag{12-65}$$

For $\mathbf{A}$ small enough (equivalently, for small enough coupling constant) we may expand

$$\delta\alpha = \delta\alpha^{[0]} + g\delta\alpha^{[1]} + \cdots$$

$$\text{and} \qquad \Delta\delta\alpha^{[0]} = 0 \qquad \Delta\delta\alpha^{[1]} - \partial_i[\delta\alpha^{[0]}, A^i] = 0 \qquad \text{etc.}$$

If α is assumed to vanish at infinity, so do all the $\alpha^{[k]}$. There is no nontrivial solution. However, Eq. (12-65) may be regarded as a Schrödinger equation. For a potential $\mathbf{A}$ large enough, it can be checked that bound states do exist, i.e., solutions of

$$\Delta\alpha + \partial_i[\alpha, A_i] = E\alpha$$

with $E < 0$. Therefore, for intermediate magnitudes of $\mathbf{A}$, there must exist zero-energy solutions of fast decrease at infinity.

This objection seems embarrassing for our program, since the comparison of the quantization in different gauges relies on the assumption of uniqueness of the transformation which relates them. However, since this phenomenon appears for large values of the field, it does not affect the construction of the perturbative series which is by essence a small-field (small-fluctuation) expansion about a given classical configuration. We shall henceforth discard this Gribov phenomenon. Whenever we talk of uniqueness of the choice of gauge (12-64), we understand it in a perturbative sense.

The auxiliary conditions (12-64) have a nontrivial Poisson bracket with the constraint

$$\{\text{div}\, \mathbf{A}_a, \Gamma_b(y)\}_{x_0=y_0} = \frac{\delta}{\delta\alpha_b(y)} \text{div}\, \mathbf{A}'_a(x)$$
$$= [-\delta_{ab}\Delta_x + C_{abc}\nabla_x \cdot \mathbf{A}_c(x)]\delta^3(\mathbf{x}-\mathbf{y}) \qquad (12\text{-}66)$$

We introduce the operator $\mathscr{M}$:

$$\mathscr{M}_{ab}(x, y) = [-\Delta_x\delta_{ab} + C_{abc}\nabla_x \cdot \mathbf{A}_c(x)]\delta^3(\mathbf{x}-\mathbf{y})$$
$$= [-\Delta_x\delta_{ab} + C_{abc}\mathbf{A}_c(x) \cdot \nabla_x]\delta^3(\mathbf{x}-\mathbf{y}) \qquad (12\text{-}67)$$

where the last expression holds when $\mathscr{M}$ is restricted to the constraint manifold. In the Coulomb gauge, we thus write the generating functional of Green functions as

$$e^{G_c(J)} = \int \mathscr{D}(A) \prod_t \left[\det \mathscr{M} \prod_x \delta(\nabla \cdot \mathbf{A})\right] \exp\left\{\frac{i}{g^2}\int d^4x\, [\mathscr{L} - 2g\, \text{tr}\,(J \cdot A)]\right\} \qquad (12\text{-}68)$$

Through a modification of the normalization in Eq. (12-68), we may replace $\det \mathscr{M}$ by $\det \mathscr{M}\mathscr{M}_0^{-1}$ where

$$\mathscr{M}_0 = -\Delta\delta_{ij}\delta^3(\mathbf{x}-\mathbf{y})$$

Using $\det \mathscr{M}\mathscr{M}_0^{-1} = \exp[\text{tr}\ln(\mathscr{M}\mathscr{M}_0^{-1})]$ and $\text{tr}\ln(1 + A) = \sum [(-1)^{n-1}/n]\, \text{tr}\, A^n$, we may derive the Feynman rules in this gauge. However, they are not covariant either.

12-2-2 Integration over the Gauge Group

The gauge transformations used so far were time independent. In order to make the Lorentz covariance manifest, it is more appropriate to consider time-dependent transformations as well. The method to implement them will appear as a by-product of a different problem. How can we show the equivalence between the axial and Coulomb gauges?

Let us drop the coupling to the external source J, replace it possibly by sources coupled to gauge-invariant quantities $O_i(x)$, and consider the functional integral (at a given time)

$$X = \int \mathscr{D}(\mathbf{A}, \mathbf{E}) \prod_x \delta(A^3)\mu(\mathbf{A}, \mathbf{E}) \qquad (12\text{-}69)$$

where μ is the gauge-invariant functional

$$\mu(\mathbf{A}, \mathbf{E}) = \exp\left\{i \int d^4x\, [\mathscr{L}(x) + J_i(x)O_i(x)]\right\}$$

In a gauge transformation

$$\mathbf{A} \to ({}^g\mathbf{A}) = -(\nabla g)g^{-1} + g\mathbf{A}g^{-1} \qquad \mathbf{E} \to ({}^g\mathbf{E}) = g\mathbf{E}g^{-1} \qquad (12\text{-}70)$$

where g is considered as a product $\prod_x g(x)$ and μ is invariant. So is the measure

$\mathscr{D}(\mathbf{A}, \mathbf{E})$, since the gauge transformations are canonical. Hence

$$X = \int \mathscr{D}(\mathbf{A}, \mathbf{E}) \prod_x \delta^3({}^gA^3)\mu(\mathbf{A}, \mathbf{E})$$

As the conditions $\nabla \cdot \mathbf{A} = 0$ or $A^3 = 0$ are two equivalent ways to define uniquely (in a perturbative sense) a point in each equivalence class, there must exist a $g_0(A)$ such that

$$^{g_0}A^3 = 0 \Rightarrow \nabla \cdot \mathbf{A} = 0 \tag{12-71}$$

Let us compute the jacobian of this transformation. We set

$$\Delta^{-1}(A) \equiv \int \mathscr{D}(g) \prod_x \delta(\nabla \cdot {}^g\mathbf{A}) \tag{12-72}$$

where $\mathscr{D}(g)$ stands for the infinite product of invariant measures on compact groups, isomorphic to the group G at every space point x, $\mathscr{D}(g) = \prod_x Dg(x)$.

The invariance of the measure $Dg' = D(gg')$ entails

$$\Delta({}^gA) = \Delta(A) \tag{12-73}$$

We then multiply Eq. (12-69) by

$$1 = \Delta(A) \int \mathscr{D}(g) \prod_x \delta(\nabla \cdot {}^g\mathbf{A}) \tag{12-74}$$

interchange the order of integrations, and use Eq. (12-73) to write

$$\begin{aligned} X &= \int \mathscr{D}(\mathbf{A}, \mathbf{E})\mu(\mathbf{A}, \mathbf{E}) \prod_x \delta[A^3(x)] \int \mathscr{D}(g) \prod_y \delta[\nabla \cdot {}^g\mathbf{A}(y)]\Delta(A) \\ &= \int \mathscr{D}(\mathbf{A}, \mathbf{E})\mu(\mathbf{A}, \mathbf{E})\Delta(A) \prod_y \delta[\nabla \cdot \mathbf{A}(y)] \int \mathscr{D}(g) \prod_x \delta[{}^{g^{-1}}A^3(x)] \end{aligned} \tag{12-75}$$

Owing to the invariance of the measure $\mathscr{D}(g)$, we may change in the last integral g^{-1} into gg_0, where g_0 is such that $^{g_0}A^3 = 0$. For simplicity, we denote $\mathbf{B} = {}^{g_0}\mathbf{A}$ and write

$$\int \mathscr{D}(g) \prod_x \delta[{}^{g^{-1}}A^3(x)] = \int \mathscr{D}(g) \prod_x \delta[{}^gB^3(x)]$$

Since $B^3 = 0$, it suffices (at least within perturbation theory) to consider only infinitesimal transformations g and to perform the group integration in the vicinity of the identity

$$g(x) = e + \alpha(x)$$

where α is infinitesimal. Under these circumstances, the measure at each point $Dg(x)$ reduces to the product $\prod_i d\alpha_i(x)$ and

$$^gB^3(x) = -\frac{\partial \alpha(x)}{\partial x^3}$$

Hence the integral

$$\int \mathscr{D}(g) \prod_x \delta[{}^gB^3(x)] = \int \mathscr{D}(\alpha) \prod_x \delta\left[\frac{\partial \alpha(x)}{\partial x^3}\right]$$

is independent of **A** and may be absorbed in the normalization. Therefore with an overall μ-independent normalization factor N, we have

$$X = N \int \mathscr{D}(\mathbf{A}, \mathbf{E}) \Delta(A) \mu(\mathbf{A}, \mathbf{E}) \prod_x \delta[\nabla \cdot \mathbf{A}(x)] \tag{12-76}$$

To recover (12-68), we have to show that $\Delta(A)$ is proportional to the determinant of the operator $\mathscr{M}$ defined by (12-67). Since $\Delta(A)$ is multiplied by $\delta(\nabla \cdot \mathbf{A})$ in (12-76), it is sufficient to compute it for transverse fields **A**. Then, only infinitesimal g contribute to the integral defining Δ:

$$\begin{aligned} \Delta^{-1}(A) &= \int \mathscr{D}(g) \prod_x \delta[\nabla \cdot {}^g\mathbf{A}(x)] \\ &= \int \mathscr{D}(\alpha) \prod_x \delta(\nabla \cdot \{-\nabla\alpha(x) + [\alpha(x), \mathbf{A}(x)]\}) \\ &= \det{}^{-1} \mathscr{M} \end{aligned} \tag{12-77}$$

with, as in (12-67),

$$\mathscr{M}_{ab}(x, y) = \frac{\delta}{\delta\alpha_b(y)} [\nabla \cdot {}^g\mathbf{A}_a(x)] \bigg|_{\alpha=0}$$

We have just shown the equivalence of the axial and Coulomb gauges, as far as the computation of gauge-invariant quantities $\mu(\mathbf{A}, \mathbf{E})$ is concerned. What happens to the source term, and hence to the Green functions, has not been examined. However, only local and canonical changes of field variables have been performed in the previous derivation. By virtue of the equivalence theorem mentioned in Chap. 9, we do not expect these transformations to modify the physical content of the theory.

The previous method is interesting, as it allows us to deal with time-dependent gauge transformations and to impose covariant auxiliary conditions. Let

$$\mathscr{F}(A) = 0 \tag{12-78}$$

be such a condition. $\mathscr{F}$ is a local functional of A, that is, a function of $A(x)$ and of its derivatives, it takes its values in the Lie algebra, and it may depend on A^0. Using the same method as above we find

$$X = \int \mathscr{D}(A) \prod_x \delta[\mathscr{F}(A)] \det \mathscr{M}_{\mathscr{F}} \mu(A)$$

up to a normalization. We are thus led to consider Green functions generated by

$$e^{G_{\mathscr{F}}(J)} = \int \mathscr{D}(A) \prod_x \delta[\mathscr{F}(A)] \det \mathscr{M}_{\mathscr{F}} \exp\left\{\frac{i}{g^2}\int d^4x \, [\mathscr{L} - 2g \operatorname{tr}(J \cdot A)]\right\} \tag{12-79}$$

with

$$\Delta_{\mathscr{F}}^{-1}(A) = (\det \mathscr{M}_{\mathscr{F}})^{-1} = \int \mathscr{D}(g) \prod_x \delta[\mathscr{F}({}^gA)] \tag{12-80}$$

computed for $\mathscr{F}(A) = 0$. Only infinitesimal transformations contribute to the integral and

$$\begin{aligned}
\mathscr{M}_{ab}(x, y) &= \frac{\delta}{\delta\alpha_b(y)} \mathscr{F}_a[{}^gA(x)]\bigg|_{g=e} \\
&= \frac{\delta}{\delta\alpha_b(y)}\left[\frac{\delta\mathscr{F}_a}{\delta A^\mu{}_c(x)} D^\mu_{cd}\alpha_d(x)\right]_{\alpha=0} \\
&= \frac{\delta\mathscr{F}_a}{\delta A^\mu{}_c(x)} D^\mu_{x,cb}\delta^4(x-y)
\end{aligned} \tag{12-81}$$

where

$$D^\mu_{x,cb} = \partial^\mu_x \delta_{cb} + C_{cbd}A^\mu{}_d(x)$$

For instance, if we use the Lorentz gauge

$$\mathscr{F}(A) \equiv \partial_\mu A^\mu = 0$$

the operator $\mathscr{M}$ reads

$$\begin{aligned}
\mathscr{M}_{L,ab}(x, y) &= \partial_\mu D^\mu{}_{ab}\delta^4(x-y) \\
&= [\Box\delta_{ab} + C_{abc}\partial^\mu_x A_{\mu c}(x)]\delta^4(x-y) \\
&= [\Box\delta_{ab} + C_{abc}A_{\mu c}(x)\partial^\mu{}_x]\delta^4(x-y)
\end{aligned} \tag{12-82}$$

Returning to the axial gauges

$$\mathscr{F} \equiv n \cdot A = 0$$

we see that $\mathscr{M}$ is independent of A, when restricted to the constraint manifold; hence $\det \mathscr{M}$ may be absorbed into the normalization.

Instead of relying on the canonical hamiltonian quantization and on the subsequent functional manipulations, we might decide to modify the ill-defined integral

$$X = \int \mathscr{D}(A)\mu(A)$$

through the insertion of the identity written as

$$1 = \int \mathscr{D}(g)\Delta(A) \prod_x \delta[\mathscr{F}({}^gA)]$$

The gauge invariance of $\mu(A)$, of $\Delta(A)$, and of the measure $\mathscr{D}(A)$ would then enable us to factorize an infinite group volume

$$X = \left[\int \mathscr{D}(g) \right] \int \mathscr{D}(A)\mu(A)\Delta(A) \prod_x \delta[\mathscr{F}(A)]$$

This crude argument has the merit of exhibiting clearly the enormous degeneracy of the degrees of freedom as the origin of the problem.

The previous developments are readily extended to auxiliary conditions of the form

$$\mathscr{F}(A) = C$$

where $\mathscr{F}$ and the given function $C(x)$ take their values in the Lie algebra. This modification does not affect the form of the operator $\mathscr{M}$ in (12-80).

The variations $(\delta/\delta g)[\mathscr{F}({}^gA) - C]$ do not depend on C, and the only dependence of $\Delta_{\mathscr{F}}$ on C comes from g_0, the gauge transformation such that $\mathscr{F}({}^{g_0}A) = C$:

$$\Delta_{\mathscr{F}}(A, C) = \Delta({}^{g_0}A)$$

We may write

$$1 = \int \mathscr{D}(g)\Delta({}^{g_0}A)\delta[\mathscr{F}({}^gA) - C] = \int \mathscr{D}(g)\Delta({}^gA)\delta[\mathscr{F}({}^gA) - C]$$

where the second equality holds because the δ function implies $g = g_0$. Any reference to C has disappeared.

Since gauge-invariant quantities should not be sensitive to changes of auxiliary conditions, it is possible to average over C with a gaussian weight, i.e., to substitute for $\delta[\mathscr{F}(A) - C]$ the quantity

$$\int \mathscr{D}(C) \exp\left[\frac{i\lambda}{g^2} \int d^4x \operatorname{tr}(C^2)\right] \delta[\mathscr{F}(A) - C] = \exp\left\{\frac{i\lambda}{g^2} \int d^4x \operatorname{tr}[\mathscr{F}^2(A)]\right\}$$

In its final form the generating function reads

$$e^{G(J)} = \int \mathscr{D}(A) \det \mathscr{M}_{\mathscr{F}} \exp\left(\frac{i}{g^2} \int d^4x \{\mathscr{L}(x) + \lambda \operatorname{tr}[\mathscr{F}^2(A)] - 2g \operatorname{tr}(J \cdot A)\}\right) \tag{12-83}$$

The perturbative expansion of det $\mathscr{M}$ leads to nonlocal interactions between gauge fields. It is more convenient to perform a final manipulation on this determinant, and to reexpress it as a local interaction of fictitious fields. The discussion of the integration on a Grassmann algebra has yielded the formula (9-76):

$$\det \mathscr{M} = \int \prod_k d\bar{\eta}_k \, d\eta_k \, e^{-\bar{\eta}\mathscr{M}\eta}$$

where $\mathscr{M}$ denotes an $n \times n$ matrix. Changing $\mathscr{M}$ into $i\mathscr{M}$ and extending this result to an infinite algebra, we may write the determinant det $\mathscr{M}_{\mathscr{F}}$ as a functional integral

$$e^{G(J)} = \int \mathscr{D}(A, \eta, \bar{\eta}) \exp\left(\frac{i}{g^2}\int d^4x \{\mathscr{L} + \lambda \operatorname{tr}[\mathscr{F}^2(A)] - \bar{\eta}\mathscr{M}\eta - 2g \operatorname{tr}(J \cdot A)\}\right) \tag{12-84}$$

The modified or effective lagrangian

$$\mathscr{L}_{\text{eff}}(A, \eta, \bar{\eta}) = \mathscr{L}(A) + \lambda \operatorname{tr} \mathscr{F}^2(A) - \bar{\eta}\mathscr{M}\eta \tag{12-85}$$

involves the gauge field A and the new anticommuting auxiliary scalar fields η and $\bar{\eta}$, the so-called Faddeev-Popov ghosts. We emphasize that these fields are unphysical and only play an algebraic role. To insure global invariance, η and $\bar{\eta}$ transform according to the adjoint representation. The ghost term in the lagrangian (12-85) reads

$$\begin{aligned}\int d^4x\, \bar{\eta}\mathscr{M}_{\mathscr{F}}\eta &= \iint d^4x\, d^4y\, \bar{\eta}_a(x) \frac{\delta \mathscr{F}_a[{}^gA(x)]}{\delta\alpha_b(y)} \eta_b(y) \\ &= \int d^4x\, \bar{\eta}_a(x) \frac{\partial \mathscr{F}_a(A)}{\partial A^\mu{}_b} D^\mu{}_{bc}\eta_c(x)\end{aligned} \tag{12-86}$$

$$(D^\mu \eta)_a = \partial^\mu \eta_a + C_{abc}\eta_b A^\mu{}_c$$

In general, the kernel $\mathscr{M}_{\mathscr{F}}$ is not hermitian; therefore the ghost lines in a Feynman diagram will have to be oriented. For instance, if $\mathscr{F}$ is the Lorentz covariant condition $\mathscr{F} = \partial_\mu A^\mu$, we find

$$\int d^4x\, \bar{\eta}\mathscr{M}\eta = \int d^4x\, \bar{\eta}_a \partial_\mu D^\mu{}_{ab}\eta_b \tag{12-87}$$

In this case, the lagrangian of Eq. (12-85) reads

$$\mathscr{L}_{\text{eff}} = \tfrac{1}{2} \operatorname{tr}(F_{\mu\nu}F^{\mu\nu}) + \lambda \operatorname{tr}(\partial \cdot A)^2 - \bar{\eta}\partial_\mu D^\mu \eta \tag{12-88}$$

The parameter λ (also denoted α^{-1} or ξ^{-1} in the literature) reminds us of the arbitrariness of the auxiliary condition. The choices $\lambda = 1$ or $\lambda^{-1} = 0$ are referred to as Feynman or Landau gauges respectively.

12-2-3 Feynman Rules

Equation (12-88) gives the desired solution to the quantization problem, since it provides a local Lorentz covariant expression for the effective lagrangian. The quadratic part in A is invertible as a consequence of the fact that the condition $\mathscr{F} = C$ has picked a single representative in each equivalence class.

We are now in a position to write the Feynman rules. We rescale the fields A and $\eta, \bar{\eta}$ by g, the coupling constant. As the Feynman diagrams also include ghost fields, it is helpful to introduce (anticommuting) sources $\bar{\xi}, \xi$ coupled to

$\eta, \bar{\eta}$. We have thus

$$e^{G(J,\xi,\bar{\xi})} = \int \mathscr{D}(A, \eta, \bar{\eta}) \exp\left\{i \int d^4x \, [\mathscr{L}_{\text{eff}}(A, \eta, \bar{\eta}; g, \lambda) + J^\mu{}_a A_{\mu a} + \bar{\xi}_a \eta_a + \bar{\eta}_a \xi_a]\right\} \tag{12-89}$$

$$\begin{aligned}
\mathscr{L}_{\text{eff}}(A, \eta, \bar{\eta}; g, \lambda) &= -\tfrac{1}{4} F_{\mu\nu a} F^{\mu\nu}{}_a - \frac{\lambda}{2}(\partial_\mu A^\mu{}_a)^2 + \partial_\mu \bar{\eta}_a (D^\mu \eta)_a \\
F^{\mu\nu}{}_a &= \partial^\mu A^\nu{}_a - \partial^\nu A^\mu{}_a - g C_{abc} A^\mu{}_b A^\nu{}_c \\
(D^\mu \eta)_a &= \partial^\mu \eta_a - g C_{abc} A^\mu{}_b \eta_c
\end{aligned} \tag{12-90}$$

The propagators of the gauge and ghost fields are respectively

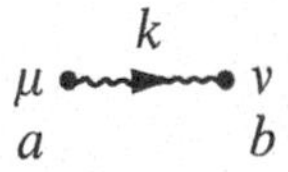

$$\{-i[\Box - (1-\lambda)\partial \otimes \partial]\}^{-1} = -i\delta_{ab}\left[\frac{g_{\mu\nu}}{k^2 + i\varepsilon} - (1 - \lambda^{-1})\frac{k_\mu k_\nu}{(k^2 + i\varepsilon)^2}\right] \tag{12-91}$$

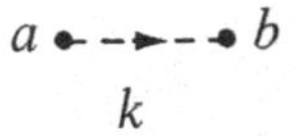

$$(i\Box)^{-1} = \frac{i\delta_{ab}}{k^2 + i\varepsilon}$$

and behave as k^{-2} at large momentum. Therefore, both fields are given the dimension one for the ultraviolet power counting. The coupling constant g is dimensionless. There are three kinds of vertices. If we orient the ghost lines from the $\bar{\eta}$ to the η (as we do for genuine fermions) and include the factor i from the expansion of $\exp[i \int d^4z \, \mathscr{L}_{\text{int}}(z)]$, we get

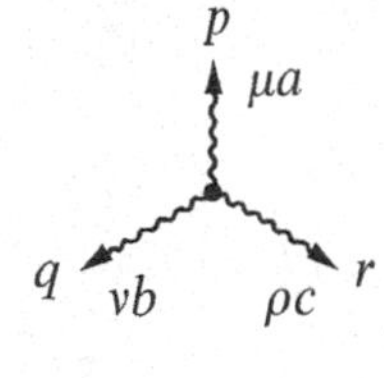

$$g C_{abc} (2\pi)^4 \delta^4(p + q + r)[g_{\mu\nu}(p - q)_\rho + g_{\nu\rho}(q - r)_\mu + g_{\rho\mu}(r - p)_\nu]$$

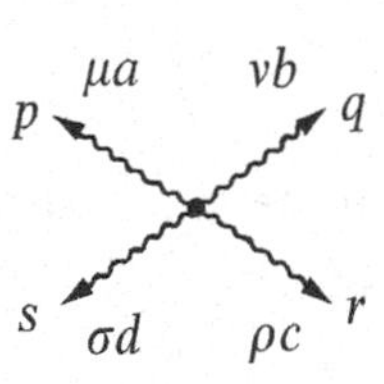

$$\begin{aligned}
&-ig^2(2\pi)^4 \delta^4(p + q + r + s) \\
&\quad \times \{C_{eab} C_{ecd}(g_{\mu\rho} g_{\nu\sigma} - g_{\mu\sigma} g_{\nu\rho}) \\
&\qquad + C_{eac} C_{edb}(g_{\mu\sigma} g_{\rho\nu} - g_{\mu\nu} g_{\rho\sigma}) \\
&\qquad + C_{ead} C_{ebc}(g_{\mu\nu} g_{\sigma\rho} - g_{\mu\rho} g_{\sigma\nu})\}
\end{aligned} \tag{12-92}$$

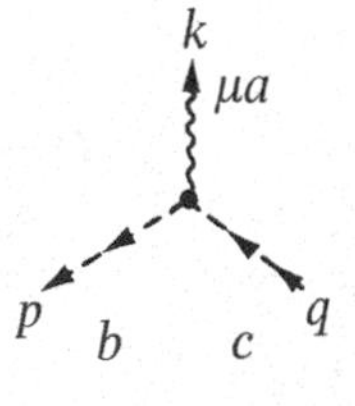

$$-g C_{abc} p_\mu (2\pi)^4 \delta^4(k + p - q)$$

Notice the (expected) asymmetric character of the last vertex. With our convention the outgoing ghost line carries the momentum arising from the differentiation. For an actual computation, the above rules have of course to be supplemented by the prescriptions derived in Chap. 6, namely, integrations with the measure $d^4k/(2\pi)^4$ of all internal momenta, factorization of the global energy momentum delta function, symmetry factors, and a factor minus one for every ghost loop.

Let us complete the Feynman rules when matter fields are coupled in a minimal way. We add to the lagrangian (12-90) the terms

$$\mathscr{L}_f = i\bar{\psi}\not{D}\psi - m\bar{\psi}\psi$$

or

$$\mathscr{L}_b = (D^\mu\phi)^\dagger(D_\mu\phi) - m^2\phi^\dagger\phi - P(\phi^\dagger\phi) \tag{12-93}$$

for fermion and boson fields respectively. Here ψ and ϕ stand for multiplets of fields, transforming according to some representation R of the group, the infinitesimal generators of which are the antihermitian matrices T^a and P is a polynomial. We recall that D^μ denotes

$$D^\mu = \partial^\mu - gA^\mu{}_a T^a$$

The additional Feynman rules are

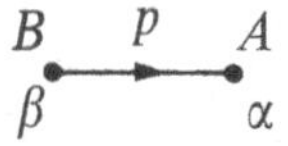

$$\left(\frac{i}{\not{p} - m}\right)_{\alpha\beta}\delta_{AB} \quad \text{and} \quad \frac{i}{p^2 - m^2}\delta_{AB}$$

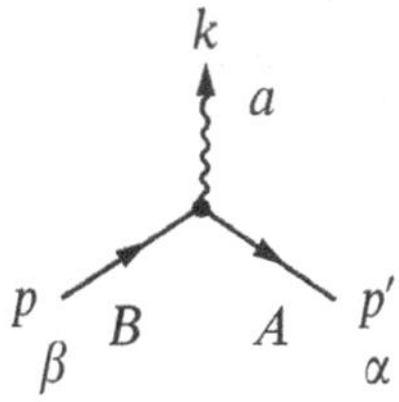

$$\begin{matrix} g(\gamma_\mu)_{\alpha\beta}T^a_{AB} \\ \times (2\pi)^4\delta^4(p - p' - k) \end{matrix} \quad \text{and} \quad \begin{matrix} gT^a_{AB}(p_\mu + p'_\mu) \\ \times (2\pi)^4\delta^4(p - p' - k) \end{matrix} \tag{12-94}$$

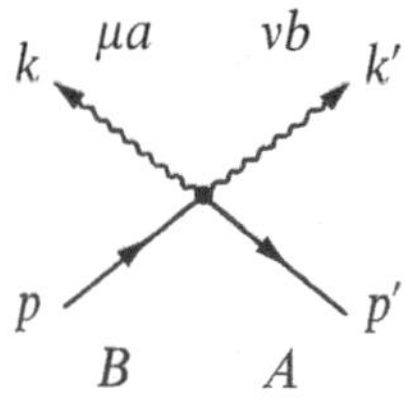

$$\begin{matrix} -ig^2 g_{\mu\nu}\{T^a, T^b\}_{AB} \\ \times (2\pi)^4\delta^4(p - p' - k - k') \end{matrix}$$

The first column refers to fermions and the second to bosons. In the latter case, additional vertices stem from the self-coupling polynomial $P(\phi^\dagger\phi)$.

For completeness, we also give here the Feynman rules in the axial gauge. To this end, we add to the lagrangian a term of the form $(n \cdot A)^2$:

$$\mathscr{L}_{\text{eff}} = \sum_a \left[-\tfrac{1}{4}F_{\mu\nu a}F^{\mu\nu}{}_a - \frac{\lambda}{2}(n_\mu A^\mu{}_a)^2\right] \tag{12-95}$$

The axial gauge is reached by letting λ^{-1} go to zero and, as shown above, no ghost term is required in this limit.

The corresponding propagator is

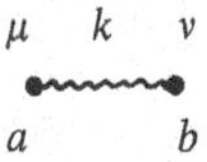

$$
\begin{aligned}
&[-i(\Box - \partial \otimes \partial - \lambda n \otimes n)]^{-1} \\
&= -i\delta_{ab}\left[\frac{g_{\mu\nu}}{k^2} + \frac{(k^2 + \lambda n^2)k_\mu k_\nu}{\lambda k^2 (k \cdot n)^2} - \frac{k_\mu n_\nu + k_\nu n_\mu}{k^2 k \cdot n}\right] \\
&\xrightarrow[\lambda^{-1} \to 0]{} -i\delta_{ab}\frac{g_{\mu\nu} - (k_\mu n_\nu + k_\nu n_\mu)(k \cdot n)^{-1} + n^2 (k \cdot n)^{-2} k_\mu k_\nu}{k^2}
\end{aligned}
\tag{12-96}
$$

The problems raised by the new type of singularity in the denominators will not be investigated here.

Notice that the propagator behaves as k^{-2} only in the limit $\lambda^{-1} \to 0$. Finally, the A^3 and A^4 vertices are the same as in Eq. (12-92).

12-3 THE EFFECTIVE ACTION AT THE ONE-LOOP ORDER

The previous Feynman rules lead to a theory renormalizable according to power counting, since all propagators behave as k^{-2} and all vertices have dimension four. However, the problem is to show that gauge invariance is preserved by the renormalization. Before embarking on a long and rather technical proof to all orders, it is instructive to perform an explicit computation of the one-loop effective action.

12-3-1 General Form

To cope with the multiplicity of indices let us use a compact functional notation. As in Eqs. (6-73) or (9-107) we obtain the effective action through a Legendre transformation

$$
i\Gamma(A, \eta, \bar{\eta}) = G(J, \xi, \bar{\xi}) - i\int d^4x\,(J \cdot A + \bar{\xi}\eta + \bar{\eta}\xi)
\tag{12-97}
$$

where
$$
A = \frac{\delta G}{i\delta J} \qquad \eta = \frac{\delta G}{i\delta \bar{\xi}} \qquad \bar{\eta} = -\frac{\delta G}{i\delta \xi}
$$

Lorentz and group indices have been omitted. The derivatives with respect to anticommuting variables are understood as left derivatives. In other words, we write

$$
\delta G(J, \xi, \bar{\xi}) = \int d^4x \left[\delta\xi(x)\frac{\delta G}{\delta\xi(x)} + \delta\bar{\xi}(x)\frac{\delta G}{\delta\bar{\xi}(x)}\right] + \int d^4x\, d^4y\, \delta\xi(x)\delta\bar{\xi}(y)\frac{\delta^2 G}{\delta\bar{\xi}(y)\delta\xi(x)} + \cdots
$$

This prescription is responsible for the minus sign in the expression (12-97) for $\bar{\eta}$.

To lowest order, Γ reduces to the action

$$
\Gamma^{[0]} = I(A, \eta, \bar{\eta}) = \int d^4x\, \mathscr{L}_{\text{eff}}(A, \eta, \bar{\eta}; g, \lambda)
\tag{12-98}
$$

while the first correction $\Gamma^{[1]}$ results from a gaussian integration over primed variables of the action expanded to second order:

$$I \approx \int d^4x\, \mathscr{L}_{\text{eff}}(A + A', \eta + \eta', \bar{\eta} + \bar{\eta}'; g, \lambda)$$

We write the quadratic form explicitly as

$$\begin{aligned} I_q = 2\int d^4x \operatorname{tr} \Big\{ &\tfrac{1}{4}([D_\mu, A'_\nu] - [D_\nu, A'_\mu])([D^\mu, A'^\nu] - [D^\nu, A'^\mu]) \\ &- \frac{g}{2} F^{\mu\nu}[A'_\mu, A'_\nu] + \frac{\lambda}{2}(\partial^\mu A'_\mu)^2 - \partial_\mu \bar{\eta}'[D^\mu, \eta'] \\ &+ g\partial_\mu \bar{\eta}[A'_\mu, \eta'] + g\partial_\mu \bar{\eta}'[A'^\mu, \eta] \Big\} \end{aligned} \tag{12-99}$$

where matrix notations are used, $\eta = \eta_a t^a$, etc. This fairly complicated form mixes commuting and anticommuting variables. We recall the formulas

$$\begin{aligned} &\int \prod_1^n \frac{dx_i}{(2\pi)^{1/2}} e^{-xQx/2 + u\cdot x} = [\det Q]^{-1/2}\, e^{uQ^{-1}u/2} \\ &\int \prod_i^n d\bar{\eta}_i\, d\eta_i\, e^{-\bar{\eta}\mathscr{M}\eta + \bar{\xi}\cdot\eta + \bar{\eta}\cdot\xi} = \det \mathscr{M}\, e^{\bar{\xi}\mathscr{M}^{-1}\xi} \end{aligned} \tag{12-100}$$

for gaussian integrals of commuting and anticommuting variables respectively. In the mixed case, we get

$$\begin{aligned} &\int \prod \frac{dx_i}{(2\pi)^{1/2}} \prod (d\bar{\eta}_i\, d\eta_i) \exp(-\tfrac{1}{2}xQx + \bar{\eta}\alpha x + x\bar{\beta}\eta - \bar{\eta}\mathscr{M}\eta + u\cdot x + \bar{\xi}\cdot\eta + \bar{\eta}\cdot\xi) \\ &\qquad = \det \mathscr{M}\, [\det \tilde{Q}]^{-1/2} \exp\{\tfrac{1}{2}[u + \bar{\xi}\mathscr{M}^{-1}\alpha \\ &\qquad\quad + (\bar{\beta}\mathscr{M}^{-1}\xi)^T]\tilde{Q}^{-1}[u + (\bar{\xi}\mathscr{M}^{-1}\alpha)^T + \bar{\beta}\mathscr{M}^{-1}\xi] + \bar{\xi}\mathscr{M}^{-1}\xi\} \end{aligned} \tag{12-101}$$

The matrices Q and $\mathscr{M}$ have commuting elements, while those of the matrices α and $\bar{\beta}$ belong to the Grassmann algebra; $\tilde{Q}$ denotes the symmetric matrix

$$\tilde{Q} = Q - \bar{\beta}\mathscr{M}^{-1}\alpha - (\bar{\beta}\mathscr{M}^{-1}\alpha)^T$$

Equation (12-101) results from the repeated application of the preceding formulas. An alternative way to recover it is to remember that the saddle-point method is exact for gaussian integrals.

These formulas are applied to the quadratic form (12-99) where

$$\begin{aligned} (\tilde{Q}A')^\mu \equiv{}& [D_\nu, [D^\nu, A'^\mu] - [D^\mu, A'^\nu]] - g[F^{\mu\nu}, A'^\nu] + \lambda\partial^\mu\partial\cdot A' \\ &+ g\{\partial^\mu\bar{\eta}, \eta'\} + g\{\partial^\mu\bar{\eta}', \eta\} \end{aligned} \tag{12-102}$$

and where η' and $\bar\eta'$ are obtained from the equations

$$\begin{aligned}\mathscr{M}\eta' \equiv \partial_\mu[D^\mu, \eta'] &= g\partial_\mu[A'^\mu, \eta]\\ [D^\mu, \partial_\mu\bar\eta'] &= -g[\partial_\mu\bar\eta, A'_\mu]\end{aligned} \tag{12-103}$$

If $g = 0$, $\mathscr{M}$ and $\tilde{Q}$ reduce to

$$\mathscr{M}_0 = \Box \qquad Q_0 = \Box - (1-\lambda)\partial \otimes \partial$$

the inverse of which are the free propagators. The one-loop effective action, normalized to $\Gamma^{[1]}(0) = 0$ is therefore

$$i\Gamma^{[1]}(A, \bar\eta, \eta) = \mathrm{Tr}\,[\ln(\mathscr{M}_0^{-1}\mathscr{M}) - \tfrac{1}{2}\ln(\tilde{Q}Q_0^{-1})] \tag{12-104}$$

The trace is performed on internal and Lorentz indices and on space-time variables.

It is not possible to obtain a more explicit expression. As the matrices Q and $\mathscr{M}$ act in the adjoint representation, it is more convenient to consider A_μ and η as matrices in that representation [compare with Eq. (12-9c)]. Explicitly,

$$\begin{aligned}(A^\mu)_{bc} &= A_a^\mu T^a{}_{bc} = C_{bac}A_a^\mu\\ (\eta)_{bc} &= C_{bac}\eta_a\end{aligned} \tag{12-105}$$

Therefore, instead of the normalization (12-26), we have

$$\mathrm{tr}\,(T^aT^b) = -\sum_{c,d} C_{cda}C_{cdb} = -C\delta^{ab} \tag{12-106}$$

For $SU(N)$, $C = N$. With these notations, we write

$$\begin{aligned}\mathscr{M}_0^{-1}\mathscr{M}(1,2) &= \delta(1,2) - ig\int \frac{d^4k}{(2\pi)^4}\,\frac{e^{-ik\cdot(x_1-x_2)}}{k^2}\,k\cdot A(x_2)\\ QQ_0^{-1}(1,2) &= \delta(1,2) + g\int \frac{d^4k}{(2\pi)^4(k^2)^2}\\ &\times \{(-ik\cdot A + \partial\cdot A - gA^2)[k^2 g_{\mu_1\mu_2} - (1-\lambda^{-1})k_{\mu_1}k_{\mu_2}]\\ &+ [k^2A_{\mu_2} - (1-\lambda^{-1})k\cdot Ak_{\mu_2}](ik_{\mu_1} + gA_{\mu_1})\\ &- [k^2\partial_{\mu_2} - (1-\lambda^{-1})k_{\mu_2}k\cdot\partial]A_{\mu_1}\\ &+ [k^2F_{\mu_1\mu_2} - (1-\lambda^{-1})F_{\mu_1\rho}k^\rho k_{\mu_2}]\}_{x_1}\,e^{-ik\cdot(x_1-x_2)}\end{aligned} \tag{12-107}$$

In the curly brackets, the argument of A or F is x_1, and the derivatives act on all x_1-dependent terms to their right. The kernel $\tilde{Q}$ differs from Q by ghost terms, not made explicit here. In the sequel, we use the same notation for a function and its Fourier transform:

$$A(x) = \int \frac{d^4k}{(2\pi)^4}\,e^{-ik\cdot x}A(k)$$

We shall now focus on the superficially divergent functions.

12-3-2 Two-Point Function

From the expansion of $\operatorname{Tr}\ln(\mathcal{M}_0^{-1}\mathcal{M}) - \frac{1}{2}\operatorname{Tr}\ln(Q_0^{-1}\tilde{Q})$ we get three contributions to the two-point function (Fig. 12-1):

$$\frac{1}{2}\int [dk] \operatorname{tr}[A(k)\Gamma(k)A(-k)]$$

$$= \frac{g^2}{2}\int [dk][dq] \frac{1}{q^2(q+k)^2} \operatorname{tr}[(q+k)\cdot A(k) q\cdot A(-k)]$$

$$+ \frac{g^2}{2}\int [dk][dq] \frac{1}{(q^2)^2} \operatorname{tr}\{q^2(d-1)A(k)\cdot A(-k)$$

$$+ (1-\lambda^{-1})[q\cdot A(k) q\cdot A(-k) - q^2 A(k)\cdot A(-k)]\}$$

$$- \frac{g^2}{4}\int [dk][dq] \frac{1}{(q^2)^2(q'^2)^2} \operatorname{tr}\{(q+q')\cdot A[q^2 - (1-\lambda^{-1})q\otimes q]$$

$$- q\otimes[q^2 A - (1-\lambda^{-1})q\cdot Aq] - A\otimes[q^2 q' - (1-\lambda^{-1})q\cdot q' q]$$

$$+ q^2(k\otimes A - A\otimes k) - (1-\lambda^{-1})(kq\cdot A - q\cdot kA)\otimes q\}_k$$

$$\times \{(q+q')\cdot A[q'^2 - (1-\lambda^{-1})q'\otimes q'] - q'\otimes[q'^2 A - (1-\lambda^{-1})q'\cdot Aq']$$

$$- A\otimes[q'^2 q - (1-\lambda^{-1})q\cdot q' q'] - q'^2(k\otimes A - A\otimes k)$$

$$+ (1-\lambda^{-1})[(kq'\cdot A - k\cdot q' A)\otimes q']\}_{-k} \tag{12-108}$$

In the last trace, $q' \equiv q + k$, and the argument in the first curly bracket is k and in the second one $-k$. Tensor notations are adopted for the sake of brevity. A Wick rotation has been carried out on the time components of both momenta and vector fields:

$$k^0 \to ik^0 \qquad q^0 \to iq^0 \qquad A^0(k) \to iA^0(k)$$

We use dimensional regularization to preserve gauge invariance and $[dk]$ stands for the measure $d^dk/(2\pi)^d$. Needless to say, the above expression may also be derived directly from the Feynman rules of Eqs. (12-91) and (12-92).

As explained in Chap. 8, it is consistent, within the dimensional regularization, to consider that the second integral of the right-hand side of (12-108) vanishes

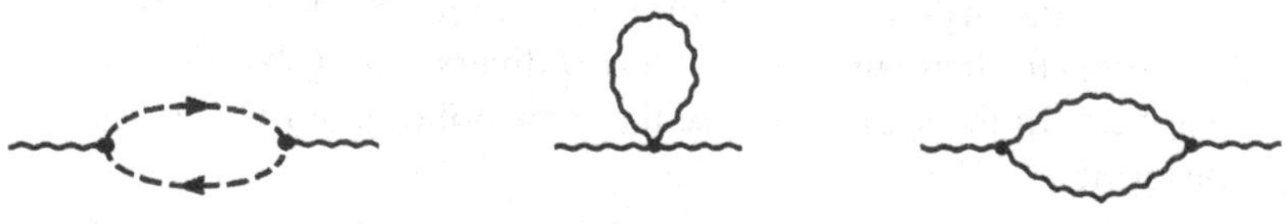

(a) (b) (c)

Figure 12-1 One-loop contributions to the vector self-energy. The broken lines stand for ghost propagators.

identically. The ghost contribution, i.e., the first term on the right-hand side of (12-108) is easy to compute in terms of Euler's functions:

$$\Gamma_{gh}(k) = \frac{g^2}{(4\pi)^{d/2}} B\left(\frac{d}{2}, \frac{d}{2}\right)(k^2)^{d/2-2}\left[\tfrac{1}{2}k^2\Gamma\left(1-\frac{d}{2}\right) - k \otimes k\Gamma\left(2-\frac{d}{2}\right)\right] \tag{12-109}$$

and after tedious algebra we obtain the total two-point function as

$$\begin{aligned}\Gamma(k) = \frac{g^2}{(4\pi)^{d/2}}(k^2)^{d/2-2}(k^2 - k \otimes k)\Bigg\{2\left[B\left(\frac{d}{2}, \frac{d}{2}\right) - B\left(\frac{d}{2}-1, \frac{d}{2}-1\right)\right]\Gamma\left(2-\frac{d}{2}\right)\\ + (1-\lambda^{-1})\left[4B\left(\frac{d}{2}-1, \frac{d}{2}\right)\Gamma\left(3-\frac{d}{2}\right) - 2B\left(\frac{d}{2}-1, \frac{d}{2}\right)\Gamma\left(2-\frac{d}{2}\right)\right]\\ -(1-\lambda^{-1})^2\tfrac{1}{4}\Gamma\left(3-\frac{d}{2}\right)B\left(\frac{d}{2}-1, \frac{d}{2}-1\right)\Bigg\}\end{aligned} \tag{12-110}$$

In these expressions, $\Gamma(k)$ must be regarded as a matrix in the adjoint representation, proportional to the unit matrix.

The ghost contribution (12-109) was crucial to achieve the transversity in k, since separate contributions do not satisfy it. The expression (12-110) is suited for the extraction of the divergent part as $d - 4 = -\varepsilon \to 0$. Recalling from Chap. 8 that g^2 should actually be written $g^2\mu^\varepsilon$, where μ is an arbitrary mass scale, we have

$$\Gamma(k) \underset{\varepsilon \to 0}{\simeq} \frac{g^2}{(4\pi)^2}(k^2 - k \otimes k)\left[\tfrac{5}{3} + \tfrac{1}{2}(1-\lambda^{-1})\right]\left(-\frac{2}{\varepsilon} + \ln\frac{k^2}{\mu^2} + \text{constant}\right) \tag{12-111}$$

where constant terms (i.e., independent of k but λ dependent) have not been computed. This expression may be rephrased in an equivalent form, by writing explicitly the group indices and returning to Minkowski space. The two-point proper function for $A^\mu{}_a(k)A^\nu{}_b(-k)$ reads

$$\Gamma^{(2)\mu\nu}_{ab} = -C\delta_{ab}\frac{g^2}{16\pi^2}(k^2 g^{\mu\nu} - k^\mu k^\nu)\left[\tfrac{5}{3} + \tfrac{1}{2}(1-\lambda^{-1})\right]\left[-\frac{2}{\varepsilon} + \ln\left(-\frac{k^2}{\mu^2}\right)\right] \tag{12-112}$$

It is now clear that the divergent term may be eliminated by the introduction of a counterterm:

$$\delta\mathscr{L}_{A^2} = (Z_3 - 1)\{\tfrac{1}{2}\operatorname{tr}[(\partial_\mu A_\nu - \partial_\nu A_\mu)(\partial^\mu A^\nu - \partial^\nu A^\mu)]\} \tag{12-113}$$

where

$$Z_3 = 1 + \frac{g^2 C}{16\pi^2}\left[\tfrac{5}{3} + \tfrac{1}{2}(1-\lambda^{-1})\right]\frac{2}{\varepsilon} \tag{12-114}$$

To this order, the term $2/\varepsilon$ may be regarded as corresponding to $\ln(\Lambda^2/\mu^2)$ in a conventional regularization. The wave-function renormalization Z_3 is gauge dependent. This may be traced to the fact that nonabelian gauge fields also play the role of charged fields. The fundamental consequence of the transversity of the function (12-110) is that no counterterm in A^2 nor $(\partial \cdot A)^2$ has been required.

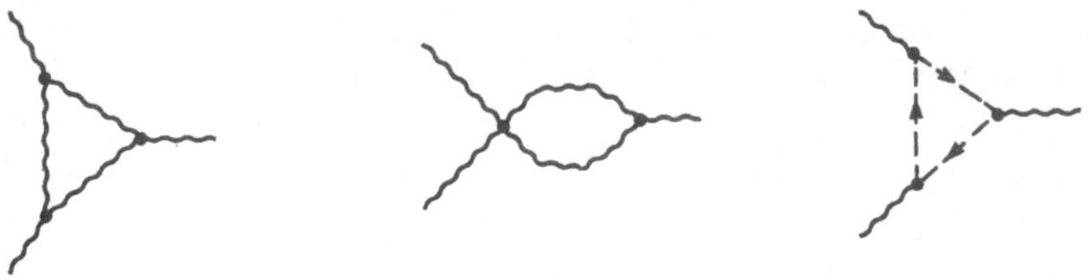

Figure 12-2 Three-point function.

The divergent terms in (12-110) come from the poles of the Γ function. However, when the computation is carried out in a different way, they may arise as singularities of the B function. This is a reflection of the fact that in a massless theory, the distinction between ultraviolet and infrared divergences becomes loose. For instance, the term of the integrand of $\Gamma^{\mu\nu}$ proportional to $(1-\lambda^{-1})\delta^{\mu\nu}$ has the form

$$\frac{(k^2-q^2)^2}{q^2(q+k)^4}$$

and leads, within the dimensional regularization, to an ultraviolet finite, but infrared divergent, integral

$$I=\int\frac{d^dq(k^2-q^2)^2}{q^2(k+q)^4}=\int\frac{d^dq(k^2)^2}{q^2(k+q)^4}\propto B\left(\frac{d}{2}-2,\frac{d}{2}-1\right)\Gamma\left(3-\frac{d}{2}\right)$$

However, after a change of variable, $q\to q'=q+k$, this integral looks ultraviolet divergent but infrared finite:

$$I=\int\frac{d^dq'[k^2-(q'-k)^2]^2}{(q'-k)^2q'^4}\propto 4B\left(\frac{d}{2}-1,\frac{d}{2}\right)\Gamma\left(3-\frac{d}{2}\right)-2B\left(\frac{d}{2},\frac{d}{2}-1\right)\Gamma\left(2-\frac{d}{2}\right)$$

Of course, the two expressions coincide, but it is important to pick all singular terms when expanding around $d=4$.

12-3-3 Other Functions

For other superficially divergent proper functions we only give the structure of the divergent term.

Three diagrams contribute to the three-point function (Fig. 12-2). The required counterterm reads

$$\delta\mathscr{L}_{A^3}=(Z_1-1)(-g\operatorname{tr}\{(\partial_\mu A_\nu-\partial_\nu A_\mu)[A^\mu,A^\nu]\})$$
$$Z_1=1+\frac{g^2C}{16\pi^2}\left[\tfrac{2}{3}+\tfrac{3}{4}(1-\lambda^{-1})\right]\frac{2}{\varepsilon}\qquad(12\text{-}115)$$

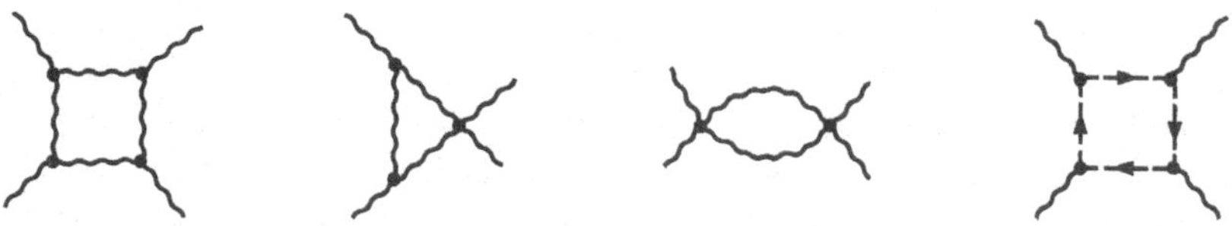

Figure 12-3 Four-point function.

Figure 12-4 Ghost self-energy to the one-loop order.

Similarly, the diagrams of Fig. 12-3 for the four-point function necessitate a counterterm

$$\begin{aligned} \delta\mathscr{L}_{A^4} &= (Z_4 - 1)\left\{\frac{g^2}{2}\operatorname{tr}([A_\mu, A_\nu][A^\mu, A^\nu])\right\} \\ Z_4 &= 1 + \frac{g^2 C}{16\pi^2}[-\tfrac{1}{3} + (1 - \lambda^{-1})]\frac{2}{\varepsilon} \end{aligned} \tag{12-116}$$

The counterterms have the same form as the initial terms of the lagrangian. It is of no wonder that this is the case for $\delta\mathscr{L}_{A^3}$, as the expression (12-115) is the only one which is Lorentz invariant, cubic in the field, of dimension four, and invariant under (global) group transformations. This is no longer true for the quartic term and the form of the counterterm (12-116) is therefore a gift.

The counterterms pertaining to functions with external ghosts must also be computed. Thanks to the structure of the operator $\tilde{Q}$ defined in (12-102) and (12-103), or equivalently to the Feynman rules (12-92), the momentum of an outgoing ghost line may always be factored out. This reduces the effective superficial degree of divergence of functions involving ghost fields, and leaves us with only two divergent functions: the ghost self-energy (Fig. 12-4) and the ghost-vector vertex (Fig. 12-5). The former does not need any mass counterterm, owing to the above property, and we find

$$\begin{aligned} \delta\mathscr{L}_{\bar{\eta}\eta} &= (\tilde{Z}_3 - 1)(-\bar{\eta}_a \partial^2 \eta_a) \\ \tilde{Z}_3 &= 1 + \frac{g^2 C}{16\pi^2}\left(\frac{1}{2} + \frac{1 - \lambda^{-1}}{4}\right)\frac{2}{\varepsilon} \end{aligned} \tag{12-117}$$

and

$$\begin{aligned} \delta\mathscr{L}_{\bar{\eta}A\eta} &= (\tilde{Z}_1 - 1)(C_{abc} A_a^\mu \partial_\mu \bar{\eta}_b \eta_c) \\ \tilde{Z}_1 &= 1 - \frac{g^2 C}{16\pi^2}\frac{\lambda^{-1}}{2}\frac{2}{\varepsilon} \end{aligned} \tag{12-118}$$

In fact $\tilde{Z}_1$ reduces to one in the Landau gauge ($\lambda^{-1} = 0$) to all orders, as a consequence of the transversity of the vector propagator and of the factorization of the incoming ghost momentum.

In practical cases, matter fields are also coupled to the gauge field. We list

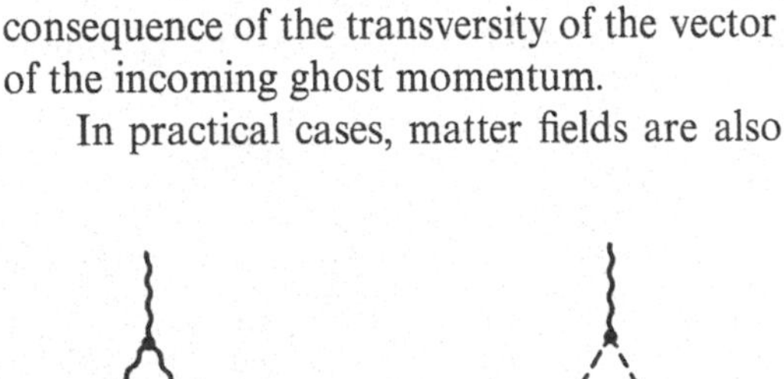

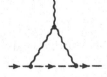

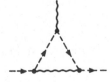

Figure 12-5 One-loop contributions to the ghost-vector vertex.

Figure 12-6 Fermion self-energy.

the counterterms involving spinor fields, as well as the extra contributions to $\delta\mathscr{L}_{A^2}$, $\delta\mathscr{L}_{A^3}$, and $\delta\mathscr{L}_{A^4}$ that they induce. The coupling has been written in Eq. (12-93). We use the following notations for the quadratic Casimir operators in the representation of the fields:

$$\begin{aligned} \operatorname{tr}(T^a T^b) &= -T_f \delta^{ab} \\ \sum_a (T^a)^2 &= -C_f I \end{aligned} \tag{12-119}$$

If r stands for the order of the group [number of generators of the Lie algebra, for example, $r = N^2 - 1$ for $SU(N)$] and n_f is the dimension of the fermion representation [$n_f = N$ for the fundamental representation of $SU(N)$], we have the relation

$$T_f r = C_f n_f \tag{12-120}$$

For the adjoint representation, $n = r$ and hence $T = C$ [$= N$ for $SU(N)$], while $T_f = \frac{1}{2}$ and $C_f = (N^2 - 1)/2N$ for the fundamental representation of $SU(N)$.

The counterterms generated by the diagrams of Figs. 12-6 and 12-7 read

$$\begin{aligned} \delta\mathscr{L}_{\bar\psi\psi} &= (Z_2 - 1)\bar\psi i\partial\!\!\!/\psi - \left(Z_2 \frac{m_0}{m} - 1\right) m\bar\psi\psi \\ Z_2 &= 1 - \frac{g^2 C_f}{16\pi^2}\lambda^{-1}\frac{2}{\varepsilon} \\ Z_2 \frac{m_0}{m} &= 1 - \frac{g^2 C_f}{16\pi^2}[4 - (1 - \lambda^{-1})]\frac{2}{\varepsilon} \end{aligned} \tag{12-121}$$

and

$$\begin{aligned} \delta\mathscr{L}_{\bar\psi A\psi} &= (Z_{1F} - 1)(-ig\bar\psi A_a T^a \psi) \\ Z_{1F} &= 1 - \frac{g^2 C_f}{16\pi^2}\left\{\left[1 - \frac{(1 - \lambda^{-1})}{4}\right]C + \lambda^{-1}C_f\right\}\frac{2}{\varepsilon} \end{aligned} \tag{12-122}$$

These expressions reduce to those computed for quantum electrodynamics in Chap. 7, if we set $C_f = 1$, $C = 0$, $g^2/4\pi = \alpha$.

Finally, the modifications induced by an internal fermion loop (Fig. 12-8) may be expressed as

$$\begin{aligned} \delta\mathscr{L}(A)\big|_{\text{fermions}} &= \delta Z^{(F)}[\tfrac{1}{2}\operatorname{tr}(F_{\mu\nu}F^{\mu\nu})] \\ \delta Z^{(F)} = \delta Z_3{}^{(F)} = \delta Z_1{}^{(F)} &= \delta Z_4{}^{(F)} = -\frac{T_f g^2}{16\pi^2}\frac{4}{3}\frac{2}{\varepsilon} \end{aligned} \tag{12-123}$$

The reader will have no difficulty in working out the case where scalar fields are coupled to the gauge vectors.

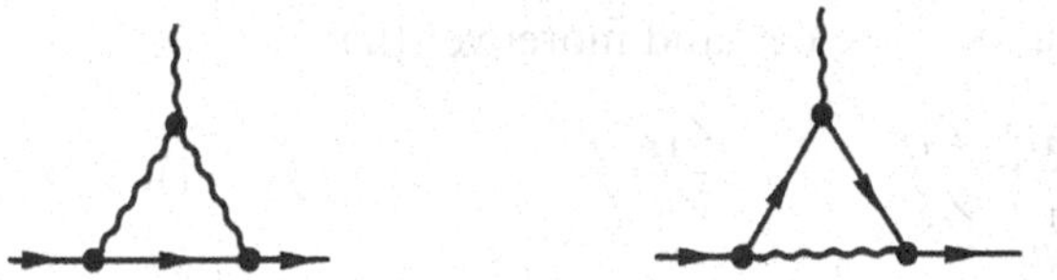

Figure 12-7 Fermion-vector vertex.

12-3-4 One-Loop Renormalization

We have discovered that all counterterms have the same structure as monomials of the initial lagrangian. This is not completely sufficient to insure that the renormalized and bare lagrangians enjoy the same symmetry properties. Omitting matter fields for a while, we found explicitly

$$\begin{aligned}\mathscr{L} + \delta\mathscr{L} = \operatorname{tr}\{\tfrac{1}{2}Z_3(\partial_\mu A_\nu - \partial_\nu A_\mu)(\partial^\mu A^\nu - \partial^\nu A^\mu) + \frac{\lambda}{2}(\partial\cdot A)^2 \\ - gZ_1(\partial_\mu A_\nu - \partial_\nu A_\mu)[A^\mu, A^\nu] + \frac{g^2}{2}Z_4[A_\mu, A_\nu][A^\mu, A^\nu]\} \\ + \tilde{Z}_3\partial_\mu\bar{\eta}\partial^\mu\eta + g\tilde{Z}_1(\partial_\mu\bar{\eta}_b A^\mu{}_a\eta_c C_{abc})\end{aligned} \tag{12-124}$$

We define the bare fields and parameters according to

$$\begin{aligned}A_0 = Z_3^{1/2}A \qquad \eta_0 = \tilde{Z}_3^{1/2}\eta \qquad \bar{\eta}_0 = \tilde{Z}_3^{1/2}\bar{\eta} \\ g_0 = Z_1 Z_3^{-3/2} g \qquad \lambda_0 = \lambda Z_3^{-1}\end{aligned} \tag{12-125}$$

Then $\mathscr{L} + \delta\mathscr{L}$ may be regarded as the initial lagrangian $\mathscr{L}(A_0, \eta_0, \bar{\eta}_0; g_0, \lambda_0)$ written in terms of bare quantities provided the following identities hold:

$$\frac{Z_4}{Z_1} = \frac{Z_1}{Z_3} = \frac{\tilde{Z}_1}{\tilde{Z}_3} \tag{12-126a}$$

A glance at the expressions (12-114) to (12-118) suffices to convince ourselves that they are satisfied to the one-loop order. These identities, which generalize the relation $Z_1 = Z_2$ of quantum electrodynamics, express the fact that the coupling constant renormalizations of the cubic, quartic, and $\eta\bar{\eta}A$ vertices coincide, i.e., that the universality of the coupling is preserved by renormalization. In the

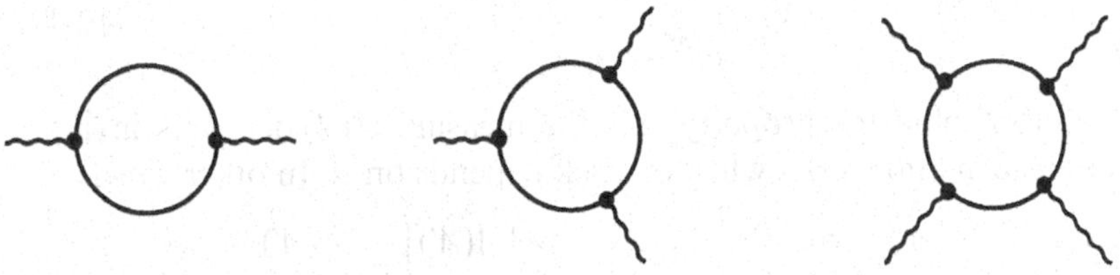

Figure 12-8 Fermion contributions to the two-, three-, and four-point functions.

presence of matter fields, fermions, say, we demand moreover that

$$\frac{Z_4}{Z_1} = \frac{Z_1}{Z_3} = \frac{\tilde{Z}_1}{\tilde{Z}_3} = \frac{Z_{1F}}{Z_2} \tag{12-126b}$$

This relation, too, is satisfied by the expressions (12-121) to (12-123).

Finally, we write the coupling constant renormalization in the presence of fermion fields:

$$\begin{aligned} g_0 &= Z_g g \\ Z_g &= 1 - \frac{g^2}{16\pi^2}\left(\tfrac{11}{6}C - \tfrac{2}{3}T_f\right) \ln \frac{\Lambda^2}{\mu^2} \end{aligned} \tag{12-127}$$

Notice that the gauge dependence has disappeared. For future use, $2/\varepsilon$ has been replaced by $\ln(\Lambda^2/\mu^2)$.

12-4 RENORMALIZATION

This section deals with the renormalization of nonabelian gauge theories with an unbroken symmetry. What happens as the local symmetry is spontaneously broken will be examined later. The issue is to know whether the remarkable features of gauge theories, in particular the universality of the coupling constant, will be preserved by the renormalization. A regularization is introduced at intermediate stages; in practice, the dimensional regularization is the most convenient. The desired properties will follow from Ward identities, first derived in this context by Slavnov and by Taylor. The reader may find the repeated use of such identities in Chaps. 8, 11, and here rather cumbersome. However, nonabelian gauge theories are more intricate and require a sophisticated analysis.

12-4-1 Slavnov-Taylor Identities

We start from the generating functional

$$e^{G(J)} = \int \mathscr{D}(A) \det \mathscr{M} \exp\left[i \int d^4x \left(\mathscr{L} - \frac{\lambda}{2}\mathscr{F}^2 + J \cdot A\right)\right]$$

where $\mathscr{M}$ is the variation of $\mathscr{F}$ with respect to a gauge transformation

$$\begin{aligned} \delta A &= D\delta\alpha \\ \delta\mathscr{F} &= \mathscr{M}\delta\alpha \end{aligned} \tag{12-128}$$

We are going to exploit the property that the measure $\mathscr{D}(A) \det \mathscr{M}$ is invariant under this transformation, even when $\delta\alpha$ itself depends on A. In other words, if

$$A = A' + \delta A \qquad \text{and} \qquad \mathscr{F}'(A') = \mathscr{F}[A(A')] = \mathscr{F}(A') + \delta\mathscr{F}$$

then

$$\mathscr{D}(A) \det \mathscr{M}_{\mathscr{F}}(A) = \mathscr{D}(A') \det \mathscr{M}_{\mathscr{F}'}(A') \tag{12-129}$$

The proof is straightforward. We write $\Delta(A)$ for det $\mathscr{M}(A)$:

$$\begin{aligned}\int \mathscr{D}(A)\Delta_{\mathscr{F}}(A) &= \int \mathscr{D}(A)\mathscr{D}(A')\Delta_{\mathscr{F}}(A)\delta(A' - A + \delta A)\\ &= \int \mathscr{D}(A)\mathscr{D}(A')\mathscr{D}(g)\Delta_{\mathscr{F}}(A)\delta(A' - {}^{g}A)\Delta_{\mathscr{F}'}(A)\delta[\mathscr{F}'({}^{g}A) - \mathscr{F}(A)]\\ &= \int \mathscr{D}(g)\mathscr{D}(A')\Delta_{\mathscr{F}'}({}^{g^{-1}}A')\Delta_{\mathscr{F}}({}^{g^{-1}}A')\delta[\mathscr{F}'(A') - \mathscr{F}({}^{g^{-1}}A')]\\ &= \int \mathscr{D}(A')\Delta_{\mathscr{F}'}(A')\end{aligned}$$

where the invariance of the measure $\mathscr{D}(A)$ and of $\Delta_{\mathscr{F}}(A)$ under a transformation $A \to {}^{g}A$ has been used.

Consequently,

$$e^{G(J)} = \int \mathscr{D}(A) \det \mathscr{M} \exp\left[i \int d^4x \left(\mathscr{L} - \frac{\lambda}{2}\mathscr{F}^2 + J\cdot A - \lambda\mathscr{F}\mathscr{M}\,\delta\alpha + J\cdot D\delta\alpha\right)\right]$$

The content of this identity is most easily explored if we take $\delta\alpha = \mathscr{M}^{-1}\delta\omega$. This corresponds to a nonlocal gauge transformation, which translates $\mathscr{F}$ by $\delta\omega$. To lowest order in $\delta\omega$, we get

$$0 = \int \mathscr{D}(A) \det \mathscr{M} \int d^4x(\lambda\mathscr{F} - J\cdot D\mathscr{M}^{-1})\delta\omega \exp\left[i\int d^4x(\mathscr{L} - \tfrac{1}{2}\mathscr{F}^2 + J\cdot A)\right]$$

or, after substitution of $\delta/i\delta J$ for A,

$$\left\{\lambda\mathscr{F}_a\left[\frac{1}{i}\frac{\delta}{\delta J(x)}\right] - \int d^4y\, J_b(y)D_{bc}\left(\frac{1}{i}\frac{\delta}{\delta J}\right)\mathscr{M}_{ca}^{-1}\left(y, x; \frac{1}{i}\frac{\delta}{\delta J}\right)\right\} e^{G(J)} = 0 \tag{12-130}$$

where summation over repeated indices is understood. The expression

$$G_{ca}(y, x) = \mathscr{M}_{ca}^{-1}\left(y, x; \frac{1}{i}\frac{\delta}{\delta J}\right)e^{G(J)} \tag{12-131}$$

may be regarded as the ghost propagator in the presence of the source J.

The Slavnov-Taylor identities finally read

$$\begin{aligned}\lambda\mathscr{F}_a\left[\frac{\delta}{i\delta J(x)}\right]e^{G(J)} &= \int d^4y\left[J_b D_{bc}\left(\frac{\delta}{i\delta J}\right)\right](y)G_{ca}(y, x)\\ \mathscr{M}_{bc}\left[\frac{\delta}{i\delta J(y)}\right]G_{ca}(y, x) &= \delta_{ba}\delta^4(x - y)\, e^{G(J)}\end{aligned} \tag{12-132}$$

It is difficult to express in a compact way the Slavnov-Taylor identities (12-132) on one-particle irreducible functions. A transformation discovered by Becchi, Rouet, and Stora enables us to reach this goal. The ghost fields are reintroduced in the action

$$I = \int d^4x\left(\mathscr{L} - \frac{\lambda}{2}\mathscr{F}^2 - \bar{\eta}\mathscr{M}\eta\right) \tag{12-133}$$

It is easy to show that this action is invariant under a combined transformation of the variables

$$\begin{cases} \delta A^\mu{}_a(x) = D^\mu{}_{ab}(x)\eta_b(x)\delta\zeta \equiv sA\delta\zeta \\ \delta\bar{\eta}_a(x) = \lambda\mathscr{F}_a[A(x)]\delta\zeta \equiv s\bar{\eta}\delta\zeta \\ \delta\eta_a(x) = -\dfrac{g}{2}C_{abc}\eta_b(x)\eta_c(x)\delta\zeta \equiv s\eta\delta\zeta \end{cases} \tag{12-134}$$

This transformation is local, in contrast to the one encountered above. It introduces an x-independent anticommuting parameter $\delta\zeta$ and mixes commuting and anticommuting variables.

The invariance of I is easy to prove. First $\mathscr{L}$ is invariant since δA is a special gauge transformation. Second, we observe that

$$\delta\left(-\frac{\lambda}{2}\mathscr{F}^2\right) - (\delta\bar{\eta})\mathscr{M}\eta = 0$$

as a consequence of the anticommutation of η and $\delta\zeta$. Finally,

$$\delta(\mathscr{M}\eta) = \delta\left[\frac{\delta\mathscr{F}}{\delta A^\mu{}_a}(D^\mu\eta)_a\right]$$

$$= \frac{-\delta^2\mathscr{F}}{\delta A^\nu{}_b\delta A^\mu{}_a}(D^\nu\eta)_b(D^\mu\eta)_a\delta\zeta + \frac{\delta\mathscr{F}}{\delta A^\mu{}_a}\delta(D^\mu\eta)_a$$

The first term vanishes because $(D^\nu\eta)_b(D^\mu\eta)_a$ is antisymmetric in the interchange of (μa) and (νb), while it is easy to see that

$$\delta(D^\mu\eta) = 0 \tag{12-135}$$

as a consequence of (12-134) and of the Jacobi identity. For later use, we also mention that similarly

$$\delta(C_{abc}\eta_b\eta_c) = 0 \tag{12-136}$$

The Becchi-Rouet-Stora transformation s is defined as the right derivative of (12-134) with respect to $\delta\zeta$, that is, $sA = D\eta$, $s\eta_a = -gC_{abc}\eta_b\eta_c/2$. Equations (12-135) and (12-136) imply that $s^2A = 0$, $s^2\eta = 0$.

This invariance implies identities between Green functions. Let us first show how the Slavnov-Taylor identity (12-130) is recovered. We start from

$$\int \mathscr{D}(A, \eta, \bar{\eta})\bar{\eta}_a(x) \exp\left[i\left(I + \int d^4y\, J\cdot A\right)\right] = 0$$

which results from the odd character in the ghost variables of the integrand. We then perform a change of integration variables of the form (12-134); such a change does not affect the integral and has a jacobian equal to one, as is easily checked. Hence

$$0 = \int \mathscr{D}(A, \eta, \bar{\eta})\left[\lambda\mathscr{F}_a(x) + \bar{\eta}_a(x)\int d^4y\, iJ_b(y)(D\eta)_b(y)\right]\exp\left[i\left(I + \int d^4y\, J\cdot A\right)\right] = 0 \tag{12-137}$$

The integration over $\eta, \bar{\eta}$ entails the substitution

$$\bar{\eta}_a(x)\eta_b(y) \to i\mathcal{M}^{-1}{}_{ba}(y, x)$$

and leads to (12-130).

12-4-2 Identities for Proper Functions

The identities are first expressed on connected functions. It is convenient to introduce sources not only for the ghost fields but also for the composite operators involved in the transformation (12-134). We write

$$e^{G(J,\xi,\bar{\xi},K,L)} = \int \mathscr{D}(A, \eta, \bar{\eta}) \exp\left[i \int d^4y \left(\mathscr{L} - \frac{\lambda}{2}\mathscr{F}^2 - \bar{\eta}\mathcal{M}\eta \right.\right.$$
$$\left.\left. + J \cdot A + \bar{\xi}\eta + \bar{\eta}\xi + KsA - Ls\eta\right)\right] \tag{12-138}$$

$$G(J) = G(J, \xi, \bar{\xi}, K, L)\big|_{\xi=\bar{\xi}=K=L=0}$$

Here $K^\mu{}_a(y)$ and $L_a(y)$ are local sources coupled to $sA_{\mu a}(y) = (D_\mu\eta)_a(y)$ and $-s\eta_a(y) = gC_{abc}\eta_b(y)\eta_c(y)/2$, and are therefore anticommuting or commuting objects respectively. More precisely, if η is given the ghost number $\gamma = -1$ (and $\bar{\eta}$, $+1$), K and L have $\gamma = 1$ and 2. For power counting sA and $s\eta$, and hence K and L, have dimension two. We assume hereafter that the function $\mathscr{F}$ is linear in A:

$$\mathscr{F}_a(y) = \phi^\mu{}_{ab}A_{\mu b}(y) \tag{12-139}$$

For instance, the generalized Feynman gauges correspond to $\phi^\mu{}_{ab} = \partial^\mu\delta_{ab}$. The use of a nonlinear gauge condition $\mathscr{F}$ would require the introduction in (12-138) of an extra source coupled to it.

After a change of variables of the form (12-134), we get

$$0 = \int \mathscr{D}(A, \eta, \bar{\eta}) \int d^4x\,(J \cdot sA + \bar{\xi}s\eta - \lambda\mathscr{F}\xi)(x) \exp\left[i \int d^4y\,(\cdots)\right]$$

since sA and $s\eta$ are also invariant. This is rewritten as

$$\int d^4x \left[J \cdot \frac{\delta}{i\delta K} - \bar{\xi}\frac{\delta}{i\delta L} - \lambda\xi\mathscr{F}\left(\frac{\delta}{i\delta J}\right)\right](x)\, e^{G(J,\xi,\bar{\xi},K,L)} = 0$$

The operator of functional differentiation is linear owing to hypothesis (12-139). Hence

$$\int d^4x \left[J\frac{\delta}{\delta K} - \bar{\xi}\frac{\delta}{\delta L} - \lambda\xi\mathscr{F}\left(\frac{\delta}{\delta J}\right)\right] G(J, \xi, \bar{\xi}, K, L) = 0 \tag{12-140}$$

In the preceding section, the identity (12-130) was supplemented by an equation of motion for the ghost propagator [Eq. (12-132)]. The analog here is obtained by expressing that the functional integral (12-138) is invariant under an infinitesimal change $\bar{\eta} \to \bar{\eta} + \delta\bar{\eta}$ where $\delta\bar{\eta}$ is arbitrary. We get the local relation

$$\int \mathscr{D}(A, \eta, \bar{\eta}) \, (-\mathscr{M}\eta + \xi)(x) \exp\left[i \int d^4y \, (\cdots)\right] = 0$$

or, since $\mathscr{M}\eta = s\mathscr{F} = \phi \cdot sA$:

$$\left(\xi - \phi \frac{\delta}{i\delta K}\right)(x) \, e^{G(J,\xi,\bar{\xi},K,L)} = 0$$

For connected functions, this reads

$$\phi \frac{\delta}{i\delta K(x)} G(J, \xi, \bar{\xi}, K, L) = \xi(x) \tag{12-141}$$

Equations (12-140) and (12-141) must now be translated in terms of proper functions. This is achieved after a Legendre transformation

$$\Gamma(A, \eta, \bar{\eta}, K, L) = -iG(J, \xi, \bar{\xi}, K, L) - \int d^4y \, (J \cdot A + \bar{\xi}\eta + \bar{\eta}\xi) \tag{12-142}$$

where

$$A = \frac{\delta G}{i\delta J} \qquad \eta = \frac{\delta G}{i\delta \bar{\xi}} \qquad \bar{\eta} = -\frac{\delta G}{i\delta \xi}$$

or, equivalently,

$$J = -\frac{\delta \Gamma}{\delta A} \qquad \bar{\xi} = \frac{\delta \Gamma}{\delta \eta} \qquad \xi = -\frac{\delta \Gamma}{\delta \bar{\eta}}$$

In this operation, the sources K and L are passive spectators:

$$\frac{\delta G}{i\delta K} = \frac{\delta \Gamma}{\delta K} \qquad \frac{\delta G}{i\delta L} = \frac{\delta \Gamma}{\delta L} \tag{12-143}$$

The identities (12-140) and (12-141), expressed on Γ, read

$$\int d^4x \left[\frac{\delta \Gamma}{\delta A}\frac{\delta \Gamma}{\delta K} + \frac{\delta \Gamma}{\delta \eta}\frac{\delta \Gamma}{\delta L} - \lambda \frac{\delta \Gamma}{\delta \bar{\eta}} \mathscr{F}(A)\right](x) = 0 \tag{12-144}$$

and

$$\phi^{\mu}{}_{ab} \frac{\delta \Gamma}{\delta K^{\mu}_{b}(x)} + \frac{\delta \Gamma}{\delta \bar{\eta}_a(x)} = 0 \tag{12-145}$$

If the modified effective action $\tilde{\Gamma}$ is introduced

$$\tilde{\Gamma} = \Gamma + \frac{\lambda}{2} \int d^4y \, \mathscr{F}^2[A(y)] \tag{12-146}$$

Equations (12-144) and (12-145) take the simpler form

$$\int d^4x \left(\frac{\delta\tilde{\Gamma}}{\delta A} \frac{\delta\tilde{\Gamma}}{\delta K} + \frac{\delta\tilde{\Gamma}}{\delta \eta} \frac{\delta\tilde{\Gamma}}{\delta L} \right)(x) = 0$$

$$\phi^\mu \frac{\delta\tilde{\Gamma}}{\delta K^\mu} + \frac{\delta\tilde{\Gamma}}{\delta\bar{\eta}} = 0 \qquad (12\text{-}147)$$

The identities (12-147) have a universal form. They no longer involve any parameter affected by the renormalization, such as the coupling constant, and make no reference to the group structure, which has been hidden in the definition of the sources. Consequently, they apply as well to the action and are therefore suited to study the structure of counterterms.

It is easy to verify that the identities (12-144) and (12-145) are satisfied to lowest order. The first one expresses the invariance of the action under the Becchi-Rouet-Stora transformation

$$\delta A = \frac{\delta\Gamma^{[0]}}{\delta K}\delta\zeta \qquad \delta\eta = -\frac{\delta\Gamma^{[0]}}{\delta L}\delta\zeta \qquad \delta\bar{\eta} = \lambda\mathscr{F}(A)\delta\zeta$$

The condition that the jacobian be equal to one reads

$$\frac{\delta^2\Gamma^{[0]}}{\delta A\delta K} = 0 \qquad \frac{\delta^2\Gamma^{[0]}}{\delta\eta\delta L} = 0 \qquad (12\text{-}148)$$

and is obviously satisfied.

It is a good exercise to perform the analogous analysis in the abelian case, in a gauge such as $\mathscr{F} = \partial_\mu A^\mu + A^2/2$ which requires the introduction of ghosts.

12-4-3 Recursive Construction of the Counterterms

In the preceding considerations, the theory was implicitly dimensionally regularized. We now want to renormalize it, in a way which preserves the previous identities. As in the one-loop computation of Sec. 12-3, it is convenient to perform the minimal renormalization defined in Sec. 8-4-4. In other words, we content ourselves with eliminating the divergent terms as the dimension d goes to 4. Order by order in $\hbar$, we will write

$$\tilde{\Gamma}^{[n]}_{\text{reg}} = \tilde{\Gamma}^{[n]}_R + \tilde{\Gamma}^{[n]}_{\text{div}} \qquad (12\text{-}149)$$

where $\Gamma^{[n]}_{\text{reg}}$ is computed by taking into account all lowest-order counterterms. More physical conditions may be substituted, provided they are consistent with the identities (12-144) and (12-145).

Equation (12-147) tells us that all functionals depend on K and $\bar{\eta}$ only through the combination $K - \bar{\eta}\phi$. We will use the compact notation

$$\Gamma_1 * \Gamma_2 \equiv \int d^4x \left(\frac{\delta\Gamma_1}{\delta A} \frac{\delta\Gamma_2}{\delta K} + \frac{\delta\Gamma_1}{\delta\eta} \frac{\delta\Gamma_2}{\delta L} \right) \qquad (12\text{-}150)$$

The identity (12-147) satisfied by the power series in $\hbar$

$$\tilde{\Gamma} = \tilde{\Gamma}^{[0]} + \tilde{\Gamma}^{[1]} + \tilde{\Gamma}^{[2]} + \cdots$$

reads to nth order

$$\sum_{p+q=n} \tilde{\Gamma}^{[p]} * \tilde{\Gamma}^{[q]} = 0 \tag{12-151}$$

and we seek counterterms such that the renormalized $\tilde{\Gamma}_R^{[p]}$ satisfy

$$\sum_{p+q=n} \tilde{\Gamma}_R^{[p]} * \tilde{\Gamma}_R^{[q]} = 0 \tag{12-152}$$

We proceed in a recursive way. To lowest order $\tilde{\Gamma}^{[0]} = \tilde{\Gamma}_R^{[0]}$ reduces to

$$\begin{aligned}\tilde{\Gamma}^0 = \tilde{I}(A, \eta, \bar{\eta}, K, L) &= \int d^4x\,[\mathscr{L}(A) - \bar{\eta}\mathscr{M}\eta + K\cdot sA - Ls\eta] \\ &= \int d^4x\,[\mathscr{L} + (K - \bar{\eta}\phi)\cdot sA - Ls\eta]\end{aligned} \tag{12-153}$$

which of course satisfies (12-152). To first order, we have

$$\begin{cases}\tilde{I} * \tilde{\Gamma}_{\text{div}}^{[1]} + \tilde{\Gamma}_{\text{div}}^{[1]} * \tilde{I} = 0 & \text{(12-154}a) \\ \tilde{I} * \tilde{\Gamma}_R^{[1]} + \tilde{\Gamma}_R^{[1]} * \tilde{I} = 0 & \text{(12-154}b)\end{cases}$$

Equation (12-154b) is simply the desired identity (12-152) for $n = 1$, while Eq. (12-154a) gives the structure of the first-order counterterm. It is suggested to modify $\tilde{I}$ into

$$\tilde{I} \to \tilde{I}_1 \stackrel{?}{=} \tilde{I} - \tilde{\Gamma}_{\text{div}}^{[1]} \tag{12-155}$$

so as to cancel the divergent part. However, the recursive procedure works only if the renormalized action I_1 satisfies itself:

$$\tilde{I}_1 * \tilde{I}_1 = 0 \tag{12-156}$$

as $\tilde{I}$ does. This is not the case for (12-155), as

$$\tilde{I}_1 * \tilde{I}_1 = \tilde{\Gamma}_{\text{div}}^{[1]} * \tilde{\Gamma}_{\text{div}}^{[1]}$$

However, the right-hand side is of order $\hbar^2$. We are thus led to define

$$\tilde{I}_1 = I - \tilde{\Gamma}_{\text{div}}^{[1]} + \Delta_1$$

where the extra piece Δ_1 is the integral of a local polynomial in the fields of degree four, of order $\hbar^2$, and defined in such a way that (12-156) is satisfied. Of course, it does not affect quantities to order one, and hence preserves the finiteness and normalization condition of $\tilde{\Gamma}_R^{[1]}$.

This complication is quite typical of symmetries involving nonlinear identities. The same phenomenon occurs, for instance, in the two-dimensional nonlinear σ model evoked at the end of Chap. 11.

The structure of Δ_1 will be deduced from the one of $\Gamma_{\text{div}}^{[1]}$. If we are able to show that

$$(\tilde{I} - \tilde{\Gamma}_{\text{div}}^{[1]})(A, \eta, \bar{\eta}, K, L; g) = \tilde{I}(A_0, \eta_0, \bar{\eta}_0, K_0, L_0; g_0) + O(\hbar^2) \tag{12-157}$$

where

$$\begin{aligned} A_0 &= Z_3^{1/2}A & \eta_0 &= \tilde{Z}_3^{1/2}\eta & \bar{\eta}_0 &= \tilde{Z}_3^{1/2}\bar{\eta} \\ K_0 &= \tilde{Z}_3^{1/2}K & L_0 &= Z_L^{1/2}L & g_0 &= Z_g g \end{aligned} \tag{12-158}$$

with $Z_3 = 1 + z_3\hbar$, etc., a natural choice will be

$$\tilde{I}_1(A, \eta, \bar{\eta}, K, L; g) = \tilde{I}(A_0, \eta_0, \bar{\eta}_0, K_0, L_0; g_0)$$

This new action $\tilde{I}_1$ will satisfy

$$\tilde{I}_1 * \tilde{I}_1 = \int d^4x \left[\tilde{Z}_3^{1/2} Z_3^{1/2} \frac{\delta\tilde{I}(A_0, \ldots; g_0)}{\delta A_0} \frac{\delta\tilde{I}(A_0, \ldots)}{\delta K_0} + \tilde{Z}_3^{1/2} Z_L^{1/2} \frac{\delta\tilde{I}(A_0, \ldots)}{\delta\eta_0} \frac{\delta\tilde{I}(A_0, \ldots)}{\delta L_0} \right]$$

If, moreover,

$$Z_L = Z_3 \tag{12-159}$$

then the condition (12-156) will be fulfilled and the recursive proof can be pursued. The aim of the forthcoming technical discussion is therefore to prove (12-157) and (12-159).

We look for the general solution of the equation

$$\sigma\tilde{\Gamma}_{\text{div}}^{[n]} \equiv \tilde{I} * \tilde{\Gamma}_{\text{div}}^{[n]} + \tilde{\Gamma}_{\text{div}}^{[n]} * \tilde{I} = 0 \tag{12-160}$$

satisfied by the divergent part $\tilde{\Gamma}_{\text{div}}^{[n]}$ to a given order n, when all counterterms of smaller order have been taken into account.

The operation σ in (12-160), which is a generalization of the Becchi-Rouet-Stora transformation s, is nilpotent

$$\sigma^2 = 0 \tag{12-161}$$

This is easily seen if σ is written as

$$\sigma = \frac{\partial\tilde{I}}{\partial x_i}\frac{\partial}{\partial\theta_i} + \frac{\partial\tilde{I}}{\partial\theta_i}\frac{\partial}{\partial x_i}$$

where the commuting and anticommuting variables have been collectively denoted x and θ respectively; in our problem $\{x_i\} = \{A, L\}$, $\{\theta_i\} = \{\eta, K\}$. Then (12-161) results from the identity

$$\frac{\partial\tilde{I}}{\partial x_i}\frac{\partial\tilde{I}}{\partial\theta_i} = 0 \tag{12-162}$$

satisfied by $\tilde{I}$. Explicitly,

$$\begin{aligned} \sigma^2 = &\left(\frac{\partial\tilde{I}}{\partial x_i}\frac{\partial\tilde{I}}{\partial x_j}\frac{\partial}{\partial\theta_i}\frac{\partial}{\partial\theta_j} + \frac{\partial\tilde{I}}{\partial\theta_i}\frac{\partial\tilde{I}}{\partial\theta_j}\frac{\partial}{\partial x_i}\frac{\partial}{\partial x_j} - \frac{\partial\tilde{I}}{\partial x_i}\frac{\partial\tilde{I}}{\partial\theta_j}\frac{\partial}{\partial\theta_i}\frac{\partial}{\partial x_j} + \frac{\partial\tilde{I}}{\partial\theta_i}\frac{\partial\tilde{I}}{\partial x_j}\frac{\partial}{\partial x_i}\frac{\partial}{\partial\theta_j} \right) \\ &+ \frac{\partial}{\partial x_j}\left(\frac{\partial\tilde{I}}{\partial x_i}\frac{\partial\tilde{I}}{\partial\theta_i}\right)\frac{\partial}{\partial\theta_j} - \frac{\partial}{\partial\theta_j}\left(\frac{\partial\tilde{I}}{\partial x_i}\frac{\partial\tilde{I}}{\partial\theta_i}\right)\frac{\partial}{\partial x_j} \end{aligned}$$

The first bracket vanishes as a consequence of anticommutativity and the two remaining terms owing to Eq. (12-162). On the other hand, any gauge-invariant functional $R_{\text{inv}}(A)$ of A only satisfies

$$\sigma R_{\text{inv}}(A) = \int d^4x \frac{\delta R_{\text{inv}}}{\delta A}\frac{\delta\tilde{I}}{\delta K} = \int d^4x \frac{\delta R_{\text{inv}}}{\delta A^\mu{}_a(x)}(D^\mu\eta)_a(x) = 0$$

Therefore an expression of the form

$$R = R_{\text{inv}}(A) + \sigma R' \tag{12-163}$$

is a solution of $\sigma R = 0$. It is possible to show that this is the general solution of the equation for $\tilde{\Gamma}^{[n]}_{\text{div}}(A)$, even when additional sources coupled to gauge-invariant composite operators are added to the initial lagrangian. In the present case, however, we can prove it by inspection, using the fact that power counting and ghost number conservation restrict the form of $\tilde{\Gamma}^{[n]}_{\text{div}}$ to

$$\tilde{\Gamma}^{[n]}_{\text{div}} = \int d^4x\,\{l(A) + [K^\mu{}_a - (\bar{\eta}\phi^\mu)_a]\Delta_{\mu ab}\eta_b + \tfrac{1}{2}d_{abc}L_a\eta_b\eta_c\}$$

In this expression, $l(A)$ is of dimension four, $\Delta_{\mu ab}$ is of dimension one, and thus at most linear in A, and $d_{abc} = -d_{acb}$ are numbers. If we assume that the global symmetry is unbroken (by the choice of gauge), then

$$\Delta_{\mu ab} = \alpha\partial_\mu\delta_{ab} + \beta g C_{abc}A_{\mu c}$$

$$d_{abc} = \gamma g C_{abc}$$

with α, β, and γ being pure numbers. Inserting these expressions into (12-160) yields

$$\beta = \gamma$$

$$D^\mu_{ab}\frac{\partial l(A)}{\partial A^\mu{}_b} + g(\beta - \alpha)C_{abc}A^\mu{}_c\frac{\partial\mathscr{L}}{\partial A^\mu{}_b} = 0$$

A particular solution to the second equation is

$$l(A) = (\beta - \alpha)A^\mu{}_a\frac{\partial\mathscr{L}}{\partial A^\mu{}_a}$$

and the general solution is obtained through the addition of a gauge-invariant functional of A, of degree four, and thus a multiple of $\mathscr{L}(A)$:

$$l(A) = a\mathscr{L}(A) + (\beta - \alpha)A^\mu{}_a\frac{\partial\mathscr{L}}{\partial A^\mu{}_a}$$

To summarize, $\tilde{\Gamma}^{[n]}_{\text{div}}$ has the form

$$\tilde{\Gamma}^{[n]}_{\text{div}} = \int d^4x\left[a\mathscr{L}(A) + (\beta - \alpha)A^\mu{}_a\frac{\partial\mathscr{L}}{\partial A^\mu{}_a} + \alpha(K - \bar{\eta}\phi)^\mu{}_a(D_\mu\eta)_a\right.$$
$$\left. + (\beta - \alpha)g(K - \bar{\eta}\phi)^\mu{}_a C_{abc}A_{\mu c}\eta_b + \beta\frac{g}{2}L_a C_{abc}\eta_b\eta_c\right] \tag{12-164}$$

where a, α, and β are of order $\hbar^n$. Simple algebra and the use of the homogeneity property of $\mathscr{L}$,

$$2\mathscr{L} = A^\mu{}_a(x)\frac{\partial\mathscr{L}}{\partial A^\mu{}_a(x)} - g\frac{\partial\mathscr{L}}{\partial g}$$

enable us to rewrite $\tilde{\Gamma}^{[n]}_{\text{div}}$ as

$$\tilde{\Gamma}^{[n]}_{\text{div}} = \left(\int d^4x\left\{\left(\beta - \alpha + \frac{a}{2}\right)\left[A^\mu{}_a(x)\frac{\delta}{\delta A^\mu{}_a(x)} + L_a(x)\frac{\delta}{\delta L_a(x)}\right]\right.\right.$$
$$\left.\left. + \frac{\alpha}{2}\left[K_a(x)\frac{\delta}{\delta K_a(x)} + \eta_a(x)\frac{\delta}{\delta\eta_a(x)} + \bar{\eta}_a(x)\frac{\delta}{\delta\bar{\eta}_a(x)}\right]\right\} - \frac{a}{2}g\frac{\partial}{\partial g}\right)\tilde{\Gamma}(A, \eta, \bar{\eta}, K, L; g) \tag{12-165}$$

which is the desired result. All counterterms arise from a renormalization of the parameters of the initial action. Moreover, A and L are renormalized in the same way. If, according to the recursion hypothesis, we write the action renormalized up to order $n-1$ as

$$\tilde{I}_{n-1} = \tilde{I}(Z_{3,n-1}^{1/2}A, \tilde{Z}_{3,n-1}^{1/2}\eta, \tilde{Z}_{3,n-1}^{1/2}\bar{\eta}, \tilde{Z}_{3,n-1}^{1/2}K, Z_{3,n-1}^{1/2}L; Z_{g,n-1}g)$$

then we have just proved that

$$\begin{aligned}\tilde{I}_n &= \tilde{I}_{n-1} - \tilde{\Gamma}_{\text{div}}^{[n]} + O(\hbar^{n+1}) \\ &= \tilde{I}(Z_{3,n}^{1/2}A, \tilde{Z}_{3,n}^{1/2}\eta, \tilde{Z}_{3,n}^{1/2}\bar{\eta}, \tilde{Z}_{3,n}^{1/2}K, Z_{3,n}^{1/2}L; Z_{g,n}g)\end{aligned} \tag{12-166}$$

with

$$Z_{3,n}^{1/2} = Z_{3,n-1}^{1/2} - \left(\beta - \alpha + \frac{a}{2}\right)$$

$$\tilde{Z}_{3,n}^{1/2} = \tilde{Z}_{3,n-1}^{1/2} - \frac{\alpha}{2}$$

$$Z_{g,n} = Z_{g,n-1} + \frac{a}{2}$$

This completes the inductive proof.

We have shown that it is possible to renormalize nonabelian gauge theories, while preserving gauge invariance expressed through the Slavnov-Taylor identities (12-144) and (12-145). Happily the whole operation boils down to wave-function and coupling constant renormalizations. We generate finite Green functions if we use the action

$$I_R(A, \eta, \bar{\eta}, K, L; g, \lambda) = I(A_0, \eta_0, \bar{\eta}_0, K_0, L_0; g_0, \lambda_0) \tag{12-167}$$

with the same notations as in (12-158) and $Z_L = Z_3$, $\lambda_0 = Z_3^{-1}\lambda$, which takes into account the nonrenormalization of the gauge term $(\lambda/2)\mathscr{F}(A)^2$ (for a linear function $\mathscr{F}$). As a consequence of (12-167), the Green functions are multiplicatively renormalized:

$$\Gamma_R(A, \eta, \bar{\eta}, K, L; g, \lambda) = \Gamma_{\text{reg}}(A_0, \eta_0, \bar{\eta}_0, K_0, L_0; g_0, \lambda_0) \tag{12-168}$$

and satisfy the identities (12-144) and (12-145). After completion of this proof, we are of course entitled to drop the auxiliary sources K and L in the two preceding equations (12-167) and (12-168).

The previous considerations extend to cases involving matter fields, coupled in a minimal way. If fermion fields are present, we assume for the time being that the couplings involve no λ_5 matrix (this point will be investigated in Sec. 12-4-5). Dimensional regularization makes the theory finite, the Becchi-Rouet-Stora transformation and the identities (12-147) can be generalized, and, as expected from our one-loop computation, equations analogous to (12-167) and (12-168) hold. The crucial feature is, of course, the universality of coupling constant renormalization.

The compact formulation of Ward identities may have obscured simple facts. We emphasize that the results found in the one-loop computation are mere consequences of the Eqs. (12-147). For instance, if the counterterms $Z_1, \tilde{Z}_1, Z_4$ are reinstated as in (12-124), it is easy to see that (12-167) means that

$$\frac{Z_1}{Z_3} = \frac{\tilde{Z}_1}{\tilde{Z}_3} = \frac{Z_4}{Z_1}$$

to all orders. Equation (12-147) may also be used to show that the corrections to the inverse propagator $\Gamma_{\mu\nu}$ are transverse to all orders.

A different regularization, investigated by Lee and Zinn-Justin, relies on the introduction of higher covariant derivatives in the initial lagrangian. This improves the large-momentum behavior of the gauge field and makes all diagrams with more than one loop finite. The one-loop diagrams must be regularized independently in a gauge-invariant way.

The reader may carry out the renormalization program in the axial gauge; it is convenient to write the condition $n \cdot A = 0$ by using a Lagrange multiplier in the functional integral. Although the ghost fields are not really coupled to the gauge field, their introduction enables us to use a slightly modified Becchi-Rouet-Stora transformation and to derive a set of identities.

12-4-4 Gauge Dependence of Green Functions

Since at the very end of our computations we are supposed to check the gauge independence of the physical quantities, it is important to control the gauge dependence of Green functions. We observed that the transformation $\delta A = D\mathscr{M}^{-1}\delta\omega$ shifts $\mathscr{F}$ into $\mathscr{F} + \delta\omega$ and modifies the action $I_{\mathscr{F}}$ of Eq. (12-133) according to

$$\begin{aligned} I_{\mathscr{F}} + \delta I &= I_{\mathscr{F}} - \int d^4x \left[\lambda \mathscr{F}_a \delta\omega_a + \bar{\eta}_a \frac{\delta(\delta\omega_a)}{\delta A^\mu{}_b} (D^\mu \eta)_b \right] \\ &= I_{\mathscr{F} + \delta\omega} \end{aligned} \tag{12-169}$$

Consequently, an infinitesimal change of the function $\mathscr{F}$ may be cancelled by a gauge transformation on the field A. The latter affects only the source term, and according to the equivalence theorem of Sec. 9-2-1 this should not modify physical quantities such as the S-matrix elements:

$$\begin{aligned} &\exp\left[G_{\mathscr{F}+\Delta\mathscr{F}}(J)\right] \\ &= \int \mathscr{D}(A) \det \mathscr{M}_{\mathscr{F}+\Delta\mathscr{F}} \exp\left\{ i \int d^4x \left[\mathscr{L} - \frac{\lambda}{2}(\mathscr{F} + \Delta\mathscr{F})^2 + J \cdot A \right] \right\} \\ &= \int \mathscr{D}(A) \det \mathscr{M}_{\mathscr{F}} \exp\left\{ i \int d^4x \left[\mathscr{L} - \frac{\lambda}{2}\mathscr{F}^2 + J \cdot (A - D\mathscr{M}^{-1}\Delta\mathscr{F}) \right] \right\} \end{aligned} \tag{12-170}$$

This discussion is quite formal at this point, since in the gauge theories studied so far the S matrix is not defined, due to severe infrared divergences.

In terms of the Becchi-Rouet-Stora transformation, the preceding property is reflected by the following structure of δI:

$$\begin{aligned} \delta I &= -\int d^4x \left[\lambda \mathscr{F}_a \Delta\mathscr{F}_a + \bar{\eta}_a \frac{\delta \Delta\mathscr{F}_a}{\delta A^\mu{}_b} (D^\mu \eta)_b \right] \\ &= s \int (-\bar{\eta}_a \Delta\mathscr{F}_a)\, d^4x \end{aligned}$$

This enables one to study the gauge dependence of proper functions and counterterms, for instance their dependence on the parameter λ. As a typical result, one may show that the coupling constant renormalization Z_g is λ independent, at least in the minimal renormalization. With other prescriptions, this may be wrong. The last remark is a hint that the physical interpretation and observation of a nonabelian gauge coupling constant may be difficult.

12-4-5 Anomalies

We now reconsider a case of physical interest set aside at the end of Sec. 12-4-3. Let us assume that fermions are coupled to the gauge field through an axial current. We have seen in the previous chapter that anomalies may occur in the conservation (or quasiconservation) of such a current, as a consequence of the impossibility of regularizing the theory while preserving chiral symmetry. In the instances studied in Chap. 11, namely, quantum electrodynamics or the σ model, this anomaly was acceptable and of physical interest to analyze the process $\pi^\circ \to 2\gamma$. If the gauge field (abelian or not) is coupled to an anomalous current, the situation is drastically different. Slavnov-Taylor identities may become invalid and renormalizability is jeopardized. In theories where the gauge field remains massless, such as those considered so far, this would mean that all possible counterterms of dimension four would be required, spoiling the universality of the coupling renormalization. The issue is much more crucial when the symmetry is spontaneously broken. As we shall see in the next section, the gauge field becomes massive and renormalizability only results from the underlying gauge invariance. Anomalies are then harmful, and it is possible to devise models where they make the theory nonrenormalizable. It is therefore important to find a criterion to rule out their appearance.

As in the previous chapter, the analysis may be restricted to one-loop diagrams. There exists a gauge-invariant regularization which preserves chiral invariance at higher orders (Sec. 12-4-3). Consider a gauge theory based on a compact group. In other words, we admit abelian factors. The lagrangian, including fermion fields, is

$$\mathscr{L} = -\tfrac{1}{4}F_{\mu\nu a}F^{\mu\nu}{}_a + \bar{\psi}(i\not{\partial} - m - ig\not{A}_a\Gamma^a)\psi \tag{12-171}$$

where Γ^a stands for some combination of T^a and $T^a\gamma_5$. It is easy to check that the anomaly of the axial current

$$\bar{\psi}\gamma_\mu\gamma_5 T^a\psi$$

as given by Eq. (11-225), is proportional to the combination

$$d_{abc} \equiv \operatorname{tr}\,(T^a\{T^b, T^c\}) \tag{12-172}$$

The vanishing of this quantity may be realized by each species of fermion (each representation) coupled to the gauge field; this is in particular what happens for real representations where the matrices T are antisymmetric:

$$d_{abc} = \operatorname{tr}\,(T^{aT}\{T^b, T^c\}^T) = -d_{abc} = 0 \tag{12-173}$$

But the condition may also result from a cancellation between different species. This will be illustrated in Sec. 12-6-4.

The harmful anomalies are those occurring in axial currents coupled to gauge fields. For instance, if the combinations $\bar{\psi}\gamma_\mu\gamma_5\lambda^a\psi$ are singlets for the internal group G, the anomalies of such currents are of no importance here. The matrices λ^a might refer to a different set of quantum numbers, e.g., flavor as opposed to color. The corresponding anomalies are proportional to

$$d_{abc} = \text{tr}\,(\lambda^a\{T^b, T^c\}) = \text{tr}\,\lambda^a\,\text{tr}\,\{T^b, T^c\}$$

which vanishes in the case of an $SU(3)$ symmetry, for instance, since $\text{tr}\,\lambda^a = 0$. Only the $U(1)$ current

$$J_{\mu 5} = \bar{\psi}\gamma_\mu\gamma_5\psi \tag{12-174}$$

is anomalous:

$$\begin{aligned}\partial^\mu J_{\mu 5} &= 2im\bar{\psi}\gamma_5\psi + Cg^2\varepsilon_{\mu\nu\rho\sigma}F^{\mu\nu}{}_a F^{\rho\sigma}{}_a \\ &= 2im\bar{\psi}\gamma_5\psi + 2Cg^2\partial^\mu\varepsilon_{\mu\nu\rho\sigma}A^\nu{}_a(\partial^\rho A^\sigma{}_a - \partial^\sigma A^\rho{}_a - \tfrac{2}{3}gC_{abc}A^\rho{}_b A^\sigma{}_c)\end{aligned} \tag{12-174a}$$

with $C = -T_f/16\pi^2$. On the other hand, there exists a conserved current

$$\tilde{J}_{\mu 5} = J_{\mu 5} - 2Cg^2\varepsilon_{\mu\nu\rho\sigma}A^\nu{}_a(\partial^\rho A^\sigma{}_a - \partial^\sigma A^\rho{}_a - \tfrac{2}{3}gC_{abc}A^\rho{}_b A^\sigma{}_c) \tag{12-175}$$

but it suffers from the lack of gauge invariance.

12-5 MASSIVE GAUGE FIELDS

12-5-1 Historical Background

The nonabelian gauge theories considered so far enjoyed an exact local symmetry, and consequently the gauge field was massless. Such theories are used nowadays to construct models of strong interactions. However, historically, after the introduction by Yang and Mills of nonabelian gauge fields, physicists have struggled for years in order to build a meaningful theory of massive gauge fields, hence breaking explicitly the local symmetry.

A strong motivation came from the study of weak interactions. We recall from Chap. 11 that the current-current Fermi theory provides a remarkable phenomenological framework. The weak interaction lagrangian (or up to a sign, the hamiltonian) was written (compare with 11-62)

$$\mathscr{L}_{\text{int}} = -\frac{G}{\sqrt{2}}J^\mu(x)J_\mu{}^\dagger(x) \tag{12-176}$$

This is, of course, a zero-range interaction.

In spite of its successes for low-energy processes, this model suffers from serious problems. As the dimensionality of the coupling constant G shows, the theory is nonrenormalizable. Alternatively, power counting assigns dimension six to the product $J^\mu J_\mu^\dagger$. At high enough energy we cannot content ourselves with the Born approximation. In order that a scattering amplitude satisfies the unitarity condition, at least perturbatively, higher-order terms must be added. These corrections, however, are plagued with ultraviolet divergences, the elimination of which introduces a growing number of arbitrary parameters. In practice, the nonrenormalizability makes this computation impossible.

Another aspect of the same problem arises when we consider the Born approximation for some cross section σ. On dimensional grounds, we expect at high energy the behavior

$$\sigma \sim \text{constant} \times G^2 s \tag{12-177}$$

where s is the total center of mass energy square, while in every partial wave the unitarity limit reads

$$\sigma \sim \frac{\text{constant}}{s} \tag{12-178}$$

Therefore, we expect a violation of unitarity to arise at energies of the order $\sqrt{s} \sim G^{-1/2} \sim 300$ GeV.

It is a good exercise to compute explicitly the constant appearing in (12-177) and (12-178) for such leptonic processes as $\nu\bar{\nu} \to \nu\bar{\nu}$, $\bar{\nu}_e e^- \to \bar{\nu}_e e^-$, and $\nu_\mu e^- \to \nu_e \mu^-$.

Both aspects, nonrenormalizability of the theory and bad high-energy behavior of the Born approximation, are manifestations of the same phenomenon. This is obvious if we use dispersion relations to compute a one-loop contribution to some elastic scattering amplitude in terms of its discontinuity, i.e., in terms of some Born cross section. The behavior of the latter results in severe divergences in the dispersion integral.

It is therefore mandatory to transform the Fermi theory into a respectable, i.e., renormalizable, field theory. A tempting hypothesis consists in introducing a charged vector field W_μ and coupling it to the current J_μ:

$$\mathcal{L}_{\text{int}} = gJ^\mu(x)W_\mu{}^\dagger(x) + \text{hc} \tag{12-179}$$

The analogy with electromagnetism is evident. The intermediate boson represented by the field W would be the quantum of weak interactions. To explain the validity of the Fermi theory at low energy, we assume W to be very massive. The implications of (12-179) would depart from those of (12-176) only at high energies. Let us consider, for instance, the μ decay. The Fermi theory gives the amplitude (Fig. 12-9*a*)

$$-\frac{iG}{\sqrt{2}}\bar{u}(p_e)\gamma_\rho(1-\gamma_5)v(p_{\bar{\nu}_e})\bar{u}(p_{\nu_\mu})\gamma^\rho(1-\gamma_5)u(p_\mu)$$

whereas the W boson contributes

$$ig^2\bar{u}(p_e)\gamma_\rho(1-\gamma_5)v(p_{\bar{\nu}_e})\frac{g^{\rho\sigma}-k^\rho k^\sigma/M_W^2}{k^2-M_W^2}\bar{u}(p_{\nu_\mu})\gamma_\sigma(1-\gamma_5)u(p_\mu)$$

Figure 12-9 Muon decay (*a*) in the Fermi theory and (*b*) as mediated by an intermediate vector boson.

The two amplitudes coincide at energies such that $k^2 \equiv (p_e + p_{\bar{\nu}_e})^2 \ll M_W^2$, provided

$$\frac{g^2}{M_W^2} = \frac{G}{\sqrt{2}} \tag{12-180}$$

Similarly the Born approximation to a scattering amplitude such as $\nu\bar{\nu} \to \nu\bar{\nu}$ would be reduced at high energy by a factor M_W^2/s with respect to the Fermi amplitude. The violations of unitarity have seemingly been eliminated. Actually, the amplitude $\nu\bar{\nu} \to W^+W^-$ still has a bad behavior. This points to the necessity of introducing more fields and more couplings.

The dynamics of these massive vector fields have now to be prescribed. Particularly troublesome is the question of renormalizability in view of the large momentum behavior of the propagator:

$$\left(g_{\mu\nu} - \frac{k_\mu k_\nu}{M_W^2}\right)(k^2 - M_W^2)^{-1} \underset{k^2 \gg M_W^2}{\sim} \text{constant}$$

However, we remember that in quantum electrodynamics the introduction of a photon mass has not spoiled renormalizability. If we adhere to the prejudice that more symmetry in a theory reduces the number of divergences, it is suggested to consider the W_μ as a member of a set of gauge fields. A candidate for a symmetry group is $SU(2)$ and the local invariance is of course explicitly broken by the mass terms of the W. This would lead to a universal coupling of the W. Since $SU(2)$ gauge fields must form a triplet, a third neutral vector field has to be introduced. Such a model, suitably amended, will be shown to be renormalizable.

For completeness, we mention a different historical reason for introducing massive vector fields. It was suggested in the early 1960s to base the theory of strong interactions on a gauge principle. The invariance group was $SU(3) \times U(1)$, corresponding to the eightfold symmetry and baryonic charge conservation. The vector bosons—massive gauge fields—were identified with the existing ρ, K^*, ω, and ϕ particles.

An interesting property of such a model is that the forces were attractive between particles of antiparallel isospins and repulsive for parallel isospins, a generalization of the electromagnetic attraction between opposite charges. Such a property is in agreement with experimental facts at low energy.

To see this, let us compute the elastic scattering amplitude of two scalar particles belonging to the real representations (1) and (2) of a simple Lie group. The lowest-order contribution (the exchange

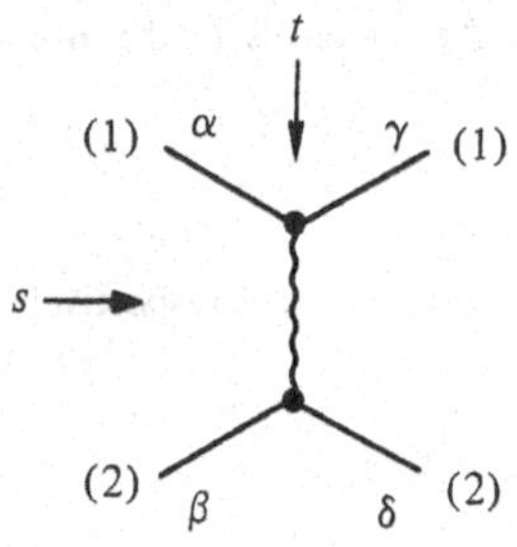

Figure 12-10 A scattering amplitude to lowest order.

of the vector field as shown in Fig. 12-10) comes from the following terms of the lagrangian:

$$\mathscr{L} = \tfrac{1}{2}(D_\mu^{(1)}\phi^{(1)})(D^{(1)\mu}\phi^{(1)}) + \tfrac{1}{2}(D_\mu^{(2)}\phi^{(2)})(D^{(2)\mu}\phi^{(2)}) \tag{12-181}$$

where $\qquad D^{(1)\mu} = \partial^\mu - gA^\mu{}_a T^{(1)a} \qquad T^{(1)a}$ antisymmetric $a = 1, \ldots, r$

and reads

$$\mathscr{T} = g^2 \frac{s-u}{M^2 - t} \sum_{a=1}^{r} T_{\alpha\gamma}^{(1)a} T_{\beta\delta}^{(2)a} \tag{12-182}$$

The quantity

$$X_{\alpha\beta\gamma\delta} = \sum_{a=1} T_{\alpha\gamma}^{(1)a} T_{\beta\delta}^{(2)a}$$

must be projected on irreducible representations. To this end, we introduce the Clebsch-Gordan matrices for the product of the representations (1) and (2). If n_1 and n_2 are the dimensions of the representations, the $n_1 \times n_2$ matrices $M^{(t)A}$ satisfy the orthogonality and completion relations

$$\begin{aligned} &\operatorname{tr}(M^{(t)A}M^{(t')B}) = \delta^{tt'}\delta^{AB} \qquad A, B = 1, \ldots, n_1 n_2 \\ &\sum_{A,t} M_{\alpha\beta}^{(t)A} M_{\alpha'\beta'}^{(t)A} = \delta_{\alpha\alpha'}\delta_{\beta\beta'} \end{aligned} \tag{12-183}$$

and transform according to the (t) representation

$$T^{(1)a}M^{(t)A} - M^{(t)A}T^{(2)a} = T_{AB}^{(t)a}M^{(t)B} \tag{12-184}$$

It is then easy to see that

$$\begin{aligned} X_{\alpha\beta\gamma\delta} &= \sum_a T_{\alpha\gamma}^{(1)a} T_{\beta\delta}^{(2)a} \\ &= -\sum_{a,A,t} M_{\alpha\beta}^{(t)A} T_{\gamma\alpha'}^{(1)a} M_{\alpha'\beta'}^{(t)A} T_{\beta'\delta}^{(2)a} \\ &= \frac{1}{2}\sum_{A,t} [C(1) + C(2) - C(t)] M_{\alpha\beta}^{(t)A} M_{\gamma\delta}^{(t)A} \end{aligned} \tag{12-185}$$

in terms of the Casimir operators for the representations (1) and (2) and (t) [see (12-119)]. The value of the Casimir operator increases with the dimension of the representation. For instance, in the case of $SU(2)$, the representation of isospin I has $C(I) = I(I+1)$. The desired property follows from the positivity of $(s-u)/(M^2-t)$ in the physical region.

This calculation of a crossing matrix, relative here to internal degrees of freedom, is analogous to the Fierz reshuffling of Chap. 3.

This model, considered by Sakurai, also predicts the widths of vector bosons. Neglect the $\phi - \omega$ mixing (which improves the computation) and couple the (π, K, η) and (ρ, K^*, ϕ) octets in a gauge-invariant way. The p-wave amplitude of the elastic $\pi\pi$, πK, and $K\bar{K}$ channels in the Born approximation reads

$$\mathcal{T}_{I=1} = \alpha \frac{g^2}{16\pi} \frac{\mathbf{q}^2}{M^2 - s}$$

where $\mathbf{q}$ is the center of mass three-momentum, and $\alpha = \frac{8}{3}$, 1, and 2 for the three channels respectively. A unitary amplitude t_1 may be constructed in the form

$$t_1 = \frac{\mathcal{T}_1}{1 - i\rho\mathcal{T}_1} \qquad \operatorname{Im} t_1 = \rho\,|t_1|^2$$

where $\rho = 2|\mathbf{q}|/\sqrt{s}$ ($\times \frac{1}{2}$ for $\pi\pi$ due to the identity of particles) is the phase space factor. This amplitude exhibits a pole at $s \simeq M^2 - i\rho\alpha\mathbf{q}^2 g^2/16\pi$ corresponding to a resonance of width $\Gamma \simeq \rho\alpha\mathbf{q}^2 g^2/16\pi M|_{s=M^2}$. Adjusting the value $g^2/16\pi = 0.63$ yields $\Gamma_\rho \simeq 130$ MeV, $\Gamma_{K^*} \simeq 38$ MeV, and $\Gamma_\phi \simeq 4.5$ MeV, which reproduces fairly well the experimental values (125, 50, 3.2 respectively).

12-5-2 Massive Gauge Theory

Is a gauge theory where mass terms are introduced by hand renormalizable?

In electrodynamics, the situation is favorable. After separation of the gauge field into transverse and longitudinal components, the longitudinal part $k_\mu k_\nu/M^2$ which gives rise to the bad behavior in the propagator does not contribute to the S matrix. This results from the noninteraction of longitudinal and transverse components and from the coupling of the field to a conserved current. In a non-abelian theory, none of these properties is satisfied. Longitudinal and transverse parts do interact, while the current to which the gauge field is coupled is not conserved. On the other hand, unexpected cancellations of divergences at the one-loop level make the theory look like renormalizable. This explains why it took some time to reach a consensus, namely, that the theory is not renormalizable. The way out of this unpleasant situation is to appeal to the mechanism of spontaneous symmetry breaking, to be explained in the next subsection.

The aim is the construction of a renormalizable theory with the requirement that the physical states be the massive vector fields only. If we use auxiliary fields, as in the Stueckelberg method for the electromagnetic field, we must check that only the three physical degrees of freedom of each massive vector boson contribute to unitarity. In the forthcoming method, where local changes of variables will be performed in the functional integral so as to improve the behavior of the propagator, this requirement will be fulfilled by virtue of the equivalence theorem. We thus consider the generating functional

$$e^{G(J)} = \int \mathcal{D}(A) \exp\left\{i \int d^4x \left[\mathcal{L} - 2\operatorname{tr}(J \cdot A)\right]\right\} \tag{12-186}$$

where we use again matrix notations

$$\mathcal{L} = \operatorname{tr}\left(\tfrac{1}{2}F_{\mu\nu}F^{\mu\nu} - M^2 A_\mu A^\mu\right) \tag{12-187}$$

The canonical quantization may be in trouble, since $\pi_{0a} = \delta\mathcal{L}/\delta\partial_0 A^0{}_a \equiv 0$. However, the presence of the mass term insures the existence of a propagator, and hence the definiteness of the previous functional integral in a perturbative sense. The Faddeev-Popov operation is not necessary, but we do it nevertheless, in order to improve the behavior of the propagator. We choose a gauge condition

$$\mathcal{F}(A) = C$$

as in Sec. 12-2-2 and insert in (12-186) the identity

$$1 = \int \mathscr{D}(g) \prod_x \delta[\mathscr{F}({}^g A) - C] \det \mathscr{M}_{\mathscr{F}}(A) \tag{12-188}$$

We obtain

$$e^{G(J)} = \int \mathscr{D}(A)\mathscr{D}(g) \prod_x \delta[\mathscr{F}({}^g A) - C] \det \mathscr{M}_{\mathscr{F}} \exp \left\{ i \int d^4x \, [\mathscr{L} - 2 \operatorname{tr}(J \cdot A)] \right\}$$

In contrast with the massless case, $\mathscr{L}$ is no longer invariant under the gauge transformation $A \to {}^g A$. If we parametrize $g(x)$ as

$$g(x) = e^{\xi(x)}$$

it is easy to show that

$$\operatorname{tr}(A^\mu A_\mu) = \operatorname{tr}\left[{}^g A^\mu \, {}^g A_\mu - \frac{2}{g} \partial_\mu \xi \, {}^g A^\mu + \frac{1}{g^2} \partial_\mu \xi P^\mu({}^g A, \xi) \right]$$

where P^μ stands for the formal series

$$P^\mu(A, \xi) = \sum_{n=0}^{\infty} \frac{2}{(n+2)!} [\cdots [D^\mu \xi, \xi], \xi], \ldots, \xi] \tag{12-189}$$

the generic term of which has n brackets. The S matrix is not affected if we replace the source term $J \cdot A$ by $J \cdot {}^g A$. After a change of variable $A \to {}^{g^{-1}} A$ and a gaussian integration over C, the new generating functional reads

$$e^{G'(J)} = \int \mathscr{D}(A, \xi, \eta, \bar\eta) \exp \left\{ i \int d^4x \, [\mathscr{L}'(A, \xi, \eta, \bar\eta) - 2 \operatorname{tr}(J \cdot A)] \right\} \tag{12-190}$$

It involves a lagrangian with the field A_μ, the Faddeev-Popov ghosts $\eta, \bar\eta$, and a new field ξ, with conventional commutation assignments

$$\mathscr{L}'(A, \xi, \eta, \bar\eta) = \operatorname{tr}\left[\tfrac{1}{2} F_{\mu\nu} F^{\mu\nu} - M^2 A_\mu A^\mu + \lambda \mathscr{F}^2(A) + 2 \frac{M^2}{g} A^\mu \partial_\mu \xi - \frac{M^2}{g^2} \partial_\mu \xi P^\mu(A, \xi) \right] - \bar\eta \mathscr{M}_{\mathscr{F}} \eta \tag{12-191}$$

If $M^2 = 0$, the field ξ disappears from the lagrangian and the ξ integration gives an (infinite) factor which does not contribute to $e^{G'(J) - G'(0)}$.

It is convenient to choose the Landau gauge, $\mathscr{F} = \partial_\mu A^\mu$, $\lambda \to \infty$, to compute the vector propagator $-i(g_{\mu\nu} - k_\mu k_\nu / k^2)(k^2 - m^2)^{-1}$. The ξ and η propagators behave as $1/k^2$. The superficial degree of divergence of an L-loop diagram is

$$\omega = 2L + 2 + \sum_i n_i(d_i - 2) \leq 2L + 2$$

where n_i is the number of vertices of type (i) and d_i is the number of field derivatives on such a vertex. For a proper diagram with E external lines (among which no ξ line), this yields

$$\omega \leq 2L + 2 - E + \sum_{\substack{\text{internal} \\ \text{vertices}}} n_i(d_i - 2) \leq 2L + 2 - E \tag{12-192}$$

to be compared with

$$\omega \leq 6L + E - 2$$

in the original theory (12-187).

Observe that for an abelian theory, $P_\mu = \partial_\mu \xi$, the fields ξ and η are not coupled to the vector

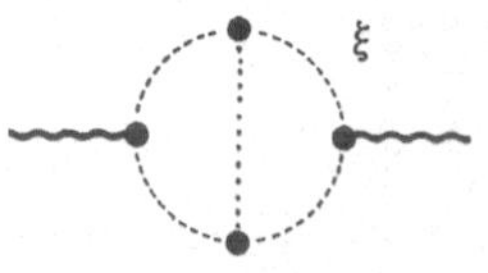

Figure 12-11 A diagram with a quartic divergence, in a massive gauge theory. The dotted lines represent auxiliary field ξ propagators.

field. We have just recovered the result of Chap. 8, namely, that massive electrodynamics is renormalizable by power counting.

In a nonabelian case, if we restrict ourselves to the one-loop approximation, Eq. (12-192) gives the same superficial degree of divergence as in a renormalizable theory, $\omega = 4 - E$. To this order an effective lagrangian for diagrams without external ξ or η lines reads in the Landau gauge

$$\mathscr{L}'_1 = \operatorname{tr}\left(\tfrac{1}{2}F_{\mu\nu}F^{\mu\nu} - M^2 A_\mu A^\mu - \frac{M^2}{g^2}\partial_\mu\xi D^\mu\xi\right) - \bar{\eta}\partial_\mu D^\mu \eta \tag{12-193}$$

The gaussian integrals over ξ and $\eta, \bar{\eta}$ may be performed. The former yields $\det^{-1/2} \mathscr{M}_{\mathscr{F}}$ while the latter gives $\det \mathscr{M}_{\mathscr{F}}$. Hence, to that order, a single auxiliary field suffices, with the prescription that a factor $-\frac{1}{2}$ is attached to each closed ghost loop. The presence of this factor $-\frac{1}{2}$ to be compared with a factor -1 in the massless case shows that the limit $M \to 0$ must be singular. We have seen in Sec. 12-3 that the ghost contribution (with a factor -1) was crucial in the massless case to maintain gauge invariance. We thus expect the modification of the prescription to modify the counterterms. For instance, it leads to mass or gauge term renormalizations, and, more seriously, to other four-point couplings. Therefore, even though the theory looks renormalizable to this order, the symmetry of the counterterms is lost and serious difficulties occur to higher orders. For instance, according to Eq. (12-192) the diagram depicted on Fig. 12-11 has a quartic divergence, $\omega = 4$.

We thus conclude that in spite of cancellations of divergences, the massive gauge theory is not renormalizable.

12-5-3 Spontaneous Symmetry Breaking

We have studied spontaneous symmetry breaking in Chap. 11, where boundary conditions allow us to choose among a set of degenerate ground states. A remarkable feature of this phenomenon, in the case of a continuous symmetry, is the appearance of massless particles. These Goldstone bosons are the zero-energy excitations connecting the possible vacua to each other. It is natural to reexamine this phenomenon in a gauge theory (abelian or nonabelian) where long-range forces are present or, alternatively, where there may exist an unphysical sector in the Hilbert space. It turns out that in the presence of a broken gauge symmetry, the long-range forces are screened. The Goldstone bosons and the gauge fields conspire to create massive excitations, and the massless excitations are unobservable.

This phenomenon was discovered and studied in the context of superconductivity. Electron pairs responsible for superconductivity may be described by a wave function $\psi = \rho e^{i\theta/\hbar}$. The charge density, proportional to $\psi^*\psi = \rho^2$, must be constant throughout the crystal to neutralize the background charge of the ions. In the presence of a magnetic field with vector potential $\mathbf{A}$, the current $\mathbf{J}$ reads

$$\mathbf{J} = \frac{1}{2m}\psi^*\left(\frac{\hbar}{i}\overleftrightarrow{\nabla} - 2q\mathbf{A}\right)\psi = \frac{\rho^2}{m}(\nabla\theta - q\mathbf{A})$$

where $q = 2e$ is the charge of the pair. The divergence of $\mathbf{J}$ vanishes, and hence in the transverse gauge $\nabla \cdot \mathbf{A} = 0$ we have $\Delta\theta = 0$. For a simple geometrical configuration, θ is constant; thus the Maxwell equation

$$\Delta \mathbf{A} = -q\mathbf{J}$$

leads to

$$\Delta \mathbf{A} = \frac{q^2 \rho^2}{m} \mathbf{A} \tag{12-194}$$

The vector potential $\mathbf{A}$ is screened on a characteristic length λ, $\lambda = \sqrt{m}/|q\rho|$. This is the Meissner effect which prevents a magnetic field from penetrating a superconductor.

Returning to field theory consider a charged field ϕ coupled to an abelian gauge field. The lagrangian is

$$\mathscr{L} = -\tfrac{1}{4}(\partial_\mu A_\nu - \partial_\nu A_\mu)(\partial^\mu A^\nu - \partial^\nu A^\mu) + (\partial_\mu - ieA_\mu)\phi^*(\partial^\mu + ieA^\mu)\phi - V(\phi) \tag{12-195}$$

The potential $V(\phi)$ is invariant under local transformations $\phi \to e^{i\omega(x)}\phi$ and its minimum occurs for a nonzero value of $\phi^*\phi$ (compare with Fig. 11-5). For instance,

$$V(\phi) = \mu^2 \phi^*\phi + \lambda(\phi^*\phi)^2 \tag{12-196}$$

with $\mu^2 < 0$, $\lambda > 0$. In the ground state $\langle\phi\rangle = v/\sqrt{2}$ with

$$\mu^2 + \lambda |v|^2 = 0 \tag{12-197}$$

A global rotation may always transform $\langle\phi\rangle$ to a real value, since it is x independent. We assume, therefore,

$$\phi = \frac{\phi_1 + i\phi_2}{\sqrt{2}} \qquad \langle\phi_1\rangle = v \quad \langle\phi_2\rangle = 0 \tag{12-198}$$

and translate ϕ:

$$\phi_1(x) = \phi'_1(x) + v \quad \langle\phi'_1\rangle = 0 \tag{12-199}$$

In terms of ϕ' (the prime will be dropped), the lagrangian reads

$$\begin{aligned}\mathscr{L} = &-\tfrac{1}{4}(\partial_\mu A_\nu - \partial_\nu A_\mu)(\partial^\mu A^\nu - \partial^\nu A^\mu) + \tfrac{1}{2}(\partial_\mu \phi_1 \partial^\mu \phi_1 + \partial_\mu \phi_2 \partial^\mu \phi_2) \\ &- \frac{\mu^2}{2}(\phi_1^2 + \phi_2^2 + 2v\phi_1) + \frac{e^2 A^2}{2}(\phi_1^2 + \phi_2^2 + 2v\phi_1 + v^2) \\ &- \frac{\lambda}{4}(\phi_1^2 + \phi_2^2 + 2v\phi_1 + v^2)^2 - eA_\mu \phi_2 \overleftrightarrow{\partial}_\mu \phi_1 + evA^\mu \partial_\mu \phi_2\end{aligned} \tag{12-200}$$

By virtue of Eq. (12-197), the coefficient of ϕ_2^2 vanishes; ϕ_2 is a Goldstone boson. But wait! a mass term for A_μ has appeared: $\frac{1}{2}e^2 v^2 A^2$, together with a mixed term $evA_\mu \partial^\mu \phi_2$. The quadratic form in A and ϕ_2 is diagonalized by the combination

$$B_\mu = A_\mu + \frac{\partial_\mu \phi_2}{ev} \tag{12-201}$$

which looks like a gauge transformation. Then, the quadratic part of $\mathscr{L}$ is

$$\mathscr{L}_q = -\tfrac{1}{4}(\partial_\mu B_\nu - \partial_\nu B_\mu)(\partial^\mu B^\nu - \partial^\nu B^\mu) + \tfrac{1}{2}\partial_\mu \phi_1 \partial^\mu \phi_1 + \tfrac{1}{2}e^2 v^2 B^2 - \tfrac{1}{2}(2\lambda v^2)\phi_1^2 \tag{12-202}$$

The net result is the following. The gauge field has acquired a mass while ϕ_2 has disappeared, at least from $\mathscr{L}_q$. In fact ϕ_2 can be eliminated altogether if we use a different parametrization of ϕ:

$$\phi(x) = e^{i\theta(x)/v} \frac{v + \rho(x)}{\sqrt{2}} \tag{12-203}$$

with hermitian fields v and ρ. After a local gauge transformation

$$\begin{aligned} \phi \to \phi' &= e^{-i\theta/v}\, \phi = \frac{v+\rho}{\sqrt{2}} \\ A_\mu \to A'_\mu &= A_\mu + \frac{1}{ev}\partial_\mu \theta \end{aligned} \tag{12-204}$$

the lagrangian takes the form

$$\begin{aligned} \mathscr{L} = &-\tfrac{1}{4}(\partial_\mu A'_\nu - \partial_\nu A'_\mu)(\partial^\mu A'^\nu - \partial^\nu A'^\mu) \\ &+ (\partial_\mu - ieA'_\mu)\phi'(\partial^\mu + ieA'^\mu)\phi' - \mu^2 \phi'^2 - \lambda \phi'^4 \end{aligned} \tag{12-205}$$

Expanding $\phi' = (\rho + v)/\sqrt{2}$, we verify that A has indeed acquired a mass $M = |e|v$ and that the Goldstone boson has disappeared. We started from a system describing a charged scalar field (two states) and a massless gauge field with two polarization states. After spontaneous symmetry breakdown, we have one real scalar field and one massive vector field with three polarizations. The number of degrees of freedom has been conserved and the Goldstone boson has been transmuted into the longitudinal polarization state of the vector field. This is the phenomenon discovered by Englert and Brout and by Higgs. The remaining massive scalar boson is referred to as the Higgs boson.

The previous mechanism may be extended to a nonabelian symmetry. We follow the analysis of Kibble and Lee and Zinn-Justin. Let G be the gauge group of order r, not necessarily semisimple, and let the gauge field be coupled to a multiplet of scalar fields ϕ, transforming according to some irreducible n-dimensional representation. The lagrangian is

$$\mathscr{L} = -\tfrac{1}{4}F_{\mu\nu a}F^{\mu\nu}{}_a + [(\partial_\mu - g_a T^a A_{\mu a})\phi]^\dagger[(\partial_\mu - g_a T^a A_{\mu a})\phi] - V(\phi) \tag{12-206}$$

The antihermitian T matrices are the infinitesimal generators of the representation and the coupling constants may depend on the simple component of the group. Finally, V is assumed G invariant and its minimum is reached for $\langle \phi \rangle = v$. Let

H be the subgroup of G (of order s) which leaves v invariant. The nature of H depends on G, on the representation chosen for ϕ, and on the form of V.

Examples

(*a*) $G = O(n)$, ϕ is the vector (n-dimensional) representation, $H = O(n-1)$.

(*b*) $G = SU(3) \times SU(3)$, $\phi = M$ is a 3×3 matrix belonging to the $(3, \bar{3})$ representation, and

$$V(M) = \alpha \operatorname{tr} (MM^\dagger)^2 + \beta[\operatorname{tr} (MM^\dagger)]^2 + \gamma(\det M + \det M^\dagger) + \delta \operatorname{tr} (MM^\dagger)$$

The reader will find that for hermitian $\langle M \rangle$, H contains at least $SU(2) \times U(1)$.

Let the infinitesimal generators of H correspond to T^a ($a = 1, \ldots, s$). The remaining generators T^a ($a = s+1, \ldots, r$) generate the coset space G/H. We parametrize $\phi(x)$ as

$$\phi(x) = \exp\left[\sum_{a=s+1}^{r} T^a \frac{\xi_a(x)}{v}\right][\rho(x) + v]/\sqrt{2} \tag{12-207}$$

where ρ and ξ have a vanishing vacuum expectation value‡. The ρ field has $n - (r - s)$ effective components. In the absence of gauge fields, the ξ would be the Goldstone bosons. However, gauge invariance enables us to eliminate them. We perform the transformation

$$\phi(x) \to \phi'(x) = \exp\left[-\sum_{a=s+1}^{r} T^a \frac{\xi_a(x)}{v}\right]\phi(x) = [\rho(x) + v]/\sqrt{2}$$

$$\sum_{a=1}^{r} g_a T^a A_{\mu a}(x) \to \sum_{a=1}^{r} g_a T^a A'_{\mu a}(x) = \exp\left[-\sum_{s+1}^{r} \frac{T^a \xi_a(x)}{v}\right] \tag{12-208}$$

$$\times \left[-\partial_\mu + \sum_{a=1}^{r} g_a T^a A_{\mu a}(x)\right] \exp\left[\sum_{s+1}^{r} \frac{T^a \xi_a(x)}{v}\right]$$

which changes (12-206) into

$$\mathscr{L}(A', \phi') = -\tfrac{1}{4} F'_{\mu\nu a} F'^{\mu\nu}{}_a + [D_\mu(A')\phi']^\dagger [D^\mu(A')\phi'] - V(\phi') \tag{12-209}$$

where
$$D_\mu = \partial_\mu - \sum_{a=1}^{r} g_a A'_{\mu a} T^a$$

Any reference to the would-be Goldstone bosons ξ has disappeared, and the hermitian nonnegative mass matrix of the vector fields is

$$(M^2)_{ab} A^\mu{}_a A_{\mu b} = g_a g_b (T^a v)^\dagger (T^b v) A^\mu{}_a A_{\mu b} \tag{12-210}$$

Since the s first generators T^a ($a = 1, \ldots, s$) map v onto zero, the matrix is block diagonal. Only the lower $(r - s) \times (r - s)$ matrix is positive definite and corresponds to massive vector bosons. The s remaining massless fields correspond to the unbroken H gauge symmetry. The total number of degrees of freedom is conserved, as in the abelian case.

‡ The factor $\sqrt{2}$ appearing on the right-hand side of Eq. (12-207) will be omitted when discussing real fields, as was the case in Chap. 11.

We may now reexamine the semiclassical picture sketched in Sec. 12-1-3 and look for solutions of the equations of motion for a theory involving gauge fields and scalar fields. Nontrivial static solutions of finite energy do not exist for gauge fields or scalar fields alone. However, theories involving both scalar and gauge fields may possess interesting classical solutions in three space dimensions.

For a static solution of finite energy, the fields must tend at spatial infinity toward one of the lowest energy configurations. Otherwise the energy density would differ from zero by a finite amount in an infinite domain. A possible way to insure stability of a nontrivial solution is the existence of a set of degenerate vacua. We may then assume that the fields tend to different vacuum configurations in different spatial directions at infinity. The solution, if it exists, will be topologically stable if it maps in a nontrivial way the S_2 sphere at spatial infinity onto the manifold of possible vacua, i.e., the coset space G/H. A sufficient condition is that the symmetry is spontaneously broken and that the homotopy group $\pi_2(G/H)$ is nontrivial.

For definiteness, let us consider the Georgi-Glashow model, which is a gauge theory of symmetry group $G = SO(3)$, where a triplet of scalar fields is coupled to the triplet of gauge fields. The symmetry is spontaneously broken into $U(1) = SO(2)$. If the remaining massless gauge field is regarded as the electromagnetic field, we have a model of quantum electrodynamics based on the group $SO(3)$. With this in mind, we call the coupling constant e in the sequel. Mathematicians teach us that $\pi_2[SO(3)/U(1)] = Z$, the group of integers, which means that the solutions are characterized by an integer topological charge n.

't Hooft and Polyakov have studied the $n = 1$ solution. This corresponds to a boundary condition such that ϕ_a points along the normal to the S_2 sphere at infinity [identity map from S_2 onto $SO(3)/U(1) \sim S_2$]:

$$\phi_a(\mathbf{x}) \underset{|\mathbf{x}|\to\infty}{\to} \frac{x_a}{|\mathbf{x}|} v \tag{12-211}$$

where v is the vacuum expectation value of the real field ϕ. If we choose the gauge $A_a^0 = 0$, imposing that $\mathbf{D}\phi$ vanishes asymptotically, leads to

$$A_a^i \to \frac{1}{e} \varepsilon_{iab} \frac{x_b}{\mathbf{x}^2} \tag{12-212}$$

It is then possible to show the existence of regular solutions $\phi_a(\mathbf{x})$, $\mathbf{A}_a(\mathbf{x})$ satisfying these boundary conditions.

The topological invariant n may be interpreted as a magnetic charge. For this time-independent solution the electric field vanishes. At infinity the magnetic field is radial:

$$B^k = -\tfrac{1}{2}\varepsilon_{ijk} F^{ij} = \frac{x^k}{er^3}$$

since it is obtained from the vacuum configuration $\langle\phi\rangle = \mathbf{v}$ through a gauge transformation.

Its flux through the surface is

$$\int \mathbf{B}\cdot d\mathbf{S} = \frac{1}{e}\int \frac{1}{r^2} r^2\, d\Omega = \frac{4\pi}{e} \equiv g \tag{12-213}$$

by definition of the magnetic charge g contained inside the sphere. Solutions of higher topological charge n carry a magnetic charge $4\pi n/e$.

As any semiclassical configuration of this kind, the energy of the configuration—i.e., the rest mass of the monopole—is proportional to the inverse square coupling constant, $1/e^2$. 't Hooft has shown that it is of order M_W/α where M_W is the vector mass acquired through spontaneous symmetry breaking. Such a monopole would be extremely heavy!

12-5-4 Renormalization of Spontaneously Broken Gauge Theories

It is clear that the gauge (12-209), the so-called unitary gauge as it involves only physical degrees of freedom, is not suited for the study of renormalization, since the propagator has a bad large-momentum behavior. We need to use the initial gauge where renormalizability is more obvious, but then we will have to show that unphysical states do not contribute to the S matrix.

The key idea is the same as in the case of the spontaneously broken σ model (Chap. 11). Renormalization is independent of whether the symmetry is exact or spontaneously broken. For simplicity of notations, we present the analysis in the case of $G = O(n)$ and for a real scalar vector multiplet ϕ. The complete lagrangian including gauge and ghost terms reads

$$\mathscr{L}_S(\mu^2) = -\tfrac{1}{4}F_{\mu\nu a}F^{\mu\nu}{}_a - \frac{\lambda}{2}(\partial_\mu A^\mu{}_a)^2 - \bar{\eta}\partial_\mu D^\mu \eta + \tfrac{1}{2}D_\mu\phi D^\mu\phi - \frac{\mu^2}{2}\phi^2 - \frac{\lambda_\phi}{4}(\phi^2)^2 \tag{12-214}$$

The ϕ^4 coupling constant has been denoted λ_ϕ to avoid confusion with the gauge parameter. For $\mu^2 < 0$, the field ϕ acquires a vacuum expectation value $\langle\phi\rangle = v$. We want to show that the counterterms of the symmetric theory ($\mu^2 > 0$) suffice to make finite the broken theory ($\mu^2 < 0$) up to a renormalization of μ^2. As in Chap. 11, we cannot blindly continue μ^2 from positive to negative values, because $\mu^2 = 0$ is not an analyticity point. A phase transition occurs at this point. It is safe, however, to introduce a small external source $\mathbf{c}$ coupled to the field $\boldsymbol{\phi}(x)$ and constant throughout space

$$\mathscr{L} = \mathscr{L}_S(\mu^2) + \mathbf{c}\cdot\boldsymbol{\phi}(x) \tag{12-215}$$

This explicit breaking induces a vacuum expectation value $\mathbf{v}$ of ϕ parallel to $\mathbf{c}$. Indeed, an identity derived for the linear σ model [Eq. (11-170)] remains true here. If $\Gamma_T^{(2)}(0) = -m^2$ denotes the inverse propagator of the transverse component ϕ_T at zero momentum, then

$$\mathbf{c} = -\mathbf{v}\Gamma_T^{(2)}(0) \tag{12-216}$$

To lowest order, this condition reads $\mathbf{c} = (\mu^2 + \lambda_\phi v^2)\mathbf{v}$ and expresses that the vacuum expectation value $\mathbf{v} = \langle\boldsymbol{\phi}\rangle$ makes the lagrangian (12-214) stationary.

The proof of renormalizability relies on two observations. First, as in the linear σ model, the generating functional of proper functions for the symmetric and the broken theories are related by

$$\Gamma(\boldsymbol{\phi}, \mathbf{c}, \mathbf{v}) = \Gamma_S(\boldsymbol{\phi} + \mathbf{v}) - \Gamma_S(\mathbf{v}) + \int d^4x\, \mathbf{c}\cdot\boldsymbol{\phi} \tag{12-217}$$

with

$$\mathbf{c} = -\frac{\delta\Gamma_S(\mathbf{v})}{\delta\mathbf{v}} \tag{12-218}$$

The other arguments of Γ and Γ_S, such as $A_\mu, \eta, \bar{\eta}, g, \ldots$, have been omitted. We stress that Γ is the generating functional of Green functions of fluctuations around the unsymmetric vacuum. This identity is important because we already know how to renormalize the symmetric theory. From Sec. 12-4-3,

$$\Gamma_{S,R}(\phi, A, \eta, \ldots; g, \lambda_\phi, \mu^2) = \Gamma_{S,\text{reg}}(Z_2^{1/2}\phi, Z_3^{1/2}A, \tilde{Z}_3^{1/2}\eta, \ldots; g_0, \lambda_{\phi_0}, \mu^2 + \delta\mu^2) \tag{12-219}$$

Consequently, the symmetric counterterms will also make the broken theory finite:

$$\Gamma_R(\phi, \mathbf{c}, \mathbf{v}, A, \ldots, \mu^2) = \Gamma_{\text{reg}}(Z_2^{1/2}\phi, \mathbf{c}_0, \mathbf{v}_0, Z_3^{1/2}A, \ldots, \mu^2 + \delta\mu^2) \tag{12-220}$$

provided $\mathbf{v}$ and $\mathbf{c}$ are renormalized so as to maintain the finiteness of $\mathbf{c}\cdot\phi$ and the validity of (12-218):

$$\mathbf{c}_0 = Z_2^{-1/2}\mathbf{c} \qquad \mathbf{v}_0 = Z_2^{1/2}\mathbf{v} \tag{12-221}$$

The second point concerns the variation of μ^2. Power counting tells us that in the unbroken theory such a variation only requires a modification of the counterterm $\delta\mu^2$. The identity (12-217) shows that this modification is also sufficient to make the broken theory finite. In this case, we might think that a variation of μ^2 also requires a modification of the mass term M_A^2 of the vector field, which would invalidate our proof. However, Eq. (12-217) says that this modification of $M^2{}_A$ comes only from the variation of v as a function of μ^2.

In conclusion we may reach any point of the (m^2, v) plane (Fig. 12-12) and renormalize the corresponding theory with the symmetric counterterms. Modifying only the μ^2 counterterm, we renormalize the spontaneously broken theory [$\mathbf{c} = 0$, $m^2 = 0$, v given (point δ on Fig. 12-12)]. See also Fig. 11-14 for plots of the (μ^2, m^2) or (μ^2, v) planes.

The corresponding normalization conditions of the various proper functions have not been explicited. They can be deduced from the identity (12-217) and from the normalization conditions of the symmetric functions. Instead of this intermediate renormalization, we may prefer more physical conditions, such as those defining the coupling constant as the value of the three-point function at some on-shell point, etc. Needless to say, these new normalization conditions must be in agreement with the identities derived from (12-217) and from those satisfied by Γ_S.

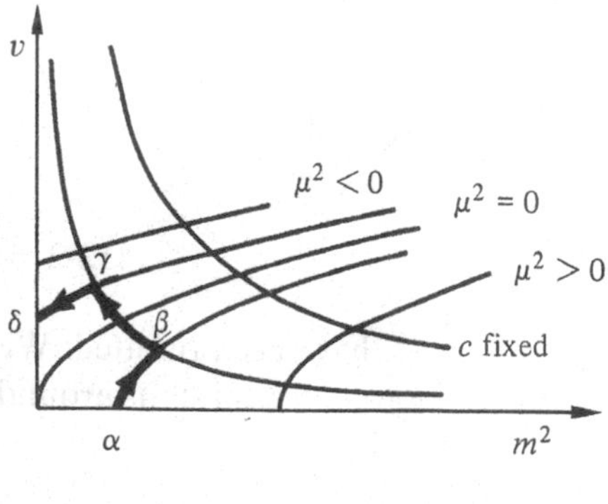

Figure 12-12 Curves of constant μ^2 and c in the (m^2, v) plane.

The method followed here is economical, since it appeals to the simpler symmetric theory in order to renormalize the spontaneously broken one. It is, however, possible to avoid any reference to the massless unbroken case. The problem then amounts to show that the Slavnov-Taylor identities satisfied by the spontaneously broken theory may be preserved by renormalization.

The previous analysis has been carried out in the renormalizable gauge where the comparison between broken and unbroken theories is obvious. This gauge is not physically satisfactory since it exhibits unphysical features such as the massless modes of ϕ_T ($m^2 \to 0$ when $\mathbf{c} \to 0$). However, the study of renormalization in the unitary gauge such as the one in Eq. (12-209) would be much more difficult, as the theory looks nonrenormalizable and contact with the symmetric case has been lost.

12-5-5 Gauge Independence and Unitarity of the S Matrix

We want to show that all unphysical states—fictitious Goldstone bosons, additional polarization states of the vector field, and Faddeev-Popov fields—do not actually contribute to S-matrix elements. All these unphysical fields have propagators with a pole at $k^2 = 0$. It is tantamount to showing that this singularity does not contribute to intermediate states.

A simple proof uses the gauge independence of the S matrix to introduce the 't Hooft gauge

$$\mathscr{F}_a = \left[\partial_\mu A^\mu{}_a - \frac{g}{\lambda}(v, T^a\phi')\right] \tag{12-222}$$

We assume that the group is simple and hence has a single coupling constant, that the scalar field ϕ belongs to a real representation, and that ϕ' is obtained after a translation $\phi = \phi' + v$, $\langle\phi'\rangle = 0$. This gauge has several merits and its invention by 't Hooft was a major step in the theoretical developments. It breaks explicitly the global invariance. Consequently, the would-be Goldstone bosons acquire a mass matrix

$$-\tfrac{1}{2}(m^2{}_\phi)_{\alpha\beta}\phi'_\alpha\phi'_\beta = -\frac{g^2}{2\lambda}(\phi'_\alpha T^a{}_{\alpha\beta}v_\beta)(\phi'_\gamma T^a{}_{\gamma\delta}v_\delta) \tag{12-223}$$

Moreover, the Faddeev-Popov ghost also acquires a mass. The operator $\mathscr{M}_{ab}$ reads

$$\mathscr{M}_{ab}(x, y) = \left\{\partial_\mu D^\mu{}_{ab} - \frac{g^2}{\lambda}(v, T^a T^b(\phi' + v))\right\}\delta^4(x - y) \tag{12-224}$$

since it results from an infinitesimal gauge transformation in $\mathscr{F}$ acting on both A and ϕ'. The ghost-mass matrix is therefore

$$(m^2)_{ab} = \frac{g^2}{\lambda}(T^a v, T^b v) \tag{12-225}$$

Finally, this choice diagonalizes the quadratic form in A and ϕ'. The crossed term in the expansion of $-(\lambda/2)\mathscr{F}^2$ just cancels $-g(\partial_\mu\phi', A^\mu{}_a T^a v)$ arising from $\frac{1}{2}(D_\mu\phi, D^\mu\phi)$. It follows that in terms of the mass matrix of Eq. (12-210), the vector propagator reads

$$\Delta^{\mu\nu}(k) = \frac{-i}{k^2 - M^2 + i\varepsilon}\left[g^{\mu\nu} - (1 - \lambda^{-1})\frac{k^\mu k^\nu}{k^2 - \lambda^{-1}M^2 + i\varepsilon}\right] \tag{12-226}$$

As $\lambda \to \infty$, one recovers the Feynman rules in the transverse (Landau) gauge, while as $\lambda \to 0$, all the unphysical masses recede to infinity. In the latter case, we do not expect these states with enormous masses to contribute to the S matrix. We proved in Sec. 12-4-4 that the S matrix does not depend on the choice of gauge. The argument which was formal due to the infrared divergences is now justified. We conclude that in any gauge, and in particular in the Landau gauge, the unphysical states do not contribute. A careful analysis should pay proper attention to renormalization. On this point the reader is referred to the literature.

Even though unphysical particles have disappeared from the physical subspace, there remains some trace of the spontaneous breaking mechanism, namely, the (massive) components of the scalar fields. Besides these scalar Higgs fields, we recall that some components of the vector field may remain massless.

We may wonder whether it is mandatory to introduce scalar fields and whether it is not possible to generate them as bound states, for instance, of a fermion-antifermion pair. Such a dynamical breakdown is illustrated by the Schwinger two-dimensional massless electrodynamics. The vacuum polarization has a pole at zero momentum, the fermions disappear from the theory, and the only remaining single particle state is a bosonic bound state of mass $e/\sqrt{\pi}$. In spite of the appeal of such a mechanism, it is not presently known how to realize it in four dimensions.

12-6 THE WEINBERG-SALAM MODEL

We present a realistic unified model of weak and electromagnetic interactions proposed independently by Weinberg and Salam and based on a spontaneously broken gauge theory. Among all the models of this type, it may be singled out because of its anteriority, its economical number of parameters, and the fact that it has received some experimental confirmation with the discovery of neutral currents and of charmed particles.

12-6-1 The Model for Leptons

The electron and its neutrino ν_e are treated on the same footing as the muon and its neutrino ν_μ. The left helicity component of the charged lepton $e_L = (1 - \gamma_5)e/2$ $[\mu_L = (1 - \gamma_5)\mu/2]$ and its neutrino $\nu_e(\nu_\mu)$ are grouped into a column matrix

$$L_e = \begin{pmatrix} \nu_e \\ e_L \end{pmatrix} \qquad L_\mu = \begin{pmatrix} \nu_\mu \\ \mu_L \end{pmatrix} \tag{12-227}$$

This suggests the introduction of a group of leptonic isospin for which L_e and L_μ are doublets, while the right components $e_R = (1 + \gamma_5)e/2 \equiv R_e$ and $\mu_R \equiv R_\mu$ are singlets. A leptonic hypercharge Y is also assigned to each of these fields in such a way that the analog of the Gell-Mann and Nishijima rule is satisfied:

$$Q = T^3 + \frac{Y}{2} \tag{12-228}$$

The left doublets have $Y = -1$ and the right singlets $Y = -2$. The weak isospin

T and hypercharge Y commute; therefore the transformation group is $SU(2) \times U(1)$.

We then construct a gauge theory with this invariance group, involving a triplet of gauge fields $\mathbf{A}_\mu$ for $SU(2)$ with a charge g and a field B_μ for $U(1)$. The $U(1)$ coupling constant will be denoted $g'/2$. Since we want a single gauge field (the photon) to remain massless after spontaneous breaking, we introduce a doublet of complex scalar fields:

$$\phi = \begin{pmatrix} \phi^+ \\ \phi_0 \end{pmatrix} \tag{12-229}$$

of hypercharge $Y = +1$. The most general renormalizable invariant potential for ϕ is

$$V(\phi^\dagger\phi) = \mu^2\phi^\dagger\phi + \lambda(\phi^\dagger\phi)^2 \tag{12-230}$$

For $\mu^2 < 0$, ϕ acquires a nonvanishing vacuum expectation value, which may be assumed real, along ϕ^0:

$$\langle\phi\rangle = \frac{1}{\sqrt{2}}\begin{pmatrix} 0 \\ v \end{pmatrix} \qquad v^2 = -\frac{\mu^2}{\lambda} + O(\hbar) \tag{12-231}$$

The symmetry $SU(2) \times U(1)_Y$ is broken but the symmetry under $U(1)_Q$ is preserved. This achieves the desired result, since one vector field coupled to the electric charge remains massless.

The lagrangian reads

$$\begin{aligned}\mathscr{L} = &-\tfrac{1}{4}\mathbf{A}_{\mu\nu}\mathbf{A}^{\mu\nu} - \tfrac{1}{4}B_{\mu\nu}B^{\mu\nu} + \Big[\bar{R}_e(i\not{\partial} - g'\not{B})R_e + \bar{L}_e\left(i\not{\partial} - \frac{g'}{2}\not{B} + g\frac{\tau_i}{2}\not{A}_i\right)L_e \\ &- G_e(\bar{L}_eR_e\phi + \phi^\dagger\bar{R}_eL_e) + e \leftrightarrow \mu\Big] \\ &+ \left(\partial_\mu\phi - i\frac{g'}{2}B_\mu\phi - \frac{ig}{2}\tau^iA_{\mu i}\phi\right)^\dagger\left(\partial^\mu\phi - i\frac{g'}{2}B^\mu\phi - \frac{ig}{2}\tau^iA^\mu{}_i\phi\right) - V(\phi^\dagger\phi)\end{aligned} \tag{12-232}$$

where $\mathbf{A}_{\mu\nu}$ and $B_{\mu\nu}$ stand for the field strength tensors

$$\mathbf{A}_{\mu\nu} = \partial_\mu\mathbf{A}_\nu - \partial_\nu\mathbf{A}_\mu + g\mathbf{A}_\mu \times \mathbf{A}_\nu \qquad B_{\mu\nu} = \partial_\mu B_\nu - \partial_\nu B_\mu$$

and τ^i $(i = 1, 2, 3)$ are the Pauli matrices. The $SU(2)$ symmetry prevents us from writing mass terms of the electron and muon, but it does not forbid the introduction of the coupling to the scalar field with coupling constants G_e and G_μ.

In order to understand the physical content of this model, let us use a unitary gauge. We use the parametrization

$$\phi(x) = e^{i\xi_i(x)\tau^i/2v}\begin{pmatrix} 0 \\ \dfrac{v + \rho(x)}{\sqrt{2}} \end{pmatrix} \tag{12-233}$$

and perform the $SU(2)$ gauge transformation

$$\begin{cases} \phi(x) \to \phi'(x) = \begin{pmatrix} 0 \\ \dfrac{v + \rho(x)}{\sqrt{2}} \end{pmatrix} \\ \dfrac{\tau^i}{2} A_{\mu i}(x) \to \dfrac{\tau^i}{2} A'_{\mu i}(x) = e^{-i\xi_i \tau^i/2v} \left(\dfrac{i}{g} \partial_\mu + \dfrac{\tau^i}{2} A_{\mu i} \right) e^{i\xi_i \tau^i/2v} \\ L \to L' = e^{-i\xi_i \tau^i/2v} L \\ B, R \text{ invariant} \end{cases} \tag{12-234}$$

Equation (12-232) is expressed as

$$\mathscr{L} = -\tfrac{1}{4}\mathbf{A}_{\mu\nu}\mathbf{A}^{\mu\nu} - \tfrac{1}{4}B_{\mu\nu}B^{\mu\nu} + \Bigg[\bar{e}_R(i\partial\!\!\!/ - g'B\!\!\!/)e_R + \bar{L}_e\left(i\partial\!\!\!/ - \frac{g'}{2}B\!\!\!/ + g\frac{\tau^i}{2}A\!\!\!/_i\right)L_e$$
$$- G_e \frac{v+\rho}{\sqrt{2}} (\bar{e}_R e_L + \bar{e}_L e_R) + e \leftrightarrow \mu \Bigg] + \tfrac{1}{2}\partial_\mu \rho \partial^\mu \rho$$
$$+ \tfrac{1}{8}(v+\rho)^2[(g' B_\mu - g A_\mu^3)^2 + g^2(A_\mu^1 A^{1\mu} + A_\mu^2 A^{2\mu})] - V\left[\frac{(v+\rho)^2}{2}\right] \tag{12-235}$$

The scalar (Higgs) field ρ has a mass $\sqrt{-2\mu^2}$. The electron and the muon have acquired masses equal to $m_e = G_e v/\sqrt{2}$ and $m_\mu = G_\mu v/\sqrt{2}$. The charged vector field

$$W_\mu^\pm = \frac{1}{\sqrt{2}}(A_\mu^1 \mp i A_\mu^2) \tag{12-236}$$

is also massive, with

$$M_W = \frac{vg}{2} \tag{12-237}$$

Finally, the quadratic form in A^3 and B is diagonalized by

$$\begin{aligned} Z_\mu &= (g^2 + g'^2)^{-1/2}(-gA_\mu^3 + g'B_\mu) \\ A_\mu &= (g^2 + g'^2)^{-1/2}(gB_\mu + g'A_\mu^3) \end{aligned} \tag{12-238}$$

so that

$$\begin{aligned} M_Z &= \frac{v}{2}(g^2 + g'^2)^{1/2} \\ M_A &= 0 \end{aligned} \tag{12-239}$$

The leptonic interaction terms of $\mathscr{L}$ may then be rewritten in terms of the physical fields $W^\pm$, Z, and A:

$$\mathscr{L}_{\text{lept}}^{\text{int}} = \frac{g}{2\sqrt{2}}[\bar{\nu}_e \gamma^\mu (1 - \gamma_5) e W_\mu^+ + \text{hc}] - \frac{e}{2}[\tan\theta_W(2\bar{e}_R\gamma^\mu e_R + \bar{\nu}_e\gamma^\mu \nu_e + \bar{e}_L\gamma^\mu e_L)$$
$$- \cot\theta_W(\bar{e}_L\gamma^\mu e_L - \bar{\nu}_e\gamma^\mu\nu_e)]Z_\mu - eA_\mu \bar{e}\gamma^\mu e + \{e \leftrightarrow \mu, \nu_e \leftrightarrow \nu_\mu\} \tag{12-240}$$

In Eq. (12-240) we have introduced the Weinberg angle θ_W, defined in such a way that

$$e = \frac{gg'}{\sqrt{g^2 + g'^2}} \qquad \tan\theta_W = \frac{g'}{g} \tag{12-241}$$

or

$$e = g' \cos\theta_W = g \sin\theta_W$$

The last term in (12-240) is the usual electromagnetic coupling to the field A_μ. The first one has the form (12-176). Its coupling is related to Fermi's constant through

$$\frac{G}{\sqrt{2}} = \frac{g^2}{8M_W^2} = \frac{1}{2v^2} \tag{12-242}$$

From the knowledge of G, we deduce lower bounds on the masses of $W^\pm$ and Z:

$$\begin{aligned} M_W &= \frac{gv}{2} = \frac{38}{\sin\theta_W}\,\text{GeV} \geq 38\ \text{GeV} \\ M_Z &= \frac{gv}{2\cos\theta_W} = \frac{76}{\sin 2\theta_W}\,\text{GeV} \geq 76\ \text{GeV} \\ M_W &< M_Z \end{aligned} \tag{12-243}$$

The coupling constants G_e and G_μ are determined from the electron and muon masses:

$$\begin{aligned} G_e &= \sqrt{2}\frac{m_e}{v} = 2^{3/4}G^{1/2}m_e \sim 2 \times 10^{-6} \\ G_\mu &= \frac{m_\mu}{m_e} G_e \sim 4 \times 10^{-4} \end{aligned} \tag{12-244}$$

This model involves a new type of coupling of Z_μ to a neutral, parity violating weak current constructed from e_L, e_R, and ν_e (and $\mu_{L,R}$, ν_μ). This is a feature of most renormalizable models of weak interactions. They introduce either a neutral current and a vector field coupled to it, or new leptons assumed to be heavy to comply with experimental facts, or both. In the former case, the explicit form of the current depends on the model, viz on the choice of representations for the various fields, etc. The Weinberg-Salam model incorporates in a natural way the electron-muon universality. Only G_e and G_μ are sensitive to the nature of the lepton. On the other hand, the model does not provide any natural explanation for the electric charge quantization.

This is not the case for other models based on simple groups such as the Georgi-Glashow model (Sec. 12-5-3). As a model for weak and electromagnetic interactions, it is now ruled out by experiment since it does not incorporate neutral currents. Among the three components of the gauge field, two become massive (the analogs of $W_\mu^\pm$ in the Salam-Weinberg model) while the last one remains massless (the photon). The benefit of dealing with a simple group is that the electric charge is quantized.

12-6-2 Electron-Neutrino Cross Sections

To expose some striking consequences of the existence of neutral leptonic currents, we will now compute the elastic $e - \nu$ cross sections to lowest order in the Weinberg-Salam model. The relevant diagrams are depicted in Fig. 12-13. In the limit where the incident neutrino energy is small as compared to the masses of the W and Z bosons, we may content ourselves with the effective lagrangian

$$\mathscr{L}_{\text{eff,int}} = -\sqrt{2}\,G\{[\tfrac{1}{2}\bar{\nu}_e\gamma^\mu(1-\gamma_5)e][\bar{e}\gamma_\mu(1-\gamma_5)\nu_e] + (\bar{\nu}_\mu\gamma^\mu\nu_\mu + \bar{\nu}_e\gamma^\mu\nu_e)(2\sin^2\theta_W\bar{e}_R\gamma_\mu e_R - \cos 2\theta_W\bar{e}_L\gamma_\mu e_L)\} \tag{12-245}$$

where the first and the second terms represent the W and Z contributions respectively. After a Fierz transformation on the first term

$$\left(\bar{\nu}_e\gamma^\mu\frac{1-\gamma_5}{2}e\right)\left(\bar{e}\gamma_\mu\frac{1-\gamma_5}{2}\nu_e\right) = \left(\bar{\nu}_e\gamma^\mu\frac{1-\gamma_5}{2}\nu_e\right)\left(\bar{e}\gamma_\mu\frac{1-\gamma_5}{2}e\right) \tag{12-246}$$

this interaction is

$$\mathscr{L}_{\text{eff,int}} = -\sqrt{2}\,G\left\{\left(\bar{\nu}_e\gamma^\mu\frac{1-\gamma_5}{2}\nu_e\right)\left[\bar{e}\gamma_\mu\left(\frac{1-\gamma_5}{2} + 2\sin^2\theta_W\right)e\right] + \left(\bar{\nu}_\mu\gamma^\mu\frac{1-\gamma_5}{2}\nu_\mu\right)\left[\bar{e}\gamma_\mu\left(-\frac{1-\gamma_5}{2} + 2\sin^2\theta_W\right)e\right]\right\} \tag{12-247}$$

A general expression of the form

$$\mathscr{L}_{\text{eff}} = -\sqrt{2}\,G\left(\bar{\nu}\gamma^\mu\frac{1-\gamma_5}{2}\nu\right)\left[\bar{e}\gamma_\mu\left(C_L\frac{1-\gamma_5}{2} + C_R\frac{1+\gamma_5}{2}\right)e\right] \tag{12-248}$$

where C_L and C_R are real coefficients, leads to a cross section

$$\frac{d\sigma}{dE'_e} = \frac{G^2}{2\pi m_e E_\nu^2}[C_L^2(p\cdot q)^2 + C_R^2(p\cdot q')^2 - C_L C_R m_e^2(q\cdot q')^2] \tag{12-249}$$

for the process $e(p) + \bar{\nu}(q) \to e(p') + \bar{\nu}(q')$. The expression for the process $e\nu \to e\nu$ is obtained by interchanging C_L and C_R. When the incident neutrino energy is much larger than the electron mass, we have

$$(p\cdot q)^2 = m_e^2 E_\nu^2$$

$$(p\cdot q')^2 = m_e^2 E_\nu^2\left(1 - \frac{E'_e}{E_\nu}\right)^2$$

$$m_e^2 q\cdot q' \simeq m_e^2 p\cdot p' = m_e^2 E_\nu^2 \frac{E'_e}{E_\nu}\frac{m_e}{E_\nu}$$

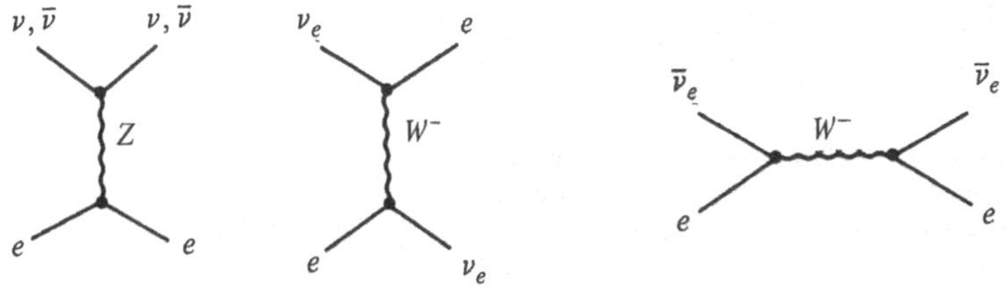

Figure 12-13 Electron-neutrino scattering to lowest order.

and the last term in (12-249) may be neglected. We finally obtain in the various channels

$$\begin{aligned}
\sigma(\bar{\nu}_e e \to \bar{\nu}_e e) &= \frac{G^2 m_e E_\nu}{2\pi}[4 \sin^4 \theta_W + \tfrac{1}{3}(1 + 2 \sin^2 \theta_W)^2] \\
\sigma(\nu_e e \to \nu_e e) &= \frac{G^2 m_e E_\nu}{2\pi}[(1 + 2 \sin^2 \theta_W)^2 + \tfrac{4}{3} \sin^4 \theta_W] \\
\sigma(\bar{\nu}_\mu e \to \bar{\nu}_\mu e) &= \frac{G^2 m_e E_\nu}{2\pi}[4 \sin^4 \theta_W + \tfrac{1}{3}(1 - 2 \sin^2 \theta_W)^2] \\
\sigma(\nu_\mu e \to \nu_\mu e) &= \frac{G^2 m_e E_\nu}{2\pi}[(1 - 2 \sin^2 \theta_W)^2 + \tfrac{4}{3} \sin^4 \theta_W]
\end{aligned} \tag{12-250}$$

These expressions are compatible with the few observed $\nu_\mu e$ and $\bar{\nu}_\mu e$ events and with the value of θ_W derived from neutrino-nucleon inclusive reactions

$$\sin^2 \theta_W \simeq 0.25 \pm 0.02 \tag{12-251}$$

Experimental investigation of the structure of neutral currents is still actively underway, in particular in atomic physics.

12-6-3 Higher-Order Corrections

A renormalizable theory of weak interactions enables us to compute higher-order corrections. In fact, in the model presented so far, only leptonic processes can be considered, such as the weak contribution to the muon anomalous magnetic moment or the radiative correction to the muon decay.

The weak contributions to $g_\mu - 2$ are shown in Fig. 12-14. For this one-loop computation, it is safer to use a renormalizable gauge of the form (12-232). The diagrams of Fig. 12-14*a*, *b*, *c* give contributions of the form

$$\begin{aligned}
&(a) \quad g^2 \int \frac{d^4k}{(k^2 - M^2)^2} \frac{O(km_\mu, m_\mu^2)}{(p-k)^2} \\
&(b) \quad g^2 \int \frac{d^4k}{k^2 - M^2} \frac{O(km_\mu, m_\mu^2)}{[(p-k)^2 - m_\mu^2]^2} \\
&(c) \quad \frac{m_\mu^2 g^2}{M_W^2} \int \frac{d^4k}{k^2 - m_{\phi_0}^2} \frac{O(km_\mu, m_\mu^2)}{[(p-k)^2 - m_\mu^2]^2}
\end{aligned} \tag{12-252}$$

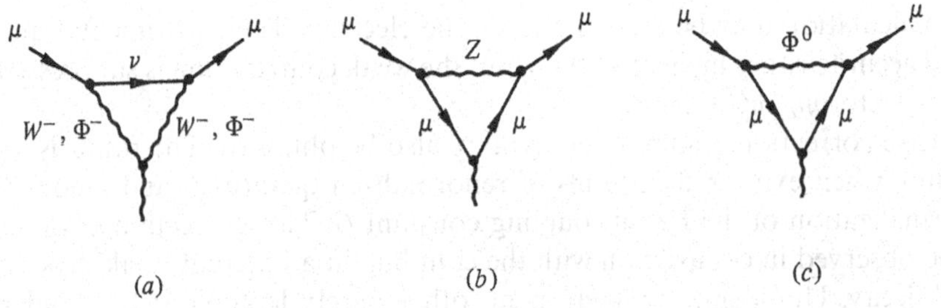

Figure 12-14 Weak corrections to the muon anomalous magnetic moment.

In cases (*a*) and (*b*), M stands for the mass of the W, of the Z, or of one of the would-be Goldstone bosons $\phi^{\pm}$. Assume that we have chosen a gauge where the latter is very large, of the order of M_W such as given by Eq. (12-223) with λ finite. The one corresponding to the combination $(\phi_0 - \phi_0^{\dagger})/\sqrt{2}$ is also large, while the physical component $(\phi_0 + \phi_0^{\dagger})/\sqrt{2}$ has an unknown mass m_{ϕ_0}. We assume, however, that m_{ϕ_0} is much larger than m_μ. In the contribution (*c*), we have taken the muon-scalar coupling

$$G_\mu = \frac{\sqrt{2}\, m_\mu}{v} = \frac{m_\mu g}{\sqrt{2}\, M_W} \tag{12-253}$$

into account. Since the F_2 form factor does not require any ultraviolet subtraction, we expect the previous integrals to behave as m_μ^2/M^2 or $m_\mu^2/m_{\phi_0}^2$ times a possible logarithmic factor. The weak corrections to the anomalous magnetic moment are thus at most of order

$$g^2 \frac{m_\mu^2}{M_W^2} \sim G m_\mu^2 \sim 10^{-7} \tag{12-254}$$

This is precisely the order of magnitude of both the experimental and theoretical uncertainties (the latter due to the hadronic contributions):

$$\begin{aligned} \left.\frac{g-2}{2}\right|_{\text{exp}} &= (11\,659.22 \pm 0.09) \times 10^{-7} \\ \left.\frac{g-2}{2}\right|_{\text{th}} &= (11\,659.19 \pm 0.10) \times 10^{-7} \end{aligned} \tag{12-255}$$

A logarithmic factor $\ln (m_\mu^2/M_W^2)$ might, however, make the weak contributions sizable. Notice that the contribution of the Higgs boson is suppressed by a factor m_μ^2/M_W^2 with respect to the two others, and is negligible. In the actual computation no logarithm appears, and the weak correction is small:

$$\begin{aligned} \left.\frac{g-2}{2}\right|_{\text{weak}} &= \frac{G m_\mu^2}{\sqrt{2}\,\pi^2}\left(\tfrac{2}{3}\sin^4\theta_W - \tfrac{1}{3}\sin^2\theta_W + \tfrac{1}{4}\right) \\ &\sim 2 \times 10^{-9} \end{aligned} \tag{12-256}$$

A similar calculation may be carried out for the electron. The experimental and theoretical accuracies are higher ($\sim 10^{-9}$), but the weak contribution is suppressed by an extra factor $m_e^2/m_\mu^2 \sim 10^{-5}$.

Radiative corrections to muon decay may also be obtained. The result is, of course, finite when expressed in terms of renormalized quantities, and amounts to a renormalization of the Fermi coupling constant G. This correction of order α might be observed in comparison with the coupling in a different weak process such as β decay. Unfortunately, there is no other purely leptonic decay, and a comparison with a hadronic system involves strong interaction corrections.

12-6-4 Incorporation of Hadrons

A natural way to incorporate hadrons in the previous scheme is to couple quark multiplets to the gauge fields. The structure of the charged hadronic current in terms of the conventional quarks u, d, s has already been given (Sec. 11-3). It contains strangeness conserving and strangeness changing components, with a Cabibbo mixing angle. This angle could perhaps be predicted in the framework of a gauge theory of weak interactions. In addition to this charged current, a neutral current coupled to the Z boson emerges in theories like the Weinberg-Salam model. The constraint that strangeness changing processes induced by this neutral current should not be in violent contradiction with experiment turns out to be a stringent demand, and requires the introduction of at least a new quark.

We start with the three quarks u, d, s and neglect for the time being their couplings to the Higgs scalars. The Cabibbo angle θ_c is assumed to be given. The usual charged hadronic current will be reproduced if we give the following assignments to the quark components:

$$N_L \equiv \begin{pmatrix} u \\ d_\theta \equiv d \cos\theta_c + s \sin\theta_c \end{pmatrix}_L \tag{12-257}$$

is a doublet of weak isospin, with $Y = \frac{1}{3}$, while u_R, d_R, s_R, and $s_{\theta L} \equiv (s \cos\theta_c - d \sin\theta_c)_L$ are isosinglets, with $Y = \frac{4}{3}, -\frac{2}{3}, -\frac{2}{3}, -\frac{2}{3}$. The interaction

$$\begin{aligned}\mathscr{L}_{\text{int}} = \bar{N}_L\left(g\not{A}\cdot\frac{\tau}{2} + \frac{1}{3}\frac{g'}{2}\not{B}\right)N_L + \left[\bar{s}_{\theta L}\left(-\frac{g'}{3}\not{B}\right)s_{\theta L} + (s_{\theta L}\to d_R) + (s_{\theta L}\to s_R)\right] \\ + \tfrac{2}{3}\bar{u}_R g'\not{B}u_R \end{aligned} \tag{12-258}$$

may be rewritten in terms of the fields W, A, and Z of Eqs. (12-236) and (12-238) as

$$\mathscr{L}_{\text{int}} = \frac{g}{2\sqrt{2}}(W_\mu^+ h^\mu + \text{hc}) + eA^\mu j_\mu^{\text{em}} - \sqrt{g^2 + g'^2}\, Z^\mu\left(\bar{N}_L\frac{\tau_3}{2}\gamma_\mu N_L - \sin^2\theta_W j_\mu^{\text{em}}\right) \tag{12-259}$$

with

$$h_\mu = \bar{u}\gamma_\mu(1-\gamma_5)(d\cos\theta_c + s\sin\theta_c)$$

$$j_\mu^{\text{em}} = \tfrac{2}{3}\bar{u}\gamma_\mu u - \tfrac{1}{3}(\bar{d}\gamma_\mu d + \bar{s}\gamma_\mu s)$$

as expected. The neutral boson Z is coupled to the electromagnetic current and to

$$\begin{aligned}\bar{N}_L\tau_3\gamma_\mu N_L = \bar{u}_L\gamma_\mu u_L - \cos^2\theta_c\bar{d}_L\gamma_\mu d_L - \sin\theta_c\cos\theta_c(\bar{d}_L\gamma_\mu s_L + \bar{s}_L\gamma_\mu d_L) \\ - \sin^2\theta_c\bar{s}_L\gamma_\mu s_L\end{aligned}$$

The terms proportional to $\sin\theta_c\cos\theta_c$ are the strangeness changing neutral currents. They are embarrassing because experimentally $\Delta S = 1$ or $\Delta S = 2$, $\Delta Q = 0$ processes are heavily suppressed. For instance,

$$\frac{\Gamma(K^\pm \to \pi^\pm\nu\bar{\nu})}{\Gamma(K^\pm \to \text{all})} = 6\times 10^{-7} \qquad \frac{\Gamma(K^0 \to \mu^+\mu^-)}{\Gamma(K^0 \to \text{all})} \simeq 10^{-8} \tag{12-260}$$

This puzzle was solved by Glashow, Iliopoulos, and Maiani (1970) through the introduction of a fourth quark, denoted c, of charge $\frac{2}{3}$ like u and carrying a new quantum number, the charm. The left-handed component c_L is supposed to form an isodoublet, together with the combination $s_{\theta L} = -d_L \sin\theta_c + s_L \cos\theta_c$. In other words, there are now two left-handed doublets

$$\begin{pmatrix} u \\ d_\theta \end{pmatrix}_L \quad \begin{pmatrix} c \\ s_\theta \end{pmatrix}_L \qquad \text{with } Y = \tfrac{1}{3} \tag{12-261}$$

and four right-handed singlets u_R, d_R, s_R, and c_R with $Y = \frac{4}{3}, -\frac{2}{3}, -\frac{2}{3}, \frac{4}{3}$ respectively.

It is easy to see that the neutral boson is now coupled to

$$(\bar{u}, \bar{d}_\theta)_L \frac{\tau_3}{2} \gamma_\mu \begin{pmatrix} u \\ d_\theta \end{pmatrix}_L + (\bar{c}, \bar{s}_\theta)_L \frac{\tau_3}{2} \gamma_\mu \begin{pmatrix} c \\ s_\theta \end{pmatrix}_L - \sin^2\theta_W j_\mu^{\text{em}} \tag{12-262}$$

Strangeness changing neutral currents occur in the first and in the second terms of (12-262) but cancel exactly. This means that the unwanted currents have been eliminated to order G. However, strangeness changing neutral transitions induced by higher-order exchanges of charged currents might still endanger the theory. This is the case of the diagrams drawn in Fig. 12-15, expected to be of order $G\alpha$ and hence in disagreement with the very low experimental rates (12-260).

However, after introduction of the c quark, the form of the charged current becomes

$$h_\mu = \bar{u}\gamma_\mu(1 - \gamma_5)d_\theta + \bar{c}\gamma_\mu(1 - \gamma_5)s_\theta \tag{12-263}$$

When all contributions are taken into account, it turns out that the dangerous amplitude is proportional to $(m_c^2 - m_u^2)/M_W^2$. For charmed quarks much lighter than the W meson, that is, $m_c \lesssim 1.5 - 2$ GeV, there remains no discrepancy with experiment.

Besides this property of suppressing the $\Delta S \neq 0$ neutral currents, the charmed quark has also an aesthetical appeal. It restores a symmetry between the four leptons and the four quarks, and, as a bonus, it removes the anomalies of the Weinberg-Salam model. Indeed, the lepton model studied in the previous sections is plagued with chiral anomalies. If we apply the analysis of Sec. 12-4 with only weak isosinglets and doublets, we readily discover that the only nonvanishing anomaly is proportional to

$$\sum_{\text{fermions}} \operatorname{tr}(T^a T^a Y) \propto \sum_{\text{doublets}} Y$$

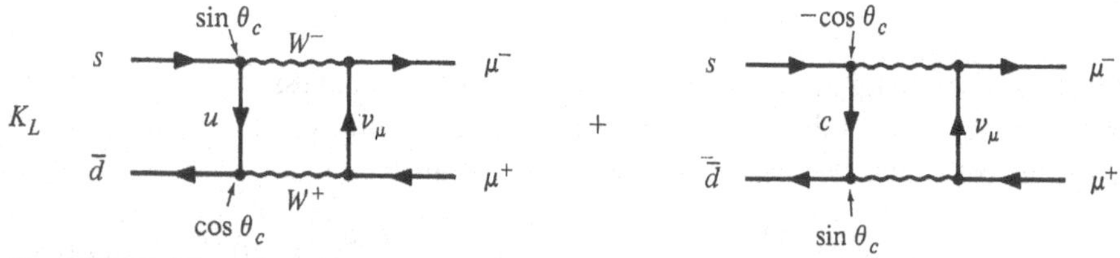

Figure 12-15 $K^0 \to \mu^+ \mu^-$ decay.

As above, T^a is some component of the weak isospin and Y denotes the weak hypercharge. This corresponds, for instance, to triangle A^aA^aB diagrams. It follows that leptons give a nonvanishing contribution to the anomaly. With the two doublets L_e and L_μ of (12-227), we have

$$\sum_{\substack{\text{leptonic}\\ \text{doublets}}} Y = 2 \times (-1) \tag{12-264}$$

However, if we incorporate hadrons according to the previous scheme, the two doublets of (12-261) contribute

$$\sum_{\substack{\text{hadronic}\\ \text{doublets}}} Y = 2 \times \tfrac{1}{3} \tag{12-265}$$

If, moreover, we assume—as in the discussion of the π° decay in Chap. 11—that quarks come in three degenerate unobservable colors, the expression in (12-265) must be supplemented by a summation over color, i.e., multiplied by a factor 3. Leptonic and hadronic contributions to the anomaly thus cancel. We emphasize the importance of the equal number of leptonic and hadronic doublets—for the previous weak isospin and charge assignments—and of the color degeneracy factor for the validity of this argument.

The computation of the π° decay is not affected by the present cancellation mechanism. Indeed, recall that π° is coupled to the divergence of the third component of the ordinary axial isocurrent, that is, $(\bar{u}, \bar{d})\gamma_\mu\gamma_5\tau_3/2\binom{u}{d}$.

We have omitted so far the Yukawa-like couplings of the scalar mesons to quarks and the quark-mass terms. Before spontaneous breakdown takes place, the theory must be $SU(2) \times U(1)$ invariant, which forbids quark-mass terms. However, after spontaneous breaking and shift of the Higgs field, the quarks acquire a mass

$$\mathscr{L}_m = -\overline{\psi}\frac{1-\gamma_5}{2}\mathscr{M}\psi - \overline{\psi}\frac{1+\gamma_5}{2}\mathscr{M}^\dagger\psi \tag{12-266}$$

Here $\mathscr{M}$ is a matrix which is a priori neither diagonal nor real, and is just restricted to commute with the charge operator Q. By independent unitary redefinitions of the left- and right-handed components, it is possible to diagonalize it:

$$\begin{aligned} \psi'_L &= U_L\psi_L \\ \psi'_R &= U_R\psi_R \\ U_R\mathscr{M}U_L^\dagger &= M \text{ diagonal} \end{aligned} \tag{12-267}$$

$$\mathscr{L}_m = -\overline{\psi'}\frac{1-\gamma_5}{2}M\psi' - \overline{\psi'}\frac{1+\gamma_5}{2}M^\dagger\psi' \tag{12-268}$$

The eigenstates ψ' of the mass matrix are the quarks u, d, s, c. On the other hand, the charged current which reads

$$h_\mu \propto \overline{\psi}\gamma_\mu(1-\gamma_5)\tau^-\psi \tag{12-269}$$

becomes

$$h_\mu \propto \bar{\psi}'\gamma_\mu(1-\gamma_5)U_{1L}^\dagger U_{2L}\psi' \tag{12-270}$$

where U_{1L} and U_{2L} are the restrictions of U_L to the upper (u, c) and lower (d, s) components respectively. By a further redefinition of the relative phases between the quarks [of course, unobservable on (12-268)] this may be cast into the form

$$h_\mu = (\bar{u}\bar{c})\gamma_\mu(1-\gamma_5)\begin{pmatrix} \cos\theta_c & \sin\theta_c \\ -\sin\theta_c & \cos\theta_c \end{pmatrix}\begin{pmatrix} d \\ s \end{pmatrix} \tag{12-271}$$

in agreement with (12-263). We conclude that the Cabibbo angle comes from the mismatch between the eigenstates of the mass matrix and the quark components entering the charged current.

A presentation of models of weak interactions should include a review of their implications—on neutrino scattering off hadrons, in particular. It is wiser to refer the reader to more competent authors for a thorough discussion.

In view of the recent experimental discoveries, the preceding theoretical framework may and must be extended to include more quarks and more leptons. We content ourselves with a simple remark. The Cabibbo mixing matrix which took the simple form (12-271) in the case of four quarks may depend on more parameters, some of which may be complex, thus introducing *CP* violations. The recent years have seen a blossoming of theoretical models, involving various groups, multiplet assignments, possible right-handed couplings, etc. Any discussion is doomed to become obsolete very soon.

How would strong interactions enter this scheme? For reasons to be discussed in the next chapter, a nonabelian gauge theory of strong interactions nowadays seems to be a good candidate. The gauge group would be a $SU(3)_c$ group, unrelated to the Gell-Mann and Ne'eman octet symmetry. The eight-gauge fields—the so-called gluons—would be coupled to the color quantum numbers of the quark triplets. In contrast with the spontaneously broken symmetry of weak and electromagnetic interactions, this local $SU(3)_c$ symmetry would be exactly implemented and the gluon would remain massless. In this gauge description of weak, electromagnetic, and strong interactions, quarks carry the two quantum numbers of color and flavor. On the other hand, vector bosons $W, Z, \ldots,$ or the photon are colorless, while the gluons have neither flavor nor charge.

Finally, it is possible to speculate that the group $G_W \times SU(3)_c$ [$G_W = SU(2) \times U(1)$ for the Weinberg-Salam model] originates from the breaking of a larger simple group. Such a superunification might even extend to gravitation.

NOTES

Nonabelian gauge fields were introduced by C. N. Yang and R. L. Mills, *Phys. Rev.*, vol. 96, p. 191, 1954, and have been considered by many authors since then. We list here only references directly related to our presentation.

The quantization was developed by R. P. Feynman, *Acta Phys. Polonica*, vol. 24, p. 697, 1963; L. D. Faddeev and V. N. Popov, *Phys. Lett.*, vol. 25, ser. B, p. 29, 1967; and B. S. DeWitt, *Phys. Rev.*, vol. 162, pp. 1195 and 1239, 1967. At about the same time, S. Weinberg, *Phys. Rev. Lett.*, vol. 19, p. 1264, 1967, and A. Salam, in "Elementary Particle Theory," edited by N. Svartholm, Almquist and Wiksells, Stockholm, 1968, proposed to base a theory of weak interactions on a spontaneously broken gauge model. This phenomenon had been discussed by F. Englert and R. Brout, *Phys. Rev. Lett.*, vol. 13, p. 321, 1964; P. W. Higgs, *Phys. Lett.*, vol. 12, p. 132, 1964, *Phys. Rev. Lett.*, vol. 13, p. 508, 1964, and *Phys. Rev.*, vol. 145, p. 1156, 1966; G. S. Guralnik, C. R. Hagen, and T. W. B. Kibble, *Phys. Rev. Lett.*, vol. 13, p. 585, 1964; and T. Kibble, *Phys. Rev.*, vol. 155, p. 1554, 1967. An early discussion in the context of superconductivity is in P. W. Anderson, *Phys. Rev.*, vol. 112, p. 1900, 1958, and vol. 130, p. 439, 1963.

This led to an intensive study of the renormalization procedure. Some crucial papers are those by G. 't Hooft, *Nucl. Phys.*, ser. B, vol. 33, p. 173, 1971, and ser. B, vol. 35, p. 167, 1971; A. A. Slavnov, *Theor. and Math. Phys.*, vol. 10, p. 99, 1972; J. C. Taylor, *Nucl. Phys.*, ser. B, vol. 33, p. 436, 1971; B. W. Lee and J. Zinn-Justin, *Phys. Rev.*, ser. D, vol. 5, pp. 3121, 3137, and 3155, ser. D, vol. 7, p. 1049, 1972; G. 't Hooft and M. T. Veltman, *Nucl. Phys.*, ser. B, vol. 50, p. 318, 1972; and C. Becchi, A. Rouet, and R. Stora, in "Renormalization Theory," edited by G. Velo and A. S. Wightman, D. Reidel, Dordrecht, Holland, and Boston, Mass., 1976.

Reviews have been written by E. S. Abers and B. W. Lee, *Phys. Rep.*, ser. C, vol. 9, p. 1, 1973; M. T. Veltman, in "Proceedings of the Sixth International Symposium on Electron and Photon Interactions at High Energies," edited by H. Rollnik and W. Pfeil, North-Holland, Amsterdam, 1974; J. Zinn-Justin, in "Trends in Elementary Particle Theory," edited by H. Rollnik and K. Dietz, Springer-Verlag, Berlin, 1975; B. W. Lee, in "Methods in Field Theory," edited by R. Balian and J. Zinn-Justin, North-Holland, Amsterdam, 1976; and W. Marciano and H. Pagels, *Phys. Rep.*, ser. C, vol. 36, p. 137, 1978. A textbook by J. C. Taylor, "Gauge Theories of Weak Interactions," Cambridge University Press, 1976, offers a general survey.

The extension of the weak interaction model to include hadrons was stimulated by the proposal of the charm quantum number by S. L. Glashow, J. Iliopoulos, and L. Maiani, *Phys. Rev.*, ser. D, vol. 2, p. 1285, 1970.

The geometry of gauge fields has been revisited by T. T. Wu and C. N. Yang, *Phys. Rev.*, ser. D, vol. 12, p. 3843, 1975. A general discussion of classical solutions is in the lecture notes by S. Coleman, in "New Phenomena in Subnuclear Physics," edited by A. Zichichi, Plenum Press, New York, 1977. The gauge monopoles were found independently by G. 't Hooft, *Nucl. Phys.*, ser. B, vol. 79, p. 276, 1974, and A. M. Polyakov, *JETP Lett.*, vol. 20, p. 194, 1974. A. A. Belavin, A. M. Polyakov, A. S. Schwartz, and Y. S. Tyupkin introduced the euclidean instantons, *Phys. Lett.*, ser. B, vol. 59, p. 85, 1975.

Gauge-fixing ambiguities are discussed by V. N. Gribov, *Nucl. Phys.*, ser. B, vol. 139, p. 1, 1978.

For the muon anomaly, see Chap. 7. The weak corrections have been computed by K. Fujikawa, B. W. Lee, and A. I. Sanda, *Phys. Rev.*, ser. D, vol. 6, p. 2923, 1972.

CHAPTER

THIRTEEN

ASYMPTOTIC BEHAVIOR

Relations between large-momentum or short-distance behavior and renormalizability properties have been discussed in the early 1950s by Stueckelberg and Peterman and by Gell-Mann and Low. In the 1970s, under the lead of Wilson, Symanzik, and Callan, this subject blossomed. We should not underestimate the flow of new ideas which resulted in the confluence with the study of critical phenomena, to which are associated the names of Kadanoff, Fisher, and Wilson. The troublesome short-distance singularities, which forced upon us the great machinery of renormalization, turn out to be the key to scaling properties of operators. A set of differential equations may be written expressing the nontrivial effect of an infinitesimal change in the scale. Upon integration they reveal that the short-distance behavior acquires, in favorable cases, a universal character, with fields and composite operators being assigned anomalous dimensions.

These ideas find successful applications to deep inelastic lepton scattering, electron-positron annihilation, and other high-energy processes. The concepts of short-distance expansion, asymptotic freedom, enrich the theorist's arsenal, and motivate hopes of developing a fundamental theory of strong interactions.

The parallel development in statistical mechanics has been tremendous, and has achieved notable successes in the comparison with experimental facts.

This chapter can only be an introduction to this vast subject. We shall avoid intricate mathematical proofs and rely mostly on heuristic arguments and examples, following the historical development at the price of some repetition.

13-1 EFFECTIVE CHARGE IN ELECTRODYNAMICS

To introduce some of the ideas we begin with the case of electrodynamics originally studied by Gell-Mann and Low.

The fundamental quantity is the electric charge or rather the fine-structure constant $\alpha = 1/137$ measured in low-energy experiments, in atomic physics, say, i.e., over distances much larger than the charged fermion Compton wavelength (for short we refer to electrons) which fixes the fundamental scale. In high-energy experiments we are rather interested in local, quasi-instantaneous properties of the interaction. We would expect naively that this regime is dictated by the bare parameters occurring in the hamiltonian. Unfortunately, as a result of renormalization the relation between bare and renormalized charges is plagued by infinities, at least perturbatively. The manner in which these infinities have to compensate imposes constraints which are reflected in the asymptotic behavior. The meaning of the word asymptotic will be clarified as we proceed.

13-1-1 The Gell-Mann and Low Function

Consider the photon's propagator and its relation with vacuum polarization

$$G_{\rho\nu}(q) = -i\,\frac{g_{\rho\nu}}{q^2[1+\bar{\omega}(q^2)]} + G^L_{\rho\nu}(q) \tag{13-1}$$

For simplicity no fictitious photon mass has been introduced. The longitudinal part $G^L_{\rho\nu}$ in $q_\rho q_\nu$ plays no role in the following. Normalization of charge is insured by the condition $\bar{\omega}(0) = 0$, and is therefore fitted to low-energy interaction or soft photon emission. The vacuum polarization is a function of q^2, α, and m the electron mass. We define the effective charge at momentum q, $d(\alpha, q^2, m^2)$, by considering the combination $\alpha G_{\rho\nu}$ associated to photon propagation

$$\alpha G_{\rho\nu}(q) = -i\,\frac{g_{\rho\nu}}{q^2}\,d(\alpha, q^2, m^2) + \alpha G^L_{\rho\nu}(q)$$

$$d(\alpha, q^2, m^2) = \frac{\alpha}{1+\bar{\omega}(\alpha, q^2, m^2)} \tag{13-2}$$

$$d(\alpha, 0, m^2) = \alpha$$

Subtractions are usually carried out at $q^2 = 0$. Consider the effect of picking a different subtraction point $q^2 = \lambda^2 < 0$ (in the euclidean region away from singularities). The perturbation series will be defined in terms of a parameter α_λ equal to

$$\alpha_\lambda = d(\alpha, \lambda^2, m^2) \tag{13-3}$$

In terms of α_λ the effective charge will be a function D such that

$$D(\alpha_\lambda, q^2, m^2, \lambda^2) = d(\alpha, q^2, m^2) \tag{13-4}$$

and

$$\alpha_\lambda = D(\alpha_\lambda, \lambda^2, m^2, \lambda^2) \tag{13-5}$$

Equation (13-4) expresses the apparently empty fact that the physical content is not modified by a mere change in the conventions. A different subtraction point leads to a modification of α_λ preserving the function D. The discussion is simplified here by the Ward identities of electrodynamics $Z_1 = Z_2$, $\alpha = Z_3\alpha_0$, implying a consistent treatment by referring only to the vacuum polarization.

Since the D function is dimensionless we may write

$$D\left(\alpha_\lambda, \frac{q^2}{\lambda^2}, -\frac{m^2}{\lambda^2}\right) = d\left(\alpha, -\frac{q^2}{m^2}\right)$$
$$\alpha_\lambda = D\left(\alpha_\lambda, 1, -\frac{m^2}{\lambda^2}\right) = d\left(\alpha, -\frac{\lambda^2}{m^2}\right) \tag{13-6}$$

and exploit these relations in the deep euclidean region where $-q^2/m^2$ becomes large. We know in principle the perturbative expansion of the vacuum polarization $(\alpha d)^{-1} - 1$. Order by order we extract its asymptotic behavior by neglecting terms behaving as $(m^2/q^2)[\ln(-q^2/m^2)]^n$. This defines a function $d^{as}(\alpha, -q^2/m^2)$ which would also result from massless quantum electrodynamics with m^2 as a scale parameter. We therefore explore the domain $-q^2/m^2 \to \infty$, $\alpha \ln(-q^2/m^2) \to 0$, hoping that neglected terms do not sum up to a nonnegligible contribution. Such an hypothesis cannot be seriously analyzed at this stage.

The right-hand side of Eq. (13-6) becomes $d^{as}(\alpha, x)$ with $x = -q^2/m^2$. If we choose the subtraction point such that $m^2 \ll -\lambda^2 \lesssim -q^2$, we can neglect the dependence on m^2 of the left-hand side, since we know from Chap. 8 that the corresponding massless limit exists. In the limit, D is replaced by D^{as} with λ^2 the scale parameter. Setting $y = -\lambda^2/m^2$ we find

$$D^{as}\left(\alpha_y, \frac{x}{y}\right) = d^{as}(\alpha, x) \tag{13-7}$$

with
$$\alpha_y = d^{as}(\alpha, y) = D^{as}(\alpha_y, 1) \tag{13-8}$$

These relations imply that d^{as} may be considered as a function of a single variable. To see this we define, following Gell-Mann and Low, the function

$$\psi(z) = \frac{\partial}{\partial x} D^{as}(z, x)\big|_{x=1} \tag{13-9}$$

differentiate Eq. (13-7) with respect to x, and set $y = x$. It follows that

$$\psi(\alpha_x) = x\frac{\partial}{\partial x} d^{as}(\alpha, x) \tag{13-10}$$

Using Eq. (13-8) this may be integrated in the form

$$\ln\frac{x_2}{x_1} = \int_{d^{as}(\alpha,x_1)}^{d^{as}(\alpha,x_2)} \frac{dz}{\psi(z)} \tag{13-11}$$

equivalent to the series

$$d^{as}(\alpha, x_2) = \sum_0^{\infty} \frac{1}{n!}\left(\ln \frac{x_2}{x_1}\right)^n \left\{\left[\psi(z)\frac{d}{dz}\right]^n z\right\}\Bigg|_{z=d^{as}(\alpha,x_1)} \tag{13-12}$$

Before giving a general discussion let us recall our earlier results (8-130) for the vacuum polarization. Remember that the power of α is given by the number of loops. To order α^2,

$$\bar{\omega}^{as}(\alpha, x) = -\frac{\alpha}{3\pi}\left(\ln x - \frac{5}{3}\right) - \frac{\alpha^2}{4\pi^2}(\ln x - C_2) + O(\alpha^3) \tag{13-13}$$

with C_2 a numerical constant. If α_1 stands for

$$\alpha_1 = d^{as}(\alpha, 1) = \frac{\alpha}{1 + 5\alpha/9\pi + C_2(\alpha^2/4\pi^2) + \cdots} \tag{13-14}$$

we have

$$d^{as}(\alpha, x) = \frac{\alpha_1}{1 - (\alpha_1/3\pi)\ln x - (\alpha_1^2/4\pi^2)\ln x + \cdots} \tag{13-15}$$

To obtain the function $\psi(\alpha_1)$, we compute the derivative $(\partial/\partial x)d^{as}(\alpha_1, x)$ for x equal to one. Let us also include the result of the three-loop calculation performed by Baker and Johnson. This gives

$$\psi(z) = \frac{z^2}{3\pi} + \frac{z^3}{4\pi^2} + \frac{z^4}{3\pi^3}\left[\xi(3) - \frac{101}{96}\right] + O(z^5) \tag{13-16}$$

where ξ is the Riemann function

$$\xi(3) = 1.202\ldots \tag{13-17}$$

The solution of the Gell-Mann and Low equation (13-10) shows that it is possible to obtain the dominant behavior of the vacuum polarization to order n, knowing d^{as} and therefore ψ to lower orders. Let us illustrate this remark by restricting ψ to the two-loop value $\psi(z) = z^2/3\pi + z^3/4\pi^2 + \cdots$ and deriving the dominant $\alpha^3(\ln x)^2$ term in d. From Eq. (13-11),

$$\begin{aligned} \ln x &= \int_{\alpha_1}^{d^{as}(\alpha,x)} \frac{dz}{\psi(z)} = f(d^{as}(\alpha, x)) - f(\alpha_1) \\ f(z) &= \text{constant} - \frac{3\pi}{z} - \frac{9}{4}\ln z + O(z) \end{aligned} \tag{13-18}$$

Upon inversion this yields

$$\begin{aligned} d^{as}(\alpha, x) &= \frac{\alpha_1}{1 - (\alpha_1/3\pi)\ln x - (\alpha_1^2/4\pi^2)\ln x - (\alpha_1^3/24\pi^3)(\ln x)^2 + O(\alpha_1^3 \ln x)} \\ \alpha_1 &= \frac{\alpha}{1 + 5\alpha/9\pi + C_2\alpha^2/4\pi^2} = \alpha - \frac{5}{9\pi}\alpha^2 + \cdots \end{aligned} \tag{13-19}$$

Consequently, we have indeed found the dominant behavior to order α^3. More modestly, with only the first term in ψ we would have found the cancellation of $\alpha^2(\ln x)^2$ terms in the two-loop computation of $\bar{\omega}$, a result which cost us considerable effort in Chap. 8. Note, however, that it would have been insufficient to obtain the coefficient of $\alpha^2 \ln x$.

If α_1 is substituted as a function of α in (13-19), we recover nonleading contributions to a given order. At any rate, the conclusion to be drawn from (13-19) is that as $-q^2$ goes to infinity the three-loop contribution to $\bar{\omega}$ behaves as $-(\alpha^3/24\pi^3)[\ln(-q^2/m^2)]^2$. The renormalization group therefore allows us to derive new results, while it seemed at first that we were only stating trivialities. We will show below that the numerical expressions obtained so far are sufficient to predict all terms of the form $\alpha^n[\ln(-q^2/m^2)]^{n-1}$.

The function ψ is only known as a power series around the origin. We have no guarantee about its convergence. On the contrary we suspect such a series to be at best asymptotic (Sec. 9-4). We recall that its range of validity requires $\alpha \ln(-q^2/m^2) \to 0$ when $-q^2/m^2 \to \infty$. This explains why the existence of a pole in d in the vicinity of the unphysical euclidean point

$$-q^2 = m^2 e^{3\pi/\alpha} \sim m^2 \times 10^{560}$$

is rather doubtful. Such a pole, sometimes referred to as the Landau ghost, occurs when keeping only the first-order contribution in $\bar{\omega}$. Since this singularity is clearly in the asymptotic region, a serious investigation would require the complete expression of d^{as}, all of its terms being of comparable size. It is therefore hazardous to draw any conclusions on the inconsistency of quantum electrodynamics at this stage.

On the other hand, assuming ψ to be meaningful, it is possible to speculate on the theory as a whole by formulating precise hypotheses on this function. It has a universal character independent of normalization conventions as a consequence of its very definition.

Let us study a few possibilities. As suggested by the first terms in its expansion, assume that $\psi(z)$ is positive in an interval $0 < z < \alpha_\infty$. To be consistent we suppose that α_1 is included in this interval.

If α_∞ is infinite, i.e., if $\psi(z)$ only vanishes at the origin, it is necessary that the integral $\int^\infty dz/\psi(z)$ diverges at its upper limit in such a way that $d^{as}(\alpha, x)$ goes to infinity as $x \to \infty$. Otherwise $d^{as}(\alpha, x)$ becomes infinite for a nonphysical value $q^2 = -m^2 \exp[\int_{\alpha_1}^\infty dz/\psi(z)]$ and the same is likely to be true for the complete function d. In other words, we have a genuine Landau ghost. Clearly the first few terms in ψ cannot give a proper indication to test this hypothesis.

The second possibility is that $\psi(z)$ has a finite positive zero α_∞. For $0 < \alpha_1 < \alpha_\infty$ the positivity of $\psi(z)$ implies that $d^{as}(x)$ is an increasing function and we require $\int^{\alpha_\infty} dz/\psi(z)$ to be divergent. Such is the case if $\psi(z)$ has a simple zero. Under these conditions $d^{as}(\alpha, x)$ tends to α_∞ when $x \to \infty$. In fact, since $\psi(z)$ decreases in the vicinity of this point, d^{as} tends to α_∞ no matter whether α_1 is smaller or larger than α_∞. This property is characterized by saying that α_∞ is an ultraviolet attractor (or stable fixed point). In the case $\alpha_1 = \alpha_\infty$, d^{as} would reduce to the constant α_∞.

In a more quantitative way let us write, close to $z = \alpha_\infty$,

$$\psi(z) = \nu(\alpha_\infty - z) + \cdots \qquad \nu > 0 \tag{13-20}$$

Integration of (13-11) yields

$$d^{as}\left(\alpha, -\frac{q^2}{m^2}\right) = \alpha_\infty - k\left(-\frac{q^2}{m^2}\right)^{-\nu} + \cdots \tag{13-21}$$

where the positive constant k depends on the precise form of ψ. The critical index $\nu[\nu = -(d\psi/dz)(\alpha_\infty) > 0]$ characterizes the approach to the fixed value α_∞ for which the photon propagator assumes its free-field value. Correspondingly, α_∞ would describe the electromagnetic coupling at short distances as if the bare coupling were finite.

13-1-2 The Callan-Symanzik Equation

Instead of comparing the effect of changing the renormalization point, we make seemingly a step backward by returning to the bare regularized Green functions using a cutoff $\Lambda \gg m$. For an appropriate choice of bare parameters we have the relation

$$d(\alpha, q^2, m^2) = \lim_{\Lambda \to \infty} d(\alpha_0, q^2, m_0^2, \Lambda^2) \tag{13-22}$$

The relation as we know was only established perturbatively and does not involve any multiplicative wave-function factor due to the Ward identity. Furthermore, we have

$$\alpha = Z_3\left(\frac{\Lambda^2}{m^2}, \alpha\right)\alpha_0 \tag{13-23}$$

The $\Lambda \to \infty$ limit will always be understood, but most of the time it is not written explicitly. The crucial observation is that, due to this limiting procedure, the renormalized amplitudes depend on one dimensional parameter less than the regularized ones.

To proceed, consider the irreducible function $d^{-1} = \alpha^{-1}(1 + \bar{\omega})$. The diagrams contributing to

$$m_0 \frac{\partial}{\partial m_0} d^{-1}(\alpha_0, q^2, m_0^2, \Lambda^2)$$

are superficially convergent. Except for the factor $g^{\mu\nu}q^2 - q^\mu q^\nu$ they correspond to the irreducible part of the Green function (Fig. 13-1)

$$\int d^4x\, e^{-iq\cdot x} \langle 0|\, Tj^\mu(x) j^\nu(0) \int d^4y\, im_0 : \bar{\psi}_0\psi_0(y): |0\rangle$$

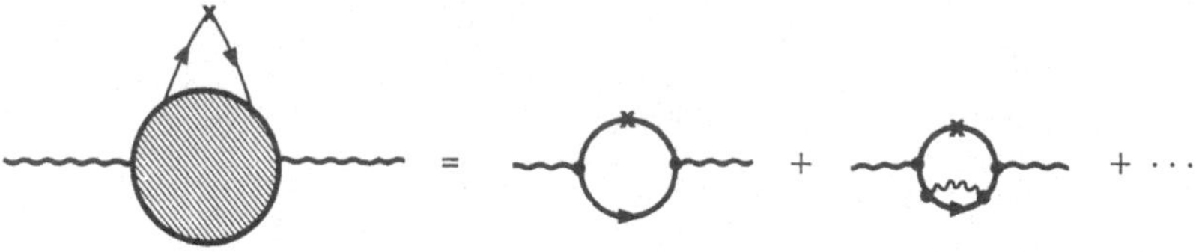

Figure 13-1 Diagrams for $m_0(\partial/\partial m_0)d^{-1}$.

The superficial convergence assumes current conservation. The insertion of the operator $im_0 \int d^4y :\bar{\psi}\psi(y):$ increases by one unit the degree of one of the denominators in the Feynman integrals. However, as discussed in Chap. 8, the subtractions of internal divergences due to the insertions of $m_0\bar{\psi}_0\psi_0$ in fermion self-energy subdiagrams require the introduction of a new counterterm or, equivalently, the multiplication of $\bar{\psi}_0\psi_0$ by $Z_{\bar{\psi}\psi}$. Ultimately

$$\Delta\left(\alpha, -\frac{q^2}{m^2}\right) = Z_{\bar{\psi}\psi} m_0 \frac{\partial}{\partial m_0} d^{-1}(\alpha_0, q^2, m_0^2, \Lambda^2) \tag{13-24}$$

will be finite in the limit $\Lambda \to \infty$. To lowest order Δ is derived from formula (7-9). It behaves as $-(m^2/q^2)\ln(-q^2/m^2)$ for large $-q^2/m^2$. At order n, Δ is $-m^2/q^2$ times a polynomial in $\ln(-q^2/m^2)$, and therefore its asymptotic limit Δ_{as} vanishes perturbatively.

We consider now a variation of m (hence also of α) for fixed α_0 and Λ. According to (13-23),

$$m\frac{\partial}{\partial m}\alpha\big|_{\alpha_0,\Lambda} = \alpha m \frac{\partial}{\partial m}\ln Z_3\big|_{\alpha_0,\Lambda}$$

If this has a limit as $\Lambda \to \infty$ with α and m fixed, the only dimensionless parameter left on which it can depend is α. From dimensional arguments we can therefore define

$$\beta(\alpha) = \lim_{\Lambda\to\infty} \alpha m \frac{\partial}{\partial m}\ln Z_3\big|_{\alpha_0,\Lambda} = -\lim_{\Lambda\to\infty} \alpha\Lambda \frac{\partial}{\partial \Lambda}\ln Z_3\left(\frac{\Lambda^2}{m^2}, \alpha_0\right) \tag{13-25}$$

We shall soon verify that $\beta(\alpha)$ exists perturbatively. For this purpose we compute the corresponding derivative of d^{-1}:

$$\lim_{\Lambda\to\infty} m\frac{\partial}{\partial m} d^{-1}\big|_{\alpha_0,\Lambda} = \left[m\frac{\partial}{\partial m} + \beta(\alpha)\frac{\partial}{\partial\alpha}\right] d^{-1}\left(\alpha, -\frac{q^2}{m^2}\right)$$

From (13-24) we can also express the left-hand side as

$$\begin{aligned}\lim_{\Lambda\to\infty} m\frac{\partial}{\partial m} d^{-1}(\alpha_0, q^2, m_0^2, \Lambda^2)\big|_{\alpha_0,\Lambda} &= \lim_{\Lambda\to\infty} \frac{m}{m_0}\frac{\partial m_0}{\partial m} \\ &\quad\times m_0\frac{\partial}{\partial m_0} d^{-1}(\alpha_0, q^2, m_0^2, \Lambda^2)\big|_{\alpha_0,\Lambda} \\ &= \lim_{\Lambda\to\infty}\left(\frac{m}{m_0}\frac{\partial m_0}{\partial m}\right) Z_{\bar{\psi}\psi}^{-1}\Delta\left(\alpha, -\frac{q^2}{m^2}\right)\end{aligned}$$

If $\beta(\alpha)$ is finite the same should be true for

$$1 + \delta(\alpha) = \lim_{\Lambda \to \infty} Z_{\bar{\psi}\psi}^{-1} \frac{m}{m_0} \frac{\partial m_0}{\partial m}\bigg|_{\alpha_0, \Lambda} \tag{13-26}$$

leading to the Callan-Symanzik equation

$$\left[m \frac{\partial}{\partial m} + \beta(\alpha) \frac{\partial}{\partial \alpha}\right] d^{-1}\left(\alpha, -\frac{q^2}{m^2}\right) = [1 + \delta(\alpha)]\, \Delta\left(\alpha, -\frac{q^2}{m^2}\right) \tag{13-27}$$

From the fact that Δ_{as} vanishes it also follows that

$$\left[-2x \frac{\partial}{\partial x} + \beta(\alpha) \frac{\partial}{\partial \alpha}\right] d_{as}^{-1}(\alpha, x) = 0 \tag{13-28}$$

Equation (13-28) implies that $\beta(\alpha)$ is finite perturbatively. It is sufficient to use the equation for some definite value of x. Conversely, we may derive renormalizability from a study of this equation and its extension to other Green's functions.

From the three-loop computation of de Rafaël and Rosner we obtain for fermionic quantum electrodynamics

$$\beta(\alpha) = \frac{2\alpha^2}{3\pi} + \frac{\alpha^3}{2\pi^2} - \frac{121}{144} \frac{\alpha^4}{\pi^3} + O(\alpha^5) \tag{13-29}$$

In contradistinction to the function ψ, the function β depends on the convention used to define the renormalized coupling α, except for its first two terms. The above expression shows that for $\alpha \to +0$, $\beta(\alpha)$ is positive.

As promised before, let us now derive the leading behavior of the vacuum polarization to an arbitrary large order as a consequence of (13-28) and (13-29). We substitute in (13-28) the formal series

$$d_{as}^{-1} = \alpha^{-1}\left[1 + \sum_1^\infty \alpha^n p_n(x)\right]$$

$$\beta(\alpha) = \sum_1^\infty b_n \alpha^{n+1} \tag{13-30}$$

where $$b_1 = \frac{2}{3\pi} \qquad b_2 = \frac{1}{2\pi^2}$$

and obtain the following equation:

$$2 \frac{d}{d \ln x} p_{n+1}(x) = \sum_{q=2}^{n} (q-1) p_q(x) b_{n-q+1} \qquad n \geq 2 \tag{13-31}$$

The solution is

$$p_1(x) = -\frac{b_1}{2} \ln x + C_1$$

$$p_n(x) = -\left(\frac{b_1}{2}\right)^{n-2} \frac{b_2}{2} \frac{(\ln x)^{n-1}}{n-1} + O(\ln x^{n-2}) \qquad n \geq 2 \tag{13-32}$$

This shows that computations up to the two-loop order yield b_1 and b_2, which in turn determine the leading behavior to order n.

We may reorganize the expansion of d_{as}^{-1} by summing successively leading logarithms, subleading ones, . . . :

$$d_{as}^{-1}(\alpha, x) = \alpha^{-1}\left(1 - \frac{\alpha}{3\pi}\ln x\right) + \frac{3}{4\pi}\ln\left(1 - \frac{\alpha}{3\pi}\ln x\right) + C_1 + \cdots \tag{13-33}$$

Unfortunately this method does not enable us to reach $d_{as}^{-1}(\alpha, x)$ for large x. For instance, if we were to keep leading logarithms we would recover the unphysical Landau ghost by setting $1 - (\alpha/3\pi)\ln x$ equal to zero. For this value the second term is logarithmically infinite, violating the assumption that it is subdominant. Stated differently, the reorganization implied by (13-33) is only useful when $|\alpha \ln x| \ll 1$. To find the true behavior for large x requires to sum up all logarithms or, equivalently, to return to the original equations (13-27) and (13-28).

It is interesting to understand diagrammatically why it is sufficient to know b_1 and b_2 alone in order to obtain the coefficient of the leading logarithm to any order. We recall an observation made in Chap. 8, according to which when summing gauge-invariant classes of diagrams the leading power of $\ln x$ is equal to the number of internal fermion loops. The reader will find it instructive to derive this property by the device of introducing N-degenerate species of fermions and studying the dependence on N. Diagrams with the largest number of fermion loops at a given order in α are of the type shown in Fig. 13-2. The resulting structure shows why a two-loop calculation yields the coefficient of the dominant term to order n as proportional to $b_1^{n-2} b_2 (\ln x)^{n-1}$.

From (13-25) it follows that the constant Z_3 satisfies the equation

$$\left[-2y\frac{\partial}{\partial y} + \frac{\beta(\alpha)}{\alpha}\left(\alpha\frac{\partial}{\partial \alpha} - 1\right)\right]Z_3(y, \alpha) = 0$$
$$y = \frac{\Lambda^2}{m^2} \tag{13-34}$$

Therefore it is possible to study the limit of infinite cutoff $y \to \infty$, provided we are willing to make some hypothesis on the function $\beta(\alpha)$.

The Gell-Mann and Low function ψ and the Callan-Symanzik coefficient $\beta(\alpha)$ are, of course, related. Take, for instance, the derivative of the expression

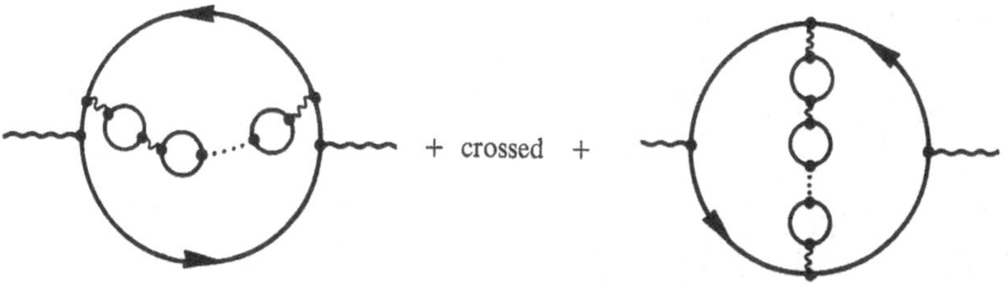

Figure 13-2 Diagrams with the maximal number $(n - 1)$ of fermion loops contributing to vacuum polarization at order n.

(13-11) with respect to x_2, and substitute in (13-28). We obtain

$$\psi[d_{as}(\alpha, x)] = \frac{1}{2}\beta(\alpha)\frac{\partial}{\partial\alpha}d_{as}(\alpha, x) \tag{13-35}$$

which may be specialized to some value of x. A possible choice is $x = 1(q^2 = -m^2)$, with $d^{as}(\alpha, 1) = \alpha_1(\alpha) = \alpha - 5\alpha^2/9\pi + \cdots$, and

$$\psi(\alpha_1) = \frac{1}{2}\beta(\alpha)\frac{d\alpha_1}{d\alpha} \tag{13-36}$$

This also shows that the zeros of the two functions are related. If α_c is such that $\beta(\alpha_c) = 0$, then ψ vanishes for $\alpha_\infty = \alpha_1(\alpha_c)$. In particular, when this value is an ultraviolet fixed point, d_{as} tends to α_∞ as x goes to infinity. Furthermore, from (13-34) Z_3 is finite and α_∞ plays the role of a finite bare square charge. Adler has given a thorough discussion of the possibility that α_c might be equal to the observed physical fine structure constant. Unfortunately, at present we lack a definite nonperturbative procedure to evaluate these functions, so that the whole matter remains speculative.

Derive for the other Green functions of electrodynamics the analogous Callan-Symanzik equations.

13-2 BROKEN SCALE INVARIANCE

As realized some time after the early investigations of Gell-Mann and Low, a more general point of view is to investigate the short-distance behavior of Green functions in a renormalizable theory when all relative distances are space-like and tend to zero simultaneously. This question may seem purely theoretical since it involves amplitudes far off-shell. Luckily this is a wrong impression. Indirect means, such as deep inelastic lepton scattering on hadronic targets, enable one to probe short-distance interactions. The results of such experiments, anticipated by theoretical considerations of Bjorken and Feynman, partly motivated these investigations by Wilson, Symanzik, and Callan.

The expectation that at large momenta when masses become negligible the theory becomes scale invariant is in fact too naive. The asymptotic behavior is given by the corresponding massless theory. Renormalization imposes the choice of an arbitrary energy scale as discussed in Chap. 8. This scale spoils dimensional analysis. Its very arbitrariness is in fact what saves the day. A change of scale may be absorbed into a modification of the coupling constants. The corresponding flow is governed by functions similar to the β coefficient encountered above, and the renormalization group transformations appear therefore as a substitute to naive dimensional analysis.

Ultraviolet fixed points in this flow (if they exist) will attract the coupling constants as λ increases to infinity. This leads to the restoration of short-distance scale invariance for specific values of the couplings largely independent of the initial data. In particular, the observed dimensions of the fields (or other composite operators) will in general depend on the dynamics.

Of special interest is the case where the origin is an ultraviolet fixed point. This situation is called asymptotic freedom. Logarithmic corrections to naive scaling will then emerge as a result of renormalization.

13-2-1 Scale and Conformal Invariance

If in a classical action all dimensional constants vanish we would expect the theory to be scale invariant. This could also be the case in a massive theory at short distance (typically $m|x| \ll 1$).

If the configuration variables are scaled down

$$x \to {}^{\lambda}x = \lambda^{-1}x \qquad \lambda > 0 \tag{13-37}$$

the fields, noted generically as φ, would transform according to

$$\varphi(x) \to {}^{\lambda}\varphi(x) = U(\lambda)\varphi(\lambda x) \tag{13-37a}$$

with $U(\lambda)$ standing for a finite dimensional representation of the abelian group of dilatations. We shall assume this representation to be fully reducible. This means that we may write

$$U(\lambda) = e^{D \ln \lambda}$$

where the matrix D can be diagonalized. In infinitesimal form, when $\ln \lambda = \delta\varepsilon$, the transformation law reads

$$\delta\varphi = \delta\varepsilon\left(x \cdot \frac{\partial}{\partial x} + D\right)\varphi \tag{13-38}$$

In a classical massless theory (13-37) or (13-38) lead to an invariance provided the eigenvalues of D are equal to $1(d/2 - 1)$ for Bose fields, and $\frac{3}{2}(d/2 - \frac{1}{2})$ for Fermi fields. The quantities in parentheses apply to the case of an arbitrary dimension d instead of four.

We can also consider the effect of such transformations in a massive theory, obtaining therefore a Ward identity reflecting the breaking of scale invariance. In this sense we differ from pure dimensional analysis, since we consider the effect of a transformation of dynamical variables (the fields) but not of the dimensional parameters such as masses. If we were to do so we would relate two different physical situations.

For our favorite example‡

$$\mathscr{L} = \frac{1}{2}(\partial\varphi)^2 - \frac{m^2}{2}\varphi^2 - g\frac{\varphi^4}{4!} \tag{13-39}$$

‡ To avoid confusion with the scale parameter λ, the φ^4 coupling constant will be denoted g throughout this chapter.

we find a variation

$$\frac{\delta \mathscr{L}}{\delta \varepsilon} = x \cdot \frac{\partial \mathscr{L}}{\partial x} + 2(D+1)\frac{(\partial \varphi)^2}{2} - 4Dg\frac{\varphi^4}{4!} - 2D\frac{m^2}{2}\varphi^2 \tag{13-40}$$

Hence, if $D = 1$,

$$\frac{\delta \mathscr{L}}{\delta \varepsilon} = \left(x \cdot \frac{\partial}{\partial x} + 4\right)\mathscr{L} + m^2\varphi^2$$

The integral $\int d^4x\, \lambda^4 \mathscr{L}(\lambda x)$ is independent of λ (positive). From a differentiation at $\lambda = 1$ it follows that

$$\int d^4x \left(x \cdot \frac{\partial}{\partial x} + 4\right)\mathscr{L}(x) = 0$$

meaning that $[x \cdot (\partial/\partial x) + 4]\mathscr{L}(x)$ is a divergence and that the variation of the action I reads

$$\delta I = \delta\varepsilon \int d^4x\, m^2\varphi^2(x) \tag{13-41}$$

Obviously, when m vanishes we have classical scale invariance.

Let us show that conformal invariance is then a consequence of scale invariance.

The conformal group is defined as the set of transformations leaving the angles invariant. This carries over to Minkowski space, where we deal with hyperbolic as well as circular angles. This group is obtained by adjoining to the Poincaré transformations an inversion with respect to an arbitrary point—the origin, for instance,

$$x^\mu \rightarrow x'^\mu = \frac{ax^\mu}{x^2} \tag{13-42}$$

For this definition to be meaningful, the usual R^4 space must be completed by a cone at infinity. Let us introduce a useful geometrical representation. We consider a six-dimensional space with a (2, 4) metric, i.e., such that

$$dz^2 = (dz_0)^2 - (dz_1)^2 - (dz_2)^2 - (dz_3)^2 - (dz_4)^2 + (dz_{-1})^2$$

The lines belonging to the isotropic cone $z^2 = 0$ are identified with the usual R^4 space completed by a cone at infinity. This may be realized, for instance, by cutting the cone $z^2 = 0$ by the hyperplane $z_{-1} = 1$ and projecting stereographically on R^4 the resulting one-sheeted hyperboloid from the point $(0, 0, 0, 0, -1)$ in R^5 (Fig. 13-3). The explicit formulas are

$$y_\mu = \frac{2x_\mu}{1 - x^2} \qquad y_4 = \frac{1 + x^2}{1 - x^2}$$

$$y_\mu y^\mu - (y_4)^2 = -1 \qquad 0 \leq \mu \leq 3 \tag{13-43}$$

$$x_\mu = \frac{y_\mu}{1 + y_4} \qquad x^2 = \frac{y_4 - 1}{y_4 + 1}$$

The action of the four-sheeted pseudoorthogonal group $O(2, 4)$ is reflected on Minkowski space as conformal transformations. In particular dilatations correspond to hyperbolic rotations in the plane

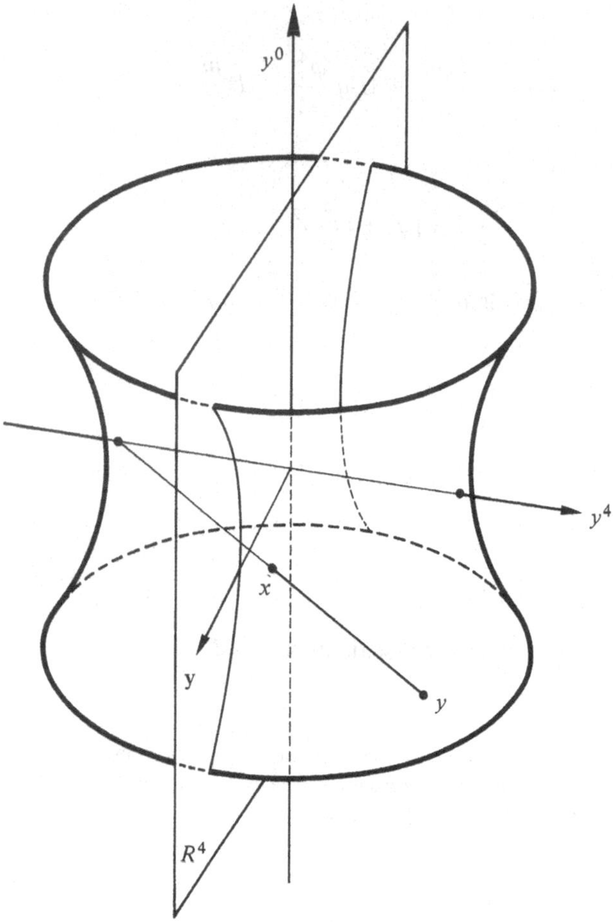

Figure 13-3 Projection of a (1, 4) hyperboloid on Minkowski space.

z_4, z_{-1}:

$$x' = e^{-\theta}x \Leftrightarrow y'_\mu = \frac{y_\mu}{\cosh\theta + y_4 \sinh\theta}, \quad y'_4 = \frac{y_4 \cosh\theta + \sinh\theta}{y_4 \sinh\theta + \cosh\theta}$$

$$\Leftrightarrow z'_\mu = z_\mu, \quad z'_4 = z_4 \cosh\theta + z_{-1} \sinh\theta, \quad z'_{-1} = z_{-1}\cosh\theta + z_4 \sinh\theta \tag{13-44}$$

As an exercise, find the four other types of conformal transformations, completing therefore the number of generators to 15. Write the corresponding transformations in Minkowski space. Perform the similar construction in the case of an euclidean four-dimensional space. The conformal group is then $O(1, 5)$ and R^4, completed by a point at infinity, may be identified with the stereographic projection of a unit sphere in five-dimensional space.

In order to prove the conformal invariance of the massless φ^4 theory it is therefore sufficient to study the effect of an inversion. In five-dimensional y space this transformation corresponds to a symmetry $y \leftrightarrow -y$ of the unit hyperboloid. We have to choose a transformation law for the field. Equation (13-37a), ${}^\lambda\varphi(x) = \lambda\varphi(\lambda x)$, suggests the definition

$$\varphi(x) \to \varphi'(x) = \frac{1}{x^2}\,\varphi\left(\frac{x}{x^2}\right) \tag{13-45}$$

It follows that

$$I' = \int d^4x \left[\frac{1}{2}(\partial\varphi')^2 - g\frac{\varphi'^4}{4!}\right] = I + \int d^4x \frac{4}{x^2}\varphi(x)\left(1 + x\cdot\frac{\partial}{\partial x}\right)\varphi(x)$$

The additional term is a four-divergence

$$\frac{4}{x^2}\varphi\left(1 + x\cdot\frac{\partial}{\partial x}\right)\varphi = 2\partial_\mu\left[\frac{x^\mu}{x^2}\varphi^2(x)\right]$$

Formally (i.e., barring possible singularities) the action, and therefore the equations of motion, are conformally invariant.

Exercises

(*a*) Express the massless φ^4 theory in five-dimensional space with dynamical variables defined on a unit hyperboloid (minkowskian case) or unit sphere (euclidean case). Write the corresponding lagrangian and equations of motion. Expand the solutions of the classical free-field equations in terms of generalized spherical harmonics.

(*b*) Show that the variation of the action of a massive theory under a dilatation [Eq. (13-41)] may be written as the integral of the four-divergence of a dilatation current. The latter is related to a modified energy momentum tensor (such that its trace vanishes in the massless case) as follows:

$$\delta I = \int d^4x\, \partial_\mu S^\mu \delta\varepsilon$$

$$S^\mu = x_\nu T^{\nu\mu} \qquad (13\text{-}46)$$

$$T^{\mu\nu} = \partial^\mu\phi\partial^\nu\phi - g^{\mu\nu}\mathscr{L} + \tfrac{1}{6}(g^{\mu\nu}\Box - \partial^\mu\partial^\nu)\varphi^2$$

For a discussion see the work of Callan, Coleman, and Jackiw.

(*c*) Investigate scale and conformal invariance in the presence of Fermi fields.

13-2-2 Modified Ward Identities

We perform a Wick rotation and study the euclidean theory. For our problem of short-distance behavior this implies no restriction. We write the normalized generating functional as

$$e^{G(j)} = \int \mathscr{D}(\varphi) \exp\left\{-I + \int d^4x\, j\varphi\right\} \qquad (13\text{-}47)$$

The euclidean action I may be split into

$$I = I_1 + I_2 \qquad I_1 = \int d^4x \left[\frac{1}{2}(\partial\varphi)^2 + g\frac{\varphi^4}{4!}\right] \qquad I_2 = \int d^4x \frac{m^2}{2}\varphi^2 \qquad (13\text{-}48)$$

A scale transformation $\varphi(x) \to {}^\lambda\varphi(x)$ may be considered here as a simple change of integration variable, under which

$$I_1 \to I_1 \qquad I_2 \to \lambda^{-2} I_2 \qquad \int d^4x\, j\varphi \to \lambda \int d^4x\, j(x)\varphi(\lambda x)$$

The change in the measure is absorbed into the normalization and we recover the naive Ward identity in the form

$$\int d^4x \left\{ m^2 \left[\frac{\delta}{\delta j(x)} \right]^2 + j(x) \left(1 + x \cdot \frac{\partial}{\partial x} \right) \frac{\delta}{\delta j(x)} \right\} e^{G(j)} - (j = 0) = 0 \qquad (13\text{-}49)$$

The term in $(\delta/\delta j)^2$ could be replaced by a derivative with respect to a source for the φ^2 operator. Of course, we expect Eq. (13-49) to be modified because of renormalization effects. In most of the cases encountered previously, such as gauge invariance or global symmetries, we were able to display a regularization and renormalization scheme preserving the symmetry and hence the Ward identities. But chiral anomalies have been the warning that modifications may appear in the quantum case.

Before displaying the aforementioned modifications let us disentangle the algebraic structure of (13-49) rewritten as

$$\int d^4x \left(j(x) \left(1 + x \cdot \frac{\partial}{\partial x} \right) \frac{\delta G}{\delta j(x)} + m^2 \left\{ \left[\frac{\delta G}{\delta j(x)} \right]^2 + \frac{\delta^2 G}{\delta j(x)^2} \right\} \right) - (j = 0) = 0 \qquad (13\text{-}50)$$

The Legendre transformation to irreducible functions reads in euclidean variables, Eq. (6-97):

$$G(j) - \Gamma(\varphi) = \int d^4x \, j(x)\varphi(x)$$

$$\varphi(x) = \frac{\delta G(j)}{\delta j(x)} \qquad (13\text{-}51)$$

$$j(x) = -\frac{\delta \Gamma(\varphi)}{\delta \varphi(x)}$$

with

$$\int d^4z \, \frac{\delta j(z)}{\delta \varphi(x)} \frac{\delta \varphi(y)}{\delta j(z)} = \delta^4(x - y)$$

implying that the kernels $-\delta^2 G(j)/\delta j(x)\delta j(y)$ and $\delta^2\Gamma(\varphi)/\delta\varphi(x)\delta\varphi(y)$ are the inverse of each other. To simplify the notations we denote them

$$\Gamma(x, y; \varphi) = \frac{\delta^2 \Gamma(\varphi)}{\delta \varphi(x) \delta \varphi(y)} \qquad \Gamma^{-1}(x, y; \varphi) = -\frac{\delta^2 G(j)}{\delta j(x) \delta j(y)} \qquad (13\text{-}52)$$

in such a way that Eq. (13-50) takes the form

$$0 = \int d^4x \left\{ \left[\left(1 + x \cdot \frac{\partial}{\partial x} \right) \varphi(x) \right] \frac{\delta \Gamma}{\delta \varphi(x)} \right.$$

$$\left. + m^2 [-\varphi^2(x) + \Gamma^{-1}(x, x; \varphi) - \Gamma^{-1}(x, x; 0)] \right\} \qquad (13\text{-}53)$$

Expand (13-53) in powers of φ to obtain the corresponding identity pertaining to the n-point irreducible function.

To zeroth order in $\hbar$, (13-53) reduces to the trivial cases encountered in the classical discussion (up to the Wick rotation):

$$\Gamma^{[0]} = -\int d^4x \left[\frac{1}{2}(\partial\varphi)^2 + \frac{m^2}{2}\varphi^2 + g\frac{\varphi^4}{4!}\right] \tag{13-54}$$

$$\int d^4x \left[\left(1 + x\cdot\frac{\partial}{\partial x}\right)\varphi(x)\right]\frac{\delta\Gamma^{[0]}}{\delta\varphi(x)} = -\int d^4x \left[\left(1 + x\cdot\frac{\partial}{\partial x}\right)\varphi(x)\right]\left(-\Box\varphi + m^2\varphi + g\frac{\varphi^3}{3!}\right)$$
$$= \int d^4x\, m^2\varphi^2(x) \tag{13-55}$$

Equation (15-53) is indeed satisfied since $[\Gamma^{-1}(x, x; \varphi) - \Gamma^{-1}(x, x; 0)]$ is of order $\hbar$, as is readily seen by restoring the factors $\hbar$ ($\Gamma \to \Gamma/\hbar$, $\delta/\delta\varphi \to \hbar\,\delta/\delta\varphi$). To order one this quantity may be written

$$-m^2 \operatorname{Tr}\left(\frac{1}{-\Box + m^2 + g\,\varphi^2/2} - \frac{1}{-\Box + m^2}\right)$$

and, formally,

$$\Gamma^{[1]} = -\frac{1}{2}\operatorname{Tr}\left\{\ln\left[\left(-\Box + m^2 + g\frac{\varphi^2}{2}\right)\frac{1}{-\Box + m^2}\right]\right\} \tag{13-56}$$

Since

$$\int d^4x \left[\left(1 + x\cdot\frac{\partial}{\partial x}\right)\varphi(x)\right]\frac{\delta\Gamma}{\delta\varphi(x)} = \lambda\frac{\partial}{\partial\lambda}\Gamma({}^\lambda\varphi)|_{\lambda=1}$$

it follows that

$$\int d^4x \left[\left(1 + x\cdot\frac{\partial}{\partial x}\right)\varphi(x)\right]\frac{\delta\Gamma^{[1]}}{\delta\varphi(x)} = -m\frac{\partial}{\partial m}\Gamma^{[1]} = m^2 \operatorname{Tr}\left(\frac{1}{-\Box + m^2 + g\,\varphi^2/2} - \frac{1}{-\Box + m^2}\right) \tag{13-57}$$

and Eq. (13-53) would be verified, if it were not for the necessary ultraviolet subtractions.

The functional

$$\Gamma_\Delta(\varphi) = \tfrac{1}{2}\int d^4x\, m^2[-\varphi^2(x) + \Gamma^{-1}(x, x; \varphi) - \Gamma^{-1}(x, x; 0)] \tag{13-58}$$

is interpreted as generating the irreducible Green functions with an insertion of the operator $-(m^2/2)\int d^4x\, \varphi^2(x)$. This was clear at the level of Eq. (13-49), where the term $m^2 \int d^4x\, [\delta/\delta j(x)]^2$ arose from the path integral over $m^2 \int d^4x\, \varphi^2(x)$. We check it at order one in Eq. (13-57). The corresponding diagrammatic expansion appears in Fig. 13-4. We call it a mass insertion.

Power counting indicates that this insertion introduces logarithmic singularities in the two-point function at order $\hbar^1$. In operator language the relevant Green function is $\int d^4x \langle 0|\, T\varphi^2(x)\varphi(y)\varphi(z)\,|0\rangle$. Let us specify our normalization conditions. Until we are faced with difficulties, it will be convenient to use inter-

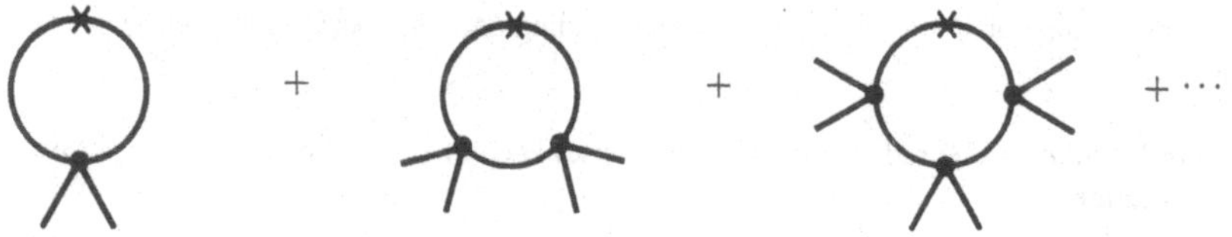

Figure 13-4 Mass insertion to the one-loop order.

mediate renormalization at zero momentum:

$$\Gamma_R^{(2)}(0) = -m^2 \qquad \left.\frac{d\Gamma_R^{(2)}(p^2)}{dp^2}\right|_{p^2=0} = -1 \qquad \Gamma_R^{(4)}(0) = -g \tag{13-59}$$

where m is proportional but not equal to the physical mass.

Using a regularization with a large cutoff Λ and adding the required counterterms to the lagrangian, we obviously modify the Ward identities of scale transformations. The relation between bare and renormalized irreducible Green's functions is

$$\Gamma_R(\varphi, m, g) = \Gamma_b(Z^{1/2}\varphi, m_0, g_0) \tag{13-60}$$

where on the right-hand side the cutoff Λ appears in the wave-function renormalization Z, the bare mass m_0, and bare coupling g_0, but the combination Γ_R has a finite limit where $\Lambda \to \infty$ with m and g held fixed. Ordinary dimensional analysis implies that the dependence on Λ of the quantities Z, g_0, m_0^2/m^2 is only through the combination Λ^2/m^2.

The bare generating functional of mass insertions is

$$\Gamma_{\Delta,b}(\varphi, m_0, g_0) = m_0 \frac{\partial}{\partial m_0} \Gamma_b(\varphi, m_0, g_0)\big|_{g_0,\Lambda} \tag{13-61}$$

We are now in a position to study scale anomalies. A variation of m_0 at fixed g_0 and Λ implies variations of the renormalized parameters g and m satisfying

$$0 = dg_0 = \frac{\partial g_0}{\partial m} dm + \frac{\partial g_0}{\partial g} dg \tag{13-62}$$

From (13-60) and (13-61),

$$2\frac{dm_0}{m_0}\Gamma_\Delta(Z^{1/2}\varphi, m_0, g_0) = \left[dm \frac{\partial}{\partial m} + dg \frac{\partial}{\partial g} - \frac{1}{2} d(\ln Z) \times \int d^4x\, \varphi(x) \frac{\delta}{\delta\varphi(x)}\right]\Gamma_R(\varphi, m, g) \tag{13-63}$$

while ordinary dimensional analysis implies

$$\left\{m\frac{\partial}{\partial m}+\int d^4x\left[\left(1+x\cdot\frac{\partial}{\partial x}\right)\varphi(x)\right]\frac{\delta}{\delta\varphi(x)}\right\}\Gamma_R(\varphi, m, g)=0 \tag{13-64}$$

This enables us to rewrite (13-63) as

$$\begin{aligned}\Big\{m\frac{\partial}{\partial m}g\left(\frac{\Lambda}{m}, g_0\right)\frac{\partial}{\partial g}-\int d^4x\Big[1+\frac{1}{2}m\frac{\partial}{\partial m}\ln Z\left(\frac{\Lambda}{m}, g_0\right)&\\ +x\cdot\frac{\partial}{\partial x}\Big]\varphi(x)\frac{\delta}{\delta\varphi(x)}\Big\}\Gamma_R(\varphi, m, g)&\\ =2\left[\frac{m}{m_0}\frac{\partial m_0(\Lambda/m, g)}{\partial m}\right]\Gamma_{\Delta,b}(Z^{1/2}\varphi, m_0, g_0)&\end{aligned} \tag{13-65}$$

Let us anticipate the fact that the dimensionless coefficients on the left-hand side have a finite limit as $\Lambda\to\infty$ with g and m fixed. Again from dimensional arguments they can only depend on g in this limit. We set

$$\beta(g)=m\frac{\partial}{\partial m}g\left(\frac{\Lambda}{m}, g_0\right)=-\Lambda\frac{\partial}{\partial\Lambda}g\left(\frac{\Lambda}{m}, g_0\right) \tag{13-66}$$

$$\gamma(g)=\frac{1}{2}m\frac{\partial}{\partial m}\ln Z\left(\frac{\Lambda}{m}, g_0\right)=-\frac{1}{2}\Lambda\frac{\partial}{\partial\Lambda}\ln Z\left(\frac{\Lambda}{m}, g_0\right) \tag{13-67}$$

As the notation indicates the derivatives are taken at fixed g_0. Of course, $\Lambda\to\infty$ is implied, which means that in perturbative calculations all terms vanishing for infinite cutoff are to be neglected.

According to the analysis of Chap. 8, the mass insertion is multiplicatively renormalizable. There exists a constant Z_{φ^2} such that

$$Z_{\varphi^2}\Gamma_{\Delta,b}(Z^{1/2}\varphi, m_0, g_0)=\Gamma_{\Delta,R}(\varphi, m, g) \tag{13-68}$$

with $\Gamma_{\Delta,R}(\varphi, m, g)$ finite. We define

$$1+\delta(g)=Z_{\varphi^2}^{-1}\frac{m}{m_0}\frac{\partial m_0(\Lambda/m, g_0)}{\partial m} \tag{13-69}$$

assuming the limit to exist, for a well-defined normalization prescription of the insertion.

Dropping the suffix R we obtain the Callan-Symanzik equation in the final form

$$\beta(g)\frac{\partial\Gamma}{\partial g}-\int d^4x\left\{\left[1+\gamma(g)+x\cdot\frac{\partial}{\partial x}\right]\varphi(x)\right\}\frac{\delta\Gamma}{\delta\varphi(x)}=2[1+\delta(g)]\Gamma_\Delta \tag{13-70}$$

Comparing this expression to the incorrect formula (13-53) which neglected renormalization, we see that it differs by the appearance of terms involving the coefficients $\beta(g)$, $\gamma(g)$, and $\delta(g)$. Before proving their finiteness, we comment on their interpretation.

First of all, they are similar to the anomalies of chiral invariance, modifying the classical Ward identity. We may look at $\gamma(g)$ as playing the role of a coupling dependent addition to the field dimension. The term in $\beta(g)\,\partial/\partial g$ is a remnant of the presence of the dimensional cutoff Λ in the relation between g and g_0. It implies that an infinitesimal scale transformation has to be accompanied by a small change in coupling constant. Finally, $\delta(g)$ could be absorbed a posteriori in a finite renormalization of the mass insertion.

To prove that these coefficients are finite we expand (13-70) in powers of φ (even powers for the φ^4 theory). In momentum space we obtain

$$-\sum_{1}^{n-1} p_k \cdot \frac{\partial}{\partial p_k}\Gamma^{(n)}(p) + \beta(g)\frac{\partial}{\partial g}\Gamma^{(n)}(p) + \{4 - n[1+\gamma(g)]\}\Gamma^{(n)}(p)$$
$$= 2[1+\delta(g)]\Gamma_{\Delta}^{(n)}(0; p) \qquad (13\text{-}71)$$

For definiteness we complete the normalization conditions (13-59) by

$$\Gamma_{\Delta}^{(2)}(0; 0) = -m^2 \qquad (13\text{-}72)$$

which is verified to order $\hbar^0$, and is sufficient to ensure the finiteness of Γ_Δ. Applying Eq. (13-71) in the vicinity of $p = 0$ for $n = 2$ we get two relations:

$$\gamma(g) + \delta(g) = 0$$
$$\gamma(g) - [1+\delta(g)]\left.\frac{\partial\Gamma_{\Delta}^{(2)}(0; p^2)}{\partial p^2}\right|_{p^2=0} = 0 \qquad (13\text{-}73)$$

which prove that $\gamma(g)$ and $\delta(g)$ are finite. From (13-71) it then follows that $\beta(g)$ is also well defined. We emphasize that in general these coefficients depend on the normalization conditions. In practical calculations it will sometimes be advantageous to use (13-66), (13-67), and (13-69), which relate them to the divergences of perturbation theory. This shows in fact that to lowest order ($\hbar^1$), β and γ do not depend on any convention.

The above presentation relies heavily on the bare theory with its infinite cutoff Λ. A scrupulous reader might suspect that it is possible to avoid such a detour, and derive Eq. (13-70) within the finite renormalized theory. The price of such an approach is the lack of intuitive appeal. On the other hand, the Callan-Symanzik equations may serve as a basis for the construction of the renormalized theory.

We have emphasized the interpretation corresponding to the modifications of broken scale invariance Ward identities. This is in fact the useful aspect of the Callan-Symanzik equations. They may, however, also be interpreted as renormalization group equations. This is achieved, for instance, in the massless theory by shifting the dilatation factor from the momenta to the arbitrary subtraction point μ with the result that Eq. (13-71) reads

$$\left[\mu\frac{\partial}{\partial\mu}+\beta(g)\frac{\partial}{\partial g}-n\gamma(g)\right]\Gamma^{(n)}(p,g,\mu)=0$$

It is also clear that the derivation generalizes to other renormalizable theories. The $\beta(g)$ function becomes a vector field in the case of several dimensionless coupling constants. In gauge theories, abelian or not, the Green functions depend in general on the gauge parameter λ, and a new term $\zeta(g,\lambda)\partial/\partial\lambda$ shows up on the left-hand side of Eq. (13-71), where

$$\zeta(g,\lambda)=m\frac{\partial}{\partial m}\lambda\left(\frac{\Lambda}{m},g_0,\lambda_0\right)\bigg|_{\Lambda,g_0,\lambda_0} \tag{13-74}$$

As λ_0 is related to the wave-function renormalization Z_3 of the gauge field through

$$\lambda=Z_3\lambda_0$$

we have

$$\zeta(g,\lambda)=2\lambda\gamma(g,\lambda)$$

in terms of the anomalous dimension $\gamma=\frac{1}{2}m(\partial/\partial m)\ln Z_3$. In nonabelian theories, the latter, as well as anomalous dimensions of other matter fields, depends on λ. This is not the case for $\beta(g)$, at least for certain normalization conditions, as noticed in Chap. 12.

Equation (13-71) may be compared to the analogous result (13-28) in electrodynamics. Note that there exists a relation between β and the anomalous dimension of the electrodynamic field γ. This is a consequence of the Ward identity: $e^2=e_0^2Z_3$.

The previous analysis also generalizes to functions involving composite operators. Each set of operators with the same dimension is multiplicatively renormalized by a renormalization matrix Z [see (8-69)]. The proper functions (of a massless theory, for simplicity) satisfy

$$\left[\mu\frac{\partial}{\partial\mu}+\beta(g)\frac{\partial}{\partial g}-n\gamma(g)\right]\Gamma^{(n)}_{O_i}+\gamma_{ij}(g)\Gamma^{(n)}_{O_j}=0$$

with

$$\gamma_{ij}(g)=-\mu\frac{\partial}{\partial\mu}Z_{ik}\big|_{g_0}(Z^{-1})_{kj} \tag{13-75}$$

The rationale for the discrepancy of sign between (13-67) and (13-75) will appear soon.

13-2-3 Callan-Symanzik Coefficients to Lowest Orders

For the φ^4 theory we could make use of the results given up to order $\hbar^2$ in Chap. 9. It is perhaps more instructive to derive β and γ from the counterterms of the lagrangian. To be specific we assume here a gaussian regularization

$$\frac{1}{p^2+m^2}=\int_0^\infty d\alpha\, e^{-\alpha(p^2+m^2)}\to\int_{1/\Lambda^2}^\infty d\alpha\, e^{-\alpha(p^2+m^2)}=\frac{e^{-(p^2+m^2)/\Lambda^2}}{p^2+m^2}$$

and compute $\Gamma^{(2)}(0)$, $(d/dp^2)\Gamma^{(2)}(0)$, and $\Gamma^{(4)}(0)$ as functions of g_0, m_0, and Λ. We will not omit the tadpole terms which must cancel at the end of the calculation.

For our purpose it is sufficient to compute $\Gamma^{(2)}(0)$ up to terms in $\hbar$; we recall that in φ^4 theory there is no wave-function renormalization to this order. We display the Feynman diagrams to the right of the corresponding expressions. As in previous chapters γ denotes Euler's constant. Terms written as constant are meant to be independent of Λ.

We find

$$
\Gamma_b^{(2)}(0) = \begin{cases} -m_0^2 \\ -\dfrac{1}{2}\dfrac{g_0}{(4\pi)^2}\left[\Lambda^2 - m_0^2 \ln \dfrac{\Lambda^2}{m_0^2} - m_0^2(1-\gamma)\right] \end{cases}
$$

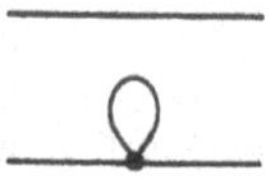

$$
\frac{d\Gamma_b^{(2)}}{dp^2}(0) = \begin{cases} -1 \\ -\dfrac{1}{12}\dfrac{g_0^2}{(4\pi)^4}\left(\ln \dfrac{\Lambda^2}{m_0^2} + \text{constant}\right) \end{cases}
$$

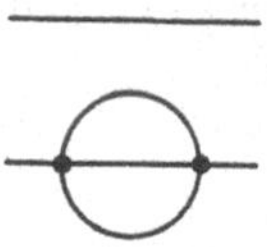

$$
\Gamma_b^{(4)}(0) = \begin{cases} -g_0 \\ +\dfrac{3}{2}\dfrac{g_0^2}{(4\pi)^2}\left(\ln \dfrac{\Lambda^2}{m_0^2} - \gamma - 1 - \ln 2\right) \\ -3\dfrac{g_0^3}{(4\pi)^4}\left[\dfrac{1}{2}\left(\ln \dfrac{\Lambda^2}{m_0^2}\right)^2 - \ln \dfrac{\Lambda^2}{m_0^2}(\gamma + \ln 2) + \text{constant}\right] \\ -\dfrac{3}{4}\dfrac{g_0^3}{(4\pi)^4}\left[\left(\ln \dfrac{\Lambda^2}{m_0^2}\right)^2 - 2(\gamma + 1 + \ln 2)\ln \dfrac{\Lambda^2}{m_0^2} \right. \\ \left. + (\gamma + 1 + \ln 2)^2\right] \\ -\dfrac{3}{4}\dfrac{g_0^3}{(4\pi)^4}\left(\dfrac{\Lambda^2}{m_0^2} - \ln \dfrac{\Lambda^2}{m_0^2} + \gamma - 1\right) \end{cases} \tag{13-76}
$$

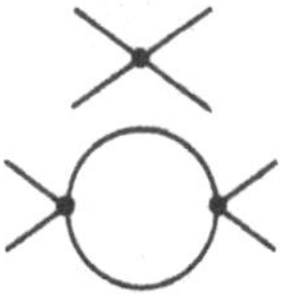

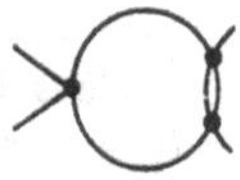

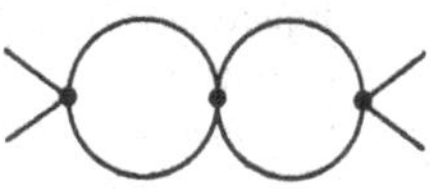

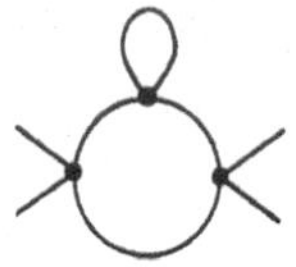

According to (13-59) we have

$$
\Gamma_b^{(2)}(0) = -m^2 Z^{-1} \qquad \frac{d\Gamma_b^{(2)}}{dp^2}(0) = -Z^{-1} \qquad \Gamma_b^{(4)}(0) = -Z^{-2} g
$$

Therefore Eq. (13-76) yields Z, g, and m as functions of m_0, g_0, and Λ. We substitute m^2 for m_0^2, keep only terms up to order g^2, and use the notations

$$
\alpha = \frac{g}{(4\pi)^2} \qquad \alpha_0 = \frac{g_0}{(4\pi)^2} \tag{13-77}
$$

It follows that

$$
Z^{-1} = 1 + \frac{\alpha_0^2}{12}\left(\ln \frac{\Lambda^2}{m^2} + \text{constant}\right) + \cdots
$$

$$
\begin{aligned} \alpha = \alpha_0 Z^2 \Bigg\{ 1 &- \frac{3}{2}\alpha_0\left(\ln \frac{\Lambda^2}{m^2} - \gamma - 1 - \ln 2\right) \\ &+ \alpha_0^2\left[\frac{9}{4}\left(\ln \frac{\Lambda^2}{m^2}\right)^2 - \ln \frac{\Lambda^2}{m^2}\left(\frac{9}{2}\gamma + \frac{3}{2} + \frac{9}{2}\ln 2\right) + \text{constant}\right] + \cdots \Bigg\} \end{aligned} \tag{13-78}
$$

Terms proportional to Λ^2 have fortunately disappeared, leaving only logarithms. Applying the definition (13-66),

$$-\frac{\beta(g)}{2(4\pi)^2}=\frac{\partial\alpha}{\partial\ln\Lambda^2}\bigg|_{m,\alpha_0}=2\alpha\frac{\partial\ln Z}{\partial\ln\Lambda^2}+Z^2\alpha_0\left[-\frac{3\alpha_0}{2}\right.$$
$$\left.+\alpha_0^2\left(\frac{9}{2}\ln\frac{\Lambda^2}{m^2}-\frac{9}{2}\gamma-\frac{3}{2}-\frac{9}{2}\ln 2\right)+\cdots\right]$$

The factor Z^2 may be replaced by unity to this order. Reexpressing α_0 in terms of α we obtain

$$\beta(g)=3\frac{g^2}{(4\pi)^2}-\frac{17}{3}\frac{g^3}{(4\pi)^4}+\cdots \tag{13-79a}$$

Similarly,

$$\gamma(g)=\frac{\partial}{\partial\ln\Lambda^2}\ln Z^{-1}\big|_{m,g_0}=\frac{g^2}{12(4\pi)^4}+\cdots \tag{13-79b}$$

$$\delta(g)=-\gamma(g)=-\frac{1}{12}\frac{g^2}{(4\pi)^4}+\cdots \tag{13-79c}$$

All coefficients are finite, as expected. Moreover, the Euler constant γ and ln 2 which appeared at intermediate stages as a result of our regularization scheme have disappeared.

Exercises

(*a*) Check that $\beta(g)$ is independent of the conventions to order $\hbar^2$.

(*b*) Study the modifications arising from an internal $O(n)$ symmetry group.

(*c*) Derive from (13-70) the Callan-Symanzik equations satisfied by the functions $V_{\text{eff}}(\varphi)$ and $Z_{\text{eff}}(\varphi)$ of Chap. 9 [Eq. (9-116)] and verify that they are obeyed perturbatively by the expressions obtained in (9-132).

We can also use the results of Chap. 12 on gauge fields to get the corresponding β function. To lowest order the relation between renormalized and bare couplings was found to be

$$g=g_0\left[1+\frac{g_0^2}{(4\pi)^2}\left(\frac{11}{3}C-\frac{4}{3}T_f\right)\ln\frac{\Lambda}{\mu}+O(g_0^4)\right] \tag{13-80}$$

The energy scale μ was an arbitrary device to secure a definition of g in the absence of a Higgs phenomenon inducing a genuine physical mass. From (13-80) we obtain $\beta(g)$ to order g^3. Let us also add the two-loop contribution computed by Caswell, Jones, Belavin, and Migdal:

$$\beta(g)=-\frac{g^3}{(4\pi)^2}\left(\frac{11}{3}C-\frac{4}{3}T_f\right)+\frac{g^5}{(4\pi)^4}\left(-\frac{34}{3}C^2+\frac{20}{3}CT_f+4C_fT_f\right)+O(g^7) \tag{13-81}$$

The interesting feature of this result is that, contrary to similar expressions for quantum electrodynamics [Eq. (13-29)] or φ^4 theory [Eq. (13-79)], $\beta(g)$ has a sign opposite to g as $g \to 0$ (provided $T_f < \frac{11}{4}C$). We shall later see the importance of this remark. The computation of anomalous dimensions is a priori of interest only for gauge invariant operators. We will also have to return to this point.

Exercises

(*a*) Show that the computation performed in quantum electrodynamics agrees with (13-81) when we set $C = 0$, $T_f = C_f = 1$, and write $\beta(\alpha)\partial/\partial\alpha \equiv \beta'(e)\partial/\partial e$.

(*b*) Discuss the case of gauge fields coupled to scalar bosons. Obtain their contribution to lowest order and verify that it is one-eighth of the fermion one, provided they correspond to a real representation.

13-3 SCALE INVARIANCE RECOVERED

13-3-1 Coupling Constant Flow

The exact equation of Callan and Symanzik finds its most interesting applications when the right-hand side involving the mass insertion may be neglected. If it were not the case we would have to face a cascade of functionals Γ, Γ_Δ, $\Gamma_{\Delta\Delta}, \ldots$ containing the full complexity of the amplitudes throughout their kinematical domains. Here we are only interested in the deep euclidean regime where all momenta become large. This is, of course, only meaningful with respect to some mass scale. The Weinberg theorem of Chap. 8 comes to the rescue here. Its application to a strictly renormalizable theory shows that perturbatively $\Gamma_\Delta^{(n)}$ in (13-71) is depressed by a power p^2 (up to logarithms) with respect to the left-hand side. Therefore the massless limit of the theory exists provided normalization conditions at nonzero momentum be chosen. This offers the choice of an energy scale. Henceforth $\Gamma_{as}^{(n)}$ will denote the corresponding massless Green functions. They satisfy homogeneous equations:

$$\left\{-\sum_{1}^{n-1} p_k \cdot \frac{\partial}{\partial p_k} + \beta(g)\frac{\partial}{\partial g} + 4 - n[1 + \gamma(g)]\right\}\Gamma_{as}^{(n)}(p_i; g) = 0 \qquad (13\text{-}82)$$

Similar relations may be derived for $\Gamma_\Delta^{(n)}$ and the coherence of our assumption checked if the solution is negligible with respect to $\Gamma^{(n)}$.

The solution of Eqs. (13-82) will exhibit a structure reminiscent of the relation between bare and renormalized theory. The reason lies clearly in the way the equations were obtained. The difference will be that infinities are no longer involved. A finite renormalization effect will accompany a change of momentum scale. To describe it let $g(\lambda)$ be the solution of the differential equation

$$\lambda \frac{d}{d\lambda} g(\lambda) = \beta[g(\lambda)] \qquad g(1) = g \qquad (13\text{-}83)$$

This generalizes to a first-order differential system in the case of several coupling constants. Let us also introduce

$$z(\lambda) = \exp\left\{\int_1^\lambda \frac{d\lambda'}{\lambda'} \gamma[g(\lambda')]\right\} = \exp\left[\int_g^{g(\lambda)} \frac{dg'}{\beta(g')} \gamma(g')\right] \qquad z(1) = 1 \qquad (13\text{-}84)$$

Equation (13-82) takes the form

$$\lambda \frac{d}{d\lambda}\left\{\lambda^{4-n} z(\lambda)^{-n} \Gamma_{as}^{(n)}[\lambda^{-1} p_i; g(\lambda)]\right\} = 0 \qquad (13\text{-}85)$$

meaning that

$$\Gamma_{as}^{(n)}(\lambda p_i; g) = \lambda^{4-n} z(\lambda)^{-n} \Gamma_{as}^{(n)}[p_i; g(\lambda)] \qquad (13\text{-}86)$$

The departure from the naive scaling factor λ^{4-n} arises as an anomalous dimension $[z(\lambda)^{-n}]$ and a change in the coupling constant $g(\lambda)$.

It remains to find what happens in the limit $\lambda \to \infty$. In particular, we have to study the limiting behavior of $g(\lambda)$. From Eq. (13-83) it is clear that the crucial issue is the location of the zeros of $\beta(g)$. It may occur exceptionally that the initial coupling $g(1)$ just coincides with such a zero, call it g_∞. In this case $g(\lambda) = g_\infty$ independently of λ.

In general, $g(\lambda)$ will vary as $\lambda \to \infty$; it grows if β is positive or decreases if $\beta < 0$. This variation is only interrupted when it meets a zero of β. A possible situation is an evasion of g at infinity when λ grows. This would happen if β is of the same sign as g for all g, and vanishes only at the origin. Such a strong coupling situation is hard to analyze.

In order that $g(\lambda)$ tends to a finite limit g_∞ as $\lambda \to \infty$ it is necessary that $\beta(g_\infty)$ vanishes and that $(g - g_\infty)\beta(g)$ be negative in the neighborhood of g_∞. This is called an ultraviolet fixed point (Fig. 13-5). It could happen that a fixed point is ultraviolet attractive on one side and ultraviolet repulsive on the other. Such is the case if two simple zeros coalesce. For a simple attractive zero $\ln \lambda$ will diverge as

$$\ln \lambda \sim \int^{g(\lambda)} \frac{dg}{g_\infty - g}$$

and $g(\lambda)$ will tend to g_∞ as an inverse power of λ. A similar analysis can be carried out for a multiple zero. On the other hand, nothing spectacular is expected

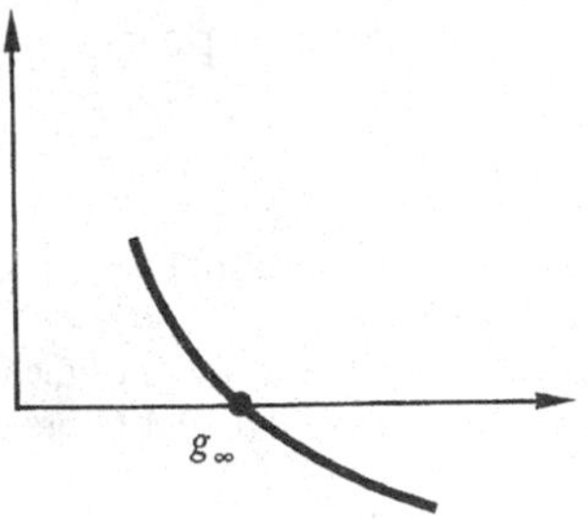

Figure 13-5 Ultraviolet fixed point.

from $\gamma(g)$ in the vicinity of g_∞. If $\gamma_\infty = \gamma(g_\infty)$, then up to a λ-independent factor

$$z(\lambda) \underset{\lambda\to\infty}{\sim} \lambda^{\gamma_\infty}$$

In a general situation, attractive and repulsive fixed points appear successively along the g axis. According to the initial value $g(1)$, the effective coupling constant $g(\lambda)$ will tend to the nearest ultraviolet attractive fixed point as $\lambda \to \infty$.

Ultraviolet unstable points, $(g - g_\infty)\beta(g) > 0$, are infrared attractors and correspond to limits of $g(\lambda)$ when $\lambda \to 0$. This is of interest when we study, for a massless theory, the limit of large distances or very small momenta with respect to an ultraviolet cutoff. This is precisely the aim of the theory of critical phenomena where $m = 0$ corresponds to the critical temperature and we investigate the long-range tail of correlations.

Returning to our original problem we conclude from the existence of a nearby ultraviolet fixed point that for large λ the asymptotic behavior of amplitudes is of the form

$$\Gamma^{(n)}(\lambda p_i; g) \underset{\lambda\to\infty}{\to} \Gamma^{(n)}_{as}(\lambda p_i; g) \sim \lambda^{4-n(1+\gamma_\infty)}\Gamma^{(n)}_{as}(p_i; g_\infty) \tag{13-87}$$

Scale invariance has been recovered, corresponding to a specific value g_∞ of the coupling and with an effective dimension for the field

$$d_{\text{eff}}(\varphi) = 1 + \gamma_\infty \tag{13-88}$$

The Green functions with a mass insertion Δ obey

$$\left(-\sum p_k \cdot \frac{\partial}{\partial p_k} + \beta(g)\frac{\partial}{\partial g} + \left\{2 - n[1 + \gamma(g)] + \gamma_\Delta(g)\right\}\right)\Gamma^{(n)}_{\Delta as}(0; p_i; g) = 0$$

with $\gamma_\Delta(g) = \partial \ln Z_{\varphi^2}/\partial \ln \Lambda$ [see Eq. (13-75)]. As long as $\gamma_\Delta(g_\infty) < 2$, the perturbative hypothesis of identifying the asymptotic regime with the massless theory is consistent. Otherwise the mass insertion would, in effect, correspond to a hard operator and the theory should be reformulated. See Sec. 13-3-3 for additional remarks on this point.

This beautiful reasoning, essentially due to Wilson, shows how the renormalization group has been an instrument in recovering a nontrivial scaling behavior. The crux is now to find whether $\beta(g)$ has such zeros. Since, however, the origin is always such a zero, as $\beta(g)$ vanishes in the absence of interactions, it is of special interest to assert perturbatively the nature of this fixed point.

Exercises

(*a*) Discuss the effect of a multiple zero on the asymptotic scaling law and find the corrections to the dominant behavior.

(*b*) Show that the positive measure in the Lehman-Källen representation for the two-point function implies $\gamma \geq 0$ and that $\gamma_\infty = 0$ implies a free theory as noted by Parisi and Callan and Gross.

13-3-2 Asymptotic Freedom

We have given examples of the computation of $\beta(g)$ for small g in the case of a self-coupled scalar field and abelian or nonabelian gauge fields. The results are represented in Fig. 13-6. For electrodynamics we have plotted the quantity $4\pi\beta(\alpha)/2e$ [with $\beta(\alpha)$ given by Eq. (13-29)] as a function of e.

If the origin is an ultraviolet fixed point we say that the theory is asymptotically free. Among the examples exhibited here, only nonabelian gauge fields with a small number of fermions possess this property. Superficially φ^4 theory would also be in this case for $g < 0$. But it is likely to be unstable for negative g. Since asymptotic freedom would mean that at least for the large momentum regime radiative corrections are calculable perturbatively, we would then find an unbounded effective potential. We conclude with Coleman that for $g < 0$ the theory is unstable.

An exhaustive examination by Coleman and Gross establishes the following result. No renormalizable field theory is asymptotically free in four dimensions, except if it contains nonabelian gauge fields.

We outline the derivation of this important result.

(*a*) *Scalar theory*
The φ^4 interaction is generalized to an arbitrary set of interacting scalar fields in the form

$$\mathscr{L}_{\text{int}} = -g_{ijkl}\varphi_i\varphi_j\varphi_k\varphi_l$$

with g_{ijkl} totally symmetric. To order $\hbar$,

$$\beta_{ijkl} = \lambda \frac{d}{d\lambda} g_{ijkl}(\lambda) = A \sum_{m,n} [g_{ilmn}(\lambda) g_{jkmn}(\lambda) + \text{permutations}]$$

with $A > 0$. Assuming the quartic form $g_{ijkl}\varphi_i\varphi_j\varphi_k\varphi_l$ positive for stability, we derive that $\beta_{iiii} > 0$.

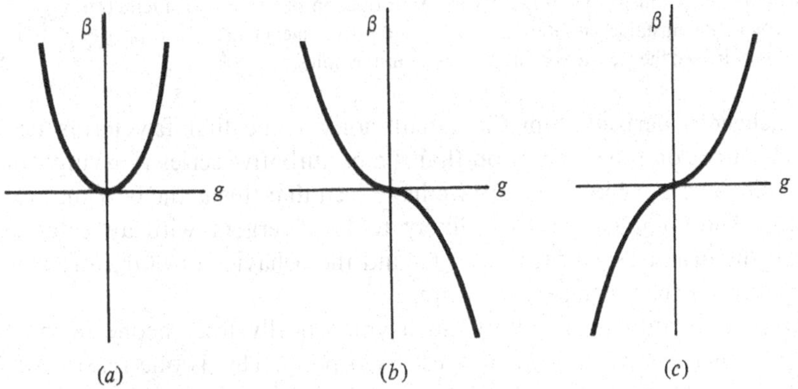

Figure 13-6 The function $\beta(g)$ near the origin: (*a*) self-coupled scalar field; (*b*) nonabelian gauge field with $11C > 4T_f$; (c) nonabelian gauge field with $11C < 4T_f$, electrodynamics $C = 0$, $T_f = 1$, or Yukawa coupling.

(*b*) *Yukawa interaction*
The interacting lagrangian is of the form

$$\mathscr{L}_I = -g_{ijkl}\varphi_i\varphi_j\varphi_k\varphi_l + \varphi_k\bar{\psi}^a(A^k_{ab} + i\gamma_5 B^k_{ab})\psi^b$$

The positivity of β_{iiii} may be spoiled, but all Yukawa couplings cannot vanish asymptotically. Indeed, if

$$g^k = A^k + i\gamma_5 B^k$$

then

$$\lambda\frac{d}{d\lambda}\operatorname{tr}\left[g^k(\lambda)g^{k\dagger}(\lambda)\right] \geq 0$$

Since abelian gauge fields are not asymptotically free either, the only possibility that remains is to include nonabelian gauge fields.

If Yang-Mills fields are coupled to fermions it is necessary for a simple gauge group that the condition $11C > 4T_f$ be fulfilled. If boson fields are also present the situation is more involved. First these fields must not be too numerous to keep $\beta(g) < 0$, where g is the gauge coupling. Furthermore, these fields have their own self-coupling. Assuming, for instance, that there exists a unique scalar field belonging to the adjoint representation, its self-coupling g_φ will obey to lowest order an equation of the type

$$\lambda\frac{d}{d\lambda}g_\varphi(\lambda) = Ag_\varphi^2(\lambda) + B'g_\varphi(\lambda)g^2(\lambda) + Cg^4(\lambda)$$

with $A > 0$. The quantity $\alpha = g_\varphi/g^2$ will then behave according to

$$\lambda\frac{d}{d\lambda}\alpha(\lambda) = A\alpha^2(\lambda) + B\alpha(\lambda) + C$$

Furthermore, we have to assume that g_φ is at least of order g^2; otherwise the positive term $A\alpha^2$ would be dominant and ruin asymptotic freedom. The right-hand side admits two roots α_1 and α_2 of order unity provided $B^2 > 4AC$. This requires a delicate balance between the number of fields, as added fermions tend to increase the quantity B. If this is satisfied, then $g_\varphi(\lambda)$ will behave as $\alpha g^2(\lambda)$ when $\lambda \to \infty$, with α one of the roots of the equation, preserving therefore the required property. The general situation with several Bose fields is even more intricate and requires a case-by-case detailed study.

The introduction of such Bose fields may be motivated by the desire of generating mass terms through the mechanism of spontaneous symmetry breaking. In practice it is difficult to combine asymptotic freedom and symmetry breaking. By the introduction of a Yukawa interaction $g_{\bar{\psi}\psi\varphi}$ we may at best achieve an unstable situation along an attractive line in the plane $(g, g_{\bar{\psi}\psi\varphi})$. For a discussion of this point see the lectures of Gross quoted in the notes.

The conclusions derived from the calculations of the first few terms near $g = 0$ rely, of course, on the assumption that the perturbative series is asymptotic. When discussing large orders (Chap. 9) we have seen that this is the best that can be hoped for. The complete series is likely to be divergent with an essential singularity at the origin. Needless to say, to find the behavior of $\beta(g)$ away from the origin remains a challenge for the future.

The large momentum behavior of an asymptotically free theory becomes calculable provided the origin is the nearest fixed point. This is one reason why it is so favored by theorists. The name asymptotic freedom is, however, slightly misleading, since even then the scaling behavior departs from the free-field case by logarithmic factors. With

$$\beta(g) = -bg^3 + O(g^5) \qquad b > 0 \tag{13-89}$$

we obtain, for $g(\lambda)$,

$$g^2(\lambda) = \frac{1}{2b \ln \lambda} + O\left[\frac{\ln(\ln \lambda)}{(\ln \lambda)^2}\right] \tag{13-90}$$

For most of the interesting operators the function $\gamma(g)$ will be of order g^2:

$$\gamma(g) = cg^2 + O(g^4) \tag{13-91}$$

Therefore the scaling factor will contain logarithms

$$z(\lambda) \sim \exp\left[-\int_g^{g(\lambda)} \frac{c}{b}\frac{dg'}{g'}\right] \sim (2bg^2 \ln \lambda)^{c/2b} \tag{13-92}$$

A Green function involving the operator O_i will then behave as $(2bg^2 \ln \lambda)^{c_i/2b}$, apart from the canonical power of λ.

The existence of asymptotically free theories has far-reaching consequences for the purpose of building models of strong interactions. They enable us to reconcile two seemingly contradictory aspects. At low and medium energies the interactions are indeed strong and complex with numerous resonances. An approximate flavor symmetry according to $SU(3)$ or $SU(4)$ gives a qualitative description of hadrons as composite bound states of quarks. Every attempt at isolating these constituents has failed so far. At higher energies, however, quarks have much weaker interactions up to the point where they seem to act as free particles. The relevant kinematical domain reached by the experimental constraints is the one of light-like separations. We shall see in Secs. 13-4 and 13-5 that it is, however, possible to extend the short-distance expansion discussed so far to this region. It would then be possible to explain this paradoxical behavior by assuming an underlying asymptotically free-field theory. The catastrophical infrared singularities would be responsible for permanent confinement of quarks. These attractive speculations require necessarily such a model to include nonabelian gauge fields coupled to an unobserved (and perhaps unobservable) color degree of freedom.

The generalization to several couplings and to the corresponding multidimensional flow uncovers a great richness of phenomena such as stable fixed points, limiting cycles, ergodic behavior, etc. The difficulty of obtaining reliable information on this flow away from the origin somehow limits the interest of such a study.

13-3-3 Mass Corrections

We return to the question of the consistency of the massless asymptotic theory. Nonleading powers in the large-momentum behavior of perturbative amplitudes have been neglected. They reflect the presence of terms in the lagrangian with dimension less than four and corresponding couplings with a positive dimension in energy. Typically these are the mass terms. We have already mentioned that we can define anomalous dimensions for composite operators O_i such as φ^2 or $\bar{\psi}\psi$, thanks to their multiplicative renormalizability. Assume to simplify that the corresponding renormalization matrix is diagonal and of the form $Z_i(\Lambda/m, g_0)$.

Define

$$\gamma_i(g) = \Lambda \frac{\partial}{\partial \Lambda} \ln Z_i\left(\frac{\Lambda}{m}, g_0\right) \tag{13-93}$$

If g_∞ is an ultraviolet fixed point, the effective dimension of the operator O_i will differ from the canonical one d_i according to

$$d_i^{\text{eff}} = d_i + \gamma_i(g_\infty) \tag{13-94}$$

Here we only deal with the case $d_i < 4$. In general, mass corrections will remain negligible as long as Wilson's criterion is satisfied:

$$d_i^{\text{eff}} < 4 \tag{13-95}$$

This will be automatically verified in an asymptotically free theory where $\gamma_i(g_\infty)$ vanishes. We recall, however, that logarithmic corrections still affect the canonical behavior.

A closely related procedure has been proposed by Weinberg. It relies on the construction of counterterms independent of the renormalized mass up to trivial dimensional factors. In other words, we do not fix the latter explicitly. A mean to achieve this result is to use dimensional regularization and renormalization. Therefore we shall no longer have to neglect the insertion of the mass term in the renormalization group equation. In φ^4 theory the solution to the corresponding equation will have the form

$$\Gamma_n(\lambda p_i; g, m^2) = \lambda^{4-n}[z(\lambda)]^{-n}\Gamma_n[p_i; g(\lambda), m^2(\lambda)] \tag{13-96}$$

where

$$m^2(\lambda) = \frac{m^2}{\lambda^2} \exp\left[\int_g^{g(\lambda)} \frac{dg'}{\beta(g')} \gamma_\Delta(g')\right]$$

Show that to lowest order the contribution of the one-loop diagram of Fig. 13-7 leads to

$$\gamma_\Delta = \frac{g}{(4\pi)^2}$$

As $\lambda \to \infty$, the condition $m^2(\lambda) \to 0$ is $\gamma_\Delta(g_\infty) < 2$, and is therefore equivalent to Wilson's criterion (13-95). Only in this case is a massless scale-invariant theory meaningful, even though this condition does not appear perturbatively. We should not conclude from the previous analysis that a soft mass insertion can be treated perturbatively. Indeed, derivatives of Green functions with respect to the mass of a sufficient high order are singular in the massless limit. This is due to the fact that after a certain point increasing the number of insertions does not improve the ultraviolet behavior (see the discussion of Weinberg's

Figure 13-7 Lowest-order mass insertion in the φ^4 theory.

theorem in Chap. 8). An intuitive argument is the following. The second (third) derivative with respect to m of a bosonic (fermionic) propagator induces in general when $m \to 0$ a logarithmic infrared singularity in a Feynman diagram. We therefore expect singularities of the type $m_B^2(\ln m_B)^a[m_F^3(\ln m_F)^a]$ for small masses.

13-4 DEEP INELASTIC LEPTON-HADRON SCATTERING AND ELECTRON-POSITRON ANNIHILATION INTO HADRONS

We consider here in more detail a subject already introduced in Chap. 11. Our aim is to discuss in the next section the implications of a field theoretic model for the description of high transfer phenomena induced by leptons.

13-4-1 Electroproduction

We begin with a discussion of the electromagnetic scattering of a charged lepton (electron or muon) off a nucleon. The process has been depicted in Fig. 11-8. The initial and final lepton momenta are l and l' and at high energy we neglect the lepton mass. The initial nucleon of momentum p (mass m) breaks into a final state X which is not observed; hence the name inclusive process. Up to radiative corrections the electromagnetic interaction acts to lowest order. A photon carries the space-like momentum transfer $q = l - l'$ from the lepton to the hadron vertex.

In practice the measured quantities are the laboratory initial and final lepton energies E and E' and the scattering angle θ. We shall not discuss here polarization effects. The kinematical invariants are

$$
\begin{aligned}
-q^2 &= 4EE' \sin^2 \frac{\theta}{2} \\
\nu &= p \cdot q = m(E - E') \\
M^2 &= (p + q)^2 = m^2 + 2m(E - E') - 4EE' \sin^2 \frac{\theta}{2}
\end{aligned}
\tag{13-97}
$$

The nucleon being the lightest state with baryonic number one, the stability condition $M^2 \geq m^2$ implies that the Bjorken variable

$$
x = \omega^{-1} = -\frac{q^2}{2\nu} \tag{13-98}
$$

satisfies the inequalities

$$
0 < x \leq 1 \tag{13-99}
$$

The upper limit corresponds to elastic scattering. Both notations x and ω are found in the literature.

Let J_μ be the hadronic component of the electromagnetic current. The scattering amplitude reads

$$S_{fi} = i(2\pi)^4\delta^4(p_X + l' - p - l)\bar{u}(l')\gamma^\mu u(l)\frac{e^2}{q^2}\langle p_X|J_\mu(0)|p\rangle \tag{13-100}$$

For unpolarized leptons and nucleons the inclusive cross section is therefore

$$d\sigma = \frac{1}{EE'}\frac{d^3l'}{(2\pi)^3}\sum_X (2\pi)^4\delta^4(p_X + l' - p - l)\left(\frac{e^2}{q^2}\right)^2 \frac{1}{2}\operatorname{tr}\left(\gamma^\mu\frac{\not{l}}{2}\gamma^\nu\frac{\not{l}'}{2}\right)$$
$$\times\frac{1}{2}\sum_{\text{pol}}\langle p|J_\mu(0)|p_X\rangle\langle p_X|J_\nu(0)|p\rangle$$

The notation implies a sum over nucleon polarizations. Explicit computation yields

$$\frac{1}{8}\operatorname{tr}(\gamma^\mu\not{l}\gamma^\nu\not{l}') = \frac{1}{2}(l^\mu l'^\nu + l^\nu l'^\mu - g^{\mu\nu}l\cdot l')$$

$$W_{\mu\nu} = \frac{1}{4\pi}\sum_X (2\pi)^4\delta^4(p_X - p - q)\sum_{\text{pol}}\langle p|J_\mu(0)|p_X\rangle\langle p_X|J_\nu(0)|p\rangle$$
$$= \frac{1}{4\pi}\sum_{\text{pol}}\int d^4x\, e^{iq\cdot x}\langle p|J_\mu(x)J_\nu(0)|p\rangle$$

In the kinematical region of interest, $W_{\mu\nu}$ may also be expressed as the Fourier transform of the current commutator

$$W_{\mu\nu} = \frac{1}{4\pi}\sum_{\text{pol}}\int d^4x\, e^{iq\cdot x}\langle p|[J_\mu(x), J_\nu(0)]|p\rangle \tag{13-101}$$

Relativistic invariance, current conservation, and parity in the electromagnetic case enable us to express $W_{\mu\nu}$ in terms of two structure functions W_1 and W_2 which generalize the elastic form factors [compare with Eq. (3-203)]

$$W_{\mu\nu} = W_1(\nu, q^2)\left(\frac{q_\mu q_\nu}{q^2} - g_{\mu\nu}\right) + W_2(\nu, q^2)\frac{1}{m^2}\left(p_\mu - \frac{\nu}{q^2}q_\mu\right)\left(p_\nu - \frac{\nu}{q^2}q_\nu\right) \tag{13-102}$$

The cross section is expressed as

$$\frac{d\sigma}{d\Omega'\,dE'} = \frac{\alpha^2}{4E^2\sin^4(\theta/2)}\left(\cos^2\frac{\theta}{2}W_2 + 2\sin^2\frac{\theta}{2}W_1\right) \tag{13-103}$$

Experimental conditions, $\sin^2(\theta/2) \ll 1$, make it difficult to extract W_1.

The amplitude $W_{\mu\nu}$ is related to the scattering of polarized virtual photons off a nucleon. Call σ_T and σ_S the transverse and longitudinal cross sections (with $\sigma_S = 0$ at $q^2 = 0$). Show that

$$W_1(\nu, q^2) = \frac{2\nu + q^2}{8\pi^2 \alpha m} \sigma_T$$
$$W_2(\nu, q^2) = \frac{2\nu + q^2}{8\pi^2 \alpha m} \frac{1}{1 - \nu^2/m^2 q^2} (\sigma_S + \sigma_T) \tag{13-104}$$

where in σ_T and σ_S the flux factor is conventionally computed as if the photons were real with the same energy.

Derive from these expressions positivity conditions for W_1 and W_2. For elastic scattering ($\omega = 1$) derive the following relations with the electric and magnetic form factors, $G_E(q^2)$ and $G_M(q^2)$:

$$W_1^{el}(\nu, q^2) = -\frac{q^2}{2m} G_M^2(q^2)\, \delta(2\nu + q^2)$$
$$W_2^{el}(\nu, q^2) = \frac{2m}{1 - q^2/4m^2} \left[G_E^2(q^2) - \frac{q^2}{4m^2} G_M^2(q^2) \right] \delta(2\nu + q^2) \tag{13-105}$$

We recall from Chap. 11 that the scattering of neutrinos involves a third structure function W_3. The cross section for neutrinos (ν) or antineutrinos ($\bar{\nu}$) reads [compare with Eq. (11-101)]

$$\frac{d\sigma^{(\nu,\bar{\nu})}}{d\Omega'\, dE'} = \frac{EE'm}{\pi} \frac{d\sigma^{(\nu,\bar{\nu})}}{d\nu\, dq^2} = G^2 \frac{E'^2}{2\pi^2} \left(\cos^2 \frac{\theta}{2} W_2^{(\nu,\bar{\nu})} + 2 \sin^2 \frac{\theta}{2} W_1^{(\nu,\bar{\nu})} \mp \frac{E + E'}{m} \sin^2 \frac{\theta}{2} W_3^{(\nu,\bar{\nu})} \right) \tag{13-106a}$$

or

$$\frac{d\sigma^{(\nu,\bar{\nu})}}{dx\, dy} = m \frac{G^2 E}{2\pi} \left[(1 - y) \frac{\nu W_2}{m} + xy^2 m W_1 \mp x \left(1 - \frac{y}{2}\right) \frac{\nu W_3}{m} \right] \tag{13-106b}$$

in terms of the variable

$$y = \frac{E - E'}{E} = \frac{\nu}{mE} \tag{13-107}$$

The extra contribution represents the interference between the vector and axial parts of the current. In the high-energy limit where the mass of the final lepton (electron or muon) has been neglected, the nonconserved part of the current has disappeared.

The experimental results show that beyond the domain of resonance excitation the cross section remains sizable in the very inelastic region, for $-q^2$ and ν very large. For fixed q^2 the integral over ν is comparable with the Mott cross section on a point nucleon. Accurate measurements at different angles allow us to separate W_1 and W_2 to obtain the ratio $R = \sigma_S/\sigma_T$ with σ_S and σ_T defined according to (13-104). Its value is small, $R \sim 0.15$. The most striking phenomenon is, however, the scaling property anticipated by Bjorken and Feynman. In the very inelastic domain the dimensionless quantities $2mW_1$ and $\nu W_2/m$ tend to nontrivial functions of the scaling variable $\omega = 2\nu/-q^2$. This is shown on Fig. 13-8 where $\nu W_2/m$ is plotted versus the modified variable $\omega' = (2\nu + m^2)/-q^2 \sim \omega$. Analogous results hold in the case of neutrino scattering.

These phenomena suggest an interpretation of the scatterer in terms of pointlike nearly noninteracting constituents, most of which have spin $\frac{1}{2}$ (which corresponds to a ratio R equal to zero). Such a phenomenological parton model, as it has been called by Bjorken and Feynman, suggests a number of sum rules in agreement with the data. An asymptotically free-field theoretic model gives a coherent framework where the constituents are the fundamental quanta: the quark and gauge gluons.

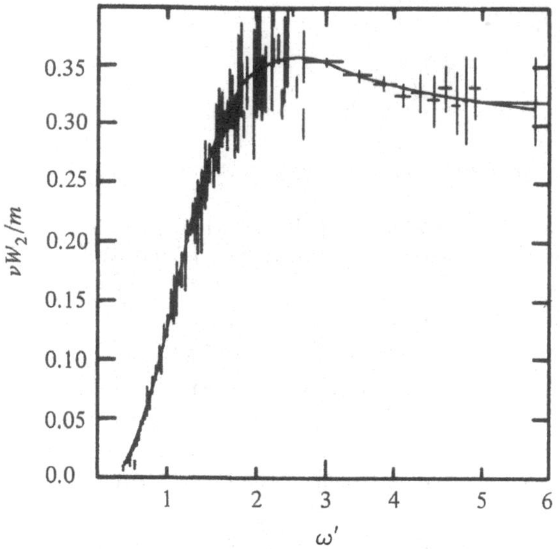

Figure 13-8 The function $\nu W_2/m$ plotted versus the modified scaling variable $\omega' = (2\nu + m^2)/-q^2$. The value of R is of the order of 0.18. The data are from SLAC and are commented in the report of R. E. Taylor at the Palerme Conference of 1975. The values of $-q^2/m^2$ being large, resonances at small ω' have been washed out.

13-4-2 Light-Cone Dynamics

Let us investigate more closely the kinematical domain which is probed in these deep inelastic experiments. At high energy, we test the singularities of the commutator $[J_\mu(x), J_\nu(0)]$. Singularities occur at short distance ($x \sim 0$) or for light-like separation ($x^2 \sim 0$). For the commutator which vanishes outside the light cone, short distances in Minkowski space are time-like. To study such a region would require us to look in momentum space for large q and small ω. The physical boundaries are, however, $\omega \geq 1$. In the experimental conditions we can write $q = \lambda n - q_0$ where n is light-like and λ large. As $\lambda \to \infty$, $-q^2 \sim 2\lambda(n \cdot q_0) \to \infty$, $\nu \sim \lambda(n \cdot p) \to \infty$, and $\omega \sim p \cdot n/q_0 \cdot n$ tends to a finite limit. Qualitatively, we expect to probe a region where $(n \cdot x) \sim 1/\lambda$ in such a way that $x^2 \sim O(1/\lambda) \sim O(1/-q^2)$. Thus we are looking at the singular structure of current commutators in the neighborhood of the light cone. It is therefore attractive to formulate the parton model hypothesis in the so-called infinite momentum frame (Fig. 13-9).

The dynamical evolution is considered starting from an hyperplane containing a light-like direction suited to the case where energies and longitudinal momenta are large and comparable. Without entering into details, let us sketch the intuitive reasoning of the parton model. Let P be the magnitude of the large longitudinal proton momentum in the center of mass frame, for instance; its four-momentum is then approximately $p \simeq (P, 0, 0, P)$. The target is thought of as an assembly of N elementary constituents, to be treated during their electromagnetic interactions as free particles with longitudinal momentum $x_i P$ and negligible transverse momentum with respect to $\sqrt{-q^2}$. The virtual photon four-momentum will be approximated in this frame by $q \simeq (\nu/2P, \sqrt{-q^2}, 0, -\nu/2P)$. The Feynman rules in the infinite momentum frame become analogous to those of nonrelativistic perturbation theory. The contribution of the ith constituent to the scattering cross section will thus be proportional to

$$Q_i^2 \frac{m}{P} \delta(E' - E - q_0)$$

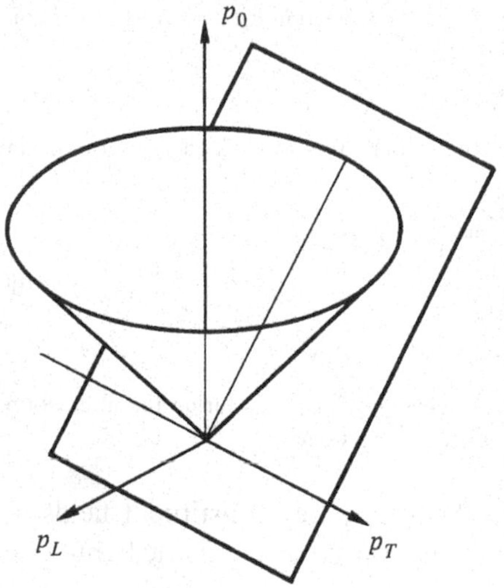

Figure 13-9 Infinite momentum frame.

with Q_i equal to its charge (Fig. 13-10). Since its initial momentum is x_iP, we have

$$E' - E - q_0 \sim \sqrt{-q^2 + \left(x_iP - \frac{\nu}{2P}\right)^2} - x_iP - \frac{\nu}{2P} \sim -\frac{q^2}{2x_iP} - \frac{\nu}{P}$$

Therefore the above contribution can also be written

$$x_iQ_i^2 \frac{m}{\nu} \delta\left(x_i + \frac{q^2}{2\nu}\right)$$

giving an interpretation of the scaling variable $x = \omega^{-1}$ as the fraction of total momentum carried by a constituent when scattering off the virtual photon. The model is completed by assuming the probability for such an occurrence to be $f(x)$ with

$$\int_0^1 dx\, f(x) = 1$$

For simplicity we have considered the case where $f(x)$ does not depend on the type i of the constituent. It then follows that the structure function W_2, say, is given by

$$\frac{\nu}{m} W_2(\nu, q^2) = F_2\left(-\frac{q^2}{2\nu}\right)$$
$$F_2(x) = \sum_i Q_i^2 x f(x) \tag{13-108}$$

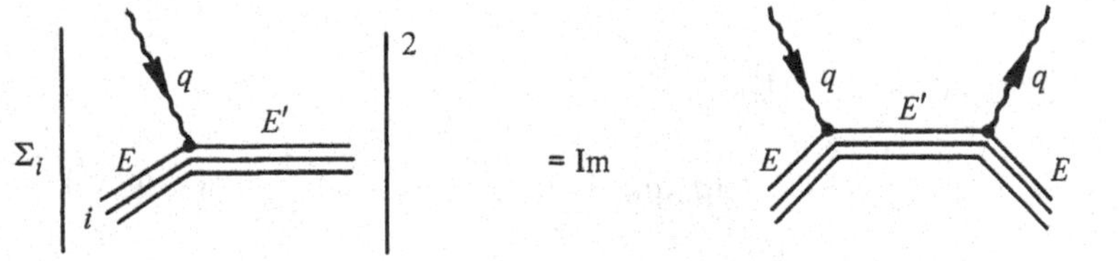

Figure 13-10 Parton contributions to the structure function.

and automatically satisfies scaling. Similarly, $mW_1(\nu, q^2)$ is equal to a function $F_1(-q^2/2\nu)$. The relation between F_1 and F_2 depends on the constituent's spin, being

$$2xF_1(x) = F_2(x) \tag{13-109}$$

for spin $\frac{1}{2}$, which corresponds to $R = \sigma_S/\sigma_T = 0$. From the conservation of total energy momentum ($\sum x_i = 1$) we derive the sum rules

$$\begin{aligned} \int_0^\infty \frac{d\nu}{\nu}\left[\frac{\nu}{m} W_2(\nu, q^2)\right] &= \int_0^1 \frac{dx}{x} F_2(x) = \sum_i Q_i^2 \\ -\frac{q^2}{2m}\int_0^\infty \frac{d\nu}{\nu} W_2(\nu, q^2) &= \int_0^1 dx\, F_2(x) = \sum_i \frac{Q_i^2}{N} \end{aligned} \tag{13-110}$$

The crucial hypothesis of this approach is the quasi free behavior of partons during the interaction with the external current and the neglect of transverse degrees of freedom.

An equivalent description amounts to substituting free constituent fields in (13-101) when estimating the contribution of the hadronic current in the light-cone region. The latter will read

$$J^\mu(x) = \overline{\psi}(x)\gamma^\mu Q \psi(x) \tag{13-111}$$

where ψ is a fermion free field and Q the charge matrix. Using the anticommutator (3-170)

$$\{\psi(x), \overline{\psi}(y)\} = \frac{1}{2\pi} \not{\partial}_x \varepsilon(x^0 - y^0)\,\delta[(x-y)^2] + O[m^2(x-y)^2]$$

we find

$$[J_\mu(x), J_\nu(y)] \underset{(x-y)^2 \to 0}{\simeq} [\overline{\psi}(x) Q^2 \gamma_\mu\gamma_\alpha\gamma_\nu \psi(y) - \overline{\psi}(y) Q^2 \gamma_\nu\gamma_\alpha\gamma_\mu \psi(x)]$$
$$\times\, \partial^\alpha \frac{\varepsilon(x^0 - y^0)}{2\pi}\, \delta[(x-y)^2]$$

When evaluated in a sum over polarizations of a diagonal element, only even terms in the $\mu \leftrightarrow \nu$ exchange contribute. We expand the product of fields in Taylor series

$$\overline{\psi}(x)\psi(-x) = \sum \frac{(-1)^n}{n!} x^{\mu_1}\cdots x^{\mu_n} \overline{\psi}(0) \overleftrightarrow{\partial}_{\mu_1} \cdots \overleftrightarrow{\partial}_{\mu_n} \psi(0)$$

and use

$$\tfrac{1}{2}(\gamma_\mu\gamma_\alpha\gamma_\nu + \gamma_\nu\gamma_\alpha\gamma_\mu) = (g_{\mu\alpha}g_{\nu\beta} + g_{\mu\beta}g_{\nu\alpha} - g_{\mu\nu}g_{\alpha\beta})\gamma^\beta$$

to obtain

$$\frac{1}{2}\{[J_\mu(x), J_\nu(-x)] + [J_\nu(x), J_\mu(-x)]\}$$
$$= i \sum_{n\,\text{odd}} x_{\mu_1}\cdots x_{\mu_n} \frac{1}{n!} O^{\beta\mu_1\cdots\mu_n}(g_{\mu\alpha}g_{\nu\beta} + g_{\mu\beta}g_{\nu\alpha} - g_{\mu\nu}g_{\alpha\beta}) \frac{1}{8\pi}\, \partial^\alpha \varepsilon(x^0)\delta(x^2)$$

$$O^{\beta\mu_1\cdots\mu_n} = i\overline{\psi}(0) Q^2 \gamma^\beta \overleftrightarrow{\partial}^{\mu_1} \cdots \overleftrightarrow{\partial}^{\mu_n} \psi(0) \tag{13-112}$$

In the vicinity of $x^2 = 0$ the commutator has been expanded in an infinite series of regular local operators multiplying the same c-number distribution. To compute the tensor $W_{\mu\nu}$ in the Bjorken limit we need the matrix elements

$$\frac{1}{2}\sum_{\text{pol}} \langle p| \frac{1}{2^n} O^{\beta\mu_1\cdots\mu_n}|p\rangle = a_{n+1}(p^\beta p^{\mu_1}\cdots p^{\mu_n} + \text{trace terms}) \qquad \text{(13-113)}$$

The trace terms involving contractions of two indices will not contribute when multiplied by a string of coordinates as in (13-112). We have

$$W_{\mu\nu} = \frac{1}{2\pi}\int d^4y\, e^{iq\cdot y}\frac{1}{2}\sum_{\text{pol}}\langle p|\left[J_\mu\left(\frac{y}{2}\right), J_\nu\left(-\frac{y}{2}\right)\right]|p\rangle$$
$$= \frac{i}{2\pi}\int d^4y\, e^{iq\cdot y}\sum_{n\,\text{odd}}\frac{1}{n!}y_{\mu_1}\cdots y_{\mu_n}\frac{1}{2}\sum_{\text{pol}}\langle p|\frac{1}{2^n}O^{\beta\mu_1\cdots\mu_n}|p\rangle$$
$$\times (g_{\mu\alpha}g_{\nu\beta} + g_{\mu\beta}g_{\nu\alpha} - g_{\mu\nu}g_{\alpha\beta})\,\partial^\alpha\frac{\varepsilon(y^0)}{2\pi}\,\delta(y^2)$$

We insert (13-113) into this expression and define a function $f(x)$, where x will turn out to be the scaling variable (not to be confused with a configuration argument), in such a way that

$$\sum_{n\,\text{odd}}\frac{(y\cdot p)^n}{n!}a_{n+1} = \frac{1}{2i}\int dx\, e^{ix(y\cdot p)}\frac{f(x)}{x} \qquad \text{(13-114)}$$

This means that the matrix elements a_{n+1} are the moments of the distribution $f(x)$. Neglecting $p^2 = m^2$ as compared to $p\cdot q$, we find

$$W_{\mu\nu} = \frac{i}{4\pi}\int d^4y\, e^{iq\cdot y}\int dx\, e^{ix(y\cdot p)}\frac{f(x)}{x}$$
$$\times [p_\mu(q+xp)_\nu + p_\nu(q+xp)_\mu - g_{\mu\nu}p\cdot q]\,\frac{\varepsilon(y^0)}{\pi}\,\delta(y^2)$$

Carrying out the integral over y we are led to the expression

$$W_{\mu\nu} = \frac{f(x)}{2x}\left(\frac{q_\mu q_\nu}{q^2} - g_{\mu\nu}\right) + \frac{f(x)}{\nu}\left(p_\mu - \frac{\nu}{q^2}q_\mu\right)\left(p_\nu - \frac{\nu}{q^2}q_\nu\right) \qquad \text{(13-115)}$$

where of course $x = \omega^{-1} = -q^2/2\nu$.

We recognize the results of the parton model with structure functions

$$\frac{\nu W_2}{m} = F_2(x) = mf(x)$$
$$mW_1 = F_1(x) = \frac{mf(x)}{2x} \qquad \text{(13-116)}$$
$$F_2(x) = 2xF_1(x)$$

Why is such a simple-minded approximation in good agreement with experiment and what are the expected corrections? The answer to these questions is given by the existence of an asymptotically free theory. The detailed implications will be discussed in Sec. 13-5.

It is possible to measure the antisymmetric part of the current commutator in experiments with polarized leptons and nucleons. Measuring spin in the direction of the incident beam, show that

$$\frac{d\sigma^{\uparrow\downarrow}}{d\Omega'\,dE'} - \frac{d\sigma^{\uparrow\uparrow}}{d\Omega'\,dE'} = \frac{4\alpha^2 E'}{-q^2 E}\frac{1}{4\pi m}[(E + E'\cos\theta)d(\nu, q^2) - (E - E'\cos\theta)(E + E')mg(\nu, q^2)] \tag{13-117}$$

where the quantities $d(\nu, q^2)$ and $g(\nu, q^2)$ are defined through

$$\frac{1}{4\pi}\int d^4x\, e^{-iq\cdot x}\{\langle p, S|[J_\mu(x), J_\nu(0)]|p, S\rangle - \langle p, -S|[J_\mu(x), J_\nu(0)]|p, -S\rangle\}$$
$$= \frac{1}{4\pi m}\varepsilon_{\mu\nu\rho\sigma}q^\rho[-S^\sigma d(\nu, q^2) + p^\sigma S\cdot q g(\nu, q^2)] \tag{13-117a}$$

and the nucleon polarization fourvector S satisfies $S\cdot p = S^2 + 1 = 0$.

13-4-3 Electron-Positron Annihilation

Electron-positron annihilation into hadrons at very high energies (several GeV) is a surprisingly fruitful domain of investigation. The discovery of narrow resonances such as the ψ and ψ' and the corresponding spectroscopy will not be discussed here because of lack of space.

The cross-sections are of the order of a few tens of nanobarns (10^{-33} cm^2) and are comparable to the rate of the electromagnetic annihilation $e^+e^- \to \mu^+\mu^-$. This process is schematized on Fig. 13-11*a*. At very high energy where the square center of mass energy q^2 is much larger than the masses it is given by

$$\sigma_{e^+e^-\to\mu^+\mu^-} = \int \frac{d^3P_+\, d^3P_-}{(2\pi)^6 P_+^0 P_-^0 p_+^0 p_-^0 v_{+-}} \frac{e^4}{(2q^2)^2}(2\pi)^4\delta^4(P_+ + P_- - p_+ - p_-)$$
$$\times \sum_{\text{pol}} |\bar{u}(P_-)\gamma^\mu v(P_+)\bar{v}(p_+)\gamma_\mu u(p_-)|^2$$
$$= \frac{4\pi\alpha^2}{3q^2} = \frac{86.9}{q^2\,(\text{GeV}^2)} \quad \text{nb} \tag{13-118}$$

The hadronic annihilation involves matrix elements of the current J_μ. For unpolarized electrons and positrons, taking into account the hermiticity of the current, it assumes the form

$$\sigma_{e^+e^-\to\text{hadrons}} = \frac{e^4}{2(q^2)^3}(g^{\mu\nu}p_+\cdot p_- - p_+^\mu p_-^\nu - p_+^\nu p_-^\mu)$$
$$\times \sum_f (2\pi)^4\delta^4(p_f - p_+ - p_-)\langle 0|J_\mu(0)|f\rangle\langle f|J_\nu(0)|0\rangle \tag{13-119}$$

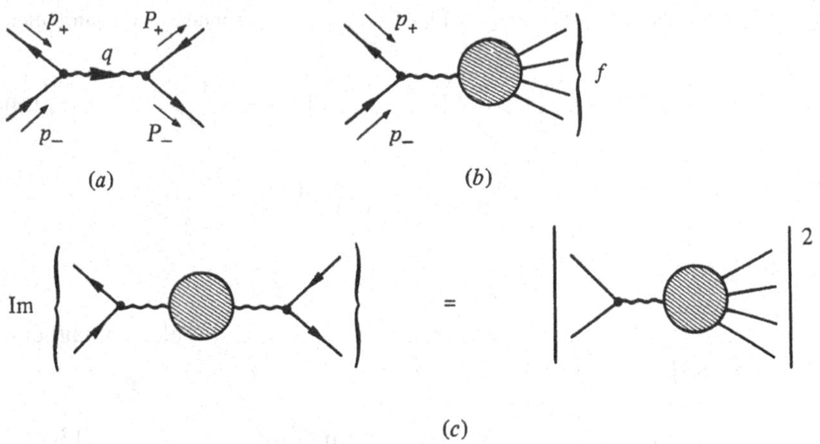

Figure 13-11 (a) Born diagram for $e^+e^- \to \mu^+\mu^-$; (b) the process $e^+e^- \to$ hadrons; (c) the cross section $\sigma_{e^+e^-\to\text{hadrons}}$ in terms of the hadronic contribution to the vacuum polarization.

Since q is time-like positive, we can also write

$$\sum_f (2\pi)^4 \delta^4(p_f - p_+ - p_-)\langle 0|J_\mu(0)|f\rangle\langle f|J_\nu(0)|0\rangle$$
$$= \int d^4x\, e^{iq\cdot x}\langle 0|J_\mu(x)J_\nu(0)|0\rangle$$
$$= \int d^4x\, e^{iq\cdot x}\langle 0|[J_\mu(x), J_\nu(0)]|0\rangle \tag{13-120}$$

Annihilation at high energy allows therefore to test the vacuum matrix element of a current commutator at short time-like distance. We are led once again to investigate the asymptotic domain. However, changing from space-like to time-like short distances may be more than an innocent modification.

The Fourier transform of the commutator in (13-120) is related to the hadronic contribution to the forward amplitude through

$$\int d^4x\, e^{iq\cdot x}\langle 0|[J_\mu(x), J_\nu(0)]|0\rangle = 2\,\mathrm{Im}\left[i\int d^4x\, e^{iq\cdot x}\langle 0|TJ_\mu(x)J_\nu(0)|0\rangle\right] \tag{13-121}$$

(Fig. 13-11c). This is also the hadronic part of the vacuum polarization tensor, up to a factor e^2:

$$i\int d^4x\, e^{iq\cdot x}\langle 0|TJ_\mu(x)J_\nu(0)|0\rangle = (g_{\mu\nu}q^2 - q_\mu q_\nu)\bar{\omega}^h(q^2) \tag{13-122}$$

Thus

$$\sigma_{e^+e^-\to\text{hadrons}} = \frac{16\pi^2\alpha^2}{q^2}\,\mathrm{Im}\,\bar{\omega}^h(q^2) \tag{13-123}$$

Note that $\sigma_{e^+e^-\to\mu^+\mu^-}$ has the same structure with $\operatorname{Im}\bar{\omega}^h$ replaced by the muon-loop contribution (7-11)

$$\operatorname{Im}\bar{\omega}_{\mu^+\mu^-}(q^2) = \frac{1}{e^2}\frac{\alpha}{3}\left(1-\frac{4m_\mu^2}{q^2}\right)^{1/2}\left(1+\frac{2m_\mu^2}{q^2}\right) \xrightarrow[m_\mu^2/q^2\to\infty]{} \frac{1}{12\pi} \tag{13-124}$$

Therefore

$$\sigma_{e^+e^-\to\mu^+\mu^-} = \frac{16\pi^2\alpha^2}{q^2}\operatorname{Im}\bar{\omega}_{\mu^+\mu^-}(q^2) = \frac{4\pi\alpha^2}{3q^2}$$

in agreement with (13-118).

It is also traditional to call R the ratio of hadronic to electromagnetic annihilation cross sections:

$$R(q^2) = \frac{\sigma_{e^+e^-\to\text{hadrons}}}{\sigma_{e^+e^-\to\mu^+\mu^-}} = 12\pi \operatorname{Im}\bar{\omega}^h(q^2) \tag{13-125}$$

An unsophisticated application of the parton model yields for the limiting value of this ratio

$$\lim_{q^2\to\infty} R(q^2) = R_\infty = \sum_i Q_i^2 \tag{13-126}$$

by simply adding the contributions of the lowest-order interactions of the charged spin $\frac{1}{2}$ elementary constituents.

Models for the internal symmetries of hadronic states yield different values for R_∞ according to the number, type, and charges of the constituents. For the octet model of quarks with charges $\frac{2}{3}$, $-\frac{1}{3}$, and $-\frac{1}{3}$ and three color indices, the predicted value is $R_\infty = 2$. If additional quantum numbers are present this ratio increases. For instance, a charmed quark of charge $\frac{2}{3}$ would contribute an extra

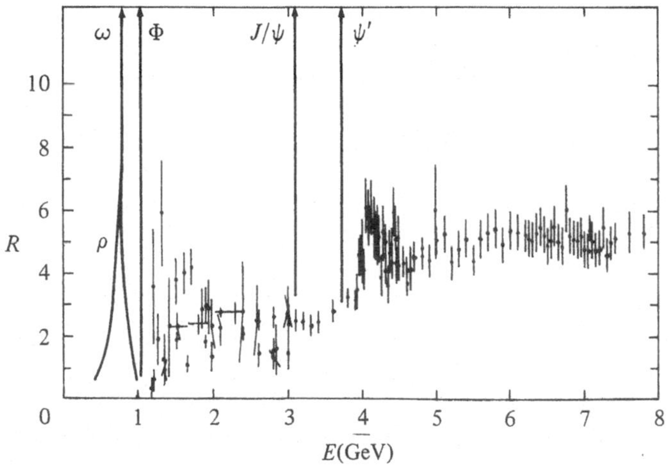

Figure 13-12 The ratio $R = \sigma_{e^+e^-\to\text{hadrons}}/\sigma_{e^+e^-\to\mu^+\mu^-}$ as a function of the total center of mass energy $E = \sqrt{q^2}$ measured at SLAC. The data were compiled by R. F. Schwitters and K. Strauch, *Ann. Rev. Nucl. Sci.*, vol. 26, p. 89, 1976.

quantity $\frac{4}{3}$, leading to $R_\infty = \frac{10}{3}$. A massive lepton around 2 GeV produced in pairs and not distinguished from the hadronic states increases again R_∞ by one unit. In any case the experimental results as of 1976 are shown in Fig. 13-12, and support the idea that $R(q^2)$ might stabilize around a value compatible with these predictions. A new generation of experiments might provide surprises.

The asymptotic property (13-126) may equivalently be stated by assuming a free-field behavior for the vacuum matrix element close to $x = 0$:

$$\langle 0|J_\mu(x)J_\nu(0)|0\rangle \underset{x\to 0}{\sim} \frac{R_\infty}{12\pi^4}(\partial_\mu\partial_\nu - g_{\mu\nu}\Box)\frac{1}{(x^2 - i\varepsilon x_0)^2} \tag{13-127}$$

We can guess the kind of corrections to R_∞ which follows from an asymptotically free-field theoretic model. From the fact that conserved currents are not renormalized (see below), we expect that for large q^2 the function R is approximately given by

$$R(\lambda^2 q^2, g) \simeq R[q^2, g(\lambda)] \simeq R_\infty\left[1 + \frac{3g^2(\lambda)}{16\pi^2}T_f + \cdots\right] \tag{13-128}$$

where $g(\lambda)$ is the running strong coupling constant. The value indicated in (13-128) uses the two-loop calculation of the vacuum polarization [Eq. (13-13)] to estimate the term in $g^2(\lambda)$ of Im $\bar{\omega}^h$. The internal fermion quantum numbers are responsible for the trace factor T_f and R_∞ is given by (13-126). For large λ, $g^2(\lambda)$ is given by

$$g^2(\lambda) \sim \frac{3(4\pi)^2}{(11C - 4T_f)\ln\lambda^2} \tag{13-129}$$

according to Eqs. (13-81) and (13-90). Such a correction predicts that R_∞ is slowly approached from above [as $(\ln q^2)^{-1}$]. In an ordinary perturbative expansion the dominant contribution would be obtained by expanding (13-124) in powers of m^2/q^2, with the result

$$R(q^2) \underset{q^2\to\infty}{\sim} R_\infty\left[1 - 8\left(\frac{m^2}{q^2}\right)^2 + \cdots\right] \tag{13-130}$$

i.e., a negative and fast varying correction.

We have poorly mistreated the beautiful and numerous experimental results on deep inelastic phenomena. Our aim was simply to illustrate an area of particle physics where modern ideas allow quantitative predictions which may be severely tested. For nonasymptotically free theories, such as quantum electrodynamics, little is known of the short-distance behavior.

13-5 OPERATOR PRODUCT EXPANSIONS

The preceding discussion illustrates the interest of studying the behavior of matrix elements of products of composite operators in several well-defined limits:

1. The space-like separation tends to zero (euclidean case).

2. The time-like separation tends to zero (minkowskian case).
3. The square separation tends to zero (light-like limit).

Problems 2 and 3 are specific of particle physics. In general, the operators involved are conserved or partially conserved currents. Problem 1 is the one that can be analyzed most thoroughly. The results can be directly applied in statistical mechanics. Crucial contributions were made by Wilson and Zimmermann.

13-5-1 Short-Distance Expansion

Consider the product of two local operators. For simplicity, we indicate only their dependence upon the configuration variable. Wilson has suggested a short-distance expansion of the form

$$A(x)B(y) \underset{x\to y}{\sim} \sum_N C_N(x-y)O_N\left(\frac{x+y}{2}\right) \tag{13-131}$$

The O_N are a sequence of local regular operators, while the c-number coefficients $C_N(x-y)$ are singular in the limit $x \to y$. Perturbatively their behavior is dictated by the canonical dimension of the corresponding operators up to logarithms

$$\begin{gathered}\lim_{x\to 0} C_N(x) \sim x^{\gamma_N}(\text{mod} \ln|x|)\\ \gamma_N = d_{O_N} - d_A - d_B\end{gathered} \tag{13-132}$$

The higher the dimension of O_N the faster the C_N go to zero. Of course, these notions will be somehow modified by renormalization group effects.

When dealing with the euclidean theory the name operator is somehow abusive. We have in mind the possibility of constructing generalized Green functions $G^{(n)}_{AB}(x, y; z_1, \ldots, z_n)$ such that the n arguments $z_1, \ldots, z_n$ refer to fundamental fields and the remaining two to A and B. The meaning of Eq. (13-131) is that when ordering the O_N according to their dimension

$$\lim_{x\to y}\left\{\left[G^{(n)}_{AB}(x, y; z_1, \ldots, z_n) - \sum^{N_{\max}} C_N(x-y)G^{(n)}_{O_N}\left(\frac{x+y}{2}; z_1, \ldots, z_n\right)\right] \times C^{-1}_{N_{\max}}(x-y)\right\} = 0 \tag{13-133}$$

The Wilson expansion is characterized by the fact that the singularities generated in the limit $x \to y$ are given by the c-number coefficients $C_N(x-y)$ independently of the arguments and types of elementary field appearing in Green functions. In Minkowski space Eq. (13-133) is understood as an asymptotic series in the weak sense of matrix elements between physical states. Clearly this is a generalization from the case considered previously, when all separations were tending simultaneously to zero. We shall see that it is possible to write for the coefficients C_N renormalization group equations.

Instead of giving a cumbersome general proof, we shall satisfy ourselves with a simple example from a scalar field theory. The operators A and B are taken as elementary fields and we look for the dominant coefficient

$$\varphi(x)\varphi(y) \underset{x\to y}{\longrightarrow} C(x-y)\varphi^2\left(\frac{x+y}{2}\right) + O[(x-y)^2] \tag{13-134}$$

We expect from (13-132) that $C(x-y)$ behaves at a given order as a polynomial in $\ln|x-y|$, up to terms of order $(x-y)^2$. To prove this, we consider the two sets of connected Green's functions $G^{(n+2)}(q_1, q_2, p_1, \ldots, p_n)$ and $G^{(n)}_{\varphi^2}(q_1+q_2; p_1, \ldots, p_n)$ and study the additional subtractions required in the construction of $G^{(n)}_{\varphi^2}$ as compared to those already necessary to define $G^{(n+2)}$. An example involving $G^{(4)}$ and $G^{(2)}_{\varphi^2}$ is shown in Fig. 13-13.

A Feynman integrand relative to $G^{(n)}_{\varphi^2}$ may be obtained by identifying two external vertices pertaining to $G^{(n+2)}$ as a unique one and adjusting the symmetry coefficient. The subtractions to be performed are, however, different. For the sake of renormalization, it suffices to consider one-particle irreducible functions $\Gamma^{(n)}_{\varphi^2}$. However, the two external lines to be contracted in $G^{(n+2)}$ need not be truncated. Let $G'^{(n+2)}(q_1, q_2; p_1, \ldots, p_n)$ denote the one-particle irreducible $(n+2)$-point function with complete propagators on the external lines of momentum q_1 and q_2. Let $\mathscr{R}_{\varphi^2}$ and $\mathscr{R}$ be the renormalized integrands of $\Gamma^{(n)}_{\varphi^2}$ and $G'^{(n+2)}$. They are related through

$$\mathscr{R}_{\varphi^2} = \mathscr{R} + \mathscr{S} \tag{13-135}$$

where $\mathscr{S}$ comes from the subtractions on the renormalization parts containing the vertex v of the φ^2 operator. The latter are irreducible Green functions with two φ external lines; they are of superficial degree zero if φ^2 is considered of dimension two. Following Chap. 8, denote by $\mathscr{F}$ the forests of renormalization parts, one of them containing v. If τ is the smallest renormalization part of $\mathscr{F}$ containing the vertex v, we write

$$\mathscr{F} = \mathscr{F}_1 \cup \tau \cup \mathscr{F}_2 \tag{13-136}$$

with $\mathscr{F}_2$ the set of renormalization parts included in τ and $\mathscr{F}_1$ those containing τ or disjoint with it. Let I be the unsubtracted integrand common to $\Gamma^{(n)}_{\varphi^2}$ and $G'^{(n+2)}$,

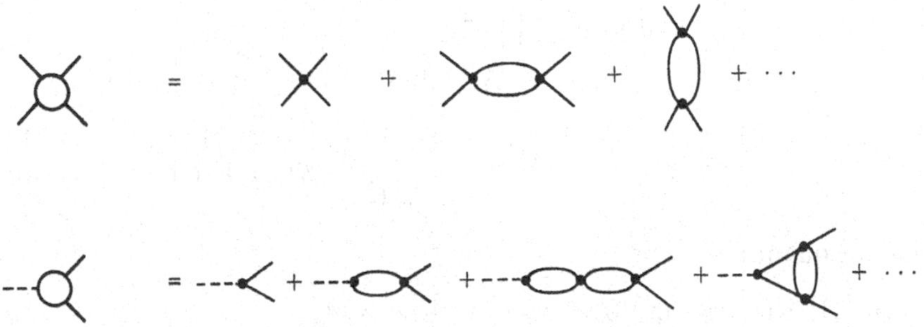

Figure 13-13 The connected Green functions $G^{(4)}$ and $G^{(2)}_{\varphi^2}$.

the explicit expression for $\mathscr{S}$ reads

$$\mathscr{S} = \sum_{\tau,\mathscr{F}_1,\mathscr{F}_2} \left[\prod_{\gamma\in\mathscr{F}_1} (-T^{\gamma}) \right] (-T^{\tau}) \left[\prod_{\gamma\in\mathscr{F}_2} (-T^{\gamma}) \right] I \tag{13-137}$$

We can organize the summation as follows. We sum over the renormalization part τ including v, the forests $\mathscr{F}_2$ of τ excluding τ itself, and over the forests $\mathscr{F}_1$ of the reduced diagram G/τ. The latter is a possible diagram for a function $\Gamma^{(n)}_{\varphi^2}$. Let σ be the part of the diagram pertaining to $G'^{(n+2)}$ corresponding to τ. It is an irreducible four-point function except for the propagators and possible self-energy insertions on the two lines which join at v in τ. We call it $\tilde{\Gamma}^{(4)}$.

For each τ the integrand I is factorized in $I = I_{G/\tau} I_{\tau}$. In an intermediate renormalization, the operators T^{γ} represent the Taylor series at zero momentum. The forests $\mathscr{F}_2$ are those of σ, $\mathscr{F}_1$ those of G/τ, and T^{τ} amounts to restricting $\mathscr{R}(\sigma)$ to its constant term $\mathscr{R}_0(\sigma)$. This operation is illustrated on Fig. 13-14 where one shows a contribution to $\tilde{\Gamma}^{(4)}$ with the two-lines candidate to be joined at v. The operation T^{τ} sets the external momenta equal to zero except for the would-be circulating momentum k in τ. Translated into symbols this means

$$\mathscr{S} = -\sum_{\tau} \mathscr{R}_{\varphi^2}(G/\tau)\mathscr{R}_0(\sigma) \tag{13-138}$$

so that, from Eq. (13-135),

$$\mathscr{R} = \mathscr{R}_{\varphi^2} + \sum_{\tau} \mathscr{R}_{\varphi^2}(G/\tau)\mathscr{R}_0(\sigma) \tag{13-139}$$

The renormalized integrand for $G'^{(n+2)}$ is expressed in a way showing the subtractions implying v which are, of course, compensated by the second term on the right-hand side. The notation $\mathscr{R}_{\varphi^2}$ is a little ambiguous since it still involves $n+2$ arguments and is relative to a Green function which can be denoted $\langle 0 | T[\varphi(x)\varphi(y)]_2 \varphi(z_1)\cdots\varphi(z_n) | 0 \rangle$.

After integration over loop momenta and summation over diagrams we may write the Green functions as

$$\begin{aligned} G^{(n+2)}(x, y, z_1, \ldots, z_n) &= \langle 0 | T\varphi(x)\varphi(y)\varphi(z_1)\cdots\varphi(z_n) | 0 \rangle \\ &= \sum_G \int \frac{d^4q_1\, d^4q_2 \prod_a d^4p_a}{(2\pi)^{4(n+2)}} (2\pi)^4 \delta^4\left(q_1 + q_2 + \sum_a p_a\right) \\ &\quad \times e^{i(q_1\cdot x + q_2\cdot y + \Sigma p_a\cdot z_a)} \int \prod_b \frac{d^4k_b}{(2\pi)^4} \mathscr{R}(q, p, k; G) \end{aligned} \tag{13-140a}$$

We define the quantities

$$\langle 0 | T[\varphi(x)\varphi(y)]_2 \varphi(z_1)\cdots\varphi(z_n) | 0 \rangle = \text{idem with } \mathscr{R} \to \mathscr{R}_{\varphi^2} \tag{13-140b}$$

$$\langle 0 | T[\varphi^2(x)]_2 \varphi(z_1)\cdots\varphi(z_n) | 0 \rangle = \text{idem with } \mathscr{R} \to \mathscr{R}_{\varphi^2} \text{ and } x = y \tag{13-140c}$$

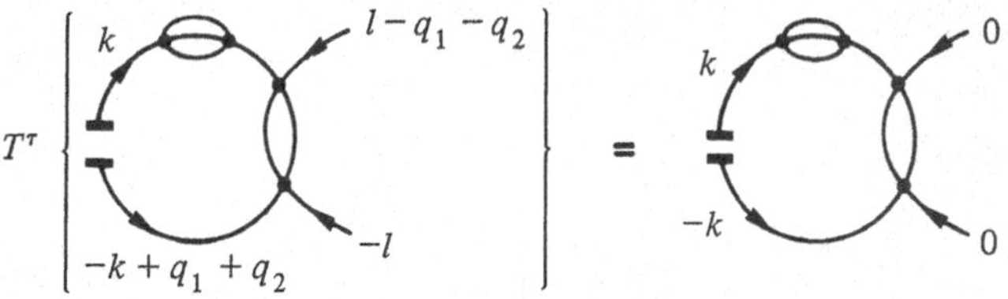

Figure 13-14 Zero-momentum subtractions on $\tilde{\Gamma}^{(4)}$.

The notation $[\ ]_2$ is to remind us that dimension two is attributed to the operator between brackets. For notational simplicity we have used the minkowskian time-ordering symbol. Setting

$$\tilde{\varphi}(0) = \int d^4z\, \varphi(z) \tag{13-141}$$

for the zero-momentum Fourier transform, we conclude from the identity (13-139) that

$$\begin{aligned}\langle 0|\,T\varphi(x)\varphi(y)\varphi(z_1)\cdots\varphi(z_n)|0\rangle &= \langle 0|\,T[\varphi(x)\varphi(y)]_2\varphi(z_1)\cdots\varphi(z_n)|0\rangle \\ &\quad + \frac{1}{2}\langle 0|\,T\varphi\left(\frac{x-y}{2}\right)\varphi\left(\frac{y-x}{2}\right)\tilde{\varphi}(0)\tilde{\varphi}(0)|0\rangle^P \\ &\quad \times \langle 0|\,T\left[\varphi^2\left(\frac{x+y}{2}\right)\right]_2\varphi(z_1)\cdots\varphi(z_n)|0\rangle\end{aligned} \tag{13-142}$$

The index P recalls that the quantity in question is obtained from the irreducible $\Gamma^{(4)}$ by adding complete propagators on two external lines. The reader might find it useful to try out the above operations on an example in order to be convinced of the presence of the symmetry factor $\frac{1}{2}$ in (13-142).

This algebraic construction is well suited to study the limit $x \to y$. Indeed, renormalization theory implies the convergence of the subtracted integral. In particular, Green functions involving $[\varphi(x)\varphi(y)]_2$ tend order by order to those of $[\varphi^2(x)]_2$ up to corrections to the type $(x-y)^2(\ln|x-y|)^\alpha$, in the euclidean region at least. Consequently, Eq. (13-142) enables us to isolate the most singular part of $G^{(n+2)}$ in the form

$$\langle 0|\,T\varphi(x)\varphi(y)\varphi(z_1)\cdots\varphi(z_n)|0\rangle \underset{x\to y}{\to} C(x-y)\langle 0|\,T\left[\varphi^2\left(\frac{x+y}{2}\right)\right]_2\varphi(z_1)\cdots\varphi(z_n)|0\rangle \tag{13-143}$$

$$C(x-y) = 1 + \frac{1}{2}\langle 0|\,T\varphi\left(\frac{x-y}{2}\right)\varphi\left(\frac{y-x}{2}\right)\tilde{\varphi}(0)\tilde{\varphi}(0)\,|0\rangle^P$$

To each order $C(x)$ behaves as a polynomial in $\ln|x|$. For instance, to lowest order $C(x)$ is proportional to

$$C(x) \sim \int \frac{d^4p}{(2\pi)^4} e^{iq\cdot x} \frac{1}{(q^2+m^2)^2} = \frac{1}{(2\pi)^3} K_0(m|x|)$$

$$\sim \frac{1}{(2\pi)^3}\left[-\ln\frac{m|x|}{2} + O(m^2x^2)\right] \tag{13-144}$$

One can refine the above derivation in order to exhibit the successive terms in the Wilson expansion. We shall skip this tedious task and, assuming the result, will study the consequences of the renormalization group on the coefficient C. According to Eq. (13-143) this implies the analysis of a Green function at exceptional momenta.

From the above example it is clear that $C(x)$ might contain subdominant terms in its perturbative behavior. We have therefore the choice to follow either the original Callan-Symanzik method with zero-momentum mass insertion or the Weinberg approach of mass-independent normalization conditions. We take for definiteness the first option and call $C_0(x)$ the asymptotic part of $C(x)$ obtained by dropping perturbative subdominant terms. We want to show that $C_0(x)$ satisfies an equation of the form

$$\left[x\cdot\frac{\partial}{\partial x} + \beta(g)\frac{\partial}{\partial g} + 2\gamma_\varphi(g) - \gamma_\Delta(g)\right]C_0(x) = 0 \tag{13-145}$$

The connected functions obey

$$\left\{x\cdot\frac{\partial}{\partial x} + y\cdot\frac{\partial}{\partial y} + z_a\cdot\frac{\partial}{\partial z_a} + \beta(g)\frac{\partial}{\partial g} + (n+2)[1+\gamma_\varphi(g)]\right\}G^{(n+2)}(x, y, z_a)$$

$$= G_\Delta^{(n+2)}(x, y, z_a) \tag{13-146}$$

We have absorbed the factor $2[1+\delta(g)]$ in the definition of Δ.

We substitute Wilson's expansion when $x \to y$ and keep only the dominant term. In this limit the left-hand side reads

$$\left\{\left[(x-y)\cdot\frac{\partial}{\partial(x-y)} + \beta(g)\frac{\partial}{\partial g} + 2\gamma_\varphi(g) - \gamma_\Delta(g)\right]C_0(x-y)\right\}G_{\varphi^2}^{(n)}\left(\frac{x+y}{2}, z_a\right)$$

$$+ C_0(x-y)\left[x\cdot\frac{\partial}{\partial x} + y\cdot\frac{\partial}{\partial y} + z_a\cdot\frac{\partial}{\partial z_a} + \beta(g)\frac{\partial}{\partial g} + n + 2 + n\gamma_\varphi(g) + \gamma_\Delta(g)\right]$$

$$\times G_{\varphi^2}^{(n)}\left(\frac{x+y}{2}, z_a\right)$$

A Callan-Symanzik equation also holds for $G_{\varphi^2}^{(n)}$ of the form

$$\left[x\cdot\frac{\partial}{\partial x} + z_a\cdot\frac{\partial}{\partial z_a} + \beta(g)\frac{\partial}{\partial g} + n + 2 + n\gamma_\varphi(g) + \gamma_\Delta(g)\right]G_{\varphi^2}^{(n)}(x, z_a)$$

$$= G_{\Delta,\varphi^2}^{(n)}(x, z_a) \tag{13-147}$$

The desired result will follow, provided that in the limit $x \to y$ the right-hand side of Eq. (13-146) can be identified with $C_0 G^{(n)}_{\Delta,\varphi^2}$:

$$G^{(n+2)}_{\Delta}(x, y, z_a) \underset{x\to y}{\to} C_0(x-y) G^{(n)}_{\Delta,\varphi^2}\left(\frac{x+y}{2}, z_a\right) \tag{13-148}$$

That this is true is a consequence of the same analysis as the one sketched above, generalized to Green functions involving a mass insertion Δ. The latter is outside the renormalization part τ which contributes to $C(x-y)$ since the only divergent function with two φ^2 insertions is a vacuum-to-vacuum amplitude. We conclude that (13-148) holds and we have therefore proved that $C_0(x)$ obeys the renormalization group equation (13-145).

All this may be extended to the successive terms of the operator expansion. Returning to a general product $A(x)B(y)$ we see that Eq. (13-132) must be corrected to read

$$\left[x \cdot \frac{\partial}{\partial x} + \beta(g)\frac{\partial}{\partial g} + d_A + \gamma_A(g) + d_B + \gamma_B(g) - d_{O_N} - \gamma_{O_N}(g)\right] C_N(x) = 0 \tag{13-149}$$

with d_A, d_B, d_{O_N} the canonical dimensions and γ_A, γ_B, γ_{O_N} the anomalous dimensions of the operators assumed to be multiplicatively renormalizable.

Strictly speaking, Eqs. (13-145) and (13-149) are only valid in the euclidean domain. They hold in the Minkowski case provided Feynman's $i\varepsilon$ is kept finite. The genuine minkowskian limit as implied in e^+e^- annihilation requires a careful discussion of possible oscillations at large momenta.

Wilson's expansion is only established in the weak sense. For instance, it holds for each function $G^{(n)}_{AB}$ with n elementary fields φ and the operators A and B. It is not guaranteed that (13-131) can be naively applied to other Green functions containing extra composite operators. For instance, it is not true that

$$\lim_{x\to y} \langle 0|T\varphi(x)\varphi(y)\varphi^2(z)|0\rangle \quad \text{is equal to} \quad C(x-y)\langle 0|T\varphi^2\left(\frac{x+y}{2}\right)\varphi^2(z)|0\rangle$$

In this case, since the global function is primitively divergent there will exist new contributions to the short-distance behavior apart from those involving two fields φ generating the coefficient $C(x-y)$. We must also account for the subtraction of the complete diagram, which will add a new function $\tilde{C}(x-y)$ independent of z:

$$\lim_{x\to y} \langle 0|T\varphi(x)\varphi(y)\varphi^2(z)|0\rangle = C(x-y)\langle 0|T\varphi^2\left(\frac{x+y}{2}\right)\varphi^2(z)|0\rangle + \tilde{C}(x-y) \tag{13-150}$$

The conclusions to be drawn from the expansion depend on the nonperturbative existence or nonexistence of ultraviolet fixed points. If such a point exists we recover the results of a modified dimensional analysis with

$$C_N(x) \sim |x|^{d_{O_N}+\gamma_{O_N}(g_\infty)-d_A-\gamma_A(g_\infty)-d_B-\gamma_B(g_\infty)} \tag{13-151}$$

In an asymptotically free theory where the functions $\gamma(g)$ are of order g^2 we write, as in Sec. 13-3,

$$\beta(g) = -bg^3 + \cdots$$
$$\gamma(g) = cg^2 + \cdots \tag{13-152}$$

Integration of (13-149) yields

$$C_N(x) \sim |x|^{d_{O_N} - d_A - d_N} \left(\ln \frac{1}{|x|} \right)^{(c_A + c_B - c_{O_N})/2b} \tag{13-153}$$

and therefore predicts logarithmic deviations from canonical scaling.

To apply the above techniques to the concrete examples we have to specify the relevant operators, study their conservation laws, and extend the analysis to light-like separations.

13-5-2 Dominant and Subdominant Operators, Operator Mixing, and Conservation Laws

In the simple example of the product $\varphi(x)\varphi(y)$ the operators O_N of the expansion are local monomials of the fields and their derivatives compatible with the symmetry properties. For instance, the next subdominant terms in a theory invariant under the change of φ into $-\varphi$ are of canonical dimension four and read

$$\varphi\left(\frac{x}{2}\right)\varphi\left(-\frac{x}{2}\right) = C_2(x)\varphi^2(0) + C_4^{(1)}(x)(\partial\varphi)^2(0) + C_4^{(2)}(x)\varphi\Box\varphi(0) + C_4^{(3)}(x)\varphi^4(0) + \cdots \tag{13-154}$$

The notation $C_2(x)$ stands here for what was previously called $C(x)$. The number of operators grows with the canonical dimension. Even when dealing with the first subdominant terms new difficulties arise. We recall that renormalization mixes operators of the same canonical dimension and the same quantum numbers. Moreover, renormalization is not exactly multiplicative since the insertion of an operator of dimension d necessitates counterterms of dimension smaller or equal to d. Special conventions are required to disentangle m^2x^2 corrections to $C_2(x)$ from contributions arising in $C_4^{(i)}(x)$. In a massless theory, where such problems do not arise, we have only to deal with a multiplicative matrix renormalization of these operators. The dominant behavior of the $C_N^{(i)}$ is therefore governed by the equation

$$\left\{\left[x\cdot\frac{\partial}{\partial x} + \beta(g)\frac{\partial}{\partial g} + d_A + \gamma_A(g) + d_B + \gamma_B(g) - d_N\right]\delta^{ij} - (\gamma_N^T)^{ij}(g)\right\} C_N^{(j)}(x) = 0 \tag{13-155}$$

where $(\gamma_N^T)^{ij}(g) = \gamma_N^{ji}(g)$ is the transposed anomalous dimension matrix of the operators $O_N^{(i)}$ mixed under renormalization [see Eq. (13-75)]. In the case of an ultraviolet fixed point g_∞, a diagonalization of $\gamma_N^{ji}(g_\infty)$ will give the observed anomalous dimensions. A similar diagonalization is necessary for the matrix c^{ij} generalizing the constant c in Eq. (13-152) for an asymptotically free theory.

It may happen that there exist relations among the operators O_N as a consequence of a specific dynamical scheme. As an example this is the case for the equations of motion.

The bare connected Green functions of the regularized φ^4 theory satisfy in euclidean space the equations of motion

$$\left[(-\Box + m_0^2)_\Lambda \frac{\delta}{\delta j(x)} + \frac{g_0}{3!}\frac{\delta^3}{\delta j(x)^3} - j(x)\right] e^{G_b(j)} = 0 \tag{13-156}$$

with $(-\Box + m_0^2)_\Lambda$ the inverse of the regularized propagator. Take a derivative with respect to $j(y)$ and let $x \to y$. We obtain an identity which translated on the proper functions reads

$$\Gamma_{\varphi(-\Box+m_0^2)_\Lambda\varphi(x),b} + \frac{g_0}{3!}\Gamma_{\varphi^4(x),b} = \varphi(x)\frac{\delta}{\delta\varphi(x)}\Gamma_b \tag{13-157}$$

The renormalized version of this relation mixes operators of dimensions two and four. As a consequence there exist functions $a_i(g)$ and $b(g)$ such that we have, identically,

$$a_1(g)\Gamma^{(n)}_{\varphi\Box\varphi}(q;p_a) + a_2(g)\Gamma^{(n)}_{\varphi^4}(q;p_a) + a_3(g)\Gamma^{(n)}_{(\partial\varphi)^2}(q;p_a) + b(g)\Gamma^{(n)}_{m^2\varphi^2}(q;p_a)$$
$$= \sum_{a=1}^{n} \Gamma^{(n)}(p_1,\ldots,p_a+q,\ldots,p_n) \tag{13-158}$$

In the massless theory, giving the dominant behavior the contribution proportional to b will vanish. From the Callan-Symanzik equations relative to the two sides of (13-158) we conclude that the matrix γ^{ij} pertaining to operators of dimension four must have a zero eigenvalue.

A similar phenomenon occurs when one of the operators O_N (or a combination of the O_N) is the generator of a continuous symmetry, conserved current, energy momentum tensor, etc. The corresponding anomalous dimension vanishes. We have already encountered this case in electrodynamics, the consequence of which was the appearance of a unique β coefficient in Eq. (13-28) for the invariant charge. A more general situation was encountered in the σ model when a symmetry is softly broken, i.e., by terms of dimension d less than four in the lagrangian. We recall the result of Symanzik showing that the counterterms of dimension higher than d could be kept symmetric. In particular, wave-function renormalization is symmetric.

Under such circumstances let J_0^μ be the bare current (of dimension three) and D_0 its divergence. The latter is of dimension smaller than four by hypothesis [see Eq. (11-3)]:

$$\partial_\mu J_0^\mu = D_0$$
$$\delta(x^0 - y^0)[J_0^0(x), \varphi_0(y)] = \tau\varphi_0(x)\delta^4(x-y) \tag{13-159}$$

with φ_0 the bare field being a vector in the internal space and τ the matrix representative of the generator. The Ward identity

$$\partial_\mu^x\langle 0|\,TJ_0^\mu(x)\varphi_0(y_1)\cdots\varphi_0(y_n)|0\rangle$$
$$= \langle 0|\,TD_0(x)\varphi_0(y_1)\cdots\varphi_0(y_n)|0\rangle$$
$$+ \sum_{a=1}^{n}\langle 0|\,T\varphi_0(y_1)\cdots\tau\varphi_0(y_a)\cdots\varphi_0(y_n)|0\rangle\,\delta^4(x-y_a) \tag{13-160}$$

becomes in renormalized form

$$
\begin{aligned}
Z_J^{-1} Z^{n/2} \partial_x^\mu \langle 0 | T J_\mu(x) \varphi(y_1) \cdots \varphi(y_n) | 0 \rangle \\
= Z_D^{-1} Z^{n/2} \langle 0 | T D(x) \varphi(y_1) \cdots \varphi(y_n) | 0 \rangle \\
+ Z^{n/2} \sum_{a=1}^{n} \langle 0 | T \varphi(y_1) \cdots \tau\varphi(y_a) \cdots \varphi(y_n) | 0 \rangle \, \delta^4(x - y_a)
\end{aligned}
\tag{13-161}
$$

if we assume no anomaly. In writing (13-161) we have taken into account the fact that wave-function renormalization is symmetric, i.e., independent of the component of φ. Since the renormalized functions are finite it follows that $Z_J = Z_D$ is finite and a proper normalization consistent with (13-159) yields

$$
Z_J = Z_D = 1 \tag{13-162}
$$

As a consequence, exact or softly broken symmetries correspond to currents for which

$$
\gamma_J = \gamma_D = 0 \tag{13-163}
$$

For the divergence D to have a dimension smaller than four it must contain an explicit dependence on the massive parameters of the theory. In a fermionic theory, for instance, where chiral invariance is broken by a mass term of dimension three, the axial current conservation is softly broken. In the absence of anomalies

$$
\partial_\mu(\bar{\psi}\gamma^\mu\gamma^5\psi) = 2mi\bar{\psi}\gamma^5\psi \tag{13-164}
$$

With a mass-independent renormalization, m depends on the dilatation factor λ as in Eq. (13-96). Let γ_m be the anomalous dimension of the operator $\bar{\psi}\psi$. According to (13-163) we shall have

$$
\begin{aligned}
\Big\{ -q \cdot \frac{\partial}{\partial q} - p_a \cdot \frac{\partial}{\partial p_a} + \gamma_m(g) m \frac{\partial}{\partial m} + \beta(g) \frac{\partial}{\partial g} + 4 + \gamma_m(g) \\
- \sum_1^n [d_a + \gamma_a(g)] \Big\} \Gamma^{(n)}_{i\bar{\psi}\gamma^5\psi}(q; p_a) = 0
\end{aligned}
\tag{13-165}
$$

The explicit dependence on m has contributed an extra term γ_m, to be interpreted as the anomalous dimension of $i\bar{\psi}\gamma^5\psi$, equal therefore to the one of $\bar{\psi}\psi$:

$$
\gamma_{i\bar{\psi}\gamma^5\psi} = \gamma_{\bar{\psi}\psi}
$$

This result is not surprising since both anomalous dimensions may be computed in the chiral limit where they obviously coincide.

Let us apply these ideas to hadronic symmetries and to their relation with the effective lagrangian of weak nonleptonic interactions.

Consider a model for strong, electromagnetic, and weak interactions based on a gauge theory for a product group $G_S \otimes G_W$ along the lines sketched at the end of Chap. 12. The group G_S will typically be the color group [for instance, $SU(3)$] and G_W will be spontaneously broken down to a $U(1)_Q$ phase group for the electric charge.

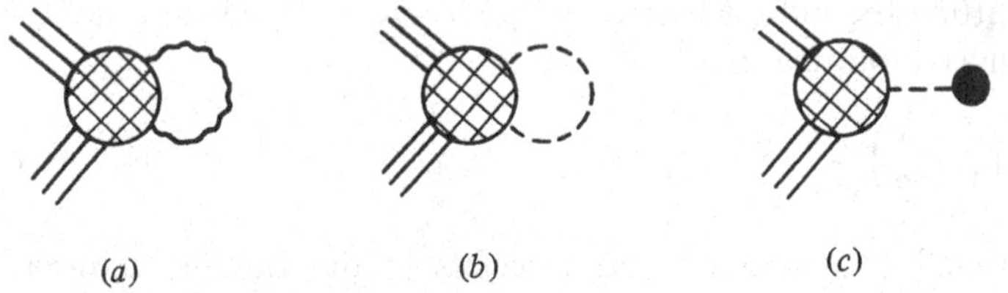

Figure 13-15 Lowest-order contributions to the effective nonleptonic lagrangian: (a) gauge boson, (b) Higgs meson, (c) tadpole term.

To order zero in α and $G_F \sim \alpha/M_W^2$ the strong interaction lagrangian will read

$$\mathscr{L}_S = i\bar{\psi}\not{D}_S\psi - \bar{\psi}M\psi - \tfrac{1}{4}F_{S,\mu\nu}F_S^{\mu\nu}$$

where M is a mass matrix, the origin of which may be related in part or totally to the breaking of G_W. Weinberg has shown that suitable redefinitions of the fields always allow to bring M to a diagonal real form without any γ_5 factor, while keeping the kinetic term invariant. In other words, to order zero in α, parity and strangeness are naturally conserved while isospin symmetry requires the additional hypothesis $M_d = M_u$ for the d and u quarks. Let us show that parity and strangeness are not violated to order α, but only to order $G_F \sim \alpha/M_W^2$.

In order to verify this point we introduce an effective lagrangian for weak nonleptonic interactions computed from lowest-order exchanges (Fig. 13-15):

$$\mathscr{L}_{\text{eff}} = -g_W^2 \int d^4x\, \Delta^{\mu\nu}(x, m^2)\langle TJ_\mu(x)J_\nu^\dagger(0)\rangle + \mathscr{L}_{(b)} + \mathscr{L}_{(c)} \tag{13-166}$$

Only the first term corresponding to gauge bosons has been written. The two additional terms represent Higgs-meson exchange $\mathscr{L}_{(b)}$ and the vacuum expectation value of Higgs bosons leading to a renormalization of the mass matrix M. We expect the contribution of Higgs bosons to be of order $\alpha(m^2/M_W^2)$ at least, with m a typical hadronic mass.

As regards the first term, apart from the massless photon contribution, the remaining part involving the propagation of heavy W and Z mesons may be analyzed using the short-distance expansion of the product of two currents. The dominant terms of order $\alpha \sim g_W^2$ lead to an additional renormalization of the mass matrix while subdominant terms contain factors $\alpha/M_W^2 \sim G$. The matrix $M + \delta M$ may again be brought to a diagonal real form. In short, to order α, parity and strangeness are still naturally preserved. This is not the case for isospin invariance. But even to this day attempts at predicting in an absolute way the supposed electromagnetic mass differences have remained without real success.

The same short-distance expansion may be used to understand the dynamical selection rules observed in nonleptonic weak transitions, such as the rule $\Delta I = \frac{1}{2}, |\Delta S| = 1$. The dominant contribution arises from dimension-six operators involving four fermion fields. Indeed, operators of dimension three only contribute to a redefinition of the mass matrix and those of dimension four may be absorbed in a wave-function renormalization. Calculations by Gaillard and Lee and by Altarelli and Maiani favour a $\Delta I = \frac{1}{2}$ enhancement with respect to the $\Delta I = \frac{3}{2}$ transitions by a logarithmic factor $[\ln(M_W^2/m^2)]^\gamma$ of the order 5 to 7 according to the models, whereas the observed enhancement factor is rather of order 20. The discussion is made difficult by the lack of absolute normalization of the matrix elements of the operators involved.

13-5-3 Light-Cone Expansion

In order to analyze the corrections to the parton model of deep inelastic phenomena we have to extend the short-distance expansion. The type of generalization required to go over to light-like separations may be anticipated from the free-field case, as in Eq. (13-112). Omitting the indices carried by the currents, we would like to show the validity of an asymptotic series

$$J\left(\frac{x}{2}\right)J\left(-\frac{x}{2}\right) \underset{x^2\to 0}{\sim} \sum_{N,\alpha} C_{N,\alpha}(x^2)x_{\mu_1}\cdots x_{\mu_N}O_{N,\alpha}^{\mu_1\cdots\mu_N}(0) \tag{13-167}$$

in terms of operators $O_{N,\alpha}^{\mu_1\cdots\mu_N}$ symmetric and traceless in the Lorentz indices. Perturbatively and up to logarithms, we expect the coefficients $C_{N,\alpha}(x)$ to scale as

$$C_{N,\alpha}(x^2) \underset{x^2\to 0}{\sim} (x^2)^{(d_{O_{N,\alpha}}-N-2d_J)/2} \tag{13-168}$$

with d_J the canonical dimension of J.

In contradistinction to the previous case, an infinite number of terms contribute to a given behavior near the light cone, in particular to the dominant one. The grouping of terms is made according to what Gross and Treiman have called twist, i.e., the difference

$$\Delta = d_{O_{N,\alpha}} - N \tag{13-169}$$

between the dimension of the operator and its spin. Strictly speaking the latter characterizes the corresponding representation of the homogeneous Lorentz group. The fact that we need such an infinity of terms is welcome since the matrix elements of this product of currents must give a scaling function $F(x)$. The knowledge of such a function is equivalent to an infinite sequence of numbers, its moments for instance. When the structure functions of lepton-hadron collisions are described as absorptive parts of Compton amplitudes the integer N will be an upper bound on the spin in the exchange t channel.

The leading contribution arises from operators of lowest twist. For a theory with spin $\frac{1}{2}$ fermions, and scalar and gauge bosons this value is two for the diagonal matrix elements of electromagnetic currents. These operators are bilinear in the fields up to covariant derivatives. They read

$$\begin{aligned}
&i^N \varphi^* \overleftrightarrow{D}_{\mu_1}\cdots \overleftrightarrow{D}_{\mu_N}\varphi \\
&i^{N-1}\overline{\psi}\gamma_{\mu_1}\overleftrightarrow{D}_{\mu_2}\cdots \overleftrightarrow{D}_{\mu_N}\psi \\
&i^{N-2}F_{\mu_1\nu}\overleftrightarrow{D}_{\mu_2}\cdots \overleftrightarrow{D}_{\mu_{N-1}}F^{\nu}{}_{\mu_N}
\end{aligned} \tag{13-170}$$

Symmetrization and extraction of traces is understood. We have only mentioned "physical" fields in this list. For specific subtleties of gauge theories, see some remarks below.

To justify Eq. (13-167) we relate it to the short-distance expansion by projecting out a given spin exchange of the crossed Compton amplitude. This quantity depends only on the variable q^2, the square virtual photon momentum, and is dominated in the limit $-q^2 \to \infty$ by the operator of corresponding spin and twist two in the Wilson expansion. This gives a definite information on the moments of the structure functions.

For short we continue to ignore the vector character of the currents and the subsequent tensor analysis of the structure functions; we therefore pretend that we deal with the scalar amplitude

$$A(q^2, \nu) = i \int d^4x \, e^{iq \cdot x} \langle p | TJ\left(\frac{x}{2}\right) J\left(-\frac{x}{2}\right) | p \rangle \qquad (13\text{-}171)$$

For finite q^2, ν it can be expanded on a basis of orthogonal polynomials in the variable $z = i\nu/m\sqrt{-q^2}$ which play the role of a cosine of the angle between the four-vectors p and q. These polynomials are orthogonal with respect to the measure $dz\sqrt{1 - z^2}$ and generalize to $O(4)$ the Legendre polynomials relative to $O(3)$ invariance. We refer to the work of Nachtmann quoted in the notes for details on this projection and the constraints arising from the positivity properties of the structure functions. We shall satisfy ourselves here with a simplified presentation.

Inserting (13-167) into the definition (13-171) we find

$$A(q^2, \nu) = i \int d^4x \, e^{iq \cdot x} \sum_{N,\alpha} C_{N,\alpha}(x^2 - i\varepsilon) x_{\mu_1} \cdots x_{\mu_N} \langle p | O_{N,\alpha}^{\mu_1 \cdots \mu_N} | p \rangle$$

Some care has been paid to the analytic properties in x space to give an infinitesimal negative imaginary part to the variable x^2. Define

$$\begin{aligned} \langle p | O_{N,\alpha}^{\mu_1 \cdots \mu_N} | p \rangle &= a_{N,\alpha}(p^{\mu_1} \cdots p^{\mu_N} + \text{trace terms}) \\ \tilde{C}_{N,\alpha}(q^2) &= i\left(-iq^2 \frac{\partial}{\partial q^2}\right)^N \int d^4x \, e^{iq \cdot x} C_{N,\alpha}(x^2 - i\varepsilon) \end{aligned} \qquad (13\text{-}172)$$

in terms of which the dominant term in $A(q^2, \nu)$ is given by

$$A(q^2, \nu) \simeq \sum_{N,\alpha} \left(\frac{2p \cdot q}{-q^2}\right)^N a_{N,\alpha} \tilde{C}_{N,\alpha}(q^2) \qquad (13\text{-}173)$$

The contributions of the operators $O_{N,\alpha}$ are related to the coefficients of the Taylor series of A in powers of the scaling variable $\omega = x^{-1} = 2p \cdot q / -q^2$, while experiments measure the absorptive part of A for ω larger than one. It is therefore necessary to isolate the coefficient of ω^N in order to use the short-distance expansion to study $\tilde{C}_{N,\alpha}(q^2)$ for large negative q^2. In this discussion the variable ω rather than its inverse x is more convenient.

Since the point $\omega = 0$ is outside the experimental reach an analytic continuation cannot be avoided. This is expressed through a forward dispersion relation for the virtual Compton amplitude A, using as discontinuity the structure function $W(q^2, \nu)$. The variable in this dispersion relation is the energy ν/m, or what amounts to the same for fixed q^2, the quantity ω. It is also necessary to use the crossing properties to define the discontinuity for ω smaller than -1. In a fictitious scalar case as here $W(q, p) = -W(-q, p)$ or $W(q^2, -\omega) = -W(q^2, -\omega)$. In the realistic vector case $W_{\mu\nu}(q, p) = -W_{\nu\mu}(-q, p)$, which means that all three structure functions W_1, W_2, W_3 may be considered as odd in ω.

Further information must be provided concerning the number of subtractions k, uniformly for large negative q^2. With $\text{Im}\, A = \pi W$, and P_{k-1} being a polynomial of degree $k-1$ in ω with q^2 dependent coefficients, we find

$$A(q^2,\omega) = P_{k-1}(q^2,\omega) + \int_{|\omega'|>1} \frac{d\omega'}{\omega'-\omega}\left(\frac{\omega}{\omega'}\right)^k W(q^2,\omega') \tag{13-174}$$

Expanding near the origin this is also

$$A(q^2,\omega) = P_{k-1}(q^2,\omega) + \sum_{N\geq k} \omega^N \int_{|\omega'|>1} \frac{d\omega'}{\omega'^{N+1}} W(q^2,\omega') \tag{13-175}$$

For $N \geq k$ we have obtained the desired relation between the moments $M_N(q^2)$ of the structure function and the Wilson coefficients

$$M^{(N)}(q^2) = \int_{|\omega|\geq 1} \frac{d\omega}{\omega^{N+1}} W(q^2,\omega) \underset{-q^2\to\infty}{\sim} \sum_\alpha a_{N,\alpha}\tilde{C}_{N,\alpha}(q^2) \tag{13-176}$$

For fixed N the leading terms on the right-hand side correspond to twist two operators. Positivity conditions on the W result in convexity properties of the moments in the variable N. The smaller the N the more sensitive are the $M^{(N)}$ to the asymptotic region since for large N we test mostly the region $\omega \sim 1$.

According to various hypothesis we have the following possibilities for $M^{(N)}(q^2)$ for large q^2:

$$M^{(N)}(q^2) \sim \begin{cases} \text{constant} & \text{naive scaling} \\ (-q^2)^{-\gamma_{O_N}(g_\infty)/2} & \text{nontrivial fixed point} \\ (\ln - q^2)^{-c_{O_N}/2b} & \text{asymptotic freedom} \end{cases} \tag{13-177}$$

The last case seems to be favored by the experimental findings.

The reader may wonder whether the observed scaling is not consistent with the existence of a non-trivial fixed point g_∞, provided that all $\gamma_{O_N}(g_\infty)$ vanish. However, this would imply that the anomalous dimension of the fundamental field itself vanishes:

$$\gamma_{O_N}(g_\infty) = 0 \qquad \forall N \Rightarrow \gamma(g_\infty) = 0$$

Together with positivity this entails that the theory is free. Therefore we are left with asymptotic freedom as the only possibility.

We return briefly to gauge theories required for asymptotic freedom in order to survey some specific intricacies arising as usual in this case. Products of physically observable operators, and hence gauge invariant, can however involve unphysical ones, in particular ghost fields, in their short-distance expansion. Such operators occur in the counterterms to Green functions and are therefore indispensible to compute physical anomalous dimensions. These additional operators are characterized using the method of Ward identities developed in Chap. 12 for Green functions without insertions. The result of Eq. (12-163) is generalized as follows. Using the same notations as in this chapter, a gauge-invariant operator O of dimension d generates counterterms of dimension smaller or equal to d with the same quantum numbers as O, which are either gauge invariant or of the form $\sigma O'$. The second class of operators is stable under renormalization since $\sigma^2 = 0$. Consequently, we can organize a computation of anomalous dimensions in a basis of the form $\{O^i_{\text{inv}}, \sigma O'^j\}$. The renormalization matrix is then upper triangular by blocks. Only the submatrix in the subspace O^i_{inv} enters the calculation of gauge-invariant physical anomalous dimensions. In favorable cases some arguments allow us to simplify the analysis by computing directly this submatrix. Finally, physical matrix elements of gauge-invariant operators, and their anomalous dimensions, do not depend on the gauge parameter. The term in $\zeta(\partial/\partial\lambda)$ introduced in Eq. (13-74) may therefore be dropped.

We close this discussion with a summary of results obtained in applying gauge theories of strong interactions to leptoproduction. They were derived by Georgi and Politzer, on the one hand, and Gross and Wilczek, on the other. For electro(or muon)-production, the previous analysis applies to the structure functions mW_1 and $\nu W_2/mx$, denoted collectively $f_a(q^2, x)$, and to their moments

$$M_a^{(N)}(q^2) = \int_0^1 dx\, x^{N-1} f_a(q^2, x) \tag{13-178}$$

In that case, the last two series of operators in (13-170) contribute. We have to face the problem of mixing of the two types at fixed N. Since we deal with an asymptotically free theory, naive scaling is violated by computable powers of logarithms

$$M^{(N)}(q^2) \sim \sum_\alpha \tilde{a}_{N,\alpha}(\ln -q^2)^{-A_{N,\alpha}} \tag{13-179}$$

where the sum runs over the eigenvector operators of the renormalization matrix. In general, the coefficients $\tilde{a}_N$ in (13-179) remain unknown.

The lowest moment $N = 2$ is the most easily handled. It implies the conserved energy momentum tensor $T_{\mu\nu}$ of zero anomalous dimension, together with another subleading operator. Moreover, the diagonal matrix element of $T_{\mu\nu}$ between proton states is known:

$$\langle p | T_{\mu\nu} | p \rangle = 2p_\mu p_\nu$$

This results in the following sum rules:

$$\lim_{-q^2\to\infty} \int_0^1 dx\, x[2mW_1(q^2, x)] = \lim_{-q^2\to\infty} \int_0^1 dx \left[\frac{\nu W_2(q^2, x)}{m}\right] = a \tag{13-180}$$

where a is given in terms of the average square charge $\langle Q^2 \rangle$ of fermion constituents through

$$a = \langle Q^2 \rangle \frac{T_f}{2C_f + T_f} = \frac{\langle Q^2 \rangle}{1 + 2(N^2 - 1)/nf} \tag{13-181}$$

The last expression applies to f flavor multiplets of fermions belonging to the same representation of dimension n of the gauge group $SU(N)$. For instance, for the three triplets of $SU(3)$ with charges $\frac{2}{3}$, $-\frac{1}{3}$, $-\frac{1}{3}$, the value of a is $\frac{2}{25}$, with four triplets of charges $\frac{2}{3}$, $-\frac{1}{3}$, $-\frac{1}{3}$, $\frac{2}{3}$, $a = \frac{5}{42} \simeq 0.12$. In the free parton model, the corresponding value was $a = \langle Q^2 \rangle$ from (13-110). In the asymptotically free model, the reduction of a is due to the fact that part of the energy momentum is carried by neutral gluons.

For N larger or equal to four, it is a reasonable approximation to retain only the eigenoperator of lowest anomalous dimension. For large N, A_N behaves as $\ln N$.

Although it is hopeless to reconstruct the structure functions from their moments without further information, some results may be derived from positivity and from the behavior (13-179). For instance, we may write, for any N,

$$f(q^2, x) \leq \frac{1}{\varepsilon}\int_x^{x+\varepsilon} dy \left(\frac{y}{x}\right)^{N-1} f(q^2, y) < \frac{1}{\varepsilon}\int_0^1 dy \left(\frac{y}{x}\right)^{N-1} f(q^2, y)$$

$$\leq \text{constant}\ \frac{1}{x^{N-1}} (\ln -q^2)^{-A_N} \tag{13-182}$$

Since $A_N \sim \ln N$ as $N \to \infty$, we expect f to decrease for fixed x faster than any power of $\ln -q^2$. This property, together with the sum rule (13-180), implies that as $-q^2$ grows the structure function has to increase in the vicinity of $x = 0$.

The asymptotic regime for the moments is slowly reached since subdominant terms are only suppressed by factors of order $\ln[\ln(-q^2/\mu^2)]/\ln(-q^2/\mu^2)$. This results from the two-loop contribution to the effective charge [Eq. (13-90)]. The rate of approach depends on the unknown scale for which the effective charge becomes small. Moreover, for finite q^2, mass effects may be sizable.

On the other hand, the moments are not easy to obtain from the experimental data, since they involve measurements at very large energy (small x), and for any finite q^2 the change of the x variable into $x' = x + O(1/q^2)$ may somehow modify the results.

In spite of these difficulties, the comparison between theory and experiment seems to encounter a reasonable success. It is convenient to rewrite the effective coupling of Eq. (13-129) in terms of a single parameter Λ which sets the scale

$$\frac{g^2(q^2)}{4\pi} = \frac{12\pi}{(33 - 2f)\ln(-q^2/\Lambda^2)} \tag{13-183}$$

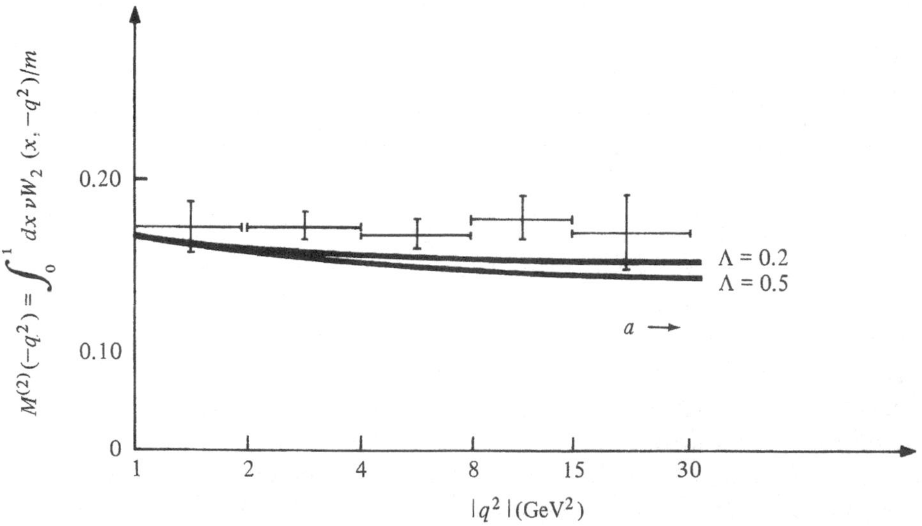

Figure 13-16 The theoretical prediction for the second moment of the muon-production structure function compared with the experimental data of H. L. Anderson et al., *Phys. Rev. Lett.*, ser. B, vol. 38, p. 1450, 1977. This drawing and the one of Fig. 13-17 were communicated by G. Altarelli.

in the case of the color group $SU(3)$ ($C = 3$) and for f triplets of quarks (f flavors). For $f = 4$, the best fits are obtained for $\Lambda \sim 400 \pm 200$ MeV, as of 1977. In Fig. 13-16, the experimental data for the muon-production moment $M^{(2)}$ are compared with the theoretical prediction. The arrow shows the limit (13-180) ($a = 0.12$ for $f = 4$). For higher moments (Fig. 13-17) only the ratio $M^{(N)}(q^2)/M^{(N)}(q_0^2)$ may be compared with experiment.

For inelastic neutrino scattering a similar analysis applies to the light-cone

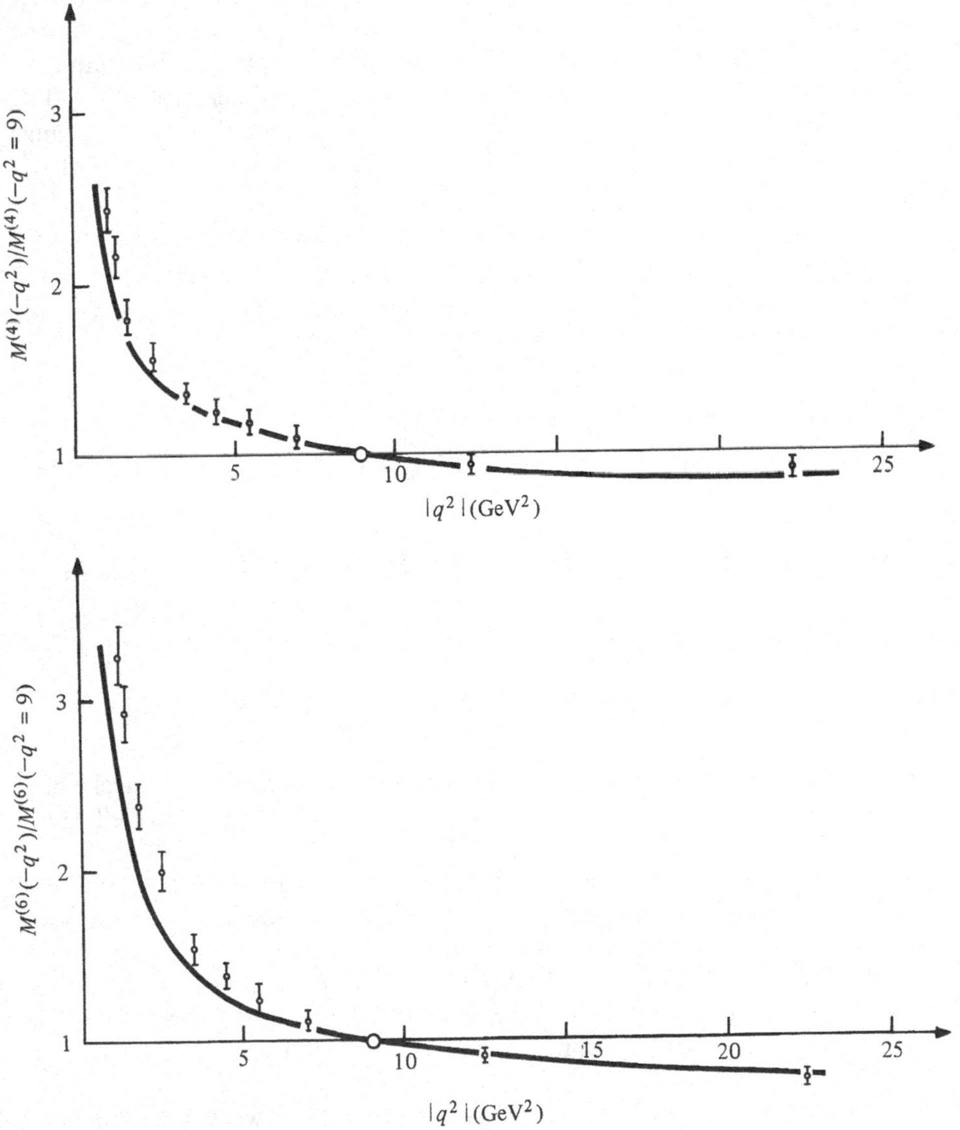

Figure 13-17 Fourth and sixth moments for muon production, normalized at $-q^2 = 9$ GeV^2. The experimental data are from Anderson et al. (Communication at the Hambourg Conference, 1977.)

expansion of the product of weak currents $(V_\mu - A_\mu)^{1+i2}(x)(V_\nu - A_\nu)^{1-i2}(0)$ in terms of the twist two operators:

$$\begin{aligned}
&(a) \qquad i^{N-1}\bar{\psi}\gamma_{\mu_1}\overleftrightarrow{D}_{\mu_2}\cdots\overleftrightarrow{D}_{\mu_N}(1 \pm \gamma_5)\psi \\
&(b) \qquad i^{N-1}\bar{\psi}\gamma_{\mu_1}\overleftrightarrow{D}_{\mu_2}\cdots\overleftrightarrow{D}_{\mu_N}(1 \pm \gamma_5)\frac{\lambda^a}{2}\psi \\
&(c) \qquad i^{N-2}F_{\mu_1\nu}\overleftrightarrow{D}_{\mu_2}\cdots\overleftrightarrow{D}_{\mu_{N-1}}F_{\mu_N}{}^{\nu}
\end{aligned} \tag{13-184}$$

A third structure function $f_3 = \nu W_3/m$ [see Eq. (13-106)] is also involved. By combining amplitudes for neutrinos and antineutrinos on proton and neutron targets, it is possible to isolate a particular channel with given quantum numbers. For instance, only the octet operator (13-184*b*) enters the difference $W^{(\nu)} - W^{(\bar{\nu})}$.

The sum rules arising from hadronic symmetries commuting with the gauge group continue to be satisfied. Thus Adler's sum rule (11-105) is verified for all q^2:

$$\int_1^\infty \frac{d\omega}{\omega}(\nu W_2{}^{(\nu n)} - \nu W_2{}^{(\nu p)}) = 2m \tag{13-185}$$

Others are only asymptotic and approached logarithmically. An example is the Callan-Gross sum rule which reads

$$\frac{\displaystyle\int_1^\infty (d\omega/\omega^{N+1})\, W_L(\omega, q^2)}{\displaystyle\int_1^\infty (d\omega/\omega^{N+1})\nu W_2(\omega, q^2)} \underset{-q^2\to\infty}{\sim} O\left(\frac{1}{\ln -q^2}\right) \tag{13-186}$$

where the quantity in the numerator $W_L = (2/\omega)W_1 - \nu W_2/m^2$ vanishes in the parton model [Eq. (13-109)].

NOTES

Early work on the renormalization group was done by E. C. G. Stueckelberg and A. Peterman, *Helv. Phys. Acta*, vol. 26, p. 499, 1953, and M. Gell-Mann and F. E. Low, *Phys. Rev.*, vol. 95, p. 1300, 1954.

Applications to quantum electrodynamics were discussed by M. Baker and K. Johnson, *Phys. Rev.*, vol. 183, p. 1292, 1969, who computed the third-order contribution to the ψ function, and by S. L. Adler, *Phys. Rev.*, ser. D, vol. 5, p. 3021, 1972. L. D. Landau, A. A. Abrikosov, and I. M. Khalatnikov studied consistency questions in *Doklady Akad. Nauk SSSR*, vol. 95, p. 1177, 1954. The β function for electrodynamics to third order is from E. de Rafaël and J. L. Rosner, *Ann. of Phys. (New York)*, vol. 82, p. 369, 1974.

Ward identities for broken-scale invariance appear in the work of K. Symanzik, *Comm. Math. Phys.*, vol. 18, p. 227, 1970, and C. G. Callan, *Phys. Rev.*, ser. D, vol. 2, p. 1541, 1970. An early discussion of the dilatation current is given in

C. G. Callan, S. Coleman, and R. Jackiw, *Ann. of Phys.* (*New York*), vol. 59, p. 42, 1970. A generalized version of the identities was given by S. Weinberg, *Phys. Rev.*, ser. D, vol. 8, p. 3497, 1973.

The one-loop computation of the β function for nonabelian gauge fields is due to D. J. Gross and F. Wilczek, *Phys. Rev. Lett.*, vol. 30, p. 1343, 1973, and H. D. Politzer, *Phys. Rev. Lett.*, vol. 30, p. 1346, 1973. The calculation to second order was carried out by W. E. Caswell, *Phys. Rev. Lett.*, vol. 33, p. 244, 1974, D. R. T. Jones, *Nucl. Phys.*, ser. B, vol. 75, p. 531, 1974, and A. A. Belavin and A. A. Migdal, *JETP Lett.*, vol. 19, p. 181, 1974.

G. Parisi, *Nuov. Cim. Lett.*, vol. 7, p. 84, 1973, and C. G. Callan and D. J. Gross, *Phys. Rev.*, ser. D, vol. 8, p. 4383, 1973, have discussed the consequences of positivity properties of the anomalous dimensions.

The asymptotic freedom theorem is due to S. Coleman and D. J. Gross, *Phys. Rev. Lett.*, vol. 31, p. 851, 1973.

For a review of deep inelastic scattering see, for instance, F. Gilman's report in "Proceedings of the XVII International Conference on High Energy Physics, London 1974," published by the Science Research Council, Rutherford Laboratory, Chilton, Didcot, U.K.

For parton dynamics in the infinite momentum frame see R. P. Feynman, *Phys. Rev. Lett.*, vol. 23, p. 1415, 1969, J. D. Bjorken and E. A. Paschos, *Phys. Rev.*, vol. 185, p. 1975, 1969, and S. D. Drell and T. M. Yan, *Ann. of Phys.* (*New York*), vol. 66, p. 578, 1971. The relation of the phenomenological model with light-cone expansions is reviewed by Y. Frishman in the "Proceedings of the Sixteenth International Conference on High Energy Physics, Batavia," vol. 4, edited by J. D. Jackson and A. Roberts, National Accelerator Laboratory, Batavia, Ill., 1972. The subject is covered in detail by R. P. Feynman in "Photon-Hadron Interactions," Benjamin, Reading, Mass., 1972.

The short-distance expansion was suggested by K. Wilson, *Phys. Rev.*, vol. 179, p. 1499, 1969, and *Phys. Rev.*, ser. D, vol. 3, p. 1818, 1971. A detailed investigation was carried out by W. Zimmermann, *Ann. of Phys.* (*New York*), vol. 77, pp. 536 and 570, 1973.

Applications to selection rules in weak interactions are found in S. Weinberg, *Phys. Rev. Lett.*, vol. 31, p. 494, 1973, *Phys. Rev.*, ser. D, vol. 8, p. 4482, 1974, and *Rev. Mod. Phys.*, vol. 46, p. 255, 1974. For the $\Delta I = \frac{1}{2}$ rule in nonleptonic decays see M. K. Gaillard and B. W. Lee, *Phys. Rev. Lett.*, vol. 33, p. 108, 1974, and G. Altarelli and L. Maiani, *Phys. Lett.*, ser. B, vol. 52, p. 351, 1974.

Extension of Wilson's expansion to light-cone infinite summation is due to R. A. Brandt and G. Preparata, *Nucl. Phys.*, ser. B, vol. 27, p. 541, 1971. O. Nachtmann, *Nucl. Phys.*, ser. B, vol. 63, p. 237, 1973, gives a discussion of group theoretic aspects and positivity constraints. Problems of gauge invariance are investigated by J. A. Dixon, *Nucl. Phys.*, ser. B, vol. 99, p. 420, 1975, H. Kluberg-Stern and J. B. Zuber, *Phys. Rev.*, ser. D, vol. 12, p. 3159, 1975, and B. W. Lee and S. Joglekar, *Ann. of Phys.* (*New York*), vol. 97, p. 160, 1976.

Applications of asymptotically free theories to the moments of the deep inelastic structure functions are considered by H. Georgi and H. D. Politzer,

Phys. Rev., ser. D, vol. 9, p. 416, 1974, and D. J. Gross and F. Wilczek, *Phys. Rev.*, ser. D, vol. 9, p. 920, 1974.

To this very incomplete list we should add the following reviews and lecture notes which were most useful in preparing this chapter: S. Coleman in "Properties of the Fundamental Interactions," edited by A. Zichichi, Editrice Compositori, Bologna, 1973; R. J. Crewther in "Weak and Electromagnetic Interactions at High Energies, Cargèse 1975," edited by M. Lévy, J. L. Basdevant, D. Speiser, and R. Gastmans, Plenum Press, New York, 1976; C. G. Callan and D. J. Gross, in "Methods in Field Theory," edited by R. Balian and J. Zinn-Justin, North-Holland, Amsterdam, 1976; and H. D. Politzer, *Phys. Rep.*, vol. 14, p. 129, 1974.

The relevance and application of the asymptotic field theory to critical phenomena is discussed by E. Brézin, J. C. Le Guillou, and J. Zinn-Justin, in "Phase Transitions and Critical Phenomena," vol. VI, edited by C. Domb and M. S. Green, Academic Press, New York, 1977, and in the textbook by D. Amit, "Field Theory, the Renormalization Group, and Critical Phenomena," McGraw-Hill, New York, 1978.

APPENDIX

A-1 METRIC

Metric tensor:

$$g_{\mu\nu} = g^{\mu\nu} = \begin{pmatrix} 1 & 0 & 0 & 0 \\ 0 & -1 & 0 & 0 \\ 0 & 0 & -1 & 0 \\ 0 & 0 & 0 & -1 \end{pmatrix} \tag{A-1}$$

Derivatives with respect to contravariant (x^μ) or covariant (x_μ) coordinates are sometimes abbreviated as

$$\partial_\mu \equiv \frac{\partial}{\partial x^\mu} \qquad \partial^\mu \equiv \frac{\partial}{\partial x_\mu} \tag{A-2}$$

Summation over repeated Lorentz (Greek) or space (Latin) indices is understood unless explicitly stated:

$$V \cdot W = V_\mu W^\mu = V^\mu W_\mu = g_{\mu\nu} V^\mu W^\nu = g^{\mu\nu} V_\mu W_\nu = V^0 W^0 - \mathbf{V} \cdot \mathbf{W} = V^0 W^0 - V^i W^i \tag{A-3}$$

A boldface letter denotes a three-vector or the three-dimensional part of a contravariant four-vector:

$$\mathbf{V} = \{V^i, i = 1, 2, 3\} = \{V_x, V_y, V_z\} \tag{A-4}$$

The only exception concerns the three-dimensional gradient

$$\boldsymbol{\nabla} = \{\nabla_x, \nabla_y, \nabla_z\} = \left(\frac{\partial}{\partial x^i} = \partial_i\right) = \left(-\partial^i = -\frac{\partial}{\partial x_i}\right) \tag{A-5}$$

The d'alembertian operator is

$$\Box = \partial^\mu \partial_\mu = \partial_0^2 - \boldsymbol{\nabla}^2 \tag{A-6}$$

and the four-momentum operator reads

$$p^\mu = i\partial^\mu = \{i\partial^0, -i\boldsymbol{\nabla}\} \tag{A-7}$$

Totally antisymmetric Levi-Civita tensor:

$$\varepsilon^{\mu\nu\rho\sigma} = \begin{cases} +1 & \text{if } \{\mu, \nu, \rho, \sigma\} \text{ is an even permutation of } \{0, 1, 2, 3\} \\ -1 & \text{if it is an odd permutation} \\ 0 & \text{otherwise} \end{cases} \tag{A-8}$$

$$\varepsilon_{\mu\nu\rho\sigma} = -\varepsilon^{\mu\nu\rho\sigma} \tag{A-9}$$

Useful identities:

$$\begin{aligned} \varepsilon^{\mu\nu\rho\sigma}\varepsilon^{\mu'\nu'\rho'\sigma'} &= -\det(g^{\alpha\alpha'}) \qquad \alpha = \mu, \nu, \rho, \sigma \\ &\qquad\qquad\qquad\quad \alpha' = \mu', \nu', \rho', \sigma' \\ \varepsilon^{\mu\nu\rho\sigma}\varepsilon_{\mu}{}^{\nu'\rho'\sigma'} &= -\det(g^{\alpha\alpha'}) \qquad \alpha = \nu, \rho, \sigma \\ &\qquad\qquad\qquad\quad \alpha' = \nu', \rho', \sigma' \\ \varepsilon^{\mu\nu\rho\sigma}\varepsilon_{\mu\nu}{}^{\rho'\sigma'} &= -2(g^{\rho\rho'}g^{\sigma\sigma'} - g^{\rho\sigma'}g^{\rho'\sigma}) \\ \varepsilon^{\mu\nu\rho\sigma}\varepsilon_{\mu\nu\rho}{}^{\sigma'} &= -6g^{\sigma\sigma'} \\ \varepsilon^{\mu\nu\rho\sigma}\varepsilon_{\mu\nu\rho\sigma} &= -24 \end{aligned} \tag{A-10}$$

Three-dimensional antisymmetric tensor:

$$\varepsilon^{ijk} = \varepsilon_{ijk} = 1 \qquad \text{if } (i, j, k) \text{ is an even permutation of } (1, 2, 3)$$

A-2 DIRAC MATRICES AND SPINORS

The γ matrices satisfy

$$\{\gamma^\mu, \gamma^\nu\} = \gamma^\mu\gamma^\nu + \gamma^\nu\gamma^\mu = 2g^{\mu\nu} \tag{A-11}$$

with γ^0 hermitian, γ^i antihermitian, and are related to the β and α matrices through

$$\gamma^0 = \beta \qquad \boldsymbol{\gamma} = \beta\boldsymbol{\alpha} \tag{A-12}$$

$$\begin{aligned} \gamma_5 = \gamma^5 = i\gamma^0\gamma^1\gamma^2\gamma^3 &= -\frac{i}{4!}\,\varepsilon_{\mu\nu\rho\sigma}\gamma^\mu\gamma^\nu\gamma^\rho\gamma^\sigma \\ &= -i\gamma_0\gamma_1\gamma_2\gamma_3 = i\gamma^3\gamma^2\gamma^1\gamma^0 = \gamma_5^\dagger \end{aligned} \tag{A-13}$$

$$\begin{aligned} \gamma_5^2 &= I \\ \{\gamma_5, \gamma^\mu\} &= 0 \end{aligned} \tag{A-14}$$

Commutator of γ matrices:

$$\sigma^{\mu\nu} = \frac{i}{2}[\gamma^\mu, \gamma^\nu] \tag{A-15}$$

$$\begin{aligned} \gamma^\mu\gamma^\nu &= g^{\mu\nu} - i\sigma^{\mu\nu} \\ [\gamma_5, \sigma^{\mu\nu}] &= 0 \end{aligned} \tag{A-16}$$

$$\begin{aligned} \gamma_5\sigma^{\mu\nu} &= \frac{i}{2}\varepsilon^{\mu\nu\rho\sigma}\sigma_{\rho\sigma} \\ \gamma_5\gamma^0\boldsymbol{\gamma} &= \boldsymbol{\Sigma} \qquad \text{where} \quad \Sigma^i \equiv \tfrac{1}{2}\varepsilon_{ijk}\sigma^{jk} \end{aligned} \tag{A-17}$$

Hermitian conjugates:

$$\begin{aligned} \gamma^0\gamma^\mu\gamma^0 &= \gamma^{\mu\dagger} \\ \gamma^0\gamma_5\gamma^0 &= -\gamma_5^\dagger = -\gamma_5 \\ \gamma^0(\gamma_5\gamma^\mu)\gamma^0 &= (\gamma_5\gamma^\mu)^\dagger \\ \gamma^0\sigma^{\mu\nu}\gamma^0 &= (\sigma^{\mu\nu})^\dagger \end{aligned} \tag{A-18}$$

For any two spinors ψ_1 and ψ_2 and any 4×4 matrix Γ,

$$(\bar{\psi}_1\Gamma\psi_2)^* = \bar{\psi}_2(\gamma_0\Gamma^\dagger\gamma_0)\psi_1 \tag{A-19}$$

while the corresponding identity for two anticommutating spin $\frac{1}{2}$ fields involves an extra minus sign.

Charge conjugation matrix:

$$\begin{aligned} C\gamma_\mu C^{-1} &= -\gamma_\mu^T \\ C\gamma_5 C^{-1} &= \gamma_5^T \\ C\sigma_{\mu\nu}C^{-1} &= -\sigma_{\mu\nu}^T \\ C(\gamma_5\gamma_\mu)C^{-1} &= (\gamma_5\gamma_\mu)^T \end{aligned} \tag{A-20}$$

Pauli matrices:

$$\sigma^1 = \begin{pmatrix} 0 & 1 \\ 1 & 0 \end{pmatrix} \qquad \sigma^2 = \begin{pmatrix} 0 & -i \\ i & 0 \end{pmatrix} \qquad \sigma^3 = \begin{pmatrix} 1 & 0 \\ 0 & -1 \end{pmatrix} \tag{A-21}$$

Dirac representation:

$$\begin{aligned} \gamma^0 &= \beta = \sigma^3 \otimes I = \begin{pmatrix} I & 0 \\ 0 & -I \end{pmatrix} \\ \boldsymbol{\alpha} &= \sigma^1 \otimes \boldsymbol{\sigma} = \begin{pmatrix} 0 & \boldsymbol{\sigma} \\ \boldsymbol{\sigma} & 0 \end{pmatrix} \\ \boldsymbol{\gamma} &= \beta\boldsymbol{\alpha} = i\sigma^2 \otimes \boldsymbol{\sigma} = \begin{pmatrix} 0 & \boldsymbol{\sigma} \\ -\boldsymbol{\sigma} & 0 \end{pmatrix} \\ \gamma_5 &= \gamma^5 = \begin{pmatrix} 0 & I \\ I & 0 \end{pmatrix} = \sigma^1 \otimes I \\ \gamma^5\gamma^0 &= -i\sigma^2 \otimes I = \begin{pmatrix} 0 & -I \\ I & 0 \end{pmatrix} \\ \gamma^5\boldsymbol{\gamma} &= -\sigma^3 \otimes \boldsymbol{\sigma} = \begin{pmatrix} -\boldsymbol{\sigma} & 0 \\ 0 & \boldsymbol{\sigma} \end{pmatrix} \\ \gamma^5\gamma^0\boldsymbol{\gamma} &= \boldsymbol{\Sigma} = I \otimes \boldsymbol{\sigma} = \begin{pmatrix} \boldsymbol{\sigma} & 0 \\ 0 & \boldsymbol{\sigma} \end{pmatrix} \\ \sigma^{0i} &= i\sigma^1 \otimes \sigma^i = i\alpha^i = i\begin{pmatrix} 0 & \sigma^i \\ \sigma^i & 0 \end{pmatrix} \\ \sigma^{ij} &= \varepsilon_{ijk} I \otimes \sigma^k = \varepsilon_{ijk}\Sigma^k = \varepsilon_{ijk}\begin{pmatrix} \sigma^k & 0 \\ 0 & \sigma^k \end{pmatrix} \end{aligned} \tag{A-22}$$

$$C = i\gamma^2\gamma^0 = -i\sigma^1 \otimes \sigma^2 = \begin{pmatrix} 0 & -i\sigma^2 \\ -i\sigma^2 & 0 \end{pmatrix}$$

$$C^T = C^\dagger = -C \qquad CC^\dagger = C^\dagger C = I \qquad C^2 = -I \tag{A-23}$$

Majorana representation:

$$\begin{aligned}
\gamma^0 &= \beta = \sigma^1 \otimes \sigma^2 = \begin{pmatrix} 0 & \sigma^2 \\ \sigma^2 & 0 \end{pmatrix} \\
\alpha^1 &= -\sigma^1 \otimes \sigma^1 = \begin{pmatrix} 0 & -\sigma^1 \\ -\sigma^1 & 0 \end{pmatrix} \\
\alpha^2 &= \sigma^3 \otimes I = \begin{pmatrix} I & 0 \\ 0 & -I \end{pmatrix} \\
\alpha^3 &= -\sigma^1 \otimes \sigma^3 = \begin{pmatrix} 0 & -\sigma^3 \\ -\sigma^3 & 0 \end{pmatrix} \\
\gamma^1 &= iI \otimes \sigma^3 = \begin{pmatrix} i\sigma^3 & 0 \\ 0 & i\sigma^3 \end{pmatrix} \\
\gamma^2 &= -i\sigma^2 \otimes \sigma^2 = \begin{pmatrix} 0 & -\sigma^2 \\ \sigma^2 & 0 \end{pmatrix} \\
\gamma^3 &= -iI \otimes \sigma^1 = \begin{pmatrix} -i\sigma^1 & 0 \\ 0 & -i\sigma^1 \end{pmatrix} \\
\gamma_5 &= \gamma^5 = \sigma^3 \otimes \sigma^2 = \begin{pmatrix} \sigma^2 & 0 \\ 0 & -\sigma^2 \end{pmatrix} \\
C &= -i\sigma^1 \otimes \sigma^2 = \begin{pmatrix} 0 & -i\sigma^2 \\ -i\sigma^2 & 0 \end{pmatrix} \qquad \text{also satisfies (A-23)}
\end{aligned} \tag{A-24}$$

Relation with the Dirac representation:

$$\gamma^\mu_{\text{Majorana}} = U\gamma^\mu_{\text{Dirac}}U^\dagger \qquad \text{with} \quad U = U^\dagger = \frac{1}{\sqrt{2}}\begin{pmatrix} I & \sigma^2 \\ \sigma^2 & -I \end{pmatrix}$$

Chiral representation:

$$\begin{aligned}
\gamma^0 &= \beta = -\sigma^1 \otimes I = \begin{pmatrix} 0 & -I \\ -I & 0 \end{pmatrix} \\
\boldsymbol{\alpha} &= \sigma^3 \otimes \boldsymbol{\sigma} = \begin{pmatrix} \boldsymbol{\sigma} & 0 \\ 0 & -\boldsymbol{\sigma} \end{pmatrix} \\
\boldsymbol{\gamma} &= i\sigma^2 \otimes \boldsymbol{\sigma} = \begin{pmatrix} 0 & \boldsymbol{\sigma} \\ -\boldsymbol{\sigma} & 0 \end{pmatrix} \\
\gamma_5 &= \gamma^5 = \begin{pmatrix} I & 0 \\ 0 & -I \end{pmatrix} \\
C &= -i\sigma^3 \otimes \sigma^2 = \begin{pmatrix} -i\sigma^2 & 0 \\ 0 & i\sigma^2 \end{pmatrix} \qquad \text{satisfies (A-23)} \\
\sigma^{0i} &= i\begin{pmatrix} \sigma^i & 0 \\ 0 & -\sigma^i \end{pmatrix} \\
\sigma^{ij} &= \varepsilon_{ijk}\begin{pmatrix} \sigma^k & 0 \\ 0 & \sigma^k \end{pmatrix}
\end{aligned} \tag{A-25}$$

Relation with the Dirac representation:

$$\gamma^\mu_{\text{chiral}} = U\gamma^\mu_{\text{Dirac}}U^\dagger \qquad \text{with} \qquad U = \frac{1}{\sqrt{2}}(1 - \gamma_5\gamma_0) = \frac{1}{\sqrt{2}}\begin{pmatrix} I & -I \\ I & I \end{pmatrix} \tag{A-26}$$

Contraction identities:

$$\begin{aligned}
\not{a}\not{b} &= a\cdot b - i\sigma_{\mu\nu}a^\mu b^\nu \\
\gamma^\lambda\gamma_\lambda &= 4 \\
\gamma^\lambda\gamma^\mu\gamma_\lambda &= -2\gamma^\mu \\
\gamma^\lambda\gamma^\mu\gamma^\nu\gamma_\lambda &= 4g^{\mu\nu} \\
\gamma^\lambda\gamma^\mu\gamma^\nu\gamma^\rho\gamma_\lambda &= -2\gamma^\rho\gamma^\nu\gamma^\mu \\
\gamma^\lambda\gamma^\mu\gamma^\nu\gamma^\rho\gamma^\sigma\gamma_\lambda &= 2(\gamma^\sigma\gamma^\mu\gamma^\nu\gamma^\rho + \gamma^\rho\gamma^\nu\gamma^\mu\gamma^\sigma) \\
\gamma^\lambda\sigma^{\mu\nu}\gamma_\lambda &= 0 \\
\gamma^\lambda\sigma^{\mu\nu}\gamma^\rho\gamma_\lambda &= 2\gamma^\rho\sigma^{\mu\nu}
\end{aligned} \tag{A-27}$$

Traces:

$$\begin{aligned}
\text{tr } I &= 4 \\
\text{tr } \gamma^\mu &= 0 \\
\text{tr } \gamma^5 &= 0
\end{aligned} \tag{A-28}$$

The trace of an odd product of γ^μ matrices vanishes:

$$\begin{aligned}
\text{tr }(\gamma^5\gamma^\mu) &= 0 \\
\text{tr }(\gamma^\mu\gamma^\nu) &= 4g^{\mu\nu} \\
\text{tr }(\sigma^{\mu\nu}) &= 0 \\
\text{tr }(\gamma^\mu\gamma^\nu\gamma^5) &= 0 \\
\text{tr }(\gamma^\mu\gamma^\nu\gamma^\rho\gamma^\sigma) &= 4(g^{\mu\nu}g^{\rho\sigma} - g^{\mu\rho}g^{\nu\sigma} + g^{\mu\sigma}g^{\nu\rho}) \\
\text{tr }(\gamma^5\gamma^\mu\gamma^\nu\gamma^\rho\gamma^\sigma) &= -4i\varepsilon^{\mu\nu\rho\sigma} = 4i\varepsilon_{\mu\nu\rho\sigma}
\end{aligned} \tag{A-29}$$

$$\begin{aligned}
\text{tr }(\not{a}_1\not{a}_2\cdots\not{a}_{2n}) &= \text{tr }(\not{a}_{2n}\cdots\not{a}_2\not{a}_1) \\
\text{tr }(\not{a}_1\cdots\not{a}_{2n}) &= a_1\cdot a_2 \text{ tr }(\not{a}_3\cdots\not{a}_{2n}) - a_1\cdot a_3 \text{ tr }(\not{a}_2\not{a}_4\cdots\not{a}_{2n}) + \cdots + a_1\cdot a_{2n} \text{ tr }(\not{a}_2\cdots\not{a}_{2n-1}) \\
&= 4\Sigma\varepsilon(a_{i_1}\cdot a_{j_1})\cdots(a_{i_n}\cdot a_{j_n})
\end{aligned} \tag{A-30}$$

ε is the signature of the permutation $i_1 j_1 \cdots i_n j_n$, and the sum runs over the $(2n)!/2^n n!$ different pairings satisfying $1 = i_1 < i_2 < \cdots < i_n$, $i_k < j_k$.

Dirac spinors u and v solutions of the Dirac equation

$$\begin{aligned}
(\not{p} - m)u^{(\alpha)}(p) &= 0 \\
(\not{p} + m)v^{(\alpha)}(p) &= 0
\end{aligned} \tag{A-31}$$

are functions of the on-shell momentum p, with $p^0 = E_p \equiv \sqrt{m^2 + \mathbf{p}^2}$ and are labeled by a polarization index $\alpha = 1, 2$.

Conjugate spinors:

$$\bar{u} = u^\dagger\gamma^0 \qquad \bar{v} = v^\dagger\gamma^0 \tag{A-32}$$

$$\begin{aligned}
\bar{u}^{(\alpha)}(p)(\not{p} - m) &= 0 \\
\bar{v}^{(\alpha)}(p)(\not{p} + m) &= 0
\end{aligned} \tag{A-33}$$

Normalization:

$$\bar{u}^{(\alpha)}(p)u^{(\beta)}(p) = \delta^{\alpha\beta}$$
$$\bar{v}^{(\alpha)}(p)v^{(\beta)}(p) = -\delta^{\alpha\beta} \tag{A-34}$$
$$\bar{v}^{(\alpha)}(p)u^{(\beta)}(p) = \bar{u}^{(\alpha)}(p)v^{(\beta)}(p) = 0$$

Density:

$$\bar{u}^{(\alpha)}(p)\gamma^0 u^{(\beta)}(p) = u^{\dagger(\alpha)}(p)u^{(\beta)}(p) = \bar{u}^{(\alpha)}(\tilde{p})u^{(\beta)}(p) = \frac{E_p}{m}\,\delta^{\alpha\beta}$$

$$\bar{v}^{(\alpha)}(p)\gamma^0 v^{(\beta)}(p) = v^{\dagger(\alpha)}(p)v^{(\beta)}(p) = -\bar{v}^{(\alpha)}(\tilde{p})v^{(\beta)}(p) = \frac{E_p}{m}\,\delta^{\alpha\beta} \tag{A-35}$$

$$\tilde{p} = (p^0, -\mathbf{p})$$

Projection operators over the positive and negative energy states:

$$\Lambda_+(p) = \frac{\not{p} + m}{2m} = \sum_{\alpha=1,2} u^{(\alpha)}(p) \otimes \bar{u}^{(\alpha)}(p)$$

$$\Lambda_-(p) = \frac{m - \not{p}}{2m} = -\sum_{\alpha=1,2} v^{(\alpha)}(p) \otimes \bar{v}^{(\alpha)}(p) \tag{A-36}$$

Projectors over a definite polarization state along a space-like four-vector n orthogonal to p, $n \cdot p = 0$:

$$u(p, n) \otimes \bar{u}(p, n) = \frac{\not{p} + m}{2m}\,\frac{1 + \gamma_5 \not{n}}{2}$$

$$-v(p, n) \otimes \bar{v}(p, n) = \frac{m - \not{p}}{2m}\,\frac{1 + \gamma_5 \not{n}}{2} \tag{A-37}$$

For comments on helicity states, see Sec. 2-2-1.

Gordon identities:

$$\bar{u}^{(\alpha)}(p)\gamma^\mu u^{(\beta)}(q) = \frac{1}{2m}\,\bar{u}^{(\alpha)}(p)[(p + q)^\mu + i\sigma^{\mu\nu}(p - q)_\nu]u^{(\beta)}(q)$$

$$\bar{u}^{(\alpha)}(p)\gamma^\mu\gamma^5 u^{(\beta)}(q) = \frac{1}{2m}\,\bar{u}^{(\alpha)}(p)[(p - q)^\mu\gamma^5 + i\sigma^{\mu\nu}(p + q)_\nu\gamma^5]u^{(\beta)}(q) \tag{A-38}$$

In particular:

$$\bar{u}^{(\alpha)}(p)\not{q}u^{(\beta)}(p) = \delta^{\alpha\beta}\,\frac{p \cdot q}{m}$$

$$u^{(\alpha)\dagger}(p)\boldsymbol{\alpha}u^{(\beta)}(p) = \delta^{\alpha\beta}\,\frac{\mathbf{p}}{m} \tag{A-39}$$

A-3 NORMALIZATION OF STATES, S MATRIX, UNITARITY, AND CROSS SECTIONS

Normalization of one-boson states:

$$\langle p \,|\, p' \rangle = 2\omega_p(2\pi^3)\delta^3(\mathbf{p} - \mathbf{p}') \tag{A-40}$$

with $\omega_p \equiv \sqrt{\mathbf{p}^2 + m^2}$ and polarization indices omitted.

One-fermion states:

$$\langle p | p' \rangle = \frac{\omega_p}{m} (2\pi)^3 \delta^3(\mathbf{p} - \mathbf{p}') \tag{A-41}$$

(For massless fermions such as neutrinos, it is safer to use a normalization of the form (A-40) in intermediate computations).

S matrix and invariant scattering amplitude:

$$\begin{aligned} S &= I + iT \\ \langle f | T | i \rangle &= (2\pi)^4 \delta^4(P_f - P_i) \mathscr{T}_{fi} \end{aligned} \tag{A-42}$$

Differential cross section for the scattering from an initial state $i = \{1, 2\}$ involving no massive fermion into a final state $f = \{3, 4, \ldots, n\}$:

$$d\sigma = \frac{1}{4[(p_1 \cdot p_2)^2 - m_1^2 m_2^2]^{1/2}} \frac{|\mathscr{T}_{fi}|^2}{S} d\tilde{p}_3 \cdots d\tilde{p}_n (2\pi)^4 \delta^4(P_i - P_f) \tag{A-43}$$

The factor S is

$$S = \prod_i k_i! \tag{A-44}$$

if there are k_i identical particles of species i in the final state. The measure $d\tilde{p}$ generally denotes

$$d\tilde{p} = \frac{d^3p}{(2\pi)^3 2\omega_p} \tag{A-45a}$$

except for massive fermions for which

$$d\tilde{p} = \frac{d^3p}{(2\pi)^3} \frac{m}{\omega_p} \tag{A-45b}$$

Accordingly, if the incident particles 1 and/or 2 are massive fermions, the expression (A-43) has to be multiplied by $2m_1$ and/or $2m_2$.

The formula (A-44) may have to be supplemented by an average over the initial polarizations and a summation over the final ones.

The decay rate $d\Gamma = d(\tau^{-1})$ of a particle of mass M into particles $3, 4, \ldots, n$ is given in its rest frame by the right-hand side of Eq. (A-43) with the flux factor $1/4[(p_1 \cdot p_2)^2 - m_1^2 m_2^2]^{1/2}$ replaced by $1/2M$. The same modifications as above are to be brought when fermions are present.

Differential cross section for two-body scattering $1 + 2 \to 3 + 4$ of nonidentical particles:

$$\frac{d\sigma}{dt} = \frac{1}{64\pi s \mathbf{q}^2} |\mathscr{T}(s, t)|^2 \tag{A-46a}$$

or

$$\frac{d\sigma}{d\Omega_{\text{cm}}} = \frac{|\mathbf{q}'|}{|\mathbf{q}|} \frac{1}{64\pi^2 s} |\mathscr{T}(s, t)|^2 \tag{A-46b}$$

in terms of the Mandelstam variables: $s = (p_1 + p_2)^2$, $t = (p_1 - p_3)^2$, and of the initial and final center of mass momenta

$$\begin{aligned} 4\mathbf{q}^2 &= \frac{\lambda(s, m_1^2, m_2^2)}{s} = \frac{[s - (m_1 + m_2)^2][s - (m_1 - m_2)^2]}{s} \\ 4\mathbf{q}'^2 &= \frac{\lambda(s, m_3^2, m_4^2)}{s} \end{aligned} \tag{A-47}$$

Optical theorem: total cross section $i \to \cdots$ in terms of the imaginary part of the forward elastic amplitude $\mathscr{T}_{ii}(s, t = 0)$:

$$\sigma_{\text{tot}}(i) = \frac{\text{Im}\, \mathscr{T}_{ii}(s, t = 0)}{\lambda^{1/2}(s, m_1^2, m_2^2)} \tag{A-48}$$

Decomposition into partial wave amplitudes for spinless particles:

$$\mathscr{T}_{fi}(s,t) = 16\pi \sum_l (2l+1) P_l(\cos\theta) \mathscr{T}^l_{fi}(s) \tag{A-49}$$

with

$$4|\mathbf{q}||\mathbf{q}'|\cos\theta = t - u + \frac{(m_1^2 - m_2^2)(m_3^2 - m_4^2)}{s}$$

$$s + t + u = m_1^2 + m_2^2 + m_3^2 + m_4^2$$

Unitarity below the inelastic threshold results in

$$\mathscr{T}^l(s) = \frac{s^{1/2}}{2|\mathbf{q}|} e^{i\delta_l(s)} \sin \delta_l(s) \tag{A-50}$$

Generalization to particles with spin has been sketched in Chap. 5.

A-4 FEYNMAN RULES

Feynman rules for the computation of a definite Green function or scattering amplitude:

1. Draw all possible topologically distinct diagrams—connected or disconnected but without vacuum-vacuum subdiagrams—contributing to the process under study, at the desired order.
2. For each diagram, and to each internal line, attach a propagator:

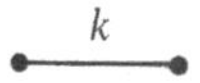

$$\frac{i}{k^2 - m^2 + i\varepsilon} \quad \text{for a spin 0 boson} \tag{A-51}$$

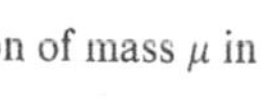

$$\left(\frac{i}{\not p - m + i\varepsilon}\right)_{\beta\alpha} \quad \text{for a spin } \tfrac{1}{2} \text{ fermion} \tag{A-52}$$

$$-i\left(\frac{g_{\rho\sigma} - k_\rho k_\sigma/\mu^2}{k^2 - \mu^2 + i\varepsilon} + \frac{k_\rho k_\sigma/\mu^2}{k^2 - \mu^2/\lambda + i\varepsilon}\right) \tag{A-53a}$$

$$= -i\left[\frac{g_{\rho\sigma}}{k^2 - \mu^2 + i\varepsilon} - \frac{(1-\lambda^{-1})k_\rho k_\sigma}{(k^2 - \mu^2 + i\varepsilon)(k^2 - \mu^2/\lambda + i\varepsilon)}\right] \tag{A-53b}$$

for a spin 1 boson of mass μ in the Stueckelberg gauge, i.e., endowed with a kinetic lagrangian

$$\mathscr{L} = -\tfrac{1}{4}(\partial_\mu A_\nu - \partial_\nu A_\mu)^2 - \frac{\lambda}{2}(\partial \cdot A)^2 + \frac{\mu^2}{2} A^2$$

3. To each vertex, assign a weight derived from the relevant monomial of the interaction lagrangian. It is composed of a factor coming from the degeneracy of identical particles in the vertex, of the coupling constant appearing in $i\mathscr{L}_{\text{int}}$, of possible tensors in internal indices, and of a momentum conservation delta function $(2\pi)^4\delta^4(\Sigma p)$. To each field derivative $\partial_\mu\phi$ is associated $-ip_\mu$ where p is the corresponding incoming momentum. Vertices for the most common theories are listed below.
4. Carry out the integration over all internal momenta with the measure $d^4k/(2\pi)^4$, possibly after a regularization.
5. Multiply the contribution of each diagram by
   (a) a symmetry factor $1/S$ where S is the order of the permutation group of the internal lines and vertices leaving the diagram unchanged when the external lines are fixed;
   (b) a factor minus one for each fermion loop; and
   (c) a global sign for the external fermion lines, coming from their permutation as compared to the arguments of the Green function at hand (see Chap. 6).

These rules yield truncated functions with no factor on the external lines. Connected functions $(2\pi)^4\delta^4(\Sigma p)G_c(p_1, \ldots, p_n)$ are obtained by retaining only connected diagrams and by putting propagators (A-51) to (A-53) on the external lines. Contributions to proper Green functions $i(2\pi)^4\delta^4(\Sigma p)\Gamma(p_1, \ldots, p_n)$ come from one-particle irreducible diagrams. Finally, the scattering amplitude $i\mathcal{T}(2\pi)^4\delta^4(P_i - P_f)$ is obtained, up to renormalization, from the previous rules by putting the external lines on their mass shell, i.e., letting $p_i^2 = m_i^2$, and providing external fermion lines with spinors $u(p)$, $v(q')$, $\bar{u}(p')$, $\bar{v}(q)$ according to whether the line enters or leaves the diagram and whether it belongs to the initial or final state ($\not{p} = \not{p}' = m$, $\not{q} = \not{q}' = -m$).

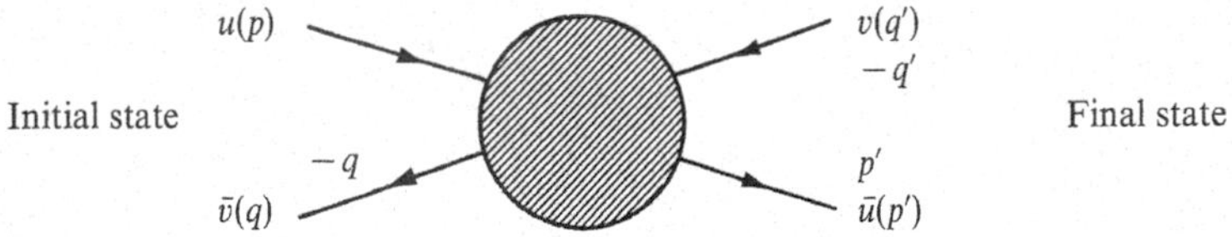

Standard theories

(*a*) φ^4 *theory*

$$\mathcal{L} = \tfrac{1}{2}(\partial\varphi)^2 - \frac{m^2}{2}\varphi^2 - \frac{\lambda\varphi^4}{4!}$$

Propagator (A-51)
Vertex $\quad -i\lambda(2\pi)^4\delta^4(\Sigma p)$

(*b*) *Quantum electrodynamics*

$$\mathcal{L} = -\tfrac{1}{4}(\partial_\mu A_\nu - \partial_\nu A_\mu)^2 - \frac{\lambda}{2}(\partial\cdot A)^2 + \bar{\psi}(i\not{\partial} - e\not{A} - m)\psi$$

Photon propagator (A-53*b*) with $\mu^2 = 0$
Fermion propagator (A-52)

Vertex 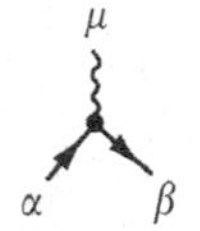 $-ie(\gamma_\mu)_{\beta\alpha}(2\pi)^4\delta^4(\Sigma p)$

(*c*) *Scalar electrodynamics*

$$\mathcal{L} = -\tfrac{1}{4}(\partial_\mu A_\nu - \partial_\nu A_\mu)^2 - \frac{\lambda}{2}(\partial\cdot A)^2 + [(\partial_\mu + ieA_\mu)\varphi]^\dagger[(\partial^\mu + ieA^\mu)\varphi] - m^2\varphi^\dagger\varphi - \frac{g}{4}(\varphi^\dagger\varphi)^2$$

Photon propagator (A-53) with $\mu^2 = 0$
Scalar propagator (A-51) oriented along the charge flow
Vertices:

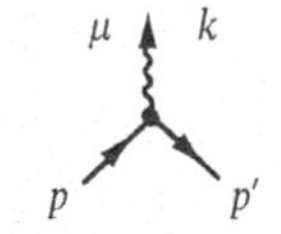

$$-ie(p_\mu + p'_\mu)(2\pi)^4\delta^4(p - p' - k)$$

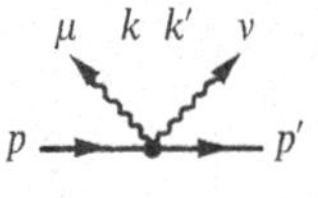

$$2ie^2 g_{\mu\nu}(2\pi)^4\delta^4(p - p' - k - k')$$

$$-ig(2\pi)^4\delta^4(p + p' - q - q')$$

(*d*) *Nonabelian gauge theory*

$$\begin{aligned}\mathscr{L} = &-\tfrac{1}{4}(\partial_\mu A_{\nu a} - \partial_\nu A_{\mu a} - gC_{abc}A_{\mu b}A_{\nu c})(\partial^\mu A^\nu{}_a - \partial^\nu A^\mu{}_a - gC_{abc}A^\mu_b A^\nu{}_c)\\ &- \frac{\lambda}{2}(\partial_\mu A^\mu{}_a)^2 - \bar{\eta}_a\partial_\mu(\partial^\mu\delta_{ac} - gC_{abc}A^\mu{}_b)\eta_c\\ &+ \bar{\psi}[i\gamma_\mu(\partial^\mu - gA^\mu{}_aT^a) - m]\psi + [(\partial^\mu - gA^\mu{}_aT^a)\phi]^\dagger[(\partial_\mu - gA_{\mu a}T_a)\phi] - m_\phi^2\phi^\dagger\phi\end{aligned}$$

Vector propagator as in (A-53*b*), with $\mu^2 = 0$
Ghost η propagator as in (A-51)
A minus sign for each ghost loop

Vertices:

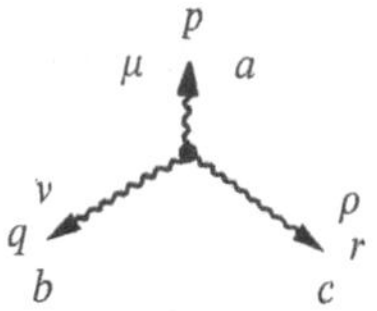

$$gC_{abc}(2\pi)^4\delta^4(p+q+r)[g_{\mu\nu}(p-q)_\rho + g_{\nu\rho}(q-r)_\mu + g_{\rho\mu}(r-p)_\nu]$$

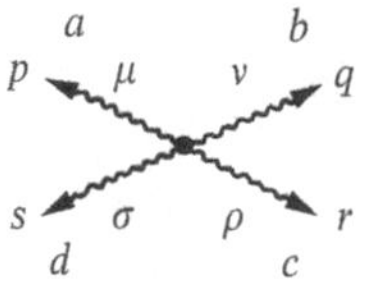

$$-ig^2(2\pi)^4\delta^4(p+q+r+s)[C_{eab}C_{ecd}(g_{\mu\rho}g_{\nu\sigma} - g_{\mu\sigma}g_{\nu\rho}) + C_{eac}C_{edb}(g_{\mu\sigma}g_{\rho\nu} - g_{\mu\nu}g_{\rho\sigma}) + C_{ead}C_{ebc}(g_{\mu\nu}g_{\sigma\rho} - g_{\mu\rho}g_{\sigma\nu})]$$

Ghost-vector vertex:

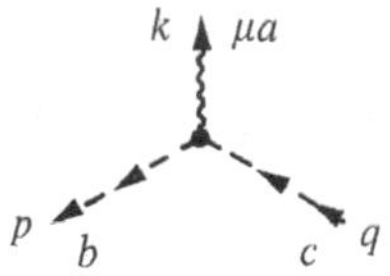

$$-gC_{abc}p_\mu(2\pi)^4\delta^4(k+p-q)$$

Fermion-vector vertex:

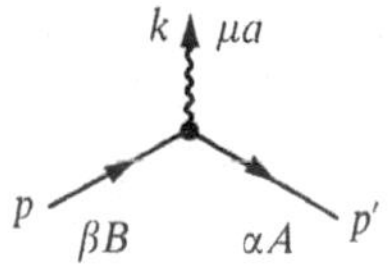

$$g(\gamma_\mu)_{\alpha\beta}T^a_{AB}(2\pi)^4\delta^4(p-p'-k)$$

Scalar-vector vertices:

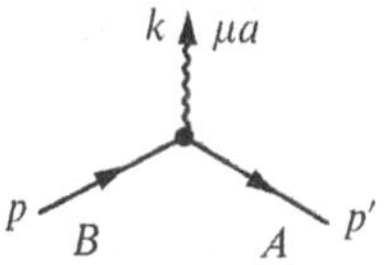

$$gT^a_{AB}(p_\mu + p'_\mu)(2\pi)^4\delta^4(p-p'-k)$$

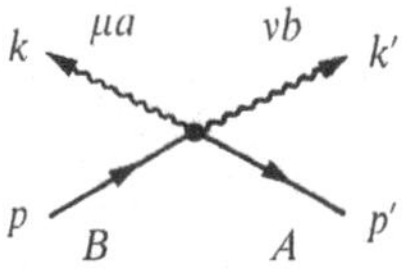

$$-ig^2g_{\mu\nu}\{T^a, T^b\}_{AB}(2\pi)^4\delta^4(p-p'-k-k')$$

with T^a antihermitian.

INDEX

A CATALOG OF SELECTED

DOVER BOOKS

IN SCIENCE AND MATHEMATICS

Physics

OPTICAL RESONANCE AND TWO-LEVEL ATOMS, L. Allen and J. H. Eberly. Clear, comprehensive introduction to basic principles behind all quantum optical resonance phenomena. 53 illustrations. Preface. Index. 256pp. 5⅜ x 8½. 0-486-65533-4

QUANTUM THEORY, David Bohm. This advanced undergraduate-level text presents the quantum theory in terms of qualitative and imaginative concepts, followed by specific applications worked out in mathematical detail. Preface. Index. 655pp. 5⅜ x 8½. 0-486-65969-0

ATOMIC PHYSICS (8th EDITION), Max Born. Nobel laureate's lucid treatment of kinetic theory of gases, elementary particles, nuclear atom, wave-corpuscles, atomic structure and spectral lines, much more. Over 40 appendices, bibliography. 495pp. 5⅜ x 8½. 0-486-65984-4

A SOPHISTICATE'S PRIMER OF RELATIVITY, P. W. Bridgman. Geared toward readers already acquainted with special relativity, this book transcends the view of theory as a working tool to answer natural questions: What is a frame of reference? What is a "law of nature"? What is the role of the "observer"? Extensive treatment, written in terms accessible to those without a scientific background. 1983 ed. xlviii+172pp. 5⅜ x 8½. 0-486-42549-5

AN INTRODUCTION TO HAMILTONIAN OPTICS, H. A. Buchdahl. Detailed account of the Hamiltonian treatment of aberration theory in geometrical optics. Many classes of optical systems defined in terms of the symmetries they possess. Problems with detailed solutions. 1970 edition. xv + 360pp. 5⅜ x 8½. 0-486-67597-1

PRIMER OF QUANTUM MECHANICS, Marvin Chester. Introductory text examines the classical quantum bead on a track: its state and representations; operator eigenvalues; harmonic oscillator and bound bead in a symmetric force field; and bead in a spherical shell. Other topics include spin, matrices, and the structure of quantum mechanics; the simplest atom; indistinguishable particles; and stationary-state perturbation theory. 1992 ed. xiv+314pp. 6⅛ x 9¼. 0-486-42878-8

LECTURES ON QUANTUM MECHANICS, Paul A. M. Dirac. Four concise, brilliant lectures on mathematical methods in quantum mechanics from Nobel Prize-winning quantum pioneer build on idea of visualizing quantum theory through the use of classical mechanics. 96pp. 5⅜ x 8½. 0-486-41713-1

THIRTY YEARS THAT SHOOK PHYSICS: THE STORY OF QUANTUM THEORY, George Gamow. Lucid, accessible introduction to influential theory of energy and matter. Careful explanations of Dirac's anti-particles, Bohr's model of the atom, much more. 12 plates. Numerous drawings. 240pp. 5⅜ x 8½. 0-486-24895-X

ELECTRONIC STRUCTURE AND THE PROPERTIES OF SOLIDS: THE PHYSICS OF THE CHEMICAL BOND, Walter A. Harrison. Innovative text offers basic understanding of the electronic structure of covalent and ionic solids, simple metals, transition metals and their compounds. Problems. 1980 edition. 582pp. 6⅛ x 9¼. 0-486-66021-4

Mathematics

FUNCTIONAL ANALYSIS (Second Corrected Edition), George Bachman and Lawrence Narici. Excellent treatment of subject geared toward students with background in linear algebra, advanced calculus, physics and engineering. Text covers introduction to inner-product spaces, normed, metric spaces, and topological spaces; complete orthonormal sets, the Hahn-Banach Theorem and its consequences, and many other related subjects. 1966 ed. 544pp. 6⅛ x 9¼. 0-486-40251-7

ASYMPTOTIC EXPANSIONS OF INTEGRALS, Norman Bleistein & Richard A. Handelsman. Best introduction to important field with applications in a variety of scientific disciplines. New preface. Problems. Diagrams. Tables. Bibliography. Index. 448pp. 5⅜ x 8½. 0-486-65082-0

VECTOR AND TENSOR ANALYSIS WITH APPLICATIONS, A. I. Borisenko and I. E. Tarapov. Concise introduction. Worked-out problems, solutions, exercises. 257pp. 5⅜ x 8¼. 0-486-63833-2

AN INTRODUCTION TO ORDINARY DIFFERENTIAL EQUATIONS, Earl A. Coddington. A thorough and systematic first course in elementary differential equations for undergraduates in mathematics and science, with many exercises and problems (with answers). Index. 304pp. 5⅜ x 8½. 0-486-65942-9

FOURIER SERIES AND ORTHOGONAL FUNCTIONS, Harry F. Davis. An incisive text combining theory and practical example to introduce Fourier series, orthogonal functions and applications of the Fourier method to boundary-value problems. 570 exercises. Answers and notes. 416pp. 5⅜ x 8½. 0-486-65973-9

COMPUTABILITY AND UNSOLVABILITY, Martin Davis. Classic graduate-level introduction to theory of computability, usually referred to as theory of recurrent functions. New preface and appendix. 288pp. 5⅜ x 8½. 0-486-61471-9

ASYMPTOTIC METHODS IN ANALYSIS, N. G. de Bruijn. An inexpensive, comprehensive guide to asymptotic methods–the pioneering work that teaches by explaining worked examples in detail. Index. 224pp. 5⅜ x 8½ 0-486-64221-6

APPLIED COMPLEX VARIABLES, John W. Dettman. Step-by-step coverage of fundamentals of analytic function theory–plus lucid exposition of five important applications: Potential Theory; Ordinary Differential Equations; Fourier Transforms; Laplace Transforms; Asymptotic Expansions. 66 figures. Exercises at chapter ends. 512pp. 5⅜ x 8½. 0-486-64670-X

INTRODUCTION TO LINEAR ALGEBRA AND DIFFERENTIAL EQUATIONS, John W. Dettman. Excellent text covers complex numbers, determinants, orthonormal bases, Laplace transforms, much more. Exercises with solutions. Undergraduate level. 416pp. 5⅜ x 8½. 0-486-65191-6

RIEMANN'S ZETA FUNCTION, H. M. Edwards. Superb, high-level study of landmark 1859 publication entitled "On the Number of Primes Less Than a Given Magnitude" traces developments in mathematical theory that it inspired. xiv+315pp. 5⅜ x 8½. 0-486-41740-9